D0793268

ON FOOD AND COOKING

The Science and Lore of the Kitchen

COMPLETELY REVISED AND UPDATED

Harold McGee

Illustrations by Patricia Dorfman, Justin Greene, and Ann McGee

SCRIBNER
New York London Toronto Sydney

SCRIBNER
1230 Avenue of the Americas
New York, NY 10020

First Scribner revised edition 2004

For information about special discounts for bulk purchases,
please contact Simon & Schuster Special Sales at
1-800-456-6798 or business@simonandschuster.com

Set in Sabon

Manufactured in the United States of America

5 7 9 10 8 6 4

Library of Congress Control Number: 2004058999

ISBN 0-684-80001-2

Page 884 constitutes a continuation of the copyright page.

To Soyoung and to my family

CONTENTS

ACKNOWLEDGMENTS

Along with many food writers today, I feel a great debt of gratitude to Alan Davidson for the way he brought new substance, scope, and playfulness to our subject. On top of that, it was Alan who informed me that I would have to revise *On Food and Cooking*—before I'd even held the first copy in my hands! At our first meeting in 1984, over lunch, he asked me what the book had to say about fish. I told him that I mentioned fish in passing as one form of animal muscle and thus of meat. And so this great fish enthusiast and renowned authority on the creatures of several seas gently suggested that, in view of the fact that fish are diverse creatures and their flesh very unlike meat, they really deserve special and extended attention. Well, yes, they really do. There are many reasons for wishing that this revision hadn't taken as long as it did, and one of the biggest is the fact that I can't show Alan the new chapter on fish. I'll always be grateful to Alan and to Jane for their encouragement and advice, and for the years of friendship which began with that lunch. This book and my life would have been much poorer without them.

I would also have liked to give this book to Nicholas Kurti—bracing myself for the discussion to come! Nicholas wrote a heartwarmingly positive review of the first edition in *Nature*, then followed it up with a Sunday-afternoon visit and an extended interrogation based on the pages of questions that he had accumulated as he wrote the review. Nicholas's energy, curiosity, and enthusiasm for good food and the telling "little experiment" were infectious, and animated the early Erice workshops. They and he are much missed.

Coming closer to home and the present, I thank my family for the affection and patient optimism that have kept me going day after day: son John and daughter Florence, who have lived with this book and experimental dinners for more than half their years, and enlivened both with their gusto and strong opinions; my father, Chuck McGee, and mother, Louise Hammersmith; brother Michael and sisters Ann and Joan; and Chuck Hammersmith, Werner Kurz, Richard Thomas, and Florence Jean and Harold Long. Throughout these last few trying years, my wife Sharon Long has been constantly caring and supportive. I'm deeply grateful to her for that gift.

Milly Marmur, my onetime publisher, longtime agent, and now great friend, has been a source of propulsive energy over the course of a marathon whose length neither of us foresaw. I've been lucky to enjoy her warmth, patience, good sense, and her skill at nudging without noodging.

I owe thanks to many people at Scribner and Simon & Schuster. Maria Guarnaschelli commissioned this revision with inspiring enthusiasm, and Scribner publisher Susan Moldow and S&S president Carolyn Reidy have been its committed advocates ever since. Beth Wareham tirelessly supervised all aspects of editing, production, and publication. Rica Buxbaum Allannic made many improvements in the manuscript with her careful editing; Mia Crowley-Hald and her team produced the book under tough time constraints with meticulous care; and Erich Hobbing welcomed my ideas about layout and designed pages that flow well and read clearly. Jeffrey Wilson kept contractual and other legal matters smooth and peaceful, and Lucy Kenyon organized some wonderful early publicity. I appreciate the marvelous team effort that has launched this book into the world.

I thank Patricia Dorfman and Justin Greene for preparing the illustrations with patience, skill, and speed, and Ann Hirsch, who produced the micrograph of a wheat kernel for this book. I'm happy to be able to include a few line drawings from the first edition by my sister Ann, who has been prevented by illness from contributing to this revision. She was a wonderful collaborator, and I've missed her sharp eye and good humor very much. I'm grateful to several food scientists for permission to share their photographs of food structure and microstructure: they are H. Douglas Goff, R. Carl Hoseney, Donald D. Kasarda, William D. Powrie, and Alastair T. Pringle. Alexandra Nickerson expertly compiled some of the most important pages in this book, the index.

Several chefs have been kind enough to invite me into their kitchens—or laboratories—to experience and talk about cooking at its most ambitious. My thanks to Fritz Blank, to Heston Blumenthal, and especially to Thomas Keller and his colleagues at The French Laundry, including Eric Ziebold, Devin Knell, Ryan Fancher, and Donald Gonzalez. I've learned a lot from them, and look forward to learning much more.

Particular sections of this book have benefited from the careful reading and comments of Anju and Hiten Bhaya, Devaki Bhaya and Arthur Grossman, Poornima and Arun Kumar, Sharon Long, Mark Pastore, Robert Steinberg, and Kathleen, Ed, and Aaron Weber. I'm very grateful for their help, and absolve them of any responsibility for what I've done with it.

I'm glad for the chance to thank my friends and my colleagues in the worlds of writing and food, all sources of stimulating questions, answers, ideas, and encouragement over the years: Shirley and Arch Corriher, the best of company on the road, at the podium, and on the phone; Lubert Stryer, who gave me the chance to see the science of pleasure advanced and immediately applied; and Kurt and Adrienne Alder, Peter Barham, Gary Beauchamp, Ed Behr, Paul Bertolli, Tony Blake, Glynn Christian, Jon Eldan, Anya Fernald, Len Fisher, Alain Harrus, Randolph Hodgson, Philip and Mary Hyman, John Paul Khoury, Kurt Koessel, Aglaia Kremezi, Anna Tasca Lanza, David Lockwood, Jean Matricon, Fritz Maytag, Jack McInerney, Alice Medrich, Marion Nestle, Ugo and Beatrice Palma, Alan Parker, Daniel Patterson, Thorvald Pedersen, Charles Perry, Maricel Presilla, P.N. Ravindran, Judy Rodgers, Nick Ruello, Helen Saberi, Mary Taylor Simeti, Melpo Skoula, Anna and Jim Spudich, Jeffrey Steingarten, Jim Tavares, Hervé This, Bob Togasaki, Rick Vargas, Despina Vokou, Ari Weinzweig, Jonathan White, Paula Wolfert, and Richard Zare.

Finally, I thank Soyoung Scanlan for sharing her understanding of cheese and of traditional forms of food production, for reading many parts of the manuscript and helping me clarify both thought and expression, and above all for reminding me, when I had forgotten, what writing and life are all about.

ON FOOD AND COOKING

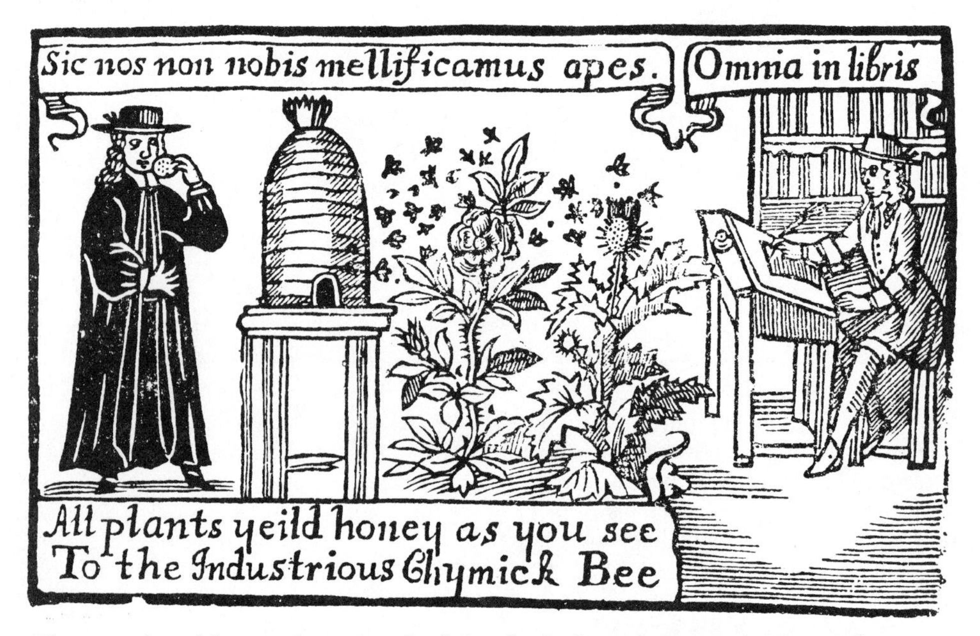

The everyday alchemy of creating food for the body and the mind. This 17th-century woodcut compares the alchemical ("chymick") work of the bee and the scholar, who transform nature's raw materials into honey and knowledge. Whenever we cook we become practical chemists, drawing on the accumulated knowledge of generations, and transforming what the Earth offers us into more concentrated forms of pleasure and nourishment. (The first Latin caption reads "Thus we bees make honey, not for ourselves"; the second, "All things in books," the library being the scholar's hive. Woodcut from the collection of the International Bee Research Association.)

INTRODUCTION

Cooking and Science, 1984 and 2004

This is the revised and expanded second edition of a book that I first published in 1984, twenty long years ago. In 1984, canola oil and the computer mouse and compact discs were all novelties. So was the idea of inviting cooks to explore the biological and chemical insides of foods. It was a time when a book like this really needed an introduction!

Twenty years ago the worlds of science and cooking were neatly compartmentalized. There were the basic sciences, physics and chemistry and biology, delving deep into the nature of matter and life. There was food science, an applied science mainly concerned with understanding the materials and processes of industrial manufacturing. And there was the world of small-scale home and restaurant cooking, traditional crafts that had never attracted much scientific attention. Nor did they really need any. Cooks had been developing their own body of practical knowledge for thousands of years, and had plenty of reliable recipes to work with.

I had been fascinated by chemistry and physics when I was growing up, experimented with electroplating and Tesla coils and telescopes, and went to Caltech planning to study astronomy. It wasn't until after I'd changed directions and moved on to English literature—and had begun to cook—that I first heard of food science. At dinner one evening in 1976 or 1977, a friend from New Orleans wondered aloud why dried beans were such a problematic food, why indulging in red beans and rice had to cost a few hours of sometimes embarrassing discomfort. Interesting question! A few days later, working in the library and needing a break from 19th-century poetry, I remembered it and the answer a biologist friend had dug up (indigestible sugars), thought I would browse in some food books, wandered over to that section, and found shelf after shelf of strange titles. *Journal of Food Science. Poultry Science. Cereal Chemistry.* I flipped through a few volumes, and among the mostly bewildering pages found hints of answers to other questions that had never occurred to me. Why do eggs solidify when we cook them? Why do fruits turn brown when we cut them? Why is bread dough bouncily alive, and why does bounciness make good bread? Which kinds of dried beans are the worst offenders, and how can a cook tame them? It was great fun to make and share these little discoveries, and I began to think that many people interested in food might enjoy them. Eventually I found time to immerse myself in food science and history and write *On Food and Cooking: The Science and Lore of the Kitchen.*

As I finished, I realized that cooks more serious than my friends and I might be skeptical about the relevance of cells and molecules to their craft. So I spent much of the introduction trying to bolster my case. I began by quoting an unlikely trio of authorities, Plato, Samuel Johnson, and Jean Anthelme Brillat-Savarin, all of whom suggested that cooking deserves detailed and serious study. I pointed out that a 19th-century German chemist still influences how many people think about cooking meat, and that around the turn of the 20th century, Fannie Farmer began her cookbook with what she called "condensed scientific knowledge" about ingredients. I noted a couple of errors in modern cookbooks by Madeleine Kamman and Julia Child, who were ahead of their time in taking chemistry seriously. And I proposed that science can make cooking more interesting by connecting it with the basic workings of the natural world.

A lot has changed in twenty years! It turned out that *On Food and Cooking* was riding a rising wave of general interest in food, a wave that grew and grew, and knocked down the barriers between science and cooking, especially in the last decade. Science has found its way into the kitchen, and cooking into laboratories and factories.

In 2004 food lovers can find the science of cooking just about everywhere. Magazines and newspaper food sections devote regular columns to it, and there are now a number of books that explore it, with Shirley Corriher's 1997 *CookWise* remaining unmatched in the way it integrates explanation and recipes. Today many writers go into the technical details of their subjects, especially such intricate things as pastry, chocolate, coffee, beer, and wine. Kitchen science has been the subject of television series aired in the United States, Canada, the United Kingdom, and France. And a number of food molecules and microbes have become familiar figures in the news, both good and bad. Anyone who follows the latest in health and nutrition knows about the benefits of antioxidants and phytoestrogens, the hazards of trans fatty acids, acrylamide, *E. coli* bacteria, and mad cow disease.

Professional cooks have also come to appreciate the value of the scientific approach to their craft. In the first few years after *On Food and Cooking* appeared, many young cooks told me of their frustration in trying to find out *why* dishes were prepared a certain way, or why ingredients behave as they do. To their traditionally trained chefs and teachers, understanding food was less important than mastering the tried and true techniques for preparing it. Today it's clearer that curiosity and understanding make their own contribution to mastery. A number of culinary schools now offer "experimental" courses that investigate the whys of cooking and encourage critical thinking. And several highly regarded chefs, most famously Ferran Adrià in Spain and Heston Blumenthal in England, experiment with industrial and laboratory tools—gelling agents from seaweeds and bacteria, non-sweet sugars, aroma extracts, pressurized gases, liquid nitrogen—to bring new forms of pleasure to the table.

As science has gradually percolated into the world of cooking, cooking has been drawn into academic and industrial science. One effective and charming force behind this movement was Nicholas Kurti, a physicist and food lover at the University of Oxford, who lamented in 1969: "I think it is a sad reflection on our civilization that while we can and do measure the temperature in the atmosphere of Venus, we do not know what goes on inside our soufflés." In 1992, at the age of 84, Nicholas nudged civilization along by organizing an International Workshop on Molecular and Physical Gastronomy at Erice, Sicily, where for the first time professional cooks, basic scientists from universities, and food scientists from industry worked together to advance gastronomy, the making and appreciation of foods of the highest quality.

The Erice meeting continues, renamed the "International Workshop on Molecular Gastronomy 'N. Kurti' " in memory of its founder. And over the last decade its focus, the understanding of culinary excellence, has taken on new economic significance. The modern industrial drive to maximize efficiency and minimize costs generally lowered the quality and distinctiveness of food products: they taste much the same, and not very good. Improvements in quality can now mean a competitive advantage; and cooks have always been the world's experts in the applied science of deliciousness. Today, the French National Institute of Agricultural Research sponsors a group in Molecular Gastronomy at the Collège de France (its leader, Hervé This, directs the Erice workshop); chemist Thorvald Pedersen is the inaugural Professor of Molecular Gastronomy at Denmark's Royal Veterinary and Agricultural University; and in the United States, the rapidly growing membership of the Research Chefs Association specializes in bringing the chef's skills and standards to the food industry.

So in 2004 there's no longer any need to explain the premise of this book. Instead, there's more for the book itself to explain! Twenty years ago, there wasn't much demand for information about extra-virgin olive oil or balsamic vinegar, farmed salmon or grass-fed beef, cappuccino or white tea, Sichuan pepper or Mexican mole, sake or well-tempered chocolate. Today there's interest in all these and much more. And so this second edition of *On Food and Cooking* is substantially longer than the first. I've expanded the text by two thirds in order to cover a broader range of ingredients and preparations, and to explore them in greater depth. To make room for new information about foods, I've dropped the separate chapters on human physiology, nutrition, and additives. Of the few sections that survive in similar form from the first edition, practically all have been rewritten to reflect fresh information, or my own fresh understanding.

This edition gives new emphasis to two particular aspects of food. The first is the diversity of ingredients and the ways in which they're prepared. These days the easy movement of products and people makes it possible for us to taste foods from all over the world. And traveling back in time through old cookbooks can turn up forgotten but intriguing ideas. I've tried throughout to give at least a brief indication of the range of possibilities offered by foods themselves and by different national traditions.

The other new emphasis is on the flavors of foods, and sometimes on the particular molecules that create flavor. Flavors are something like chemical chords, composite sensations built up from notes provided by different molecules, some of which are found in many foods. I give the chemical names of flavor molecules when I think that being specific can help us notice flavor relationships and echoes. The names may seem strange and intimidating at first, but they're just names and they'll become more familiar. Of course people have made and enjoyed well seasoned dishes for thousands of years with no knowledge of molecules. But a dash of flavor chemistry can help us make fuller use of our senses of taste and smell, and experience more—and find more pleasure—in what we cook and eat.

Now a few words about the scientific approach to food and cooking and the organization of this book. Like everything on earth, foods are mixtures of different chemicals, and the qualities that we aim to influence in the kitchen—taste, aroma, texture, color, nutritiousness—are all manifestations of chemical properties. Nearly two hundred years ago, the eminent gastronome Jean Anthelme Brillat-Savarin lectured his cook on this point, tongue partly in cheek, in *The Physiology of Taste*:

> You are a little opinionated, and I have had some trouble in making you understand that the phenomena which take place in your laboratory are nothing other than the execution of the eternal laws of nature, and that certain things which you do without thinking, and only because you have seen others do them, derive nonetheless from the highest scientific principles.

The great virtue of the cook's time-tested, thought-less recipes is that they free us from the distraction of having to guess or experiment or analyze as we prepare a meal. On the other hand, the great virtue of thought and analysis is that they free us from the necessity of following recipes, and help us deal with the unexpected, including the inspiration to try something new. Thoughtful cooking means paying attention to what our senses tell us as we prepare it, connecting that information with past experience and with an understanding of what's happening to the food's inner substance, and adjusting the preparation accordingly.

To understand what's happening within a food as we cook it, we need to be familiar with the world of invisibly small molecules and their reactions with each other. That idea may seem daunting. There are a hundred-plus chemical elements, many more combinations of those elements into molecules, and several different forces that rule their behavior. But scientists always simplify reality in order to understand it, and we can do the same. Foods are mostly built out of just four kinds of molecules—water, proteins, carbohydrates, and fats. And their behavior can be pretty well described with a few simple principles. If you know that heat is a manifestation of the movements of molecules, and that sufficiently energetic collisions disrupt the structures of molecules and eventually break them apart, then you're very close to understanding why heat solidifies eggs and makes foods tastier.

Most readers today have at least a vague idea of proteins and fats, molecules and energy, and a vague idea is enough to follow most of the explanations in the first 13 chapters, which cover common foods and ways of preparing them. Chapters 14 and 15 then describe in some detail the molecules and basic chemical processes involved in all cooking; and the Appendix gives a brief refresher course in the basic vocabulary of science. You can refer to these final sections occasionally, to clarify the meaning of pH or protein coagulation as you're reading about cheese or meat or bread, or else read through them on their own to get a general introduction to the science of cooking.

Finally, a request. In this book I've sifted through and synthesized a great deal of information, and have tried hard to double-check both facts and my interpretations of them. I'm greatly indebted to the many scientists, historians, linguists, culinary professionals, and food lovers on whose learning I've been able to draw. I will also appreciate the help of readers who notice errors that I've made and missed, and who let me know so that I can correct them. My thanks in advance.

As I finish this revision and think about the endless work of correcting and perfecting, my mind returns to the first Erice workshop and a saying shared by Jean-Pierre Philippe, a chef from Les Mesnuls, near Versailles. The subject of the moment was egg foams. Chef Philippe told us that he had thought he knew everything there was to know about meringues, until one day a phone call distracted him and he left his mixer running for half an hour. Thanks to the excellent result and to other surprises throughout his career, he said, *Je sais, je sais que je sais jamais:* "I know, I know that I never know." Food is an infinitely rich subject, and there's always something about it to understand better, something new to discover, a fresh source of interest, ideas, and delight.

A Note About Units of Measurement, and About the Drawings of Molecules

Throughout this book, temperatures are given in both degrees Fahrenheit (°F), the standard units in the United States, and degrees Celsius or Centigrade (°C), the units used by most other countries. The Fahrenheit temperatures shown in several charts can be converted to Celsius by using the formula °C = (°F−32) x 0.56. Volumes and weights are given in both U.S. kitchen units—teaspoons, quarts, pounds—and metric units—milliliters, liters, grams, and kilograms. Lengths are generally given in millimeters (mm); 1 mm is about the diameter of the degree symbol °. Very small lengths are given in microns (μ). One micron is 1 micrometer, or 1 thousandth of a millimeter.

Single molecules are so small, a tiny fraction of a micron, that they can seem abstract, hard to imagine. But they are real and concrete, and have particular structures that determine how they—and the foods made out of them—behave in the kitchen. The better we can visualize what they're like and what happens to them, the easier it is to understand what happens in cooking. And in cooking it's generally a molecule's overall shape that matters, not the precise placement of each atom. In most of the drawings of molecules in this book, only the overall shapes are shown, and they're represented in different ways—as long thin lines, long thick lines, honeycomb-like rings with some atoms indicated by letters—depending on what behavior needs to be explained. Many food molecules are built from a backbone of interconnected carbon atoms, with a few other kinds of atoms (mainly hydrogen and oxygen) projecting from the backbone. The carbon backbone is what creates the overall structure, so often it is drawn with no indications of the atoms themselves, just lines that show the bonds between atoms.

CHAPTER 1

MILK AND DAIRY PRODUCTS

What better subject for the first chapter than the food with which we all begin our lives? Humans are mammals, a word that means "creatures of the breast," and the first food that any mammal tastes is milk. Milk is food for the beginning eater, a gulpable essence distilled by the mother from her own more variable and challenging diet. When our ancestors took up dairying, they adopted the cow, the ewe, and the goat as surrogate mothers. These creatures accomplish the miracle of turning meadow and straw into buckets of human nourishment. And their milk turned out to be an elemental fluid rich in possibility, just a step or two away from luxurious cream, fragrant golden butter, and a multitude of flavorful foods concocted by friendly microbes.

No wonder that milk captured the imaginations of many cultures. The ancient Indo-Europeans were cattle herders who moved out from the Caucasian steppes to settle vast areas of Eurasia around 3000 BCE; and milk and butter are prominent in the creation myths of their descendents, from India to Scandinavia. Peoples of the Mediterranean and Middle East relied on the oil of their olive tree rather than butter, but milk and cheese still figure in the Old Testament as symbols of abundance and creation.

The modern imagination holds a very different view of milk! Mass production turned it and its products from precious, marvelous resources into ordinary commodities, and medical science stigmatized them for their fat content. Fortunately a more balanced view of dietary fat is developing; and traditional versions of dairy foods survive. It's still possible to savor the remarkable foods that millennia of human ingenuity have teased from milk. A sip of milk itself or a scoop of ice cream can be a Proustian draft of youth's innocence and energy and possibility, while a morsel of fine cheese is a rich meditation on maturity, the fulfillment of possibility, the way of all flesh.

MAMMALS AND MILK

The Evolution of Milk

How and why did such a thing as milk ever come to be? It came along with warm-bloodedness, hair, and skin glands, all of which distinguish mammals from reptiles. Milk may have begun around 300 million years ago as a protective and nourishing skin secretion for hatchlings being incubated on their mother's skin, as is true for the platypus today. Once it evolved, milk contributed to the success of the mammalian family. It gives newborn animals the advantage of ideally formulated food from the mother even after birth, and therefore the opportunity to continue their physical development outside the womb. The human species has taken full advantage of this opportunity: we are completely helpless for months after birth, while our brains finish growing to a size that would be difficult to accommodate in the womb and birth canal. In this sense, milk helped make possible the evolution of our large brain, and so helped make us the unusual animals we are.

Milk and Butter: Primal Fluids

When the gods performed the sacrifice, with the first Man as the offering, spring was the melted butter, summer the fuel, autumn the offering. They anointed that Man, born at the beginning, as a sacrifice on the straw. . . . From that full sacrifice they gathered the grains of butter, and made it into the creatures of the air, the forest, and the village . . . cattle were born from it, and sheep and goats were born from it.

—The *Rg Veda,* Book 10, ca. 1200 BCE

. . . I am come down to deliver [my people] out of the hands of the Egyptians, and to bring them up out of that land unto a good land and a large, unto a land flowing with milk and honey. . . .

—God to Moses on Mount Horeb (Exodus 3:8)

Hast thou not poured me out as milk, and curdled me like cheese?

—Job to God (Job 10:10)

The Rise of the Ruminants

All mammals produce milk for their young, but only a closely related handful have been exploited by humans. Cattle, water buffalo, sheep, goats, camels, yaks: these suppliers of plenty were created by a scarcity of food. Around 30 million years ago, the earth's warm, moist climate became seasonally arid. This shift favored plants that could grow quickly and produce seeds to survive the dry period, and caused a great expansion of grasslands, which in the dry seasons became a sea of desiccated, fibrous stalks and leaves. So began the gradual decline of the horses and the expansion of the deer family, the *ruminants,* which evolved the ability to survive on dry grass. Cattle, sheep, goats, and their relatives are all ruminants.

The key to the rise of the ruminants is their highly specialized, multichamber stomach, which accounts for a fifth of their body weight and houses trillions of fiber-digesting microbes, most of them in the first chamber, or *rumen.* Their unique plumbing, together with the habit of regurgitating and rechewing partly digested food, allows ruminants to extract nourishment from high-fiber, poor-quality plant material. Ruminants produce milk copiously on feed that is otherwise useless to humans and that can be stockpiled as straw or silage. Without them there would be no dairying.

Dairy Animals of the World

Only a small handful of animal species contributes significantly to the world's milk supply.

The Cow, European and Indian The immediate ancestor of *Bos taurus,* the common dairy cow, was *Bos primigenius,* the long-horned wild aurochs. This massive animal, standing 6 ft/180 cm at the shoulder and with horns 6.5 in/17 cm in diameter, roamed Asia, Europe, and North Africa in the form of two overlapping races: a humpless European-African form, and a humped central Asian form, the zebu. The European race was domesticated in the Middle East around 8000 BCE, the heat- and parasite-tolerant zebu in south-central Asia around the same time, and an African variant of the European race in the Sahara, probably somewhat later.

In its principal homeland, central and south India, the zebu has been valued as much for its muscle power as its milk, and remains rangy and long-horned. The European dairy cow has been highly selected for milk production at least since 3000 BCE, when confinement to stalls in urban Mesopotamia and poor winter feed led to a reduction in body and horn size. To this day, the prized dairy breeds—Jerseys, Guernseys, Brown Swiss, Holsteins—are short-horned cattle that put their energy into making milk rather than muscle and bone. The modern zebu is not as copious a producer as the European breeds, but its milk is 25% richer in butterfat.

The Buffalo The water buffalo is relatively unfamiliar in the West but the most important bovine in tropical Asia. *Bubalus bubalis* was domesticated as a draft animal in Mesopotamia around 3000 BCE, then taken to the Indus civilizations of present-day Pakistan, and eventually through India and China. This tropical animal is sensitive to heat (it wallows in water to cool down), so it proved adaptable to milder climates. The Arabs brought buffalo to the Middle East around 700 CE, and in the Middle Ages they were introduced throughout Europe. The most notable vestige of that introduction is a population approaching 100,000 in the Campagna region south of Rome, which supplies the milk for true mozzarella cheese, *mozzarella di bufala.* Buffalo milk is much richer than cow's milk, so mozzarella and Indian milk dishes are very different when the traditional buffalo milk is replaced with cow's milk.

The Yak The third important dairy bovine is the yak, *Bos grunniens.* This long-haired, bushy-tailed cousin of the common cow is beautifully adapted to the thin, cold, dry air and sparse vegetation of the Tibetan plateau and mountains of central Asia. It was domesticated around the same time as lowland cattle. Yak milk is substantially richer in fat and protein than cow milk. Tibetans in particular make elaborate use of yak butter and various fermented products.

The Goat The goat and sheep belong to the "ovicaprid" branch of the ruminant family, smaller animals that are especially at home in mountainous country. The goat, *Capra hircus,* comes from a denizen of the mountains and semidesert regions of central Asia, and was probably the first animal after the dog to be domesticated, between 8000 and 9000 BCE in present-day Iran and Iraq. It is the hardiest of the Eurasian dairy animals, and will browse just about any sort of vegetation, including woody scrub. Its omnivorous nature, small size, and good yield of distinctively flavored milk—the highest of any dairy animal for its body weight—have made it a versatile milk and meat animal in marginal agricultural areas.

The Sheep The sheep, *Ovis aries,* was domesticated in the same region and period as its close cousin the goat, and came to be valued and bred for meat, milk, wool, and fat. Sheep were originally grazers on grassy foothills and are somewhat more fastidious than goats, but less so than cattle. Sheep's milk is as rich as the buffalo's in fat, and even richer in protein; it has long been valued in the Eastern Mediterranean for making yogurt and feta cheese, and elsewhere in Europe for such cheeses as Roquefort and pecorino.

The Camel The camel family is fairly far removed from both the bovids and ovicaprids, and may have developed the habit of rumination independently during its early evolution in North America. Camels are well adapted to arid climates, and were domesticated around 2500 BCE in central Asia, primarily as pack animals. Their milk, which is roughly comparable to cow's milk, is collected in many countries, and in northeast Africa is a staple food.

The Origins of Dairying

When and why did humans extend our biological heritage as milk drinkers to the cultural practice of drinking the milk of *other* animals? Archaeological evidence suggests that sheep and goats were domesticated in the grasslands and open forest of present-day Iran and Iraq between 8000 and 9000 BCE, a thousand years before the far larger, fiercer cattle. At first these animals would have been kept for meat and skins, but the discovery of milking was a significant advance. Dairy animals could produce the nutritional equivalent of a slaughtered meat animal or more each year for several years, and in manageable daily increments. Dairying is the most efficient means of obtaining nourishment from uncultivated land, and may have been especially important as farming communities spread outward from Southwest Asia.

Small ruminants and then cattle were almost surely first milked into containers fashioned from skins or animal stomachs. The earliest hard evidence of dairying to date consists of clay sieves, which have been found in the settlements of the earliest northern European farmers, from around 5000 BCE. Rock drawings of milking scenes were made a thousand years later in the Sahara, and what appear to be the remains of cheese have been found in Egyptian tombs of 2300 BCE.

Diverse Traditions

Early shepherds would have discovered the major transformations of milk in their first containers. When milk is left to stand, fat-enriched cream naturally forms at the top, and if agitated, the cream becomes butter. The remaining milk naturally turns acid and curdles into thick yogurt, which draining separates into solid curd and liquid whey. Salting the fresh curd produces a simple, long-keeping cheese. As dairyers became more adept and harvested greater quantities of milk, they found new ways to concentrate and preserve its nourishment, and developed distinctive dairy products in the different climatic regions of the Old World.

In arid southwest Asia, goat and sheep milk was lightly fermented into yogurt that could be kept for several days, sun-dried, or kept under oil; or curdled into cheese that could be eaten fresh or preserved by drying or brining. Lacking the settled life that makes it possible to brew beer from grain or wine from grapes, the nomadic Tartars even fermented mare's milk into lightly alcoholic *koumiss,* which Marco Polo described as having "the qualities and flavor of white wine." In the high country of Mongolia and Tibet, cow, camel, and yak milk was churned to butter for use as a high-energy staple food.

In semitropical India, most zebu and buffalo milk was allowed to sour overnight into a yogurt, then churned to yield buttermilk and butter, which when clarified into *ghee* (p. 37) would keep for months. Some milk was repeatedly boiled to keep it sweet, and then preserved not with salt, but by the combination of sugar and long, dehydrating cooking (see box, p. 26).

The Mediterranean world of Greece and Rome used economical olive oil rather than butter, but esteemed cheese. The Roman Pliny praised cheeses from distant provinces that are now parts of France and Switzerland. And indeed cheese making reached its zenith in continental and northern Europe, thanks to abundant pastureland ideal for cattle, and a temperate climate that allowed long, gradual fermentations.

The one major region of the Old World not to embrace dairying was China, perhaps because Chinese agriculture began where the natural vegetation runs to often toxic relatives of wormwood and epazote rather than ruminant-friendly grasses. Even so, frequent contact with central Asian nomads introduced a variety of dairy products to China, whose elite long enjoyed yogurt, koumiss, butter, acid-set curds, and, around 1300 and thanks to the Mongols, even milk in their tea!

Dairying was unknown in the New World. On his second voyage in 1493, Columbus brought sheep, goats, and the first of the Spanish longhorn cattle that would proliferate in Mexico and Texas.

Milk in Europe and America: From Farmhouse to Factory

Preindustrial Europe In Europe, dairying took hold on land that supported abundant pasturage but was less suited to the cultivation of wheat and other grains: wet Dutch lowlands, the heavy soils of western France and its high, rocky central massif, the cool, moist British Isles and Scandinavia, alpine valleys in Switzerland and Austria. With time, livestock were selected for the climate and needs of different regions, and diversified into hundreds of distinctive local breeds (the rugged Brown Swiss cow for cheesemaking in the mountains, the diminutive Jersey and Guernsey for making butter in the Channel Islands). Summer milk was preserved in equally distinctive local cheeses. By medieval times, fame had come to French Roquefort and Brie, Swiss Appenzeller, and Italian Parmesan. In the Renaissance, the Low Countries were renowned for their butter and exported their productive Friesian cattle throughout Europe.

Until industrial times, dairying was done on the farm, and in many countries mainly by women, who milked the animals in early morning and after noon and then worked for hours to churn butter or make cheese. Country people could enjoy good fresh milk, but in the cities, with confined cattle fed inadequately on spent brewers' grain, most people saw only watered-down, adulterated, contaminated milk hauled in open containers through the streets. Tainted milk was a major cause of child mortality in early Victorian times.

Industrial and Scientific Innovations Beginning around 1830, industrialization transformed European and American dairying. The railroads made it possible to get fresh country milk to the cities, where rising urban populations and incomes fueled demand, and new laws regulated milk quality. Steam-powered farm machinery meant that cattle could be bred and raised for milk production alone, not for a compromise between milk and hauling, so milk production boomed, and more than ever was drunk fresh. With the invention of machines for milking, cream separation, and churning, dairying gradually moved out the hands of milkmaids and off the farms, which increasingly supplied milk to factories for mass production of cream, butter, and cheese.

From the end of the 19th century, chemical and biological innovations have helped make dairy products at once more hygienic, more predictable, and more uniform. The great French chemist Louis Pasteur inspired two fundamental changes in dairy practice: *pasteurization,* the pathogen-killing heat treatment that bears his name; and the use of standard, purified microbial cultures to make cheeses and other fermented foods. Most traditional cattle breeds have been abandoned in favor of high-yielding black-and-white Friesian (Holstein) cows, which now account for 90% of all American dairy cattle and 85% of British. The cows are farmed in ever larger herds and fed an optimized diet that seldom includes fresh pasturage, so most modern milk lacks the color, flavor, and seasonal variation of preindustrial milk.

Dairy Products Today Today dairying is split into several big businesses with nothing of the dairymaid left about them. Butter and cheese, once prized, delicate concentrates of milk's goodness, have become inexpensive, mass-produced, uninspiring commodities piling up in government warehouses. Manufacturers now remove much of what makes milk, cheese, ice cream, and butter distinctive and pleasurable: they remove milk fat, which suddenly became undesirable when medical scientists found that saturated milk fat tends to raise blood cholesterol levels and can contribute to heart disease. Happily the last few years have brought a correction in the view of saturated fat, a reaction to the juggernaut of mass production, and a resurgent interest in full-flavored dairy products crafted on a small scale from traditional breeds that graze seasonally on green pastures.

MILK AND HEALTH

Milk has long been synonymous with wholesome, fundamental nutrition, and for good reason: unlike most of our foods, it is actually designed to be a food. As the sole sustaining food of the calf at the beginning of its life, it's a rich source of many essen-

Food Words: *Milk* and *Dairy*

In their roots, both *milk* and *dairy* recall the physical effort it once took to obtain milk and transform it by hand. *Milk* comes from an Indo-European root that meant both "milk" and "to rub off," the connection perhaps being the stroking necessary to squeeze milk from the teat. In medieval times, *dairy* was originally *dey-ery,* meaning the room in which the *dey,* or woman servant, made milk into butter and cheese. *Dey* in turn came from a root meaning "to knead bread" (*lady* shares this root)—perhaps a reflection not only of the servant's several duties, but also of the kneading required to squeeze buttermilk out of butter (p. 34) and sometimes the whey out of cheese.

tial body-building nutrients, particularly protein, sugars and fat, vitamin A, the B vitamins, and calcium.

Over the last few decades, however, the idealized portrait of milk has become more shaded. We've learned that the balance of nutrients in cow's milk doesn't meet the needs of human infants, that most adult humans on the planet can't digest the milk sugar called lactose, that the best route to calcium balance may not be massive milk intake. These complications help remind us that milk was designed to be a food for the young and rapidly growing calf, not for the young or mature human.

MILK NUTRIENTS

Nearly all milks contain the same battery of nutrients, the relative proportions of which vary greatly from species to species. Generally, animals that grow rapidly are fed with milk high in protein and minerals. A calf doubles its weight at birth in 50 days, a human infant in 100; sure enough, cow's milk contains more than double the protein and minerals of mother's milk. Of the major nutrients, ruminant milk is seriously lacking only in iron and in vitamin C. Thanks to the rumen microbes, which convert the unsaturated fatty acids of grass and grain into saturated fatty acids, the milk fat of ruminant animals is the most highly saturated of our common foods. Only coconut oil beats it. Saturated fat does raise blood cholesterol levels, and high blood cholesterol is associated with an increased risk of heart disease; but the other foods in a balanced diet can compensate for this disadvantage (p. 253).

The box below shows the nutrient contents of both familiar and unfamiliar milks. These figures are only a rough guide, as the breakdown by breed indicates; there's also much variation from animal to animal, and in a given animal's milk as its lactation period progresses.

The Compositions of Various Milks

The figures in the following table are the percent of the milk's weight accounted for by its major components.

Milk	Fat	Protein	Lactose	Minerals	Water
Human	4.0	1.1	6.8	0.2	88
Cow	3.7	3.4	4.8	0.7	87
Holstein/Friesian	3.6	3.4	4.9	0.7	87
Brown Swiss	4.0	3.6	4.7	0.7	87
Jersey	5.2	3.9	4.9	0.7	85
Zebu	4.7	3.3	4.9	0.7	86
Buffalo	6.9	3.8	5.1	0.8	83
Yak	6.5	5.8	4.6	0.8	82
Goat	4.0	3.4	4.5	0.8	88
Sheep	7.5	6.0	4.8	1.0	80
Camel	2.9	3.9	5.4	0.8	87
Reindeer	17	11	2.8	1.5	68
Horse	1.2	2.0	6.3	0.3	90
Fin whale	42	12	1.3	1.4	43

Milk in Infancy and Childhood: Nutrition and Allergies

In the middle of the 20th century, when nutrition was thought to be a simple matter of protein, calories, vitamins, and minerals, cow's milk seemed a good substitute for mother's milk: more than half of all six-month-olds in the United States drank it. Now that figure is down to less than 10%. Physicians now recommend that plain cow's milk not be fed to children younger than one year. One reason is that it provides too much protein, and not enough iron and highly unsaturated fats, for the human infant's needs. (Carefully prepared formula milks are better approximations of breast milk.) Another disadvantage to the early use of cow's milk is that it can trigger an allergy. The infant's digestive system is not fully formed, and can allow some food protein and protein fragments to pass directly into the blood. These foreign molecules then provoke a defensive response from the immune system, and that response is strengthened each time the infant eats. Somewhere between 1% and 10% of American infants suffer from an allergy to the abundant protein in cow's milk, whose symptoms may range from mild discomfort to intestinal damage to shock. Most children eventually grow out of milk allergy.

Milk after Infancy: Dealing with Lactose

In the animal world, humans are exceptional for consuming milk of any kind after they have started eating solid food. And people who drink milk after infancy are the exception within the human species. The obstacle is the milk sugar lactose, which can't be absorbed and used by the body as is: it must first be broken down into its component sugars by digestive enzymes in the small intestine. The lactose-digesting enzyme, *lactase,* reaches its maximum levels in the human intestinal lining shortly after birth, and then slowly declines, with a steady minimum level commencing at between two and five years of age and continuing through adulthood.

The logic of this trend is obvious: it's a waste of its resources for the body to produce an enzyme when it's no longer needed; and once most mammals are weaned, they never encounter lactose in their food again. But if an adult without much lactase activity does ingest a substantial amount of milk, then the lactose passes through the small intestine and reaches the large intestine, where bacteria metabolize it, and in the process produce carbon dioxide, hydrogen, and methane: all discomforting gases. Sugar also draws water from the intestinal walls, and this causes a bloated feeling or diarrhea.

Low lactase activity and its symptoms are called *lactose intolerance.* It turns out that adult lactose intolerance is the rule rather than the exception: lactose-tolerant adults are a distinct minority on the planet. Several thousand years ago, peoples in northern Europe and a few other regions underwent a genetic change that allowed them to produce lactase throughout life, probably because milk was an exceptionally important resource in colder climates. About 98% of Scandinavians are lactose-tolerant, 90% of French and Germans, but only 40% of southern Europeans and North Africans, and 30% of African Americans.

Coping with Lactose Intolerance Fortunately, lactose intolerance is not the same as milk intolerance. Lactase-less adults can consume about a cup/250 ml of milk per day without severe symptoms, and even more of other dairy products. Cheese contains little or no lactose (most of it is drawn off in the whey, and what little remains in the curd is fermented by bacteria and molds). The bacteria in yogurt generate lactose-digesting enzymes that remain active in the human small intestine and work for us there. And lactose-intolerant milk fans can now buy the lactose-digesting enzyme itself in liquid form (it's manufactured from a fungus, *Aspergillus*), and add a few drops to any dairy product just before they consume it.

New Questions about Milk

Milk has been especially valued for two nutritional characteristics: its richness in calcium, and both the quantity and quality of its protein. Recent research has raised some fascinating questions about each of these.

Perplexity about Calcium and Osteoporosis Our bones are constructed from two primary materials: proteins, which form a kind of scaffolding, and calcium phosphate, which acts as a hard, mineralized, strengthening filler. Bone tissue is constantly being deconstructed and rebuilt throughout our adult lives, so healthy bones require adequate protein and calcium supplies from our diet. Many women in industrialized countries lose so much bone mass after menopause that they're at high risk for serious fractures. Dietary calcium clearly helps prevent this potentially dangerous loss, or *osteoporosis*. Milk and dairy products are the major source of calcium in dairying countries, and U.S. government panels have recommended that adults consume the equivalent of a quart (liter) of milk daily to prevent osteoporosis.

This recommendation represents an extraordinary concentration of a single food, and an unnatural one—remember that the ability to drink milk in adulthood, and the habit of doing so, is an aberration limited to people of northern European descent. A quart of milk supplies two-thirds of a day's recommended protein, and would displace from the diet other foods—vegetables, fruits, grains, meats, and fish—that provide their own important nutritional benefits. And there clearly must be other ways of maintaining healthy bones. Other countries, including China and

The Many Influences on Bone Health

Good bone health results from a proper balance between the two ongoing processes of bone deconstruction and reconstruction. These processes depend not only on calcium levels in the body, but also on physical activity that stimulates bone-building; hormones and other controlling signals; trace nutrients (including vitamin C, magnesium, potassium, and zinc); and other as yet unidentified substances. There appear to be factors in tea and in onions and parsley that slow bone deconstruction significantly. Vitamin D is essential for the efficient absorption of calcium from our foods, and also influences bone building. It's added to milk, and other sources include eggs, fish and shellfish, and our own skin, where ultraviolet light from the sun activates a precursor molecule.

The amount of calcium we have available for bone building is importantly affected by how much we excrete in our urine. The more we lose, the more we have to take in from our foods. Various aspects of modern eating increase calcium excretion and so boost our calcium requirement. A high intake of salt is one, and another is a high intake of animal protein, the metabolism of whose sulfur-containing amino acids acidifies our urine, and pulls neutralizing calcium salts from bone.

The best insurance against osteoporosis appears to be frequent exercise of the bones that we want to keep strong, and a well-rounded diet that is rich in vitamins and minerals, moderate in salt and meat, and includes a variety of calcium-containing foods. Milk is certainly a valuable one, but so are dried beans, nuts, corn tortillas and tofu (both processed with calcium salts), and several greens—kale, collards, mustard greens.

Japan, suffer much lower fracture rates than the United States and milk-loving Scandinavia, despite the fact that their people drink little or no milk. So it seems prudent to investigate the many other factors that influence bone strength, especially those that slow the deconstruction process (see box, p. 15). The best answer is likely to be not a single large white bullet, but the familiar balanced diet and regular exercise.

Milk Proteins Become Something More We used to think that one of the major proteins in milk, casein (p. 19), was mainly a nutritional reservoir of amino acids with which the infant builds its own body. But this protein now appears to be a complex, subtle orchestrator of the infant's metabolism. When it's digested, its long amino-acid chains are first broken down into smaller fragments, or peptides. It turns out that many hormones and drugs are also peptides, and a number of casein peptides do affect the body in hormone-like ways. One reduces breathing and heart rates, another triggers insulin release into the blood, and a third stimulates the scavenging activity of white blood cells. Do the peptides from cow's milk affect the metabolism of human children or adults in significant ways? We don't yet know.

MILK BIOLOGY AND CHEMISTRY

How the Cow Makes Milk

Milk is food for the newborn, and so dairy animals must give birth before they will produce significant quantities of milk. The mammary glands are activated by changes in the balance of hormones toward the end of pregnancy, and are stimulated to continue secreting milk by regular removal of milk from the gland. The optimum sequence for milk production is to breed the cow again 90 days after it calves, milk it for 10 months, and let it go dry for the two months before the next calving. In intensive operations, cows aren't allowed to waste energy on grazing in variable pastures; they're given hay or silage (whole corn or other plants, partly dried and then preserved by fermentation in airtight silos) in confined lots, and are milked only during their two or three most productive years. The combination of breeding and optimal feed formulation has led to per-animal yields of a hundred pounds or 15 gallons/58 liters per day, though the American average is about half that. Dairy breeds of sheep and goats give about one gallon per day.

The first fluid secreted by the mammary gland is colostrum, a creamy, yellow solution of concentrated fat, vitamins, and proteins, especially immunoglobulins and antibodies. After a few days, when the colostrum flow has ceased and the milk is saleable, the calf is put on a diet of reconstituted and soy milks, and the cow is milked two or three times daily to keep the secretory cells working at full capacity.

The Milk Factory The mammary gland is an astonishing biological factory, with many different cells and structures working together to create, store, and dispense milk. Some components of milk come directly from the cow's blood and collect in the udder. The principal nutrients, however—fats, sugar, and proteins—are assembled by the gland's secretory cells, and then released into the udder.

A Living Fluid Milk's blank appearance belies its tremendous complexity and vitality. It's alive in the sense that, fresh from the udder, it contains living white blood cells, some mammary-gland cells, and various bacteria; and it teems with active enzymes, some floating free, some embedded in the membranes of the fat globules. Pasteurization (p. 22) greatly reduces this vitality; in fact residual enzyme activity is taken as a sign that the heat treatment was insufficient. Pasteurized milk contains very few living cells or active enzyme molecules, so it is more predictably free of bacteria that could cause food poisoning, and more sta-

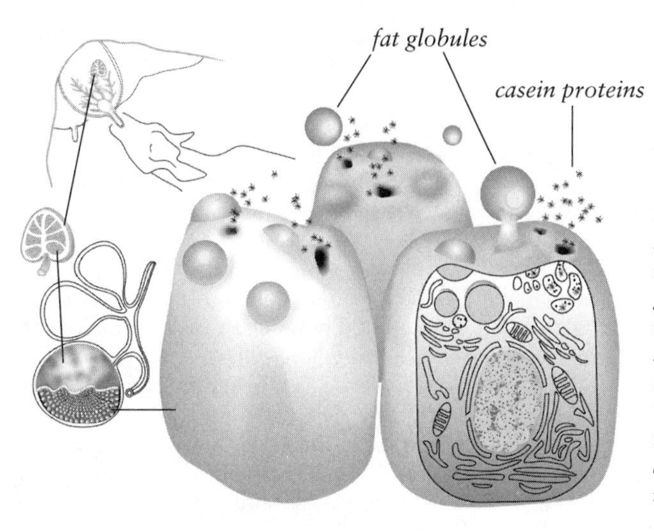

The making of milk. Cells in the cow's mammary gland synthesize the components of milk, including proteins and globules of milk fat, and release them into many thousands of small compartments that drain toward the teat. The fat globules pass through the cells' outer membranes, and carry parts of the cell membrane on their surface.

ble; it develops off-flavors more slowly than raw milk. But the dynamism of raw milk is prized in traditional cheese making, where it contributes to the ripening process and deepens flavor.

Milk owes its milky opalescence to microscopic fat globules and protein bundles, which are just large enough to deflect light rays as they pass through the liquid. Dissolved salts and milk sugar, vitamins, other proteins, and traces of many other compounds also swim in the water that accounts for the bulk of the fluid. The sugar, fat, and proteins are by far the most important components, and we'll look at them in detail in a moment.

First a few words about the remaining components. Milk is slightly acidic, with a pH between 6.5 and 6.7, and both acidity and salt concentrations strongly affect the behavior of the proteins, as we'll see. The fat globules carry colorless vitamin A and its yellow-orange precursors the carotenes, which are found in green feed and give milk and undyed butter whatever color they have. Breeds differ in the amount of carotene they convert into vitamin A; Guernsey and Jersey cows convert little and give especially golden milk, while at the other extreme sheep, goats, and water buffalo process nearly all of their carotene, so their milk and butter are nutritious but white. Riboflavin, which has a greenish color, can sometimes be seen in skim milk or in the watery translucent whey that drains from the curdled proteins of yogurt.

Milk Sugar: Lactose

The only carbohydrate found in any quantity in milk is also peculiar to milk (and a handful of plants), and so was named *lactose,* or "milk sugar." (*Lac-* is a prefix based on the Greek word for "milk"; we'll encounter it again in the names of milk proteins, acids, and bacteria.) Lactose is a composite of the two simple sugars glucose and galactose, which are joined together in the secretory cell of the mammary gland, and nowhere else in the animal body. It provides nearly half of the calories in human milk, and 40% in cow's milk, and gives milk its sweet taste.

The uniqueness of lactose has two major practical consequences. First, we need a special enzyme to digest lactose; and many adults lack that enzyme and have to be careful about what dairy products they consume (p. 14). Second, most microbes take some time to make their own lactose-digesting enzyme before they can grow well in milk, but one group has enzymes at the ready and can get a head start on all the others. The bacteria known as *Lactobacilli* and *Lactococci* not only grow on lactose immediately, they also convert it into lactic acid ("milk acid"). They thus acidify the milk, and in so doing, make it less habitable by other microbes, including many that would make the milk unpalatable or cause disease. Lactose and the lactic-acid bacteria therefore turn milk sour, but help prevent it from spoiling, or becoming undrinkable.

Lactose is one-fifth as sweet as table sugar, and only one-tenth as soluble in water (200 vs. 2,000 gm/l), so lactose crystals readily form in such products as condensed milk and ice cream and can give them a sandy texture.

Milk Fat

Milk fat accounts for much of the body, nutritional value, and economic value of milk. The milk-fat globules carry the fat-soluble vitamins (A, D, E, K), and about half the calories of whole milk. The higher the fat content of milk, the more cream or butter can be made from it, and so the higher the price it will bring. Most cows secrete more fat in winter, due mainly to concentrated winter feed and the approaching end of their lactation period. Certain breeds, notably Guernseys and Jerseys from the Channel Islands between Britain and France, produce especially rich milk and large fat globules. Sheep and buffalo milks contain up to twice the butterfat of whole cow's milk (p. 13).

The way the fat is packaged into globules accounts for much of milk's behavior in the kitchen. The membrane that surrounds each fat globule is made up of phospholipids (fatty acid emulsifiers, p. 802) and proteins, and plays two major roles. It separates the droplets of fat from each other and prevents them from pooling together into one large mass; and it protects the fat molecules from fat-digesting enzymes in the milk that would otherwise attack them and break them down into rancid-smelling and bitter fatty acids.

Creaming When milk fresh from the udder is allowed to stand and cool for some hours, many of its fat globules rise and form a fat-rich layer at the top of the container. This phenomenon is called *creaming*, and for millennia it was the natural first step toward obtaining fat-enriched cream and butter from milk. In the 19th century, centrifuges were developed to concentrate the fat globules more rapidly and thoroughly, and homogenization was invented to prevent whole milk from separating in this way (p. 23). The globules rise because their fat is lighter than water, but they rise much faster than their buoyancy alone can account for. It turns out that a number of minor milk proteins attach themselves loosely to the fat globules and knit together clusters of about a million globules that have a stronger lift than single globules do. Heat denatures these proteins and prevents the globule clustering, so that the fat globules in unhomogenized but pasteurized milk rise more slowly into a shallower, less distinct layer. Because of their small globules and low clustering activity, the milks of goats, sheep, and water buffalo are very slow to separate.

Milk Fat Globules Tolerate Heat . . . Interactions between fat globules and milk proteins are also responsible for the remarkable tolerance of milk and cream to heat. Milk and cream can be boiled and reduced for hours, until they're nearly dry, without breaching the globule membranes enough to release their fat. The globule membranes are robust to begin with, and it turns out that heating unfolds many of the milk proteins and makes them more prone to stick to the globule surface and to each other—so the globule armor actually gets progressively thicker as heating proceeds. Without this stability to heat, it would be impossible to make many cream-enriched sauces and reduced-milk sauces and sweets.

. . . But Are Sensitive to Cold Freezing is a different story. It is fatal to the fat globule membrane. Cold milk fat and freezing water both form large, solid, jagged crystals that pierce, crush, and rend the thin veil of phospholipids and proteins around the globule, just a few molecules thick. If you freeze milk or cream and then thaw it, much of the membrane material ends up floating free in the liquid, and many of the fat globules get stuck to each other in grains of butter. Make the mistake of heating thawed milk or cream, and the butter grains melt into puddles of oil.

Milk Proteins: Coagulation by Acid and Enzymes

Two Protein Classes: Curd and Whey

There are dozens of different proteins floating around in milk. When it comes to cooking behavior, fortunately, we can reduce the protein population to two basic groups: Little Miss Muffet's curds and whey. The two groups are distinguished by their reaction to acids. The handful of curd proteins, the *caseins,* clump together in acid conditions and form a solid mass, or *coagulate,* while all the rest, the whey proteins, remain suspended in the liquid. It's the clumping nature of the caseins that makes possible most thickened milk products, from yogurt to cheese. The whey proteins play a more minor role; they influence the texture of casein curds, and stabilize the milk foams on specialty coffees. The caseins usually outweigh the whey proteins, as they do in cow's milk by 4 to 1.

Both caseins and whey proteins are unusual among food proteins in being largely tolerant of heat. Where cooking coagulates the proteins in eggs and meat into solid masses, it does not coagulate the proteins in milk and cream—unless the milk or cream has become acidic. Fresh milk and cream can be boiled down to a fraction of their volume without curdling.

The Caseins The casein family includes four different kinds of proteins that gather together into microscopic family units called *micelles.* Each casein micelle contains a few thousand individual protein molecules, and measures about a ten-thousandth of a millimeter across, about one-fiftieth the size of a fat globule. Around a tenth of the volume of milk is taken up by casein micelles. Much of the calcium in milk is in the micelles, where it acts as a kind of glue holding the protein molecules together. One portion of calcium binds individual protein molecules together into small clusters of 15 to 25. Another portion then helps pull several hundred of the clusters together to form the micelle (which is also held together by the water-avoiding hydrophobic portions of the proteins bonding to each other).

Keeping Micelles Separate . . . One member of the casein family is especially influential in these gatherings. That is kappa-casein, which caps the micelles once they reach a

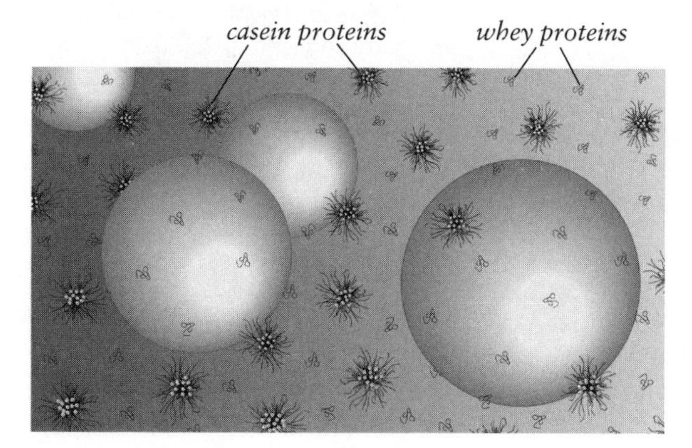

A close-up view of milk. Fat globules are suspended in a fluid made up of water, individual molecules of whey protein, bundles of casein protein molecules, and dissolved sugars and minerals.

certain size, prevents them from growing larger, and keeps them dispersed and separate. One end of the capping-casein molecule extends from the micelle out into the surrounding liquid, and forms a "hairy layer" with a negative electrical charge that repels other micelles.

. . . And Knitting Them Together in Curds The intricate structure of casein micelles can be disturbed in several ways that cause the micelles to flock together and the milk to curdle. One way is souring. Milk's normal pH is about 6.5, or just slightly acidic. If it gets acid enough to approach pH 5.5, the capping-casein's negative charge is neutralized, the micelles no longer repel each other, and they therefore gather in loose clusters. At the same acidity, the calcium glue that holds the micelles together dissolves, the micelles begin to fall apart, and their individual proteins scatter. Beginning around pH 4.7, the scattered casein proteins lose their negative charge, bond to each other again and form a continuous, fine network: and the milk solidifies, or curdles. This is what happens when milk gets old and sour, or when it's intentionally cultured with acid-producing bacteria to make yogurt or sour cream.

Another way to cause the caseins to curdle is the basis of cheese making. Chymosin, a digestive enzyme from the stomach of a milk-fed calf, is exquisitely designed to give the casein micelles a haircut (p. 57). It clips off just the part of the capping-casein that extends into the surrounding liquid and shields the micelles from each other. Shorn of their hairy layer, the micelles all clump together—without the milk being noticeably sour.

The Whey Proteins Subtract the four caseins from the milk proteins, and the remainder, numbering in the dozens, are the whey proteins. Where the caseins are mainly nutritive, supplying amino acids and calcium for the calf, the whey proteins include defensive proteins, molecules that bind to and transport other nutrients, and enzymes. The most abundant one by far is lactoglobulin, whose biological function remains a mystery. It's a highly structured protein that is readily denatured by cooking. It unfolds at 172°F/78°C, when its sulfur atoms are exposed to the surrounding liquid and react with hydrogen ions to form hydrogen sulfide gas, whose powerful aroma contributes to the characteristic flavor of cooked milk (and many other animal foods).

In boiling milk, unfolded lactoglobulin binds not to itself but to the capping-casein

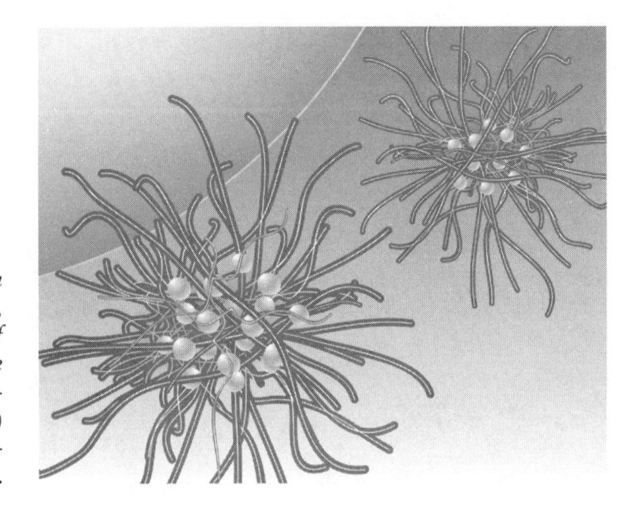

A model of the milk protein casein, which occurs in micelles, or small bundles a fraction of the size of a fat globule. A single micelle consists of many individual protein molecules (lines) *held together by particles of calcium phosphate* (small spheres).

on the casein micelles, which remain separate; so denatured lactoglobulin doesn't coagulate. When denatured in acid conditions with relatively little casein around, as in cheese whey, lactoglobulin molecules do bind to each other and coagulate into little clots, which can be made into whey cheeses like true ricotta. Heat-denatured whey proteins are better than their native forms at stabilizing air bubbles in milk foams and ice crystals in ice creams; this is why milks and creams are usually cooked for these preparations (pp. 26, 43).

Milk Flavor

The flavor of fresh milk is balanced and subtle. It's distinctly sweet from the lactose, slightly salty from its complement of minerals, and very slightly acid. Its mild, pleasant aroma is due in large measure to short-chain fatty acids (including butyric and capric acids), which help keep highly saturated milk fat fluid at body temperature, and which are small enough that they can evaporate into the air and reach our nose. Normally, free fatty acids give an undesirable, soapy flavor to foods. But in sparing quantities, the 4- to 12-carbon rumen fatty acids, branched versions of these, and acid-alcohol combinations called esters, provide milk with its fundamental blend of animal and fruity notes. The distinctive smells of goat and sheep milks are due to two particular branched 8-carbon fatty acids (4-ethyl-octanoic, 4-methyl-octanoic) that are absent in cow's milk. Buffalo milk, from which traditional mozzarella cheese is made, has a characteristic blend of modified fatty acids reminiscent of mushrooms and freshly cut grass, together with a barnyardy nitrogen compound (indole).

The basic flavor of fresh milk is affected by the animals' feed. Dry hay and silage are relatively poor in fat and protein and produce a less complicated, mildly cheesy aroma, while lush pasturage provides raw material for sweet, raspberry-like notes (derivatives of unsaturated long-chain fatty acids), as well as barnyardy indoles.

Flavors from Cooking Low-temperature pasteurization (p. 22) slightly modifies milk flavor by driving off some of the more delicate aromas, but stabilizes it by inactivating enzymes and bacteria, and adds slightly sulfury and green-leaf notes (dimethyl sulfide, hexanal). High-temperature pasteurization or brief cooking—heating milk above 170°F/76°C—generates traces of many flavorful substances, including those characteristic of vanilla, almonds, and cultured butter, as well as eggy hydrogen sulfide. Prolonged boiling encourages browning or Maillard reactions between lactose and milk proteins, and generates molecules that combine to give the flavor of butterscotch.

The Development of Off-Flavors The flavor of good fresh milk can deteriorate in several different ways. Simple contact with oxygen or exposure to strong light will cause the oxidation of phospholipids in the globule membrane and a chain of reactions that slowly generate stale cardboard, metallic, fishy, paint-like aromas. If milk is kept long enough to sour, it also typically develops fruity, vinegary, malty, and more unpleasant notes.

Exposure to sunlight or fluorescent lights also generates a distinctive cabbage-like, burnt odor, which appears to result from a reaction between the vitamin riboflavin and the sulfur-containing amino acid methionine. Clear glass and plastic containers and supermarket lighting cause this problem; opaque cartons prevent it.

UNFERMENTED DAIRY PRODUCTS

Fresh milk, cream, and butter may not be as prominent in European and American cooking as they once were, but they are still essential ingredients. Milk has bubbled up to new prominence atop the coffee craze of the 1980s and '90s.

Milks

Milk has become the most standardized of our basic foods. Once upon a time, people lucky enough to live near a farm could taste the pasture and the seasons in milk fresh from the cow. City life, mass production, and stricter notions of hygiene have now put that experience out of reach. Today nearly all of our milk comes from cows of one breed, the black-and-white Holstein, kept in sheds and fed year-round on a uniform diet. Large dairies pool the milk of hundreds, even thousands of cows, then pasteurize it to eliminate microbes and homogenize it to prevent the fat from separating. The result is processed milk of no particular animal or farm or season, and therefore of no particular character. Some small dairies persist in milking other breeds, allowing their herds out to pasture, pasteurizing mildly, and not homogenizing. Their milk can have a more distinctive flavor, a rare reminder of what milk used to taste like.

Raw Milk Careful milking of healthy cows yields sound raw milk, which has its own fresh taste and physical behavior. But if it's contaminated by a diseased cow or careless handling—the udder hangs right next to the tail—this nutritious fluid soon teems with potentially dangerous microbes. The importance of strict hygiene in the dairy has been understood at least since the Middle Ages, but life far from the farms made contamination and even adulteration all too common in cities of the 18th and 19th centuries, where many children were killed by tuberculosis, undulant fever, and simple food poisoning contracted from tainted milk. In the 1820s, long before anyone knew about microbes, some books on domestic economy advocated boiling all milk before use. Early in the 20th century, national and local governments began to regulate the dairy industry and require that it heat milk to kill disease microbes.

Today very few U.S. dairies sell raw milk. They must be certified by the state and inspected frequently, and the milk carries a warning label. Raw milk is also rare in Europe.

Pasteurization and UHT Treatments In the 1860s, the French chemist Louis Pasteur studied the spoilage of wine and beer and developed a moderate heat treatment that preserved them while minimizing changes in their flavor. It took several decades for pasteurization to catch on in the dairy. Nowadays, in industrial-scale production, it's a practical necessity. Collecting and pooling milk from many different farms increases the risk that a given batch will be contaminated; and the plumbing and machinery required for the various stages of processing afford many more opportunities for contamination. Pasteurization extends the shelf life of milk by killing pathogenic and spoilage microbes and by inactivating milk enzymes, especially the fat splitters, whose slow but steady activity can make it unpalatable. Pasteurized milk stored below 40°F/5°C should remain drinkable for 10 to 18 days.

There are three basic methods for pasteurizing milk. The simplest is *batch* pasteurization, in which a fixed volume of milk, perhaps a few hundred gallons, is slowly agitated in a heated vat at a minimum of 145°F/62°C for 30 to 35 minutes. Industrial-scale operations use the *high-temperature, short-time* (HTST) method, in which milk is pumped continuously through a heat exchanger and held at a minimum of 162°F/72°C for 15 seconds. The batch process has a relatively mild effect on flavor, while the HTST method is hot enough to denature around 10% of the whey proteins and generate the strongly aromatic gas hydrogen sulfide (p. 87). Though this "cooked" flavor was considered a defect in the early days, U.S. consumers have come to expect it, and dairies now often intensify it by pasteurizing at well above the minimum temperature; 171°F/77°C is commonly used.

The third method of pasteurizing milk is the *ultra-high temperature* (UHT) method, which involves heating milk at 265–300°F/

130–150°C either instantaneously or for 1 to 3 seconds, and produces milk that, if packaged under strictly sterile conditions, can be stored for months without refrigeration. The longer UHT treatment imparts a cooked flavor and slightly brown color to milk; cream contains less lactose and protein, so its color and flavor are less affected.

Sterilized milk has been heated at 230–250°F/110–121°C for 8 to 30 minutes; it is even darker and stronger in flavor, and keeps indefinitely at room temperature.

Homogenization Left to itself, fresh whole milk naturally separates into two phases: fat globules clump together and rise to form the cream layer, leaving a fat-depleted phase below (p. 18). The treatment called *homogenization* was developed in France around 1900 to prevent creaming and keep the milk fat evenly—homogeneously—dispersed. It involves pumping hot milk at high pressure through very small nozzles, where the turbulence tears the fat globules apart into smaller ones; their average diameter falls from 4 micrometers to about 1. The sudden increase in globule numbers causes a proportional increase in their surface area, which the original globule membranes are insufficient to cover. The naked fat surface attracts casein particles, which stick and create an artificial coat (nearly a third of the milk's casein ends up on the globules). The casein particles both weigh the fat globules down and interfere with their usual clumping: and so the fat remains evenly dispersed in the milk. Milk is always pasteurized just before or simultaneously with homogenization to prevent its enzymes from attacking the momentarily unprotected fat globules and producing rancid flavors.

Homogenization affects milk's flavor and appearance. Though it makes milk taste blander—probably because flavor molecules get stuck to the new fat-globule surfaces—it also makes it more resistant to developing most off-flavors. Homogenized milk feels creamier in the mouth thanks to its increased population (around sixty-fold) of fat globules, and it's whiter, because the carotenoid pigments in the fat are scattered into smaller and more numerous particles.

Nutritional Alteration; Low-Fat Milks One nutritional alteration of milk is as old as dairying itself: skimming off the cream layer substantially reduces the fat content of the remaining milk. Today, low-fat milks are made more efficiently by centrifuging off some of the globules before homogenization. Whole milk is about 3.5% fat, low-fat milks usually 2% or 1%, and skim milks can range between 0.1 and 0.5%.

More recent is the practice of supplementing milk with various substances. Nearly all milks are fortified with the fat-soluble vitamins A and D. Low-fat milks have a thin body and appearance and are usually filled out with dried milk proteins, which can lend them a slightly stale flavor.

Powdered Milk in 13th-Century Asia

[The Tartar armies] make provisions also of milk, thickened or dried to the state of a hard paste, which they prepare in the following manner. They boil the milk, and skimming off the rich or creamy part as it rises to the top, put it into a separate vessel as butter; for so long as that remains in the milk, it will not become hard. The milk is then exposed to the sun until it dries. [When it is to be used] some is put into a bottle with as much water as is thought necessary. By their motion in riding, the contents are violently shaken, and a thin porridge is produced, upon which they make their dinner.

—Marco Polo, *Travels*

"Acidophilus" milk contains *Lactobacillus acidophilus,* a bacterium that metabolizes lactose to lactic acid and that can take up residence in the intestine (p. 47). More helpful to milk lovers who can't digest lactose is milk treated with the purified digestive enzyme lactase, which breaks lactose down into simple, absorbable sugars.

Storage Milk is a highly perishable food. Even Grade A pasteurized milk contains millions of bacteria in every glassful, and will spoil quickly unless refrigerated. Freezing is a bad idea because it disrupts milk fat globules and protein particles, which clump and separate when thawed.

Concentrated Milks A number of cultures have traditionally cooked milk down for long keeping and ease of transport. According to business legend, the American Gail Borden reinvented evaporated milk around 1853 after a rough transatlantic crossing that sickened the ship's cows. Borden added large amounts of sugar to keep his concentrated milk from spoiling. The idea of sterilizing unsweetened milk in the can came in 1884 from John Meyenberg, whose Swiss company merged with Nestlé around the turn of the century. Dried milk didn't appear until around the turn of the 20th century. Today, concentrated milk products are valued because they keep for months and supply milk's characteristic contribution to the texture and flavor of baked goods and confectionery, but without milk's water.

Condensed or *evaporated milk* is made by heating raw milk under reduced pressure (a partial vacuum), so that it boils between 110 and 140°F/43–60°C, until it has lost about half its water. The resulting creamy, mild-flavored liquid is homogenized, then canned and sterilized. The cooking and concentration of lactose and protein cause some browning, and this gives evaporated milk its characteristic tan color and note of caramel. Browning continues slowly during storage, and in old cans can produce a dark, acidic, tired-tasting fluid.

For *sweetened condensed milk,* the milk is first concentrated by evaporation, and then table sugar is added to give a total sugar concentration of about 55%. Microbes can't grow at this osmotic pressure, so sterilization is unnecessary. The high concentration of sugars causes the milk's lactose to crystallize, and this is controlled by seeding the milk with preformed lactose crystals to keep the crystals small and inconspicuous on the tongue (large, sandy lactose crystals are sometimes encountered as a quality defect). Sweetened condensed milk has a milder, less "cooked" flavor than evaporated milk, a lighter color, and the consistency of a thick syrup.

Powdered or *dry milk* is the result of

The Composition of Concentrated Milks

The figures are the percentages of each milk's weight accounted for by its major components.

Kind of Milk	Protein	Fat	Sugar	Minerals	Water
Evaporated milk	7	8	10	1.4	73
Evaporated skim milk	8	0.3	11	1.5	79
Sweetened condensed milk	8	9	55	2	27
Dry milk, full fat	26	27	38	6	2.5
Dry milk, nonfat	36	1	52	8	3
Fresh milk	3.4	3.7	4.8	1	87

taking evaporation to the extreme. Milk is pasteurized at a high temperature; then about 90% of its water is removed by vacuum evaporation, and the remaining 10% in a spray drier (the concentrated milk is misted into a chamber of hot air, where the milk droplets quickly dry into tiny particles of milk solids). Some milk is also freeze-dried. With most of its water removed, powdered milk is safe from microbial attack. Most powdered milk is made from low-fat milk because milk fat quickly goes rancid when exposed to concentrated milk salts and atmospheric oxygen, and because it tends to coat the particles of protein and makes subsequent remixing with water difficult. Powdered milk will keep for several months in dry, cool conditions.

Cooking with Milk Much of the milk that we use in the kitchen disappears into a mixture—a batter or dough, a custard mix or a pudding—whose behavior is largely determined by the other ingredients. The milk serves primarily as a source of moisture, but also contributes flavor, body, sugar that encourages browning, and salts that encourage protein coagulation.

When milk itself is a prominent ingredient—in cream soups, sauces, and scalloped potatoes, or added to hot chocolate, coffee, and tea—it most often calls attention to itself when its proteins coagulate. The skin that forms on the surface of scalded milk, soups, and sauces is a complex of casein, calcium, whey proteins, and trapped fat globules, and results from evaporation of water at the surface and the progressive concentration of proteins there. Skin formation can be minimized by covering the pan or whipping up some foam, both of which minimize evaporation. Meanwhile, at the bottom of the pan, the high, dehydrating temperature transmitted from the burner causes a similar concentration of proteins, which stick to the metal and eventually scorch. Wetting the pan with water before adding milk will reduce protein adhesion to the metal; a heavy, evenly conducting pan and a moderate flame help minimize scorching, and a double boiler will prevent it (though it's more trouble).

Between the pan bottom and the surface, particles of other ingredients can cause curdling by providing surfaces to which the milk proteins can stick and clump together. And acid in the juices of all fruits and vegetables and in coffee, and astringent tannins in potatoes, coffee, and tea, make milk proteins especially sensitive to coagulation and curdling. Because bacteria slowly sour milk, old milk may be acidic enough to curdle instantly when added to hot coffee or tea. The best insurance against curdling is fresh milk and careful control of the burner.

Cooking Sweetened Condensed Milk Because it contains concentrated protein

Intentionally Curdled Milk

For most cooks most of the time, curdled milk betokens crisis: the dish has lost its smoothness. But there are plenty of dishes in which the cook intentionally causes the milk proteins to clot precisely for the textural interest this creates. The English *syllabub* was sometimes made by squirting warm milk directly from the udder into acidic wine or juice; and in the 17th century, the French writer Pierre de Lune described a reduced milk "marbled" by the addition of currant juice. More contemporary examples include roast pork braised in milk, which reduces to moist brown nuggets; the Kashmiri practice of cooking milk down to resemble browned ground meat; and eastern European summertime cold milk soups like the Polish *chlodnik*, thickened by the addition of "sour salt," or citric acid.

and sugar, sweetened condensed milk will "caramelize" (actually, undergo the Maillard browning reaction, p. 778) at temperatures as low as the boiling point of water. This has made cans of sweetened condensed milk a favorite shortcut to a creamy caramel sauce: many people simply put the can in a pot of boiling water or a warm oven and let it brown inside. While this does work, it is potentially dangerous, since any trapped air will expand on heating and may cause the can to burst open. It's safer to empty the can into an open utensil and then heat it on the stovetop, in the oven, or in the microwave.

Milk Foams A foam is a portion of liquid filled with air bubbles, a moist, light mass that holds its shape. A meringue is a foam of egg whites, and whipped cream is a foam of cream. Milk foams are more fragile than egg foams and whipped cream, and are generally made immediately before serving, usually as a topping for coffee drinks. They prevent a skin from forming on the drink, and keep it hot by insulating it and preventing evaporative cooling.

Milk owes its foaming power to its proteins, which collect in a thin layer around the pockets of air, isolate them, and prevent the water's strong cohesive forces from popping the bubbles. Egg foams are also stabilized by proteins (p. 101), while the foam formed by whipping cream is stabilized by fat (below, p. 31). Milk foams are more fragile and short-lived than egg foams because milk's proteins are sparse—just 3% of the milk's weight, where egg white is 10% protein—and two-thirds of the milk proteins are resistant to being unfolded and coagulated into a solid network, while most of the egg proteins readily do so. However, heat around 160°F/70°C does unfold the whey proteins (barely 1% of milk's weight). And if they unfold at the air-water boundary of a bubble wall, then the force imbalance does cause the proteins to bond to each other and briefly stabilize the foam.

Milks and Their Foams Some milks are better suited to foaming than others. Because the whey proteins are the critical stabilizers, milks that are fortified with added protein—usually reduced-fat and skim milks—are most easily foamed. Full-fat foams, on the other hand, are fuller in texture and flavor. Milk should always be as fresh as possible, since milk that has begun to sour can curdle when heated.

India's Galaxy of Cooked Milks

For sheer inventiveness with milk itself as the primary ingredient, no country on earth can match India. Its dozens of variations on the theme of cooked-down milk, many of them dating back a thousand years, stem from a simple fact of life in that warm country: the simplest way to keep milk from souring is to boil it repeatedly. Eventually it cooks down to a brown, solid paste with about 10% moisture, 25% lactose, 20% protein and 20% butterfat. Even without added sugar, *khoa* is almost a candy, so it makes sense that over time, it and the intermediate concentrations that precede it became the basis for the most widely made Indian milk sweets. Doughnut-like fried *gulabjamun* and fudge-like *burfi* are rich in lactose, calcium, and protein: a glass of milk distilled into a morsel.

A second, separate constellation of Indian milk sweets is based on concentrating the milk solids by curdling them with heat and either lime juice or sour whey. The drained curds form a soft, moist mass known as *chhanna,* which then becomes the base for a broad range of sweets, notably porous, springy cakes soaked in sweetened milk or syrup (*rasmalai, rasagollah*).

Espresso Steamers: Simultaneous Bubbles and Heat Milk foams are usually made with the help of the steam nozzle on an espresso coffee machine. Steaming milk accomplishes two essential things simultaneously: it introduces bubbles into the milk, and it heats the bubbles enough to unfold and coagulate the whey proteins into a stabilizing web. Steam itself does not make bubbles: it is water vapor, and simply condenses into the colder water of the milk. Steam makes bubbles by splashing milk and air together, and it does this most efficiently when the nozzle is just below the milk surface.

One factor that makes steaming tricky is that very hot milk doesn't hold its foam well. A foam collapses when gravity pulls the liquid out of the bubble walls, and the hotter the liquid, the faster it drains. So you have to use a large enough volume of cold milk—at least ⅔ cup/150 ml—to make sure that the milk doesn't heat up too fast and become too runny before the foam forms.

Cream

Cream is a special portion of milk that is greatly enriched with fat. This enrichment occurs naturally thanks to the force of gravity, which exerts more of a pull on the milk's water than on the less dense fat globules. Leave a container of milk fresh from the udder to stand undisturbed, and the globules slowly rise through the water and crowd together at the top. The concentrated cream layer can then be skimmed off from the fat-depleted "skim" milk below. Milk with 3.5% fat will naturally yield cream that is about 20%.

We value cream above all for its feel. *Creaminess* is a remarkable consistency, perfectly balanced between solidity and fluidity, between persistence and evanescence. It's substantial, yet smooth and seamless. It lingers in the mouth, yet offers no resistance to teeth or tongue, nor becomes merely greasy. This luxurious sensation results from the crowding of the fat glob-

Keys to Foaming Milk

To get a good volume of milk foam from the steam attachment on an espresso machine:

- Use fresh milk right out of the refrigerator, or even chilled for a few minutes in the freezer.
- Start with at least ⅔ cup/150 ml of milk in a container that will hold at least double the initial volume.
- Keep the nozzle at or just under the milk surface so that it froths continuously with a moderate flow of steam.

To foam a small volume of milk without steam, separate the foaming and heating steps:

- Pour cold, fresh milk into a jar, tighten the lid, and shake it vigorously for 20 seconds or until the contents have doubled in volume. (Or froth in a plunger-style coffee maker, whose fine screen produces an especially thick, creamy foam.)
- Then stabilize the foam: remove the lid, place the jar in the microwave, and heat on high for about 30 seconds, or until the foam rises to the top of the jar.

ules, which are far too small for our senses to distinguish, into a small volume of water, whose free movement is thus impeded and slowed.

In addition to its appealing texture, cream has distinctive "fatty" aroma notes from molecules also found in coconut and peach (lactones). And it offers the virtue of being a robust, forgiving ingredient. Milk contains roughly equal weights of protein and fat, while in cream fat outweighs protein by at least 10 to 1. Thanks to this dilution of the protein, cream is less likely to curdle. And thanks to its concentration of fat globules, it can be inflated into whipped cream: a far more substantial and stable foam than milk alone can make.

Though it has certainly been appreciated since the beginning of dairying, cream spoils faster than the butter that could be made from it, and so it played a minor role in all but farmhouse kitchens until fairly recently. By the 17th century, French and English cooks were frothing cream into imitation snow; the English exploited its layering nature to pile cream skins in the form of a cabbage, and used long, gentle heating to produce solid, nutty "clouted" cream. Cream's heyday arrived in the 18th century, when it went into cakes, puddings, and such savory dishes as fricassees, stews, and boiled vegetables, and became popular in frozen form as ice cream. The popularity of cream declined in the 20th century with the nutritional condemnation of saturated fats, so much so that in many parts of the United States it's only available in the long-keeping ultrapasteurized form.

Making Cream The natural separation of cream from milk by means of gravity takes 12 to 24 hours, and was superseded late in the 19th century by the merry-go-round forces of the French centrifugal separator. Once separated, the cream is pasteurized. In the United States, the minimum temperatures for pasteurizing cream are higher than the milk minimum (for 20% fat or less, 30 minutes at 155°F/68°C; otherwise at 165°F/74°C). "Ultrapasteurized" cream is heated for 2 seconds at 280°F/140°C (like UHT-treated milk, p. 22; however the cream is not packaged under strictly sterile conditions, and so is kept refrigerated). Under refrigeration, ordinary pasteurized cream keeps for about 15 days before bacterial activity turns it bitter and rancid; ultrapasteurized cream, which has a stronger cooked flavor, keeps for several weeks. Normally cream is not homogenized because this makes it harder to whip, but long-keeping ultrapasteurized cream and relatively thin half-and-half are usually homogenized to prevent continuing slow separation in the carton.

The Importance of Fat Content Cream is manufactured with a number of different fat levels and consistencies, each for particular purposes. Light creams are poured into coffee or onto fruit; heavy creams are whipped or used to thicken sauces; clotted

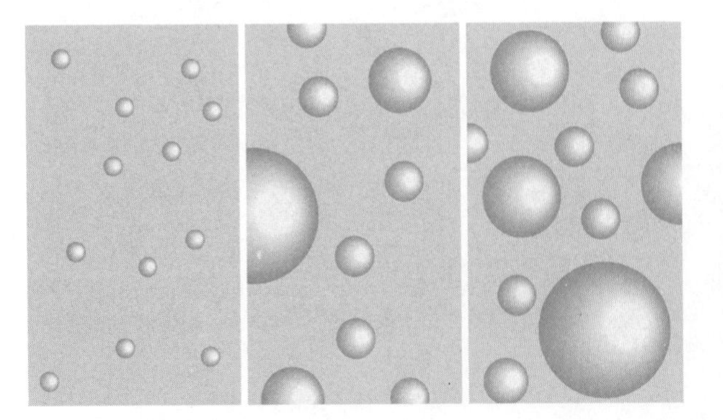

Fat globules in milk and cream. Left to right: *Fat globules in homogenized milk (3.5% fat), and in unhomogenized light cream (20% fat), and in heavy cream (40% fat). The more numerous fat globules in cream interfere with the flow of the surrounding fluid and give cream its full-bodied consistency.*

or "plastic" creams are spread onto breads, pastries, or fruit. The proportion of fat determines both a cream's consistency and its versatility. Heavy cream can be diluted with milk to approximate light cream, or whipped to make a spreadable semisolid. Light cream and half-and-half contain insufficient numbers of fat globules to stabilize a whipped foam (p. 32), or to resist curdling in a sauce. Whipping cream, at between 30 and 40% fat, is the most versatile formulation.

Stability in Cooking How does a high fat content permit the cook to boil a mixture of heavy cream and salty or acidic ingredients without curdling it, as when dissolving pan solids or thickening a sauce? The key seems to be the ability of the fat globule's surface membrane to latch onto a certain amount of the major milk protein, casein, when milk is heated. If the fat globules account for 25% or more of the cream's weight, then there's a sufficient area of globule surface to take most of the casein out of circulation, and no casein curds can form. At lower fat levels, there's both a smaller globule surface area *and* a greater proportion of the casein-carrying water phase. Now the globule surfaces can only

Kinds of Cream

U.S. Term	European Term	Fat Content, %	Use
Half-and-half		12 (10.5–18)	Coffee, pouring
	*Crème légère**	12–30	Coffee, pouring, enriching sauces, soups, etc., whipping
	Single cream	18+	Coffee, pouring
Light cream		20 (18–30)	Coffee, pouring (seldom available)
	Coffee cream	25	Coffee, pouring
Light whipping cream		30–36	Pouring, enriching, whipping
	Crème fraîche† (*fleurette* or *épaisse*)*	30–40	Pouring, enriching, whipping (if rich, spreading)
Whipping cream		35+	Pouring, enriching, whipping
Heavy whipping cream		38 (36+)	Pouring, enriching, whipping
	Double cream	48+	Spreading
	Clotted cream	55+	Spreading
Plastic cream		65–85	Spreading

**légère*: "light"; *fleurette*: "liquid"; *épaisse*: "thick" due to bacterial culture

†*fraîche*: "fresh, cool, new." In France, *crème fraîche* may be either "sweet" or cultured with lactic acid bacteria; in the United States, the term always means cultured, tart, thick cream. See p. 49.

absorb a small fraction of the casein, and the rest bonds together and coagulates when heated. (This is why acid-curdled mascarpone cheese can be made from light cream, but not from heavy cream.)

Problems with Cream: Separation A common problem with unhomogenized cream is that it continues to separate in the carton: the fat globules slowly rise and concentrate further into a semisolid layer at the top. At refrigerator temperatures, the fat inside the globules forms solid crystals whose edges break through the protective globule membrane, and these slightly protruding fat crystals get stuck to each other and form microscopic butter grains.

Clotted Creams These days cooks generally consider the separation and solidification of cream a nuisance. In the past, and in present-day England and the Middle East, congealed cream has been and is appreciated for its own sake. The cooks of 17th-century England would patiently lift the skins from shallow dishes of cream and arrange them in wrinkled mounds to imitate the appearance of a cabbage. Cabbage cream is now a mere curiosity. But the 16th-century English invention called *clotted* cream (and its Turkish and Afghan relatives *kaymak* and *qymaq*) remain vital traditions.

Old-fashioned clotted cream is made by heating cream just short of the boil in shallow pans for several hours, then letting it cool and stand for a day or so, and removing the thick solid layer. Heat accelerates the rise of the fat globules, evaporates some of the water, melts some of the aggregated globules into pockets of butterfat, and creates a cooked flavor. The result is a mix of thick, granular, fatty areas and thin, creamy ones, with a rich, nutty flavor and a straw-colored surface. Clotted cream is around 60% fat, and is spread onto scones and biscuits and eaten with fruit.

Whipped Cream The miraculous thing about whipped cream is that simple physical agitation can transform a luscious but

Food Words: *Cream, Crème, Panna*

The English name for the fat-rich portion of milk, like the French word from which it derives, has associations that are startling but appropriate to its status as a textural ideal.

Before the Norman Conquest, and to this day in some northern dialects, the English word for cream was *ream,* a simple offshoot of the Indo-European root that also gave the modern German *Rahm.* But the French connection introduced a remarkable hybrid term. In 6th-century Gaul, fatty milk was called *crama,* from the Latin *cremor lactis,* or "heat-thickened substance of milk." Then in the next few centuries it somehow became crossed with a religious term: *chreme,* or "consecrated oil," which stems from the Greek word *chriein,* "to anoint," that gave us *Christ,* "the anointed one." So in France *crama* became *crème,* and in England *ream* gave way to *cream.*

Why this confusion of ancient ritual with rich food? Linguistic accident or error, perhaps. On the other hand, anointing oil and butterfat *are* essentially the same substance, so perhaps it was inspiration. In the monastic or farm kitchens of Normandy, the addition of cream to other foods may have been considered not just an enrichment, but a kind of blessing.

The Italian word for cream, *panna,* has been traced back to the Latin *pannus,* or "cloth." This is apparently a homely allusion to the thin covering that cream provides for the milk surface.

unmanageable liquid into an equally luscious but shapeable "solid." Like foamed milk, whipped cream is an intimate intermingling of liquid and air, with the air divided into tiny bubbles and the cream spread out and immobilized in the microscopically thin bubble walls. Common as it is today, this luxurious, velvety foam was very laborious to make until 1900. Before then, cooks whipped naturally separated cream for an hour or more, periodically skimming off the foam and setting it aside to drain. The key to a stable foam of the whole mass of cream is enough fat globules to hold all the fluid and air together, and naturally separated cream seldom reaches that fat concentration, which is about 30%. It took the invention of the centrifugal separator to produce easily whipped cream.

How Fat Stabilizes Foamed Cream Unlike the protein foams of egg white, egg yolk, and milk, the cream foam is stabilized by fat. Initially, the whisk introduces short-lived air bubbles into the cream. After the first half-minute or so, the bubble walls begin to be stabilized by the *de*stabilization of the fat globules. As the globules are knocked all around and into each other by the whipping, parts of their protective membranes are stripped away by the shearing action of the whisk, and by the force imbalance in the air bubble walls. The patches of naked fat, which by their nature avoid contact with water, settle in one of two regions in the cream: either facing the air pocket in the bubble walls, or stuck to a patch of naked fat on another globule. The fat globules thus form walls around the air bubbles, and connections between neighboring walls: and so a continuous network develops. This network of solid fat spheres not only holds the air bubbles in place, but also prevents the intervening pockets of fluid from moving very far. And so the foam as a whole takes on a definite, persistent structure.

If the beating continues past the point at which a fat network has just barely formed, the gathering of the fat globules continues also, but this process now *de*stabilizes the foam. The fine globule clusters coalesce with each other into ever coarser masses of butterfat, and the pockets of air and fluid that they hold in place coarsen as well. The foam loses volume and weeps, and the vel-

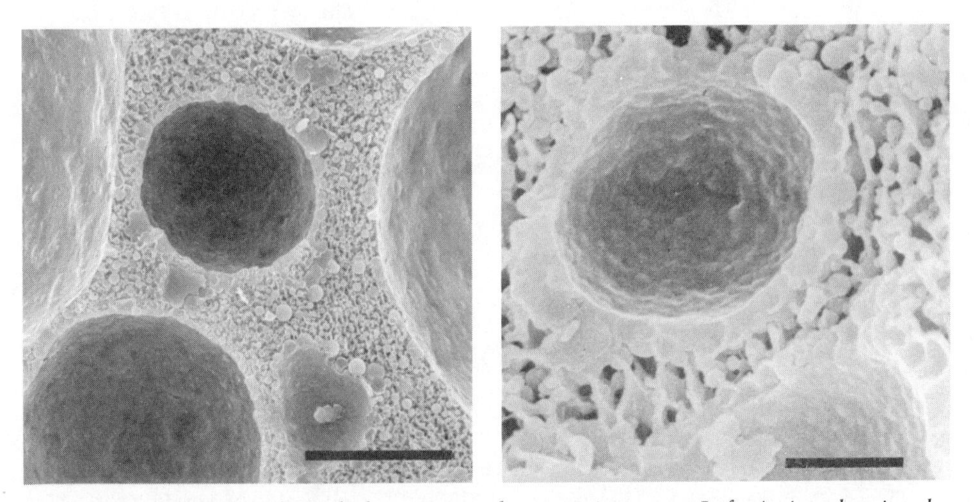

Whipped cream as seen through the scanning electron microscope. Left: *A view showing the large cavity-like air bubbles and smaller spherical fat globules (the black bar represents 0.03 mm).* Right: *Close-up of an air bubble, showing the layer of partly coalesced fat that has stabilized the bubble (the bar represents 0.005 mm).*

vety texture of the perfectly whipped cream becomes granular. The butter grains in overwhipped cream leave a greasy residue in the mouth.

The Importance of Cold Because even mild warmth softens the butterfat skeleton of a cream foam, and liquid fat will collapse the air bubbles, it's essential to keep cream cold while it's whipped. It should start out at the low end of 40–50°F/5–10°C, and bowl and beaters should be chilled as well, since both air and beating will quickly warm everything. Ideally, the cream is "aged" in the refrigerator for 12 hours or more before whipping. Prolonged chilling causes some of the butterfat to form crystalline needles that hasten the membrane stripping and immobilize the small portion of fat that's liquid even in cold cream. Cream that has been left at room temperature and chilled just before use leaks bubble-deflating liquid fat from the beginning of whipping, never rises very high, and more easily becomes granular and watery.

How Different Creams Behave When Whipped Cream for whipping must be sufficiently rich in fat to form a continuous skeleton of globules. The minimum fat concentration is 30%, the equivalent of "single" or "light whipping" cream. "Heavy" cream, at 38 to 40% fat, will whip faster than light cream, and forms a stiffer, denser, less voluminous foam. It also leaks less fluid, and so is valued for use in pastries and baked goods, and for piping into decorative shapes. For other purposes, heavy cream is usually diluted with a quarter of its volume of milk to make 30% cream and a lighter, softer foam.

The fat globules in homogenized cream are smaller and more thickly covered with milk proteins. Homogenized cream therefore forms a finer-textured foam, and takes at least twice as long to whip (it's also harder to overwhip to the granular stage). The cook can cut the whipping time of any cream by slightly acidifying it (1 teaspoon/5 ml lemon juice per cup/250 ml), which makes the proteins in its globule membranes easier to strip away.

Methods: Hand, Machine, Pressurized Gas Cream can be foamed by several different methods. Whisking by hand takes more time and physical exertion than an electric beater, but incorporates more air and produces a greater volume. The lightest, fluffiest whipped cream is produced with the help of pressurized gas, usually nitrous oxide (N_2O). The most familiar

Early Whipped Cream

My Lord of S. Alban's Cresme Fouettee

Put as much as you please to make, of sweet thick cream into a dish, and whip it with a bundle of white hard rushes, (of such as they make whisks to brush cloaks) tied together, till it come to be very thick, and near a buttery substance. If you whip it too long, it will become butter. About a good hour will serve in winter. In summer it will require an hour and a half. Do not put in the dish you will serve it up in, till it be almost time to set it upon the table. Then strew some powdered fine sugar into the bottom of the dish it is to go in, and with a broad spatule lay your cream upon it: when half is laid in, strew some more fine sugar upon it, and then lay in the rest of the cream (leaving behind some whey that will be in the bottom) and strew some more sugar upon that.

—Sir Kenelm Digby, *The Closet Opened,* 1669

gas-powered device is the aerosol can, which contains a pressurized mixture of ultrapasteurized cream and dissolved gas. When the nozzle is opened and the mixture released, the gas expands instantly and explodes the cream into a very light froth. There is also a device that aerates ordinary fresh cream with a replaceable canister of nitrous oxide, which is released in the nozzle and causes great turbulence as it mixes with the cream.

Butter and Margarine

These days, if a cook actually manages to *make* butter in the kitchen, it's most likely a disaster: a cream dish has been mishandled and the fat separates from the other ingredients. That's a shame: all cooks should relax now and then and intentionally overwhip some cream! The coming of butter is an everyday miracle, an occasion for delighted wonder at what the Irish poet Seamus Heaney called "coagulated sunlight" "heaped up like gilded gravel in the bowl." Milkfat is indeed a portion of the sun's energy, captured by the grasses of the field and repackaged by the cow in scattered, microscopic globules. Churning milk or cream damages the globules and frees their fat to stick together in ever larger masses, which we eventually sieve into the golden hoard that imparts a warm, sweet richness to many foods.

Ancient, Once Unfashionable All it takes to separate the fat from milk is 30 seconds of sloshing, so butter was no doubt discovered in the earliest days of dairying. It has long been important from Scandinavia to India, where nearly half of all milk production goes to making butter for both cooking and ceremonial purposes. Its heyday came much later in northern Europe, where throughout the Middle Ages it was eaten mainly by peasants. Butter slowly infiltrated noble kitchens as the only animal fat allowed by Rome on days of abstention from meat. In the early 16th century it was also permitted during Lent, and the rising middle classes adopted the rustic coupling of bread and butter. Soon the English were notorious for serving meats and vegetables swimming in melted butter, and cooks throughout Europe exploited butter in a host of fine foods, from sauces to pastries.

Normandy and Brittany in northwest France, Holland, and Ireland became especially renowned for the quality of their butter. Most of it was made on small farms using cream that was pooled from several milkings, and was therefore a day or two old and somewhat soured by lactic acid bacteria. Continental Europe still prefers the flavor of this lightly fermented "cultured" butter to the "sweet cream" butter made common in the 19th century by the use of ice, the development of refrigeration, and the mechanical cream separator.

Around 1870, a shortage of butter in France led to the invention of an imitation, *margarine,* which could be made from a variety of cheap animal fats and vegetable oils. More margarine than butter is now consumed in the United States and parts of Europe.

Making Butter Butter making is in essence a simple but laborious operation: you agitate a container of cream until the fat globules are damaged and their fat leaks out and comes together into masses large enough to gather.

Preparing the Cream For butter making, cream is concentrated to 36–44% fat. The cream is then pasteurized, in the United States usually at 185°F/85°C, a high temperature that develops a distinct cooked, custardy aroma. After cooling, the cream for cultured butter may be inoculated with lactic acid bacteria (see p. 35). The sweet or cultured cream is then cooled to about 40°F/5°C and "aged" at that temperature for at least eight hours so that about half of the milk fat in the globules forms solid crystals. The number and size of these crystals help determine the how quickly and completely the milk fat separates, as well as the final texture of the butter. The properly aged cream is then warmed a few degrees Fahrenheit and churned.

Churning Churning is accomplished by a variety of mechanical devices that may take 15 minutes or a few seconds to damage the fat globules and form the initial grains of butter. The fat crystals formed during aging distort and weaken the globule membranes so that they rupture easily. When damaged globules collide with each other, the liquid portion of their fat flows together to make a continuous mass, and these grow as churning continues.

Working Once churning generates the desired size of butter grains, often the size of a wheat seed, the water phase of the cream is drained off. This is the original buttermilk, rich in free globule membrane material and with about 0.5% fat (p. 50). The solid butter grains may be washed with cold water to remove the buttermilk on their surfaces. The grains are then "worked," or kneaded together to consolidate the semisolid fat phase and to break up the embedded pockets of buttermilk (or water) into droplets around 10 micrometers in diameter, or about the size of a large fat globule. Cows that get little fresh pasturage and its orange carotene pigments produce pale milk fat; the butter maker can compensate for this by adding a dye such as annatto (p. 423) or pure carotene during the working. If the butter is to be salted, either fine granular salt or a strong brine goes in at this stage as well. The butter is then stored, blended, or immediately shaped and packaged.

Kinds of Butter Butter is made in several distinct styles, each with its own particular qualities. It's necessary to read labels carefully to learn whether a given brand has been made with plain cream, fermented cream, or cream flavored to taste like fermented cream.

Raw cream butter, whether sweet or cultured, is now nearly extinct in the United States and a rarity even in Europe. It is prized for its pure cream flavor, without the cooked-milk note due to pasteurization. The flavor is fragile; it deteriorates after about 10 days unless the butter is frozen.

Sweet cream butter is the most basic, and the commonest in Britain and North America. It's made from pasteurized fresh cream, and in the United States must be at least 80% fat and no more than 16% water; the remaining 4% is protein, lactose, and salts contained in the buttermilk droplets.

Salted sweet cream butter contains between 1 and 2% added salt (the equivalent of 1–2 teaspoons per pound/5–10 gm

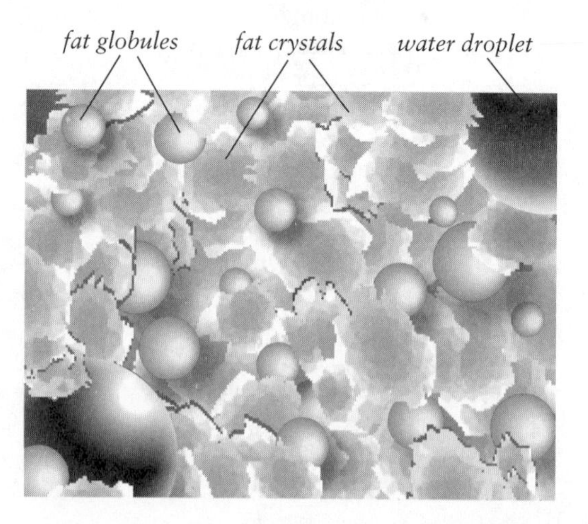

The structure of butter, which is about 80% milk fat and 15% water. The fat globules, solid crystals, and water droplets are embedded in a continuous mass of semisolid "free" fat that coats them all. A high proportion of ordered crystals imparts a stiff firmness to cold butter, while free fat lends spreadability and the tendency to leak liquid fat as it warms and softens.

per 500 gm). Originally salt was added as a preservative, and at 2%, the equivalent of about 12% in the water droplets, it still is an effective antimicrobial agent.

Cultured cream butter, the standard in Europe, is the modern, controlled version of the commonest preindustrial butter, whose raw cream had been slightly soured by the action of lactic acid bacteria while it slowly separated in the pan before churning. Cultured butter tastes different: the bacteria produce both acids and aroma compounds, so the butter is noticeably fuller in flavor. One particular aroma compound, diacetyl, greatly intensifies the basic butter flavor itself.

There are several different methods for manufacturing cultured butter or something like it. The most straightforward is to ferment pasteurized cream with cream-culture bacteria (p. 49) for 12 to 18 hours at cool room temperature before churning. In the more efficient method developed in the Netherlands in the 1970s and also used in France, sweet cream is churned into butter, and then the bacterial cultures and preformed lactic acid are added; flavor develops during cold storage. Finally, the manufacturer can simply add pure lactic acid and flavor compounds to sweet cream butter. This is an artificially flavored butter, not a cultured butter.

European-style butter, an American emulation of French butter, is a cultured butter with a fat content higher than the standard 80%. France specifies a minimum fat content of 82% for its butter, and some American producers aim for 85%. These butters contain 10–20% less water, which can be an advantage when making flaky pastries (p. 563).

Whipped butter is a modern form meant to be more spreadable. Ordinary sweet butter is softened and then injected with about a third its volume of nitrogen gas (air would encourage oxidation and rancidity). Both the physical stress and the gas pockets weaken the butter structure and make it easier to spread, though it remains brittle at refrigerator temperature.

Specialty butters are made in France for professional bakers and pastry chefs. *Beurre cuisinier, beurre pâtissier,* and *beurre concentré* are almost pure butterfat, and are made from ordinary butter by gently melting it and centrifuging the fat off of the water and milk solids. It can then be recooled as is, or slowly crystallized and separated into fractions that melt at temperatures from 80°F/27°C to 104°F/40°C, depending on the chef's needs.

Butter Consistency and Structure Well made butters can have noticeably different consistencies. In France, for example, butter from Normandy is relatively soft and favored for spreading and making sauces—Elizabeth David said, "When you get melted butter with a trout in Normandy it is difficult to believe that it is not cream"—while butter from the Charentes is firmer, and preferred for making pastries. Many dairies will often produce softer butter in the summer than they do in the winter. The consistency of butter reflects its microscopic structure, and this is strongly influenced by two factors: what the cows eat, and how the butter maker handles their milk. Feeds rich in polyunsaturated fats, especially fresh pasturage, produce softer butters; hay and grain harder ones. The butter maker also influences consistency by the rate and degree of cooling to which he subjects the cream during the aging period, and by how extensively he works the new butter. These conditions control the relative proportions of firming crystalline fat and softening globular and free fat.

Keeping Butter Because its scant water is dispersed in tiny droplets, properly made butter resists gross contamination by microbes, and keeps well for some days at room temperature. However, its delicate flavor is easily coarsened by simple exposure to the air and to bright light, which break fat molecules into smaller fragments that smell stale and rancid. Butter also readily absorbs strong odors from its surroundings. Keep reserves in the freezer, and daily butter in the cold and dark as much as possible. Rewrap remainders airtight, preferably with the original foiled paper and not with aluminum foil; direct contact with metal can hasten fat oxidation, particularly in salted butter. Translucent, dark yellow patches on the surface of a butter stick are areas where the butter has been exposed to the air and dried out; they taste rancid and should be scraped off.

Cooking with Butter Cooks use butter for many different purposes, from greasing cake pans and soufflé molds to flavoring butterscotch candies. Here are notes on some of its more prominent roles. The important role of butter in baking is covered in chapter 10.

Butter as Garnish: Spreads, Whipped Butters Good plain bread spread with good plain butter is one of the simplest pleasures. We owe butter's buttery consistency to the peculiar melting behavior of milk fat, which softens and becomes spreadable around 60°F/15°C, but doesn't begin to melt until 85°F/30°C.

This workable consistency also means that it's easy to incorporate other ingredients into the butter, which then carries their flavor and color and helps apply them evenly to other foods. Composed butters are masses of room-temperature butter into which some flavoring and/or coloring has been kneaded; these can include herbs, spices, stock, a wine reduction, cheese, and pounded seafood. The mixture can then be spread on another food, or refrigerated, sliced, and melted into a butter sauce when put onto a hot meat or vegetable. And whipped butter prepared by the cook is butter lightened by the incorporation of some air, and flavored with about half its volume of stock, a puree, or some other liquid, which becomes dispersed into the butter fat in small droplets.

***Butter as Sauce: Melted Butter,* Beurre Noisette, *and* Beurre Noir** Perhaps the simplest of sauces is the pat of butter dropped on a heap of hot vegetables, or stirred into rice or noodles, or drawn across the surface of an omelet or steak to give a sheen. Melted butter can be enlivened with lemon juice, or "clarified" to remove the milk solids (see below). *Beurre noisette* and *beurre noir,* "hazel" and "black" butter, are melted butter sauces that the French have used since medieval times to enrich fish, brains, and vegetables. Their flavor is deepened by heating the butter to about 250°F/120°C until its water boils off and the molecules in the white residue, milk sugar and protein, react with each other to form brown pigments and new aromas (the browning reaction, p. 777). Hazel butter is cooked until it's golden brown, black butter until it's dark brown (truly black butter is acrid). They're often balanced with vinegar or lemon juice, which should be added only after the butter has cooled below the boiling point; otherwise the cold liquid will cause spattering and the lemon solids may brown. On their own, they lend a rich nutty flavor to baked goods.

The emulsified butter sauces—*beurre blanc,* hollandaise, and their relatives—are described in chapter 11.

Clarified Butter *Clarified* butter is butter whose water and milk solids have been removed, leaving essentially pure milk fat that looks beautifully clear when melted and that is better suited for frying (the milk solids scorch at relatively low frying temperatures). When butter is gently heated to the boiling point of water, the water bubbles to the top, where the whey proteins form a froth. Eventually all the water evaporates, the bubbling stops, and the froth dehydrates. This leaves a skin of dry whey protein on top, and dry casein particles at the bottom. Lift off the whey skin, pour the liquid fat off of the casein residue, and the purification is complete.

Frying with Butter Butter is sometimes used for frying and sautéing. It has the advantage that its largely saturated fats are resistant to being broken down by heat, and so don't become gummy the way unsaturated oils do. It has the disadvantage that its milk solids brown and then burn around 250°F, 150° below the smoke point of many vegetable oils. Adding oil to butter does not improve its heat tolerance. Clarifying does; butter free of milk solids can be heated to 400°F/200°C before burning.

Margarine and Other Dairy Spreads

Margarine has been called "a creation of political intuition and scientific research." It was invented by a French chemist in 1869, three years after Napoleon III had offered funds for the development of an inexpensive food fat to supplement the inadequate butter supply for his poorly nourished but growing urban populace. Others before Hippolyte Mège-Mouriès had modified solid animal fats, but he had the novel idea of flavoring beef tallow with milk and working the mixture like butter.

Margarine caught on quickly in the major European butter producers and exporters—Holland, Denmark, and Germany—in part because they had surplus skim milk from butter making that could be used to flavor margarine. In the United States large-scale production was underway by 1880. Here, the dairy industry and its allies in government put up fierce resistance in the form of discriminatory taxes that persisted into the 1970s. Today, basic margarine remains cheap compared to butter, and Americans consume more than twice as much margarine as butter. Scandinavia and northern Europe also favor margarine, while France and Britain still give a substantial edge to butter.

The Rise of Vegetable Margarine Modern margarine is now made not from solid animal fats, but from normally liquid vegetable oils. This shift was made possible around 1900 by German and French chemists who developed the process of *hydrogenation,* which hardens liquid oils by altering the structures of their fatty acids (p. 801). Hydrogenation allowed manufacturers to make a butter substitute that spreads easily even at refrigerator temperature, where butter is unusably hard. An unanticipated bonus for the shift to vegetable oils was the medical discovery after World War II that the saturated fats typical of meats and dairy products raise blood cholesterol levels and the risk of heart disease. The ratio of saturated to unsaturated fat in hard stick margarine is only 1 to 3, where in butter it is 2 to 1. Recently, however, scientists have found

Indian Clarified Butter: *Ghee*

In India, clarified butter is the most eminent of all foods. In addition to being used as an ingredient and frying oil, it is an emblem of purity, an ancient offering to the gods, the fuel of holy lamps and funeral pyres. Ghee (from the Sanskrit for "bright") was born of necessity. Ordinary butter spoils in only ten days in much of the country, while the clarified fat keeps six to eight months. Traditionally, ghee has been made from whole cow or buffalo milk that is soured by lactic acid bacteria into yogurt-like *dahi,* then churned to obtain butter. Today, industrial manufacturers usually start with cream. The preliminary souring improves both the quantity of butter obtained and its flavor; ghee made from sweet cream is said to taste flat. The butter is heated to 190°F/90°C to evaporate its water, then the temperature is raised to 250°F/120°C to brown the milk solids, which flavors the ghee and generates antioxidant compounds that delay the onset of rancidity. The brown residue is then filtered off (and mixed with sugar to make sweets), leaving the clear liquid ghee.

that *trans fatty acids* produced by hydrogenation actually raise blood cholesterol levels (see box). There are other methods for hardening vegetable oils that don't produce trans fatty acids, and manufacturers are already producing "trans free" margarines and shortenings.

Making Margarine The gross composition of margarine is the same as butter's: a minimum of 80% fat, a maximum of 16% water. The water phase is either fresh or cultured skim milk, or skim milk reconstituted from powder. Salt is added for flavor, to reduce spattering during frying, and as an antimicrobial agent. In the United States, the fat phase is blended from soybean, corn, cottonseed, sunflower, canola, and other oils. In Europe, lard and refined fish oils are also used. The emulsifier lecithin is added (0.2%) to stabilize the water droplets and reduce spattering in the frying pan; coloring agents, flavor extracts, and vitamins A and D are also incorporated. Nitrogen gas may be pumped in to make a whipped, softer spread.

Kinds of Margarine and Related Spreads
Stick and *tub margarines* are the two most common kinds. They are formulated to approximate the spreadable consistency of butter at room temperature, and to melt in the mouth. Stick margarine is only slightly softer than butter in the refrigerator, and like butter can be creamed with sugar to make icings. Tub margarine is substantially less saturated and easily spreadable even at 40ºF/5ºC, but too soft to cream or to use in layered pastries.

Reduced-fat spreads contain less oil and more water than standard margarines, rely on carbohydrate and protein stabilizers, and aren't suited to cooking. The stabilizers can scorch in the frying pan. If used to replace butter or margarine in baking, high-moisture spreads throw liquid-solid proportions badly out of balance. Very-low-fat and no-fat spreads contain so much starch, gum, and/or protein that there's nothing there to melt when heated: they dry out and eventually burn.

Specialty margarines are generally available only to professional bakers. Like the original French oleomargarine, they some-

Hydrogenation By-Products: Trans Fatty Acids

Trans fatty acids are unsaturated fatty acids that nevertheless behave more like saturated fatty acids (p. 801). They're formed in the hydrogenation process, and are the reason that margarines can be as solid as butter and yet contain half the saturated fat; the trans unsaturated fats contribute a great deal to margarine firmness. Trans unsaturated fats are also less prone to oxidation or heat damage and make cooking oils more stable.

Trans fatty acids have come under scrutiny due to the likelihood that they may contribute to human heart disease. Research has shown that they not only raise undesirable LDL cholesterol levels in the blood as saturated fats do, they also lower desirable HDL levels. Manufacturers are now modifying their processing methods to lower trans fatty acids levels in U.S. margarines and cooking oils from the present levels, which reach 20–50% of total fatty acids in hard margarines (less in softer products).

Margarine manufacturers are not the only producers of trans fatty acids: the microbes in animal rumens are too! Thanks to their activity, the fat in milk, butter, and cheese averages 5% trans fatty acids, and the meat fat of ruminant animals—beef and lamb—ranges from 1 to 5%.

times contain beef tallow. They're formulated to have a firm but spreadable consistency over a much broader temperature range than butter (p. 562).

Ice Cream

Ice cream is a dish that manages to heighten the already remarkable qualities of cream. By freezing it, we make it possible to taste the birth of creaminess, the tantalizing transition from solidity to fluidity. But it was no simple matter to freeze cream in a way that does it justice.

The Invention and Evolution of Ice Cream Plain frozen cream is hard as a rock. Sugar makes it softer, but also lowers its freezing point (the dissolved sugar molecules get in the way as the water molecules settle into ordered crystals). So sweetened cream freezes well below the freezing point of pure water, and can't freeze in the slush that forms when a warm object is placed in snow or ice. What made ice cream possible was a sprinkling of chemical ingenuity. If salts are added to the ice, the salts dissolve in the slush, lower *its* freezing point, and allow it to get cold enough to freeze the sugared cream.

The effect of salts on freezing was known in the 13th-century Arab world, and that knowledge eventually made its way to Italy, where ices made from fruit were described in the early 17th century. The English term "ice cream" first appears in a 1672 document from the court of Charles II, and the first printed recipes for frozen waters and creams appear in France and Naples in the 1680s and 1690s. By the time of the American Revolution, the French had discovered that frequent stirring of the freezing mix gave a finer, less crystalline texture. They had also developed super-rich versions with 20 egg yolks per pint of cream (*glace au beurre,* "ice butter"!), and ice creams flavored with various nuts and spices, orange blossoms,

The First Recipes for Ice Cream

Neige de fleurs d'orange ("Snow of Orange Flowers")

You must take sweet cream, and put thereto two handfuls of powdered sugar, and take petals of orange flowers and mince them small, and put them in your cream . . . and put all into a pot, and put your pot in a wine cooler; and you must take ice, crush it well and put a bed of it with a handful of salt at the bottom of the cooler before putting in the pot. . . . And you must continue putting a layer of ice and a handful of salt, until the cooler is full and the pot covered, and you must put it in the coolest place you can find, and you must shake it from time to time for fear it will freeze into a solid lump of ice. It will take about two hours.

—*Nouveau confiturier,* 1682

Fromage à l'angloise (English Cheese)

Take a *chopine* [16 oz] of sweet cream and the same of milk, half a pound of powdered sugar, stir in three egg yolks and boil until it becomes like a thin porridge; take it from the fire and pour it into your ice mould, and put it in the ice for three hours; and when it is firm, withdraw the mold, and warm it a little, in order more easily to turn out your cheese, or else dip your mould for a moment in hot water, then serve it in a compôtier.

—François Massialot, *La Nouvelle instruction pour les confitures* (1692)

caramel, chocolate, tea, coffee, and even rye bread.

In America, a Food for the Masses America transformed this delicacy into a food for the masses. Ice cream making was an awkward, small-batch procedure until 1843, when a Nancy Johnson of Philadelphia patented a freezer consisting of a large bucket for the brine and a sealed cylinder containing the ice-cream mix and a mixing blade, whose shaft protruded from the top and could be cranked continuously. Five years later, William G. Young of Baltimore modified Johnson's design to make the mix container rotate in the brine for more efficient cooling. The Johnson-Young freezer allowed large quantities of fine-textured ice cream to be made with a simple, steady mechanical action.

The second fateful advance toward mass production came in the early 1850s, when a Baltimore milk dealer by the name of Jacob Fussell decided to use his seasonal surplus of cream to make ice cream, was able to charge half the going price in specialty shops, and enjoyed great success as the first large-scale manufacturer. His example caught on, so that by 1900 an English visitor was struck by the "enormous quantities" of ice cream eaten in America. Today Americans still eat substantially more ice cream than Europeans do, nearly 20 quarts/liters per person every year.

The Industrialization of Ice Cream Once ice cream became an industrial product, industry redefined it. Manufacturers could freeze their ice cream faster and colder than the handmade version, and so could produce very fine ice crystals. Smoothness of texture became the hallmark of industrial ice cream, and manufacturers accentuated it by replacing traditional ingredients with gelatin and concentrated milk solids. After World War II, they dosed ice cream with greater amounts of stabilizers to preserve its smoothness in the new and unpredictable home freezers. And price competition led to the increasing use of additives, powdered milk from surplus production, and artificial flavors and colors. So an ice cream hierarchy developed. At the top is traditional but relatively expensive ice cream; at the bottom, a lower-quality but more stable and affordable version.

The Structure and Consistency of Ice Cream

Ice Crystals, Concentrated Cream, Air Ice cream consists of three basic elements: ice crystals made of pure water, the concentrated cream that the crystals leave behind as they form from the prepared mix, and tiny air cells formed as the mix is churned during the freezing.

- The ice crystals form from water molecules as the mix freezes, and give ice cream its solidity; they're its backbone. And their size determines whether it is fine and smooth or coarse and grainy. But they account for only a fraction of its volume.
- The concentrated cream is what is left of the mix when the ice crystals form. Thanks to all the dissolved sugar, about a fifth of the water in the mix remains unfrozen even at 0°F/–18°C. The result is a very thick fluid that's about equal portions of liquid water, milk fat, milk proteins, and sugar. This fluid coats each of the many millions of ice crystals, and sticks them together—but not too strongly.
- Air cells are trapped in the ice cream mix when it's agitated during the freezing. They interrupt and weaken the matrix of ice crystals and cream, making that matrix lighter and easier to scoop and bite into. The air cells inflate the volume of the ice cream over the volume of the original mix. The increase is called *overrun,* and in a fluffy ice cream can be as much as 100%: that is, the final ice cream volume is half mix and half air. The lower the overrun, the denser the ice cream.

Balance The key to making a good ice cream is to formulate a mix that will freeze into a balanced structure of ice crystals, concentrated cream, and air. The consistency of a balanced, well made ice cream is creamy, smooth, firm, almost chewy. The smaller the proportion of water in the mix, the easier it is to make small crystals and a smooth texture. However, too much sugar and milk solids gives a heavy, soggy, syrupy result, and too much fat can end up churning into butter. Most good ice cream recipes produce a mix with a water content around 60%, a sugar content around 15%, and a milk-fat content between 10%—the minimum for commercial U.S. ice cream—and 20%.

Styles of Ice Cream Flavorings apart, there are two major styles of ice cream, and several minor ones.

- *Standard* or *Philadelphia-style* ice cream is made from cream and milk, sugar, and a few other minor ingredients. Its appeal is the richness and delicate flavor of cream itself, complemented by vanilla or by fruits or nuts.
- *French* or *custard* ice cream contains an additional ingredient: egg yolks, as many as 12 per quart/liter. The proteins and emulsifiers in egg yolk can help keep ice crystals small and the texture smooth even at relatively low milk-fat and high water levels; some traditional French ice cream mixes are a *crème anglaise* (p. 98) made with milk, not cream. A mix that contains yolks must be cooked to disperse the proteins and emulsifiers (and kill any bacteria in the raw yolks), and the resulting thickened, custard-like mix makes an ice cream with a characteristic cooked, eggy flavor.

  A distinct style of custard ice cream is the Italian *gelato,* which is typically high in butterfat as well as egg yolks, and frozen with little overrun into a very rich, dense cream. (The name simply means "frozen," and in Italy is applied to a range of frozen preparations.)
- *Reduced-fat, low-fat, and nonfat ice creams* contain progressively less fat than the 10% minimum specified in the commercial American definition of ice cream. They keep their ice crys-

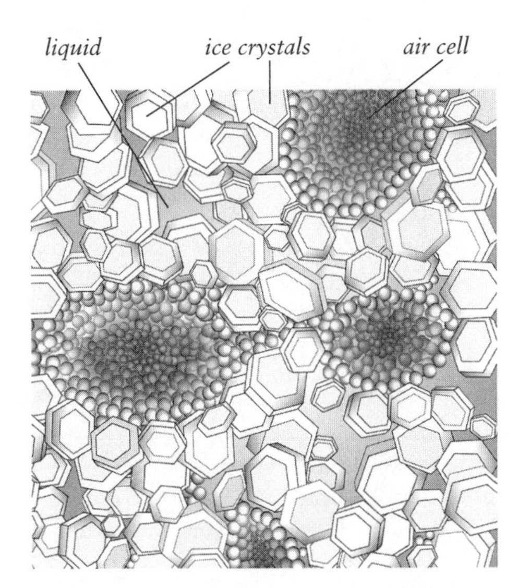

Ice cream, a semisolid foam. The process of freezing the ice-cream mix forms ice crystals—solid masses of pure water—and concentrates the remaining mix into a liquid rich in sugar and milk proteins. Churning fills the mix with air bubbles, which are stabilized by layers of clustered fat globules.

The Typical Compositions of Ice Creams

With the exception of overrun and calories, the percentages shown are the percentages by weight of the ice cream.

Style	% Milk Fat	% Other Milk Solids	% Sugar	% Yolk Solids (Stabilizers)	% Water	Overrun (% volume of original mix)	Calories per ½ c/125 ml
Premium standard	16–20	7–8	13–16	(0.3)	64–56	20–40	240–360
Name-brand standard	12–14	8–11	13–15	(0.3)	67–60	60–90	130–250
Economy standard	10	11	15	(0.3)	64	90–100	120–150
"French" (commercial)	10–14	8–11	13–15	2	67–58	60–90	130–250
French (handmade)	3–10	7–8	15–20	6–8	69–54	0–20	150–270
Gelato	18	7–8	16	4–8	55–50	0–10	300–370
Soft-serve	3–10	11–14	13–16	(0.4)	73–60	30–60	175–190
Low-fat	2–4	12–14	18–21	(0.8)	68–61	75–90	80–135
Sherbet	1–3	1–3	26–35	(0.5)	72–59	25–50	95–140
Kulfi	7	18	5–15	—	70–60	0–20	170–230

tals small with a variety of additives, including corn syrup, powdered milk, and vegetable gums. "Soft-serve" ice cream is a reduced-fat preparation whose softness comes from being dispensed at a relatively high temperature (20–22°F/–6°C).
- *Kulfi,* the Indian version of ice cream that may go back to the 16th century, is made without stirring from milk boiled down to a fraction of its original volume, and therefore concentrated in texture-smoothing milk proteins and sugar. It has a strong cooked-milk, butterscotch flavor.

Generally, premium-quality ice creams are made with more cream and egg yolks than less expensive types. They also contain less air. Hefting cartons is a quick way to estimate value; there can be as much cream and sugar in an expensive pint as there is in a cheap quart, which may be up to half empty space.

Making Ice Cream There are three basic steps in making ice cream: preparing the mix, freezing it, and hardening it.

Preparing the Mix The first step is to choose the ingredients and combine them. The basic ingredients are fresh cream and milk and table sugar. A mix made of up to 17% milk fat (equal volumes of whole milk and heavy cream) and 15% table sugar (¾ cup per quart/180 gm per liter of liquid) will be smooth when frozen quickly in kitchen ice cream makers. A smooth but lower-fat ice cream can be made by making a custard-style mix with egg yolks; or by replacing some of the cream with high-protein evaporated, condensed, or powdered milk; or replacing some of the sugar with thickening corn syrup.

In commercial practice, most or all of the mix ingredients are combined, then pasteurized, a step that also helps dissolve and hydrate the ingredients. If carried out at a high enough temperature (above 170°F/76°C), cooking can improve the body and smoothness of the ice cream by denaturing the whey proteins, which helps minimize the size of the ice crystals. Mixes that include egg yolks are always cooked until they just thicken. Simple home mixtures of cream and sugar can be frozen uncooked and have their own fresh flavor.

Freezing Once the mix has been prepared, it's prechilled to speed the subsequent freezing. It's then frozen as rapidly as possible in a container with coolant-chilled walls. The

Freezing Ice Cream
with Flying Fortresses and Liquid Nitrogen

On March 13, 1943, the *New York Times* reported that American fliers stationed in Britain had discovered an ingenious way of making ice cream while on duty. A story titled "Flying Fortresses Double as Ice-Cream Freezers" disclosed that the airmen "place prepared ice-cream mixture in a large can and anchor it to the rear gunner's compartment of a Flying Fortress. It is well shaken up and nicely frozen by flying over enemy territory at high altitudes."

These days, a popular, spectacular, and effective method among chemistry teachers is to freeze the mix in an open bowl with a gallon or two/8–10 liters of liquid nitrogen, whose boiling point is –320°F/–196°C. When the liquid nitrogen is stirred in, it boils, bubbles, and chills the mix almost instantly throughout, a combination that makes a very smooth—and initially very cold!—ice cream.

mix is stirred to expose it evenly to the cold walls, to incorporate some air, and above all to produce a smooth texture. Slow cooling of an unstirred mix—"quiescent cooling"—causes the formation of relatively few ice crystals that grow to a large size, grow together into clumps, and give a coarse, icy texture. Rapid cooling with stirring causes the quick production of many "seed" crystals which, because they share the available water molecules among themselves, cannot grow as large as a smaller population could; the agitation also helps prevent several crystals from growing into each other and forming a cluster that the tongue might notice. And many small crystals give a smooth, velvety consistency.

Hardening Hardening is the last stage in making ice cream. When the mix becomes thick and difficult to stir, only about half of its water has frozen into ice crystals. Agitation is then stopped, and the ice cream is finished with a period of quiescent freezing, during which another 40% of its water migrates onto existing ice crystals, leaving the various solid components less lubricated. If hardening is slow, some ice crystals take up more water than others and coarsen the texture. Hardening can be accelerated by dividing the newly frozen ice cream into several small containers whose greater surface area will release heat faster than one large container.

Storing and Serving Ice Cream Ice cream is best stored as cold as possible, at 0°F/–18°C or below, to preserve its smoothness. The inevitable coarsening during storage is due to repeated partial thawings and freezings, which melt the smallest ice crystals completely and deposit their water molecules on ever fewer, ever larger crystals. The lower the storage temperature, the slower this coarsening process.

The ice cream surface suffers in two ways during storage: its fat absorbs odors from the rest of the freezer compartment, and can be damaged and go rancid when dried out by the freezer air. These problems can be prevented simply by pressing plastic wrap directly into the surface, being careful not to leave air pockets.

Ideally, ice cream should be allowed to warm up from 0°F before being served. At 8–10°F/–13°C, it doesn't numb the tongue and taste buds as much, and it contains more liquid water, which softens the texture. At 22°F/–6°C—the typical temperature of soft-serve ice cream—half of the water is in liquid form.

FRESH FERMENTED MILKS AND CREAMS

One of the remarkable qualities of milk is that it invites its own preservation. It can spontaneously foster a particular group of microbes that convert its sugar into acid, and thereby preserve it for some time from spoiling or harboring disease. At the same time, the microbes also change the milk's texture and flavor in desirable ways. This benign transformation, or *fermentation,* doesn't happen all the time, but it happened often enough that milks fermented by bacteria became important among all dairying peoples. Yogurt and soured creams remain widely popular to this day.

Why this fortunate fermentation? It's a combination of milk's unique chemistry, and a group of microbes that were ready to exploit this chemistry long before mammals and milk arrived on earth. The *lactic acid bacteria* are what make possible the variety of fermented dairy products.

Lactic Acid Bacteria

Milk is rich in nutrients, but its most readily tapped energy source, lactose, is a sugar found almost nowhere else in nature. This means that not many microbes have the necessary digestive enzymes at the ready. The elegantly simple key to the success of the milk bacteria is that they specialize in digesting lactose, and they extract energy from lactose by breaking it down to lactic acid. Then they release the lactic acid into the milk, where it accumulates and retards the growth of most other microbes, including those that cause human disease. They also make some antibacterial substances, but their main defense is a pleasantly puckery tartness, one that also causes the casein proteins to gather together in semisolid curds (p. 20) and thicken the milk.

There are two major groups of lactic acid bacteria. The small genus *Lactococcus* (a combination of the Latin for "milk" and "sphere") is found primarily on plants (but it's a close relative of *Streptococcus*, whose members live mainly on animals and cause a number of human diseases!). The 50-odd members of the genus *Lactobacillus* ("milk" and "rod") are more widespread in nature. They're found both on plants and in animals, including the stomach of milk-fed calves and the human mouth, digestive tract, and vagina; and their clean living generally benefits our insides (see box, p. 47).

The bacteria responsible for the major fermented products were identified around 1900, and pure cultures of individual strains became available then. Nowadays, few dairies leave their fermentations to chance. Where traditional spontaneously fermented products may involve a dozen or more different microbes, the industrial versions are usually limited to two or three. This biological narrowing may affect flavor, consistency, and health value.

Families of Fresh Fermented Milks

Unlike most cheeses (p. 51), which undergo several stages of manipulation and continue to evolve for weeks or months, fresh fermented milks are usually finished and ready for eating within hours or days. A recent encyclopedia catalogued several hundred different kinds! Most of them originated in western Asia, eastern Europe, and Scandinavia, and have been carried across the globe by countless emigrants, many of whom dipped a cloth in their family's culture, dried it gently, and guarded it until they could moisten it in the milk of their new home.

The handful of fresh fermented milks familiar in the West, yogurt and soured creams and buttermilk, represent two major families that developed from the dairying habits of peoples in two very different climates.

Yogurt and its relatives are native to a broad and climatically warm area of central and southwest Asia and the Middle East, an

The curdling of milk by lactic acid bacteria. As the bacteria ferment lactose and produce lactic acid, the increasingly acid conditions cause the normal bundled micelles of casein proteins (left) *to fall apart into separate casein molecules, and then rebond to each other* (right). *This general rebonding forms a continuous meshwork of protein molecules that traps the liquid and fat globules in small pockets, and turns the fluid milk into a fragile solid.*

Traditional Fresh Fermented Milks and Creams

Product	Region	Microbes	Fermentation Temperature, Time	Acidity	Characteristics
Yogurt	Middle East to India	*Lactobacillus delbrueckii, Streptococcus salivarius* (in rural areas, assorted lactococci and lactobacilli)	106–114°F/41–45°C, 2–5 hrs, or 86°F/30°C, 6–12 hrs	1–4%	Tart, semisolid, smooth; green aroma
Buttermilk	Eurasia	*Lactococcus lactis, Leuconostoc mesenteroides*	72°F/22°C, 14–16 hrs	0.8–1.1%	Tart, thickened liquid; buttery aroma
Crème fraîche	Europe	*Lactococcus lactis, Leuconostoc mesenteroides*	68°F/20°C, 15–20 hrs	0.2–0.8%	Mildly tart and thickened; buttery aroma
Sour cream	Europe	*Lactococcus lactis, Leuconostoc mesenteroides*	72°F/22°C, 16 hrs	0.8%	Mildly tart, semisolid; buttery aroma
Ropy milks	Scandinavia	*Lactococcus lactis, Leuconostoc mesenteroides* (*Geotrichum* mold)	68°F/20°C, 18 hrs	0.8%	Mildly tart, semisolid, slimy; buttery aroma
Koumiss	Central Asia	Lactobacilli, yeasts	80°F/27°C, 2–5 hrs plus cool aging	0.5–1%	Mildly tart, thickened liquid; effervescent, 0.7–2.5% alcohol
Kefir	Central Asia	Lactococci, Lactobacilli, *Acetobacter,* yeasts	68°F/20°C, 24 hrs	1%	Tart, thickened liquid; effervescent, 0.1% alcohol

area that includes the probable home of dairying, and where some peoples still store milk in animal stomachs and skins. The lactobacilli and streptococci that produce yogurt are "thermophilic," or heat-loving species that may have come from the cattle themselves. They're distinguished by their ability to grow rapidly and synergistically at temperatures up to 113°F/45°C, and to generate high levels of preservative lactic acid. They can set milk into a very tart gel in just two or three hours.

Sour cream, crème fraîche, and buttermilk are indigenous to relatively cool western and northern Europe, where milk spoils more slowly and was often left overnight to separate into cream for buttermaking. The lactococci and *Leuconostoc* species that produce them are "mesophilic," or moderate-temperature lovers that probably first got into milk from particles of pasturage on the cows' udders. They prefer temperatures around 85°F/30°C but will work well below that range, and develop moderate levels of lactic acid during a slow fermentation lasting 12 to 24 hours.

Yogurt

Yogurt is the Turkish word for milk that has been fermented into a tart, semisolid mass; it comes from a root meaning

The Health Benefits of Fermented Milks

The bacteria in dairy products may do more for us than just predigest lactose and create flavor. Recent research findings lend some support to the ancient and widespread belief that yogurt and other cultured milks can actively promote good health. Early in the 20th century, the Russian Nobelist Ilya Metchnikov (who discovered that white blood cells fight bacterial infection) gave a scientific rationale to this belief, when he proposed that the lactic acid bacteria in fermented milks eliminate toxic microbes in our digestive system that otherwise shorten our lives. Hence Dr. James Empringham's charming title of 1926: *Intestinal Gardening for the Prolongation of Youth.*

Metchnikov was prescient. Research over the last couple of decades has established that certain lactic acid bacteria, the Bifidobacteria, are fostered by breast milk, do colonize the infant intestine, and help keep it healthy by acidifying it and by producing various antibacterial substances. Once we're weaned onto a mixed diet, the Bifidobacterial majority in the intestine recedes in favor of a mixed population of *Streptococcus, Staphylococcus, E. coli,* and yeasts. The standard industrial yogurt and buttermilk bacteria are specialized to grow well in milk and can't survive inside the human body. But other bacteria found in traditional, spontaneously fermented milks—*Lactobacillus fermentum, L. casei,* and *L. brevis,* for example—as well as *L. plantarum* from pickled vegetables, and the intestinal native *L. acidophilus,* do take up residence in us. Particular strains of these bacteria variously adhere to and shield the intestinal wall, secrete antibacterial compounds, boost the body's immune response to particular disease microbes, dismantle cholesterol and cholesterol-consuming bile acids, and reduce the production of potential carcinogens.

These activities may not amount to prolonging our youth, but they're certainly desirable! Increasingly, manufacturers are adding "probiotic" Lactobacilli and even Bifidobacteria to their cultured milk products, and note that fact on the label. Such products, approximations of the original fermented milks that contained an even more diverse bacterial flora, allow us to plant our inner gardens with the most companionable microbes we're currently aware of.

"thick." Essentially the same product has been made for millennia from eastern Europe and North Africa across central Asia to India, where it goes by a variety of names and is used for a variety of purposes: it's eaten on its own, diluted into drinks, mixed into dressings, and used as an ingredient in soups, baked goods, and sweets.

Yogurt remained an exotic curiosity in Europe until early in the 20th century, when the Nobel Prize–winning immunologist Ilya Metchnikov connected the longevity of certain groups in Bulgaria, Russia, France, and the United States with their consumption of fermented milks, which he theorized would acidify the digestive tract and prevent pathogenic bacteria from growing (see box, p. 47). Factory-scale production and milder yogurts flavored with fruit were developed in the late 1920s, and broader popularity came in the 1960s with Swiss improvements in the inclusion of flavors and fruits and the French development of a stable, creamy stirred version.

The Yogurt Symbiosis By contrast to the complex and variable flora of traditional yogurts, the industrial version is reduced to the essentials. Standard yogurt contains just two kinds of bacteria, *Lactobacillus delbrueckii* subspecies *bulgaricus*, and *Streptococcus salivarius* subspecies *thermophilus*. Each bacterium stimulates the growth of the other, and the combination acidifies the milk more rapidly than either partner on its own. Initially the streptococci are most active. Then as the acidity exceeds 0.5%, the acid-sensitive streptococci slow down, and the hardier lactobacilli take over and bring the final acidity to 1% or more. The flavor compounds produced by the bacteria are dominated by acetaldehyde, which provides the characteristic refreshing impression of green apples.

Making Yogurt There are two basic stages in yogurt making: preparing the milk by heating and partly cooling it; and fermenting the warm milk.

The Milk Yogurt is made from all sorts of milk; sheep and goat were probably the first. Reduced-fat milks make especially firm yogurt because manufacturers mask their lack of fat by adding extra milk proteins, which add density to the acid-coagulated protein network. (Manufacturers may also add gelatin, starch, and other stabilizers to help prevent separation of whey and curd from physical shocks during transportation and handling.)

Heating the Milk Traditionally the milk for yogurt was given a prolonged boiling to concentrate the proteins and give a firmer texture. Today, manufacturers can boost protein content by adding dry milk powder, but they still cook the milk, for 30 minutes at 185°F/85°C or at 195°F/90°C for 10 minutes. These treatments improve the consistency of the yogurt by denaturing the whey protein lactoglobulin, whose otherwise unreactive molecules then participate by clustering on the surfaces of the casein particles (p. 20). With the helpful interference of the lactoglobulins, the casein particles can only bond to each other at a few spots, and so gather not in clusters but in a fine matrix of chains that is much better at retaining liquid in its small interstices.

The Fermentation Once the milk has been heated, it's cooled down to the desired fermentation temperature, the bacteria are added (often in a portion of the previous batch), and the milk kept warm until it sets. The fermentation temperature has a strong influence on yogurt consistency. At the maximum temperature well tolerated by the bacteria, 104–113°F/40–45°C, the bacteria grow and produce lactic acid rapidly, and the milk proteins gel in just two or three hours; at 86°F/30°C, the bacteria work far more slowly, and the milk takes up to 18 hours to set. Rapid gelling produces a relatively coarse protein network whose few thick strands give it firmness but also readily leak whey; slow gelling produces a finer, more delicate, more intricately branched network whose individual strands are weaker but whose smaller pores are better at retaining the whey.

Frozen Yogurt Frozen yogurt became popular in the 1970s and '80s as a low-fat, "healthy" alternative to ice cream. In fact, frozen yogurt is essentially ice milk whose mix includes a small dose of yogurt; the standard proportion is 4 to 1. Depending on the mixing procedure, the yogurt bacteria may survive in large numbers or be largely eliminated.

Soured Creams and Buttermilk, Including Crème Fraîche

Before the advent of the centrifugal separator, butter was made in western Europe by allowing raw milk to stand overnight or longer, skimming off the cream that rose to the top, and churning the cream. During the hours of gravity separation, bacteria would grow spontaneously in the milk and give the cream and the butter made from it a characteristic aroma and tartness.

"Cream cultures" is a convenient shorthand for products that are now intentionally seeded with these same bacteria, which are various species of *Lactococcus* and *Leuconostoc,* and have three important characteristics. They grow best at moderate temperatures, well below the typical temperature of yogurt fermentation; they're only moderate acid-producers, so the milks and creams they ferment never get extremely sour; and certain strains have the ability to convert a minor milk component, citrate, into a warmly aromatic compound called diacetyl that miraculously complements the flavor of butterfat. It's fascinating that this single bacterial product is so closely associated with the flavor of butter that all by itself, diacetyl makes foods taste buttery: even chardonnay wines (p. 730). To accentuate this flavor note, manufacturers sometimes add citrate to the milk or cream before fermentation, and they ferment in the cool conditions that favor diacetyl production.

Crème Fraîche Crème fraîche is a versatile preparation. Thick, tart, and with an aroma that can be delicately nutty or buttery, it is a wonderful complement to fresh fruit, to caviar, and to certain pastries. And thanks to its high fat and correspondingly low protein content, it can be cooked in a sauce or even boiled down without curdling.

In France today, crème fraîche means cream with 30% fat that has been pasteurized at moderate temperatures, not UHT pasteurized (p. 22) or sterilized. (*Fraîche* means "cool" or "fresh.") It may, however, be either liquid (*liquide, fleurette*) or thick (*épaisse*). The liquid version is unfermented and has an official refrigerated shelf life of 15 days. The thick version is fermented with the typical cream culture for 15 to 20 hours, and has a shelf life of 30 days. As with all fermented milks, the thickening is an indication that the product has reached a certain acidity (0.8%, pH 4.6) and so a distinct tartness. Commercial American crème fraîche is made essentially as the French fermented version is, though some manufacturers add a small amount of rennet for a thicker consistency. A distinctly buttery flavor is found in products made with Jersey and Guernsey milks (rich in citrate) and with diacetyl-producing strains of bacteria.

Making Crème Fraîche in the Kitchen A home version of crème fraîche can be made by adding some cultured buttermilk or sour cream, which contain cream-culture bacteria, to heavy cream (1 tablespoon per cup/15 ml per 250 ml), and letting it stand at a cool room temperature for 12 to 18 hours or until thick.

Sour Cream Sour cream is essentially a leaner, firmer, less versatile version of crème fraîche. At around 20% milk fat, it contains enough protein that cooking temperatures will curdle it. Unless it is used to enrich a dish just before serving, then, it will give a slightly grainy appearance and texture. Sour cream is especially prominent in central and eastern Europe, where it has traditionally been added to soups and stews (goulash, borscht). Immigrants brought a taste for it to American cities in the 19th century, and by the middle of the 20th it had become fully naturalized as a base for dips and salad dressings, a topping for baked potatoes, and an ingredient in cakes. American sour cream is heavier-bodied than the European original thanks to the practice of passing the cream through a homogenizer twice before culturing it. A small dose of rennet is sometimes added with the bacteria; this enzyme causes the casein proteins to coagulate into a firmer gel.

A nonfermented imitation called "acidified sour cream" is made by coagulating the cream with pure acid. "Sour creams" labeled "low-fat" and "nonfat" replace butterfat with starch, plant gums, and dried milk protein.

Buttermilk Most "buttermilk" sold in the United States is not buttermilk at all. True buttermilk is the low-fat portion of milk or cream remaining after it has been churned to make butter. Traditionally, that milk or cream would have begun to ferment before churning, and afterwards the buttermilk would continue to thicken and develop flavor. With the advent of centrifugal cream separators in the 19th century, buttermaking produced "sweet" unfermented buttermilk, which could be sold as such or cultured with lactic bacteria to develop the traditional flavor and consistency. In the United States, a shortage of true buttermilk shortly after World War II led to the success of an imitation, "cultured buttermilk," made from ordinary skim milk and fermented until acid and thick.

What's the difference? True buttermilk is less acid, subtler and more complex in flavor, and more prone to off-flavors and spoilage. Its remnants of fat globule membranes are rich in emulsifiers like lecithin, and make it especially valuable for preparing smooth, fine-textured foods of all kinds, from ice cream to baked goods. (Its excellence for emulsifying led to the Pennsylvania Dutch using it as a base for red barn paint!) Cultured buttermilk is useful too; it imparts a rich, tangy flavor and tenderness to griddle cakes and many baked goods.

U.S. "cultured buttermilk" is made by giving skim or low-fat milk the standard yogurt heat treatment to produce a finer protein gel, then cooling it and fermenting it with cream cultures until it gels. The gelled milk is cooled to stop the fermentation and gently agitated to break the curd into a thick but smooth liquid. "Bulgarian buttermilk" is a version of cultured buttermilk in which the cream cultures are supplemented or replaced by yogurt cultures, and fermented at a higher temperature to a higher acidity. It's noticeably more tart and gelatinous, with the apple-like sharpness typical of yogurt.

Ropy Scandinavian Milks

A distinctive subfamily among the cream cultures are the "ropy" milks of Scandinavia, so-called because they're more than stringy: lift a spoonful of Finnish *viili,* Swedish *långfil,* or Norwegian *tättemjölk,* and the rest of the bowl follows it into the air. Some ropy milks are so cohesive that they're cut with a knife. This consistency is created by particular strains of cream culture bacteria that produce long strands of starch-like carbohydrate. The stretchy carbohydrate absorbs water and sticks to casein particles, so manufacturers are using ropy strains of *Streptococcus salivarius* as natural stabilizers of yogurt and other cultured products.

Cooking with Fermented Milks

Most cultured milk products are especially susceptible to curdling when made into sauces or added to other hot foods. Fresh milk and cream are relatively stable, but the extended heat treatment and high acidity characteristic of cultured products have already caused some protein coagulation. Anything the cook does to push this coagulation further will cause the protein network to shrink and squeeze out some of the whey and produce distinct white particles—protein curds—floating in the thinned liquid. Heat, salt, additional acid, and vigorous stirring can all cause curdling. The key to maintaining a smooth texture is gentleness. Heat gradually and moderately, and stir slowly.

There is a common misconception that crème fraîche is uniquely immune to curdling. It's true that while yogurt, sour cream, and buttermilk all will curdle if they get anywhere near the boil, crème fraîche can be boiled with impunity. But this versatility has nothing to do with fermentation: it's a simple matter of fat content. Heavy cream, at 38 to 40% fat, has so little protein that it doesn't form noticeable curds (p. 29).

CHEESE

Cheese is one of the great achievements of humankind. Not any cheese in particular, but cheese in its astonishing multiplicity, created anew every day in the dairies of the world. Cheese began as a simple way of concentrating and preserving the bounty of the milking season. Then the attentiveness and ingenuity of its makers slowly transformed it into something more than mere physical nourishment: into an intense, concentrated expression of pastures and animals, of microbes and time.

The Evolution of Cheese

Cheese is a modified form of milk that is more concentrated, more durable, and more flavorful food than milk is. It's made more concentrated by curdling milk and removing much of its water. The nutritious curds of protein and fat are made more durable

Unusual Fermented Milks: Koumiss and Kefir

Because milk contains an appreciable amount of the sugar lactose, it can be fermented like grape juice and other sugary fluids into an alcoholic liquid. This fermentation requires unusual lactose-fermenting yeasts (species of *Saccharomyces, Torula, Candida,* and *Kluyveromyces*). For thousands of years, the nomads of central Asia have made *koumiss* from mare's milk, which is especially rich in lactose, and this tart, effervescent drink, with 1–2% alcohol and 0.5–1% acid, remains very popular there and in Russia. Other European and Scandinavian peoples have made alcoholic products from other milks, as well as sparkling "wine" from whey.

Another remarkable fermented milk little known in the West is *kefir,* which is most popular in the Caucasus and may well have originated there. Unlike other fermented milks, in which the fermenting microbes are evenly dispersed, kefir is made by large, complex particles known as kefir grains, which house a dozen or more kinds of microbes, including lactobacilli, lactococci, yeasts, and vinegar bacteria. This symbiotic association grows at cool room temperatures to produce a tart, slightly alcoholic, effervescent, creamy product.

by the addition of acid and salt, which discourage the growth of spoilage microbes. And they're made more flavorful by the controlled activity of milk and microbe enzymes, which break the protein and fat molecules apart into small flavorful fragments.

The long evolution of cheese probably began around 5,000 years ago, when people in warm central Asia and the Middle East learned that they could preserve naturally soured, curdled milk by draining off the watery whey and salting the concentrated curds. At some point they also discovered that the texture of the curd became more pliable and more cohesive if the curdling took place in an animal stomach or with pieces of stomach in the same container. These first cheeses may have resembled modern brine-cured feta, which is still an important cheese type in the eastern Mediterranean and the Balkans. The earliest good evidence of cheesemaking known to date, a residue found in an Egyptian pot, dates from around 2300 BCE.

The Ingredient Essential to Diverse Cheeses: Time This basic technique of curdling milk with the help of the stomach extract now called rennet, then draining and brining the curds, was eventually carried west and north into Europe. Here people gradually discovered that curds would keep well enough in these cooler regions with much milder treatments: a less puckery souring and only a modest brining or salting. This was the discovery that opened the door to the great diversification of cheeses, because it introduced a fifth ingredient after milk, milk bacteria, rennet, and salt: time. In the presence of moderate acidity and salt, cheese became a hospitable medium for the continuing growth and activity of a variety of microbes and their enzymes. In a sense, cheese came to life. It became capable of pronounced development and change; it entered the cyclical world of birth, maturation, and decline.

When were modern cheeses born? We don't really know, but it was well before Roman times. In his *Rei rusticae* ("On Rustic Matters," about 65 CE), Columella describes at length what amounts to standard cheesemaking practice. The curdling was done with rennet or various plant fluids. The whey was pressed out, the curds sprinkled with salt, and the fresh cheese put in a shady place to harden. Salting and hardening were repeated, and the ripe cheese was then washed, dried, and packed for storage and shipping. Pliny, who also wrote in the first century, said that Rome most esteemed cheeses from its provincial outposts, especially Nîmes in southern France, and the French and Dalmatian Alps.

The Growth of Diversity During the 10 or 12 centuries after Rome's strong rule, the art of cheesemaking progressed in the feudal estates and monasteries, which worked

Cheeses as Artifacts

Behind every cheese there is a pasture of a different green under a different sky: meadows encrusted with salt that the tides of Normandy deposit every evening; meadows perfumed with aromas in the windy sunlight of Provence; there are different herds, with their shelters and their movements across the countryside; there are secret methods handed down over the centuries. This shop is a museum: Mr. Palomar, visiting it, feels as he does in the Louvre, behind every displayed object the presence of the civilization that gave it form and takes form from it.

—Italo Calvino, *Palomar,* 1983

steadily at settling in forested areas or mountain meadows and clearing the land for grazing. These widely dispersed communities developed their cheesemaking techniques independently to suit their local landscape, climate, materials, and markets. Small, perishable soft cheeses, often made from the milk of a few household animals, were consumed locally and quickly and could only be sent to nearby towns. Large hard cheeses required the milk of many animals and were often made by cooperatives (the Gruyère *fruiteries* began around 1200); they kept indefinitely and could be transported to market from distant regions. The result was a remarkable diversity of traditional cheeses, which number from 20 to 50 in most countries and several hundred in France alone, thanks to its size and range of climates.

Charlemagne Learns to Eat Moldy Cheese

During the Middle Ages, when cheese was evolving into a finely crafted food, even an emperor of France had to learn a thing or two about how to appreciate it. About 50 years after Charlemagne's death in 814, an anonymous monk at the monastery of Saint Gall wrote a biography of him that includes this fascinating anecdote (slightly modified from *Early Lives of Charlemagne*, transl. A. J. Grant, 1922). Charlemagne was traveling, and found himself at a bishop's residence at dinnertime.

> Now on that day, being the sixth day of the week, he was not willing to eat the flesh of beast or bird. The bishop, being by reason of the nature of the place unable to procure fish immediately, ordered some excellent cheese, white with fat, to be placed before him. Charles . . . required nothing else, but taking up his knife and throwing away the mold, which seemed to him abominable, he ate the white of the cheese. Then the bishop, who was standing nearby like a servant, drew close and said "Why do you do that, lord Emperor? You are throwing away the best part." On the persuasion of the bishop, Charles . . . put a piece of the mold in his mouth, and slowly ate it and swallowed it like butter. Then, approving the bishop's advice, he said "Very true, my good host," and he added, "Be sure to send me every year to Aix two cartloads of such cheeses."

The word I've translated as "mold" is *aerugo* in the Latin: literally, "the rust of copper." The cheese isn't named, and some writers have deduced that it was a Brie, which then had an external coat of gray-green mold, much the same color as weathered copper. But I think it was probably more like Roquefort, a sheep's-milk cheese veined internally with blue-green mold. The rest of the anecdote fits a large, firm, internally ripened cheese better than a thin, soft Brie. It also marks what may have been the first appointment of an official cheese affineur!

> The bishop was alarmed at the impossibility of the task and . . . rejoined: "My lord, I can procure the cheeses, but I cannot tell which are of this quality and which of another. . . ." Then Charles . . . spoke thus to the bishop, who from childhood had known such cheeses and yet could not test them. "Cut them in two," he said, "then fasten together with a skewer those that you find to be of the right quality and keep them in your cellar for a time and then send them to me. The rest you may keep for yourself and your clergy and family."

Cheeses of Reputation The art of cheesemaking had progressed enough by late medieval times to inspire connoisseurship. The French court received shipments from Brie, Roquefort, Comté, Maroilles, and Geromé (Münster). Cheeses made near Parma in Italy and near Appenzell in Switzerland were renowned throughout Europe. In Britain, Cheshire cheese was famous by Elizabethan times, and Cheddar and Stilton by the 18th century. Cheese played two roles: for the poor, fresh or briefly ripened types were staple food, sometimes called "white meat," while the rich enjoyed a variety of aged cheeses as one course of their multicourse feasts. By the early 19th century, the French gastronome Brillat-Savarin found cheese to be an aesthetic necessity: he wrote that "a dessert without cheese is like a beautiful woman who is missing an eye." The golden age of cheese was probably the late 19th and early 20th centuries, when the art was fully developed, local styles had developed and matured, and the railroads brought country products to the city while they were still at their best.

Modern Decline The modern decline of cheesemaking has its roots in that same golden age. Cheese and butter factories were born in the United States, a country with no cheesemaking tradition, just 70 years after the Revolution. In 1851, an upstate New York dairy farmer named Jesse Williams agreed to make cheese for neighboring farms, and by the end of the Civil War there were hundreds of such "associated" dairies, whose economic advantages brought them success throughout the industrialized world. In the 1860s and '70s, pharmacies and then pharmaceutical companies began mass-producing rennet. At the turn of the century scientists in Denmark, the United States, and France brought more standardization in the form of pure microbial cultures for curdling and ripening cheese, which had once been accomplished by the local, complex flora of each cheesemaker's dairy.

The crowning blow to cheese diversity and quality was World War II. In continental Europe, agricultural lands became battlefields, and dairying was devastated. During the prolonged recovery, quality standards were suspended, factory production was favored for its economies of scale and ease of regulation, and consumers were grateful for any approximation of the prewar good life. Inexpensive standardized cheese rose to dominance. Ever since, most cheese in Europe and the United States has been made in factories. Even in France, which in 1973 established a certification program ("*Fromage appellation d'origine contrôlée*") to indicate that a cheese has been made by traditional methods and in the traditional area of production, less than 20% of the total national production qualifies. In the United States, the market for *process cheese,* a mixture of aged and fresh cheeses blended with emulsifiers and repasteurized, is now larger than the market for "natural" cheese, which itself is almost exclusively factory-made.

At the beginning of the 21st century, most cheese is an industrial product, an expression not of diverse natural and human particulars, but of the monolithic imperatives of standardization and efficient mass production. Industrial cheese also requires great ingenuity, has its economic merits, and suits its primary role as an ingredient in fast-food sandwiches, snacks, and prepared foods (a role that doubled U.S. per capita cheese consumption between 1975 and 2001). But in its own way, industrial cheese is a throwback to primitive cheese, a simplified food that could be and is made anywhere, and that tastes of nowhere in particular.

The Revival of Tradition and Quality Though finely crafted cheeses will always be a minor part of modern dairy production, recent years have brought modest signs of hope. The postwar era and its economic limitations have faded. Some European countries have seen a revival of appreciation for traditional cheeses, and air travel has brought them to the attention of an ever-growing number of food lovers. Once "white meat" for the rural poor, they are now pricey treats for the urban middle class. In the United States, a few small producers blend respect for tradition with 21st-century understanding, and make superb cheeses of their own. For enthusiasts willing to seek them out, the world still offers delightful expressions of this ancient craft.

The Ingredients of Cheese

The three principal ingredients of cheese are milk, rennet enzymes that curdle the milk, and microbes that acidify and flavor the milk. Each strongly influences the character and quality of the final cheese.

Milks Cheese is milk concentrated five- to tenfold by the removal of water; so the basic character of the milk defines the basic character of the cheese. Milk character is in turn determined by the kind of animal that produces it, what the animal eats, the microbes that inhabit the milk, and whether it is raw or pasteurized.

Species The milks of cows, sheep, and goats taste different from each other (p. 21), and their cheeses do too. Cow's milk is more neutral than other milks. Sheep and buffalo milk have relatively high fat and protein contents and therefore make richer cheeses. Goat's milk has a relatively low proportion of curdle-able casein and usually produces a crumbly, less cohesive curd compared to other milks.

Breed During the spread of cheesemaking in the Middle Ages, hundreds of different dairy animal varieties were bred to make the best use of local pasturage. The Brown Swiss is thought to go back several thousand years. Today, most of these locally adapted breeds have been replaced by the omnipresent black-and-white Holstein or Friesian, bred to maximize the milk it yields on standardized feed. Traditional breeds produce a lower volume of milk, but a milk richer in protein, fat, and other desirable cheese constituents.

Feed: The Influence of the Seasons Today most dairy animals are fed year-round on silage and hay made from just a few fodder crops (alfalfa, maize). This standard regimen produces a standard, neutral milk that can be made into very good cheese. However, herds let out to pasture to eat fresh greenery and flowers give milk of greater aromatic complexity that can make extraordinary cheese. Thanks to newly sensitive analytical instruments, dairy chemists have recently verified what connoisseurs have known for centuries: an animal's diet influences its milk and the cheese made from it. French studies of alpine Gruyère found a larger number of flavor compounds in cheeses made during summer pasturage compared to winter stable feeding, and more herbaceous and floral terpenes and other aromatics (p. 273) in mountain cheeses than cheeses from the high plateaus, which in turn have more than cheeses from the plains (alpine meadows have more diverse vegetation than the grassy lowlands).

Like fruits, cheeses made from pasture-fed animals are seasonal. The season depends on the local climate—the summer is green in the Alps, the winter in California—and how long it takes a particular cheese to mature. Cheeses made from pasturage are generally recognizable by their deeper yellow color, due to the greater content of carotenoid pigments in fresh vegetation (p. 267). (Bright orange cheeses have been dyed.)

Pasteurized and Raw Milks In modern cheese production, the milk is almost always pasteurized to eliminate disease and spoilage bacteria. This is really a practical necessity in industrial cheesemaking, which requires that milk be pooled and stored from many farms and thousands of animals. The risk of contamination—which only takes one diseased cow or dirty udder—is too great. Since the late 1940s, the U.S. Food and Drug Administration has required that any cheese made from unpasteurized, "raw" milk must be aged a minimum of 60 days at a temperature above 35°F/2°C, conditions that are thought to eliminate whatever pathogens might have been in the milk; and since the early 1950s it has also banned the import of raw-milk cheeses aged less than 60 days. This means that soft cheeses made with raw milk are essentially contraband in the United States. The World Health Organization has considered recommending a complete ban on the production of raw-milk cheeses.

Of course until barely a century ago, nearly all cheeses were made in small batches with raw milk, fresh from the udders of small herds whose health was more easily monitored. And French, Swiss, and Italian regulations actually *forbid* the use of pasteurized milk for the traditional production of a number of the world's greatest cheeses, including Brie, Camembert, Comté, Emmental, Gruyère, and Parmesan. The reason is that pasteurization kills useful milk bacteria, and inactivates many of the milk's own enzymes. It thus eliminates two of the four or five sources of flavor development during ripening, and prevents traditional cheeses from living up to their own standards of excellence.

Pasteurization is no guarantee of safety, because the milk or cheese can be contaminated during later processing. Nearly all outbreaks of food poisoning from milk or cheese in recent decades have involved pasteurized products. It will be genuine progress when public health officials help ambitious cheesemakers to ensure the safety of raw-milk cheeses, rather than making rules that restrict consumer choice without significantly reducing risk.

The Key Catalyst: Rennet The making and use of rennet was humankind's first venture in biotechnology. At least 2,500 years ago, shepherds began to use pieces of the first stomach of a young calf, lamb, or goat to curdle milk for cheese; and sometime later they began to make a brine extract from the stomach. That extract was the world's first semipurified enzyme. Now, by means of genetic engineering, modern biotechnology produces a pure version of the same calf enzyme, called *chymosin,* in a bacterium, a mold, and a yeast. Today, most cheese in the United States is made with these engineered "vegetable rennets," and less than a quarter with traditional rennet from calf stomach (which is often required for traditional European cheeses).

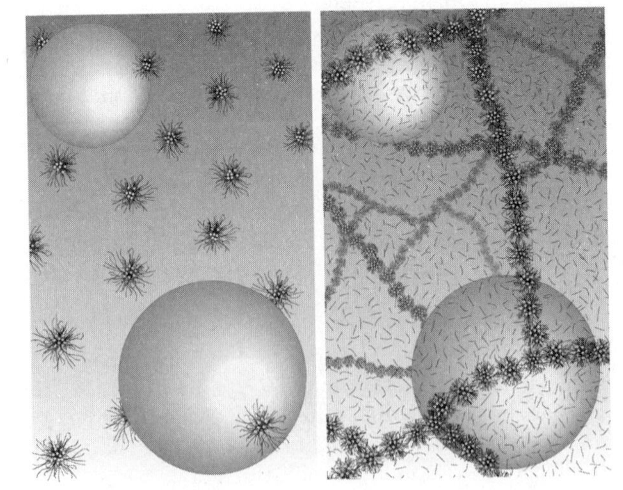

The curdling of milk by the rennet enzyme chymosin. The bundled micelles of casein in milk are kept separate from each other by electrically charged micelle components that repel each other (left). *Chymosin selectively trims away these charged kappa-caseins, and the now uncharged micelles bond to each other to form a continuous meshwork* (right). *The liquid milk coagulates into a moist solid.*

The Curdling Specialist Traditional rennet is made from the fourth stomach or abomasum of a milk-fed calf less than 30 days old, before chymosin is replaced by other protein-digesting enzymes. The key to rennet's importance in cheesemaking is chymosin's specific activity. Where other enzymes attack most proteins at many points and break them into many pieces, chymosin effectively attacks only one milk protein, and at just one point. Its target is the negatively charged kappa-casein (p. 19) that repels individual casein particles from each other. By clipping these pieces off, chymosin allows the casein particles to bond to each other and form a continuous solid gel, the curd.

Since plain acidity alone causes milk to curdle, why do cheesemakers need rennet at all? There are two reasons. First, acid disperses the casein micelle proteins and their calcium glue before it allows the proteins to come together, so some casein and most of the calcium are lost in the whey, and the remainder forms a weak, brittle curd. By contrast, rennet leaves the micelles mostly intact and causes each to bond to several others into a firm, elastic curd. Second, the acidity required to curdle casein is so high that flavor-producing enzymes in the cheese work very slowly or not at all.

Cheese Microbes Cheeses are decomposed and recomposed by a colorful cast of microbes, perhaps a handful in most modern cheeses made with purified cultures, but dozens in some traditional cheeses made with a portion of the previous batch's starter.

Starter Bacteria First there are the lactic acid bacteria that initially acidify the milk, persist in the drained curd, and generate much of the flavor during the ripening of many semihard and hard cheeses, including Cheddar, Gouda, and Parmesan. The numbers of live *starter* bacteria in the curd often drop drastically during cheesemaking, but their enzymes survive and continue to work for months, breaking down proteins into savory amino acids and aromatic by-products (see box, p. 62). There are two broad groups of starters: the moderate-temperature lactococci that are also used to make cultured creams, and the heat-loving lactobacilli and streptococci that are also used to make yogurt (p. 48). Most cheeses are acidified by the mesophilic group, while the few that undergo a cooking step—mozzarella, the alpine and Italian hard cheeses—are acidified by thermophiles that can survive and continue to contribute flavor. Many Swiss and Italian starters are still only semidefined mixtures of heat-loving milk bacteria, and are made the old-fashioned way, from the whey of the previous batch.

True "Vegetable Rennets" from Thistle Flowers

It has been known at least since Roman times that some plant materials can curdle milk. Two have been used for centuries to make a distinctive group of cheeses. In Portugal and Spain, flowers of the wild cardoon thistles (*Cynara cardunculus* and *C. humilis*) have long been collected and dried in the summer, and then soaked in warm water in the winter to make sheep and goat cheeses (Portuguese Serra, Serpa, Azeitão; Spanish Serena, Torta del Casar, Pedroches). The cardoon rennets are unsuited to cow's milk, which they curdle but also turn bitter. Recent research has revealed that Iberian shepherds had indeed found a close biochemical relative of calf chymosin, which the thistle flower happens to concentrate in its stigmas.

The Propionibacteria An important bacterium in Swiss starter cultures is *Propionibacter shermanii,* the hole-maker. The propionibacteria consume the cheese's lactic acid during ripening, and convert it to a combination of propionic and acetic acids and carbon dioxide gas. The acids' aromatic sharpness, together with buttery diacetyl, contributes to the distinctive flavor of Emmental, and the carbon dioxide forms bubbles, or the characteristic "holes." The propionibacteria grow slowly, and the cheesemaker must coddle them along by ripening the cheese at an unusually high temperature—around 75°F/24°C—for several weeks. This need for warmth may reflect the cheese propionibacteria's original home, which was probably animal skin. (At least three other species of propionibacteria inhabit moist or oily areas of human skin, and *P. acnes* takes advantage of plugged oil glands.)

The Smear Bacteria The bacterium that gives Münster, Epoisses, Limburger, and other strong cheeses their pronounced stink, and contributes more subtly to the flavor of many other cheeses, is *Brevibacterium linens.* As a group, the brevibacteria appear to be natives of two salty environments: the seashore and human skin. Brevibacteria grow at salt concentrations that inhibit most other microbes, up to 15% (seawater is just 3%). Unlike the starter species, the brevibacteria don't tolerate acid and need oxygen, and grow only on the cheese surface, not inside. The cheesemaker encourages them by wiping the cheese periodically with brine, which causes a characteristic sticky, orange-red "smear" of brevibacteria to develop. (The color comes from a carotene-related pigment; exposure to light usually intensifies the color.) They contribute a more subtle complexity to cheeses that are wiped for only part of the ripening (Gruyère) or are ripened in humid conditions (Camembert). Smear cheeses are so reminiscent of cloistered human skin because both *B. linens* and its human cousin, *B. epidermidis,* are very active at breaking down protein

Why Some People Can't Stand Cheese

The flavor of cheese can provoke ecstasy in some people and disgust in others. The 17th century saw the publication of at least two learned European treatises "*de aversatione casei,*" or "on the aversion to cheese." And the author of "*Fromage*" in the 18th-century *Encyclopédie* noted that "cheese is one of those foods for which certain people have a natural repugnance, of which the cause is difficult to determine." Today the cause is clearer. The fermentation of milk, like that of grains or grapes, is essentially a process of limited, controlled spoilage. We allow certain microbes and their enzymes to decompose the original food, but not beyond the point of edibility. In cheese, animal fats and proteins are broken down into highly odorous molecules. Many of the same molecules are also produced during uncontrolled spoilage, as well as by microbial activity in the digestive tract and on moist, warm, sheltered areas of human skin.

An aversion to the odor of decay has the obvious biological value of steering us away from possible food poisoning, so it's no wonder that an animal food that gives off whiffs of shoes and soil and the stable takes some getting used to. Once acquired, however, the taste for partial spoilage can become a passion, an embrace of the earthy side of life that expresses itself best in paradoxes. The French call a particular plant fungus the *pourriture noble,* or "noble rot," for its influence on the character of certain wines, and the Surrealist poet Leon-Paul Fargue is said to have honored Camembert cheese with the title *les pieds de Dieu*—the feet of God.

into molecules with fishy, sweaty, and garlicky aromas (amines, isovaleric acid, sulfur compounds). These small molecules can diffuse into the cheese and affect both flavor and texture deep inside.

The Molds, Especially Penicillium Molds are microbes that require oxygen to grow, can tolerate drier conditions than bacteria, and produce powerful protein- and fat-digesting enzymes that improve the texture and flavor of certain cheeses. Molds readily develop on the rind of almost any cheese that is not regularly wiped to prevent it. The French St.-Nectaire develops a surface as variegated as lichen-covered rocks in the fields, with spots of bright yellow or orange standing out from a complex, muted background. Some cheeses are gardened to allow a diverse flora to develop, while others are seeded with one particular desired mold. The standard garden variety molds come from the large and various genus *Penicillium,* which also gave us the antibiotic penicillin.

Blue Molds *Penicillium roqueforti,* as its name suggests, is what gives sheep's milk Roquefort cheese its veins of blue. It and its cousin *P. glaucum* also color the interior of Stilton and Gorgonzola and the surface of many aged goat cheeses with the complex pigment produced in their fruiting structures. The blue penicillia are apparently unique in their ability to grow in the low-oxygen (5%, compared to 21% in the air) conditions in small fissures and cavities within cheese, a habitat that echoes the place that gave Roquefort its mold in the first place: the fissured limestone caves of the Larzac. The typical flavor of blue cheese comes from the mold's metabolism of milk fat, of which *P. roqueforti* breaks up 10 to 25%, liberating short-chain fatty acids that give the peppery feel to sheep's milk and goat milk blues, and breaking the longer chains and converting them into substances (methyl ketones and alcohols) that give the characteristic blue aroma.

White Molds In addition to the blue penicillia, there are the white ones, all strains of *P. camemberti,* which make the small, milder surface-ripened soft cow's milk cheeses of northern France, Camembert and Brie and Neufchâtel. The white penicillia create their effects mainly by protein breakdown, which contributes to the creamy texture and provides flavor notes of mushrooms, garlic, and ammonia.

Making Cheese

There are three stages in the transformation of milk into cheese. In the first stage, lactic acid bacteria convert milk sugar into lactic acid. In the second stage, while the acidifying bacteria are still at work, the cheesemaker adds the rennet, curdles the casein proteins, and drains the watery whey from the concentrated curds. In the last stage, ripening, a host of enzymes work together to create the unique texture and flavor of each cheese. These are mainly protein- and fat-digesting enzymes, and they come from the milk, from bacteria originally present in the milk, the acidifying bacteria, the rennet, and any bacteria or molds enlisted especially for the ripening process.

Cheese is certainly an expression of the milk, enzymes, and microbes that are its major ingredients. But it is also—perhaps above all—an expression of the skill and care of the cheesemaker, who chooses the ingredients and orchestrates their many chemical and physical transformations. Here is a brief summary of the cheesemaker's work.

Curdling With the exception of some fresh cheeses, the cheesemaker curdles nearly all cheeses with a combination of starter-bacteria acid and rennet. Acid and rennet form very different kinds of curd structures—acid a fine, fragile gel, rennet a coarse but robust, rubbery one—so their relative contributions, and how quickly they act, help determine the ultimate texture of the cheese. In a predominantly acid coagulation, the curd forms over the course

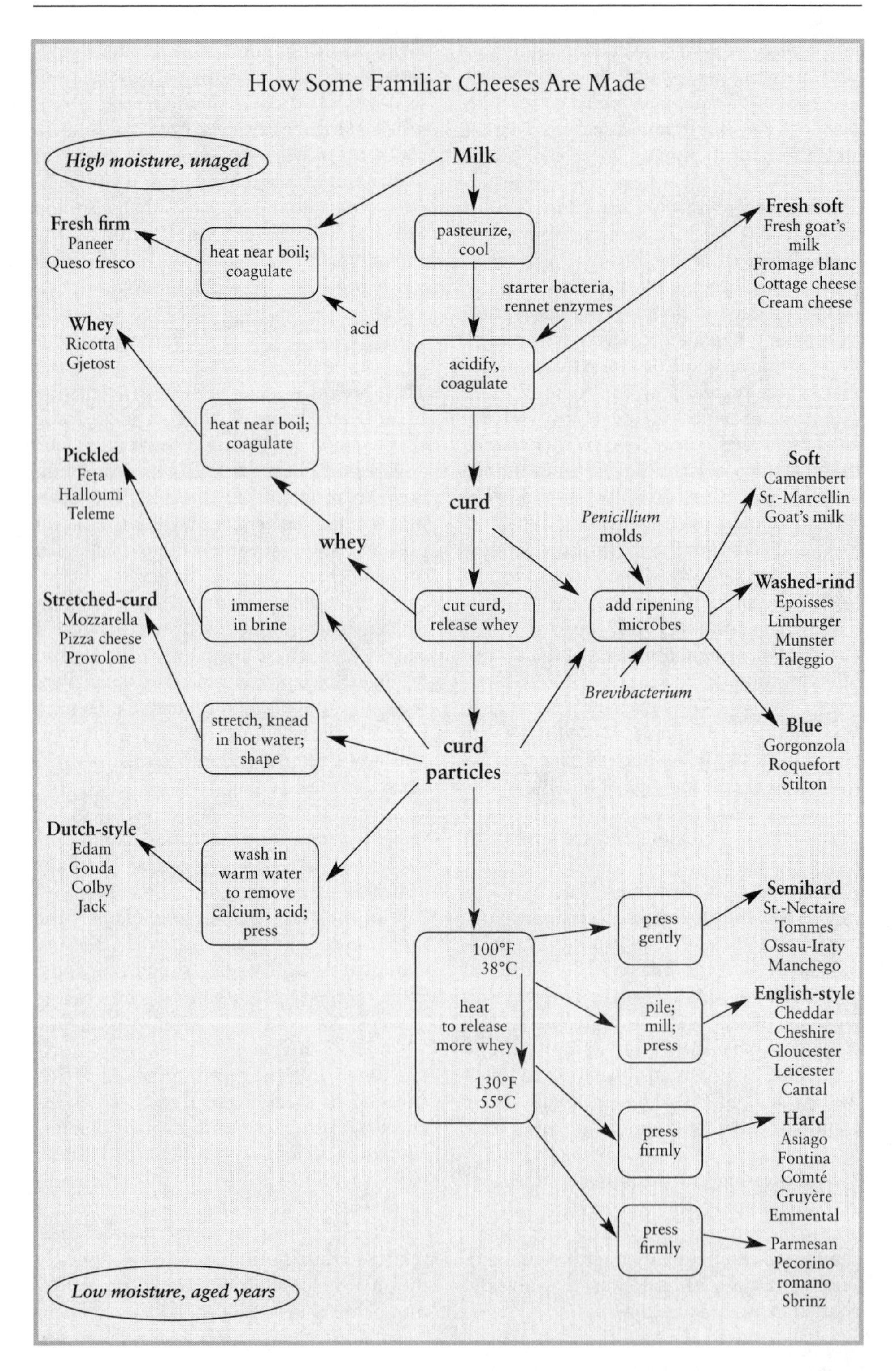
How Some Familiar Cheeses Are Made
High moisture, unaged
Milk
pasteurize, cool
heat near boil; coagulate
Fresh firm
Paneer
Queso fresco
acid
starter bacteria, rennet enzymes
Fresh soft
Fresh goat's milk
Fromage blanc
Cottage cheese
Cream cheese
acidify, coagulate
Whey
Ricotta
Gjetost
heat near boil; coagulate
Pickled
Feta
Halloumi
Teleme
curd
whey
Penicillium molds
Soft
Camembert
St.-Marcellin
Goat's milk
Stretched-curd
Mozzarella
Pizza cheese
Provolone
immerse in brine
cut curd, release whey
add ripening microbes
Washed-rind
Epoisses
Limburger
Munster
Taleggio
Brevibacterium
stretch, knead in hot water; shape
curd particles
Blue
Gorgonzola
Roquefort
Stilton
Dutch-style
Edam
Gouda
Colby
Jack
wash in warm water to remove calcium, acid; press
press gently
Semihard
St.-Nectaire
Tommes
Ossau-Iraty
Manchego
100°F 38°C
heat to release more whey
130°F 55°C
pile; mill; press
English-style
Cheddar
Cheshire
Gloucester
Leicester
Cantal
press firmly
Hard
Asiago
Fontina
Comté
Gruyère
Emmental
press firmly
Parmesan
Pecorino romano
Sbrinz
Low moisture, aged years

of many hours, is relatively soft and weak, and has to be handled gently, so it retains much of its moisture. This is how fresh cheeses and small, surface-ripened goat cheeses begin. In a predominantly rennet coagulation, the curd forms in less than an hour, is quite firm, and can be cut into pieces the size of a wheat grain to extract large amounts of whey. This is how large semihard and hard cheeses begin, from Cheddar and Gouda to Emmental and Parmesan. Cheeses of moderate size and moisture content are curdled with a moderate amount of rennet.

Draining, Shaping, and Salting the Curds The curds can be drained of their whey in several ways, depending on how much moisture the cheesemaker wants to remove from the curd. For some soft cheeses, the whole curd is carefully ladled into molds and allowed to drain by force of gravity alone, for many hours. For firmer cheeses, the curd is precut into pieces to provide more surface area from which the whey can drain or be actively pressed. The cut curd of large hard cheeses may also be "cooked" in its whey to 130°F/55°C, a temperature that not only expels whey from the curd particles, but also affects bacteria and enzymes, and encourages flavor-producing chemical reactions among some milk components. Once the curd pieces are placed in the mold that gives the cheese its final shape, they may be pressed to squeeze out yet more moisture.

The cheesemaker always adds salt to the new cheese, either by mixing dry salt with the curd pieces or by applying dry salt or brine to whole cheese. The salt provides more than its own taste. It inhibits the growth of spoilage microbes, and it's an essential regulator of cheese structure and the ripening process. It draws moisture out of the curds, firms the protein structure, slows the growth of ripening microbes, and alters the activity of ripening enzymes. Most cheeses contain between 1.5 and 2% salt by weight; Emmental is the least salty traditional cheese at about 0.7%, while feta, Roquefort, and pecorino may approach 5%.

Ripening, or *Affinage* Ripening is the stage during which microbes and milk enzymes transform the salty, rubbery, or crumbly curd into a delicious cheese. The French term for ripening, *affinage,* comes from the Latin *finus,* meaning "end" or "ultimate point," and was used in medieval alchemy to describe the refining of impure materials. For at least 200 years it has also meant bringing cheeses to the point at which flavor and texture are at their best. Cheeses have lives: they begin young and bland, they mature into fullness of character, and they eventually decay into harshness and coarseness. The life of a moist cheese like Camembert is meteoric, its prime come and gone in weeks, while the majority of cheeses peak at a few months, and a dry Comté or Parmesan slowly improves for a year or more.

The cheesemaker initiates and manages this maturation process by controlling the temperature and humidity at which the cheese is stored, conditions that determine the moisture content of the cheese, the growth of microbes, the activity of enzymes, and the development of flavor and texture. Specialist cheese merchants in France and elsewhere are also *affineurs*: they buy cheeses before they have fully matured, and carefully finish the process on their premises, so that they can sell the cheeses at their prime.

Industrial producers usually ripen their cheeses only partly, then refrigerate them to

OPPOSITE: *Principal cheese families. Only distinctive processing steps are shown; most cheeses are also salted, shaped in molds, and aged for some time. Curdling the milk, cutting the curd, heating the curd particles, and pressing are methods of removing progressively more moisture from a cheese, slowing its aging, and extending its edible lifetime.*

suspend their development before shipping. This practice maximizes the cheeses' stability and shelf life at the expense of quality.

The Sources of Cheese Diversity

So these are the ingredients that have generated the great diversity of our traditional cheeses: hundreds of plants, from scrubland herbs to alpine flowers; dozens of animal breeds that fed on those plants and transformed them into milk; protein-cutting enzymes from young animals and thistles; microbes recruited from meadow and cave, from the oceans, from animal insides and skins; and the careful observation, ingenuity, and good taste of generations of cheesemakers and cheese lovers. This remarkable heritage underlies even today's simplified industrial cheeses.

The usual way of organizing the diversity of cheeses into a comprehensible system is to group them by their moisture content and the microbes that ripen them. The more moisture removed from the curd, the harder the cheese's eventual texture, and the longer its lifespan. A fresh cheese with 80% water lasts a few days, while a soft cheese (45–55%) reaches its prime in a few weeks, a semihard cheese (40–45%) in a few months, a hard cheese (30–40%) after a year or more. And ripening microbes create distinctive flavors. The box on p. 60 shows how cheesemakers create such different cheeses from the same basic materials.

Choosing, Storing, and Serving Cheese

It has always been a challenge to choose a good cheese, as Charlemagne's instructor admitted (p. 53). A late medieval compendium of maxims and recipes for the

Cheese Flavors from Proteins and Fats

The flavor of a good cheese seems to fill the mouth, and that's because enzymes from the milk and rennet and microbes break down the concentrated protein and fat into a wide range of flavor compounds.

The long, chain-like casein proteins are first broken into medium-sized pieces called peptides, some still tasteless, some bitter. Usually these are eventually broken down by microbial enzymes into the 20 individual protein building blocks, the amino acids, a number of which are sweet or savory. The amino acids can in turn be broken into various amines, some of which are reminiscent of ocean fish (trimethylamine), others of spoiling meat (putrescine); into strong sulfur compounds (a specialty of smear bacteria), or into simple ammonia, a powerful aroma that in overripened cheeses is harsh, like household cleaner. Though few of these sound appetizing, bare hints of them together build the complexity and richness of cheese flavor.

Then there are the fats, which are broken down into fatty acids by blue-cheese *Penicillium roqueforti* and by special enzymes added to Pecorino and Provolone cheeses. Some fatty acids (short-chain) have a peppery effect on the tongue and an intensified sheepy or goaty aroma. The blue molds further transform some fatty acids into molecules (methyl ketones) that create the characteristic aroma of blue cheese. And the copper cauldrons in which the Swiss cheeses and Parmesan are made damage some milk fat directly, and the fatty acids thus liberated are further modified to create molecules with the exotic aromas of pineapple and coconut (esters, lactones).

The more diverse the cast of ripening enzymes, the more complex the resulting collection of protein and fat fragments, and the richer the flavor.

middle-class household, known as *Le Ménagier de Paris,* includes this formula "To recognize good cheese":

Not at all white like Helen,
Nor weeping, like Magdalene.
Not Argus, but completely blind,
And heavy, like a buffalo.
Let it rebel against the thumb,
And have an old moth-eaten coat.
Without eyes, without tears, not at all white,
Moth-eaten, rebellious, of good weight.

But these rules wouldn't work for young goat cheeses (white and coatless), Roquefort (with its pockets of whey), Emmental (eyefull and light), or Camembert (which should give when thumbed). As always, the proof is in the tasting.

These days, the most important thing is to understand that bulk supermarket cheeses are only pale (or dyed) imitations of their more flavorful, distinctive originals. The way to find good cheeses is to buy from a specialist who loves and knows them, chooses the best and takes good care of them, and offers samples for tasting.

Cut to Order Whenever possible, buy portions that are cut while you watch. Precut portions may be days or weeks old, and their large exposed surfaces inevitably develop rancid flavors from contact with air and plastic wrap. Exposure to light in the dairy case also damages lipids and causes off-flavors in as little as two days; in addition it bleaches the annatto in orange-dyed cheeses, turning it pink. Pregrated cheese has a tremendous surface area, and while it is often carefully wrapped, it loses much of its aroma and its carbon dioxide, which also contributes the impression of staleness.

Cool, Not Cold If cheese must be kept for more than a few days, it's usually easiest to refrigerate it. Unfortunately, the ideal conditions for holding cheese—a humid 55–60°F/12–15°C, simply a continuation of its ripening conditions—is warmer than most refrigerators, and cooler and moister than most rooms. Refrigeration essentially puts cheese into suspended animation, so if you want an immature soft cheese to ripen further, you'll need to keep it warmer.

Cheeses should never be served direct from the refrigerator. At such low temperatures the milk fat is congealed and as hard as refrigerated butter, the protein network unnaturally stiff, the flavor molecules imprisoned, and the cheese will seem rubbery and flavorless. Room temperature is best, unless it's so warm (above about 80°F/26°C) that the milk fat will melt and sweat out of the cheese.

Cheese Crystals

Cheeses usually have such a smooth, luscious texture, either from the beginning or as a hard cheese melts in the mouth, that an occasional crunch comes as a surprise. In fact a number of cheeses develop hard, salt-like crystals of various kinds. The white crystals often visible against the blue mold of a Roquefort, or detectable in the rind of a Camembert, are calcium phosphate, deposited because the *Penicillium* molds have made the cheese less acid and calcium salts less soluble. In aged Cheddar there are often crystals of calcium lactate, formed when ripening bacteria convert the usual form of lactic acid into its less soluble mirror ("D") image. In Parmesan, Gruyère, and aged Gouda, the crystals may be calcium lactate or else tyrosine, an amino acid produced by protein breakdown that has limited solubility in these low-moisture cheeses.

Loose Wrapping Tight wrapping in plastic film is inadvisable for three reasons: trapped moisture and restricted oxygen encourages the growth of bacteria and molds, not always the cheese's own; strong volatiles such as ammonia that would otherwise diffuse from the cheese instead impregnate it; and trace volatile compounds and plastic chemicals migrate into the cheese. Whole, still-developing cheeses should be stored unwrapped or very loosely wrapped, other cheeses loosely wrapped in wax paper. Stand them on a wire rack or turn them frequently to prevent the bottom from getting soggy. It can be fun to play the role of *affineur* and encourage surface or blue mold from a good Camembert or Roquefort to grow on a fresh goat cheese or in a piece of standard Cheddar. But there's some risk that other microbes will join in. If a piece of cheese develops an unusual surface mold or sliminess or an unusual odor, the safest thing is to discard it. Simply trimming the surface will not remove mold filaments, which can penetrate some distance and may carry toxins (p. 67).

Rinds Should cheese rinds be eaten? It depends on the cheese and the eater. The rinds of long-aged cheeses are generally tough and slightly rancid, and are best avoided. With softer cheeses it's largely a matter of taste. The rind can offer an interesting contrast to the interior in both flavor and texture. But if safety is a concern, then consider the rind a protective coating and trim it off.

Cooking with Cheese

When used as an ingredient in cooking, cheese can add both flavor and texture: either unctuousness or crispness, depending on circumstances. In most cases, we want the cheese to melt and either mix evenly with other ingredients or spread over a surface. A certain giving cohesiveness is part of the pleasure of melted cheese. Stringy cheese can be enjoyable on pizzas, but a nuisance in more formal dishes. To understand cheese cooking, we need to understand the chemistry of melting.

Cheese Melting What is going on when we melt a piece of cheese? Essentially two things. First, at around 90ºF, the milk fat melts, which makes the cheese more supple, and often brings little beads of melted fat to the surface. Then at higher temperatures—around 130ºF/55ºC for soft cheeses, 150ºF/65ºC for Cheddar and Swiss types, 180ºF/82ºC for Parmesan and pecorino—enough of the bonds holding the casein proteins together are broken that the protein matrix collapses, and the piece sags and flows as a thick liquid. Melting behavior is largely determined by water content. Low-moisture hard cheeses require more heat to melt because their protein molecules are more concentrated and so more intimately bonded to each other; and when melted, they flow relatively little. Separate pieces of grated moist mozzarella will melt together, while flecks of Parmesan remain separate. With continued exposure to high heat, moisture will evaporate from the liquefied cheese, which gets progressively stiffer and eventually resolidifies. Most cheeses will leak some melted fat, and extensive breakdown of the protein fabric accentuates this in high-fat cheeses. The ratio of fat to surrounding protein is just 0.7 in part-skim Parmesan, around 1 in mozzarella and the alpine cheeses, but 1.3 in Roquefort and Cheddar, which are especially prone to exuding fat when melted.

Nonmelting Cheeses There are several kinds of cheese that do not melt on heating: they simply get drier and stiffer. These include Indian paneer and Latin queso blanco, Italian ricotta, and most fresh goat cheeses; all of them are curdled exclusively or primarily by means of acid, not rennet. Rennet creates a malleable structure of large casein micelles held together by relatively few calcium atoms and hydrophobic bonds, so this structure is readily weakened by heat. Acid, on the other hand, dissolves the calcium glue that holds the casein proteins together in micelles (p. 20), and it eliminates each protein's negative electrical charge, which would otherwise cause the proteins to repel each other. The proteins are free to flock together and bond extensively into microscopic clumps. So when an acid curd is heated, the first thing to be shaken loose is not the proteins, but water: the water boils away, and this simply dries out and concentrates the protein even further. This is why firm paneer and queso blanco can be simmered or fried like meat, and goat cheeses and ricotta maintain their shape on pizzas or in pasta stuffings.

Stringiness Melted cheese becomes stringy when mostly intact casein molecules are cross-linked together by calcium into long, rope-like fibers that can stretch but get stuck to each other. If the casein has been attacked extensively by ripening enzymes, then the pieces are too small to form fibers; so well-aged grating cheeses don't get stringy. The degree of cross-linking also matters: a lot and the casein molecules are so tightly bound to each other that they can't give with pulling, and simply snap apart; a little and they pull apart right away. The cross-linking is determined by how the cheese was made: high acidity removes calcium from the curd, and high moisture, high fat, and high salt help separate casein molecules from each other. So the stringiest cheeses are moderate in acidity, moisture, salt, and age. The most common stringy cheeses are intentionally fibrous mozzarella, elastic Emmental, and Cheddar. Crumbly cheeses like Cheshire and Leicester, and moist ones like Caerphilly, Colby, and Jack are preferred for making such melted preparations as Welsh rarebit, stewed cheese, and grilled-cheese sandwiches. Similarly, Emmental's alpine cousin Gruyère is preferred in fondues because it's moister, fatter, and saltier. And the Italian grating cheeses—Parmesan, grana Padano, the pecorinos—have had their protein fabric sufficiently broken that its pieces readily disperse in sauces, soups, risottos, polenta, and pasta dishes.

Cheeses are at their stringiest right around their melting point—which usually means right about the point that a piping-hot dish gets cool enough to eat—and get more so the more they are stirred and stretched. One French country dish, *aligot* from the Auvergne, calls for unripened Cantal cheese to be sliced, mixed with just-boiled potatoes, and sweepingly stirred until it forms an elastic cord that can stretch for 6 to 10 feet/2–3 meters!

Cheese Sauces and Soups When cheese is used to bring flavor and richness to a sauce (Gruyère or Parmesan in French *sauce Mornay,* Fontina in Italian *fonduta*) or a soup, the aim is to integrate the cheese evenly into the liquid. There are several ways to avoid the stringiness, lumps, and fat separation that result when the cheese proteins are allowed to coagulate.

- Avoid using a cheese that is prone to stringiness in the first place. Moist or well-aged grating cheeses blend better.
- Grate the cheese finely so that you can disperse it evenly throughout the dish from the beginning.
- Heat the dish as little as possible after the cheese has been added. Simmer the other ingredients together first, let the pot cool a bit, and *then* add the cheese. Remember that temperatures above the cheese's melting point will tend to tighten the protein patches into hard clumps and squeeze out their fat. On the other hand, don't let the dish cool down too much before serving. Cheese gets stringier and tougher as it cools down and congeals.
- Minimize stirring, which can push the dispersed patches of cheese protein back together into a big sticky mass.
- Include starchy ingredients that will coat the protein patches and fat pockets and keep them apart. These stabilizing ingredients include flour, cornstarch, and arrowroot.

- If the flavor of the dish permits, include some wine or lemon juice—a preventive or emergency measure well known to fans of the ultimate cheese sauce, *fondue*.

Cheese Fondue In the Swiss Alps, where for centuries cheese has been melted in a communal pot at the table and kept hot over a flame for dipping bread, it's well known that wine can help keep melted cheese from getting stringy or seizing up. The ingredients in a classic fondue, in fact, are just alpine cheese—usually Gruyère—a *tart* white wine, some kirsch, and sometimes (for added insurance) starch. The combination of cheese and wine is delicious but also savvy. The wine contributes two essential ingredients for a smooth sauce: water, which keeps the casein proteins moist and dilute, and tartaric acid, which pulls the cross-linking calcium off of the casein proteins and binds tightly to it, leaving them glueless and happily separate. (Alcohol has nothing to do with fondue stability.) The citric acid in lemon juice will do the same thing. If it's not too far gone, you can sometimes rescue a tightening cheese sauce with a squeeze of lemon juice or a splash of white wine.

Toppings, Gratins When a thin layer of cheese is heated in the oven or under a broiler—on a gratin, a pizza, or bruschetta—the intense heat can quickly dehydrate the casein fabric, toughen it, and cause its fat to separate. To avoid this, watch the dish carefully and remove it as soon as the cheese melts. On the other hand, browned, crisp cheese is quite delicious: the *religieuse* at the bottom of the fondue pot crowns the meal. If you want a cheese topping to brown, then pick a robust cheese that resists fat loss and stringiness. The grating cheeses are especially versatile; Parmesan can be formed into a thin disk and melted and lightly browned in a frying pan or the oven, then molded into cups or other shapes.

Process and Low-Fat Cheeses

Process cheese is an industrial version of cheese that makes use of surplus, scrap, and unripened materials. It began as a kind of resolidified, long-keeping fondue made from trimmings of genuine cheeses that were unsaleable due to partial defects or damage. The first industrial attempts to melt together a blend of shredded cheeses were made at the end of the 19th century. The key insight—the necessity of "melting salts" analogous to the tartaric acid and citric acid in a fondue's wine or lemon juice—came in Switzerland in 1912. Five years later, the American company Kraft patented a combination of citric acid and phosphates, and a decade after that it brought out the popular cheddar look-alike Velveeta.

Today, manufacturers use a mixture of sodium citrate, sodium phosphates, and sodium polyphosphates, and a blend of new, partly ripened, and fully ripened cheeses. The polyphosphates (negatively charged chains of phosphorus and oxygen atoms that attract a cloud of water molecules) not only remove calcium from the casein matrix, but also bind to the casein themselves, bringing moisture with them and thus further loosening the protein matrix. The same salts that melt the component cheeses into a homogeneous mass also help the resulting blended cheese melt nicely when cooked. This characteristic, together with its low cost, has made process cheese a popular ingredient in fast-food sandwiches.

Low- and no-fat "cheese products" replace fat with various carbohydrates or proteins. When heated, such products don't melt; they soften and then dry out.

Cheese and Health

Cheese and the Heart As a food that is essentially a concentrated version of milk, cheese shares many of milk's nutritional advantages and disadvantages. It's a rich source of protein, calcium, and energy. Its abundant fat is highly saturated and therefore tends to raise blood cholesterol levels. However, France and Greece lead the world in per capita cheese consumption, at better than 2 oz/60 gm per day, about double the U.S. figures, yet they're remarkable among Western countries for their relatively low rates of heart disease, probably thanks to their high consumption of heart-protective vegetables, fruits, and wine (p. 253). Eating cheese as part of a balanced diet is fully compatible with good health.

Food Poisoning

Cheeses Made from Raw and Pasteurized Milks Government concerns about the danger of the various pathogens that can grow in milk led to the U.S. requirement (originating in 1944, reaffirmed in 1949, and extended to imports in 1951) that all cheeses aged less than 60 days be made with pasteurized milk. Since 1948 there have been only a handful of outbreaks of food poisoning in the United States caused by cheese, nearly all involving contamination of the milk or cheese after pasteurization. In Europe, where young raw-milk cheeses are still legal in some countries, most outbreaks have also been caused by pasteurized cheeses. Cheeses in general present a relatively low risk of food poisoning. Because any soft cheese contains enough moisture to permit the survival of various human pathogens, both pasteurized and unpasteurized versions are probably best avoided by people who may be especially vulnerable to infection (pregnant women, the elderly and chronically ill). Hard cheeses are inhospitable to disease microbes and very seldom cause food poisoning.

Storage Molds In addition to the usual disease microbes, the molds that can grow on cheese are of some concern. When cheeses are held in storage for some time, toxin-producing foreign molds (*Aspergillus versicolor, Penicillium viridicatum* and *P. cyclopium*) may occasionally develop on their rinds and contaminate them to the depth of up to an inch/2 cm. This problem appears to be very rare, but does make it advisable to discard cheeses overgrown with unusual mold.

Amines There is one normal microbial product that can cause discomfort to some people. In a strongly ripened cheese, the casein proteins are broken down to amino acids, and the amino acids can be broken down into amines, small molecules that can serve as chemical signals in the human body. Histamine and tyramine are found in large quantities in Cheddar, blue, Swiss, and Dutch-style cheeses, and can cause a rise in blood pressure, headaches, and rashes in people who are especially sensitive to them.

Tooth Decay Finally, it has been recognized for decades that eating cheese slows tooth decay, which is caused by acid secretion from relatives of a yogurt bacterium (especially *Streptococcus mutans*) that adhere to the teeth. Just why is still not entirely clear, but it appears that eaten at the end of a meal, when streptococcal acid production is on the rise, calcium and phosphate from the cheese diffuse into the bacterial colonies and blunt the acid rise.

CHAPTER 2

EGGS

The egg is one of the kitchen's marvels, and one of nature's. Its simple, placid shape houses an everyday miracle: the transformation of a bland bag of nutrients into a living, breathing, vigorous creature. The egg has loomed large as a symbol for the enigmatic origins of animals, of humans, of gods, of the earth, of the entire cosmos. The Egyptian Book of the Dead, the Indian Rg Veda, Greek Orphic mysteries, and creation myths throughout the world have been inspired by the eruption of life from within a lifeless, blank shell.

Humpty Dumpty has had a great fall! If eggs inspire any notable feeling today, it's boredom tinged by wariness. The chicken egg is now an industrial product, so familiar that it would be almost invisible—except that it was stigmatized by the cholesterol phobia of the 1970s and 1980s.

Neither familiarity nor fear should obscure eggs' great versatility. Their contents are primal, the unstructured stuff of life. This is why they are protean, why the cook can use them to generate such a variety of structures, from a light, insubstantial meringue to a dense, lingeringly rich custard. Eggs reconcile oil and water in a host of smooth sauces; they refine the texture of candies and ice creams; they give flavor, substance, and nutritiousness to soups, drinks, breads, pastas, and cakes; they put a shine on pastries; they clarify meat stocks and wines. On their own, they're amenable to being boiled, fried, deep-fried, baked, roasted, pickled, and fermented.

Meanwhile modern science has only deepened the egg's aptness as an emblem of creation. The yolk is a stockpile of fuel obtained by the hen from seeds and leaves, which are in turn stockpiles of the sun's radiant energy. The yellow pigments that gave the yolk its name also come directly from plants, where they protect the chemical machinery of photosynthesis from being overwhelmed by the sun. So the egg does embody the chain of creation, from the developing chick back through the hen to the plants that fed her, and then to the ultimate source of life's fire, the yellow sphere of the sky. An egg is the sun's light refracted into life.

Many animals lay eggs, and humans exploit a number of them, from pigeons and turkeys to wild birds, penguins, turtles, and crocodiles. The chicken egg is by far the most commonly eaten in most countries, so I'll concentrate on it, with occasional asides on duck eggs.

THE CHICKEN AND THE EGG

Over the centuries there have been several clever answers to the conundrum, Which came first: the chicken or the egg? The Church Fathers sided with the chicken, pointing out that according to Genesis, God first created the creatures, not their reproductive apparatus. The Victorian Samuel Butler awarded the egg overall priority when he said that a chicken is just an egg's way of making another egg. About one point, however, there is no dispute: eggs existed long before chickens did. Ultimately, we owe our soufflés and sunny-sides-up to the invention of sex.

THE EVOLUTION OF THE EGG

Sharing DNA Defined broadly, the egg is a kind of cell that is specialized for the process of sexual reproduction, in which two parents contribute genes to the making of a new individual. The first living things were single cells and reproduced on their own, each cell simply making a copy of its DNA and then dividing itself into two cells.

The World Egg

In the beginning this world was nonexistent. It became existent. It developed. It turned into an egg. It lay for the period of a year. It split apart. One of the parts became silver, one gold.

That which was silver is this earth. That which was gold is the sky. That which was the outer membrane is the mountains. That which was the inner membrane is clouds and mist. What were the veins are the rivers. What was the fluid within is the ocean.

What was born from the egg is the sun. When it was born, shouts and hurrahs and all beings and all desires rose up toward it. Therefore at its rise and at its every return, shouts and hurrahs and all beings and all desires rise up toward it.

—*Chandogya Upanishad*, ca. 800 BCE

The first sexual organisms, probably single-celled algae, paired up and exchanged DNA with each other before dividing—a mixing that greatly facilitated genetic change. Specialized egg and sperm cells became necessary around a billion years ago, when many-celled organisms evolved and this simple transfer of DNA was no longer possible.

What makes an egg an egg? Of the two reproductive cells, it's the larger, less mobile one. It receives the sperm cell, accommodates the joining of the two gene sets, and then divides and differentiates into the embryonic organism. It also provides food for at least the initial stages of this growth. This is why eggs are so nutritious: Like milk and like plant seeds, they are actually designed to be foods, to support new creatures until they are able to fend for themselves.

Improving the Package The first animal eggs were released into the equable oceans, where their outer membrane could be simple and their food supply minimal. Some 300 million years ago, the earliest fully land-dwelling animals, the reptiles, developed a self-contained egg with a leathery skin that slowed fatal water loss, and with enough food to support prolonged embryonic development into a fully formed animal. The eggs of birds, animals that arose some 100 million years later, are a refined version of the primitive reptile egg. Their hard, mineralized shell is impermeable enough that the embryo can develop in the driest habitats; and they contain an array of antimicrobial defenses. These developments made the bird egg into an ideal human food. It contains a sizeable and balanced portion of animal nutrients; and it's so well packaged that it keeps for weeks with little or no care.

The Chicken, from Jungle to Barnyard

Eggs, then, are nearly a billion years older than the oldest birds. The genus *Gallus,* to which the chicken belongs, is a mere 8 million years old, and *Gallus gallus,* the chicken species, has been around only for the last 3 to 4 million years.

For a barnyard commoner, the chicken has a surprisingly exotic background. Its immediate ancestors were jungle fowl native to tropical and subtropical Southeast Asia and India. The chicken more or less as we know it was probably domesticated in Southeast Asia before 7500 BCE, which is when larger-than-wild bones date from in Chinese finds far north of the jungle fowl's current range. By 1500 BCE chickens had found their way to Sumer and Egypt, and they arrived around 800 BCE in Greece, where they became known as "Persian birds," and where quail were the primary source of eggs.

The Domestic Egg We'll never know exactly why chickens were domesticated, but they may well have been valued more for their prolific egg production than for their meat. Some birds will lay only a set number of eggs at a time, no matter what happens to the eggs. Others, including the

Food Words: *Egg* and *Yolk*

Egg comes from an Indo-European root meaning "bird."

The brusque-sounding *yolk* is rich in overtones of light and life. It comes from the Old English for "yellow," whose Greek cousin meant "yellow-green," the color of new plant growth. Both the Old English and the Greek derive ultimately from an Indo-European root meaning "to gleam, to glimmer." The same root gave us *glow* and *gold.*

chicken, will lay until they accumulate a certain number in the nest. If an egg is taken by a predator, the hen will lay another to replace it—and may do so indefinitely. Over a lifetime, these "indeterminate layers" will produce many more eggs than the "determinate" layers. Wild Indian jungle fowl lay clutches of about twelve glossy, brown eggs a few times each year. In industrial production—the ecological equivalent of unlimited food resources combined with unrelenting predation—their domesticated cousins will lay an egg a day for a year or more.

Cooked Eggs Doubtless bird eggs have been roasted ever since humans mastered fire; in *As You Like It* Shakespeare has Touchstone call Corin "damned, like an ill-roasted egg, all on one side." Salting and pickling eggs are ancient treatments that preserved the spring's bounty for use throughout the year. We know from the recipes of Apicius that the Romans ate *ova frixa, elixa, et hapala*—fried, boiled, and "soft" eggs—and the *patina,* which could be a savory quiche or a sweet custard. By medieval times, the French were sophisticated omelet makers and the English were dressing poached eggs with the sauce that would come to be called *crème anglaise.* Savory yolk-based sauces and egg-white foams developed over the next three centuries. By around 1900, Escoffier had a repertoire of more than 300 egg dishes, and in his *Gastronomie Pratique,* Ali Bab gave a playful recipe for a "Symphony of Eggs"—a four-egg omelet containing two chopped hard-cooked and six whole poached eggs.

The Industrial Egg

Hen Fever The chicken underwent more evolutionary change between 1850 and 1900 than it had in its entire lifetime as a species, and under an unusual selection pressure: the fascination of Europeans and Americans with the exotic East. A political opening between England and China brought specimens of previously unknown Chinese breeds, the large, showy Cochins, to the West. These spectacular birds, so different from the run of the barnyard, touched off a chicken-breeding craze comparable to the Dutch tulip mania of the 17th century. During this "hen fever," as one observer of the American scene called

Roman Custards, Savory and Sweet

Patina of Soles

Beat and clean the soles and put in patina [a shallow pan]. Throw in oil, liquamen [fish sauce], wine. While the dish cooks, pound and rub pepper, lovage, oregano; pour in some of the cooking liquid, add raw eggs, and make into one mass. Pour over the soles and cook on a slow fire. When the dish has come together, sprinkle with pepper and serve.

"Cheese" Patina

Measure out enough milk for your pan, mix with honey as for other milk dishes, add five eggs for [a pint], three for [a half-pint]. Mix them in the milk until they make one mass, strain into a dish from Cuma, and cook over a slow fire. When it is ready, sprinkle with pepper and serve.

—from Apicius, first few centuries CE

it, poultry shows were very popular and hundreds of new breeds were developed.

Ordinary farm stock was also improved. Just a few decades after its arrival in the United States from Tuscany around 1830, descendents of the White Leghorn emerged as the champion layers. Versions of the Cornish, itself the offshoot of Asiatic fighting breeds, were deemed the best meat birds; and the Plymouth Rock and Rhode Island Red, whose eggs are brown, were bred as dual-purpose chickens. As interest in the show birds faded, the egg and meat breeds became ever more dominant. Today, an egg or meat chicken is usually the product of four purebred grandparents. Nearly all of the diversity generated in the 1800s has disappeared. Among industrialized countries, only France and Australia have remained independent of the handful of multinational corporations that provide laying stock to the egg industry.

Mass Production The 20th century saw the general farm lose its poultry shed to the poultry farm or ranch, which has in turn been split up into separate hatcheries and meat and egg factories. Economies of scale dictate that production units be as large as possible—one caretaker can manage a flock of 100,000, and many ranches now have a million or more laying hens. Today's typical layer is born in an incubator, eats a diet that originates largely in the laboratory, lives and lays on wire and under lights for about a year, and produces between 250 and 290 eggs. As Page Smith and Charles Daniel put it in their *Chicken Book,* the chicken is no longer "a lively creature but merely an element in an industrial process whose product [is] the egg."

A Medieval Omelet and English Cream

Arboulastre *(An Omelet)*

[First prepare mixed herbs, including rue, tansy, mint, sage, marjoram, fennel, parsley, violet leaves, spinach, lettuce, clary, ginger.] Then have seven eggs well beaten together, yolks and whites, and mix with the herbs. Then divide in two and make two *allumelles,* which are fried in the following manner. First you heat your frying pan well with oil, butter, or whatever fat you like. When it is well heated, especially toward the handle, mix and cast your eggs upon the pan, and turn frequently with a paddle over and under; then throw some good grated cheese on top. Know that it is done thus because if you mix the cheese with the eggs and herbs, when you fry the *allumelle,* the cheese that is underneath sticks to the pan. . . . And when your herbs are fried in the pan, shape your *arboulastre* into a square or round form, and eat it neither too hot nor too cold.

—*Le Ménagier de Paris,* ca. 1390

Poche to Potage (Poached Eggs in Crème Anglaise)

Take eggs and break them into boiling water, and let them seethe, and when they are done take them out, and take milk and yolks of eggs, and beat them well together, and put them in a pot; and add sugar or honey, and color it with saffron, and let it seethe; and at the first boil take it off, and cast therein powder of ginger, and dress the cooked eggs in dishes, and pour the pottage above, and serve it forth.

—from a manuscript published in *Antiquitates Culinariae,* 1791 (ca. 1400)

Benefits and Costs The industrialization of the chicken has brought benefits, and these shouldn't be underestimated. A pound of broiler can now be produced from less than two pounds of feed, a pound of eggs from less than three, so both chickens and eggs are bargains among animal foods. Egg quality has also improved. City and country dwellers alike enjoy fresher, more uniform eggs than formerly, when small-farm hens ran free and laid in odd places, and when spring eggs were stored until winter in limewater or waterglass (see p. 115). Refrigeration alone has made a tremendous difference. Year-round laying (made possible by controlled lighting and temperature), prompt gathering and cooling, and daily shipping by rapid, refrigerated transport mean that good eggs deteriorate much less between hen and cook than they did in the more relaxed, more humane past.

There are drawbacks to the industrial egg. While average quality has improved, people who pay close attention to eggs say that flavor has suffered: that the chicken's natural, varied diet of grains, leaves, and bugs provides a richness that the commercial soy and fish meals don't. (This difference has proven hard to document in taste tests; see p. 87.) In addition, mass husbandry has played a role in the rising incidence of salmonella contamination. "Spent" hens are often recycled into feed for the next generation of layers, so that salmonella infection is readily spread by careless processing. Finally, there is a more difficult question: whether we can enjoy good, cheap eggs more humanely, without reducing descendents of the spirited jungle fowl to biological machines that never see the sun, scratch in the dust, or have more than an inch or two to move.

Freer Range? Enough people have become uncomfortable with the excesses of industrialization, and willing to pay a substantial premium for their eggs, that smaller-scale, "free-range" and "organically fed" laying flocks have made a comeback in the United States and Europe. Swiss law now requires that all hens in that country have free access to the outdoors. The term "free-range" can be misleading; it sometimes means only that the chickens live in a slightly larger cage than usual, or have brief access to the outdoors. Still, with people eating fewer eggs in the home, spending so little on those eggs, and paying more attention to what they eat, the odds are good that this modest de-industrialization of the egg will continue.

EGG BIOLOGY AND CHEMISTRY

How the Hen Makes an Egg

The egg is so familiar that we seldom remember to marvel at its making. All animals work hard at the business of reproduction, but the hen does more than most. Her "reproductive effort," defined as the fraction of body weight that an animal deposits daily in her potential offspring, is 100 times greater than a human's. Each egg is about 3% of the hen's weight, so in a year of laying, she converts about eight times her body weight into eggs. A quarter of her daily energy expenditure goes toward egg-making; a duck puts in half.

The chicken egg begins with the pinhead-sized white disc that we see riding atop the yellow yolk. This is the business end of the egg, the living germ cell that contains the hen's chromosomes. A hen is born with several thousand microscopic germ cells in her single ovary.

Making the Yolk As the hen grows, her germ cells gradually reach a few millimeters in diameter, and after two or three months accumulate a white, primordial form of yolk inside their thin surrounding membrane. (The white yolk can be seen in a hard-cooked egg; see box, p. 74.) When the hen reaches laying age at between four and six months, the egg cells begin to mature, with different cells at different stages at any given time. Full maturation takes about ten weeks. During the tenth, the germ cell rapidly accumulates yellow yolk, mostly fats and proteins, which is synthesized in the hen's liver. Its color depends on the pigments in the hen's feed; a diet rich in corn or alfalfa makes a deeper yellow. If the hen feeds only once or twice a day, her yolk will show distinct layers of dark and light. In the end, the yolk comes to dwarf the germ cell, containing as it must the provisions for 21 days during which the chick will develop on its own.

Making the White The rest of the egg provides both nourishment and protective housing for the germ cell. Its construction takes about 25 hours and begins when the ovary releases the completed yolk. The yolk is then gripped by the funnel-shaped opening of the oviduct, a tube 2–3 feet/0.6–0.9 meter long. If the hen has mated in recent days, there will be sperm stored in a "nest" at the upper end of the oviduct, and one will fuse with the egg cell. Fertilized or not—and most eggs are not—the yolk spends two to three hours slowly passing down the upper end of the oviduct. Protein-secreting cells in the oviduct lining add a thickening layer to its membrane, and then coat it with about half the final volume of the egg white, or *albumen* (from the Latin *albus,* meaning "white"). They apply this portion of albumen in four layers that are alternately thick and thin in consistency.

The first thick layer of albumen protein is twisted by spiraling grooves in the oviduct wall to form the *chalazae* (from the Greek for "small lump," "hailstone"), two dense, slightly elastic cords which anchor the yolk to the ends of the shell and allow it to rotate while suspending it in the middle of the egg. This system keeps as much cushioning albumen as possible between the embryo and the shell, and prevents premature contact between shell and embryo, which could distort the embryo's development.

Membranes, Water, and Shell Once the albumen proteins have been applied to the yolk, it spends an hour in the next section of the oviduct being loosely enclosed in two tough, antimicrobial protein membranes that are attached to each other everywhere except for one end, where the air pocket will later develop to supply the hatching chick with its first gulps of air. Then comes a long stretch—19 or 20 hours—in the 2-inch-/5-cm-long uterus, or shell gland. For five hours, cells in the uterus wall pump water and salts through the membranes and into the albumen and "plump" the egg to its full volume. When the membranes are taut, the uterine lining secretes calcium carbonate and protein to form the shell, a process that takes about 14 hours. Since the embryo needs air, the shell is riddled (especially at the blunt end) with some 10,000 pores that add up to a hole about 2 mm in diameter.

Germ-Side Up: Primordial Yolk

Have you ever noticed that when you crack open a raw egg, the germ cell—the pinhead-sized white disc that carries the hen's DNA—usually comes to the top of the yolk? It does so because the channel of primordial white yolk below it is less dense than the yellow yolk—so the egg cell's side of the yolk is lighter, and rises. In the intact egg, the chalazae allow the germ cell to return to the top whenever the hen rearranges her eggs.

That persistent bit of uncoagulated yolk at the center of a hard-cooked egg is primordial white yolk, especially rich in iron, which the hen deposits in its eggs when they're barely a quarter-inch/6 mm in diameter.

Cuticle and Color The hen's finishing touch on her egg is a thin proteinaceous cuticle. This coating initially plugs up the pores to slow water loss and block the entry of bacteria, but gradually fractures to allow the chick to get enough oxygen. Along with the cuticle comes color, in the form of chemical relatives of hemoglobin. Egg color is determined by the hen's genetic background, and has no relation to the egg's taste or nutritional value. Leghorns lay very lightly pigmented "white" eggs. Brown eggs are produced by breeds that were originally dual-purpose egg and meat birds, including Rhode Island Reds and Plymouth Rocks; New Hampshire and Australorps hens were bred for intensive brown-egg production. Chinese Cochin hens paint their eggs with fine yellow dots. Thanks to a dominant trait unknown in any other wild or domestic chickens, the rare Chilean Araucana lays blue eggs. Crosses between Araucanas and brown-egg breeds make both blue and brown pigments and thus green shells.

The completed egg is expelled blunt end first about 25 hours after leaving the ovary. As the egg cools down from the hen's high body temperature (106°F/41°C), its contents shrink slightly. This contraction pulls the inner shell membrane away from its outer partner at the blunt end and thereby forms the air space, whose size is an indicator of egg freshness (p. 81).

THE YOLK

The yolk accounts for just over a third of a shelled egg's weight, and its biological purpose is almost exclusively nutritive. It carries three-quarters of the calories and most of the iron, thiamin, and vitamin A of the egg as a whole. The yolk's yellow color comes not from the vitamin-A precursor beta-carotene, the orange pigment in carrots and other plant foods, but from plant pigments called xanthophylls (p. 267), which the hen obtains mainly from alfalfa and corn feeds. Producers may supplement the feeds with marigold petals and other additives to deepen the color. Duck yolks owe their deeper orange color both to beta-carotene and to the reddish pigment canthaxanthin, which wild ducks obtain from small water insects and crustaceans, egg-laying ducks from feed supplements. One minor component of the yolk that can cause a major culinary disaster is the starch-digesting enzyme amylase, which has liquefied many a normal-looking pie filling from within (see p. 98).

Spheres Within Spheres That's a yolk by numbers and nutrients. But there's a lot more to this concentrated pool of the sun's rays. Its structure is intricate, much like a Chinese set of nested spheres carved from a single block of jade. We see the first layer of structure whenever we cut into a hard-cooked egg. Where heat gels the white into

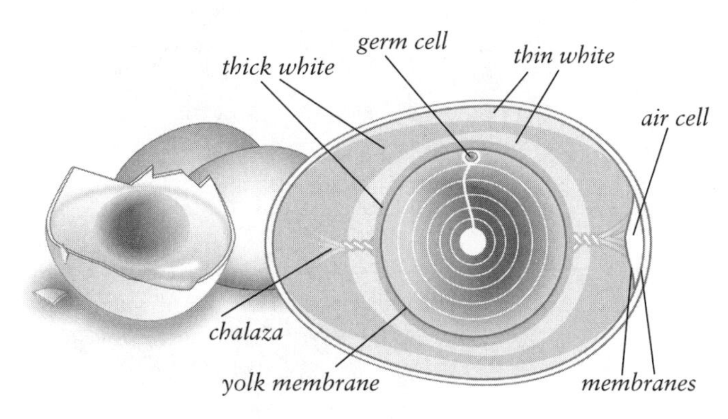

The structure of the hen's egg. The egg white provides physical and chemical protection for the living germ cell, and protein and water for its development into a chick. The yolk is rich in fats, proteins, vitamins and minerals. The layering of color in the yolk is caused by the hen's periodic ingestion of grain and its fat-soluble pigments.

a smooth, continuous mass, the yolk sets into a crumbly mass of separate particles. The intact yolk turns out to consist of spherical compartments about a tenth of a millimeter across, each contained within a flexible membrane, and so tightly packed that they're distorted into flat-sided shapes (much like the oil droplets that egg yolk stabilizes in mayonnaise; see p. 626). When a yolk is cooked intact, these spheres harden into individual particles and give the yolk its characteristic crumbly texture. But break the yolk out before you cook it so that the spheres can move freely, and it becomes less granular.

What's inside these large yolk spheres? Though we think of the yolk as rich and fatty, in fact its chambers are filled mostly with water. Floating in that water are subspheres about one hundredth the size of the spheres. The subspheres are too small to see with the naked eye or to be broken up by a kitchen beating. But they can be seen indirectly, and disrupted chemically. Yolk is cloudy because these subspheres are large enough to deflect light and prevent it from passing through the yolk directly. Add a pinch of salt to a yolk (as you do when making mayonnaise) and you'll see the yolk become simultaneously clearer and thicker. Salt breaks apart the light-deflecting subspheres into components that are too small to deflect light—and so the yolk clears up.

And what do the subspheres contain? A mixture similar to the liquid that surrounds them in the spheres. First, water. Dissolved in the water, proteins: hen blood proteins outside the subspheres; inside, phosphorus-rich proteins that bind up most of the egg's iron supply. And suspended in the water, *sub*subspheres about 40 times smaller than the subspheres, some of which turn out to be familiar from the human body. The subsubspheres are aggregates of four different kinds of molecules: a core of fat surrounded by a protective shell of protein, cholesterol, and phospholipid, a hybrid fat-water mediator which in the egg is mainly lecithin. Most of these subsubspheres are "low-density lipoproteins," or LDLs—essentially the same particles that we keep track of in our own blood to monitor our cholesterol levels.

Stand back from all these spheres within spheres and the picture becomes less dizzying. The yolk is a bag of water that contains free-floating proteins and protein-fat-cholesterol-lecithin aggregates—and these

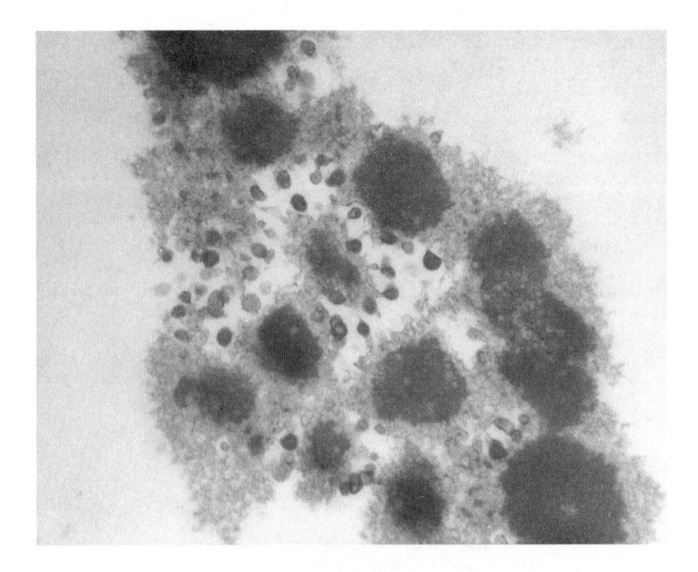

An egg yolk granule as seen through the electron microscope. It has fallen apart after immersion in a salt solution, and is an intricate assembly of proteins, fats, phospholipids, and cholesterol.

lipoprotein aggregates are what give the yolk its remarkable capacities for emulsifying and enriching.

The White

Next to the yolk's riches, the white seems colorless and bland. It accounts for nearly two thirds of the egg's shelled weight, but nearly 90% of that is water. The rest is protein, with only traces of minerals, fatty material, vitamins (riboflavin gives the raw white a slightly yellow-green cast) and glucose. The quarter-gram of glucose, which is essential for the embryo's early growth, isn't enough to sweeten the white, though in such preparations as long-cooked eggs (p. 89) and thousand-year preserved eggs (p. 116) it's sufficient to turn the white a dramatic brown. The white's structural interest is limited to the fact that it comes in two consistencies, thick and thin, with the yolk cords being a twisted version of the thick.

Protective Proteins Pale though it is, the egg white has surprising depths. Of course it supplies the developing embryo with essential water and protein. But biochemical studies have revealed that the albumen proteins are not mere baby food. At least four of the proteins block the action of digestive enzymes. At least three bind tightly to vitamins, which prevents them from being useful to other creatures, and one does the same for iron, an essential mineral for bacteria and animals alike. One protein inhibits the reproduction of viruses, and another digests the cell walls of bacteria. In sum, the egg white is first of all a chemical shield against infection and pre-

The Proteins in Egg White

Protein	Percent of Total Albumen Protein	Natural Functions	Culinary Properties
Ovalbumin	54	Nourishment; blocks digestive enzymes?	Sets when heated to 180°F/80°C
Ovotransferrin	12	Binds iron	Sets when heated to 140°F/60°C or foamed
Ovomucoid	11	Blocks digestive enzymes	?
Globulins	8	Plug defects in membranes, shell?	Foam readily
Lysozyme	3.5	Enzyme that digests bacterial cell walls	Sets when heated to 170°F/75°C; stabilizes foam
Ovomucin	1.5	Thickens albumen; inhibits viruses	Stabilizes foam
Avidin	0.06	Binds vitamin (biotin)	?
Others	10	Bind vitamins (2+); block digestive enzymes (3+) . . .	?

dation, forged during millions of years of battling between the nourishing egg and a world of hungry microbes and animals.

A few of the dozen or so egg-white proteins are especially important for the cook and worth knowing by name.

- *Ovomucin* accounts for less than 2% of the total albumen protein, but has by far the greatest influence on the fresh egg's commercial and culinary value. It makes fried and poached eggs compact and attractive by making the thick white thick—40 times more so than the thin white. Ovomucin somehow pulls together the otherwise soupy protein solution into an organized structure; gently tear a piece of hard-cooked white and you can see its laminations along the edge of the tear. This structure is thought to help cushion the yolk and slow the penetration of microbes through the white. It gradually disintegrates with age in the raw egg, which may make the white more digestible for the developing chick, and certainly makes the egg less useful for the cook.
- *Ovalbumin*, the most plentiful egg protein, was the first protein ever crystallized in the laboratory (in 1890), yet its natural function remains unclear. It seems related to a family of proteins that inhibit protein-digesting enzymes, and may be a mainly nutritional relic of ancient battles against a now-extinct microbe. It is the only egg protein to have reactive sulfur groups, which make decisive contributions to the flavor, texture, and color of cooked eggs. Interestingly for the cook, ovalbumin's heat resistance increases for several days after laying, so that very fresh eggs need less cooking for a given consistency than eggs a few days old.
- *Ovotransferrin* holds tightly onto iron atoms to prevent bacteria from using them, and to transport iron in the developing chick's body. It is the first protein to coagulate when an egg is heated, and so determines the temperature at which eggs set. The setting temperature is higher for whole eggs than for egg white, because ovotransferrin becomes more stable and resistant to coagulation when it binds the abundant iron in the yolk. The color of ovotransferrin changes when it latches onto metals, which is why egg whites whipped in a copper bowl turn golden; you can also make a pink meringue by dosing the whites with a pinch of ground-up iron supplement.

The Nutritional Value of Eggs

An egg contains everything you need to make a chick, all the ingredients and chemical machinery and fuel. That fact is its strength as a food. Cooked—to neutralize the protective antinutritional proteins—the egg is one of the most nutritious foods we have. (Raw, it causes laboratory animals to *lose* weight.) It's unmatched as a balanced source of the amino acids necessary for animal life; it includes a plentiful supply of linoleic acid, a polyunsaturated fatty acid that's essential in the human diet, as well as of several minerals, most vitamins, and two plant pigments, lutein and zeaxanthin, that are especially valuable antioxidants (p. 255). The egg is a rich package.

Cholesterol in Eggs *Too* rich for our blood, it's been thought: a belief that contributed to the steep drop in U.S. egg consumption beginning around 1950. Among our common foods, the egg is the richest source of cholesterol. One large egg contains around 215 milligrams, while an equivalent portion of meat has about 50.

Why is there so much cholesterol in the egg? Because it's an essential component of animal cell membranes, of which the chicken embryo must construct many millions before it hatches. There is some variability in the cholesterol contents of different breeds, and the hen's diet has some effect—a feed high in sitosterol, a vegetable relative of cholesterol, brings egg cholesterol down by a third. But these reductions still leave egg yolk way ahead of most other foods.

Since high blood cholesterol does increase the risk of heart disease, many medical associations have long recommended limiting our yolk consumption to two or three per week. However, recent studies of moderate eaters have shown that egg consumption has little influence on blood cholesterol. This is partly because blood cholesterol is raised far more powerfully by saturated fats in the diet than by cholesterol itself, and most of the fat in egg yolk is unsaturated. It also appears that other fatty substances in the yolk, the phospholipids, interfere with our absorption of yolk cholesterol. So there no longer seems to be any reason to bother counting our weekly yolks. Of course, eggs shouldn't displace positively heart-protective fruits and vegetables from the diet; and on a strict regimen to deal with serious heart disease or obesity, it may make sense to avoid egg yolks along with similarly fatty animal foods. Better than 60% of the calories in a whole egg come from fat, a third of them from saturated fat.

Egg Substitutes Largely impelled by the public desire for cholesterol-free eggs, food manufacturers have come up with formulations that imitate whole beaten eggs, and can be cooked into scrambled eggs or omelets or used in baking. These products consist of genuine egg whites mixed with an imitation of the yolk, which is usually made from vegetable oil, milk solids, gums that provide a thick consistency, as well as colorings, flavorings, and vitamin and mineral supplements.

Fertilized Eggs Despite folklore to the contrary, there is no detectable nutritional difference between unfertilized and fertilized eggs. By the time a fertilized egg is laid, the single germ cell has divided into tens of thousands of cells, but its diameter has only grown from 3.5 millimeters to 4.5, and any

The Composition of a U.S. Large Egg

A shelled U.S. Large egg weighs 2 ounces, or 55 grams. In the following table, all weights are given in grams (g) or thousandths of a gram (mg). Fat accounts for about 60% of the calories in an egg, saturated fat around 20%.

	Whole Egg	Egg White	Egg Yolk
Weight	55 g	38 g	17 g
Protein	6.6 g	3.9 g	2.7 g
Carbohydrate	0.6 g	0.3 g	0.3 g
Fat	6 g	0	6 g
Monounsaturated	2.5 g	0	2.5 g
Polyunsaturated	0.7 g	0	0.7 g
Saturated	2 g	0	2 g
Cholesterol	213 mg	0	213 mg
Sodium	71 mg	62 mg	9 mg
Calories	84	20	64

biochemical changes are negligible. Refrigerated storage prevents any further growth or development. In the U.S. grading system, any significant development of the egg—from minute blood vessels (which appear after two to three days of incubation) to a recognizable embryo—is considered a major defect, and automatically puts it in the "inedible" category. Of course this is a cultural judgment. In China and the Philippines, for example, duck eggs containing two- to three-week embryos are boiled and eaten, in part for their supposed contribution to virility. Because embryos obtain some nourishment from the shell, these duck embryos do contain more calcium than the eggs that they developed from.

Egg Allergies Eggs are one of the commonest foods to which people develop food allergies. Portions of the major egg-white protein ovalbumin appear to be the usual culprits. The immune system of sensitive people interprets these parts of ovalbumin to be a threat, and mounts a massive and self-destructive defense that can take the form of fatal shock. Since a sensitivity to egg white often forms in early life, pediatricians commonly recommend that children not eat egg whites until after the age of one. Egg yolks are far less allergenic and can safely be eaten by nearly all infants.

EGG QUALITY, HANDLING, AND SAFETY

What is a good egg? An intact, uncontaminated egg with a strong shell; a firm yolk and yolk membrane, which prevents the yolk from breaking and mixing with the white; and a high proportion of cohesive, jellylike thick white compared to runny thin white.

And what makes a good egg? Above all, a good hen: a hen of a select laying breed that is healthy and not approaching the end of a laying year, when shells and whites deteriorate (this stage is shortened by restricting the hen's food, which induces her to molt and reset her biological clock). A nutritious feed, free of contaminants, and without ingredients (fish meal, raw soy meal) that impart off-flavors. And careful evaluation and handling once the egg leaves the hen.

In order to determine egg quality without actually breaking them, producers *candle* their eggs, or place them in front of a light bright enough to pass through them and illuminate their contents. (Candle and eye were the original equipment; today electric lights and scanners do the work automatically.) Candling readily detects cracks in the shell, harmless but unappealing blood spots on the yolk (from burst capillaries in the hen's ovary or yolk sac), and "meat spots" in the whites (either brown blood spots or tiny bits of tissue sloughed off from the oviduct wall), and large air cells, all characteristics that relegate an egg to the lower grades. To determine the condition of the yolk and white, the egg is quickly twirled. The yolk's shadow will remain indistinct if its membrane is strong enough and the white thick enough to have kept it from getting close to the shell. If the yolk is easy to see, then it's too easily deformed or mobile, and the egg is of lower quality.

Egg Grades

Eggs sold in stores are usually (but not mandatorily) classified by United States Department of Agriculture (USDA) grades. Egg grade has nothing to do with either freshness or size, and is not a guarantee of egg quality in your kitchen. It's an approximate indication of the quality of the egg back at the ranch, at the time it was collected. Because candling isn't foolproof, USDA definitions allow several eggs per carton to be below grade at the time of packing. Once the eggs have arrived in stores, the below-grade allowance doubles, because egg quality naturally declines with time, and jostling and vibration during transport can cause the white to thin out.

Generally, only the two top grades, AA

and A, are seen in stores. If you're going to use eggs fairly soon and will be scrambling them or making a custard or pancakes, then the higher grade isn't worth the higher price. But if you go through eggs slowly, or like your hard-boiled yolks well centered and your poached and fried eggs neat and compact, or are planning to make a meringue, soufflé, or egg-leavened cake, then you may be better off with the premium grade, with its thicker white and a yolk membrane less likely to leak foam-lowering yolk into the white.

In any case, the quality of a carton of eggs depends mainly on how old they are. Even Grade AA eggs eventually develop flat yolks and thin whites. So be sure to check the sell-by date stamped on the carton (usually four weeks from the packing date; sometimes the pack date itself is indicated by a single number from 1 to 365), and choose the carton with the latest date. Fresh grade A eggs can be a better buy than old grade AA.

Deterioration in Egg Quality

Designed as it was to protect itself for the duration of the chick's development, the egg is unique among our raw animal foods in its ability to remain edible for weeks, as long as it's kept intact and cool. Even so, the moment the egg leaves the hen, it begins to deteriorate in important ways. There is a fundamental chemical change: both the yolk and the white get more alkaline (less acidic) with time. This is because the egg contains carbon dioxide, which takes the form of carbonic acid when it's dissolved in the white and yolk, but is slowly lost in its gaseous form through the pores in the shell. The pH scale provides a measure of acidity and alkalinity (p. 795). On the pH scale, the yolk rises from a slightly acidic pH of 6.0 to a nearly neutral 6.6, while the albumen goes from a somewhat alkaline 7.7 to a very alkaline 9.2 and sometimes higher.

This alkalinization of the white has highly visible consequences. Because albumen proteins at the pH of a fresh egg tend to cluster in masses large enough to deflect light rays, the white of a fresh egg is indeed cloudily white. In more alkaline conditions these proteins repel each other rather than cluster, so the white of an older egg tends to be clear, not cloudy. And the white gets progressively more runny with time: the proportion of thick albumen to thin, initially about 60% to 40%, falls below 50–50.

The relatively minor change in yolk acidity is less important than a simple physical change. The yolk starts out with more dissolved molecules than the white, and this osmotic imbalance creates a natural pressure for water in the white to migrate across the yolk membrane. At refrigerator temperatures, about 5 milligrams of water cross into the yolk each day. This influx causes the yolk to swell, which stretches and weakens the yolk membrane. And the added water thins the yolk dramatically.

A Home Test Finally, the egg as a whole also loses moisture through its porous shell, so the contents of the egg shrink, and the air cell at the wide end expands. Even an oil-coated egg in a humid refrigerator loses 4 milligrams of water to evaporation each

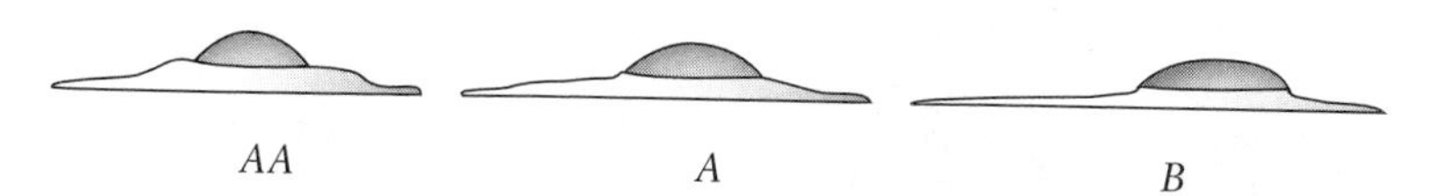

Three different grades of eggs. The AA egg has a high proportion of thick white and a firm, rounded yolk. The A egg has a less thick albumen and a weaker yolk membrane, so it spreads more when cracked into a pan. The B egg spreads even further, and its yolk membrane is easily broken.

day. The cook can use this moisture loss to estimate the freshness of an egg. A new egg with an air space less than ⅛ inch/3 mm deep is denser than water and will sink to the bottom of a bowl of water. As an egg ages and its air cell expands, it gets progressively less dense, and the wide end of the egg rises higher and higher in the water. An egg that actually floats is very old and should be discarded. Around 1750, the English cookbook author Hannah Glasse gave two ways of determining the freshness of an egg, an important talent at a time when it might have been sitting for some time in an odd corner of the yard. One is to feel how warm it is—probably less than reliable—but the second indirectly assays the air cell: "[Another way] to know a good egg, is to put the egg into a pan of cold water; the fresher the egg the sooner it will fall to the bottom; if rotten, it will swim at the top."

All of these trends are probably part of the normal development of the egg. The increase in alkalinity makes the albumen even less hospitable to invading bacteria and molds. The thinning of the albumen allows the yolk to rise and the embryo to approach the shell, its early source of oxygen, and may make it easier for the embryo to tap the shell's calcium stores. A weaker yolk membrane could mean an easier attachment to the shell membranes. And the larger air cell gives the chick more oxygen for its first few breaths.

These changes may be good for the chick, but they're mostly bad for the cook. A thinner white is runnier in the pan; a flabby yolk membrane is more likely to break when the egg is cracked open; and a large air cell means an irregular shape for a whole hard-cooked egg. The only culinary benefit to an older egg is that it's easier to peel.

Handling and Storing Eggs

Producers handle eggs in ways that are meant to slow down the inevitable deterioration in quality. Eggs are gathered as shortly after laying as possible and immediately cooled. In the United States, they are then washed in warm water and detergent to remove the thousands of bacteria deposited on the shell during its passage through the hen's cloacal opening. In the past, the washed eggs were given a fresh coat of mineral oil to retard the loss of both CO_2 and moisture; today, with most eggs getting to market just two days after laying and refrigerated during shipping as well as storage, oiling is limited to long-haul delivery routes.

Egg Storage at Home: Cold, Still, Sealed Egg quality deteriorates as much in a day at room temperature as in four days under refrigeration, and salmonella bacteria (p. 83) multiply much faster at room temperature. So it's best to buy your eggs cold—out of the cooler, not off an open shelf—and keep them cold. Agitation thins

Storage Position

Does it make a difference what posture we store our eggs in? Studies in the 1950s found albumen quality to decline more slowly in eggs stored blunt end up, and many states adopted this as the official position for packing egg cartons. Studies in the 1960s and '70s, when retailers began to stack the cartons on their side to display the top label, found that posture doesn't affect albumen quality. Eggs that are stored on their sides give somewhat better-centered yolks when hard-cooked, perhaps because both yolk cords fight equally against gravity.

the white, so an inner refrigerator shelf is preferable to the door. An airtight container is better than the standard loose carton at slowing moisture loss and the absorption of odors from other foods, although it accentuates the stale flavor that gradually develops in the eggs themselves. Bought fresh and treated with care, eggs should keep for several weeks in the shell. Once broken open, they're far more susceptible to spoilage and should be used promptly or frozen.

Freezing Eggs Eggs can be stored frozen for several months in airtight containers. Remove them from the shell, which would shatter, as its contents expand during freezing. Allow some room for expansion in the containers, and press plastic wrap onto the surface to prevent freezer burn (see p. 146) before covering with a lid. Whites freeze fairly well; they lose only a modest amount of their foaming power. Yolks and blended whole eggs, however, require special treatment. Frozen as is, they thaw to a pasty consistency and can no longer be readily combined with other ingredients. Thoroughly mixing the yolks with either salt, sugar, or acid will prevent the yolk proteins from aggregating, and leaves the thawed mixture fluid enough to mix. Yolks require 1 teaspoon salt, 1 tablespoon sugar, or 4 tablespoons lemon juice per pint (respectively 5 gm, 15 gm, or 60 ml per half liter), and whole eggs half these amounts. The equivalent of a U.S. Large egg is 3 tablespoons whole egg, or 2 tablespoons white and 1 tablespoon yolk.

Egg Safety: The Salmonella Problem

Beginning around 1985, a hitherto minor bacterium called *Salmonella enteritidis* was identified as the culprit in growing numbers of food poisonings in continental Europe, Scandinavia, Great Britain, and North America. Salmonella can cause diarrhea or more serious chronic infection of other body organs. Most of these outbreaks were associated with the consumption of raw or lightly cooked eggs. Further investigation demonstrated that even intact, clean, Grade A eggs can harbor large numbers of salmonella. In the early 1990s, U.S. health authorities estimated that perhaps one egg in 10,000 carried this particularly virulent form of salmonella. Thanks to a variety of preventive measures, the prevalence of contaminated eggs is now much lower—but it's not zero.

Precautions Until the day of the certified salmonella-free egg, all cooks should know how to minimize the risk to themselves and to others, particularly the very young and very old and people with weakened immune systems. The best way to reduce the already small chance of using a badly contaminated egg is to buy only refrigerated eggs and to speed them into your own refrigerator. Cook all egg dishes sufficiently to kill any bacteria that might be present. This generally means holding a temperature of at least 140°F/60°C for 5 minutes, or 160°F/70°C for 1 minute. Egg yolks will remain runny at the first temperature, but will harden at the second. For many lightly cooked egg dishes—soft-boiled and poached eggs, for example, and the yolk-based sauces—it's possible to modify traditional recipes so as to eliminate any salmonella that might be present (see box, p. 91).

Pasteurized Eggs Three safer alternatives to fresh eggs are eggs pasteurized in the shell, liquid eggs, and dried egg whites, all of which are available in supermarkets. Intact eggs, blended whole eggs, or separated yolks and whites can all be pasteurized by careful heating to temperatures between 130 and 140°F/55–60°C, just below the range in which the egg proteins begin to coagulate. Dried egg whites, which are reconstituted in water to make lightly cooked meringues, can be pasteurized either before or after the drying. For most uses, these products do an adequate job of replacing fresh eggs, though there is usually some loss in foaming or emulsifying power and in stability to further heating; and heating and drying do alter the mild egg flavor.

THE CHEMISTRY OF EGG COOKING: HOW EGGS GET HARD AND CUSTARDS THICKEN

The most commonplace procedures involving eggs are also some of the most astonishing kitchen magic. You begin with a slippery, runny liquid, do nothing more than add heat, and presto: the liquid rapidly stiffens into a solid that you can cut with a knife. No other ingredient is as readily and drastically transformed as is the egg. This is the key to its great versatility, both on its own and as a structure builder in complex mixtures.

To what does the egg owe its constructive powers? The answer is simple: to its proteins and their innate capacity to bond to each other.

Protein Coagulation

Pulling Proteins Together . . . The raw egg begins as a liquid because both yolk and white are essentially bags of water containing dispersed protein molecules, with water molecules outnumbering proteins 1,000 to 1. As molecules go, a single protein is huge. It consists of thousands of atoms bonded together into a long chain. The chain is folded up into a compact wad whose shape is maintained by bonds between neighboring folds of the chain. In the chemical environment of the egg white, most of the protein molecules accumulate a negative electrical charge and repel each other, while in the yolk, some proteins repel each other and some are bound up in fat-protein packages. So the proteins in a raw egg mostly remain compact and separate from one another as they float in the water.

When we heat the egg, all its molecules move faster and faster, collide with each other harder and harder, and eventually begin to break the bonds that hold the long protein chains in their compact, folded shape. The proteins unfold, tangle with each other, and bond to each other into a kind of three-dimensional network. There's still much more water than protein, but the water is now divided up among countless little pockets in the continuous protein network, so it can't flow together any more. The liquid egg thus becomes a moist solid. And because the large protein molecules have clustered together densely enough to deflect light rays, the initially transparent egg albumen becomes opaque.

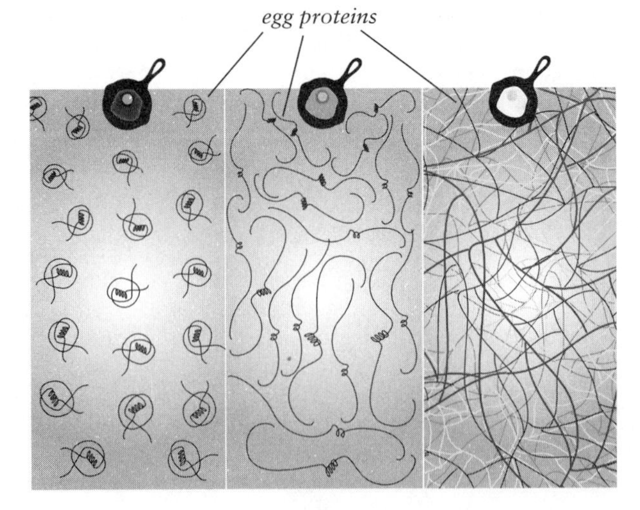

How heat solidifies a liquid egg. Egg proteins begin as folded chains of amino acids (left). *As they're heated, their increased motion breaks some bonds, and the chains unfold* (center). *The unfolded proteins then begin to bond to each other. This results in a continuous meshwork of long molecules* (right), *and a moist but solid egg.*

The other treatments that cause egg to firm up—pickling them in acid or salt, beating them into a foam—work in the same basic way, by overcoming the proteins' aloofness and encouraging them to bond to each other. When you combine treatments—adding both acid and heat, for example—you can achieve a whole range of consistencies and appearances, depending on the degree of protein unfolding and bonding: from tough to delicate, dry to moist, lumpy to jellylike, opaque to clear.

. . . But Not *Too* Close In nearly every egg dish we make, we want to bond a liquid—the egg alone or a mixture of eggs and other liquids—into a moist, delicate solid. Overcooking either gives the dish a rubbery texture or else curdles it into a mixture of hard lumps and watery liquid. Why? Because it bonds the proteins too exclusively to each other and squeezes out the water from the protein network. This is why it is that boiled or fried eggs lose water in the form of steam and get rubbery, while mixtures of eggs and other liquids separate into two phases, the added water and the solid lumps of protein.

The key to cooking egg dishes, then, is to avoid overcooking them and carrying coagulation too far. Above all, this means temperature control. For tender, succulent results, egg dishes should be cooked only just to the temperature at which their proteins coagulate, which is always well below the boiling point, 212°F/100°C. The exact temperature depends on the mixture of ingredients, but is usually higher than the temperature needed to kill bacteria and make the dish safe. (Warm but still liquid yolk is another story; see p. 91). Generally, plain undiluted eggs coagulate at the lowest temperatures. Egg white begins to thicken at 145°F/63°C and becomes a tender solid when it reaches 150°F/65°C. This solidification is due mainly to the most heat-sensitive protein, ovotransferrin, even though it's only 12% of the total protein. The major albumen protein, ovalbumin, doesn't coagulate until about 180°F/80°C, at which temperature the tender white gets much firmer. (The last albumen protein to coagulate is heat-resistant ovomucin, which is why the ovomucin-rich yolk cords remain liquid in scrambled eggs long after the rest has set.) The yolk proteins begin to thicken at 150°F and set at 158°F/70°C, and whole egg—the yolk and white mixed together—sets around 165°F/73°C.

The Effects of Added Ingredients Eggs are often combined with other ingredients, from a sprinkling of salt or lemon juice, to spoonsful of sugar or cream, to cups of

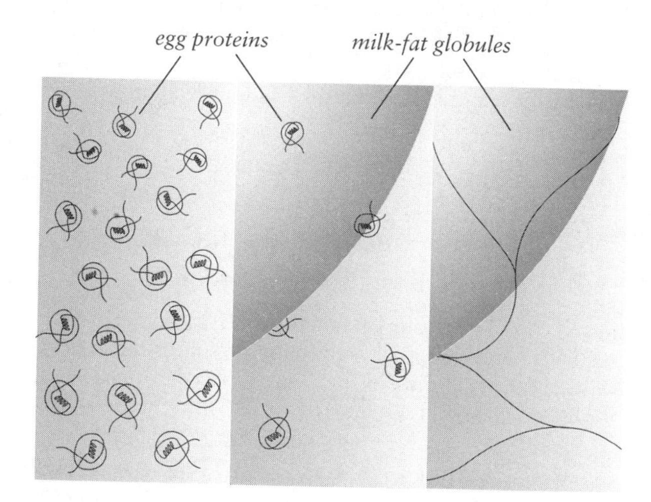

The dilution of egg proteins in a custard. Left: *An egg is rich in proteins; when unfolded by cooking, they are numerous enough to form a firm solid network.* Center: *When mixed with milk or cream, whose proteins don't coagulate with heat, the egg proteins are greatly diluted.* Right: *When a custard mix is cooked, the egg proteins unfold and form a solid meshwork, but that meshwork is open and fragile, and the custard's consistency is delicate.*

milk or brandy. Each of these additions affects egg-protein coagulation and the dish's consistency.

Milk, Cream, and Sugar Dilute, Delay, and Tenderize When we dilute eggs with other liquids, we raise the temperature at which thickening begins. Dilution surrounds the protein molecules with many more water molecules, and the proteins must be hotter and moving around more rapidly in order to find and bond to each other at a noticeable rate. Sugar also raises the thickening temperature, and for the same reason: its molecules dilute the proteins. A tablespoon of sugar surrounds each protein molecule in a one-egg dish with a screen of several thousand sucrose molecules. Combine the diluting effects of water, sugar, and milk fat, and a custard mix containing a cup of milk, a tablespoon of sugar, and an egg begins to thicken not at 160°F/70°C, but at 175 or 180°F/78–80°C. And because the protein network is stretched out into such a large volume—in a custard, the proteins from a single egg have to embrace not three tablespoons of liquid but 18 or 20!—the coagulum is far more delicate, and easily disrupted by overheating. At the extreme, in a concoction like eggnog or the Dutch brandy drink advocaat, the egg proteins are so diluted that they can't possibly accommodate all the liquid, and instead merely give it some body.

Acids and Salt Tenderize There's no truth to the common saying that acidity and salt "toughen" egg proteins. Acids and salt do pretty much the same thing to egg proteins. They get the proteins together sooner, but they don't let them get as *close* together. That is, acids and salt make eggs thicken and coagulate at a lower cooking temperature, but actually produce a more tender texture.

The key to this seeming paradox is the negative electrical charge that most of the egg proteins carry, and that tends to keep them at a distance from each other. Acids—cream of tartar, lemon juice, or the juice of any fruit or vegetable—lower the pH of the egg, and thus diminish the proteins' mutually repelling negative charge. Similarly, salt dissolves into positively and negatively charged ions that cluster around the charged portions of the proteins and effectively neutralize them. In both cases, the proteins no longer repel each other as strongly, and therefore approach each other and bond together earlier in the cooking and unfolding process, when they're still mostly balled up and can't intertwine and bond with each other as tightly. In addition, coagulation of the yolk proteins and of some albumen proteins depends on sulfur chemistry that is suppressed in acidic conditions (see the discussion of egg foams, p. 103). So eggs end up more tender when salted, and especially when acidified.

Cooks have known this for a long time. In Morocco, Paula Wolfert found that eggs are often beaten with lemon juice before long cooking to prevent them from becoming leathery; and Claudia Roden gives an

Early Acid-Tenderized Eggs

Marmelades *or Scrambled Eggs and Verjus, Without Butter*

Break four eggs, beat them, adjust with salt and four spoonsful of verjus [sour grape juice], put the mix on the fire, and stir gently with a silver spoon just until the eggs thicken enough, and then take them off the fire and stir them a bit more as they thicken. One can make scrambled eggs in the same way with lemon or orange juice . . .

—*Le Patissier françois,* ca. 1690

Arab recipe for scrambled eggs made unusually creamy with vinegar (the eggs' alkalinity reduces the amount of free, odorous acetic acid, so the flavor is surprisingly subtle). Eggs scrambled with tart fruit juices were popular in 17th-century France, and may have been the ancestors of lemon curd.

The Chemistry of Egg Flavor

Fresh eggs have a mild flavor that has proven difficult to analyze. The white contributes the main sulfury note, the yolk a sweet, buttery quality. The aroma produced by a given egg is slightest immediately after laying, and gets stronger the longer it's stored before cooking. In general, egg age and storage conditions have a greater influence on flavor than the hen's diet and freedom to range. However, both diet and pedigree can have noticeable effects. Brown-egg breeds are unable to metabolize an odorless component of rapeseed and soy meals (choline), and their intestinal microbes then transform it into a fishy-tasting molecule (triethylamine) that ends up in the eggs. Fish-meal feeds and certain feed pesticides cause off-flavors. The unpredictable diet of truly free-range hens will produce unpredictable eggs.

Something between 100 and 200 compounds have been identified in the aroma of cooked eggs. The most characteristic is hydrogen sulfide, H_2S. In large doses—in a spoiled egg or industrial pollution—H_2S is very unpleasant. In a cooked egg it contributes the distinctively eggy note. It's formed predominantly in the white, when the albumen proteins begin to unfold and free their sulfur atoms for reaction with other molecules, at temperatures above 140°F/60°C. The longer the albumen spends at these temperatures, the stronger the sulfury aroma. Greater quantities of H_2S are produced when the egg is older and the pH higher (the highly alkaline conditions in Chinese preserving methods, p. 116, also liberate copious amounts of H_2S). Added lemon juice or vinegar reduces H_2S production and its aroma. Because hydrogen sulfide is volatile, it escapes from cooked eggs during storage, so they get milder with time. Small quantities of ammonia are also created during cooking and make a subliminal contribution to egg flavor (but an overpowering one in Chinese preserved eggs).

BASIC EGG DISHES

Eggs Cooked in the Shell

"Boiling an egg" is often taken as a measure of minimal competence in cooking, since you leave the egg safe in its shell and have only to keep track of the water temperature and the time. Though we commonly speak of hard- and soft-boiled eggs, boiling is not a good way to cook eggs. Turbulent water knocks the eggs around and cracks shells, which allows albumen to leak out and overcook; and for hard-cooked eggs, a water temperature way above the protein coagulation temperature means that the outer layers of the white get rubbery while the yolk cooks through. Soft-cooked eggs aren't cooked long enough to suffer in the same way, and should be cooked in barely bubbling water, just short

Telling Cooked Eggs from Raw

It's easy to tell whether an intact egg is raw or already cooked. Give it a spin on its side. If it spins fast and smoothly, it's cooked. If it seems balky and wobbly, it's raw—the liquid insides slip and slosh and resist the movement of the solid shell.

of the boil. Hard-cooked eggs should be cooked at a bubble-less simmer, between 180 and 190°F/80–85°C. Eggs in the shell can also be steamed, a technique that requires the least water and the least energy and time to heat the water. Leaving the lid slightly ajar on a gently bubbling steamer will reduce the effective cooking temperature to something below the boil and produce a tenderer white.

Times and Textures Cooking times for in-shell eggs are determined by the desired texture (they also depend on egg size, starting temperature, and cooking temperature; the times here are rough averages). There's a continuum of eggs cooked in the shell for different periods of time. The French *oeuf à la coque* ("from the shell") is cooked for only two or three minutes and remains semi-liquid throughout. Coddled or "soft-boiled" eggs, cooked 3 to 5 minutes, have a barely solid outer white, a milky inner white, and a warm yolk, and are spooned from the shell. The less familiar mollet eggs (from the French *molle,* "soft"), cooked for 5 or 6 minutes, have a semi-liquid yolk but a sufficiently firm outer white that they can be peeled and served whole.

Hard-cooked eggs are firm throughout after cooking for 10 to 15 minutes. At 10 minutes the yolk is still dark yellow, moist, and somewhat pasty; at 15, it's light yellow, dry, and granular. Hard-cooking is sometimes prolonged for hours to heighten color and flavor (p. 89). Chinese tea eggs, for example, are simmered until set, then gently cracked, and simmered for another hour or two in a mixture of tea, salt, sugar, and flavorings to produce a marbled, aromatic, very firm white.

Hard-Cooked Eggs A properly prepared hard-cooked egg is solid but tender, not rubbery; its shell intact and easy to peel; its yolk well centered and not discolored; its flavor delicate, not sulfurous. Good texture and flavor are obtained by taking care not to overcook the eggs, which overcoagulates their proteins and generates too much hydrogen sulfide. Any method that keeps the cooking temperature well below the boil will help avoid overcooking, as will plunging the cooked eggs into ice water. Gentle cooking also takes care of most shell and yolk problems—but not all.

Easily Cracked and Not So Easily Peeled Shells A shell that cracks during hard cooking makes a mess and a sulfurous stink, while a shell that doesn't peel away cleanly makes an ugly, pockmarked egg. A traditional preventative measure for both problems is to poke a pinhole in the wide end of the shell, but studies have found that this doesn't make much difference. The best way to avoid cracking is to heat fresh eggs gently, without the turbulence of boiling water. On the other hand, the best guarantee of easy peeling is to use old eggs! Difficult peeling is characteristic of fresh eggs with a relatively low albumen pH, which somehow causes the albumen to adhere to the inner shell membrane more strongly than it coheres to itself. At the pH typical after several days of refrigeration, around 9.2, the shell peels easily. If you end up with a carton of very fresh eggs and need to cook them right away, you can add a half teaspoon of baking soda to a quart of water to make the cooking water alkaline (though this intensifies the sulfury flavor). It also helps to cook fresh eggs somewhat longer to make the white more cohesive, and to allow the white to firm up in the refrigerator before peeling.

Off-Center Yolks and Flat-Bottomed Whites Well-centered yolks for attractive slices or stuffed halves are easiest to obtain from fresh, high-grade eggs with small air cells and plenty of thick albumen. As eggs age, the albumen loses water and becomes more dense, which makes the yolk rise. Industry studies have found that you can increase the proportion of centered yolks somewhat by storing eggs on their sides instead of their ends. Various cooking strategies have also been suggested, including rotating the eggs around their long axis during the first several minutes in the pot, and standing them on end. None of these is completely reliable.

Green Yolks The occasional green-gray discoloration on the surface of hard-cooked yolks is a harmless compound of iron and sulfur, ferrous sulfide. It forms at the interface of white and yolk because that's where reactive sulfur from the former comes into contact with the iron from the latter. The alkaline conditions in the white favor the stripping of sulfur atoms from the albumen proteins when heat unfolds them, and the sulfur reacts with iron in the surface layer of yolk to form ferrous sulfide. The older the egg, the more alkaline the white, and the more rapidly this reaction occurs. High temperatures and prolonged cooking produce more ferrous sulfide.

Yolk greening can be minimized by using fresh eggs, by cooking them as briefly as possible, and by cooling them rapidly after cooking.

Long-*Cooked Eggs* An intriguing alternative to the standard hard-cooked egg is the Middle Eastern *hamindas* (Hebrew) or *beid hamine* (Arabic), which are cooked for anywhere from 6 to 18 hours. They derive from the Sephardic Sabbath mixed stew (called *hamin,* from the Hebrew for "hot"), which was put together on Friday, cooked slowly in the oven overnight, and served as a midday Sabbath meal. Eggs included in the stew shell and all, or alternatively long-simmered in water, come out with a stronger flavor and a striking, tan-colored white. During prolonged heating in alkaline conditions, the quarter-gram of glucose sugar in the white reacts with albumen protein to generate flavors and pigments typical of browned foods (see the explanation of the Maillard reaction on p. 778). The white will be very tender and the yolk creamy if the cooking temperature is kept in a very narrow range, between 160 and 165°F/71–74°C.

Eggs Cooked Out of the Shell

Baked, Shirred, en Cocotte There are several ways of soft-cooking eggs that are broken out of the shell and into a container, which might be a dish or a hollowed-out fruit or vegetable. As is true of in-shell soft-cooked eggs, timing is of the essence to avoid overcoagulation of the white and yolk proteins, and depends on the nature and placement of the heat source. In the case of baked or shirred eggs, the dish should be set on the middle rack to avoid overcooking the top or bottom while the rest cooks through. Eggs *en*

Eggs and Fire

Another Way with Eggs (Roasting)

Turn fresh eggs carefully in warm ashes near the fire so that they cook on all sides. When they begin to leak they are thought to be freshly done, and so are served to guests. These are the best and are most agreeably served.

Eggs on a Spit

Pierce eggs lengthwise with a well-heated spit and parch them over the fire as if they were meat. They should be eaten hot. This is a stupid invention and unsuitable and a cooks' joke.

—Platina, *De honesta voluptate et valetudine,* 1475

cocotte ("in the casserole") are cooked in dishes set in a pan of simmering water, either on the stovetop or in the oven. Here the eggs are well buffered from the heat source, yet cook just as quickly as baked eggs because water transfers heat more rapidly than the oven air.

Poached Eggs A poached egg is a containerless, soft-cooked egg that generates its own skin of coagulated protein in the first moments of cooking. Slid raw into a pan of already simmering water—or cream, milk, wine, stock, soup, sauce, or butter—it cooks for three to five minutes, until the white has set, but before the yolk does.

The Problem of Untidy Whites The tricky thing about poached eggs is getting them to set into a smooth, compact shape. Usually the outer layer of thin white spreads irregularly before it solidifies. It's helpful to use fresh Grade AA eggs shelled just before cooking, which have the largest proportion of thick white and will spread the least, and water close to but not at the boil, which will coagulate the outer white as quickly as possible without turbulence that would tease the thin albumen all over the pan. Other conventional cookbook tips are not very effective. Adding salt and vinegar to the cooking water, for example, does speed coagulation, but it also produces shreds and an irregular film over the egg surface. An unconventional but effective way to improve the appearance of poached eggs is simply to remove the runny white from the egg *before* poaching. Crack the egg into a dish, then slide it into a large perforated spoon and let the thin white drain away for a few seconds before sliding the egg into the pan.

Timing Poached Eggs by Levitation There's a professional method for poaching eggs that also makes great amateur entertainment. This is the restaurant technique in which eggs are cracked into boiling water in a tall stockpot, disappear into the depths, and—as if by magic!—bob up to the surface again just when they're done: a handy way indeed to keep track of many eggs being cooked at once. The trick is the use of vinegar and salt (at about ½ and 1 tablespoon respectively for each quart of cooking water, 8 and 15g per liter) and keeping the water at the boil. The vinegar reacts with bicarbonate in the thin white to form tiny buoyant bubbles of carbon dioxide, which get trapped at the egg surface as its proteins coagulate. The salt increases the density of the cooking liquid just enough that the egg and three minutes' worth of bubbles will float.

Fried Eggs The containerless fried egg is even more prone to spreading than the poached egg because it is heated only from below, so its white is slower to coagulate. Fresh, high-grade eggs give the most compact shape, and straining off the thin white also helps. The ideal pan temperature for a pale, tender fried egg is around 250°F/120°C, when butter has finished sizzling but hasn't yet browned, or oil to which a drop of water has been added has stopped sputtering. At higher temperatures, you lose tenderness but gain a more flavorsome, browned and crisp surface. The top of the egg can be cooked by turning the egg over

Poached Threads

A kind of poached egg that was enjoyed in 17th-century France and England, and still is in modern China and Portugal, is egg yolk trailed in a thin stream into hot syrup, then lifted out as sweet, delicate threads.

after a minute or so, or by adding a teaspoon of water to the pan and covering it to trap the resulting steam, or—as in the browned Chinese "coin-purse" egg—the egg can be folded over onto itself when barely set, so that top and bottom are crisped but the yolk remains protected and creamy.

Scrambled Eggs Scrambled eggs and omelets are made from yolks and whites mixed together, and are therefore a good fate for fragile, runny lower-quality eggs. These dishes frequently include other ingredients. Cream, butter, milk, water, or oil (used in China) will dilute the egg proteins and produce a tenderer mass when the eggs are carefully cooked; overheating, however, will cause some of the added liquid to separate. Watery vegetables like mushrooms should be precooked to prevent them from weeping into the eggs. Chopped herbs, vegetables, or meats should be warm—not hot or cold—to avoid uneven heating of adjacent egg proteins.

The Key to Scrambled Eggs: Slow Cooking Scrambled eggs made in the usual quick, offhand way are usually hard and forgettable. The key to moist scrambled eggs is low heat and patience; they will take several minutes to cook. The eggs should be added to the pan just as butter begins to bubble, or oil makes a water drop dance gently. Texture is determined by how and when the eggs are disturbed. Large, irregular curds result if the cook lets the bottom layer set for some time before scraping to distribute the heat. Constant scraping and stirring prevents the egg proteins at the bottom from setting into a separate, firm layer, and produces a creamy, even mass of yolk and thin white punctuated with very fine curds of thick white. Scrambled eggs should be removed from the pan while still slightly underdone, since they will continue to thicken for some time with their residual heat.

Omelets If good scrambled eggs demand patience, a good omelet takes panache—a two- or three-egg omelet cooks in less than a minute. Escoffier described the omelet as scrambled eggs held together in a coagulated envelope, a skin of egg heated past the moist, tender stage to the dry and tough, so that it has the strength to contain and shape the rest. Its formation requires a hotter pan than do evenly tender scrambled eggs. But a hot pan means fast cooking to avoid overcooking.

An important key to a successful omelet is contained in the name of the dish, which since the Middle Ages has gone through various forms—*alemette, homelaicte, omelette* (the standard French)—and comes ultimately from the Latin *lamella,* "thin plate." The volume of eggs and the pan diameter should be balanced so that the mix forms a relatively *thin* layer; otherwise the scrambled mass will take too long to cook and be hard to hold together. The usual recommendation is three eggs in a medium-sized frying pan, which should

Safe Poached Eggs

The runny yolk in ordinary poached eggs hasn't been heated enough to eliminate any salmonella bacteria that might be present. To eliminate bacteria while keeping the yolk soft, transfer the finished egg to a second large pan full of water at 150°F/65°C, cover, and let sit for 15 minutes. Check the thermometer every few minutes; if the water drops below 145°F/63°C, put it back on the heat. If you want to cook the eggs a short time before serving them, this hot-water bath is a useful alternative to chilling and then reheating.

have a well-seasoned or nonstick surface so that the skin will come away from it cleanly.

The skin of an omelet can be formed either just at the end of the cooking, or right from the beginning. The fastest technique is to scramble the eggs vigorously with a spoon or fork in a hot pan until they begin to set, then push the curds into a rough disk, let the bottom consolidate for a few seconds, shake the pan to release the disk, and fold it onto itself. A more substantial and more uniform-looking skin results if the eggs are left undisturbed for a while to allow the bottom surface to set. The pan is then shaken periodically to free the skin from the pan while the still-liquid portion alone is stirred until creamy, and the disk finally folded and slipped onto a plate. Yet another way is to let the bottom of the mix set, then lift an edge with the fork and tip the pan to let more of the liquid egg run underneath. This is repeated until the top is no longer runny, and the mass then folded over.

An omelet with an especially light texture (*omelette soufflée*) is made by whipping the eggs until they're full of bubbles, or by whipping the separated whites into a foam and folding them gently back into the mixture of yolks and flavorings. The mix is poured into a heated pan and cooked in a moderate oven.

EGG-LIQUID MIXTURES: CUSTARDS AND CREAMS

DEFINITIONS

Eggs are mixed with other liquids across a tremendous range of proportions. One tablespoon of cream will enrich a scrambled egg, while one beaten egg will slightly thicken a pint of milk into an eggnog. Just about in the middle of this range—at around 4 parts liquid to 1 part egg, or 1 cup/250 ml to 1 or 2 eggs—are the custards and creams, dishes in which the egg proteins give substantial body to otherwise thin liquids. These terms are often used interchangeably, which obscures a useful distinction.

In this section I'll use *custard* to mean a dish prepared and served in the same container, often baked and therefore unstirred, so that it sets into a solid gel. The custard family includes savory quiches and timbales as well as sweet flans, crèmes caramels, pots de crème, crèmes brûlées, and cheesecakes. *Creams*, by contrast, are auxiliary preparations, made from essentially the same mix as custards but stirred continuously during stovetop cooking to produce a thickened but malleable, even pourable mass. Pastry cooks in particular use crème anglaise (so-called "custard

Classically Smooth Scrambled Eggs

Oeufs brouillés au jus *(Scrambled Eggs with Meat Demiglace)*

Break a dozen fresh eggs into a dish, beat them thoroughly, pass them through a strainer into a casserole dish, add six ounces of Isigny butter cut into small pieces, season with salt, white pepper, and grated nutmeg; place on a moderate stove and whip gently with a little egg-white whip. As soon as they begin to thicken remove the casserole from the flame and continue to whip until the eggs form a light, smooth cream. Then add a little chicken demiglace, as big as a nut of butter, cut into pieces, return to the stove to finish the cooking, pour into a silver casserole and garnish with croutons passed through nicely colored butter.

—Antonin Carême, *L'Art de la cuisine française au 19ième siècle*, 1835

cream"), pastry cream (crème pâtissière), and their relatives to coat or fill or underlie a great variety of baked sweets.

Dilution Demands Delicacy

Nearly all the problems that arise in custard and cream making come from the fact that the egg proteins are spread very thin by the other ingredients. Take the nearly identical recipes for a typical sweet milk custard or a crème anglaise: 1 whole egg, 1 cup/250 ml milk, 2 tablespoons/30 gm sugar. The milk alone increases the volume of the mix—which the proteins must span and knit together—by a factor of 6! And each tablespoon of sugar surrounds every protein molecule in the egg with several thousand sucrose molecules. Because the egg proteins are so outnumbered by water and sugar molecules, the coagulation temperature in a custard is between 10 and 20°F higher than in the undiluted egg, between 175 and 185°F/79–83°C. And the protein network that does form is tender, tenuous, and fragile. Exceed the coagulation range by just 5 or 10°F and the network begins to collapse, forming water-filled tunnels in the custard, grainy curds in the cream.

Gentle Heat Many cooks have known the temptation to crank up the heat after a custard has been in the oven for an hour with no sign of setting, or a cream has been stirred and stirred with no sign of thickening. But there's good reason to resist. The gentler these dishes are heated, the greater the safety margin between thickening and curdling. Turning up the heat is like accelerating on a wet road while you're looking for an unfamiliar driveway. You get to your destination faster, but you may not be able to brake in time to avoid skidding past it. Chemical reactions like coagulation develop momentum, and don't stop the second you turn off the heat. If the thickening proceeds too fast, you may not be able to detect and stop it before it overshoots done and hits curdled. A curdled cream can often be salvaged by straining out the lumps, but an overcooked custard is a loss.

Always Add Hot Ingredients to Cold Careful heating is also important during preparation of the mix. Most custard and cream mixes are made by scalding milk or cream—quickly heating it just to the boil—and then stirring it into the combined eggs and sugar. This technique heats the eggs gently but quickly to 140 or 150°F, just 30 to 40°F short of the setting temperature. Doing the reverse—adding cold eggs to the hot milk—would immediately heat the first dribbles of egg close to the boil and cause premature coagulation and curdling.

Though scalding was a form of insurance in times when milk quality was uncertain, it can now be dispensed with in custard making—unless you need to flavor

Green Eggs in the Chafing Dish

Scrambled eggs and omelets kept hot in a chafing dish or on a steam table will sometimes develop green patches. This discoloration results from the same reaction that turns hard-cooked yolks green (p. 89), and is encouraged by the persistent high temperature and the increased alkalinity of the cooked eggs (the rise is about half a pH unit). It can be prevented by including an acidic ingredient in the egg mixture, around a half teaspoon/2 gm lemon juice or vinegar per egg; half that amount will slow the discoloration and affect the flavor less.

the milk by infusing it with vanilla or coffee beans, citrus peel, or another solid flavoring. A custard mixed cold has just as even a texture and sets almost as quickly as a pre-scalded one. Pre-scalding the milk remains handy in making creams because milk (or cream) can be boiled quickly with little attention from the cook, while heating the milk-egg mix from room temperature requires a low flame and constant stirring to prevent coagulation at the pan bottom.

Curdling Insurance: Starch in Custards and Creams Flour or cornstarch can protect against curdling in custards and creams, even if they're cooked quickly over direct heat and actually boil. (The same is true for egg-based sauces like hollandaise; see p. 628.) The key is the gelation of the solid starch granules in these materials. When heated to 175°F/77°C and above—right around the temperature at which the egg proteins are bonding to each other—the granules absorb water, swell up, and begin to leak their long starch molecules into the liquid. The swelling granules slow protein binding by absorbing heat energy themselves, and the dissolved starch molecules get in the proteins' way and prevent them from bonding to each other too intimately. Because they contain starch, both chocolate and cocoa can also help stabilize custards and creams.

A full tablespoon/8g of flour per cup/250ml liquid (or 2 teaspoons/5g pure starch in the form of cornstarch or arrowroot) is required to prevent curdling. The disadvantage is that this proportion of starch also turns a creamily smooth dish into a coarser, thicker one, and diminishes its flavor.

Custard Theory and Practice

In the West, custards are almost always made with milk or cream, but just about any liquid will do as long as it contains some dissolved minerals. Mix an egg with a cup of plain water and you get curdled egg floating in water; include a pinch of salt and you get a coherent gel. Without minerals, the negatively charged, mutually repelling protein molecules avoid each other as they unfold in the heat, and each forms only a few bonds with a few others. With minerals, positively charged ions cluster around the negatively charged proteins and provide a neutralizing shield, which makes it possible for the proteins to unfold near each other and bond extensively into a fine network. Meats are rich in minerals, and the Japanese make savory custards, *chawan-mushi* (soft) and *tamago dofu* (firm), from both bonito and chicken broths. Vegetable stocks also work.

Proportions The consistency of a custard can be firm or soft, slick or creamy, depending on its egg content. The greater the proportion of whole eggs or whites, the firmer and glossier the custard. Extra yolks, or using yolks alone, will produce a more tender, creamier effect. A custard to be served

> Food Words: *Custard, Cream, Flan*
>
> The nomenclature for egg-milk mixtures has always been loose. The English "custard" began as "croustade" in medieval times, and meant dishes served in a crust—thus, for egg-milk combinations, usually baked and unstirred, and so solid. Early English creams could be either liquid or solid, as could the French *crèmes*. Those congealed past the point of creaminess became known as *crèmes prises*, or "set creams."
>
> *Flan,* a French word, comes from the late Latin for "flat cake."

in the container it was cooked in can be as soft as the cook desires. Those that are to be turned out of a container for serving must be firm enough to stand on their own, which means that they must contain either some egg whites or at least 3 yolks per cup/250 ml of liquid (the LDL-bound yolk proteins are less efficient networkers than the free-floating albumen proteins, so we need more of them to make a firm gel). The replacement of some or all of the milk with cream reduces the proportion of eggs required for a given firmness, since cream contains 20 to 40% less water and the egg proteins are proportionally less diluted. Unmolding is easiest from a buttered ramekin, and when the custards have been allowed to cool thoroughly; cooling firms protein gels.

Custards that contain fruits or vegetables can turn out very uneven, with pockets of fluid and curdling. (Usually this is undesirable, though the Japanese expect *chawan-mushi* to weep and treat it as a combination of custard and soup.) The culprits are juices that leak out of the plant tissue, and fibrous particles, which cause local overcoagulation of egg proteins. The juice leakage can be reduced by precooking the fruit or vegetable, and including some flour in the mix to help bind excess liquid and minimize overcoagulation. These dishes are best cooked very gently and only until barely done.

Cooking Cooks have known for thousands of years that a low cooking temperature provides the greatest safety margin for making custards: that is, it gives us more time to recognize that the dish is properly done and remove it from the heat, before it toughens and tunnels. Custards are usually baked in a moderate oven with the protection of a water bath, which keeps the effective cooking temperature below the boiling point. The actual temperature depends on the pan material and whether and how the water bath is covered (see box). It's a mistake to cover the whole water bath, since this forces the water to the boil and makes it more likely that the custards will be overcooked. The most gentle heating results when the individual molds are set covered on a rack in an open, thin metal pan of hot water.

Custard doneness can be judged by bumping the dish—the contents should move only sluggishly—or by probing the interior with a toothpick or knife, which should return without any mix clinging to it. When the proteins have coagulated enough that the mix clings mostly to itself, the dish is done. Unless the custard needs to be firm enough to unmold, it's best taken from the oven while the center is still slightly underdone and jiggly. The egg proteins continue to set somewhat with the residual heat, and the custard will in any case be firmer once cooled to serving temperature.

"Ribboning" Yolks with Sugar

Cookbooks often assert the importance of beating yolks with sugar until they lighten in color and thicken sufficiently to form a ribbon when trailed from a spoon. This stage does not mark any critical change in the yolk components. It's simply a sign that much of the sugar has dissolved in the limited yolk water (about half the volume of the yolks themselves), which makes the mix viscous enough to pour thickly and retain air bubbles (the cause of the whitening). Sugar grains are a convenient means for mixing the yolks and albumen remnants thoroughly, but the quality of a cream or custard will not suffer if you mix the yolks and sugar thoroughly but stop short of the ribbon.

Savory Custards: The Quiche The quiche (a French version of the German *Kuchen,* "little cake") can be thought of either as a savory custard or a close relative of the omelet. It is a pie-shaped mixture of eggs and cream or milk that contains small pieces of a vegetable, meat, or cheese. To make it firm enough to be cut into wedges for serving, a quiche normally contains 2 whole eggs per cup/250 ml of liquid, and is baked unprotected by a water bath, either alone or in a precooked crust. The Italian *frittata* and Egyptian *eggah* are similar preparations that omit any milk or cream.

Crème Caramel and Crème Brûlée Crème caramel is a freestanding sweet custard with a layer of moist caramel on top. It's made by coating the bottom of the dish with a layer of caramelized sugar (see p. 656) before the custard mix is poured in and cooked. The caramel does harden and stick to the dish, but moisture from the custard mix softens it, and the two layers become partly integrated. The custard is turned out of the dish while still slightly warm and the caramel soft. If the custard must be refrigerated before serving, leave it in the mold; the caramel can be softened again by placing the dish in a shallow pan of hot water for a minute or two before unmolding.

Crème brûlée ("burned cream") is also a custard topped with caramel, but here the caramel should be hard enough to shatter when rapped with a spoon. The trick is to harden and brown the sugar topping without overcooking the custard. The standard modern method is to bake the custard and then chill it for several hours, so that the subsequent caramelizing step won't overcook the egg proteins. The hard crust is then made by coating the custard surface with granulated sugar, and then melting and browning the sugar, either with a propane torch or by placing the dish right under the broiler. The dishes are sometimes immersed in an ice-water bath to protect the custard from a second cooking. From the time of its invention in

The Surprising Science of Water Baths

Most cooks know that oven heat can be moderated with a water bath. Though the oven may be at 350°F, the liquid water can't exceed 212°F/100°C, the temperature at which it boils and turns from liquid into vapor. Less well known is the fact that the water temperature can vary over a range of 40°F depending on the pan containing the water and whether it's covered. A pan of water is heated by the oven, but it's simultaneously cooled as water molecules evaporate from the surface. The actual water temperature is determined by the balance between heating of the water mass through the pan, and evaporative cooling at the water surface. More heat accumulates in a thick cast iron pan or passes through infrared-transparent glass than is transmitted by thin stainless steel. So in a moderate oven, a cast-iron water bath may reach 195°F/87°C, a glass bath 185°F/ 83°C, and a stainless one 180°F/80°C. If the pans are covered with foil, then evaporative cooling is prevented, and all of them will come to a full boil.

Custards are tenderest when heated gently, and so are best cooked in an open water bath—one, however, that is sure to reach at least 185°F; otherwise the mix may never completely set. Many cooks take the precaution of folding a kitchen towel in the bottom of the water bath so that the custard cups or dish won't be in direct contact with the hot pan, but this can backfire: the towel prevents the water from circulating under the cups, so the water trapped there reaches the boil and rocks the cups around. A wire rack works better.

the 17th century until early in the 20th, crème brûlée was a stirred cream, prepared on the stovetop by making a crème anglaise, pouring it into dishes, and caramelizing the sugar topping with a red-hot metal plate, or "salamander."

Cheesecake We don't ordinarily think of cheesecake as a custard, probably because the presence of eggs is masked by the richness of the filling they bind together, which is some combination of ricotta cheese, cream cheese, sour cream, heavy cream, and butter. The proportions for cheesecake are similar to those for other custards, approaching 1 egg per cup/250 ml of filling, though the greater richness and tartness of the filling demand more sugar for balance, around 4 tablespoons per cup (60 gm per 250 ml) instead of 2. Flour or cornstarch is sometimes included to stabilize the gel and, in the case of ricotta cheesecakes, to absorb water that may be released from the fresh cheese.

The thick texture and high fat content of cheesecake filling necessitate more delicate treatment than a standard custard. Instead of a preliminary cooking on the stovetop, the sugar is first mixed into the cream ingredients, and the eggs then incorporated along with other flavorings. The cool mix is poured into the pan (often preceded by a crumb crust) and baked at a gentle 325°F/163°C, often in a water bath. The

First Recipes for Crème Brûlée, Crème Anglaise, and Crème Caramel

Massialot's recipe for crème brûlée is the first I know of. The identical recipe in the 1731 edition of his book is renamed "Crême a l'Angloise," which may well be the origin of that basic stirred cream. An English model for "English cream" hasn't yet been unearthed.

Crème brûlée

> Take four or five egg yolks, according to the size of your platter. You mix them well in a casserole with a good pinch of flour; and little by little you pour in some milk, about [3 cups/750 ml]. Add a little stick cinnamon, and chopped green citron peel. . . . Put on the stovetop and stir continuously, taking care that your cream doesn't stick to the bottom. When it is well cooked, place a platter on the stove, pour the cream onto it, and cook it again until you see it stick to the platter rim. Then remove from the heat and sugar it well: take the fire iron, good and red, and burn the cream so that it takes on a fine gold color.
>
> —F. Massialot, *Le Cuisinier roial et bourgeois,* 1692

A few decades later, Vincent La Chapelle plagiarized Massialot's recipe for his own version of crème brûlée, which comes close to the modern crème caramel. La Chapelle copies Massialot word for word up to the point that the cream is cooked on the stovetop. Then . . .

> When the cream is well cooked, put a silver platter onto the hot stove with some powdered sugar and a little water to dissolve it; and when your sugar has colored, pour the cream on top; turn the sugar along the platter rim onto the top of your cream, and serve at once.
>
> —V. La Chapelle, *Le Cuisinier moderne,* 1742

last phase of cooking may take place with the heat off and the oven door ajar, which smooths the transition between cooking and cooling.

The most common problem with cheesecakes is the development of depressions and cracks in the surface, which result when the mix expands and rises during the cooking, then shrinks and falls as it cools down. Rising is essential for soufflés and sponge cakes, but it is antithetical to the dense richness of cheesecake. Four basic strategies will minimize it. First, beat the ingredients slowly, gently, and only long enough to obtain an even mix. Vigorous or long beating incorporates more air bubbles that will fill with steam and expand during baking. Second, bake the cheesecake slowly in a low oven. This will allow trapped air and steam to disperse gradually and evenly. Third, don't overbake. This will dry the filling and cause it to shrink from moisture loss. Finally, cool the cheesecake gradually in the open oven. Cooling causes any trapped air or steam to contract, and the more gradually this happens, the more gently the cheesecake surface is pulled in.

Cream Theory and Practice

Creams are easier to make than custards in two respects. They're heated on the stovetop, so the cook doesn't have to consider the fine points of heat transfer in the oven. And because they're not served as is, in the container they're cooked in, some curdling can be tolerated and remedied by putting the cream through a strainer before it's served.

Pourable and Stiff Creams There are two broad classes of creams, and they demand entirely different handling by the cook. The *pourable creams,* crème anglaise for example, are meant to have the consistency of heavy cream at serving temperature. They contain the standard eggs, milk, and sugar (sugar is omitted for a savory cream), and are cooked only until they just begin to thicken, far below the boil. The *cream fillings*—crème pâtissière, banana cream, and so on—are meant to stay put in a dish and hold their shape. They are therefore stiffened with a substantial dose of flour or cornstarch; and this means not only that they *can* be heated to the boil, they *must* be boiled. Egg yolks contain a starch-digesting enzyme, amylase, that is remarkably resistant to heat. Unless a starch-egg mix is brought to a full boil, the yolk amylase will survive, digest the starch, and turn the stiff cream into a pourable one.

When stored for any time, creams should be protected against the formation of the leathery skin that results from evaporation, which concentrates and toughens the surface layer of protein and starch. Butter can be dotted onto the warm surface, where the milkfat will melt and spread into a protective layer; and sprinkled sugar will form a layer of concentrated syrup that resists evaporation. The most straightforward solution is to press waxed paper or

A Medieval Cheesecake

Tart de bry

Take raw yolks of eggs, and good fat cheese, and dress it, and mix it well together; and add powder of ginger, and of cinnamon, and sugar, and saffron, and put it in a crust, and bake it, and serve it forth.

—from a manuscript published in *Antiquitates Culinariae,* 1791 (ca. 1400)

buttered parchment directly onto the cream. Avoid plastic wrap; its plasticizing chemicals tend to migrate into fat-rich foods.

Crème Anglaise and Other Pourable Creams The mix for a stirred cream is made much as baked-custard mixes are. An especially rich cream may call for yolks only, as many as 4 or 5 per cup/250 ml milk. The eggs and sugar are mixed with scalded milk or cream, and the mixture is then stirred constantly on the stovetop until it thickens enough to cling to the spoon, at around 180°F/80°C. The gentle heat of a double boiler minimizes the possibility of curdling, but it takes longer than direct heat. The thickened cream is then strained of any coagulated egg or other solid particles, and cooled, with occasional stirring to prevent the proteins from setting into a solid gel. An ice bath will cool the cream quickly, but demands more frequent stirring to maintain an even texture. Fruit purees are generally added after the cooling, because their acidity and fibrous particles can cause curdling during the cooking.

Pastry Cream, *Bouillie*, and Cream-Pie Fillings Along with crème anglaise, pastry cream is one of the most versatile of the dessert maker's stock preparations. It's used mainly to fill and decorate cakes and pastries, and is a common reinforcing base for sweet soufflés; in Italy and France it's even cut into pieces and fried on its own. It must therefore be thick enough to hold its shape at room temperature, and so is stiffened with between 1 and 2 tablespoons flour (or about half that amount of pure starch) per cup liquid/10–20 gm per 250 ml.

Pastry cream is made by adding scalded milk to the mixture of sugar, eggs, and flour, whose protective action allows the mix to be brought to a full boil over direct heat without curdling. After a minute or so of boiling (and constant stirring) to thoroughly inactivate the yolk amylase enzyme and to

The First Recipe for Pastry Cream

Pastry cream has been a standard professional preparation for more than three centuries.

The Manner of Making Cresme de Pâstissier

Take for example a *chopine* [3 cups/750 ml] of good milk. . . . Put the milk in a pot on the fire: you must also have four eggs, and while the milk heats up, break two eggs, and mix the white and yolk with about a half *litron* [7 oz/185 gm] flour, as if for making porridge, and a little milk. And when the flour is well diluted so that it has no more lumps, you throw in the other two eggs to mix them well with this preparation.

And when the milk begins to boil you pour in little by little this mix of eggs and flour and milk, and boil together on a low flame that is clear and without smoke; stir with a spoon as you would a porridge. You must also add salt at your discretion as it cooks, and a quarteron [a quarter-pound/125 gm] good fresh butter.

This cream should be cooked for 20 to 25 minutes, then pour it into a bowl and set aside this preparation, which pastry cooks call cream and use in many baked goods.

—*Le Pâtissier françois,* ca. 1690

extract starch from its granules, and to improve the flavor, the thickened cream is scraped into a bowl and allowed to cool with minimal stirring (stirring breaks the developing starch network and thins it out). Once cool, pastry cream is sometimes enriched with cream or butter, or lightened with foamed egg whites, or simultaneously enriched and lightened with whipped cream.

A traditional French variant on pastry cream is the *bouillie* (literally "boiled"; the word means a plain porridge-like cereal paste), which is made at the last minute, and primarily to reinforce soufflés. For a *bouillie,* milk, sugar, and flour are heated together to the boil, removed from the heat, and the eggs beaten in as the mix cools. Because the egg proteins are not as thoroughly heated and coagulated as they are in the technique for pastry cream, the consistency of a *bouillie* is lighter and smoother. Some yolk amylase enzyme survives in a *bouillie,* but this doesn't matter if the dish is to be made and served immediately; the enzyme takes hours to digest a noticeable amount of starch.

However, the survival of yolk amylase can spell disaster in the fillings for American cream pies, which are often made in the fashion of a *bouillie* rather than a pastry cream, and are held for hours or days before serving, enough time for a perfect cream pie to disintegrate into a soupy mess. No matter what a recipe may say, always be sure that the egg yolks in a starch-thickened pie filling are heated all the way to the boil.

Fruit Curds Fruit curds—lemon curd is the most common—can be thought of as a kind of cream in which the place of milk is taken by fruit juice, usually enriched with butter. (They may have begun as a sweetened version of creamy eggs scrambled with fruit juice; see p. 86.) Fruit curds are meant to have a spoonable consistency that works well as a filling for small pastries or a breakfast spread, and must be sweet enough to balance the acidity of the juice. They therefore contain no flour, more sugar, and more eggs than do milk creams, typically 4 eggs (or 8 yolks) and a cup or more of sugar for a half-cup of butter and a half-cup of juice (375 gm sugar per 125 ml each of butter and juice).

EGG FOAMS: COOKING WITH THE WRIST

If the transformation of eggs by heat seems remarkable, consider what beating can do! Physical agitation normally breaks down and destroys structure. But beat eggs and you *create* structure. Begin with a single dense, sticky egg white, work it with a whisk, and in a few minutes you have a cupful of snowy white foam, a cohesive structure that clings to the bowl when you turn it upside down, and holds its own when mixed and cooked. Thanks to egg whites we're able to harvest the air, and make it an integral part of meringues and mousses, gin fizzes and soufflés and sabayons.

The full foaming power of egg white seems to have burst forth in the early 17th century. Cooks had noticed the egg's readiness to foam long before then, and by Renaissance times were exploiting it in two fanciful dishes: imitation snow and the confectioner's miniature loaves and biscuits. But in those days the fork was still a novelty, and twigs, shreds of dried fruits, and sponges could deliver only a coarse froth at best (see box, p. 101). Sometime around 1650, cooks began to use more efficient whisks of bundled straw, and meringues and soufflés start to appear in cookbooks.

Like the head on a beer or a cappuccino, an egg foam is a liquid—the white—filled with a gas—air—in such a way that the mixture of liquid and gas keeps its shape, like a solid. It's a mass of bubbles, with air inside each bubble, and the white spread out into a thin film to form the bubble walls. And the makeup of those liquid walls determines how long a foam can stand up. Pure water has such a strong surface tension—such strong attractive forces among its molecules—that it immediately starts to pull itself together into a compact puddle; and it's so runny that it puddles almost immediately. The many nonwater molecules in egg white both reduce the surface tension of the water they float in, and make it less runny, and thus allow the bubbles to survive long enough to accumulate into a sizeable mass. What gives the mass of foam a useful kitchen lifetime is the white's team of proteins.

How the Egg Proteins Stabilize Foams

Stress Builds Protein Solidarity As is true for the setting of heated eggs and custards, the key to the stable egg foam is the tendency of the proteins to unfold and bond to each other when they're subjected to physical stress. In a foam this creates a kind of reinforcement for the bubble walls, the culinary equivalent of quick-setting cement. Whipping exerts two kinds of physical stress on the proteins. First, as we force the whisk through the white, the whisk wires drag some of the liquid with them, and create a pulling force that unfolds the compacted protein molecules. And second, because water and air are very different physical environments, the simple mixing of air into the whites creates an imbalance of forces that also tugs the proteins out of their usual folded shape. All these unfolded proteins (mainly the globulins and ovotransferrin) tend to gather where air and water meet, with their water-

Early Egg-White Foams: "Snow" and Biscuits

How to Break Whites of Eggs Speedily

A fig or two shred in pieces and then beaten amongst the whites of eggs will bring them into an oil speedily: some break them with a stubbed rod, and some by wringing them often through a sponge.

—Sir Hugh Platt, *Delightes for Ladies,* 1605

Eggs in Snow

Break the eggs, separate the whites from the yolks, place the eggs on a plate with some butter, season them with salt, place on hot coals. Beat and whip the whites well, and just before serving throw them on the yolks with a drop of rosewater, the fire iron underneath: sugar, then serve.

Another way: You may put the yolks in the middle of the snow that is made with your whipped whites, and then cook them before the fire on a plate.

—François Pierre de La Varenne, *Le Cuisinier françois,* 1651

To Make Italian Biskets

Take a quarter of a pound of searsed [sieved] Sugar, and beat it in an Alabaster Mortar with the white of an Egg, and a little Gum Dragon [gum tragacanth] steept in Rose water to bring it to a perfect Paste, then mould it up with a little Anniseed and a grain of Musk; then make it up like Dutch bread, and bake it on a Pye-plate in a warm Oven, till they rise somewhat high and white, take them out, but handle them not till they be throughly dry and cold.

—*Queen's Closet Open'd,* 1655

loving portions immersed in the liquid and their water-avoiding portions projecting into the air. Thus disturbed and concentrated, they readily form bonds with each other. So a continuous, solid network of proteins pervades the bubble walls, holding both water and air in place.

Permanent Reinforcement A raw egg-white foam will eventually coarsen, settle, and separate. It must therefore be reinforced when it is turned into a final dish. This may be done by adding other thickening ingredients—such things as flour, cornstarch, chocolate, or gelatin. But if the foam is to be used relatively pure, as in a meringue or a flourless soufflé, the egg proteins have to do the job themselves. With the help of heat, they do beautifully.

Ovalbumin, the major protein in egg white, is relatively immune to beating and doesn't contribute much to the raw foam. But it is sensitive to heat, which causes it to unfold and coagulate. So when the raw foam is cooked, ovalbumin more than doubles the amount of solid protein reinforcement in the bubble walls. At the same time, much of the free water in the foam evaporates. Heat thus allows the cook to transform a transient semiliquid foam into a permanent solid one.

How Proteins *Destabilize* Foams

The very same forces that make egg foams also break egg foams. Often just as the foam is reaching its optimum texture, it will get grainy, lose volume, and separate into a dry froth and a runny liquid. As the proteins bond to each other to support the foam, they embrace each other too tightly, and squeeze out the water they had held between them. There are several different kinds of bonds by which the long, unfolding egg proteins are joined to each other in a reinforcing network: bonds between positively and negatively charged parts of molecules, between water-like parts, between fat-like parts, and between sulfur groups. The protein network begins to collapse when too many of these bonds accumulate and the proteins cluster together too tightly. Fortunately, there are simple ways for the cook to limit the accumulation of bonds and prevent the collapse of albumen foams.

Blocking Sulfur Bonds with Copper Bowls . . . Long before anyone knew about egg proteins or their chemical bonds, cooks had come up with a way of controlling them. The French tradition has long specified the use of copper utensils for

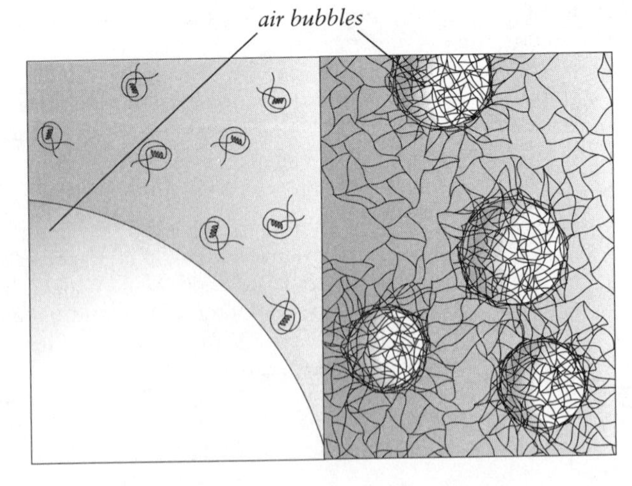

Foamed egg whites. The folded proteins in egg white (left) *produce a light, long-lived foam by unfolding at the interface between liquid and air, the walls of the air bubbles. The unfolded proteins then bond to each other, and form a solid meshwork of reinforcement around the bubbles* (right).

making egg foams. One early trace of this tradition is a 1771 illustration in the French *Encyclopédie* that shows a boy in a pastry kitchen working with a straw whisk and what the accompanying key identifies as "a copper bowl for beating egg whites." It turns out that along with a very few other metals, copper has the useful tendency to form extremely tight bonds with reactive sulfur groups: so tight that the sulfur is essentially prevented from reacting with anything else. So the presence of copper in foaming egg whites essentially eliminates the strongest kind of protein bond that can form, and makes it harder for the proteins to embrace each other too tightly. Sure enough, if you whip egg whites in a copper bowl—or in a glass bowl to which you've added a pinch of a powdered copper supplement from a health food store—the foam stays glossy and never develops grains. A silver-plated bowl will do the same thing.

. . . And Acids There are disadvantages to the traditional copper bowl: it's expensive, and a nuisance to keep clean. (Copper contamination is negligible; a cup of foam contains a tenth of our normal daily intake.) Fortunately there's a nonmetallic alternative for controlling reactive sulfur groups. The sulfur bonds form when the sulfur-hydrogen (S-H) groups on two different protein molecules shed their hydrogens and form a sulfur-sulfur (S-S) connection with each other. The addition of an acid boosts the number of free-floating hydrogen (H) ions in the egg white, which makes it much harder for the S-H groups to shed their own H, and so slows the sulfur bonding down to a crawl. A good dose is ⅛ teaspoon/0.5g cream of tartar or ½ teaspoon/2ml lemon juice per egg white, added at the beginning of the beating.

The Enemies of Egg Foams

There are three enemies to the successful mounting of a foam which the cook should be careful to exclude from the bowl: egg yolk, oil or fat, and detergent. All are chem-

Copper bowls and eggs in the 18th century. This is a detail of "Pâtissier," *or* "The Pastry-cook," *from the* Encyclopédie, *an engraving first published in 1771. The boy at right wields what the accompanying key calls "a copper bowl for beating egg whites and mixing them with the dough from which biscuits are made."*

ical relatives, and interfere with foaming in the same ways: by competing with the proteins for a place at the air-water interface without offering any structural reinforcement; and by interfering with the bonding of the protein molecules. Traces of these troublemakers won't absolutely prevent you from making a foam, but they'll make you work harder and longer, and the foam won't be as light or stable. Of course yolk and fat can safely be mixed with a finished foam, as happens in many recipes for soufflés and egg-leavened batters.

The Effects of Other Ingredients

Egg white foams are almost always made with other ingredients, and these can influence the beating process and the final consistency.

Salt Salt increases the whipping time and decreases the foam's stability. Salt crystals dissolve into positively charged sodium and negatively charged chloride ions, and these probably compete for bonding sites on the unfolded protein molecules, thereby reducing the number of protein-protein bonds and so weakening the overall structure. It's therefore best to add salt to the other components of a dish—the base of a soufflé, for example—rather than to the foam itself.

Sugar Sugar both hinders and helps foam making. Added early in the process, it delays foaming, and it reduces the foam's ultimate volume and lightness. The delay comes from sugar's interference with the unfolding and bonding of the proteins. And the reduction in volume and lightness is caused by the syrupy sugar-egg mixture being harder to spread into thin bubble walls. Slow foaming is a real disadvantage when the whites are whipped by hand—at standard soft-meringue levels, it doubles the work—but less so if you're using a stand mixer.

The helpful thing about sugar is that it improves the foam's stability. By making the liquid thick and cohesive, sugar greatly slows drainage from the bubble walls and coarsening of texture. In the oven, the dissolved sugar hangs onto the water molecules and so delays their evaporation in the high heat until after ovalbumin has had time to coagulate and reinforce the raw foam. And it eventually contributes reinforcement of its own in the form of fine but solid, cotton-candy-like strands of dry sugar.

Sugar is usually incorporated into the egg whites after the foam has begun to

A Silver Bullet for the Copper Theory

Why do copper bowls make more stable egg foams? I've wondered about this for many years. In 1984 I did some experiments with the help of Stanford University biologists, and then published a theory in the British science journal *Nature* and in the first edition of this book. The experiments suggested that one of the albumen proteins, ovotransferrin, takes up copper from the bowl surface and is thereby rendered resistant to unfolding—which could make the foam as a whole resistant to overcoagulating. That theory stood up for ten years, until one day on a whim I tried whipping egg whites in a silver-plated bowl. Ovotransferrin doesn't bind silver, so the foam should have turned grainy. It didn't. It remained light and glossy. I resumed my frothy investigations, and learned that both copper and silver do block sulfur reactions between proteins. Hence the revised edition of the copper theory outlined here.

form, when many proteins are already unfolded. For some purposes, cooks will mix sugar and whites at the outset, in order to obtain a very firm, dense foam.

Water Water is seldom called for, but in small amounts it increases the volume and lightness of the foam. Because water thins the whites, however, it's more likely that some liquid will drain from the foam. Albumen diluted by 40% or more of its volume in water cannot produce a stable foam.

Basic Egg-Beating Techniques

Beating egg whites into a foam is one of those techniques about which cooks and cookbooks wax stern and stringent. In fact it's not all that sensitive to details. Just about any egg and bowl and whisk can give you a good foam.

Choosing the Eggs An egg foam begins with the eggs. Old eggs at room temperature are often recommended on the grounds that the whites are thinner and therefore foam more rapidly. This is true, and *very* fresh eggs are said to be almost impossible to foam by hand. But fresh eggs are less alkaline and so make a more stable foam; the older thin white also drains from the foam more easily, and old eggs are more likely to leave traces of yolk in the white. Cold yolks are less likely to break as you separate them from the whites, and the whipping process quickly warms cold eggs anyway. Fresh eggs right out of the refrigerator will work fine, especially if you're using an electric mixer. Egg foams can also be made with dried egg whites. Powdered egg whites are pure, pasteurized, freeze-dried egg whites. "Meringue powder" contains more sugar than egg, and includes gums to stabilize the foam.

Bowl and Whisk The bowl in which you beat the whites should be large enough to accommodate an eightfold expansion of their volume. It's often recommended that the cook avoid making egg foams in plastic bowls, because plastics are hydrocarbon relatives of fats, and tend to retain traces of fats and soaps. While this is true, the bowl is also unlikely to release such traces into a mass of egg white. Ordinary cleaning is adequate to make a plastic bowl suitable for foaming eggs.

If you're beating by hand, a large "balloon whisk" aerates a greater volume of the egg whites at a time and will speed your work. If you have a choice of machines, a stand mixer whose beater both spins on its shaft and traces a curlicue path from the center to the edge of the bowl (a "hypocycloidal" or planetary motion) beats the whites more evenly and leaves less unfoamed. Less efficient beaters produce a denser texture.

Interpreting the Foam's Appearance There are various ways to judge when the foam is at its optimum, from seeing whether the foam will support the weight of a coin or an egg, to seeing how it supports itself, in soft mounds or sharply defined peaks, to seeing whether it clings to the bowl or slips along its surface, whether its surface looks glossy or dry. All these tests tell us how crowded the air bubbles are, and how much lubrication they have between them in the way of liquid from the egg white. And different dishes will define an optimum foam differently. The lightening power of an egg foam depends not just on the foam's volume, but also on how easily it can be mixed with other ingredients, and how well it can accommodate bubble expansion in the oven. Soufflés and cakes require the lubrication and expansion tolerance of a somewhat underbeaten foam, while in meringues and related pastries volume is less important than shape-holding stiffness.

Glossy Soft Peaks and Stiff Peaks At the "soft peak" stage, when glossy foam edges retain some shape but droop, and when the foam doesn't yet cling to the bowl, the somewhat coarse bubbles are still lubricated by plenty of liquid, which would quickly drain to the bottom of the bowl. At the "stiff peak" stage, where the foam is still glossy but now retains a well-defined edge and clings to the bowl, the foam is approaching 90% air, and the egg liquid has been spread so thin that the protein webs in adjacent bubble walls begin to catch on each other and on the bowl surface. There's just enough lubrication left for the foam to be creamy and easily mixed with other ingredients. This stage, or perhaps just before it, is the optimum for making mousses, soufflés, sponge cakes, and similar dishes that involve mixing and further rising in the oven. Further beating gains little additional volume.

Dry Peaks and Beyond Just past the stiff-peak stage, the foam is even firmer, takes on a dull, dry appearance and crumbly consistency, and begins to leak some liquid, so that it slips away from the bowl again. At this "slip-and-streak" stage, as pastry chef Bruce Healy describes it, the protein webs in adjacent bubble walls are bonding to each other and squeezing out what little liquid once separated them. Pastry makers look for this stage to give them the firmest foam for a meringue or cookie batter; they stop the incipient overcoagulation and weeping by immediately adding sugar, which separates the proteins and absorbs the water. They also start the beating with about half the cream of tartar per egg that a cake or soufflé maker will, so that the foam will in fact progress to this somewhat overwhipped condition. Past the slip-and-streak stage, the foam begins to lose volume and get grainy.

Egg foams can be used on their own or as the aerating ingredient in a variety of complicated mixtures.

Meringues: Sweet Foams on Their Own

Though they're sometimes folded into cake or cookie batters or fillings, meringues—sweetened egg foams—generally stand by themselves as a discrete element in a dish: as a frothy topping, for example, or a creamy icing, or a hard edible container, or melt-in-the-mouth decoration. A meringue foam must therefore be stiff and stable enough to hold its shape. The cook obtains both stiffness and stability by the addition of sugar and/or of heat. Meringues are often baked very slowly in a low oven (200°F/93°C) to dry them out into a brittle, pristinely white morsel or container. (The door of electric ovens should be left slightly ajar to allow the meringue's moisture to escape; gas ovens are already vented.) When quickly browned in a hot oven or under the broiler—atop a pie, for example—the surface gets crisp while the interior remains moist. Poached in milk for the dish called Floating Islands, they are firm yet moist throughout.

Sugar in Meringues The addition of sugar is what makes a fragile egg-white foam into a stable, glossy meringue. The more sugar added, the more body the meringue will have, and the crisper it will be when baked. The proportion (by either volume or weight) of sugar to egg white ranges from about 1 to 1 to about 2 to 1, the equivalent of a 50% and a 67% sugar solution, respectively. The higher is typical of jams and jellies—and also the room-temperature limit of sugar's solubility in water. Ordinary granulated sugar won't dissolve completely in a "hard" meringue, and will leave a gritty texture and weeping syrup drops. Superfine and powdered "confectioner's" sugar, or a premade syrup, are better choices. (Powdered sugar, which weighs half as much as the other sugars cup for cup, contains 10% cornstarch to help prevent caking, which some cooks dislike and others value as moisture-absorbing insurance.)

Meringue Types The traditional meringue terminology—French, Italian, Swiss, and so on—is unclear and used inconsistently. These foams are best classified according to the method of preparation and resulting texture. Meringues can be either *uncooked* or *cooked*. If the sugar is added after the egg whites have been whipped on their own, the meringue will be relatively *light*; if the sugar is added early in the whipping, the meringue will be relatively *dense*.

Uncooked Meringues Uncooked meringues are the simplest and most common, and provide a broad range of textures, from frothy to creamy to dense and stiff. The lightest possible consistency is obtained by first beating the whites to a firm foam and then gently folding in the sugar with a spatula. The sugar dissolves into the existing bubble walls and adds both bulk and cohesiveness to them. The added bulk gives the bubbles more room to slide past each other and creates a soft, frothy consistency suitable for a spread pie topping or for folding into a mousse or chiffon mix, but too fragile to shape. A creamier, firmer consistency results when the sugar is not merely folded in, but beaten in. In this case, the sugar's added bulk is spread out as the beating further subdivides the bubbles, and the cohesiveness of the sugar-water mixture noticeably tightens the foam's texture. The longer you beat the egg-sugar mixture, the stiffer it will get and the more finely it can be shaped.

These standard methods take only a few minutes but require the cook's attention. Some professionals, particularly in France, make firm meringues suitable for the pastry pipe on the kitchen equivalent of autopilot. They place all the sugar in the bowl of a stand mixer, add a portion of the egg whites with some lemon juice to prevent graining, mix for several minutes—the timing is not critical—then add more whites, mix a while, and so on. The result is a fine-textured, stiff, supple meringue. Beating the eggs gradually into the sugar rather than the other way around does slow the foaming, but requires little supervision. Such "automatic" meringues are denser than usual and less brittle when dried down.

Food Words: *Meringue*

Thanks to the *Larousse Gastronomique*, it's widely believed that the meringue was invented by a pastry chef in the Swiss town of Mieringen around 1720, and brought to France a couple of decades later by the Polish father-in-law of Louis XV. Sounds suitably colorful: except that the French writer Massialot had already published a recipe for "Meringues" in 1691.

The linguist Otto Jänicke has traced the word *meringue* back to an alteration of the Latin word *merenda*, meaning "light evening meal," into *meringa*, a form that was found in the Artois and Picardie near what is now Belgium. Jänicke cites many variations on *merenda* that variously meant "evening bread," "shepherd's loaf," "food taken to the field and forest," "traveler's snack."

What do breads and road food have to do with whipped egg whites? Early baked sugar-egg pastes were called "biscuits," "breads," and "loaves" because they were miniature imitations of these baked goods (biscuits, being thoroughly dried and therefore light and durable, were standard traveler's fare). Perhaps such a confection was called *meringa* in northeast France. Then, when cooks from that region discovered the advantage of beating the eggs thoroughly with the new straw whisk before adding sugar, the local term spread with their invention, and in the rest of France served to distinguish this delicate foam from its dense predecessors.

Between the two extremes—adding all the sugar after the foam has been made, or adding it all at the start of foaming—are a host of methods that call for adding certain portions of the sugar along the way. There's plenty of latitude in meringue making! Just remember: the earlier the sugar is added in the course of beating, the firmer and finer-textured the meringue. Sugar folded in after the beating stops will soften the texture.

Cooked Meringues Cooked meringues are more trouble to make than uncooked meringues, and are generally denser because heat sets the albumen proteins and prematurely limits the trapping of air. However, they offer several advantages. Because sugar is more soluble in hot liquid than in cold, they more readily absorb a large proportion of sugar. Like the dense automatic meringue (above), they're less brittle when dried down. Partial coagulation of the egg proteins stabilizes these foams enough to sit without separating for a day or more. And for cooks concerned about the safety of raw eggs, some cooked meringues get hot enough to kill salmonella bacteria.

There are two basic kinds of cooked meringues. The first ("Italian") is the *syrup-cooked meringue*. Sugar is boiled separately with some water to 240 or 250°F/115–120°C (the "soft-ball" stage, around 90% sugar, at which fudge and fondant are made), the whites whipped to stiff peaks, and the syrup then streamed and beaten into the whites. The result is a fluffy yet fine-textured, stiff foam. It has enough body to decorate pastries and to hold for a day or two before use, but is also light enough to blend into batters and creams. Because much of the syrup's heat is lost to the bowl, whisk, and air, the foam mass normally gets no hotter than 130 or 135°F/55–58°C, which is insufficient to kill salmonella.

The second sort of cooked meringue ("Swiss") is most clearly described as a *cooked meringue* plain and simple (the French *meringue cuite*). To make it, eggs, acid, and sugar are heated in a hot-water bath and beaten until a stiff foam forms. The bowl is then removed from the heat and the foam beaten until it cools. This preparation can pasteurize the egg whites. Thanks to the protective effects of sugar, cream of tartar, and constant agitation, you can heat the meringue mixture to 170 or 175°F/75–78°C and still end up with a stable though dense foam. The cooked meringue can be refrigerated for several days, and is usually piped into decorative shapes.

Meringue Problems: Weeping, Grittiness, Stickiness Meringues can go wrong in a number of ways. Under- or overbeaten foams may weep syrup into unsightly beads or puddles. Beads also form when the sugar hasn't been completely dissolved; residual crystals attract water from their surroundings and make pockets of concentrated syrup. Undissolved sugar (including invisibly small particles present in an undercooked syrup that then slowly grow at room temperature) will give a gritty texture to a meringue. Too high an oven tem-

Royal Icing

A given weight of egg whites can't dissolve more than about double that weight in sugar. Yet royal icing, a traditional decorative material in pastry work, is made by whipping a 4 to 1 mixture of powdered sugar and egg white for 10 or 15 minutes. Royal icing is not a simple foam—it's a combination of a very dense foam and a paste. Much sugar remains undissolved, but it's so fine that we can't feel it on the tongue.

perature can squeeze water from the coagulating proteins faster than it can evaporate and produce syrup beads; it can also cause the foam to rise and crack, and turn its surface an unappealing yellow.

A common problem with meringue pie toppings is that they weep syrup onto the base and don't adhere properly. This can be caused both by relative undercooking of the foam bottom when the pie base is cold and the oven hot, or by relative overcooking on a hot pie base in a moderate oven. Preventive measures include covering the pie base with a syrup-absorbing layer of crumbs before adding the meringue topping, and including starch or gelatin in the foam to help it retain moisture.

Humid weather is bad for meringues. Their sugary surface absorbs moisture from the air and gets soft and sticky. It's best to transfer dried meringues directly from the oven to an airtight container, and serve as soon as possible after removing from the container.

Cold Mousses and Soufflés: Reinforcement from Fat and Gelatin In addition to being served as is in the form of a sugar- and heat-stabilized meringue, an egg foam can also be enrobed in a mixture of other ingredients, for which the foam serves as a hidden scaffolding. The cold mousse and cold soufflé (essentially a mousse molded to look like a hot soufflé that has risen above its dish) hold well for hours, even days, and require only minimal cooking. Instead of being stabilized when heat *coagulates* egg proteins, these mixtures are stabilized when cold *congeals* fats and gelatin protein.

The classic dish of this kind is chocolate mousse. In its purest form, it is made by melting chocolate—a blend of cocoa butter, starchy cocoa particles, and finely ground sugar—at around 100°F/38°C, combining it with raw egg yolks, and combining this mixture with 3 to 4 times its volume of stiffly beaten egg whites (see p. 112). The watery foam walls are thus augmented with the thick, yolky chocolate, and much of the egg moisture is absorbed by the cocoa solids and sugar, which further thickens the bubble walls. While still warm, the mousse is spooned into serving dishes, and these are then refrigerated for several hours. As the mousse cools, the cocoa butter congeals, and the bubble walls become rigid enough to maintain the foam structure indefinitely. The chocolate thus strengthens the egg foam, and the foam spreads the stodgy chocolate mass into a gossamer structure that melts on the tongue.

Soufflés: A Breath of Hot Air

Soufflés—savory and sweet mixes lightened with an egg-white foam, then dramatically inflated above their dish by oven heat—have the reputation for being difficult preparations. Certainly they can be among the most delicate, as their name—French for "puffed," "breathed," "whispered"—suggests. In fact, soufflés are reliable and

Edible Insulation

Egg foams are often used to cover and conceal the heart of a dish. Among the most entertaining of these constructions is the hot, browned meringue enclosing a mass of chilly ice cream: the baked Alaska, which derives from the French *omelette surprise.* This thermal contrast is made possible by the excellent insulating properties of cellular structures like foams. For the same reason, a cup of cappuccino cools more slowly than a cup of regular coffee.

resilient. Many soufflé mixes can be prepared hours, even days in advance, and refrigerated or frozen until needed. If you manage to get *any* air into the mix, an inexorable law of nature will raise it in the oven, and opening the door for a few seconds won't do it any harm. The inevitable post-oven deflation can be minimized by your choice of ingredients and cooking method, and can even be reversed.

The basic idea of the soufflé—and of egg-leavened cakes as well—dates back at least to the 17th century, when confectioners noticed that a "biscuit" paste of egg whites and sugar worked in a mortar would rise in the oven like a loaf of bread. Sometime around 1700, French cooks began to incorporate foamed whites into the yolks to make a puffy *omelette soufflée.* At mid-century, Vincent La Chapelle could offer five *omelettes soufflées* and—under the names *timbale* and *tourte*—the first recorded soufflés as we now know them, their foams reinforced with pastry cream, which came to displace the *omelette soufflée* in restaurants. The great 19th-century chef Antonin Carême called the reinforced soufflé "the queen of hot pastries," but also saw its success as the triumph of convenience and stability over the *omelette soufflée*'s incomparable delicacy of texture and flavor. Carême wrote, "The *omelette soufflée* must be free of the concoction that goes into the soufflé, whether it be rice flour or starch. The gourmet must have the patience to wait if he wishes to eat the *omelette soufflée* in all its perfection."

Convenience is certainly one reason for the soufflé's popularity among cooks. It can be largely prepared in advance, even

Early Recipes for the Omelette Soufflée and Soufflé

This 18th-century recipe for the omelette soufflée is an interesting mix of savory and sweet ingredients, while the timbales are soufflés reinforced with pastry cream.

Omelette Soufflée *with Veal Kidney*

Take a roasted veal kidney, chop with its fat; put in a casserole and cook for a moment to break apart. Then off the fire add a large spoonful of sweet cream and a dozen egg yolks, whose whites you will whip; season the mixture with salt, minced parsley, minced candied lemon peel. Whip your egg whites into snow, mix with the rest, and beat well. Then put a piece of butter in a pan, and when it has melted pour in your mixture, and cook gently. Hold a red-hot fire iron above it. Then invert it onto the serving platter and put it on a small stove, so that it will rise up; when risen to a handsome enough height, powder with sugar and glaze with the fire iron without touching the omelette. Serve hot as an entremet.

Timbales of Cream

You will have a good pastry cream, bitter-almond biscuits, candied lemon peel, orange flower; add to these egg whites whipped into snow. You will have little timbale dishes greased with good fresh butter: you powder them with bread crumbs; then you fill them with your cream, and cook them in the oven. When they are done, turn them out and serve as a small hot entremet.

—Vincent La Chapelle, *Le Cuisinier moderne,* 1742

precooked and reheated. Versatility is another. Soufflés can be made from practically every sort of food—pureed fruits and vegetables and fish; cheese, chocolate, liqueurs—and in a broad range of textures, from the puddinglike to the meltingly fragile *soufflé à la minute,* which is Carême's starch-free *omelette soufflée* barely altered.

The Soufflé Principle, Up Side: It *Must* Rise The physical law that animates the soufflé was discovered a few decades after its invention by—appropriately—a French scientist and balloonist, J. A. C. Charles. Charles's law is this: all else equal, the volume occupied by a given weight of gas is proportional to its temperature. Heat an inflated balloon and the air will take up more space, so the balloon expands. Similarly, put a soufflé in the oven and its air bubbles heat up and swell, so the mix expands in the only direction it can: out the top of the dish.

Charles's law is part of the story, but not the whole story—it accounts for about a quarter of the typical soufflé rise. The rest comes from the continuous evaporation of water from the bubble walls into the bubbles. As portions of the soufflé approach the boiling point, more liquid water becomes water vapor and adds to the *quantity* of gas molecules in the bubbles, which increases the pressure on the bubble walls, which causes the walls to stretch and the bubbles to expand.

Down Side: It Must Fall Charles's law also means that what must go up in the oven must come down at the table. A balloon expands as its temperature rises, but shrinks again if its temperature falls. Of course a soufflé must be taken out of the oven to be served, and from that moment on it loses heat. As the soufflé bubbles cool, the air they contain contracts in volume, and the vapor that came from liquid water in the mix condenses back into liquid.

Rules of Thumb Several basic facts follow from the nature of the driving forces behind the soufflé. First, the higher the cooking temperature, the higher a soufflé will rise: the plain heat expansion will be greater, and more mix moisture will be vaporized. At the same time, a higher cooking temperature also means a greater subsequent overpressure and swifter fall. Then there's the effect of consistency. A thick soufflé mix can't rise as easily as a thin mix, but it also won't fall as easily. A stiff foam can resist the overpressure.

So the two critical factors that determine the behavior of a soufflé are the cooking temperature and the consistency of the

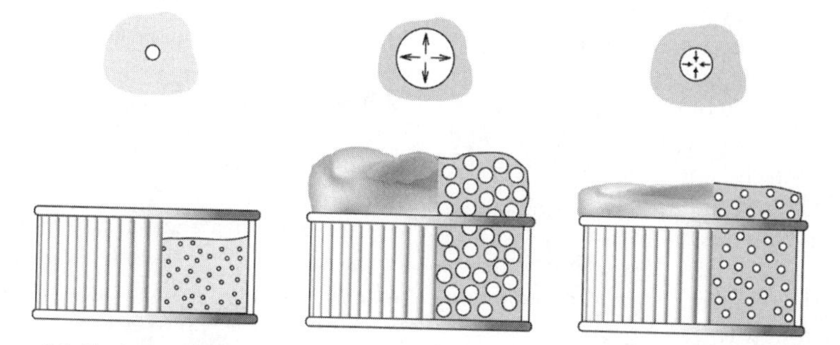

The rise and fall of a soufflé. Left: *The soufflé mix begins filled with small air bubbles.* Center: *Heat causes gases to expand and water to vaporize into steam, so the bubbles expand and raise the mix.* Right: *After the soufflé has been cooked, cooling causes the bubble gases to contract and the steam to condense into liquid water, so that the bubbles contract and the soufflé shrinks.*

soufflé base. A hot oven and thin mix create a more dramatic rise than a moderate oven (or water bath) and a thick mix, but also a more dramatic collapse at the table.

Finally, a fact that follows from both the up and the down sides of the soufflé principle: a fallen soufflé will rise again if put back into the oven. Those air bubbles are still in there, as is most of the moisture; and both air and moisture will expand again as the temperature goes up. You won't get as high a rise the second or third time around, because the soufflé mix has stiffened and there's less water available. But you can resurrect leftovers, or cook the soufflé once to set it and unmold it, then again to serve it.

The Soufflé Base The soufflé base, the preparation into which the foamed egg whites are incorporated, serves two essential purposes. The first is to provide the soufflé's flavor (the base must be *over*flavored to compensate for its dilution by tasteless egg white and air). The second purpose is to contribute a reservoir of moisture for the soufflé's rise, and starch and protein to make the bubble walls viscous enough that they won't ooze down again. Usually the base is precooked and can't actually thicken during the soufflé's rise. The bubble walls are set by the egg white proteins, which can be effective only if they're not excessively diluted by the base material. The usual rule is to allow at least one white or one cup whipped white per half-cup/125 ml base.

The consistency of the base has a strong influence on soufflé quality. Too liquid, and the soufflé will rise and spill over before the egg proteins have a chance to set. Too stiff, and it won't mix evenly with the foamed whites or rise much. A common rule of thumb is that the base should be cohesive yet soft enough to fall of its own weight from a spoon.

Many Formulas Soufflé bases are made from a broad range of ingredients. Those that contain just egg yolks, sugar, and flavoring are the lightest and most delicate and produce the equivalent of the *omelette soufflée*, often called *soufflé à la minute* because it can be made quickly with no advance preparation. A concentrated sugar syrup will make the bubble walls more viscous and stable, as will the various carbohydrates (cellulose, pectin, starch) in pureed fruits and vegetables, and the proteins in a puree of cooked meat, fish, or poultry. If the pureed flesh is raw, then its proteins will coagulate during the cooking along with the egg whites and provide substantial reinforcement to the foam. The starchy brown particles in cocoa and chocolate stiffen the bubble walls by both absorbing moisture and getting sticky and swollen as they do so.

The most versatile kind of soufflé base is thickened with cooked starch in the form of stock preparations like pastry cream or béchamel sauce, or a *panade* (like pastry cream, but without sugar and including butter) or *bouillie* (p. 99). The standard consistency of a starchy base is that of a medium-thick sauce, and produces a moist, fairly light soufflé. Double the flour and you get a drier, denser soufflé that is robust enough to be unmolded, placed in a dish with a hot sauce, and raised again in the oven or under the broiler (Escoffier's *soufflé à la suissesse*). Triple the flour and you get a so-called "pudding soufflé"—with the bready texture you would expect from the name—that won't fall no matter what you do to it. (Increase the flour 15-fold and you have a sponge cake.)

Whipping and Folding the Egg Whites The best consistency for egg whites in a soufflé preparation is stiff yet moist, glossy peaks. A stiff but dry foam is harder to mix evenly with the base, while a softer foam is still coarse—so the soufflé texture will be the same—and may leave the mix so runny that it will overflow before it sets.

The trick is to mix the two materials as evenly as possible while losing as little air as possible. Typically, between a quarter and half of the foam volume is lost at this stage. The traditional method of mixing base and foam is to vigorously stir a quarter of the foam into the base to lighten it, then use a spatula to "fold" the two together by repeatedly scooping some base, cutting vertically through the foam, and depositing the base along the cut surface.

Why laboriously fold rather than quickly stir? Because the rough mass of starch, fats, and other foreign matter in the base pops bubbles, and the more you rub the bubbles against such a mass, the more bubbles you lose. Simply stirring continuously grates the two phases together and causes a substantial loss of air. Folding has the advantage of disturbing the foam only along the surface where the base is being deposited, and that surface is only disturbed for a single stroke. The result is minimal grating of bubbles against mix, and maximal bubble survival.

Despite the usual cookbook direction to fold the whites and base together quickly, it's best to fold *slowly*. The disruptive shear force felt by a given bubble is proportional to the velocity at which it's being pushed along the base. The slower your spatula moves, the less damage it will do to the foam.

The one exception to the folding rules is the soufflé made with a fruit puree or juice cooked with sugar to a thick syrup. Such a base can be poured onto the foam as it's beaten—a soufflé version of the Italian meringue—and will actually increase the mix volume.

Preparing and Filling the Soufflé Dish Ever since La Chapelle's timbale of cream, soufflé dishes have been prepared in two steps: first the interior is buttered, and then coated with sugar for a sweet soufflé, with breadcrumbs or grated cheese for a savory one. The butter supposedly helps the soufflé mix slide up the side as it expands, while the particles give the mix something to cling to as it climbs. Contradictory claims, and not true! Soufflés made in unbuttered or uncrumbed dishes rise just as high. The butter simply makes the soufflé surface easier to detach from the dish, and sugar, breadcrumbs, and cheese make a nice crunchy, brown crust for the otherwise soft interior.

Once put in its dish, a reasonably stiff soufflé mix can be held for several hours in the refrigerator before the foam deteriorates. It will keep indefinitely in the freezer.

Cooking Soufflés Baking soufflés is not a perilous enterprise. Put a room-temperature soufflé mix in a hot oven and it will rise. Don't worry about opening the oven door. The mix can't fall unless it actually begins to cool down, and even if that did happen, it will rise again when it heats up again.

Most soufflés are placed directly on a rack or baking sheet in the oven, but small individual soufflés are often light enough that they can be blasted halfway out of their dish by the steam generated at the oven-hot dish bottom, so the dish ends up half empty. A baking pan filled with water, or individual foil cups of water on a baking sheet, will moderate the bottom temperature and keep a small soufflé in its dish.

A soufflé's appearance and consistency are strongly affected by the oven temperature. At temperatures above 400°F/200°C, the mix rises the fastest, and the surface can brown while the interior is still moist and creamy. At 325 to 350°F/160 to 180°C, the rising is more modest, and surface browning coincides with a firming of the interior. A slow oven may coagulate the surfaces so gradually that the expanding mix spills out of its dish rather than rising vertically. Doneness can be determined by probing the interior with a toothpick, and is a matter of taste; some people like a creamy interior that still clings to the toothpick, others prefer a more fully cooked consistency, which clings to itself and leaves the toothpick clean.

Yolk Foams: Zabaglione and Sabayons

Yolks Can't Foam Without Help Beat an egg white for two minutes and it will expand eightfold into a semisolid foam.

Beat an egg yolk for ten minutes and you'll be lucky to double its volume. Yolks are richer in protein than whites, and have the added advantage of emulsifying phospholipids that do a fine job of coating fat droplets: so why can't they stabilize air bubbles and make a decent foam?

One clue is what happens when you wash out your yolky bowl: the moment you pour in some water, it foams! It turns out that the protein-rich, emulsifier-rich yolk is deficient in water. Not only does it contain about half the water that the white does, but nearly all of it is tightly bound to all the other materials. In one tablespoon/15 ml of yolk, the volume typical of a large egg, there's about a third of a teaspoon/2 ml of free, foamable water. Add two teaspoons to give it the same free water as a white, and it foams enthusiastically.

Enthusiastically but fleetingly. Put your ear to the foam and you'll hear the bubbles popping. The other deficiency of the egg yolk is that its proteins are too stable. Neither the physical abuse of whipping nor the presence of air bubbles causes the yolk proteins to unfold and bond with each other into a reinforcing matrix. Of course heat will, as we know from hard-boiled yolks and custards. So supplement the yolk with liquid, and the whipping with careful cooking, and the mixture will rise to four or more times its original volume. Exactly this procedure is the principle of zabaglione and sabayon sauces.

Medieval Precursors of Zabaglione and Sabayon

Our modern Italian and French versions of foamed egg yolks began in medieval times as yolk-thickened wine, simply flavored in France and Italy, highly spiced in England.

Chaudeau flament ("Flemish Hot Drink," for the Sick)

Set a little water to boil; then beat egg yolks without the whites, mix them with white wine and pour gradually into your water stirring it well to keep it from setting; add salt when it is off the fire. Some people add a very little verjuice.

—Taillevent, *Le Viandier*, ca. 1375

Cawdell Ferry

Take raw yolks of eggs separated from the whites; then take good wine, and warm it in a pot on a fair fire, and throw in the yolks, and stir it well, but let it not boil, till it be thick; and throw in sugar, saffron, and salt, mace, gillyflowers and galingale [a relative of ginger] ground small, and powdered cinnamon; and when you serve it, sprinkle with powdered ginger, cinnamon, and nutmeg.

—Harleian MS 279, ca. 1425

Zabaglone

For four cups of Zabaglone get twelve fresh egg yolks, three ounces of sugar, half an ounce of good cinnamon and a beaker of good sweet wine. Cook this until it is as thick as a broth, then take it out and set it on a plate in front of the boys. And if you like you can add a bit of fresh butter.

—*Cuoco Napoletano,* ca. 1475, transl. Terence Scully

From Zabaglione to Sabayon The recipe trail for yolk foams is spotty. Zabaglione—from a root meaning "mixed," "confused"—was an Italian yolk-thickened spiced wine in the 15th century, and by 1800 was sometimes foamy and sometimes not. (Even some modern zabaglione recipes are not whipped but stirred, and come out more like a winey crème anglaise.) The French discovered zabaglione around 1800, and by 1850 had incorporated it into their system of sauces as a dessert cream with the more refined-sounding name *sabayon.* In the 20th century they extended the principle to savory cooking broths and stocks, and to lighten classical yolk-based butter and oil sauces, including hollandaise and mayonnaise. (For the sauces, see p. 639.)

Zabaglione Technique The standard method for making zabaglione is to mix equal volumes of sugar and yolks, add the wine—usually Marsala, and anywhere from the same to four times the volume of yolks—set the bowl above a pan of simmering water, and whip for several minutes until the mix becomes foamy and thick. During the mixing and initial foaming, the elaborately nested spheres of yolk proteins are unpacked for action. Dilution, the wine's acidity and alcohol, and air bubbles all disrupt the yolk granules and lipoprotein complexes into their component molecules so that those molecules can coat the air bubbles and stabilize them. When the temperature reaches 120°F/50°C, high enough to unfold some of the yolk proteins, the mix thickens, traps air more efficiently, and begins to expand. As the proteins continue to unfold and then bond to each other, the foam rises into fluffy mounds. The key to a maximally light zabaglione is to stop the heating just when the foam teeters on the cusp between liquid and solid. Further cooking will produce a stiffer, denser, eventually tough sponge as the proteins overcoagulate.

Zabaglione is traditionally made in a copper bowl over a water bath; the mix thickens at such a low temperature that direct heat can quickly overcook it. In professional kitchens, where experience is long and time is short, zabaglione and sabayons are sometimes cooked right over a flame. The advantage of a copper bowl in making yolk foams is not chemical, but physical: its excellent heat conductivity makes it quickly responsive to the cook's adjustments. However, copper does impart a distinct metallic flavor to the foam, and some cooks prefer stainless steel for this reason.

The ideal zabaglione or sweet sabayon is soft and meltingly evanescent, yet stable enough that it can be refrigerated and served cold. Savory sabayons may be cooked short of maximal fluffiness so that they remain easily pourable, but the lubricating liquid in the bubble walls will eventually drain out and separate. Fortunately, a separated sabayon can be rebeaten to its original consistency.

PICKLED AND PRESERVED EGGS

Until the recent developments in breeding and artificial lighting, domesticated birds produced eggs seasonally: they would begin laying in the spring, continue through the summer, and then cease in the fall. So, just as they did for milk and for fruits and vegetables, our ancestors developed methods for preserving eggs so that they could be eaten year-round. Many of these methods simply isolated the eggs from the air and left them largely unchanged. Water saturated with lime, or calcium hydroxide, is alkaline enough to discourage bacteria, and coats the egg shell with a thin layer of calcium carbonate that partly seals the shell pores. Oiling with linseed oil apparently began on Dutch farms around 1800. The early 20th century brought the use of waterglass, or a solution of sodium silicate, which again seals the shell pores and is bactericidal. These treatments were rendered obsolete by the advent of refrigeration and year-round egg production.

Still vital 500 years after their first known description are Chinese preservation methods that maintain the nutritional value of the egg but drastically change its flavor, consistency, and appearance. The closest Western counterpart to this ovo-alchemy is cheesemaking, which transforms milk into an entirely different food. Ordinary vinegar-pickled eggs offer only a hint of the possibilities; they are to Chinese preserved eggs as yogurt is to Stilton.

Pickled Eggs

Common pickled eggs are made by first boiling the eggs and then immersing them in a solution of vinegar, salt, spices, and often a coloring like beet juice, for 1 to 3 weeks. Over that time the vinegar's acetic acid dissolves much of the shell's calcium carbonate, penetrates the eggs, and lowers their pH sufficiently to prevent the growth of spoilage microbes. (The vinegar in Easter-egg dyes etches the shell surface and helps the dye penetrate.) Pickled eggs will keep for a year or more without refrigeration.

Pickled eggs can be eaten shell—or its remains—and all. In addition to being tart, they are firmer than freshly boiled eggs; the white is sometimes described as rubbery. A more tender consistency can be obtained by including ample salt in the pickling liquid and having the liquid at the boil when the eggs are immersed. Though the eggs won't spoil at room temperature, they will suffer less from swollen yolks and split whites (which result when the egg absorbs the pickling liquid too rapidly) if stored in the cold.

Chinese Preserved Eggs

Though the average Chinese consumes only a third as many eggs as the average American, and though most of those eggs are chicken eggs, China is renowned for its preserved duck eggs, including the "thousand-year-old" eggs. These and plain salt-preserved eggs come from the duck-rich southern provinces, where they made it possible to transport eggs to distant markets and store them for months during the off season. The proteins and membranes of chicken eggs are less suited to some of these treatments.

Salted Eggs The simplest method for preserving eggs is to treat them with salt, which draws the water out of bacteria and molds and inhibits their growth. The eggs are immersed in a 35% salt solution, or coated individually with a paste of salt, water, and clay or mud. After 20 or 30 days, the egg stops absorbing salt and reaches chemical equilibrium. Strangely, the white remains liquid, but the yolk at its center becomes solid. The high levels of positive sodium and negative chloride ions actually shield the albumen proteins from each other, but cause the yolk particles to agglomerate into a grainy mass. Salted eggs, which are variously called *hulidan* and *xiandan*, are boiled before they're eaten.

Fermented Eggs A second kind of preserved egg, little seen in the West, is made by covering gently cracked eggs in a fermenting mass of cooked rice or other grains mixed with salt: in essence a concentrated and salty version of sake or beer. *Zaodan* mature in four to six months and take on the aromatic, sweet, alcoholic flavor of their surroundings. Both white and yolk coagulate and fall out of the softened shell. Such eggs can be eaten as is or cooked first.

***Pidan*: "Thousand-Year-Old" Alkali-Cured Eggs** The most famous of preserved eggs are the so-called "thousand-year-old" duck eggs, which actually have only been made for about 500 years, take between one and six months to mature, and keep for a year or so. They owe their popular name—the Chinese term is *pidan*, or "coated eggs"—to their startlingly decrepit appearance: the shell encrusted with mud, the white a transparent brown jelly, and the yolk a semisolid, somber jade. The flavor too is earthy and elemental, eggy in the extreme, salty, stonily alkaline, with strong accents of sulfur and ammonia. *Pidan* are toned down by rinsing the shelled egg and allowing it time to "breathe" before serving. They are a delicacy in China, and are usually served as an appetizer.

There are only two essential ingredients for making *pidan,* in addition to the eggs: salt, and a strongly alkaline material, which can be wood ash, lime, sodium carbonate, lye (sodium hydroxide), or some combination of these. Tea is often used for flavor, and mud to create a paste that dries to a protective crust, though the eggs can also be immersed in a water solution of the curing ingredients (this gives a faster cure but also a coarser alkaline flavor). A mild, soft-yolked version of *pidan* is sometimes made by adding some lead oxide to the cure. The lead reacts with sulfur from the egg white to form a fine black powder of lead sulfide, which blocks the shell pores and slows the further movement of salt and alkaline ingredients into the egg. (Lead is a potent nerve toxin, so such eggs should be avoided; look for packages clearly labeled "no lead oxide." A similar effect can be obtained by replacing lead with zinc.)

Creating Clarity, Color, and Flavor The real transforming agent in *pidan* is the alkaline material, which gradually raises the already alkaline egg from a pH of around 9 to 12 or more. This chemical stress causes what might be thought of as an inorganic version of fermentation: that is, it denatures the egg proteins, and breaks down some of the complex, flavorless proteins and fats into simpler, highly flavorful components. The disruptively high pH forces the egg proteins to unfold, and at the same time confers on them a strongly repelling negative charge. The dissolved salt, with its positive and negative ions, moderates the repulsion enough that the fine strands of widely dispersed albumen proteins are able to bond into a solid yet transparent gel. In the yolk, the same extreme conditions destroy the organized structure of the yolk spheres, and with it the usual graininess; the yolk proteins coagulate into a creamy mass. The extreme alkalinity also browns the albumen by accelerating the reaction between the proteins and the trace of glucose (see p. 89), and it greens the yolk by encouraging the formation of ferrous sulfide throughout the yolk, not just at its surface (as in hard-cooked eggs; see p. 89). Finally, the alkalinity intensifies the egg's flavor by breaking down both proteins and phospholipids into hydrogen sulfide, distinctly animal fatty acids, and pungent ammonia (the fumes from a freshly opened egg will turn litmus paper blue).

Nouveaux *Pidan* Recently, two Taiwanese food scientists devised a method for making a striking, toned-down version of *pidan.* They minimized the chemical stress, and thus the alteration of color and flavor, by limiting the alkaline treatment to eight days in a solution of 5% salt and 4.2% lye. Such eggs don't solidify on their own. But when the unfolding and bonding are supplemented by gentle heating at 160°F/70°C for 10 minutes, these eggs set to a golden yolk and a colorless, clear white!

Pine-Blossom Eggs An especially prized variant of *pidan* is one in which the aspic-colored white is marked throughout with tiny, pale, snowflake traceries. Such eggs are known as *songhuadan,* or "pine-blossom" eggs. The "blossoms" turn out to be crystals of modified amino acids, which the high alkalinity has broken off from from the albumen proteins. They're thus an index of protein breakdown and flavor generation, a delicate inscription of the mineral world on the blank orb of the animal, and an example of the unexpected delight that can lie hidden in the crudest of preparations.

CHAPTER 3

MEAT

Of all the foods that we obtain from animals and plants, meat has always been the most highly prized. The sources of that prestige lie deep in human nature. Our primate ancestors lived almost exclusively on plant foods until 2 million years ago, when the changing African climate and diminishing vegetation led them to scavenge animal carcasses. Animal flesh and fatty bone marrow are more concentrated sources of food energy and tissue-building protein than nearly any plant food. They helped feed the physical enlargement of the brain that marked the evolution of early hominids into humans. Later, meat was the food that made it possible for humans to migrate from Africa and thrive in cold regions of Europe and Asia, where plant foods were seasonally scarce or even absent. Humans became active hunters around 100,000 years ago, and it's vividly clear from cave paintings of wild cattle and horses that they saw their prey as embodiments of strength and vitality. These same qualities came to be attributed to meat as well, and a successful hunt has long been the occasion for pride, gratitude, and celebratory feasting. Though we no longer depend on the hunt for meat, or on meat for survival, animal flesh remains the centerpiece of meals throughout much of the world.

Paradoxically, meat is also the most widely avoided of major foods. In order to eat meat, we necessarily cause the death

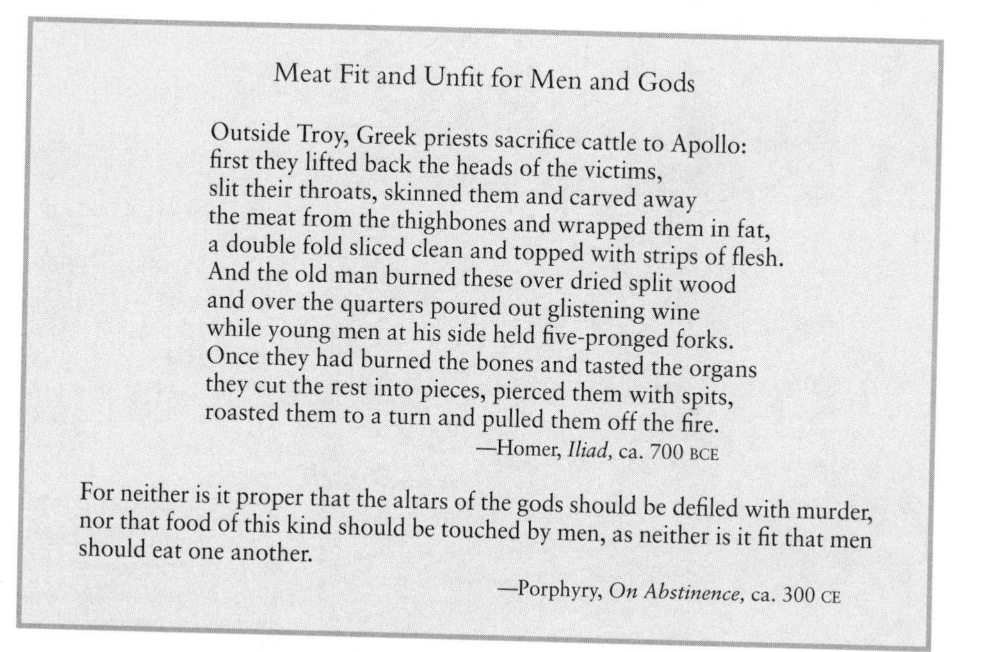

Meat Fit and Unfit for Men and Gods

Outside Troy, Greek priests sacrifice cattle to Apollo:
first they lifted back the heads of the victims,
slit their throats, skinned them and carved away
the meat from the thighbones and wrapped them in fat,
a double fold sliced clean and topped with strips of flesh.
And the old man burned these over dried split wood
and over the quarters poured out glistening wine
while young men at his side held five-pronged forks.
Once they had burned the bones and tasted the organs
they cut the rest into pieces, pierced them with spits,
roasted them to a turn and pulled them off the fire.

—Homer, *Iliad*, ca. 700 BCE

For neither is it proper that the altars of the gods should be defiled with murder, nor that food of this kind should be touched by men, as neither is it fit that men should eat one another.

—Porphyry, *On Abstinence*, ca. 300 CE

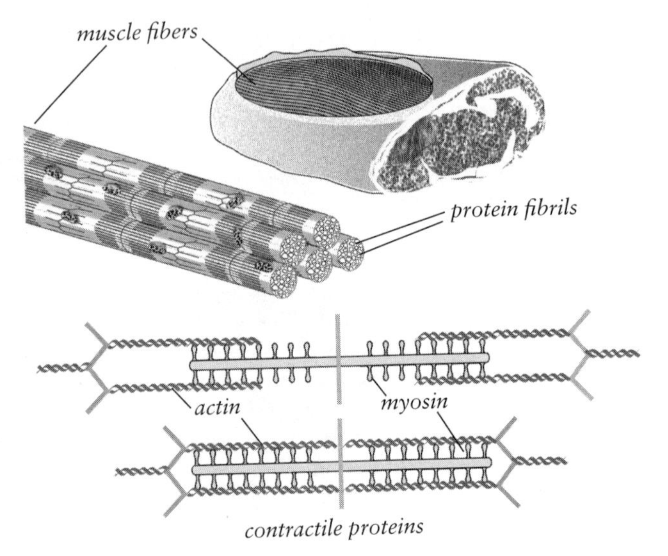

The structure of muscle tissue and meat. A piece of meat is composed of many individual muscle cells, or fibers. The fibers are in turn filled with many fibrils, which are assemblies of actin and myosin, the proteins of motion. When a muscle contracts, the filaments of actin and myosin slide past each other and decrease the overall length of the complex.

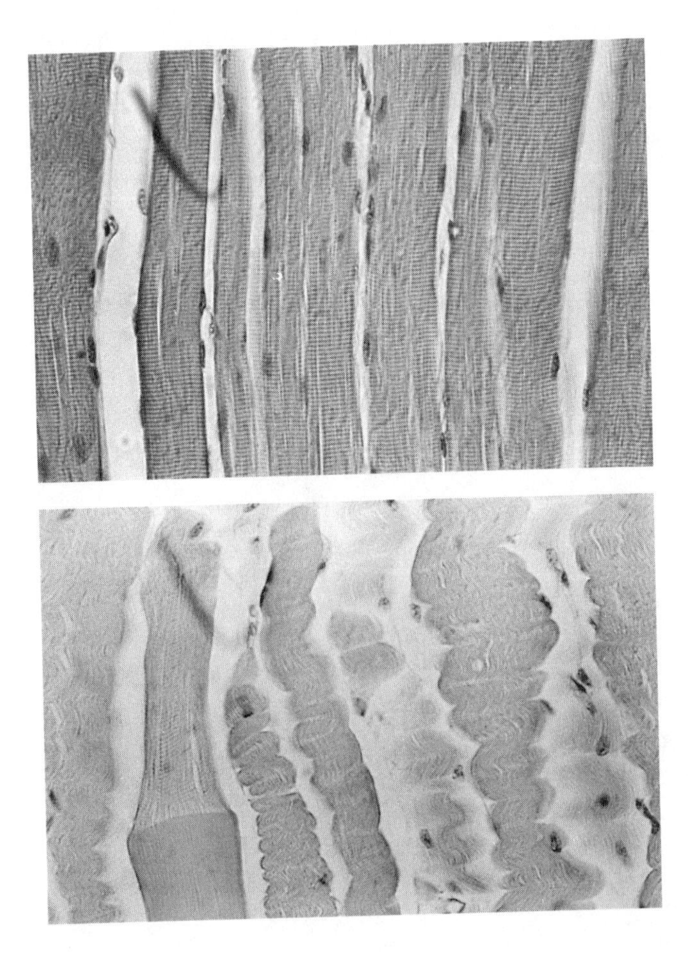

Muscle contraction. The view through a light microscope of rabbit muscle fibers, relaxed (above) *and contracted* (below).

of other creatures that feel fear and pain, and whose flesh resembles our own. Many people throughout history have found this a morally unacceptable price for our own nourishment and pleasure. The ethical argument against eating meat suggests that the same food that fueled the biological evolution of modern humans now holds us back from full humaneness. But the biological and historical influences on our eating habits have their own force. However culturally sophisticated we may be, humans are still omnivorous animals, and meat is a satisfying and nourishing food, an integral part of most food traditions.

Less philosophical questions, but more immediate ones for the cook, have been raised by the changing quality of meat over the last few decades. Thanks to the industrial drive toward greater efficiency, and consumer worries about animal fats, meat has been getting younger and leaner, and therefore more prone to end up dry and flavorless. Traditional cooking methods don't always serve modern meat well, and cooks need to know how to adjust them.

Our species eats just about everything that moves, from insects and snails to horses and whales. This chapter gives details for only the more common meats of the developed world, but the general principles apply to the flesh of all animals. Though fish and shellfish are as much flesh foods as meat and poultry, their flesh is unusual in several ways. They are the subject of chapter 4.

EATING ANIMALS

By the word *meat* we mean the body tissues of animals that can be eaten as food, anything from frog legs to calf brains. We usually make a distinction between meats proper, muscle tissue whose function is to move some part of the animal, and *organ* meats, such innards as the liver, kidneys, intestine, and so on.

The Essence of the Animal: Mobility from Muscle

What is it that makes a creature an *animal*? The word comes from an Indo-European root meaning "to breathe," to move air in and out of the body. The definitive characteristic of animals is the power to move the body and nearby parts of the world. Most of our meats are muscles, the propulsive machinery that moves an animal across a meadow, or through the sky or sea.

The job of any muscle is to shorten itself, or contract, when it receives the appropriate signal from the nervous system. A muscle is made up of long, thin cells, the muscle fibers, each of which is filled with two kinds of specialized, contractile protein filaments intertwined with each other. This packing of protein filaments is what makes meat such a rich nutritional source of protein. An electrical impulse from the nerve associated with the muscle causes the protein filaments to slide past each other, and then lock together by means of *cross-bridging*, or forming bonds with each other. The change in relative position of the filaments shortens the muscle cell as a whole, and the cross bridges maintain the contraction by holding the filaments in place.

Portable Energy: Fat Like any machine, the muscle protein machine requires energy to run. Almost as important to animals as their propulsive machinery is an energy supply compact enough that it doesn't weigh them down and impede their movement. It turns out that fat packs twice as many calories into a given weight as carbohydrates do. This is why mobile animals store up energy almost exclusively in fat, and unlike stationary plants, are rich rather than starchy.

Because fat is critical to animal life, most animals are able to take advantage of abundant food by laying down large stores of fat. Many species, from insects to fish to birds to mammals, gorge themselves in preparation for migration, breeding, or surviving seasonal scarcity. Some migratory birds put on 50% of their lean weight in fat in just a few weeks, then fly 3,000 to 4,000 kilometers from the northeast United States to South America without refueling. In seasonally cold parts of the world, fattening has been part of the resonance of autumn, the time when wild game animals are at their plumpest and most appealing, and when humans practice their cultural version of fattening, the harvest and storing of crops that will see them through winter's scarcity. Humans have long exploited the fattening ability of our meat animals by overfeeding them before slaughter, to make them more succulent and flavorful (p. 135).

Humans as Meat Eaters

Meat became a predictable part of the human diet beginning around 9,000 years ago, when early peoples in the Middle East managed to tame a handful of wild animals—first dogs, then goats and sheep, then pigs and cattle and horses—to live alongside them. Livestock not only transformed inedible grass and scraps into nutritious meat, but constituted a walking larder, a store of concentrated nourishment that could be harvested whenever it was needed. Because they were adaptable enough to submit to human control, our meat animals have flourished and now number in the billions, while many wild animals are being squeezed by the growth of cities and farmlands into ever smaller habitats, and their populations are declining.

The History of Meat Consumption

The Scarcity of Meat in Agricultural Societies Around the time that our ancestors domesticated animals, they also began to cultivate a number of grasses, plants that grow in extensive stands and produce large numbers of nutritious seeds. This was the beginning of agriculture. With the arrival of domesticated barley and wheat, rice and maize, nomadic peoples settled down to farm the land and produce food, populations boomed—and most people ate very little meat. Grain crops are simply a far more efficient form of nourishment than animals grazing on the same land, so meat became relatively expensive, a luxury reserved for the rulers. From the prehistoric invention of agriculture to the Industrial Revolution, the great majority of people on the planet lived on cereal gruels and breads. Beginning with Europe and the Americas in the 19th century, industrialization has generally made meat less expensive and more widely available thanks to the development of managed pastures and formulated feeds,

Food Words: *Meat*

The English word *meat* has not always meant animal flesh, and its evolution indicates a shift in the eating habits of English-speaking people. In the *Oxford English Dictionary*'s first citation for *meat,* from the year 900, the word meant solid food in general, in contrast to drink. A vestige of this sense survives today in the habit of referring to the meat of nuts. It wasn't until 1300 that *meat* was used for the flesh of animals, and not until even later that this definition displaced the earlier one as animal flesh became preeminent in the English diet, in preference if not in quantity. (The same transformation can be traced in the French word *viande.*) One sign of this preference is Charles Carter's 1732 *Compleat City and Country Cook,* which devotes 50 pages to meat dishes, 25 to poultry, and 40 to fish, but only 25 to vegetables and a handful to breads and pastries.

the intensive breeding of animals for efficient meat production, and improved transportation from farms to cities. But in less developed parts of the world, meat is still a luxury reserved for the wealthy few.

Abundant Meat in North America

From the beginning, Americans have enjoyed an abundance of meat made possible by the size and richness of the continent. In the 19th century, as the country became urbanized and more people lived away from the farm, meats were barreled in salt to preserve them in transit and in the shops; salt pork was as much a staple food as bread (hence such phrases as "scraping the bottom of the barrel" and "pork-barrel politics"). In the 1870s a wider distribution of fresh meat, especially beef, was made possible by several advances, including the growth of the cattle industry in the West, the introduction of cattle cars on the railroads, and the development of the refrigerated railroad car by Gustavus Swift and Philip Armour.

Today, with one fifteenth of the world's population, the United States eats one third of the world's meat. Meat consumption on this scale is possible only in wealthy societies like our own, because animal flesh remains a much less efficient source of nourishment than plant protein. It takes much less grain to feed a person than it does to feed a steer or chicken in order to feed a person. Even today, with advanced methods of production, it takes 2 pounds of grain to get 1 pound of chicken meat, and the ratios are 4 to 1 for pork, 8 to 1 for beef. We can afford to depend on animals as a major source of food only because we have a surplus of seed proteins.

Why Do People Love Meat?

If meat eating helped our species survive and then thrive across the globe, then it's understandable why many peoples fell into the habit, and why meat would have a significant place in human culture and tradition. But the deepest satisfaction in eating meat probably comes from instinct and biology. Before we became creatures of culture, nutritional wisdom was built into our sensory system, our taste buds, odor receptors, and brain. Our taste buds in particular are designed to help us recognize and pursue important nutrients: we have receptors for essential salts, for energy-rich sugars, for amino acids, the building blocks of proteins, for energy-bearing molecules called nucleotides. Raw meat triggers all these tastes, because muscle cells are relatively fragile, and because they're biochemically very active. The cells in a plant leaf or seed, by contrast, are protected by tough cell walls that prevent much of their contents from being freed by chewing, and their protein and starch are locked up in inert

Food Words: Animals and Their Meats

As the novelist Walter Scott and others pointed out long ago, the Norman Conquest of Britain in 1066 caused a split in the English vocabulary for common meats. The Saxons had their own Germanic names for the animals—*ox, steer, cow, heifer,* and *calf; sheep, ram, wether, ewe,* and *lamb; swine, hog, gilt, sow,* and *pig*—and named their flesh by attaching "meat of" to the animal name. When French became the language of the English nobility in the centuries following the Conquest, the animal names survived in the countryside, but the prepared meats were rechristened in the fashion of the court cooks: the first recipe books in English call for *beef* (from the French *boeuf*), *veal* (*veau*), *mutton* (*mouton*), and *pork* (*porc*).

storage granules. Meat is thus mouth-filling in a way that few plant foods are. Its rich aroma when cooked comes from the same biochemical complexity.

MEAT AND HEALTH

Meat's Ancient and Immediate Nutritional Advantages . . .

The meat of wild animals was by far the most concentrated natural source of protein and iron in the diet of our earliest human ancestors, and along with oily nuts, the most concentrated source of energy. (It's also unsurpassed for several B vitamins.) Thanks to the combination of meat, calcium-rich leaf foods, and a vigorous life, the early hunter-gatherers were robust, with strong skeletons, jaws, and teeth. When agriculture and settled life developed in the Middle East beginning 10,000 years ago, human diet and activity narrowed considerably. Meats and vegetables were displaced from the diet of early farmers by easily grown starchy grains that are relatively poor in calcium, iron, and protein. With this and the higher prevalence of infectious disease caused by population growth and crowding, the rise of agriculture brought about a general decline in human stature, bone strength, and dental health.

A return to something like the robustness of the hunter-gatherers came to the industrialized world beginning late in the 19th century. This broad improvement in stature and life expectancy owed a great deal to improvements in medicine and especially public hygiene (water quality, waste treatment), but the growing nutritional contribution of meat and milk also played an essential role.

. . . And Modern, Long-Term Disadvantages

By the middle of the 20th century, we had a pretty good understanding of the nutritional requirements for day-to-day good health. Most people in the West had plenty of food, and life expectancy had risen to seven or eight decades. Medical research then began to concentrate on the role of nutrition in the diseases that cut the good life short, mainly heart disease and cancer. And here meat and its strong appeal turned out to have a significant disadvantage: a diet high in meat is associated with a higher risk of developing heart disease and cancer. In our postindustrial life of physical inactivity and essentially unlimited ability to indulge our taste for meat, meat's otherwise valuable endowment of energy contributes to obesity, which increases the risk of various diseases. The saturated fats typical of meats raise blood cholesterol levels and can contribute to heart disease. And to the extent that meat displaces from our diet the vegetables and fruits that help fight heart disease and cancer (p. 255), it increases our vulnerability to both.

It's prudent, then, to temper our species' infatuation with meat. It helped make us what we are, but now it can help unmake us. We should eat meat in moderation, and accompany it with the vegetables and fruits that complement its nutritional strengths and limitations.

Minimizing Toxic By-Products in Cooked Meats We should also prepare meat with care. Scientists have identified three families of chemicals created during meat preparation that damage DNA and cause cancers in laboratory animals, and that may increase our risk of developing cancer of the large intestine.

Heterocyclic Amines HCAs are formed at high temperatures by the reaction of minor meat components (creatine and creatinine) with amino acids. HCA production is generally greatest at the meat surface where the temperature is highest and the meat juices collect, and on meats that are grilled, broiled, or fried well done. Oven roasting leaves relatively few HCAs on the meat but large amounts in the pan drippings. Acid marinades reduce HCA production, as does cooking gently and aiming for a rare or medium doneness. Vegetables, fruits, and acidophilus bacteria (p. 47) appear to bind HCAs in the digestive tract and prevent them from causing damage.

Polycyclic Aromatic Hydrocarbons PAHs are created when nearly any organic material, including wood and fat, is heated to the point that it begins to burn (p. 448). Cooking over a smoky wood fire therefore deposits PAHs from the wood on meat. A charcoal fire is largely smokeless, but will create PAHs from fat if the fat is allowed to fall and burn on the coals, or if the fat ignites on the meat surface itself. Small quantities of PAHs can also be formed during high-temperature frying. The PAH hazard can be minimized by grilling over wood only when it has been reduced to coals, by leaving the grill uncovered so that soot and vapors can dissipate, by avoiding fat flareups, and by eating smoked meats only rarely.

Nitrosamines Nitrosamines are formed when nitrogen-containing groups on amino acids and related compounds combine with nitrite, a chemical that has been used for millennia in salt-cured meats, and that suppresses the bacterium that causes botulism (p. 174). This reaction between amino acids and nitrites takes place both in our digestive system and in very hot frying pans. Nitrosamines are known to be powerful DNA-damaging chemicals, yet at present there's no clear evidence that the nitrites in cured meats increase the risk of developing cancer. Still, it's probably prudent to eat cured meats in moderation and cook them gently.

Meat and Food-Borne Infections

Beyond the possibility that it may chip away at our longevity by contributing to heart disease and cancer, meat can also pose the much more immediate hazard of causing infection by disease microbes. This problem remains all too common.

Bacterial Infection Exactly because it is a nutritious material, meat is especially vulnerable to colonization by microbes, mainly bacteria. And because animal skins and digestive tracts are rich reservoirs of bacteria, it's inevitable that initially clean meat surfaces will be contaminated during slaughter and the removal of skin, feathers, and innards. The problem is magnified in standard mechanized operations, where carcasses are handled less carefully than they would be by skilled butchers, and where a single infected carcass is more likely to contaminate others. Most bacteria are harmless and simply spoil the meat by consuming its nutrients and eventually generating unpleasant smells and a slimy surface. A number, however, can invade the cells of our digestive system, and produce toxins to destroy the host cells and defenses and to speed their getaway from the body. The two most prominent causes of serious meat-borne illness are *Salmonella* and *E. coli*.

Salmonella, a genus that includes more than 2,000 distinct bacterial types, causes more serious food-borne disease in Europe and North America than any other microbe, and appears to be on the rise. It's a resilient group, adaptable to extremes of temperature, acidity, and moisture, and found in most if not all animals, including fish. In the United States it's especially prevalent in poultry and eggs, apparently thanks to the practices of industrial-scale poultry farming: recycling animal by-products (feathers, viscera) as feed for the next generation of animals, and crowding the animals together in very close confinement, both of which favor the spread of the bacteria. Salmonella often have no obvious effect on the animal carriers, but in humans can cause diarrhea and chronic infection in other parts of the body.

Escherichia coli is the collective name for many related strains of bacteria that are normal residents of the intestines of warm-blooded animals, including humans. But several strains are aliens, and if ingested will invade the cells of the digestive tract and cause illness. The most notorious *E. coli,* and the most dangerous, is a special strain called O157:H7 that causes bloody diarrhea and sometimes kidney failure, especially in children. In the United States, about a third of people diagnosed with *E. coli* O157:H7 need to be hospitalized, and about 5% die. *E. coli* O157:H7 is harbored in cattle, especially calves, and other animals, but has little if any effect on them. Ground beef is by far the most common source of *E. coli* O157:H7 infection. Grinding mixes and spreads what may be only a small contaminated portion throughout the entire mass of meat.

Prevention Prevention of bacterial infection begins with the well warranted assumption that all meat has been contaminated with at least some disease bacteria. It requires measures to ensure that those bacteria are not spread to other foods, and are eliminated from the meats during cooking. Hands, knives, cutting boards, and countertops used to prepare meats should be cleaned with hot soapy water before being used to prepare other foods. *E. coli* are killed at 155°F/68°C, so ground meats are safest if their center gets at least this hot. Salmonella and other bacteria can multiply at significant rates between 40 and 140°F/5–60°C, so meats should not be left in this range for more than two hours. Buffet dishes should be kept hot, and leftovers promptly refrigerated and reheated at least to 160°F/70°C.

Trichinosis Trichinosis is a disease caused by infection with the cysts of a small parasitic worm, *Trichina spiralis.* In the United States, trichinosis was long associated with undercooked pork from pigs fed garbage that sometimes included infected rodents or other animals. Uncooked garbage was banned as pork feed in 1980, and since then the incidence of trichinosis in the United States has declined to fewer than ten cases annually. Most of these are not from pork, but from such game meats as bear, boar, and walrus.

For many years it was recommended that pork be cooked past well done to ensure the elimination of trichinae. It's now known that a temperature of 137°F/58°C, a medium doneness, is sufficient to kill the parasite in meat; aiming for 150°F/65°C gives reasonable safety margin. Trichinae can also be eliminated by frozen storage for a period of at least 20 days at or lower than 5°F/–15°C.

"Mad Cow Disease"

"Mad cow disease" is the common name for bovine spongiform encephalopathy, or BSE, a disease that slowly destroys the brains of cattle. It's an especially worrisome disease because the agent of infection is a nonliving protein particle that cannot be destroyed by cooking, and that appears to cause a similar and fatal disease in people who eat infected beef. We still have a lot to learn about it.

BSE originated in the early 1980s when cattle were fed by-products from sheep suffering from a brain disease called scrapie, whose cause appears to be a chemically stable protein aggregate called a prion. The sheep prions somehow adapted to their new host and began to cause brain disease in the cattle.

Humans are not susceptible to sheep scrapie. But there's a mainly hereditary human brain disease similar to scrapie and caused by a similar prion; it is called Creutzfeldt-Jakob disease (CJD), typically strikes old people with loss of coordination and then dementia, and eventually kills them. In 1995 and 1996, ten relatively young Britons died from a new variant of CJD, and the prion agent found in their bodies was closely related to the BSE prion. This strongly suggests that humans can contract a devastating disease by eating meat from BSE-infected cattle. The cattle brain, spinal cord, and retina are thought to be the tissues in which prions are concentrated, but a 2004 report suggests that they may also be found in muscles and thus in common cuts of beef.

BSE appears to have been eliminated in Britain thanks to the culling of affected herds, changes in feeding, and surveillance. But diseased cattle have turned up elsewhere in Europe, as well as in the United States, Canada, and Japan. As a precautionary measure, a number of countries have suspended some traditional practices at least temporarily. These include eating flavorful meat from older animals (which are more likely to carry BSE), as well as beef brains, sweetbreads and spleen (immune-system organs), and intestines (which contain immune-system tissues). Some countries also forbid the use of "mechanically recovered meat"—tiny scraps removed from the skeleton by machine and incorporated into ground beef—from the head and spinal column. These rules will probably be modified as rapid tests for the animal disease are developed and implemented, and as we learn more about how it is transmitted to people.

To date, the known human death toll from BSE-infected beef numbers in the low hundreds, and the overall risk of contracting the prion disease from beef appears to be very small.

CONTROVERSIES IN MODERN MEAT PRODUCTION

Meat production is big business. In the United States just a few decades ago, it was second only to automobile manufacturing. Both industry and government have long underwritten research on innovative ways to control meat production and its costs. The result has been a reliable supply of relatively inexpensive meat, but also a production system increasingly distant from its origins in the family farmer's pasture, pigsty, and chicken coop, and troubling in various ways. Many innovations involve the use of chemicals to manipulate animal metabolism. These chemicals act as drugs in the animals, and raise worries that they may influence human health as well. Other innovations involve the animals' living conditions, which have become increasingly artificial and crowded, and their feed, which often includes reprocessed waste materials from various agricultural industries, and which contributed to the origin of mad cow disease and the persistence of salmonella in chickens. The scale and con-

Invisible Animals

Historian William Cronon has written eloquently about the disappearance of our food animals as the system of meat production changed in the 19th century:

> Formerly, a person could not easily have forgotten that pork and beef were the creation of an intricate, symbiotic partnership between animals and human beings. One was not likely to forget that pigs and cattle had died so that people might eat, for one saw them grazing in familiar pastures, and regularly visited the barnyards and butcher shops where they gave up their lives in the service of one's daily meal. . . . As time went on, fewer of those who ate meat could say that they had ever seen the living creature whose flesh they were chewing; fewer still could say that they had actually killed the animal themselves. In the packers' world, it was easy not to remember that eating was a moral act inextricably bound to killing. . . . Meat was a neatly wrapped package one bought at the market. Nature did not have much to do with it.
>
> —William Cronon, *Nature's Metropolis: Chicago and the Great West,* 1991

centration of modern meat production, with hundreds of thousands of animals confined in a single facility, have caused significant water, soil, and air pollution. Enough consumers and producers have become uneasy about these developments that there is now a modest segment of the industry devoted to meats raised more traditionally, on a smaller scale, and with more attention to the quality of the animals' life and meat.

Hormones

The manipulation of animal hormones is an ancient technology. Farmers have castrated male animals for thousands of years to make them more docile. Testicle removal not only prevents the production of sex hormones that stimulate aggressive behavior, but also turns out to favor the production of fat tissue over muscle. This is why steers and capons have long been preferred as meat animals over bulls and cocks. The modern preference for lean meat has led some producers to raise uncastrated animals, or to replace certain hormones in castrates. Several natural and synthetic hormones, including estrogen and testosterone, produce leaner, more muscular cattle more rapidly and on less feed. There is ongoing research into a variety of growth factors and other drugs that would help producers fine-tune the growth and proportions of fat to lean in cattle and other meat animals.

Currently, beef producers are allowed to treat meat cattle with six hormones in the United States, Canada, Australia, and New Zealand, but not in Europe. Hormone treatments were outlawed in the European Economic Community in 1989 in response to well-publicized abuses; a few Italian veal producers injected their calves with large quantities of the banned steroid DES, which ended up in bottled baby food and caused changes in the sexual organs of some infants. Laboratory studies indicate that meat from animals treated with allowed hormone levels contains only minute hormone residues, and that these residues are harmless when ingested by humans.

Antibiotics

Efficient industrial-scale meat production requires that large numbers of animals be raised in close confinement, a situation that favors the rapid spread of disease. In order to control animal pathogens, many producers routinely add antibiotics to their feed. This practice turns out to have the additional advantage of increasing growth rate and feed efficiency.

Antibiotic residues in meat are minute and apparently insignificant. However, there's good evidence that the use of antibiotics in livestock has encouraged the evolution of antibiotic-resistant campylobacter and salmonella bacteria, and that these bacteria have caused illness in U.S. consumers. Because resistant bacteria are more difficult to control, Europe and Japan restrict the use of antibiotics in animals.

Humane Meat Production

To many people, the mass production of livestock is itself undesirable. In a series of legislative acts and executive orders dating back to 1978, Switzerland has mandated that producers accommodate the needs of their animals for such things as living space, access to the outdoors, and natural light, and limit the size of herds and flocks. The European Union is also adopting animal welfare guidelines for meat production, and producers in a number of countries have grouped together to establish and monitor their own voluntary guidelines.

Mass production has certainly made meat a more affordable food than it would be otherwise. But because we raise meat animals in order to eat them, it seems only just that we try to make their brief lives as satisfying as possible. It would certainly be a challenge to raise meat animals economically while taking their nature and instincts into account and allowing them the opportunity to roam, nest, and nurture their young. But it's a challenge at least as worthy as finding a way to trim another 1% from production costs.

THE STRUCTURE AND QUALITIES OF MEAT

Lean meat is made up of three basic materials: it's about 75% water, 20% protein, and 3% fat. These materials are woven into three kinds of tissue. The main tissue is the mass of muscle cells, the long fibers that cause movement when they contract and relax. Surrounding the muscle fibers is the connective tissue, a kind of living glue that harnesses the fibers together and to the bones that they move. And interspersed among the fibers and connective tissue are groups of fat cells, which store fat as a source of energy for the muscle fibers. The qualities of meat—its texture, color, and flavor—are determined to a large extent by the arrangement and relative proportions of the muscle fibers, connective tissue, and fat tissue.

MUSCLE TISSUES AND MEAT TEXTURE

Muscle Fibers When we look at a piece of meat, most of what we see are bundles of muscle cells, the fibers that do the moving. A single fiber is very thin, around the thickness of a human hair (a tenth to a hundredth of a millimeter in diameter), but it can be as long as the whole muscle. The muscle fibers are organized in bundles, the larger fibers that we can easily see and tease apart in well-cooked meat.

The basic texture of meat, dense and firm, comes from the mass of muscle fibers, which cooking makes denser, dryer, and tougher. And their elongated arrangement accounts for the "grain" of meat. Cut parallel to the bundles and you see them from the side, lined up like the logs of a cabin wall; cut across the bundles and you see just their ends. It's easier to push fiber bundles apart from each other than to break the bundles themselves, so it's easier to chew along the direction of the fibers than across them. We usually carve meat *across* the grain, so that we can chew *with* the grain.

Muscle fibers are small in diameter when the animal is young and its muscles little used. As it grows and exercises, its muscles get stronger by enlarging—not by increasing the number of fibers, but by increasing the number of contractile protein fibrils within the individual fibers. That is, the number of muscle cells stays the same, but they get thicker. The more protein fibrils there are packed together in the cells, the harder it is to cut across them. So the meat of older, well exercised animals is tougher than the meat of young animals.

Connective Tissue Connective tissue is the physical harness for all the other tissues in the body, muscle included. It connects individual cells and tissues to each other, thus organizing and coordinating their actions. Invisibly thin layers of connective tissue surround each muscle fiber and hold neighboring fibers together in bundles, then merge to form the large, silver-white sheets that organize fiber bundles into muscles, and the translucent tendons that join muscles to bones. When the fibers contract, they pull this harness of connective tissue with them, and the harness pulls the bones. The more force that a muscle exerts, the more connective tissue it needs for reinforcement, and the stronger the tissue needs to be. So as an animal's growth and exercise bulk up the muscle fibers, they also bulk up and toughen the connective tissue.

Connective tissue includes some living cells, but consists mainly of molecules that the cells secrete into the large spaces between them. The most important of these molecules for the cook are the protein filaments that run throughout the tissue and reinforce it. One, a protein called *elastin* for its stretchiness, is the main component of blood vessel walls and ligaments, and is especially tough; its cross-links cannot be broken by the heat of cooking. Fortunately there isn't much of it in most muscle tissue.

The major connective-tissue filament is the protein called *collagen,* which makes up about a third of all the protein in the animal body, and is concentrated in skin, tendons, and bones. The name comes from the Greek for "glue producing," because when it's heated in water, solid, tough collagen partly dissolves into sticky *gelatin* (p. 597). So unlike the muscle fibers, which become tougher with cooking, the connective tissue becomes softer. An animal starts out life with a large amount of collagen that's easily dissolved into gelatin. As it grows and its muscles work, its total collagen supply declines, but the filaments that remain are more highly crosslinked and less soluble in hot water. This is why cooked veal seems gelatinous and tender, mature beef less gelatinous and tougher.

Fat Tissue Fat tissue is a special form of connective tissue, one in which some of the cells take on the role of storing energy. Animals form fat tissue in three different parts of the body: just under the skin, where it can provide insulation as well as energy; in well-defined deposits in the body cavity, often around the kidneys, intestine, and heart; and in the connective tissue separating muscles and the bundles within muscles. The term "marbling" is used to describe the pattern of white splotches in the red matrix of muscle.

Tissues and Textures The texture of tender meat is as distinctive and satisfying as its flavor: a "meaty" food is something you can sink your teeth into, dense and substantial, initially resistant to the tooth but soon giving way as it liberates its flavor. Toughness is a resistance to chewing that persists long enough to become unpleasant. Toughness can come from the muscle fibers, the connective tissue surrounding them, and from the lack of marbling fat.

Generally, the toughness of a cut of meat is determined by where it comes from in the animal's body, and by the animal's age and activity. Get down on all fours and "graze," and you'll notice that the neck, shoulders, chest, and front limbs all work hard, while the back is more relaxed. Shoulders and legs are used continually in walking and standing, and include a number of different muscles and their connective-tissue sheaths. They are therefore relatively tough. The tenderloin is appropriately named because it is a single muscle with little internal connective tissue that runs along the back and gets little action; it's tender. Bird legs are tougher than breasts for the same reasons; the protein in chicken legs is 5–8% collagen compared to 2% in the breast. Younger animals—veal, lamb, pork, and chicken all come from younger animals than beef does—have tenderer muscle fibers because they are smaller and less exercised; and the

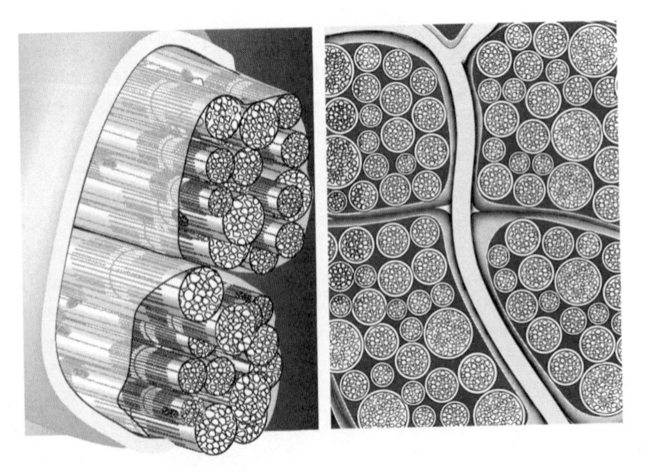

Connective tissue. Muscle fibers are bundled, held in place, and reinforced by sheets of connective tissue. The more connective tissue in a given piece of meat, the tougher its texture.

collagen in their connective tissue is more rapidly and completely converted to gelatin than older, more cross-linked collagen.

Fat contributes to the apparent tenderness of meat in three ways: fat cells interrupt and weaken the sheet of connective tissue and the mass of muscle fibers; fat melts when heated rather than drying out and stiffening as the fibers do; and it lubricates the tissue, helping to separate fiber from fiber. Without much fat, otherwise tender meat becomes compacted, dry, and tough. Beef shoulder muscles contain more connective tissue than the leg muscles, but they also include more fat, and therefore make more succulent dishes.

Muscle Fiber Types: Meat Color

Why do chickens have both white and dark meat, and why do the two kinds of meat taste different? Why is veal pale and delicate, beef red and robust? The key is the muscle fiber. There are several different kinds of muscle fiber, each designed for a particular kind of work, and each with its own color and flavor.

White and Red Fibers Animals move in two basic ways. They move suddenly, rapidly, and briefly, for example when a startled pheasant explodes into the air and lands a few hundred yards away. And they move deliberately and persistently, for example when the same pheasant supports its body weight on its legs as it stands and walks; or a steer stands and chews its cud. There are two basic kinds of muscle fibers that execute these movements, the white fibers of pheasant and chicken breasts, and the red fibers of bird and steer legs. The two types differ in many biochemical details, but the most significant difference is the energy supply each uses.

White Muscle Fibers White muscle fibers specialize in exerting force rapidly and briefly. They are fueled by a small store of a carbohydrate called glycogen, which is already in the fibers, and is rapidly converted into energy by enzymes right in the cell fluids. White cells use oxygen to burn glycogen, but if necessary they can generate their energy faster than the blood can deliver oxygen. When they do so, a waste product, lactic acid, accumulates until more

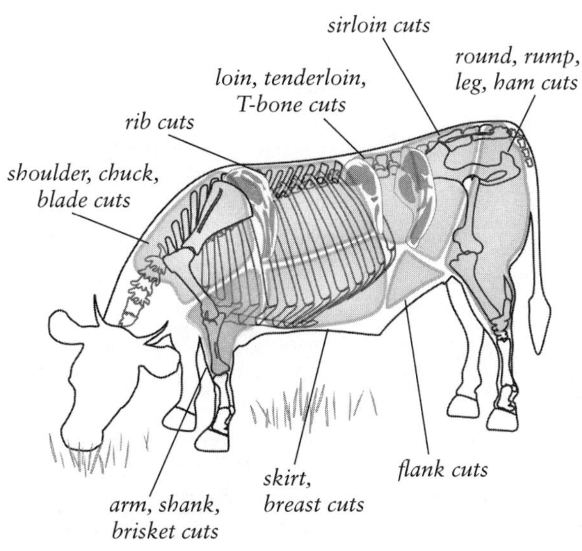

Steer anatomy and cuts of beef. The shoulder, arm, and leg do most of the work of supporting the animal. They therefore contain a large proportion of reinforcing connective tissue, are tough, and best cooked thoroughly for an hour or more to dissolve the connective-tissue collagen into gelatin. The rib, short loin, and sirloin do less work, are generally the tenderest cuts, and are tender even when cooked briefly to a medium doneness.

oxygen arrives. This accumulation of lactic acid limits the cells' endurance, as does their limited fuel supply. This is why white cells work best in short intermittent bursts with long rest periods in between, during which the lactic acid can be removed and glycogen replaced.

Red Muscle Fibers Red muscle fibers are used for prolonged efforts. They are fueled primarily by fat, whose metabolism absolutely requires oxygen, and obtain both fat (in the form of fatty acids) and oxygen from the blood. Red fibers are relatively thin, so that fatty acids and oxygen can diffuse into them from the blood more easily. They also contain their own droplets of fat, and the biochemical machinery necessary to convert it into energy. This machinery includes two proteins that give red cells their color. *Myoglobin,* a relative of the oxygen-carrying hemoglobin that makes blood red, receives oxygen from the blood, temporarily stores it, and then passes it to the fat-oxidizing proteins. And among the fat oxidizers are the *cytochromes,* which like hemoglobin and myoglobin contain iron and are dark in color. The greater the oxygen needs of the fiber, and the more it's exercised, the more myoglobin and cytochromes it will contain. The muscles of young cattle and sheep are typically 0.3% myoglobin by weight and relatively pale, but the muscles of the constantly moving whale, which must store large amounts of oxygen during its prolonged dives, have 25 times more myoglobin in their cells, and are nearly black.

Fiber Proportions: White Meat and Dark Meat Because most animal muscles are used for both rapid and slow movements, they contain both white and red muscle fibers, as well as hybrid fibers that combine some characteristics of the other two. The proportions of the different fibers in a given muscle depend on the inherited genetic design for that muscle and the actual patterns of muscle use. Frogs and rabbits, which make quick, sporadic movements and use very few of their skeletal muscles continuously, have very pale flesh consisting mainly of white fast fibers, while the cheek muscles of ruminating, perpetually cud-chewing steers are exclusively red slow fibers. Chickens and turkeys fly only when startled, run occasionally, and mostly stand and walk; so their breast muscles consist predominantly of white fibers, while their leg muscles are on average half white fibers, half red. The breast muscles of such migratory birds as ducks and pigeons are predominantly red fibers because they're designed to help the birds fly for hundreds of miles at a time.

Muscle Pigments The principal pigment in meat is the oxygen-storing protein *myoglobin,* which can assume several different forms and hues depending on its chemical environment. Myoglobin consists of two connected structures: a kind of molecular cage with an iron atom at the center, and an attached protein. When the iron is holding onto a molecule of oxygen, myoglobin is bright red. When the oxygen is pulled away by enzymes in the muscle cell that need it, the myoglobin becomes dark purple. (Similarly, hemoglobin is red in our arteries because it's fresh from our lungs, and blue in our veins because it has unloaded oxygen

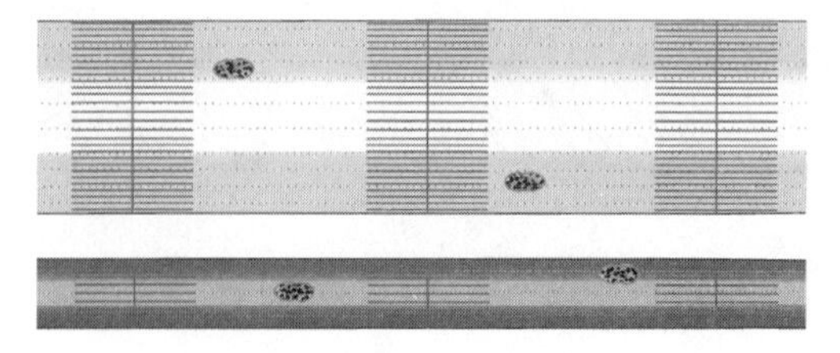

White and red muscle fibers. Fast muscle cells are thicker than slow cells, contain little oxygen-storing myoglobin pigment and few fat-burning mitochondria. The thinness of slow, red muscle fibers speeds the diffusion of oxygen from the external blood supply to the center of the fibers.

into our cells.) And when oxygen manages to rob the iron atom of an electron and then escape, the iron atom loses its ability to hold oxygen at all, has to settle for a water molecule, and the myoglobin becomes brown.

Each of these myoglobins—the red, the purple, and the brown—is present in red meat. Their relative proportions, and so the meat's appearance, are determined by several factors: the amount of oxygen available, the activity of oxygen-consuming enzymes in the muscle tissue, and the activity of enzymes that can resupply brown myoglobin with an electron, which turns it purple again. Acidity, temperature, and salt concentration also matter; if any is high enough to destabilize the attached protein, myoglobin is more likely to lose an electron and turn brown. Generally, fresh red meat with active enzyme systems will be red on the surface, where oxygen is abundant, and purple inside, where the little oxygen that diffuses through is consumed by enzymes. When we cut into raw meat or into a rare steak, the initially purple interior quickly "blooms," or reddens, thanks to its direct exposure to the air. Similarly, vacuum-packed meat appears purple due to the absence of oxygen, and reddens only when removed from the package.

The pink color of salt-cured meats comes from yet another alteration of the myoglobin molecule (p. 148).

Muscle Fibers, Tissues, and Meat Flavor

The main source of meat's great appeal is its flavor. Meat flavor has two aspects: what might be called generic meatiness, and the special aromas that characterize meats from different animals. Meatiness is largely provided by the muscle fibers, character aromas by the fat tissue.

Muscle Fibers: The Flavor of Action

Meaty flavor is a combination of mouth-filling taste sensations and a characteristic, rich aroma. Both arise from the proteins and energy-generating machinery of the muscle fibers—after they have been broken down into small pieces by the muscle's enzymes and by the heat of cooking. Some of these pieces—single amino acids

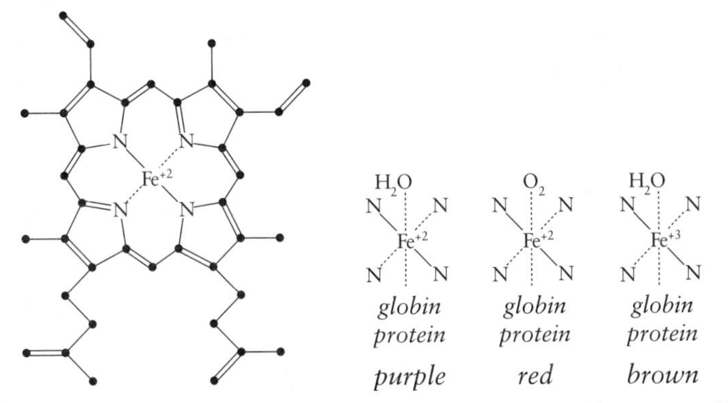

Meat pigments. Left: *The heme group, a carbon-ring structure at the center of both hemoglobin and myoglobin molecules that holds oxygen for use by the animal body's cells. The protein portion of these molecules, the globin, is a long, folded chain of amino acids, and is not shown here.* Right: *Three different states of the heme group in uncooked meat. In the absence of oxygen, myoglobin is purple. Myoglobin that has bound a molecule of oxygen gas is red. When little oxygen is available for some time, the iron atom in the heme group is readily oxidized—robbed of an electron—and the resulting pigment molecule is brownish* (right).

and short chains of them, sugars, fatty acids, nucleotides, and salts—are what stimulate the tongue with sweet, sour, salty, and savory sensations. And when they're heated, they react with each other to form hundreds of aromatic compounds. In general, well-exercised muscle with a high proportion of red fibers (chicken leg, beef) makes more flavorful meat than less exercised, predominantly white-fibered muscle (chicken breast, veal). Red fibers contain more materials with the potential for generating flavor, in particular fat droplets and fat-like components of the membranes that house the cytochromes. They also have more substances that help break these flavor precursors down into flavorful pieces, including the iron atoms in myoglobin and cytochromes, the oxygen that those molecules hold, and the enzymes that convert fat into energy and recycle the cell's proteins.

This connection between exercise and flavor has been known for a very long time. Nearly 200 years ago, Brillat-Savarin made fun of "those gastronomes who pretend to have discovered the special flavor of the leg upon which a sleeping pheasant rests his weight."

Fat: The Flavor of the Tribe The machinery of the red or white muscle fiber is much the same no matter what the animal, because it has the specific job of generating movement. Fat cells, on the other hand, are essentially storage tissue, and any sort of fat-soluble material can end up in them. So the contents of fat tissue vary from species to species, and are also affected by the animal's diet and resident gastrointestinal microbes. It's largely the contents of the fat tissue that give beef, lamb, pork, and chicken their distinctive flavors, which are composites of many different kinds of aroma molecules. The fat molecules themselves can be transformed by heat and oxygen into molecules that smell fruity or floral, nutty or "green," with the relative proportions depending on the nature of the fat. Compounds from forage plants contribute to the "cowy" flavor of beef. Lambs and sheep store a number of unusual molecules, including branched-chain fatty acids that their livers produce from a compound generated by the microbes in their rumen, and thymol, the same molecule that gives thyme its aroma. The "piggy" flavor of pork and gamy flavor of duck are thought to come from intestinal microbes and their fat-soluble products of amino-acid metabolism, while the "sweetness" in pork aroma comes from a kind of molecule that also gives coconut and peach their character (lactones).

Grass versus Grain In general, grass or forage feeding results in stronger-tasting meat than grain or concentrate feeding, thanks to the plants' high content of various odorous substances, reactive polyunsaturated fatty acids, and chlorophyll, which rumen microbes convert into chemicals called terpenes, relatives of the aroma compounds in many herbs and spices (p. 273). Another important contributor

Meat Pigments Are a Good Source of Iron

One of meat's nutritional strong points is that the body absorbs its iron more efficiently than it does iron from vegetable sources. The reason for this is not well understood, but it's possible that the pigment proteins hold onto iron and prevent it from being bound up with indigestible plant compounds. Meat color is a good indicator of its iron content; red beef and lamb contain on average two to three times as much as pale pork; relatively dark pork shoulder contains twice as much as the loin.

to grass-fed flavor is skatole, which on its own smells like manure! The deep "beefy" flavor of beef, however, is more prominent in grain-fed animals. And the flavor carried in fat gets stronger as animals get older, as more of the flavor compounds are put into storage. This is why lamb is generally more popular than mutton from mature sheep.

Production Methods and Meat Quality

Full-flavored meat comes from animals that have led a full life. However, exercise and age also increase muscle fiber diameter and the cross-linking of connective tissue: so a full life also means tougher meat. In centuries past, most people ate mature, tough, strongly flavored meat, and developed long-cooked recipes to soften it. Today, most of us eat young, tender, mild meat that is at its best quickly cooked; long-cooking often dries it out. This shift in meat quality has resulted from a shift in the way the animals are raised.

Rural and Urban Styles of Meat There are two traditional, indeed ancient ways of obtaining meat from animals, and they produce meats of distinctive qualities.

One method is to raise animals primarily for their value as living companions—oxen and horses for their work in the fields, laying hens for their eggs, cows and sheep and goats for their milk and for wool—and turn them into meat only when they are no longer productive. In this system, slaughtering animals for meat is the last use of a resource that is more valuable when alive. The meat comes from mature animals, and is therefore well exercised and relatively tough, lean but flavorful. This method was by far the most common one from prehistoric times until the 19th century.

The second way of obtaining meat from animals is to raise the animals exclusively for that purpose. This means feeding the animals well, sparing them unnecessary exercise, and slaughtering them young to obtain tender, mild, fatty flesh. This method also goes back to prehistory, when it was applied to pigs and to the otherwise useless male offspring of hens and dairy animals. With the rise of cities, meat animals were confined and fattened exclusively for the urban elite who could afford such a luxury, an art represented in Egyptian murals and described by Roman writers.

For many centuries, rural and urban meats coexisted, and inspired the development of two distinct styles of meat preparation: roasting for the tender, fattened meats of the wealthy, and stewing for the tough, lean meats of the peasants.

The Rural Style Disappears With the Industrial Revolution, draft animals were slowly replaced by machines. City populations and the middle class grew, and along with them the demand for meat, which encouraged the rise of large-scale specialized meat production. In 1927 the U.S. Department of Agriculture enshrined the identification of quality with urban-style fattiness when it based its beef grading system on the amount of "marbling" fat deposited within the muscles (see box). Meat from mature animals began to disappear in North America, and ever more efficient industrial production took the urban style to new extremes.

Mass Production Favors Immaturity Today nearly all meat comes from animals raised exclusively for that purpose. Mass production methods are dictated by a simple economic imperative: the meat should be produced at minimum cost, which generally means in the shortest possible time. Animals are now confined to minimize the expenditure of feed on unnecessary movement, and they're slaughtered before they reach adulthood, when the growth of their muscles slows down. Rapid, confined growth favors the production of white muscle fibers, so modern meats are relatively pale. They're also tender, because the animals get little exercise, because rapid growth means that their connective-tissue collagen is continuously taken apart and rebuilt and develops fewer strong cross-links, and because rapid growth means high levels of the protein-breaking enzymes that tenderize meat during aging (p. 143). But many meat lovers feel that meat has gotten less flavorful in recent decades. Life intensifies flavor, and modern meat animals are living less and less.

Changing Tastes for Fat: The Modern Style In the early 1960s, American consumers began to abandon well marbled beef and pork for less fatty cuts and for lean poultry. Since marbling develops only after the animals' rapid muscle growth slows, the meat industry was happy to minimize fattening and improve its production efficiency. Consumer and producer preferences for lean beef led the USDA to reduce its marbling requirements for the top grades in 1965 and 1975.

The modern style of meat, then, combines elements of the traditional styles: young like the city meats, lean like the country meats, and therefore both mild and easily dried out during cooking. Cooks now face the challenge of adapting hearty country traditions to these finicky ingredients.

Quality Production: A French Example There have been small but significant exceptions to this general trend toward producing meat as cheaply as possible. In the 1960s, the French poultry industry found that many consumers were dissatisfied with the standard chicken's bland flavor and tendency to shrink and fall off the bone when cooked. Some producers then developed a production scheme guided by considerations of quality as well as efficiency. The result was the popular *label rouge,* or "red label," which identifies chickens that have been produced according to specific standards: they are slow-growing varieties, fed primarily on grain rather than artificially concentrated feeds, raised in flocks of moderate size and with access to the outdoors, and slaughtered at 80 or more days of age rather than 40 to 50. Red-label chickens are leaner and more

USDA Beef Grades: The Triumph of Fat over Lean

As economist V. James Rhodes recounts, the USDA grading system for beef did not arise from an objective government analysis of meat quality. Instead, it was conceived and pushed during an agricultural recession in the early 1920s by cattlemen in the Midwest and East, who wanted to boost demand for their purebred, fat, corn-fed animals at the expense of lean dairy and "scrub" cattle. The chief propagandist was Alvin H. Sanders, editor of the *Breeder's Gazette,* who colorfully denigrated the slow cooking of economical cuts as "the same old continental European story of how to make a banquet out of a few bones and a dash of 'cat-meat.'"

Sanders and his colleagues set out to convince the country that "the muscular tissues of animals are made tender and fully flavored only by the presence of plenty of fat." In the summer of 1926, a well-placed breeder and New York financier named Oakleigh Thorne personally tutored the Secretary of Agriculture, who soon offered to begin free quality grading—based on the amount of visible fat marbling—at all packing houses subject to federal health supervision. U.S. "Prime" beef was born in 1927. A few years later, government-funded studies found that heavy marbling does *not* guarantee either tender or flavorful beef. But the prestige of heavily marbled Prime beef persisted, and the United States became one of only three countries—the others being Japan and Korea—to make fat content a major criterion for meat quality.

muscular than their standard industrial equivalent, lose a third less of their moisture during cooking, are firmer in texture, and have a more pronounced flavor. Similar quality-based meat production schemes exist today in a number of countries.

So economic forces have conspired to make mild, tender meat the modern norm, but small producers of more mature, flavorful meats, sometimes from rare "heirloom" breeds, are finding their own profitable market among consumers willing to pay a premium for quality.

MEAT ANIMALS AND THEIR CHARACTERISTICS

Each of the animals that we raise for food has its own biological nature, and its own history of being shaped by humans to meet their changing needs and tastes. This section sketches the distinctive qualities of our more common meats, and the main styles in which they're now produced.

Domestic Meat Animals

Cattle Cattle are descendents of the wild ox or aurochs, *Bos primigenius,* which browsed and grazed in forests and plains all across temperate Eurasia. Cattle are our largest meat animals and take the longest to reach adulthood, about two years, so their meat is relatively dark and flavorful. Breeders began to develop specialized meat animals in the 18th century. Britain produced the compact, fat-carcassed English Hereford and Shorthorn and Scots Angus, while continental meat breeds remained closer to the rangy, lean draft type; these include the French Charolais and Limousin, and the Italian Chianina, which is probably the largest breed in the world (4,000-pound bulls, double the size of the English breeds).

American Beef The United States developed a uniform national style when federal grading standards for beef were introduced in 1927 (see box, p. 136), with the highest "Prime" grade reserved for young, fine-textured meat with abundant marbling. Purebred Angus and Hereford beef were the model for three decades. The shift in consumer preference to lower-fat meat brought revisions of the USDA grades to allow leaner meat to qualify for the Prime and Choice grades (see box below). Nowadays, U.S. beef comes mainly from steers (males castrated as calves) and heifers (females that have never calved) between 15 and 24 months old, and fed on grain for the last

U.S. Beef Quality and Grades Today

Despite the prestige of Prime beef, the current consensus among meat scientists is that fat marbling accounts for no more than a third of the variation in the overall tenderness, juiciness, and flavor of cooked beef. The other important factors include breed, exercise and feed, animal age, conditions during slaughter, extent of post-slaughter aging (p. 143), and storage conditions before sale. Most of these are impossible for the consumer to evaluate, though there is a movement toward store and producer "brands" that may provide greater information about and consistency of production. Potentially more flavorful beef from older animals can be recognized by its darker color and coarser muscle fibers.

Most graded supermarket beef today is graded "Choice," with 4–10% fat, or "Select," with 2–4% fat. Prime beef is now around 10–13% fat. Ground beef, which may be all lean meat or a mixture of lean and fat, ranges from 5 to 30% fat content.

four to eight months. Recent years have brought a new interest in beef from cattle raised exclusively on grass, which is leaner and stronger in flavor (p. 134) than mainstream beef.

European Beef Other beef-loving countries raise their cattle differently and have produced distinctive beefs. Italy prefers young meat from animals slaughtered at 16–18 months. Until the advent of BSE, much French and British beef came from dairy stock several years old. According to a standard French handbook, *Technologie Culinaire* (1995), the meat of an animal less than two years old is "completely insipid," while meat "at the summit of quality" comes from a steer three to four years old. But because the risk of an animal having BSE rises as it grows older, a number of countries now require that meat cattle be slaughtered at less than three years of age. In 2004, most French and British beef came from animals no older than 30 months.

Japanese Beef Japan prizes its *shimofuri*, or highly marbled beef, of which the best known comes from the Kobe region. Steers of the native Wagyu draft breed are slaughtered at 24–30 months. High-quality heifers (and some steers) are identified and then fattened for a further year or more on grain. (Currently Japan tests all meat cattle for BSE.) This process produces beef that is mature, flavorful, tender, and very rich, with as much as 40% marbling fat. The best cuts are usually sliced very thin, in 1.5–2 mm sheets, and simmered in broth for a few seconds in the one-pot dishes called *sukiyaki* and *shabu shabu*.

Veal Veal is the meat of young male offspring of dairy cows. Veal has traditionally been valued for being as different as possible from beef: pale, delicate in flavor, with a softer fat, and succulently tender thanks to its soluble collagen, which readily dissolves into gelatin when cooked. Calf flesh becomes more like beef with every day of ordinary life, so most veal calves aren't allowed an ordinary life: they're confined so that exercise won't darken, flavor, and toughen their muscles, and fed a low-iron diet with no grass to minimize the production of myoglobin pigment and prevent rumen development (p. 13), which would saturate and thus harden the fat. In the United States, veal generally comes from confined animals fed a soy or milk formula and slaughtered between 5 and 16 weeks old, when they weigh 150 to 500 lb/70–230 kg. "Bob" or "drop" veal comes from unconfined, milk-fed animals three weeks old or less. "Free-range" and "grain-fed" veal have become increasingly common as more humane alternatives, but are more like beef in the color and flavor of their meat.

Sheep Along with goats, sheep were probably the first animals to be domesticated after the dog, thanks to their small size, around a tenth that of cattle, and their herding instinct. Most European breeds of sheep are specialized for milk or wool; there are relatively few specialized meat breeds.

Lamb and Mutton Lamb and sheep meat is finer grained and more tender than beef, but well endowed with red myoglobin and with flavor, including a characteristic odor (p. 134) that becomes more pronounced with age. Pasture-feeding, particularly on alfalfa and clover, increases the levels of a compound called skatole, which also contributes a barnyardy element to pork flavor, while lambs finished on grain for a month before slaughter are milder. In the United States, lambs are sold in a range of ages and weights, from 1 to 12 months and 20–100 lb/9–45 kg, under a variety of names, including "milk" and "hothouse" lamb for younger animals, "spring" and "Easter" lamb for the rest (though production is no longer truly seasonal). New Zealand lamb is pasture-fed but slaughtered at four months, younger than most American lamb, and remains mild. In France, older lambs (*mouton*) and young female sheep (*brebis*) are aged for a week or more after slaughter, and develop an especially rich flavor.

Pigs Pigs are descendents of the Eurasian wild boar, *Sus scrofa.* If beef has been the most esteemed of meats in Europe and the Americas, pork has fed far more people, both there and in the rest of the world: in China the word for "pork" is also the generic word for "meat." The pig has the virtues of being a relatively small, voracious omnivore that grows rapidly and bears large litters. Its indiscriminate appetite means that it can turn otherwise useless scraps into meat, but that meat can harbor and transmit parasites from infected animals and their remains (see p. 126 on trichinosis). Perhaps in part for this reason, and because pigs are difficult to herd and will devour field crops, pork eating has been forbidden among various peoples, notably Middle Eastern Jews and Muslims.

There are several specialized styles of pigs, including lard breeds, bacon breeds, and meat breeds, some large-boned and massive, some (Iberian and Basque ham pigs) relatively lean, slow-growing, small and dark-fleshed, much like their wild south-Europe ancestors. Today most of the specialized breeds have been displaced by the fast-growing descendents of a few European bacon and meat breeds.

Pork Like modern beef, modern pork comes from much younger and leaner animals than was true a century ago. Pigs are typically slaughtered at six months and 220 lb/100 kg, just as they reach sexual maturity, when the connective tissue is still relatively soluble and the meat tender. Individual cuts of American and European pork generally contain half to a fifth of the fat they did in 1980. Pork is a pale meat because the pig uses its muscles more intermittently than do cattle and sheep, and therefore has a lower proportion of red muscle fibers (around 15%). Some small Chinese and European breeds have darker and significantly more flavorful flesh.

Domestic Meat Birds

Chickens Chickens are descendents of the aggressive, pugnacious red jungle fowl of northern India and southern China. *Gallus gallus* is a member of the pheasant family or Phasianidae, a large, originally Eurasian group of birds that tend to colonize open forest or the edge between field and wood. Chickens seem to have been domesticated in the vicinity of Thailand before 7500 BCE, and arrived in the Mediterranean around 500 BCE. In the West, they were largely unpampered farmyard scavengers until the 19th-century importation of large Chinese birds created a veritable chicken-breeding craze in Europe and North America. Mass production began in the 20th century, when much of the genetic diversity in meat chickens evaporated in favor of a fast-growing cross between the broad-breasted Cornish (developed in Britain from Asian fighting stock) and the U.S. White Plymouth Rock.

Chicken Styles The modern chicken is a product of the drive to breed fast-growing animals and raise them as rapidly and on as little feed as possible. It's an impressive feat of agricultural engineering to produce a 4-pound bird on 8 pounds of feed in six weeks! Because such a bird grows very fast and lives very little, its meat is fairly bland, and that of the younger "game hen" or "poussin" even more so.

Largely in reaction to the image of industrial chicken, so-called "free range" chickens are now sold in the United States, but the term only means that the birds have access to an outdoor pen. "Roasting" chickens and capons (castrated males) are raised to double or more the age of the standard broiler, are heavier, and so have given their leg muscles more exercise; the capon may also be more succulent thanks to the infiltration of marbling fat.

Turkeys Turkeys are also members of the sedentary pheasant family. *Meleagris gallopavo* descended from ancestors that once ranged through North America and Asia. The modern colossal turkey dates from 1927–1930, when a breeder in British Columbia developed a 40 lb/18 kg bird with oversized flight and thigh muscles, and breeders in the U.S. northwest used his stock to perfect the Broad-Breasted Bronze. The little-used breast muscle is tender, mild, and lean; the leg muscles that support the breast are well-exercised, dark, and flavorful.

Today, industrial facilities produce 14–20 lb/6–9 kg birds year-round in 12–18 weeks; some small American farms extend the period to 24 weeks, while the name-controlled French Bresse turkey is raised for 32 weeks or more, confined and fattened for the last several weeks on corn and milk.

Ducks and Squab Ducks and squab are notable for having dark, flavorful breast meat, abundantly endowed with myoglobin-rich red muscle fibers, thanks to their ability to fly hundreds of miles in a day with few stops. The most common breeds of duck in China, much of Europe, and the United States are descendents of the wild green-headed mallard, *Anas platyrhynchos,* an aquatic migratory bird that puts on as much as a third of its carcass weight in fat for fuel and under-skin insulation. Ducks are eaten at two ages: in the egg as 15–20-day embryos (the Philippine boiled delicacy *balut*), and at 6 to 16 weeks. The Muscovy duck is an entirely different bird: *Cairina moschata,* the greater wood duck, which is native to the west coast of Central and northern South America, differs in three important ways from mallard varieties: it lays down about a third less body fat, grows significantly larger, and has a more pronounced flavor.

Squab, dove, and pigeon are various names for the European rock dove, *Columba livia,* a species that includes the common city pigeon; "squab" means a bird young enough that it has never flown. Its flying muscles weigh five times as much as its leg muscles. Today, domestic squab are raised for four weeks and slaughtered at about 1 lb/450 g, just before they're mature enough to fly.

Game Animals and Birds

Wild animals—sometimes called *game* or *venison*—have always been especially prized in the autumn, when they fatten themselves for the coming winter. While the autumn game season is still celebrated in many European restaurants with wild duck, hare, pheasant, partridge, deer, and boar, in the United States wild meats are banned from commerce (only inspected meat can be sold legally, and hunted meat is

Food Words: *Turkey*

Ornithological and geographical confusion appear to be responsible for the common names of this bird, which came late to Europe. The turkey was first seen by the Spanish in Mexico around 1518, and they named it with variants on the word *pavo,* "pea fowl." In most other European languages its early names referred to India: French *dinde, dindon* (*d'Inde,* "of India"), German *Kalikutische Hahn* ("hen of Calicut," an Indian port), Italian *pollo d'India* ("fowl of India"). The turkey was indeed in India by 1615, so it could well have been introduced to much of Europe via Asia. The English connection with Turkey goes back quite early, to around 1540, and is more obscure. It may reflect a vague impression that the bird came from some outpost of the exotic Ottoman Empire, which originated in and was identified with Turkey.

not inspected). Most "game" meats available to the U.S. consumer these days come from animals raised on farms and ranches. They're perhaps better described as "semi-domestic" meats. Some of these animals have been raised in captivity since Roman times, but haven't been as intensively bred as the domesticated animals, and so are still much like their wild counterparts.

Today Americans are buying more venison (various species of deer and antelope), buffalo, and other game meats thanks to their distinctive flavors and leanness. The very low fat content of game meat causes it to conduct heat and cook faster than standard meats, and to dry out more easily. Cooks often shield it from direct oven heat by "barding" it with a sheet of fat or fatty bacon, and baste it during cooking, which cools the meat surface by evaporation and slows the movement of heat into the meat (p. 158).

Gaminess True wild game has the appeal of rich, variable flavor, thanks to its mature age, free exercise, and mixed diet. Carried

Some Characteristics of Meat Birds

In general, older and larger birds and those with more red fibers have a more pronounced flavor.

Bird	Age, weeks	Weight, lb/kg	Red Fibers in Breast Muscle, %
Chickens			10
Industrial broiler, fryer	6–8	1.5–3.5/0.7–1.6	
Roaster	12–20	3.5–5/1.6–2.3	
French *label rouge*	11.5	2–3.5/1–1.6	
French *appellation contrôlée*	16	2–3.5/1–1.6	
Game hen	5–6	1–2/0.5–1	
Capon	<32	5–8/2.3–3.6	
Stewing fowl	>40	3.5–6/1.6–2.7	
Turkeys		8–30/3.6–14	10
Industrial	12–18		
French *fermière*, U.S. premium brands	24		
French *appellation contrôlée*	32		
Duck	6–16	3.5–7/1.6–3.2	80
Goose	24–28	7–20/3.2–9	85
Quail (wild)	6–10	0.25–0.33/0.1–0.15	75
Squab	4–5	0.75–1.3/0.3–0.6	85
Guinea hen	10–15	2–3.5/1–1.6	25
Pheasant	13–24	2–3/1–1.4	35

to excess, this interesting wild flavor becomes "gamy." In the time of Brillat-Savarin, game was typically allowed to hang for days or weeks until it began to rot. This treatment was called *mortification* or *faisandage* (after the pheasant, *faisan*), and had two purposes: it tenderized the meat, and further heightened its "wild" flavor. Gamy game is no longer the style. Modern farmed animals are often relatively sedentary, eat a uniform diet, and are slaughtered before they reach sexual maturity, so they're usually milder in flavor and more tender than their wild counterparts. Since distinctive meat flavors reside in the fat, they can be minimized by careful trimming.

THE TRANSFORMATION OF MUSCLE INTO MEAT

The first step in meat production is to raise a healthy animal. The second step is to transform the living animal into useful portions of its flesh. The ways in which this transformation occurs affect the quality of the meat, and can explain why the same cut of meat from the same store can be moist and tender one week and dry and tough the next. So it's useful to know what goes on in the slaughterhouse and packing plant.

SLAUGHTER

The Importance of Avoiding Stress By a fortunate coincidence, the methods of slaughter that result in good-quality meat are also the most humane. It has been recognized for centuries that stress just before an animal's death—whether physical work, hunger, duress in transport, fighting, or simple fear—has an adverse effect on meat quality. When an animal is killed, its muscle cells continue to live for some time and consume their energy supply (glycogen, an animal version of starch). In the process they accumulate lactic acid, which reduces enzyme activity, slows microbial spoilage, and causes some fluid loss, which makes the meat seem moist. Stress depletes the muscles of their energy supply before slaughter, so that after slaughter they accumulate less lactic acid and produce readily spoiled "dark, firm, dry" or "dark-cutting" meat, a condition first described in the 18th century. So it pays to treat animals well. In November 1979, the *New York Times* reported that a Finnish slaughterhouse had evicted a group of young musicians from a nearby building because their practice sessions were resulting in dark-cutting meat.

Procedures Meat animals are generally slaughtered as untraumatically as possible. Each animal is stunned, usually with a blow or electrical discharge to the head,

---

Food Words: *Game* and *Venison*

The word *game* is Germanic in origin. Its original meaning in Old English was "amusement," "sport," and after some centuries was applied to hunted animals by people wealthy enough to consider hunting as entertainment. (*Hunt* originally meant "to seize.") The term *venison* comes from the Latin verb *venari,* "to hunt," but ultimately from an Indo-European root meaning "to desire, to strive for," which also gave us the words *win, wish, venerate, Venus,* and *venom* (originally a love potion). It once meant all hunted animals, but now refers mainly to deer and antelope, both ruminants, like cattle and sheep, that can eat weeds and brush and thrive on poorer land than their domesticated relatives.

and then is hung up by the legs. One or two of the major blood vessels in the neck are cut, and the animal bleeds to death while unconscious. As much blood as possible (about half) is removed to decrease the risk of spoilage. (Rarely, as in the French Rouen duck, blood is retained in the animal to deepen the meat's flavor and color.) After bleeding, cattle and lamb heads are removed, the hides stripped off, the carcasses cut open, and the inner organs removed. Pig carcasses remain intact until they have been scalded, scraped and singed to remove bristles; the head and innards are then removed, but the skin is left in place.

Chickens, turkeys, and other fowl must be plucked. The slaughtered birds are usually immersed in a bath of hot water to loosen the feathers, plucked by machine, and cooled in a cold-water bath or cold-air blast. Prolonged water-chilling can add a significant amount of water to the carcass: U.S. regulations allow 5–12% of chicken weight to be absorbed water, or several ounces in a 4-pound bird. By contrast, air-chilling, which is standard in much of Europe and Scandinavia, actually removes water, so that the flesh becomes more concentrated and the skin will brown more readily.

Kosher and halal meats are processed according to Jewish and Muslim religious laws respectively, which among other things require a brief period of salting. These practices don't allow meat birds to be scalded before plucking, so their skin is often torn. The plucked carcasses are then salted for 30–60 minutes and briefly rinsed in cold water; like air-chilled birds, they absorb little if any extraneous moisture. Salting makes meat fats more prone to oxidation and the development of off flavors, so kosher and halal meats don't keep as long as conventionally processed meats.

Rigor Mortis

The Importance of Timing, Posture, and Temperature For a brief period after the animal's death its muscles are relaxed, and if immediately cut and cooked will make especially tender meat. Soon, however, the muscles clench in the condition called *rigor mortis* ("stiffness of death"). If cooked in this state, they make very tough meat. Rigor sets in (after about 2.5 hours in the steer, 1 hour or less in lamb, pork, and chicken) when the muscle fibers run out of energy, their control systems fail and trigger a contracting movement of the protein filaments, and the filaments lock in place. Carcasses are hung up in such a way that most of their muscles are stretched by gravity, so that the protein filaments can't contract and overlap very much; otherwise the filaments bunch up and bond very tightly and the meat becomes exceptionally tough. Eventually, protein-digesting enzymes within the muscle fibers begin to eat away the framework that holds the actin and myosin filaments in place. The filaments are still locked together, and the muscles cannot be stretched, but the overall muscle structure weakens, and the meat texture softens. This is the beginning of the aging process. It becomes noticeable after about a day in beef, after several hours in pork and chicken.

The inevitable toughening during rigor mortis can be worsened by poor temperature control, and may be the source of excessive toughness in retail meats.

Aging

Like cheese and wine, meat benefits from a certain period of aging, or slow chemical change, during which it gets progressively more flavorful. Meat also becomes more tender. In the 19th century, beef and mutton joints would be kept at room temperature for days or weeks, until the outside was literally rotten. The French called this *mortification,* and the great chef Antonin Carême said that it should proceed "as far as possible." The modern taste is for somewhat less mortified flesh! In fact most meat in the United States is aged only incidentally, during the few days it takes to be shipped from packing plant to market. This is enough for chicken, which benefits from a day or two of aging, and for pork and lamb, which benefit from a week. (The unsaturated fats of pork and poultry go rancid relatively quickly.) But the flavor and texture of beef keeps improving for up to a month, especially when whole, unwrapped sides are *dry-aged* at 34–38°F/1–3°C and at a relative humidity of 70–80%. The cool temperature limits the growth of microbes, while the moderate humidity causes the meat to lose moisture gradually, and thus become denser and more concentrated.

Muscle Enzymes Generate Flavor . . . The aging of meat is mainly the work of the muscle enzymes. Once the animal is slaughtered and the control systems in its cells stop functioning, the enzymes begin attacking other cell molecules indiscriminately, turning large flavorless molecules into smaller, flavorful fragments. They break proteins into savory amino acids; glycogen into sweet glucose; the energy currency ATP into savory IMP (inosine monophosphate); fats and fat-like membrane molecules into aromatic fatty acids. All of these breakdown products contribute to the intensely meaty, nutty flavor of aged meat. During cooking, the same products also react with each other to form new molecules that enrich the aroma further.

. . . And Diminish Toughness Uncontrolled enzyme activity also tenderizes meat. Enzymes called *calpains* mainly weaken the supporting proteins that hold the contracting filaments in place. Others called *cathepsins* break apart a variety of proteins, including the contracting filaments and the supporting molecules. The cathepsins also weaken the collagen in connective tissue, by breaking some of the strong cross-links between mature collagen fibers. This has two important effects: it causes more collagen to dissolve into gelatin during cooking, thus making the meat more tender and succulent; and it reduces the squeezing pressure that the connective tissue exerts during heating (p. 150), which means that the meat loses less moisture during cooking.

Enzyme activity depends on temperature. The calpains begin to denature and lose activity around 105°F/40°C, the cathepsins around 122°F/50°C. But below this critical range, the higher the temperature, the faster the enzymes work. Some accelerated "aging" can take place during cooking. If meat is quickly seared or blanched in boiling water to eliminate microbes on its surface, and then heated up slowly during the cooking—for example, by braising or roasting in a slow oven—then the aging enzymes within the meat can be very active for several hours before they denature. Large 50 lb/23 kg slow-roasted "steamship" rounds of beef spend 10 hours or more rising to 120–130°F/50–55°C, and come out more tender than small portions of the same cut cooked quickly.

Aging Meat in Plastic and in the Kitchen Despite the contribution that aging can make to meat quality, the modern meat industry generally avoids it, since it means tying up its assets in cold storage and losing about 20% of the meat's original weight to evaporation and laborious trimming of the dried, rancid, sometimes moldy surface. Most meat is now butchered into retail cuts at the packing plant shortly after slaughter, wrapped in plastic, and shipped to market immediately, with an average of 4 to 10 days between slaughter and sale. Such meat is sometimes *wet-aged*, or kept in its plastic wrap for some days or weeks, where it's shielded from oxygen and retains moisture while its enzymes work. Wet-aged meat can develop some of the flavor and tenderness of dry-aged meat, but not the same concentration of flavor.

Cooks can age meat in the kitchen. Simply buying meat several days before it's needed will mean some informal aging in the refrigerator, where it can be kept tightly wrapped, or uncovered to allow some evaporation and concentration. (Loose or no wrapping may cause dry spots, the absorption of undesirable odors, and the necessity of some trimming; this works best with large roasts, not steaks and chops.) And as we've seen, slow cooking gives the aging enzymes a chance to do in a few hours what would otherwise take weeks.

Cutting and Packaging

In the traditional butchering practice that prevailed until the late 20th century but is now rare, animal carcasses are divided at the slaughterhouse into large pieces—halves or quarters—which are then delivered to retail butchers, who break them down into roasts, steaks, chops, and the other standard cuts. The meat might not be wrapped at all until sale, and then only loosely in "butcher's paper." Such meat is continuously exposed to the air, so it tends to be fully oxygenated and red, and it slowly dries out, which concentrates its flavor at the same time that it leaves some surface areas discolored and off-flavored and in need of trimming.

The modern tendency in butchering is to break meat down into the retail cuts at the packing house, vacuum-wrap them in plastic precisely to avoid exposure to the air, and deliver these prepackaged cuts to the supermarket. Vacuum-packed meat has the economic advantage of assembly-line efficiency, and keeps for weeks (as much as 12 for beef, 6 to 8 for pork and lamb) without any weight loss due to drying or trimming. Once repackaged, meat has a display-case life of a few days.

Carefully handled, well packaged meat will be firm to the touch, moist and even-colored in appearance, and mild and fresh in smell.

MEAT SPOILAGE AND STORAGE

Fresh meat is an unstable food. Once a living muscle has been transformed into a piece of meat it begins to change, both chemically and biologically. The changes that we associate with aging—the generation of flavor and tenderness by enzymes throughout the meat—are desirable. But the changes that take place at the meat surface are generally undesirable. Oxygen in the air and energetic rays of light generate off-flavors and dull color. And meat is a nourishing food for microbes as well as for humans. Given the chance, bacteria will feast on meat surfaces and multiply. The result is both unappetizing and unsafe, since some microbial digesters of dead flesh can also poison or invade the living.

Meat Spoilage

Fat Oxidation and Rancidity The most important chemical damage suffered by meats is the breakdown of their fats by both oxygen and light into small, odorous fragments that define the smell of *rancidity.* Rancid fat won't necessarily make us sick, but it's unpleasant, so its development limits how long we can age and store meat. Unsaturated fats are most susceptible to rancidity, which means that fish, poultry, and game birds go bad most quickly. Beef has the most saturated and stable of all meat fats, and keeps the longest.

Fat oxidation in meats can't be prevented, but it can be delayed by careful handling. Wrap raw meat tightly in oxygen-impermeable plastic wrap (saran, or polyvinylidene chloride; polyethylene is permeable), overwrap it with foil or paper to keep it in the dark, store it in the coldest corner of the refrigerator or freezer, and use it as soon as possible. When cooking with ground meat, grind the meat fresh, just before cooking, since dividing the meat into many small pieces exposes a very large surface area to the air. The development of rancidity in cooked meats can be delayed by minimizing the use of salt, which encourages fat oxidation, and by using ingredients with antioxidant activity: for example the Mediterranean herbs, especially rosemary (p. 395). Browning the meat surface in a hot pan also generates antioxidant molecules that can delay fat oxidation.

Spoilage by Bacteria and Molds The intact muscles of healthy livestock are generally free of microbes. The bacteria and molds that spoil meat are introduced during processing, usually from the animal's hide or the packing-plant machinery. Poultry and fish are especially prone to spoilage because they're sold with their skin intact, and many bacteria persist despite washing. Most of these are harmless but unpleasant. Bacteria and molds break down cells at the meat surface and digest proteins and amino acids into molecules that smell fishy, skunky, and like rotten eggs. Spoiled meat smells more disgusting than other rotten foods exactly because it contains the proteins that generate these stinky compounds.

Refrigeration

In the developed world, the most common domestic method for preserving meat is simply to keep it cool. Refrigeration has two great advantages: it requires little or no preparation time, and it leaves the meat relatively unchanged from its fresh state. Cooling meat extends its useful life because both bacteria and meat enzymes become less active as the temperature drops. Even so, spoilage does continue. Meats keep best at temperatures approaching or below the freezing point, 32°F/0°C.

Freezing Freezing greatly extends the storage life of meat and other foods because it halts all biological processes. Life requires liquid water, and freezing immobilizes the food's liquid water in solid crystals of ice. Well-frozen meat will keep for millennia, as has been demonstrated by the discovery of mammoth flesh frozen 15,000 years ago in the ice of northern Siberia. It's best to keep meat as cold as possible. The usual recommendation for home freezers is 0°F/–18°C (many operate at 10–15°F/–12 to –9°C).

Freezing will preserve meat indefinitely from biological decay. However, it's a drastic physical treatment that inevitably causes damage to the muscle tissue, and therefore diminishes meat quality in several ways.

Cell Damage and Fluid Loss As raw meat freezes, the growing crystals protrude into the soft cell membranes and puncture them. When the meat is thawed, the ice crystals melt and unplug the holes they've made in the muscle cells, and the tissue as a whole readily leaks a fluid rich in salts, vitamins, proteins, and pigments. Then when the meat is cooked, it loses more fluid than usual (p. 150), and more readily ends up dry, dense, and tough. Cooked meat suffers less from freezing because its tissue has already been damaged and lost fluid when it was heated.

Cell damage and fluid loss are minimized by freezing the meat as rapidly as possible and keeping it as cold as possible. The faster the meat moisture freezes, the smaller the crystals that it forms, and the less they protrude into the cell membranes; and the colder the meat is kept, the less enlargement of existing crystals will occur. Freezing can be accelerated by setting the freezer at its coldest temperature, dividing the meat into small pieces, and leaving it unwrapped until after it has solidified (wrapping acts as insulation and can double the freezing time).

Fat Oxidation and Rancidity In addition to inflicting physical damage, freezing causes chemical changes that limit the storage life of frozen meats. When ice crystals form and remove liquid water from the muscle fluids, the increasing concentration of salts and trace metals promotes the oxidation of unsaturated fats, and rancid flavors accumulate. This inexorable process means that quality declines noticeably for fresh fish and poultry after only a few months in the freezer, for pork after about six months, for lamb and veal after about nine months, and for beef after about a year. The flavors of ground meats, cured meats, and cooked meats deteriorate even faster.

Freezer Burn A last side effect of freezing is *freezer burn,* that familiar brownish-white discoloration of the meat surface that develops after some weeks or months of storage. This is caused by water "sublimation"—the equivalent of evaporation at below-freezing temperatures—from ice crystals at the meat surface into the dry freezer air. The departure of the water leaves tiny cavities in the meat surface which scatter light and so appear white. The meat surface is now in effect a thin layer of freeze-dried meat where oxidation of fat and pigment is accelerated, so texture, flavor, and color all suffer.

Freezer burn can be minimized by covering the meat as tightly as possible with water-impermeable plastic wrap.

Thawing Meats Frozen meats are usually thawed before cooking. The simplest method—leaving the meat on the kitchen counter—is neither safe nor efficient. The surface can rise to microbe-friendly temperatures long before the interior thaws, and air transfers heat to the meat very slowly, at about one-twentieth the rate that water does. A much faster and safer method is to immerse the wrapped meat in a bath of ice water, which keeps the surface safely cold, but still transfers heat into the meat efficiently. If the piece of meat is too large for a water bath, or isn't needed right away, then it's also safe to thaw it in the refrigerator. But cold air is an especially inefficient purveyor of warmth, so it can take days for a large roast to thaw.

Cooking Unthawed Meats Frozen meats can be cooked without thawing them first, particularly with relatively slow methods such as oven roasting, which give the heat time to penetrate to the center without drastically overcooking the outer portions of the meat. Cooking times are generally 30–50% longer than for fresh cuts.

Irradiation

Because ionizing radiation (p. 782) damages delicate biological machinery like DNA and proteins, it kills spoilage and disease microbes in food, thus extending its shelf life and making it safer to eat. Tests have shown that low doses of radiation can kill most microbes and more than double the shelf life of carefully wrapped refrigerated meats. There is, however, a characteristic radiation flavor, described as metallic, sulfurous, and goaty, which may be barely noticeable or unpleasantly strong.

Beginning in 1985, the U.S. Food and Drug Administration has approved irradiation to control a number of pathogens in meat: first trichinosis in pork, then salmonella in chickens, and *E. coli* in beef. A treatment like irradiation is an especially valuable form of insurance for the mass production of ground meats, in which a single infected carcass can contaminate thousands of pounds of meat, and affect thousands of consumers. But its use remains limited due to consumer wariness. Decades of testing indicate that irradiated meat is safe to eat. But one other objection is quite reasonable. If meat has been contaminated with enough fecal matter to cause infection with *E. coli,* then irradiation will kill the bacteria and leave the meat edible for three months. However, it will still be adulterated meat. Many consumers set a higher standard than the absence of living pathogens and months of shelf life for the food from which they take daily nourishment and pleasure. People who care about food quality will seek out meat produced locally, carefully, and recently, and for enjoyment within a few days, when it's at its best.

COOKING FRESH MEAT: THE PRINCIPLES

We cook meat for four basic reasons: to make it safe to eat, easier to chew and to digest (denatured proteins are more vulnerable to our digestive enzymes), and to make it more flavorful. The issue of safety is detailed beginning on p. 124. Here I'll describe the physical and chemical transformations of meat during cooking, their effects on flavor and texture, and the challenge of cooking meat well. These changes are summarized in the box on p. 152.

Heat and Meat Flavor

Raw meat is tasty rather than flavorful. It provides salts, savory amino acids, and a slight acidity to the tongue, but offers little in the way of aroma. Cooking intensifies the taste of meat and creates its aroma. Simple physical damage to the muscle fibers causes them to release more of their fluids and therefore more stimulating substances for the tongue. This fluid release is at its maximum when meat is only lightly cooked, or done "rare." As the temperature increases and the meat dries out, physical change gives way to chemical change, and to the development of aroma as cell molecules break apart and recombine with each other to form new molecules that not only smell meaty, but also fruity and floral, nutty and grassy (esters, ketones, aldehydes).

Surface Browning at High Temperatures If fresh meat never gets hotter than the boiling point of water, then its flavor is largely determined by the breakdown products of proteins and fats. However, roasted, broiled, and fried meats develop a crust that is much more intensely flavored, because the meat surface dries out and gets hot enough to trigger the Maillard or browning reactions (p. 778). Meat aromas generated in the browning reactions are generally small rings of carbon atoms with additions of nitrogen, oxygen, and sulfur. Many of these have a generic "roasted" character, but some are grassy, floral, oniony or spicy, and earthy. Several hundred aromatic compounds have been found in roasted meats!

Heat and Meat Color

The appearance of meat changes in two different ways during cooking. Initially it's somewhat translucent because its cells are filled with a relatively loose meshwork of proteins suspended in water. When heated to about 120°F/50°C, it develops a white opacity as heat-sensitive myosin denatures and coagulates into clumps large enough to scatter light. This change causes red meat color to lighten from red to pink, long before the red pigments themselves are affected. Then, around 140°F/60°C, red myoglobin begins to denature into a tan-colored version called hemichrome. As this change proceeds, meat color shifts from pink to brown-gray.

The denaturation of myoglobin parallels the denaturation of fiber proteins, and this makes it possible to judge the doneness of fresh meat by color. Little-cooked meat and its juices are red, moderately cooked meat and its juices are pink, thoroughly cooked meat is brown-gray and its juices clear. (Intact red myoglobin can escape in the meat juices; denatured brown myoglobin has bonded to other coagulated proteins in the cells and stays there.) However, there

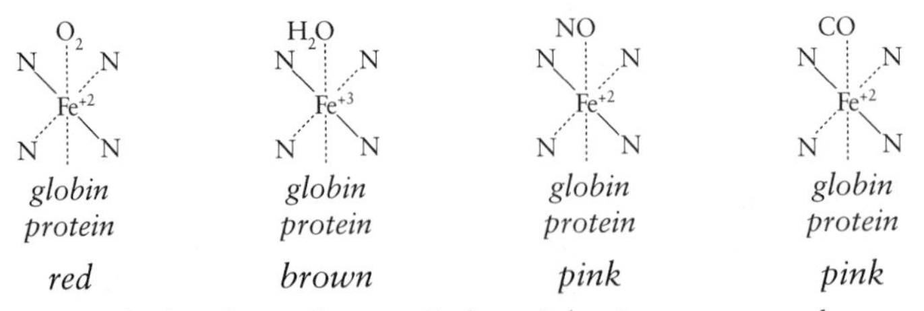

The pigments in cooked and cured meats. Left to right: *In raw meat, the oxygen-carrying myoglobin is red; in cooked meat, the oxidized, denatured form of myoglobin is brown; in meats cured with nitrite, including corned beef and ham, the myoglobin assumes a stable pink form (NO is nitric oxide, a product of nitrite); and in uncured meats cooked in a charcoal grill or gas oven, traces of carbon monoxide (CO) accumulate and produce another stable pink form.*

are a number of oddities about myoglobin that can lead to misleading redness or pinkness even in well-cooked meat (see box). And it's also possible for undercooked meat to look brown and well-done, if its myoglobin has already been denatured by prolonged exposure to light or to freezing temperatures. If it's essential that meat be cooked to microbe-destroying temperatures, then the cook should use an accurate thermometer to confirm that it has reached a minimum of 160°F/70°C. Meat color can be misleading.

Heat and Meat Texture

The texture of a food is created by its physical structure: the way it feels to the touch, the balance of solid and liquid components, and the ease or difficulty with which our teeth break it down into manageable pieces. The key textural components in meat are its moisture, around 75% of its weight, and the fiber proteins and connective tissue that either contain and confine that moisture, or release it.

Raw and Cooked Textures The texture of raw meat is a kind of slick, resistant mushiness. The meat is chewy yet soft, so that chewing compresses it instead of cutting through it. And its moisture manifests itself in slipperiness; chewing doesn't manage to liberate much juice.

Heat changes meat texture drastically. As it cooks, meat develops a firmness and

Persistent Colors in Cooked Meats

Thoroughly cooked meat is usually a dull, brownish-gray in appearance due to the denaturing of its myoglobin and cytochrome pigments. But two cooking methods can leave well-done meat attractively red or pink.

- Barbecued meat, stew meat, a pot roast, or a confit can be surprisingly pink or red inside—if it was heated very gradually and gently. Myoglobin and cytochromes can survive somewhat higher temperatures than the other muscle proteins. When meat is heated quickly, its temperature rises quickly, and some of the muscle proteins are still unfolding and denaturing when the pigments begin to do the same. The other proteins are therefore able to react with the pigments and turn them brown. But when meat is heated slowly, so that it takes an hour or two to reach the denaturing temperature for myoglobin and cytochromes, the other proteins finish denaturing first, and react with each other. By the time that the pigments become vulnerable, there are few other proteins left to react with them, so they stay intact and the meat stays red. The preliminary salting for making a confit (p. 177) greatly accentuates this effect in duck meat.
- Meats cooked over wood, charcoal, or gas flames—barbecued pork or beef, for example, or even poultry cooked in a gas oven—often develop "pink ring," which reaches from the surface to a depth of 8–10 mm. This is caused by nitrogen dioxide (NO_2) gas, which is generated in trace amounts (parts per million) by the burning of these organic fuels. It appears that NO_2 dissolves at the meat surface to form nitrous acid (HNO_2), which diffuses into the muscle tissue and is converted to nitric oxide (NO). NO in turn reacts with myoglobin to form a stable pink molecule, like the molecule found in nitrite-cured meats (p. 174).

resilience that make it easier to chew. It begins to leak fluid, and becomes juicy. With longer cooking, the juices dry up, and resilience gives way to a dry stiffness. And when the cooking goes on for hours, the fiber bundles fray away from each other, and even tough meat begins to fall apart. All of these textures are stages in the denaturation of the fiber and connective-tissue proteins.

Early Juiciness: Fibers Coagulate One of the two major contracting filaments, the protein myosin, begins to coagulate at about 120°F/50°C; this lends each cell some solidity and the meat some firmness. As the myosin molecules bond to each other, they squeeze out some of the water molecules that had separated them. This water collects around the solidifying protein core, and is actively squeezed out of the cell by its thin, elastic sheath of connective tissue. In intact muscles, juices break through weak spots in the fiber sheaths. In chops and steaks, which are thin slices of whole muscles, it also escapes out the cut ends of the fibers. Meat served at this stage, the equivalent of rare, is firm and juicy.

Final Juiciness: Collagen Shrinks As the meat's temperature rises to 140°F/60°C, more of the proteins inside its cells coagulate and the cells become more segregated into a solid core of coagulated protein and a surrounding tube of liquid: so the meat gets progressively firmer and moister. Then between 140 and 150°F/60–65°C, the meat suddenly releases lots of juice, shrinks noticeably, and becomes chewier. These changes are caused by the denaturing of collagen in the cells' connective-tissue sheaths, which shrink and exert new pressure on the fluid-filled cells inside them. The fluid flows copiously, the piece of meat loses a sixth or more of its volume, and its protein fibers become more densely packed and so harder to cut through. Meat served in this temperature range, the equivalent of medium-rare, is changing from juicy to dry.

Falling-Apart Tenderness: Collagen Becomes Gelatin If the cooking continues, the meat will get progressively dryer, more compacted, and stiff. Then around 160°F/70°C, connective-tissue collagen begins to dissolve into gelatin. With time, the connective tissue softens to a jelly-like consistency, and the muscle fibers that it had held tightly together are more easily pushed apart. The fibers are still stiff and dry, but they no longer form a monolithic mass, so the meat seems more tender. And the gelatin provides a succulence of its own. This is the delightful texture of slow-cooked meats, long braises, and stews and barbecues.

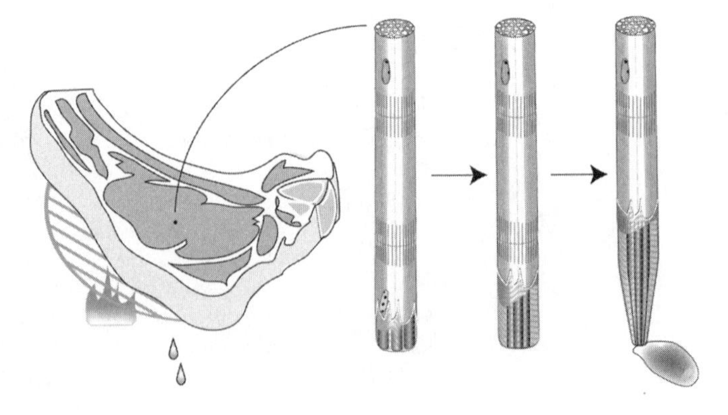

How cooking forces moisture from meat. Water molecules are bound up in the protein fibrils that fill each muscle cell. As the meat is heated, the proteins coagulate, the fibrils squeeze out some of the water they had contained and shrink. The thin elastic sheet of connective tissue around each muscle cell then squeezes the unbound water out the cut ends of the cells.

The Challenge of Cooking Meat: The Right Texture

Generally, we like meat to be tender and juicy rather than tough and dry. The ideal method for cooking meat would therefore minimize moisture loss and compacting of the meat fibers, while maximizing the conversion of tough connective-tissue collagen to fluid gelatin. Unfortunately, these two aims contradict each other. Minimizing fiber firming and moisture loss means cooking meat quickly to no hotter than 130–140°F/55–60°C. But turning collagen to gelatin requires prolonged cooking at 160°F/70°C and above. So there is no ideal cooking method for all meats. The method must be tailored to the meat's toughness. Tender cuts are best heated rapidly and just to the point that their juices are in full flow. Grilling, frying, and roasting are the usual fast methods. Tough cuts are best heated for a prolonged period at temperatures approaching the boil, usually by stewing, braising, or slow-roasting.

It's Easy to Overcook Tender Meat

Cooking tender meat to perfection—so that its internal temperature is just what we want—is a real challenge. Imagine that we grill a thick steak just to medium rare, 140°F/60°C, at the center. Its surface will have dried out enough to get hotter than the boiling point, and in between the center and surface, the meat temperature spans the range between 140°F/60°C—medium rare—and 212°F/100°C—cooked dry. In fact the bulk of the meat is overcooked. And it only takes a minute or two to overshoot medium rare at the center and dry out the whole steak, because meat is cooked but juicy in only a narrow temperature range, just 30°F/15°C. When we grill or fry an inch-thick steak or chop, the rate of temperature increase at the center can exceed 10°F/5°C per minute.

Solutions: Two-Stage Cooking, Insulation, Anticipation There are several ways to give the cook a larger window of time for stopping the cooking, and to obtain meat that is more evenly done.

The most common method is to divide the cooking into two stages, an initial high-temperature surface browning, and a subsequent cooking through at a much lower temperature. The low cooking temperature means a smaller temperature difference between center and surface, so that more of the meat is within a few degrees of the center temperature. It also means that the meat cooks more slowly, with a larger window of time during which the interior is properly done.

Another trick is to cover the meat surface with another food, such as strips of fat or bacon, batters and breadings, pastry and bread dough. These materials insulate the meat surface from direct cooking heat and slow the heat penetration.

The Nature of Juiciness

Food scientists who have studied the subjective sensation of juiciness find that it consists of two phases: the initial impression of moisture as you bite into the food, and the continued release of moisture as you chew. Juiciness at first bite comes directly from the meat's own free water, while continued juiciness comes from the meat's fat and flavor, both of which stimulate the flow of our own saliva. This is probably why well-seared meat is often credited with greater juiciness despite the fact that searing squeezes more of the meat's own juice out. Above all else, searing intensifies flavor by means of the browning reactions, and intense flavor gets *our* juices flowing.

The Effects of Heat on Meat Proteins, Color, and Texture							
Meat Temperature	Doneness	Meat Qualities	Fiber-Weakening Enzymes	Fiber Proteins	Connective-Tissue Collagen	Protein-Bound Water	Myoglobin Pigment
100°F 40°C	Raw	•Soft to touch •Slick, smooth •Translucent, deep red	Active	Beginning to unfold	Intact	Begins to escape from proteins, accumulate within cells	Normal
110°F 45°C	*Bleu*						
120°F 50°C	Rare, 120–130°	•Becoming firmer •Becoming opaque	Very active	Myosin begins to denature, coagulate		Escape and accumulation accelerate	
130°F 55°C	Medium rare, 130–135°	•Resilient to touch •Less slick, more fibrous •Releases juice when cut •Opaque, lighter red	Denature, become inactive, coagulate	Myosin coagulated			
140°F 60°C	Medium 135–145° (USDA "rare")	•Begins to shrink •Losing resilience •Exudes juice •Red fades to pink		Other fiber proteins denature, coagulate	Collagen sheaths shrink, squeeze cells	Flows from cells under collagen pressure	Begins to denature
150°F 65°C	Medium well, 145–155° (USDA "medium rare")	•Continues to shrink •Little resilience •Less free juice •Pink fades to gray-brown					
160°F 70°C	Well, 155° and above (USDA "medium")	•Continues to shrink •Stiff •Little free juice •Gray-brown			Begins to dissolve	Flow ceasing	Mostly denatured, coagulated
170°F 75°C	(USDA "well")	•Stiff •Dry •Gray-brown					
180°F 80°C				Actin denatures, coagulates; cell contents densely compacted			
190°F 85°C							
200°F 90°C		•Fibers more easily separated from each other			Dissolving rapidly		

Cooks can also avoid zooming through the zone of ideal doneness by removing the meat from the oven or pan before it's completely done, and relying on lingering afterheat to finish the cooking more gradually, until the surface cools enough to draw the heat back out of the meat interior. The extent of afterheating depends on the meat's weight, shape, and center temperature, and the cooking temperature, and can range from a negligible few degrees in a thin cut to 20°F/10°C in a large roast.

Knowing When to Stop Cooking The key to cooking meat properly is knowing when to stop. Cookbooks are full of formulas for obtaining a given doneness—so many minutes per pound or per inch thickness—but these are at best rough approximations. There are a number of unpredictable and significant factors that they just can't take into account. Cooking time is affected by the meat's starting temperature, the true temperatures of frying pans and ovens, and the number of times the meat is flipped or the oven door opened. The meat's fat content matters, because fat is less conductive than the muscle fibers: fatty cuts cook more slowly than lean ones. Bones make a difference too. The ceramic-like minerals in bone give it double the heat conductivity of meat, but its frequently honeycombed, hollow structure generally slows its transfer of heat and turns bone into an insulator. This is why meat is often said to be "tender at the bone," more succulent there because less thoroughly cooked. Finally, cooking time depends on how the meat's surface is treated. Naked or basted meat evaporates moisture from its surface, which cools the meat and slows cooking, but a layer of fat or a film of oil forms a barrier to such evaporation and can cut cooking times by a fifth.

With so many variables affecting cooking time, it's clear that no formula or recipe can predict it infallibly. It's up to the cook to monitor the cooking and decide when it should stop.

Judging Doneness The best instruments for monitoring the doneness of meat remain the cook's eye and finger. Measuring internal temperature with a thermometer works well for roasts but not for smaller cuts. (Standard kitchen thermometers register temperature along an inch span of their thick metal shaft, not just at the tip. Dial thermometers also require frequent recalibration to maintain their accuracy.) The simplest way to be sure is to cut into the meat and check its color (the loss of fluid is local and minor).

Most professional cooks still evaluate meats by their "feel" and by the way their juices flow:

- *Bleu* meat, cooked at the surface but just warmed within, remains rela-

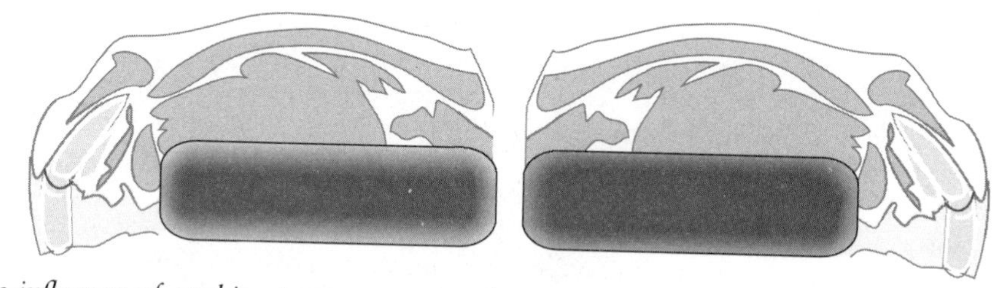

The influence of cooking temperature on the evenness of cooking. Left: *In meat cooked through over high heat, the outer layer gets overcooked while the center reaches the desired temperature.* Right: *In meat cooked through over low heat, the outer layers get less overcooked, and the meat is more evenly done.*

tively unchanged—soft to the touch, like the muscle between thumb and forefinger when it's completely relaxed, with little or no colored juice (some colorless fat may melt out).

- *Rare* meat, some of whose protein has coagulated, is more resilient when poked with the finger—like the thumb-forefinger muscle when the two digits are stretched apart—and red juice begins to appear at the surface. To some people this is meat at its most succulent; to others it is still raw, "bloody" (though the juices are not blood), and potentially hazardous.
- *Medium*-done meat, whose connective-tissue collagen has shrunk, is more firm—like the thumb-forefinger muscle when the two digits are squeezed together—and squeezes droplets of red juice to steak and chop surfaces, while the interior pales to pink. Most but not all microbes are killed in this range.
- *Well-done* meat, nearly all its proteins denatured, is frankly stiff to the touch, little juice is apparent, and both juice and interior are a dull tan or gray. Microbes are dead, and many meat lovers would say that the meat is too. However, prolonged, gentle cooking will loosen the connective-tissue harness and bring back a degree of tenderness.

Meat Doneness and Safety

As we've seen, meats inevitably harbor bacteria, and it takes temperatures of 160°F/70°C or higher to guarantee the rapid destruction of the bacteria that can cause human disease—temperatures at which meat is well-done and has lost much of its moisture. So is eating juicy, pink-red meat risky? Not if the cut is an intact piece of healthy muscle tissue, a steak or chop, and its surface has been thoroughly cooked: bacteria are on the meat surfaces, not inside. Ground meats are riskier, because the contaminated meat surface is broken into small fragments and spread throughout the mass. The interior of a raw hamburger usually does contain bacteria, and is safest if cooked well done. Raw meat dishes—steak tartare and carpaccio—should be prepared only at the last minute from cuts carefully trimmed of their surfaces.

Making a Safer Rare Hamburger One way to enjoy a less risky rare hamburger is to grind the meat yourself after a quick treatment that will kill surface bacteria. Bring a large pot of water to a rolling boil, immerse the pieces of meat in the water for 30–60 seconds, then remove, drain and pat dry, and grind in a scrupulously clean meat grinder. The blanching kills surface bacteria while overcooking only the outer 1–2 millimeters, which grinding then disperses invisibly throughout the rest of the meat.

Now that we understand the basic nature of heat and how it moves into and through meat, let's survey the common methods of cooking meat and how to make the best of them.

COOKING FRESH MEAT: THE METHODS

Many traditional meat recipes were developed at a time when meats came from mature, fatty animals, and so were fairly tolerant of overcooking. Fat coats and lubricates meat fibers during cooking, and stimulates the flow of saliva and creates the sensation of juiciness no matter how dry the meat fibers themselves have become. Recipes for hours-long braising or stewing were developed for mature animals with substantially cross-linked collagen that took a long time to dissolve into gelatin. However, today's industrially produced meats come from relatively young animals with more soluble collagen and with far less fat; they cook quickly, and suffer more from overcooking. Grilled chops and steaks may be just right at the center but dry elsewhere; long-braised pot roasts and stews are often dry throughout.

The cook's margin of error in cooking meat is narrower than it used to be. So it's more useful than ever to understand how the various methods for cooking meat work, and how best to apply them to the meat of the 21st century.

Modifying Texture Before and After Cooking

There are a number of traditional techniques that tenderize tough meat before cooking, so that the cooking itself, and the drying of the muscle fibers, can be minimized. The most straightforward of these is to damage the meat structure physically, to fragment the muscle fibers and connective-tissue sheets by pounding, cutting, or grinding. Pieces of veal pounded into sheets (*escalopes, scallopini*) are both tenderized and made so thin that they cook through in a moisture-sparing minute or two. Grinding the meat into small pieces creates an entirely different sort of texture: the gently gathered ground beef in a good hamburger has a delicate quality quite unlike even a tender steak.

A traditional and labor-intensive French method for modifying tough meat is *larding*, the insertion of slivers of pork fat into the meat by means of hollow needles. In addition to augmenting the meat's fat content, larding also breaks some fibers and connective-tissue sheets.

Marinades Marinades are acidic liquids, originally vinegar and now including such ingredients as wine, fruit juices, buttermilk, and yogurt, in which the cook immerses meat for hours to days before cooking. They have been used since Renaissance times, when their primary function was to slow spoilage and to provide flavor. Today, meats are marinated primarily to flavor them and to make them more moist and tender. Perhaps the most common marinated meat dish is a stew, for which the meat is immersed in a mixture of wine and herbs and then cooked in it.

The acid in marinades does weaken muscle tissue and increase its ability to retain moisture. But marinades penetrate slowly, and can give the meat surface an overly sour flavor while they do so. The penetration time can be reduced by cutting meat into thin pieces or by using a cooking syringe to inject the marinade into larger pieces.

Meat Tenderizers Meat tenderizers are protein-digesting enzymes extracted from a number of plants, including papaya, pineapple, fig, kiwi, and ginger. They are available either in the original fruit or leaf, or purified and powdered for the shaker, diluted in salt and sugar. (Despite lore to the contrary, wine corks do not contain active enzymes and don't tenderize octopus or other tough meats!) The enzymes act slowly at refrigerator or room temperature, and some five times faster between 140 and 160°F/60–70°C, so nearly all the tenderizing action takes place during cooking. The problem with tenderizers is that they penetrate into meat even more slowly than acids, a few millimeters per day, so that the meat surface tends to accumulate too much and get overly mealy, while the interior remains unaffected. The distribution can be improved by injecting the tenderizer into the meat.

Brining The tendency of modern meats to dry out led cooks to rediscover light brining, a traditional method in Scandinavia and elsewhere. The meats, typically poultry or pork, are immersed in a brine containing 3 to 6% salt by weight for anywhere from a few hours to two days (depending on thickness) before being cooked as usual. They come out noticeably juicier.

Brining has two initial effects. First, salt disrupts the structure of the muscle filaments. A 3% salt solution (2 tablespoons per quart/30 gm per liter) dissolves parts of the protein structure that supports the contracting filaments, and a 5.5% solution (4 tablespoons per quart/60 gm per liter) partly dissolves the filaments themselves. Second, the interactions of salt and proteins result in a greater water-holding capacity in the muscle cells, which then absorb water from the brine. (The inward movement of salt and water and disruptions of the muscle filaments into the meat also increase its absorption of aromatic molecules from any herbs and spices in the brine.) The meat's weight increases by 10% or more. When cooked, the meat still loses around 20% of its weight in moisture, but this loss is counterbalanced by the brine absorbed, so the moisture loss is effectively cut in half. In addition, the dissolved protein filaments can't coagulate into normally dense aggregates, so the cooked meat seems more tender. Because the brine works its way in from the outside, it has its earliest and strongest effects on the meat region most likely to be overcooked, so even a brief, incomplete soaking can make a difference.

The obvious disadvantage of brining is that it makes both the meat and its drippings quite salty. Some recipes balance the saltiness by including sugar or such ingredients as fruit juice or buttermilk, which provide both sweetness and sourness.

Shredding Even if a tough roast has been cooked to the point that it has become tender but unpleasantly dry, the cook can restore a certain succulence to the meat by pulling it apart into small shreds and pouring over them the meat's collected juices, or a sauce. A film of liquid clings to the surface of each shred and thus coats many fibers with some of their lost moisture. The finer the shredding, the greater the surface that can take up liquid, and the moister the meat will seem. When "pulled" meat and sauce are very hot, the sauce is more fluid and tends to run off the shreds; when cooler, the sauce becomes thicker and clings more tenaciously to the meat.

Flames, Glowing Coals, and Coils

Fire and red-hot coals were probably the first heat sources used to cook meat, and thanks to temperatures high enough to generate browning-reaction aromas, they can produce the most flavorful results. But

Aids to Successful Grilling and Frying: Warm Meat and Frequent Flips

Because grilling and frying involve high heat, they tend to overcook the outer portions of meat while the interior cooks through. This overcooking can be minimized in two ways: prewarming the meat, and flipping it frequently.

- The warmer the meat starts out, the less time it takes to cook through, and so the less time the outer layers are exposed to high heat. The cooking and overcooking time can be reduced by a third or more by wrapping steaks and chops, immersing them for 30–60 minutes in warm water, so that they approach body temperature, 100°F/40°C, and then cooking immediately (bacteria grow quickly on warm meat).
- How often should the cook turn a steak or hamburger when grilling or frying? If perfect grill marks are necessary, once or twice. If texture and moistness are more important, then flip every minute. Frequent turns mean that neither side has the time either to absorb or to release large amounts of heat. The meat cooks faster, and its outer layers end up less overdone.

this "primitive" method takes some care to get a juicy interior underneath the delicious crust.

Grilling and Broiling The term "grilling" is generally used to mean cooking meat on a metal grate directly over the heat source, while "broiling" means cooking meat in a pan below the heat source. The heat source may be glowing coals, an open gas flame, or ceramic blocks heated by a gas flame, or a glowing electrical element. The primary means of heat transfer is infrared radiation, the direct emission of energy in the form of light: hence the glow of coals, flames, and heating elements (p. 781). The meat surface is only a few inches away from the heat, which is very hot indeed: gas burns at around 3,000°F/1,650°C, coals and electrical elements glow at 2,000°F/1,100°C. Because these temperatures can blacken food surfaces before the inside is cooked through, grilling is limited to such relatively thin and tender cuts as chops, steaks, poultry parts, and fish.

The most flexible grill arrangement is a dense bed of glowing coals or high gas flame under one area for surface browning, sparser coals or a lower gas flame under another for cooking through, and the distance between meat and fire an inch or two. The meat is cooked over high heat to brown each side well but as briefly as possible, in two or three minutes, and then moved to the cooler area to heat through gently and evenly.

Spit-Roasting Spit-roasting—impaling meat on a metal or wood spike and turning it continuously near the radiating heat source—is best suited to large, bulky cuts, including roasts and whole animals. It exposes the meat surface to browning temperatures, but it does so both evenly and intermittently. Each area receives an intense, browning blast of infrared radiation, but only for a few seconds. During the many seconds when it faces away from the heat, the hot surface gives up much of its heat to the air, so only a fraction of each blast penetrates into the meat, and the interior therefore cooks through relatively gently. In addition, the constant rotation causes the juices to cling to and travel around the meat surface, basting and coating it with proteins and sugars for the browning reactions.

The full advantages of spit roasting are obtained when the roasting is done in the open air, or in an oven with the door ajar. A closed oven quickly heats up to baking temperatures, and the meat will accordingly heat through less gently.

Barbecuing This distinctively American cooking method took its modern form about a century ago. Barbecuing is the low-temperature, slow heating of meat in a closed chamber by means of hot air from smoldering wood coals. It's an outdoor cousin to the slow oven roast, and produces smoky, fall-apart tender meat.

Modern barbecuing devices allow the

Food Words: *Barbecue*

The term *barbecue* comes via the Spanish *barbacoa* from the West Indies, and a Taino word that meant a framework of green sticks suspended on corner posts, on which meat, fish, and other foods were laid and cooked in the open over fire and coals. Both the height and the fire were adjustable, so food could either be quickly grilled or slowly smoked and dried. In American colonial times the *barbecue* was a popular and festive bout of mass outdoor meat cooking. By the beginning of the 20th century it had evolved into the familiar slow cooking of highly flavored meat.

cook to control the amount of heat and smoke produced, and facilitate periodic basting with a wide range of sauces, most of them spicy and vinegary, to intensify flavor, moisten the meat surface, and further slow the cooking. In the best devices, the wood is burned in one chamber and the meat cooked in a second connected chamber, so that there's no direct radiation from the coals, and only the relatively cool smoke (around 200°F/90°C) transfers heat, inefficiently and therefore gently. It takes several hours to bring large cuts of meat—slabs of ribs, pork shoulders and legs, beef briskets—to an internal temperature of 165–70°F/75°C, and a whole hog will take 18 hours or more. These are ideal conditions for the tenderizing of tough, inexpensive cuts.

Many barbecued meats end up with a "smoke ring," a permanent pink or red zone under the surface (p. 149).

Hot Air and Walls: Oven "Roasting"

In contrast to the grill, the oven is an indirect and more uniform means of cooking. The primary heat source, whether flame, coil, or coal, heats the oven, and the oven then heats the food from all sides, by means of convection currents of hot air and infrared radiation from the oven walls (p. 784). Oven heating is a relatively slow method, well suited to large cuts of meat that take time to heat through. Its efficiency is especially influenced by the cooking temperature, which can be anything from 200 up to 500°F/95–260°C and above. Cooking times range from 60 to 10 (or fewer) minutes per pound/500 gm.

Low Oven Temperatures At low oven temperatures, below 250°F/125°C, the moist meat surface dries very slowly. As moisture evaporates, it actually cools the surface, so despite the oven temperature, the surface temperature of the meat may be as low as 160°F/70°C. This means relatively little surface browning and long cooking times, but also very gentle heating of the interior, minimal moisture loss, a relatively uniform doneness within the meat, and a large window of time in which the meat is properly done. In addition, a slow inner temperature rise to 140°F/60°C—over the course of several hours in a large roast—allows the meat's own protein-breaking enzymes to do some tenderizing (p. 144). Ovens equipped with fans to force the hot air over the meat ("forced convection") improve surface browning at low roasting temperatures. Low-temperature roasting is suited to both tender cuts, whose moistness it preserves, and tough cuts that benefit from long cooking to dissolve collagen into gelatin.

High Oven Temperatures At high oven temperatures, 400°F/200°C and above, the meat surface quickly browns and develops the characteristic roasted flavor, and cooking times are short. On the other hand, the meat loses a lot of moisture, its outer portions end up much hotter than the center, and the center can go from done to overdone in just a few minutes.

High-temperature roasting is ideal for tender and relatively small cuts of meat that cook through quickly, and whose surface wouldn't have time to brown without the exposure to high heat.

Moderate Oven Temperatures Moderate temperatures, around 350°F/175°C, offer a compromise that produces acceptable results with many cuts of meat. So does a two-stage cooking: for example, starting the oven at a high temperature for an initial browning (or browning the meat on the stovetop in a hot pan), and then turning the thermostat down to cook the meat through more gently.

The Effects of Shielding and Basting At moderate and hot oven temperatures, the oven walls, ceiling, and floor radiate heat energy in significant amounts. This means that if an object lies between the food and one of the oven surfaces, the food will receive less heat from that direction and will cook more slowly. This shielding effect can be both a nuisance and a useful tool. The pan underneath a roast slows the heating of the roast bottom, and the cook should turn the roast periodically to make sure that top and bottom receive equal amounts of heat. But a sheet of aluminum foil deliberately placed over the meat will deflect a substantial portion of heat energy and thus slow the cooking of the whole roast. So will basting with a water-containing liquid, which cools the meat surface as it evaporates.

The Challenge of Whole Birds Chickens, turkeys, and other meat birds are difficult to roast whole, because their two kinds of meat are best cooked differently. The tender breast meat gets dry and tough if heated much above 155°F/68°C. The leg meat is full of connective tissue, and is chewy if cooked to less than 165°F/73°C. So usually the cook must choose: either the leg meat is sufficiently cooked and the breast meat dry, or the breast meat succulent and the leg meat gristly.

Cooks try to overcome this dilemma in many ways. They turn the bird in various routines to expose the thigh joint to more heat. They cover the breast with foil, or with wet cheesecloth, or strips of pork fat ("barding"), or baste it, all to slow its cooking. They cover the breast with an ice pack and let the bird sit at room temperature for an hour, so that the legs start the cooking warmer than the breast. They brine the bird to juice up its breast. Perfectionists cut the bird up and roast legs and breasts separately.

Hot Metal: Frying, or Sautéing

Simple frying, or sautéing, cooks by the direct conduction of heat energy from hot metal pan to meat, usually via a thin layer of oil that prevents the meat from sticking and conducts heat evenly across minute gaps between meat and pan. Metals are the best heat conductors known, and frying therefore cooks the meat surface rapidly. Its distinctive characteristic is the ability to brown and flavor the meat surface in a matter of seconds. This searing action requires a combination of heat source and pan that can maintain a high temperature even while the leaking meat juices are being vaporized. If the pan gets cool enough to let moisture accumulate—for example because it was insufficiently preheated, or overloaded with cold wet meat—then the meat stews in its own juices until they boil off, and its surface doesn't brown well. (The same thing will happen if the pan is covered, so that the water vapor is trapped and falls back into the pan.) The appetizing sizzle of frying meat is actually the sound of moisture from the meat being vaporized

Predicting Roasting Times

A number of different guidelines have been proposed for predicting how much time it should take to roast a given piece of meat. Minutes per inch thickness and minutes per pound are the usual approximations. However, the mathematics of heat transfer show that cooking times are actually proportional to the thickness *squared,* or to the weight *to the ⅔ power.* And the cooking time also depends on many other factors. There is no simple and accurate equation that can tell us how long to cook a particular piece of meat in our particular kitchen. The best we can do is monitor the actual cooking, and anticipate when we should stop by following the temperature rise at the center of the meat.

as it hits the hot metal pan, and cooks use this sound to judge the pan temperature. A continuous strong hiss indicates the immediate conversion of moisture to steam by a hot pan, and efficient surface browning; weak and irregular sputtering indicates that the moisture is collecting in distinct droplets, and the pan is barely hot enough to boil it off.

Because frying is a rapid cooking method, it's applied mainly to the same thin, tender cuts best suited for grilling and broiling. As with grilling, frying will be both faster and gentler if the meat starts at room temperature or above and is turned frequently (see box, p. 156). Cooks make frying even more efficient by pressing down on the meat—with the spatula or a heavy pan or brick—to improve the thermal contact between meat and pan. For thicker cuts whose insides take time to heat through, the cook slows heat transfer after the initial browning to prevent the outer portions from being overcooked. This can be done simply by lowering the burner heat, or by shifting the pan to the oven, which continues the heating from all sides and frees the cook from the necessity of turning the meat. Restaurant cooks often "finish" fried meats by putting the pan in the oven as soon as the first side has been browned and the meat turned.

Hot Oil: Shallow and Deep Frying

Fats and oils are a useful cooking medium because they can be heated to temperatures well above the boiling point of water, and can therefore dry, crisp, and brown the food surface. In shallow-fat frying, pieces of meat are cooked in enough melted fat or oil to bathe the bottom and sides of the meat; in deep-fat frying, there's enough oil to immerse the meat completely. Heat is transferred from the pan to the meat by way of convection currents in the fat or oil. These materials are less efficient than both metal and water at transferring heat, and yet more than twice as efficient as an oven. This thermal moderation, together with the ability to contact the meat evenly and intimately, makes fat frying an especially versatile technique. It's used primarily for poultry and fish, everything from thin fillets and chicken breasts to whole 15 lb/7 kg turkeys, which take something over an hour to cook (compared to two or three hours in the oven). The usual cooking temperature ranges between 300 and 350°F/150 and 175°C. The oil starts out near 350°, cools

The Keys to Crisp Skin

One of the special pleasures of a well-cooked bird is its crisp, rich skin. The skin of birds and other animals is mainly water (about 50%), fat (40%), and connective-tissue collagen (3%). In order to crisp the skin, the cook must dissolve the leathery collagen into tender gelatin in the skin's water, and then vaporize the water out of the skin. The high heat of a hot oven or frying pan does this most effectively; slow cooking at a low oven temperature can desiccate the skin while its collagen is intact, and preserve its leatheriness. A crisp skin is easier to obtain with a dry-processed bird—kosher or halal, for example—whose skin hasn't been plumped with added water (p. 143). It also helps to let the bird air-dry uncovered in the refrigerator for a day or two, and to oil the skin before roasting. (Oiling improves heat transfer from hot oven air to moist meat.) The cooked bird should be served promptly, since crisp skin quickly reabsorbs moisture from the hot meat beneath, and becomes flabby as it sits on the plate.

when the meat is introduced and its moisture begins to boil and bubble away, then heats up again as moisture flow slows and the burner heat catches up. The temperature is high enough to dehydrate, brown, and crisp the surface, while the gradual movement of heat into the meat gives the cook a reasonable window of time in which to stop the cooking while the meat is still moist.

For some purposes, meats may be partly precooked at a relatively low oil temperature, and then cooked through and browned at a higher temperature just before serving. Fast-food fried chicken is prepared in special pressure cookers (p. 785), which fry at the usual oil temperatures, but raise the boiling point of water, so that less of the moisture in the meat vaporizes during the cooking. The result is more rapid cooking (less cooling by evaporation) and moister meat.

The Searing Question

The best-known explanation of a cooking method is probably this catchy phrase: "Sear the meat to seal in the juices." The eminent German chemist Justus von Liebig came up with this idea around 1850. It was disproved a few decades later. Yet this myth lives on, even among professional cooks.

Before Liebig, most cooks in Europe cooked roasts through at some distance from the fire, or protected by a layer of greased paper, and then browned them quickly at the end. Juice retention was not a concern. But Liebig thought that the water-soluble components of meat were nutritionally important, so it was worth minimizing their loss. In his book *Researches on the Chemistry of Food,* he said that this could be done by heating the meat quickly enough that the juices are immediately sealed inside. He explained what happens when a piece of meat is plunged into boiling water, and then the temperature reduced to a simmer:

> When it is introduced into the boiling water, the albumen immediately coagulates from the surface inwards, and in this state forms a crust or shell, which no longer permits the external water to penetrate into the interior of the mass of flesh. . . . The flesh retains its juiciness, and is quite as agreeable to the taste as it can be made by roasting; for the chief part of the sapid [flavorful] constituents of the mass is retained, under these circumstances, in the flesh.

And if the crust can keep water out during boiling, it can keep the juices in during roasting, so it's best to sear the roast immediately, and then continue at a lower temperature to finish the insides.

Liebig's ideas caught on very quickly among cooks and cookbook writers, including the eminent French chef Auguste Escoffier. But simple experiments in the 1930s showed that Liebig was wrong. The crust that forms around the surface of the meat is not waterproof, as any cook has experienced: the continuing sizzle of meat in the pan or oven or on the grill is the sound of moisture continually escaping and vaporizing. In fact, moisture loss is proportional to meat temperature, so the high heat of searing actually dries out the meat surface more than moderate heat does. But searing does flavor the meat surface with products of the browning reactions (p. 777), and flavor gets our juices flowing. Liebig and his followers were wrong about meat juices, but they were right that searing makes delicious meat.

Breadings and Batters Nearly all meats that are shallow- or deep-fried are coated with a layer of dry breading or flour-based batter before they're cooked. These coatings do not "seal in" moisture. Instead, they provide a thin but critical layer of insulation that buffers the meat surface from direct contact with the oil. The coating, not the meat, quickly dries out into a pleasingly crisp surface, and forms a poorly conducting matrix of dry starch with pockets of steam or immobilized oil. Because rare meat that still exudes juice would quickly make the crisp crust soggy, oil-fried meats are generally cooked until bubbling in the oil ceases, a sign that their juices have ceased to flow.

Hot Water: Braising, Stewing, Poaching, Simmering

As a medium for cooking meat, water has several advantages. It transmits heat rapidly and evenly; its own temperature is easily adjusted to the cook's needs, and it can carry and impart flavor and become a sauce. Unlike oil, it can't get hot enough to generate browning flavors at the meat surface; but meats can be prebrowned and then finished in water-based liquids.

There are several names for the simple and versatile method of heating meat in these liquids, which may be meat or vegetable stock, milk, wine or beer, pureed fruits or vegetables. The many variations involve differences in the cooking liquid used, the size of the meat pieces, the relative proportions of meat and liquid, and initial precooking. (Braises and pot roasts involve larger cuts and less liquid than do stews.) In all of them, however, the key variable is temperature, which should be kept well below the boil, around 180°F/80°C, so that the outer portions don't overcook badly. Many slow braises and stews are cooked in a low oven, but the usual temperatures specified—325–350°F/165–175°C—are high enough that they'll eventually raise the contents of a covered pot to the boil. Unless the pot is left uncovered, which allows cooling evaporation (and concentrates and creates flavor at the liquid surface), the oven temperature should be kept below 200°F. (The original *braisier* in France was a closed pot sitting on and topped with a few live coals.)

Meats cooked in liquid should be allowed to cool in that liquid, and are best served at temperatures well below the cooking temperature, around 120°F/50°C. The capacity of the meat tissue to hold water increases as it cools, so it will actually reabsorb some of the liquid it lost during the cooking.

Tender Meats: Surprisingly Quick Cooking Hot water is such an effective heat transmitter that it cooks flat tender cuts of meat very quickly. Chops, chicken breasts, fish steaks and filets will all be

Food Words: *Poach, Simmer, Braise, Stew*

These various terms for the same basic process have wildly different origins. *Poach* is a medieval word from the French for the "pouch" of gently cooked egg white that forms around the yolk. The original 16th-century form of *simmer* was *simper,* an affected, conceited facial expression, the connection possibly being the coy blinking of the bubbles as they begin to break at the surface. *Braise* and *stew* are both 18th-century borrowings from the French, the first coming from a word for "coal," and referring to the practice of putting coals under and atop the cooking pot, the second from *étuve,* meaning stove or heated room and so a hot enclosure.

done in just a few minutes. If they're browned first in a frying pan to develop flavor, they may need only a minute or two to finish cooking through. For the most consistent results with tender meats, bring the braising liquid to the boil, add the meat to destroy surface bacteria, and after a few seconds add some cold liquid to cool the pan to 180°F/80°C, so that the outer portions of the meat won't overheat and there is a broader window of time during which the center is properly done. If the liquid needs to be boiled down to concentrate flavor or to create a thicker consistency for a sauce, remove the meat first.

Tough and Large Cuts: Slower Means Moister Meats with a significant amount of tough connective tissue must be cooked to a minimum of 160–180°F/70–80°C to dissolve their collagen into gelatin, but that temperature range is well above the 140–150°F/60–65°C at which the muscle fibers lose their juices. So it's a challenge to make tough meats succulent. The key is to cook slowly, at or just above the collagen-dissolving minimum, to minimize the drying-out of the fibers. The meat should be checked regularly and taken off the heat as soon as its fibers are easily pushed apart ("fork tender"). The connective tissue itself can help, because once dissolved, its gelatin holds onto some of the juice squeezed from the muscle fibers and thus imparts a kind of succulence to the meat. The shanks, shoulders, and cheeks of young animals are rich in collagen and so make fairly forgiving, gelatin-thickened braises.

One useful ingredient in long-cooked braises and stews can be a prolonged time—an hour or two—during which the cook carefully manages the meat's temperature rise up to the simmer. The time that the meat spends below 120°F/50°C amounts to a period of accelerated aging that weakens the connective tissue and reduces the time needed at fiber-drying temperatures. One sign that braised or

Guidelines for Succulent Braises and Stews

A moist, tender braise or stew results from the cook's cumulative attention to several details of procedure. The most important rule: never let the meat interior get anywhere near the boil.

- Keep the meat as intact as possible to minimize cut surfaces through which fluids can escape.
- If the meat must be cut, cut it into relatively large pieces, at least an inch/2.5 cm on a side.
- Brown the meat very quickly in a hot pan so that the inside of the meat warms only slightly. This kills microbes on the meat surfaces, and creates flavor.
- Start the pot with meat and cooking liquid in a cold oven, the pot lid ajar to allow some evaporation, and set the thermostat to 200°F/93°C, so that it heats the stew to around 120°F/50°C slowly, over two hours.
- Raise the oven temperature to 250°F/120°C so that the stew slowly warms from 120°F to 180°F/80°C.
- After an hour, check the meat every half hour, and stop the cooking when it is easily penetrated by the tines of a fork. Let the meat cool in the stew, where it will reabsorb some liquid.
- The liquid will probably need to be reduced by boiling to improve flavor and consistency. Remove the meat first.

stewed meat has been heated very gently and gradually is a distinct red color throughout the meat, even though it's well done: the same slow heating that allows meat enzymes to tenderize and flavor the meat also allows more of the myoglobin pigment to remain intact (p. 149).

Water Vapor: Steaming

Steaming is by far the fastest method for pouring heat into food, thanks to the large amount of energy that water vapor releases when it condenses into droplets on the food surface. However, it works rapidly only as long as the meat surface is cooler than the boiling point. Because heat moves through meat more slowly than steam deposits it on the surface, heat accumulates at the surface, which soon reaches the boiling point, and the heat transfer rate falls to a level just sufficient to keep the surface at the boil. Though it heats meat by means of moisture, steaming does not guarantee moist meat. Muscle fibers heated to the boiling point shrink and squeeze out much of their moisture, and the steamy atmosphere can't replace it.

Because steaming brings the meat surface to the boil so quickly, it's a method best suited to thin, tender cuts of meat that will cook through quickly in just a few minutes, before their outer portions become badly overcooked and dried out. Meats are often wrapped—in an edible lettuce or cabbage leaf, an inedible but flavorsome banana leaf or corn husk, or in parchment or foil—to protect the surface from the harsh steam heat and cook it more gradually. The meat must be arranged on an open rack in a single layer or else in separate tiers; any surface not exposed directly to the atmosphere inside the pot will cook much more slowly than the rest. The pot should contain enough water that it won't cook dry as steam escapes around the lid. Herbs and spices are often included in the water to aromatize the meat.

Low-Temperature Steam When steaming, the cook usually takes care to keep the lid tight on the pot and the heat high, to make sure that the pot atmosphere is saturated with vapor. However steaming can also be done at reduced temperatures and therefore more gently. Water at a 180°F/80°C simmer in a covered pot will keep the pot atmosphere around 180° as well, and leave the outer portions of the meat less overdone. In China, some dishes are steamed in open pots, where the water vapor mixes with ambient air and the temperature is well below the boil. Commercial convection steamers can produce saturated vapor all the way from body temperature to the boil. They make it possible for restaurant cooks to prepare moist meats and fish with very little attention and keep them at serving temperature until needed.

High-Pressure and Low-Pressure Cooking While conventional cooking is limited to an effective maximum temperature of the boiling point of water (p. 784), the pressure cooker allows us to raise that maximum from 212 to 250°F/100 to 120°C. It does so by tightly sealing the meat and cooking liquid in the pan and allowing the vaporizing water to build up the pressure to about double the normal air pressure at sea level. This increased pressure increases the boiling point, and high pressure and temperature put together produce an overall doubling or tripling of the heat transfer rate into the meat, as well as an extremely efficient conversion of collagen into gelatin. Pot roasts cook in less than an hour instead of two or three. Of course the proteins get very hot and therefore squeeze out much of their moisture; meat must be well endowed with fat and collagen to end up anything but dry.

At the other end of the pressure scale is cooking at high altitude, where the atmospheric pressure is significantly lower than it is at sea level. The boiling point of water is also lower (203°F/95°C at mile-high Denver, 194°F/90°C at 10,000 feet/3,000 meters), and meat cooking more gentle—and more time-consuming.

Microwave Cooking

Microwave cooking is neither dry nor a moist technique, but electromagnetic (p. 786). High-frequency radio waves generated in the oven cause electrically asymmetrical water molecules to vibrate, and these molecules in turn heat up the rest of the tissue. Because radio waves penetrate organic matter, the meat is cooked directly to a depth of an inch or so. Microwave cooking is thus very fast, but it also tends to result in greater fluid loss than conventional means. Generally, large cuts of meats "roasted" in the microwave oven get badly overcooked in the outer inch while the interior cooks through; they end up dryer and tougher than standard roasts. Since the air in the oven is not heated, microwave ovens can't brown meat surfaces unless they're assisted by special packaging or a broiling element. (An exception to this rule is cured meats like bacon, which get so dry when cooked that they can brown.)

More reliable results can be obtained in the microwave oven when the meat is immersed in some liquid, cooked in a loosely covered container, and checked carefully for signs of proper doneness. There's some evidence that microwaves are unusually effective at dissolving collagen into gelatin.

After the Cooking: Resting, Carving, and Serving

A meat dish can be cooked perfectly and yet disappoint if it's mishandled on the way to the table. Large oven roasts should be allowed to rest on the countertop for at least a half hour before carving, not only to allow the "afterheat" to finish cooking the center (p. 153), but also to allow the meat to cool down, ideally to 120°F/50°C or so. (This may take well over an hour; some chefs allow for a rest period equal to the roasting time.) As the temperature drops, the meat structure becomes firmer and more resistant to deformation, and its water-holding capacity increases. Cooling therefore makes the meat easier to carve and reduces the amount of fluid lost during carving.

Whenever possible, meat is carved across the grain of the muscle fibers to reduce the impression of fibrousness in the mouth and make the meat easier to chew. Carving knives should be kept sharp. Sawing away with a dull blade compresses the tissue and squeezes its delicious liquid away.

Finally, remember that the saturated fats of beef, lamb, and pork are solid at room temperature, which means that they rapidly congeal on the plate. Also, gelatinized collagen begins to set around body temperature and makes the meat seem noticeably stiffer. Preheated platters and plates prolong the table appeal of any hot meat dish.

Leftovers

Warmed-Over Flavor At the same time that cooking develops the characteristic flavors of meat, it also promotes chemical changes that lead to characteristic, stale, cardboard-like "warmed-over flavors" when the meat is stored and reheated. (Complex or strongly flavored dishes may actually improve with time and reheating; warmed-over flavor develops within the meat itself.) The principal source of off-flavors is unsaturated fatty acids, which are damaged by oxygen and iron from myoglobin. This damage occurs slowly in the refrigerator and more rapidly during reheating. Meats with a greater proportion of unsaturated fat in their fat tissue—poultry and pork—are more susceptible to warmed-over flavor than beef and lamb. Cured meats suffer less because their nitrite acts as an antioxidant.

There are several ways to minimize the development of off-flavors in leftovers. Season the food with herbs and spices that contain antioxidant compounds (chapter 8). Use low-permeability plastic wraps to cover the meat (saran or polyvinyl chloride; polyethylene is surprisingly permeable to oxygen), and eliminate air pockets in the package. Eat the leftovers as soon as possible, and with the minimum degree of reheating consistent with safety. Leftover roast chicken, for example, tastes fresher when served cold.

Maintaining Moistness If you've taken the trouble to cook a meat dish gently, then apply the same care to reheating: it only takes moments at the boil to dry out a good stew. Bring the liquid alone to the boil, return the meat to it so that its surfaces are exposed to the boil very briefly, and then reduce the heat and stir so that the liquid quickly comes down to 150°F/65°C. Then let the meat warm through at this gentle temperature.

Safety As a general rule, leftover meats are safest when refrigerated or frozen within two hours of the end of cooking, and reheated quickly to at least 150°F/65°C before serving a second time. To be served cold, the meat should be well cooked to begin with, refrigerated quickly, and served within a day or two, fresh out of the refrigerator. If in doubt, it's best to heat the meat thoroughly, and compensate for the adverse effects on taste and texture by shredding the meat and moistening it with a flavorful liquid.

OFFAL, OR ORGAN MEATS

Animals have muscles because they nourish themselves on other living things and must move around to find them. And they have innards—livers, kidneys, intestines, and

Composition of Organ Meats

Organ meats are generally similar to skeletal muscle in their chemical composition, but often contain substantially more iron and vitamins thanks to their special tasks. (Poultry heart and liver and veal liver are especially rich in folate, a vitamin that is associated with a significantly reduced risk of heart disease.) Their higher cholesterol levels reflect the fact that their cells are much smaller than muscle cells and therefore include proportionally more cell membrane, of which cholesterol is an essential component. The chart below lists broad ranges of nutrient content for organs of various animals. Cholesterol and iron levels are given in milligrams per 100 grams/3.6 oz; folate in micrograms per 100 grams.

Meat	Protein, %	Fat, %	Cholesterol, milligrams	Iron, milligrams	Folate, micrograms
Standard cuts	24–36	5–20	70–160	1–4	5–20
Heart	24–30	5–8	180–250	4–9	3–80
Tongue	21–26	10–21	110–190	2–5	3–8
Gizzard	25–30	3–4	190–230	4–6	50–55
Tripe, beef (stomach)	15	4	95	2	2
Liver	21–31	5–9	360–630	3–18	70–770
Sweetbreads	12–33	3–23	220–500	1–2	3
Kidney	16–26	3–6	340–800	3–12	20–100
Brain	12–13	10–16	2,000–3,100	2–3	4–6

other organs—to break down these complex foods and separate the useful building blocks from waste materials, to distribute nourishment throughout the body, and to coordinate the body's activities.

The word *meat* is used most commonly to mean the limb-moving *skeletal* muscles of animals. But skeletal muscle only accounts for about half of the animal body. The various other organs and tissues are also nutritious and offer their own diverse, often pronounced flavors and textures. The nonskeletal muscles—stomach, intestines, heart, tongue—generally contain much more connective tissue than ordinary meats—up to 3 times as much—and benefit from slow, moist cooking to dissolve the collagen. The liver contains relatively little collagen: it is an agglomeration of specialized cells held together by a network of connective tissue that, because it experiences little mechanical stress, is unusually fine and delicate. Liver is thus tender if minimally cooked, crumbly and dry if overcooked.

Unlike standard meats cut from discrete and largely sterile skeletal muscles, many organ meats carry extraneous matter. Before cooking, they're often trimmed and cleaned, then "blanched," or covered with cold water that is slowly brought to a simmer. The slow heating first washes proteins and microbes off the meat, then coagulates them and floats them to the water surface where they can be skimmed off. Blanching also moderates strong odors on the meat surface.

Liver

The liver is the biochemical powerhouse of the animal body. Most of the nutrients that the body absorbs from food go here first and are either stored or processed for distribution to other organs. All this work takes a lot of energy, and this is why the liver is dark red with fat-burning mitochondria and their cytochrome pigments. It also requires direct access of the liver cells to the blood, and accordingly there is very little connective tissue between the minute hexagonal columns of cells. It's a delicate organ that is best briefly cooked; long cooking simply dries it out. The characteristic flavor of liver has been little investigated, but seems to depend importantly on sulfur compounds (thiazoles and thiazolines), and gets stronger with prolonged cooking. Generally, both flavor and texture coarsen with age. The occasionally milky appearance of chicken livers is due to an unusual but harmless accumulation of fat, about double the amount in a normal red liver (8% instead of 4%).

Foie Gras

Of the various animal innards that cooks have put to good use, one deserves special mention, because it is in a way the ultimate meat, the epitome of animal flesh and its essential appeal. *Foie gras* is the "fat liver" of force-fed geese and ducks. It has been made and appreciated since Roman times and probably long before; the force-feeding of geese is clearly represented in Egyptian art from 2500 BCE. It's a kind of living pâté, ingeniously prepared in the growing bird before it's slaughtered. Constant overnourishment causes the normally small, lean, red organ to grow to 10 times its normal size and reach a fat content of 50 to 65%. The fat is dispersed in insensibly fine droplets within the liver cells, and creates an incomparably integrated, delicate blend of smoothness, richness, and savoriness.

Preparing Foie Gras A good-quality liver is recognized by its unblemished appearance, pale thanks to the minute fat droplets, and by its consistency. The liver tissue itself is firm but pliable (like chicken liver), while the fat is only semisolid at cool room temperature. When cool and pressed with the finger, a good foie gras will give, retain the imprint, and feel somewhat supple and unctuous, while an underfattened liver will feel elastic, hard, and wet. An overfattened, weakened liver feels soft and frankly oily.

Foie gras is at its best fresh out of the bird. Apart from its use in pâtés, it is generally prepared in two ways. One is to slice it fairly thick, briefly saute in a hot, dry pan until the surface is browned and the interior just warmed through, and serve it immediately. The sensation of warm, firm, flavorful flesh melting away between tongue and palate is unparalleled. Liver quality is especially important in this preparation, since high pan heat will release a flood of fat from an overfattened or otherwise weakened organ, and the texture is unpleasantly flabby.

A second preparation is to cook the liver whole, chill it, and slice and serve it cool. This is more forgiving of second-quality livers, and offers its own kind of lusciousness. To make a terrine, the livers are pressed gently into a container and cooked in a water bath; to prepare a torchon of foie gras, they're wrapped in a cloth and poached in stock or in duck or goose fat. Fat loss is minimized by gentle, gradual heating just to the desired doneness (from 110 to 160°F/45–70°C, lower temperatures giving a creamier texture), the liquid kept only a few degrees above the target temperature. Cooling partly solidifies the fat, which allows the terrine or torchon to be sliced cleanly, and then contributes a melting firmness to the dish's texture as it's eaten.

Skin, Cartilage, and Bones

Usually cooks don't welcome large amounts of toughening connective tissue in meat. But taken on their own, animal skin, cartilage, and bones are valuable exactly because they're mostly connective tissue and therefore full of collagen (skin also provides flavorful fat). Connective tissue has two uses. First, in long-cooked stocks, soups, and stews, it dissolves out of bones or skin to provide large quantities of gelatin and a substantial body. And second, it can be turned into a delicious dish itself, with either a succulent gelatinous texture or a crisp, crunchy one, depending on the cut and the cooking method. Long moist cooking gives tender veal ears, cheeks, and muzzle for tête de veau, or Chinese beef tendon or fatty pork skin. A briefer cooking produces crunchy or chewy cartilaginous pig's ears, snouts, and tails; and rapid frying gives crisp pork rinds.

Fat

Solid fat tissue is seldom prepared as such: instead we usually extract the fat from its storage cells, and then use it as both a cooking medium and an ingredient. There are two major exceptions to this rule. The first is *caul fat,* a thin membrane of connective tissue with a lacework of small fat deposits embedded in it. This membrane is the omentum or peritoneum, usually from the pig or sheep, which covers the organs of the abdominal cavity. Caul fat has been used at least since Roman times as a wrap to hold foods together and protect and moisten their surface while they are cooked. During the cooking, much of the fat is rendered from the membrane and the membrane itself is softened, so that it all but disappears into the food.

The second fat tissue frequently used as is is mild, soft-textured pork fat, especially the thick deposits lying immediately under the skin of belly and back. Bacon is largely fat tissue from the belly, while back fat is the preferred fat for making sausages (p. 170). Italian *lardo* is pork fat cured in salt, flavorings, and wine, eaten as is or used to flavor other dishes. In classic French cooking, pork fat is used to provide both flavor and succulence to lean meats, applied either in a thin sheet that protects the surface during roasting, or in thin splinters inserted into the meat by means of larding needles.

Rendered Fats Pure fat is rendered from fat tissue by cutting the tissue into small pieces and gently heating them. Some fat melts out of the tissue, and more is squeezed out by applying pressure. Rendered beef fat is called *tallow,* and pork fat *lard.* The fats from different animals differ in flavor and in consistency. Fats from ruminant cattle and sheep are more saturated and therefore harder than pig or bird fats (due to their rumen microbes; see p. 13); and fats stored just under the skin are less saturated and therefore softer than fats stored in the body core, because their environment is cooler. Beef suet, from around the kidneys, is the hardest culinary fat, followed by subcutaneous beef fat, then leaf lard from pig kidneys, and lard from back and belly fat. Chicken, duck, and goose fat are still less saturated and so semiliquid at room temperature.

MEAT MIXTURES

The transformation of a steer or pig into the standard roasts, steaks, and chops generates a large assortment of scraps and byproducts. These remainders have always been put to use, reassembled into everything from the "goat sausage bubbling fat and blood" that the disguised Odysseus wins in a warm-up fight before his battle with Penelope's suitors, to the Scots haggis of sheep's liver, heart, and lung stuffed into its stomach, to the modern canned mixture of ham, pork shoulder, and flavorings called Spam. Chopped or ground up, mixed with other ingredients, and pressed together, meat scraps can provide one of the heartiest parts of a meal—and even one of the most luxurious.

SAUSAGES

The word *sausage* comes from the Latin for "salt," and names a mixture of chopped meat and salt stuffed into an edible tube. Salt plays two important roles in the sausage: it controls the growth of microbes, and it dissolves one of the fiber filament proteins (myosin) out of the muscle fibers and onto the meat surfaces, where it acts as a glue to bind the pieces together. Traditionally the edible container was the animal's stomach or intestine, and fat accounted for at least a third of the mixture. Today many sausages are housed in artificial casings and contain far less fat.

There are an infinite number of variations on the sausage theme, but most of them fall into a handful of families. Sausages may be sold raw and eaten freshly cooked; they may be fermented; they may be air-dried, cooked, and/or smoked to

Early Sausage Recipes

Lucanians

Pound pepper, cumin, savory, rue, parsley, seasoning, bay berries, and *liquamen* [salted fish sauce], and mix with well-pounded flesh, grinding both together. Mix in *liquamen,* whole peppercorns, plenty of fat and pine-nuts, force into an intestine stretched thinly, and hang in smoke.

—Apicius, first few centuries CE

Liver Sausage (Esicium ex Iecore)

Grind pork or other livers after they have boiled a little. Then cut up pork belly to the amount of liver, and mix with two eggs, sufficient aged cheese, marjoram, parsley, raisins, and ground spices. When these form a mass make balls the size of a nut, wrap in caul fat, and fry in a pan with lard. They require slow and low heat.

—Platina, *De honesta voluptate et valetudine,* 1475

varying degrees in order to keep for a few days or indefinitely. The meat and fat may be chopped into discrete pieces of varying size, or they may be disintegrated, blended together, and cooked into a homogeneous mass. And the sausage may either be mostly meat and fat, or it may include a substantial proportion of other ingredients.

Fermented sausages are a form of preserved meat, and are described on p. 176.

Fresh and Cooked Sausages Fresh sausages are just that: freshly made, unfermented and uncooked, and therefore highly perishable. They should be cooked within a day or two of being made or purchased.

Cooked sausages are heated as part of their production, and can be bought and eaten without further cooking for several days, or longer if they've been partly dried or smoked. But they're often cooked again just before eating. They can be made from the usual mixture of meat and fat, or from a number of other materials that thicken on cooking. The French white sausage, *boudin blanc,* is made from various white meats bound together with milk, eggs, bread crumbs, or flour, while the black *boudin noir* contains no meat at all: it's around one-third pork fat, one-third onions, apples, or chestnuts, and one-third pork blood, which coagulates during poaching to help provide a solid matrix. Liver sausage is made by cooking a blend of finely ground liver and fat. Manufacturers often use soy protein and nonfat milk solids to help thicken and retain moisture.

Emulsified Sausages Emulsified sausages are a special kind of cooked sausage, best known in the form of frankfurters or wieners and so called for their presumed origins in Germany (Frankfurt) or Austria (Wien). Italian mortadella ("bologna") is similar. These sausages have a very fine-textured, homogeneous, tender interior, and a relatively mild flavor. They're made by combining pork, beef, or poultry with fat, salt, nitrite, flavorings, and usually additional water, and shearing the ingredients together in a large blender until they form a smooth "batter," which is similar to an emulsified sauce like mayonnaise (p. 625): the fat is evenly dispersed in small droplets, which are surrounded and stabilized by fragments of the muscle cells and by salt-dissolved muscle proteins. The temperature during blending is critical: if it rises above 60°F/16°C in a pork batter, 70°F/21°C in beef, the emulsion will be unstable and leak fat. The batter is then extruded into a casing and cooked to about 160°F/70°C. Heat coagulates the meat proteins and turns the batter into a cohesive, solid mass from which the casing can be removed. Due to their relatively high water content, around 50–55%, emulsified sausages are perishable and must be refrigerated.

Sausage Ingredients: Fat and Casings The fat for sausage making is generally pork fat from under the skin of the animal's back. Pork fat has the advantage of being relatively neutral in flavor, and back fat in particular has just the right consistency: hard enough not to melt and separate as the meat is ground or stored at warm room temperatures, but soft enough that it's not granular and pasty when eaten cool. Belly fat is softer than ideal, kidney fat and beef and lamb fat harder; poultry fats are too soft. In standard nonemulsified sausages, the 30%+ fat content helps separate the meat fragments and provides tenderness and moistness. The coarser the meat fragments, the lower the surface area that fat must lubricate, and so the less fat required for an appealing texture (as little as 15%).

Sausage casings were traditionally various parts of the animal digestive tube. Today, most "natural" casings are the thin connective-tissue layers of hog or sheep intestine, stripped of their inner lining and outer muscular layers by heat and pressure, partly dried and packed in salt until they're filled. (Beef casings include some muscle.) There are also manufactured sausage containers made from animal collagen, plant cellulose, and paper.

Cooking Fresh Sausages Since their fragmented interior guarantees a certain kind of tenderness, sausages are often cooked very casually. But they benefit from being heated as carefully as other fresh meats. Five centuries ago, Platina remarked on the need to cook liver sausage gently (see box, p. 169), and said that another sausage was called *mortadella* "because it is surely more pleasant a little raw than overcooked." Fresh sausages should be thoroughly cooked to kill microbes, but no hotter than well-done meat, or 160°F/70°C. Gentle cooking prevents the interior from reaching the boil, at which point the skin will burst and leak moisture and flavor, and which hardens the texture. Intentionally piercing the skin will release moisture throughout the cooking, but provides insurance against more disfiguring splitting toward the end.

Pâtés and Terrines

Most medieval European cookbooks offer several recipes for meat pies, in which chopped meat and fat are cooked inside a pastry crust or in a well-greased earthenware pot. Over the centuries, French cooks refined this preparation, while in other countries it survived in rustic forms. And so England has pasties and patties, France the pâté and the terrine. These last two terms are largely synonymous, though today "pâté" usually suggests a fairly uniform and fine-textured mixture based on liver, "terrine" a coarser, often patterned one. Pâtés and terrines thus span a wonderful range, from coarse, rustic massings of pork innards and head in the French pâté de campagne, to luxurious layerings of brandy-scented foie gras and truffles.

Modern pâtés and terrines often contain little fat, but traditional mixes were based on a meat to fat ratio of around 2 to 1 to give a rich, melt-in-the-mouth consistency. Pork and veal, an immature meat with relatively little tough connective tissue and an abundant producer of gelatin, are the usual main ingredients. They are ground together with the fat—usually pork for its ideal consistency—to mix protein and fat intimately. Hand chopping is less likely to heat the mixture or damage intact fat cells, which would cause more liquid fat to separate from the mix during cooking. The mix is

Pâtés and Terrines: Early Recipes

As these medieval recipes demonstrate, even early pâtés were made in pots and dishes without the pastry that originally gave them their name.

Pastez de beuf

Take good young beef, and remove all fat. Cut the lean into pieces and boil, and afterwards take to the pastry cook to be chopped, and fatten it with beef marrow.

—*Le Ménagier de Paris,* ca. 1390

Pastilli di carne

Take as much lean meat as you want and cut it up fine with small knives. Mix veal fat and spices into this meat. Wrap in crusts and bake in an oven. . . . This can even be made in a well-greased dish without a crust.

—Maestro Martino, ca. 1450

seasoned more strongly than many foods both because it's rich in flavor-binding proteins and fats, and because it's generally served cool, which reduces the aroma. The mix is placed in a mold, covered, and cooked gently in a water bath until the juices run clear and the internal temperature reaches 160°F/70°C. (Terrines of foie gras are often cooked to a much lower temperature, perhaps 120°F/55°C, especially if intact lobes are layered together; they come out rosy pink.) The proteins have coagulated into a solid matrix, trapping much of the fat in place. The pâté is then topped with a weight to compact it, and refrigerated for several days to firm and allow the flavors to blend. The cooked mixture keeps for about a week.

PRESERVED MEATS

The preservation of meat from biological spoilage has been a major challenge throughout human history. The earliest methods, which go back at least 4,000 years, were physical and chemical treatments that make meat inhospitable to microbes. Drying meat in the sun and wind or by the fire removes enough water to halt bacterial growth. A smoky fire deposits cell-killing chemicals on the meat surface. Heavy salting—with partly evaporated seawater, or rock salt, or the ashes of salt-concentrating plants—also draws vital moisture from cells. Moderate salting permits the growth of a few hardy and harmless microbes that help exclude harmful ones. Out of these crude methods to stave off spoilage have come some of our most complex and interesting foods, the dry-cured hams and fermented sausages.

The Industrial Revolution brought a new approach: preserve meat not by changing the meat itself, but by controlling its environment. Canning encloses cooked meat in a sterile container hermetically sealed against the entry of microbes. Mechanical refrigeration and freezing keep meat cold enough to slow microbial growth or suspend it altogether. And irradiation of prepackaged meat kills any microbes in the package while leaving the meat itself relatively unchanged.

DRIED MEATS: JERKY

Microbes need water to survive and grow, so one simple and ancient preservation technique has been to dry meat, originally by exposing it to the wind and sun. Nowadays, meat is dried by briefly salting it to inhibit surface microbes and then heating it in low-temperature convection ovens to remove at least two-thirds of its weight and 75% of its moisture (more than 10% moisture may allow *Penicillium* and *Aspergillus* molds to grow). Because its flavor has been concentrated and its texture is interesting, dried meat remains popular. Modern examples include American jerky, Latin American *carne seca,* Norwegian *fenalår* and southern African *biltong,* whose textures can range from chewy to brittle. Two refined versions are Italian *bresaola* and Swiss *Buendnerfleisch,* which are beef salted and sometimes flavored with wine and herbs before a slow, cool drying period of up to several months. They're served in paper-thin slices.

Freeze-Drying Freeze-drying is the technique originally used by Andean peoples to make *charqui*; they took advantage of the thin dry air to evaporate moisture from meat during sunny days and sublimate it from ice crystals during freezing nights. The result was an uncooked, honeycombed tissue that would readily reabsorb water during later cooking. In the industrial version, the meat is rapidly frozen under vacuum, then mildly heated to sublimate its water. Because this kind of desiccation doesn't cause cooking and compaction of the tissue, relatively thick pieces can be dried and reconstituted.

Salted Meats: Hams, Bacon, Corned Beef

Like drying, salting preserves meat by depriving bacteria and molds of water. The addition of salt—sodium chloride—to meat creates such a high concentration of dissolved sodium and chloride ions outside the microbes that water inside their cells is drawn out, salt is drawn in, and their cellular machinery is disrupted. The microbes either die or slow down drastically. The muscle cells too are partly dehydrated and absorb salt. Traditional cured meats, made by dry-salting or brining large cuts for several days, are about 60% moisture and 5–7% salt by weight. The resulting hams (from pig legs), bacon (from pig sides), corned beef ("corn" coming from the English word for grains, including salt grains), and similar products keep uncooked for many months.

Useful Impurities: Nitrates and Nitrites

Sodium chloride is not the only salt with an important role in salt-curing. The others were unpredictable mineral impurities in the rock, sea, and vegetable salts originally used for curing. One of these, potassium nitrate (KNO_3), was discovered during the Middle Ages and named *saltpeter* because it was found as a salt-like crystalline outgrowth on rocks. In the 16th or 17th century, it was found to brighten meat color and improve its flavor, safety, and storage life. Around 1900, German chemists discovered that during the cure certain salt-tolerant bacteria transform a small portion of the nitrate into nitrite (NO_2), and that nitrite rather than nitrate is the true active ingredient. Once this was known, producers could eliminate saltpeter from the curing mixture and replace it with much smaller doses of pure nitrite. This is now the rule except in the production of traditional dry-cured hams and bacons, where prolonged ripening benefits from the ongoing bacterial production of nitrite from nitrate.

We now know that nitrite does several important things for cured meats. It con-

---

Traditional Versions of Cured Pork

Of curing hams: This is the way to cure hams in jars or tubs. . . . Cover the bottom of the jar or tub with salt and put in a ham, skin down. Cover the whole with salt and put another ham on top, and cover this in the same manner. Be careful that meat does not touch meat. So proceed, and when you have packed all the hams, cover the top with salt so that no meat can be seen, and smooth it out even. When the hams have been in salt five days, take them all out with the salt and repack them, putting those which were on top at the bottom. . . .

After the twelfth day remove the hams, brush off the salt, and hang them for two days in the wind. On the third day wipe them off clean with a sponge and rub them with oil. Then hang them in smoke for two days, and on the third day rub them with a mixture of vinegar and oil.

Then hang them in the meat house, and neither bats nor worms will touch them.

—Cato, *On Agriculture*, 50 BCE

Bacon, to dry: Cut the Leg with a piece of the Loin (of a young Hog) then with Salt-peter, in fine Pouder and brown Sugar mix'd together, rub it well daily for 2 or 3 days, after which salt it well; so will it look red: let it lye for 6 or 8 Weeks, then hang it up (in a drying-place) to dry.

—William Salmon, *The Family Dictionary: Or, Household Companion*, London, 1710

---

tributes its own sharp, piquant flavor. It reacts in the meat to form nitric oxide (NO), which retards the development of rancid flavors in the fat by preemptively binding to the iron atom in myoglobin, thus preventing the iron from causing fat oxidation. The same iron binding produces the characteristic bright pink-red color of cured meat. Finally, nitrite suppresses the growth of various bacteria, most importantly the spores of the oxygen-intolerant bacterium that causes deadly botulism. *Clostridium botulinum* can grow inside sausages that have been insufficiently or unevenly salted; German scientists first named the poisoning it causes *Wurstvergiftung*, or sausage disease (*botulus* is Latin for sausage). Nitrite apparently inhibits important bacterial enzymes and interferes with energy production.

Nitrate and nitrite can react with other food components to form possible cancer-causing nitrosamines. This risk now appears to be small (p. 125). Nevertheless, residual nitrate and nitrite in cured meats is limited to 200 parts per million (0.02%) in the United States, and is usually well below this limit.

Sublime Hams The many months that salted meats keep turned out to transform pig flesh into some of the great foods of the world! First among them are the dry-cured hams, which go back at least to classical times. The modern versions, which include Italian prosciutto di Parma, Spanish serrano, French Bayonne, and American country hams, may be aged for a year or more. Though they can be cooked, dry-cured hams are at their best when eaten in paper-thin raw slices. With their vivid, rose-colored translucency, silken texture, and a flavor at once meaty and fruity, they are to fresh pork what long-aged cheeses are to fresh milk: a distillation, an expression of the transforming powers of salt, enzymes, and time.

The Effects of Salt In addition to protecting hams from spoilage as they mature, salt contributes to their appearance and texture. High salt concentrations cause the normally tightly bunched protein filaments in the muscle cells to separate into individual filaments, which are too small to scatter light: so the normally opaque muscle tissue becomes translucent. The same unbunching also weakens the muscle fibers, while at the same time dehydration makes the tissue denser and more concentrated: hence the close but tender texture.

The Enigma of Hams Cured Without Nitrite

Though most traditional long-cured hams are treated with saltpeter to provide a steady supply of nitrite, a few are not. The eminent prosciuttos of Parma and San Daniele are cured with sea salt only, yet somehow still develop the characteristic rosy color of nitrite-stabilized myoglobin. Sea salt does contain nitrate and nitrite impurities, but not enough to affect ham color. Recently, Japanese scientists found that the stable red pigment of these hams is not nitrosomyoglobin, and its formation seems associated with the presence of particular ripening bacteria (*Staphylococcus carnosus* and *caseolyticus*). And it may be that the absence of nitrite is one of the keys to the exceptional quality of these hams. Nitrite protects meat fats from oxidation and the development of off-flavors. But fat breakdown is also one of the sources of desirable ham flavor, and nitrite-free Parma hams have been found to contain more fruity esters than nitrite-cured Spanish and French hams.

The Alchemy of Dry-Cured Flavor Some of the muscles' biochemical machinery survives intact, in particular the enzymes that break flavorless proteins down into savory peptides and amino acids, which over the course of months may convert a third or more of the meat protein to flavor molecules. The concentration of mouth-filling, meaty glutamic acid rises ten- to twenty-fold, and as in cheese, so much of the amino acid tyrosine is freed that it may form small white crystals. In addition, the unsaturated fats in pig muscle break apart and react to form hundreds of volatile compounds, some of them characteristic of the aroma of melon (a traditional and chemically fitting accompaniment to ham!), apple, citrus, flowers, freshly cut grass, and butter. Other compounds react with the products of protein breakdown to give nutty, caramel flavors normally found only in cooked meats (concentration compensates for the subcooking temperature). In sum, the flavor of dry-cured ham is astonishingly complex and evocative.

Modern Wet-Cured Meats Salted meats continue to be popular even in the age of refrigeration, when salting is no longer essential. But because we now salt meats for taste, not to extend storage life, industrial versions are treated with milder cures, and generally must be refrigerated and/or cooked. And they're made very quickly, which means that their flavor is less complex than dry-cured meats. Industrial bacon is made by injecting brine (typically about 15% salt, 10% sugar) into the pork side with arrays of fine needles, or else cutting it into slices, then immersing the slices in a brine for 10 or 15 minutes. In either method the "maturing" period has shrunk to a few hours, and the bacon is packed the same day. Hams are injected with brine, then "tumbled" in large rotating drums for a day to massage the brine evenly through the meat and make it more supple, and finally pressed into shape, partly or fully cooked, chilled, and sold with no maturing period. For some boneless "hams," pork pieces are tumbled with salt to draw out the muscle protein myosin, which forms a sticky layer that holds the pieces together. Most corned beef is now injected with brine as well; the briskets never touch any actual salt grains.

Modern ham and bacon contain more moisture than the dry-cured versions (sometimes more than the original raw meat!) and about half the salt—3–4% instead of 5–7%. Where slices of traditional ham and bacon fry easily and retain 75% of their weight, the wetter modern versions spatter, shrink, and curl as they give up their water, and retain only a third of their initial weight.

Smoked Meats

Smoke from burning plant materials, usually wood, has helped to preserve food ever since our ancestors mastered fire. Smoke's usefulness results from its chemical complexity (p. 448). It contains many hundreds of compounds, some of which kill or inhibit the growth of microbes, some of which retard fat oxidation and the development of rancid flavors, and some of which add an appealing flavor of their own. Because smoke only affects the surface of food, it has long been used in conjunction with salting and drying—a happy combination because salted meats are especially prone to developing rancidity. American country hams and bacons are examples of smoked salted foods. Because there are now other ways to store meat, and because some smoke components are known to be health hazards (p. 449), smoke is now used less frequently as a full-strength preservative, and more often as a lightly applied flavoring.

Hot and Cold Smoking Meat can be smoked in two different ways. When *hot-smoked,* the meat is held directly above or in the same enclosure as the wood, and therefore cooks while it's smoked. This will give it a more or less firm, dry texture, depending on the temperature (usually between 130 and 180°F/55–80°C) and time involved, and can kill microbes throughout the meat, not just on the surface. (Barbecuing is a form of hot smoking; see p. 157.) When it is *cold-smoked,* the meat is held in an unheated chamber through which smoke is passed from a separate firebox. The texture of the meat, and any microbes within it, are relatively unaffected. The cold-smoking chamber may be as low as 32°F/0°C but more usually ranges between 60 and 80°F/15–25°C. Smoke vapors are deposited onto the meat surface as much as seven times faster in hot smoking; however, cold-smoked meats tend to accumulate higher concentrations of the sweet-spicy phenolic components and so may have a finer flavor. (They also tend to accumulate more possible carcinogens.) The humidity of the air also makes a difference; smoke vapors are deposited most efficiently onto moist surfaces, so "wet" smoking has a stronger effect in a shorter time.

Fermented Meats: Cured Sausages

Milk is transformed into long-keeping and flavorful cheese by removing some of its moisture, salting it, and encouraging harmless microbes to grow in and acidify it: and meat can be be treated in much the same way to the same effect. There are many different kinds of *sausage,* or re-formed masses of chopped, salted meat (p. 169). Fermented sausages are the most flavorful thanks to bacteria that break down bland proteins and fats into smaller, intensely savory and aromatic molecules.

Fermented sausages probably developed in prehistoric times from the practice of salting and drying meat scraps to preserve them. When salted scraps are squeezed together, microbe-laden surfaces end up inside the moist mass, and salt-tolerant bacteria that can grow without oxygen thrive there. For the most part, these bacteria turn out to be the same ones that can grow in salty, air-poor cheese: namely the Lactobacilli and Leuconostocs (and such relatives as the Micrococci, Pediococci, and Carnobacteria). They produce lactic and acetic acids, which lower the meat pH from 6 to 4.5–5 and make it even less hospitable to spoilage microbes. Then, as the sausage slowly dries out with time, the salt and acidity become more concentrated, and the sausage increasingly resistant to spoilage.

Southern and Northern Styles of Sausage Fermented sausages come in two general styles. One is the dry, salty, well-spiced sausage typical of the warm, dry Mediterranean. Italian salami and Spanish and Portuguese chorizos are 25–35% water, contain more than 4% salt, and can be stored at room temperature. The other style is the moister, less salty, usually smoked and/or cooked sausage typical of northern Europe, whose cool, humid climate made drying difficult. These "summer" sausages and German cervelats are 40–50% water, around 3.5% salt, and must be refrigerated. Both can be eaten uncooked.

Making Fermented Sausages These days, nitrates (Europe) or nitrites (U.S.) to suppress botulism bacteria are added to the mix of meat, fat, bacterial culture, salt, and spices, as is some sugar, at least part of which the bacteria transform into lactic acid. Fermentation lasts from 18 hours to three days, depending on temperature (60–100°F/15–38°C, with dry sausages at the low end) and sausage size, until the acidity reaches 1%, the pH 4.5–5. High-temperature fermentation tends to produce volatile acids (acetic, butanoic) with a sharp aroma, while low-temperature fermentation produces a more complex blend of nutty aldehydes and fruity esters (the traditional salami flavor). The sausage may then be cooked and/or smoked, and finally is dried for two to three weeks to the desired final moisture content. A powdery white coat of harmless molds and yeasts (species of *Penicillium, Candida, Debaromyces*) may develop on the casing during drying; these microbes contribute to flavor and prevent the growth of spoilage microbes.

Fermented sausages develop a dense, chewy texture thanks to the salt extraction of the meat proteins, their denaturation by the bacterial acids, and to the general drying of the meat mass. Their tangy, aromatic flavor comes from the bacterial acids and volatile molecules, and from fragments of protein and fat generated by enzymes from both the microbes and the meat.

Confits

In ancient times, cooks from central Asia to western Europe learned that cooked meat could be preserved by burying it under a thick, airtight seal of fat. Today the best known version is the southwest French *confit* of goose and duck legs, which became fashionable in the 19th century on the coattails of foie gras—which may in turn have been an accidental by-product of cramming geese to get the fat for unfashionable farmhouse confits! The French confit probably began as a household method for preserving pork in its own lard through the year following the autumn slaughter. The confit of goose and duck seems to have been developed by makers of salted meats around Bayonne in the 18th century, when local maize production made it economical to force-feed fowl and generate the necessary fat. In the age of canning and refrigeration, confits are still made as a convenient, long-keeping ingredient that lends its distinctive flavor to salads, stews, and soups.

The traditional French confit is made by salting pieces of meat for a day, sometimes along with herbs and spices, then drying them, immersing them in fat, and heating very gradually and gently for several hours. The meat, often still pink or red inside (p. 149), is then drained, placed in a sterilized container over an additional sprinkling of salt, the fat skimmed from any spoilage-prone meat juices, reheated, and then poured over the meat. Sealed and stored in a cool place, the confit keeps for several months, and can be reheated periodically to extend its useful life.

The small but real risk that botulism bacteria could grow in this low-oxygen environment is reduced by the second dose of salt, by storage temperatures below 40°F/4°C, and by the addition of nitrate or nitrite to the salt. Most modern versions of the confit are either canned or are refrigerated for safety and made to be eaten within a few days, so they're salted mildly, more for flavor and color than for preservation.

The flavor of a traditional confit is said

Food Words: *Confit*

These days the word *confit* is used loosely to describe just about anything cooked slowly and gently to a rich, succulent consistency: onions in olive oil, for example, or shrimp cooked and stored under clarified butter. In fact the term is a fairly inclusive one. It comes via the French verb *confire*, from the Latin *conficere,* meaning "to do, to produce, to make, to prepare." The French verb was first applied in medieval times to fruits cooked and preserved in sugar syrup or honey (hence French *confiture* and English *confection*) or in alcohol. Later it was applied to vegetables pickled in vinegar, olives in oil, various foods in salt, and meats under fat. The general sense has been to immerse a food in and often impregnate it with a substance that both flavors it and preserves it. In modern usage of the term *confit,* the connotations of immersion, impregnation, flavoring, and slow, deliberate preparation survive, while the idea of preservation—and the special flavors that develop over weeks and months—has faded away.

to improve over the course of several months. Though the cooking presumably kills bacteria and inactivates all enzymes in the meat, there will certainly be biochemical changes in the meat over time, and the fat will oxidize. A slight rancidity is part of the flavor of a traditional confit.

Canned Meats

Around 1800, a French brewer and confectioner named Nicolas Appert discovered that if he sealed food in a glass container and then heated the container in boiling water, the food would keep indefinitely without spoiling. This was the beginning of canning, a form of preservation in which the food is first isolated from air and external contamination by microbes, and then heated sufficiently to destroy any microbes already in the food. (Pasteur hadn't yet proven the existence of microbes; Appert simply observed that all "ferments" were destroyed in his process.) When done properly, canning is quite effective: canned meat a century old has been eaten without harm, if also without much pleasure. The canning of meats is almost exclusively an industrial process today, in part because it offers the cook little in the way of desirable flavors or textures.

CHAPTER 4

FISH AND SHELLFISH

Fish and shellfish are foods from the earth's other world, its vast water underworld. Dry land makes up less than a third of the planet's surface, and it's a tissue-thin home compared to the oceans, whose floor plunges as much as 7 miles below the waves. The oceans are voluminous and ancient, the "primordial soup" in which all life began, and in which the human imagination has found rich inspiration for myths of destruction and creation, of metamorphosis and rebirth. The creatures that live in this cold, dark, dense, airless place are unmatched among our food animals in their variety and their strangeness.

Our species has long nourished itself on fish and shellfish, and it built nations on them as well. The world's coastlines are dotted with massive piles of oyster and mussel shells that commemorate feasts going back 300,000 years. By 40,000 years ago the hunters of prehistoric Europe were carving salmon images and making the first hooks to catch river fish; and not long afterward, they ventured onto the ocean in boats. From the late Middle Ages on, the seagoing nations of Europe and Scandinavia exploited the Atlantic's abundant stocks of cod and herring, drying and salting them into commodities that were the foundation of their modern prosperity.

Five hundred years later, at the beginning of the 21st century, the oceans' productivity is giving out. It has been exhausted by feeding a tenfold increase in the human population, and by constant advances in fishing technology and efficiency. With the help of faster and larger ships, sonar to see into the depths, miles-long nets and lines, and the mechanization of all aspects of the harvest, we've managed to fish many important food species to the verge of commercial extinction. Formerly common fish—cod and herring, Atlantic salmon and swordfish and sole, sturgeon and shark—are increasingly rare. Others—orange roughy, Chilean sea bass, monkfish—come and go from the market, temporarily abundant until they too are overfished.

The decline in the populations of wild fish has encouraged the widespread revival and modernization of aquaculture. Fish farms are now our nearly exclusive source for freshwater fish, for Atlantic salmon, and for mussels. Many of these operations effectively spare wild populations, but others further deplete them and cause environmental damage of their own. It takes some effort these days to find and choose fish and shellfish that have been produced in environmentally responsible, sustainable ways.

Yet it's a good time to be eating from the waters. More fish of excellent quality are available more widely than ever before, and they come from all over the globe,

Brillat-Savarin on Fish

Fish are an endless source of meditation and astonishment. The varied forms of these strange creatures, their diverse means of existence, the influence upon this of the places in which they must live and breathe and move about. . . .

—*Physiology of Taste*, 1825

offering the opportunity to discover new ingredients and new pleasures. At the same time, their variety and variability make it challenging to choose and prepare them well. Fish and shellfish are more fragile and less predictable than ordinary meats. This chapter will take a close look at their special nature, and how they're best handled and prepared.

FISHERIES AND AQUACULTURE

Of all our foods, fish and shellfish are the only ones that we still harvest in significant quantities from the wild. The history of the world's fisheries is the saga of human ingenuity, bravery, hunger, and wastefulness evolving into a maw that now swallows much of the oceans' tremendous productivity. In 1883, the eminent biologist T. H. Huxley expressed his belief that "the cod fishery, the herring fishery, the pilchard fishery, the mackerel fishery, and probably all the great sea fisheries are inexhaustible; that is to say that nothing we do seriously affects the numbers of fish." Just over a century later, cod and herring stocks on both sides of the North Atlantic have collapsed, many other fish are in decline, and the U.N. Food and Agriculture Organization estimates that we are harvesting two-thirds of the major commercial fish in the world at or beyond the level at which they can sustain themselves.

In addition to dangerously depleting its target fish populations, modern fishing causes collateral damage to other species, the "bycatch" of undiscriminating nets and lines that is simply discarded, and it can damage ocean-bottom habitats. Fishing is also an unpredictable, dangerous job, subject to the uncertainties of weather and the hazards of working at sea with heavy equipment. To this highly problematic system of production, there is an increasingly important alternative: aquaculture, or fish farming, which in many parts of the world goes back thousands of years. Today in the United States, all of the rainbow trout and nearly all of the catfish sold are farmed on land in various kinds of ponds and tanks. Norway pioneered the ocean farming of Atlantic salmon in large offshore pens in the 1960s; and today more than a third of the salmon eaten in the world is farmed in Europe and North and South America. About a third of the world warm-water shrimp harvest is cultured, mainly in Asia. In all, about 70 species are now farmed worldwide.

Advantages and Drawbacks of Aquaculture

There are several distinct advantages to aquaculture. Above all, it allows the pro-

The Oceans' Silver Streams

Fish . . . may seem a mean and a base commodity; yet who will but truly take the pains and consider the sequel, I think will allow it well worth the labour. . . . The poor Hollanders chiefly by fishing at a great charge and labour in all weathers in the open sea, . . . are made so mighty, strong, and rich, as no state but Venice of twice their magnitude is so well furnished, with so many fair cities, goodly towns, strong fortresses. . . . The sea [is] the source of those silver streams of all their virtue, which hath made them now the very miracle of industry, the only pattern of perfection for these affairs . . .

—Capt. John Smith, *The Generall Historie of Virginia, New England, and the Summer Isles,* London, 1624

ducer unequaled control over the condition of the fish and the circumstances of the harvest, both of which can result in better quality in the market. Farmed fish can be carefully selected for rapid growth and other desirable characteristics, and raised to a uniform and ideal stage for eating. By adjusting water temperature and flow rate and light levels, fish can be induced to grow far more rapidly than in the wild, and a balance can be struck between energy consumption and muscle-toning exercise. Farmed fish are often fattier and so more succulent. They can be slaughtered without suffering the stress and physical damage of being hooked, netted, or dumped en masse on deck; and they can be processed and chilled immediately and cleanly, thus prolonging their period of maximum quality.

However, aquaculture is not a perfect solution to the problems of ocean fishing, and has itself created a number of serious problems. Farming in offshore pens contaminates surrounding waters with wastes, antibiotics, and unconsumed food, and allows genetically uniform fish to escape and dilute the diversity of already endangered wild populations. The feed for carnivorous and scavenger species (salmon, shrimp) is mainly protein-rich fish meal, so some aquaculture operations actually consume wild fish rather than sparing them. And very recent studies have found that some environmental toxins (PCBs, p. 184) become concentrated in fish meal and are deposited in the flesh of farmed salmon.

A less serious problem, but one that makes a difference in the kitchen, is that the combination of limited water flow, limited exercise, and artificial feeds can affect the texture and flavor of farmed fish. In taste tests, farmed trout, salmon, and catfish are perceived to be blander and softer than their wild counterparts.

Modern aquaculture is still young, and ongoing research and regulation will certainly solve some of these problems. In the meantime, the most environmentally benign products of aquaculture are freshwater fish and a few saltwater fish (sturgeon, turbot) farmed on land, and molluscs farmed on

Farmed Fish and Shellfish

These are some commonly available fish and shellfish that are being farmed on a commercial scale at the beginning of the 21st century.

Freshwater Fish	Saltwater Fish	Molluscs	Crustaceans
Carp	Salmon	Abalone	Shrimp
Tilapia	Sea Bass	Mussel	Crayfish
Catfish	Sturgeon	Oyster	
Trout (rainbow)	Trout (steelhead)	Clam	
Nile perch	Char	Scallop	
Eel	Turbot		
Striped bass (hybrid)	Mahimahi		
	Milkfish		
	Yellowtail		
	Amberjack		
	Breams		
	Fugu		
	Tuna		

seacoasts. Concerned cooks and consumers can get up-to-date information about the health of fisheries and aquacultural practices from a number of public interest groups, including the Monterey Bay Aquarium in California.

SEAFOOD AND HEALTH

Fish is good for us: this belief is one important reason for the growing consumption of seafood in the developed world. There is indeed good evidence that fish oils can contribute significantly to our long-term health. On the other hand, of all our foods, fish and shellfish are the source of the broadest range of immediate health hazards, from bacteria and viruses to parasites, pollutants, and strange toxins. Cooks and consumers should be aware of these hazards, and of how to minimize them. The simplest rule is to buy from knowledgeable seafood specialists whose stock turns over quickly, and to cook fish and shellfish promptly and thoroughly. Raw and lightly cooked preparations are delicious but carry the risk of several kinds of food-borne disease. They are best indulged in at established restaurants that have access to the best fish and the expertise to prepare it.

HEALTH BENEFITS

Like meats, fish and shellfish are good sources of protein, the B vitamins, and various minerals. Iodine and calcium are special strengths. Many fish are very lean, and so offer these nutrients along with relatively few calories. But the fat of ocean fish turns out to be especially valuable in its own right. Like other fats that are liquid at room temperature, fish fats are usually referred to as "oils."

The Benefits of Fish Oils As we'll see (p. 189), life in cold water has endowed sea creatures with fats rich in unusual, highly unsaturated *omega-3 fatty acids.* (The name means that the first kink in the long chain of carbon atoms is at the third link from the end; see p. 801.) The human body can't make these fatty acids very efficiently from other fatty acids, so our diet supplies most of them. A growing body of evidence indicates that they happen to have a number of beneficial influences on our metabolism.

One benefit is quite direct, the others indirect. Omega-3 fatty acids are essential to the development and function of the brain and the retina, and it appears that an abundance in our diet helps ensure the health of the central nervous system in infancy and throughout life. But the body also transforms omega-3 fatty acids into a special set of calming immune-system signals (eicosanoids). The immune system responds to various kinds of injuries by generating an inflammation, which kills cells in the vicinity of the injury in preparation for repairing it. But some inflammations can become self-perpetuating, and do more harm than good: most importantly, they can damage arteries and contribute to heart disease, and they can contribute to the development of some cancers. A diet rich in omega-3 fatty acids helps limit the inflammatory response, and thus lowers the incidence of heart disease and cancer. By reducing the body's readiness to form blood clots, it also lowers the incidence of stroke. And it lowers the artery-damaging form of blood cholesterol.

In sum, it looks as though a moderate and regular consumption of fatty ocean fish is good for us in several ways. Fish obtain their omega-3 fatty acids directly or indirectly from tiny oceanic plants called phytoplankton. Farmed fish generally have lower levels of the omega-3s in their formulated feed, and so less in their meat. Freshwater fish don't have access to the oceanic plankton, and so provide negligible amounts of omega-3s. However, all fish contain low amounts of cholesterol-raising saturated fats, so to the extent that they replace meat in the diet, they lower artery-damaging blood cholesterol and reduce the risk of heart disease.

Health Hazards

There are three general kinds of hazardous materials that contaminate fish and shellfish: industrial toxins, biological toxins, and disease-causing microbes and parasites.

Toxic Metals and Pollutants Because rain washes chemical pollution from the air to the ground, and rain and irrigation wash it from the ground, almost every kind of chemical produced on the planet ends up in the rivers and oceans, where they can be accumulated by fish and shellfish. Of the potentially hazardous substances found in fish, the most significant are heavy metals and organic (carbon-containing) pollutants, preeminently dioxins and polychlorinated biphenyls, or PCBs. The heavy metals, including mercury, lead, cadmium, and copper, interfere with oxygen absorption and the transmission of signals in the nervous system; they're known to cause brain damage in humans. Organic pollutants cause liver damage, cancer, and hormonal disturbances in laboratory animals, and they accumulate in body fat. Fatty coho salmon and trout in the Great Lakes carry such high levels of these pollutants that government agencies advise against eating them.

Cooking doesn't eliminate chemical toxins, and there's no direct way for consumers to know whether fish contain unhealthy levels of them. In general, they concentrate in filter-feeding shellfish like oysters, which strain suspended particles from large volumes of water, and in large predatory fish at the top of the food chain, which are long-lived and eat other creatures that accumulate toxins. In recent years, common ocean fish have been found

Fat Contents of Common Fish

Low-Fat Fish (0.5–3%)	Moderately Fatty Fish (3–7%)	High-Fat Fish (8–20%)
Cod	Anchovy	Arctic char
Flounder	Bluefish	Carp
Halibut	Catfish	Chilean sea bass
Monkfish	Salmon: pink, coho	(Patagonian toothfish)
Rockfish	Shark	Eel
Skate	Smelt	Herring
Snapper	Sole: Dover	Mackerel
Tuna: bigeye,	Striped bass	Pompano
yellowfin, skipjack	Sturgeon	Sablefish
Turbot	Swordfish	Salmon: Atlantic,
	Tilapia	king, sockeye
Escolar*	Trout	Shad
Orange roughy*	Tuna: bluefin, albacore	
Ruvettus/walu*	Whitefish	

*These fish contain oil-like wax esters (p. 187) that the human body can't digest; they therefore seem rich but are really low-fat fish.

to contain so much mercury that the U.S. Food and Drug Administration advises children and pregnant women not to eat any swordfish, shark, tilefish, and king mackerel, and to limit their overall fish consumption to 12 ounces/335 grams per week. Even tuna, currently the most popular seafood in the United States after shrimp, may join the list of fish that are best eaten only occasionally. The fish least likely to accumulate mercury and other toxins are smaller, short-lived fish from the open ocean and from farms with a controlled water supply. They include Pacific salmon and soles, common mackerel, sardines, and farmed trout, striped bass, catfish, and tilapia. Sport fishing in freshwater or near large coastal cities is more likely to land an unwholesome catch contaminated by runoff or industrial discharge.

Infectious and Toxin-Producing Microbes Seafoods carry about the same risk of bacterial infections and poisonings as other meats (p. 125). The riskiest seafoods are raw or undercooked shellfish, particularly bivalves, which trap bacteria and viruses as they filter the water for food, and which we eat digestive tract and all, sometimes raw. As early as the 19th century, public health officials connected outbreaks of cholera and typhoid fever with shellfish from polluted waters. Government monitoring of water quality and regulation of shellfish harvest and sales have greatly reduced these problems in many countries. And scrupulous restaurant owners make sure to buy shellfish for the summer raw bar from monitored sources, or from less risky cold-water sources. But lovers of raw or lightly cooked seafood should be aware of the possibility of infection.

As a general rule, infections by bacteria and parasites can be prevented by cooking seafood to a minimum of 140°F/60°C. Temperatures above 185°F/82°C are required to eliminate some viruses. Some chemical toxins produced by microbes survive cooking, and can cause food poisoning even though the microbes themselves are destroyed.

Among the most important microbes in fish and shellfish are the following:

- *Vibrio* bacteria, natural inhabitants of estuary waters that thrive in warm summer months. One species causes cholera, another a milder diarrheal disease, and a third (*V. vulnificus*), usually contracted from raw oysters and the deadliest of the seafood-related diseases, causes high fever, a drop in blood pressure, and damage to skin and flesh, and kills more than half of its victims.
- Botulism bacteria, which grow in the digestive system of unchilled fish and produce a deadly nerve toxin. Most cases of fish-borne botulism are caused by improperly cold-smoked, salt-cured, or fermented products.
- Intestinal viruses, the "Norwalk" viruses, which attack the lining of the small intestine and cause vomiting and diarrhea.
- Hepatitis viruses A and E, which can cause long-lasting liver damage.

Scombroid Poisoning Scombroid poisoning is unusual in that it is caused by a number of otherwise harmless microbes when they grow on insufficiently chilled mackerels of the genus *Scomber* and other similarly active swimmers, including tuna, mahimahi, bluefish, herring, sardine, and anchovy. Within half an hour of eating one of these contaminated fish, even fully cooked, the victim suffers from temporary headache, rash, itching, nausea, and diarrhea. The symptoms are apparently caused by a number of toxins including histamine, a substance that our cells use to signal each other in response to damage; antihistamine drugs give some relief.

Shellfish and Ciguatera Poisonings Fish and shellfish share the waters with many thousands of animal and plant species, some of which engage in nasty chemical warfare with each other. At least 60 species of one-celled algae called dinoflagellates produce defensive toxins that also poison the human digestive and nervous systems. Several of these toxins can kill.

We don't consume dinoflagellates directly, but we do eat animals that eat them. Bivalve filter feeders—mussels, clams, scallops, oysters—concentrate algal toxins in their gills and/or digestive organs, and then transmit the poisons to other shellfish—usually crabs and whelks—or to humans. Accordingly, most dinoflagellate poisonings are called "shellfish poisonings." Many countries now routinely monitor waters for the algae and shellfish for the toxins, so the greatest risk is from shellfish gathered privately.

There are several distinct types of shellfish poisoning, each caused by a different toxin and each with somewhat different symptoms (see box below), though all but one are marked by tingling, numbness, and weakness within minutes to hours after eating. Dinoflagellate toxins are not destroyed by ordinary cooking, and some actually become more toxic when heated. Suspect shellfish should therefore be avoided altogether.

Finfish generally don't accumulate toxins from algae. The exceptions are a group of tropical reef fish—barracuda, groupers, jacks, king mackerel, mahimahi, mullets, porgies, snappers, wahoo—that prey on an algae-eating snail (*cigua*) and can cause ciguatera poisoning.

Parasites Parasites are not bacteria or viruses: they're animals, from single-celled protozoa to large worms, that take up residence in one or more animal "hosts" and use them for both shelter and nourishment during parts of their life cycle. There are more than 50 that can be transmitted to people who eat fish raw or undercooked, a handful of which are relatively common, and may require surgery to remove. Thanks to their more complex biological organization, parasites are sensitive to freezing (bacteria generally aren't). So there's a simple rule for eliminating parasites in fish and shellfish: either cook the food to a minimum of 140°F/60°C, or prefreeze it. The U.S. FDA recommends freezing at –31°F/–35°C for 15 hours, or

Poisonings Caused by Toxic Algae

Type of Poisoning	Usual Regions	Usual Sources	Toxin
Diarrhetic shellfish poisoning	Japan, Europe, Canada	Mussels, scallops	Okadaic acid
Amnesic shellfish poisoning	U.S. Pacific coast, New England	Mussels, clams, Dungeness crab	Domoic acid
Neurotoxic shellfish poisoning	Gulf of Mexico, Florida	Clams, oysters	Brevetoxins
Paralytic shellfish poisoning	U.S. Pacific coast, New England	Clams, mussels, oysters, scallops, cockles	Saxitoxins
Ciguatera poisoning	Caribbean, Hawaii, South Pacific	Barracuda, grouper, snapper, other reef fish	Ciguatoxin

–10°F/–23°C for seven days, treatments that are not feasible in home freezers, which seldom dip below 0°F.

Anisakid and Cod Worms These species of *Anisakis* and *Pseudoterranova* can be an inch/2.5 centimeters or more long, with a diameter of a few human hairs. Both often cause only a harmless tingling in the throat, but they sometimes invade the lining of the stomach or small intestine and cause pain, nausea, and diarrhea. They're commonly found in herring, mackerel, cod, halibut, salmon, rockfish, and squid, and can be contracted from sushi or lightly marinated, salted, or cold-smoked preparations. Farmed salmon are much less likely to be infected than wild salmon.

Tapeworms and Flukes Larvae of the tapeworm *Diphyllobothrium latum,* which can grow in the human intestine to as long as 27 feet/9 meters, are found in freshwater fish of temperate regions worldwide. Notable among these is the whitefish, which caused many infections when home cooks made the traditional Jewish dish gefilte fish and tasted the raw mix to correct the seasoning.

More serious hazards are a number of flukes, or flatworms, which are carried by fresh- and brackish-water crayfish, crabs, and fish. They damage the human liver and lungs after being consumed in such live Asian delicacies as "jumping salad" and "drunken crabs."

Potential Carcinogens Formed During Fish Preparation Certain cooking processes transform the proteins and related molecules in meat and fish into highly reactive products that damage DNA and may thereby initiate the development of cancers (p. 124). So the rule for cooking meat also holds for cooking fish: to minimize the creation of potential carcinogens, steam, braise, and poach fish rather than grilling, broiling, or frying it. If you do use high, dry heat, then consider applying a marinade, whose moisture, acidity, and other chemical qualities reduce carcinogen production.

LIFE IN WATER AND THE SPECIAL NATURE OF FISH

As a home for living things, the earth's waters are a world apart. The house rules are very different than they are for our cattle and pigs and chickens. The adaptations of fish and shellfish to life in water are the source of their distinctive qualities as foods.

The Paleness and Tenderness of Fish Flesh

Fish owe their small, light bones, delicate connective tissue, and large, pale muscle

A Health Inconvenience: *Waxy Fish*

There's an unusual digestive consequence to eating the fish called escolar and walu (*Lepidocybium flavobrunneum* and *Ruvettus pretiotus*). They, and to a lesser extent the orange roughy, accumulate substances called "wax esters," which are an oil-like combination of a long-chain fatty acid and a long-chain alcohol. Humans lack the digestive enzymes necessary to break these molecules into their smaller, absorbable parts. The wax esters therefore pass intact and oily from the small intestine into the colon, where a sufficient quantity will cause diarrhea. Restaurants are the best place to experience these luscious fish—the flesh is as much as 20% calorie-free "oil"—because they usually limit the serving size to a tolerable amount.

masses to the fact that water is much denser than air. Fish can attain a neutral buoyancy—can be almost weightless—simply by storing some lighter-than-water oils or gas in their bodies. This means that they don't need the heavy skeletons or the tough connective tissues that land animals have developed in order to support themselves against the force of gravity.

The paleness of fish flesh results from water's buoyancy and its resistance to movement. Continuous cruising requires long-term stamina and is therefore performed by slow-twitch red fibers, well supplied with the oxygen-storing pigment myoglobin and fat for fuel (p. 132). Since cruising in buoyant water is relatively effortless, fish devote between a tenth and a third of their muscle to that task, usually a thin dark layer just under the skin. But water's resistance to movement increases exponentially with the fish's speed. This means that fish must develop very high power very quickly when accelerating. And so they devote most of their muscle mass to an emergency powerpack of fast-twitch white cells that are used only for occasional bursts of rapid movement.

In addition to red and white muscle fibers, fish in the tuna family and some others have intermediate "pink" fibers, which are white fibers modified for more continuous work with oxygen-storing pigments.

The Flavor of Fish and Shellfish

The flavors of ocean and freshwater creatures are very different. Because ocean fish breathe and swallow salty water, they had to develop a way of maintaining their body fluids at the right concentration of dissolved substances. Water in the open ocean is about 3% salt by weight, while the optimum level of dissolved minerals inside animal cells, sodium chloride included, is less than 1%. Most ocean creatures balance the saltiness of seawater by filling their cells with amino acids and their relatives the amines. The amino acid glycine is sweet; glutamic acid in the form of monosodium glutamate is savory and mouthfilling. Shellfish are especially rich in these and other tasty amino acids. Finfish contain some, but also rely on a largely tasteless amine called TMAO (trimethylamine oxide). And sharks, skates, and rays use a different substance: slightly salty and bitter urea, which is what animals generally turn protein waste into in order to excrete it. The problem with TMAO and urea is that once the fish are killed, bacteria and fish enzymes convert the former into stinky TMA (trimethylamine) and the latter into kitchen-cleanser ammonia. They're thus responsible for the powerfully bad smell of old fish.

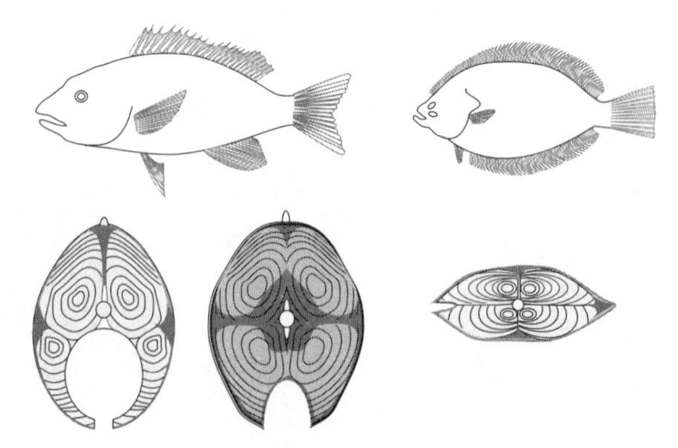

Fish muscle tissues, shown in cross-section. Below left: *Most fish swim intermittently, so their muscle mass consists mainly of fast white fibers, with isolated regions of slow red fibers.* Center: *Tuna swim more continuously and contain larger masses of dark fibers, while even their white fibers contain some myoglobin.* Right: *Soles, halibuts, and other bottom-hugging flatfish swim on their side.*

Freshwater fish are a different story. Their environment is actually less salty than their cells, so they have no need to accumulate amino acids, amines, or urea. Their flesh is therefore relatively mild, both when it's fresh and when it's old.

The Healthfulness of Fish Oils

Why should fish and not Angus steers provide the highly unsaturated fats that turn out to be good for us? Because oceanic waters are colder than pastures and barns, and most fish are cold-blooded. Throw a beefsteak in the ocean and it congeals; its cells are designed to operate at the animal's usual body temperature, around 100°F/40°C. The cell membranes and energy stores of ocean fish and the plankton they eat must remain fluid and workable at temperatures that approach 32°F/0°C. Their fatty acids are therefore very long and irregular in structure (p. 801), and don't solidify into orderly crystals until the temperature gets very low indeed.

The Perishability of Fish and Shellfish

The cold aquatic environment is also responsible for the notorious tendency of fish and shellfish to spoil faster than other meats. The cold has two different effects. First, it requires fish to rely on the highly unsaturated fatty acids that remain fluid at low temperatures: and these molecules are highly susceptible to being broken by oxygen into stale-smelling, cardboardy fragments. More importantly, cold water requires fish to have enzymes that work well in the cold, and the bacteria that live in and on the fish also thrive at low temperatures. The enzymes and bacteria typical of our warm-blooded meat animals normally work at 100°F/40°C, and are slowed to a crawl in a refrigerator at 40°F/5°C. But the same refrigerator feels perfectly balmy to deep-water fish enzymes and spoilage bacteria. And among fishes, cold-water species, especially fatty ones, spoil faster than tropical ones. Where refrigerated beef will keep and even improve for weeks, mackerel and herring remain in good condition on ice for only five days, cod and salmon for eight, trout for 15, carp and tilapia (a freshwater African native) for 20 days.

The Sensitivity and Fragility of Fish in the Pan

Most fish pose a double challenge in the kitchen. They are more easily overcooked to a dry fibrousness than ordinary meats. And even when they're perfectly done, their flesh is very fragile and tends to fall apart when moved from pan or grill to plate. The sensitivity of fish to heat is related to their perishability: muscle fibers that are specialized to work well in the cold not only spoil at lower temperatures, they become cooked at lower temperatures. The muscle proteins of ocean fish begin to unfold and coagulate at room temperature!

Though overcooked fish gets dry, it never gets tough. The fragility of cooked fish results from its relatively small amounts connective-tissue collagen, and from the low temperature at which that collagen is dissolved into gelatin.

The Unpredictability of Fish Quality

The quality of many fish and shellfish can vary drastically from season to season. This is because they live out life cycles that typically include one phase during which they grow and mature, accumulating energy reserves and reaching their peak of culinary quality, and a subsequent phase during which they expend those reserves to migrate and create masses of eggs or sperm for the next generation. And most fish don't store their reserves in layers of fat, as land animals do. Instead they use the proteins of their muscle mass as their energy pack. During migrations and spawning, they accumulate protein-digesting enzymes in their muscle and literally transform their own flesh into the next generation. Then and afterward, their muscle is meager and spent, and makes a spongy, mushy dish.

Because different fish have different cycles, and can be in different phases depending on the part of the world in which they've been caught, it's often hard to know whether a given wild fish in the market is at its prime.

THE ANATOMY AND QUALITIES OF FISH

Fish and shellfish have many things in common, but anatomy is not one of them. Fish are vertebrates, animals with backbones; shellfish are boneless invertebrates. Their muscles and organs are organized differently, and as a result they can have very different textures. The anatomy and special qualities of shellfish are described separately, beginning on p. 218.

Fish Anatomy

For about 400 million years, beginning well before reptiles or birds or mammals had even made an appearance, fish have had the same basic body plan: a streamlined bullet shape that minimizes the water's resistance to their movement. There are exceptions, but most fish can be thought of as sheets of muscle tissue anchored with connective tissue and the backbone to a propulsive tail. The animals push water behind them, developing thrust by undulations of the whole body and flexing of the tail.

Skin and Scales Fish skin consists of two layers, a thin outer epidermis and a thicker underlying dermis. A variety of gland cells in the epidermis secrete protective chemicals, the most evident of which is mucus, a proteinaceous substance much like egg white. The skin is often richer than the flesh, averaging 5–10% fat. The thick dermis layer of the skin is especially rich in connective tissue. It's generally about one-third collagen by weight, and therefore can contribute much more thickening gelatin to stocks and stews than the fish's flesh (0.3–3% collagen) or bones. Moist heating will turn the skin into a slick gelatinous sheet, while frying or grilling enough to desiccate it will make it crisp.

Scales are another evident form of protection for the fish skin. They are made up of the same hard, tough calcareous minerals as teeth, and are removed by scraping against their grain with a knife blade.

Bones The main skeleton of a small or moderate-size fish, consisting of the backbone and attached rib cage, can often be separated from the meat in one piece. However, there are usually also bones projecting into the fins, and fish in the herring, salmon, and other families have small "floating" or "pin" bones unattached to the main skeleton, which help stiffen some of the connective-tissue sheets and direct the muscular forces along them. Because fish bones are smaller, lighter, and less mineralized with calcium than land-animal bones, and because their collagen is less tough, they can be softened and even dissolved by a relatively short period near the boil (hence the high calcium content of canned salmon). Fish skeletons are even eaten on their own: in Catalonia, Japan, and India they're deep-fried until crunchy.

Fish Innards The innards of fish and shellfish offer their own special pleasures. Fish eggs are described below (p. 239). Many fish livers are prized, including those of the goatfish ("red mullet"), monkfish, mackerel, ray, and cod, as is the comparable organ in crustaceans, the hepatopancreas (p. 219). The "tongues" of cod and carp are actually throat muscles and associated connective tissue that softens with long cooking. Fish heads can be 20% fatty material and are stuffed and slow-cooked until the bones soften. And then there are "sounds," or swim bladders, balloons of connective tissue that such fish as cod, carp, catfish, and sturgeon fill with air to adjust their buoyancy. In Asia, fish sounds are dried, fried until they puff up, and slowly cooked in a savory sauce.

Fish Muscle and Its Delicate Texture

Fish have a more delicate texture than the flesh of our land animals. The reasons for this are the layered structure of fish muscle, and the sparseness and weakness of fish connective tissue.

Muscle Structure In land animals, individual muscles and muscle fibers can be quite long, on the order of several inches, and the muscles taper down at the ends into a tough tendon that connects them to bone. In fish, by contrast, muscle fibers are arranged in sheets a fraction of an inch thick ("myotomes"), and each short fiber merges into very thin layers of connective tissue ("myosepta"), which are a loose mesh of collagen fibers that run from the backbone to the skin. The muscle sheets are folded and nested in complex W-like shapes that apparently orient the fibers for greatest efficiency of force transmission to the backbone. There are about 50 muscle sheets or "flakes" along the length of a cod.

Connective Tissue Fish connective tissue is weak because its collagen contains less structure-reinforcing amino acids than beef collagen does, and because the muscle tissue also serves as an energy store that's repeatedly built up and broken down, whereas in land animals it is progressively reinforced with age. Meat collagen is tough and must be cooked for some time near the boil to be dissolved into gelatin, but in most fish it dissolves at 120 or 130°F/50–55°C, at which point the muscle layers separate into distinct flakes.

Succulence from Gelatin and Fat Both gelatin and fat can contribute an impression of moistness to fish texture. Fish with little collagen—trout, bass—seem drier when cooked than those with more—halibut, shark. Because the motion for steady swimming comes mostly from the back end of the fish, the tail region contains more connective tissue than the head end, and seems more succulent. Red muscle fibers are thinner than white fibers and require more connective tissue to join them with each other, so dark meat has a noticeably finer, more gelatinous texture.

The fat content of fish muscle runs a tremendous range, from 0.5% in cod and other white fish to 20% in well-fed herring

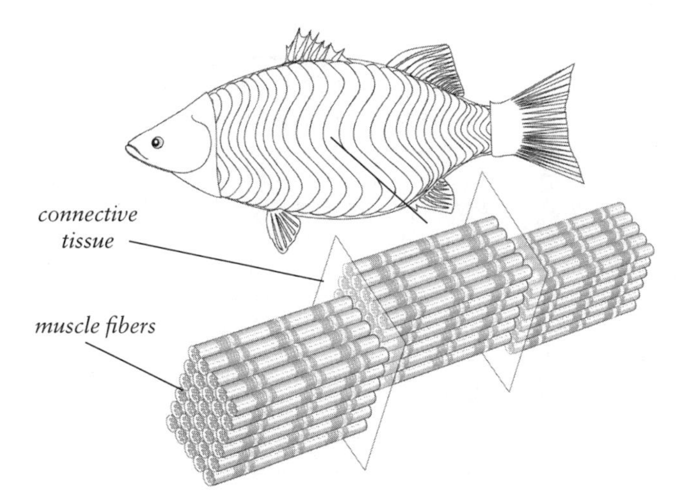

Fish anatomy. Unlike the muscles of land animals (p. 120), fish muscles are arranged in layers of short fibers, and organized and separated by sheets of connective tissue that are thin and delicate.

and their relatives (p. 184). Fat storage cells are found primarily in a distinct layer under the skin, and then in the visible sheets of connective tissue that separate the myotomes. Within a given fish, the belly region is usually the fattiest, while muscle segments get progressively leaner toward the back and tail. A center-cut salmon steak may have twice the fat content of a slice from the tail.

Softness Certain conditions can lead to fish flesh becoming unpleasantly soft. When fish flesh is depleted by migration or by spawning, their sparse muscle proteins bond to each other only very loosely, and the overall texture is soft and flabby. In extreme cases, such as "sloppy" cod or "jellied" sole, the muscle proteins are so tenuously bonded that the muscle seems almost liquefied. Some fish come out mushy when thawed after frozen storage, because freezing disrupts the cells' compartments and liberates enzymes that then attack the muscle fibers. And enzyme activity during cooking can turn firm fish mushy in the pan; see p. 211.

Fish Flavor

The flavor of fish may well be the most variable and changeable among our basic foods. It depends on the kind of fish, the salinity of its home waters, the food it eats, and the way it is harvested and handled.

Fish Taste In general, seafood is more full-tasting than meats or freshwater fish, because ocean creatures accumulate amino acids to counterbalance the salinity of seawater (p. 188). The flesh of ocean fish generally contains about the same amount of salty sodium as beef or trout, but three to ten times more free amino acids, notably sweet glycine and savory glutamate. Shellfish, sharks and rays, and members of the herring and mackerel family are especially rich in these amino acids. Because the salt content of seawater varies substantially—it's high in the open ocean, lower near river mouths—the amino-acid content and therefore taste intensity of fish varies according to the waters they're caught in.

An additional element of fish taste is contributed indirectly by the energy-carrying compound ATP (adenosine triphosphate). When a cell extracts energy from ATP, it is transformed into a series of smaller molecules, one of which, IMP (inosine monophosphate), has a savory taste similar to that of glutamate. However, IMP is a transient substance. So the savoriness of fish increases for some time after its death as IMP levels rise, then declines again as IMP disappears.

Fish Aroma

Fresh and Plant-like Few of us get the chance to enjoy the experience, but very fresh fish smell surprisingly like crushed plant leaves! The fatty materials of both plants and fish are highly unsaturated, and both leaves and fish skin have enzymes (lipoxygenases) that break these large smell-less molecules down into the same small, aromatic fragments. Nearly all fish emit fragments (8 carbon atoms long) that have a heavy green, geranium-leaf, slightly metallic smell. Freshwater fish also produce fragments that are typical of freshly cut grass (6 carbons), and earthy fragments also found in mushrooms (8 carbons). Some freshwater and migratory species, especially the smelts, produce fragments characteristic of melons and cucumbers (9 carbons).

Smell of the Seacoast Ocean fish often have an additional, characteristic aroma of the seacoast. This ocean aroma appears to be provided by compounds called bromophenols, which are synthesized by algae and some primitive animals from bromine, an abundant element in seawater. Bromophenols are propelled into the seacoast air by wave action, where we smell them directly. Fish also accumulate them, either by eating algae or by eating algae eaters, and the fish can thus remind us of the sea air. Farmed saltwater fish lack the oceanic aroma unless their artificial feed is supplemented with bromophenols.

Muddiness Freshwater fish sometimes carry an unpleasant muddy aroma. It's most often encountered in bottom-feeding fish, especially catfish and carp that are raised in ponds dug directly in the earth. The chemical culprits are two compounds that are produced by blue-green algae, especially in warm weather (geosmin and methylisoborneol). These chemicals appear to concentrate in the skin and the dark muscle tissue, which can be cut away to make the fish more palatable. Geosmin breaks down in acid conditions, so there is a good chemical reason for traditional recipes that include vinegar and other acidic ingredients.

Fishiness The moment fish are caught and killed, other aromas begin to develop. The strong smell that we readily identify as "fishy" is largely due to the saltwater-balancing compound TMAO (p. 188), which bacteria on the fish surfaces slowly break down to smelly TMA. Freshwater fish generally don't accumulate TMAO, and crustaceans accumulate relatively little, so they don't get as fishy as ocean fish. In addition, the unsaturated fats and fresh-smelling fragments (aldehydes) produced from them slowly react to produce other molecules with stale, cheesy characters, some of which accentuate the fishiness of TMA. And during frozen storage, the fish's own enzymes also convert some TMA to DMA (dimethylamine), which smells weakly of ammonia.

Fortunately, the fishiness of fish past its prime can be greatly reduced a couple of simple treatments. TMA on the surface can be rinsed off with tap water. And acidic ingredients—lemon juice, vinegar, tomatoes—help in two ways. They encourage the stale fragments to react with water and become less volatile; and they contribute a hydrogen ion to TMA and DMA, which thereby take on a positive electrical charge, bond with water and other nearby molecules, and never escape the fish surface to enter our nose.

The aromas of cooked fish are discussed on p. 208.

Flavor Compounds in Raw Fish and Shellfish

The basic flavors of fish and shellfish arise from their different combinations of taste and aroma molecules.

Source	Amino acids: sweet, savory	Salts: salty	IMP: savory	TMA: fishy	Bromophenol: sea-air	Ammonia (from urea)	Geosmin, borneol: muddy
Terrestrial meats	+	+	+	–	–	–	–
Freshwater fish	+	+	+	–	–	–	+
Saltwater fish	+++	+	+++	+++	+	–	–
Sharks and rays	+++	++	++	+++	+	+++	–
Molluscs	+++	+++	+	++	+	–	–
Crustaceans	++++	+++	+	+	+	–	–

Fish Color

Pale Translucence Most of the muscle in most raw fish is white or off-white and delicately translucent compared to raw beef or pork, whose cells are surrounded by more light-scattering connective tissue and fat cells. Especially fatty portions of fish, such as salmon and tuna bellies, look distinctly milky compared to flesh from just a few inches away. The translucence of fish muscle is turned into opacity by cooking treatments that cause the muscle proteins to unfold and bond to each other into large, light-scattering masses. Both heat and marination in acid unfold proteins and turn fish flesh opaque.

Red Tunas The meaty color of certain tunas is caused by the oxygen-storing pigment myoglobin (p. 132), which these fish need for their nonstop, high-velocity life (p. 201). Fish myoglobin is especially prone to being oxidized to brownish metmyoglobin, especially at freezer temperatures down to –22°F/–30°C; tuna must be frozen well below this to keep its color. During cooking, fish myoglobins denature and turn gray-brown at around the same temperature as beef myoglobin, between 140 and 160°F/60 and 70°C. Because they are often present in small quantities, their color change can be masked by the general milkiness caused when all the other cell proteins unfold and bond to each other. This is why fish with distinctly pink raw flesh (albacore tuna, mahimahi) will turn as white as any white fish when cooked.

Orange-Pink Salmons and Trouts The characteristic color of the salmons is due to a chemical relative of the carotene pigment that colors carrots. This compound, astaxanthin, comes from the salmons' small crustacean prey, which create it from the beta-carotene they obtain from algae. Many fish store astaxanthin in their skin and ovaries, but only the salmon family stores it in muscle. Because farmed salmon and trout don't have access to the wild crustaceans, they have paler flesh unless their feed is supplemented (usually with crustacean shell by-products or an industrially produced carotenoid called canthaxanthin).

THE FISH WE EAT

The number of different kinds of fish in the world is staggering. Of all the animals that have backbones, fish account for more than half, something approaching 29,000 species. Our species regularly eats hundreds of these. Perhaps two dozen are at least occasionally available in U.S. supermarkets, and another several dozen in upscale and ethnic restaurants, often under a variety of names. The box beginning on p. 195 surveys the family relations of some commonly eaten fish, and the paragraphs that follow provide a few details about the more important families.

Shellfish are also a diverse group of animals. They lack backbones and differ from finfish in important ways, so they're described separately, p. 218.

The Herring Family: Anchovy, Sardine, Sprat, Shad

The herring family is an ancient, successful, and highly productive one, and for centuries was the animal food on which much of northern Europe subsisted. Its various species school throughout the world's oceans in large, easily netted numbers and are relatively small, often just a few inches long but sometimes reaching 16 in/40 cm and 1.5 lb/0.75 kg.

Members of the herring family feed by constantly swimming and straining tiny zooplankton from the seawater. They thus have very active muscle and digestive enzymes that can soften their flesh and generate strong flavors soon after they're harvested. Their high fat content, upwards of 20% as they approach spawning, also makes them vulnerable to the off-flavors of easily oxidized polyunsaturated fats.

Names and Family Relations of Commonly Eaten Fishes

Closely related families are grouped together, and neighboring groups in the chart are more closely related than widely separated groups. Saltwater families are listed without special indication; "f" means a freshwater family and "f&s" a family that includes both freshwater and saltwater species.

Family	Number of Species	Examples
Shark (several)	350	Blue (*Prionace*), thresher (*Alopias*), hammerhead (*Sphyrna*), black-tipped (*Carcharinchus*), dogfish (*Squalus*), porbeagle (*Lamna*), smooth hound (*Mustelus*)
Skate	200	Skates (*Raja*)
Ray	50	Rays (*Dasyatis, Myliobatis*)
Sturgeon	24	Beluga, kaluga (*Huso*); osetra, sevruga, Atlantic, lake, green, white (all *Acipenser*)
Paddlefish (f)	2	American, Chinese paddlefish (*Polyodon, Psephurus*)
Gar	7	Gar (*Lepisosteus*)
Tarpon	2	Tarpon (*Tarpon*)
Bonefish	2	Bonefish (*Albula*)
Eel, Common (f&s)	15	European, North American, Japanese eel (all *Anguilla*)
Eel, Moray	200	Moray eel (*Muraena*)
Eel, Conger	150	Conger eel (*Conger*), pike conger eel (*Muraenesox*)
Anchovy	140	Anchovy (*Engraulis, Anchoa, Anchovia, Stolephorus*)
Herring	180	Herring (*Clupea*), sardine, pilchard (*Sardina pilchardus*); sprat (*Sprattus*), shad (*Alosa*), hilsa (*Hilsa*)
Milkfish	1	Milkfish (*Chanos*)
Carp (f)	2,000	Carp (*Cyprinus, Carassius, Hypophthalmichthys,* etc.), minnow (*Notropis, Barbus*), tench (*Tinca*)
Catfish (f)	50	North American catfish (*Ictalurus*), bullhead (*Ameirus*)
Sheatfish (f)	70	Wels (*Silurus*), Eastern European
Catfish, Sea	120	Sea catfish (*Arius, Ariopsis*)
Pike (f)	5	Pike, pickerel (*Esox*)
Smelt	13	Smelt (*Osmerus, Thaleichthys*), capelin (*Mallotus*), ayu (*Plecoglossus*)

Family	Number of Species	Examples
Salmon (s&f)	65	Salmons (*Salmo, Oncorhynchus*), trouts (*Salmo, Oncorhynchus, Salvelinus*), char (*Salvelinus*), whitefish & cisco (*Coregonus*), grayling (*Thymallus*), huchen (*Hucho*)
Lizardfish	55	Lizardfish (*Synodus*), Bombay duck (*Harpadon*)
Moonfish	2	Moonfish, opah (*Lampris*)
Cod	60	Cod (*Gadus*), haddock (*Melanogrammus*), saithe and pollock (*Pollachius*), pollack (*Pollachius, Theragra*), ling (*Molva*), whiting (*Merlangus, Merluccius*), burbot (*Lota*) (f)
Hake	20	Hake (*Merluccius, Urophycis*)
Southern Hake	7	Hoki (*Macruronus*)
Grenadier	300	Grenadier (*Coelorhynchus, Coryphaenoides*)
Goosefish	25	Monkfish (*Lophius*)
Mullet	80	Grey mullet (*Mugil*)
Silversides	160	Silversides, grunion (*Leuresthes*)
Needlefish	30	Needlefish, belone (*Belone*)
Saury	4	Saury (*Scomberesox*)
Flying Fish	50	Flying fish (*Cypselurus, Hirundichthys, Exocoetus*)
Roughies	30	Orange roughy (*Hoplostethus*)
Alfonsino	10	Alfonsino (*Beryx, Centroberyx*)
Dory	10	John Dory, St. Pierre (*Zeus*)
Oreo	10	Oreos (*Allocyttus, Neocyttus*)
Rockfish	300	Rockfish, "ocean perch," U.S. coastal "snappers" (*Sebastes*); scorpionfish (*Scorpaena*)
Searobin	90	Gurnard (*Trigla*)
Sablefish	2	"Black cod" (*Anoplopoma*)
Greenling	10	Greenling (*Hexagrammos*), "ling cod" (*Ophiodon*)
Sculpin	300	Sculpin (*Cottus, Myoxocephalus*), cabezon (*Scorpaenichthys*)
Lumpfish	30	Lumpfish (*Cyclopterus*)
Snook (f&s)	40	Nile perch, Australian barramundi (*Lates*); snook (*Centropomus*)
Bass, Temperate (f&s)	6	European sea bass (*Dicentrarchus*), American striped, white, yellow bass (all *Morone*)
Bass, Sea	450	Black sea bass (*Centropristis*), groupers (*Epinephelus, Mycteroperca*)
Sunfish (f)	30	Sunfish, bluegill (*Lepomis*); small- & large-mouth bass (*Micropterus*), crappies (*Pomoxis*)

Family	Number of Species	Examples
Perch (f)	160	Perches (*Perca*), walleye (*Stizostedion*)
Tilefish	35	Tilefish (*Lopholatilus*)
Bluefish	3	Bluefish (*Pomatomus*)
Dolphin Fish	2	Dolphin fish, mahimahi (*Coryphaena*)
Jack	150	Jack (*Caranx*), amberjack & yellowtail (*Seriola*), horse mackerel (*Trachurus*), scad (*Decapterus*), pompanos (*Trachinotus*)
Butterfish	20	Pomfrets (*Pampus, Peprilus, Stromateus*)
Snapper	200	Snappers (*Lutjanus, Ocyurus, Rhomboplites*), Hawaiian onaga (*Etelis*), uku (*Aprion*), opakapaka (*Pristipomoides*)
Porgy	100	Porgies (*Calamus, Stenotomus, Pagrus*), tai (*Pagrosomus*), sea breams (*Sparus*), dentex (*Dentex*), sheepshead (*Archosargus*)
Drum/Croaker	200	Redfish (*Sciaenops*), Atlantic croaker (*Micropogonias*)
Goatfish	60	Red mullets, rouget (*Mullus*)
Cichlid (f)	700	Tilapia (*Oreochromis = Tilapia*)
Cod Icefish	50	"Chilean sea bass" (*Dissostichus*)
Barracuda	20	Barracudas (*Sphyraena*)
Snake Mackerel	25	Escolar (*Lepidocybium*), waloo, ruvettus (*Ruvettus*)
Cutlassfish	20	Cutlassfish (*Trichiurus*)
Tuna and Mackerel	50	Tunas (*Thunnus, Euthynnus, Katsuwonus, Auxis*), Atlantic, chub mackerels (*Scomber*); Spanish, sierra, cero mackerel (*Scomberomorus*); wahoo/ono (*Acanthocybium*), bonitos (*Sarda*)
Billfish	10	Sailfish (*Istiophorus*), spearfish (*Tetrapturus*), marlin (*Makaira*), swordfish (*Xiphias*)
Flounder, Lefteye	115	Turbot (*Psetta*), brill (*Scophthalmus*)
Flounder, Righteye	90	Halibuts (*Hippoglossus, Reinhardtius*), plaice (*Pleuronectes*), flounders (*Platichthys, Pseudopleuronectes*)
Sole	120	True soles (*Solea, Pegusa*)
Puffer	120	Pufferfish, fugu (*Fugu*); blowfish (*Sphoeroides, Tetraodon*)
Sunfish	3	Mola (*Mola*)

Adapted from J. S. Nelson, *Fishes of the World,* 3d ed. (New York: Wiley, 1994).

Thanks to this fragility most of these fish are preserved by smoking, salting, or canning.

Carp and Catfish

The freshwater carp family arose in east Europe and west Asia, and is now the largest family of fish on the planet. Some of the same characteristics that have made them so successful—the ability to tolerate stagnant water, low oxygen levels, and temperatures from just above freezing to 100°F/38°C—have also made them ideal candidates for aquaculture, which China pioneered three millennia ago. Carp themselves can reach 60 lb/30 kg or more, but are generally harvested between one and three years when they weigh a few pounds. They're relatively bony fish, with a coarse texture and a low to moderate fat content.

The mostly freshwater catfish family is also well adapted to an omnivorous life in stagnant waters, and therefore to the fish farm. Its most familiar member is the North American channel catfish (*Ictalurus*), which is harvested when about 1 ft/30 cm long and 1 lb/450 gm, but can reach 4 ft/1.2 m in the wild. Catfish have the advantage over the carps of a simpler skeleton that makes it easy to produce boneless fillets; they keep well, as much as three weeks when vacuum-packed on ice. Both carp and catfish can suffer from a muddy flavor (p. 193), particularly in the heat of late summer and fall.

Salmons, Trouts, and Relatives

The salmons and trouts are among the most familiar of our food fishes—and among the most remarkable. The family is one of the oldest among the fishes, going back more than 100 million years. The salmons are carnivores that are born in freshwater, go to the sea to mature, and return to their home streams to spawn. The freshwater trouts evolved from several landlocked groups of Atlantic and Pacific salmon.

Salmons Salmon develop their muscle mass and fat stores in order to fuel their egg production and nonstop upstream migra-

Salmons and Their Characteristics

	Fat Content, %	Size, lb/kg	Major Uses
Atlantic			
Atlantic: *Salmo salar*	14	100/45; 6–12/3–5 farmed	Fresh, smoked
Pacific			
King, Chinook: *Oncorhynchus tshawytscha*	12	30+/14	Fresh, smoked
Sockeye, Red: *O. nerka*	10	8/4	Fresh, canned
Coho, Silver: *O. kisutch*	7	30/14	Fresh, canned
Chum, Dog: *O. keta*	4	10–12/4–5	Roe, pet food
Pink: *O. gorbuscha*	4	5–10/2–4	Canned
Cherry, Amago (Japan and Korea): *O. masou*	7	4–6/2–3	Fresh

tion, processes that consume nearly half of their weight and leave their flesh mushy and pale. Salmon quality is thus at its peak as the fish approach the mouth of their home river, which is where commercial fishermen take them. The stocks of Atlantic salmon have been depleted by centuries of overfishing and damage to their home rivers, so nowadays most market fish come from farms in Scandinavia and North and South America. The wild Alaska fishery is still healthy. Opinions vary on the relative qualities of wild and farmed salmon. Some professional cooks prefer the fattiness and more consistent quality of farm fish, while others prefer the stronger flavor and firmer texture of wild fish at their best.

The Atlantic and the Pacific king salmons are well supplied with moistening fat, and yet don't develop the strong flavor that similarly fatty herring and mackerel do. The distinctive salmon aroma may be due in part to the stores of pink astaxanthin pigment, which the fish accumulate from ocean crustaceans (p. 194), and which when heated gives rise to volatile molecules found in and reminiscent of fruits and flowers.

Trouts and Chars These mainly freshwater offshoots of the salmons are excellent sport fish and so have been transplanted from their home waters to lakes and streams all over the world. Their flesh lacks the salmon coloration because their diet doesn't include the pigmented ocean crustaceans. Today, the trout found in U.S. markets and restaurants are almost all farmed rainbows. On a diet of fish and animal meal and vitamins, rainbow trout take just a year from egg to mild, single-portion (0.5–1 lb/225–450 gm) fish. The Norwegians and Japanese raise exactly the same species in saltwater to produce a farmed version of the steelhead trout, which can reach 50 lb/23 kg, and has the pink-red flesh and flavor of a small Atlantic salmon. Arctic char, which can grow to 30 lb/14 kg as migratory fish, are farmed in Iceland, Canada, and elsewhere to about 4 lb/2 kg, and can be as fatty as salmon.

The Cod Family

Along with the herring and tuna families, the cod family has been one of the most important fisheries in history. Cod, haddock, hake, whiting, pollack, and pollock

Trouts, Chars, & Relatives

Trout family relations are complicated. Here's a list of the more common species and the part of the world they came from.

Common Name	Scientific Name	Original Home
Brown, salmon trout	*Salmo trutta*	Europe
Rainbow trout; Steelhead (seagoing)	*Oncorhynchus mykiss*	W. North America, Asia
Brook trout	*Salvelinus fontinalis*	E. North America
Lake trout	*Salvelinus namaycush*	N. North America
Arctic char	*Salvelinus alpinus*	N. Europe and Asia, N. North America
Whitefish	*Coregonus* species	N. Europe, North America

are medium-sized predators that stay close to the ocean bottom along the continental shelves, where they swim relatively little—and thus have relatively inactive enzyme systems and stable flavor and texture. Cod set the European standard for white fish, with its mild flavor and bright, firm, large-flaked flesh, nearly free of both red muscle and fat.

Members of the cod family mature in two to six years, and once provided about a third the tonnage of the herring-family catch. Many populations have been exhausted by intensive fishing; but the northern Pacific pollock fishery is still highly productive (it's used mostly in such prepared foods as surimi and breaded or battered frozen fish). Some cod are farmed in Norway in offshore pens.

Nile Perch and Tilapia

The mainly freshwater family of true perches are fairly minor foodfish in both Europe and North America. More prominent today are several farmed relatives that provide alternatives to scarce cod and flatfish fillets. The Nile or Lake Victoria perch can grow to 300 lb/135 kg on a diet of other fish, and is farmed in many regions of the world. The herbivorous tilapia is also a widely farmed native of Africa; it's hardy and grows well at 60–90°F/20–35°C in both fresh and brackish water. A number of different species and hybrids are sold under the name tilapia, and have different qualities. *Oreochromis nilotica* is said to have been cultured the longest and to have the best flesh. The Nile perch and tilapia are among the few freshwater fish to produce TMAO, which breaks down into fishy-smelling TMA (p. 193).

Basses

The freshwater basses and sunfish of North America are mostly sport fish, but one has become an important product of aquaculture: the hybrid striped bass, a cross between the freshwater white bass of the eastern United States and the seagoing striped bass. The hybrid grows faster than either parent, is more robust, and yields more meat, which can remain edible for up to two weeks. Compared to the wild striped bass, the hybrid has a more fragile texture and bland flavor. Occasionally muddy aroma can be reduced by removing the skin.

The ocean basses—the American striped bass and European sea bass (French *loup de mer,* Italian *branzino*) are prized for their firm, fine-flavored flesh and simple skeletons; the sea bass is now farmed in the Mediterranean and Scandinavia.

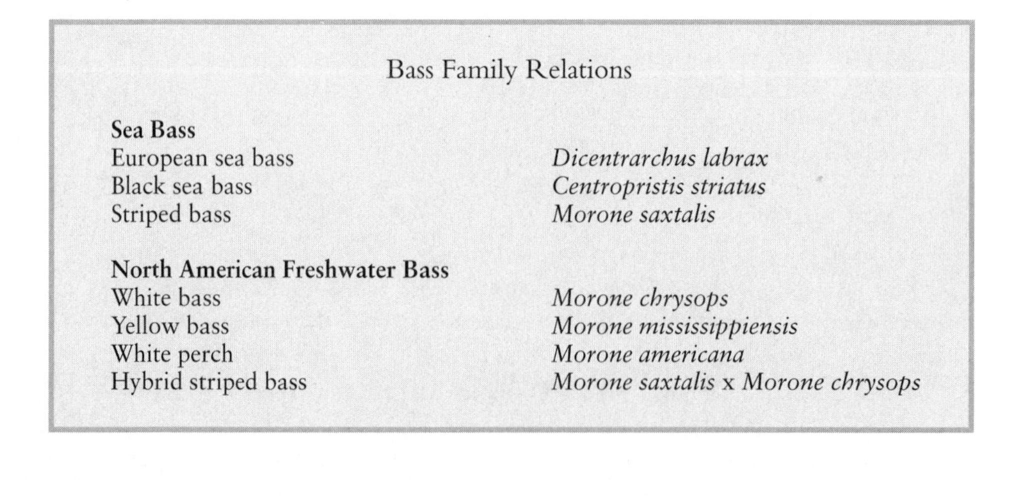

Bass Family Relations

Sea Bass	
European sea bass	*Dicentrarchus labrax*
Black sea bass	*Centropristis striatus*
Striped bass	*Morone saxtalis*
North American Freshwater Bass	
White bass	*Morone chrysops*
Yellow bass	*Morone mississippiensis*
White perch	*Morone americana*
Hybrid striped bass	*Morone saxtalis* x *Morone chrysops*

Icefish

The "cod icefish" family is a group of large, sedentary plankton-eaters that live in the cold deep waters off Antarctica. The best known of them is the fatty "Chilean sea bass," an inaccurate but more palatable commercial name for the Patagonian toothfish (*Dissostichus eleginoides*), which can reach 150 lb/70 kg. Its fat is located in a layer under the skin, in the chambered bones, and dispersed among the muscle fibers: toothfish flesh can be nearly 15% fat. It wasn't until the mid-1980s that cooks came to know and appreciate this lusciously rich, large-flaked fish, which is unusually tolerant of overcooking. Like the orange roughy and other deepwater creatures, the toothfish is slow to reproduce, and there are already signs that its numbers have been dangerously depleted by overfishing.

Tunas and Mackerel

Who would know from looking at a cheap can of tuna that it was made from one of the most remarkable fish on earth? The tunas are large predators of the open ocean, reaching 1,500 lb/680 kg and swimming constantly at speeds up to 40 miles/70 km per hour. Even their fast-twitch muscle fibers, which are normally white and bland, contribute to the nonstop cruising, and have a high capacity for using oxygen, a high content of oxygen-storing myoglobin pigment, and active enzymes for generating energy from both fat and protein. This is why tuna flesh can look as dark red as beef, and has a similarly rich, savory flavor. The meaty aroma of cooked and canned tuna comes in part from a reaction between the sugar ribose and the sulfur-containing amino acid cysteine, probably from the myoglobin pigment, which produces an aroma compound that's also typical of cooked beef.

Tuna has been the subject of connoisseurship at least since classical times. Pliny tells us that the Romans prized the fatty belly (the modern Italian *ventresca*) and neck the most, as do the Japanese today. Tuna belly, or *toro,* can have ten times the fat content of the back muscle on the same fish, and commands a large premium for its velvety texture. Because the bluefin and bigeye tunas live longest, grow largest, and prefer deep, cold waters, they accumulate more fat for fuel and insulation than other species, and their meat can fetch hundreds of dollars per pound.

The Tuna Family

These major oceangoing tuna species are found worldwide.

Common Name	Scientific Name	Abundance	Size	Fat Content, %
Bluefin	*Thunnus thynnus* (northern); *T. maccoyii* (southern)	very rare	to 1500 lb/675 kg	15
Bigeye, ahi	*T. obesus*	rare	20–200 lb/9–90 kg	8
Yellowfin, ahi	*T. albacares*	abundant	3–200 lb/1–90 kg	2
Albacore	*T. alalunga*	abundant	20–45 lb/9–20 kg	7
Skipjack	*Katsuwonus pelamis*	abundant	4–40 lb/2–20 kg	2.5

These days, most tuna are harvested in the Pacific and Indian oceans. By far the largest catches are of skipjack and yellowfin tuna, small and medium-sized lean fish that reproduce rapidly and can be netted in schools near the surface. They also provide most of the world's canned tuna, with the solitary light-fleshed albacore (Hawaiian tombo) giving "white" tuna. (Italian canned tuna is often made from the darker, stronger bluefin and from the dark portions of skipjack.)

Mackerels The mackerels are small relatives of the tunas. The mackerel proper is a native of the North Atlantic and Mediterranean, typically 18 inches/45cm long and 1–2 lb/0.5–1 kg. Like the tuna, it's an energetic predator, with a large complement of red fibers, active enzymes, and an assertive flavor. It is usually netted in large numbers and sold whole, and deteriorates rapidly unless immediately and thoroughly iced.

Swordfish

The billfish are a family of large (to 13 ft/4 m and 2,000 lb/900 kg), active predators of the open oceans, with a spear-like projection from their upper jaw and dense, meaty, nearly boneless flesh that has been sought after for thousands of years. The preeminent billfish is the swordfish, whose Atlantic stock is thought to be down to less than a tenth of its original size and in need of protection. Swordfish have a dense, meaty texture and keep unusually well on ice, as long as three weeks.

Flatfish: Soles, Turbot, Halibuts, Flounders

Flatfish are bottom-dwelling fish whose bodies have been compressed from the sides into a bottom-hugging shape. Most flatfish are relatively sedentary, and therefore are only modestly endowed with the enzyme systems that generate energy for the fish and flavor for us. Their mild flesh generally keeps well for several days after harvest.

The most prized flatfish is Dover or English sole, the principal member of a family found mainly in European waters (lesser U.S. flatfish are often misleadingly called sole). It has a fine-textured, succulent flesh said to be best two or three days after harvest, a trait that makes it an ideal fish for air-shipping to distant markets. The other eminent flatfish, the turbot, is a more active hunter. It can be double the size of the sole, with a firmer flesh that is said to be sweetest in a freshly killed fish. Thanks to their ability to absorb some oxygen through the skin, small turbot are farmed in Europe and shipped live in cold, moist containers to restaurants worldwide.

The halibut is the largest of the flatfish and a voracious hunter. The Atlantic and Pacific halibuts (both species of *Hippoglossus*) can reach 10 ft/3 m and 650 lb/300 kg, and their firm, lean flesh is said to retain good quality for a week or more. The distantly related "Greenland halibut" is softer and fattier, and the small "California halibut" is actually a flounder.

FROM THE WATERS TO THE KITCHEN

The quality of the fish we cook is largely determined by how it is harvested and handled by fishermen, wholesalers, and retail markets.

The Harvest

As we've seen, fish and shellfish are a more delicate and sensitive material than meat. They're the animal equivalent of ripe fruit, and ideally they would be handled with corresponding care. The reality is otherwise. In a slaughterhouse it's possible to kill each animal in a controlled way, minimize the physical stress and fear that adversely affect meat quality, and process the carcass immediately, before it begins to deteriorate. The fisherman has no such mastery over the circumstances of the catch, though the fish farmer has some.

Harvest from the Ocean There are several common ways of harvesting fish from the wild, none of them ideal. In the most controlled and least efficient method, a few fisherman catch a few fish, ice them immediately, and deliver them to shore within hours. This method can produce very fresh and high-quality fish—*if* they are caught quickly with minimal struggle, expertly killed and cleaned, quickly and thoroughly iced, and promptly delivered to market. But if the fish are exhausted, processing is less than ideal, or cold storage is interrupted, quality will suffer. Far more common are fish caught and processed by the thousands and delivered to port every few days or weeks. Their quality often suffers from physical damage caused by the sheer mass of the catch, delays in processing, and storage in less than ideal conditions. Factory-scale trawlers and longliners also harvest huge numbers of fish, but they do their own processing on board, and often clean, vacuum-pack, and freeze their catch within hours. Such fish can be superior in quality to unfrozen fish caught locally and recently but handled carelessly.

Harvest in Aquaculture By contrast to the logistical challenge posed by fishing, consider the care with which salmon are harvested in the best aquaculture operations. First, the fish are starved for seven to ten days to reduce the levels of bacteria and digestive enzymes in the gut that may otherwise accelerate spoilage. The fish are

Flatfish Family Relations

There are many flatfish, and even more names for them; this list includes only the more common. The names are often misleading: American waters don't harbor true soles; some halibuts aren't halibuts or turbots turbots.

True European soles	
Dover, English sole	*Solea solea*
French sole	*Pegusa lascaris*
Other European flatfish	
Turbot	*Psetta maxima*
Atlantic halibut	*Hippoglossus hippoglossus*
Plaice	*Pleuronectes platessa*
Flounder	*Platichthys flesus*
West Atlantic flatfish	
Halibut	*Hippoglossus hippoglossus*
Winter, common flounder, lemon sole	*Pseudopleuronectes americanus*
Summer flounder	*Paralichthys dentatus*
Greenland halibut or turbot	*Reinhardtius hippoglossoides*
East Pacific flatfish	
Petrale sole	*Eopsetta jordani*
Rex sole	*Glyptocephalus zachirus*
Pacific sand dab	*Citharichthys sordidus*
Pacific halibut	*Hippoglossus stenolepsis*
California halibut	*Paralichthys californicus*

anesthetized in chilled water saturated with carbon dioxide, then killed either with a blow to the head or by bleeding with a cut through the blood vessels of the gill and tail. Because the blood contains both enzymes and reactive hemoglobin iron, bleeding improves the fish's flavor, texture, color, and market life. Workers then clean the fish while it's still cold, and may wrap it in plastic to protect it from direct contact with ice or air.

The Effects of Rigor Mortis and Time

We sometimes eat fish and shellfish very fresh indeed, just minutes or hours after their death, and before they pass through the chemical and physical changes of rigor mortis (p. 143). This stiffening of the muscles may begin immediately after death in a fish already depleted by struggling, or many hours later in a fat-farmed salmon. It "resolves" after a few hours or days when the muscle fibers begin to separate from each other and from the connective-tissue sheets. Fish and shellfish cooked and eaten before rigor has set in are therefore somewhat chewier than those that have passed through rigor. Some Japanese enjoy slices of raw fish that are so fresh that they're still twitching (*ikizukuri*); Norwegians prize cod held in tanks at the market and killed to order just before cooking (*blodfersk*, or "blood-fresh"); Chinese restaurants often have tanks of live fish at the ready; the French prepare freshly killed "blue" trout; and many shellfish are cooked alive.

In general, delaying and extending the period of rigor will slow the eventual deterioration of texture and flavor. This can be done by icing most fish immediately after harvest, before rigor sets in. However, early icing can actually toughen some fish—sardine, mackerel, and warm-water fish such as tilapia—by disrupting their contraction control system. Fish are generally at their prime just when rigor has passed, perhaps 8 to 24 hours after death, and begin to deteriorate soon after that.

Recognizing Fresh Fish

Nowadays, consumers often have no idea where a given piece of fish in the market has come from, when and how it was harvested, how long it has been in transit, or how it has been handled. So it's important to be able to recognize good-quality fish when we see it. But looks and smell can be deceiving. Even perfectly fresh fish may not be of the best quality if it has been caught in a depleted state after spawning. So the ideal solution is to find a knowledgeable and reliable fish merchant who knows the seasonality of fish quality, and buys accordingly. Such a merchant is also more likely to be selective about his suppliers, and less likely to sell seafood that's past its prime.

It's preferable to have fillets and steaks

Handling Freshly Killed Fish

Sport fishermen may not get around to cooking their catch until it has already begun to stiffen. Fortunately, fish in rigor aren't as tough as beef or pork would be. It's a mistake, however, to cut up a freshly killed, pre-rigor fish into steaks or fillets, and not either cook or freeze the pieces immediately. If rigor develops in the pieces, the severed muscle fibers are free to contract, and they will shorten by as much as half into a corrugated, rubbery mass. If instead the pieces are quickly frozen, and then allowed to thaw gradually so that the muscle energy stores slowly run down while the piece shapes are maintained by some ice crystals, this contraction can be mostly avoided.

cut to order from a whole fish, because cutting immediately exposes new surfaces to microbes and the air. Old cut surfaces will be stale and smelly.

In the case of a whole fish:

- The skin should be glossy and taut. On less fresh fish it will be dull and wrinkled. Color is not a helpful guide because many skin colors fade quickly after the fish dies.
- If present, the natural proteinaceous mucus covering the skin should be transparent and glossy. With time it dries out and dulls, the proteins coagulate to give a milky appearance, and the color goes from off-white to yellow to brown. The mucus is often washed off when the fish is cleaned.
- The eyes should be bright, black, and convex. With time the transparent surface becomes opaque and gray and the orb flattens out.
- The belly of an intact fish should not be swollen or soft or broken, all signs that digestive enzymes and bacteria have eaten through the gut into the abdominal cavity and muscle. In a dressed fish, all traces of the viscera should have been removed, including the long red kidney that runs along the backbone.

If the fish has already been cut up, then:

- The steaks and fillets should have a full, glossy appearance. With time, the surfaces dry out and the proteins coagulate into a dull film. There should be no brown edges, which are a sign of drying, oxidation of oils, and off-flavors.
- Whether the fish is precut or whole, its odor should resemble fresh sea air or crushed green leaves, and be only slightly fishy. Strong fishiness comes from prolonged bacterial activity. More advanced age and spoilage are indicated by musty, stale, fruity, sulfurous, or rotten odors.

Storing Fresh Fish and Shellfish: Refrigeration and Freezing

Once we've obtained good fish, the challenge is to keep it in good condition until we use it. The initial stages of inevitable deterioration are caused by fish enzymes and oxygen, which conspire to dull colors, turn flavor stale and flat, and soften the texture. They don't really make the fish inedible. That change is caused by microbes, especially bacteria, with which fish slime and gills come well stocked—particularly *Pseudomonas* and its cold-tolerant ilk. They make fish inedible in a fraction of the time they take to spoil beef or pork, by consuming the savory free amino acids and then proteins and turning them into obnoxious nitrogen-containing substances (ammonia, trimethylamine, indole, skatole, putrescine, cadaverine) and sulfur compounds (hydrogen sulfide, skunky methanethiol).

The first defense against incipient spoilage is rinsing. Bacteria live and do their damage on the fish surface, and thor-

Shellfish That Glow in the Dark

Some ocean bacteria (species of *Photobacterium* and *Vibrio*) produce light by way of a particular chemical reaction that releases photons, and can cause shrimp and crab to glow in the dark! So far, these luminescent bacteria appear to be harmless to humans, though some can cause disease in the crustaceans. Their glow indicates that the crustaceans are laden with bacteria and thus not pristinely fresh.

ough washing can remove most of them and their smelly by-products. Once the fish is washed and blotted dry, a close wrapping in wax paper or plastic film will limit exposure to oxygen.

But by far the most important defense against spoilage is temperature control. The colder the fish, the slower enzymes and bacteria do their damage.

Refrigeration: The Importance of Ice For most of the foods that we want to store fresh for a few days, the ordinary refrigerator is quite adequate. The exception to the rule is fresh fish, whose enzymes and microbes are accustomed to cold waters (p. 189). The key to maintaining the quality of fresh fish is ice. Fish lasts nearly twice as long in a 32°F/0°C slush as it does at typical refrigerator temperatures of 40–45°F/5–7°C. It's desirable to keep fish on ice as continuously as possible: in the market display case, the shopping cart, the car, and in the refrigerator. Fine flake or chopped ice will make more even contact than larger cubes or slabs. Wrapping will prevent direct contact with water that leaches away flavor.

In general, well iced fatty saltwater fish—salmon, herring, mackerel, sardine—will remain edible for about a week, lean cold-water fish—cod, sole, tuna, trout—about two weeks, and lean warm-water fish—snappers, catfish, carp, tilapia, mullets—about three weeks. A large portion of these ice-lives may already have elapsed before the fish appear in the market.

Freezing To keep fish in edible condition for more than a few days, it's necessary to lower its temperature below the freezing point. This effectively stops spoilage by bacteria, but it doesn't stop chemical changes in the fish tissues that produce stale flavors. And the proteins in fish muscle (especially cod and its relatives) turn out to be unusually susceptible to "freeze denaturation," in which the loss of their normal environment of liquid water breaks some of the bonds holding the proteins in their intricately folded structure. The unfolded proteins are then free to bond to each other. The result is tough, spongy network that can't hold onto its moisture when it's cooked, and in the mouth becomes a dry, fibrous wad of protein.

So once you've brought frozen fish home, it's best to use it as soon as possible. In general, the storage life of fish in ordinary freezers, wrapped tightly and/or glazed with water to prevent freezer burn (freeze the fish, then dip in water, refreeze, and repeat to build up a protective ice layer) is about four months for fatty fish such as salmon, six months for most lean white fish and shrimp. Like frozen meats, frozen fish should be thawed in the refrigerator or in a bath of ice water (p. 147).

IRRADIATION

Irradiation preserves food by way of high-energy particles that damage the DNA and proteins of spoilage microbes (p. 782). Pilot studies have found that irradiation can extend the refrigerated shelf life of fresh fish by as much as two weeks. However, the initial deterioration of fish quality is caused by the action of fish enzymes and oxygen, and this action proceeds despite irradiation. Also, irradiation can produce off-flavors of its own. It's unclear whether irradiation will become an important means of preservation for fish.

UNHEATED PREPARATIONS OF FISH AND SHELLFISH

People in many parts of the world enjoy eating ocean fish and shellfish raw. Unlike meats, fish have the advantage of relatively tender muscle and a naturally savory taste, and are easier and more interesting to eat raw. They offer the experience of a kind of primal freshness. The cook may simply provide a few accompanying ingredients with complementary flavors and textures, or firm the fish's texture by means of light acidification (*ceviche*), salting (*poke*), or both (anchovies briefly cured in salt and lemon juice). And raw preparations don't require the use of fuel, which is often scarce on islands and coastlines.

All uncooked fresh fish pose the risk of carrying a number of microbes and parasites that can cause food poisoning or infection (p. 185). Only very fresh fish of the highest quality should be prepared for consumption raw, and they should be handled very carefully in the kitchen to avoid contamination by other foods. Because parasitic worms are often found in otherwise high-quality fish, the U.S. Food Code specifies that fish sold for raw consumption should be frozen throughout for a minimum of 15 hours at –31°F/–35°C, or for seven days at –4°F/–20°C. The exceptions to this rule are the tuna species commonly served in Japanese sushi and sashimi (bluefin, yellowfin, bigeye, albacore), which are rarely infected with parasites. Despite this exception, most tuna are blast-frozen at sea so that the boats can stay out for several days at a time. Sushi connoisseurs say that the texture of properly frozen tuna is acceptable, but that the flavor suffers.

Sushi and Sashimi

Probably the commonest form of raw fish is sushi, whose popularity spread remarkably in the late 20th century from its home in Japan. The original sushi seems to have been the fermented preparation *narezushi* (p. 235); *sushi* means "salted" and now applies more to the flavored rice, not the fish. The familiar bite-sized morsels of raw fish and lightly salted and acidified rice are *nigiri* sushi, meaning "grasped" or "squeezed," since the rice portion is usually molded by hand. The mass-produced version of sushi found in supermarkets is formed by industrial robots.

Sushi chefs take great care to avoid contamination of the fish. They use a solution of cold water and chlorine bleach to clean surfaces between preparations, and they change cleaning solutions and cloths frequently during service.

Tart Ceviche and Kinilaw

Ceviche is an ancient dish from the northern coast of South America, in which small cubes or thin slices of raw fish are "cooked" by immersing them in citrus juice or another acidic liquid, usually with onion, chilli peppers, and other seasonings. This period of marination changes both the appearance and texture of the fish: in a thin surface layer if it lasts 15–45 minutes, throughout if it lasts a few hours. The high acidity denatures and coagulates the proteins in the muscle tissue, so that the gel-like translucent tissue becomes opaque and firm: but more delicately than it does when heated, and with none of the flavor changes caused by high temperatures.

Kinilaw is the indigenous Philippine version of acid marination. Morsels of fish or shellfish are dipped for only a few seconds into an acidic liquid, often vinegar made from the coconut, nipa palm, or sugarcane, to which condiments have been added. In the case of "jumping salad," tiny shrimp or crabs are sprinkled with salt, doused with lime juice, and eaten alive and moving.

Salty Poke and Lomi

To the world's repertoire of raw fish dishes, the Hawaiian islands have contributed *poke* ("slice," "cut") and *lomi* ("rub," "press," "squeeze"). These are small pieces of tuna, marlin, and other fish, coated with salt for varying periods (until the fish stiffens, if it's to be kept for some time), and mixed with other flavorful ingredients, traditionally seaweed and roasted candlenuts. Lomi is unusual in that the piece of fish is first worked between the thumb and fingers before salting, to break some of the muscle sheets and fibers apart from each other and soften the texture.

COOKING FISH AND SHELLFISH

The muscle tissues of fish and shellfish react to heat much as beef and pork do, becoming opaque, firm, and more flavorful. However, fish and shellfish are distinctive in a few important ways, above all in the delicacy and activity of their proteins. They therefore pose some special challenges to the cook who wants to obtain a tender, succulent texture. Shellfish in turn have some special qualities of their own; they're described beginning on p. 218.

If it's more important to produce the safest possible dish than the most delicious one, then the task is simpler: cook all fish and shellfish to an internal temperature between 185°F/83°C and the boil. This will kill both bacteria and viruses.

How Heat Transforms Raw Fish

Heat and Fish Flavor The mild flavor of raw fish gets stronger and more complex as its temperature rises during cooking. At first, moderate heat speeds the activity of muscle enzymes, which generate more amino acids and reinforce the sweet-savory taste, and the volatile aroma compounds already present become more volatile and more noticeable. As the fish cooks through, its taste becomes somewhat muted as amino acids and IMP combine with other molecules, while the aroma grows yet stronger and more complex as fatty-acid fragments, oxygen, amino acids, and other substances react with each other to produce a host of new volatile molecules. If the surface temperature exceeds the boiling point, as it does during grilling and frying, the Maillard reactions produce typical roasted, browned aromas (p. 778).

Shellfish have their own distinctive cooked flavors (pp. 221, 225). Cooked fish fall into four broad flavor families.

- Saltwater white fish are the mildest.
- Freshwater white fish have a stronger aroma thanks to their larger repertoire of fatty-acid fragments and traces of earthiness from ponds and tanks. Freshwater trout have characteristic sweet and mushroomy aromas.
- Salmon and sea-run trout, thanks to the carotenoid pigments that they accumulate from ocean crustaceans, develop fruity, flowery aromas and a distinctive family note (from an oxygen-containing carbon ring).
- Tuna, mackerel, and their relatives have a meaty, beefy aroma.

Fishiness and How to Fight It The house-permeating "fishy" aroma of cooked fish appears to involve a group of volatile molecules formed by fatty-acid fragments reacting with TMAO (p. 193). Japanese scientists have found that certain ingredients help reduce the odor, apparently by limiting fatty-acid oxidation or preemptively reacting with TMAO: these include green tea and such aromatics as onion, bay, sage, clove, ginger, and cinnamon, which may also mask the fishy smell with their own. Acidity—whether in a poaching liquid, or in a buttermilk dip before frying—also mutes the volatility of fishy amines

Preparing Fish in Ancient Rome

In summer in their lower rooms they often had clear fresh water run in open channels underneath, in which there were a lot of live fish, which the guests would select and catch in their hands to be prepared to the taste of each. Fish has always had this privilege, as it still does, that the great have pretensions of knowing how to prepare it. Indeed its taste is much more exquisite than that of flesh, at least to me.

—Michel de Montaigne, "Of Ancient Customs," ca. 1580

and aldehydes, and helps break down muddy-smelling geosmin that farmed freshwater fish (catfish, carp) sometimes accumulate from blue-green algae.

Simple physical treatments can also minimize fishy odors. Start with very fresh fish and wash it well to remove oxidized fats and bacteria-generated amines from the surface. Enclose the fish in a covered pan, or pastry crust, or parchment or foil envelope, or poaching liquid, to reduce the exposure of its surface to the air; frying, broiling, and baking all propel fishy vapors into the kitchen. And let the fish cool down to some extent before removing it from its enclosure; this will reduce the volatility of the vapors that do escape.

Heat and Fish Texture The real challenge in cooking both fish and meat is to get the texture right. And the key to fish and meat texture is the transformation of muscle proteins (p. 149). The cook's challenge is to control the process of coagulation so that it doesn't proceed too far, to the point that the muscle fibers become hard and the juice flow dries up completely.

Target Temperatures In meat cooking, the critical temperature is 140°F/60°C, when the connective-tissue collagen sheath around each muscle cell collapses, shrinks, and puts the squeeze on the fluid-filled insides, forcing juice out of the meat. But fish collagen doesn't play the same critical role, because its squeezing power is relatively weak and it collapses before coagulation and fluid flow are well underway. Instead, it's mainly the fiber protein myosin and its coagulation that determine fish texture. Fish myosin and its fellow fiber proteins are more sensitive to heat than their land-animal counterparts. Where meats begin to shrink from coagulation and major fluid loss at 140°F/60°C and are dry by 160°F/70°C, most fish shrink at 120°F/50°C and begin to become dry around 140°F/60°C. (Compare the behaviors of meat and fish proteins in the boxes on pp. 152 and 210).

In general, fish and shellfish are firm but still moist when cooked to 130–140°F/55–60°C. Some dense-fleshed fish, including tuna and salmon, are especially succulent at 120°F, when still slightly translucent and jelly-like. Creatures with a large proportion of connective-tissue collagen—

Why Some Fish Seem to Dry Out Faster Than Others

One puzzling aspect of fish cooking is the fact that different fish can have surprisingly different tolerances for overcooking, despite similar protein and fat contents. Rockfish, snappers, and mahimahi, for example, seem more moist and forgiving than tuna or swordfish, which tend to become firm and dry very quickly. Japanese researchers have peered through the microscope and identified the likely culprits: the enzymes and other proteins in muscle cells that are not locked in the contracting fibrils, but float free in the cell to perform other functions. These proteins generally coagulate at a higher temperature than the main contractile protein myosin. So when myosin coagulates and squeezes cell fluids out, these other proteins flow out with the fluid. Some of them then coagulate in the spaces between the muscle cells, where they glue the cells together and prevent them from sliding easily apart when we chew. Highly active swimmers like tunas and billfish require more enzymes than sedentary bottom fish like snappers and cod, so their fibers get glued more firmly to each other if they are cooked to 130°F/55°C and above.

The Effects of Heat on Fish Proteins and Texture

Fish Temperature	Fish Qualities	Fiber-Weakening Enzymes	Fiber Proteins	Connective-Tissue Collagen	Protein-Bound Water
70°F 20°C	•Soft to touch •Slick, smooth •Translucent	Active	Beginning to unfold	Beginning to weaken	Beginning to escape
. . .					
100°F 40°C	•Soft to touch •Slick, smooth •Translucent •Wet surface	Active	Myosin begins to denature, coagulate	Collagen sheaths shrink, rupture	Escape accelerates; leaks from cells
110°F 45°C	•Begins to shrink •Becoming firmer •Becoming opaque •Exudes juice				
120°F 50°C	•Continues to shrink •Resilient •Less slick, more fibrous •Opaque •Exudes juice when cut or chewed	Very active	Myosin coagulated	Thick myocommata sheets begin to shrink, rupture	Leakage at maximum
130°F 55°C	•Sheets of muscle begin to separate •Becoming flaky	Most denature, become inactive	Other cell proteins denature, coagulate		
140°F 60°C	•Continues to shrink •Firm •Fibrous •Fragile •Little free juice	Some become very active and may badly fragment muscle fibers		Sheath collagen dissolves into gelatin	Leakage ceasing
150°F 65°C	•Getting progressively more firm, dry, flaky, fragile		Heat-resistant enzymes denature, coagulate	Thick myocommata sheets dissolve into gelatin	
160°F 70°C	•Stiff •Dry		Actin denatures, coagulates		
170°F 75°C		All now denatured, inactive			
180°F 80°C	•Stiffness at maximum				
190°F 85°C					
200°F 90°C	•Fibers begin to disintegrate				

notably the cartilagenous sharks and skates—benefit from higher temperatures and longer cooking to turn it into gelatin, and can be chewy unless cooked to 140°F/60°C or higher. Some molluscs are also rich in collagen and benefit from long cooking (p. 225).

Gentle Heat and Close Attention In practice, it's all too easy to overshoot the ideal temperature range for fish. It takes only a matter of seconds to overcook a thin fillet. Two characteristics of fish add to the trickiness of cooking them well. First, whole fish and fillets are thick at the center and taper down to nothing at the edges: so thin areas overcook while the thick areas cook through. And second, fish vary widely in their chemical and physical condition, and therefore in their response to heat. The fillets of cod, bluefish, and other species often suffer from some degree of gaping, separations of muscle layers through which heat penetrates more rapidly. Such fish as tuna, swordfish, and shark have very dense flesh, crammed full of protein (around 25%), which absorbs a lot of heat before its temperature rises; less active members of the cod family get by with less protein (15–16%) in their muscle, and cook more rapidly. Fat transfers heat more slowly than protein, so fatty fish take longer to cook than lean fish of the same size. And the very same species of fish can be protein- or fat-rich one month, depleted and quickly heated the next.

There are several ways to work around these inherent obstacles and uncertainties:

- Cook the fish through with the gentlest possible heat, so that the outer portions aren't badly overcooked. Oven baking and poaching well below the boil are two good ways to do this, after an initial and brief high-temperature treatment to brown and/or sterilize the surfaces.
- Compensate for uneven thickness by cutting slashes in the thick areas every 1–2 cm. This effectively divides the thick areas into smaller portions and allows heat to penetrate more rapidly. Another strategy for relatively large portions is to cover thin areas loosely with aluminum foil, which blocks radiant heat and slows their cooking.
- Check the fish early and often for doneness. Simple formulas—10 minutes to the inch is a popular one—and past experience can get you in the vicinity of the correct time, but there's no substitute for checking the particular piece. This can be done by measuring the internal temperature with a reliable thermometer, peering into a small incision to see whether the interior is still translucent or already opaque, pulling on a small bone to see whether the connective tissue has dissolved enough to release it, or pushing a small skewer or toothpick into the flesh to see whether it encounters resistance from coagulated muscle fibers.

Why Careful Cooking Sometimes Makes Fish Mushy Slow and gentle heating has an important place in meat cooking, and some fish—Atlantic salmon, for example—can develop an almost custard-like texture if heated gently to 120°F /50°C. In fish cooking, however, slow cooking can sometimes produce an unpleasant, mushy texture. This is caused by protein-digesting enzymes in the muscle cells of active fish and shellfish that help convert muscle mass into energy (p. 189). Some of these enzymes become increasingly active as the temperature rises during cooking, until they're inactivated at 130–140°F/55–60°C. Mush-prone fish (see box, p. 212) are best either cooked quickly to an enzyme-killing but somewhat drying 160°F/70°C, or else cooked to a lower temperature and served immediately.

Preparations for Cooking

Cleaning and Cutting Most fish in U.S. markets are sold precleaned and precut.

This is certainly convenient, but it also means that the scaled and cut surfaces have been exposed to the air and bacteria for hours or days, drying out and developing off-flavors. Preparing fish at the last minute can give fresher results. Both whole fish and pieces should be rinsed thoroughly in cold water to remove fragments of inner organs, the accumulation of odorous TMA, other bacterial by-products, and bacteria themselves.

Presalting Japanese cooks briefly presalt most fish and shrimp to remove surface moisture and odor and firm the outer layers. This is especially useful for getting fish skin to crisp and brown quickly when fried. As is true for meats, presoaking fish and shellfish in a 3–5% salt brine will cause the flesh to absorb both water and salt, with moisturizing and tenderizing results (p. 155).

Techniques for Cooking Fish and Shellfish

The many methods for heating meats and fish are described in detail in the previous chapter, pp. 156–65. Briefly, "dry" heating methods—grilling, frying, baking—produce surface temperatures high enough to produce the colors and flavors of the browning reactions, while "moist" techniques—steaming, poaching—fail to trigger browning, but heat foods more rapidly and can supply flavors from other ingredients. (Chinese cooks often get the best of both methods by first frying a fish and then finishing it with a brief braise in a flavorful sauce.) Fish don't require long cooking to dissolve their connective tissue and become tender. The purpose of any given technique is to get the center of the fish promptly to the proper temperature without overcooking the outer portions.

Handling Delicate Flesh Its delicate and sparse connective tissue means that most cooked fish is troublesomely fragile to work with. It's best to manipulate fish as little as possible during and after cooking, and to support the whole piece when moving it, small ones with a spatula, large ones on a rack or a stretcher of foil or cheesecloth. Neat individual portions should be cut before cooking, when the tissue is still cohesive; after cooking, even a sharp knife pulls flakes and shreds from the weakened matrix.

Grilling and Broiling Grilling and broiling are high-temperature techniques that cook mainly by radiant heat, and are well suited to relatively thin whole fish, fillets, and steaks. For successful results, the thickness of the fish and the distance from the heat must be balanced so that the fish can be cooked through at the center without

Mush-Prone Fish and Shellfish

Japanese studies have found that the following fish and shellfish have especially active protein-digesting enzymes in their muscle, and tend to become mushy when cooked slowly or held at temperatures around 130–140°F/55–60°C.

Sardine	Chum salmon	Shrimp
Herring	Whiting	Lobster
Mackerel	Pollack	
Tunas	Tilapia	

the outer portions becoming badly overcooked and dry. The fish must either be firm enough to hold together when turned with a spatula—tuna, swordfish, and halibut do well—or be supported in a closed wire rack that can be turned without disturbing the fish. Thin fillets of sole and other flatfish are sometimes put on a preheated buttered plate or aromatic cedar board and broiled without turning.

Baking Oven baking is a versatile method for cooking fish. Because it transfers heat to the fish mainly by hot air, which is an inefficient method (p. 784), it's relatively slow and gentle, and makes it easier to avoid overcooking. This is true as long as the container remains open to the oven air, when the fish moisture evaporates and cools the surface to well below the thermostat temperature. If the container is closed, water vapor builds up inside and the fish quickly steams rather than bakes. The dry oven air is also useful for concentrating the fish juices and any moist flavoring ingredients—wine, or a bed of aromatic vegetables, for example—and it can also trigger aroma-producing browning reactions.

Low-Temperature Baking In one extreme version of baking, the oven is set for temperatures as low as 200 or 225°F/95–110°C, and the cooking is gentle indeed. Because the fish surface is simultaneously warmed by the oven air and cooled by evaporation of its moisture, the actual maximum temperature of the fish surface in such an oven may be just 120–130°F, the internal temperature even lower, and the fish ends up with a barely cooked, almost custard-like texture. The appearance of fish cooked this way is often marred by the off-white globs of solidified cell fluid, which is able to leak out of the tissue before it gets hot enough for its dissolved proteins to coagulate (normally these proteins, which constitute as much as 25% of the total, coagulate within the muscle).

High-Temperature Baking At the other extreme, a very hot oven is often used in restaurant kitchens to finish cooking through a portion of fish whose skin side has been browned in a hot frying pan; pan and fish together are then slipped into the oven, and the fish cooks through in a few minutes with heat from all directions, without the necessity of turning it. A 500°F/260°C oven can also be used to "oven fry" pieces of fish that have been breaded, spread out on a baking sheet, and moistened with oil.

Cooking Under Wraps: Crusts, Envelopes, and Others An ancient way of cooking fish is to enclose it in a layer of some material—clay, coarse salt, leaves—to shield it from direct heat, and then cook the whole package (see box below). The fish inside will be more evenly and gently cooked, though checking the temperature is still essential to avoid overcooking. Showy preparations with an edible crust of pastry

Roman Fish in Parchment

Stuffed Bonito

Bone the bonito. Pound together pennyroyal, cumin, pepper, mint, nuts, and honey. Stuff the fish with this mixture and sew it up. Wrap the fish in paper and place it in a covered pan over steam. Season with oil, reduced wine, and fermented fish paste.

—Apicius, first few centuries CE

or brioche (French *en croûte*) are baked in the oven. A more versatile technique is the use of a thin envelope of parchment (*en papillote*), or aluminum foil, or a leaf, either neutral (lettuce) or flavorful (cabbage, fig, banana, lotus, *hoja santa*), which can be used with almost any heat source, from grill to steamer. But once the contents get hot enough, nearly all the heating is done by the juices of the fish and vegetables themselves, which surround the food and steam it. The envelope can be served intact and opened by the diner, releasing aromas that would otherwise have been left behind in the kitchen.

Frying Fish is fried in hot metal pans in two different ways: with just enough oil to lubricate the fish surface in contact with the pan, or with enough oil to surround and cover most or all of the fish. Either way, the fish is exposed to temperatures sufficient to dry out and brown its surfaces, and therefore develops a contrastingly crisp outside and characteristic, rich aroma. Because high heat also makes the lean flesh fibrous and chewy, fish to be fried is often given a protective coating of starchy and/or proteinaceous material, so that the coating can crisp while the fish remains moist. Common coatings include flour and flour-based batters; cornmeal or breadcrumbs; ground spices or nuts or shredded coconut; thin shreds, strings, or sheets of potato or another starchy root (sometimes cut and arranged to look like fish scales); and rice paper. The adhesion of coating to fish can be improved by first lightly salting the fish, which draws some protein-rich, sticky fluid to the surface.

Frying is also an excellent way to crisp the skin on a whole fish or fillet. The skin will dry out more rapidly and thoroughly if it's first salted to remove moisture.

Fried surfaces stay crispest when they're exposed to the air; confined between the moist fish and plate, a crisp skin or coating soon reabsorbs moisture and softens. Serve crunchy-skinned fillets skin-side up, or at least give the skin room to breathe.

Sautéing When frying in a small amount of oil, it's best to heat the pan before adding the oil (this reduces oil breakdown into sticky polymers), or lightly oil the fish surfaces instead. If an especially crisp skin or crust is desired, the fish should be started on that side, pressed gently to maximize contact between hot pan and skin, and left long enough on high heat to develop the desired texture, then turned once and allowed to finish cooking through on lower heat. Thin fillets cook in just a few minutes per side, and require a hotter pan in order to brown quickly.

Deep Frying In deep frying, the fish is usually protected with a batter or breading, and more or less immersed in oil, a relatively inefficient conductor of heat, at a temperature around 350°F/175°C, well above the boiling point of water. The surface dries out and gets hot enough to brown and to develop a characteristic rich aroma and a crisp crust that acts as a layer of insulation and slows subsequent heating. The fish is therefore heated evenly from all directions, but fairly gently, giving the cook some leeway in removing it while it's still moist inside.

Japanese Tempura The classic Japanese version of fried fish is fish *tempura,* a preparation and term that were borrowed in the late 16th century from Portuguese and Spanish missionaries who cooked fish during fasting seasons (*tempora* means "period of time"). Tempura—which now means a batter-fried food of any sort—is characterized by relatively small pieces that cook in just a few minutes, and a fresh, barely mixed batter made from an egg yolk, 1 cup/120 gm flour, and 1 cup/250 ml ice water stirred together with chopsticks just before the frying. As in all batters, cold water makes the mixture more viscous and thus better retained on the fish surface. The freshness of the batter means that the flour particles have little time to soak up water, so the moisture is rapidly removed from their surfaces during frying to produce a crisp crust. And the minimal mixing means an uneven batter consistency and therefore an uneven, lacy coating on the fish, rather than a monolithic sheet.

Simmering, Poaching, Stewing Immersing fish in hot liquid is a simple, flexible method that offers the cook unmatched control over the heating. The liquid can be very hot for cooking thin pieces in a matter of seconds, moderately hot for thicker pieces, or start out cold for gently cooking a whole fish through; it can be flavored in many different ways; and it can be turned into a sauce. When a fish or shellfish is served in a generous quantity of its cooking liquid, however supplemented with other ingredients, the French fittingly call it a preparation *à la nage,* or "aswim."

The Cooking Liquids Because fish don't require prolonged cooking, there's little time for fish and cooking liquid to exchange flavors and mellow together. Cooking liquids for fish are therefore either fairly neutral and discarded—salted water, or a mixture of water and milk—or are prepared ahead of time to develop their flavor. In the French tradition, there are two classic liquids for poaching fish: a tart, light infusion of vegetables and herbs, and a richer stock made from fish and vegetables.

Court bouillon, or "briefly boiled liquid," is a mixture of water, salt, wine or vinegar, and vegetable aromatics, cooked together for 30–60 minutes into a medium that will lightly flavor the fish. The vegetables soften and release flavor more rapidly if the acid ingredient is added toward the end; black or white pepper is also added in the last 10 minutes to avoid overextraction of its bitter components. A whole fish poached in court bouillon will contribute both flavor and gelatin to the liquid, which can then be boiled down to a succulent sauce, or else kept as a fish stock and used later.

Fish stocks, or *fumets* (from the French for "aroma"), are also generally prepared in an hour or less, since longer simmering of fragile fish bones can dissolve calcium salts that then cloud the liquid and give it a chalky taste. Stocks are made with fish bones, skins, trimmings, and heads, which are an especially rich source of gelatin and flavor. (Gills are omitted because their flavor deteriorates quickly.) The higher the proportion of fish, the more flavorful the stock; equal weights of water and fish work well (e.g. 2 lb/1 kg per quart/liter). The pot is left uncovered to prevent accidental boiling and clouding, and to allow slow evaporation and concentration. To make a clear consommé, the resulting strained stock can be clarified with a whipped mixture of egg whites and pureed raw fish, whose massed proteins trap the tiny protein particles that cloud the liquid (p. 601) into a solid, easily removed mass.

Fish are also poached in a variety of other liquids, including oil, butter, and such emulsions as beurre blanc and beurre monté (p. 632). These offer the advantage

Fish Aspics

Ordinary fish consommés are seldom concentrated enough in gelatin to set into the firm, stable gel of an aspic (p. 607). For giving a glossy, aspic-like coating to a cold fish preparation, cooks may supplement their simple consommé with a small amount of commercial gelatin, or cook a second batch of fish in the consommé. Fish gelatin melts at a lower temperature than pig and beef gelatin—around 77°F/25°C, instead of 86°F/30°C—so a true fish aspic melts more readily in the mouth, seems more delicate, and releases its flavor faster.

of slower, more gentle heat conduction and a more stable temperature thanks to reduced evaporative cooling.

Poaching Temperatures The great advantage of poaching fish is the ease of controlling the heat to obtain a moist, succulent result. Moderate-sized fillets and steaks should be started in liquid just below the boil and so hot enough to kill surface microbes instantly. The pot should then be taken off the heat, cool liquid added to bring its temperature down more quickly to around 150–160°F/65–70°C, and the fish cooked through gently. Allowing the cooked fish to cool while immersed in its liquid will leave it moister, since a hot piece of fish exposed to the air evaporates its surface moisture away.

Poaching at the Table Fish and shellfish cook so quickly in hot liquid that some cooks make poaching part of the presentation at the table. Pour steaming consommé into a bowl containing raw scallops or small cubes of fish, and the diner can witness their instant opacification and savor the evolution of their texture.

Soups and Stews; Bouillabaisse Fish stews and soups are dishes in which small pieces of fish, sometimes several different fish, are served in their cooking liquid, often with vegetables. The basic rules for simmering apply. The soup or stew base is prepared ahead of time, and the fish pieces added at the end and cooked just long enough to heat through: thick and dense pieces first, thin and delicate last. Combinations of fish and shellfish are a nice acknowledgment of the sea's bounteous variety.

A gentle simmer is usually preferred to a rolling boil so as to avoid breaking up delicate morsels. A partial exception to this rule is the *bouillabaisse* of southern France, whose name includes the idea of boiling, and whose unique character depends on the vigorous agitation that boiling provides. A bouillabaisse starts with a stock made from scraps and small bony fish to provide gelatin and flavor, tomatoes and aromatics for flavor and color, and a large dollop of olive oil—perhaps a third of a cup/75 ml per quart/liter of liquid—which a fierce 10-minute boil emulsifies into fine droplets throughout the soup. The dissolved fish gelatin and suspended proteins coat the oil droplets and slow their coalescence (p. 628). The other pieces of fish are added last and simmered to cook through, and the soup is served immediately, before the oil has a chance to separate.

Steaming Steaming is a rapid way to cook fish and is especially appropriate for thin fillets, which can cook through quickly (thick pieces would overcook on the surface while their interior cooks through). Subtle aromas are contributed by herbs and spices, vegetables, and even seaweed, if they're included in the steaming water or provide a bed on which the fish sits.

Even cooking requires that the fish pieces be the same thickness, and that the steam have equal access to all surfaces. If fillets taper down to a very thin end, fold the thin layers over or interleave them with each other. More than one layer's worth of fish should be cooked in batches or divided among separate levels (as in stackable Chinese bamboo steamers). Relatively thick steaks or whole fish are best steamed below the boil, at an effective temperature of 180°F/80°C, to minimize overcooking of the surface. This can be achieved by lowering the heat on the pot and/or leaving the pot lid ajar. An even gentler effect is achieved by the Chinese method of steaming fish without a lid, in which steam and room air combine to give an effective cooking temperature of 150–160°F/65–70°C.

Microwaving Microwave versions of simmered or steamed fish can be quite successful thanks to the relatively thin dimensions of fillets and steaks, which the electromagnetic waves can penetrate fully and cook quickly. To prevent especially thin portions from overcooking, cover them with radiation-blocking pieces of aluminum foil (p. 787), or overlap them with each other to a consistent thickness. As in most microwave cooking, the food should be enclosed so that the surface doesn't dry out and toughen: wrap the fish pieces in parchment or the cooking dish with plastic wrap, or simply place the fish between two inverted plates. Waiting for the fish to cool down some before uncovering the dish will mean less likelihood of a steam burn, a smaller billow of fishy aromas into the air, and less moisture loss from the fish surface.

Stovetop Smoking Smoking whole fish is a time-consuming and elaborate process, and cold-smoking requires an appliance with separate chambers for the smoke source and the fish (p. 236). But it's a simple matter to flavor a few portions with smoke on the backyard grill, or even indoors. Line the interior of an ordinary saucepan and its lid with aluminum foil, scatter smokeable materials—small dry wood chips or sawdust, sugar, tea leaves, spices—on the bottom, place presalted fish pieces on a rack, turn the heat on high until smoke appears, then reduce the heat to medium, cover the pot tightly, and allow the fish to "bake" in this 400–500°F/ 200–250°C stovetop oven until barely cooked through.

Fish Mixtures

Like meats, fish can be chopped or pounded or ground up and mixed with other ingredients to make balls, cakes, sausages, pâtés, terrines, and so on. This is an excellent way to use small scraps or cooked leftovers, or fish that are bony or otherwise unsuited to serving in large pieces. While meat mixtures are often tenderized and enriched by chunks of fat, and firmed by conversion of the meat's connective tissue into gelatin, fish contain little connective tissue and no fat that is solid at room temperature. Instead, many fish mixtures aim for a distinctive lightness, and have for many centuries, as is clear from Anthimus's early version of the classic French dish *quenelles de brochet* (see box below).

Mousselines, Quenelles The basic preparation for many refined fish mixtures is the *mousseline,* from the French *mousse,* or "foam," a term that describes the airy, delicate consistency aimed for. Chilled raw fish is very finely chopped or pureed (with care to avoid overheating in high-speed processors), then whisked with one or more of several binding and enriching ingredients. The whisking also incorporates air, which lightens the mixture. If the fish is very fresh, then it can be enriched and tenderized with cream and bound simply with salt, which extracts some myosin protein from the muscle fibers to help them stick together. With less pristine fish—weeks in the freezer can cause premature protein aggregation and a wet, crumbly puree—egg whites help hold the particles of fish muscle together, as do various starchy materials, including bread crumbs, flour-based béchamel and velouté sauces, pastry doughs, and mashed rice or potatoes. The mousseline mixture is firmed by refrigeration, then shaped into dumpling-like *quenelles,* or wrapped inside thin fish fillets

Ancient *Quenelles*

Pike is good too. Egg white should be mixed into the dish called spumeum which is made with pike, so that this dish may be quite soft rather than hard, and wholesome when mixed together.

—Anthimus, *On the Observance of Foods,* ca. 600 CE

(*paupiettes*), and gently poached; or it's put in ramekins or a pan and cooked in a water bath to make pâtés and terrines. The target temperature at the center is 140–150°F/60–65°C; higher temperatures give a harder, heavier result.

Fish Balls and Cakes Quenelles are essentially refined fish balls, a genre of which there are many regional variations. Chinese fish balls are bound with egg and cornstarch, lightened with water; Norwegian fish balls are enriched with butter and cream, bound with potato flour; Jewish gefilte fish (thought to derive via eastern Europe from French *quenelles*), bound with eggs and matzoh meal, and aerated by chopping. Less delicate and tricky fish mixtures include coarse cakes and croquettes bound with eggs and starchy particles like bread crumbs, and mousses made from cooked fish and held together with starchy sauces or gelatin.

Fingers and Burgers, Surimi Commercial "minced" fish products are made from a variety of white ocean fish that would otherwise be discarded as too small or bony. They run the range from coarse-textured fish sticks and fishburgers to finer patties and paste-like spreads. Imitation fish fillets and shellfish meats are made by extruding highly processed mixtures of fish paste and other structure-reinforcing ingredients, including seaweed-derived alginate gums and textured vegetable proteins.

The most widely consumed form of processed fish is *surimi,* the Japanese term for "minced fish," which is nearly 1,000 years old and is now made into many imitation shellfish products. Surimi is made by finely mincing fish scraps (today, usually pollack), washing them, pressing them to remove the wash water, salting and seasoning the mince, shaping it, and boiling it until it solidifies. Washing the mince removes nearly everything from the muscle except the muscle fiber membranes and contracting proteins. Salting then dissolves the protein myosin out of the muscle fibers, so that when it's heated, the myosin will coagulate into a continuous, solid, elastic gel in which the other fiber materials are embedded. The result is a flavorless, colorless, homogeneous matrix that can be flavored, colored, and formed to imitate nearly any seafood.

SHELLFISH AND THEIR SPECIAL QUALITIES

Though shellfish have much in common with finfish and are cooked in many of the same ways, they're also distinctive. Most of the shellfish we eat are creatures from one of two groups, the crustaceans and the molluscs. Unlike true fish, these creatures are invertebrates: they don't have a backbone or internal skeleton; and most of them don't swim much. Their body tissues are therefore organized differently, undergo different kinds of seasonal changes, and require special treatment from the cook.

Crustaceans: Shrimps, Lobsters, Crabs, and Relatives

Crustaceans are the shellfish that have legs and sometimes claws: shrimps and prawns, lobsters and crayfish, and crabs. Like the molluscs, the crustaceans are an ancient and successful group of animals. There were primitive shrimps 200 million years ago; today there are some 38,000 crustacean species, the largest with a claw span of 12 feet/4 meters! The crustaceans are members of the large animal group known as *arthropods,* and are relatives of the insects. Like the insects, they have a body made up of several segments, a hard outer cuticle, or exoskeleton, that protects and supports the muscles and organs within, and many rigid appendages that are adapted to a variety of purposes, including swimming, crawling, and attacking prey. Most edible crustaceans are "decapods," meaning they have five pairs of legs, one of which is sometimes greatly enlarged into claws. The meat of crustaceans is mainly skeletal muscle like that of fish and our land livestock. (Notable exceptions are the immobile barnacles, prized in Spain and South America.)

Because they're mobile, carnivorous, and often cannibalistic, crustaceans aren't as easy to farm as molluscs. The greatest success has come with the shrimps, thanks to their ability to grow rapidly on both plant feeds and very small animals.

Crustacean Anatomy All crustaceans share the same basic body plan, which can be divided roughly into two parts. The forward portion, or *cephalothorax,* often called the "head" in shrimp, is the equivalent of our head and trunk put together. It includes the mouth, sensing antennae and eyes, five pairs of manipulating and crawling appendages, and the main organs of the digestive, circulatory, respiratory, and reproductive systems. The rear portion, or *abdomen,* usually called the "tail," is mostly a large, meaty block of swimming muscle that moves the fin-like plates at the back end. The major exception to this body plan is the crab, which seldom swims; its abdomen is a thin plate folded up underneath a greatly enlarged cephalothorax.

The most important organ in the crustacean is what biologists call the *midgut gland* or *hepatopancreas,* and what the rest of us usually call the "liver." This is the source of enzymes that flow into the digestive tube and break down ingested foods; it's also the organ in which fatty materials are absorbed and stored to provide energy during molting (below). It's thus one of the richest, most flavorful parts of the body, and is especially prized in lobsters and crabs. But it's also what makes crustaceans spoil so readily. The gland is made up of tiny fragile tubes; and when the animal is killed, the tubules are readily attacked and damaged by their own enzymes, which then spread into the muscle tissue and break it down into mush. There are several ways to avoid this spoilage. Lobsters and crabs are sold either live, their digestive system intact, or fully cooked, their enzymes inactivated by the cooking. Because the shrimp liver is relatively small, processors often remove the "head" that contains it, and sell only the tail meat. Raw shrimp that are sold "head-on" must be handled with greater care (iced immediately and continuously) and don't keep as long.

The Crustacean Cuticle, Molting, and Seasonal Quality Another defining characteristic of the crustaceans is a "shell" or cuticle made up of *chitin,* a network of molecules that are something of a hybrid between carbohydrates and proteins. In

Crustacean anatomy. The forward part of the crustacean body, the cephalothorax or "head," contains the digestive and reproductive organs. The rear part, the abdomen or "tail," is mainly fast muscle tissue that moves the rear fins and propels shrimp (top) *and lobsters* (center) *in brief swimming maneuvers. The crab* (bottom) *has only a vestigial abdomen tucked under its massive cephalothorax.*

shrimp, the cuticle is thin and transparent; in larger animals it's thick and opaque, hardened to a rock-like mass with calcium minerals filling the space between chitin fibers.

As a crustacean grows, it must periodically cast off the old cuticle and create a larger new one. This process is called *molting*. The animal constructs a new, flexible cuticle under the old one from its body's protein and energy reserves. It squeezes its shrunken body through weakened joints in the old shell, then pumps up itself with water—from 50 to 100% of its original weight—to stretch the new cuticle to its maximum volume. It then hardens the new cuticle by cross-linking and mineralizing it, and gradually replaces its body water with muscle and other tissues.

Molting means that the quality of crustacean flesh is highly variable, and this is why wild harvests are seasonal, with the seasons depending on the particular animal and location. An actively growing animal has dense, abundant muscle, while an animal preparing to molt loses muscle and liver mass, and a newly molted animal may be as much water as muscle.

Crustacean Color Crustacean shells and eggs provide some of the table's most vivid colors. They are generally a dark green-blue-red-brown that helps them blend in with the sea bottom, but turn a bright orange-red when cooked. The animals create their protective coloration by attaching bright carotenoid pigments derived from their planktonic diet (astaxanthin, canthaxanthin, beta-carotene and others) to protein molecules, thus muting and altering their color. Cooking denatures the proteins and frees the carotenoids to reveal their own true colors.

The shells of lobsters, crayfish, and some crabs are often cooked to extract both flavor and color for sauces (the French sauce Nantua), soups, and aspics. Because carotenoid pigments are much more soluble in fat than in water, more color will be extracted if the cooking liquid is mainly fat or oil—butter, for example—or contains some.

Crustacean Texture Like fish flesh, most crustacean flesh consists of white, fast muscle fibers (p. 131). Its connective-tissue collagen is both more abundant than fish collagen and less easily dissolved by heat, so crustacean meat is less delicate and easily dried out than fish. But the protein-breaking enzymes in the muscle are very active, and can turn the meat mushy if they aren't rapidly inactivated by the heat of cooking. These enzymes work fastest when the temperature hits 130–140°F/55–60°C,

Food Words: *Shrimp, Prawn, Crab, Crayfish, Lobster, Crustacean*

Most of our words for crustaceans go back to prehistoric times. *Shrimp* comes from the Indo-European root *skerbh,* meaning to turn, bend, or shrink, perhaps reflecting the curled shape of these creatures. The near-synonym *prawn* first appears in medieval times, and its origins are unknown. *Crab* and *crayfish* both derive from the Indo-European *gerbh,* meaning to scratch or carve, something that crustacean claws readily do to human skin. Finally, *lobster* shares with *locust* the Indo-European root *lek,* meaning to leap or fly: a remarkably early recognition of the family resemblance of crustaceans and insects.

Crustacean itself comes from an Indo-European root meaning to freeze, to form a crust, and describes the hard outer skeleton of these creatures. It shares this root with *crystal.*

so the cook should either heat the flesh well above this range as quickly as possible, or get it just into this range (for maximum moistness) and then serve it immediately. Boiling and steaming are the most rapid heating methods, and the usual treatments for shrimp, lobster, and crab.

Crustacean texture is also more tolerant of freezing than most fish; frozen shrimp in particular can be quite good. However, domestic freezers are warmer than commercial freezers and allow undesirable chemical changes and general toughening to occur (p. 206), so frozen crustacean meats should be used as quickly as possible.

Crustacean Flavor The aromas of boiled shrimp, lobster, crayfish, and crab are remarkable for their nutty, popcorn-like qualities, quite distinct from either mollusc or fish aromas. Even meats don't develop these notes unless they're actually roasted rather than boiled. They're due to an abundance of molecules (pyrazines, thiazoles) that are normally produced when amino acids and sugars react at high temperatures (the Maillard reactions, p. 778). These reactions evidently take place at lower temperatures in crustaceans, perhaps thanks to the unusual concentration of free amino acids and sugars in their muscle tissue. Among the amino acids that sea creatures accumulate in their cells to balance the salt in the water, crustaceans favor glycine, which has a sweet taste and lends sweetness to their meat.

The distinctive iodine-like flavor found frequently in gulf brown shrimps and occasionally in other crustaceans originates in bromine compounds that the animals accumulate from algae and other foods, and then convert to unusual and more odorous compounds (bromophenols) in their gut.

It's often observed that crustaceans are more flavorful when cooked in their shells. The cuticle reduces the leaching of flavor compounds from the flesh, and is itself a concentrated mass of proteins, sugars, and pigment molecules that can flavor the outer layer of flesh.

Choosing and Handling Crustaceans

Because their flesh is so easily damaged by their own enzymes once they're dead, crustaceans are generally sold to consumers either frozen, cooked, or alive. Most "fresh" raw shrimp have been obtained frozen and thawed by the store. Ask for a sniff of one, and don't buy if you smell ammonia or other off-odors. Cook them the same day.

The larger crustaceans, lobsters and crabs, are generally sold either precooked or alive. Live crustaceans should come from a clean-looking tank, and should be active. They can be kept alive in a moist wrapping in the refrigerator for a day or two. Relatively small lobsters and crabs will have finer muscle fibers and so a finer texture.

Traditional recipes often treat lobsters, crayfish, and crabs as if they were insensible to pain, calling for the cook either to cut them up or drop them in boiling water while they're still alive. These creatures don't really have a central nervous system. The "brain" in the head region receives input only from the antennae and eyes, and each body segment has its own nerve cluster, so it's hard to know whether or how pain can be minimized. The most sensible-sounding advice comes from marine biologists: anaesthetize the animal in iced salt water for 30 minutes just before cutting up or boiling.

Shrimps and Prawns Shrimp and prawns are the most commonly available shellfish in the world. Their predominance stems from their delicious flavor, conveniently small size, rapid reproduction in the wild and in aquaculture, and the tolerance of their flesh to freezing. The two terms are often used for the very same animals; in the United States, "prawn" usually means a larger variety of shrimp. There are some 300 species of shrimp and close relatives exploited for food around the world, but the most common belong to one semitropical and tropical genus, *Penaeus*. Species of *Penaeus* can mature in a year or less and grow as long as

9 in/24 cm. Temperate-water shrimp belong to a slower-growing group and are usually smaller (a maximum of 6 in/15 cm). Today about a third of world production is cultivated, mainly in Asia.

Shrimp Quality Shrimp flavor declines in just a few days on ice due to the slow loss of amino acids and other tasty small molecules. But thanks to their protective cuticle, shrimp can remain edible for as much as 14 days. Shrimpers often treat them with a bleaching solution of bisulfite to prevent discoloration, and like scallops, with a sodium polyphosphate solution to keep them moist; these practices can cause off-flavors.

The mainly muscular "tail" of the shrimp amounts to about two-thirds of its body weight, so producers often separate it from the flavorful "head" and its midgut enzymes, which can accelerate spoilage. The dark "vein" along the outside curve of the abdomen is the end of the digestive tube, and can be gritty with the sand from which the animals glean bacteria and debris; it's easily pulled away from the surrounding muscle. Though peeled, cooked shrimp are widely available and convenient, serious shrimp lovers seek out fresh whole shrimp and cook them in the shell, rapidly and briefly.

Lobsters and Crayfish Saltwater lobsters (species of *Homarus* and *Nephrops*) and freshwater crayfish (*Astacus, Procambarus,* and others) are generally the largest crustaceans in their neighborhood. The American lobster once weighed as much as 40 lb/19 kg, while today it's typically 1–3 lb/450–1,350 gm. And more than 500 species of crayfish have evolved in the fresh waters of isolated rivers and streams, especially in North America and Australia. Most are relatively small, but the Australian marrons and "Murray lobsters" can exceed 10 lb/4.5 kg. Crayfish are the most easily cultured of the crustaceans, and have been raised in natural ponds in the Atchafalaya Basin of Louisiana for better than two centuries. They're also prized in Sweden.

The main attraction of all these creatures is their white "tail" meat. Three European and American lobster species and their crayfish cousins have large claws, which in the American lobster can amount to half the total body weight. A larger group of more distant relatives, the spiny and rock lobsters (*Palinurus, Panuliris, Jasus* and others), are less impressively endowed, and are called "clawless"; they supply much frozen lobster tail, because their meat freezes better than clawed lobster meat. The claw meat is noticeably different from the main body and tail meats. Because they require more stamina, claw muscles include a substantial proportion of slow red fibers (p. 132), and have a distinctive, richer flavor.

Lobsters and crayfish are often sold live to consumers. The prime season for Louisiana crayfish is generally local winter through spring, when the animals are heaviest and firm-fleshed. The lobster body contains the flavorful digestive gland known as the liver or tomalley, a pale mass that turns green when cooked. Females may also contain an ovary, a mass containing thousands

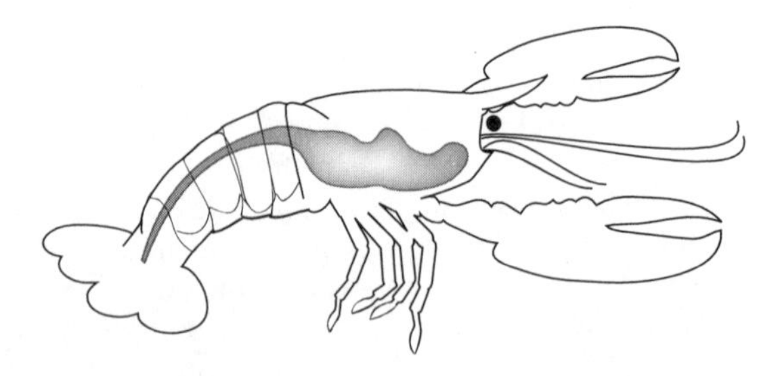

Crustacean innards. The cephalothorax of crustaceans contains a large, flavorful digestive gland, the hepatopancreas, whose enzymes can also damage the surrounding muscle. The dark, sometimes gritty "vein" along the tail muscle is actually the end of the digestive tube.

of 1–2 mm eggs, which turns red-pink when cooked; hence its name "coral." Lobster liver and coral are sometimes removed before cooking, and then crushed to a paste and added to hot sauces at the last minute to contribute their color and flavor.

Crabs Crabs are tailless. Instead they have a massive cephalothorax, whose musculature enables these creatures to live in the deepest sea, burrow on land, and climb trees. Most crabs have one or two powerful claws for holding, cutting, and crushing their prey. Crab claw meat is flavorful but coarser and harder to get at than the body meat, and generally not as prized. Exceptions are the massive and flavorsome single claws of the Florida stone crab and European fiddler crab. The legs of the north Pacific king crabs, which can span 4–6 ft/1.2–1.8 m, provide large cylinders of meat that are often sold frozen.

Most commercial crabs (species of *Callinectes, Carcinus, Cancer,* and others) are still caught alive in baited traps or dredges. They may be sold live, or cooked and whole, or cooked and processed into shell-less meat. This meat is then sold fresh, or pasteurized, or frozen for longer keeping. In addition to the muscle tissue, the crab's large digestive gland, its "mustard" or "butter," is prized for a rich, intense flavor and creamy texture, which it lends to sauces or to crab pastes. Crab liver can accumulate the toxins from algae that cause shellfish poisoning (p. 186), so state regulators monitor toxin levels and restrict crabbing when they become significant.

Soft-shell crabs Because freshly molted crustaceans have just spent much of their protein and fat reserves and are absorbing water to fill out their new shell, eaters generally disdain them. The major exceptions to this rule are the soft-shell shore crab of Venice, and the soft-shell blue crab of the U.S. Atlantic coast, which is fried and eaten whole. Animals that are about to molt are watched carefully and removed from salt water as soon as they shed their old shell, since their new cuticle would otherwise become leathery within hours and calcified hard in two or three days.

Molluscs: Clams, Mussels, Oysters, Scallops, Squid, and Relatives

Molluscs are the strangest creatures we eat. Take a close look sometime at an intact abalone or oyster or squid! But strange or not, molluscs are plentiful and delicious. Judging by the massive prehistoric piles of oyster, clam, and mussel shells that dot the planet's seacoasts, humans have feasted on these conveniently sluggish creatures from the earliest days. This highly successful and diverse branch of the animal kingdom got its start half a billion years ago and currently includes 100,000 species, double the number of fish and animal species with backbones, from snails just a millimeter across to giant clams and squids.

The secret to the molluscs' success—and their strangeness—is their adaptable body plan. It includes three major parts: a muscular "foot" for moving; an intricate assembly that includes the circulatory, digestive, and sexual organs; and enveloping this assembly, a versatile sheet-like "mantle" that takes on such jobs as secreting materials for a shell, supporting eyes and small tentacles that detect food or danger, and contracting and relaxing to control water flow into the interior. The molluscan shellfish that we eat have combined these parts in very different ways.

- Abalones, the most primitive, have one cup-like shell for protection, and a massive, tough muscular foot for moving along and clinging to the seaweed on which their rasping mouths feed.
- Clams are enclosed in two shells, and burrow into the sand with their foot. Modifications of the mantle have provided them with two pegs of muscle for closing the shells, and with the muscular tube—the siphon or

"neck"—that they extend to the sand surface and use to draw in passing food particles. All the bivalves—clams, mussels, oysters—have comb-like gills for filtering food particles from the water that the mantle draws in and expels.
- Mussels are also two-shelled filter feeders, but they attach their foot permanently to intertidal and subtidal rocks. They have no need for a siphon, and one of their tough shell-closing muscles is much reduced.
- Oysters cement themselves to inter- and sub-tidal rocks. Their two heavy shells are closed by a single large muscle at their center, around which the mantle and other organs are organized. The bulk of their body is the tender mantle and food-trapping gills.
- Scallops neither attach nor bury themselves. They lie free on the ocean floor, and escape predators by swimming. Their massive central muscle claps their shells shut and forces water out one end, thus propelling them in the other direction.
- Squids and octopuses are molluscs turned inside out and transformed into highly mobile, streamlined carnivores with large eyes and arms. The remnants of a shell provide an internal support, and the mantle is now a specialized muscular sheet that expands and contracts to provide jet propulsion through a small funnel derived from the foot muscle.

The immobile molluscs do very well in aquaculture. They can be raised in large numbers in the water's three dimensions, suspended in nets or on ropes, and grow rapidly thanks to the good circulation of oxygen and nutrients.

Bivalve Adductor Muscles The two-shelled or "bivalve" molluscs must spread their shells apart to allow water and food particles in, and pull their shells together to protect their soft innards against predators or—in the case of intertidal mussels and oysters—the drying air. To do this work they have evolved a special muscle system, one that poses some challenges to the cook but is mostly a boon, since these prepackaged animals can survive for many days in the refrigerator covered only with a moist towel.

Bivalve shells are normally held open mechanically, by means of a spring-like ligament that connects and pulls them together at the hinge end, and thus pulls the opposite wide ends apart. To close the shells, the animal must power a muscle, called an "adductor" (from the Latin *adducere,* "to bring together"), which extends between the broad ends of the shell and contracts to overcome the spring force of the ligament.

Tender Quick, Tough Catch The adductor muscle has to perform two very different kinds of work. One is to close the shell quickly to expel sediment, accumulated wastes, or eggs, or to slam the door on predators. The other is to keep the shell tightly closed for hours, sometimes even days, until the danger passes. These two jobs are performed by adjoining parts of the muscle. The fast-contracting "quick" portion is quite similar to the fast muscles of fish and crustaceans; it's white, translucent, and relatively tender. But the slow, tension-maintaining "catch" portion is among the strongest muscles known, and can maintain its contraction with very little expenditure of energy, thanks to biochemical tricks that lock the muscle fibers in place once they've shortened, and reinforcement with large amounts of connective-tissue collagen. Catch muscles have an opalescent appearance, much like the tough tendons in a chicken leg or leg of lamb, and they are tough to eat as well unless cooked for a long time. In the scallop, the small catch portion would detract from the large quick portion's tenderness, and so is usually cut away.

Mollusc Texture The adductor muscles largely determine the texture of several bivalves—especially the scallop, whose large and tender "swimming" muscle is often the only portion served. The other bivalve bodies are eaten whole, and include one or two adductors together with miscellaneous innards; small tubes and thin sheets of muscle and connective tissue; soft masses of eggs, sperm, and food particles; and a general proteinaceous mucus that lubricates and binds food particles. Clams, mussels, and oysters are thus slick and both crunchy and tender when raw, chewy when cooked. The greater the proportion of muscle tissue, the chewier the mollusc.

Mollusc texture is also strongly affected by the animals' reproductive stage. And as they approach spawning and their bodies fill with eggs and/or sperm, the bivalves develop a soft creaminess that cooking sets to a custard-like texture. Immediately after spawning, the depleted tissues are thin and flabby.

Abalone, octopus, and squid meats are mainly muscle tissue with a lot of connective-tissue collagen and a complex fiber arrangement. They're chewy when lightly cooked, tough when cooked to the denaturation temperature of their collagen, around 120–130°F/50–55°C, and become tender with long cooking.

Mollusc Flavor Oysters, clams, and mussels are prized for their rich, mouth-filling taste, especially when eaten raw. We owe this savoriness to their accumulation of internal taste-active substances as an energy reserve and to balance the external salinity of their home waters. For osmotic balance, marine fish (and squid and octopus) use tasteless TMAO and relatively small amounts of amino acids, while most molluscs rely almost entirely on amino acids: in the bivalves, especially brothy glutamic acid. And instead of storing energy in the form of fat, molluscs accumulate other amino acids—proline, arginine, alanine, and some combined forms—as well as glycogen, the animal version of starch, which is itself tasteless, though it probably provides a sense of viscosity and substance, and is slowly transformed to sweet molecules (sugar phosphates).

Because shellfish use amino acids to counteract salt concentration, the saltier the water, the more savory the shellfish. This fact accounts for at least some of the flavor differences among shellfish from different waters, and it is part of the rationale for "finishing" oysters for a few weeks or months in particular locations. Because shellfish use up their energy stores as they prepare for spawning, they become noticeably less tasty as spawning approaches.

When molluscs are cooked, their savoriness is somewhat diminished because heat traps some of the amino acids in the web of coagulated protein and so withholds them from the tongue. However, heating alters and intensifies the aroma, which is generally dominated by dimethyl sulfide, a compound formed from an odd sulfur-containing substance (dimethyl-β-propiothetin) that molluscs accumulate from the algae on which they feed. DMS is also a prominent aroma in canned corn and in heated milk: one reason that oysters and clams go so well with these ingredients in seafood soups and stews.

Choosing and Handling Molluscs

Unless they've already been removed from their shell, fresh bivalves should be alive and healthy: otherwise they are likely to have begun spoiling. A healthy bivalve has an intact shell, and its adductor muscle is active and holds the shells tightly together, especially when sharply tapped. Molluscs keep best on ice covered with a damp cloth, and should not be allowed to sit in a puddle of meltwater, which is saltless and therefore fatal to sea creatures. Clams and relatives often benefit from several hours' immersion in a bucket of cold salt water (⅓ cup salt per gallon, or 20 gm/l) to clean themselves of residual sand and grit.

When the cook wants to "shuck" an oyster or clam, or open the shell and remove the raw meat, it's the hinge ligament and adductor muscles that must be dealt with. The usual technique is to wedge the blade of a small, strong knife between the shells near the hinge, then cut through the elastic ligament. Then run the knife along the inner surface of one shell to sever the adductor muscle(s) (clams and mussels have two, oysters and scallops one). Remove the loose shell, and cut the other end of the adductor(s) to free the body from the remaining shell.

Heat causes the adductor muscle to relax, which is why mollusc shells open during cooking. Shells that don't open may not contain a live animal and should be discarded.

Abalone There are about 100 species in the abalone genus *Haliotis*; they have a single low-slung shell, and the largest grow to 12 in/30 cm and 8 lb/4 kg. In the United States, the red abalone, *Haliotis rufescens*, is now farmed in offshore cages and onshore tanks, reaching 3.5 in/9 cm across and yielding 0.25 lb/100g meat in about three years. Abalone meat can be quite tough, in part because they apparently accumulate connective-tissue collagen as an energy reserve! Either very gentle or prolonged heating is essential; the meat toughens badly when it exceeds 120°F/50°C, and the collagen shrinks and compacts the tissue. Once this happens, continued simmering will eventually dissolve the collagen into gelatin and make the meat densely silken. Japanese cooks simmer abalone for several hours to obtain a more savory flavor (free amino acids apparently react to form taste-active peptides).

Clams Clams are the burrowing bivalves. They dig themselves into ocean or river sediments by extending a foot muscle downward, expanding its end into an anchor, and then contracting the foot while squirting water and rocking the shell. In order to reach the water from their burrow to breathe and feed, they have a pair of muscular tubes or "siphons," one for inhaling and the other for exhaling, which may be separate or else joined together into a single "neck."

The U.S. term "hard shell" is applied to sturdy clams that close completely (littleneck, quahog), while "soft shell" clams have siphons much longer than the shell, which is thin and always gapes (steamer, longneck). The Japanese or Manila hard-shell clam (*Ruditapes philippinarum*) is the only one to be cultivated on a large scale worldwide, thanks to its robustness and preference for shallow burial. The other dozen or so common clam species are mainly regional products. Some species of

Food Words: *Mollusc, Abalone, Clam, Oyster, Scallop, Squid*

The general term for these hard-shelled creatures, *mollusc,* comes from the Indo-European root *mel,* meaning "soft," which the inner body parts indeed are. *Abalone* entered English via Spanish from the Monterey Indian word for this streamlined snail, *aulun. Clam* began in the Indo-European *gel,* a compact mass: *cloud, cling,* and *clamp* are its linguistic relatives. *Mussel* derives from the Indo-European *mus,* meaning both "mouse" and "muscle," which moves quickly like a mouse under the skin. Since mussels hardly move at all, their dark, oblong shapes must have suggested the comparison. *Oyster,* from the Indo-European *ost,* "bone," names the mollusc with the heavy and bone-colored shell. *Scallop,* with its unusually symmetrical and patterned valves, comes via the Middle French *escalope*, from a Germanic word for "shell." And *squid*? To date, the linguists are stumped. It appeared out of nowhere in the 17th century.

the large surf clam (*Mactromeris* species) absorb plankton pigments and have a striking red layer on several muscles. The largest and most grotesque of the temperate commercial clams is the deep-burrowing geoduck of the Pacific Northwest subtidal mudflats (*Panope generosa*), whose neck looks like a small elephant's trunk. Though most are 3 lb/1.5 kg, geoducks can reach 15 lb/8 kg with a neck 3 ft/1 m long!

Their burrowing and siphoning musculature makes clams fairly chewy creatures. The tenderer portions of large clams (mantle, quick muscle) may be cut out and prepared separately. The large geoduck neck is usually scalded and the tough protective skin removed before the meat is sliced and/or pounded very thin for eating raw or either gentle or prolonged cooking.

Mussels The handful of mussel species we usually eat have become cosmopolitan: they have hitched rides or been intentionally introduced to various parts of the world, where they both grow naturally and are farmed and marketed at 2.5 in/6 cm in less than two years. The Mediterranean and Atlantic species of *Mytilus* have complementary habits; the Atlantic is in its prime in the spring and spawns in the summer; the Mediterranean is best in summer and spawns in winter.

Mussels anchor themselves in the intertidal zone by means of a thatch of tough proteinaceous fibers called the *byssus*, or "beard." Where the clams have two similar adductor muscles to close and hold the shells tightly shut, the mussel has one large adductor at the wide end and a small one at the narrow end. The rest of the mussel body comprises the respiratory and digestive systems and the mantle. Sexual tissues develop throughout these systems. Coloration depends on sex, diet, and species; orange pigments from algae and crustaceans accumulate more in female and Atlantic mussels.

Mussels are the easiest molluscs to prepare; they tolerate some overcooking and readily come off the shell. Both characteristics reflect the relatively small amount of muscle tissue. Because the beard is attached to the body inside, tugging on it can injure the animal. Beard removal should be put off until just before cooking. To avoid toughening mussels, it's best to cook them in a broad, shallow pan in essentially a single layer: this allows the cook to remove the early openers so that they don't overcook while the others finish.

Oysters Oysters are the most prized of the bivalves. They are the sea's tenderest morsels, the marine equivalent of penned veal or the fattened chicken, which just sit and eat. Their shell-closing adductor amounts to just a tenth of the body weight, the thin, delicate sheets of all-enclosing mantle and gills account for more than half, and the visceral mass for a third. The oyster is a special delicacy when cut from the shell and eaten raw. It's big enough to make a generous morsel, has a full, complex flavor and suggestively slippery moistness; and its delicacy is a striking contrast to the encrusted, rocky shell.

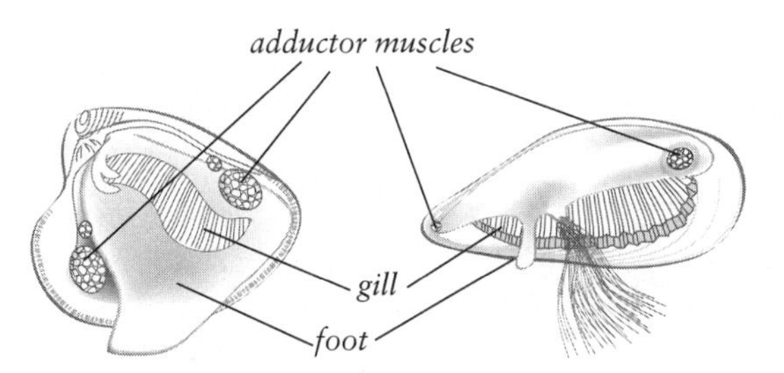

Clam and mussel anatomy. The bulk of the clam body (left) *is the muscular foot, while the mussel body* (right) *is mainly a nonmuscular mantle and the digestive and reproductive organs it encloses. The shell-closing adductor muscles are relatively minor parts. The mussel's "beard" is a thatch of tough protein fibers that anchor it to a rock or other support.*

Oyster Types Oysters became scarce as early as the 17th century, and are now largely farmed. A handful of the two dozen oyster species are commercially important; they have different shapes and subtly different flavors. European flat oysters (*Ostrea edulis*) are relatively mild with a metallic taste; Asian cupped oysters (*Crassostrea gigas*) have melon and cucumber aromas; and Virginia cupped oysters (*Crassostrea virginica*) smell like green leaves. Though there are exceptions, most oysters produced in Europe are the native flat, "Portuguese," and Asian; on the east and Gulf coasts of North America, the Virginia; and on the west coast, the Asian and the Pacific (*Ostrea lurida*). The "Portuguese" oyster is almost certainly a race of the Asian oyster that hitched a ride from China or Taiwan to the Iberian peninsula in the ships of early explorers, four or five centuries ago.

Oyster Waters The flavor of an oyster also depends on its home waters, which is why it makes sense to give geographical designations to oysters. The greater the salinity of the water, the more taste-active amino acids the oyster's cells must contain to balance the dissolved salt outside, and so the more savory its flavor. The local plankton and dissolved minerals will leave distinctive traces in the animal; and predators, currents, and exposure in the tidal zone will exercise and enlarge its adductor muscle. Water temperature determines how rapidly the oyster grows, and even its sex: warmth and plentiful food usually mean fast growth and development into a plump female creamy with millions of tiny eggs; cold water means slow growth, an indefinitely postponed sexual maturity, and a leaner, crisper texture.

Handling and Preparing Oysters Live oysters can survive for a week or more under moist wraps in the refrigerator, cupped shell down. Up to a point, this holding period can heighten their flavor, since metabolism without oxygen causes savory succinic acid to accumulate in their tissues. Preshucked oysters are rinsed with cold fresh water and then bottled in their subsequent secretions, which should be mostly clear; pronounced cloudiness indicates that the oyster tissues are breaking. Bottled oysters are often sub-pasteurized (heated to around 120°F/50°C) to delay spoilage while mostly retaining the fresh texture and flavor.

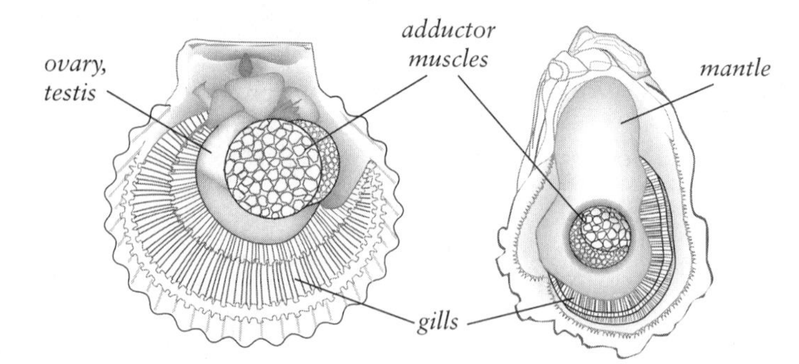

Scallop and oyster anatomy. The prized portion of the scallop (left) *is the large main adductor muscle, a tender bundle of fast muscle fibers that claps the shells together to propel the scallop away from danger. The crescent of "catch" muscle alongside it holds the shell closed. It is rich in connective tissue and tough, and is usually cut away from the adductor. The pink and tan reproductive tissues are prized in Europe but not in the United States. The oyster body* (right) *is mainly digestive and reproductive organs enclosed in a fleshy mantle; it's usually eaten whole, the adductor and catch muscles providing a crunchy chewiness.*

Scallops The scallop family includes about 400 species that range from a few millimeters to a yard across. Most food scallops are still harvested from the ocean floor. Large "sea scallops" (species of *Pecten* and *Placopecten*) are dredged from deep, cold waters year-round on trips that may last weeks, while smaller "bay" and "calico" scallops (*Argopecten*) are either dredged or hand-gathered by divers closer to shore during a defined season.

Unlike all the other molluscs, the scallop is mostly delectably tender, sweet muscle! This is because it's the only bivalve that swims. It defends itself from predators by clapping its shells together and forcing water out the hinge end, using a central striated muscle that can be an inch/2 cm or more across and long. This adductor muscle makes up such a large portion of the scallop's body that it also serves as protein and energy storage. Its sweet taste comes from large amounts of the amino acid glycine and of glycogen, a portion of which is gradually converted by enzymes into glucose and a related molecule (glucose 6 phosphate) when the animal is killed.

Because their shells don't close tightly, scallops are usually shucked soon after harvest, with only the adductor muscle kept for the U.S. market, the adductor and yellow and pink reproductive organs for Europe. This means that meat quality usually begins to deteriorate long before it gets to market. On boats that go out for more than a day, the catch may therefore be frozen and/or dipped in a solution of polyphosphates, which the adductors absorb and retain, becoming plump and glossy white. However, such scallops have less flavor and lose large amounts of liquid when heated. Untreated scallops have a duller, off-white appearance with pink or orange tones.

In the kitchen, the cook sometimes needs to separate the large, tender swimming muscle from the adjoining, smaller, tough catch muscle that holds the two shells shut. When sautéed, scallops quickly develop a rich brown crust thanks to their combination of free amino acids and sugars, which undergo Maillard reactions.

Squid, Cuttlefish, Octopus The *cephalopod* group are the most advanced of the molluscs, with their mantle turned into a muscular body wall and the remnants of their shell within (the term means "head-foot": the foot muscle is near the head). The octopus, species of *Octopus* and *Cistopus,* has eight arms clustered around its mouth with which it clambers along the bottom and seizes prey; the coastal-bottom cuttlefish (species of *Sepia*) and open-ocean squid (species of *Loligo, Todarodes, Ilex*) have short arms and two long tentacles.

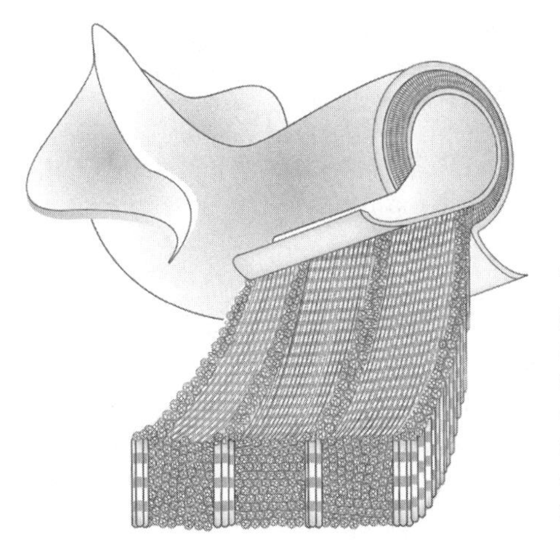

The anatomy of the squid mantle. This main portion of the squid body consists of an envelope of muscle that propels the animal by contracting and squeezing water through a small opening. The mantle muscle is built up from tough connective tissue and alternating rings of muscle fibers, some oriented across the mantle wall and some along it.

Cephalopod Texture The muscle fibers of squid and octopus are extremely thin—less than a tenth the diameter of a typical fiber in a fish or steer (0.004 mm, vs. 0.05–0.1 mm)—which makes the flesh dense and fine-textured. They're arrayed in multiple layers, and greatly reinforced with strengthening and toughening connective-tissue collagen, some three to five times more than fish muscle has. Unlike the fragile collagen of fish, squid and octopus collagen is extensively cross-linked and behaves more like the collagen of meat animals.

Like the abalone and clam, squid and octopus must be cooked either barely and briefly to prevent the muscle fibers from toughening, or for a long time to break down the collagen. Cooked quickly to 130–135°F/55–57°C, their flesh is moist and almost crisp. At 140°F/60°C it curls and shrinks as the collagen layers contract and squeeze moisture from the muscle fibers. Continued gentle simmering for an hour or more will dissolve the tough, contracted collagen into gelatin and give the flesh a silken succulence. Pounding can also help disorganize and thus tenderize mantles and arms.

Cephalopod Flavor and Ink Like finfish, squid and octopus maintain their osmotic balance largely with the tasteless TMAO (p. 188) rather than with free amino acids. Their flesh is therefore less sweet and savory than that of the other molluscs, and can turn fishy when bacteria convert TMAO to TMA.

Cephalopod ink is a bag of pigment that the animal can squirt into the water when endangered. It's a heat-stable mix of phenolic compounds (animal cousins of the phenolic complexes that discolor cut fruits and vegetables; p. 269), and cooks use it to color stews and pastas a dark brown.

Other Invertebrates: Sea Urchins

Spiny sea urchins are members of the animal group called *echinoderms* (Greek for "prickly skin"), which may account for 90% of the biomass on deep-sea floors. There are about a half dozen commercial species of sea urchins with average diameters of 2.5–5 in/6–12 cm. They're almost entirely enclosed in a sphere of mineralized plates covered with protective spines, and are collected mainly for their golden, creamy, richly flavored reproductive tissues, which can account for up to two-thirds of the internal tissues. Both testes and ovaries are prized, and are hard to tell apart. Sea-urchin gonads average 15–25% fat and 2–3% savory amino acids, peptides, and IMP. In Japan, sea urchins are eaten raw in sushi or salted and fermented into a savory paste; in France they're added to scrambled eggs, soufflés, fish soups and sauces, and sometimes poached whole.

PRESERVED FISH AND SHELLFISH

Few foods go bad faster than fish. And until recently, few people in the world had the chance to eat fresh fish. Before refrigeration and motorized transportation became common, fish were harvested in such numbers and spoiled so rapidly that most had to be preserved by drying, salting, smoking, fermenting, or some combination of these antimicrobial treatments. Preserved forms of fish are still important and appreciated in most parts of the world, especially in Europe and Asia. It's true that their flavor is much more assertive than the mild fresh fish that are now the U.S. standard. But preserved fish aren't just an inferior relic of preindustrial necessity. They can be a delicious alternative, and they offer a taste of history.

Dried Fish

Drying foods in the sun and wind is an ancient method of preservation. Fresh fish is about 80% water; below 25%, bacteria have trouble growing, and below 15% molds do too. Happily, dehydration also intensifies and alters flavor by disrupting cellular structure and so promoting enzyme action, and by concentrating flavorsome molecules to the point that they begin to react with each other to form additional layers of flavor. Very lean fish and shellfish are the usual choice, since air-drying will inevitably cause fat oxidation and some development of rancid flavors. Fatty fish are usually smoked, or salt-cured in closed containers to minimize rancidity. Often drying is preceded by salting and/or cooking, which draw moisture from the fish and make their surfaces less hospitable to spoilage microbes during the drying proper.

China and Southeast Asia are the largest producers and consumers of dried fish and shellfish. Cooks there use dried shrimp as is, either whole or ground, to season various dishes; they steam and shred dried scallops before adding them to soups; they reconstitute tough abalone, octopus, squid, jellyfish, and sea cucumber by soaking in water, then simmer them until tender. They do the same with shark fins, which give a gelatinous thickness to soups.

Stockfish Perhaps the best known dried fish in the West is the Scandinavian stockfish, which traditionally has been cod, ling, or their relatives, freeze-dried for several weeks on rocky beaches along the cold, windy coasts of Norway, Iceland, and Sweden. The result is a hard, light slab that's nearly all protein and has a pronounced, almost gamy flavor when cooked. Today, stockfish is mechanically air-dried for two to three months at 40–50°F/5–10°C. Stockfish fanciers in Scandinavia and the Mediterranean region reconstitute the woody mass in water for from one to several days, with frequent changes to prevent bacterial growth. The skin is then removed and the fish gently simmered, then served in pieces, in boneless flakes, or else pounded into a paste, and with a variety of enrichments and flavorings: in the north, often butter and mustard; in the Mediterranean, olive oil and garlic.

Salted Fish

Preservation by natural drying works well in cold and hot climates. Temperate Europe, where fish generally spoil before they can dry sufficiently, developed the habit of salting fish first, or instead. A day's salting would preserve many fish for several days more, long enough to be carried inland, while saturating the fish with around 25% salt keeps it stable for a year. Lean cod and relatives were salted and then

Alkaline Fish: *Lutefisk*

Distinctly alkaline foods are rare and have a slippery, soapy quality that takes getting used to. (Alkalinity is the chemical opposite of acidity.) Egg white is one such food, and another is *lutefisk*, a peculiar Norwegian and Swedish way of preparing stockfish that probably began in late medieval times, and that gives it a jiggly, jelly-like consistency. *Lutefisk* is made by soaking the partly reconstituted dry cod for a day or more in a water solution that is strongly alkaline, originally from the addition of potash (the carbonate- and mineral-rich ashes from a wood fire), sometimes lime (calcium carbonate), and later lye (pure sodium hydroxide, at the rate of about 5 grams per liter water). These strong alkaline substances cause the proteins in the muscle fibers to accumulate a positive electrical charge and repel each other. When the fish is then simmered in the usual way (after several days of rinsing to remove excess lye), the fiber proteins bind to each other only weakly.

air-dried, while fatty herring and their ilk were guarded from air-induced rancidity by immersing them in barrels of brine, or by subsequent smoking. The best of these are the piscatory equivalent of salt-cured hams. In both, salt buys time for transformation: it preserves them long and gently enough for enzymes of both fish and harmless salt-tolerant bacteria to break down flavorless proteins and fats into savory fragments, which then react further to create flavors of great complexity.

It's hard to draw a clear distinction between salted and fermented fish. Bacteria play some role even in hard-cured cod; and most fish fermentations start with a salting to control the bacterial population and activity. Most salted cod, herring, and anchovy products are not generally thought of as fermented, so I'll describe them in this section.

Salt Cod Bountiful cod was one resource that attracted Europeans to the New World, where the standard treatment was to split and salt the fish, and lay them out on rocks or racks to dry for several weeks. Nowadays cod may be hard-cured for 15 days to saturate the flesh with salt (25%), then held without drying for months. During that time, *Micrococcus* bacteria generate flavor by producing free amino acids and TMA; and oxygen breaks up to half the very small amount of fatty substances into free fatty acids and then into a range of smaller molecules that also contribute to aroma. The final artificial drying takes less than three days.

Salt cod remains a popular food around the Mediterranean as well as in the Caribbean and Africa, where it was introduced during the slave trade. Scandinavia and Canada are still the largest producers. White pieces are preferred to yellowish or reddish ones, the colors being indicators of oxidized or microbial off-flavors. Cooks first reconstitute and desalt it by soaking it for hours to days in several changes of water. Perhaps the best-known preparation is the Provençal *brandade,* a paste made by pounding the shredded poached fish along with olive oil, milk, garlic, and sometimes potato.

Salt Herring Herring and their relatives may be up to 20% fat by weight, and are therefore susceptible to becoming rancid when exposed to the air. Medieval fishermen solved this problem by barreling the fish in brine, where they would keep for as much as a year. Then sometime around 1300, the Dutch and northern Germans developed a quick gutting technique that left in place a portion of the intestine rich in digestive enzymes (the pyloric caecum). During one to four months of curing in a moderate brine (16–20% salt), these enzymes circulate and supplement the activity of both muscle and skin enzymes, breaking down proteins to create a tender, luscious texture and a wonderfully complex flavor, at once fishy, meaty, and cheesy. Such herring are eaten as is, without desalting or cooking.

Two particularly prized types of cured herring are the lightly salted Dutch *groen* and *maatjes,* or "green" and "maiden" herring, which traditionally broke the winter-long diet of hard-cured beef and fish. Because all lightly cured fish must now be prefrozen to rid them of parasites (p. 186), these formerly seasonal delicacies are now made and enjoyed year-round.

Cured Anchovies Anchovies, smaller and more southerly relatives of the herring, are cured in and around the Mediterranean to make that region's version of flavor-enhancing fish sauce (see box, p. 235). The fish are headed and gutted, then layered with enough salt to saturate their tissue. This mass is then weighted down and held for six to ten months at a relatively high temperature, between 60 and 86°F/15–30°C. The fish can then be sold as is, or the fillets repacked in cans or bottles, or ground and mixed with oil or butter into a paste. Enzymes from the muscle, skin, blood cells, and bacteria generate many flavor components; and their concentration, together with the warm curing temperature, encourages early stages of the browning reactions, which generate another range of aromatic molecules. The result is a remarkably full flavor that includes fruity, fatty, fried, cucumbery, floral, sweet, buttery, meaty, popcorn, mushroom, and malty notes. This concentrated complexity, together with the way that the cured flesh readily disintegrates in a dish, has led cooks from the 16th century on to use anchovies as a general flavor enhancer in sauces and other dishes.

Gravlax and Lox Gravlax originated in medieval Scandinavia as a lightly salted, pressed form of salmon that was preserved by fermentation (p. 235) and had a strong smell. By the 18th century, it had evolved into a lightly salted and pressed but unfermented dish. This new gravlax had a subtle flavor, a dense, silken texture that makes it possible to cut very thin slices, and a glistening, translucent appearance. This refined version of gravlax has become popular in many countries.

Modern recipes for gravlax call for widely varying amounts of salt, sugar, and time. Fresh dill is now the standard flavoring, probably a domestic replacement for the original pine needles, which are a delightful alternative. The salt, sugar, and flavoring are sprinkled evenly over all surfaces of salmon fillets, the fillets are weighted down, and the container refrigerated for one to four days. The weighting provides intimate contact between flesh and flavorings, presses excess fluid from the fish, and compacts the flesh. Salt dissolves the major contracting protein myosin in the muscle fibers, and thus gives the flesh its compact tenderness.

Lox, most familiar as a delicatessen accompaniment to the bagel, is a heavily brined form of salmon. It's usually soaked to remove some salt before being sliced for sale.

Fermented Fish

Many cultures from the Arctic to the tropics have recruited microbes to grow on fish and transform their texture and flavor. But the world center of fish fermentation is eastern Asia, where it has served two important purposes: to preserve and put to use the large numbers of small fish that inhabit the coastal and inland waters; and to provide a concentrated source of appetite-stimulating flavors—above all the savory monosodium glutamate and other amino acids—for a diet dominated by bland rice.

Fish fermentation apparently arose several thousand years ago in the freshwaters of southwest China and the Mekong River region. It then spread to the coastal deltas and was applied to ocean fish. Two broadly different techniques evolved: simply salting a mass of small fish or fish parts and allowing it to ferment; and salting larger fish lightly, then embedding them in a fermenting mass of rice or other grains, vegetables, or fruits. In the simple fermentation, the proportion of salt is usually enough by itself to preserve the fish from spoilage, and bacteria are important mainly as flavor modifiers. But in the mixed fermentation, a smaller dose of salt preserves the fish for just a few weeks while the plant-based ingredients feed the same microbes that sour milk or turn grape juice into wine. The fish is then preserved by the microbes' acids or alcohol, and flavored by the many by-products of their growth.

From these simple principles, Asian peoples have developed dozens of distinctive fermented fish products, and Europeans a handful. These include the original sushi, which was not a pristinely fresh piece of fish on mildly vinegared rice! Here I'll describe some of the more common ones.

Asian Fish Pastes and Sauces Asian fermented fish pastes and sauces are vital manifestations of a preparation that has mostly disappeared in Europe but was once well known as *garum* or *liquamen,* the fish sauce of Rome (see box, p. 235). (Modern ketchup, a sweet-sour tomato condiment, owes its name to *kecap,* an Indonesian salty fish condiment.) Fish sauces play the same role that soy sauces do in regions where soy doesn't grow well, and were probably the original model for soy sauce.

Fish pastes and sauces are two phases of the same simple preparation. A mass of fish or shellfish is mixed with salt to give an overall salt concentration between 10% and 30%, and sealed in a closed container for from one month (for pastes) to 24 months (for sauces). Fish pastes tend to have relatively strong fish and cheese notes, while the more thoroughly transformed fish sauces are more meaty and savory. The most prized fish sauces come from the first tapping of the mass; after boiling, flavoring, and/or aging, they play the lead role in dipping sauces. Second-quality sauces from re-extraction of the mass may be supplemented with caramel, molasses, or roasted rice, and are used in cooking to add depth to the flavor of a complex dish.

Sour Fish: The Original Sushi and Gravlax There are remarkable parallel traditions in Asia and Scandinavia in which fish are stored with carbohydrate-rich foods that bacteria ferment to produce acids that

Some Asian Fermented Fish Products

This chart gives an idea of the great variety of fermented fish condiments made in Asia.

Country	Fish Pieces or Paste	Fish Sauce	Sour-Fermented (carbohydrate source)
Thailand	Kapi (usually shrimp)	Nam-plaa	Plaa-som (cooked rice)
			Plaa-raa (roasted rice)
			Plaa-chao (fermented rice)
			Plaa-mum (papaya, galangal)
			Khem-bak-nad (pineapple)
Vietnam	Mam	Nuoc mam	
Korea	Jeot-kal	Jeot-kuk	Sikhae (millet, malt, chilli, garlic)
Japan	Shiokara (squid, fish viscera)	Shottsuru	Narezushi (cooked rice)
			Kasuzuke (cooked rice, sake wine sediments)
		Ika-shoyu (squid viscera)	
Philippines	Bagoong	Patis	Burong isda (cooked rice)
Indonesia	Pedah		Bekasam (roasted rice)
	Trassi (shrimp)		Makassar (rice fermented with red-pigmented yeast)
Malaysia	Belacan (shrimp)	Budu (anchovy)	Pekasam (roasted rice, tamarind)
		Kecap ikan (other fish)	Cincaluk (shrimp, cooked rice)

preserve the fish. These traditions have given birth to more widely popular but unfermented preparations: *sushi* and *gravlax*.

Asian Mixtures of Rice and Fish Of the many Asian fermentations that mix fish and grains, one of the most influential has been the Japanese *narezushi,* the original form of modern sushi (p. 207). The best-known version is *funa-zushi,* made with rice and goldfish carp (*Carassius auratus*) from Lake Biwa, north of Kyoto. Various bacteria consume the rice carbohydrates and produce a range of organic acids that protect against spoilage, soften the head and backbone, and contribute to the characteristic tart and rich flavor, which has vinegary, buttery, and cheesy notes. In modern sushi, made with pristinely fresh raw fish, the tartness of *narezushi* survives through the addition of vinegar to the rice.

Scandinavian Buried Fish: Gravlax According to food ethnologist Astri Riddervold, Scandinavian fermented fishes—the original *gravlax,* Swedish *surlax* and *sursild,* Norwegian *rakefisk* and *rakørret*—were probably the result of a simple dilemma facing medieval fisherman at remote rivers, lakes, and coastlines, who landed many fish but had little salt and few barrels. The solution was to salt the cleaned fish lightly and bury them where they had been caught, in a hole in the ground, perhaps wrapped in birch bark: *gravlax* means "buried salmon." The low summer temperature of the far northern earth, the airlessness, minimal salt, and added carbohydrates (from the bark, or from whey, malted barley, or flour), all conspired to encourage a lactic fermentation that acidified the fish surface. And enzymes from the fish muscle and the bacteria broke protein and fish oil down to produce a buttery texture and powerful, sharp, cheesy smell: the *sur* in *sursild* and *surlax* means "sour."

Modern, unfermented gravlax is made by dry-salting salmon fillets for a few days at refrigerator temperatures (p. 233).

Smoked Fish

The smoking of fish may have begun with fishermen drying their catch over a fire

Garum: The Original Anchovy Paste

One of the defining flavors of the ancient world was a fermented fish sauce variously called *garos* (Greece), *garum,* and *liquamen* (Rome). According to the Roman natural historian Pliny, "garum consists of the guts of fish and other parts that would otherwise be considered refuse, so that garum is really the liquor from putrefaction." Despite its origins and no doubt powerful aroma, Pliny noted that "scarcely any other liquid except perfume has become more highly valued"; the best, from mackerel only, came from Roman outposts in Spain. Garum was made by salting the fish innards, letting the mixture ferment in the sun for several months until the flesh had mostly fallen apart, and then straining the brown liquid. It was used as an ingredient in cooked dishes and as a sauce at the table, sometimes mixed with wine or vinegar (*oenogarum, oxygarum*). Some form of garum is called for in nearly every savory recipe in the late Roman recipe collection attributed to Apicius.

Preparations like garum persisted in the Mediterranean through the 16th century, then died out as the modern-day, solid version of garum rose to prominence: salt-cured but innard-free anchovies.

when sun, wind, and salt were inadequate. Certainly many familiar smoked fishes come from cool northern nations: smoked herring from Germany, Holland, and Britain, cod and haddock from Britain, sturgeon from Russia, salmon from Norway, Scotland, and Nova Scotia (the origin of the "Nova" salmon found in delicatessens), and smoked skipjack from Japan. It turned out that smoke imparts a flavor that can mask stale fishiness, and it helps preserve both the fish and its own flavor; the many chemicals generated by burning wood have both antimicrobial and antioxidant properties (p. 449). Traditional smoking treatments were extreme; the medieval Yarmouth red herring was left ungutted, saturated with salt and then smoked for several weeks, leaving it capable of lasting as long as a year, but also odiferous enough to become a byword for establishing—or covering up—a scent trail. When rail transport reduced the time from production to market in the 19th century, both salt and smoke cures became much milder. Today salt contents are kept around or under 3%, the salinity of seawater, and smoking is limited to a few hours, contributing flavor and extending the shelf life of refrigerated fish for a matter of days or weeks. Much modern smoked fish and shellfish is preserved in cans!

Preliminary Salting and Drying Nowadays, fish destined for the smoker are generally soaked in a strong brine for a few hours to days, long enough to pick up a little salt (a few percent, not enough to inhibit microbial spoilage). This also draws to the surface some of the proteins in the muscle fiber, notably myosin. When the fish is hung and allowed to drip dry, the sticky layer of dissolved myosin on the surface forms a shiny gel or *pellicle* that will give the smoked fish an attractive golden sheen. (The gold color is created by browning reactions between aldehydes in the smoke and amino acids in the pellicle, as well as condensation of dark resins from the smoke vapor.)

Cold and Hot Smoking The initial smoking (often using sawdust, which can produce more smoke at a lower temperature than intact wood) takes place at a relatively cool temperature around 85°F/30°C, which avoids hardening the surface and forming a barrier to moisture movement from the interior. This also allows the fish flesh to lose some moisture and become denser without being cooked, which would denature connective-tissue collagen and cause the fish to fall apart. Finally, the fish is smoked for several hours in one of two temperature ranges. In cold smoking, the temperature remains below

Lightly Salted, Strong-Smelling Fish: *Surstrømming*

Fish pastes and sauces are cured with enough salt to limit the growth and activity of microbes. There are also fish fermentations that involve far less salt, so that bacteria thrive and have a far more powerful influence on flavor. One notorious example is Swedish *Surstrømming*. Herring are fermented in barrels for one to two months, then sealed in cans and allowed to continue for as much as another year. The cans swell, which is normally a warning sign for the growth of botulism bacteria, but for *surstrømming* a sign of promising flavor development. The unusual bacteria responsible for ripening in the can are species of *Haloanaerobium,* which produce hydrogen and carbon dioxide gases, hydrogen sulfide, and butyric, propionic, and acetic acids: in effect a combination of rotten eggs, rancid Swiss cheese, and vinegar, overlaid onto the basic fish flavor!

90°F/32°C, and the fish retains its delicate raw texture. In hot smoking, the fish is essentially cooked in air at temperatures that gradually rise and approach the boiling point; it reaches an internal temperature of 150–170°F/65–75°C fairly quickly, and has a cohesive yet dry, flaky texture. Fish smoked cold and long can keep for as long as a couple of months in the refrigerator, while a light smoking, hot or cold, will only keep the fish for a few days or weeks.

Fine smoked salmon may be treated with salt and sometimes sugar for a few hours to a few days, then rinsed, air-dried, and cold-smoked for anywhere from five to 36 hours, with the temperature rising from 85° to 100°F/30° to 40°C toward the end to bring some glossy oil to the surface.

Four-Way Preservation: Japanese Katsuobushi

The most remarkable preserved fish is *katsuobushi,* a cornerstone of Japanese cooking, which dates from around 1700 and is made most often from one fish, the skipjack tuna *Katsuwonus pelamis.* The fish's musculature is cut away from the body in several pieces, which are gently boiled in salt water for about an hour, and their skin removed. Next, they undergo a routine of daily hot-smoking above a hardwood fire until they have fully hardened. This stage lasts 10 to 20 days. Then the pieces are inoculated with one or more of several different molds (species of *Aspergillus, Eurotium, Penicillium*), sealed in a box, and allowed to ferment on their surface for about two weeks. After a day or two of sun-drying, the mold is scraped off; this molding process is repeated three or four times. At the end, after a total of three to five months, the meat has turned light brown and dense; when struck, it's said to sound like a resonant piece of wood.

Why go to all this trouble? Because it accumulates a spectrum of flavor molecules whose breadth is approached only in the finest cured meats and cheeses. From the fish muscle itself and its enzymes come lactic acid and savory amino acids, peptides, and nucleotides; from the smoking come pungent phenolic compounds; from the boiling, smoking, and sun-drying come the roasted, meaty aromas of nitrogen- and sulfur-containing carbon rings; and from the mold's attack on fish fat come many flowery, fruity, green notes.

Katsuobushi is to the Japanese tradition what a concentrated veal stock is to the French: a convenient flavor base for many soups and sauces. It contributes its months of flavor-making in a matter of moments in

Smoked Fish Terminology

Kippered herring	Herring, gutted and split, cold-smoked
Bloater, bokking	Herring, whole, cold-smoked
Buckling	Herring, whole, hot-smoked
Sild	Herring, immature, whole, hot-smoked
Red herring	Herring, gutted, unsplit, cold-smoked
Brisling	Sprat, immature, whole, hot-smoked
Finnan haddie	Haddock, gutted, split, cold-smoked (peat)
Norwegian/Scotch smoked salmon; "Nova"	Salmon fillets, cold-smoked

the form of fine shavings. For the basic broth called *dashi*, cold water is brought just to the boil with a piece of kombu seaweed, which is then removed. The katsuobushi shavings are added, the liquid brought again to the boil, and poured off the shavings the moment they absorb enough water to fall to the bottom. The broth's delicate flavor is spoiled by prolonged steeping or pressing the shavings.

Marinated Fish

In chemical terms, an acid is a substance that readily releases free protons, the small reactive nuclei of hydrogen atoms. Water is a weak acid, and living cells are designed to operate while bathed in it. But strong acids flood living cells with more protons than they can handle, and cripple their chemical machinery. This is why acids are good at preserving foods: they cripple microbes. In the case of acidifying fish, a happy side benefit is that it leaves the fish with a distinctive, almost fresh aroma. Acid conditions cause heavy-smelling aldehydes, which accentuate the fishiness of TMA, to react with water molecules and become nonvolatile, so that lighter alcohols dominate the aroma. Pickled herring and other fish can be surprisingly delicate.

As the recipe from Apicius shows (see box below), inhabitants of the Mediterranean region have been marinating fish for thousands of years. The common modern term, *escabeche* and variants on it, derives from the Arabic *sikbaj*, which in the 13th century named meat and fish dishes with vinegar (acetic acid, p. 772) added toward the end of the preparation. Other acidic liquids were also used, including wine and verjuice, the juice of unripe grapes.

Fish and shellfish can be marinated in acid either raw or after an initial salting or cooking. In northern Europe, for example, raw herring are immersed in marinade (3 parts fish to 2 parts of a 10% salt, 6% acetic acid mixture) for up to a week, at a temperature around 50°F/10°C; while for marinated Japanese mackerel (*shimesaba*) the fillets are first dry-salted for a day, then immersed in vinegar for a day. In the case of precooked fish, the initial heat treatment kills bacteria and firms texture, so the subsequent marination is a milder one, and there is less development of texture and flavor.

Canned Fish

Because canned fish keep indefinitely without refrigeration and in a handy package, this is the preserved fish that most of us eat most often. In the United States, it is the most popular of all fish products: we consume more than a billion cans of tuna every year. Fish and shellfish were first heated in a hermetically sealed container around 1810 by Nicholas Appert, principal inventor of the process. Fellow Frenchman Joseph Colin started canning sardines a little over a decade later; American fishermen canned oysters in Delaware around 1840 and Pacific salmon around 1865, and Italian immigrants founded the canned tuna industry around San Diego in 1903. Today, salmon, tuna, and sardines are the most popular canned seafoods worldwide.

Most canned fish are heated twice: once before the cans are sealed, to cause the

Ancient Escabeche

To make fried fish keep longer. The moment that they are fried and lifted from the pan, pour hot vinegar over them.

—Apicius, first few centuries CE

inevitable cooking losses and allow the moisture (as well as flavor and healthful oils) to be drained away, so that the can contents won't be watery; and once after the cans are sealed to sterilize the contents, usually under pressurized steam at about 240°F/115°C. This second treatment is sufficient to soften fish bones, so fish canned with its bones is an excellent source of calcium (fresh fish contains about 5 milligrams of calcium per 4 oz/100 g; canned salmon contains 200 to 250). A number of additives are permitted in canned fish, particularly tuna, to improve flavor and appearance. These include monosodium glutamate and various forms of hydrolyzed protein, which are proteins broken down into savory amino acids (including glutamate). Premium canned fish is cooked only once, in the container, retaining its juices, and needs no improvement by additives.

FISH EGGS

Of all foods from the waters, the most expensive and luxurious are fish eggs. Caviar, the salted roe of the sturgeon, is the animal kingdom's truffle: a remarkable food that has become increasingly rare as civilization has encroached on its wild source. Happily, sturgeon farms are now producing good caviar, and a variety of other fish eggs are available as affordable and interesting alternatives.

The ovaries or "roes" of fish accumulate vast numbers of eggs in preparation for spawning: as many as 20,000 in a single salmon, and several million in a sturgeon, carp, or shad. Because fish eggs contain all the nutrients that one cell will need to grow into a hatchling, they're often a more concentrated form of nourishment than the fish itself, with more fat (between 10 and 20% in sturgeon and salmon caviars) and large quantities of savory building-block amino acids and nucleic acids. They often contain attractive pigments, sometimes bright pink or yellow carotenoids, sometimes camouflaging brown-black melanins.

The best roes for both cooking and salting are neither very immature nor fully ripe: immature eggs are small and hard and have little flavor; eggs ready for spawning are soft, easily crushed, and quick to develop off-flavors. Roes consist of separate eggs barely held together in a dilute protein solution and enclosed in a thin, fragile membrane. They can be easier to handle in the kitchen if they're first briefly poached to coagulate the protein solution and give them a firmer consistency.

Male fish accumulate sperm to release into the water when the females release

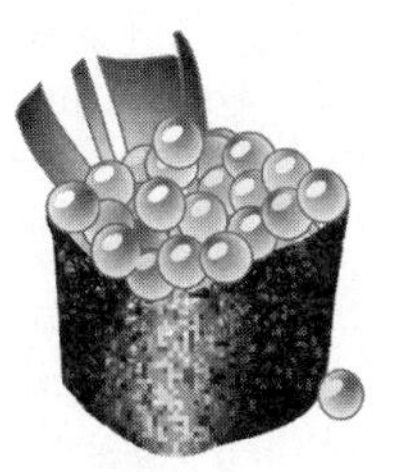

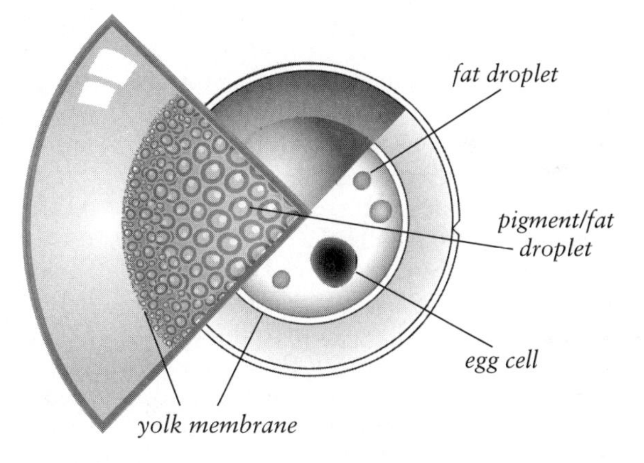

A salmon egg. Like the chicken egg, the inner yolk is surrounded by a protein-rich fluid, and contains fatty materials, including fat-soluble carotenoid pigments, and the living egg cell.

their eggs. The sperm mass is called white roe, milt, or *laitance,* and is creamy rather than granular (the sperm cells suspended in the proteinaceous fluid are microscopic). Sea bream and cod milts are prized in Japan, where they're cooked gently to a delicate custard-like consistency.

Salt Transforms Egg Flavor and Texture

Heavy Salting: *Bottarga* Fish eggs are more frequently consumed salted than they are fresh. Originally, salting was simply a means of preserving the eggs. For millennia in the Mediterranean, whole mullet and tuna ovaries have been dry-salted, pressed, and dried to make what's now best known as *bottarga* (there are almost identical Asian versions). The salting and drying cause a concentration of amino acids, fatty materials, and sugars, which react with each other in the complex browning reactions to darken the color to a deep red-brown and generate rich, fascinating flavors reminiscent of parmesan cheese and even tropical fruits! Bottarga is now a delicacy, sliced paper-thin and served as an antipasto, or grated onto plain hot pasta.

Light Salting: Caviar It turned out that salting has even more to offer when applied sparingly to loose, moist fish eggs. A small dose of salt triggers the action of protein-digesting enzymes in the egg, which boost the levels of taste-stimulating free amino acids. It also triggers another enzyme (transglutaminase) that cross-links proteins in the outer egg membrane and helps toughen it, thus giving the egg more texture. By generating a brine that gets drawn into the space between the outer and yolk membranes, salt plumps the egg, making it rounder and firmer. And by changing the distribution of electrical charges on the proteins within, it causes the proteins to bond to each other and thicken the watery egg fluids to a honey-like luxuriousness.

In sum, a light salting transforms fish eggs from a mere pleasant mouthful into the remarkable food known as caviar: a fleeting taste of the primordial brine and the savory molecules from which all life springs.

Caviar

Caviar appears to have arisen in Russia sometime around 1200 CE as a more palatable alternative to the traditional preserved sturgeon ovaries. Though the term *caviar* is now widely used to describe any sort of lightly salted loose fish eggs, for many centuries it referred only to loose sturgeon eggs. The most sought-after caviar still comes from a handful of sturgeon species mainly harvested by Russian and Iranian fishermen as the fish enter the rivers that drain into the Caspian Sea.

Just 150 years ago, sturgeon were common in many large rivers the northern hemisphere, and caviar was plentiful enough in Russia that Elena Molokhovets suggested using it to clarify bouillons and to decorate sauerkraut "so that it appears as if it were strewn with poppy seeds"! But overfishing, dams and hydroelectric plants, and industrial pollution have since put many sturgeon species in danger of extinction. Around 1900, sturgeon roe became rare, expensive, and therefore a sought-after luxury—and so even more expensive. The trend has continued, with Caspian sturgeon populations plummeting and U.N. organizations considering an export ban on caviar from the region. In recent decades, caviar production has been growing further east, along the Amur River in both Russia and China, and on sturgeon farms in the United States and elsewhere.

Making Caviar In traditional caviar-making, sturgeon are captured alive in nets, stunned, and their roe sacs removed before they are killed and butchered. The caviar maker passes the roe through screens to loosen the eggs and separate them from the ovary membrane, sorts and grades the eggs, and then dry-salts and mixes them by hand for two to four minutes to obtain a final salt concentration between 3 and 10%. (Small amounts of alkaline borax [sodium borate] have been used since the 1870s to replace part of the salt, making the caviar taste sweeter and improving its shelf life, but the United States and some other countries forbid borax in their imports.) The eggs are allowed to drain for 5 to 15 minutes, filled into large cans, and chilled to 26°F/–3°C (the salt prevents freezing at this temperature).

The most highly prized caviar is the most perishable. It goes by the Russian term *malossol,* which means "little salt," and ranges from 2.5–3.5% salt. The classic Caspian caviars have distinctive sizes, colors, and flavors. Beluga is the rarest, largest, and most expensive. Osetra, the most common wild caviar, comes mainly from the Black and Azov seas, is tinged with brown, and has a flavor reminiscent of oysters. Sevruga caviar is dark and has a less complicated flavor. "Pressed caviar" is a relatively inexpensive, saltier (to 7%), strong-tasting paste made from overmature eggs, and can be frozen.

Salmon and Other Caviars Russia pioneered the development of salmon caviar in the 1830s, and it's a delicious and affordable alternative, with its striking red-pink translucence and large grains. The sepa-

Commonly Eaten Fish Eggs

Source	Qualities, Names
Carp	Very small, light pink; sometimes salted Greece: *tarama*
Cod, pollack	Very small, pink, sometimes salted, pressed, dried, smoked Japan: *ajitsuki, tarako, momijiko*
Flying fish	Small, yellow, often dyed orange or black, crunchy Japan: *tobiko*
Grey mullet	Small; often salted; pressed and dried for *bottarga* Italy: *bottarga,* Greece: *tarama*; Japan: *karasumi*
Herring	Medium, yellow-gold, sometimes salt-cured; prized in Japan when attached to kelp Japan: *kazunoko*
Lumpfish	Small, fish common in North Atlantic and Baltic; greenish eggs often dyed red or black, heavily salted, pasteurized, bottled
Salmon	Large (4–5 mm) red-orange eggs mainly from chum salmon (*Oncorhynchus keta*), usually lightly brined and sold fresh Japan: whole ovary *sujiko,* separated eggs *ikura*
Shad	Small, from herring relative
Sturgeon	Medium-sized; lightly salted to make caviar
Trout	Large yellow eggs from Great Lakes trout
Tuna	Small; often salted; pressed and dried for *bottarga* Italy: *bottarga*
Whitefish	Small, golden, crunchy, from freshwater cousins of salmon in Northern Hemisphere; often flavored or smoked

rated eggs of chum and pink salmons are soaked in saturated brine for 2 to 20 minutes to achieve a final salt level of 3.5–4%, then drained and dried for up to 12 hours. Lumpfish caviar dates from the 1930s, when the sevruga-sized eggs of this otherwise little-used fish were salted and dyed to imitate the real thing. Whitefish eggs are similar in size and left undyed to retain their golden color. In recent years, the roe of herring, anchovy, and even lobster have been used to make caviars. Caviars may be pasteurized (120–160°F/50–70°C for 1–2 hours) to prolong their shelf life, but this can produce a rubbery off-aroma and chewy texture.

CHAPTER 5

EDIBLE PLANTS

An Introduction to Fruits and Vegetables, Herbs and Spices

We turn now from milk, eggs, meats, and fish, all expressions of animating protein and energizing fat, and enter the very different world that sustains them and us alike. The plant world encompasses earthy roots, bitter and pungent and refreshing leaves, perfumed flowers, mouth-filling fruits, nutty seeds, sweetness and tartness and astringency and pleasing pain, and aromas by the thousands! It turns out that this exuberantly diverse world was born of simple, harsh necessity. Plants can't move as animals do. In order to survive their immobile, exposed condition, they became virtuosic chemists. They construct themselves from the simplest materials of the earth itself, water and rock and air and light, and thus transform the earth into food on which all animal life depends. Plants deter enemies and attract friends with colors, tastes, and scents, all chemical inventions that have shaped our ideas of beauty and deliciousness. And they protect themselves from the common chemical stresses of living with substances that protect us as well. So when we eat vegetables and fruits and grains and spices, we eat the foods that made us possible, and that opened our life to a kaleidoscopic world of sensation and delight.

Human beings have always been plant eaters. For a million years and more, our omnivorous ancestors foraged and lived on a wide range of wild fruits, leaves, and seeds. Beginning around 10,000 years ago they domesticated a few grains, seed legumes, and tubers, which are among the richest sources of energy and protein in the plant world, and can be grown and stored in large quantities. This control over the food supply made it possible for many people to be fed reliably from a small patch of land: so cultivation of the fields led to settlement, the first cities, and cultivation of the human mind. On the other hand, agriculture drastically reduced the variety of plant foods in the human diet. Millennia later, industrialization reduced it even further. Fruits and vegetables became accessory, even marginal foods in the modern Western diet. Only recently have we begun to understand how the human body still depends for its long-term health on a various diet rich in fruits and vegetables, herbs and spices. Happily, modern technologies now give us unprecedented access to the world's cornucopia of edible plants. The time is ripe to explore this fascinating—and still evolving—legacy of natural and human inventiveness.

This chapter is a general introduction to the foods that we obtain from plants. Because there are so many of them, particular fruits and vegetables, herbs and spices are described in subsequent chapters. Foods derived from seeds—grains, legumes, nuts—have special properties, and are described separately in chapter 9.

The Original Food

The idea that plants are our original and therefore only proper food has deep cultural roots. In the Golden Age described by Greek and Roman mythology, the earth gave of itself freely, without cultivation, and humans ate only nuts and fruit. And in Genesis, Adam and Eve spend their brief innocence as gardeners:

> And the Lord God planted a garden eastward in Eden; and there he put the man whom he had formed. And out of the ground made the Lord God to grow every tree that is pleasant to the sight, and good for food . . . And the Lord God took the man, and put him into the Garden of Eden to dress it and to keep it.

The Bible doesn't mention meat as food until after it records the first killing, Cain's murder of his brother Abel. Many individuals and groups from Pythagoras to the present have chosen to eat only plant foods to avoid taking the life of another creature capable of feeling pain. And most people throughout history have had no choice, because meat is far more costly to produce than grains and tubers.

PLANTS AS FOOD

The Nature of Plants

Plants and animals are very different kinds of living things, and this is because they have evolved different solutions to a single basic challenge: how to obtain the energy and substance necessary to grow and reproduce. Plants essentially nourish themselves. They build their tissues out of water, minerals, and air, and run them on the energy in sunlight. Animals, on the other hand, can't extract energy and construct complex molecules from such primitive materials. They must obtain them premade, and they do so by consuming other living things. Plants are independent *autotrophs,* while animals are parasitic *heterotrophs.* (Parasitism may not sound especially admirable, but without it there would be no need to eat and so none of the pleasures of eating and cooking!)

There are various ways of being an autotroph. Some archaic bacteria, which are microbes consisting of a single cell, manipulate sulfur, nitrogen, and iron compounds to produce energy. The most important development for the future of eating came more than 3 billion years ago with the evolution of a bacterium that could tap the energy in sunlight and store it in carbohydrate molecules (molecules built from carbon, hydrogen, and oxygen). Chlorophyll, the green pigment we see in vegetation all around us, is a molecule that captures sunlight and initiates this process of *photosynthesis,* which culminates in the creation of the simple sugar glucose.

$$6CO_2 + 6H_2O + \text{light energy} \rightarrow C_6H_{12}O_6 + 6O_2$$

carbon dioxide + water + light energy →
glucose + oxygen

The bacteria that managed to "invent" chlorophyll gave rise to algae and all green land plants—and indirectly to land animals

The challenging life of the plant. Plants are rooted to one spot in the earth, where they absorb water and minerals from the soil, carbon dioxide and oxygen from the air, and light energy from the sun, and transform these inorganic materials into plant tissues—and into nourishment for insects and other animals. Plants defend themselves against predators with a variety of chemical weapons, some of which also make them flavorful, healthful, or both. In order to spread their offspring far and wide, some plants surround their seeds with tasty and nourishing fruits that animals carry away and eat, often spilling some seeds in the process.

as well. Before photosynthesis, the earth's atmosphere contained little oxygen, and the sun's killing ultraviolet rays penetrated all the way to the ground and several feet into the oceans. Living organisms could therefore survive only in deeper waters. When photosynthetic bacteria and early algae burgeoned, they liberated vast quantities of oxygen (O_2), which radiation in the upper atmosphere converted to ozone (O_3), which in turn absorbed ultraviolet light and prevented much of it from reaching the earth's surface. Land life was now possible.

So we owe our very existence as oxygen-breathing, land-dwelling animals to the greenery we walk through and cultivate and consume every day of our lives.

Why Plants Aren't Meaty Land-dwelling plants that can nourish themselves still need access to the soil for minerals and trapped water, to the atmosphere for carbon dioxide and oxygen, and to the sun for energy. All of these sources are pretty reliable, and plants have developed an economical structure that takes advantage of this reliability. Roots penetrate the soil to reach stable supplies of water and minerals; leaves maximize their surface area to capture sunlight and exchange gases with the air; and stalks support leaves and connect them with roots. Plants are essentially stationary chemical factories, made up of chambers for carbohydrate synthesis and carbohydrate storage, and tubes to transfer chemicals from one part of the factory to another, with structural reinforcement—also mainly carbohydrates—to provide mechanical rigidity and strength. Parasitic animals, by contrast, must find and feed on other organisms, so they are constructed mainly of muscle proteins that transform chemical energy into physical motion (p. 121).

Why Plants Have Strong Flavors and Effects Animals can also use their mobility to avoid becoming another creature's meal, by fleeing or fighting. But stationary plants? They compensate for their immobility with a remarkable ability for chemical synthesis. These master alchemists produce thousands of strong-tasting, sometimes poisonous warning signals that discourage bacteria, fungi, insects, and us from attacking them. A partial list of their chemical warfare agents would include irritating compounds like mustard oil, hot-chilli capsaicin, and the tear-inducing factor in onions; bitter and toxic alkaloids like caffeine in coffee and solanine in potatoes; the cyanide compounds found in lima beans and many fruit seeds; and substances that interfere with the digestive process, including astringent tannins and inhibitors of digestive enzymes.

If plants are so well endowed with their own natural pesticides, then why isn't the world littered with the corpses of their victims? Because animals have learned to recognize and avoid potentially harmful plants with the help of their senses of smell and taste, which can detect chemical compounds in very small concentrations. Animals have developed appropriate innate responses to significant tastes—aversion to the bitterness typical of alkaloids and cyanide, attraction to the sweetness of nutritionally important sugars. And some animals have developed specific detoxifying enzymes that enable them to exploit an otherwise toxic plant. The koala bear can eat eucalyptus leaves, and monarch butterfly caterpillars milkweed. Humans invented their own ingenious detoxifying methods, including plant selection and breeding and cooking. Cultivated varieties of such vegetables as cabbage, lima beans, potatoes, and lettuce are less toxic than their wild ancestors. And many toxins can be destroyed by heat or leached away in boiling water.

A fascinating wrinkle in this story is that humans actually prize and seek out certain plant toxins! We've managed to learn which irritating warning signals are relatively harmless, and have come to enjoy sensations whose actual purpose is to repel us. Hence our seemingly perverse love of mustard and pepper and onions. This is the essential appeal of herbs and spices, as we'll see in chapter 8.

Why Ripe Fruits Are Especially Delicious The higher plants and animals reproduce by fusing genetic material from male and female sex organs, usually from different individuals. Animals have the advantage of being mobile: male and female can sense each other's presence and move toward each other. Plants can't move, and instead have to depend on mobile go-betweens. The male pollen of most land plants is carried to the female ovule by the wind or by animals. To encourage animals to help out, advanced plants evolved the flower, an organ whose shape, color, and scent are designed to attract a particular assistant, usually an insect. As it flies around and collects nutritious nectar or pollen for food, the insect spreads the pollen from one plant to another.

Once male and female cells have come together and developed into offspring, they must be given a good start. The animal mother can search out a promising location and deposit her young there. But plants need help. If the seeds simply dropped from the plant to the ground, they would have to compete with each other and with their overshadowing parent for sunlight and soil minerals. So successful plant families have developed mechanisms for dispersing their seeds far and wide. These mechanisms include seed containers that pop open and propel their contents in all directions, seed appendages that catch the wind or the fur of a passing animal—and structures that hitch a ride *inside* passersby. Fruits are plant organs that actually invite animals to eat them, so that the animals will carry their seeds away, and often pass them through their digestive system and deposit them in a nourishing pile of manure. (The seeds escape destruction in various ways, among them by being large and armored, or tiny and easily spilled, or poisonous.)

So, unlike the rest of the plant, fruit is *meant* to be eaten. This is why its taste, odor, and texture are so appealing to our animal senses. But the invitation to eat must be delayed until the seeds are mature and viable. This is the purpose of the changes in color, texture, and flavor that we call ripening. Leaves, roots, stalks can be eaten at any time, generally the earlier the tenderer. But we must wait for fruit to signal that it is ready to be eaten. The details of ripening are described in chapter 7 (p. 350).

Our Evolutionary Partners Like us, most of our food plants are relative newcomers to the earth. Life arose about 4 billion years ago, but flowering plants have been around for only about 200 million years, and dominant for the last 50 million. An even more recent development is the "herbaceous" habit of life. Most food plants are not long-lived trees, but relatively small, delicate plants that produce their seeds and die in one growing season. This herbaceous habit gives plants greater flexibility in adapting to changing conditions, and it has worked to our advantage as well. It allows us to grow crops to maturity in a few months, change plantings from year to year, rapidly breed new varieties, and eat plant parts that would be inedible were they toughened to endure for years. Herbaceous plants became widespread only in the last few million years, just as the human species was emerging. They made possible our rapid cultural development, and we in turn have used selection and breeding to direct their biological development. We and our food plants have been partners in each other's evolution.

Definitions

We group the foods we obtain from plants into several loose categories.

Fruit and Vegetable Apart from such plant seeds as wheat and rice, which are described in chapter 9, the most prominent plant foods in our diet are fruits and vegetables. *Vegetable* took on its current sense just a few centuries ago, and essentially means a plant material that is neither fruit nor seed. So what is a fruit? The word has both a technical and a common meaning. Beginning in the 17th century, botanists defined it as the organ that develops from the flower's ovary and surrounds the plant's seeds. But in common usage, seed-surrounding green beans, eggplants, cucumbers, and corn kernels are called vegetables, not fruits. Even the United States Supreme Court has preferred the cook's definition over the botanist's. In the 1890s, a New York food importer claimed duty-free status for a shipment of tomatoes, arguing that tomatoes were fruits, and so under the regulations of the time, not subject to import fees. The customs agent ruled that tomatoes were vegetables and imposed a duty. A majority of the Supreme Court decided that tomatoes were "usually served at a dinner in, with, or after the soup, fish, or meat, which constitute the principal part of the repast, and not, like fruits, generally as dessert." Ergo tomatoes were vegetables, and the importer had to pay.

The Key Distinction: Flavor Why do we customarily prepare vegetables as side dishes to the main course, and make fruits the centerpiece of the meal's climax? Culinary fruits are distinguished from vegetables by one important characteristic: they're among the few things we eat that we're meant to eat. Many plants have engineered their fruits to appeal to the animal senses, so that animals will eat them and disperse the seeds within. These fruits are the natural world's soft drinks and candies, flashily packaged in bright colors, and test-marketed through millions of years of natural selection. They tend to have a high sugar content, to satisfy the innate liking for sweetness shared by all animals. They have a pronounced and complex aroma, which may involve several hundred different chemicals, far more than any other natural ingredient. And they soften themselves to an appealingly tender, moist consistency. By contrast, the plant foods that we treat as vegetables remain firm, have either a very mild flavor—green beans and potatoes—or else an excessively strong one—onions and cabbage—and therefore require the craft of the cook to make them palatable.

The very words *fruit* and *vegetable* reflect these differences. *Vegetable* comes from the Latin verb *vegere,* meaning to invigorate or enliven. *Fruit,* on the other hand, comes from Latin *fructus,* whose cluster of related meanings includes gratification, pleasure, satisfaction, enjoyment. It's the nature of fruit to taste good, to appeal to our basic biological interests, while vegetables stimulate us to find and create more subtle and diverse pleasures than fruits have to offer.

Herb and Spice The terms *herb* and *spice* are more straightforward. Both are categories of plant materials used primarily as flavorings, and in relatively small amounts. Herbs come from green parts of plants, usually leaves—parsley, thyme, basil—while spices are generally seeds, bark, underground stems—black pepper, cinnamon, ginger—and other robust materials that were well suited to international trade in early times. The word *spice* came from the medieval Latin *species,* which meant "kind of merchandise."

Despite the fact that we consider them vegetables, capsicum "peppers," pea pods, cucumbers, and even corn kernels are actually fruits: plant parts that originate in the flower's ovary and surround one or more seeds.

Plant Foods Through History

How long has the Western world been eating the plant foods we eat today, and in the way that we eat them? Only a very few common vegetables have *not* been eaten since before recorded history (the relative newcomers include broccoli, cauliflower, brussels sprouts, celery). But it was only with the age of exploration in the 16th century that the variety of foods we now know became available to any single culture. In the Western world, fruit has been eaten as dessert at least since the Greeks; recognizable salads go back to the Middle Ages, and boiled vegetables in delicate sauces to 17th-century France.

Prehistory and Early Civilizations Many plants came under human cultivation by the unsophisticated but slowly effective means of gathering useful plants and leaving a few seeds in fertile refuse heaps. Judging from archaeological evidence, early Europeans seem to have relied on wheat, fava beans, peas, turnips, onions, radishes, and cabbage. In Central America, corn, beans, hard squashes, tomatoes, and avocados were staples around 3500 BCE, while Peruvian settlements relied heavily on the potato. Northern Asia started with millets, cabbage relatives, soybeans, and tree fruits in the apple and peach families; southern Asia had rice, bananas, coconuts, yams, cabbage relatives, and citrus fruits. Indigenous African crops included related but distinct millets, sorghum, rice, and bananas, as well as yams and cowpeas. Mustard seed flavored foods in Europe and in Asia, where ginger may also have been used. Chilli "pepper" was probably the chief spice in the Americas.

By the time of the earliest civilizations in Sumer and Egypt about 5,000 years ago, most of the plants native to that area and eaten today were already in use (see box, p. 250). Trade between the Middle East and Asia is also ancient. Egyptian records of around 1200 BCE document huge offerings of cinnamon, a product of Sri Lanka.

Greece, Rome, and the Middle Ages With the Greeks and Romans we begin to see the outlines of modern Western cuisine. The Greeks were fond of lettuce, and habitually ate fruit at the end of meals. Pepper from the Far East was in use around 500 BCE and quickly became the most popular spice of the ancient world. In Rome, lettuce was served at both the beginning and end of meals, and fruit as dessert. Thanks to the art of grafting growing shoots from desirable trees onto other trees, there were about 25 named apple varieties and 35 pears. Fruits were preserved whole by immersing them, stems and all, in honey, and the gastronome Apicius gave a recipe for pickled peaches. From the Roman recipes that survive, it would seem that few foods were served without the application of several strong flavors.

When the Romans conquered Europe they brought along tree fruits, the vine, and cultivated cabbage, as well as their heavy spice habit. Sauce recipes from the 14th century resemble those of Apicius, and the English lettuce-free salad would also have been quite pungent (see box, p. 251). Medieval recipe collections include relatively few vegetable dishes.

New World, New Foods Plants—and especially the spice plants—helped shape world history in the last five centuries. The ancient European hunger for Asian spices was an important driving force in the development of Italy, Portugal, Spain, Holland, and England into major sea powers during the Renaissance. Columbus, Vasco da Gama, John Cabot, and Magellan were looking for a new route to the Indies in order to break the monopoly of Venice and southern Arabia on the ancient trade in cinnamon, cloves, nutmeg, and black pepper. They failed in that quest, but succeeded in opening the "West Indies" to European exploitation. The New World was initially disappointing in its yield of sought-for spices. But vanilla and chillis quickly became popular; and its wealth of new vegetables was largely adaptable to

Vegetables, Fruits, and Spices Used in the West

Vegetables		Fruits	Herbs and Spices	
Mediterranean Area Natives, Used BCE				
Mushroom Beet Radish Turnip Carrot Parsnip Asparagus Leek	Onion Cabbage Lettuce Artichoke Cucumber Broad bean Pea Olive	Apple Pear Cherry Grape Fig Date Strawberry	Basil Marjoram Fennel Mint Rosemary Sage Savory Thyme Anise Caraway Coriander Cumin	Dill Parsley Oregano Bay Caper Fenugreek Garlic Mustard Poppy Sesame Saffron
Later Additions				
Spinach Celery Rhubarb Cauliflower Broccoli Brussels sprouts				
Asian Natives, Brought to the West BCE				
		Citron Apricot Peach	Cardamom Ginger Cinnamon Turmeric Black pepper	
Imported Later				
Yam Water chestnut Bamboo Eggplant		Lemon Lime Orange Melon	Tarragon Mace	Clove Nutmeg
New World Natives, Imported 15th–16th Centuries				
Potato Sweet potato Pumpkin Squashes Tomato	Kidney bean Lima bean Capsicum pepper Avocado	Pineapple	Allspice Chillis Vanilla	

Europe's climate: so the common bean, corn, squashes, tomatoes, potatoes, and sweet chillis eventually became staple ingredients in the new cuisines of the Old World.

The 17th and 18th centuries were a time of assimilating the new foods and advancing the art of cooking them. Cultivation and breeding received new attention; Louis XIV's orchards and plantings at Versailles were legendary. And cooks took a greater interest in vegetables, and handled them with greater refinement, in part to make the meatless diet of Lent and other Catholic fasts more interesting. France's first great culinary writer, Pierre François de La Varenne, chef to Henri IV, included meatless recipes for peas, turnips, lettuce, spinach, cucumbers, cabbage (five ways), chicory, celery, carrots, cardoons, and beets,

Plant Ingredients in Rome and Medieval Europe

A Roman Sauce for Shellfish

Cumin Sauce, for Shellfish: Pepper, lovage, parsley, mint, aromatic leaf [e.g., bay], malabathrum [a Middle Eastern leaf], plenty of cumin, honey, vinegar, liquamen [a fermented fish paste similar to our anchovy paste].

—from Apicius, first few centuries CE

*Medieval Sauces, French (Taillevent, ca. 1375) and English (*The Forme of Cury*, ca. 1390)*

Sauce Cameline, for Meats:
France: Ginger, mace, cinnamon, cloves, grain of paradise, pepper, vinegar, bread [to thicken].
England: Ginger, cloves, cinnamon, currants, nuts, vinegar, bread crusts.

Verde Sauce:
France: Parsley, ginger, vinegar, bread.
England: Parsley, ginger, vinegar, bread, mint, garlic, thyme, sage, cinnamon, pepper, saffron, salt, wine.

*Salad and a Vegetable Compote (*The Forme of Cury*, ca. 1390)*

Salat: Take parsley, sage, garlic, scallions, onions, leeks, borage, mints, young leeks, fennel, cress, new rosemary, purslane; wash them clean; pick them and pluck them small with your hands, and mix them well with raw oil. Lay on vinegar and salt, and serve it forth.

Compost: Take root of parsley and parsnip, scrape them and wash them clean. Take turnips and cabbages pared and cut. Take an earthen pan with clean water, and set it on the fire. Cast all these things in. When they are boiled, add pears and parboil them well. Take these things out and let them cool on a fair cloth. Put in a vessel and add salt when it is cold. Take vinegar and powder and saffron and add. And let all these things lie there all night or day. Take Greek wine and honey clarified together, Lombardy mustard, and raisins, whole currants, and grind sweet powder and whole anise, and fennel seed. Take all these things and cast them together in a pot of earth, and take some when you wish, and serve it forth.

as well as ordinary dishes of artichokes, asparagus, mushrooms, and cauliflower. And the recipes leave a major role for the vegetables' own flavors. Similarly, the Englishman John Evelyn wrote a book-length disquisition on salads, once again firmly based on the lettuces, and emphasized the importance of balance.

With the 19th century, English vegetable cooking became ever simpler until it almost always meant boiled and buttered, a quick and simple method for homes and restaurants alike, while in France the elaborate professional style reached its apogee. The influential chef Antonin Carême declared in his *Art of French Cooking in the 19th Century* (1835) that "it is in the confection of the Lenten cuisine that the chef's science must shine with new luster." Carême's enlarged repertoire included broccoli, truffles, eggplant, sweet potatoes, and potatoes, these last fixed *à l'anglaise, dites, Mache-Potetesse* ("in the English style, that is, mashed"). Of course, such luster tends to undermine the whole point of Lent. In his *366 Menus* (1872), Baron Brisse asked: "Are the meatless meals of our Lenten enthusiasts really meals of abstinence?"

The Influence of Modern Technology

The age of exploration and the advancement of fine cooking brought a new prominence to fruits and vegetables in Europe. Then the social and technical innovations of the industrial age conspired to make them both less available and less desirable. Beginning early in the 19th century, as industrialization drew people from the agricultural countryside to the cities, fruits and vegetables became progressively rarer in the diets of Europe and North America. Urban supplies did improve with the development of rail transportation in the 1820s, then canning at mid-century, and refrigeration a few decades later. Around the turn of the 20th century, vitamins and their nutritional significance were discovered, and fruits and vegetables were soon officially canonized as one of the four food groups that should be eaten at every meal.

Refinements of 17th-Century Vegetable Cooking

Choose the largest asparagus, scrape them at the bottom, and wash. Cook them in some water, salt them well, and do not let them overcook. When done, let them drain, and make a sauce with some good fresh butter, a little vinegar, salt, and nutmeg, and an egg yolk to bind the sauce; take care that it doesn't curdle. Serve the asparagus well garnished with whatever you like.

—La Varenne, *Le Cuisinier françois,* 1655

. . . by reason of its soporifous quality, lettuce ever was, and still continues the principal foundation of the universal tribe of Sallets, which is to cool and refresh, besides its other properties [which included beneficial influences on "morals, temperance, and chastity"]. We have said how necessary it is that in the composure of a sallet, every plant should come in to bear its part, without being overpower'd by some herb of a stronger taste, so as to endanger the native sapor and virtue of the rest; but fall into their places, like the notes in music, in which there should be nothing harsh or grating: And though admitting some discords (to distinguish and illustrate the rest) striking in all the more sprightly, and sometimes gentler notes, reconcile all dissonancies, and melt them into an agreeable composition.

—John Evelyn, *Acetaria: A Discourse of Sallets,* 1699

Still, the consumption of fresh produce continued to decline through much of the 20th century, at least in part because its quality and variety were also declining. In the modern system of food production, with crops being handled in massive quantities and shipped thousands of miles, the most important crop characteristics became productivity, uniformity, and durability. Rather than being bred for flavor and harvested at flavor's peak, fruits and vegetables were bred to withstand the rigors of mechanical harvesting, transport, and storage, and were harvested while still hard, often weeks or months before they would be sold and eaten. A few mediocre varieties came to dominate the market, while thousands of others, the legacy of centuries of breeding, disappeared or survived only in backyard gardens.

At the end of the 20th century, several developments in the industrialized world brought renewed attention to plant foods, to their diversity and quality. One was a new appreciation of their importance for human health, thanks to the discovery of trace "phytochemicals" that appear to help fight cancer and heart disease (p. 255). Another was the growing interest in exotic and unfamiliar cuisines and ingredients, and their increasing availability in ethnic markets. Yet another, at the opposite extreme, was the rediscovery of the traditional system of food production and its pleasures: eating locally grown foods, often forgotten "heirloom" or other unusual varieties, that were harvested a matter of hours beforehand, then sold at farmers' markets by the people who grew them. Allied to this trend was the growing interest in "organic" foods, produced without relying on the modern array of chemicals for controlling pests and disease. Organic practices mean different things to different people, and don't guarantee either safer or more nutritious foods—agriculture is more complicated than that. But they represent an essential, prominent alternative to industrial farming, one that encourages attention to the quality of agricultural produce and the sustainability of agricultural practices.

These are good times for curious and adventurous eaters. There are many forgotten varieties of familiar fruits and vegetables to revive, and many new foods to taste. It's estimated that there are 300,000 edible plant species on earth, and perhaps 2,000 that are cultivated to some extent. We have plenty of exploring to do!

PLANT FOODS AND HEALTH

Plant foods can provide us all the nourishment we need in order to live and thrive. Our primate ancestors started out eating little else, and many cultures still do. But meat and other animal foods became important to our species at its birth, when their concentrated energy and protein probably helped accelerate our evolution (p. 119). Meat continued to have a deep biological appeal for us, and in societies that could afford to feed livestock on staple grains and roots, it became the most prized of foods. In the industrialized world, meat's prestige and availability pushed grains, vegetables, and fruits to the side of the plate and the end of the meal. And for decades, nutritional science affirmed their accessory status. Fruits and vegetables in particular were considered to be the source of a few nutrients that we need only in small amounts, and of mechanically useful roughage. In recent years, though, we've begun to realize just how many valuable substances plant foods have always held for us. And we're still learning.

Essential Nutrients in Fruits and Vegetables: Vitamins

Most fruits and vegetables contribute only modestly to our intake of proteins and calories, but they're our major source for several vitamins. They provide nearly all of our vitamin C, much of our folic acid, and half of our vitamin A. Each of these plays a

Genetic Engineering and Food

The most far-reaching development in 20th-century agriculture was the introduction in the 1980s of genetic engineering, the technology that makes it possible to alter our food plants and animals by surgically precise manipulation of the DNA that makes up their genes. This manipulation bypasses the natural barriers between species, so theoretically a gene from any living thing, plant or animal or microbe, can be introduced into any other.

Genetic engineering is still in its infancy, and to date has had a limited impact on the foods we eat. In the United States, an estimated 75% of all processed foods now contain genetically modified ingredients. But this remarkable figure is due to just three agricultural commodities—soybean, canola, and corn—all of them modified for improved resistance to insect pests or herbicides. As I write in 2004, the only other significant engineered U.S. crop is Hawaiian papaya, which is now resistant to a formerly devastating virus disease. A few other foods are processed with enzymes made in engineered microbes—for example, much cheese is coagulated with rennet made by microbes into which the cattle gene for the enzyme has been inserted. But in general, our raw ingredients remain relatively untouched by genetic engineering.

This will certainly change in coming years, and not just in the West: China also has a very active program in agricultural biotechnology. Genetic engineering is the modern fruit of agriculture itself, an outgrowth of the ancient human realization that living things can be shaped to human desires. That shaping began when the first farmers selectively cultivated plants and animals that grew larger or tasted better or looked more interesting. In its own way, this simple process of observation and selection became a powerful biological technology. It gradually revealed the hidden potential for diversity within individual species, and made that potential real in the form of hundreds of distinct varieties of wheat and cattle, citrus fruits and chillis, many of which had never before existed in nature. Today, genetic engineers are exploring the hidden potential for improving a given food plant or animal not just within that species, but among all species, in the entire living world's cornucopia of DNA and its possible modifications.

Genetic engineering holds the promise of bringing great improvements to the production and quality of our foods. However, like any powerful new technology, it also has the potential to cause unintended and far-reaching consequences. And as the instrument of industrial agriculture, it's likely to contribute to the ongoing erosion of traditional, decentralized, small-scale food production and its ancient heritage of biological and cultural diversity. It's important that these environmental, social, and economic issues be considered by all concerned—by the biotechnology and agriculture industries, the governments that regulate them, the farmers who plant and raise their products, the cooks and manufacturers who turn the products into something edible, and the consumers who support the whole system by buying and eating food—so that in the long run this new agricultural revolution will benefit the common good as much as possible.

number of roles in the metabolism of our cells. For example, vitamin C refreshes the chemical state of metal components in many enzymes, and helps with the synthesis of connective-tissue collagen. Vitamin A, which our bodies make from a precursor molecule in plants called beta-carotene (p. 267), helps regulate the growth of several different kinds of cells, and helps our eyes detect light. Folic acid, named from the Latin word for "leaf," converts a by-product of our cells' metabolism, homocysteine, into the amino acid methionine. This prevents homocysteine levels from rising, causing damage to blood vessels, and possibly contributing to heart disease and stroke.

Vitamins A, C, and E are also antioxidants (see below).

Phytochemicals

The first edition of this book reflected the prevailing nutritional wisdom circa 1980: we should eat enough fruits and vegetables to avoid vitamin and mineral deficiencies, and to keep our digestive system moving. Period.

What a difference 20 years makes!

Nutritional science has undergone a profound revolution in that time. For most of the 20th century it aimed to define an *adequate* diet. It determined our body's minimal requirements for chemical building blocks (protein, minerals, fatty acids), for essential cogs in its machinery (vitamins), and for the energy it needs to run and maintain itself from day to day. Toward the end of the century, it became clear from laboratory studies and comparisons of health statistics in different countries that the major diseases of the adequately nourished developed world—cancer and heart disease—are influenced by what we eat. Nutritional science then began to focus on defining the elements of an *optimal* diet. So we discovered that minor, nonessential food components can have a cumulative effect on our long-term health. And plants, the planet's biochemical virtuosos, turn out to be teeming with trace *phytochemicals*—from the Greek *phyton*, meaning "leaf"—that modulate our metabolism.

Antioxidants

Oxidative Damage: The Price of Living

One major theme in modern nutrition is the body's need to cope with the chemical wear and tear of life itself. Breathing is essential to human life because our cells use oxygen to react with sugars and fats and generate the chemical energy that keeps the cellular machinery functioning. Unfortunately, it turns out that energy generation and other essential processes involving oxygen generate chemical by-products called "free radicals," very unstable chemicals that react with and damage our own complex and delicate chemical machinery. This damage is called *oxidative* because it usually originates in reactions involving oxygen. It can affect different parts of the cell, and different organs in the body. For example, oxidative damage to a cell's DNA can cause that cell to multiply uncontrollably and grow into a tumor. Oxidative damage to the cholesterol-carrying particles in our blood can irritate the lining of our arteries, and initiate damage that leads to a heart attack or stroke. The high-energy ultraviolet rays in sunlight create free radicals in the eye that damage proteins in the lens and retina, and cause cataracts, macular degeneration, and blindness.

Our bodies stave off such drastic consequences by means of *antioxidant* molecules, which react harmlessly with free radicals before they have a chance to do any damage to the cells' chemical machinery. We need a continuous and abundant supply of antioxidants to maintain our good health. The body does make a few important antioxidant molecules of its own, including some powerful enzymes. But the more help it gets, the better it's able to defend itself from the constant onslaught of free radicals. And plants turn out to be a goldmine of antioxidants.

Some Beneficial Effects of Chemicals in Fruits and Vegetables, Herbs and Spices

This is a very broad survey of a rich and complex subject. It's meant to give a general idea of how a variety of plant chemicals can affect various aspects of our health by a variety of means. Certain phenolic compounds, for example, appear capable of helping us fight cancer by preventing oxidative damage to DNA in healthy cells, by preventing the body from forming its own DNA-damaging chemicals, and by inhibiting the growth of already cancerous cells.

Prevent oxidative damage to important molecules in body: antioxidants

- **Eye:** slow cataracts and macular degeneration
  - Kale, many dark green vegetables (carotenoids: lutein)
  - Citrus fruits, corn (carotenoids: zeaxanthin)
- **Blood lipids:** slow development of heart disease
  - Grapes, other berries (phenolics: anthocyanidins)
  - Tea (phenolics)
- **General:** reduce DNA damage, development of cancer
  - Tomatoes (carotenoids: lycopene)
  - Carrots, other orange and green vegetables (carotenoids)
  - Tea (phenolics)
  - Green vegetables (chlorophyll)
  - Broccoli, daikon, cabbage family (glucosinolates, thiocyanates)

Moderate the body's inflammatory response

- **General:** slow development of heart disease, cancer
  - Raisins, dates, chillis, tomatoes (salicylates)

Reduce the body's own production of DNA-damaging chemicals

- Many fruits, vegetables (phenolics: flavonoids)
- Broccoli, daikon, cabbage family (glucosinolates, thiocyanates)
- Citrus fruits (terpenes)

Inhibit the growth of cancer cells and tumors

- Many fruits, vegetables (phenolics: flavonoids)
- Soybeans (phenolics: isoflavones)
- Grapes, berries (phenolics: ellagic acid)
- Rye, flaxseed (phenolics: lignans)
- Citrus fruits (terpenes)
- Mushrooms (carbohydrates)

Slow the body's removal of calcium from bones

- Onions, parsley (responsible agents not yet identified)

Encourage the growth of beneficial bacteria in the intestine

- Onion family, sunchokes (inulin)

Prevent the adhesion of infectious bacteria to walls of urinary tract

- Cranberries, grapes (phenolics: proanthocyanidins)

Antioxidants in Plants Nowhere in living things is oxidative stress greater than in the photosynthesizing leaf of a green plant, which harvests energetic particles of sunlight, and uses them to split water molecules apart into hydrogen and oxygen atoms in order to make sugars. Leaves and other exposed plant parts are accordingly chock-full of antioxidant molecules that keep these high-energy reactions from damaging essential DNA and proteins. Among these plant antioxidants are the carotenoid pigments, including orange beta-carotene, yellow lutein and zeaxanthin, and the red lycopene that colors tomato fruits. Green chlorophyll itself is an antioxidant, as are vitamins C and E. Then there are thousands of different "phenolic" compounds built from rings of 6 carbon atoms, which play several roles in plant life, from pigmentation to antimicrobial duty to attracting and repelling animals. All fruits, vegetables, and grains probably contain at least a few kinds of phenolic compounds; and the more pigmented and astringent they are, the more they're likely to be rich in phenolic antioxidants.

Each plant part, each fruit and vegetable, has its own characteristic cluster of antioxidants. And each kind of antioxidant generally protects against a certain kind of molecular damage, or helps regenerate certain other protective molecules. No single molecule can protect against all kinds of damage. Unusually high concentrations of single types can actually tip the balance the wrong way and *cause* damage. So the best way to reap the full benefits of the antioxidant powers of plants is not to take manufactured supplements of a few prominent chemicals: it is to eat lots of different vegetables and fruits.

Other Beneficial Phytochemicals

Antioxidants may be the most important group of ingredients for maintaining long-term health, but they're not the only one. Trace chemicals in plants, including herbs and spices, are turning out to have helpful effects on many other processes that affect the balance between health and disease. For example, some act like aspirin (originally found in plants) to prevent the body from overreacting to minor damage with an inflammation that can lead to heart disease or cancer; some prevent the body from turning mildly toxic chemicals into more powerful toxins that damage DNA and cause cancer; some inhibit the growth of cells that are already cancerous. Others slow the loss of calcium from our bones, encourage the growth of beneficial bacteria in our system, and discourage the growth of disease bacteria.

The box on p. 256 lists some of these effects, and the chemicals and plants that cause them. Our knowledge of this aspect of nutrition is still in its infancy, but we know enough right now for at least one conclusion to be evident: no single fruit or vegetable offers the many kinds of protections that a varied diet can provide.

So today's provisional nutritional wisdom goes like this: fruits and vegetables, herbs and spices supply us with many different beneficial substances. Within an otherwise adequate diet, we should eat as much of them as we can, and as great a variety as we can.

Estimating Healthfulness by Eye There's a useful guideline for estimating the relative healthfulness of vegetables and fruits: the deeper its color, the more healthful the food is likely to be. The more light a leaf gets, the more pigments and antioxidants it needs to handle the energy input, and so the darker the coloration of the leaf. For example, the light-colored inner leaves of lettuce and cabbage varieties that form tight heads contain a fraction of the carotene found in the darker outer leaves and in the leaves of more open varieties. Similarly, the dark leaves of open romaine lettuce contain nearly 10 times the eye-protecting lutein and zeaxanthin of the pale, tight heads of iceberg lettuce. Other deeply colored fruits and vegetables also contain more beneficial carotenoids and phenolic compounds than their pale counterparts. Their skins are especially rich sources. Among the fruits highest in antioxidant content are cherries, red grapes, blueberries, and strawberries; among vegetables, garlic, red and yellow onions, asparagus, green beans, and beets.

Fiber

Fiber is defined as the material in our plant foods that our digestive enzymes can't break down into absorbable nutrients. These substances therefore aren't absorbed in the small intestine, and pass intact into the large intestine, where some are broken down by intestinal bacteria, and the rest are excreted. The four main components of fiber come from plant cell walls (p. 265). Cellulose and lignin form solid fibers that don't dissolve in our watery digestive fluids, while pectins and hemicelluloses do dissolve into their individual molecules. Minor components of fiber include uncooked starch and various gums, mucilages, and other unusual carbohydrates (e.g., mushroom chitin, seaweed agar and carrageenan, inulin in onions, artichokes, and sunchokes). Particular foods offer particular kinds of fiber. Wheat bran—the dry outer coat of the grain—is a rich source of insoluble cellulose, while oat bran is a rich source of soluble glucan (a carbohydrate), and juicy ripe fruits are a relatively dilute source of soluble pectins.

The different fiber components contribute to health in different ways. Insoluble cellulose and lignin mainly provide bulk to the intestinal contents, and thus increase the rate and ease with which they pass through the large intestine. It's thought that rapid excretion may help minimize our exposure to DNA-damaging chemicals and other toxins in our foods, and the fiber materials may bind some of these toxins and prevent them from being absorbed by our cells. Soluble fiber components make the intestinal contents thicker, so that there is slower mixing and movement of both nutrients and toxins. They, too, probably bind certain chemicals and prevent their absorption. Soluble fiber has been shown to lower blood cholesterol and slow the rise of blood sugar after a meal. Inulin in particular encourages the growth of beneficial intestinal bacteria, while reducing the numbers of potential troublemakers. The details are complex, but overall it appears that soluble fiber helps protect against heart disease and diabetes.

In sum, the indigestible portion of fruits and vegetables does us good. It's a mistake to think that a juiced orange or carrot is as valuable as the whole fruit or vegetable.

Toxins in Some Fruits and Vegetables

Many plants, perhaps all plants, contain chemicals meant to discourage animals from eating them. The fruits and vegetables that we eat are no exception. While domestication and breeding have reduced their toxin contents to the point that they're not generally hazardous, unusual preparations or serving sizes can cause problems. The following plant toxins are worth being aware of.

Alkaloids Alkaloids are bitter-tasting toxins that appeared in plants about the time that mammals evolved, and seem especially effective at deterring our branch of the animal family by both taste and aftereffects. Almost all known alkaloids are poisonous at high doses, and most alter animal metabolism at lower doses: hence the attractions of caffeine and nicotine. Among familiar foods, only the potato accumulates potentially troublesome alkaloid levels, which make greened potatoes and potato sprouts bitter and toxic (p. 302).

Cyanogens Cyanogens are molecules that warn and poison animals with bitter hydrogen cyanide, a deadly poison of the enzymes that animals use to generate energy. When the plant's tissue is damaged by chewing, the cyanogens are mixed with the plant enzyme that breaks them apart and releases hydrogen cyanide (HCN). Cyanogen-rich foods, including manioc, bamboo shoots, and tropical varieties of lima beans, are made safe for consumption by open boiling, leaching in water, and fermentation. The seeds of citrus, stone, and pome fruits generate cyanide, and stone-fruit seeds are prized because their cyanogens also produce benzaldehyde, the characteristic odor of almond extract (p. 506).

Hydrazines Hydrazines are nitrogen-containing substances that are found in relatively large amounts (500 parts per million) in the common white mushroom and other mushroom varieties, and that persist after cooking. Mushroom hydrazines cause liver damage and cancer when fed to laboratory mice, but have no effect in rats. It's not yet clear whether they pose a significant hazard to humans. Until we know, it's best to eat mushrooms in moderation.

Protease Inhibitors and Lectins These are proteins that interfere with digestion: inhibitors block the action of protein-digesting enzymes, and lectins bind to intestinal cells and prevent them from absorbing nutrients. Lectins can also enter the bloodstream and bind red blood cells to each other. They're found mainly in soy, kidney, and lima beans. Both inhibitors and lectins are inactivated by prolonged boiling. But they can survive in beans that are eaten raw or undercooked, and cause symptoms similar to food poisoning.

Flavor Chemicals Flavor chemicals are generally consumed in only tiny amounts, but a few may cause problems when overindulged in. Safrole, the main aromatic in oil of sassafras and therefore of traditional root beer, causes DNA damage and was banned as an additive in 1960 (root beer is now made with safe sarsaparilla or artificial flavorings). Myristicin, the major flavor contributor in nutmeg, seems largely responsible for intoxication and hallucinations that result from ingesting large amounts. Glycyrrhizin, an intensely sweet-tasting substance in true licorice root, induces high blood pressure. Coumarin, which gives sweet clover its sweet aroma and is also found in lavender and vanilla-like tonka beans (*Dipteryx odorata*), interferes with blood clotting.

Toxic Amino Acids Toxic amino acids are unusual versions of the building blocks for our proteins that interfere with proper protein functioning. Canavanine interferes with several cell functions and has been associated with the development of lupus; it's found in large quantities in alfalfa sprouts and the jack bean. Vicine and convicine in the fava bean cause a blood-cell-destroying anemia, favism, in susceptible people (p. 490).

Oxalates Oxalates are various salts of oxalic acid, a waste product of plant metabolism found in a number of foods, notably spinach, chard, beets, amaranth, and rhubarb. The sodium and potassium salts are soluble, while the calcium salts are insoluble and form crystals that irritate the mouth and digestive system. Soluble oxalates can combine with calcium in the human kidney to form painful kidney stones. In very large doses—a few grams—oxalic acid is corrosive and can be fatal.

Bracken-Fern Toxins Bracken-fern toxins cause several blood disorders and cancer in animals that graze on this common fern (*Pteridium*), which is sometimes collected in the young "fiddlehead" stage for human consumption. Ostrich ferns, *Matteuccia* species, are thought to be a safer source of fiddleheads, but there's little solid information about the safety of eating ferns. It's prudent to eat fiddleheads in moderation, and to avoid bracken ferns by checking labels and asking produce sellers.

Psoralens Psoralens are chemicals that damage DNA and cause blistering skin inflammations. They're found occasionally in badly handled celery and celery root, parsley, and parsnips, when these vegetables have been stressed by near-freezing temperatures, intense light, or infection by mold. Psoralens are absorbed through the skin during handling, or by being ingested with the vegetable, either raw or cooked. They lie dormant in skin cells until they're struck by ultraviolet rays in sunlight, which causes them to bind to and damage DNA and important cell proteins. The psoralen-generating vegetables should be bought as fresh as possible and used quickly.

In addition to their own chemical defenses, fruits and vegetables can carry other toxins that come from contaminating molds (patulin in apple juice, from a *Penicillium* mold growing on damaged fruit), agricultural chemicals (pesticides, herbicides, fungicides), and soil and air pollutants (dioxins, polycyclic aromatic hydrocarbons). In general, it's thought that the usual levels of these contaminants do not constitute an immediate health hazard. On the other hand, they are toxins, and therefore undesirable additions to our diet. We can reduce our intake of them by washing produce, by peeling off surface layers, and by buying certified organic produce, which is grown in relatively clean soil without the use of most agricultural chemicals.

Fresh Produce and Food Poisoning

Though we generally associate outbreaks of food poisoning with foods derived from animals, fruits and vegetables are also a significant source. They have caused outbreaks of nearly every major food pathogen known (see box below). There are several reasons for this. Fruits and vegetables are grown in the soil, a vast reservoir of microbes. Field facilities for the harvesting crew (toilets, wash water) and for processing and packing may not be hygienic, so the produce is easily contaminated by people, containers, and machinery. And produce is often eaten raw. Salad bars in restaurants and cafeterias can collect and grow bacteria for hours, and have been associated with many outbreaks of food poisoning. Fruit juices, often made by crushing whole fruits, are readily contaminated by a

Disease Outbreaks Caused by Raw Fruits and Vegetables

This selected list demonstrates that raw produce is capable of causing a wide range of food-borne illnesses. These disease outbreaks are not common or a cause for great concern, but they do mean that produce should be prepared carefully, and ideally should be cooked for people with weak immune systems—the very young and very old and people suffering from other illnesses.

Microbe	Food
Clostridium botulinum	Garlic in oil
E. coli	Salad bars, alfalfa and radish sprouts, melons, apple juice
Listeria	Cabbage (long cold storage)
Salmonella	Salad bars, alfalfa sprouts, orange juice, melons, tomatoes
Shigella	Parsley, lettuce
Staphylococcus	Prepared salads
Vibrio cholerae	Fruits and vegetables contaminated by water
Yersinia	Sprouts contaminated by water
Cyclospora (protozoa)	Berries, lettuce
Hepatitis viruses	Strawberries, scallions

small number of infected pieces; so fresh cider has become hard to find. Nearly all juice production in the United States is now pasteurized.

The prudent consumer will thoroughly wash all produce, including fruits whose skins will be discarded (knives and fingers can introduce surface bacteria to the flesh). Soapy water and commercial produce washes are more effective than water alone. Washing can reduce microbial populations a hundredfold, but it's impossible to eliminate all microbes from uncooked lettuce and other produce—they can evade even heavily chlorinated water by hiding in microscopic pores and cracks in the plant tissue. Raw salads are therefore not advised for people who are especially vulnerable to infections. Once fruits and vegetables have been cut up, they should be kept refrigerated and used as soon as possible.

THE COMPOSITION AND QUALITIES OF FRUITS AND VEGETABLES

What makes a vegetable tender or tough? Why do leafy greens shrink so much when cooked? Why do apples and avocados turn brown when cut open? Why are green potatoes dangerous? Why do some fruits get sweet in the bowl, and others just older? The key to understanding these and other characteristics is a familiarity with the structural and chemical makeup of plant tissues.

Plant Structure: Cells, Tissues, and Organs

The Plant Cell Like animals, plants are built up out of innumerable microscopic chambers called cells. Each cell is surrounded and contained by a thin, balloon-like cell membrane constructed from certain fat-like molecules and proteins, and permeable to water and other small molecules. Immediately inside the membrane is a fluid called the *cytoplasm,* which is filled with much of the complex chemical machinery necessary to the cell's growth and function. Then within the cytoplasm float a variety of other membrane-contained bags, each with its own chemical nature. Nearly all plant cells contain a large watery *vacuole,* which may be filled with enzymes, sugars, acids, proteins, water-soluble pigments, and waste or defensive compounds. Often one large vacuole will fill 90% of the cell volume and squeeze the cytoplasm and *nucleus* (the body that contains most of the cell's DNA) up against the cell membrane. Leaf cells contain dozens to hundreds of *chloroplasts,* bags filled with green chlorophyll and other molecules that do the work of photosynthesis. The cells of fruits often contain *chromoplasts,* which concentrate yellow, orange, and red pig-

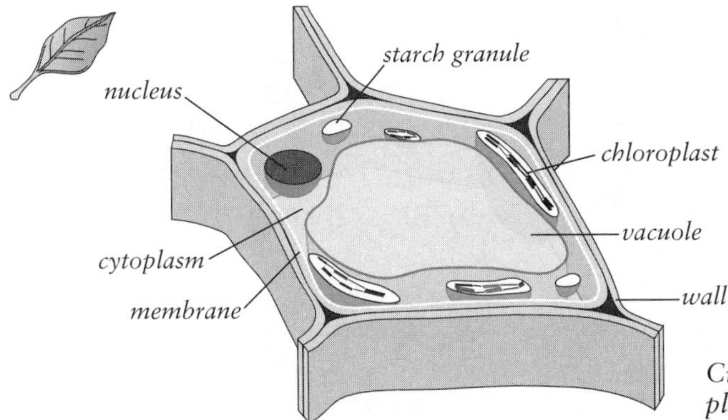

Cross-section through a typical plant cell.

ments that are soluble in fat. And storage cells are often filled with *amyloplasts,* which hold many granules of the long sugar chains called starch.

The Cell Wall One last and very important component of the plant cell is its *cell wall,* something that animal cells lack entirely. The plant cell wall surrounds the membrane and is strong and rigid. Its purpose is to lend structural support to the cell and the tissue of which it is a part. Neighboring cells are held together by the outer, glue-like layers of their cell walls. Some specialized strengthening cells become mostly cell wall and do their job even after their death. The gritty grains in pear flesh, the fibers in celery stalks, the stone that surrounds a peach seed, and the seed coats of beans and peas are all mainly the cell-wall material of strengthening cells.

Broadly speaking, the texture of plant foods is determined by the fullness of the storage vacuole, the strength of the cell walls, and the absence or presence of starch granules. Color is determined by the chloroplasts and chromoplasts, and sometimes by water-soluble pigments in vacuoles. Flavor comes from the contents of the storage vacuoles.

Plant Tissues Tissues are groups of cells organized to perform a common function. Plants have four basic tissues.

Ground tissue is the primary mass of cells. Its purpose depends on its location in the plant. In leaves the ground tissue performs photosynthesis; elsewhere it stores nutrients and water. Cells in the ground tissue usually have thin cell walls, so the tissue is generally tender. Most of our fruits and vegetables are mainly ground tissue.

Vascular tissue runs through the ground tissue, and resembles our veins and arteries. It is the system of microscopic tubes that transport nutrients throughout the plant. The work is divided between two subsystems: *xylem,* which takes water and minerals from the roots to the rest of the plant, and *phloem,* which conducts sugars down from the leaves. Vascular tissue usually provides mechanical support as well, and is often tough and fibrous compared to the surrounding tissue.

Dermal tissue forms the outer surface of the plant, the layer that protects it and helps it retain its moisture. It may take the form of either *epidermis* or *periderm.* The epidermis is usually a single layer of cells that secretes several surface coatings, including a fatty material called cutin, and wax (long molecules made by joining fatty acids with alcohols), which is what makes many fruits naturally take a shine. Periderm is found instead of epidermis on underground organs and older tissues, and has a dull, corky appearance. Our culinary experience of periderm is usually limited to the skins of potatoes, beets, and so on.

Secretory tissue usually occurs as iso-

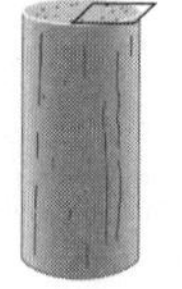

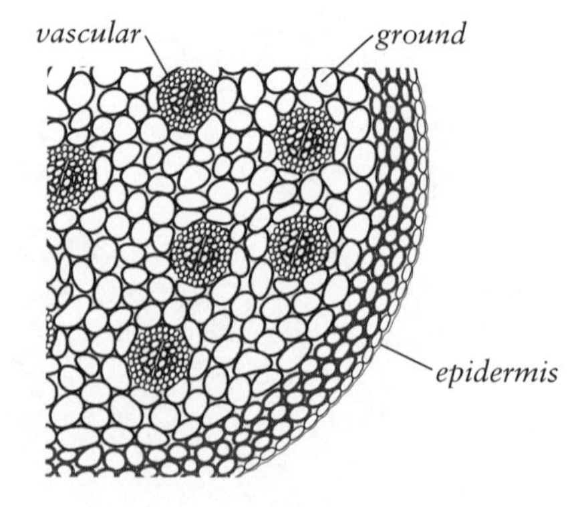

The three kinds of plant tissue in a stem. Fibrous vascular tissue and thick dermal layers are common causes of toughness in vegetables.

lated cells on the surface or within the plant. These cells correspond to the oil and sweat glands in our skin, and produce and store various aroma compounds, often to attract or repel animals. The large mint family, which includes other common herbs like thyme and basil, is characterized by glandular hairs on stems and leaves that contain aromatic oils. Vegetables in the carrot family concentrate their aromatic substances in inner secretory cells.

Plant Organs There are six major plant organs: the root, the stem, the leaf, the flower, the fruit, and the seed. We'll take a closer look at seeds in chapter 9.

Roots Roots anchor the plant in the ground, and absorb and conduct moisture and nutrients to the rest of the plant. Most roots are tough, fibrous, and barely edible. The exceptions are roots that swell up with nonfibrous storage cells; they allow plants to survive temperate-zone winter to flower in their second year (carrots, parsnips, radishes) or seasonal dryness in the tropics (sweet potatoes, manioc). Root vegetables develop this storage area in different ways, and so have different anatomies. In the carrot, storage tissue forms around the central vascular core, which is less flavorful. The beet produces concentric layers of storage and vascular tissue, and in some varieties these accumulate different pigments, so their slices appear striped.

Stems, Stalks, Tubers, and Rhizomes Stems and stalks have the main function of conducting nutrients between the root and leaves, and providing support for the aboveground organs. They therefore tend to become fibrous, which is why asparagus and broccoli stems often need to be peeled before cooking, celery and cardoon stalks deveined. The junction between stem and root, which is called the *hypocotl,* can swell into a storage organ; turnips, celery "root," and beets are actually part stem, part root. And some plants, including the potato, yam, sunchoke, and ginger, have developed special underground stem structures for nonsexual reproduction: they "clone" themselves by forming a storage organ that can produce its own roots and stem and become an independent—but genetically identical—plant. The common potato and true yam are such swollen underground stem tips called *tubers,* while the sunchoke and ginger "root" are horizontal underground stems called *rhizomes.*

Leaves Leaves specialize in the production of high-energy sugar molecules via photosynthesis, a process that requires exposure to sunlight and a good supply of carbon dioxide. They therefore contain very little storage or strengthening tissue that would

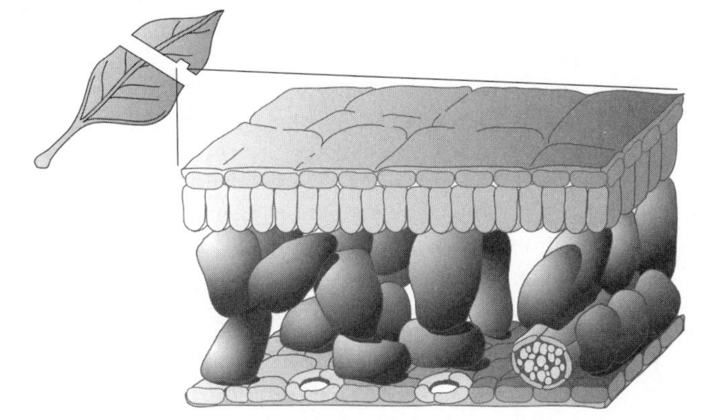

Cross section of a leaf. Because photosynthesis requires a continuous supply of carbon dioxide, leaf tissue often has a spongy structure that directly exposes many inner cells to the air.

interfere with access to light or air, and are the most fragile and short-lived parts of the plant. To maximize light capture, the leaf is flattened out into a thin sheet with a large surface area, and the photosynthetic cells are heavily populated with chloroplasts. To promote gas exchange, the leaf interior is filled with thousands of tiny air pockets, which further increase the area of cells exposed to the air. Some leaves are as much as 70% air by volume. This structure helps explain why leafy vegetables shrink so much when cooked: heat collapses the spongy interior. (It also wilts the leaves so that they pack together more compactly.)

An exception to the rule against storage tissue in leaves is the onion family (tulips and other bulb ornamentals are exceptions as well). The many layers of the onion (and the single layer of a garlic clove) surrounding the small inner stem are the swollen bases of leaves whose tops die and fall off. The leaf bases store water and carbohydrates during the plant's first year of growth so that they can be used during the second, when it will flower and produce seed.

Flowers Flowers are the plant's reproductive organs. Here the male pollen and female ovules are formed; here too they unite in the chamber that contains the ovules, the ovary, and develop into embryos and seeds. Flowers are often brilliantly colored and aromatic to attract pollinating insects, and can be a striking ingredient. However, some familiar plants protect their flowers from animal predators with toxins, so their edibility should be checked before use (p. 326). We also eat a few flowers or their supporting tissues before they mature; broccoli, cauliflower, and artichokes are examples.

Fruits The fruit is the organ derived from the flower's ovary (or adjacent stem tissue). It contains the seeds, and promotes their dispersal away from the mother plant. Some fruits are inedible—they're designed to catch the wind, or the fur of a passing animal—but the fruits that we eat were made by the plant to be eaten, so that an animal would intentionally take it and the seeds away. The fruit has no support, nutrition, or transport responsibilities to the other organs. It therefore consists almost entirely of storage tissue filled with appealing and useful substances for animals. When ready and ripe, it's usually the most flavorful and tenderest part of the plant.

Texture

The texture of raw fruits and vegetables can be crisp and juicy, soft and melting, mealy and dry, or flabby and chewy. These qualities are a reflection of the way the plant tissues break apart as we chew. And their breaking behavior depends on two main factors: the construction of their cell walls, and the amount of water held in by those walls.

The cell walls of our fruits and vegetables have two structural materials: tough fibers of cellulose that act as a kind of framework, and a semisolid, flexible mixture of water, carbohydrates, minerals, and proteins that cross-link the fibers and fill the space between them. We can think of the semisolid mixture as a kind of cement whose stiffness varies according to the proportions of its ingredients. The cellulose fibers act as reinforcing bars in that cement. Neighboring cells are held together by the cement where their walls meet.

Crisp Tenderness: The Roles of Water Pressure and Temperature Cell walls are thus firm but flexible containers. The cells that they contain are mostly water. When water is abundant and a cell approaches its maximum storage capacity, the vacuole swells and presses the surrounding cytoplasm (p. 261) against the cell membrane, which in turn presses against the cell wall. The flexible wall bulges to accommodate the swollen cell. The pressure exerted against each other by many bulging cells—which can reach 50 times the pressure of the surrounding air—results in a full, firm, turgid fruit or vegetable. But if the cells are low on water, the mutually supporting pressure disappears, the flexible cell walls sag, and the tissue becomes limp and flaccid.

Water and walls determine texture. A vegetable that is fully moist and firm will seem both crisp and more tender than the same vegetable limp from water loss. When we bite down on a vegetable turgid with water, the already-stressed cell walls readily break and the cells burst open; in a limp vegetable, chewing compresses the walls together, and we have to exert much more pressure to break through them. The moist vegetable is crisp and juicy, the limp one chewy and less juicy. Fortunately, water loss is largely reversible: soak a limp vegetable in water for a few hours and its cells will absorb water and reinflate. Crispness can also be enhanced by making sure that the vegetable is icy cold. This makes the cell-wall cement stiff, so that when it breaks under pressure, it seems brittle.

Mealiness and Meltingness: The Role of Cell Walls Fruits and vegetables can sometimes have a mealy, grainy, dry texture. This results when the cement between neighboring cells is weak, so that chewing breaks the cells apart from each other rather than breaking them open, and we end up with lots of tiny separate cells in our mouth. Then there's the soft, melting texture of a ripe peach or melon. This too is a manifestation of weakened cell walls, but here the weakening is so extreme that the walls have practically disintegrated, and the watery cell interior oozes out under the least pressure. The contents of the cells also have an effect: a ripe fruit's vacuole full of sugar solution will give a melting, succulent impression, while a potato's solid starch grains will contribute a firm chalkiness. Because starch absorbs water when heated, cooked starchy tissue becomes moist but mealy or pasty, never juicy.

The changes in texture that occur during ripening and cooking result from changes in the cell-wall materials, in particular the cement carbohydrates. One group is the hemicelluloses, which form strengthening cross-links between celluloses. They are built up from glucose and xylose sugars, and can be partly dissolved and removed from cell walls during cooking (p. 282). The other important component is the pectic substances, large branched chains of a sugar-like molecule called galacturonic acid, which bond together into a gel that fills the spaces between cellulose fibers. Pectins can be either dissolved or consolidated by cooking, and their gel-like consistency is exploited in the making of fruit jellies and jams (p. 296). When fruits soften during ripening, their enzymes weaken the cell walls by modifying the pectins.

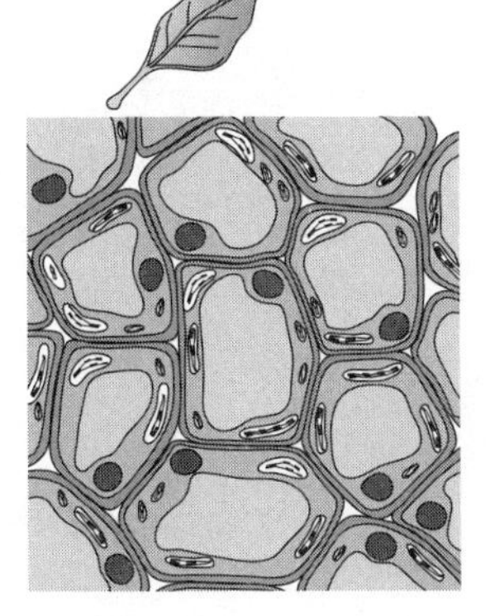

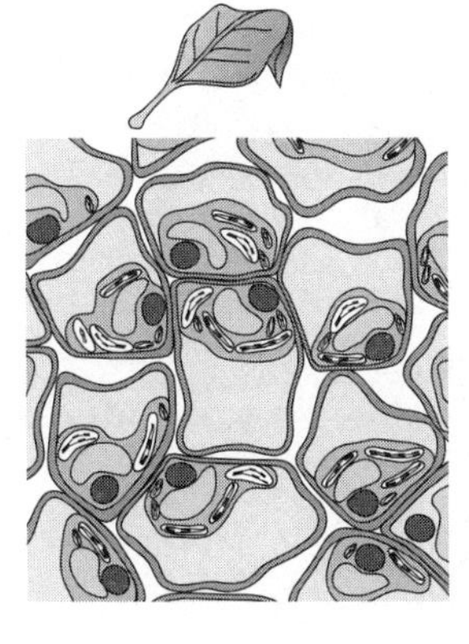

Wilting in vegetables. Plant tissue that is well supplied with water is filled with fluids and mechanically rigid (left). *Loss of water causes cell vacuoles to shrink. The cells become partly empty, the cell walls sag, and the tissue weakens* (right).

Tough Cellulose and Lignin Cellulose, the other major cell-wall component, is very resistant to change, and this is one reason that it's the most abundant plant product on earth. Like starch, cellulose consists of a chain of glucose sugar molecules. But a difference in the way they're linked to each other allows neighboring chains to bond tightly together into fibers that are invulnerable to human digestive enzymes and all but extreme heat or chemical treatment. Cellulose becomes most visible to us in the winter as hay, a stubble field, or the fine skeletons of weeds. This remarkable stability makes cellulose valuable to long-lived trees and to the human species as well. Wood is one-third cellulose, and cotton and linen fibers are almost pure cellulose. However, cellulose is a problem for the cook: it simply can't be softened by normal kitchen techniques. Sometimes, as in the gritty "stone cells" of pears, quince, and guava, this is a relatively minor distraction. But when it's concentrated to provide structural support in stems and stalks—in celery and cardoons, for example—cellulose makes vegetables permanently stringy, and the only remedy is to pull the fibers from the tissue.

One last cell wall component is seldom significant in food. Lignin is also a strengthening agent and very resistant to breakdown; it's the defining ingredient of wood. Most vegetables are harvested well in advance of appreciable lignin formation, but occasionally we do deal with woody asparagus and broccoli stems. The only remedy for this kind of toughness is to peel away the lignified areas.

Color

Plant pigments are one of life's glories! The various greens of forest and field, the purples and yellows and reds of fruits and flowers—these colors speak to us of vitality, renewal, and the sheer pleasure of sensation. Some pigments are designed to catch our eye, some actually become part of our eye, and some made possible the very existence of us and our eyes (see box, p. 271). Many turn out to have beneficial effects on our health. The cook's challenge is to preserve the vividness and appeal of these remarkable molecules.

There are four families of plant pigments, each with different functions in the plant's life and different behaviors in the

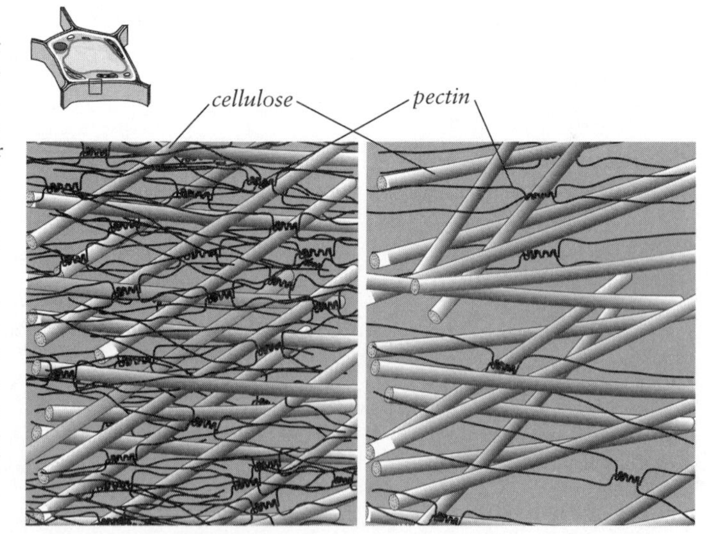

The softening of plant cell walls. The walls are made up of a framework of cellulose fibers embedded in a mass of amorphous materials, including the pectins (left). *When cooked in boiling water, the cellulose fibers remain intact, but the amorphous materials are partly extracted into fluids from within the cells, thus weakening the walls* (right) *and tenderizing the vegetable or fruit.*

kitchen. All of them are large molecules that appear to be a certain color because they absorb certain wavelengths of light, and thus leave only parts of the spectrum for our eyes to detect. Chlorophylls are green, for example, because they absorb red and blue wavelengths.

Green Chlorophylls The earth is painted green with chlorophylls, the molecules that harvest solar energy and funnel it into the photosynthetic system that converts it into sugar molecules. Chlorophyll *a* is bright blue-green, chlorophyll *b* a more muted olive color. The *a* form dominates the *b* by 3 to 1 in most leaves, but the balance is evener in plants that grow in the shade, and in aging tissues, where the *a* form is degraded faster. The chlorophylls are concentrated in cell bodies called chloroplasts, where they're embedded in the many folds of a membrane along with the other molecules of the photosynthetic system. Each chlorophyll molecule is made up of two parts. One is a ring of carbon and nitrogen atoms with a magnesium atom at the center, quite similar to the heme ring in the meat myoglobin pigment (p. 133). This ring portion is soluble in water, and does the work of absorbing light. The second part is a fat-soluble tail of 16 carbon atoms, which anchors the whole molecule in the chloroplast membrane. This part is colorless.

These complex molecules are readily altered when their membrane home is disrupted during cooking. This is why the bright green of fresh vegetables is fragile. Ironically, prolonged exposure to intense light also damages chlorophylls. Attention to cooking times, temperatures, and acidities are thus essential to serving bright green vegetables (p. 280).

Yellow, Orange, Red Carotenoids *Carotenoids* are so named because the first member of this large family to be chemically isolated came from carrots. These pigments absorb blue and green wavelengths and are responsible for most of the yellow and orange colors in fruits and vegetables (beta-carotene, xanthophylls, zeaxanthin), as well as the red of tomatoes, watermelons, and chillis (lycopene, capsanthin, and capsorubin; most red colors in plants are caused by anthocyanins). Carotenoids are zigzag chains of around 40 carbon atoms and thus resemble fat molecules (p. 797). They're generally soluble in fats and oils and are relatively stable, so they tend to stay bright and stay put when a food is cooked in water. Carotenoids are found in two different places in plant cells. One is in special pigment bodies, or chromoplasts, which signal animals that a flower is open for business or a fruit is ripe. Their other home is the photosynthetic membranes of chloroplasts, where there is one carotenoid molecule for every five or so chlorophylls. Their main role there is to protect chlorophyll and other parts of the photosynthetic system. They absorb potentially damaging wavelengths in the light spectrum, and act as antioxidants by soaking up the many high-energy chemical by-products generated in photosynthesis. They can do the same in the human body, particularly in the eye (p. 256). Chloroplast carotenoids are usually invisible, their presence masked by green chlorophyll, but it's a good rule of thumb that the darker green the vegetable, the more chloroplasts and chlorophyll it contains, and the more carotenoids as well.

About ten carotenoids have a nutritional as well as aesthetic significance: they are converted to vitamin A in the human intestinal wall. Of these the most common and active is beta-carotene. Strictly speaking, only animals and animal-derived foods contain vitamin A itself; fruits and vegetables contain only its precursors. But without these pigment precursors there would be no vitamin A in animals either. In the eye, vitamin A becomes part of the receptor molecule that detects light and allows us to see. Elsewhere in the body it has a number of other important roles.

Red and Purple Anthocyanins, Pale Yellow Anthoxanthins Anthocyanins (from the Greek for "blue flower") are responsible for most of the red, purple, and blue colors in plants, including many berries, apples, cabbage, radishes, and potatoes. A related group, the anthoxanthins ("yellow flower") are pale yellow compounds found in potatoes, onions, and cauliflower. This third major class of plant pigments is a subgroup of the huge phenolic family, which is based on rings of 6 carbon atoms with two-thirds of a water molecule (OH) attached to some of them, which makes phenolics soluble in water. The anthocyanins have 3 rings. There are about 300 known anthocyanins, and a given fruit or vegetable will usually contain a mixture of a dozen or more. Like many other phenolic compounds, they are valuable antioxidants (p. 255).

Anthocyanins and anthoxanthins reside in the storage vacuole of plant cells, and readily bleed into surrounding tissues and ingredients when cell structures are damaged by cooking. This is why the lovely color of purple-tinted asparagus, beans, and other vegetables often disappears with cooking: the pigment is stored in just the outer layers of cells, and gets diluted to invisibility when the cooked cells break open. The main function of the anthocyanins is to provide signaling colors in flowers and fruits, though they may have begun their career as light-absorbing protection for the photosynthetic systems in young leaves (see box, p. 271). Anthocyanins are very sensitive to the acid-alkaline balance of foods—alkalinity shifts their color to the blue—and they're altered by traces of metals, so they are often the source of strange off-colors in cooked foods (p. 281).

Red and Yellow Betains A fourth group of plant pigments is the betains, which are only found in a handful of distantly related species. However, these include three popular and vividly colored vegetables: beets and chard (both varieties of the

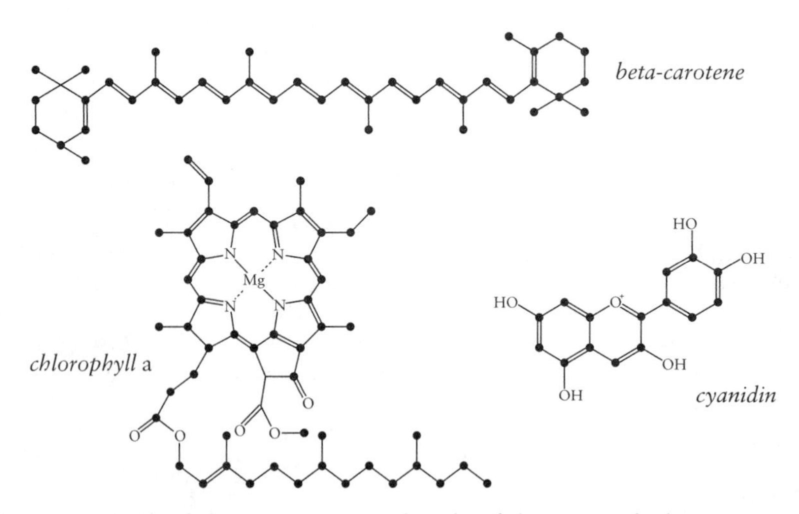

The three major kinds of plant pigments. For the sake of clarity, most hydrogen atoms are left unlabeled; dots indicate carbon atoms. Top: *Beta-carotene, the most common carotenoid pigment, and the source of the orange color of carrots. The long fat-like carbon chain makes these pigments much more soluble in fats and oils than in water.* Bottom left: *Chlorophyll* a, *the main source of the green in vegetables and fruits, with a heme-like region (p. 133) and a long carbon tail that makes chlorophyll more soluble in fats and oils than in water.* Bottom right: *Cyanidin, a blue pigment in the anthocyanin family. Thanks to their several hydroxyl (OH) groups, anthocyanins are water-soluble, and readily leak out of boiled vegetables.*

same species), amaranth, and the prickly pear, the fruit of a cactus. The betains (sometimes called betalains) are complex nitrogen-containing molecules that are otherwise similar to anthocyanins: they are water-soluble, sensitive to heat and light, and tend toward the blue in alkaline conditions. There are about 50 red betains and 20 yellow betaxanthins, combinations of which produce the almost fluorescent-looking stem and vein colors of novelty chards. The human body has a limited ability to metabolize these molecules, so a large dose of red beets or prickly pears can give a startling but harmless tinge to the urine. The red betains contain a phenolic group and are good antioxidants; yellow betaxanthins don't and aren't.

Discoloration: Enzymatic Browning

Many fruits and vegetables—for example apples, bananas, mushrooms, potatoes—quickly develop a brown, red, or gray discoloration when cut or bruised. This discoloration is caused by three chemical ingredients: 1- and 2-ring phenolic compounds, certain plant enzymes, and oxygen. In the intact fruit or vegetable, the phenolic compounds are kept in the storage vacuole, the enzymes in the surrounding cytoplasm. When the cell structure is damaged and phenolics are mixed with enzymes and oxygen, the enzymes oxidize the phenolics, forming molecules that eventually react with each other and bond together into light-absorbing clusters. This system is one of the plant's chemical defenses: when insects or microbes damage its cells, the plant releases reactive phenolics that attack the invaders' own enzymes and membranes. The brown pigments that we see are essentially masses of spent weapons. (A similar kind of enzyme acting on a similar compound is responsible for the "browning" of humans in the sun; here the pigment itself is the protective agent.)

Minimizing Brown Discoloration Enzymatic browning can be discouraged by several means. The single handiest method for the cook is to coat cut surfaces with lemon juice: the browning enzymes work very slowly in acidic conditions. Chilling the food below about 40°F/4°C will also slow the enzymes down somewhat, as will immersing the cut pieces in cold water, which limits the availability of oxygen. In the case of precut lettuce for salads, enzyme activity and browning can be reduced by immersing the freshly cut leaves in a pot of water at 115°F/47°C for three minutes before chilling and bagging them. Boiling temperatures will destroy the enzyme, so cooking will eliminate the problem. However, high temperatures can encourage phenolic oxidation in the absence of enzymes: this is why the water in which vegetables have been cooked sometimes turns brown on standing. Various sulfur compounds will combine with the phenolic substances and block their reaction with the enzyme, and these are

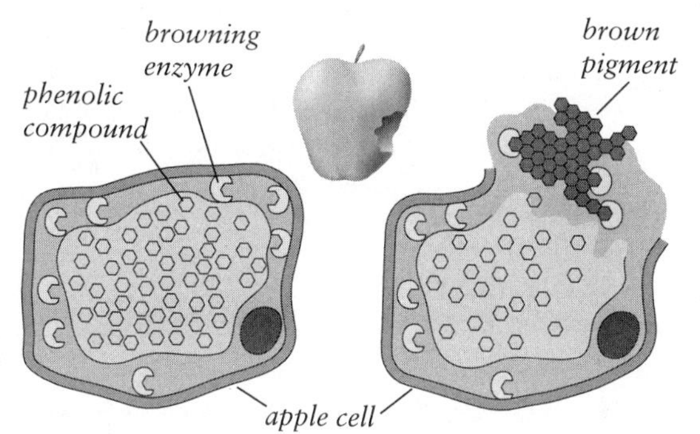

Brown discoloration caused by plant enzymes. When the cells in certain fruits and vegetables are damaged by cutting, bruising, or biting, browning enzymes in the cell cytoplasm come into contact with small, colorless phenolic molecules from the storage vacuole. With the help of oxygen from the air, the enzymes bind the phenolic molecules together into large, colored assemblies that turn the damaged area brown.

often applied commercially to dried fruits. Sulfured apples and apricots retain their natural color and flavor, while unsulfured dried fruits turn brown and develop a more cooked flavor.

Another acid that inhibits browning by virtue of its antioxidant properties is ascorbic acid, or vitamin C. It was first identified around 1925 when the Hungarian biochemist Albert Szent-Györgyi found that the juice of some nonbrowning plants, including the chillis grown for paprika, could delay the discoloration of plants that do brown, and he isolated the responsible substance.

FLAVOR

The overall flavor of a fruit or vegetable is a composite of several distinct sensations. From the taste buds on our tongues, we register salts, sweet sugars, sour acids, savory amino acids, and bitter alkaloids. From the cells in our mouth sensitive to touch, we notice the presence of astringent, puckery tannins. A variety of cells in and near the mouth are irritated by the pungent compounds in peppers, mustard, and members of the onion family. Finally, the olfactory receptors in our nasal passages can detect many hundreds of volatile molecules that are small and chemically repelled by water, and therefore fly out of the food and into the air in our mouth. The sensations from our mouth give us an idea of a food's basic composition and qualities, while our sense of smell allows us to make much finer discriminations.

Taste: Salty, Sweet, Sour, Savory, Bitter

Of the five generally recognized tastes, three are especially prominent in fruits and vegetables. Sugar is the main product of photosynthesis, and its sweetness is the main attraction provided by fruits for their animal seed dispersers. The average sugar content of ripe fruit is 10 to 15% by weight. Often the unripe fruit stores its sugar as tasteless starch, which is then converted back into sugar during ripening to make the fruit more appealing. At the same time, the fruit's acid content usually drops, a development that makes the fruit seem even sweeter. There are several organic acids—citric, malic, tartaric, oxalic—that plants can accumulate in their vacuoles and variously use as alternative energy stores, chemical defenses, or metabolic wastes, and that account for the acidity of most fruits and vegetables (all are acid to some degree). The sweet-sour balance is especially important in fruits.

Most vegetables contain only moderate amounts of sugar and acid, and these are

Browning Enzymes, Breath Fresheners, and the Order of the Meal

The browning enzymes are normally considered a nuisance, because they discolor foods as we prepare them. Recently a group of Japanese scientists found a constructive use for their oxidizing activities: they can help clear our breath of persistent garlic, onion, and other sulfurous odors! The reactive phenolic chemicals produced by the enzymes combine with sulfhydryl groups to form new and odorless molecules. (Phenolic catechins in green tea do the same.) Many raw fruits and vegetables are effective at this, notably pome and stone fruits, grapes, blueberries, mushrooms, lettuces, burdock, basil, and peppermint. This may be one of the benefits of ending a meal with fruit, and one of the reasons that some cultures serve a salad after the main course, not before.

quickly used up by the plant cells after harvest. This is why vegetables picked just before cooking are more full-flavored than store-bought produce, which is usually days to weeks from the field.

Bitter tastes are generally encountered only among vegetables and seeds (for example, coffee and cocoa beans), which contain alkaloids and other chemical defenses meant to discourage animals from eating them. Farmers and plant breeders have worked for thousands of years to reduce the bitterness of such crops as lettuce, cucumbers, eggplants, and cabbage, but chicory and radicchio, various cabbage relatives, and the Asian bitter gourd are actually prized for their bitterness. In many cultures, bitterness is thought to be a manifestation of medicinal value and therefore of healthfulness, and there may be some truth to this association (p. 334).

Though savory, mouth-filling amino acids are more characteristic of protein-rich animal foods, some fruits and vegetables do contain significant quantities of glutamic acid, the active portion of MSG. Notable among them are tomatoes, oranges, and many seaweeds. The glutamic acid in tomatoes, together with its balanced sweetness and acidity, may help explain why this fruit is so successfully used as a vegetable, both with meats and without.

Touch: Astringency Astringency is neither a taste nor an aroma, but a tactile sensation: that dry, puckery, rough feeling that follows a sip of strong tea or red wine, or a bite into an unripe banana or peach. It is caused by a group of phenolic compounds consisting of 3 to 5 carbon rings, which are just the right size to span two or more normally separate protein molecules, bond to them, and hold them together. These phenolics are called *tannins* because they have been used since prehistory to tan animal hides into tough leather by bonding with the skin proteins. The sensation of astringency is caused when tannins bond to proteins in our saliva, which normally provide lubrication and help food particles slide smoothly along the mouth surfaces. Tannins cause the proteins to clump together and stick to particles and surfaces, increasing the friction between them. Tannins are another of the plant kingdom's chemical defenses. They counteract bacteria and fungi by interfering with their

Leaves and Fruits Shaped Our Vision

We can distinguish and enjoy the many hues of anthocyanin- and carotenoid-rich plants—as well as the same hues in paintings and clothing, makeup and warning signs—because our eyes are designed to see well in this color range of yellow to orange to red. It now looks as though we owe this ability to leaves and fruits! It turns out that we are among a small handful of animal species with eyes that can distinguish red from green. The other species are tropical forest–dwelling primates like our probable ancestors, and they have in common a need to detect their foods against the green of the forest canopy. The young leaves of many tropical plants are red with anthocyanins, which apparently absorb excess solar energy during the momentary shafts of direct sunlight in an otherwise shaded life; and young leaves are more tender than the older, green, fibrous leaves, more easily digested and nutritious, and more sought after by monkeys. Without good red vision, it would be hard to find them—or carotenoid-colored fruits—among the green leaves. So leaves and fruits shaped our vision. The pleasure we take today in their colors was made possible by our ancestors' hunger, and the sustenance they found in red-tinged leaves and yellow-orange fruits.

surface proteins, and deter plant-eating animals by their astringency and by interfering with digestive enzymes. Tannins are most often found in immature fruit (to prevent their consumption before the seeds are viable), in the skins of nuts, and in plant parts strongly pigmented with anthocyanins, phenolic molecules that turn out to be the right size to cross-link proteins. Red-leaf lettuces, for example, are noticeably more astringent than green.

Though a degree of astringency can be desirable in a dish or drink—it contributes a feeling of substantialness—it often becomes tiresome. The problem is that the sensation becomes stronger with each dose of tannins (whereas most flavors become less prominent), and it lingers, with the duration also increasing with each exposure. So it's worth knowing how to control astringency (p. 284).

Irritation: Pungency The sensations caused by "hot" spices and vegetables—chillis, black pepper, ginger, mustard, horseradish, onions, and garlic—are most accurately described as irritation and pain (for why we can enjoy such sensations, see p. 394). The active ingredients in all of these are chemical defenses that are meant to annoy and repel animal attackers. Very reactive sulfur compounds in the mustard and onion families apparently do mild damage to the unprotected cell membranes in our mouth and nasal passages, and thus cause pain. The pungent principles of the peppers and ginger, and some of the mustard compounds, work differently; they bind to a specific receptor on the cell membranes, and the receptor then triggers reactions in the cell that cause it to send a pain signal to the brain. The mustard and onion defenses are created only when tissue damage mixes together normally separate enzymes and their targets. Because enzymes are inactivated by cooking temperatures, cooking will moderate the pungency of these foods. By contrast, the peppers and ginger stockpile their defenses ahead of time, and cooking doesn't reduce their pungency.

The nature and use of pungent ingredients are described in greater detail in the next several chapters, in entries on particular vegetables and spices.

Aroma: Variety and Complexity The subject of aroma is both daunting and endlessly fascinating! Daunting because it involves many hundreds of different chemicals and sensations for which we don't have a good everyday vocabulary; fascinating because it helps us perceive more, and find more to enjoy, in the most familiar foods. There are two basic facts to keep in mind when thinking about the aroma of any food. First, the distinctive aromas of particular foods are created by specific volatile chemicals that are characteristic of those foods. And second, nearly all food aromas are composites of many different volatile molecules. In the case of vegetables, herbs, and spices, the number may be a dozen or two, while fruits typically emit several hundred volatile molecules. Usually just a handful create the dominant element of an aroma, while the others supply background, supporting, enriching notes. This combination of specificity and complexity helps explain why we find echoes of one food in another, or find that two foods go well together. Some affinities result when the foods happen to share some of the same aroma molecules.

One way to approach the richness of plant flavors is to taste actively and with other people. Rather than simply recognizing a familiar flavor as what you expect, try to dissect that flavor into some of its component sensations, just as a musical chord can be broken down into its component notes. Run through a checklist of the possibilities, and ask: Is there a green-grass note in this aroma? A fruity note? A spicy or nutty or earthy note? If so, which kind of fruit or spice or nut? Chapters 6–8 give interesting facts about the aromas of particular fruits, vegetables, herbs, and spices.

Aroma Families The box on pp. 274–275 identifies some of the more prominent aromas to be found in plant foods. Though I've divided them by type of food, this division is arbitrary. Fruits may have green-leaf aromas; vegetables may contain chemicals more characteristic of fruits or spices; spices and herbs share many aromatics with fruits. Some examples: cherries and bananas contain the dominant element of cloves; coriander contains aromatics that are prominent in citrus flowers and fruits; carrots share piney aromatics with Mediterranean herbs. While a given plant does usually specialize in the production of a certain kind of aromatic, plants in general are biochemical virtuosos, and may operate a number of different aromatic production lines at once. Some of the most important production lines are these:

- "Green," cucumber/melon, and mushroom aromas, produced from unsaturated fatty acids in cell membranes when tissue damage mixes an oxidizing enzyme (lipoxygenase) with unsaturated fatty acids in cell membranes. This enzyme breaks the long fatty acid chains into small, volatile pieces, and other enzymes then modify the pieces.
- "Fruity" aromas, produced when enzymes in the intact fruit combine an acid molecule with an alcohol molecule to produce an *ester.*
- "Terpene" aromas, produced by a long series of enzymes from small building blocks that also get turned into carotenoid pigments and other important molecules. They range from flowery to citrusy, minty, herbaceous, and piney (p. 391).
- "Phenolic" aromas, produced by a series of enzymes from an amino acid with a 6-carbon ring. These are offshoots of the biochemical pathway that makes woody lignin (p. 266), and include many spicy, warming, and pungent molecules (p. 391).
- "Sulfur" aromatics, usually produced when tissue damage mixes enzymes with nonaromatic aroma precursors. Most sulfur aromatics are pungent chemical defenses, though some give a more subtle depth to a number of fruits and vegetables.

Fascinating and useful as it is to analyze the flavors of the plant world, the greatest pleasure still comes from savoring them whole. This is one of the great gifts of life in the natural world, as Henry David Thoreau reminded us:

> Some gnarly apple which I pick up in the road reminds by its fragrance of all the wealth of Pomona. There is thus about all natural products a certain volatile and ethereal quality which represents their highest value. . . . For nectar and ambrosia are only those fine flavors of every earthly fruit which our coarse palates fail to perceive,—just as we occupy the heaven of the gods without knowing it.

HANDLING AND STORING FRUITS AND VEGETABLES

Post-Harvest Deterioration

There's no match for the flavor of a vegetable picked one minute and cooked the next. Once a vegetable is harvested it begins to change, and that change is almost always for the worse. (Exceptions include plant parts designed to hibernate, for example onions and potatoes.) Plant cells are hardier than animal cells, and may survive for weeks or even months. But cut off from their source of renovating nutrients, they consume themselves and accumulate waste products, and their flavor and texture suffer. Many varieties of corn and peas lose half their sugar in a few hours at room temperature, either by converting it to starch or using it for energy to stay alive. Bean pods, asparagus, and broccoli begin to use their sugar to make tough lignified fibers. As crisp, crunchy lettuce and celery use up their water, their cells lose turgor pressure and they become limp and chewy (p. 265).

Some of the Aromas in Foods from Plants

This table provides a quick overview of the kinds of aromas found in plant foods, where they come from, and how they behave when the food is cooked.

Aroma	Examples	Chemicals responsible	Origin	Characteristics
		Vegetables		
"Green leaf": fresh-cut leaves, grass	Most green vegetables; also tomatoes, apples, other fruits	Alcohols, aldehydes (6-carbon)	Cutting or crushing; enzyme action on unsaturated cell membrane lipids	Delicate, reduced by cooking (stops enzymes, alters chemicals)
Cucumber	Cucumbers, melons	Alcohols, aldehydes (9-carbon)	Cutting or crushing; enzyme action on unsaturated cell membranes	Delicate, reduced by cooking (stops enzymes, alters chemicals)
"Green vegetable"	Bell peppers, fresh peas	Pyrazines	Preformed	Strong, persistent
Earthy	Potatoes, beets	Pyrazines, geosmin	Preformed	Strong, persistent
Fresh mushroom	Mushrooms	Alcohols, aldehydes (8-carbon)	Cutting or crushing; enzyme action on unsaturated cell membrane lipids	Delicate, reduced by cooking (stops enzymes, alters chemicals)
Cabbage-like	Cabbage family	Sulfur compounds	Cutting or crushing; enzyme action on sulfur-containing precursors	Strong, persistent, altered and strengthened by cooking

Aroma	Examples	Chemicals responsible	Origin	Characteristics
Onion-like, mustard-like	Onion family	Sulfur compounds	Cutting or crushing; enzyme action on sulfur-containing precursors	Strong, persistent, altered and strengthened by cooking
Floral	Edible flowers	Alcohols, terpenes, esters	Preformed	Delicate, altered by cooking
		Fruits		
"Fruity"	Apple, pear, banana, pineapple, strawberry	Esters (acid+ alcohol)	Preformed	Delicate, altered by cooking
Citrus	Citrus family	Terpenes	Preformed	Persistent
"Fatty," "creamy"	Peach, coconut	Lactones	Preformed	Persistent
Caramel, nutty	Strawberry, pineapple	Furanones	Preformed	Persistent
Tropical fruit, "exotic," musky	Grapefruit, passion fruit, mango, pineapple; melon; tomato	Sulfur compounds, complex	Preformed	Persistent
		Herbs & Spices		
Pine-like, mint-like, herbaceous	Sage, thyme, rosemary, mint, nutmeg	Terpenes	Preformed	Strong, persistent
Spicy, warming	Cinnamon, clove, anise, basil, vanilla	Phenolic compounds	Preformed	Strong, persistent

Fruits are a different story. Some fruits may actually get better after harvest because they continue to ripen. But ripening soon runs its course, and then fruits also deteriorate. Eventually fruit and vegetable cells alike run out of energy and die, their complex biochemical organization and machinery break down, their enzymes act at random, and the tissue eats itself away.

The spoilage of fruits and vegetables is hastened by microbes, which are always present on their surfaces and in the air. Bacteria, molds, and yeasts all attack weakened or damaged plant tissue, break down its cell walls, consume the cell contents, and leave behind their distinctive and often unpleasant waste products. Vegetables are mainly attacked by bacteria, which grow faster than other microbes. Species of *Erwinia* and *Pseudomonas* cause familiar "soft rot." Fruits are more acidic than vegetables, so they're resistant to many bacteria but more readily attacked by yeasts and molds (*Penicillium, Botrytis*).

Precut fruits and vegetables are convenient but especially susceptible to deterioration and spoilage. Cutting has two important effects. The tissue damage induces nearby cells to boost their defensive activity, which depletes their remaining nutrients and may cause such changes as toughening, browning, and the development of bitter and astringent flavors. And it exposes the normally protected nutrient-rich interior to infection by microbes. So precut produce requires special care.

Handling Fresh Produce

The aim in storing fruits and vegetables is to slow their inevitable deterioration. This begins with choosing and handling the produce. Mushrooms as well as some ripe fruits—berries, apricots, figs, avocados, papayas—have a naturally high metabolism and deteriorate faster than lethargic apples, pears, kiwi fruits, cabbages, carrots, and other good keepers. "One rotten apple spoils the barrel": moldy fruit or vegetables should be discarded and refrigerator drawers and fruit bowls should be cleaned regularly to reduce the microbial population. Produce shouldn't be subjected to physical stress, whether dropping apples on the floor or packing tomatoes tightly into a confined space. Even rinsing in water can make delicate berries more susceptible to infection by abrading their protective epidermal layer with clinging dirt particles. On the other hand, soil harbors large numbers of microbes, and should be removed from the surfaces of sturdier fruits and vegetables before storing them.

The Storage Atmosphere

The storage life of fresh produce is strongly affected by the atmosphere that surrounds it. All plant tissues are mostly water, and require a humid atmosphere to avoid drying out, losing turgidity, and damaging their internal systems. Practically, this means it's best to keep plant foods in restricted spaces—plastic bags, or drawers within a refrigerator—to slow down moisture loss to the compartment as a whole and to the outside. At the same time, living produce exhales carbon dioxide and water, so moisture can accumulate and condense on the food surfaces, which encourages microbial attack. Lining the container with an absorbent material—a paper towel or bag—will delay condensation.

The metabolic activity of the cells can also be slowed by limiting their access to oxygen. Commercial packers fill their bags of produce with a well-defined mixture of nitrogen, carbon dioxide, and just enough oxygen (8% or less) to keep the plant cells functioning normally; and they use bags whose gas permeability matches the respiration rate of the produce. (Too little oxygen and fruits and vegetables switch to anaerobic metabolism, which generates alcohol and other odorous molecules characteristic of fermentation, and causes internal tissue damage and browning.)

Home and restaurant cooks can approximate such a controlled atmosphere by packing their produce in closed plastic bags with most of the air squeezed out of them. The plant cells consume oxygen and create carbon dioxide, so the oxygen levels in the bags slowly decline. However, a major disadvantage of a closed plastic bag is that it traps the gas ethylene, a plant hormone that advances ripening in fruits and induces defensive activity and accelerated aging in other tissues. This means that bagged fruits may pass from ripe to overripe too quickly, and one damaged lettuce leaf can speed the decline of a whole head. Recently, manufacturers have introduced produce containers with inserts that destroy ethylene and extend storage life (the inserts contain permanganate).

A very common commercial treatment that slows both water loss and oxygen uptake in whole fruits and fruit-vegetables—apples, oranges, cucumbers, tomatoes—is to coat them at the packing facility with a layer of edible wax or oil. A number of different materials are used, including natural beeswax and carnauba, candellila, and rice-bran waxes and vegetable oils, and such petrochemical by-products as paraffin, polyethylene waxes, and mineral oil. These treatments are harmless, but they can make produce surfaces unpleasantly waxy or hard.

Temperature Control: Refrigeration

The most effective way to prolong the storage life of fresh produce is to control its temperature. Cooling slows chemical reactions in general, so it slows the metabolic activity of the plant cells themselves, and the growth of the microbes that attack them. A reduction of just 10°F/5°C can nearly double storage life. However, the ideal storage temperature is different for different fruits and vegetables. Those native to temperate climates are best kept at or near the freezing point, and apples may keep for nearly a year if the storage atmosphere is also controlled. But fruits and vegetables native to warmer regions are actually injured by temperatures that low. Their cells begin to malfunction, and uncontrolled enzyme action causes damage to cell walls, the development of off-flavors, and discoloration. Chilling injury may become apparent during storage, or only after the produce is brought back to room temperature. Banana skins turn black in the refrigerator; avocados darken and fail to soften further; citrus fruits develop spotted skins. Foods of tropical and subtropical origin keep best at the relatively high temperature of 50°F/10°C, and are often better off at room temperature than in the refrigerator. Among them are melons, eggplants, squash, tomatoes, cucumbers, peppers, and beans.

Temperature Control: Freezing

The most drastic form of temperature control is freezing, which stops cold the overall metabolism of fruits, vegetables, and spoilage microbes. It causes most of the water in the cells to crystallize, thus immobilizing other molecules and suspending most chemical activity. The microbes are hardy, and most of them revive on warming. But freezing kills plant tissues, which suffer two kinds of damage. One is chemical: as the water crystallizes, enzymes and other reactive molecules become unusually concentrated and react abnormally. The other damage is physical disruption caused by the water crystals, whose edges puncture cell walls and membranes. When the food is thawed, the cell fluids leak out of the cells, and the food loses crispness and becomes limp and wet. Producers of frozen foods minimize the size of the ice crystals, and so the amount of damage done, by freezing the food as quickly as possible to as low a temperature as possible, often –40°F/–40°C. Under these conditions, many small ice crystals form; at higher temperatures fewer and larger crystals form, and do more damage. Home and restaurant freezers are warmer than commercial freezers and their temperatures fluctuate, so during storage some water melts and refreezes into larger crystals, and the food's texture suffers.

Although freezing temperatures generally reduce enzymatic and other chemical activity, some reactions are actually enhanced by the concentrating effects of ice formation, including enzymatic breakdown of vitamins and pigments. The solution to this problem is *blanching*. In this process the food is immersed in rapidly boiling water for a minute or two, just enough time to inactivate the enzymes, and then just as rapidly immersed in cold water to stop further cooking and softening of the cell walls. If vegetables are to be frozen for more than a few days, they should be blanched first. Fruits are less commonly blanched because their cooked flavor and texture are less appealing. Enzymatic browning in frozen fruit can be prevented by packing it in a sugar syrup supplemented with ascorbic acid (between ¼ and ¾ teaspoon per quart, 750–2,250 mg per liter, depending on the fruit's susceptibility to browning). Sugar syrup (usually around 40%, or 1.5 lb sugar per quart water, 680 gm per liter) can also improve the texture of frozen fruit by being absorbed into the cell-wall cement, which becomes stiffer. Frozen produce should be wrapped as air- and watertight as possible. Surfaces left exposed to the relatively dry atmosphere of the freezer will develop freezer burn, the slow, patchy drying out caused by the evaporation of frozen water molecules directly into vapor (this is called "sublimation"). Freezer-burned patches develop a tough texture and stale flavor.

COOKING FRESH FRUITS AND VEGETABLES

Compared to meats, eggs, and dairy products, vegetables and fruits are easy to cook. Animal tissues and secretions are mainly protein, and proteins are sensitive molecules; moderate heat (140°F/60°C) causes them to cling tightly to each other and expel water, and they quickly become hard and dry. Vegetables and fruits are mainly carbohydrates, and carbohydrates are robust molecules; even boiling temperatures simply disperse them more evenly in the tissue moisture, so the texture becomes soft and succulent. However, the cooking of vegetables and fruits does have its fine points. Plant pigments, flavor compounds, and nutrients are sensitive to heat and to the chemical environment. And even carbohydrates sometimes behave curiously! The challenge of cooking vegetables and fruits is to create an appealing texture without compromising color, flavor, and nutrition.

How Heat Affects the Qualities of Fruits and Vegetables

Color Many plant pigments are altered by cooking, which is why we can often judge by their color how carefully vegetables have been prepared. The one partial exception to this rule is the yellow-orange-red carotenoid group, which is more soluble in fat than in water, so the colors don't readily leak out of the tissue, and are fairly stable. However, even carotenoids are changed by cooking. When we heat carrots, their beta-carotene

> **Aromas from Altered Carotenoid Pigments**
>
> Both drying and cooking break some of the pigment molecules in carotenoid-rich fruits and vegetables into small, volatile fragments that contribute to their characteristic aromas. These fragments provide notes reminiscent of black tea, hay, honey, and violets.

shifts structure and hue, from red-orange toward the yellow. Apricots and tomato paste dried in the sun lose much of their intact carotenoids unless they're treated with antioxidant sulfur dioxide (p. 291). But compared to the green chlorophylls and multihued anthocyanins, the carotenoids are the model of steadfastness.

Green Chlorophyll One change in the color of green vegetables as they are cooked has nothing to do with the pigment itself. That wonderfully intense, bright green that develops within a few seconds of throwing vegetables into boiling water is a result of the sudden expansion and escape of gases trapped in the spaces between cells. Ordinarily, these microscopic air pockets cloud the color of the chloroplasts. When they collapse, we can see the pigments much more directly.

The Enemy of Green: Acids Green chlorophyll is susceptible to two chemical changes during cooking. One is the loss of its long carbon-hydrogen tail, which leaves the pigment water-soluble—so that it leaks out into the cooking liquid—and more susceptible to further change. This loss is encouraged by both acid and alkaline conditions and by an enzyme called chlorophyllase, which is most active between 150–170°F/66–77°C and only destroyed near the boiling point. The second and more noticeable change in chlorophyll is the dulling of its color, which is caused when either heat or an enzyme nudge the magnesium atom from the center of the molecule. The replacement of magnesium by hydrogen is by far the most common cause of color change in cooked vegetables. In even slightly acidic water, the plentiful hydrogen ions displace the magnesium, a change that turns chlorophyll *a* into grayish-green pheophytin *a*, chlorophyll *b* into yellowish pheophytin *b*. Cooking vegetables without water—stir-frying, for example—will also cause a color change, because when the temperature of the plant tissue rises above 140°F/60°C, the organizing membranes in and around the chloroplast are damaged,

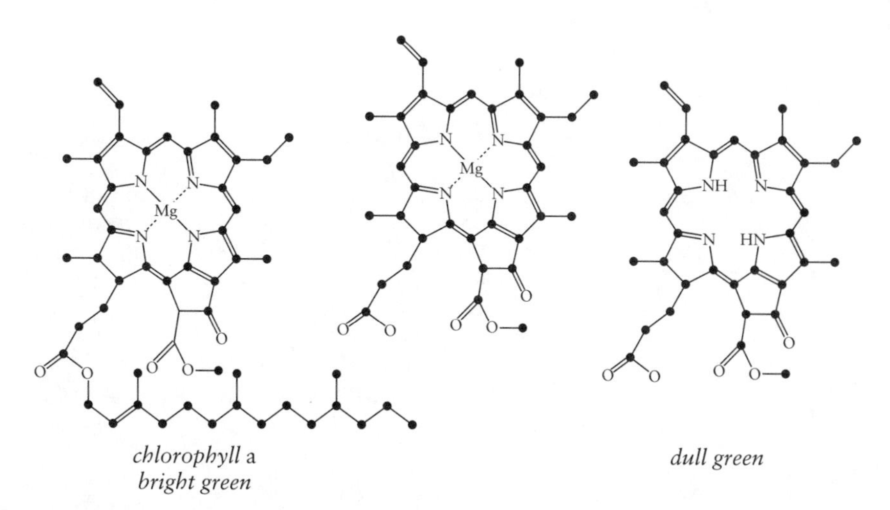

Changes in chlorophyll during cooking. Left: *The normal chlorophyll molecule is bright green and has a fat-like tail that makes it soluble in fats and oils.* Center: *Enzymes in the plant cells can remove the fat-like tail, producing a tailless form that is water-soluble and readily leaks into cooking liquids.* Right: *In acid conditions, the central magnesium atom is replaced by hydrogens, and the resulting chlorophyll molecule is a dull olive green.*

and chlorophyll is exposed to the plant's own natural acids. Freezing, pickling, dehydration, and simple aging also damage chloroplasts and chlorophyll. This is why dull, olive-green vegetables are so common.

Traditional Aids: Soda and Metals There are two chemical tricks that can help keep green vegetables bright, and cooks have known about them for hundreds and even thousands of years. One is to cook them in alkaline water, which has very few hydrogen ions that are free to displace the magnesium in chlorophyll. The great 19th-century French chef Antonin Carême de-acidified his cooking water with wood ash; today baking soda (sodium bicarbonate) is the easiest. The other chemical trick is to add to the cooking water other metals—copper and zinc—that can replace magnesium in the chlorophyll molecule, and resist displacement by hydrogen. However, both tricks have disadvantages. Copper and zinc are essential trace nutrients, but in doses of more than a few milligrams they can be toxic. And while there's nothing toxic about sodium bicarbonate, excessively alkaline conditions can turn vegetable texture to mush (p. 282), speed the destruction of vitamins, and leave a soapy off-taste.

Watch the Water, Time, and Sauce Dulling of the greens can be minimized by keeping cooking times short, between five and seven minutes, and protecting chlorophyll from acid conditions. Stir-frying and microwaving can be very quick, but they expose chlorophyll fully to the cells' own acids. Ordinary boiling in copious water has the advantage of diluting the cells' acids. Most city tap water is kept slightly alkaline to minimize pipe corrosion, and slightly alkaline water is ideal for preserving chlorophyll's color. Check the pH of your water: if it's acid, its pH below 7, then experiment with adding small amounts of baking soda (start with a small pinch per gallon/4 liters) to adjust it to neutral or slightly alkaline. Once the vegetables are cooked, either serve them immediately or plunge them briefly in ice water so that they don't continue to cook and get dull. Don't dress the vegetables with acidic ingredients like lemon juice until the last minute, and consider protecting them first with a thin layer of oil (as in a vinaigrette) or butter.

Old Tricks for Green Vegetables

Cooks had worked out the practical chemistry of chlorophyll long before it had a name. The Roman recipe collection of Apicius advises, "*omne holus smaragdinum fit, si cum nitro coquatur.*" "All green vegetables will be made emerald colored, if they are cooked with nitrum." Nitrum was a natural soda, and alkaline like our baking soda. In her English cookbook of 1751, Hannah Glasse directed readers to "Boil all your Greens in a Copper Sauce-pan by themselves, with a great Quantity of Water. Use no iron pans, etc., for they are not proper; but let them be Copper, Brass, or Silver." Cookbooks of the early 19th century suggest cooking vegetables and making cucumber pickles with a copper ha'penny coin thrown in to improve the color. All of these practices survived in some form until the beginning of the 20th century, though Sweden outlawed the use of copper cooking pots in its armed services in the 18th century due to the toxicity of copper in large, cumulative doses. And "Tabitha Tickletooth" wrote in *The Dinner Question* (1860): "Never, under any circumstances, unless you wish entirely to destroy all flavor, and reduce your peas to pulp, boil them with soda. This favorite atrocity of the English kitchen cannot be too strongly condemned."

Red-Purple Anthocyanins and Pale Anthoxanthins The usually reddish anthocyanins and their pale yellow cousins, the anthoxanthins, are chlorophyll's opposites. They're naturally water-soluble, so they always bleed into the cooking water. They too are sensitive to pH and to the presence of metal ions, but acidity is good for them, metals bad. And where chlorophyll just gets duller or brighter according to these conditions, the anthocyanins change color completely! This is why we occasionally see red cabbage turn blue when braised, blueberries turn green in pancakes and muffins, and garlic turn green or blue when pickled. (The betacyanins and betaxanthins in beets and chard are different compounds and somewhat more stable.)

The Enemies: Dilution, Alkalinity, and Metals Anthocyanins and anthoxanthins are concentrated in cell vacuoles, and sometimes (as in purple beans and asparagus) just in a superficial layer of cells. So when the food is cooked and the vacuoles damaged, the pigments escape and can become so diluted that their color fades or disappears—especially if they're cooked in a pot of water. The pigments that remain are affected by the new chemical environment of the cooked plant tissue. The vacuoles in which anthocyanins are stored are generally acid, while the rest of the cell fluids are less so. Cooking water is often somewhat alkaline, and quick breads include distinctly alkaline baking soda. In acid conditions, anthocyanins tend toward the red; around neutral pH, they're colorless or light violet; and in alkaline conditions, bluish. And pale anthoxanthins become more deeply yellow as alkalinity rises. So red fruits and vegetables can fade and even turn blue when cooked, while pale yellow ones darken. And traces of metals in the cooking liquid can generate very peculiar colors: some anthocyanins and anthoxanthins form grayish, green, blue, red, or brown complexes with iron, aluminum, and tin.

The Aid: Acids The key to maintaining natural anthocyanin coloration is to keep fruits and vegetables sufficiently acidic, and avoid supplying trace metals. Lemon juice in the cooking water or sprinkled on the food can help with both aims: its citric acid binds up metal ions. Cooking red cabbage with acidic apples or vinegar keeps it from turning purple; dispersing baking soda evenly in batters, and using as little as possible to keep the batter slightly acidic, will keep blueberries from turning green.

Creating Color from Tannins On rare and wonderful occasions, cooking can actually create anthocyanins: in fact, it transforms touch into color! Colorless quince slices

Turning Red Wine into White

The sensitivity of anthocyanin pigments to pH is the basis for a remarkable recipe in the late Roman collection attributed to Apicius:

> To make white wine out of red wine. Put bean-meal or three egg whites into the flask and stir for a very long time. The next day the wine will be white. The ashes of white grape vines have the same effect.

Both vine ashes and egg whites are alkaline substances and do transform the wine's color—though when I've tried this with eggs, the result is not so much a white wine as a gray one.

cooked in a sugar syrup lose their astringency and develop a ruby-like color and translucency. Quinces and certain varieties of pear are especially rich in phenolic chemicals, including aggregates (proanthocyanidins) of from 2 to 20 anthocyanin-like subunits. The aggregates are the right size to cross-link and coagulate proteins, so they feel astringent in our mouth. When these fruits are cooked for a long time, the combination of heat and acidity causes the subunits to break off one by one; and then oxygen from the air reacts with the subunits to form true anthocyanins: so the tannic, pale fruits become more gentle-tasting and anything from pale pink to deep red. (Interestingly, the similar development of pinkness in canned pears is considered discoloration. It's accentuated by tin in unenameled cans.)

Texture We've seen that the texture of vegetables and fruits is determined by two factors: the inner water pressure of the tissue's cells, and the structure of the cell walls (p. 265). Cooking softens plant tissues by releasing the water pressure and dismantling the cell walls. When the tissue reaches 140°F/60°C, the cell membranes are damaged, the cells lose water and deflate, and the tissue as a whole goes from firm and crisp to limp and flabby. (Even vegetables surrounded by boiling water lose water during cooking, as weighings before and after will prove.) At this stage, vegetables often squeak against the teeth: they've lost the crunch of turgid tissue, but the cell walls are still strong and resist chewing. Then as the tissue temperature approaches the boiling point, the cell walls begin to weaken.The cellulose framework remains mostly unchanged, but the pectin and hemicellulose "cement" softens, gradually breaks down into shorter chains, and dissolves. Teeth now easily push adjacent cells apart from each other, and the texture becomes tender. Prolonged boiling will remove nearly all of the cell-wall cement and cause the tissue to disintegrate, thus transforming it into a puree.

Acid and Hard Water Maintain Firmness; Salt and Alkalinity Speed Softening The wall-dissolving, tenderizing phase of fruit and vegetable cooking is strongly influenced by the cooking environment. Hemicelluloses are not very soluble in acid conditions, and readily soluble in alkaline conditions. This means that fruits and vegetables cooked in an acid liquid—a tomato sauce for example, or other fruit juices and purees—may remain firm during hours of cooking, while in neutral boiling water, neither acid nor alkaline, the same vegetables soften in 10 or 15 minutes. In distinctly alkaline water, fruits and vegetables quickly become mushy. Table salt in neutral cooking water speeds vegetable soft-

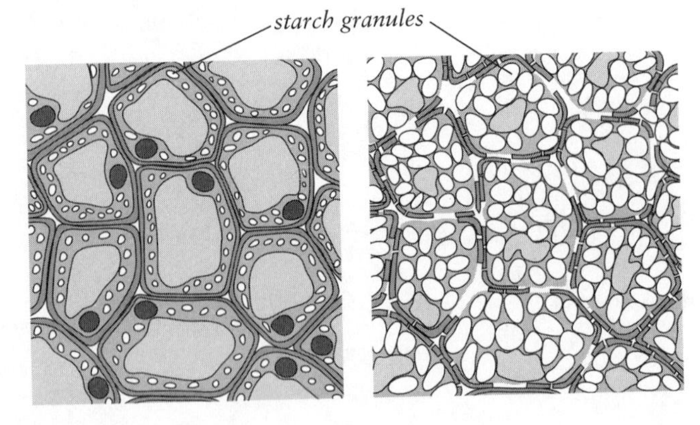

Cooking starchy vegetables. Left: *Before cooking, the plant cells are intact, the starch granules compact and hard.* Right: *Cooking causes the starch granules to absorb water from the cell fluids, swell, and soften.*

ening, apparently because its sodium ions displace the calcium ions that cross-link and anchor the cement molecules in the fruit and vegetable cell walls, thus breaking the cross-links and helping to dissolve the hemicelluloses. On the other hand, the dissolved calcium in hard tap water slows softening by reinforcing the cement cross-links. When vegetables are cooked without immersion in water—when they're steamed or fried or baked—the cell walls are exposed only to the more or less acid cell fluids (steam itself is also a somewhat acidic pH 6), and a given cooking time often produces a firmer result than boiling.

The cook can make use of these influences to diagnose the cause of excessively rapid or slow softening and adjust the preparation—for example, precooking vegetables in plain water before adding them to a tomato sauce, or compensating for hard water with a softening pinch of alkaline baking soda. In the case of green vegetables, shortening the softening time with the help of salt and a discreet dose of baking soda helps preserve the bright green of the chlorophyll (p. 280).

Starchy Vegetables Potatoes, sweet potatoes, winter squashes, and other starchy vegetables owe their distinctive cooked texture to their starch granules. In the raw vegetables, starch granules are hard, closely packed, microscopic agglomerations of starch molecules, and give a chalky feeling when chewed out of the cells. They begin to soften at about the same temperature at which the membrane proteins denature, the "gelation range," which in the potato is from 137–150°F/58–66°C (it varies from plant to plant). In this range the starch granules begin to absorb water molecules, which disrupt their compact structure, and the granules swell up to many times their original size, forming a soft gel, or sponge-like network of long chains holding water in the pockets between chains. The overall result is a tender but somewhat dry texture, because the tissue moisture has been soaked up into the starch. (Think of the textural difference between cooked high-starch potatoes and low-starch carrots.) In starchy vegetables with relatively weak cell walls, the gel-filled cells may be cohesive enough to pull away from each other as separate little particles, giving a mealy impression. This water absorption and the large surface area of separate cells are the reasons that mashed potatoes and other cooked starchy purees benefit from and accommodate large amounts of lubricating fat.

Precooking Can Give a Persistent Firmness to Some Vegetables and Fruits It turns out that in certain vegetables and fruits—including potatoes, sweet potatoes, beets, carrots, beans, cauliflower, tomatoes, cherries, apples—the usual softening during cooking can be reduced by a low-temperature precooking step. If preheated to 130–140°F/55–60°C for 20–30 minutes, these foods develop a persistent firmness that survives prolonged final cooking. This can be valuable for vegetables meant to hold their shape in a long-cooked meat dish, or potatoes in a potato salad, or for foods to be preserved by canning. It's also valuable for boiled whole potatoes and beets, whose outer regions are inevitably over-softened and may begin to disintegrate while the centers cook through. These and other long-cooked root vegetables are usually started in cold water, so that the outer regions will firm up during the slow temperature rise. Firm-able vegetables and fruits have an enzyme in their cell walls that becomes activated at around 120°F/50°C (and inactivated above 160°F/70°C), and alters the cell-wall pectins so that they're more easily cross-linked by calcium ions. At the same time, calcium ions are being released as the cell contents leak through damaged membranes, and they cross-link the pectin so that it will be much more resistant to removal or breakdown at boiling temperatures.

Persistently Crisp Vegetables A few underground stem vegetables are notable for retaining some crunchiness after prolonged cooking and even canning. These include the Chinese water chestnut, lotus root, bamboo shoots, and beets. Their textural robustness comes from particular phenolic compounds in their cell walls (ferulic acids) that form bonds with the cell-wall carbohydrates and prevent them from being dissolved away during cooking.

Flavor The relatively mild flavor of most vegetables and fruits is intensified by cooking. Heating makes taste molecules—sweet sugars, sour acids—more prominent by breaking down cell walls and making it easier for the cell contents to escape and reach our taste buds. Carrots, for example, taste far sweeter when cooked. Heat also makes the food's aromatic molecules more volatile and so more noticeable, and it creates new molecules by causing increased enzyme activity, mixing of cell contents, and general chemical reactivity. The more prolonged or intense the heating, the more the food's original aroma molecules are modified and supplemented, and so the more complex and "cooked" the flavor. If the cooking temperature exceeds the boiling point—in frying and baking, for example—then these carbohydrate-rich materials will begin to undergo browning reactions, which produce characteristic roasted and caramelized flavors. Cooks can create several layers of flavor in a dish by combining well-cooked, lightly cooked, and even raw batches of the same vegetables or herbs.

One sensory quality unique to plants is astringency (p. 271), and it can make such foods as artichokes, unripe fruits, and nuts less than entirely pleasant to eat. There are ways to control the influence of tannins in these foods. Acids and salt increase the perception of astringency, while sugar reduces it. Adding milk, gelatin, or other proteins to a dish will reduce its astringency by inducing the tannins to bind to food proteins before they can affect salivary proteins. Ingredients rich in pectin or gums will also take some tannins out of circulation, and fats and oils will slow the initial binding of tannins and proteins.

Nutritional Value Cooking destroys some of the nutrients in food, but makes many nutrients more easily absorbed. It's a good idea to include both raw and cooked fruits and vegetables in our daily diet.

Some Diminishment of Nutritional Value . . . Cooking generally reduces the nutritional content of fruits and vegetables. There are some important exceptions to this rule, but the levels of most vitamins, antioxidants, and other beneficial substances are diminished by the combination of high temperatures, uncontrolled enzyme activity, and exposure to oxygen and to light. They and minerals can also be drawn out of plant tissues by cooking water. These losses can be minimized by rapid and brief cooking. Baked potatoes, for example, heat up relatively slowly and lose much more vitamin C to enzyme action than do boiled potatoes. However, some techniques that speed cooking—cutting vegetables into small pieces, and boiling in a large volume of water, which maintains its temperature—can result in increased leaching of water-soluble nutrients, including minerals and the B and C vitamins. To maximize the retention of vitamins and minerals, cook small batches of vegetables and fruits in the microwave oven, in a minimal amount of added water.

. . . And Some Enhancement Cooking has several general nutritional benefits. It eliminates potentially harmful microbes. By softening and concentrating foods, it also makes them easier to eat in significant quantities. And it actually improves the availability of some nutrients. Two of the most important are starch and the carotenoid pigments. Starch consists of long chains of sugar molecules crammed into masses called granules. Our digestive enzymes can't penetrate past the outer layer of raw starch granules, but cooking unpacks the starch chains and lets our enzymes break them down. Then there are beta-carotene, the precursor to vitamin A, its chemical relative lycopene, an important antioxidant, and other valuable carotenoid pigments.

Because they're not very soluble in water, we simply don't extract these chemicals very efficiently by just chewing and swallowing. Cooking disrupts the plant tissues more thoroughly and allows us to extract much more of them. (Added fat also significantly improves our absorption of fat-soluble nutrients.)

There are many different ways of cooking vegetables and fruits. What follows is a brief outline of the most common methods and their general effects. They can be divided into three groups: moist methods that transfer heat by means of water; dry methods that transfer heat by means of air, oil, or infrared radiation; and a more miscellaneous group that includes ways of restructuring the food, either turning it into a fluid version of itself, or extracting the essence of its flavor or color.

Hot Water: Boiling, Steaming, Pressure-Cooking

Boiling and steaming are the simplest methods for cooking vegetables, because they require no judgment of cooking temperature: whether water is boiling on a high flame or low, its temperature is 212°F/100°C (near sea level, with predictably lower temperatures at higher elevations). And because hot water and steam are excellent carriers of heat, these are efficient methods as well, ideal for the rapid cooking of green vegetables that minimizes their loss of color (p. 280). One important difference is that hot water dissolves and extracts some pectin and calcium from cell walls, while steaming leaves them in place: so boiling will soften vegetables faster and more thoroughly.

Boiling In the case of boiling green vegetables, it's good to know the pH and dissolved mineral content of your cooking water. Ideally it should be neutral or just slightly alkaline (pH 7–8), and not too hard, because acidity dulls chlorophyll, and acidity and calcium both slow softening and so prolong the cooking. A large volume of rapidly boiling water will maintain a boil even after the cold vegetables are added, cut into pieces small enough to cook through in about five minutes. Salt in the cooking water at about the concentration of seawater (3%, or 2 tablespoons/30 gm per quart/liter) will speed softening (p. 282) and also minimize the loss of cell contents to the water (cooking water without its own dissolved salt will draw salts and sugars from the plant cells). When just tender enough, the vegetables should be removed and either served immediately or scooped briefly into ice water to stop the cooking and prevent further dulling of the color.

Starchy vegetables, especially potatoes cooked whole or in large pieces, benefit from a different treatment. Their vulnerability is a tendency for the outer portions to soften excessively and fall apart while the interiors cook through. Hard and slightly acid water can help them maintain their surface firmness, as will starting them in cold water and raising the temperature only gradually to reinforce their cell walls (p. 283). Salt is best omitted from the water, since it encourages early softening of the vulnerable exterior. Nor is it necessarily best to cook them at the boiling point: 180–190°F/80–85°C is sufficient to soften starch and cell walls and won't overcook the exterior as badly, though the cooking through will take longer.

When vegetables are included in a meat braise or stew and are expected to have a tender integrity, their cooking needs as much attention as the meat's. A very low cooking temperature that keeps the meat tender may leave the vegetables hard, while repeated bouts of simmering to dissolve a tough cut's connective tissue may turn them to mush. The vegetables can be precooked separately, either to soften them for a low-temperature braise or firm them for long simmering; or they can be removed from a long-simmered dish when they reach the desired texture and added back when the meat is done.

Steaming Steaming is a good method for cooking vegetables at the boiling point, but without the necessity of heating a whole pot of water, exposing the food directly to turbulent water, and leaching out flavor or color or nutrients. It doesn't allow the cook to control saltiness, calcium cross-linking, or acidity (steam itself is a slightly acid pH 6, and plant cells and vacuoles are also more acid than is ideal for chlorophyll); and evenness of cooking requires that the pieces be arranged in a single layer, or that the pile be very loose to allow the steam access to all food surfaces. Steaming leaves the food tasting exclusively of its cooked self, though the steam can also be aromatized by the inclusion of herbs and spices.

Pressure Cooking Pressure cooking is sometimes applied to vegetables, especially in the canning of low-acid foods. It is essentially cooking by a combination of boiling water and steam, except that both are at about 250°F/120°C rather than 212°F/100°C. (Enclosing the water in an airtight container traps the water vapor, which in turn raises the boiling point of the water.) Pressure cooking heats foods very rapidly, which means that it's also very easy to overcook fresh vegetables. It's best to follow specialized recipes closely.

Hot Air, Oil, and Radiation: Baking, Frying, and Grilling

These "dry" cooking methods remove moisture from the food surface, thus concentrating and intensifying flavor, and can heat it above the boiling point, to temperatures that generate the typical flavors and colors of the browning reactions (p. 777).

Baking The hot air in an oven cooks vegetables and fruits relatively slowly, for several reasons. First, air is not as dense a medium as water or oil, so air molecules collide with the food less often, and take longer to impart energy to it. Second, a cool object in a hot oven develops a stagnant "boundary layer" of air molecules and water vapor that slows the collision rate even further. (A convection fan speeds cooking by circulating the air more rapidly and disrupting the boundary layer.) Third, in a dry atmosphere the food's moisture evaporates from the surface, and this evaporation absorbs most of the incoming energy, only a fraction of which gets to the center. So baking is much less efficient than boiling or frying.

Of course, the oven's thin medium is why the oven is a good means for drying foods, either partly—for example, to concentrate the flavor of watery tomatoes—or almost fully, to preserve and create a chewy or crisp texture. And once the surface has dried and its temperature rises close to the oven's, then carbohydrates and proteins can undergo the browning reactions, which generate hundreds of new taste and aroma molecules and so a greater depth of flavor.

Often vegetables are coated with oil before baking, and this simple pretreatment has two important consequences. The thin surface layer of oil doesn't evaporate the way the food moisture does, so all the heat the oil absorbs from the oven air goes to raising its and the food's temperature. The surface therefore gets hotter than it would without the oil, and the food is significantly quicker both to brown and to cook through. Second, some of the oil molecules participate in the surface browning reactions and change the balance of reaction products that are formed; they create a distinctly richer flavor.

Frying and Sautéing Baking oiled vegetables is sometimes called "oven frying," and indeed true frying in oil also desiccates the food surface, browns it, and enriches the flavor with the characteristic notes contributed by the oil itself. A food may be fried partly or fully immersed in oil, or just well lubricated with it (sautéing); and typical oil temperatures range from 325–375°F/160–190°C. True frying is faster than oven frying because oil is much denser than air, so energetic oil molecules collide with the food much more frequently. The key to successful frying is getting the piece size and frying temperature right, so that the pieces cook through in the time that the surfaces require to be properly browned. Starchy vegetables are the most commonly fried plant foods, and I describe the important example of potatoes in detail in chapter 6 (p. 303). Many more delicate vegetables and even fruits are fried with a protective surface coating of batter (p. 553) or breading, which browns and crisps while the food inside is insulated from direct contact with the high heat.

Stir-Frying and Sweating Two important variations on frying exploit opposite ends of the temperature scale. One is high-temperature stir-frying. The vegetables are cut into pieces sufficiently small that they heat through in about a minute, and they're cooked on a smoking-hot metal surface with just enough oil to lubricate them, and with constant stirring to ensure even heating and prevent burning. In stir-frying it's important to preheat the pan alone and add the oil just a few seconds before the vegetables; otherwise the high heat will damage the oil and make it unpalatable, viscous, and sticky. The rapidity of stir-frying makes it a good method for retaining pigments and nutrients. At the other extreme is a technique sometimes called "sweating" (Italian *soffrito* or Catalan *soffregit,* both meaning "underfrying"): the very slow cooking over low heat of finely chopped vegetables coated with oil, to develop a flavor base for a dish featuring other ingredients. Often the cook wants to avoid browning, or to minimize it; here the low heat and oil function to soften the vegetables, develop and concentrate their flavors, and blend those flavors together. Vegetables cooked in a version of the *confit* (p. 177) are immersed in oil and slowly cooked to soften them and infuse them with the oil's flavor and richness.

Grilling Grilling and broiling cook by means of the intense infrared radiation emitted from burning coals, flames, and glowing electrical elements. This radiation can desiccate, brown, and burn in rapid succession, so it's important to adjust the distance between heat source and food to make sure that the food can heat through before the surface chars. As in baking, a coating of oil speeds the cooking and improves flavor. Enclosing the food in a wrapper—fresh corn in its husk, plantains in their skin, potatoes in aluminum foil—can give some protection to the surface and essentially steam the food in its own moisture, while allowing in some of the smoky aroma from the heat source and smoldering wrapper. And some foods actually benefit from charring. Large sweet and hot chillis have a thick, tough cuticle or "skin" that is tedious to peel away. Because it's relatively dry compared to the underlying flesh, and made up in part of flammable waxes, it can be burned to a crisp before the flesh gets soft. Once burned, the skin can be scraped or rinsed off with ease. Similarly, the flesh of eggplants is smokily perfumed and easily scraped from the skin when the whole vegetable is grilled until the flesh softens and the skin dries and toughens.

Microwave Cooking

Microwave radiation selectively energizes the water molecules in fruits and vegetables, and the water molecules then heat up the cell wall, starch, and other plant molecules (p. 786). Because radiation penetrates into food an inch/2 cm or so, it can be a fairly rapid method, and is an excellent one for retaining vitamins and minerals. However, it has several quirks that the cook must anticipate and compensate for. Because the microwaves penetrate a limited distance into the food, they will cook evenly only if the food is cut into similar-sized thin pieces, and the pieces arranged in a single layer or very loose pile. Energetic water molecules turn into water vapor and escape from the food: so microwaves tend to dry foods out. Vegetables should be enclosed in an almost steam-tight container, and often benefit from starting out with a small amount of added water so that their surfaces don't lose too much moisture and shrivel. And because the foods must be enclosed, they retain some volatile chemicals that would otherwise escape—so their flavor can seem strong and odd. The inclusion of other aromatics can help mask this effect.

Cooks can exploit the drying quality of microwave radiation to crisp thin slices of fruits and vegetables. This is best done at a low power setting so that the heating is gentle and even, and doesn't rapidly progress to browning or burning. When there's little water left in a patch of tissue, it takes more energy to break it free, so the local boiling point rises to a temperature high enough to break apart carbohydrates and proteins, and this causes browning and then blackening.

Pulverizing and Extracting

In addition to preparing fruits and vegetables more or less as is, with their tissue structure intact, cooks often deconstruct them completely. In some preparations, we blend the contents of the plant cells with the walls that normally separate and contain them. In others, we separate the food's flavor or color from its flavorless, colorless cell-wall fibers or abundant water, and produce a concentrated extract of that food's essence.

Purees The simplest deconstructed version of fruits and vegetables is the puree, which includes such preparations as tomato and apple sauces, mashed potatoes, carrot soup, and guacamole. We make purees by applying enough physical force to crush the tissue, break apart and break open its cells, and mix cell innards with fragments of the cells' walls. Thanks to the high water content of the cells, most purees are fluid versions of the original tissue. And thanks to the thickening powers of the cell-wall carbohydrates, which bind up water molecules and get entangled with each other, they also have a considerable, velvety body—or can develop such a body when we boil off excess water and concentrate the carbohydrates. (Potatoes and other starchy vegetables are a major exception: starch granules in the cells absorb all the free moisture in the tissue, and are best left intact in unbroken cells so the solid puree doesn't become gluey. See the discussion of mashed potatoes on p. 303.) Purees are made into sauces and soups, frozen into ices, and dried into "leathers." For purees as sauces, see p. 620.

Many ripe fruits have sufficiently weakened cell walls that they are easily pureed raw, while most vegetables are first cooked to soften the cell walls. Precooking has the additional advantage of inactivating cell enzymes which, when cellular organization is disrupted, would otherwise destroy vitamins and pigments, alter flavor, and cause unsightly browning (p. 269). The size of solid particles in the puree, and so its textural fineness, is determined by how thoroughly ripening or cooking have dismantled the cell walls, and by the method used to crush the tissue. Mashing by hand leaves large cell aggregates intact; the screens in food mills and strainers produce smaller pieces; machine-powered food processor blades chop very finely, and blender blades, working in a more confined space, chop and shear more finely still. Persistent cellulose-rich fibers can be removed only by passing the puree through a strainer.

Juices Juices are refined versions of the puree: they are mainly the fluid contents of fruit and vegetable cells, made by crushing the raw food and separating off most of the solid cell-wall materials. Some of these materials inevitably end up in the juice—for example, the pulp in orange juice—and can cause both desirable and undesirable haziness and body. Because juicing mixes together the contents of living cells, including active enzymes and various reactive and oxygen-sensitive substances, fresh juices are unstable and change rapidly. Apple and pear juices turn brown, for example, thanks to the action of browning enzymes and oxygen (p. 269). If not used immediately, they're best kept chilled or frozen, perhaps after a heat treatment just short of the boil to inactivate enzymes and kill microbes. Modern juicing machines can apply very strong forces, and make it possible to extract juice from any fruit or vegetable, not just the traditional ones.

Foams and Emulsions The cell-wall carbohydrates in purees and juices can be used to stabilize two otherwise fleeting physical structures, a foam of air bubbles and an emulsion of oil droplets (pp. 638, 625), which are especially easy to prepare with modern electrical blenders and mixers. If a puree or juice is whipped to fill it with air bubbles, the cell-wall carbohydrates slow the flow of water out of the bubble walls, so the bubbles take longer to collapse. This allows the cook to make a foam or mousse that lasts long enough to be savored; foams from juice are especially ethereal. Similarly, when oil is whisked into a puree or juice, the plant carbohydrates insulate the oil droplets from each other, and the oil and water phases separate more slowly. The cook can therefore incorporate oil into a puree or juice to form a temporary emulsion, with richer dimensions of flavor and texture than the puree alone. The thicker the puree, the more stable and less delicate the foam or emulsion. The consistency of a thick preparation can be lightened by adding liquid (water, juice, stock).

Frozen Purees and Juices: Ices, Sorbets, Sherbets When purees and juices are frozen, they form a refreshing semisolid mass that's known by a variety of names, including ice, sorbet, granita, and sherbet. This kind of preparation was first refined in 17th-century Italy, which gave us the term *sorbet* (via *sorbetto* from the Arabic *sharab*, or "syrup"). Its flavor is essentially that of the fruit (sometimes an herb, spice, flower, coffee, or tea), usually heightened with added sugar and acid (to 25–35% and 0.5% respectively), and with an overall sugar-acid ratio similar to that of the melons (30–60:1; see p. 382). The puree or juice is often diluted with some water as well, sometimes to reduce the acidity (lemon and lime juices), sometimes to stretch an ingredient in short supply, and sometimes to improve the flavor, which is interestingly affected by the very cold serving temperature: for example, undiluted melon can taste too much like its close relative the cucumber, and thinned pear puree tastes less like frozen fruit, more delicate and perfumed. In the United States, "sherbet" is the term applied to fruit ices with milk solids included (3–5%) to fill out the flavor and help soften the texture.

Though traditional ices are made with fruits, vegetable ices can be refreshing too, as a cool mouthful and as a surprise.

The Texture of Frozen Purees and Juices Ice texture can vary from rocky to coarse to creamy, depending on the proportions of ingredients, how the ice is made, and the temperature at which it's served. During the freezing process, water in the mix solidifies into millions of tiny ice crystals, which are surrounded by all the other substances in the mix: mainly leftover liquid water that forms a syrup with dissolved sugars, both from the fruit and added by the cook, as well as contents of the plant cells and cell walls. The more syrup and plant debris there are, the more the solid crystals are lubricated, the more easily they slide past each other when we press with spoon or tongue, and the softer the ice's texture. Most ices are made with about double the sugar of ice cream (whose substantial fat and protein content helps soften the texture, p. 40), between 25 and 35% by weight. Sweet fruits require less added sugar to reach this proportion, and purees rich in pectins and other plant debris (pineapple, raspberry) require less total sugar for softening. Many cooks replace a quarter to a third of the added table sugar (sucrose) with corn syrup or glucose, which helps soften without adding as much perceptible sweetness. The size of the ice crystals, and so the ice's coarseness or creaminess, is determined by the content of sugar and plant solids, and by agitation during freezing. Sugar and solids encourage the formation of many small crystals rather than a few large ones, and so do stirring and churning (p. 44). Ices served right from the freezer are relatively hard and crystalline; allowing them to warm and thus partly melt produces a softer, smoother consistency.

Vegetable Stocks A vegetable stock is a water extract of several vegetables and herbs that can serve as a flavorful base for soups, sauces, and other preparations. By simmering the vegetables until soft, the cook breaks down their cell walls and releases the cell contents into the water. These contents include salts, sugars, acids, and savory amino acids, as well as aromatic molecules. Carrots, celery, and onions are almost always included for their aromatics, and mushrooms and tomatoes are the richest source of savory amino acids. The vegetables are finely chopped to maximize their surface area for extraction. Precooking some or all of the vegetables in a small amount of fat or oil has two advantages: it adds new flavors, and the fat it contributes is a better solvent than water for many aromatic molecules. It's important not to dilute the extracted flavors in too much water; good proportions by weight (volume varies by piece size) are 1 part vegetables to 1.5 or 2 parts water. The vegetables and water are simmered uncovered (to allow evaporation and concentration) for no more than an hour, after which it's generally agreed that the stock flavor ceases to improve and even deteriorates. Once the vegetables are strained out, the stock can be concentrated by boiling it down.

Flavored Oils, Vinegars, Syrups, Alcohols Cooks extract the characteristic aroma chemicals of fruits and vegetables, herbs and spices, into a variety of liquids that then serve as convenient ready-made flavorings for sauces, dressings, and other preparations. In general, the freshest-tasting extracts come from slowly steeping intact raw fruits or herbs at room or refrigerator temperature for days or weeks. The flavors of dried herbs and spices are less altered by heat, and can be extracted more rapidly in hot liquids.

The growth of microbes that cause spoilage or illness is inhibited by the acidity of vinegar, the concentrated sugar in syrups, and the alcohol in vodka (whose own neutral flavor makes it a good medium for flavor extraction), so flavored vinegars, syrups, and alcohols are relatively trouble-free preparations. However, flavored oils require special care. The air-free environment within the oil can encourage the growth of botulism bacteria, which live in the soil, are found on most field-grown foods, and have spores that survive ordinary cooking temperatures. Cold temperatures inhibit their growth. Uncooked oils flavored with garlic or herbs are safest when made in the refrigerator, and both uncooked and cooked flavored oils should be stored in the refrigerator.

"Chlorophyll" A somewhat arcane but fascinating vegetable extract is culinary chlorophyll, an intensely green coloring agent that is not identical to biochemical chlorophyll, but is certainly a concentrated source of it. Culinary chlorophyll is made by finely grinding dark green leaf vegetables to isolate and break open cells; soaking them in water to dilute pigment-damaging enzymes and acids, and separate off solid fibers and cell-wall debris; gently simmering the water to inactivate enzymes and cause the cells and free chloroplasts to rise to the surface; and straining off and draining the green mass. Though the chemical chlorophyll in culinary chlorophyll will still turn drab when heated in an acid food, it can be added at the last minute to acid and other sauces and maintain its vibrant green through the meal.

PRESERVING FRUITS AND VEGETABLES

Fruits and vegetables can be preserved indefinitely by killing the living tissue and thus inactivating its enzymes, and then making it either inhospitable or unavailable to microbes. Some of these techniques are ancient, some a product of the industrial age.

DRYING AND FREEZE-DRYING

Drying Drying preserves foods by reducing the tissue's water content from around 90% to between 5 and 35%, a range in which very little can grow on it. This is one of the oldest preservative techniques; the sun, fire, and mounds of hot sand have been used to dry foods since prehistory. Fruits and vegetables usually benefit from treatments to inactivate the enzymes that cause vitamin and color damage. Commercially dried vegetables are usually blanched; and fruits are dipped or sprayed with a number of sulfur compounds that prevent oxidation and thereby enzymatic browning and the loss of antioxidant phenolic compounds, vitamins, and flavor. While sun-drying used to be the most common treatment for prunes, raisins, apricots, and figs, forced hot air-drying is now widely used because it is more predictable. Home and restaurant cooks can use the oven or small electric driers whose temperature is easier to control. Fruits and vegetables are dried at relatively low temperatures, 130–160°F/55–70°C, to minimize the loss of flavor and color and prevent the surface from drying too fast and impeding moisture loss from within. Pureed fruits are spread out into thin sheets to make "fruit leather." Relatively moist dried fruits and vegetables are nicely soft, but they're also vulnerable to some hardy yeasts and molds, and therefore are best stored in the refrigerator.

Freeze-Drying Freeze-drying is a controlled version of freezer burn: it removes moisture not by evaporation but by sublimation, the transformation of ice directly into water vapor. Although we think of freeze-drying as a recent industrial innovation, the natives of Peru have been freeze-drying potatoes in the Andes for millennia. To make *chuño,* which can be stored indefinitely, they trample potatoes to break down their structure and expose them constantly to the dry, cold mountain air, so that they freeze at night and lose some moisture by sublimation, then thaw during the day and lose more water by evaporation. *Chuño* develops a strong flavor from the disruption of the potato tissues and long exposure to the air and sun, and is reconstituted in water to make stews.

In modern industrial freeze-drying, foods are quickly chilled to as low as –70°F/–57°C, then slightly warmed and subjected to a vacuum, which pulls their water molecules out and dries them. Because the foods aren't heated or exposed to oxygen, their flavor and color remain relatively fresh. Many fruits and vegetables are freeze-dried today and used as is for snack foods, or reconstituted with water in instant soup mixes, emergency rations, and camping foods.

FERMENTATION AND PICKLING: SAUERKRAUT AND KIMCHI, CUCUMBER PICKLES, OLIVES

Fermentation is one of the oldest and simplest means of preserving foods. It requires no particular kind of climate, no cooking, and so no expenditure of fuel: just a container, which can be a mere hole in the ground, and perhaps some salt or seawater. Olives and sauerkraut—fermented cabbage—are familiar examples of fermented fruits and vegetables. An overlapping category is the *pickle,* a food preserved by immersion in brine or a strong acid such as vinegar. Brines often encourage fermentation, and fermentation generates preservative acids, so the term "pickle" is applied to both fermented and unfermented preparations of cucumbers and other foods. Less familiar but intriguing relatives of sauerkraut and olives include North African preserved lemons, the pickled plums, radishes, and other vegetables of Japan, and the highly spiced, multifarious pickled fruits and vegetables of India.

The Nature of Fermentation Preserving fruits and vegetables by fermentation is based on the fact that plants are the natural home of certain benign microbes which in the right conditions—primarily the absence of air—will flourish and suppress the growth of other microbes that cause spoilage and disease. They accomplish this suppression by being the first to consume the plant material's readily metabolized sugars, and by producing a variety of antimicrobial substances, including lactic and other acids, carbon dioxide, and alcohol. At the same time, they leave most of the plant material intact, including its vitamin C (protected from oxidation by the carbon dioxide they generate); they often add significant amounts of B vitamins; and they generate new volatile substances that enrich the food's aroma. These benign "lactic acid bacteria" apparently evolved eons ago in oxygen-poor piles of decaying vegetation, and now transform our carefully gathered harvests into dozens of different foods across the globe (see box, p. 308), as well as turning milk into yogurt and cheese and chopped meat into tangy sausages (pp. 44 and 176).

Fermentation Conditions and Results While some fruits and vegetables are fermented alone in tightly covered pits or jars, most are either dry-salted or brined to help draw water, sugars, and other nutrients out of the plant tissues, and to provide a liquid to cover the food and limit its exposure to oxygen. The characteristics of the pickle depend on the salt concentration and the fermentation temperature, which determine which microbes dominate and the substances they produce. Low salt concentrations and temperatures favor *Leuconostoc mesenteroides,* which generates a mild but complex mixture of acids, alcohol, and aroma compounds; higher temperatures favor *Lactobacillus plantarum,* which produces lactic acid almost exclusively. Many pickles undergo a microbial succession, with *Leuconostoc* dominating early and then giving way to *Lactobacillus* as the acidity rises. Some Asian pickles are made not by spontaneous lactic fermentations, but by the addition of another fermented "starter" material, the by-products of producing wine or miso or soy sauce. Japanese *nukazuke* are unique in employing rice bran, whose abundant B vitamins end up enriching the pickled daikon and other vegetables.

Problems Problems in vegetable fermentations are generally caused by inadequate or excessive salt concentrations or temperatures, or exposure to the air, all conditions that favor the growth of undesirable microbes. In particular, if the vegetables are not weighted down to keep them below the brine surface, or if the brine surface is itself not tightly covered, a film of yeasts, molds, and air-requiring bacteria will form, lower the brine acidity by consuming its lactic acid, and encourage the growth of spoilage microbes. The results may include discoloration, softening, and rotten smells from the breakdown of fats and proteins. Even the helpful *Lactobacillus plantarum* can generate an undesirably harsh acidity if the fermentation is too vigorous or prolonged.

Unfermented, Directly Acidified Pickles There are also a host of fruit and vegetable products that are pickled not by fermentation, but by the direct addition of acid in the form of wine or vinegar, which inhibits the growth of spoilage microbes. This ancient technique is much faster than fermentation and allows greater control over texture and salt content, but it produces a simpler flavor. Today, the usual method is to add enough hot vinegar to produce a final acetic acid concentration of around 2.5% (half that of standard vinegar) in such materials as beans, carrots, okra, pumpkin, mushrooms, watermelon rind, pears, and peaches. Nonfermented pickles are usually heat-treated (185°F/85°C for 30 minutes) to prevent spoilage. The simple flavor of directly acidified pickles is often augmented by the addition of spices and/or sugar.

Pickle Texture Most pickled fruits and vegetables are eaten raw as a condiment, and are preferred crisp. The use of unrefined sea salt improves crispness thanks to its calcium and magnesium impurities, which help cross-link and reinforce cell-wall pectins. Especially crisp cucumber and watermelon-rind pickles are made by adding alum (aluminum hydroxide), whose aluminum ions cross-link cell-wall pectins,

Some Fermented Vegetables and Fruits

Method	Material	Microbes	Region	Example
Bury in leaf-lined pit	Banana	Lactic acid bacteria	Africa	Kocho
	Breadfruit, root vegetables		South Pacific	Poi (taro)
Enclose in jars	Mustard and related greens	Lactic acid bacteria	Nepal	Gundruk
	Radish roots		Nepal, India	Sinki
Salt, 1–2%	Cabbage	Lactic acid bacteria	Europe	Sauerkraut
Salt, 2–3%	Carrots (purple) grated in water	Lactic acid bacteria	Pakistan, North India	Kanji
Salt, 3–4%	Cabbage, radish bacteria	Lactic acid	Asia	Kimchi
Salt, 4–10% (sometimes with rice bran)	Radish, cabbage, eggplant, cucumbers	Lactic acid bacteria, yeasts	Asia	Tsukemono (nukazuke)
Salt, 5–8%	Cucumbers	Lactic acid bacteria	Europe, Asia	Pickles
Salt, 5–10%	Lemons	Yeasts	West Asia, North Africa	Lamoun makbous, preserved lemons
Salt, 6–10%	Olives	Lactic acid bacteria, yeasts	Europe	Olives
Salt, 20%	Lemons, limes, unripe mangoes	Bacteria, yeasts	India	Achars, pickles

Adapted from G. Campbell Platt, *Fermented Foods of the World—A Dictionary and Guide* (London: Butterworth, 1987).

or by presoaking the raw materials in a solution of "pickling lime," or calcium hydroxide, whose calcium ions do the same. (Lime is strongly alkaline and its excess must be washed from the ingredients before pickling to avoid neutralizing the pickles' acidity.) When subsequently cooked, pickles may not soften because their acidity stabilizes cell walls (p. 282). Tender pickles are produced by precooking the vegetable until soft.

Fermented Cabbage: Sauerkraut and Kimchi Two popular styles of cabbage pickles illustrate the kind of distinctiveness that can be achieved with slight variations in the fermentation process. European sauerkraut is a refreshing side dish for rich meats, and Korean kimchi is a strong accompaniment to bland rice. Sauerkraut—the word is German for "sour cabbage"—is made by fermenting finely shredded head cabbage with a small amount of salt at a cool room temperature; it's allowed to become quite tart and develops a remarkable, almost flowery aroma thanks to some yeast growth. Kimchi is made by fermenting intact stems and leaves of Chinese cabbage together with hot peppers and garlic, and sometimes other vegetables, fruits (apple, pear, melon), and fish sauce. More salt is used, and the fermentation temperature is significantly lower, a reflection of its original production in pots partly buried in the cold earth of late autumn and winter. The result is a crunchy, pungent pickle that is noticeably less acid but saltier than sauerkraut, and may even be fizzy due to the dominance of gas-producing bacteria below about 58°F/14°C.

Cucumber Pickles Today there are three different styles of cucumber pickle in the United States, and the two most common are really flavored cucumbers; they don't keep unless refrigerated. True fermented cucumbers have become relatively hard to find.

All cucumber pickles start with thin-skinned varieties that are harvested while immature so that the seed region hasn't yet begun to liquefy, and cleaned of flower remnants that harbor microbes with enzymes that cause softening. Fermented cucumbers are cured in a 5–8% brine at 64–68°F/18–20°C for two to three weeks, and accumulate 2–3% salt and 1–1.5% lactic acid: so they're relatively strong. Such pickles are sometimes moderated before bottling by soaking out some salt and lactic acid, and adding acetic acid. The most common style of cucumber pickle, crisper and

Fermented Cabbage Two Ways

The German and Korean versions of fermented cabbage are made differently and develop distinctive qualities.

	Sauerkraut	Kimchi
Piece size	1 mm shreds	Small leaves and stems
Ingredients other than cabbage and salt	None	Chillis, garlic, fish sauce
Fermentation temperature	64–76°F/18–24°C	41–57°F/5–14°C
Fermentation time	1–6 weeks	1–3 weeks
Final salt content	1–2%	3%
Final acidity	1–1.5%	0.4–0.8%
Qualities	Tart, aromatic	Strong flavor, crunchy, tingly

more gentle in flavor, is made by soaking the cucumbers briefly in vinegar and salt until they reach 0.5% acetic acid and 0–3% salt, and then pasteurizing them before bottling. Such pickles need to be refrigerated after opening. Finally, there are the freshest-tasting but very perishable pickles, which are soaked in vinegar and salt but not pasteurized. They are kept refrigerated from the moment they're packaged.

Common problems in home-pickled cucumbers include cheesy and rancid off-flavors, which come from the growth of undesirable bacteria when there's not enough salt or acidity to inhibit them, and hollow "bloaters," which are pickles swollen with carbon dioxide produced by yeasts (or sometimes by *Lactobacillus brevis* or *mesentericus*) when the salt level is too high.

Olives Fresh olives are practically inedible thanks to their ample endowment of a bitter phenolic substance, *oleuropein,* and its relatives. The olive tree was first cultivated in the eastern Mediterranean around 5,000 years ago, probably as a source of oil. Olive fermentation may have been discovered when early peoples learned to remove the bitterness by soaking the fruit in changes of water. By Roman times, the soaking water was often supplemented with alkaline wood ashes, which cut the debittering period from weeks to hours. (The modern industrial treatment is a 1–3% solution of sodium hydroxide, or lye.) Alkaline conditions actually break bitter oleuropein down, and also breach the waxy outer cuticle and dissolve cell-wall materials. These effects make the fruit as a whole more permeable to the salt brine that follows (after a wash and acid treatment to neutralize the alkalinity), and help the fermentation proceed faster. Lactic acid bacteria are the main fermenters, though some yeasts also grow and contribute to the aroma. Olives may be debittered and fermented while still green ("Spanish" style, the major commercial type) or once their skin has turned dark with purplish anthocyanins, when they are less bitter.

Olives are also fermented without any preliminary leaching or alkaline treatment, but this results in a different kind of fermentation. Nutrients for the microbes in the brine diffuse very slowly from the flesh through the waxy cuticle, and the intact phenolic materials inhibit microbial growth. So the temperature is kept low (55–64°F/13–18°C), and yeasts rather than lactic acid bacteria dominate in a slow alcoholic fermentation that takes as long as a year. This method is usually applied to black ripe olives (Greek, Italian Gaeta, French Niçoise). They turn out more bitter and less tart than the pretreated kinds (an acidity of 0.3–0.5% rather than 1%), and have a distinctively winey, fruity aroma.

Unfermented "ripe black olives" are an invention of the California canning industry. They're made from unripe green olives, which may undergo an incidental and partial fermentation while being stored in brine before processing. But their unique character is determined by repeated brief lye treatments to leach out and break down oleuropein, and the addition of an iron solution and dissolved oxygen to react with phenolic compounds and turn the skin black. Olives so treated are then packed in a light 3% brine, canned, and sterilized. They have a bland, cooked flavor and often some residual alkalinity, which gives them a slippery quality.

Unusual Fermentations: Poi, Citron, Preserved Lemons Poi is a Hawaiian preparation of taro root (p. 306). The starchy taro is cooked, mashed, thinned with water, and then allowed to stand for one to three days. Lactic acid bacteria sour it, and produce some volatile acids as well (vinegary acetic, cheesy propionic). In longer fermentations, yeasts and *Geotrichum* molds also grow and contribute fruity and mushroomy notes.

Citron peel, candied from a relative of the lemon, owes its traditionally complex flavor to fermentation. Originally the citron fruits were preserved for some weeks in seawater or a 5 to 10% brine while they were shipped from Asia and the Middle East to Europe; now they're brined to develop flavor. Yeasts grow on the peel and produce alcohol, which then supports acetic acid bacteria. The result is the production of volatile esters that deepen the aroma of the peel. The preserved lemons of Morocco and other north African countries have a similar character; they're made by packing cut lemons with salt and fermenting for days to weeks.

Sugar Preserves

Another venerable technique for preserving fruits is to boost their sugar content. Like salt, sugar makes the fruit inhospitable to microbes: it dissolves, binds up water molecules, and draws moisture out of living cells, thus crippling them. Sugar molecules are quite heavy compared to the sodium and chloride ions in salt, so it takes a larger mass of sugar to do the same job of preserving. The usual proportion by weight of added sugar to fruit is about 55 to 45, with sugar accounting for nearly two-thirds of the final cooked mixture. Of course sugar preserves are very sweet, and this is a large part of their appeal. But they also develop an intriguing consistency otherwise found only in meat jellies—a firm yet moist solidity that can range from stiff and chewy to quiveringly tender. And they can delight the eye with a crystalline clarity: in the 16th century, Nostradamus described a quince jelly whose color "is so diaphanous that it resembles an oriental ruby." These remarkable qualities arise from the nature of pectin, one of the components of the plant cell wall, and its fortuitous interaction with the fruit's acids and the cook's added sugar.

The Evolution of Sugar Preserves The earliest sugar preserves were probably fruit pieces immersed in syrupy honey (the Greek term for quinces packed in honey, *melimelon*, gave us the word *marmalade*) or in the boiled-down juice of wine grapes. The first step toward jams and jellies was the discovery that when they were cooked together, sugar and fruit developed a texture that neither could achieve on its own. In the 4th century CE, Palladius gave directions for cooking down shredded quince in honey until its volume was reduced by half, which would have made a stiff, opaque paste similar to today's "fruit cheese" (spreadable "fruit butter" is less reduced). By the 7th century there were recipes for what were probably clear and delicate jellies made by boiling the juice of quince with honey. A second important innovation was the introduction from Asia of cane sugar, which unlike honey is nearly pure sugar, with no moisture that needs boiling off, and no strong flavor that competes with the flavor of the fruit. The Arab world was using cane sugar by the Middle Ages, and brought it to Europe in the 13th century, where it soon became the preferred sweetener for fruit preserves. However, jams and jellies didn't become common fare until the 19th century, when sugar had become cheap enough to use in large quantities.

Pectin Gels Fruit preserves are a kind of physical structure called a *gel:* a mixture of water and other molecules that is solid because the other molecules bond together into a continuous, sponge-like network that traps the water in many separate little pockets. The key to creating a fruit gel is pectin, long chains of several hundred sugar-like subunits, which seems to have been designed to help form a highly concentrated, organized gel in plant cell walls (p. 265). When fruit is cut up and heated near the boil, the pectin chains are shaken loose from the cell walls and dissolve into the released cell fluids and any added water. They can't simply re-form their gel for a couple of reasons. Pectin molecules in water accumulate a negative electrical charge, so they repel each other rather than bond to each other; and they're now so diluted by water molecules that even if they did bond, they couldn't form a continuous network. They need help to find each other again.

The cook does three things to cooked fruit to bring pectin molecules back together into a continuous gel. First, he adds a large dose of sugar, whose molecules attract water molecules to themselves, thus pulling the water away from the pectin chains and leaving them more exposed to each other. Second, he boils the mixture of fruit and sugar to evaporate some of the water away and bring the pectin chains even closer together. Finally, he increases the acidity, which neutralizes the electrical charge and allows the aloof pectin chains to bond to each other into a gel. Food scientists have found that the optimal conditions for pectin gelation are a pH between 2.8 and 3.5—about the acidity of orange juice, and 0.5% acid by weight—a pectin concentration of 0.5 to 1.0%, and a sugar concentration of 60 to 65%.

Preparing Preserves Preserve making begins with cooking the fruit to extract its pectin. Quince, apples, and citrus fruits are especially rich in pectin and often included to supplement other pectin-poor fruits, including most berries. The combination of heat and acid will eventually break pectin chains into pieces too small to form a network, so this preliminary cooking should be as brief and gentle as possible. (If a sparkling, clear jelly is desired, then the cooked fruit is gently strained to remove all solid particles of cell debris.) Then sugar is added, supplemental pectin if necessary, and the mixture rapidly brought to the boil to remove water and concentrate the other ingredients. The boiling is continued until the temperature of the mix reaches 217–221°F/103–105°C (at sea level; 2°F/1°C lower for every 500 ft/165 m elevation), which indicates that the sugar concentration has reached 65% (for the relationship between sugar content and boiling point, see p. 680). A fresher flavor results when this cooking is done at a gentle simmer in a wide pot with a large surface area for evaporation. (Industrial manufacturers cook the water out under a vacuum at much lower temperatures, 100–140°F/38–60°C, to maintain as much fresh flavor and color as possible.) Now supplemental acid is added (late in the process, to avoid breaking down the pectin chains), and the readiness of the mix is tested by placing a drop on a cold spoon or saucer to see whether it gels. Finally, the mix is poured into sterilized jars. The mix sets as it cools below about 180°F/80°C, but firms most rapidly at 86°F/30°C and

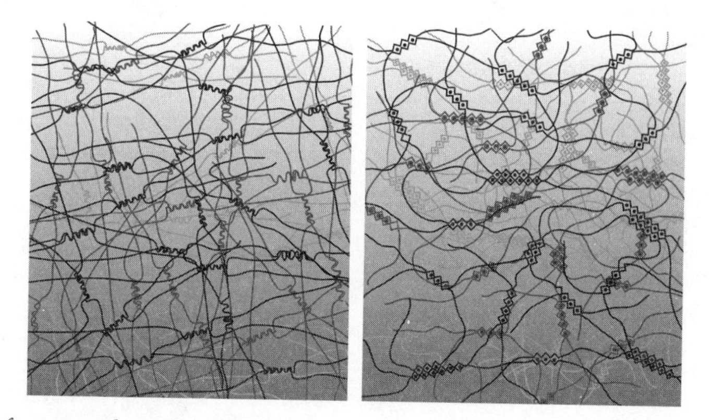

Two kinds of pectin gels. Left: *In ordinary fruit preserves, the cook causes pectin molecules to bond directly to each other and form a continuous meshwork by carefully adjusting acidity and sugar content.* Right: *A modified form of pectin (low methoxy) can be bonded into a continuous meshwork by means of added calcium ions (the black dots), no matter what the sugar content. This is how low-sugar preserves are made.*

continues to get firmer for some days or weeks.

The usual problem with preserve making is failure of the mix to set even at the proper boiling temperature and sugar concentration. This can be caused by three different factors: inadequate amounts of either acid or good-quality pectin, or prolonged cooking that damages the pectin. Failures can sometimes be rescued by the addition of a commercial liquid pectin preparation and/or cream of tartar or lemon juice, and a brief reboiling. Too much acid can cause weeping of fluid from an overfirm gel.

Uncooked and Unsweetened "Preserves" Modern preserve making has been transformed by the availability of concentrated pectin, extracted and purified from citrus and apple wastes, which can be added to any crushed fruit, cooked or not, to guarantee a firm gel. "Freezer jams" are made by loading up crushed fresh fruit with supplemental pectin and sugar, letting them sit for a day while the pectin molecules slowly form their network and form the gel, and then "preserving" them in the refrigerator or freezer (the uncooked fruit would otherwise soon be spoiled by sugar-tolerant molds and yeasts). Pectin is also used to make clear jelly candies and other confections.

Food chemists have developed several different versions of pectin for special commercial applications. The most notable of these is a pectin that sets without the need for any added sugar to pull water molecules away from the long pectin chains; instead, the chains bond to each other strongly by means of cross-linking calcium, which is added after the fruit and pectin mixture has been cooked. This pectin is what makes it possible to produce low-calorie "preserves" with artificial sweeteners.

Candied Fruits Candied fruits are small whole fruits or pieces that are impregnated with a saturated sugar syrup, then drained, dried, and stored at room temperature as separate pieces. Fruit cooked in a sugar syrup remains relatively firm and maintains its shape thanks to the interaction of sugar molecules with the cell-wall hemicelluloses and pectins. Candying can be a tedious process because it takes time for sugar to diffuse from the syrup evenly into the fruit. Typically the fruit is gently cooked to soften it and make its tissues more permeable, then soaked for several days at room temperature in a syrup that starts out at 15–20% sugar, and is made more concentrated each day until it reaches 70–74%.

Canning

Canning was a cause for wonder when it was invented by Nicolas Appert around 1810: contemporaries said that it preserved fruits and vegetables almost as if fresh! True, it preserves them without the desiccated texture of drying, the salt and sourness of fermentation, or the sweetness of sugar preserves; but there's no mistaking that canned foods have been cooked. Canning is essentially the heating of food that has been isolated in hermetically sealed containers. The heat deactivates plant enzymes and destroys harmful microbes, and the tight seal prevents recontamination by microbes in the environment. The food can then be stored at room temperature without spoiling.

The arch villain of the canning process is the bacterium *Clostridium botulinum,* which thrives in low-acid, airless conditions—oxygen is toxic to it—and produces a deadly nerve toxin. The botulism toxin is easily destroyed by boiling, but the dormant bacterial spores are very hardy and can survive prolonged boiling. Unless they are killed by the extreme condition of higher-than-boiling temperatures (which require a pressure cooker), the spores will proliferate into active bacteria when the can cools down, and the toxin will accumulate. One precautionary measure is to boil any canned produce after opening to destroy any toxin that may be there. But all suspect cans, especially those bulging from the pressure of gases produced by bacterial growth, should be discarded.

The low pH (high acidity) of tomatoes and many common fruits inhibits the growth of botulism bacteria, so these foods require the least severe canning treatment, usually about 30 minutes in a bath of boiling water to heat the contents to 185–195°F/85–90°C. Most vegetables, however, are only slightly acid, with a pH of 5 or 6, and are much more vulnerable to bacteria and molds. They're typically heated in a pressure cooker at 240°F/116°C for 30 to 90 minutes.

CHAPTER 6

A SURVEY OF COMMON VEGETABLES

Chapter 5 described the general nature of plant foods and their behavior in the kitchen. This chapter and the next two survey some familiar vegetables, fruits, and flavorings. Because we eat hundreds of different plants, and countless varieties of them, these surveys can only be selective and sketchy. They're meant to highlight the distinctive qualities of these foods, to help the food lover appreciate those qualities more fully and make the best use of them.

These chapters give special attention to two features of our plant foods. One is *family relationships,* which tell us which plants are related to each other, and conversely how varied a given species can be. Such information helps us make sense of similarities and differences among particular foods, and may suggest ideas for interesting combinations and themes.

The second feature emphasized in the following pages is *flavor chemistry.* Fruits and vegetables, herbs and spices are the most complex foods we eat. If we know even a little bit about which substances create their flavor, then we become more attuned to how the flavor is built, and better able to perceive echoes and harmonies among different ingredients. Such perceptions enrich the experience of eating, and can help us become better cooks. All aromas come from particular volatile chemicals, and I sometimes name those chemicals to be as specific as possible about the qualities of a given food. The names may look foreign and incomprehensible, but they're simply names—and sometimes make more sense than the names of the foods they're in!

This survey of vegetables begins underground, with the plant parts that sustain much of the earth's population. It then moves up the plant, from stem to leaf to flower and fruit, and finishes with water plants and those delicious nonplants, the mushrooms.

ROOTS AND TUBERS

Potatoes, sweet potatoes, yams, cassava—these roots and tubers are staple foods for billions of people. They are subterranean organs in which plants store starch, large molecular aggregates of the sugars they create during photosynthesis. They are therefore a concentrated and long-lived package of nourishment for us as well. Some anthropologists theorize that roots and tubers may have helped fuel human evolution, when the climate of the African savanna cooled about 2 million years ago and fruits became scarce. Because tubers were plentiful and far more nutritious when cooked—raw starch granules resist our digestive enzymes, while gelated starch does not—they may have offered a significant advantage to early humans who learned to dig for them and roast them in the embers of a fire.

Though some underground vegetables are a third or more starch by weight, many others—carrots, turnips, beets—contain little or no starch. Because starch granules absorb moisture from their cells as they cook, starchy vegetables tend to have a dry, mealy texture, while nonstarchy vegetables remain moist and cohesive.

Food Words: *Root, Radish, Tuber, Truffle*

Our word *root* comes from an Indo-European word that meant both "root" and "branch." *Radish* and *licorice* share that same ancestor. *Tuber* comes from an Indo-European (linguistic) root meaning "to swell," as many plant storage organs do. The same root gave us *truffle,* the swollen underground fungus, as well as *thigh, thumb, tumor,* and *thousand.*

Potatoes

There are more than 200 species of potato, relatives of the tomato, chilli, and tobacco that are indigenous to moist, cool regions of Central and South America. Some were cultivated 8,000 years ago. Spanish explorers brought one species, *Solanum tuberosum,* from Peru or Colombia to Europe around 1570. Because it was hardy and easy to grow, the potato was inexpensive and the poor were its principal consumers. (An Irish peasant ate 5–10 pounds per day at the time of the 1845 blight.) It now leads all other vegetables in worldwide production. More potatoes are consumed in the United States than any other vegetable, around a third of a pound/150 gm per person per day.

The potato is a tuber, the tip of an underground stem that swells with stored starch and water and bears primordial buds, the "eyes," that generate the stem and roots of a new plant. It is sometimes a little sweet, with a slight but characteristic bitterness, and has a mild earthy flavor from a compound (a pyrazine) produced by soil microbes, but also apparently within the tuber itself.

Harvest and Storage True "new" potatoes are immature tubers, harvested from green vines during the late spring and throughout the summer. They are moist and sweet, relatively low in starch, and perishable. Mature potatoes are harvested in the fall. The vines are killed by cutting or drying, and the tubers are left in the soil for several weeks to "cure" and toughen their skin. Potatoes can be stored in the dark for months, during which their flavor intensifies; slow enzyme action generates fatty, fruity, and flowery notes from cell-membrane lipids. The ideal storage temperature is 45–50°F/7–10°C. At warmer temperatures they may sprout or decay, and at colder temperatures their metabolism shifts in a complicated way that results in the breakdown of some starch to sugars. Makers of potato chips must "recondition" cold-stored potatoes at room temperature for several weeks to reduce their levels of glucose and fructose, which otherwise cause the chips to brown too rapidly and develop a bitter taste. Internal black spots in potatoes are essentially bruises, formed when an impact during handling damages cells and causes the browning enzymes to create dark complexes of the amino acid tyrosine (alkaloid formation and therefore bitterness often rise also).

Nutritional Qualities Potatoes are a good source of energy and vitamin C. Yellow-fleshed varieties owe their color to fat-soluble carotenoids (lutein, zeaxanthin), purple and blue ones to water-soluble and antioxidant anthocyanins. Potatoes are notable for containing significant levels of the toxic alkaloids solanine and chaconine, a hint of whose bitterness is part of their true flavor. Most commercial varieties contain 2 to 15 milligrams of solanine and chaconine per quarter-pound (100 grams) of potato. Progressively higher levels result in a distinctly bitter taste, a burning sensation in the throat, digestive and neurological problems, and even death. Stressful growing conditions and exposure to light can double or triple the normal levels. Because light also induces chlorophyll formation, a green cast to the surface is a sign of abnormally high alkaloid levels. Greened potatoes should either be peeled deeply or discarded, and strongly bitter potatoes should not be eaten.

Cooking Types and Behavior There are two general cooking categories of potato, called the "mealy" and the "waxy" for their textures when cooked. Mealy types (russets, blue and purple varieties, Russian and banana fingerlings) concentrate more dry starch in their cells, so they're denser than waxy types. When cooked, the cells tend to swell and separate from each other, producing a fine, dry, fluffy texture that works well in fried potatoes and in baked and mashed potatoes, which are moistened with butter or cream. In waxy types (true new potatoes and common U.S. red- and white-skinned varieties), neighboring cells cohere even when cooked, which gives them a solid, dense, moist texture, and holds them together in intact pieces for gratins, potato cakes, and salads. Both types can be made firmer and more coherent, less prone to the "sloughing" of outer layers when boiled, by treating them to the low-temperature precooking that strengthens cell walls (p. 283).

Cooked potatoes sometimes develop a large internal region of bluish-gray discoloration. This "after-cooking darkening" is caused by the combination of iron ions, a phenolic substance (chlorogenic acid), and oxygen, which react to form a pigmented complex. This problem can be minimized in boiled potatoes by making the pH of the water distinctly acidic with cream of tartar or lemon juice after the potatoes are half-cooked.

The flavor of boiled potatoes is dominated by the intensified earthy and fatty, fruity, and flowery notes of the raw tuber. Baked potatoes develop another layer of flavor from the browning reactions (p. 777), including malty and "sweet" aromas (methylbutanal, methional). Leftover potatoes often suffer from a stale, cardboard-like flavor that develops over several days in the refrigerator, but within a few hours if the potatoes are kept hot for prolonged service. It turns out that the aromatic fragments of membrane lipids are temporarily stabilized by the tuber's antioxidant vitamin C; but with time the vitamin C is used up and the fragments become oxidized to a series of less pleasant aldehydes.

Potatoes are prepared in many ways, and used as an ingredient in many dishes. Here are brief notes on a few starring roles.

Mashed and Pureed Potatoes There are many different styles of mashed potatoes, but all of them involve cooking the potatoes whole or in pieces, crushing them to a more or less fine particle size, and lubricating and enriching the particles with a combination of water and fat, usually in the form of butter and milk or cream. Some luxurious versions may be almost as much butter as potato, or include eggs or egg yolks. Mealy types fall apart into individual cells and small aggregates, so they offer a large surface area for coating by the added ingredients, and readily produce a fine, creamy consistency. Waxy potatoes require more mashing to obtain a smooth texture, exude more gelated starch, and don't absorb enrichment as easily. The classic French *pommes purées*, pureed potatoes, are made from waxy potatoes, pieces of which are pushed through a fine sieve or food mill and then worked hard—to the point of having what an eminent French cookbook writer, Mme Ste-Ange, called a "dead arm"—first alone and then with butter, to incorporate air and obtain the lightness of whipped cream. American recipes take a more gentle approach, sieving mealy varieties and carefully stirring in liquid and fat to avoid excessive cell damage, starch release, and glueyness.

Fried Potatoes Fried potatoes are some of the world's favorite foods. Deep-fried potato sticks and slices and the technique of double-frying were all well known in Europe by the middle of the 19th century, and in England were attributed mainly to the French: hence the term "French fry" for what the French simply call fried potatoes (*pommes frites*). These products happily turned out to be among the few foods whose quality need not be compromised by mass production. Of course they're rich: the frying oil in which they're immersed coats their surface and is drawn into the tiny pores created when the surface dries out. The proportion of oil to potato depends on the surface area. Chips, which are almost all surface, average about 35% oil, while thick fries are more like 10–15%.

French Fries "French fries" may first have been made in significant quantities by Parisian street vendors early in the 19th century. They are potato sticks cut with a square cross section, 5–10 mm on a side, deep-fried in oil, with a crisp gold exterior and a moist interior that's fluffy if the potatoes are high-starch russets, creamy otherwise. Simple quick frying doesn't work very well; it gives a thin, delicate crust that's quickly softened by the interior's moisture. A crisp crust requires an initial period of gentle frying, so that starch in the surface cells has time to dissolve from the granules and reinforce and glue together the outer cell walls into a thicker, more robust layer.

Good fries can be made by starting the potato strips in relatively cool oil, 250–325°F/120–163°C, cooking for 8–10 minutes, then raising the oil temperature to 350–375°F/175–190°C and cooking for 3–4 minutes to brown and crisp the outside. The most efficient production method is to pre-fry all the potato strips at the lower temperature ahead of time, set them aside at room temperature, and then do the brief high-temperature frying at the last minute.

Potato Chips Potato chips are essentially french fries that are all crust and no interior. The potatoes are cut into thin cross sections around 1.5 mm thick, the equivalent of just 10–12 potato cells, then deep-fried until dry and crisp. There are two basic ways of frying chips, and they produce two different textures. Cooking at a fairly constant and high oil temperature, around 350°F/175°C, heats the slices so rapidly that the starch granules and cell walls have little chance to absorb any moisture before they're desiccated and done, in 3–4 minutes. The texture is therefore delicately crisp and fine-grained. Most packaged chips have this texture because they're made in a continuous processor whose oil temperature stays high. On the other hand, cooking at an initially low and slowly increasing temperature, beginning around 250°F/120°C and reaching 350°F/175°C in 8–10 minutes, gives the starch granules time to absorb water, exude dissolved starch into the potato cell walls, and reinforce and glue them together. The result is a much harder, crunchier chip. This is the texture created by "kettle frying," or cooking the slices by the batch in a vessel like an ordinary pot. The temperature of the preheated kettle drops immediately when a batch of cold potatoes is dumped in, so the potatoes cook in oil whose temperature starts low and rises slowly as the potatoes' moisture is cooked out and the heater catches up.

Soufflée Potatoes Soufflée potatoes are a kind of hybrid French fry-chip in which the potato slices puff up into delicate brown balloons. They are made by cutting potato slices around 3 mm (⅛ in) thick, and deep-frying them at a moderate temperature, 350°F/175°C, until their surfaces become leathery and just begin to brown. The slices are cooled, then fried a second time at a high temperature, around 380°F/195°C. Now when the interior moisture is heated to the boil and vaporized, the stiffened surfaces resist the pressure, and the vapor pushes the two surfaces apart, leaving a hollow center.

Sweet Potatoes

The sweet potato is the true storage root of *Ipomoea batatas,* a member of the morning glory family. It is native to northern South America, and may have reached Polynesia in prehistoric times. Columbus brought the sweet potato to Europe, and by the end of the 15th century it was established in China and the Philippines. China now produces and consumes far more sweet potatoes than the Americas, enough to make it the second most important vegetable worldwide. There are many different varieties, ranging from dry and starchy varieties common in tropical regions, some pale and others red or purple with anthocyanins, to the moist, sweet version, dark orange with beta-carotene, that is popular in the United States and was confusingly named a "yam" in 1930s marketing campaigns (for true yams, see p. 306). The bulk of the U.S. crop is grown in the Southeast and cured for several days at 86°F/30°C to heal damaged skin and encourage sugar development. True to their subtropical heritage, sweet potatoes store best at 55–60°F/13–16°C. Chilling injury can contribute to "hardcore," a condition in which the root center remains hard even when cooked.

Most sweet potato varieties sweeten during cooking thanks to the action of an enzyme that attacks starch and breaks it down to maltose, a sugar made up of two glucose molecules that's about a third as sweet as table sugar. Moist or "soggy" varieties convert as much as 75% of their starch to maltose, so they seem permeated with syrup! The enzyme starts to make maltose when the tightly packed starch granules absorb moisture and expand, beginning around 135°F/57°C, and it stops when the rising heat denatures it, at around 170°F/75°C. Slow baking therefore gives the enzyme a longer time to work than does rapid cooking in steam, boiling water, or a microwave, and produces a sweeter result. Freshly harvested "green" roots available in the autumn have less enzyme activity and so don't become as sweet or moist.

Pale and red-purple sweet potatoes have a delicate, nutty aroma, while orange types have the heavier, pumpkin-like quality created by fragments of the carotenoid pigments. Some varieties (e.g., red-skinned Garnet) suffer from after-cooking darkening (p. 303) due to their abundant phenolic compounds.

Tropical Roots and Tubers

Root and tuber vegetables that come from the tropics generally contain less water than common potatoes, and as much as double the starch (potatoes are 18% carbohydrate by weight, cassavas 36%). They therefore become very floury when baked, dense and waxy when boiled or steamed, and help thicken soups and stews in which they're included. They have a relatively short storage life and suffer chilling damage if refrigerated, but can be frozen after preliminary peeling and cutting.

Cassava, Manioc, and Yuca These are all names for the elongated root of a tropical plant in the spurge family, *Manihot esculenta,* which has the very useful habit of lasting in the ground for as much as three years. It was domesticated in northern South America, and has spread through the lowland tropics of Africa and Asia in the last century or so. It's often made into flatbreads or fermented as well as cooked on its own. There are two general groups of cassava varieties: potentially toxic "bitter" varieties that are used in the producing countries, and safer "sweet" varieties that are exported and found in our ethnic markets. Bitter varieties, which are highly productive crop plants, have defensive cells that generate bitter cyanide throughout the root, and must be thoroughly treated—for example, by shredding, pressing, and washing—to become safe and palatable. They're mainly processed in the producing countries into flour and tapioca, small balls of dried cassava starch that become pleasantly jelly-like when remoistened in desserts and drinks. Sweet cassava varieties are less productive crop plants, but have cyanide defenses only near their surface, and are

Food Words: *Potato, Yam*

Potato came into English via the Spanish *patata,* a version of the word used by the Taino peoples of the Caribbean for the sweet potato, *batata.* The Peruvian Quechua word for the true potato of the Andes was *papa. Yam* comes via Portuguese from a West African word meaning "to eat."

safe to eat after peeling and normal cooking. The root flesh is snow-white and dense, with a bark-like skin and a fibrous core usually removed before cooking. Cassava benefits from cooking in water to moisten the starch before being fried or baked.

Taro and Dasheen Taro and dasheen are two of many names for tubers of a water-loving plant native to eastern Asia and the Pacific islands, *Colocasia esculenta,* which is in the arum family (as are calla lilies and philodendrons). Like other arums, taro contains protective crystalline needles of calcium oxalate (40–160 mg per 100 gm), and deposits them near stores of protein-digesting enzymes. The result is an arsenal of something resembling poison-tipped darts: when the tuber is eaten raw, the crystals puncture the skin and then the enzymes eat away at the wound, producing considerable irritation. Cooking overcomes this defensive system by denaturing the enzymes and dissolving the crystals.

Taro is commonly found in two sizes, one the main tuberous growth which may be several pounds, the other smaller side-growths, each a few ounces, and with a moister texture. The flesh is mottled by vessels purplish with phenolic compounds; during cooking the phenolics and color diffuse into and tinge the cream-colored flesh. Taro retains its shape when simmered, and it becomes waxy on cooling. It has a pronounced aroma that reminds some of chestnuts, others of egg yolk. In Hawaii taro is boiled, mashed, and fermented into poi, one element in the luau (p. 295).

Taro is sometimes confused with malanga, yautia, and cocoyam, tubers of a number of New World tropical species in the genus *Xanthosoma,* which are also arums protected by oxalate crystals. Malanga grows in drier soils than taro, is more elongated, has an earthier flavor, and more readily falls apart when simmered in soups and stews.

Yam True yams are starchy tubers of tropical plants that are related to the grasses and lilies, a dozen or so cultivated species of *Dioscorea* from Africa, South America, and the Pacific with varying sizes, textures, colors, and flavors. They are seldom seen in mainstream American markets, where "yam" means a sugary orange sweet potato (p. 304). True yams can grow to 100 lb/50 kg and more, and in the Pacific islands have been honored with their own little houses. They appear to have been cultivated as early as 8000 BCE in Asia. Many yams contain oxalate crystals just under the skin, as well as soap-like saponins, which give a slippery, frothy quality to their juices. Some varieties contain a toxic alkaloid called dioscorine that must be removed by grating and leaching in water. Yam tubers help their plants survive drought, and they have a longer pantry life than cassava or taro.

The Carrot Family: Carrots, Parsnips, and Others

Root vegetables in the carrot family share the family habit of containing distinctive aromatic molecules, so they're often used to lend complexity to stocks, stews, soups, and other preparations. Carrots and parsnips contain less starch than potatoes and are notably sweet; they may be 5% sugars, a mixture of sucrose, glucose, and fructose. Carrots have found their way into cakes and sugar preserves in the West, are shredded and sweetened for rice dishes in Iran, and in India are cooked down in milk to make a vegetable kind of fudge (halwa).

Carrot Cultivated carrots are swollen taproots of the species *Daucus carota,* which arose in the Mediterranean region. There are two main groups of cultivated carrots. The eastern anthocyanin carrot developed in central Asia, and has reddish-purple to purple-black outer layers and a yellow core of conducting vessels. It's eaten in its home region and can also be found in Spain. The Western carotene carrot appears to be a hybrid among three different groups of ancestors: yellow carrots cultivated in Europe and the Mediterranean since medieval times; white carrots that had been cultivated since classical times; and some wild carrot populations. The familiar orange carrot, the richest vegetable source of the vitamin A precursor beta-carotene, appears to have been developed in Holland in the 17th century. There are also Asian carrot varieties whose roots are red with lycopene, the tomato carotenoid. Carotene carrots have the practical advantage of retaining their oil-soluble pigments in water-based dishes, while anthocyanin carrots bleed their water-soluble colors into soups and stews.

The distinctive aroma of carrots is due largely to terpenes (p. 273), and is a composite of pine, wood, oil, citrus, and turpentine notes; cooking adds a violet-like note from fragmented carotene. White varieties tend to be the most strongly aromatic. Exposure to sunlight, high temperatures, or physical damage can cause the roots to generate alcohol, which adds to the solvent-like aroma, as well as a bitter defensive chemical. Peeling the thin outer layer removes most of the bitterness as well as phenolic compounds that cause brown discoloration. The sweetness is most noticeable when the roots are cooked, which weakens the strong cell walls and frees the sugars to be tasted. The carrot core carries water from the root to the leaves and has less flavor than the outer storage layers.

Pre-peeled "baby" carrots, actually cut from mature ones, often have a harmless white fuzz on their surface due to damaged outer cell layers that dehydrate within hours of processing.

Parsnip *Pastinaca sativa,* along with its aromatic taproot, is native to Eurasia, was known to the Greeks and Romans, and like the turnip was an important staple food before the introduction of the potato. The version known to us today was developed in the Middle Ages. The parsnip accumulates more starch than the carrot, but converts it to sugars when exposed to cold temperatures; so winter roots are sweeter than autumn roots, and before sugar became cheap were used to make cakes and jams in Britain. Its pale, somewhat dry tissue softens faster during cooking than either the potato's or carrot's.

Parsley Root Parsley root is the taproot of a particular variety of parsley, *Petroselinum crispum* var. *tuberosum,* is also flavored by a mixture of terpenoids, and is more complex and pungent than parsley leaves. Parsley is a Eurasian native (p. 408).

Arracacha Arracacha is the root of a South American member of the carrot family, *Arracacia xanthorhiza.* It has smooth roots of various colors, and a rich flavor that the eminent plant explorer David Fairchild called much superior to carrots.

The Lettuce Family: Sunchoke, Salsify, Scorzonera, Burdock

Roots and tubers from northerly members of the lettuce family share three characteristics: an abundance of fructose-based carbohydrates, little starch, and a mild flavor reminiscent of the true artichoke (also a lettuce relative). The fructose carbohydrates (small-chain fructosans and starch-like inulin) provide both an energy store and an antifreeze mechanism for the overwintering plants. Humans don't have the enzymes necessary to digest fructose chains, so beneficial bacteria in our intestines feed on them instead, in the process generating carbon dioxide and other gases that can cause abdominal discomfort if we've eaten a large portion of these vegetables.

The sunchoke is the nonfibrous, plump tuber of a North American sunflower (*Helianthus tuberosus*), whose traditional and obscure name is "Jerusalem artichoke." It's pleasantly moist, crunchy, and sweet when raw, and becomes soft and sweet after brief cooking. When cooked for 12–24 hours at a low temperature, around 200°F/93°C, sunchoke carbohydrates are largely converted to digestible fructose, and the flesh becomes sweet and translucently brown, like a vegetable aspic.

Salsify (*Tragopogon porrifolius*), sometimes called "oyster plant" for the supposed flavor resemblance, and black salsify or scorzonera *(Scorzonera hispanica)* are Mediterranean natives. Their Eurasian relative burdock (*Arctium lappa*) is most appreciated in Japan as *gobo*. All three of these elongated taproots become undesirably fibrous with size and age, are rich in phenolic compounds (those in *gobo* are potent antioxidants), and therefore readily turn grayish-brown—at the surface when cut and peeled, throughout when cooked.

Other Common Roots and Tubers

Chinese Water Chestnut and Tiger Nut The Chinese water chestnut and the tiger nut, or chufa, are both members of the sedge family, a group of water grasses that includes papyrus. The water chestnut is a swollen underwater stem tip of *Eleocharis dulcis*, a native of the Far East cultivated primarily in China and Japan. (Horned water chestnuts or caltrops are the seeds of species of *Trapa*, water plants native to Africa, central Europe, and Asia.) Tiger nuts are small tubers of *Cyperus esculentus*, a native of northern Africa and the Mediterranean that was cultivated in ancient Egypt. Both taste slightly sweet and nutty, and both are remarkable for retaining their crispness when cooked and even when canned, thanks to phenolic compounds in their cell walls that cross-link and strengthen them. The Spanish make the sweet drink *horchata de chufa* from dried tiger nuts by soaking them in water, grinding and resoaking, straining, and adding sugar.

In Asia, where Chinese and horned water chestnuts are sometimes cultivated in contaminated water, these foods have been known to transmit cysts of a parasitic intestinal fluke to people who shell them with their teeth. Fresh versions should be washed and scrubbed thoroughly before trimming away their tough outer layer, then washed again. A brief immersion in boiling water will guarantee their safety.

Crosnes, or Chinese Artichokes Crosnes are small tubers of several species of *Stachys*, an Asian member of the mint family; they were brought from China to France in the late 19th century. Crosnes are crisp and taste nutty and sweet, something like a sunchoke. They're notable for containing an unusual carbohydrate, stachyose, a combination of two galactoses and one sucrose. We can't digest stachyose, so a large serving of crosnes can cause gassy discomfort. Crosnes contain little starch, and turn mushy when even slightly overcooked.

Jicama Jicama is the swollen storage root of *Pachyrhizus erosus*, a South American member of the bean family. Its main virtue is its sturdy crispness: it keeps well, is slow to discolor, and retains some crunch when cooked. Jicama is often eaten raw, in salads or dipped into a sauce, and is sometimes used as a fresh replacement for Chinese water chestnuts, though it doesn't have the same sweet and nutty character.

Lotus Root Lotus root is the mud-dwelling rhizome of *Nelumbo nucifera*, a water lily native to Asia that has North American and Egyptian relatives. The lily is an important image in Buddhism and other systems of thought—a stalk rising from the mire to bear a beautiful flower over its floating leaves—so lotus root can carry extraculinary connotations. The rhizome contains large void spaces, so cross-sectional slices have a characteristic lacy pattern. It is crisp and remains so after cooking, for the same reason that water chestnuts do. It has a mild aroma and slight astringency, and discolors rapidly when cut due to phenolic compounds. Lotus root is cooked in many different ways, after an initial peeling (and blanching in the case of salads), from rapid stir-frying to braising and candying. Its modest store of starch is also extracted.

Oca Oca is the small tuber of a South American relative of wood sorrel, *Oxalis* *tuberosa*. It is variably starchy or juicy, comes in a number of anthocyanin-based skin colors, from yellow to red to purple, and is unusual in being distinctly tart, thanks to the oxalic acid typical of the family. In Peru and Bolivia it's usually cooked in stews and soups.

LOWER STEMS AND BULBS: BEET, TURNIP, RADISH, ONION, AND OTHERS

The members of this mixed category of vegetables sit at or just below ground level, and have one characteristic in common: they store little starch compared to most roots and tubers. They're therefore usually less dense, cook more rapidly, and retain a moist texture.

Beets

Beet "roots" are mainly the lower stem of *Beta vulgaris*, a native of the Mediterranean and Western Europe. People have eaten this plant since prehistory, initially its leaves (chard, p. 325), then the underground part of specialized varieties (subspecies *vulgaris)*. In Greek times beet roots were long, either white or red, and sweet; Theophrastus reported around 300 BCE that they were pleasant enough to eat raw. The fat red type is first depicted in the 16th century. Table beets are about 3% sugar and some large animal-feed varieties are 8%; in the 18th century, selection for sugar production led to sugar beets with 20% sucrose.

Colored beets owe their red, orange, and yellow hues to betain pigments (p. 268), which are water-soluble and stain other ingredients. There are variegated varieties with alternating red layers of phloem tissue and unpigmented layers of xylem (p. 262); they look their best in raw slices because cooking causes cell damage and pigment leakage. When we eat beets, the red pigment is usually decolorized by high stomach acidity and reaction with iron in the large intestine, but people sometimes excrete the intact pigment, a startling but harmless event. The persistent firmness of cooked beets is caused by phenolic reinforcement of the cell walls, as in bamboo shoots and water chestnuts (p. 283).

Beet aroma comes largely from an earthy-smelling molecule called geosmin, which was long thought to originate with soil microbes, but now appears also to be produced by the beet root itself. The sugariness of beets is sometimes put to use in chocolate cakes, syrups, and other sweets.

Celery Root

Celery root or celeriac is the swollen lower portion of the main stem of a special variety of celery, *Apium graveolens* var. *rapaceum*. Roots project from a knobbly surface that requires deep peeling. Celeriac tastes much like celery thanks to the same oxygen-ring aromatics, and contains a moderate amount of starch (5–6% by weight). It's usually cooked like other root vegetables, but also finely shredded to make a crunchy raw salad.

The Cabbage Family: Turnip, Radish

The turnip, *Brassica rapa*, has been under cultivation for about 4,000 years in Eurasia as a staple, fast-growing food. It consists of both lower stem and taproot, can have a number of different shapes and colors, and has the sulfury aroma typical of the family (p. 321). Small, mild varieties may be eaten raw and crunchy like radishes, larger ones cooked until soft: but not too long, or the overcooked cabbage flavor dominates and the texture becomes mushy. Turnips are also pickled.

The crisp, sometimes pungent radish is a different species, *Raphanus sativus*, a native of western Asia, and had reached the Mediterranean by the time of the ancient Egyptians and Greeks. Like the turnip it's mainly a swollen lower stem, and has been shaped by human selection into many distinctive forms and striking colors (for example, green at the surface and red inside). Most familiar in the United States are small, early-maturing spring varieties, usually with a bright red skin, which take only a few weeks to grow, and become harsh and woody in summer heat. These are usually eaten raw in salads. But there are also large Spanish and German varieties, some with black skins and some white, that reach several inches in diameter and mature over several months for harvest in the autumn. These types are firm and dry, and take well to braising and roasting. And there are the large, long white Asian radishes, best known by the Japanese term *daikon,* which can be more than a foot/25 cm long and weigh 6 lb/3 kg. They are relatively mild and used both raw and cooked, sometimes almost as a crisp pear might be. Radish pungency is created by an enzyme reaction that forms a volatile mustard oil (p. 321). Much of that enzyme is found in the skin, so peeling will moderate the pepperiness. Though most often eaten raw or pickled, radishes can be cooked like turnips, a treatment that minimizes their pungency (the enzyme is inactivated) and brings out their sweetness.

An unusual radish species, *R. caudatus,* is known as the "rat-tailed radish" because it bears long edible seedpods.

THE ONION FAMILY: ONIONS, GARLIC, LEEKS

There are around 500 species in the genus *Allium,* a group of plants in the lily family that are native to northern temperate regions. About 20 are important human foods, and a handful have been prized for thousands of years, as is attested by the well-known lament of the exiled Israelites in the Old Testament: "We remember the fish, which we did eat in Egypt freely; the cucumbers, and the melons, and the leeks, and the onions, and the garlic." Onions, garlic, and most of their relatives are grown primarily for their underground bulbs, which are made up of swollen leaf bases or "scales" that store energy for the beginning of the next growing season, and which naturally keep well for months. Like the sunchoke and its relatives, the onion family accumulates energy stores not in starch, but in chains of fructose sugars (p. 805), which long, slow cooking breaks down to produce a marked sweetness. Of course the fresh green leaves of bulb-forming alliums are also eaten, and nonbulbing kinds, including leeks, chives, and some onions, give only their leaves.

The key to the onion family's appeal is a strong, often pungent, sulfury flavor whose original purpose was to deter animals from eating the plants. Cooking transforms this chemical defense into a deliciously savory, almost meaty quality that adds depth to many dishes in many cultures.

The Flavors and Sting of Raw Alliums

The distinctive flavors of the onion family come from its defensive use of the element sulfur. The growing plants take up sulfur from the soil and incorporate it into four different kinds of chemical ammunition, which float in the cell fluids while their enzyme trigger is held separately in a storage vacuole (p. 261). When the cell is dam-

Onion and garlic bulbs. The bulbs in the onion family consist of a central stem bud and surrounding leaf bases, which swell with stored nutrients during one growing season and then supply them to the bud in the next.

aged by chopping or chewing, the enzyme escapes and breaks the ammunition molecules in half to produce irritating, strong-smelling sulfurous molecules. Some of these are very reactive and unstable, so they continue to evolve into other compounds. The mixture of molecules produced creates the food's raw flavor, and depends on the initial ammunition, how thoroughly the tissue is damaged, how much oxygen gets into the reactions, and how long the reactions go on. Onion flavor typically includes apple-like, burning, rubbery, and bitter notes; leek flavor has cabbage-like, creamy, and meaty aspects, while garlic seems especially potent because it produces a hundredfold higher concentration of initial reaction products than do other alliums. Chopping, pounding in a mortar, and pureeing in a food processor all give distinctive results. Chopped alliums to be eaten raw—as a garnish or in an uncooked sauce—are best rinsed to remove all the sulfur compounds from the damaged surfaces, since these tend to become harsher with time and exposure to the air.

One sulfur product is produced in significant quantities only in the onion, shallot, leek, chive, and rakkyo: the "lacrimator," which causes our eyes to water. This volatile chemical escapes from the damaged onion into the air, and lands in the onion cutter's eyes and nose, where it apparently attacks nerve endings directly, then breaks down into hydrogen sulfide, sulfur dioxide, and sulfuric acid. A very effective molecular bomb! Its effects can be minimized by prechilling the onions for 30–60 minutes in ice water. This treatment slows the ammunition-breaking enzyme down to a crawl, and gives all the volatile molecules less energy to launch themselves into the air. It also hydrates the papery onion skin, which makes it tougher and less brittle, and so easier to peel off.

The Flavors of Cooked Alliums When onions and their relatives are heated, the various sulfur compounds react with each other and with other substances to produce a range of characteristic flavor molecules. The cooking method, temperature, and medium strongly affect the flavor balance. Baking, drying, and microwaving tend to generate trisulfides, the characteristic notes of overcooked cabbage. Cooking at high temperatures in fat produces more volatiles and a stronger flavor than do other techniques. Relatively mild garlic compounds persist in butter but are changed to rubbery, pungent notes in more reactive unsaturated vegetable oils. Blanching whole garlic apparently inactivates the flavor-generating enzyme and limits its action, so the flavor of garlic cooked whole is only slightly pungent, and sweet, nutty notes come to the fore. Similarly, pickled garlic and onions are relatively mild.

The sugar and sugar-chain content of

Food Words: *Onion, Garlic, Shallot, Scallion*

Vegetable names in the onion family come from diverse sources. *Onion* itself comes from the Latin for "one," "oneness," "unity," and was the name given by Roman farmers to a variety of onion (*cepa*) that grew singly, without forming multiple bulbs as garlic and shallots do. *Garlic* is an Anglo-Saxon word that meant "spear-leek": a leek with a slim, pointed leaf blade rather than a broad, open one. And both *shallot* and *scallion* come via Latin from Ashqelon, the Hebrew name for a city in what in classical times was southwest Palestine.

onions and garlic is largely responsible for their readiness to brown when fried, and contributes a caramel note to the cooked flavor.

Onions and Shallots Onions are plants of the species *Allium cepa,* which originated in central Asia but has spread across the globe in hundreds of different varieties. There are two major categories of market onions in the United States, defined not by variety but by season and harvesting practice. Spring or short-day onions are planted as seedlings in the late fall, and harvested before they're fully mature in the next spring and early summer. They're relatively mild and moist and perishable, and best kept in the refrigerator. A special category of spring onion is the "sweet" onion—"mild" is more accurate—which is usually a standard yellow spring onion grown in sulfur-poor soils, and therefore endowed with half or less of the usual amounts of sulfur-containing defensive chemicals. The second major kind of market onion is the storage onion, grown through the summer and harvested when mature in the fall, rich in sulfur compounds, drier, and easily stored in cool conditions for several months.

White onion varieties are somewhat moister and don't keep quite as well as yellow onions, which owe their color to phenolic flavonoid compounds. Red onions are pigmented by water-soluble anthocyanins, but only in the surface layers of each leaf scale, so cooking dilutes and dulls their color.

Green onions, or scallions, can be either bulb-forming onion varieties harvested quite young, or special varieties that never do form bulbs. Shallots are a distinctive, clustering variety of onion whose bulbs are smaller, finer-textured, and somewhat milder and sweeter, often with a purple coloration. They're especially valued in France and southeast Asia.

Garlic Garlic is the central Asian native *Allium sativum,* which produces a tight head of a dozen or more bulbs, or "cloves." "Elephant garlic" is actually a bulbing variety of leek, with a milder flavor, and "wild garlic" yet another species, *A. ursinum.* Unlike multilayered onion bulbs, garlic cloves consist of a single swollen storage leaf surrounding the young shoot. That leaf contains much less water than onion scales do—under 60% of its weight, compared to 90% for onions—and a much higher concentration of fruc-

Important Members of the Onion Family

Onions, scallions	*Allium cepa*
Shallots	*Allium cepa* var. *ascalonicum*
Garlic	*Allium sativum*
Leeks, wild	*Allium ampeloprasum*
Leeks, cultivated	*Allium ampeloprasum* var. *porrum*
Great-headed (elephant) garlic	*Allium ampeloprasum* var. *gigante*
Leeks, Egyptian	*Allium kurrat*
Ramps, ramson (broad-leaf leek)	*Allium tricoccum*
Chives	*Allium schoenoprasum*
Chives, "garlic" or "Chinese"	*Allium tuberosum*
Japanese long onion	*Allium ramosum*
Japanese bunching onion	*Allium fistulosum*
Rakkyo	*Allium chinense*

tose and fructose chains, so during frying or roasting it dries out and browns much quicker than onions do.

There are many different garlic varieties, with different proportions of sulfur compounds and so different flavors and pungencies. The main commercial varieties are grown for their yield and storage life, not their flavor. Cold growing conditions produce a more intense garlic flavor. Garlic is at its moistest soon after harvest, from late summer to late fall, and becomes more concentrated as it slowly dries out during storage. Refrigerated storage causes a decline in distinctively garlicky flavor, and an increase in more generic onion flavors.

Because the peeling and chopping of the small cloves is tedious work, garlic is sometimes prepared in quantity and then stored under oil for later use. This procedure turns out to encourage the growth of deadly botulism bacteria, which thrive in the absence of air. Bacterial growth can be prevented by soaking the garlic in acidic vinegar or lemon juice for several hours before putting it under oil, and by storing the container in the refrigerator. Occasionally, acid-pickled garlic turns a strange shade of bluish-green, a reaction that apparently involves one of the sulfurous flavor precursors. This discoloration can be minimized by blanching the garlic before pickling.

Leeks Unlike onions and garlic, leeks don't form useful storage bulbs, and are grown instead for their scallion-like mass of fresh leaves. (There's one exception to this rule: the leek variety confusingly named "elephant garlic" because it produces a garlic-like bulb cluster that can reach 1 lb/450 gm.) Leeks are very tolerant of cold and in many regions can be harvested throughout the winter. They grow to a large size, and the prized white base portion of their leaves is often increased (to as much as 1 ft/3 m long and 3 in/7.5 cm thick) by hilling soil up around the growing plant to shield more of it from the sun. This practice also fills the spaces between leaves with grit, and necessitates careful washing. The inner leaves (and seldom-used roots) have the strongest flavor. The upper green portion of each leek leaf is edible, but tends to be tougher and to have a less oniony, more cabbage-like flavor than the lower white portion. It's also rich in long-chain carbohydrates that give the cooked vegetable a slippery texture, will gel when chilled, and can lend body to soups and stews.

STEMS AND STALKS: ASPARAGUS, CELERY, AND OTHERS

Vegetables derived from plant stems and stalks often present a particular challenge

Garlic on the Breath

Does the chemistry of garlic flavor offer any help when it comes to dealing with garlic breath? One major component of garlic breath appears to be various chemical relatives of skunk spray (e.g., methanethiol) that persist in the mouth. Another component (methyl allyl sulfide) is apparently generated from garlic as it passes through the digestive system, and peaks in the breath between 6 and 18 hours after the meal. Residual thiols in the mouth can be transformed into odorless molecules by the browning enzymes in many raw fruits and vegetables (p. 269), so eating a salad or an apple will help. Mouthwashes that contain strong oxidizing agents (e.g., chloramine) are also effective. Sulfides from the digestive system are probably beyond our reach!

to the cook. Stems and stalks support other plant parts and conduct essential nutrients to and from them, so they consist in large part of fibrous vascular tissue and special stiffening fibers—for example, the ridges along the outer edge of celery and cardoons—that are from 2 to 10 times tougher than the vascular fibers themselves. These fibrous materials become increasingly reinforced with insoluble cellulose as the stem or stalk matures. Sometimes there's nothing to do except to strip away the fibers, or cut the vegetable into thin pieces to minimize their fibrousness, or puree them and strain off the fibers. The keys to tender celery, cardoons, and rhubarb are on the farm rather than in the kitchen: choosing the right variety, providing plenty of water so that the stalks can support themselves with turgor pressure (p. 264), and providing mechanical support by hilling with soil or tying the stalks together, so that mechanical stress doesn't induce fiber growth.

One group of stem vegetables is inherently tender: the tips of such crop plants as peas, melons and squashes, grapevines, and hops, which grow rapidly in the spring, and have long been enjoyed as among the first fresh vegetables of the new season.

Asparagus

Asparagus is the main stalk of a plant in the lily family, *Asparagus officinalis,* a native of Eurasia that was a delicacy in Greek and Roman times. The stalk doesn't support ordinary leaves; the small projections from the stem are leaf-like bracts that shield immature clusters of feathery photosynthetic branches. The stalks grow up from long-lived underground rhizomes, and have been widely prized as a tender manifestation of spring. Many other vegetables have been called "poor man's asparagus," including young leeks, blackberry shoots, and hop shoots. It remains expensive today because the shoots grow at different rates and must be harvested by hand. In Europe, the even more labor-intensive white version, blanched by being covered with soil and cut from underground, has been popular since the 18th century. It has a more delicate aroma than green asparagus (which is rich in dimethyl sulfide and other sulfur volatiles), and some bitterness toward the stem end. Exposed to light after harvest, white asparagus will turn yellow or red. Purple asparagus varieties are colored with anthocyanins, whose color generally fades during cooking, leaving the green of the chlorophyll.

Harvested early and fresh from the soil, asparagus is very juicy and noticeably sweet (perhaps 4% sugar). As the season progresses, the rhizomes become depleted of stored energy, and sugar levels in the shoots decline. Once harvested, the actively growing shoot continues to consume its sugars, and does so more rapidly than any other common vegetable. Its flavor flattens out; it loses its juiciness, and it becomes increasingly fibrous from the base up. These changes are especially rapid in the first 24 hours after harvest, and are accelerated by warmth and light. Moisture and sugar losses can be partly remedied by soaking the spears in dilute sugar water (5–10%, or 5–10 gm per 100 ml/1–2 teaspoons per half-cup) before cooking. White asparagus is always more fibrous than green and toughens faster in storage. It and some

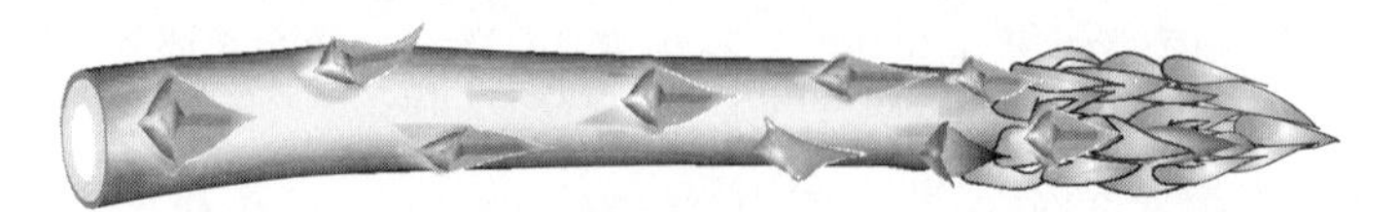

Asparagus and its peculiar branches, the phylloclades, which are clustered near the tip of the immature stem.

green asparagus may be peeled to remove some unsoftenable tissue, but woody lignin formation also takes place deeper in the stem. Cooks have dealt with this internal toughening in the same way for 500 years: they bend the stalk, and allow mechanical stress to find the border between tough and tender and break the two apart.

Asparagus has long been notorious for an unusual side effect on those who eat it: it gives a strong odor to their urine. Apparently the body metabolizes a sulfur-containing substance, asparagusic acid, a close chemical relative of the essence of skunk spray called methanethiol. In part because some people claim to be immune to this effect, biochemists have studied this phenomenon in some detail. It now appears that thanks to genetic differences, most but not all people do produce methanethiol after eating asparagus, and most but not all are able to smell it.

The Carrot Family: Celery and Fennel

The carrot family has provided two aromatic stalk vegetables.

Celery Celery, *Apium graveolens,* is the mild, enlarged version of a bitter, thin-stalked Eurasian herb called smallage. Chinese celery (var. *secalinum*) is closer in form and flavor to smallage, while Asian water celery is a more distant relative (*Oenanthe javanica*) with a distinctive flavor. Our familiar celery was apparently bred in 15th-century Italy, and remained a delicacy well into the 19th. It consists of greatly enlarged, pleasantly crunchy leaf stalks, or petioles, and has a distinctive but subtle aroma due to unusual compounds called phthalides that it shares with walnuts (hence their successful pairing in Waldorf salads), and terpenes that provide light pine and citrus notes. Celery is often combined with carrots and onions in gently fried aromatic base preparations for other dishes (French *mirepoix,* Italian *soffrito,* Spanish *sofregit*; in the Louisiana Cajun "trinity" of aromatics the carrots are replaced by green capsicums). In parts of Europe, celery has been preferred in a more delicately flavored blanched form, originally produced by covering the growing stalks with soil, then by growing pale green "self-blanching" varieties. Celery is often served raw, and its crispness is maximized by presoaking in cold water (p. 265). Both celery and celery root contain defensive chemicals that can cause skin and other reactions in sensitive people (p. 259).

Fennel Bulb or Florence or finocchio fennel is a vegetable variety (var. *azoricum*) of

Asparagus Aromatics

Not only can people differ in their ability to detect the product of asparagus metabolism, those who can detect it differ in their assessment.

> [Asparagus] cause a filthy and disagreeable smell in the urine, as everybody knows.
>
> —Louis Lemery, *Treatise of All Sorts of Foods*, 1702

> . . . all night long after a dinner at which I had partaken of [asparagus], they played (lyrical and coarse in their jesting as the fairies in Shakespeare's *Dream*) at transforming my chamber pot into a vase of aromatic perfume.
>
> —Marcel Proust, *In Search of Lost Time,* 1913

Foeniculum vulgare, the plant that produces fennel seeds (p. 415). Its enlarged leaf-stalk bases form a tight, bulb-like cluster. (The rest of the leaf stalk, the equivalent of celery stalk, remains tough and fibrous.) Fennel has a strong anise aroma thanks to the same chemical (anethole) that flavors anise seeds and star anise, and this makes fennel a more dominating, less versatile ingredient than celery and carrots. It also has a distinct citrus note (from the terpene limonene) that's especially prominent in the sparse foliage. Fennel is eaten both raw, thinly sliced and crunchy, and cooked, often braised or in a gratin.

The Cabbage Family: Kohlrabi and Rutabaga

Kohlrabi Kohlrabi is a version of the basic cabbage plant (*Brassica oleracea* var. *gongylodes*) in which the main stem swells to several inches in diameter. It has the moist texture and mild flavor of broccoli stalk. The name comes from the German for "cabbage turnip," and indeed kohlrabi resembles a turnip in its rounded appearance. Young kohlrabi are tender enough to eat raw for their crisp moistness or cook briefly; overmature stems are woody.

Rutabaga The rutabaga, or swede, is the result of a cross between the turnip and cabbage species, and is thought to have been born sometime before 1600 in Eastern Europe, perhaps in gardens where kale and turnips were growing side by side. Like kohlrabi, it's a swollen portion of the main stem; like the turnip, it may be white or yellow. It is sweeter and starchier than the turnip or kohlrabi, though still with only half the carbohydrate content of a potato; it's often boiled and mashed.

Tropical Stems: Bamboo Shoots and Hearts of Palm

Bamboo Shoots Bamboo shoots are the very young stems of several tropical Asian bamboos (species of *Phyllostachys* and others), which are woody members of the grass family. As the new shoots begin to break the soil surface, additional soil is heaped on to minimize their exposure to light and thereby their production of bitter cyanide-generating compounds (p. 258). Cooks and food manufacturers then eliminate all cyanide compounds from the fresh shoots by boiling them in water until they're no longer bitter. Along with Chinese water chestnuts and lotus root, bamboo shoots are valued for the ability to retain their firm, crisp, meaty texture during and after cooking, and even after the extreme heat treatment of canning (p. 283). Their flavor has an unusual medicinal or barnyard note thanks to cresol, as well as more common bready and brothy aromas from simple sulfur compounds (methional, dimethyl sulfide).

Hearts of Palm Hearts of palm are the growing stem tips of various palm trees, especially the South American peach palm *Bactris gasipaes,* which readily resprouts after its tip is cut. The tissue is fine-grained and crisp, with a sweet and slightly nutty taste, and is eaten both raw and cooked. Hearts of other palms may have a bitter edge and a tendency to brown discoloration; and their harvest often results in the wasteful death of the entire tree.

Other Stem and Stalk Vegetables

Cactus Pads or Nopales Cactus pads, nopales, and nopalitos are all names for the flattened stem segments of the prickly-pear cactus *Opuntia ficus-indica* (p. 369), a native of the arid regions of Mexico and the U.S. Southwest. They're eaten raw in salads or salsas, baked, fried, pickled, and added to stews. Nopalitos are remarkable for two things: a mucilage that probably helps them retain water, and that can give them a somewhat slimy consistency (dry cooking methods minimize this), and a startling tartness thanks to their malic acid content. Cactuses, purslane, and other plants that live in hot, dry environments have developed a special form of photosynthesis in which they keep their pores closed during the day to conserve water, then open them at night to take in carbon dioxide, which they then store in the form of malic acid. During the day, they use the energy from sunlight to convert the malic acid to glucose. Pads harvested in the early morning therefore contain as much as 10 times more malic acid than pads harvested in the afternoon. The acid levels in the pads slowly drop after harvest, so the difference is less apparent after a few days.

Cardoons Cardoons are the leaf stalks of *Cynara cardunculus,* the Mediterranean plant from which the artichoke (*C. scolymus*) apparently descends; the stalks are often covered for several weeks before harvest to protect them from sunlight, or blanch them. Cardoons have a flavor quite similar to the artichoke's, and are abundantly endowed with astringent, bitter phenolic compounds that quickly form brown complexes when the tissue is cut or damaged. They're often cooked in milk, whose proteins bind phenolic compounds and can reduce astringency (as in tea, p. 440). Phenolics can also cause a toughening of cell walls, and cardoon fibers are often remarkably resistant to softening. Bringing them to a gradual boil in several changes of water can help leach out phenolics and soften them, though flavor is leached out as well. Sometimes it's necessary to peel the reinforcing fibers from the cardoon stalk, or cut it into thin cross sections to keep the fibers relatively short and inconspicuous.

Fiddleheads Fiddleheads are the immature leaf stalk and fronds of ferns, named for the resemblance of their furled tips to the scroll of a violin. Long a traditional springtime delicacy, harvested as the fern fronds begin to elongate and unroll but before they toughen, fiddleheads are now cause for some caution. A common species especially enjoyed in Japan and Korea, the bracken fern *Pteridium aequilinum,* has been found to contain a potent DNA-damaging chemical (p. 259). It should be avoided. Stalks of the ostrich fern, various species of *Matteuccia,* are thought to be safer to eat.

Rhubarb Rhubarb, the leaf stalks of a large perennial herb, is unusual for containing a high concentration of oxalic acid. Its main use in the West is as a tart stand-in for fruit, so I describe it in the next chapter (p. 367).

Sea Beans Sea beans are the small, fleshy stems and branches of salt-tolerant and salty seacoast plants in the genus *Salicornia,* which is in the beet family. They are known under many other names, including samphire (a name they share with a seacoast plant in the carrot family), glasswort, pickleweed, and poussepierre. Young plants are crisp and tender and can be eaten raw or briefly blanched for their fresh, briny flavor; older ones can be cooked or steamed with seafood to intensify the sea aroma.

Sprouts Sprouts are seedlings, newborn plants just an inch or so long, and are mainly stem, which elongates to push the first set of leaves aboveground into the sunlight. Of course these infantile stems are tender and not at all fibrous; they're usually eaten raw or very briefly cooked. Many different plants are germinated to make edible sprouts, but most of them come from a handful of families: the beans (mung and soy, alfalfa), the grains (wheat, corn), the cabbage family (cress, broccoli, mustard, radish), the onion family (onions, chives). Because seedlings are so vulnerable, they're sometimes protected with strong chemical defenses. In alfalfa sprouts, the defenses include the toxic amino acid canavanine (p. 259); in broccoli sprouts, the defenses are sulforaphanes, a kind of isothiocyanate (p. 321) that appears to help prevent the development of cancer. Because the wet, warm conditions of sprout production also favor the growth of microbes, raw sprouts are frequently a cause of food poisoning. They should be bought as fresh as possible and kept refrigerated, and are safest when cooked thoroughly.

LEAVES: LETTUCES, CABBAGES, AND OTHERS

Leaves are the quintessential vegetable. They're usually the most prominent and abundant parts of a plant, and they're nutritious enough that many of our primate relatives eat little else. The salad of raw greens is a truly primeval dish! People also cook and eat the leaves of many different plants, from weeds to root and fruit crops. In temperate regions, nearly all the tender leaves of spring are edible, and were traditionally a welcome harbinger of the new crops to come after winter's scarcity; in northeast Italy, for example, *pistic* is a springtime collection of more than 50 different wild greens boiled and then sauteed together.

Because they're thin and broad, leaves both edible (lettuce, cabbage, grape) and inedible (banana, fig, bamboo) are used as wrappers to contain, protect, and aromatize fillings of meat, fish, grains, and other foods. They're often blanched first to make the leaves flaccid and pliable.

Though many leaf vegetables have distinctive flavors, most of them share a common, fresh aroma note referred to as "green" or "grassy." This comes from particular molecules that are 6 carbon atoms long—"leaf alcohol" (hexanol) and "leaf aldehyde" (hexanal)—and that are produced when leaves are cut or crushed. The cell damage frees enzymes that break up the long fatty-acid carbon chains in the membranes of the chloroplasts (p. 261). Cooking inactivates the enzymes and causes their products to react with other molecules, so the fresh green note fades and other aromas become more prominent.

The Lettuce Family: Lettuces, Chicories, Dandelion Greens

The lettuce family, or the Compositae, is the second largest family of flowering plants,

Preparing Salads

Even though salads of raw greens don't require cooking, they do require care in the preparation. Start with good ingredients: fresh young leaves are the least fibrous and have the most delicate flavor, while old overgrown lettuces can taste almost rubbery. If the leaves need to be divided into smaller pieces, this should be done with the least possible physical pressure, which can crush cells and initiate the development of off-flavors and darkened patches. Cutting with a sharp knife is generally the most effective method; tearing by hand requires squeezing, which may damage tender leaves. The greens should be well rinsed in several changes of water to remove grit, soil, and other surface contamination. Soaking them for a while in ice water will fill their cells with any water they've lost, making them turgid and crisp. Dry the greens as thoroughly as possible, so that the dressing will coat the leaves without being diluted. Thick, viscous dressings are slower to run off the leaf surface than thin, runny ones. A simple vinaigrette dressing (p. 637) can be made more viscous by chilling it in the freezer.

Oil-based dressings like vinaigrettes should be added only at the last minute, because oil readily wets the waxy leaf cuticle, percolates through the empty spaces within, and soon makes the leaf dark and sodden. Water-based cream dressings are better for greens that need to be dressed well in advance of eating.

and yet it contributes only a few food plants. The most prominent are lettuce and its relatives, the primary components of our raw salads.

Lettuces: Nonbitter Greens Today's mild, widely popular lettuces, varieties of the species *Lactuca sativa,* derive from an inedibly bitter weedy ancestor, *L. serriola,* that grew in Asia and the Mediterranean and has been under cultivation and improvement for 5,000 years. Lettuce seems to be represented in some ancient Egyptian art, and was certainly enjoyed by the Greeks and by the Romans, who had several varieties and ate them cooked as well as raw in salads at the beginning or end of the meal. The first syllable of its Latin name, *lac,* means "milk," and refers to the defensive white latex that oozes from the freshly cut base. Though lettuce is now mostly eaten raw in the West, in Asia it's often shredded and cooked. This can be a good use for the older, tougher leaves sometimes found on supermarket lettuces.

There are several broad groups of lettuce varieties, each with a characteristic growth form and texture (see box below). Most lettuces have a similar taste, though some red-leaf lettuces are noticeably astringent thanks to their anthocyanin pigments. Generally, varieties whose leaves are shielded from sunlight by the formation of a head offer much lower levels of vitamins and antioxidants. The pale, watery crisphead variety known as iceberg triumphed in the United States due to a combination of its durability in shipping and storage—it brought lettuce to the American table year-round in the 1920s—and its refreshing, crunchy-wet texture. Head lettuces respire more slowly and so keep better than leaf lettuces; both keep significantly longer at 32°F/0°C than at 40°F/4°C. There's also a type known as stem lettuce, or celtuce; it's especially popular in Asia for its prominent and crisp stalk, which is stripped of its small leaves, peeled, sliced, and cooked. Stem lettuce and the solid core of leaf lettuces are sometimes even candied.

Chicories and Endives: Bitterness Under Control The original intense bitterness of lettuce, which came from a terpene compound called lactucin and its relatives, has been bred out of the culti-

Greens from the Lettuce Family

***Lactuca sativa:* nonbitter lettuces**
- Loose-leaf varieties: open cluster of leaves
- Butter varieties: open cluster of soft, tender leaves, small midribs
- Batavian varieties: semi-open cluster of crisp, dense leaves
- Cos, Romaine varieties: loose head of elongated large leaves, prominent midribs
- Crisphead varieties: large, tightly wrapped heads of brittle, crunchy leaves

***Cichorium intybus:* bitter chicories**
- Chicory: open cluster of prominent stems and leaves
- Belgian "endive," witloof: tight elongated head of blanched crisp leaves
- Radicchio: tight round to elongated head of red leaves
- Puntarelle: open cluster of prominent narrow stems and leaves

***Cichorium endivia:* bitter endives**
- Curly endive: open cluster of curly leaves
- Frisée: open cluster of finely cut, frizzy leaves
- Escarole: open cluster of moderately broad leaves

The Cabbage Family: Relationships and Pungencies

Botanical nomenclature is constantly evolving, especially in complicated groups like the cabbage family. Though particular names may change, the broad relationships shown here seem secure.

Mediterranean origins

Brassica oleracea
- Cabbage (var. *capitata*)
- Portuguese cabbage (var. *tronchuda*)
- Kale, collards (var. *acephala*)
- Broccoli (var. *italica*)
- Cauliflower (var. *botrytis*)
- Brussels sprouts (var. *gemmifera*)
- Kohlrabi (var. *gongylodes*)

Mustard, black: *Brassica nigra*
Mustard, white: *Sinapis alba*
Rocket; arugula: *Eruca sativa, Diplotaxis* species
Watercress: *Nasturtium* species
Garden cress: *Lepidium* species
Upland, winter cress: *Barbarea* species
Nasturtium: *Tropaeolum* species
Garlic mustard: *Alliaria* species

Central Asian origins

Brassica rapa
- Turnip (var. *rapifera*)
- Broccoli rabe, broccoletti di rape (var. *rapifera*)
- Chinese cabbage, bok choy (var. *chinensis*)
- Chinese cabbage, napa (var. *pekinensis*)
- Tatsoi (var. *narinosa*)
- Mizuna, mibuna (var. *nipposinica*)

Chinese kale/broccoli, gai lan: *Brassica oleracea* var. *alboglabra*
Radish: *Raphanus sativus*
Horseradish: *Armoracia rusticana*

Recent Hybrids

Accidental
- Rutabaga, canola: *Brassica napus* (*rapa* x *oleracea*)
- Brown mustard, mustard greens: *Brassica juncea* (*rapa* x *nigra*)
- Ethiopian mustard: *Brassica carinata* (*oleracea* x *nigra*)

Intentional
- Broccolini: *Brassica oleracea* x *alboglabra*

Relative Amounts of Sulfur Pungency Precursors

Brussels sprouts	35	White cabbage	15	Radish	7
Green cabbage	26	Horseradish	11	Chinese cabbage	3
Broccoli	17	Red cabbage	10	Cauliflower	2

vated forms. But a number of close lettuce relatives are cultivated and included in salads or cooked on their own especially to provide a civilized dose of bitterness. These are plants in the genus *Cichorium,* which include endive, escarole, chicory, and radicchio. Growers go to a lot of trouble to control their bitterness. Open rosettes of escaroles and endives are often tied into an artificial head to keep the inner leaves in the dark and relatively mild. And popular "Belgian endive," also known as *witloof* ("white-head"), is a double-grown, slightly bitter version of an otherwise very bitter chicory. The plant is grown from seed in the spring, defoliated and dug up in the fall, and the taproot with its nutrient reserves kept in cold storage. The root is then either replanted indoors and kept covered with soil and sand as it leafs out, or else it's grown hydroponically in the dark. The root takes about a month to develop a fist-sized head of white to pale green leaves, with a delicate flavor and crunchy yet tender texture. This delicacy is easily lost. Exposing the heads to light in the market will induce greening and bitterness in the outer leaves, and the flavor becomes harsh.

Salads of bitter greens are often accompanied by a salty dressing or other ingredients; salt not only balances the bitterness, but actually suppresses our perception of bitterness.

Dandelion Greens The dandelion (*Taraxicum officinale*) seems to be found naturally on all continents, although most cultivated varieties are native to Eurasia. It's occasionally grown on a small scale, and has been gathered from the wild (or the backyard) since prehistory. The plant is a perennial, so if its taproot is left intact, it will give a rosette of leaves repeatedly. The bitter leaves are often blanched before eating to make them more palatable.

The Cabbage Family: Cabbage, Kale, Brussels Sprouts, and Others

Like the onion family, the cabbage family is a group of formidable chemical warriors with strong flavors. It's also a uniquely protean family. From two weedy natives of the Mediterranean and central Asia, we have managed to develop more than a dozen major crops of very different kinds: some leaves, some flowers, some stems, some seeds. Then there are a dozen or more relatives, notably the radishes and mustards (see chapter 8 for the spicy ones), and crosses between species: altogether a rich and ongoing collaboration between nature's inventiveness and our own! Beyond the cabbage family itself, some of its distant botanical relatives share elements of its biochemistry and therefore its flavor; these include capers and papayas (pp. 409, 381).

The Flavor Chemistry of the Cabbage Family Like onions, cabbages and their relatives stockpile two kinds of defensive chemicals in their tissues: flavor precursors, and enzymes that act on the precursors to liberate the reactive flavors. When the plant's cells are damaged, the two stockpiles are mixed, and the enzymes start a chain of reactions that generates bitter, pungent, and strong-smelling compounds. The special cabbage-family system is effective enough to have inspired a notorious manmade version, the mustard gas of World War I. And the cabbage family turns out to have parts of the onion defensive system (p. 310) as well; these contribute some sulfur aromatics to the overall family flavor.

The stockpiled defensive precursors in the cabbage family are called *glucosinolates.* They differ from the onion precursors in containing not only sulfur, but also nitrogen, so they and their immediate flavor products, mainly the *isothiocyanates,* have distinctive qualities. Some of the flavor precursors and products are very bitter, and some have significant effects on our metabolism. Particular isothiocyanates interfere with the proper function of the thyroid gland and can cause it to enlarge if the diet is poor in iodine. But others help protect against the development of cancer by fine-tuning our system for disposing of foreign chemicals. This is the case for substances in broccoli and broccoli sprouts.

A given vegetable will contain a number of different precursor glucosinolates, and the combinations are characteristic. This is why cabbage, brussels sprouts, broccoli, and mustard greens have similar but distinctive flavors. The chemical defensive system is most active—and the flavor strongest—in young, actively growing tissues: the center of brussels sprouts, for example, and portions of the cabbage core, which are twice as active as the outer leaves. Growing conditions have a strong influence on the amounts of flavor precursors the plant stockpiles. Summer temperatures and drought stress increase them, while the cold, wetness, and dim sunlight of autumn and winter reduce them. Autumn and winter vegetables are usually milder.

The Effects of Chopping Different preparation and cooking methods give different flavor balances in cabbage relatives. It's been found, for example, that simply chopping cabbage—for making coleslaw, for example—increases not only the liberation of flavor compounds from precursors, but also increases the production of the precursors! And if the chopped cabbage is then dressed with an acidic sauce, some pungent products increase sixfold. (Soaking the chopped cabbage in cold water will leach out most of the flavor compounds formed by chopping, at the same time that it hydrates the leaves and makes them crisper.) When cabbages and their relatives are fermented to make sauerkraut and other pickles, nearly all of the flavor precursors and their products are transformed into less bitter, less pungent substances.

The Effects of Heat Heating cabbages and their friends has two different effects. Initially the temperature rise within the tissue speeds the enzyme activity and flavor generation, with maximum activity at around 140°F/60°C. The enzymes stop working altogether somewhere short of the boiling point. If the enzymes are quickly inactivated by plunging the vegetables into abundant boiling water, then many of the flavor precursor molecules will be left intact. This isn't always desirable: cooking some mustard greens quickly, for example, minimizes their hot pungency but preserves the intense bitterness of their pungency precursors. Boiling in a large excess of water leaches flavor molecules out into the water, and produces a milder flavor than does stir-frying or steaming. If the cooking period is prolonged, then the constant heat gradually transforms the flavor molecules. Eventually the sulfur compounds end up forming trisulfides, which accumulate and are mainly responsible for the strong and lingering smell of overcooked cabbage. Prolonged cooking makes members of the onion family more sweet and mellow, but the cabbage family gets more overbearing and unpleasant.

Food Words: *Cabbage, Kale, Collards, Cauliflower*

Several vegetables in the cabbage family, *kale, collards,* and *cauliflower,* have names that derive from the Latin word *caulis,* meaning "stem" or "stalk," the part of the plant from which the edible portions emerge. *Cabbage* itself comes from the Latin *caput,* meaning "head": it's the one form in which the stem is reduced to a short stub, and the leaves form a head around it.

Thanks to the fact that they share some enzyme systems, mixtures of the onion and cabbage families can produce surprising effects. Add bits of raw scallion to some cooked and therefore nonpungent mustard greens, and the scallion enzymes will transform heat-stable mustard precursors into pungent products: so the bits of scallion taste more mustardy than the greens themselves!

Cabbage, Kale, Collards, Brussels Sprouts The original wild cabbage is native to the Mediterranean seaboard, and this salty, sunny habitat accounts for the thick, succulent, waxy leaves and stalks that help make these plants so hardy. It was domesticated around 2,500 years ago, and thanks to its tolerance of cold climates, it became an important staple vegetable in Eastern Europe. The practice of pickling it appears to have originated in China.

Collards, kale, and Portuguese tronchuda cabbage resemble wild cabbage in bearing separate leaves along a fairly short main stalk; tronchuda has especially massive midribs. Cultivated cabbage forms a large head of closely nested leaves around the tip of the main stalk. There are many varieties, some dark green, some nearly white, some red with anthocyanin pigments, some deeply ridged, and some smooth. In general, open-leaved plants accumulate more vitamins C and A and antioxidant carotenoids than heading varieties whose inner leaves never see the light of day. Heading cabbages often contain more sugar, and store well for months after harvest.

Brussels sprouts come from a cabbage variant that develops small, numerous heads along a greatly elongated central stalk. It may have been developed in northern Europe in the 15th century, but clear evidence for its existence only dates from the 18th. For many people who are sensitive to bitter tastes, brussels sprouts are simply too bitter to eat. They contain very high levels of glucosinolates. One of the major types (sinigrin, also the major mustard precursor) tastes bitter itself but produces a nonbitter thiocyanate, while the other (progoitrin) is nonbitter but produces a bitter thiocyanate. So whether we cook sprouts rapidly to minimize the production of thiocyanates, or slowly to transform all of the glucosinolates, the result is still bitter. Since these flavor components are concentrated in the center of the sprout,

Some vegetables in the remarkably various cabbage family. Center: *Kale leaf.* To its right and clockwise: *The swollen stem of kohlrabi, the terminal bud of head cabbage, the lateral buds of brussels sprouts, the flowering stalk of broccoli, and cauliflower, a mass of undeveloped flower stalks.*

it helps to halve the sprouts and cook them in a large pot of boiling water, which will leach out both precursors and products.

Rocket, Cress, Mustard Greens, Ethiopian Mustard Rocket (or Italian arugula, both from the Latin root *roc* meaning "harsh, rough") is the name given to several different plants and their leaves, all small, weedy cabbage relatives from the Mediterranean region that are especially pungent, with a full, almost meaty flavor built from various aldehydes, including almond-essence benzaldehyde. They're frequently used to enliven a salad of mixed greens, but are also pureed into a brilliant green sauce or put on pizzas. Even the briefest cooking will inactivate their protection-generating enzymes and turn them into tame greens. Some large-leaved varieties are quite mild. Like rocket, the different forms of cress—water, garden, winter—are small-leaved and pungent, usually used as a garnish or as a refreshing counterpoint to a rich meat. Their close relative, the window box nasturtium plant, sometimes lends its slightly pungent flower for a garnish; the more peppery flower buds are also used.

Mustard greens are leaves from varieties of brown mustard (*B. juncea*) that have been selected for their foliage rather than their seeds. Their texture is more delicate than that of cabbage. They're often quite pungent in the fashion of seed mustard, but are usually cooked, which may leave them mild and cabbage-like or intensely bitter, depending on the variety. Ethiopian mustard, a natural hybrid between the cabbage and mustard groups, probably arose in northeastern Africa, where the fast-growing young leaves are eaten both raw and lightly cooked. An improved variety developed in the United States is called texsel greens.

Asian Cabbages and Relatives The several very different forms of Chinese cabbage, including bok choy, napa, and tatsoi, all stem from the same species of *Brassica* that gave us the turnip. *B. rapa* is one of the oldest cultivated plants, possibly bred first for its seeds, and now among the most important vegetables in Asia. The larger modern forms are mainly elongated heads weighing up to 10 lb/4.5 kg, and are distinguished from European cabbages by their prominent white midribs, less prominent light green leaves, and mildness. Their smaller relatives mizuna and mibuna form low, spreading clumps of long, narrow leaves, those of mizuna being finely divided and feathery. Tatsoi makes a rosette of rounded leaves. These small leaves do well as additions to Western salads; they tolerate storage and dressings better than more delicate lettuces.

Spinach and Chard

Spinach Spinach (*Spinacia oleracea*) is a member of the beet family that was domesticated in central Asia, and is most productive in the cool seasons (heat and long days cause it to go to seed while it has relatively few leaves). In the late Middle Ages the Arabs brought it to Europe, where it soon displaced its smaller-leaved relatives orache and lamb's-quarters, as well as amaranth and sorrel. In the classic cuisine of France, spinach was likened to *cire-vierge,* or virgin beeswax, capable of receiving any impression or effect, while most other vegetables imposed their taste upon the dish. Today it's the most important leaf vegetable apart from lettuce, valued for its rapid growth, mild flavor, and tender texture when briefly cooked. (Some varieties are tender when raw, while thick-leaved varieties are chewy and less suitable for salads.) When cooked, its volume is reduced by about three-quarters. Spinach has a high content of potentially troublesome oxalates (p. 259), but it remains an excellent source of vitamin A as well as of phenolic antioxidants and compounds that reduce potential cancer-causing damage to our DNA. Folic acid was first purified from spinach, which is our richest source of this important vitamin (p. 255).

A number of unrelated but tender-leafed

plants are called spinach. Malabar spinach is an Asian climber, *Basella alba,* notable for its heat tolerance and the mucilaginous texture of its leaves, which may be green or red. New Zealand spinach is a relative of the succulent ice plant (also eaten!), *Tetragonia tetragonioides,* productive in hot weather but thick-leaved and best when cooked. Water spinach is an Asian relative of the sweet potato, *Ipomoea aquatica,* with elongated leaves and crunchy, hollow stems that are good at taking up sauce.

Chard Chard is the name given to varieties of the beet, *Beta vulgaris,* that have been selected for thick, meaty leaf stalks (subspecies *cicla*) rather than their roots. The beet is a distant relative of spinach, and its leaves—including ordinary, thin-midribbed beet greens—also contain oxalates. Chard stalks and leaf veins can be colored brilliant yellow, orange, and crimson by the same betain pigments that color the roots, which are water-soluble and stain cooking liquids and sauces. Some of the recently revived colored varieties are heirlooms that go back to the 16th century.

Miscellaneous Leafy Greens

Here are notes on a select handful of the many other greens that find their way to the table.

Amaranth Amaranth (species of *Amaranthus*), sometimes called Chinese spinach and other names, has been enjoyed since ancient times in both Europe and Asia. Its tender, earthy-flavored leaves are rich in vitamin A, but also in oxalates: two to three times more than spinach, for example. Boiling in copious water will remove some of them.

Grape Leaves Grape leaves are best known in pickled form as the wrappers for Greek dolmades. They are more delicate and delicious when blanched fresh. Grape leaves are distinctly tart from their large store of malic and tartaric acids.

Mâche Mâche, also known as lamb's lettuce or corn salad (*Valerianella locusta* and *V. eriocarpa*), has small, tender, slightly mucilaginous leaves and a distinctive, complex, fruity-flowery aroma (from various esters, linalool, mushroomy octenol, and lemony citronellol), which make it a popular addition or alternative to a lettuce salad in Europe.

Nettles Nettles (*Urtica dioica*) are a common Eurasian weed that has now spread throughout the Northern Hemisphere. They're notorious for their stinging hairs, which have a brittle silicate tip and a gland that supplies a cocktail of irritant chemicals, including histamine, for injection when skin meets needle. The hairs can be disarmed by a quick blanch in boiling water, which releases and dilutes the chemicals. But the harvest and washing require protective gloves. Nettles are made into soup, stewed, and mixed with cheese to stuff pasta.

Purslane Purslane (*Portulaca oleracea*) is a low-lying weed with fat stems and small thick leaves, which thrives in midsummer heat on neglected ground. It's a European native that has spread throughout the world. One nickname for purslane is pigweed, and the 19th-century Englishman William Cobbett said it was suitable only for pigs and the French. But people in many countries enjoy its combination of tartness and soothing, mucilaginous smoothness, both raw in salads and added to meat and vegetable dishes during the last few minutes of cooking. There are now cultivated varieties with larger leaves shaded yellow and pink. Its qualities are similar to those of the cactus pad because both have adapted in similar ways to hot, dry habitats (p. 316). Purslane is notable for its content of calcium, several vitamins, and an omega-3 fatty acid, linolenic acid (p. 801).

FLOWERS: ARTICHOKES, BROCCOLI, CAULIFLOWER, AND OTHERS

FLOWERS AS FOODS

Flowers are plant organs that attract pollinating animals with a strong scent, bright colors, or both; so they can add both aromatic and visual appeal to our foods. But the most important edible flowers in the West are neither colorful nor flowery! Broccoli and cauliflower are immature or developmentally arrested flower structures, and artichokes are eaten before they have a chance to open. Aromatic flowers have played a more prominent role in the Middle East and Asia. In the Middle East, the distilled essence of indigenous roses, and later of bitter orange flowers from China, has long been used to embellish the flavors of many dishes, rosewater in baklava and "Turkish delight," for example, and orange-flower water in Moroccan salads and stews and in Turkish coffee. Food historian Charles Perry has called these extracts "the vanilla of the Middle East." They were commonly used in the West as well until vanilla displaced them in the middle of the 19th century.

Many flowers can be and are used as edible garnishes, or cooked into aromatic fritters, or infused to make a tea or sorbet. The petals are the main source of volatile chemicals, which are held in surface cells or specialized oil glands. Both the petals and their flavors are delicate, so they should be cooked very briefly or added at the last minute. Flower petals are candied by brief cooking in a strong sugar syrup, or by brushing them gently with egg white or a solution of gum arabic, dusting them with sugar, and allowing them to dry. In

Some Edible and Inedible Flowers

Edible Flowers	Inedible Flowers
Herbs (chive, rosemary, lavender)	Lily of the valley
Rose	Hydrangea
Violet, pansy	Narcissus, daffodil
Daylily	Oleander
Begonia	Poinsettia
Jasmine	Rhododendron
Geranium (many herb and fruit scents)	Sweet pea
Lilac	Wisteria
Orchids	
Chrysanthemum, marigold	
Lotus	
Nasturtium	
Elderflower	
Citrus	
Apple, pear	
Tulip	
Gardenia	
Peony	
Linden (*tilleul*)	
Redbud	

the second technique, the egg white provides both antimicrobial proteins (p. 77) and a sticky liquid for the sugar to dissolve into, and the concentrated sugar pulls water out of any surviving microbes. When working with flowers, the cook should observe two cautions: avoid flowers that are known to contain defensive plant toxins, or that may have been treated with pesticides or fungicides in the greenhouse or garden.

Banana Flowers Banana flowers are the large male portion of the tropical banana tree's flower and its protective layers. They're somewhat astringent from the presence of tannins, and are cooked as a vegetable.

Daylily Buds Daylily buds, mostly from species of *Hemerocallis,* are eaten both fresh and dried in Asia—the dried form is sometimes called "golden needles"—and provide a valuable supplement of carotenoid and phenolic antioxidants.

Roselle, Hibiscus, and Jamaica These are all names for the bright red, tart, aromatic fleshy flower covering (the *calyx,* more familiar as the leaf-like stubs at the base of a strawberry) of a kind of hibiscus. *Hibiscus sabdariffa* is a native of Africa and a relative of okra. It's much used in Mexico and the Caribbean, sometimes fresh, sometimes dried and infused to make drinks, sometimes rehydrated and cooked with other ingredients. In the United States it's most familiar as an ingredient in Hawaiian punch and many red herbal teas (the pigments are anthocyanins). Jamaica ("ha-MY-ka") is remarkable for being a concentrated source of vitamin C, phenolic antioxidants, and gel-forming pectin.

Squash Blossoms The large flowers of zucchini and its relatives (p. 332) are sometimes stuffed, and variously deep-fried or chopped and added to soups or egg dishes. Their aroma is musky and complex, with green, almond, spicy, violet, and barnyard notes.

Artichokes

The artichoke is the large flower bud of a kind of thistle, *Cynara scolymus,* native to the Mediterranean region. It was probably developed from the cardoon, *C. cardunculus,* which has small and meager buds whose base and stem were eaten in ancient Greece. Artichokes were a delicacy in Rome, a fact of which Pliny professed to be ashamed: "thus we turn into a corrupt feast the earth's monstrosities, those which even the animals instinctively avoid" (Book 19). The name is a corruption, via Italian, of the Arabic *al'qarshuf,* meaning "little cardoon"; food historian Charles Perry

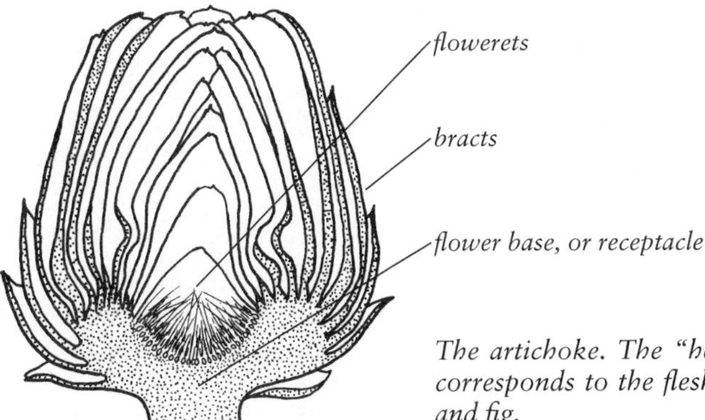

The artichoke. The "heart" is the flower base and corresponds to the fleshy portion of the strawberry and fig.

suggests that the large buds we know today, several inches in diameter, were developed in the late Middle Ages in Moorish Spain.

Thistles are members of the lettuce family and so relatives of salsify and sunchokes, all of which share a similar flavor. The edible parts of the artichoke are the fleshy bases of the bracts, or protective leaves, and the heart, which is actually the base of the flower structure, the upper part of the stalk. The "choke" is made up of the actual flowerets, which if allowed to bloom turn a deep violet-blue. The small artichokes sometimes seen in markets or in jars come from flowering stalks low on the plant rather than the main stalk. They grow very slowly, and so are picked at an immature stage when there's little or no choke inside.

The qualities of the artichoke are largely determined by its copious content of phenolic substances, which manifest themselves immediately when the flesh is cut or tasted raw. Cut surfaces turn brown very quickly as the phenolics react with oxygen to form colored complexes, and raw slices are noticeably astringent thanks to the reaction of phenolics with our salivary proteins. Cooking minimizes both effects. By disrupting the cells, it causes the phenolics to bind to a host of different molecules including each other; this gives the flesh an evenly dark tone, and leaves few phenolics free to cause astringency. Some artichoke phenolics have antioxidant and cholesterol-lowering effects, and one in particular, a compound dubbed cynarin, has the unusual effect of making foods eaten after a bite of artichoke taste sweet. Cynarin apparently inhibits the sweet receptors on our taste buds, so when it's swept off the tongue by the next bite, the receptors start up again, and we notice the contrast. Because they therefore distort the flavor of other foods, artichokes are thought to be an inappropriate accompaniment to fine wines.

The Cabbage Family: Broccoli, Cauliflower, Romanesco

These vegetables are all varieties of cabbage in which the normal development of flower stalks and flowers is arrested, so that the immature flowering tissues proliferate and accumulate into large masses. Based on recent genetic and geographic analysis, it appears that broccoli arose in Italy and in turn gave rise to cauliflower, which was known in Europe by the 16th century.

In the case of broccoli, extra flower-stalk tissue develops, fuses into thick "spears," and then goes on to produce clusters of small green flower buds. In cauliflower and its interestingly angular, green variant, romanesco, the stalk-production stage is extended indefinitely and forms a dense mass or "curd" of immature flower-stalk branches. Because the curd is developmentally immature, it remains relatively unfibrous and rich in cell-wall pectins and hemicelluloses (p. 265), and so can be pureed to a very fine, creamy consistency (and if overcooked whole, it readily turns to mush). To get as white a cauliflower curd as possible, growers usually tie the leaves over it to protect it from sunlight, which induces the production of yellowish pigments.

Broccoli rabe, slender stalks topped with a small cluster of flower buds, is unrelated to true broccoli. The name is a corruption of "broccoletti di rape," or "little sprouts of turnip," and refers to a variety of turnip that bears somewhat thickened flower stalks along its main stalk. Broccoli rabe is notably more bitter than true broccoli. Broccolini, similar to broccoli rabe but less bitter, is a recent hybrid between European and Asian brassicas.

FRUITS USED AS VEGETABLES

The botanical fruits that cooks treat as vegetables generally require cooking to make them interesting or soft enough to eat. The two familiar exceptions to this rule are tomatoes and cucumbers, frequently served raw in salads.

The Nightshade Family: Tomato, Capsicums, Eggplant, and Others

This remarkable plant family includes several of the world's most popular vegetables as well as tobacco and deadly nightshade; in fact it was the tomato's resemblance to nightshade that slowed its acceptance in Europe. Members of the nightshade family share the habit of stockpiling chemical defenses, usually bitter alkaloids. Many generations of selection and breeding have reduced these defenses in most edible nightshade fruits, though their leaves are often still toxic. There's one nightshade defense that humans have fallen in love with: the pungent capsaicins of the chilli "peppers." Chillis are the most popular spice in the world; their pungency is discussed in chapter 8. In this section I'll describe the milder capsicums that are eaten as vegetables.

Tomatoes Tomatoes started out as small, bitter berries growing on bushes in the west coast deserts of South America. Today, after their domestication in Mexico (their name comes from the Aztec term for "plump fruit," *tomatl*), and a period of European suspicion that lasted into the 19th century, they're eaten all over the world in a great variety of sizes, shapes, and carotenoid-painted colors. In the United States they're second in vegetable popularity only to the potato, a starchy staple.

What accounts for their great appeal? And why are these sweet-tart fruits treated as a vegetable? I think that the answers lie in their unique flavor. In addition to a relatively low sugar content for a fruit (3%), similar to that of cabbage and brussels sprouts, ripe tomatoes have an unusually large amount of savory glutamic acid (as much as 0.3% of their weight), as well as aromatic sulfur compounds. Glutamic acid and sulfur aromas are more common in meats than fruits, and so predispose them to complement the flavor of meats, even to replace that flavor, and certainly to add depth and complexity to sauces and other mixed preparations. (This may also be why although many rotten fruits smell pleasantly fermented, rotten tomatoes are absolutely foul!) In any case, tomatoes are

Vegetables in the Nightshade Family

Potato	*Solanum tuberosum*
Eggplant	*Solanum melongena*; *S. aethiopicum, macrocarpon*
Tomato	*Lycopersicon esculentum*
Capsicums, chillis	*Capsicum* species
Bell, pimiento, paprika, jalapeño, serrano, poblano . . .	C. *annuum*
Tabasco	C. *frutescens*
Scotch bonnet, habanero	C. *chinense*
Aji	C. *baccatum*
Manzano	C. *pubescens*
Tomatillo	*Physalis ixocarpa, P. philadelphica*
Tree tomato	*Cyphomandra betacea*

a good thing to like. They're rich in vitamin C, and the standard red varieties give us an excellent dose of the antioxidant carotenoid lycopene, which is especially concentrated in tomato paste and ketchup.

Tomato Anatomy and Flavor Aside from the relatively dry paste varieties, most tomatoes have four different kinds of tissue: a thin, tough cuticle, or skin, which is sometimes removed; the outer fruit wall; the central pith; and a semiliquid jelly and juice surrounding the seeds. The wall tissue contains most of the sugars and amino acids, while the concentration of acid in the jelly and juice is double that of the wall. And most of the aroma compounds are found in the cuticle and wall. The flavor of a tomato slice thus depends on the relative proportions of these tissues. Many cooks prepare tomatoes for cooking by first removing the skin, seedy jelly, and juice. This practice makes the tomato flesh more refined and less watery, but it changes the flavor balance in favor of sweetness, and sacrifices aroma. The tomato's citric and malic acids aren't volatile and don't cook away, so acidity and some aroma can be restored by cooking the skins, jelly, and juice together until much of the liquid has evaporated, then straining the remainder into the cooking tomato flesh. As cooks have long known and flavor chemists have verified, the overall flavor of tomatoes can be intensified by the addition of both sugar and acidity.

Tomatoes that are allowed to ripen fully on the vine accumulate more sugar, acid, and aroma compounds, and have the fullest flavor. An important element of ripe-tomato flavor is provided by the aroma compound furaneol, which resembles sweet-savory caramel (it also contributes to the flavors of ripe strawberries and pineapples). Most supermarket tomatoes are picked and shipped while still green and artificially stimulated to redden by treatment with ethylene gas (p. 351), so they have little ripe-fruit flavor, and in fact have become a byword for flavorless produce. However, parts of Europe and Latin America prefer to make salads with less fruity, more vegetable-like mature green tomatoes, and people in many regions cook (or pickle) and enjoy green tomatoes for their own kind of savoriness. And in rural Peru, the prized varieties of both tomato and tomatillo are frankly bitter.

Cooked Tomatoes When fresh tomatoes are cooked down to make a thick sauce, they gain some flavors—notably rose- and violet-like fragments of the carotenoid pigments—but they lose the fresh "green" notes provided by unstable fragments of fatty acids and by a particular sulfur compound (a thiazole). Because tomato leaves have a pronounced fresh-tomato aroma thanks to their leaf enzymes (p. 273) and prominent aromatic oil glands, some cooks add a few leaves to a tomato sauce toward the end of the cooking, to restore its fresh

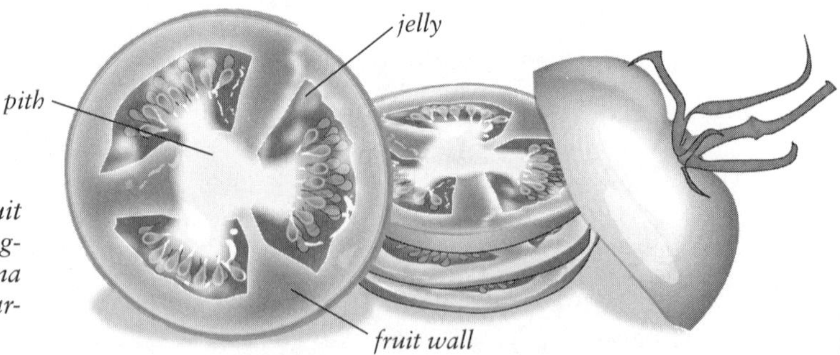

Tomato anatomy. The fruit wall is especially rich in sugars, amino acids, and aroma molecules, the jelly in sugar-balancing acids.

notes. Tomato leaves have long been considered potentially toxic because they contain a defensive alkaloid, tomatine, but recent research has found that tomatine binds tightly to cholesterol molecules in our digestive system, so that the body absorbs neither the alkaloid nor its bound partner. It thus reduces our net intake of cholesterol! (Green tomatoes also contain tomatine and have the same effect.) It's fine, then, to freshen the flavor of tomato sauces with the leaves.

Fresh tomatoes readily cook down to a smooth puree, but many canned tomatoes don't. Canners frequently add calcium salts to firm the cell walls and keep the pieces intact, and this can interfere with their disintegration during cooking. If you want to make a fine-textured dish from canned tomatoes, check the labels and buy a brand that doesn't list calcium among its ingredients.

Storage Tomatoes came originally from a warm climate, and should be stored at room temperature. Their fresh flavor readily suffers from refrigeration. Tomatoes at the mature-green stage are especially sensitive to chilling at temperatures below about 55°F/13°C, and suffer damage to their membranes that results in minimal flavor development, blotchy coloration, and a soft, mealy texture when they're brought back to room temperature. Fully ripe tomatoes are less sensitive, but lose flavor due to the loss of flavor-producing enzyme activity. Some of this activity can come back, so refrigerated tomatoes should be allowed to recover at room temperature for a day or two before eating.

The "tree tomato" is the vaguely tomato-like fruit of a woody plant in the nightshade family. It comes in red and yellow types and has a tough peel and bland flavor.

Tomatillos Tomatillos are the fruits of *Physalis ixocarpa,* a tomato relative that in fact was cultivated before the tomato in Mexico and Guatemala, to whose cool highlands it is better adapted. The tomatillo fruit is smaller than a standard tomato but similar in structure, and is borne on the plant enclosed in a papery husk. Its skin is thick and tough, sticky with a water-soluble secretion (the plant's species name, *ixocarpa,* means "sticky fruit"), that helps it keep well for several weeks. The tomatillo remains green when ripe and has a tart but otherwise mild, green flavor, a relatively firm and dry texture. It's usually cooked and/or pureed into sauces, with other ingredients adding depth or strength of flavor. A related species, *P. philadelphica,* provides an esteemed purple version called *miltomate.*

Capsicums or Sweet Peppers Capsicums, like tomatoes, are fruits of the New World that conquered the Old. They were domesticated in South America, and are now a defining element of the cuisines of Mexico, Spain, Hungary, and many countries in Asia (the countries with the highest per capita consumption are Mexico and Korea). This triumph is due largely to the defensive chemical capsaicin, which activates pain and heat receptors in our mouths, and which many human cultures have perversely come to love. This spicy aspect of the chillis is what inspired Columbus to call them peppers, though they're not at all related to true black pepper. (*Chilli* was the Aztec term.) For chillis as spices, see chapter 8.

Capsicums are essentially hollow berries, with a relatively thin, crisp wall of storage cells (spice types have been selected for very thin, easily dried fruits; vegetable types such as the pimiento have been bred for meatier walls). There are five domesticated species in the chilli genus *Capsicum,* with most vegetable types coming from *C. annuum.* Many varieties have been developed that are mild enough to be eaten as vegetables rather than condiments, and with a range of colors, shapes, sweetnesses, and aromas. Capsicums ripen to shades of yellow, brown, purple, or red, depending on the mix of pigments (purple comes from anthocyanins, brown from the combination of red carotenoids and green chlorophyll), but all can be picked and eaten green. The familiar green bell pepper has a strong, distinctive aroma thanks to a particular compound (isobutyl methoxypyrazine) carried in oil droplets within its cells; the very same compound occasionally pops up in cabernet sauvignon and sauvignon blanc wines and gives them a usually unwelcome green-vegetable note. Green fruits and mature yellow varieties are also rich in the carotenoid lutein, which helps prevent oxidative damage in the eye (p. 256). In red varieties, both lutein and the green aroma disappear during ripening along with chlorophyll, and other carotenoid pigments accumulate, the main ones being capsanthin, capsorubin, as well as beta-carotene, the precursor of vitamin A. Mature red capsicums are among the richest carotenoid sources we have; paprika powder may be more than 1% pigment by weight. They're also rich in vitamin C. And thanks to their content of cell-wall pectins, both fresh and rehydrated dried capsicums develop a thick, smooth consistency when cooked and pureed for a soup or sauce.

Eggplants, or Aubergines Eggplants are the only major vegetable in the nightshade family that came from the Old World. An early ancestor may have floated from Africa to India or Southeast Asia, where it was domesticated, and where small, bitter varieties are still appreciated as a condiment. Arab traders brought it to Spain and north Africa in the Middle Ages, and it was eaten in Italy in the 15th century, in France by the 18th. (The etymology of *aubergine* mirrors this history; it comes via Spanish and Arabic from the Sanskrit name.) Thanks to its tropical origins, the eggplant doesn't keep well in the refrigerator; internal chilling damage leads to browning and off-flavors in a few days.

There are many varieties of eggplant, white- and orange- and purple-skinned, pea- and cucumber- and melon-sized, very mild and intensely bitter. Most market types are colored with purple anthocyanins, while a different species (*S. aethiopicum*) provides the orange carotenoid types. All eggplants have a spongy interior, with many tiny air pockets between cells. When cooked, the air pockets collapse and the flesh consolidates into a fine-textured mass, sometimes creamy (most Asian varieties) and sometimes meaty (most European varieties) depending on the variety, maturity, and preparation. In baked casseroles—the Greek moussaka and Italian eggplant parmigiana—eggplant slices retain some structure; in the Middle Eastern dip baba ghanoush, grilled pureed eggplant provides the smooth, melting body that carries the flavors of sesame paste, lemon juice, and garlic.

Eggplant's spongy structure has two notable consequences for the cook. One is that eggplants shrink down to a relatively small volume when cooked. The other is that when fried, raw eggplant pieces soak up oil, leaving little on the pan for lubrication and making the vegetable very rich. In some preparations—such as the famous Arab dish *Imam bayaldi,* "the priest fainted," in which halved eggplants are stuffed and baked in copious olive oil—this richness is desired and maximized. Otherwise, the absorptiveness of eggplant can be reduced by collapsing its spongy structure before frying. This is accomplished by precooking it—microwaving works well—or by salting slices to draw out moisture from the cells and into the air pockets. Salting is often recommended as a way to remove the bitterness sometimes found in older eggplants grown in dry conditions, but it probably just reduces our perception of the alkaloids (p. 640); the bulk of the cell fluids remains in the cells.

The Squash and Cucumber Family

The squash or cucurbit family, the Cucurbitaceae, has made three broad contributions to human pleasure and nutrition. These are the sweet, moist melons described in the next chapter, the sweet, starchy, nutritious "winter" squashes, which are harvested fully mature and hard and keep for months, and the not-so-sweet, moist cucumber and "summer" squashes, which are harvested while immature and tender, and keep for a few weeks. ("Squash" comes from a Narragansett Indian word meaning "a green thing eaten raw.") When cooked, winter squashes develop a consistency and flavor something like those of a sweet potato, while the summer squashes and immature Asian gourds develop a mild but distinctive aroma and a translucent, slick, almost gelatinous texture. Fruits of *Cucurbita maxima,* the Hubbard and other winter squashes, can reach 300 lb/135 kg, and are the largest fruits of any plant. Most cucurbits produce a particular form of berry called a pepo, with a protective rind and a mass of storage tissue containing many seeds. All of them are native to warm climates, so they suffer from chill injury if stored at standard refrigerator temperatures. In addition to the flesh of their fruits, cucurbits also offer edible vines, flowers, and seeds.

Winter Squashes Winter squashes were domesticated in the Americas beginning around 5000 BCE. They are both nutritious—many are rich in beta-carotene and other carotenoids as well as starch—and versatile. The flesh of most varieties is firm enough to sauté or stew in chunks (fibrous spaghetti squash is an exception), but once cooked it also can be pureed to a very fine consistency; and its moderate sweetness makes it suitable for both savory and sweet preparations, from soups or side dishes to pies and custards. Their tough, dry skin

The Squash Family

Asian and African Species	
Cucumber	*Cucumis sativus*
Gherkin	*Cucumis anguria*
Melons: cantaloupe, honeydew, etc.	*Cucumis melo*
Watermelon	*Citrullus lanatus*
Winter/fuzzy melon, wax gourd	*Benincasa hispida*
Luffa, ridged gourd	*Luffa acutangula*
Bottle gourd, cucuzza, calabash	*Lagenaria siceraria*
Bitter gourd, bitter melon	*Momordica charantia*
South and North American Species	
Summer and acorn squash, zucchini, pumpkin, spaghetti squash	*Cucurbita pepo*
Winter squash: butternut, cheese, kabocha	*Cucurbita moschata*
Winter squash: Hubbard, turban, banana, kabocha	*Cucurbita maxima*
Cushaw	*Cucurbita mixta*
Chayote, mirliton	*Sechium edule*

and hollow structure encourage their use as edible containers; they can be filled with sweet or savory liquids, then baked, and eaten along with their contents. Winter squashes can be stored for months and many are available year-round, but they're at their prime shortly after harvest in late fall. They keep best at a temperature around 55°F/15°C and in relatively dry conditions (50–70% relative humidity).

Summer Squashes Summer squashes have been bred into a delightful variety of shapes. There are scallops or pattypans, flat and scalloped around the edges; crooknecks and straight necks, with constricted stem ends; elongated vegetable marrows, or zucchinis; acorn squash; and distinctive Middle Eastern and Asian types as well. Some are green-skinned, some vibrantly yellow with carotenoid pigments, some variegated, and all have a pale, delicately spongy flesh that softens quickly when cooked. They're sweetest when picked young, and keep for a few weeks at 45–50°F/7–10°C.

Cucumbers The cucumber was domesticated in India around 1500 BCE, arrived in the Mediterranean region about a thousand years later, and is now the second most important cucurbit worldwide after the watermelon. Like the watermelon, cucumbers are notable for their crisp, moist, mild, refreshing character. They're mainly consumed raw or pickled, and sometimes juiced to make a delicately flavored liquid for use in salad dressings, poaching fish, and other procedures. The distinctive yet melon-like aroma of cucumbers develops when the flesh is cut or chewed, and comes from the action of enzymes that break long membrane fatty-acid molecules into smaller chains that are 9 carbon atoms long; the characteristic melon fragments are alcohols, the cucumber fragments aldehydes. The larger a cucumber grows, the lower its acidity and the higher its modest sugar content (1–2%).

Cucumber Types There are five broad groups of cucumber varieties. Middle Eastern and Asian types are relatively small and thin. American pickling varieties are either smaller or slower-growing than "slicing" varieties, and have a thin skin to ease the penetration of brine. The standard American slicing cucumbers have been bred for the rigors of field production and long-distance distribution. They tend to be short and thick, with a relatively tough skin, dry flesh, prominent seeds, a strong cucumber flavor, and some bitterness at the stem end and just under the skin, thanks to the presence of defensive chemicals called cucurbitacins that discourage pests. European varieties, which are mainly grown in the controlled environment of greenhouses, are typically long and slender, with a thin tender skin, moist flesh, unformed seeds due to an absence of pollinating insects, a milder cucumber flavor, and no bitterness (breeders have eliminated the cucurbitacins). American cucumbers are often waxed to slow moisture loss, and so are almost always peeled before use, while European varieties are wrapped in plastic to accomplish the same purpose without compromising the edibility of their skin.

So-called "Armenian cucumbers" are actually an elongated African melon. True gherkins are the abundant prickly fruits, round and about 1 in/2 cm long, of yet another African relative.

Bitter Gourds Bitter gourds have long been prized in Asia for a trait that's considered a defect in cucumbers, the presence of bitter cucurbitacins. There may be good reasons to cultivate a taste for cucurbitacins, because recent studies have found that they may help slow the development of cancers. Bitter gourds are pale green, with an irregular, warty surface. They're usually eaten while immature, sometimes after an initial blanching to remove some of the water-soluble cucurbitacins, and are either stuffed or combined with other ingredients, which moderate the bitterness. Mature fruits contain a red, sticky, sweet material that covers the seeds and is sometimes eaten.

Bottle Gourds Bottle gourds or calabashes are most often allowed to mature and then dried to make containers and ornaments. Italians call the immature fruits *cucuzze,* and peel and cook them to produce a relatively bland version of a summer squash.

Luffa or Angled Gourds Luffa gourds, sometimes called Chinese okra, are elongated fruits with prominent ridges, and a mild taste and delicate texture when eaten immature. (A different species is used to make fibrous luffa "sponges"; true sponges are sea creatures.)

Winter or Wax or Fuzzy Melons Winter melons accumulate enough protective wax on their rinds that it can be scraped off and made into candles. On young fruits the wax-producing glands are more prominent than the wax itself, so they are known as hairy or fuzzy gourds or melons. These are cooked like summer squash, and their flesh becomes almost translucent. Winter melons keep well, and in Chinese cooking are used as an edible container for a festive soup.

Chayote or Mirliton Chayote is the squash that seems least like a squash. It's the fruit of a Central American vine, and looks something like a large pear some 5–8 in/12–20 cm long, with a single large seed at the center. Its flesh is finer-textured than the flesh of summer squashes, takes longer to cook, but is otherwise similar in its mild aroma and moist texture. The seed cavity is sometimes stuffed, and the seed sometimes eaten.

The Bean Family: Fresh Beans and Peas

Unlike most edible fruits, the fruits of the bean family were not designed to appeal to seed-dispersing animals. This group of plants is often called the *legumes,* "legume" being the name for their particular kind of fruit, a thin-walled pod, dry and brittle when mature, that encloses several seeds and disperses them by popping open when they're disturbed. It's in their dry form that we harvest most legume crops, since they can be stored indefinitely and are a concentrated source of nourishment (see chapter 9). Green beans and peas are immature pods and/or seeds, harvested before they begin to dry out, and are both very ancient and very recent foods. Early humans likely first ate the green pods and seeds, since dried seeds required cooking. However, the dried forms were so much more useful that varieties with pods specialized for eating green—with no tough inner "parchment" layer and reduced fiber throughout—have only been bred for a few hundred years.

Green legume seeds are tasty and nourishing because they're collecting sugars, amino acids, and other nutrients from the rest of the plant, but haven't yet packed them all into compact and tasteless starch and proteins. The green pods are tasty and nourishing because they serve as a temporary storage depot for the seeds' supplies. The pods also generate their own sugars by photosynthesis, making use of carbon dioxide that its enclosed seeds give off as they grow. After they're harvested, the green pods continue to send sugars to the seeds, so they lose their sweetness. We eat the green seeds of many legumes, notably lima beans, fava beans, and soybeans (chapter 9), but the pods of only a few: the common bean, long bean, and pea.

Green Beans Green beans come from a climbing plant native to Central America and the Andes region of northern South America. Though the peoples who domesticated them have probably always eaten some immature pods, the breeding of specialized vegetable bean varieties is less than 200 years old. There are now chlorophyll-free, yellowish "wax" varieties, and purple, chlorophyll-masking anthocyanin varieties that turn green when cooked (p. 281). The fibrous "strings" that normally join the two walls of the pod and are stripped away with the stem during preparation—hence the name "string beans"—were eliminated by a New York breeder in the late 19th century; these days only heirloom varieties tend to have strings. There are two general forms of green bean, one with round and thin pods, the other with flat and broad pods. Flat varieties have been found to have a more intense flavor. The flavor of cooked green beans is interestingly complex; it includes a number of sulfur and "green" compounds, but also the essence of fresh mushroom (octenol) and a flowery terpene (linalool).

Good-quality green beans can be hard to find, because they're one of the most fragile vegetables. Their tissue is very active, so they quickly consume their sugars and lose sweetness even in cold storage. And thanks to their subtropical origins, they don't keep well at refrigerator temperatures; their cells become damaged and lose chlorophyll. Once they're picked, tender low-fiber varieties quickly become wrinkled and limp as they lose moisture and sugars. Commercial varieties have been bred with more fiber to help them survive shipping and marketing with a better shape.

Long or Yard-Long Beans Long beans, sometimes called asparagus beans, do sometimes reach a yard/meter in length. They're the thin, small-seeded pods of a subspecies of black-eyed pea, a native of Africa that was taken to Asia more than 2,000 years ago. Asian cultures already had a number of excellent seed legumes but no warm-climate vegetable legume, and it was India or China that developed the yard-long version of the black-eyed pea. These beans have a higher fiber content than common green beans and therefore a drier, firmer texture when cooked. They're also sensitive to chilling (they keep best in the cold but deteriorate quickly if then left at room temperature).

Peas Peas come from a climbing plant native to the Mediterranean area, and are eaten immature both in the pods and as shelled green seeds (their tender shoots, stems, and leaves are also popular vegetables in Asia). Vegetable varieties of pea were first developed in the 17th century, beginning in Holland and then in England, and for a long time they remained luxury foods. There are several different types of pod peas, including the traditional English or European, round and thin-walled, the very recent "sugar snap" pea, round and thick-walled and quite crunchy, and "snow" or Asian peas, flat and broad, thin-walled, with small seeds. Peas and green capsicum "bell peppers" contain similar and very potent "green" aroma compounds (isobutyl methoxypyrazines).

Some Vegetable Beans and Their Origins

Green bean	*Phaseolus vulgaris*	Central America
Lima bean	*Phaseolus lunatus*	South America
Green peas, sugar peas, snow peas, pea shoots	*Pisum sativum*	Western Asia
Fava bean	*Vicia faba*	Western Asia
Long bean	*Vigna unguiculata*	Africa
Soybean	*Glycine max*	East Asia
Winged, asparagus bean	*Tetragonolobus purpureus*	Africa

Other Fruits Used as Vegetables

Avocado The avocado tree *Persea americana* is a native of Central America and a member of the laurel family, a relative of the bay laurel, California bay, and sassafras. Like its relatives, it has aromatic leaves that are used as flavorings (p. 408). Avocado fruits are remarkable for containing little or no sugar or starch, and for being as much as 30% oil, the equivalent of well-marbled meat (but marbled with olive oil; avocado oil is largely monounsaturated). They apparently evolved to appeal to large animals with a high calorie requirement. The name comes from the Nahuatl word *ahuacatl,* which was apparently inspired by the fruit's pear-like shape and irregular surface; it means "testicle."

There are three geographical groups of avocado. The Mexican group evolved in relatively cool subtropical highlands, so the trees are the most cold-tolerant; they produce small and smooth-fleshed fruit that are high in oil and can take relatively low storage temperatures, around 40°F/4°C. The lowland group evolved on the semitropical west coast of Guatemala and are the least cold-tolerant; their fruits tend to be large and coarse-fleshed and suffer chilling injury below about 54°F/12°C. And the Guatemalan group, from the semitropical highlands, is intermediate in most respects; the flesh of its fruits is least fibrous and the proportion of seed the lowest. In the United States, where most avocados are grown in southern California, commercial varieties are of mixed background. The most common variety, and one of the best, is the black, pebbly-skinned Hass, which is mainly Guatemalan. The smooth, green-skinned Fuerte, Pinkerton, and Reed are also relatively rich, while green Bacon and Zutano, and Florida's Booth and Lula, have more lowland ancestry, tend to remain firm, and have half or less the fat content of the Hass avocado.

Avocados don't begin to ripen until after they've been picked, so they're stored on the tree. All types ripen from the broad end toward the stem within about a week of harvest, and develop the best quality at temperatures between 60–75°F/15–24°C. Ripening can be accelerated by enclosing the fruit in a paper bag with an ethylene-emitting banana. If these warm-climate fruits are refrigerated while unripe, their cellular machinery is damaged and they will never ripen; once ripe, however, they can be refrigerated for several days and retain their quality. The aroma of avocado comes mainly from a group of mildly spicy terpenes including woody caryophyllene, as well as unusual 10- and 7-carbon fragments of fatty acids.

The flesh of rich avocado varieties readily turns into an unctuous puree without the necessity of any cooking, while leaner varieties retain some crispness and hold up well in sliced form for salads. Avocado flesh is well known for browning rapidly once cut or mashed (p. 269), a problem that can be remedied by adding an acidic ingredient (often lime juice) or by airtight wrapping with a plastic film that blocks oxygen effectively (polyvinylidene chloride, alias saran, is far more effective than polyethylene or PVC). In the case of mashed avocado, this means pressing the wrap directly into the surface. Though not usually cooked—heat generates a bitter compound and brings out an odd, eggy quality—avocados are sometimes added at the last minute to thicken and flavor soups, sauces, and stews.

Sweet Corn The corn that we eat as a vegetable is a fresh version of the same grain that gives us dry, starchy popcorn and cornmeal (chapter 9). Each individual grain on an ear of corn is a miniature fruit that's mainly seed, a combination of a small embryonic plant and its relatively large food supply of storage proteins and starch. We eat fresh corn about three weeks after pollination, while the fruits are immature, their storage tissues still sweet and juicy. Corn owes its typical yellow color to carotenoid pigments, including zeaxanthin (which derives its name from corn, *Zea mays,* and is one of the two main eye-protecting antioxidants). There are also white varieties with low carotenoid levels, as well as anthocyanin-colored red and blue varieties, and green ones too.

Fresh Corn Carbohydrates and Qualities Fresh corn contains three different forms of carbohydrate that contribute different qualities, and that are present in different proportions depending on the variety. The corn plant produces sugars and sends them to the seed, where they are temporarily held as is and impart sweetness until the cells string them together into large storage molecules. Very large sugar chains get packed into starch granules, which have no taste and lend a chalky texture to uncooked corn. Then there are medium-sized, tasteless sugar aggregates called "water-soluble polysaccharides," which have many short branches of sugar molecules. These bushy structures are small enough to float in dissolved form in the cell fluids, yet large enough that they bind up a lot of water molecules and get in each other's way, and thus thicken the fluid to a creamy, smooth consistency.

Traditional sweet corns derived from a genetic trait that arose in the cultivated fields of pre-Columbian South America, and that reduced starch levels in the maturing fruits while raising both sugar and soluble polysaccharide levels. The fresh kernels were thus sweeter and creamier than standard corns. In the early 1960s, breeders in the United States released new "supersweet" varieties, with very high sugar levels, little starch, but also fewer soluble polysaccharides: so their kernel fluids are less creamy and more watery (see box). Supersweet varieties have the advantage of losing less sweetness during shipping and storage—in three days, traditional sweet corn converts half of its sweet sugar into tasteless chains—but some corn lovers consider them too sweet, their flavor one-dimensional.

Preparing Corn While we usually prepare and consume the kernels whole, most of the flavor comes from the inner tissues, so some cooks grate, blend, or juice the raw kernels and separate the fluids from the seed coats, which get increasingly thick and tough with age. Because the fluids contain some starch, they will thicken like a sauce if heated above about 150°F/65°C. Heating also intensifies the characteristic aroma of corn, which is largely due to dimethyl and hydrogen sulfides and other sulfur volatiles (methane- and ethanethiols). Dimethyl sul-

Carbohydrates and the Qualities of Fresh Corn

This table gives the proportions in different corns of the carbohydrates that make cooked fresh corn taste sweet and feel creamy or dry in the mouth. The figures are percentages of the corn's fresh weight, when harvested 18–21 days after pollination.

	Sugars (sweetness)	**Water-soluble polysaccharides (creaminess)**	**Starch (dryness)**
Standard corn	6	3	66
Sweet corn	16	23	28
Supersweet corn	40	5	20

Source: A. R. Hallauer, ed., *Specialty Corns*. 2nd ed. (2001).

fide is also prominent in the aroma of cooked milk and molluscs, which is one reason why corn works so well in chowders. Sweet corn is also dried, which gives it a toasted, light caramel note. The hard, inedible support structure called the cob can lend flavor to vegetable stocks; that flavor is nuttier if the cob is first browned in the oven.

Baby Corn Miniature or "baby" corn consists of immature, unpollinated ears from full-sized corn varieties, picked two to four days after the silks emerge from the ear, when the cob is still edible, crisp, and sweet. (The rest of the plant becomes animal feed.) The ear may be 2–4 in/5–10 cm long and contains 2–3% sugar. Miniature corn production was developed in Taiwan and advanced in Thailand; Central America has recently become a major source.

Okra Okra comes from the annual plant *Hibiscus (Abelmoschus) esculentus,* a member of the hibiscus family, and a relative of roselle (p. 327) and cotton. It originated in either southwest Asia or eastern Africa, and came to the southern United States with the slave trade. The portion that we eat is the immature seedpod or capsule, with its distinctive five-cornered shape, star-like in cross section, and its notoriously slimy mucilage. Plant mucilage is a complex mixture of long, entangled carbohydrate molecules and proteins that helps plants and their seeds retain water. (Cactus and purslane are similarly slimy; the seeds of basil, fenugreek, and flax exude water-trapping mucilage when soaked, and are therefore used as thickeners or to add texture to drinks.) Okra mucilage can be exploited as a thickener in soups and stews (as it is in Louisiana gumbo, either to replace or augment powdered sassafras leaf), or its qualities can be minimized by using dry cooking methods (frying, baking). In Africa, slices of the pod are sun-dried. Okra has a mild flavor (though a relative, *A. moschatus,* produces aromatic seeds from which perfumers extract the musky ingredient ambrette).

Okra fruits can be hairy and sometimes even spiny, and their inner layers bear bundles of fibers that thicken and toughen as they mature. Small young fruits three to five days old are the most tender. These subtropical natives are damaged by storage temperatures below about 45°F/7°C.

Olives Olives are the small fruits of *Olea europaea,* a remarkably hardy, drought-tolerant tree that's native to the eastern Mediterranean region, and that can live and bear for a thousand years. In addition to sustenance, the olive has given us an everyday word: its ancient Greek name *elaia* is the source of the English *oil* (and Italian *olio,* French *huile*). The pulp layer surrounding the large central seed can be as much as 30% oil, which prehistoric peoples could extract by simple grinding and draining, and used in cooking and lamps, and for cosmetic purposes. Olives are also unusual among our commonly eaten fruits for being extremely unpalatable! They are

Olive Oil

Among food oils, olive oil is unique for being extracted not from a dry grain or nut, but from a fleshy fruit, and for carrying the prominent flavors of that fruit. The most prized olive oils are sold unrefined and shortly after harvest, as fresh as possible, and are used more as a delicious, delicate flavoring in their own right, not as a medium in which to cook other ingredients. Italy, France, and other Mediterranean countries are the largest producers and consumers.

richly endowed with bitter phenolic compounds, which offer some protection from both microbes and mammals. (Wild olives are eaten and their seeds dispersed mainly by birds, which swallow them whole; mammals chew and damage seeds.) Their bitterness has long been moderated or removed by various curing techniques (p. 295). The dark color of ripe olives comes from purplish anthocyanin pigments in the outer layer of the fruit.

Today, about 90% of the large worldwide crop goes to making olive oil.

Making Olive Oil Olive oil is made from the olive fruits when they are six to eight months old, mature and approaching their maximum oil content, and just beginning to turn color from green to purple; fully ripe fruits develop less of the valued green aroma. The olives are cleaned, coarsely crushed, pit and all (sometimes along with some leaves from the tree), and finely ground into a paste to break the fruit cells open and free their oil. The paste is mixed for 20–40 minutes to give the oil droplets a chance to separate from the watery mass of olive flesh and coalesce with each other (this step is called "malaxation"). Then the paste is pressed to squeeze both oil and watery liquid from the solids. More oil, but of lesser quality, is extracted by pressing repeatedly and by heating the paste; oil extracted in the "first cold pressing" is the most delicate and stable, and most likely to yield "extra virgin" oil (below). Finally, the oil is separated from the liquid by centrifuge or other means, and filtered.

The Color and Flavor of Olive Oil The result is an oil that's green-gold from the presence of chlorophyll and carotenoid pigments (beta-carotene and lutein), more or less pungent from a variety of phenolic compounds and certain products of fat breakdown (hexanol), and aromatic from dozens of volatile molecules. These include flowery and citrusy terpenes, fruity esters, nutty and earthy and almondy and hay-like molecules; but above all there are the grassy, "green"-smelling fragments of fatty acids that are also characteristic of leaf and other green vegetables (artichokes), herbs, and apples. Most of these molecules are generated during the grinding and malaxation, when active enzymes from the damaged fruit cells come into contact with vulnerable polyunsaturated fatty acids in the green chloroplasts. (Leaves are sometimes included in the grinding to supply more chloroplasts.) The oil itself is predominantly monounsaturated (oleic acid) and less vulnerable to oxidation.

The Quality of Olive Oil Olive oil quality is judged by its overall flavor and by its content of "free fatty acids," or fatty carbon chains that should be bound up in intact oil molecules but instead are floating free, and that are evidence that the oil is damaged and unstable. Under the regulations of the European Economic Community, "extra virgin" olive oil must contain less than 0.8% free fatty acids, "virgin" oil less than 2%. (To date, quality labeling of U.S. oils is not regulated.) Oils with higher free fatty acid levels are usually processed to remove nearly all impurities from the remaining intact oil molecules—including desirable flavor molecules. Producers usually blend such refined or "pure" oil with some virgin oil to give it flavor.

Storing Olive Oil The fact that virgin olive oils are unrefined has both desirable and undesirable consequences. Of course the beautiful color and rich flavor are great assets. The oils also contain significant quantities of antioxidant substances—phenolic compounds, carotenoids, and tocopherols (vitamin E and relatives)—which make them more resistant than other oils to damage by oxygen in the air. However, the same chlorophyll that colors them also makes them especially vulnerable to damage by light, whose energy the chlorophyll is designed to collect. To prevent "photooxidation" and the development of stale, harsh aromas, olive oil is best stored in the dark—in opaque cans, for example—and in cool conditions, which slow all chemical reactions.

Plantains Plantains are varieties of banana that retain much of their starchiness even when ripe, and are treated like other starchy vegetables. They're described along with their sweeter cousins on p. 378.

SEAWEEDS

Seaweed is a very general term for large plants that inhabit the oceans. Nearly all ocean plants are *algae,* a biological group that has dominated the waters for close to a billion years, and that gave rise to all land plants, including those that feed us. There are more than 20,000 species of algae, and humans have enjoyed many hundreds of them. They've been especially important foods throughout coastal Asia, in the British Isles, and in places as different as Iceland and Hawaii, where they're among very few native edibles. The Japanese use seaweeds as wrappers and to make salads and soups; in China they serve as a vegetable; in Ireland they're mashed up in porridge and thicken desserts. Most seaweeds have a richly savory taste and a fresh

Some Prominent Edible Seaweeds

	Scientific Name	Uses
Green Algae		
Sea lettuce	*Ulva lactuca*	Raw salads, soups
Sea grapes	*Caulerpa racemosa*	Peppery; eaten fresh or sugarcoated (Indonesia)
Awonori	*Enteromorpha, Monostrema* species	Powdered condiment (Japan)
Red Algae		
Nori, laver	*Porphyra* species	Oatmeal mush (Ireland); sushi wrappers or fried sheets (Japan)
Agar, tengusa	*Gracilaria* species	Branching stems; raw, salted, pickled, gelling agent for molded sweets (agar-agar, Japanese kanten)
Irish moss, carrageen	*Chondrus crispus*	Thickening agent for desserts (carrageenan)
Dulse, sea parsley	*Palmaria palmata*	With potatoes, milk, soup, breads (Ireland)
Brown Algae		
Kelp, kombu	*Laminaria* species	Soup base (dashi), salads, fried (Japan)
Wakame	*Undaria* species	Miso soup, salads (Japan)
Hiziki	*Hizikia fusiformis*	Vegetable, soups, "tea" (Japan, China)

Seaweed and the Original MSG

It was a seaweed that provoked a breakthrough in the understanding of human taste—and also brought the world the controversial food additive known as MSG. For better than a thousand years, the Japanese have been using the brown alga kombu as the base for soup stocks. In 1908 a Japanese chemist, Kikunae Ikeda, found that kombu is an especially rich source of monosodium glutamate—in fact it forms crystals on the surface of dried kombu. He also found that MSG provides a unique, savory taste sensation, different from the standard sweet, sour, salty, and bitter. He named this sensation *umami* (a rough translation is "delicious"), and pointed out that other foods, including meats and cheese, also provide it. For decades, Western scientists were skeptical that umami was a genuine taste sensation of its own, and not just a general taste enhancer. Finally, in 2001, biologist Charles Zuker at the University of California, San Diego, and colleagues demonstrated conclusively that humans and other animals do have a specific taste receptor for MSG.

A few years after Ikeda's observations, a colleague of his discovered a different umami substance (inosine monophosphate, or IMP) in cured skipjack tuna, another soup-base ingredient (p. 237). Then in 1960, Akira Kuninaka reported an umami substance in shiitake mushrooms (guanosine monophosphate, or GMP). Kuninaka also discovered that these different substances were synergistic with each other and with MSG: a very small amount of each strengthens the other's taste. Sensory scientists are still working to understand the nature of these effects.

A year after Ikeda's discovery, the Japanese company Ajinomoto began selling pure MSG as a seasoning, extracting it from the wheat gluten proteins that are a rich source of glutamate and in fact gave it its name. It caught on quickly, first with cooks in Japan and China and then with food manufacturers throughout the world. Ajinomoto is now a large multinational corporation; it and other companies produce MSG by the ton using bacteria that synthesize large amounts and excrete it into the liquid they grow in.

Beginning in the late 1960s, MSG was blamed for the "Chinese restaurant syndrome," in which distressing sensations of burning, pressure, and chest pain suddenly strike susceptible people who begin a Chinese meal with MSG-laden soup. Many studies later, toxicologists have concluded that MSG is a harmless ingredient for most people, even in large amounts. The most unfortunate aspect of the MSG saga is how it has been exploited to provide a cheap, one-dimensional substitute for real and remarkable foods. As Fuchsia Dunlop writes in her book on Sichuan cooking, *Land of Plenty*,

> It is a bitter irony that in China of all places, where chefs have spent centuries developing the most sophisticated culinary techniques, this mass-produced white powder should have been given the name *wei jing*, "the essence of flavor."

aroma reminiscent of the seacoast, which in fact they help to perfume. Many are good sources of vitamins A, B, C, and E, of iodine and other minerals, and when dried may be a third protein. Seaweeds are abundant, renew themselves rapidly over a life span of one or two years, and are easily preserved by drying. In Japan, where they've been cultivated since the 17th century, the farmed production of the seaweed used to wrap sushi is more valuable than the harvest of any other aquacultural product, including fish and shellfish.

The watery home of the seaweeds has shaped their nature in several ways that matter to the cook:

- Their buoyancy in water has allowed free-floating algae to minimize tough structural supports and maximize photosynthetic tissue. Some algae (e.g., nori, sea lettuce) are essentially all leaf, just one or two cells thick, very tender and delicate.
- Their immersion in salt waters of varying concentration has led algae to accumulate various molecules to keep their cells in osmotic balance. Some of these molecules contribute to their characteristic flavor. Mannitol, a sugar alcohol, is sweet (and, since our bodies can't metabolize it, low-calorie; see p. 662); glutamic acid is savory; and certain complex sulfur compounds give rise to aromatic, oceany dimethyl sulfide.
- Because water selectively absorbs red wavelengths from sunlight, some algae supplement their chlorophyll with special pigments for capturing the remaining wavelengths. Many seaweeds are either brown or reddish-purple, and change color when cooked.
- The many physical stresses of ocean life have encouraged some seaweeds to fill their cell walls with large quantities of jelly-like material that gives their tissues strength and flexibility, can be sloughed from their surface, and can help keep coastal species moist when they're exposed to the air at low tide. These special carbohydrates turn out to be useful for making jellies (agar) and for thickening various foods (algin, carrageenan). (More on thickeners in chapter 11.)

Green, Red, and Brown Algae

Nearly all edible seaweeds belong to one of three broad groups: the green algae, the red algae, and the brown algae.

- Green algae—sea lettuce, awonori—are the most like the land plants to which they gave rise. Their primary photosynthetic pigments are chlorophylls, with smaller amounts of carotenoids, and they store energy in the form of starch.
- Red algae—nori, dulse—are most common in tropical and subtropical waters. They owe their color to special pigment-protein complexes that are soluble in water and sensitive to heat: so during cooking their color can change quite strikingly from red to green. Red algae store their energy in a distinctive form of starch, and also produce large quantities of the sugar galactose and chains made up from it, which give us the gelling agents agar-agar and carrageenan.
- Brown algae—kelp, wakame—dominate temperate waters and supplement their chlorophyll with a group of carotenoid pigments, notably brownish fucoxanthin. They store some of their energy in the sweet sugar alcohol mannitol, which can amount to a quarter of the dry weight of fall-harvested kelp, and their typical mucilaginous material is algin.

Some freshwater algae are also collected from rivers and ponds: for example, species of *Cladophora,* which in Southeast Asia are pressed into nori-like sheets and used similarly (Laotian kaipen). Two alga-like creatures that are actually blue-green bacteria sometimes figure in the kitchen: the nutritional supplement spirulina and the Chinese "hair vegetable" or "hair moss," species of *Nostoc,* which grows in mountain springs in the Mongolian desert.

Seaweed Flavors

When it comes to flavor, the three seaweed families share a basic salty-savory taste from concentrated minerals and amino acids, especially glutamic acid, which is one of the molecules used to transport energy from one part of the seaweed to another. Seaweeds also share the aroma of dimethyl sulfide, which is found in cooked milk, corn, and shellfish as well as in seacoast air. There are also fragments of highly unsaturated fatty acids (mainly aldehydes) that contribute green-tea-like and fishy overtones. Against this common background, the three families do have distinctive characters. Dried, the red seaweeds tend to develop a deeper sulfury aroma from hydrogen sulfide and methanethiol, as well as flowery, black-tea-like notes from breakdown of their carotene pigments. When fried, dulse develops a distinct aroma of bacon. Some red algae, including the limu kohu of Hawaii (*Asparagopsis*), accumulate compounds of bromine and iodine, and can have a strong iodine flavor. The generally mild brown seaweeds have a characteristic iodine note (iodooctane) as well as a hay-like one (from the terpene cubenol). A few, notably species of *Dictyopteris* used as flavorings in Hawaii, have spicy aroma compounds that are apparently reproductive signals. Some browns are noticeably astringent thanks to the presence of tannin-like phenolic compounds, which in the dried seaweed form brown-black complexes (phycophaeins).

Prolonged cooking in liquid tends to accentuate fishy aromas, so often seaweeds are cooked only briefly. The first step in making the Japanese soup base dashi, for example, is to start dried brown kelp kombu in cold water, bring it just to the boil, then remove the kombu, leaving behind mainly its savory soluble minerals and amino acids. Because flavorsome minerals and amino acids crystallize on and in dried seaweeds, they contribute more when they're left unwashed, and if thick, when scored with a knife to release substances from within.

MUSHROOMS, TRUFFLES, AND RELATIVES

Mushrooms and their relatives are not true plants. They belong to a separate biological kingdom, the Fungi (the plural form of *fungus*), which they share with molds and yeasts.

Creatures of Symbiosis and Decay

Unlike plants, the fungi have no chlorophyll and cannot harvest energy from sunlight. They therefore live off the substance of other living things, including plants and plant remains. Different mushrooms do this in different ways. Some mushrooms, including boletes and truffles, form a symbiosis with living trees, a relationship in which both partners benefit: the mushrooms gather soil minerals and share them with the tree roots, which in turn share the tree's sugars with the mushrooms. Some fungi are parasites on living plants and cause disease; we eat the plant parasite that infects corncobs (corn smut, or huitlacoche). And some, including the world's most popular mushrooms, live off the decaying remains of dead plants. White and brown mushrooms apparently evolved along with plant-eating mammals to take advantage of the animals' partly digested but nutrient-rich dung! They now thrive in artificial piles of compost and manure.

Mushrooms that live on decaying plants have been relatively easy to cultivate. The Chinese were raising shiitake mushrooms on oak logs in the 13th century. The cultivation of the common white mushroom began in 17th-century France and boomed during the Napoleonic era in quarry tunnels near Paris. Today, *Agaricus bisporus* (or *A. brunnescens*) is grown on a mixture of manure, straw, and soil in dark buildings with carefully controlled humidity and temperature. The tropical zone's version of the button mushroom is the (paddy-)straw mushroom, which grows on composted rice straw. On the other hand, cultivation of symbiotic species is difficult, because the mushrooms need living trees, and intensive production requires a forest. This is why boletes, chanterelles, and truffles are relatively rare and expensive: they're still largely gathered from the wild. Of an estimated 1,000 edible mushroom species, only a few dozen have been successfully cultivated.

The Structure and Qualities of Mushrooms

Mushrooms differ from plants in several important ways. The part we eat is only one small portion of the organism, most of which lives invisibly underground as a fine, cottony network of fibers, or hyphae, which ramify through the soil to gather nutrients. A single cubic centimeter of soil—a small fraction of a cubic inch—can contain as much as 2,000 meters/yards of hyphae! When the underground mass of fibers has accumulated enough material and energy, it organizes a new, dense growth of interwoven hyphae into a fruiting body, which it pumps up with water to break above the soil surface and release its offspring spores into the air. The mushrooms that we eat are these fruiting bodies. (Morels form unusual hollow fruiting bodies with a distinctive honeycombed cap; the depressions bear the spores.)

Because the fruiting body is critical to the mushroom's reproduction and survival, it's often protected from animal attack by defensive poisons. Some mushroom poisons are deadly. This is why wild mushrooms should be gathered only by experts in mushroom identification. One mushroom traditionally gathered and eaten in Europe is now thought to present an unpredictable but real risk of potentially fatal hydrazine poisoning; this is the gyromitre or false morel (species of *Gyromitra*).

Inflated by water as they are, mushrooms are 80–90% water, with a thin outer cuticle that allows rapid moisture loss and gain. Their cell walls are reinforced not by plant cellulose, but by *chitin*, the carbohydrate-amine complex that also makes up the outer skeleton of insects and crustaceans. Mushrooms are notable for containing much more protein and vitamin B_{12} than other fresh produce. A number of

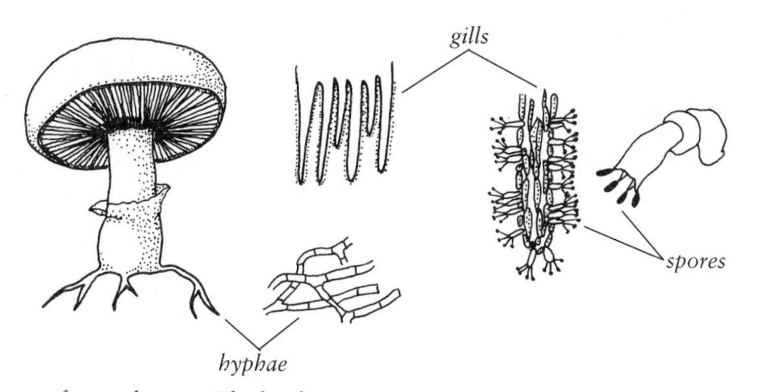

The anatomy of a mushroom. The hyphae are nutrient-gathering threads that grow through the soil. The main body of the mushroom is a fruiting body that the hyphae push up through the soil surface, which disperses spores from its gills.

mushrooms have been used in traditional medicines, and there is scientific evidence that some of the unusual cell-wall carbohydrates in shiitakes, matsutakes, and the interestingly crunchy-gelatinous ear mushrooms contain substances that inhibit tumor growth. Another factor in shiitakes may limit the production of mutagenic nitrosamines in our digestive systems.

The Distinctive Flavors of Mushrooms

We prize fungi for their rich, almost meaty flavor and their ability to intensify the flavor of many dishes. These qualities are largely due to a high content of free amino acids, including glutamic acid, which makes mushrooms—like seaweeds—a concentrated natural source of monosodium glutamate. Another taste enhancer that's synergistic with glutamate, GMP (guanosine monophosphate), was first discovered in shiitake mushrooms, and contributes to the rich taste.

The characteristic aroma of fresh common mushrooms is mainly due to octenol (an 8-carbon alcohol), which is produced by enzymes from polyunsaturated fats when the tissue is damaged, and which helps deter attack by some snails and insects. More octenol is generated from the gill tissues than from other parts, and this is one reason that common mushrooms with immature, unopened caps are less flavorful than the mature version with prominent gills. Brown and field mushrooms have more flavor than the white mushroom, and the "portobello," a brown mushroom allowed to mature for an additional five or six days until it's about 6 in/15 cm across, is especially intense.

Other mushrooms offer a wide range of aromas. A close relative of the common mushroom produces the essence of almond extract, while more exotic species are valued for such flavor notes as cinnamon, pepper, garlic, pine needles, butterscotch, and shellfish. Shiitake mushrooms owe their distinctive aroma to an unusual molecule called lenthionine, a ring of carbon and sulfur atoms, which is created by enzymes when the tissue is damaged. Lenthionine production is maximized by the common practice of drying and then rehydrating shiitakes in warm water (it's minimized by rapid cooking of the fresh or dried mushroom, which destroys enzymes before they have a chance to act). With a few exceptions (chanterelles, oysters, matsutakes), drying intensifies mushroom flavor by a combination of heightened enzyme activities and browning reactions between amino acids and sugars. Shiitakes and boletes, or porcini, are familiar examples, and especially flavorful because they're endowed with sulfur compounds that generate meaty aromas. Even home-dried button mushrooms are far more flavorful than the fresh originals, though they lose their fresh-mushroom octenol.

Storing and Handling Mushrooms

Mushrooms remain very active after harvest compared to most produce, and may even continue to grow. During four days' storage at room temperature, they lose about half of their energy reserves to the formation of cell-wall chitin. At the same time, they lose some of the enzyme activities that generate their fresh flavor, while protein-digesting enzymes become active in the stalk and turn the stalk proteins into amino acids for the cap and gills; so these parts become slightly more savory. Refrigeration at 40–45°F/4–6°C will slow mushroom metabolism, but they should be loosely wrapped in moisture-absorbing packaging to avoid having the moisture they exhale wet their surfaces and encourage spoilage. Mushrooms should be used as quickly as possible after purchase.

Cookbooks often advise against washing mushrooms so as not to make them soggy or dilute their flavor. However, they're already mostly water, and lose little if any flavor from a brief rinse. They should be cooked immediately, however, since washing can damage the surface cells and cause general discoloration.

Cooking Mushrooms

Mushrooms can be cooked in many different ways. Their flavor is generally most developed and intense when they are cooked slowly with dry heat to allow enzymes some time to work before being inactivated, and to cook out some of their abundant water and concentrate the amino acids, sugars, and aromas. Heat also collapses air pockets and consolidates the texture. (The combination of water and air loss means that mushrooms shrink considerably when cooked.) Like cellulose, chitin and some other cell-wall materials are not soluble in water, so mushrooms don't get mushy with prolonged cooking. The jelly and ear fungi, which are popular in Asian cuisines, contain an unusual amount of soluble carbohydrates, and this is why they develop a gelatinous texture.

Many mushrooms, and especially their gills, are rich in browning enzymes and blacken rapidly when cut or crushed. The dark pigments are water-soluble and can stain other ingredients in a dish, which may or may not be desirable.

Truffles

Truffles are the fruiting bodies of species in the genus *Tuber,* of which there are a handful of commercially important ones. They're typically a dense, knobby mass, ranging from walnut- to fist-sized or larger. Unlike mushrooms, truffles remain hidden underground. They spread their spores by emitting a scent to attract animals—including beetles, squirrels, rabbits, and deer—which find and eat them and spread the spores in their dung. This is why truffles have a musky, persistent aroma—to attract their spore spreaders—and why they're still gathered with the help of trained dogs or pigs or by spotting truffle "flies," insects that hover over truffled ground and lay their eggs there so that the larvae can burrow down and feed on the fungus.

Truffles grow only in symbiosis with trees, usually oaks, hazels, or lindens, so cultivation means finding or planting a forest, with significant harvests coming only after a decade or more. The Périgord region in France remains renowned for its black winter truffles, *Tuber melanosporum,* and northern and central Italy for its white truffle, *Tuber magnatum Pico.* Both are in great demand, in limited supply, and so quite expensive. Their flavor can be bought more reasonably in the form of cooked whole truffles or truffle paste, or truffle-infused oils, butter, and flours, though some of these may be flavored artificially. There are a number of other truffle species harvested in Europe, Asia, and North America, but they're not as flavorful. Unripe truffles of any species will have little flavor.

The flavors of black and white truffles are quite distinct. Black truffles are relatively subtle and earthy, with a mix of a dozen or so alcohols and aldehydes, and

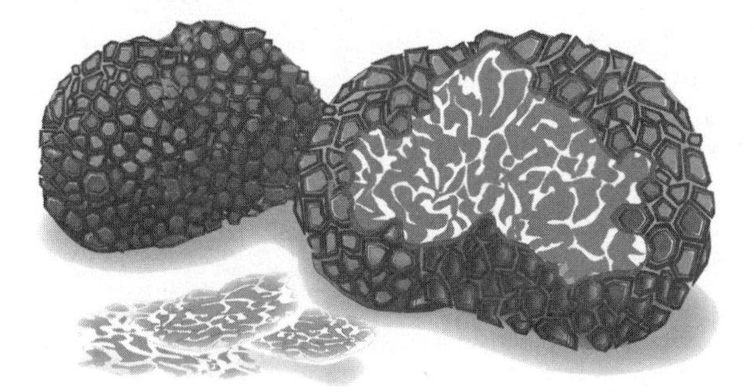

Truffle anatomy. Like mushrooms, the truffle is the fruiting body of a fungus; unlike mushrooms, it remains underground. The spores are contained in the thick masses of tissue between the vein-like folds.

Mushroom Types and Qualities

Mushrooms are grouped by broad family relationships. Most edible mushrooms bear their spores on gills.

Mushrooms with Gills		
Common mushrooms White, button Brown, cremini, portobello Champignon, field Almond	Cultivated leaf and dung decomposers Almond flavor	*Agaricus* species *A. bisporus* var. *alba* *A. bisporus* var. *avellanea* *Agaricus campestris* *Agaricus subrufescens*
Oyster, tree oyster	Cultivated wood decomposers	*Pleurotus* species
Shiitake	Cultivated oak decomposers	*Lentinus edodes*
Matsutake	Wild red pine decomposers; pine, cinnamon flavors	*Tricholoma* species
Honey	Wild wood decomposers	*Armillariella* species
Mousseron, fairy ring	Wild leaf decomposers	*Marasmius* species
Winter, enokitake	Cultivated wood decomposers; grows at 32°F/0°C	*Flammulina velutipes*
Blewit	Wild leaf decomposers; purple, blue colors	*Clitocybe nuda*
Straw mushroom	Rice straw decomposers	*Volvariella volvacea*
Parasol	Wild, cultivated leaf decomposers	*Lepiota* species
Ink cap	Wild compost decomposers	*Coprinus* species
Nameko, cinnamon cap	Cultivated wood decomposers; gelatinous cap	*Pholiota* species
Bolete, porcini, cèpe	Wild tree symbionts	*Boletus* species
Candy cap	Wild wood decomposers	*Lactarius rubidus*
Mushrooms without Gills		
Hen of the woods, maitake	Wild oak decomposers	*Grifola frondosa*
Chicken of the woods, sulfur shelf	Wild tree decomposers	*Laetiporus sulphureus*
Hedgehog, lion's mane	Wild tree symbionts	*Hydnum* species
Cauliflower	Wild tree parasites	*Sparassis crispa*
Chanterelle	Wild tree symbionts; gold, white, red	*Cantharellus* species
Black chanterelle, horn of plenty	Wild tree symbionts	*Craterellus* species
Ears: wood ear, tree ear, cloud ear	Cultivated wood decomposers; gelatinous; protease	*Auricularia* species
Snow, silver ears, white jelly	Cultivated wood decomposers; gelatinous; desserts!	*Tremella fuciformis*
Puffballs	Wild compost decomposers	*Calvatia, Lycoperdon* sp.
Morels	Wild tree decomposers	*Morchella* species
Truffles	Wild tree symbionts	*Tuber* species
Lobster	Wild mushroom decomposers	*Hypomyces lactifluorum*

some dimethyl sulfide. (They also contain small amounts of androstenone, a steroid compound also found in men's underarm sweat and secreted in the saliva of the male pig, where it prompts mating behavior in the sow. Some people are unable to smell androstenone, while others can and may find it unappetizing.) White truffles have a stronger, pungent, somewhat garlicky aroma thanks to a number of unusual sulfur compounds. The flavor of black truffles is generally thought to be enhanced by gentle cooking, while the flavor of white truffles, though strong, is fragile, and best enjoyed by shaving paper-thin slices onto a dish just before serving. Such cross sections of truffle reveal its inner structure: a network of fine veins running between masses of spore-bearing cells.

Fresh truffles are very perishable and emit their aroma in storage. They're best kept refrigerated in a closed container with some material—often rice—to absorb moisture and keep their surface from getting wet and spoiled by microbes.

Huitlacoche, or Corn Smut

Huitlacoche is a parasitic fungus, *Ustilago maydis,* that attacks corn plants, and that has been eaten in Mexico and Central America since Aztec times. It infects various plant parts, including the kernels in the growing ear, and develops into irregular spongy masses or "galls" that are a combination of greatly enlarged plant cells, nutrient-absorbing fungal threads, and blue-black spores. Fully mature galls are dry, black spore bags. The optimum stage for harvest is two to three weeks after infection, when the galls on a single ear can weigh as much as a pound/500 gm and are about three-quarters black inside. When cooked, these immature galls develop a sweet, savory, woody flavor thanks to glucose, sotolon, and vanillin. In the United States, corn smut was simply a disease until the 1990s, when growing interest in Mexican food led some farmers to cultivate it intentionally.

A related smut, *U. esculenta,* is eaten in China and Japan. An Asian wild rice, *Zizania latifolia,* develops the infection in its upper stem, which swells with hyphae. The stems are cooked and eaten as a vegetable (Chinese kah-peh-sung, Japanese makomotake) whose flavor is said to resemble bamboo shoots.

Mycoprotein, or Quorn

Mycoprotein is a 20th-century invention, an edible form of the normally useless underground hyphae of a common fungus, *Fusarium venenatum.* A strain of the fungus originally taken from a field in Buckinghamshire, England, is grown in a liquid medium in a factory-scale fermenter. The resulting mass of hyphae is harvested, washed, and rapidly heated. This produces microscopic fibers that are rich in protein and about 0.5 mm long and 0.003–0.005 mm in diameter, or about the dimensions of muscle fibers in meat. This essentially tasteless mycoprotein (from *myco-,* "related to fungi") can then be manufactured into meat substitutes and a variety of other food products.

CHAPTER 7

A SURVEY OF COMMON FRUITS

The vegetables described in chapter 6 are mainly plant parts with either mild, accidental flavors (roots, leaves, stalks) or strong defensive ones (the onion and cabbage families). We usually cook them, because cooking improves their flavor and makes them softer and easier to eat. The fruits described in this chapter are parts that the plant creates in order to attract animals to eat them and disperse the seeds within them. So the plant fills these fruits with a mouthwatering mixture of sugars and acids, endows them with pleasant aromas and eye-catching colors, and softens them for us: they're delicious and beautiful even when raw. The box on pp. 382–383 summarizes the essential flavor elements of some common fruits, especially the balance between sweet and sour that provides their taste foundation.

THE MAKING OF FRUIT: RIPENING

Among all our foods, fruits are unique in the way that they progress from inedibility to deliciousness. Immature vegetables and young meat animals are at their tenderest and most delicate, but immature fruits are usually at their least appealing. We may still eat and enjoy them—green tomatoes, green papayas, green mangoes—but we treat them as vegetables, cut them small for a salad or cook or pickle them. In order to graduate from vegetablehood, fruits must undergo the process called ripening, which creates their distinctive character.

Before Ripening: Growth and Expansion

A fruit is a distinct organ that develops from the flower, and in particular from the flower's female tissue, the ovary, which encloses the plant's maturing seeds. Most fruits are simply the thickened ovary wall, or else they incorporate nearby tissues as well. Apples and pears, for example, are made up mainly of the stem tip in which the flower parts are embedded. The fruit usually develops into three distinct layers: a thin outer protective skin, a thin inner protective coat around the central mass of seeds, and a thick, succulent, flavorful layer in between.

Fruit goes through four distinct stages of development. The first is usually fertilization of the female ovule by male pollen, which initiates the production of growth-promoting hormones and so leads to the expansion of the flower's ovary wall. Some conveniently seedless fruits, including bananas, navel oranges, and some grapes, manage to develop without fertilization. The second, relatively brief stage of fruit development is the multiplication of cells in the ovary wall, which in the tomato is virtually complete at the moment of fertilization (you can see the fully formed but tiny fruit at the base of the flower as soon as it opens).

Most of the noticeable growth during fruit development takes place during the third stage, the expansion of the storage cells. This growth can be remarkable. Melons at their most active put on better than 5 cubic inches/80 cc a day. Most of this expansion is due to the accumulation of water-based sap in the cell vacuoles. Mature fruit storage cells are among the largest in the plant kingdom, in watermelons approaching a millimeter in diameter. During this growth stage, sugar is stored in the cell vacuole as is or in more compact granules of starch. Defensive compounds, among them poisonous alkaloids and astringent tannins, accumulate in the cell vacuoles to deter infection or predation, and various enzyme systems are readied for action. When the seeds become capable of growing on their own and the fruit is ready to attract animals to disperse them, the fruit is said to be mature.

The Work of Ethylene and Enzymes

The final stage of fruit development is ripening, a drastic change in the life of the fruit that leads to its death. It consists of several simultaneous events. Starch and acid levels decrease, and sugars increase. The texture softens; defensive compounds disappear. A characteristic aroma develops. Skin color changes, usually from green to a shade of yellow or red. The fruit thus becomes sweeter, softer, and tastier, and it advertises these improvements visually. Because ripening soon gives way to rot-

Food Words: *Ripe, Climacteric*

Our word *ripe* began as an Old English word meaning "ready for reaping," and like *reap* comes ultimately from an Indo-European root meaning "to cut." *River, rope, row,* and *rigatoni* are all relatives. *Climacteric* can be traced back to a root meaning "to lean," which led to the Greek *climax,* "ladder," then offshoots meaning "rung" and therefore "a dangerous place," and finally *climacteric* itself, meaning a critical stage in life—whether a human's or a fruit's.

Fruits: Their Potential for Improvement after Harvest, and Optimal Storage Temperatures

Fruit	Improvements after Harvest	Store at 32°F/0°C	Store at 45°F/7°C	Store at 55°F/13°C
Pome				
Apple	Sweetness, aroma, softness	+		
Pear	Sweetness, aroma, softness	+		
Stone				
Apricot	Aroma, softness	+		
Cherry	–	+		
Peach	Aroma, softness	+		
Plum	Aroma, softness	+		
Citrus				
Orange	–		+	
Grapefruit	–			+
Lemon	–			+
Lime	–			+
Berries				
Blackberry	–	+		
Black currant	–	+		
Blueberry	Aroma, softness	+		
Cranberry	–	+		
Gooseberry	–	+		
Grape	–	+		
Raspberry	Aroma, softness	+		
Red currant	–	+		
Strawberry	–	+		
Melons				
Cantaloupe	Aroma, softness		+	
Honeydew	Aroma, softness		+	
Watermelon	–			+
Tropical				
Banana	Sweetness, aroma, softness			+
Cherimoya	Aroma, softness			+
Guava	Aroma, softness			+
Lychee	–	+		
Mango	Sweetness, aroma, softness			+
Papaya	Aroma, softness			+
Passion fruit	Aroma, softness		+	
Pineapple	–		Ripe	Unripe
Others				
Avocado	Aroma, softness	Ripe	Unripe	
Date	–			+
Fig	–	+		
Kiwi	Sweetness, aroma, softness	+		
Persimmon	Aroma, softness	+		
Pomegranate	–	+		
Tomato	Aroma, softness			+

ting, ripening was long considered to be an early stage in the fruit's general disintegration. But it's now clear that ripening is a last, intense phase of life. As it ripens, the fruit actively prepares for its end, organizing itself into a feast for our eye and palate.

Most of the changes in ripening are caused by a host of enzymes, which break down complex molecules into simpler ones, and also generate new molecules just for this moment in the fruit's life. There is a single trigger that sets the ripening enzymes into action. The first clues to its identity came around 1910. From the Caribbean islands came a report that bananas stored near some oranges had ripened earlier than the other bunches. Then California citrus growers noticed that green fruit kept near a kerosene stove changed color faster than the rest. What secret ripening agent did stove and fruit have in common? The answer came two decades later: ethylene, a simple hydrocarbon gas produced by both plants and kerosene combustion, which triggers ripening in mature but unripe fruit. Much later, scientists found that fruits themselves produce ethylene well in advance of ripening. It is thus a hormone that initiates this process in an organized way.

Two Styles of Ripening, Two Ways of Handling

There are two different styles of ripening among fruits. One is dramatic. When triggered by ethylene, the fruit stimulates itself by producing more ethylene, and begins to *respire*—to use up oxygen and produce carbon dioxide—from two to five times faster than before. Its flavor, texture, and color change rapidly, and afterwards they often decline rapidly as well. Such "climacteric" fruits can be harvested while mature but still green, and will ripen well on their own, especially if nudged by an artificial dose of ethylene. They often store their sugars in the form of starch, which enzymes convert back into sweetness during the post-harvest ripening.

The other style of ripening is undramatic. "Nonclimacteric" fruits don't respond to ethylene with their own escalating ethylene production. They ripen gradually, usually don't store sugars as starch, and so depend on their connection to the parent plant for continued sweetening. Once harvested, they get no sweeter, though other enzyme actions may continue to soften cell walls and generate aroma molecules.

These basic styles of ripening determine how fruits are handled in commerce and in the kitchen. Climacteric fruits like bananas and avocados, pears and tomatoes can be picked mature but still hard to minimize physical damage, packed and shipped to their destinations, then gassed with ethylene to ripen them for the produce bin. Consumers can hasten the process by enclosing the fruits in a paper bag with a ripe fruit (plastic traps too much moisture) to expose them to an active ethylene emitter and concentrate the ethylene gas in the air around them. Nonclimacteric fruits like pineapples, citrus fruits, most berries, and melons don't store starch or improve markedly after harvest, so their quality depends mainly on how far they had ripened on the plant. They're best when picked and shipped as ripe as possible, and there's nothing consumers can do to influence their quality: we simply have to choose good ones in the first place.

With just a few exceptions (pears, avocados, kiwis, bananas), even climacteric fruits will be much better if they're allowed to ripen on the plant, from which they can continue to accumulate the raw materials of flavor until the harvest.

COMMON FRUITS OF TEMPERATE CLIMATES: APPLE AND PEAR, STONE FRUITS, BERRIES

Pome Fruits: Apple, Pear, and Relatives

Apples, pears, and quinces are closely related members of the rose family, natives of Eurasia that were domesticated in prehistoric times. They are a kind of fruit known as a *pome* (from the Latin for "fruit"). The fleshy portion of a pome fruit is the greatly enlarged tip of the flower stem. The remains of the flower project from the bottom of the fruit, and the few small seeds are protected in a tough-walled core. Apples and their relatives are climacteric fruit, and contain starch stores that can be turned into sugar after harvest. They generally keep well in cold storage, though late-harvested fruit tend to develop brown cores. Apples are generally sold ripe and keep best if immediately wrapped and refrigerated; pears are sold unripe and are best ripened at relatively cool room temperatures, then refrigerated without close wrapping.

The reddish colors of pome fruits (usually in the skin but sometimes in the flesh) are due mainly to water-soluble anthocyanin pigments, their yellow and cream colors to fat-soluble carotenoids, including beta-carotene and lutein (pp. 257, 267). These fruits are good sources of phenolic antioxidant compounds (p. 267), particularly simple ones (chlorogenic acid, also found in coffee), which are especially concentrated in the skin. Some apples have an antioxidant activity equivalent to the vitamin C in 30 equal portions of orange!

Apples and pears owe their primary flavors to characteristic esters (see box, p. 355). The flavors of pome fruits differ among varieties, among fruits on different parts of a single tree, and even within a single fruit, from top to bottom and from the core outwards. Pears are often noticeably more flavorful at the flower end than at the stem end. Both apples and pears contain an indigestible, slightly sweet sugar alcohol, sorbitol (0.5%), so a large helping of cider can cause the same discomfort as do inulin-rich foods (p. 307).

Apples Apple trees are especially hardy and are probably the most widely distributed fruit trees on the planet. There are 35 species in the genus *Malus.* The species that gives us most of our eating apples, *Malus* x *domestica,* seems to have originated in the mountains of Kazakhstan from crossings of an Asian species (*Malus sieversii*) with several cousins. The domesticated apple spread very early through the Middle East. It was known in the Mediterranean region by the time of the Greek epics, and the Romans introduced it to the rest of Europe. These days apple production is an international enterprise, with southern hemisphere countries supplementing northern stored apples during the off-season, and common varieties as likely to have come from Asia (e.g., Fuji, from Japan) as from the West. There are several thousand named apple varieties, which can be divided into four general groups.

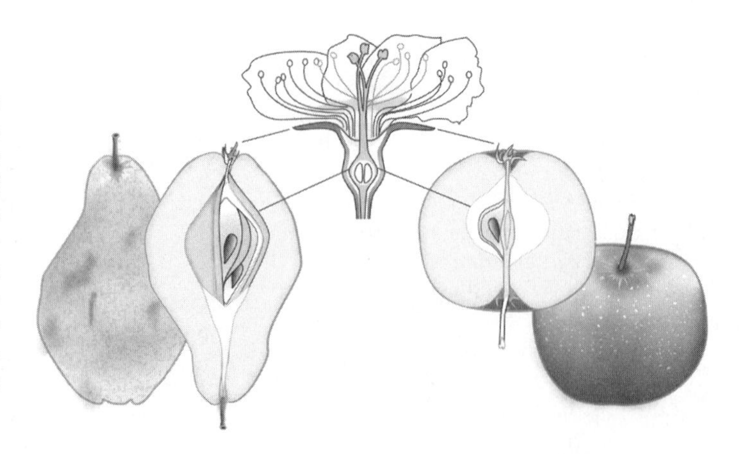

Pome fruits and the flower from which they arise. The edible portion of apples and pears derives from the flower base, or receptacle. Because the ovary is set below most of the flower parts, the flower remains a scar at the bottom of the fruit.

- *Cider apples* (mainly the European native *Malus sylvestris*) are high-acid fruits rich in astringent tannins, qualities that help control the alcoholic fermentation and clarify the liquid (tannins cross-link protein and cell-wall particles and cause them to precipitate). They're used only in cider making.
- *Dessert or eating apples* are crisp and juicy, and have a pleasing balance between sour and sweet when eaten raw (pH 3.4, 15% sugar), but become relatively bland when cooked. Most of the apples in supermarkets and produce markets are dessert apples.
- *Cooking apples* are distinctly tart when raw (a pH around 3 and sugar content around 12%), well-balanced when cooked, and have a firm flesh that tends to maintain its structure when heated in pies or tarts, rather than falling immediately into a puree or—as in some early "codling" varieties—into a fluffy froth. Many countries have traditionally had their standard cooking apples (France its *Calville blanc d'hiver,* England its Bramley's Seedling, Germany its *Glockenapfel* for strudel). But these are being replaced by dual-purpose varieties.
- *Dual-purpose apples* are adequate either raw or cooked (e.g., Golden Delicious, Granny Smith). These are usually at their best for cooking when young and tart, best for eating when older and more mellow.

An apple's potential for cooking can be tested by wrapping a few slices in aluminum foil and baking in a hot oven for 15 minutes, or microwaving a few slices wrapped in plastic film until the film balloons with steam.

Apple Flavor Apple varieties can have very distinctive flavors, and these evolve even after the fruit are picked from the tree. The English were great connoisseurs a century ago, and Edward Bunyard wrote that by storing apples properly in a cool place and tasting them periodically, the apple lover could "catch the volatile ethers at their maximum development, and the acids and sugars at their most grateful balance." Apples do become more mellow with time because they consume some of their malic acid for energy. Much of their aroma comes

Fruit Aroma Compounds: Esters

Many fruits owe their characteristic aroma to chemicals called *esters.* An ester molecule is a combination of two other molecules, an acid and an alcohol. A typical plant cell contains many different kinds of acids, and several different kinds of alcohol. The acids may be either tart substances in the cell fluids or vacuole—acetic acid, cinnamic acid—or fatty-acid portions of oil molecules and the molecules that make up cell membranes: hexanoic acid, butyric acid. The alcohols are usually by-products of cell metabolism. Fruits have enzymes that join these basic cell materials into aromatic esters. A single fruit will emit many esters, but one or two account for most of its characteristic aroma. Some examples:

ethyl alcohol + acetic acid = ethyl acetate, a characteristic note in apples
hexyl alcohol + acetic acid = hexyl acetate in pears
ethyl alcohol + butyric acid = ethyl butyrate in pineapple
isoamyl alcohol + acetic acid = isoamyl acetate in bananas

from the skin, where volatile-creating enzymes are concentrated. The distinctive aroma of cooked apple pulp comes largely from a floral-smelling fragment of the carotenoid pigments (damascenone).

Apple Air and Texture Apples differ from pears in having as much as a quarter of their volume occupied by air, thanks to open spaces between cells in the fruit. (Pears are less than 5% air.) The air spaces contribute to the typical mealiness of an overripe apple: as the cell walls soften and the cell interiors dry out, biting into the apple simply pushes the largely separated cells apart from each other rather than breaking the cells and releasing pent-up juices. Air cells become a factor in baking whole apples; they fill with steam and expand as the apple cooks, and the skin will split unless a strip is removed from the top to release the pressure.

Both apples and crabapples are good sources of cell-wall pectins (p. 265) and make excellent jellies. For the same reason, a simple puree of apples has a thick, satisfying consistency when briefly cooked into applesauce, or slowly reduced to "apple butter."

Apple Juice and Cider Apple juice can be either opalescent or clear depending on whether its pectins and proteins are left intact to deflect light rays. Made fresh, it will stay pale and retain its fresh flavor for about an hour, after which the darkening and aroma-modifying influences of enzymes and oxygen become evident. Browning can be minimized by heating the juice rapidly to the boil to inactivate the browning enzymes, but of course this lends a cooked flavor to the juice. Pasteurized apple juice was first manufactured around 1900 in Switzerland, and is now one of the most important commercial fruit products in the United States. Cider is still an important product in northwest Spain, western France, and England, where the traditional method was to let the fruit pulp ferment slowly through the cold winter, reaching an alcohol content around 4%.

Pears Pears are fruit of the genus *Pyrus,* more temperamental to grow than the apple and less common, but called by some "the queen of fruit" for their refinement of flavor, texture, and shape. Pears are less tart than apples, and denser. The familiar elongated European pears, with mostly smooth flesh, are varieties of the west Asian *Pyrus communis.* "Asian pears" are varieties of two species native to China but intensively improved in Japan, *P. pyrifolia* and *P. ussuriensis.* They have a juicy but

Some Distinctive Apple Flavors and Varieties

Flavor	Varieties
Simple, refreshing	Gravenstein, Granny Smith
Strawberry, raspberry	Northern Spy, Spitzenburg
Winey	McIntosh (well matured)
Aromatic and flowery	Cox's Orange and Ribston Pippins
Honey	Golden Delicious (well matured), Fuji, Gala
Anise or tarragon	Ellison's Orange, Fenouillet
Pineapple	Newtown Pippin, Ananas Reinette
Banana	Dodds
Nutty	Blenheim Orange
Nutmeg	D'Arcy Spice

crisp flesh, more or less gritty with cellulose-rich "stone cells," and may be elongated or apple-shaped. The characteristic aroma of pears comes from several esters, including the "pear ester" (ethyl decadienoate).

In general, pears have a higher respiratory rate than apples and don't store as well. They're unique among temperate fruits in being of the highest quality when picked mature but still hard and ripened off the tree; picked after ripening begins, their texture becomes mushy and their core breaks down. They will also develop a mealy core if excessively warmed after cold storage. They're best ripened slowly over several days at between 65–68°F/18–20°C. Pears are sensitive to carbon dioxide, so they shouldn't be enclosed in plastic bags at any stage. Asian pears are especially prone to bruising and are often marketed in protective sleeves.

Pear Varieties Originally, all pears were gritty "sand pears" and hard-fleshed. Centuries of breeding greatly reduced the prominence of gritty stone cells (but not in varieties for making perry, the pear version of cider, where they're valued for helping to grind the flesh before fermentation). The soft "butter" texture characteristic of many European pears was developed in the 18th century by Belgian and French breeders. European pears are classified in three groups according to when they're harvested and their traditional storage life (now extended by controlled atmospheres and temperatures). Summer pears like the Bartlett (also called Williams or Bon chretien) are harvested in July and August and keep for one to three months; autumn pears like the Bosc and Comice are harvested in September and October and keep two to four months; and winter pears like the Anjou and Winter Nellis are harvested in October and November and keep six to seven months.

Quince Quinces, fruit of the central Asian tree *Cydonia oblonga,* give us a taste of what apples and pears might have been like in their primitive form. They are gritty with stone cells, astringent, and hard even when ripe. But they have a distinctive, flowery aroma (thanks to lactones and violet-like ionones, all derived from carotenoid molecules) that's especially concentrated in the fuzzy yellow skin. And cooking domesticates them: heat breaks down and softens their pectin-rich cell walls, and the astringent tannins become bound up in the debris, so the taste softens as well. Quince paste firm enough to slice is a traditional product of Spain (*membrillo*) and Italy (*cotognata*), and in Portugal a quince preserve was the original marmalade (*marmalada*). The 16th-century alchemist and confectioner Nostradamus gave several recipes for quince preserves and observed that cooks "who peel them [before cooking] don't know why they do this, for the skin augments the odor." (The same is true for apples.)

Quinces have another enchanting quality: when slices are slowly cooked in sugar over several hours, they turn from a pale off-white to pink to a translucent, deep ruby red. The key to this transformation is the fruit's store of colorless phenolic compounds, some of which cooking turns into anthocyanin pigments (p. 281). Pears contain the same compounds, but in smaller quantities (common Bartletts about a twenty-fifth, Packhams a tenth to a half), and so usually get pink at best.

Medlar Medlars are small fruits of an apple relative (*Mespilus germanica*) native to central Asia, now rare but once commonly grown in Europe as a winter fruit. Like the quince, the medlar remains hard and astringent even when ripe, so it keeps well and even improves if left on the tree through early frosts. It was made into preserves, but more often it was "bletted" (a 19th-century coinage from the French *blessé,* "bruised"), or picked from the tree and kept in a cool, dry place for several weeks until the enzymes in its own cells digest it from within, and its flesh turns soft and brown. The astringency disappears, the malic acid is used up, and the aroma develops strong overtones of spice, baked apples, wine, and gentle decay, what D. H. Lawrence described as an "exquisite odour of leave-taking."

Loquat Loquats bear little resemblance to their cousin pomes. They are small, elongated fruits of a Chinese tree, *Eriobotrya japonica,* which was greatly improved by the Japanese and taken to many subtropical regions in the 19th century, notably Sicily, where they are called *nespole.* They usually ripen early, before cherries. They have a mild, delicate flavor and a wall of carotenoid-containing flesh that runs from white to orange, surrounding several large seeds. U.S. varieties are mainly ornamental and produce small fruit, while European and Asian fruits may approach a half pound/250 gm. They're eaten fresh, made into jellies and jams, and cooked in a spicy syrup in the manner of "pickled" peaches. Loquats are neither climacteric nor chill-sensitive, and so keep well.

Stone Fruits: Apricot, Cherry, Peach, and Plum

Stone fruits are all species of the genus *Prunus,* members of the large rose family and relatives of the pome fruits. They owe their name to the stone-hard "shell" that surrounds a single large seed at their center. Though the 15 species of *Prunus* are found throughout the northern hemisphere, the important stone fruits mostly come from Asia. They do not store starch and so get no sweeter after harvest, though they do soften and develop aroma. Their internal tissues tend to become mealy or break down in prolonged cold storage, so fresh stone fruits are more seasonal than hardier apples and pears. Like some of the pome fruits, stone fruits accumulate the indigestible sugar alcohol sorbitol (a frequent ingredient in sugar-free gums and candies, p. 662); they're also rich in antioxidant phenolic compounds. The seeds of stone fruits are protected by a cyanide-generating enzyme that also produces the characteristic aroma of almond extract (almonds are seeds of *Prunus amygdalus*). They thus lend an almond character when included in sugar and alcohol preserves, and can replace "bitter almonds" in European pastries and sweets (p. 506).

Apricot The apricots that are most familiar in the West are fruits of *Prunus armeniaca,* a native of China that was taken to the Mediterranean during Roman times. There are now thousands of different varieties, white and red (from lycopene) as well as orange, and most of them adapted to specific climates; apricots flower and fruit early (the name comes from the Latin *praecox,* "precocious"), and therefore bear best in areas with mild, predictable winters. Several other species are grown in Asia, including *P. mume,* whose fruits the Japanese salt-pickle and color red to make the condiment *umeboshi.* The distinctive aroma of

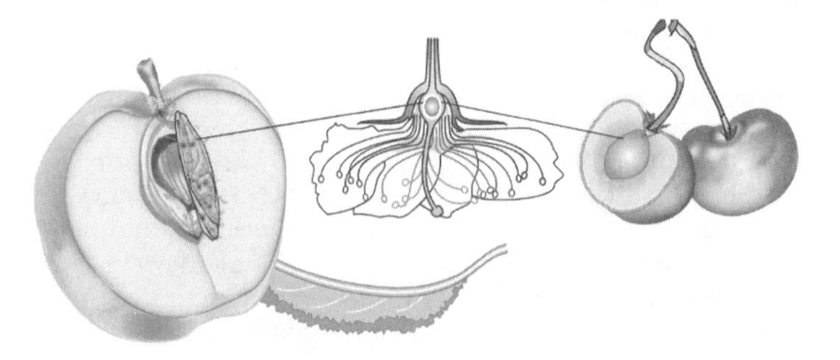

Stone fruits and the flower from which they arise. Peaches and cherries derive from an ovary that sits above the base of the flower parts, so the fruits show no remnants of the flower.

fresh apricots comes from a rich mixture of terpenes that provide citrus, herbal, and floral notes, and from peach-like compounds (lactones). They are rich in pectin, which gives them a luscious texture when fully ripe, a meaty texture when dried.

Apricots are delicate fruits that don't travel well, so most are processed. They're especially suited to drying, which concentrates their sweet-tart flavor well even when they're overripe. Most dried apricots in the United States come either from a few western states or from Turkey, which exports a relatively pale, bland variety with half the carotenoid pigments and acidity of the Blenheim and Patterson varieties of California. The fruits are dried in the sun in early summer for one or two weeks, until they reach a moisture content of 15–20%. Usually they're treated with sulfur dioxide to preserve the abundant beta-carotene and other carotenoids, vitamin C, and fresh flavor. Unsulfured apricots are brown and have a flatter, cooked taste.

Cherry Cherries come in two basic types from two different species, both of which are native to western Asia and southeast Europe. Sweet cherries are the fruits of *Prunus avium,* which is probably one of the parents of the sour cherry, *Prunus cerasus.* Sweet and sour cherries differ mainly in their maximum sugar content, with sweet cherries accumulating significantly more. Cherries don't improve once they're harvested, so they must be picked ripe and fragile. Most sweet cherries grown in the United States are sold fresh, but far more sour cherries are grown, and most of these are processed. Cherries are prized not only for their flavor, but for their color, which may range from very deep red (rich in anthocyanins) to a pale yellow. The red varieties are an excellent source of phenolic antioxidants.

Cherry flavor comes mainly from almondy benzaldehyde, a flowery terpene (linalool), and essence of clove (eugenol). Heating increases both the almond and flowery notes, especially if the pits are left in the fruit. This is why the classic French cherry clafoutis, a custardy tart, is intensely flavored but requires care in eating!

The familiar "maraschino" cherry originated several centuries ago in northeastern Italy and the neighboring Balkans, where the local marasca cherry was preserved in its own liqueur for winter eating. In the modern industrial version, light-fleshed varieties are bleached with sulfur dioxide and stored in brine until needed, then infused with sugar syrup, dyed cherry red, flavored with almond extract, and pasteurized. After all that, what's left of the original cherry is mainly its skeleton, the cell walls and skin.

Peach and Nectarine Peaches and nectarines are both fruits of the species *Prunus persica.* Nectarines are varieties with a smooth skin that are usually also smaller, firmer, and more aromatic than their fuzzy siblings. The words "peach" and "persica" come from "Persia," by way of which the fruit reached the Mediterranean world from China by about 300 BCE.

Modern peach and nectarine varieties fall into a handful of categories. Their flesh may be white or yellow, and either firm or melting, strongly attached to the large central stone (clingstone) or easily detached (freestone). The genetically dominant characteristics are white, melting, freestone flesh. Yellow varieties were developed mainly after 1850, and firm clingstone varieties have been bred mainly for drying, canning, and improved tolerance of shipping and handling. The yellow coloration comes from a handful of carotenoid pigments, beta-carotene among them; rarer red varieties contain anthocyanins (as the skin often does). Peaches begin to ripen at the stem end and along the groove, or "suture," and are said to continue their flavor development even after harvest. The distinctively aromatic flavor of peaches and nectarines comes largely from compounds called lactones, which are also responsible for the aroma of coconut; some varieties also contain clove-like eugenol.

The most frequent problem with peaches is mealy flesh, apparently due to impaired pectin breakdown when the fruit has been temporarily stored in the cold, at temperatures below about 45°F/8°C. This is especially common in supermarket fruits.

Plum and Plum Hybrids Most plums are the fruits of two species of *Prunus*. A Eurasian species, *P. domestica,* gave rise to the European plums, which include French and Italian prune plums, the greengage and Reine Claude, the yellow-egg and imperatrice. The most common of these are the prune types, purplish-blue ovals with a meaty, semimelting, semifreestone flesh. The second, Asian species, *P. salicina,* originated in China, was improved in Japan, and further bred by Luther Burbank and others in the United States after 1875. Varieties of the Asian species (Santa Rosa, elephant heart, and many others) tend to be larger, rounder, from yellow to red to purple, clingstone, and often melting. European plums are usually dried or made into preserves, Asian plums eaten fresh. Plums are climacteric fruit, so they can be harvested before ripening, stored at 32°F/0°C for up to 10 days, and then allowed to mature slowly at 55°F/13°C. Their aroma varies from kind to kind, but generally includes almondy benzaldehyde, flowery linalool, peachy lactones, and spicy methyl cinnamate.

Plum-apricot hybrids, known as pluots (more plum parentage) or plumcots (equal parentage), are generally sweeter than plums and more complex in aroma. There are also a number of minor plums, among them the English damson and sloe (*P. insititia* and *spinosa*), the latter small, astringent fruits that are steeped to make sloe gin.

Prunes The firm-fleshed prune plums dry well in the sun or during 18–24 hours in a dehydrator at around 175°F/79°C. They develop a rich flavor thanks to the concentration of sugars and acids—nearly 50% and 5% of their weight respectively—and to browning reactions that generate caramel and roasted notes as well as their color, a brown-black deep enough to be attractive rather than drab. This richness is the reason that prunes work well in many savory meat dishes. Prunes are such a concentrated source of antioxidant phenolic compounds (up to 150 mg per 100 gm) that they make an excellent natural flavor stabilizer: they prevent the development of warmed-over flavor in ground meats when

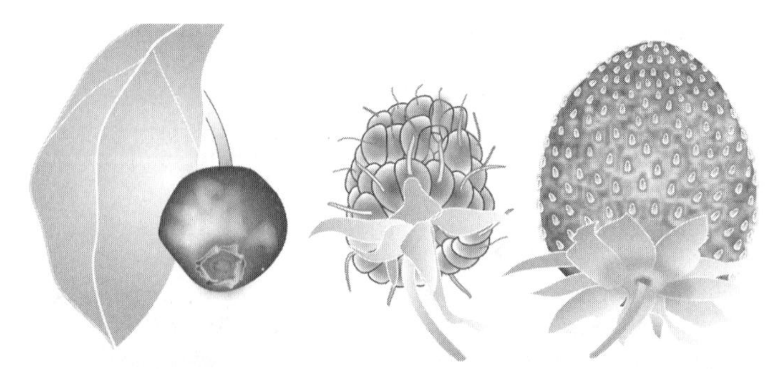

Common berries. Blueberries (left) *are true berries, or single fruits derived from the plant's ovaries. Caneberries and strawberries are not true berries, but multiple fruits that develop from many ovaries set in the same flower receptacle. Each little segment of a raspberry or blackberry* (center) *is a complete stone fruit. The strawberry* (right) *is a "false fruit": the small "seeds" borne on the surface of the swollen flower base are dry but entire fruits that correspond to the caneberry segments.*

included at the rate of just a few percent (1 tablespoon per pound). They're also rich in moisture-retaining fiber and sorbitol and so are used to replace fat in hamburgers and a variety of baked foods. (Dried cherries have many similar properties and uses.) Their well-known laxative action on the human digestive tract is not entirely understood but probably involves the sugar alcohol sorbitol (p. 662), which accounts for up to 15% of the weight of both prune and juice. We can't digest sorbitol, so it passes into our intestines where it may have a number of stimulating effects.

Berries, Including Grapes and Kiwi Fruit

Though the term *berry* has a precise botanical definition, in common usage it generally refers to small fruits borne on bushes and low plants, not trees. Most of our familiar berries are natives of northern woodlands.

Caneberries: Blackberries, Raspberries, and Relatives Caneberries are fruits of the genus *Rubus*, which grows naturally across most of the temperate northern hemisphere in the form of long thin, thorny stalks, or canes. There are hundreds of species of blackberry native to both Europe and the Americas, but just a few species of raspberry. Serious caneberry cultivation probably began around 1500, and a number of blackberry-raspberry hybrids have been created, including boysenberry, loganberry, youngberry, and tayberry from American species, the Bedford giant from European species. Less familiar caneberries include cloudberries, a yellow-orange Scandinavian fruit, and dark red, intensely aromatic Arctic bramble fruits.

Caneberries are composite fruits: a single flower has from 50 to 150 ovaries, and each ovary makes a separate small fruitlet, like a miniature plum with a stony seed. The fruitlets are nourished through contact with the flower base and held together by the entanglement of small hairs borne on their surface (the original inspiration for Velcro). When they ripen, blackberries separate from the cane at the bottom of the flower base, so the base comes with the fruit; raspberries instead separate from the

Some Caneberry Relationships

All caneberries are members of the prolific genus *Rubus*, a member of the rose family.

Raspberry, European	*Rubus idaeus vulgatus*
Raspberry, American	*R. idaeus strigosus*
Raspberry, black (American)	*R. occidentalis*
Blackberry, European	*R. fruticosus*
Blackberry, American	*R. ursinus, laciniatus, vitifolius,* etc.
Dewberry, European	*R. caesius*
Dewberry, American	*R. flagellaris, trivialis*
Boysenberry, loganberry, marionberry, olallieberry, youngberry	Various blackberry-raspberry crosses
Cloudberry	*R. chamaemorus*
Salmonberry	*R. spectabilis*
Arctic bramble	*R. arcticus*

base itself, and so have an inner cavity. Caneberries are climacteric fruit, and have one of the highest respiration rates of any fruit; thanks to this and their thin skin, they're extremely fragile and perishable.

Raspberries have a distinctive flavor due to a compound dubbed the raspberry ketone, and also have a violet-like note (from carotenoid fragments called ionones). The flavor of wild berries has been found to be by far the most intense. Blackberries vary in their flavor, the European varieties relatively mild, the American more intense, with spicy aroma notes (from terpenes). Most caneberry colors are provided by anthocyanin pigments, whose sensitivity to pH can cause dark purple blackberries to turn red when frozen (p. 281). These fruits are a good source of phenolic antioxidants, at least one of which (ellagic acid) actually increases during jam making. When made into preserves, the numerous caneberry seeds (several thousand per quarter pound/100 gm) can sometimes absorb syrup, become translucent, and give the normally deep-colored jam a milky dullness.

Blueberry, Cranberry, and Relatives

These berries are borne by several different species in the genus *Vaccinium,* which ranges across northern Europe and North America.

Blueberry Blueberries are the small fruits of a bushy North American species of the genus *Vaccinium,* which ranges from the tropics to the arctic. *V. angustifolium* and *corymbosum* are weedy pioneers in burned fields, and their fruits were gathered from the wild until the 1920s, when the first selected "highbush" (*corymbosum*) plants were developed in New Jersey. The bilberry, *V. myrtillus,* is a European relative, and the rabbit-eye blueberry, *V. ashei,* a similar but less flavorful native of the southern United States. The huckleberry, various species of *Vaccinium,* has a few large hard seeds, while blueberries have many small ones.

Blueberries have a distinctive, spicy aroma apparently due to several terpenes, and are rich in phenolic antioxidants and in anthocyanin pigments, especially in the skin. These small berries freeze well, and retain their shape and substance when baked. The pigments can turn odd shades of green if cooked with alkaline ingredients (for example, baking soda in muffins).

Cranberry and Relatives Cranberries are fruits of the North American perennial vine *Vaccinium macrocarpon,* which is native to low, swampy areas of northern states from New England to the Midwest. Cultivation and efforts at improvement began in the 19th century, and the familiar jelly-like cranberry sauce was born early in the 20th century when a large producer decided to process his damaged berries into a canned puree.

Cranberries can be harvested dry, with a comb-like machine, or wet, by flooding the bog. Dry-harvested berries keep better, for several months. Cranberries store well for a couple of reasons. One is their high acidity, exceeded only by lemons and limes, and the main obstacle to eating them straight. The other is their very high content of phenolic compounds (up to 200 milligrams per 100 grams), some of which are antimicrobial and probably protect the fruit in its damp habitat. Many of these phenolic materials are also useful to us, some as antioxidants and others as antimicrobials. One example is benzoic acid, now a common preservative in prepared foods. A particular pigment precursor in cranberries (also found in blueberries) prevents bacteria from adhering to various tissues in the human body, and so helps prevent urinary tract infections.

The spicy aroma of cranberries is created by a combination of terpenes and spicy phenolic derivatives (cinnamates, benzoates, vanillin, almondy benzaldehyde). Some of its phenolic compounds contribute a notable astringency. Cranberries are rich in pectin, which is why a barely cooked puree thickens immediately into a sauce; it's also why cranberries macerated in alcohol may cause the alcohol to gel.

Lingonberries or cowberries are the fruit of a European relative of the cranberry, *V. vitis-idaea*; they have a distinctive, complex flavor. The European cranberry, *V. oxycoccus*, has a stronger and more grassy, herbaceous flavor than the American species.

Currants and Gooseberries Currants and gooseberries are all species of the genus *Ribes*, which is found in northern Europe and North America. These small berries don't appear to have been cultivated until around 1500. (Their cultivation in the United States has been retarded by federal and state restrictions because they can harbor a disease that attacks white pines.) There are white and red currants, *R. sativum* and *R. rubrum*, and hybrids between the two. The black currant *R. nigrum*, is higher in acid than the others, and has a distinctively intense aroma made up of many spicy terpenes, fruity esters, and a musky, "catty" sulfur compound that is also found in sauvignon blanc wines. Black currants are also notably rich in vitamin C and in antioxidant phenolic compounds—as much as 1% of their weight—about a third of which are anthocyanin pigments. Currants are mainly made into preserves, and the French make black currants into a liqueur, *crème de cassis*.

The gooseberry, *R. grossularia*, is larger than the currant, and is often picked unripe for cooking in tarts and sauces. The jostaberry is a black currant–gooseberry cross.

Grapes Grapes are the berries of woody vines in the genus *Vitis*. *V. vinifera*, the major source of wine and table grapes, is native to Eurasia (p. 772). There are also about 10 grape species native to temperate Asia, and 25 to North America, including the *V. labrusca* that gives us Concord and Catawba grapes. About two-thirds of the world's grape production goes to make wine; of the rest, about two-thirds are eaten fresh and a third are made into raisins. There are many thousands of grape varieties. Most wine varieties originated in Europe, while varieties for eating fresh or making into raisins can often be traced back to western Asian parents. Wine grapes come in relatively small clusters and are acidic enough to help control the yeast fermentation; table grapes come in large clusters and are less tart; raisin varieties have a thin skin, high sugar content, and loose cluster structure to facilitate drying. The commonest table and raisin grape in the United States, the Thompson seedless or sultana, is a variant of an ancient Middle Eastern all-purpose variety, the Kishmish.

Table grapes are quite diverse. They may be seedy or seedless, deep purple with anthocyanins or pale yellow; their sugar content may range from 14 to 25%, their acidity from 0.4 to 1.2%. They may have a fairly neutral, green aroma (Thompson seedless), or be flowery and citrusy from terpenes (muscat), or musky with anthranilate and other esters (Concord and other American varieties). Most commercial varieties today have been bred to be seedless, crisp, tart, and sweet, with a long storage life. Thompson seedless grapes picked in the cool of the morning and treated with antimicrobial sulfur dioxide can be held for as long as two months at 32ºF/0ºC.

Raisins Grapes are easily preserved by sun-drying to make raisins. In the United States this is usually done by laying the grapes on paper between rows in the vineyard for about three weeks. Raisins are naturally brown and have caramel flavor notes due to a combination of browning-enzyme oxidation of phenolic compounds and direct browning reactions between sugars and amino acids (pp. 269, 778). Both of these processes are accelerated by high temperatures, so a lighter color can be obtained by drying the grapes in the shade. Golden raisins are made by treating the grapes with antioxidant sulfur dioxide and drying them mechanically at controlled temperatures and humidities; the result is a much fruitier, lighter flavor. Zante "currants" are made from the small black Corinth grape, and are tarter than ordinary raisins thanks to their higher proportion of skin to pulp.

Verjus and Saba Two ancient grape preparations make versatile ingredients in the kitchen. Fruits thinned six to eight weeks before the main harvest are crushed and filtered to produce *verjus,* a tart alternative to vinegar or lemon juice, slightly sweet, with a delicate green aroma. And ripe grapes are cooked down to a thick, sweet-tart, aromatic syrup (Roman *sapa,* Italian *saba* or *mosto cotto,* Turkish *pekmez,* Arab *dibs*). Like syrups from other fruit (pomegranates), grape syrup was an important sweetener in the times before cheap table sugar, but provides tartness and aroma as well as sweetness. It's thought that balsamic vinegar may have evolved from grape syrup that was kept long enough to ferment (p. 775).

Kiwi "Kiwi" fruit is the name that New Zealand producers came up with for the striking, tart berry of a Chinese vine, *Actinidia deliciosa,* when they pioneered its international marketing in the 1970s. Several other species of *Actinidia* are now also cultivated, including the yellow-to-red-fleshed *A. chinensis.* Kiwi fruit are unusual in appearance and ripening behavior. Their thin, hairy skin doesn't change color during ripening, and the translucent inner flesh is green with chlorophyll, with as many as 1,500 small black seeds embedded in a ring and connected to the core by white rays of vascular tissue. (There are also chlorophyll-free varieties with yellow, red, and purple flesh.) Cross-sectional slices of kiwi are thus very attractive. When harvested, kiwi fruits contain a large amount of starch. During months of storage at 32°F/0°C, the starch is slowly converted to sweet sugars. Then at room temperature they undergo a climacteric ripening that takes about 10 days. The flesh softens and the aroma becomes more pronounced, with strongly fruity esters (benzoates, butanoates) coming to dominate more delicate, grassy alcohols and aldehydes. Some kiwi varieties are rich in vitamin C and in carotenoids.

Kiwi fruits present a couple of challenges to the cook. They contain a strong protein-digesting enzyme, actinidin, which can damage other ingredients in a mixture and irritate sensitive skin. Heat inactivates enzymes, but it also muddies the fruit's delicate color and translucency. Kiwi fruits also contain crystals of calcium oxalate (p. 259), which pureeing, juicing, and drying can make more apparent and irritating in the mouth and throat.

Mulberry Mulberries are the surprisingly small, fragile, composite fruits of trees of the genus *Morus.* They resemble the blackberry, but each small fruitlet actually arises from a separate flower on a short flowering stalk. The white mulberry, *M. alba,* is native to China, where its leaves have long been used to feed silkworms. Its color ranges from white to purple, and it is relatively bland; often it's dried, which helps intensify its flavor. The Persian or black species, *M. nigra,* comes from west Asia; it's always a dark purple and is more flavorful. The North American red mulberry, *M. rubra,* is mainly tart. Mulberries are used to make preserves, syrups, and sorbets.

Strawberry Strawberries come from small perennial plants of the genus *Fragaria,* whose 20 species range across the northern hemisphere. The plants are easy to grow and therefore are grown widely, from subarctic Finland to tropical Ecuador. The strawberry is unusual in bearing its "seeds" on the surface of the fleshy portion, not inside. The "seeds" are actually miniature dry fruits (achenes), similar to buckwheat and sunflower "seeds," and the fleshy portion is the flower's swollen base, not its ovary. During ripening, the cells of the strawberry interior enlarge and pull apart from each other. The berry is therefore filled with tiny air pockets, and its shape is maintained by the pressure of the cell contents pushing each cell onto its neighbors. When this pressure is released, by water loss from drying out or from freezing that punctures the cell walls, the structure weakens and the fruit becomes soft and mushy. Strawberries don't improve once picked, so they must be picked ripe. Thanks to their thin skin and fragile structure, they only last a few days, even in cold storage.

The pineapple note in standard strawberries comes from the presence of ethyl esters. Some sulfur compounds and a complex caramel-like oxygen-containing ring, furaneol (also characteristic of pineapple), round out strawberry aroma. The smaller European woodland strawberries have a flavor of Concord grapes thanks to anthranilates, and a clove-like spicy note (from the phenolic eugenol). Strawberries are rich in ascorbic acid and in phenolic antioxidants, including its red anthocyanin pigments. They are poor in pectic materials, so strawberry preserves are often supplemented with prepared pectin or pectin-rich fruits.

The Domestication of the Strawberry Most of the strawberries grown today derive from two American species which were brought together and hybridized less than 300 years ago—and in Europe, not in the Americas!

Europe had its own native strawberry (*F. vesca* and *F. moschata*), which is now called the "wild" strawberry or *fraise de bois* ("woodland strawberry"), even though it's cultivated. This strawberry was mentioned in Roman literature, subsequently cultivated, and by the 15th century had a wonderful fragrance but was still small, pithy and unproductive. Early European visitors to North America were impressed by the size and vigor of an American species, *F. virginiana*, and brought it back to Europe. Then a Frenchman by the stunningly appropriate name of Frézier found the walnut-sized fruits of another New World species, *F. chiloensis*, growing in Chile, and took that species to France in 1712. Around 1750, in the strawberry-producing area around Plougastel in Britanny, an accidental hybrid between the two American species arose. Then across the Channel in England, a natural mutant of the Chilean species arose, large and pink, with a shape and aroma reminiscent of pineapple. Modern strawberry varieties, large and red and flavorful, derive from these two all-American ancestors. They have been given the scientific name *F. x ananassa* to indicate their hybrid origins (*x*) and distinctive pineapple aroma (*ananassa*).

Elderberry and Barberry Elderberries and barberries are minor fruits worth rediscovering. Elderberries are pleasantly aromatic fruits borne by trees of the genus *Sambucus*, which are found across the northern hemisphere. They're usually cooked or made into wine because they are too tart to be enjoyed raw, and contain antinutritional lectins (p. 259) that require heating to be inactivated. Elderberries are rich in anthocyanin pigments and antioxidant phenolic compounds. So is the barberry, from shrubby species of the northern-hemisphere

Food Words: *Berry, Strawberry*

Berry comes from an Indo-European root meaning "to shine," perhaps for the bright colors of many small fruits. The *straw* in *strawberry* comes from a root for "to spread, to strew." Straw is the dried stalks strewn about the field after the grain harvest; perhaps strawberries were named for the plant's habit of spreading by means of runners. A related cooking term is *streusel,* an informally scattered topping on baked goods.

Berberis, which is something like a miniature cranberry and dries well. Barberries are much used in Persian cooking, where they provide the tart rubies in a dish called jeweled rice.

Other Temperate Fruits

Ground Cherry This is one name for a couple of related fruits from low-lying plants in the nightshade family, close relatives of the tomatillo (p. 331). The Peruvian or Cape ground cherry or gooseberry, *Physalis peruviana*, came from South America, while the ordinary ground cherry, *P. pubescens*, is native to both North and South America. Both fruits resemble miniature, thick-skinned yellowish tomatoes, are enclosed in papery husks (thus another name, husk tomato), and keep well at room temperature. The Peruvian ground cherry has floral and caramel aroma notes in addition to generically fruity esters. These fruits are made into preserves and pies.

Persimmon Persimmons are fruits of trees in the genus *Diospyros*, which is native to both Asia and North America. There's a plum-sized native American persimmon, *D. virginiana*, and a Mexican species known as the black sapote (*D. digyna*), but the most important persimmon species worldwide is *D. kaki*, a tree with apple-sized fruits native to China and adopted by Japan; it's sometimes said that persimmons are to the Japanese what apples are to Americans. Japanese persimmons are sweet, low-acid, mild fruits, with a few brown seeds surrounded by flesh that's bright orange from various carotenoid pigments, including beta-carotene and lycopene. They have a very mild aroma reminiscent of winter squash that probably derives from breakdown products of carotenoids.

Japanese persimmons come in two general kinds, astringent and nonastringent. Astringent varieties, including the tapered Hachiya, have such high levels of tannins that they're edible only when completely ripe, with translucent and almost liquid flesh. Nonastringent types, including the flat-bottomed Fuyu or Jiro, are not tannic, and can be eaten while underripe and crisp (they also don't get as soft as the astringent types). Centuries ago, the Chinese figured out a way to remove the astringency from unripe persimmons before they ripen. This method may be the first example of controlled-atmosphere storage! They simply buried the fruit in mud for several days. It turns out that when the fruits are deprived of oxygen, they shift their metabolism in a way that results in the accumulation of an alcohol derivative called acetaldehyde, and this substance binds with tannins in the cells, thus preempting them from binding to our tongues. Modern cooks can accomplish the same thing by wrapping persimmons snugly in a truly airtight plastic film, polyvinylidene chloride (saran).

Persimmons are commonly eaten raw, frozen whole into a natural sorbet, and made into pudding. Traditional American persimmon pudding owes its distinctive black-brown color to the combination of the fruit's glucose and fructose, flour and egg proteins, alkaline baking soda, and hours of cooking, which encourage exten-

Food Words: *Rhubarb*

Rhubarb is a medieval Latin coinage, a combination of Greek *rha* and *barbarum*: "rhubarb" and "foreign." *Rha* also meant the Volga River, so the plant may have been named after it: it came from foreign lands to the east of the Volga.

sive browning reactions (p. 778; replace the baking soda with neutral baking powder or shorten the cooking time and you get a light orange pudding). Persimmon flesh can be whipped into a long-lived foam thanks to its tannins, which help bind fragments of cell walls together to stabilize the air pockets. In Japan, most Hachiya persimmons are dried, massaged every few days to even out the moisture and break down some of the fibrousness to a soft, doughy consistency.

Rhubarb Rhubarb is a vegetable that often masquerades as a fruit. It is the startlingly sour leaf stalks of a large herb, *Rheum rhabarbarum,* that is native to temperate Eurasia and became popular in early 19th-century England as one of the first fruit-like produce items to appear in the early spring. The rhubarb root had long been used as a cathartic in Chinese medicine, and traded widely as a medicinal. The stalks were also used as a vegetable in Iran and Afghanistan (in stews, with spinach) and in Poland (with potatoes). By the 18th century the English were using them to make sweet pies and tarts. The 19th century brought better varieties and techniques for digging up mature roots and forcing rapid stalk growth in warm dark sheds, which produced sweeter, tenderer stalks. These improvements, cheaper sugar, and a growing supply resulted in a rhubarb boom, which peaked between the world wars.

Rhubarb stalks may be red with anthocyanin pigments, green, or an intermediate shade, depending on the variety and production techniques. Their acidity is due to a number of organic acids, notably oxalic acid, which contributes about a tenth of the total acidity of 2–2.5%. (This is double or triple the oxalate content of spinach and beets.) Rhubarb leaves are said to be toxic in part due to their high oxalate content, as much as 1% of the leaf weight, but other chemicals are probably also responsible. Today rhubarb is available much of the year thanks to greenhouse production, though some cooks prefer the more intense flavor and color of the field-grown crop of late spring. The color of red stalks is best preserved by minimizing both the cooking time and the quantity of added liquid, which dilutes the pigments.

FRUITS FROM WARM CLIMATES: MELONS, CITRUS FRUITS, TROPICAL FRUITS, AND OTHERS

Melons

Except for the watermelon, melons are fruits of *Cucumis melo,* a close relative of the cucumber (*C. sativus*) and a native of the semiarid subtropics of Asia. The melon plant was domesticated in central Asia or India and arrived in the Mediterranean at the beginning of the 1st century CE, where their large size and rapid growth made them a common symbol of fertility, abundance, and luxury. There are many melon varieties with distinctive rinds, flesh colors (orange types are an excellent source of beta-carotene), textures, aromas, sizes, and keeping qualities.

Melons are generally used fresh, either sliced or pureed. They contain a protein-

Food Words: *Melon*

In Greek, *melon* meant "apple," but also other fruits containing seeds. The Greeks called our melon *melopepon,* or "apple-gourd," and this became shortened to *melon.*

digesting enzyme, cucumisin, and thus will prevent gelatin gels from setting unless the enzyme is denatured by cooking or an excess of gelatin is used. The melon surface can become contaminated with microbes in the field and cause food poisoning when the microbes are introduced into the flesh during cutting; it's now recommended that melons be thoroughly washed in hot soapy water before preparing them.

Melon Families and Qualities The most common Western melons fall into two families:

- Summer melons are highly aromatic and perishable, and usually have rough rinds. They include true cantaloupes and muskmelons.
- Winter melons are less aromatic and less perishable, and usually have smooth or wrinkled rinds. They include honeydews, casabas, and canaries.

The differences between the two melon families are caused by differences in their physiology. The aromatic summer melons are generally climacteric fruits that (with the exception of cantaloupes) separate from their stems when ripe; and they contain active enzymes that generate more than 200 different esters from amino acid precursors, and thus help create their characteristically rich aroma. The winter melons are generally nonclimacteric fruit like their relatives the cucumbers and squashes, and have low ester-enzyme activities and therefore a milder flavor.

Vine-ripening is important for all melons because they don't store starch and so get no sweeter after harvest. A remnant of the stem on an aromatic melon indicates that it was harvested before becoming fully ripe, while all winter melons (and true cantaloupes), even ripe ones, carry a piece of stem. The aroma of melons may continue to develop off the vine, but will not be the same as the aroma of vine-ripened

Some Melon Varieties

Summer Melons: very aromatic, keep one to two weeks
- **Cantaloupe:** smooth or lightly netted, orange flesh, rich flavor (Charentais, Cavaillon)
- **Muskmelon:** deeply netted (most U.S. varieties, sometimes misnamed "cantaloupe")
  - **Galia, Ha Ogen, Rocky Ford:** green flesh, sweet and aromatic
  - **Ambrosia, Sierra Gold:** orange flesh
  - **Persian:** large, orange, mild
  - **Sharlyn/Ananas:** translucent pale flesh
- **Pancha (charentais x muskmelon):** netted and ribbed, orange, very aromatic

Winter Melons: less aromatic, keep weeks to months
- **Honeydew:** smooth rind, green or orange flesh, sweet, mild aroma (many varieties)
- **Casaba, Santa Claus:** wrinkled or smooth rind, white flesh, less sweet or aromatic than honeydew
- **Canary:** slightly wrinkled rind, white flesh, crisp, aromatic

Hybrids
- **Crenshaw (Persian x casaba):** green-yellow wrinkled rind, orange flesh, juicy, aromatic

fruit. In addition to fruity esters, melons contain some of the same green, grassy compounds that give cucumbers their distinctive flavor, as well as sulfur compounds that provide a deeper, more savory dimension.

Minor Melons In addition to the Western melons, there are several groups of Asian melons, including Japanese pickling or tea melons, many of them crisp-fleshed, and the *flexuosus* group, long and twisted like a snake, which includes the "Armenian cucumber." There is also the *dudaim* group of small, especially musky melons used in the U.S. South and elsewhere for preserves and simply to scent the air (pocket melon, pomegranate melon, smell melon); *dudaim* is Hebrew for "love-plants." The horned melon, also called jelly melon and kiwano, is the fruit of *Cucumis metuliferus,* a native of Africa with a spiky yellow skin and a relatively scant amount of emerald-green, translucent gel surrounding its seeds. The gel has a sweet cucumbery flavor and is used in drinks, fresh sauces, and sorbets. The hollowed-out skin makes a decorative container.

Watermelon The watermelon is a distant relative of the other melons, the fruit of an African vine, *Citrullus lanatus,* whose wild relatives are very bitter. The Egyptians were eating it 5,000 years ago, and the Greeks knew it by the 4th century BCE. World production of watermelons is now double the production of all other melons combined. Watermelons are notable for the large size both of their cells, which are easily seen with the naked eye, and their fruits, which can reach 60 lb/30 kg and more. Unlike other melons, the watermelon consists of seed-bearing placental tissue rather than seed-surrounding—thus seed-free—ovary wall. "Seedless melons," which actually contain small undeveloped seeds, were first bred in Japan in the 1930s. The classic watermelon is dark red with the carotenoid pigment lycopene, and in fact is a much richer source of this antioxidant than tomatoes! Recent years have brought yellow-orange varieties. A good watermelon has a crunchy, crisp, yet tender consistency, a moderately sweet taste, and a delicate, almost green aroma. External signs of quality are a substantial heaviness for the melon's size, yellow skin undertones indicative of chlorophyll loss and thus ripeness, and a solid resonance when thumped.

In addition to being eaten fresh, watermelon flesh is pickled and candied (often after a preliminary drying), and cooked down into a syrup or thick puree. The dense rind is often made into sour or sweet preserves. There is a subgroup of watermelons, *C. lanatus citroides,* known as citron or preserving melons, with inedible flesh but abundant rind for these preparations. Both melon and watermelon seeds are used in several regions, roasted or ground and infused to make beverages.

Fruits from Arid Climates: Fig, Date, and Others

Cactus Pear "Cactus pear" is the modern marketing term for "prickly pear" (Spanish *tuna*), the fruit of the American cactus *Opuntia ficus-indica.* The species name comes from the early European idea that the dried fruit was an "Indian fig." The cactus arrived in the Old World in the 16th century and spread like a weed in the southern Mediterranean and Middle East. While both stem pads and fruits are eaten in the Americas, Europeans concentrated on the fruits, which ripen in the summer and fall and have a thick skin, green to red or purple, and many hard seeds embedded in a reddish, sometimes magenta flesh. The main pigment is not an anthocyanin but a beet-like betain (p. 268). The aroma is mild, reminiscent of melons thanks to similar alcohols and aldehydes. Like the pineapple and kiwi, cactus pears contain a protein-digesting enzyme that can affect gelatin gels unless it's inactivated by cooking. The pulp is removed and generally eaten fresh as juice or in salsas, or boiled down to a syrup or further to a pasty consistency. The paste is made into candies and cakes with flour and nuts.

Dates Dates are the sweet, easily dried fruits of a desert palm, *Phoenix dactylifera,* that can tolerate some cold and thrives as long as it has a source of water. Their original home was Middle Eastern and African oases, where they were being cultivated with artificial irrigation and pollination more than 5,000 years ago; they're now also grown in Asia and California. Though we usually see only two or three dried versions, there are thousands of different date varieties that differ in size, shape, color, flavor, and ripening schedule.

Growers and aficionados distinguish four stages in date development: green and immature; mature but unripe, when they're yellow or red and hard, crunchy, and astringent; ripe (Arabic *rhutab*), when they're soft, golden brown, and delicate; and finally dried, when they're brown and wrinkled and powerfully sweet. Drying is usually done on the tree. Dates are moist and succulent when fresh, from 50 to 90% water, and chewy and concentrated when dry, with less than 20% moisture. Dry dates are 60 to 80% sugar, together with some texture-providing pectins and other cell-wall materials, and a few percent fatty materials, including the surface wax. They're ground into a coarse powder to make "date sugar."

The drying process causes dates to develop a brown color and browned flavor thanks both to the action of browning enzymes on phenolic materials, and to the browning reactions between concentrated sugars and amino acids. Some varieties rich in phenolic materials, notably the Deglet Noor, can develop an increased astringency and red coloration when heated. Phenolic and other compounds give dates notable antioxidant and antimutagenic activities.

Fig Figs are the fruits of *Ficus carica,* a tree native to the Mediterranean and Middle East, and a relative of the mulberry. Because like the date they readily dry in the sun to a long-keeping, concentrated source of nourishment, they have been an important human food for many thousands of years. The fig is the fruit mentioned most often in the Bible, and was said to grow in the Garden of Eden. Spanish explorers brought it to the Americas via Mexico, and it now grows in many dry subtropical regions. There are many varieties, some green-skinned and some purple, some with bright red interiors. Ripe fresh figs are 80% water, very fragile and perishable. The vast bulk of the world crop is preserved by drying, a process that normally begins on the tree and then concludes on the orchard floor or in mechanical dryers.

The fig is unusual in being more flower than fruit. The main body is a fleshy flower base folded in on itself, with an open pore opposite the stem, and inner female florets that develop into small, individual dry fruits that crunch like "seeds." The florets are pollinated by tiny wasps that enter through the pore. Many fig varieties will set fruit

Food Words: *Date, Pomegranate*

Date comes from the Greek word for "finger," *daktulos,* which the elongated fruits resemble. *Pomegranate* comes from medieval French, and is a combination of Latin roots meaning "apple" and "grainy" or "seedy."

without pollination and produce "seeds" with no embryo inside, but fig experts say that fertilization and seed development seem to generate different flavors. (Wasps carry microbes into the fig interior, so fertilized fruits also suffer greater spoilage.) Smyrna figs and their descendants ("Calimyrna" is the California version) will not set fruit unless fertilized. They must be grown alongside a separate and inedible crop of "caprifigs," from which the wasps obtain fig pollen and lay their eggs.

The walls of the fig fruit contain latex vessels that carry a protein-digesting enzyme, ficin, and tannin cells that contribute astringency. Figs are remarkable for containing very large amounts of phenolic compounds, some of them antioxidants, and large amounts of calcium for a fruit. When ripe, figs have a unique aroma that comes mainly from spicy phenolic compounds and a flowery terpene (linalool).

Jujube Jujubes, also known as Chinese dates, are the fruits of *Ziziphus jujuba,* a tree native to central Asia. They do bear some resemblance to dates, as do the version known in India as the ber (*Z. mauritania*). Both trees tolerate heat and drought and are now grown in arid regions throughout the world. Jujubes are small, somewhat dry and spongy, more sweet than tart. They're an excellent source of vitamin C, containing more than double the amount in an equal weight of oranges. They're eaten fresh, dried, pickled, in rice-based cakes, and fermented into alcoholic drinks.

Pomegranate Pomegranates are fruits of the shrubby tree *Punica granatum,* a native of the arid and semiarid regions of the Mediterranean and western Asia; the finest varieties are said to grow in Iran. With their dull, dry rind surrounding two layered chambers of translucent, ruby-like fruitlets (there are also pale and yellow varieties), they figured very early in mythology and art. Pomegranate-shaped goblets have been found in prehistoric Troy, and in Greek myth it was a pomegranate that tempted Persephone and kept her in the underworld. Pomegranates are very sweet, fairly tart, and often astringent thanks to their strongly pigmented juice well-stuffed with anthocyanins and related phenolic antioxidants. Juice manufactured by crushing whole fruits is much more tannic than the fruitlets themselves; the rind is so rich in tannins that it was once used for tanning leather! Because each fruitlet contains a prominent seed, pomegranates are usually processed into juice, which then can be used as is or cooked down to make a syrup or "molasses," or fermented into a wine. True grenadine is pomegranate juice mixed with a hot sugar syrup. Today most commercial grenadines are synthetic. In northern India, pomegranate fruitlets are dried and ground for use as an acidifying powder.

The Citrus Family: Orange, Lemon, Grapefruit, and Relatives

Citrus fruits are among the most important of all tree fruits. From their birthplace in

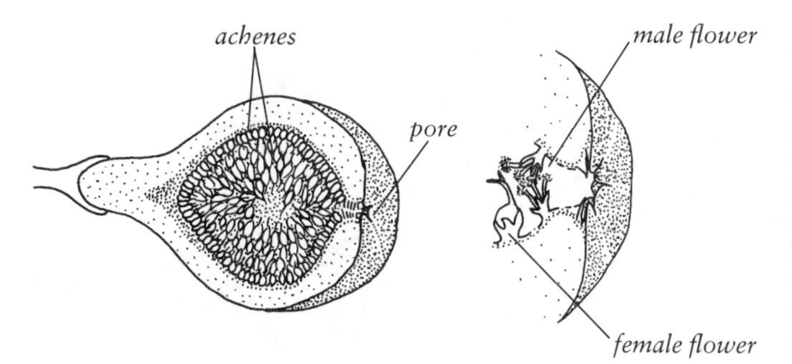

The fig. It contains small flowers within the fleshy "fruit," which is actually the swollen flower base. The fig is an inverted version of the strawberry: it surrounds rather than underlies the true tiny fruits, or achenes.

southern China, northern India, and Southeast Asia they spread throughout the subtropics and mild temperate zones of the world. Ancient trade took the citron to western Asia and the Middle East before 500 BCE, and medieval crusaders brought sour oranges back with them to Europe; Genoese and Portuguese traders introduced sweet oranges around 1500, and Spanish explorers carried them to the Americas. Today, Brazil and the United States produce most of the world's oranges. Barely a century ago, oranges were special holiday treats; now much of the Western world starts its day with orange juice.

Why are citrus fruits so popular? They offer an unusual set of virtues. Above all, their peels have distinctive and strong aromas, and these may have been their original attraction, well before human selection developed varieties with sweet juices. The improved varieties do have a refreshing, tart to sweet-tart juice that can be extracted with little pulp. The peel is rich in gel-making pectins. And citrus fruits are also fairly robust. They're nonclimacteric, so they retain their quality for some time after harvest, and the meaty peel offers good protection against physical damage and attack by spoilage microbes.

Citrus Anatomy Each segment of a citrus fruit is a compartment of the ovary, and is stuffed with small, elongated bags called *vesicles,* each of which contains many individual microscopic juice cells that fill with water and dissolved substances as the fruit develops. Surrounding the segments is a thick, white, spongy layer called the *albedo,* usually rich in both bitter substances and in pectin. And riding atop the albedo is the skin, a thin, pigmented layer with tiny spherical glands that create and store volatile oils. Flexing a piece of citrus peel will burst the oil glands and send a visible, aromatic—and flammable!—spray into the air.

Citrus Color and Flavor Citrus fruits owe their yellow and orange colors (*orange* comes ultimately from the Sanskrit word for the fruit) to a complex mixture of carotenoids, only a small portion of which has vitamin A activity. The fruit peels start out green, and in the tropics often stay that way even when the fruit ripens. In other regions, cold temperatures trigger destruction of chlorophyll in the peel, and the carotenoids become visible. Fruits of commerce are often picked green and treated with ethylene to improve their color, and coated with an edible wax to slow moisture loss. Pink and red grapefruit are colored by lycopene, and red sweet oranges by a mixture of lycopene and beta-carotene and by cryptoxanthin. The purple-red of blood oranges comes from anthocyanins.

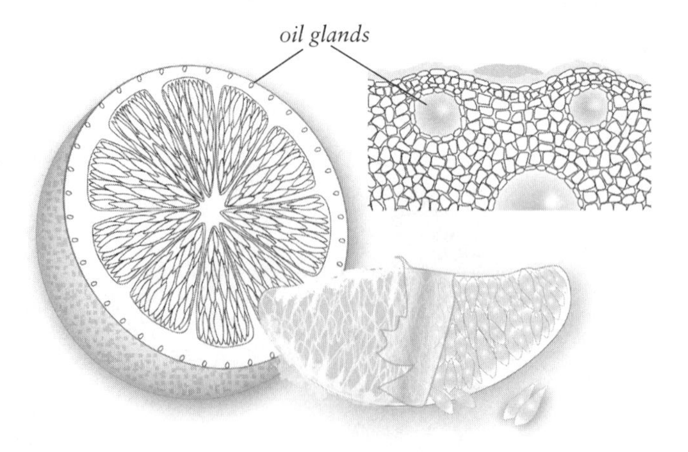

Citrus anatomy. The protective outer rind includes aromatic oil glands embedded in a bitter white pith, the albedo. Each segment, or carpel, *of a citrus fruit consists of many delicate juice sacs contained within a strong membrane.*

The taste of citrus fruits is created by a handful of substances, including citric acid (so named because it is typical of the family), sugars, and certain bitter phenolic compounds, which are usually concentrated in the albedo and peel. Citrus fruits are surprisingly rich in the savory amino acid glutamate, sometimes rivaling the tomato (oranges reach 70 milligrams per 100 grams, grapefruits 250). They contain little starch and therefore don't sweeten much after picking. Usually the blossom end of the fruit contains both more acid and more sugars, and so has a more intense taste than the stem end. Neighboring segments can also vary significantly in taste.

Citrus aroma is produced both by the oil glands in the skin and oil droplets in the juice vesicles—and these two sources are usually quite distinct. Generally the vesicle oils contain more fruity esters, and peel oil more green aldehydes and citrusy/spicy terpenes (p. 274). A few aroma compounds are shared by most citrus fruit, including generically citrusy limonene and small amounts of eggy hydrogen sulfide. In freshly made juice, the sac oil droplets gradually aggregate with the pulpy materials, and this aggregation reduces the aroma available to the taster, especially if some of the pulp is strained off.

Citrus Peel The intensely flavored citrus peel has long been used to flavor dishes (for example, dried orange peel in Sichuan cooking), and as a preparation in itself in the form of candied rind. The outer epidermis contains the aromatic oil glands, while the underlying white, spongy, pectin-rich albedo usually contains protective bitter phenolic substances. Both the oil with its terpenes and the antioxidant phenolics are valuable phytochemicals (pp. 256, 257). The bitters are water-soluble, while the oils are not. Cooks can therefore leach the peel repeatedly with hot (rapid) or cold (slow) water to remove the bitter compounds, then gently cook the peel if still necessary to soften the albedo, and finally infuse it with a concentrated sugar syrup. Through all this processing the water-insoluble oils stay largely in the rind. *Marmalade,* a sugar preserve that includes citrus peel, was originally a Portuguese fruit paste made with quince, but by the 18th century the high-pectin, readily gelled sour orange had begun to replace the quince. Marmalade made with sweet oranges doesn't gel as readily and lacks the characteristic flavor, including a bitterness that helps balance the sugar.

As is true for most fruits, the skin of citrus fruits is more easily removed from the underlying tissue by briefly immersing the fruit in simmering water. Thick citrus rinds require an immersion of several minutes. Heat softens the cell-wall cement that attaches rind to fruit, and may also encourage some enzymes to begin to dissolve the cement.

Kinds of Citrus Fruits Trees in the genus *Citrus* are wonderfully variable and prone to form hybrids with each other, which makes it a challenge for scientists to work out family relationships. Currently it's thought that the common domesticated citrus fruits all derive from just three parents: the citron *Citrus medica,* the mandarin orange *Citrus reticulata,* and the pummelo *Citrus grandis.* At least one offspring is relatively young: the grapefruit apparently originated in the West Indies in the 18th century as a cross between the pummelo and the sweet orange.

Citron Perhaps the first citrus fruit to reach the Middle East around 700 BCE and the Mediterranean around 300 BCE, citrons are native to the Himalayan foothills. They gave their name to the genus, and their name came in turn from their resemblance to the cone of a Mediterranean evergreen cedar (Greek *kedros*). The several varieties have little juice, but an intensely aromatic rind that can perfume a room—citrons are used in both Asian and Jewish religious ceremonies—and that has long been candied (p. 295). In China's Sichuan province, the rind is made into a hot pickle.

Flavor Notes in Some Citrus Fruits

The chemicals listed in the first five flavor headings are *terpenes,* which are especially characteristic of citrus fruits and some herbs and spices (p. 390).

Fruit	Citrus (limonene)	Pine (pinene)	Herbaceous (terpinene)	Lemony (neral/ geranial)	Flowery (linalool etc.)	Waxy, peel (decanal, octenol)	Musky (sulfur compounds)	Thyme (thymol)	Spicy (other terpenoids)	Others
Orange, sweet	+	+		+	+	+	+		+	cooked marmalade (valencene, sinensal)
Orange, sour	+				+	+				
Orange, blood	+	+							+	(valencene)
Mandarin (tangerine)	+	+	+	+	+			+		
Lemon	+	+	+							
Lemon, Meyer	+	+						+		
Lime	+	+	+						+	
Citron	+		+	+						
Grapefruit	+			+	+	+	+			(sinensal)
Yuzu	+	+	+		+		+		+	
Makrut lime leaf				+						
Bergamot	+	+	+		+				+	

Mandarin, or Tangerine Mandarin oranges were cultivated in ancient India and China at least 3,000 years ago. The well-known Japanese type, the satsuma, appeared by the 16th century, and Mediterranean types ("tangerines," from the Moroccan city of Tangier) in the 19th century. Mandarins tend to be relatively small and flat, with a reddish, easily peeled rind and a distinctive, rich aroma that has notes of thyme and Concord grape (thymol, methyl anthranilate). They're the most cold-hardy of citrus trees, yet the fruits are fairly fragile. Satsumas are seedless and commonly processed into canned segments.

Pummelo Pummelos require the warmest growing conditions of the common citrus fruits and have been slow to spread from their home in tropical Asia, where they were cultivated in ancient times. They're large, 10 in/25 cm or more in diameter, with a relatively thick albedo layer; large, easily separated juice vesicles that burst in the mouth; thick and tough segment membranes; and an absence of the bitterness that flavors its offspring the grapefruit. Some varieties have pink-red vesicles.

Orange Nearly three-fourths of all citrus fruit produced in the world are oranges, whose juiciness and moderate size, sweetness, and acidity make them especially versatile. They're probably an ancient hybrid between the mandarin and the pummelo, and have in turn been developed into several very different kinds of fruit.

Navel oranges probably originated in China, but became a major commodity worldwide when a Brazilian variety arrived in the United States in 1870. The navel-like appearance of their blossom end is caused by the development of a small secondary set of segments. Navels are the ideal orange for eating fresh because they're seedless and easily peeled. However, the trees are finicky to grow, and their juice contains fewer fruity esters than the best juice varieties. And juice made from navel oranges will become noticeably bitter after about 30

Citrus Family Relationships

The Parent Species	
Citron	*Citrus medica*
Mandarin, tangerine	*Citrus reticulata*
Pummelo	*Citrus grandis*
Their Offspring	
Sour orange	*Citrus aurantium*
Sweet orange	*Citrus sinensis,* pummelo x mandarin(?)
Grapefruit	*Citrus paradisi,* pummelo x sweet orange
Sour lime	*Citrus aurantifolia*
Persian/Tahiti lime	*Citrus latifolia,* sour lime x citron(?)
Lemon	*Citrus limon,* citron x sour lime x pummelo(?)
Meyer lemon	*Citrus limon,* lemon x mandarin or sweet orange?
Modern Hybrids	
Tangelo	*Citrus* x *tangelo,* tangerine x grapefruit
Tangor	*Citrus* x *nobilis,* tangerine x sweet orange

minutes. This happens because when the juice cells are broken and their contents mixed, acids and enzymes convert a tasteless precursor molecule into an intensely bitter terpene compound called limonin.

Common or *juice* oranges have a smooth blossom end, generally contain seeds, and have a more adherent skin than navel oranges. Commercial orange juice is made from juice varieties with little tendency to develop limonin bitterness. The mild flavor of the juice is usually augmented by the addition of peel oils.

Blood oranges have been grown in the southern Mediterranean at least since the 18th century, and may have originated there or in China. They're now the major type of orange grown in Italy. Blood oranges owe the deep maroon color of their juice to anthocyanin pigments, which develop only when night temperatures are low, in the Mediterranean autumn and winter. The pigments tend to accumulate at the blossom end and in vesicles immediately next to the segment walls, and continue to accumulate after harvest when the fruits are held in cold storage. The pigments and their phenolic precursors give blood oranges a higher antioxidant value than other oranges. The unique flavor of blood oranges combines citrus notes with a distinct raspberry-like aroma.

Acidless oranges are grown in small numbers in north Africa, Europe, and South America, and have about a tenth the acidity—and less orange aroma—of common and navel oranges.

Sour oranges come from a different species than the kinds described above, and are both sour and bitter (thanks not to limonin but a related compound, neohesperidin), with an intense and distinctive peel aroma. They arrived in Spain and Portugal in the 12th century, and soon displaced quince as the main ingredient in marmalade. Sour-orange flowers are used to make orange flower water.

Grapefruit The grapefruit originated as a hybrid of the sweet orange and pummelo in the Caribbean in the 18th century, and is still mainly grown in the Americas. The red types owe their color to lycopene, and first appeared as chance mutations in Florida and in Texas early in the 20th century (the more recent and popular Star Ruby and Rio Red varieties were created by intentionally inducing mutations with radiation). Unlike the anthocyanin coloration of blood oranges, grapefruit lycopene requires consistent high growing temperatures to develop well, appears evenly through all the juice vesicles, and is stable to heat. The characteristic moderate bitterness is caused by a phenolic substance called naringin, whose concentration declines as the fruit ripens. Like navel oranges, grapefruits also contain a precursor of limonin, and its juice becomes bitter on standing. Some grapefruit phenolic compounds turn out to interfere with our metabolism of certain drugs, cause the drugs to persist longer in the body, and thus cause the equivalent of an overdose, so medicine labels sometimes warn against consuming grapefruit or its juice along with the medicine. (These same phenolics are now being developed into activity-boosting drug ingredients.) Grapefruits have an especially complex aroma, which includes meaty and musky sulfur compounds.

Lime Limes are the most acidic of the citrus fruits, as much as 8% of their weight coming from citric acid. The small, seedy Mexican or Key lime, *C. aurantifolia,* is the standard sour citrus fruit in the tropics, where lemons don't grow well. In western Asia it's sun-dried whole, then ground and used as an aromatic, somewhat musty acidifier. The larger, seedless, more cold-tolerant Persian or Tahiti or Bearss lime, *C. latifolia,* may be a hybrid between the true lime and citron, and is more common in the United States and Europe. Despite the general impression that limes are characteristically "lime-green," both turn pale yellow when fully ripe. They owe their distinctive limeness to pine, floral, and spicy aroma notes (from terpenes).

Lemon Lemons may have originated as a two-step hybrid, the first (citron crossed with lime) arising in the area of northwest India and Pakistan, the second ([citron crossed with lime] crossed with pummelo) in the Middle East. Lemons arrived in the Mediterranean around 100 CE, were planted in orchards in Moorish Spain by 400, and are now mainly cultivated in subtropical regions. They're valued for their acidity, often 5% of the juice, and their fresh, bright aroma, which is the base for many popular fresh and bottled drinks. There are many varieties of true lemon, and also a couple of further hybrids. The large, coarse Ponderosa variety is probably a lemon-citron cross, and the Meyer lemon, a thin-skinned, less acid version brought to California in the early 20th century, is probably a cross between the lemon and either orange or mandarin, with a distinctive flavor due in part to a thyme note (from thymol). Lemons are generally "cured" for better shelf life; they're picked green and held in controlled conditions for several weeks, where their skin yellows, thins, and develops a waxy surface, and the juice vesicles enlarge.

The preserved lemons of northern Africa have recently become more widely appreciated as a condiment. They're made by cutting and salting lemons and letting them ferment for some weeks. The growth of bacteria and yeasts softens the rind and changes their aroma from bright and sharp to rich and rounded. Short versions of the process—for example, freezing and thawing the lemons to speed salt penetration, then salting for a few hours or days—will bring about some chemical changes as the oil glands are disrupted and their contents mix with other substances, but not the full flavor development of fermentation.

Other Citrus Fruits Lesser-known citrus fruits that are worth knowing about include the following:

- Bergamot, *C. bergamia,* possibly a cross between the sour orange and sweet lime (*C. limettoides*), is grown mainly in Italy for its floral-scented rind oil. It was one of the components of the original eau de cologne developed in 17th-century Germany, and is mainly used in perfumes, tobaccos, and Earl Grey tea.
- Kumquats, species of the genus *Fortunella,* are bite-size fruits that are eaten whole, thin rind and all. They are generally tart but not bitter. The calamondin or calamansi, also a diminutive citrus, is probably derived in part from the kumquat.
- The finger lime, *Microcitrus australasica,* is a small, elongated citrus relative native to Australia. Its fruit has robust, decorative round juice vesicles that can be pale or pink-red, and a distinctive aroma.
- The makrut or kaffir lime, *Citrus hystrix,* is common throughout Southeast Asia. Its rough green peel has a lime-like aroma with general citrus and pine notes (from limonene, pinene), and is used to flavor various prepared dishes, as are its intensely lemon-scented leaves (p. 410).

Food Words: *Orange, Lemon, Lime*

The orange fruit got its modern name from its ancient Sanskrit name, *naranga,* and the fruit then lent its name to the vivid color that characterizes it. *Lemon* and *lime* both come via Arabic from a Persian word, reflecting the route by which these Asian fruits were brought to the West.

- The tangelo and tangor are modern hybrids between the tangerine and grapefruit, and tangerine and orange, with hybrid flavors as well, and mostly eaten as fresh fruit.
- The yuzu, *Citrus junos,* possibly a mandarin hybrid, came from China but was developed in Japan beginning around a thousand years ago. The rind of the small yellow-orange fruit is used to flavor various dishes, and to make vinegar, teas, and preserves. It has a complex flavor that includes musky sulfur compounds, and clove and oregano notes (from the phenolics eugenol and carvacrol).

Some Common Tropical Fruits

A century ago, only a handful of tropical fruits were available to Europe and North America, and they were luxuries. Nowadays the banana is a common breakfast food and new fruits appear in the market every year. Here are the most familiar.

Banana and Plantain Thanks to their productivity and starchy nutritiousness, bananas and plantains top the roster of world fruit production and trade. The worldwide annual per capita consumption is almost 30 lb/14 kg, and in regions where they're a staple food, individuals consume several hundred kilograms per year. Bananas and plantains are the seedless berries of a tree-sized herb related to the grasses, *Musa sapientum,* which originated in the tropics of Southeast Asia. A banana plant produces a single flower structure with from 1 to 20 "hands" or fruit clusters, as many as 300 "fingers" of individual fruit, each fruit weighing from a couple of ounces to 2 lb/50–900 gm. The characteristic curve of long fruits develops because the fruit tip grows upward, against the downward pull of gravity. Bananas and plantains are climacteric fruits, store their energy as starch, and convert some or most of that starch to sugar during ripening. In the dramatic case of the banana, a starch-to-sugar ratio of 25 to 1 in the mature but unripe fruit becomes 1 to 20 in the ripe fruit.

The terms *banana* and *plantain* are used for two broad and overlapping categories for the many varieties of these fruits. Bananas are generally sweet dessert varieties, and plantains are starchy cooking varieties. Bananas are very sweet when ripe, their nearly 20% sugar content exceeded only by dates and jujubes, while ripe plantains may be only 6% sugar and 25% starch. Both are picked green and ripened in storage, and are very perishable once ripe thanks to their active metabolism. Bananas develop a meltingly smooth consistency, and a distinctive aroma due primarily to amyl acetate and other esters, and green, floral, and clove (eugenol) notes. Banana acidity also increases during ripening, sometimes twofold, so the flavor becomes more full in several dimensions. Ripe plantains generally retain a dry, starchy texture and can be treated as potatoes are, fried or mashed or cooked in chunks.

The pulp of these fruits is colored by carotenoids—plantain varieties often more noticeably—and tannins often make the unripe pulp astringent. Bananas and plantains are very prone to brown-black discoloration thanks to browning enzymes and phenolic substances in defensive latex-bearing vessels associated with the vascular system. These substances slowly decline by about half during ripening, so once a fruit is ripe, it can be refrigerated with relatively little discoloration of the flesh (the peel will still turn black).

Though a small handful of banana varieties (Grand Nain, Gros Michel, Cavendish) dominate in international trade, there are many interesting Latin American and Asian varieties to be found in ethnic markets, usually shorter, with tinted skin and flesh and intriguingly different flavors.

Cherimoya and Atemoya The cherimoya and atemoya are tree fruits of species in the genus *Annona,* a native of tropical and subtropical South America (the soursop or guanabana and the custard apple belong to the same genus). They are medium-sized masses of fused ovaries with their seeds, enclosed in inedible green or tan skins. Like pears, they can contain gritty stone cells. Cherimoyas and atemoyas are climacteric fruit that store starch and convert it to sugar during ripening; the result is a soft, sweet, low-acid flesh with about double the calories of common temperate fruits. They owe a vaguely banana-like note to esters, and flowery and citrus notes to a number of terpenes. They must be kept warmer than 55°F/13°C until they ripen, after which they can be refrigerated for a few days. Cherimoyas and atemoyas are eaten with a spoon both chilled and frozen, and are pulped and made into drinks and sorbets.

Durian Durian is the large, thorn-covered fruit of a tree, *Durio zibethinus,* that's a native of Southeast Asia and cultivated mainly in Thailand, Vietnam, and Malaysia. Durian is notorious for its very unfruit-like aroma, a powerful smell that can be reminiscent of onions, cheese, and meat at various stages of decay! At the same time many people prize it for its delicious flavor and creamy, custard-like texture. The armored mass of fused ovaries, each containing a seed, can weigh more than 13 lb/6 kg, and apparently evolved to appeal to elephants, tigers, pigs, and other large jungle creatures, which are drawn to it by its battery of powerful sulfur compounds, including some found in onions, garlic, overripe cheese, skunk spray, and rotten eggs. These compounds are mainly found in the outer rind, while the fleshy segments surrounding the seeds are more conventionally fruity and savory, with an especially high content of sugars and other dissolved solids (36%). Durian is eaten as is, made into drinks, candies, and cakes, and incorporated into rice and vegetable dishes. It's also fermented to make it even stronger-tasting (Malaysian *tempoyak*).

Guava and Feijoa Guavas are the large berries of a bush or small tree in the genus *Psidium,* a native of the tropical Americas, and a member of the myrtle family, which includes the clove, cinnamon, nutmeg, and allspice trees. True to their family background, they have a strong, spicy/musky aroma (from cinnamate esters and some sulfur compounds). Their flesh contains hundreds of small seeds and many gritty stone cells, so guavas are most often used to make purees, juices, syrups, and preserves. The Spanish colonizers exploited their high pectin content to make a New World version of quince paste. Guavas are remarkable for a vitamin C content that can reach 1 gram per 100 grams, with much of it concentrated near and in the thin, fragile peel.

The so-called pineapple guava, or feijoa, comes from the shrub *Feijoa sellowiana,* also a South American member of the myrtle family. It shares a similar size and structure and some flavor elements with guava, but its strong aroma is distinctive and less complex, dominated by a particular group of esters (from benzoic acid). It too is usually pulped and strained for use in liquid preparations.

Breadfruit and Jackfruit Breadfruit and jackfruit are fruits of two species of the Asian genus *Artocarpus,* a relative of the mulberry and fig, and resemble each other in structure. They're very large assemblies of fused ovaries and their seeds; breadfruits may reach 9 lb/4 kg, and jackfruits 10 times that weight. Jackfruit, a native of India, has a conventional composition for a fruit—mostly water, with 8% sugar and 4% starch—and develops a strong, complex aroma with musky, berry, pineapple, and caramel notes. It's eaten raw and in ice creams, as well as dried, preserved, and pickled. Breadfruit, whose origins in the Pacific islands remain unclear, gets its name from its very high starch content, as much as 65% by weight (with 18% sugar and just 10% water) when mature but unripe, and when cooked into a dry, absorbent mass. It's a staple food in the South Pacific and in the Caribbean, where it was taken by Captain Bligh of the notorious *Bounty* mutiny. It may be boiled, roasted, fried, or fermented into a sour paste, then dried and ground into flour. Ripe breadfruit is sweet and soft, even semiliquid, and made into desserts.

Lychee Lychees are subtropical Asian tree fruits (from *Litchi chinensis*) the size of a small plum, with a dry, loose skin and a large seed. The edible portion is its fleshy seed covering, or aril, which is pale white, sweet, and distinctively floral due to the presence of a number of terpenes (rose oxide, linalool, geraniol; Gewürztraminer grapes and wine share many of the same notes). Lychees with small, undeveloped seeds are called "chicken-tongue" fruit and are prized because there's more flesh than seed. Lychees don't improve in flavor once taken from the tree. A common problem with lychees is a brown discoloration of the flesh due to drying out or chilling injury. They're best held at cool room temperatures in a loose plastic bag. When cooked, fresh lychees sometimes develop a pink undertone as phenolic aggregates are broken apart and converted into anthocyanin pigments (p. 281). They're eaten fresh, canned in syrup, made into drinks, sauces, and preserves, cooked briefly and served with meats and seafood, and frozen into sorbets and ice creams. "Lychee nuts" are the dried fruits, not the seeds.

Rambutans, longans, and pulasans are all arils of Asian fruits in the same family as the lychee (the Sapindaceae), and have similar qualities.

Mango Mangoes are the succulent, aromatic fruits of an Asian tree, *Mangifera indica,* a distant relative of the pistachio and cashew trees, that has been cultivated for many thousands of years. There are hundreds of varieties with very different qualities, including flavor and degrees of fibrousness and astringency. The mango skin contains an irritant and allergenic phenolic compound similar to that in the cashew. Their deep orange color comes from carotenoid pigments, mainly beta-carotene. Mangoes are climacteric fruits that accumulate starch, so they can be picked green and will sweeten and soften as they ripen, from the seed outwards. Their flavor is especially complex, and may be dominated by the compounds that characterize peaches and coconuts (lactones), generically fruity esters, medicinal or even turpentiny terpenes, and caramel notes. Green mangoes are very tart, and are made into pickles as well as dried and ground to make an acidifying powder (Hindi *amchur*). Mango pickles were so admired in 18th-century England that the fruit lent its name to the preparation and to other suitable materials: hence "mango peppers."

Mangosteen The mangosteen is the medium-sized, leathery-skinned fruit of an Asian tree, *Garcinia mangostana.* Its white flesh consists of the arils around several seeds, and is moist and has a pleasing sweet-tart balance, with a delicate, fruity and flowery aroma, something like a lychee. It's usually eaten fresh or in preserves, and is also canned.

Food Words: Tropical Fruits

Many of our words for tropical fruits come from the peoples among whom Western travelers first encountered them. *Banana* comes from several West African languages, *mango* from south Indian Tamil, *papaya* from Carib, *durian* from Malay (a word meaning "thorn").

Papaya Papayas are species of the genus *Carica,* a native of the American tropics that looks like a small tree but is actually a large herbaceous plant. The common papaya, *C. papaya,* consists of a thickened ovary wall, orange to orange-red with carotenoid pigments, and a few dark seeds in a large central cavity. It is a climacteric fruit that doesn't store any starch. Ripening begins at the center and progresses outward, and causes a manyfold increase in carotenoid pigments and aroma molecules, as well as a marked softening. Softening causes the apparent sweetness to increase even though the actual sugar content doesn't change (the sugars are more readily released from the softened tissue). A ripe papaya is a low-acid fruit with a delicate, flowery aroma thanks to terpenes, and a touch of cabbage-like pungency due to the surprising presence of isothiocyanates (p. 321). These compounds are especially concentrated in the seeds, which can be dried and used as a mildly mustardy seasoning.

Unripe, crisp green papaya is made into salads and pickles. The green fruit contains vessels of milky latex rich in the protein-digesting enzyme papain, which is found in some meat tenderizers. Papain levels drop during ripening, but can still cause texture and taste problems like those caused by the pineapple enzyme bromelain (p. 384).

Two other papaya species can be found in markets. The large cool-climate mountain papaya, *C. pubescens,* is less sweet than the lowland papaya, but richer in papain and in carotenoid pigments, often including lycopene, which gives its flesh a reddish tinge. The babaco, *C. pentagona,* is apparently a natural hybrid, and has cream-colored, tart, seedless flesh.

Passion Fruit, Granadilla Passion fruits and the granadilla come from about a dozen species of vines of the genera *Passiflora* and *Tacsonia,* natives of tropical lowlands and subtropical highlands in South America. They consist of a brittle (*Passiflora*) or soft (*Tacsonia*) outer husk, with a mass of hard seeds embedded in pulpy seed coverings, or arils. The arils are the only edible portion, and make up barely a third of the fruit weight. Though the pulp is sparse, its flavor is concentrated and actually benefits from dilution. Passion fruits are unusual for their relatively high acid content, mainly citric—more than 2% of the pulp weight in purple-skinned types, and double that in most yellow ones—and their strong, penetrating aroma, which appears

Meat-Eating Fruits: The Puzzle of Plant Proteases

At first glance, it seems strange that fruits should contain meat- and gelatin-digesting enzymes, the molecules that prevent cooks from making jellies with those raw fruits. Of course there are a few carnivorous plants that trap insects and other small creatures in digestive juices. And in some plant parts, similar enzymes probably provide protection against attack by insects and larger animals, whose insides they can irritate or injure. But fruits are meant to be eaten by animals so that the animals will disperse the plant's seeds. So why fill them with proteases?

In the cases of papaya, pineapple, melon, fig, and kiwi, the enzymes may limit the number of fruits that any one animal eats: too many and the animal's digestive system suffers. Another intriguing suggestion is that in moderation, the enzymes actually benefit seed-dispersing animals by ridding them of intestinal parasites. Some tropical peoples use fig and papaya latex for this purpose, and it's known that the enzymes do indeed dissolve live tapeworms.

Flavor Elements in Some Common Fruits

The sugar and acid contents of fruits vary, and depend largely on ripeness. The figures below generally represent commercial reality rather than the ideal, and are meant to provide a rough way to compare the qualities of different fruits. Generally, the sweeter a fruit is, the tastier it is; but even a sweet fruit will seem one-dimensional without some counterbalancing acidity. The aroma notes listed represent volatile chemicals that flavor chemists have found in the fruit but that don't smell simply like the fruit; they have qualities of their own that contribute to the overall flavor. Blank entries indicate a lack of information, not a lack of interesting aroma!

Fruit	Sugar Content, % by weight	Acid Content, % by weight	Sugar/ Acid Ratio	Contributing Aroma Notes
Pome				
Apple	10	0.8	13	Many; depend on variety (p. 356)
Pear	10	0.2	50	
Stone				
Apricot	8	1.7	5	Citrus, floral, almond
Cherry	12	0.5	24	Almond, clove, floral
Peach	10	0.4	25	Creamy, almond
Plum	10	0.6	17	Almond, spicy, floral
Citrus				
Orange	10	1.2	8	Floral, musky (sulfurous), spicy
Grapefruit	6	2	3	Musky, green, meaty, metallic
Lemon	2	5	0.4	Floral, pine
Lime	1	7	0.1	Pine, spicy, floral
Berries				
Blackberry	6	1.5	4	Spicy
Black currant	7	3	2	Spicy, musky
Blueberry	11	0.3	37	Spicy
Cranberry	4	3	1	Spicy, almond, vanilla
Gooseberry	9	1.8	5	Spicy, musky
Grape	16	0.2	80	Many; depend on variety (p. 363)
Raspberry	6	1.6	4	Floral (violet)
Red currant	4	1.8	2	
Strawberry	6	1	6	Green, caramel, pineapple; clove, grape (wild)

Fruit	Sugar Content, % by weight	Acid Content, % by weight	Sugar/ Acid Ratio	Contributing Aroma Notes
Melon				
Cantaloupe	8	0.2	40	Green, cucumber, musky
Honeydew	10	0.2	50	Green, musky
Watermelon	9	0.2	45	Green, cucumber
Tropical				
Banana	18	0.3	60	Green, floral, clove
Cherimoya	14	0.2	70	Banana, citrus, floral
Guava	7	1	7	Spicy, musky
Lychee	17	0.3	57	Floral
Mango	14	0.5	28	Coconut, peach, caramel, turpentine
Papaya	8	0.1	80	Floral
Passion fruit	8	3	3	Floral, musky
Pineapple	12	2	6	Caramel, meaty, clove, vanilla, basil, sherry
Others				
Avocado	1	0.2	5	Spicy, woody
Cactus pear	11	0.1	110	Melon
Date (semidry)	60			Caramel
Fig	15	0.4	38	Floral, spicy
Kiwi	11	3	4	Green
Persimmon	14	0.2	70	Pumpkin
Pomegranate	12	1.2	10	
Tomato	3	0.5	6	Green, musky, caramel

to be a complex mixture of fruity and flowery notes (esters, peach-like lactones, violet-like ionone), and unusual musky notes (from sulfur compounds like those in black currants and sauvignon blanc wines). Passion fruit pulp is used mainly to make beverages, ices, and sauces, with the milder purple *P. edulis* generally consumed fresh and the stronger yellow *P. edulis var. flavicarpa* processed (an early commercial application was Hawaiian punch).

Pineapple Pineapples are the large, pinecone-like fruit of *Ananas comosus,* a member of the bromeliad family (which includes bromeliad houseplants) and native to tropical but arid South America. (*Ananas* comes from a Guarani Indian word for the fruit; *pineapple* from the Spanish *piña* due to its resemblance to the similarly composite pinecone.) The plant had already spread to the Caribbean before Columbus saw it there in 1493, and modern breeding efforts began shortly thereafter in French and Dutch glasshouses.

Pineapples consist of spirals of separate seedless fruitlets, between 100 and 200 of which fuse together and become joined to a central core. During the fusing process, bacteria and yeasts become incorporated in the interior and may later cause hidden spoilage. The fruit doesn't store starch, is not a climacteric fruit, and will not sweeten or improve in flavor once picked, though it will soften. Fully ripe pineapples don't ship well, so exported pineapples are harvested early, with as little as half the sugar content that they're capable of developing, and a fraction of the aroma. Brown or black regions in the interior are caused by chilling injury during shipment or storage; translucent areas seem to be caused by growing conditions that load the fruit cell walls with sugars. The quality of pineapples from the subtropics is less reliable than that of fruit from near the equator, where seasonal and climatic variation is minimal.

Pineapple Flavor Pineapples are remarkable for the intensity of their flavor, the experience of which the 19th-century English writer Charles Lamb described as "almost too transcendent . . . a pleasure bordering on pain, from the fierceness and insanity of her relish." At their best they are both very sweet and quite tart (from citric acid), and with a rich aroma provided by a complex mixture of fruity esters, pungent sulfur compounds, essences of vanilla and clove (vanillin, eugenol), and several oxygen-containing carbon rings with caramel and sherry overtones. A given pineapple has many different flavor zones. The fruitlets near the base form first and are therefore the oldest and sweetest, and the acidity of the flesh doubles from the core to the surface. Thanks to their assertive flavor and firm, somewhat fibrous flesh, pineapples can be cut into chunks and baked, grilled, or fried. They have an affinity for the flavors of butter and caramel and work well in baked goods, as well as various raw preparations (salsas, drinks, sorbets).

Pineapple Enzymes Pineapples contain several active protein-digesting enzymes that are used in meat tenderizers, but can cause problems in other prepared dishes. (In medicine they have been exploited as a means of cleaning burns and other wounds, and they help control inflammatory diseases in animals.) Bromelain, the major enzyme, will break down gelatin, so pineapple for gelatin-based desserts must be cooked first, to inactivate the enzyme. And if incorporated into a mixture containing milk or cream, bromelain will break down the casein proteins and produce bitter-tasting protein fragments. Again, precooking the pineapple will prevent this.

Star Fruit Star fruit or carambola come from the small Southeast Asian tree *Averrhoa carambola,* a member of the wood sorrel (*Oxalis*) family. These medium-sized, yellowish fruits are notable for their starlike cross section, a decorative touch in salads and garnishes, for an aroma with notes of Concord grapes and quince, and for the presence of oxalic acid, mainly in the five ridges. When unripe and especially rich in oxalic acid, star fruit are sourly reminiscent of similarly endowed sorrel (p. 411) and are used to clean and polish metal! Star fruit are colored by carotenoid pigments, including beta-carotene. A relative, the bilimbi, is too tart to eat fresh and in the tropics is made into preserves and drinks.

CHAPTER 8

FLAVORINGS FROM PLANTS

Herbs and Spices, Tea and Coffee

Herbs and spices are ingredients that we use to add flavor to foods and drinks. Herbs are plant leaves, fresh or dried, and spices are bits of dry seed, bark, and root. We consume them in tiny amounts, and they have practically no nutritional value. Yet from earliest times, these aromatic bits have been among the most highly prized and costly of all ingredients. In the ancient world they were more than mere foods: they were thought to have medicinal and even transcendental properties. Sacrificial fires wafted the fumes of aromatics upward to please the gods, and at the same time offered earthbound humans a whiff of heaven. Spices came from the ends of the earth, from Arabia and legendary lands to the east. The growing hunger for the aromas of paradise helped drive the European exploration of the globe, the discovery of the Americas, and the biological and cultural exchange that helped shape the modern world.

Few people today think of herbs and spices as emissaries from paradise or to heaven. Yet they're more popular than ever: because herbs and spices do indeed bring other worlds to our table. They mark the foods of different cultures with distinctive flavors, and provide us a taste of Morocco at one meal and Thailand the next. They help us recapture the kind of sensory variety that our ancestors enjoyed in foods before agriculture made eating both more reliable and more monotonous. And because smell is one of the senses through which we experience our immediate surroundings, herbs and spices delight by lending our foods hints reminiscent of the forest, the meadow, the flower garden, the seacoast. They can conjure a familiar part of the natural world in a bite or sip.

This chapter surveys herbs, spices, and three other important flavorings derived from plants. Tea and coffee are such promi-

A Brief History of Spices

The story of spices is a colorful one, and has been told many times. It turned out that tropical Asia was especially rich in spice plants. To the peoples of the Mediterranean and Europe, who depended on Arab traders for both the spices and information about them, this meant that cinnamon and pepper and ginger were rare treasures from fabled lands.

The Romans knew a number of Eastern spices but in cooking used mainly pepper. A thousand years later, Arab cultural influence introduced other spices to wealthy medieval tables throughout Europe, and demand for them grew with the middle classes. Medieval sauces often call for a half-dozen spices, usually beginning with cinnamon, ginger, and grains of paradise. The Turkish control of supply routes and prices impelled Portugal and Spain to search for a new sea route to Asia; Columbus reached the Americas, the home of chillis and vanilla, in 1492, and Vasco da Gama reached India in 1498. The Portuguese and then the Spanish controlled the Spice Islands and the trade in nutmeg and cloves until around 1600, when the Dutch embarked on two centuries of brutally efficient control.

As spices were planted in other tropical countries and became cheaper and more commonly available, they slowly faded from their former prominence in European dishes, persisting mainly in sweets. But at the end of the 20th century, the consumption of herbs and spices rose sharply in the West. In the United States it tripled between 1965 and 2000 (to about 4 grams per day per person), thanks to a growing appreciation of Asian and Latin American foods, and especially the spiciness of "hot" chillis.

nent ingredients in their own right that we don't think of them as herb or spice, but that's essentially what they are: tea is a dried leaf and coffee a roasted seed, and we use them to flavor water (and infuse it with a useful drug, caffeine). And wood smoke is a flavoring created when intense heat breaks plant tissues down into some of the same aromatics found in true spices.

THE NATURE OF FLAVOR AND FLAVORINGS

Flavor Is Part Taste, Mostly Smell

The function of herbs and spices is to add flavor to our foods. Flavor is a composite quality, a combination of sensations from the taste buds in our mouth and the odor receptors in the upper reaches of our nose. And these sensations are chemical in nature: we taste tastes and smell odors when our receptors are triggered by specific chemicals in foods. There are only a handful of different tastes—sweet, sour, salty, bitter, and savory or umami (p. 342), while there are many thousands of different odors. It's odor molecules that make an apple "taste" like an apple, not like a pear or radish. If our nose is blocked by a cold or pinching fingers, it's hard to tell the difference between an apple and a pear. So most of what we experience as flavor is odor, or aroma. Herbs and spices heighten flavor by adding their characteristic aroma molecules. (The exceptions to this rule are the pungent spices and herbs, which stimulate and irritate nerves in the mouth; see p. 394.)

Odors and the Suggestiveness of Volatility The aroma chemicals of herbs and spices are *volatile*: that is, they're small and light enough to evaporate from their source and fly through the air, which allows them to rise with our breath into the nose, where we can detect them. High temperatures make volatile chemicals more volatile, so heating herbs and spices liberates more of their aroma molecules and fills the air

---

The Aromas of Holiness and Paradise

In the religions of the ancient world, spices were a means by which spiritual fulfillment could be symbolized and experienced as a form of sensuous delight.

> A garden inclosed is my sister, my spouse; a spring shut up, a fountain sealed.
> Thy plants are an orchard of pomegranates, with pleasant fruits; camphire, with spikenard, Spikenard and saffron, calamus and cinnamon, with all trees of frankincense; myrrh and aloes, with all the chief spices: A fountain of gardens, a well of living waters, and streams from Lebanon. Awake, O north wind; and come, thou south; blow upon my garden, that the spices thereof may flow out.
> —Song of Solomon, 4:12–15

> Allah will deliver them from the evil of that day and make their faces shine with joy. He will reward them for their steadfastness with robes of silk and the delights of Paradise. . . . They shall be served with silver dishes, and beakers as large as goblets; silver goblets which they themselves shall measure: and cups brim-full with ginger-flavored water from the Fount of Selsabil.
> —The Koran, "Man," 76: 11–15

with their odor. Unlike most of the objects that we sense around us, which we see or touch or hear, aromas are an invisible, intangible presence. To cultures that knew nothing of molecules and odor receptors, this ethereal, penetrating quality suggested a realm of invisible beings and powers. So herbs and spices became important in the sacrificial fires and incense of religious ceremonies; they were offerings to the gods, a way of evoking their presence and imagining their heaven. Perfumes—a word that comes from the Latin for "through fire"—have long been the source of a similarly mysterious appeal.

The Evolving World of Taste and Smell

We humans are animals, and for all animals, the sense of smell does far more than provide information about a mouthful of food. Smell detects whatever volatile molecules are in the air. It therefore tells an animal about its surroundings: the air, the ground, the plants growing in the ground, other animals moving nearby that might be enemies, mates, or a meal. This more general role explains why we're sensitive to aroma notes in foods that are reminiscent of the world: wood, stone, soil, air, animals, flowers, dry grass, the seacoast and the forest. It's also essential for animals to learn from experience, and therefore to associate particular sensations with the situations they accompany. This may be why odors are so evocative of memories and the emotions associated with them.

The Variety of Gathered Foods, the Monotony of Agriculture Our earliest human ancestors were omnivores: they ate whatever they could find worth eating on the African savanna, from meat scraps on an animal carcass to nuts, fruits, leaves, and tubers. They relied on taste and smell to judge whether a new object was edible—sweetness meant nourishing sugars, bitterness toxic alkaloids, foulness dangerous decay—and to help identify and recall the effects of objects they had encountered before. And they ate a varied diet that probably included several hundred different kinds of foods. They had a lot of flavors to keep track of.

When humans developed agriculture around 10,000 years ago, they traded their diverse but chancy diet for a more predictable and monotonous one. Now they lived largely on wheat, barley, rice, and corn, all concentrated sources of energy

Spices Haven't Always Gone with Foods

In the time of classical Greece and Rome, when spices were much used in religious ceremonies and in perfumes, not everyone thought that they also belong in foods!

> The question may be raised, why aromatics and other fragrant things, while they give a pleasant taste to wine, do not have this effect on any other article of food. In all cases, they spoil food, whether it be cooked or not.
>
> —Theophrastus, *De causis plantarum*, 3rd century BCE

> Today we need "supplements" for meat. We mix oil, wine, honey, fish paste, vinegar, with Syrian and Arabian spices, as though we were really embalming a corpse for burial.
>
> —Plutarch, *Moralia*, 2nd century CE

and protein, and all relatively bland. They had very few flavors to keep track of. But they still had their senses of taste and smell.

Flavorings Provide Stimulation and Play One distinctly human characteristic is a drive to explore and manipulate the world of natural materials around us, to change those materials to suit our needs and interests. And these needs and interests include the stimulation of our senses, the creation of sensory patterns that engage our brains. After the development of agriculture and its radically simplified diet, our ancestors found ways to give our taste buds and nose more to experience again. One way was to make use of plant parts that are especially concentrated sources of flavor. Herbs and spices made it possible not only to give bland foods more flavor, but to give them more varied flavors, to ornament foods and highlight flavor for flavor's sake.

Flavorings Are Chemical Weapons

And why are some plants' parts especially potent, intense sources of flavor? What role do the chemicals that give them their flavor play in the lives of the plants themselves?

One simple clue is their very potency. Try the experiment of chewing on an oregano leaf, or a clove, or a vanilla bean. The result is far from pleasurable! When eaten as is, most spices and herbs are acrid, irritating, numbing. And the chemicals responsible for these sensations are actually toxic. The purified essence of oregano and of thyme can be bought from chemical supply companies, and come with bright warning labels: these chemicals damage skin and lungs, so don't touch or inhale. This is precisely the primary function of these chemicals: to make the plants that produce them obnoxious and therefore resistant to attack by animals or microbes. The flavors of herbs and spices are defensive chemical weapons that are released from plant cells when the plant is chewed on. Their volatility gives them the advantage of counterattacking through the air, not just on direct contact, and of being a warning signal that can train some animals to be deterred by smell alone.

Turning Weapons into Pleasures: Just Add Food

And yet humans have come to prize these weapons that are meant to repel us. What makes herbs and spices not only nontoxic and edible but delicious is a simple principle of cooking: dilution. If we bite into an intact leaf of oregano or a peppercorn, the concentrated dose of defensive chemicals overwhelms and irritates our senses; but those same chemicals diffused throughout a dish of other foods—a few milligrams in a pound or two—stimulate without overwhelming. They add flavors that our grains and meats don't have, and make those foods more complex and appealing.

THE CHEMISTRY AND QUALITIES OF HERBS AND SPICES

Most Flavorings Resemble Oils

The flavorful material in an herb or spice is traditionally called its *essential oil*. The term reflects an important practical fact: aroma chemicals are more similar to oils and fats than they are to water, and are therefore more soluble in oil than they are in water (p. 797). This is why cooks make prepared flavor extracts by infusing herbs and spices in oil, not water. They do infuse herbs in watery vinegar and in alcohols, but both alcohol and vinegar's acetic acid are small cousins of fat molecules, and help dissolve more aromatics than plain water could.

The defensive aroma chemicals can have disruptive effects on a plant's own cells as well as on predators, so plants take care to isolate them from their inner workings. Herbs and spices stockpile their aroma chemicals in specialized oil-storage cells, in glands on the surfaces of leaves, or in channels that open up between cells. Some dry spices are as much as 15% essential oil by weight, and many are 5–10%. Fresh and dry herbs generally contain much less, around 1%, fresh herbs because their water content is much higher, and dry herbs because they lose aroma chemicals in the drying process.

The Flavor of an Herb or Spice Is Several Flavors Combined

As we've seen many times and in many foods, flavor is a composite quality. A ripe fruit may contain hundreds of different aromatic compounds; and the same goes for a roast. Though we tend to think of a particular herb or spice as having its own distinctive flavor, it too is always a composite of several different aroma compounds. Sometimes one of those compounds predominates and provides the main character—as in cloves, cinnamon, anise, thyme—but often it's the mixture that creates the character, and that makes a spice well suited to serve as a unifying bridge among several different ingredients. Coriander seed, for example, is simultaneously flowery and lemony; bay leaf combines eucalyptus, clove, pine, and flowery notes. It can be fascinating—and useful—to taste spices analytically, trying to perceive the separate components and how their flavors are built. Terms from perfumery can be helpful: there are "top notes," perceived right away, ethereal and quick to fade; there are "mid-notes," the main flavors; and there are "bottom notes," which are slow to develop and which persist. The charts on pp. 392 and 393 list the prominent aroma components in a selection of herbs and spices. There are two particular chemical families that contribute many of the aroma compounds in herbs and spices.

Flavor Families: The Terpenes

Terpene compounds are constructed from a zigzag building block of five carbon atoms, which turns out to be amazingly versatile and can be combined, twisted, and decorated into tens of thousands of different molecules. Plants usually produce a mixture of defensive terpenes. They are characteristic of the needles and bark of coniferous trees, of citrus fruits (p. 374), and of flowers, and provide pine-like, citrusy, floral, leaf-like, and "fresh" notes to the overall flavor of many herbs and spices. As a family, terpenes tend to be especially volatile and reactive. This means that they're often the first molecules to reach the nose, and provide the initial impression of these lighter, more ethereal notes. It also means that they're readily boiled off or modified by even brief cooking, which is why these fresh, light notes disappear. If desired, they can be restored to a cooked dish by adding a new dose of the herb or spice just before serving.

Flavor Families: The Phenolics

Phenolic compounds are constructed from a simple closed ring of six carbon atoms and at least one fragment of a water mole-

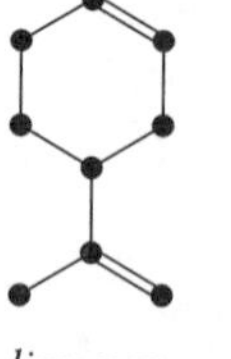

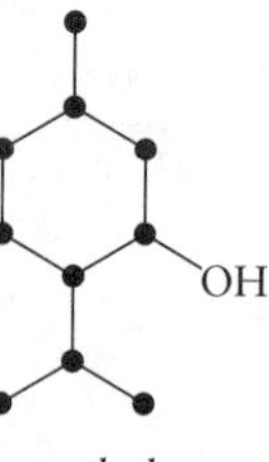

limonene (citrus) *menthol (mint)* *myrcene ("resinous")*

Examples of terpene aroma compounds. The black dots show the backbone of carbon atoms. Limonene and menthol are distinctive, while myrcene provides a background note in a number of spices and herbs.

cule (an oxygen-hydrogen combination). Single rings can then be modified by adding other atoms to one or more of the carbons, and two or more rings can be linked together to form polyphenolic compounds, including anthocyanin pigments and lignin. Unlike the terpene aromatics, which often have a generic quality to them, the phenolic aromatics are distinctive and define the flavor of such spices as cloves, cinnamon, anise, and vanilla, as well as the herbs thyme and oregano. The pungent components of chillis, black pepper, and ginger are also synthesized from a phenolic base.

Thanks to the water fragment on the carbon ring, phenolic compounds are somewhat more soluble in water than most terpenes. They tend to be more persistent in foods and in the mouth as we eat and taste.

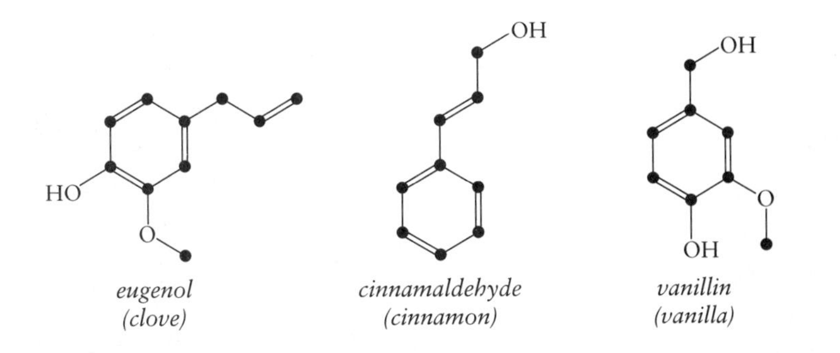

Examples of phenolic aroma compounds.

Flavor Families: Important Terpenes and Phenolics and Their Aromas

Chemical Compound	**Aroma**
Terpenes	
Pinenes	Pine needles and bark
Limonene, terpinene, citral	Citrus fruits
Geraniol	Roses
Linalool	Lily of the valley
Cineole	Eucalyptus
Menthol and menthone	Peppermint
L-carvone	Spearmint
D-carvone	Caraway
Phenolics	
Eugenol	Clove
Cinnamaldehyde	Cinnamon and cassia
Anethole	Anise
Vanillin	Vanilla
Thymol	Thyme
Carvacrol	Oregano
Estragole	Tarragon

The Flavor Components in Common Herbs

This chart and the chart opposite are a guide to experiencing flavorings from plants as the flavor mixtures they truly are. The charts identify some of the important flavor notes in individual herbs and spices, both by general sensory quality and by the names of the chemicals that contribute them. This information can help us perceive more in the flavor of a given herb or spice, and understand its affinities for other ingredients.

The lists of flavor qualities and chemicals are selective, and the groupings informal. The "light" category comprises mainly terpene compounds, the "warm" and "penetrating" categories mainly phenolic compounds. "Distinctive" compounds are those that are found almost exclusively in one herb or spice and contribute much of its character.

	LIGHT				WARM, SWEET			OTHER QUALITIES		
	Fresh	Pine	Citrus	Floral	Woody	Warm, "sweet"	Anise	Penetrating	Pungent	Distinctive
Angelica	phellandrene	pinene	limonene			angelica lactone				
Avocado leaf							estragole, anethole			
Basil	cineole			linalool		methyl eugenol	estragole	cineole, eugenol		
Bay	cineole	pinene		linalool		methyl eugenol		cineole, eugenol		
Bay, Calif.	cineole, sabinene	pinene			sabinene			cineole		
Borage										cucumber aldehyde
Celery										phthalides
Chervil							estragole			
Coriander			decenal							
Curry leaf	phellandrene	pinene		terpineol	caryophyllene					
Dill	phellandrene	pinene	limonene		myristicin					dill ether
Epazote		pinene	limonene							ascaridol
Fennel					myristicin		anethole			
Hoja santa										safrole
Hyssop	pinocamphone	pinene						camphor		
Juniper berry		pinene			sabinene	myrcene				
Lavender	lavandulyl acetate, cineole			linalool	terpinenol	ocimene		cineole		linalyl acetate
Lemongrass			citral	geraniol, linalool						
Lemon verbena			citral	linalool						
Lovage				terpineol				carvacrol		phthalides
Makrut lime			citronellal							
Marjoram	sabinene		terpinene	linalool	sabinene					
Mint, pepper-		pinene								menthol
Mint, spear-	cineole	pinene	limonene			myrcene		cineole		L-carvone, pyridines
Oregano								carvacrol		carvacrol
Parsley	phellandrene				myristicin	myrcene				menthatriene
Perilla			limonene							perillaldehyde
Rosemary	cineole	pinene		terpineol	borneol	myrcene		cineole, camphor		
Sage	cineole	pinene						cineole, camphor		thujone
Sassafras	phellandrene	pinene	limonene	linalool		myrcene				
Savory								carvacrol, thymol		
Screwpine										pyrroline
Tarragon	phellandrene	pinene	limonene			myrcene	estragole			
Thyme		pinene	cymene	linalool				thymol		thymol
Wintergreen										methyl salicylate

The Flavor Components in Common Spices

	LIGHT				WARM, SWEET			OTHER QUALITIES		
	Fresh	Pine	Citrus	Floral	Woody	Warm, “sweet”	Anise	Penetrating	Pungent	Distinctive
Ajwain		pinene	terpinene					thymol		thymol
Allspice	cineole				caryophyllene			cineole, eugenol		
Anise							anethole			anethole
Annatto		pinene	limonene		humulene	myrcene				
Asafoetida										di-, tri-, tetrasulfides
Caraway			limonene							D-carvone
Cardamom	sabinene, cineole	pinene	limonene	terpineol, linalool	sabinene	terpenyl acetate		cineole		
Cardamom, large	cineole							cineole, camphor		
Cassia						cinnamyl acetate		methoxy-cinnamate		cinnamaldehyde
Celery seed			limonene							phthalide, sedanolide
Chilli									capsaicin	
Cinnamon	cineole			linalool	caryophyllene		cinnamyl acetate	cineole, eugenol		cinnamaldehyde
Clove					caryophyllene	eugenyl acetate		eugenol		eugenol
Coriander		pinene	citral	linalool				camphor		
Cumin	phellandrene	pinene								cuminaldehyde
Dill seed	phellandrene	pinene	limonene							D-carvone
Fennel seed		pinene	limonene				anethole	fenchone		anethole
Fenugreek						sotolon				sotolon
Galangal	cineole	pinene		geranyl acetate		methyl cinnamate		cineole, camphor, eugenol		galangal acetate
Ginger	phellandrene, cineole		citral	linalool	zingiberene			cineole	gingerol, shogaol	
Grains of paradise				linalool	humulene, caryophyllene				gingerol, shogaol	
Horseradish									thiocyanates	
Licorice					paeonol					ambrettolide
Mace	sabinene	pinene			myristicin	methyl eugenol				
Mastic		pinene				myrcene				
Mustard									thiocyanates	
Nigella		pinene						carvacrol		
Nutmeg	sabinene, cineole	pinene	limonene	geraniol	myristicin	myrcene, methyl eugenol		cineole		safrole
Pepper, black	sabinene	pinene	limonene		caryophyllene				piperine	
Pepper, cubeb	sabinene			terpineol				cineole		
Pepper, long					caryophyllene				piperine	
Pepper, pink	phellandrene	pinene	limonene			carene			cardanol	
Pepper, Sichuan	phellandrene	pinene	citronellol	geraniol, linalool					sanshool	
Saffron										safranal
Sansho			citronellal	geraniol, linalool					sanshool	
Star anise			limonene	linalool			estragole, anethole			anethole
Sumac		pinene	limonene							
Turmeric	phellandrene, cineole				turmerone, curcumene			cineole		
Vanilla				linalool		vanillin		eugenol, cresol, guaiacol		vanillin
Wasabi									thiocyanates	

Flavor Families: Pungent Chemicals

There is a major exception to the rule that herbs and spices provide aroma. The two most popular spices in the world are chillis and black pepper. They and a handful of other spices—ginger, mustard, horseradish, wasabi—are especially valued for a quality often called hotness, but best called *pungency*: neither a taste nor a smell, but a general feeling of irritation that verges on pain. Pungency is caused by two general groups of chemicals. One group, the thiocyanates, are formed in mustard plants and their relatives, horseradish and wasabi, when the plant cells are damaged. Most thiocyanates are small, light, water-repelling molecules—a dozen or two atoms—that readily escape from the food into the air in our mouth, and up our nasal passages. In both the mouth and nose they stimulate nerve endings that then send a pain message to the brain. The second group of pungent chemicals, the alkylamides, are found pre-formed in a number of unrelated plants, including the chilli, black pepper, ginger, and Sichuan pepper. These molecules are larger and heavier—40 or 50 atoms—and therefore less prone to escape the food and get up our nose; they mostly affect the mouth. And their action turns out to be very specific. They bind to particular receptors on certain sensory nerves and essentially cause those nerves to become hypersensitive to ordinary sensations—and thus to register the sensation of irritation or pain. The mustard thiocyanates appear to act in a similar way in the mouth and nose.

Why Pain Can Be Pleasurable

Why should irritating spices be our favorites? Psychologist Paul Rozin has proposed a couple of different explanations. Perhaps spicy foods are the edible equivalent of riding a rollercoaster or jumping into Lake Michigan in January, an example of "constrained risk" that sets off uncomfortable warning signals in the body. But since the situations are not truly dangerous, we can ignore the normal meaning of these sensations and savor the vertigo, shock, and pain for their own sakes. The sensation of pain may also cause the brain to release natural pain-relieving body chemicals that leave a pleasant glow when the burning fades.

Stimulation and Sensitizing We may also enjoy spicy food because irritation adds a new dimension to the experience of eating. Recent research has found that, at least in the case of the pepper and chilli irritants, there's a lot more to pungency than a simple burn. These compounds induce a

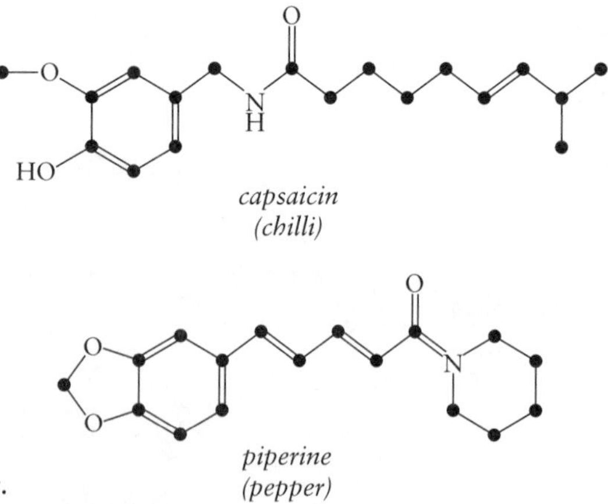

Examples of pungent flavor compounds.

temporary inflammation in the mouth, transforming it into an organ that is more "tender," more sensitive to other sensations. Those heightened sensations include touch, temperature, and the irritating aspects of various other ingredients, including salt, acids, carbonation (which becomes carbonic acid), and alcohol. It's the pepper that makes Chinese hot and sour soup, which is hot and acidic and salty, into such an intense experience. A few mouthfuls and we become conscious of simply breathing: our mouth becomes so sensitive that exhaling body-temperature air feels like a textured hot bath, inhaling room-temperature air like a refreshingly cool breeze.

Strong pungency actually diminishes our sensitivity to true tastes—to sweetness, tartness, saltiness, bitterness—and to aroma, in part because it usurps some of the attention our brains would normally pay to these other sensations. Our sensitivity to pungent flavorings also declines with exposure to it, and that desensitization lasts for 2–4 days. This is part of the reason that regular chilli-eaters can tolerate hotter dishes than people who enjoy pungent food only occasionally.

Herbs, Spices, and Health

Herbs and Spices as General Medicinals The idea that herbs and spices have medicinal value is an ancient one, and grounded in fact: plants are virtuosos of biochemical invention, and have been the original source of many important drugs (aspirin, digitalis, quinine, taxol are just a few). The health effects of plant foods in general are discussed above, p. 253. Herbs and spices, with their specialization in phenolic and terpene compounds, are notable for three broadly helpful tendencies. Phenolic compounds often have antioxidant activity; oregano, bay leaf, dill, rosemary, and turmeric are among the most effective. Antioxidants are useful both in the body, to prevent damage to DNA, cholesterol particles, and other important materials, but also in foods, to slow the deterioration in their flavor. Terpenes don't prevent oxidation, but they do help reduce the body's production of DNA-damaging molecules that can cause cancer, and help control the growth of tumors. And some phenolic compounds and terpenes are antiinflam-

The Relative Strengths of Pungent Chemicals in Black Pepper, Chillis, and Ginger

In this list, the pungency of piperine, the active ingredient in black pepper, is arbitrarily set at 1. The ingredients in ginger and grains of paradise are similar in strength, while the capsaicins in chillis are much stronger. The actual pungency of a given spice depends both on the identity of the active ingredient and its concentration in the spice.

Pungent Compound	Spice	Relative Pungency
Piperine	Black pepper	1
Gingerol	Ginger, fresh	0.8
Shogaol	Ginger, old (from gingerol)	1.5
Zingerone	Ginger, heated (from gingerol)	0.5
Paradol	Grains of paradise	1
Capsaicin	Chilli	150–300
Capsaicin variants	Chilli	85–90

matory agents; they moderate the body's overreaction to cell damage, which can otherwise contribute to the development of both heart disease and cancer.

We don't yet know whether the consumption of herbs and spices can significantly reduce the risk of any disease; but it's a real possibility.

Herbs, Spices, and Food Poisoning It has been suggested that people first began to use herbs and spices, particularly in tropical countries, because their defensive chemicals help control the microbes that cause food poisoning, and thus made food safer to eat. While some—garlic, cinnamon, cloves, oregano, thyme—are fairly effective at killing important disease microbes, most are not. And many, especially black pepper and others that take several days to dry in tropical climates, carry millions of microbes in every pinch, sometimes including *E. coli* and disease-causing species of *Salmonella, Bacillus,* and *Aspergillus.* This is why spices are often fumigated with various chemicals (ethylene or propylene oxide in the United States) or steamed. About 10% of imported spices are irradiated to eradicate microbes.

HANDLING AND STORING HERBS AND SPICES

Preserving Aroma Compounds

The aim in handling herbs and spices is to retain their characteristic aroma compounds. The volatility of these compounds means that they readily evaporate, and their reactive nature means that they are likely to be altered if they're exposed to oxygen and moisture in the air, or to reaction-causing heat or light. To preserve herbs and spices, their tissues must be killed and dried out, so that they don't rot, but as gently as possible, so that water is removed without removing all the flavor. Then the dried material must be kept in closed containers, in a dark, cool place. As a general rule, herbs and spices keep best in opaque glass containers in the freezer (the container should be warmed to room temperature before opening to prevent moisture in the air from condensing onto cold flavorings). In practice, most cooks keep their flavorings at room temperature. As long as they aren't regularly exposed to strong light, whole spices keep well for a year, and ground spices for a few months. The fine particles of ground spices have a large surface area and lose their aroma molecules to the air more rapidly, while whole spices retain the aromas within intact cells.

Storing Fresh Herbs

Many herbs are young, delicate stems and leaves, and so more fragile than other produce. Because their stems have been cut, they're likely to be producing the wound hormone ethylene, which in a closed container will accumulate and trigger general deterioration. Most are best stored in the refrigerator in partly open plastic bags, loosely wrapped in cloth or paper that will absorb moisture and prevent microbes from growing rapidly on wet leaves. Because they come from warm climates, basil and perilla suffer chilling injury in the refrigerator and so are best kept at room temperature, with freshly cut stems immersed in water.

The flavor of many herbs is well preserved by freezing, though the tissues suffer damage from ice crystals and become unattractively dark and limp when thawed. Immersion in oil, which protects the tissue from oxygen, also works for a few weeks, after which much of the flavor has oozed into the oil. Herbs in oil should always be kept in the refrigerator, because the same absence of oxygen that's good for flavor preservation is also good for the growth of botulism bacteria. The bacteria don't grow or produce toxins at refrigerator temperatures.

Drying Fresh Herbs

Drying is a process that removes most of the water in an herb, which when fresh may be more than 90% water. The basic dilemma is that many aroma chemicals are more volatile than water, so any process that evaporates most of the water will also evaporate most of the flavor. This is why many dried herbs don't taste anything like the fresh version, and instead have a generic dried-leaf, hay aroma. There are a few exceptions to this rule, mainly Mediterranean herbs in the mint family that are native to hot, arid areas and have aromatics that persist in drying conditions (oregano, thyme, rosemary, as well as bay laurel from the laurel family). Though sun-drying sounds appealing, its high temperatures and strong dose of visible and ultraviolet light mean that it generally removes and alters flavor. Air-drying over the course of a few days in the shade is much preferable. Herbs can be dried in just a few hours in a low oven or a dehydrator, but the higher temperature usually causes greater flavor losses than air-drying. Some commercial herbs are freeze-dried, which often preserves more of the original flavor.

The microwave oven turns out to work fairly well for drying small quantities of herbs, thanks to the selective and rapid effects of its radiation. Microwave energy excites water molecules while leaving nonpolar oil molecules relatively unaffected, and it penetrates instantly throughout thin leaves and stems (p. 786). This means that all the water molecules in a batch of herbs reach the boiling point within a few seconds and begin to escape from the leaves, while the structures containing the oil-like flavor compounds (glands and canals, pp. 402 and 407) heat up more gradually and indirectly, via the heat of the water molecules. The herbs dry in a matter of a few minutes, with less drastic flavor losses than result from ordinary oven drying.

COOKING WITH HERBS AND SPICES

Herbs and spices are generally cooked along with other ingredients, and as a relatively minor component of the mixture, 1% or less of the total weight. This section explores the extraction and transformation of flavor in such dishes. But some preparations rely on flavorings to provide more than just flavor (p. 401). And a number of herbs—parsley, sage, basil—are delicious on their own, deep-fried just long enough to become crisp and mellow their flavor.

Flavor Extraction

For herbs and spices to provide us with flavor, cooks must find ways to liberate the flavor chemicals from within their tissue and convey them to our taste and odor receptors. For fragile herbs, this may be as simple as strewing the fresh leaves on the dish, as in Vietnamese soups: the eater liberates the aromas by chewing on the leaves, and enjoys them at their freshest. But if the flavors are to be incorporated into the dish, then the flavor compounds must somehow escape from the herb or spice. The cook may leave the flavoring intact and use liquids and heat to encourage the flavors to seep out gradually, or he may break it into particles—chopping fresh herbs, crumbling dry ones, grinding spices—to expose the flavor molecules directly to the dish. The finer the crumble or grind, the greater the surface area from which flavor molecules can escape, and so the more rapidly the flavor moves from flavoring into dish.

Rapid extraction may or may not be desirable. In a briefly cooked dish, it's essential. In a long-cooked stew, however, a slower release from coarser particles or whole leaves and seeds may be preferable. In pickles and preserves, whole spices provide flavor without clouding the liquid. Once flavor molecules have been extracted into a preparation, they begin to react with oxygen and with other food molecules, and their original flavor is transformed, however subtly. Larger particles release the original flavors over a longer time. Another way to assure some fresh flavors in a long-cooked dish is to add the herb or spice—either altogether or a supplemental dose—toward the end of cooking, or even after the cooking is done.

Prepared flavor extracts such as vanilla extract are handy because the flavor molecules have already been dissolved in a liquid and immediately permeate the dish. Because cooking will only cause their flavor to evolve or to evaporate, extracts are best added toward the end of a preparation.

Grinding, Crushing, Chopping There are several ways of crushing herbs and spices, and they have different effects on flavor. Grinders, choppers, and mortars all generate heat. The hotter the aroma mole-

Some Classic Mixtures of Herbs and Spices

	France
Bouquet garni	Bay, thyme, parsley
Fines herbes	Tarragon, chervil, chive
Quatre épices	Black pepper, nutmeg, clove, cinnamon
Herbes de Provence	Thyme, marjoram, fennel, basil, rosemary, lavender
	Morocco
Chermoula	Onion, garlic, coriander leaf, chilli, cumin, black pepper, saffron
Ras el hanout	20+, including cardamom, cassia, mace, clove, cumin, chilli, rose petals
	Middle East
Za'atar	Marjoram, oregano, thyme, sesame, sumac
Zhug	Cumin, cardamom, garlic, chilli
	India
Garam masala	Cumin, coriander, cardamom, black pepper, clove, mace, cinnamon
Panch phoran	cumin, fennel, nigella, fenugreek, mustard
	China
Five-spice	Star anise, Sichuan pepper, cassia, clove, fennel
	Japan
Shichimi	Sansho, mustard, poppyseed, sesame seed, mustard, dried mandarin peel
	Mexico
Recado rojo	Annatto, Mexican oregano, cumin, clove, cinnamon, black pepper, allspice, garlic, salt

cules get, the more volatile they become and the more readily they escape, and the more reactive and changeable they become. The original flavors are best preserved by prechilling both spice and grinder to keep the aromatics as cool as possible. Food processors slice into herbs and introduce a lot of air and therefore aroma-altering oxygen, while a pestle pounding in a mortar crushes herbs and minimizes aeration. Careful chopping with a sharp knife leaves much of the herb structure intact to provide fresh flavor while minimizing cell damage to the cut edges; by contrast, a dull knife crushes rather than cuts, bruises a wide swath of cells, and can result in rapid brown-black discoloration.

One positive effect of oxygen on finely ground spices is manifested in the aging of ground blended spices, which are said to mellow over the course of several days or weeks.

The Influence of Other Ingredients

Because aroma chemicals are generally more soluble in oils, fats, and alcohol than in water, the ingredients in the dish will also influence the speed and degree of flavor extraction, as well as the release of flavor during eating. Oils and fats dissolve more aroma molecules than water during cooking, but also hang on to them during eating, so that their flavor appears more gradually and persists longer. Alcohol also extracts aromas more efficiently, but because it too is volatile, it releases them relatively quickly.

Two methods of flavor extraction that take advantage of the volatility of aroma molecules are steaming and smoking. Herbs and spices can be immersed in the steaming water, or can form a bed on which the food sits above the steam; either way, heat drives aroma molecules into the steam, which then condenses onto the cooler surface of the food, and flavors it. If placed on smoldering coals, or onto a heated pan, herbs and spices will emit not only their usual aromatics, but aromatics transformed by the high heat.

Marinades and Rubs

In the case of large solid chunks of meat or fish, it's easy to get herb and spice flavors onto the food surface, but not so easy to get them inside. Water- and oil-based marinades coat the meat with flavorful liquid, while pastes and dry rubs put the solid aromatics in more direct contact with the meat surface. Because flavors are mainly fat-soluble molecules, and meat is 75% water, flavor molecules can't move very far inside. A distinctly salty marinade or rub can help somewhat by disrupting the meat tissue (p. 155) and making it easier for some slightly water-soluble aromas to penetrate it. A more efficient method is to use a cooking syringe, and inject small portions of the flavorful liquid into many different parts of the meat interior.

Herbs and Spices as Coatings

A useful side effect of coating meat and fish with a paste or rub of herbs and spices is that such a coating acts as a protective layer—like the skin on poultry—that insulates the meat itself from the direct high heat of oven or grill. This means that the outer layers of meat end up less overcooked, and so moister. Coarsely cracked spices, coriander in particular, can provide a crunchy counterpoint to the softer insides. The flavor of a spice crust is improved if the coating contains some oil, which essentially causes the crust to fry rather than simply dry out.

Extracts: Flavored Oils, Vinegars, Alcohols

A special case of flavor extraction is the making of flavor extracts themselves: preparations that serve as instant sources of flavor for other dishes. The most common materials used for extraction are oils, vinegars, sugar syrups (especially for flowers), and alcohols (for example, a neutral vodka for flavoring with citrus peel). The herb and/or spice is usually bruised to damage the cellular structure and make it easier for the liquid to penetrate and aromas to escape. Oils, vinegars, and syrups are often heated before the herb or spice is added to kill bacteria and facilitate their initial penetration into the tissue, then are allowed to cool to avoid changing the flavor. Delicate flowers may require less than an hour to flavor a syrup, while leaves and seeds are usually infused in the extracting liquids for weeks at a cool room temperature. When the extract has reached its desired strength, the liquid is strained off and then stored in a cool, dark place.

Because alcohol, acetic acid, and concentrated sugar all kill bacteria or inhibit their growth, flavored alcohols, vinegars, and syrups pose few safety problems. Oils, however, actually encourage the growth of deadly *Clostridium botulinum*, whose spores can survive brief boiling and germinate when protected from the air. Most herbs and spices don't provide enough nutrients for botulism bacteria to grow on, but garlic does. Infused oils are safest when they're made and stored at refrigerator temperatures, which do slow extraction, but also prevent bacterial growth and slow deterioration.

Commercial Extracts Commercial flavor extracts, unlike kitchen-made extracts, are highly concentrated and are added to foods in tiny quantities, a few drops or a fraction of a spoonful in a whole dish. Vanilla, almond, mint, and anise are common examples. Some extracts and oils are prepared from actual herbs and spices, while others are prepared from one or a few synthetic chemicals that capture the essence of the flavoring, but don't match it in complexity and mellowness (artificial extracts often taste harsh and off). The advantage of synthetic extracts is their low price.

Flavor Evolution

Once the aroma molecules in herbs and spices are released into a preparation and exposed to other ingredients, the air, and heat, they begin to undergo a host of chemical reactions. Some fraction of the original aroma chemicals becomes altered into a variety of other chemicals, so the initially strong, characteristic notes become more subdued, and the general complexity of the mixture increases. This maturing can be a simple side effect of cooking the flavorings with the other ingredients, but it often constitutes a separate preparation step. When cumin or coriander are toasted on their own, for example, their sugars and amino acids undergo browning reactions and generate savory aroma molecules typical of roasted and toasted foods (pyrazines), thus developing a new layer of flavor that complements the original raw aroma.

Maturing Spice Flavors: The Indian System The use of spices is especially ancient and sophisticated in India and Southeast Asia. Indian cooks have several different ways of maturing spice flavors before their incorporation into a dish.

- The toasting on a hot pan of whole dry spices, typically mustard, cumin, or fenugreek, for a minute or two until the seeds begin to pop, the point at which their inner moisture has vaporized and they are just beginning to brown. Spices cooked in this way are mellowed, but individually; they retain their own identities.
- The frying in oil or ghee of mixed powdered spices, often including turmeric, cumin, and coriander. This step allows the different aroma chemicals to react with each other so that the flavors become more integrated, and is usually followed by the sequential addition of garlic, ginger, onions, and other fresh components of what will become the sauce-like phase of the dish.
- The slow frying of a paste of powdered and fresh spices, with constant stirring until much of the moisture evaporates, the oil separates from the

paste, and the spice mixture begins to darken. Mexican cooks treat their pureed chilli mixtures in much the same way. This technique yields its own unique flavors, since dried and fresh ingredients (including active enzymes from the latter) can interact from the beginning, and moisture from the fresh spices prevents the dried spices from being as affected by the heat as they are when fried on their own.
- The brief frying in ghee of whole spices, which are then sprinkled on top of a just-cooked dish as a final garnish.

Indian cooks also aromatize some dishes with a remarkable combination of smoking and spicing called *dhungar.* They put the dish into a pot along with a hollowed onion or small bowl that contains a live coal, sprinkle the coal with ghee and sometimes spices, and cover the pot tightly to infuse the dish with the fumes.

In sum, herbs and spices are remarkably diverse ingredients in themselves, and are capable of producing a remarkable diversity of effects. Combinations, proportions, particle sizes, the temperature and duration of cooking, all have an influence on the flavor of a dish.

Herbs and Spices as Thickeners

Some herbs and spices are used to provide the substance of a dish as well as its aromatic essence. A puree of fresh herbs, as in the Italian pesto sauce made from basil, is thick because the herb's own moisture is already bound up with various cell materials. And thanks to the abundance of those cell materials—mainly cell walls and membranes—such purees also do a good job of coating oil droplets and so creating a stable, luxurious emulsion (p. 628). Fresh chillis, which are fruits, produce a watery puree, but one that cooks down to a wonderful smoothness thanks to its abundant cell-wall pectins. Many Mexican sauces are made from a backbone of dried chillis, which are easily rehydrated to produce the same smooth puree; and Hungarian paprikashes are thickened with powdered chillis.

Indian and Southeast Asian dishes often owe their thickness to a combination of dried and fresh spices. Ground coriander absorbs a lot of water thanks to its thick dry husk; ginger, turmeric, and galangal are starchy root-like rhizomes, and their starch dissolves during prolonged simmering to provide a thickening tangle of long molecular chains. Ground dried sassafras leaves, or filé powder, similarly thickens Louisianan gumbo. And fenugreek is remarkable for its high content of a mucilaginous carbohydrate called galactomannan, which is released simply by soaking the ground seeds.

A SURVEY OF COMMON HERBS

Most of the herbs used in traditional European cooking are members of two plant groups, the mint family and the carrot family. The family members resemble each other to varying degrees, so in this survey I've grouped them together. The remaining herbs then follow in mostly alphabetical order.

Fresh herbs are usually harvested from mature plants, often as they're beginning to flower, when their defensive essential oil content is at its peak. The oil content of Mediterranean herbs is higher on the side of the plant facing the sun. An interesting variation is to harvest them as young sprouts with just a few leaves, when their essential oil content can be very different. Fennel sprouts, for example, contain relatively little anise-like anethole, which dominates the flavor of the mature plant.

The Mint Family

The mint family is a large one, with around 180 genera, and it provides more of our familiar kitchen herbs than any other family. Why such generosity? A fortunate combination of several factors. Members of the mint family dominate the dry, rocky Mediterranean scrublands where few other plants grow, and they cope with their exposed situation with a vigorous chemical defense. Their chemical defenses are located mainly in small glands that project from their leaves, external and therefore expandable storage tanks that can make up as much as 10% of the leaf's weight. And members of the mint family are both promiscuous chemists and promiscuous breeders: individual species make a broad range of aromatic chemicals, and they readily hybridize with each other. The result is a great variety of plants and aromas.

Basil Basils are a large and fascinating group of herbs. They're members of the tropical genus *Ocimum,* which probably originated in Africa, and was domesticated in India. There are around 165 species in the genus *Ocimum,* several of which are eaten. Basil was known to the Greeks and Romans, took firmest root in Liguria and Provence, inventors of the popular basil purees called pesto and pistou, and was hardly known in the United States until the 1970s. The standard "sweet basil" of Europe and North America, *Ocimum basilicum,* is among the more virtuosic of the herbs, and has been developed into several different flavor varieties, including lemon, lime, cinnamon, anise, and camphor. Most varieties of sweet basil are dominated by flowery and tarragon notes, though the variety used in Genoa to make the classic sauce *pesto genovese* is apparently dominated by mildly spicy methyleugenol and clove-like eugenol, with no tarragon aroma at all. Thai basil (*O. basilicum* and *tenuiflorum*) tends toward the anise-like and camphoraceous; Indian holy basil (*O. tenuiflorum*) is dominated by eugenol.

The flavor of basil depends not only on the variety, but on growing conditions and the stage at which it is harvested. Generally, aroma compounds make up a larger proportion of young sweet basil leaves than old, by as much as five times. In leaves that are still growing, the relative proportions of the different compounds actually vary along the length of the leaf, with the older tip richer in tarragon and clove notes, the younger base in eucalyptus and floral notes.

Bergamot This herb, also known as bee balm and Oswego tea, comes from a North American member of the mint family, *Monarda didyma,* whose aroma is somewhat lemony. The same name is given to a member of the citrus family whose essential oil is rich in flowery linalyl acetate, and which is the distinctive addition to Earl Grey tea. (Confusingly, European watermint [p. 404] is also sometimes called bergamot.)

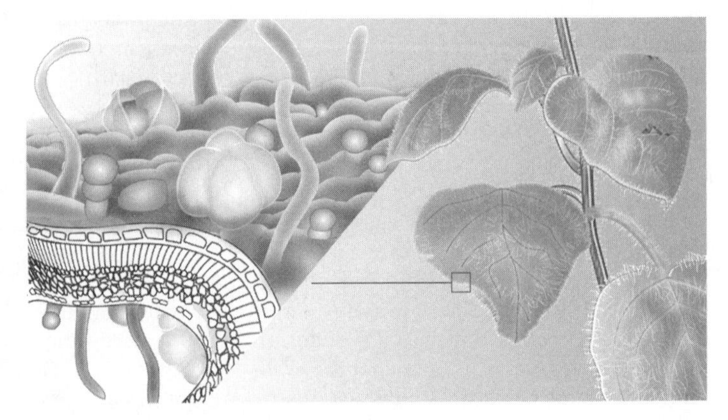

Mint family anatomy. A leaf of oregano, showing the microscopic oil glands that cover the surfaces of herbs in the mint family. The fragile, exposed glands filled with pungent essential oil offer a first line of defense against predators.

Horehound Horehound, so called for its hairy white (hoary) leaves, is a Eurasian species, *Marrubium vulgare,* with a musky and bitter flavor, more often used in candies than in cooking.

Hyssop Hyssop is an ambiguous name. It is sometimes applied to a kind of plant mentioned in the Bible and much used in the Middle East, a cluster of species characterized by the penetrating quality of true oregano (see below). Hyssop proper, *Hyssopus officinalis,* is a milder-mannered European herb with fresh-spicy and camphor notes. It was enjoyed in ancient Rome but is now more commonly used in Thai and Vietnamese cooking. Hyssop contributes to the flavors of several alcohols, including Pernod, Ricard, and Chartreuse.

Lavender Lavender is a Mediterranean plant long and widely valued for its tenacious floral-woody perfume (from a mix of flowery linalyl acetate and linalool, plus eucalyptus-like cineole), but more familiar in soaps and candles than in foods; its name comes from the Latin for "wash." Still, the dried blossoms of *Lavandula dentata* are a traditional ingredient in the mixture *herbes de provence* (along with basil, rosemary, marjoram, thyme, and fennel). They and the blossoms of English lavender, *L. angustifolia,* are also useful alone when used discreetly as a garnish or to infuse their qualities in sauces and sweets. Spanish lavender (*L. stoechas*) has a complex scent reminiscent of Indian chutneys.

Lemon Balm Lemon or bee balm is an Old World species, *Melissa officinalis,* distinguished by its mixture of citrusy and floral terpenes (citronellal and -ol, citral, geraniol). Lemon balm is usually paired with fruit dishes and other sweets.

Marjoram Marjoram was once classified as coming from a sister genus of oregano, but now is officially a species of oregano itself, *Origanum majorana.* Whatever the precise family relationship, marjoram differs from the oreganos in having a milder flavor, fresh and green and floral, with little of their penetrating quality. It therefore works well as one component in many herb blends and dishes.

Mints The true mints are mainly small natives of damp habitats in Europe and

Herbs of the Mint Family

Basil	*Ocimum basilicum*
Bergamot	*Monarda didyma*
Horehound	*Marrubium vulgare*
Hyssop	*Hyssopus officinalis*
Lavender	*Lavendula dentata, L. angustifolia*
Lemon balm	*Melissa officinalis*
Marjoram	*Origanum majorana*
Mints	*Mentha* species
Oregano	*Origanum* species
Perilla	*Perilla frutescens*
Rosemary	*Rosmarinus officinalis*
Sage	*Salvia officinalis*
Savory	*Satureja* species
Thyme	*Thymus vulgaris*

Asia. There are about 25 species in the genus *Mentha* and some 600 varieties, though the family tendency to hybridizing and chemical variation confuses the picture. The mints of most interest to the cook are spearmint (*Mentha spicata*) and peppermint (*M. piperata*), which is an ancient hybrid between spearmint and watermint (*M. aquatica*).

Both of the major culinary mints have a refreshing quality, but they are quite different. Spearmint has a distinctive aroma thanks to a particular terpene, L-carvone, and a richness and complexity thanks to pyridines, nitrogen-containing compounds more typical of roasted foods than raw ones. Spearmint is widely used in the Eastern Mediterranean as well as in India and Southeast Asia, in large quantities, both fresh and cooked, and in both sweet and savory contexts. Simpler, clearer-tasting peppermint contains little or no carvone or pyridines; instead it makes a terpene called *menthol*, which gives it a uniquely cooling quality. In addition to having its own aroma, menthol actually binds to receptors on temperature-sensing nerve cells in the mouth, and causes those cells to signal the brain that they are cooler than they really are by 7–13ºF/4–7ºC. Menthol is a reactive chemical that rapidly degenerates when heated, so peppermint is usually not cooked. Its concentration increases with the age of the leaf, so older leaves taste more cooling; hot and dry growing conditions cause menthol to be transformed into a noncooling, somewhat harsh by-product (pulegone, the characteristic volatile in pennyroyal).

A handful of other mints are worth knowing about. Watermint, one of the parents of peppermint and sometimes called bergamot or orange mint, has a strong aroma and used to be much cultivated in Europe, but now is more popular in Southeast Asia. Pennyroyal (*M. pulegium*) is an especially pungent, peppery minor mint, apple or pineapple mint (M. *suaveolens*) a sweet, apple-like one, and *Mentha x piperata* "citrata" the perfumy lemon or eau de cologne mint. *Nepitella* is the Italian name for *Calamintha nepeta*, a sometimes minty, sometimes pungent herb native to the southern Mediterranean, used in Tuscany to flavor pork, mushroom, and artichoke dishes. "Korean mint" comes from an anise-flavored Asian member of the mint family, *Agastache rugosa*.

Oregano There are about 40 species in the Mediterranean genus *Origanum*, most of them low, shrubby inhabitants of rocky places. The name comes from the Greek for "joy (or ornament) of the mountains," though we have no evidence about how the Greeks enjoyed it. Oregano was little known in the United States until the rise of the pizza after World War II. Oregano species easily form hybrids with each other, so it's not easy to sort out identities. The important thing for the cook is that they come in a range of flavors, from mild to strong and penetrating. The penetrating quality comes from the phenolic compound carvacrol. Greek oreganos are typically rich in carvacrol, while milder Italian, Turkish, and Spanish oreganos contain more thyme-like thymol and fresh, green, floral, and woody terpenes.

Mexican oregano is an entirely different plant, various species of the Mexican genus *Lippia*, a member of the verbena family. Some varieties do have a high carvacrol content, some more resemble thyme, and some are more woody and piney. They all have a substantially higher essential oil content than true oregano (3–4% in the dry leaf, vs. 1%), and therefore seem stronger.

Despite its name, Cuban oregano is an Asian member of the mint family, *Plectranthus amboinicus*, with fuzzy succulent leaves and a good dose of carvacrol. It's now widely cultivated throughout the tropics; in India the fresh leaves are battered and fried.

Perilla or Shiso Perilla is the leaf of *Perilla frutescens*, a mint relative native to China and India. It was taken to Japan in the 8th or 9th century and named shiso; many

Westerners get their first taste of it in sushi restaurants. The distinctive aroma of perilla is due to a terpene called perillaldehyde, which has a fatty, herbaceous, spicy character. There are several different perilla varieties, some green, some red to purple with anthocyanins, some with no perillaldehyde and instead tasting of dill or lemon. The Japanese eat the leaves and flower heads with seafood and grilled meats, and use a red variety to color and flavor the popular pickled plum, umeboshi. Koreans obtain both flavor and cooking oil from perilla seeds.

Rosemary Rosemary is a distinctive woody shrub, *Rosmarinus officinalis*, that grows in the dry Mediterranean scrublands, with leaves so narrow and tightly rolled that they look like pine needles. It has a strong, composite scent, made up of woody, pine, floral, eucalyptus, and clove notes. In southern France and Italy it traditionally flavors grilled meats, but it can also complement sweet dishes. Rosemary aroma is unusually well preserved by drying.

Sage The genus *Salvia* is the largest in the mint family, with around a thousand species that are rich in unusual chemicals, and have been used in many different folk medicines. The genus name comes from a Latin root meaning "health." Sage extracts have been found to be excellent antimicrobial and antioxidant materials. However, common garden sage, *S. officinalis*, is rich in two terpene derivatives, thujone and camphor, that are toxic to the nervous system, so its use as anything but an occasional flavoring is not a good idea.

Common or Dalmatian sage has a penetrating, warm quality from thujone, the note of camphor, and a eucalyptus note from cineole. Greek sage (*S. fruticosa*) has more cineole, while clary sage (*S. sclarea*) is very different,with a tea-like quality and floral and sweet notes from a number of other terpenes (linalool, geraniol, terpineol). Spanish sage, *S. lavandulaefolia*, is fresher-smelling and less distinctive, with pine, eucalyptus, citrus, and other notes partly replacing thujone. Pineapple sage, *S. elegans (rutilans)*, comes from Mexico and is said to have a sweet, fruity aroma.

Sage is especially prominent in northern Italian cooking, and in the U.S. flavors poultry stuffings and seasonings and pork sausages; it seems to have an affinity for fat. Most dried sage used to be "Dalmatian" sage, from the Balkan coast; today Albania and other Mediterranean countries are the largest producers. "Rubbed" sage is minimally ground and coarsely sieved leaves; it loses its aroma more slowly than finely ground sage.

Savory Savory comes in two types, which are two species of the northern-hemisphere genus *Satureja*. Both summer savory (*S. hortensis*) and winter savory (*S. montana*) taste like a mixture of oregano and thyme; they contain both carvacrol and thymol. Summer savory is often the milder of the two. It's thought that savory may well be the parent genus of the various oreganos and marjoram. A native of western North America, *S. douglasii*, is known as yerba buena in California, and has a mild, mint-like flavor.

Thyme Thyme got its name from the Greeks, who used it as an aromatic in their burnt sacrifices; it shares its root with the words for "spirit" and "smoke." There are a lot of thymes: 60–70 species in the shrubby, tiny-leaved, mainly Mediterranean genus *Thymus*, and as many or more varieties of the common thyme, *Thymus vulgaris*. There are also many flavors of thyme, including lemon, mint, pineapple, caraway, and nutmeg. A number of thyme species and varieties taste much like oregano because they contain carvacrol. Distinctive thyme species and varieties are rich in the phenolic compound called thymol. Thymol is a kinder, gentler version of carvacrol, penetrating and spicy, but not as aggressively so. It's this moderate quality that probably endeared thymol thyme to the French, and that makes it a more versatile flavoring than oregano and savory; European cooks have long used it in meat and vegetable dishes of all kinds. Despite its gentler aroma, thymol is as powerful a chemical as carvacrol, which is why thyme oil has long been used as an antimicrobial agent in mouthwashes and skin creams.

The Carrot Family

Though the carrot family gave fewer flavoring plants to Europe than the mint family, it is remarkable for including several that provide aromatic interest as both herb and spice, and some even as vegetables. Members of the carrot family grow in less extreme conditions than the Mediterranean mints, are generally tender biennials rather than shrubby or woody perennials, and have flavors that are generally milder, sometimes even sweet. The seeds (actually small dry fruits) may have chemical defenses—and therefore are spices—because they're fairly large and tempting to insects and birds. One terpene, myristicin, shared by dill, parsley, fennel, and carrots, and giving them a common woody, warm note, is thought to be a defense against molds. The aromatic compounds are stored in oil canals within the leaves, under large and small veins, and are generally found in smaller quantities than the externally stored defenses of the mint family.

Angelica Angelica is a large, rangy plant of northern Europe, *Angelica archangelica,* that has fresh, pine, and citrus notes, but is dominated by a sweet-smelling compound called the angelica lactone. Its candied stems were a popular delicacy from medieval times through the 19th century, but they're seldom seen in the kitchen nowadays. Various parts of the plant now flavor gins, vermouths, liqueurs, candies, perfumes, and other manufactured products.

Celery Celery was a thin-stalked, aromatic but bitter herb called smallage before gardeners developed the mild, thick-stalked vegetable. *Apium graveolens* is a native of damp European habitats near the sea. The distinctive flavor of its leaves and stalks comes from compounds called phthalides, which it shares with lovage and with walnuts. It also has citrus and fresh notes. Celery is often simmered or sautéed with onions and carrots to provide a broad aromatic base for sauces and braises.

Chervil Chervil (*Anthriscus cerefolium*) has small, pale, finely divided leaves, and a delicate flavor that comes from relatively small amounts of the tarragon aromatic estragole; it's best used raw or barely warmed, since heat drives away its flavor. Chervil is a component of the French mixture *fines herbes.*

Herbs of the Carrot Family

Angelica	*Angelica archangelica*
Celery	*Apium graveolens*
Chervil	*Anthriscus cerefolium*
Coriander leaf	*Coriandrum sativum*
Dill	*Anethum graveolens*
Fennel	*Foeniculum vulgare*
Lovage	*Levisticum officinale*
Mitsuba	*Cryptotaenia japonica*
Parsley	*Petroselinum crispum*
Saw-leaf herb	*Eryngium foetidum*

Coriander Coriander or cilantro is said to be the most world's most widely consumed fresh herb. *Coriandrum sativum* is a native of the Middle East. Its seed has been found in Bronze Age settlements and in the tomb of King Tut; it was taken early to China, India, and Southeast Asia, and later to Latin America, and its rounded, notched, tender leaves are popular in all these regions. In Central and South America they came to replace culantro (p. 408), an indigenous relative with very similar flavor, but with large, tough leaves. Coriander herb is not very popular in the Mediterranean and Europe, where its aroma is sometimes described as "soapy." The main component of the aroma is a fatty aldehyde, decenal, which also provides the "waxy" note in orange peel. Decenal is very reactive, so coriander leaf quickly loses its aroma when heated. It's therefore used most often as a garnish or in uncooked preparations. In Thailand, the root of the herb is an ingredient in some pounded spice pastes; the root contains no decenal and instead contributes woody and green notes, something like parsley.

Dill Dill (*Anethum graveolens*) is a native of southwest Asia and India with tough stalks but very delicate, feather-like leaves. Dill was known in ancient Egypt, and became popular in northern Europe, perhaps thanks to its affinity with the local native caraway. Dillweed blends the distinctive flavor of its seed with pleasant green, fresh notes and a unique, characteristic note of its own (dill ether), and in Western cooking is most often used with fish. It is prepared in large amounts, almost as a vegetable and often with rice, in Greece and in Asia. India has its own distinctive variety, *A. graveolens* var. *sowa,* which is used as a vegetable as well as for its seeds.

Fennel Fennel is a native of the Mediterranean and southwest Asia; like dill, it has fibrous leaf stalks but feathery, tender leaves. There is one species of fennel, *Foeniculum vulgare,* and it comes in three different forms. The wild subspecies, *piperitum,* is sometimes collected from the countryside in southern Italy and Sicily, where it's known as carosella and valued for its sharpness in meat and fish cooking. (Fennel now grows wild throughout central California as well.) The cultivated subspecies *vulgare* is known as sweet fennel thanks to its far richer content of the phenolic compound anethole, which is 13 times sweeter than table sugar, and also gives the characteristic sweet aroma of anise. And a specialized variety of sweet fennel, var. *azoricum,* develops the enlarged leaf-stalk bases of bulb or Florence fennel, which is used as an aromatic vegetable.

Lovage Lovage (*Levisticum officinale*) is a large western Asian plant that has aromat-

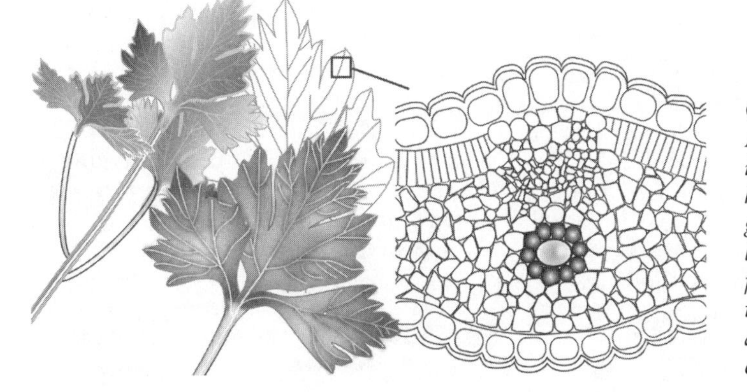

Carrot family anatomy. A leaf of parsley. Herbs in the carrot family have defensive oil glands within their leaves, not on the surfaces. The glands cluster around long canals and fill them with the essential oil.

ics in common with celery and oregano, along with a sweet, flowery note. It was used in ancient Greece and Rome and known as Ligurian celery. Today its large lobed leaves flavor beef dishes in central Europe, and tomato sauces in Liguria. Elsewhere it's little known.

Mitsuba Mitsuba, sometimes called Japanese parsley, is a native of both Asia and North America, *Cryptotaenia japonica* (or *canadensis*), whose mild, large leaves are used by the Japanese in soups and salads. They're flavored mainly by a mixture of minor, woody-resinous terpenes (germacrene, selinene, farnesene, elemene).

Parsley Parsley is a native of southeast Europe and west Asia; its name comes from the Greek and means "rock celery." *Petroselinum crispum* is one of the most important herbs in European cooking, perhaps because its distinctive flavor (from menthatriene) is accompanied by fresh, green, woody notes that are somewhat generic and therefore complement many foods. When parsley is chopped, its distinctive note fades, the green notes become dominant, and a faintly fruity note develops. There are both curly- and flat-leaf varieties with different characteristics; the flat leaves have a strong parsley flavor when young and later develop a woody note, while curly leaves start out mild and woody and develop the parsley character when more mature. The curly leaves are smaller and more incised and therefore crisp faster when fried.

Saw-Leaf Herb Saw-leaf herb or culantro is the New World's version of coriander leaf (cilantro), still used in the Caribbean but now most commonly found in Asian cooking. There are more than a hundred species of *Eryngium,* some of them in Europe, but *E. foetidum* comes from subtropical South America, and is easier to grow in hot climates. Culantro has almost the same flavor as fresh coriander leaf, the main aromatic component being a slightly longer fatty aldehyde than coriander's (dodecanal). Its leaves are large and elongated, with a serrated edge, and thicker and tougher than coriander leaves. They're frequently used in Vietnamese dishes, often torn and strewn on just before eating.

The Laurel Family

The ancient laurel family, mostly large tropical trees, is most notable for giving us cinnamon; but it does give us one well-known herb and three less familiar but interesting ones. The leaves of various cinnamon species are also used as herbs in Asia, but are seldom seen in the West.

Avocado Leaf Mexican races of the avocado tree (*Persea americana*) have leaves with a distinct tarragon aroma, thanks to the same volatiles that flavor tarragon and anise (estragole, anethole). More tropical avocado races (p. 337) lack this aroma. In Mexico, avocado leaves are dried, then crumbled or ground to flavor chicken, fish, and bean dishes.

Bay Laurel
Bay leaves, one of the most useful European herbs, come from an evergreen tree or shrub native to the hot Mediterranean, *Laurus nobilis.* The medium-sized, tough, dry leaf accumulates oils in spherical glands in the leaf interior, and has a well-rounded mixture of woody, floral, eucalyptus, and clove notes. The leaves are generally dried in the shade. Laurel branches were made into fragrant crowns in the ancient world; today the leaves are a standard ingredient in many savory dishes.

California Bay California bay leaves come from an entirely different tree, the California native *Umbellularia californica.* Their aroma bears some resemblance to bay laurel, though it is distinctly stronger, with a dominant eucalyptus note (from cineole).

Sassafras or Filé Sassafras leaves come from a North American tree, *Sassafras* *albidum.* The Choctaw Indians introduced them to French settlers in Louisiana, and they are still most commonly encountered as the dry filé powder used to thicken and flavor Louisiana gumbos. They carry woody, floral, and green notes, and contain little or no safrole, a compound that's prominent in the tree's roots and bark, and that used to give root beer its characteristic flavor until it was found to be a likely carcinogen (see hoja santa below).

Other Common Herbs

Borage Borage is a medium-sized Mediterranean native, *Borago officinalis,* with bright blue flowers and large, fuzzy leaves that have the distinct flavor of cucumber, thanks to enzymes that convert its fatty acids into the same nine-carbon chain (nonanal) produced by cucumber enzymes. It was once a common ingredient of mixed salads (see the recipe on p. 251). Plants in the borage family accumulate potentially toxic alkaloids, so borage should be eaten in moderation.

Capers Capers are the unopened flower buds of a Mediterranean bush, *Capparis spinosa,* which have been gathered from the wild and pickled for thousands of years, though cultivated only for a couple of centuries. The caper bush is distantly related to the cabbage family and has its pungent sulfur compounds, which dominate in the raw flower bud. The bud is preserved in various ways—in brine, in vinegar, dry-salted—and used as a sour-salty accent in sauces and dishes, especially with fish. When dry-salted, the caper bud undergoes an astonishing transformation: its radish and onion notes are displaced by the distinct aroma (from ionone and raspberry ketone) of violets and raspberries!

Curry Leaf Curry leaf is the leaf of a small tree in the citrus family, *Murraya koenigii,* a native of south Asia. It is primarily used in south India and Malaysia, where households often have their own plant and add it to many dishes. Despite its name, the curry leaf doesn't taste like Indian curries; it is mild and subtle, with woody, fresh notes. Curry leaves are added to stews or other simmered dishes, or briefly sautéed to flavor a cooking oil. They are also remarkable for containing unusual alkaloids (carbazoles) with antioxidant and antiinflammatory properties.

Curry Plant Curry plant is a Mediterranean member of the lettuce family, *Helichrysum italicum,* that has been said to be reminiscent of Indian curries. It does contain a number of terpenes that give it a vaguely spicy, pleasant aroma; it's used to flavor egg dishes, teas, and sweets.

Epazote Epazote is a scented member of the large "goosefoot" family, which also provides us with spinach, beets, and the grain quinoa. *Chenopodium ambrosioides* is a weedy native of temperate central America that has spread throughout much of the world, and lends a characteristic aroma to Mexican beans, soups, and stews. That aroma is variously described as fatty, herbaceous, and penetrating, and is due to a terpene called ascaridole. Ascaridole is also responsible for the use of epazote in folk medicine; it is toxic to intestinal worms.

Hoja Santa This herb's name is Spanish for "holy leaf," and refers to the large leaf of New World relatives of black pepper, *Piper auritum* and *P. sanctum.* It's used from southern Mexico to northern South America to wrap foods and flavor them while they're cooked, and is also added directly to dishes as a flavoring. The main aromatic in hoja santa is safrole, the characteristic note of sassafras familiar from root beers, and a suspected carcinogen.

Houttuynia Houttuynia is a small perennial Asian plant, *Houttuynia cordata,* in the primitive lizard's-tail family (Saururaceae), a relative of black pepper. Its leaves are used in Vietnamese and Thai salads, stews, and other dishes. There are two main varieties, one with a citrus aroma, and the other with an unusual scent said to resemble a mixture of meat, fish, and coriander.

Juniper Berries Juniper berries are not leaves, but their essence is the aroma of pine needles, so I include them here, along with the observation that pine and other evergreen needles are often used as flavorings. The Chinese steam fish over pine needles, and the original flavoring for salt-cured salmon (gravlax) was probably pine needles rather than dill. Pineyness is also one element in the aromas of many herbs and spices (see charts, pp. 392, 393).

There are about 10 species of *Juniperus,* a distant relative of the pine, all native to the northern hemisphere. They make small cone-like reproductive structures, about a third of an inch or 10 mm across, but the scales remain fleshy and coalesce to form a "berry" that surrounds the seeds. The berries take from one to three years to mature, during which they turn from green to purplish black. When immature, their aroma is dominated by the terpene pinene; when mature, they carry a mix of pine, green-fresh, and citrus notes. After two years in a spice bottle there is very little aroma left: so juniper berries are best when foraged and fresh. They are much used in northern Europe and Scandinavia to flavor meats, especially game, and cabbage dishes. Juniper is the distinguishing flavor in gin, and gave gin its name (originally Dutch *genever*).

Lemongrass Lemongrass is one of a small group of aromatic species in the grass family. *Cymbopogon citratus* accumulates the lemony terpene citral (a mixture of two compounds, neral and geranial), as well as flowery geraniol and linalool, in special oil cells in the middle of its leaves. It is a native of seasonally dry southern Asia, including the foothills of the Himalayas, and is important in the cooking of Southeast Asia. Lemongrass forms clumps of thick shoots; all parts are aromatic, but only the lower stalk is tender enough to be edible itself. The older outer leaves can be used to infuse a dish with their flavor, or be made into an herbal tea. In Thailand the tender stalk is a standard component of pounded spice pastes, and it's also eaten fresh in salads.

Lemon Verbena Lemon verbena is a South American plant, *Aloysia triphylla,* a relative of Mexican oregano. The lemony flavor of its leaves comes from the same terpenes, collectively called citral, that flavor lemongrass; other terpenes lend a flowery note.

Lolot Lolot is the large, heart-shaped leaf of a black-pepper relative, *Piper lolot,* a native of Southeast Asia, and used there as a flavoring wrap for grilling meats.

Orange Flowers Orange flowers come from the bitter or Seville orange, *Citrus aurantium,* and they have been used for millennia to flavor sweets and other dishes in the Middle East, usually in the extract called orange-flower water. The distinctive perfume results from a mixture of terpenes also found in roses and lavender, with an important contribution from the same compound that flavors concord grapes (methyl anthranilate).

Makrut or Kaffir Lime *Ma krut* is the Thai name for the tree also known as kaffir lime ("kaffir" is Arabic for "unbeliever" and has derogatory connotations). This Southeast Asian member of the citrus family, *Citrus hystrix,* has distinctively aromatic leaves and fruit rinds that are an important ingredient in Thai and Laotian cooking, especially soups, stews, and fish dishes. The rind has an unremarkable mix of citrus, pine, and fresh notes, but the tough leaves are richly endowed with citronellal, which gives them an intense, fresh, lingering lemon-green character distinct from that of sweeter citral-flavored lemongrass (with which it's often cooked). Citronellal is named for citronella, its original and main source, which is a sister species to lemongrass (*Cymbopogon winterianus*).

Nasturtium The flowers, leaves, and immature fruits of the familiar South American native *Tropaeolum major* all have a pungency like that of watercress, and enliven salads.

Nightshade Nightshade is a relative of the potato, *Solanum torvum,* which grows into a small and short-lived tree. It's a native of the West Indies, but is now found throughout tropical Asia. Its small berry-like fruits are intensely bitter, and are used in Thailand, Malaysia, and Indonesia to contribute just that quality to sauces and salads.

Rau Ram Rau ram is the Vietnamese name for a sprawling herb in the buckwheat family known in English as fragrant knotweed, and now as Vietnamese cilantro. *Polygonum odoratum* is a native of Southeast Asia whose leaves mix the aromas of coriander and lemon with a slightly peppery taste. It's often paired with mint, and eaten fresh with many foods.

Rice-Paddy Herb Rice-paddy herb is an aquatic plant in the snapdragon family, *Limnophila chinensis* ssp. *aromatica,* a native of Asia and the Pacific islands, whose small leaves are used in Southeast Asian fish dishes, soups, and curries, especially in Vietnam. It has a lemony but complex aroma produced by small quantities of a citrusy terpene and the main ingredient of perilla (perillaldehyde).

Rose Flowers Roses, mainly from the Eurasian hybrid *Rosa x damascena*, have been used for millennia from the Middle East through Asia, usually dried or in the form of an extract (rose water). Their aroma comes mainly from the terpene geraniol. Roses are most often used in sweets, but are also found in the savory Moroccan spice mixture ras el hanout, and in North African sausages.

Screwpine or Pandan Screwpine is the aromatic, strap-like leaf of shrubs related to the lily family that are native to Indonesia (species of *Pandanus*). Screwpine leaves are used in India and southeast Asia to flavor rice dishes and sweets, and to wrap meats and fish. Their primary volatile compound is the same one that gives basmati and jasmine rices their distinctive nutty aroma (2-acetyl-1-pyrroline, which is also prominent in the aroma of popcorn and of crabmeat). The screwpine flower is also aromatic and the source of a more perfume-like extract called kewra, which flavors many Indian milk sweets.

Sorrel Sorrel is the startlingly sour leaf of several European relatives of rhubarb and buckwheat that are rich in oxalic acid: *Rumex acetosa, scutatus,* and *acetosella.* Cooks use them mainly as a source of acidity, and they also provide a more generic green aroma. Sorrel readily disintegrates with a little cooking into a sauce-like puree that complements fish, but whose chlorophyll turns drab olive from the acidity. The color can be brightened by pureeing some raw sorrel and adding that to the sauce just before serving.

Tarragon Tarragon is the small, narrow leaf of a native of western and northern Asia, *Artemisia dracunculus,* a member of the lettuce family. The robust wild tarragon, often sold in plant nurseries as Russian tarragon, has a harsh and uninteresting flavor, while the relatively fragile cultivated form, "French" tarragon, has a distinctive aroma thanks to the presence of a phenolic compound called estragole (from the French name for the plant, *estragon*) in oil cavities alongside the leaf veins. Estragole is a close chemical relative of the anise aromatic anethole, and does have an anise-like character. Tarragon is a component of the French mix *fines herbes,* is the primary flavoring in béarnaise sauce, and is often used to flavor vinegars.

"Mexican tarragon" is a marigold-like New World native, *Tagetes lucida,* whose leaves do indeed contain a mixture of anise-like anethole and tarragon's estragole.

Tobacco Tobacco is used occasionally as a food flavoring, and its curing resembles the making of tea (p. 437). The leaves of the notorious North American native *Nicotiana tabacum,* a relative of the potato and tomato, are harvested when they begin to turn yellow and develop resinous secretions, then either sun-cured or fermented in heaps for several weeks, and dried by contact with hot metal. These treatments develop a complex aroma with woody, leathery, earthy, and spicy notes, and these are sometimes augmented with the addition of various essential oils (vanilla, cinnamon, clove, rose, and others). Tobacco leaves contain astringent tannins and bitter nicotine, so they are usually infused only lightly into sauces, syrups, and creams. Sometimes whole leaves are used as a disposable wrapper to flavor a food during cooking.

Water Pepper Water pepper is a relative of rau ram, or Vietnamese coriander. *Polygonum hydropiper* is a widespread native of wet areas in the Northern Hemisphere. Its leaves have been used as a pepper substitute in Europe, and are now used mainly in Japan to provide a peppery, somewhat numbing pungency (from polygodial). Water pepper also has woody, pine, and eucalyptus notes.

Wintergreen Wintergreen is the leaf of *Gaultheria procumbens* or *fragrantissima,* a North American bush in the blueberry and cranberry family, whose distinctive, refreshing aroma is created mainly by methyl salicylate.

A SURVEY OF TEMPERATE-CLIMATE SPICES

As is true of herbs, many of our temperate-climate spices come from just a few plant families. In the following survey, I group together spices that are botanically related, and list the rest in alphabetical order. Tropical spices are surveyed separately in the next section.

The Carrot Family

In addition to many leafy herbs, the carrot family also gives us a number of our favorite spices. The aromatic "seeds" of plants in the carrot family are actually small but complete fruits that are dry rather than fleshy. They are borne in pairs, enclosed in a protective husk, and are usually sold husked and separated. The individual fruits have a characteristic ridged surface, and their aromatic oil is contained in canals that lie under the ridges.

Ajwain Ajwain (*Trachyspermum ammi*) is a close relative of caraway, used in North Africa and Asia, especially India, and can

Carrot family anatomy. Fennel seeds. Seeds of plants in the carrot family carry their essential oil in hollow chambers beneath their outer ridges.

be thought of as a seed version of thyme: it carries the essence of thyme, thymol, in a caraway-like seed.

Anise Anise is the seed of a small central Asian plant, *Pimpinella anisum,* which has been prized since ancient time. It's remarkable for its high content of the phenolic compound anethole, which is both distinctively aromatic and sweet-tasting, and has been mainly used to flavor sweets and alcohols (Pernod, pastis, ouzo), though the Greeks also use it in meat dishes and tomato sauces.

Asafoetida Asafoetida is one of the strangest and strongest of all spices. It comes from a perennial plant in the carrot family native to the mountains of Central Asia, from Turkey through Iran and Afghanistan to Kashmir; India and Iran are major producers. *Ferula asafoetida, F. alliacea, F. foetida,* and *F. narthex* look something like giant carrot plants, growing to 5 feet/1.5 m and developing massive carrot-like roots 6 in/15 cm in diameter, from which new sprouts arise every spring. The spice is obtained after the new foliage begins to turn yellow. The top of the root is exposed, the foliage pulled out, and the root surface periodically scraped to wound it and gather the protective sap that collects in the wound. The sap slowly hardens and develops a strong, sulfurous aroma reminiscent of human sweat and washed-rind cheese (p. 58). Sometimes the resin is aged in fresh goat or sheep skin to augment its aroma, which is so strong that the resin is commonly ground and diluted for sale with gum arabic and flour. The aroma of asafoetida is due to a complex mixture of sulfur compounds, a dozen identical with volatiles in the onion family, and a number of less common di-, tri-, and tetrasulfides. Asafoetida can give the impression of onions, garlic, eggs, meat, and white truffles, and in India is a prominent ingredient in the cooking of the Jains, who avoid animal foods and also onion and garlic (because they contain a bud that would otherwise grow into a new plant).

Caraway Caraway comes from the small herb *Carum carvi.* There are annual and biennial forms, the first native to central Europe, the second to the Eastern Mediterranean and Middle East. The biennial form develops a taproot the first summer, then flowers and fruits the second; the taproots are sometimes cooked like carrots in northern Europe. Caraway may have been among the first spice plants cultivated in Europe; its seeds were found in ancient Swiss lake dwellings, and have continued to be an important ingredient in Eastern Europe. The distinctive flavor of caraway

Spices of the Carrot Family

Ajwain	*Trachyspermum ammi*
Anise	*Pimpinella anisum*
Asafoetida	*Ferula asafoetida*
Caraway	*Carum carvi*
Celery seed	*Apium graveolens*
Coriander	*Coriandrum sativum*
Cumin	*Cuminum cyminum*
Cumin, black	*Cuminum nigrum*
Dill seed	*Anethum graveolens*
Fennel seed	*Foeniculum vulgare*

comes from the terpene D-carvone (which it shares with dill), with citrusy limonene the only other major volatile. Caraway is used in cabbage, potato, and pork dishes, in breads and cheeses, and in the Scandinavian alcohol aquavit.

Celery Seed

Celery seed is essentially a concentrated, dried version of the same aromas found in fresh celery (*Apium graveolens*), though of course it lacks the fresh green notes. The main aromas are a distinctive celery note from unusual compounds called phthalides, together with citrus and sweet notes. Celery seed was used in the ancient Mediterranean and is still common in European and American sausages, pickling mixes, and salad dressings. "Celery salt" is a mixture of salt and ground celery seeds.

Coriander Coriander (*Coriandrum sativum*) has been valued and cultivated since ancient times, more for its dried fruits than its leaves, which have entirely different flavors. The flavor of the fruit oil is startlingly floral and lemony, and makes coriander unique and irreplaceable in the cook's arsenal of aromas. It's generally used in combination with other spices, as a component of a pickling or sausage mix, in gin and other alcohols, or as half of the coriander-cumin backbone of many Indian dishes. Coriander is also one of the distinguishing flavors in American hot dogs.

There are two common types of coriander. The European type has small fruits (1.5–3 mm), a relatively high essential oil content, and a large proportion of flowery linalool; the Indian type has larger fruits (to 5 mm), a lower oil content, less linalool, and several aromatics not found in the European type.

Coriander seed is generally supplied whole, with the two dry fruits still enclosed in their husk. When it's ground along with the aromatic fruits, the brittle, fibrous husk makes a good water absorber and thickener for sauces (the liquid portion of a curry, for example). Coarsely ground coriander is also used to coat meats and fish and provide flavor, crunch, and insulation at the same time.

Cumin Cumin comes from a small annual plant (*Cuminum cyminum*) native to southwest Asia, and was enjoyed by the Greeks and Romans; the Greeks kept it at the table in its own box, much as pepper is treated

The Flavor of Anise

The volatile chemical that creates the typical aroma of anise—as well as of fennel, star anise, the Central American pepper relative *Piper marginatum,* and the herb sweet cicely (*Myrrhis odorata*)—is called trans anethole. It is one of a group of compounds that are not only distinctively aromatic, but also intensely sweet—13 times sweeter than table sugar, weight for weight. Star anise is chewed in China, and fennel seed in India, to "sweeten the breath," and they are also literally sweet for the person chewing. A related sweet aromatic is estragole (methyl chavicol), which is most prominent in sweet basil and tarragon.

Anethole is unusual among phenolic flavor compounds for remaining pleasant to the taste at high concentrations. Its very high concentration in anise-flavored liquors is the reason for the dramatic clouding that results when these liquors are diluted with water: anethole dissolves in alcohol but not in water, so when the added water dilutes the alcohol, the anethole molecules cluster together in bunches big enough to scatter light.

today. For some reason cumin largely disappeared from European cooking during the Middle Ages, though the Spanish kept it long enough to help it take root in Mexican cooking. The Dutch still make a cumin-flavored cheese, and the Savoie French a cumin bread, but cumin now mainly marks the foods of North Africa, western Asia, India, and Mexico. Its distinctive aroma comes from an unusual chemical (cuminaldehyde) that is related to the essence of bitter almond (benzaldehyde). It also has fresh and pine notes.

Black cumin is the seed of a different species (*Cuminum nigrum*), darker and smaller, with less cuminaldehyde and a more complex aroma. It is much used in savory dishes of North Africa, the Middle East, and North India.

Dill Seed Dill seed has a stronger flavor than dill weed, the feathery leaves of the same plant (*Anethum graveolens*). It's mildly reminiscent of caraway thanks to its content of the caraway terpene carvone, but also has fresh, spicy, and citrus notes. It's mainly used in central and northern Europe in cucumber pickles (the combination goes back at least to the 17th century), sausages, condiments, cheeses, and baked goods. Indian dill, var. *sowa*, produces a larger seed with a somewhat different balance of aromas; it's used in spice mixtures of northern India.

Fennel Seed and Pollen Fennel seed has the same anise-like aroma and sweet taste as the stalk and leaves of the plant that bears it, *Foeniculum vulgare*. Its dominant volatile is the phenolic compound anethole (see anise, above), supported by citrus, fresh, and pine notes. Most fennel seed comes from sweet fennel varieties (p. 407) and tastes sweet; seed from the less cultivated types are also bitter due to the presence of a particular terpene (fenchone). Fennel seed is a distinctive ingredient in Italian sausages and in Indian spice mixes, and in India is chewed as an after-meal breath freshener.

The fine yellow pollen of the fennel flower is also collected and used as a spice. Fennel pollen combines anise and floral aromas, and in Italy is sprinkled on dishes at the last minute.

The Cabbage Family: Pungent Mustards, Horseradish, Wasabi

Of the various spices that manage to please us by causing irritation and pain, the mustards and their relatives are unique in providing a volatile pungency, one that travels from the food through the air to irritate our nasal passages as well as our mouth. The active ingredients of chillis and black pepper become significantly volatile only at high temperatures, above about 140°F/60°C, which is why roasting hot chillis or toasting peppercorns can cause everyone in the kitchen to start sneezing. But mustard and horseradish and wasabi can get into the nose even at room or mouth temperature. Theirs is a head-filling hotness.

The pungency of mustards and their relatives arises from the same chemical-defense system used by their vegetable relatives in the cabbage family (p. 321). The plants store their irritant defenses, the isothiocyanates, by combining them with a sugar molecule. The storage form is not irritating, but it does taste bitter. When their cells are damaged, special enzymes reach the storage form and break it apart, liberating the irritant molecules (and at the same time eliminating the bitterness). Mustard seeds and horseradish roots are pungent because we grind them up raw and encourage their enzymes to liberate the irritant molecules. When mustard seeds are cooked—for many Indian dishes they are toasted or fried until they pop—the liberating enzymes are inactivated, no irritants are liberated, and their flavor ends up nutty and bitter rather than pungent.

Mustards Mustard seed has been found in prehistoric sites from Europe to China, and was the first and only native pungent spice available to early Europe. It has been made into the familiar European condiment at least since Roman times; its name in most European languages comes not from the Latin name of the seed or plant (*sinapis*), but from the condiment, which was made with freshly fermented wine (*mustum*), and the hot (*ardens*) seeds. Different nations have distinctive prepared mustards whose roots go back as far as the Middle Ages. Mustard is also widely used as the whole seed, especially in Indian cooking, and flavors a broad range of dishes, including fruits preserved in sugar (Italian *mostarda di frutta*).

Black, Brown, and White Mustards There are three main kinds of mustard plants and seed, each with its own character.

- Black mustard, *Brassica nigra,* is a Eurasian native, small and dark-hulled, with a high content of the defensive storage compound sinigrin and therefore a high pungency potential. It was long important in Europe and still is in India, but is an inconvenient crop and in many countries has been replaced by brown mustard.
- Brown mustard, *B. juncea,* is a hybrid of black mustard and the turnip (*B. rapa*) that is easier to cultivate and harvest. It has large, brown seeds that contain somewhat less sinigrin than black mustard and therefore less potential pungency. Most European prepared mustards are made with brown mustard.
- White or yellow mustard, *Sinapis alba* (or *Brassica hirta*), is a European native with large pale seeds and a different defensive storage compound, sinalbin. The irritating portion of sinalbin is much less volatile than the irritant in sinigrin, so little of white mustard's pungency rises into the nose. It mostly affects the mouth, and generally seems milder than black or brown mustard. White mustard is used mainly in the United States, in prepared mustards as well as whole in pickle mixes.

Making and Using Mustard Prepared mustard condiments can be made either from whole seeds or from powdered mustard, also called mustard flour, which has been ground and sieved to remove the seed coat. Dry mustard seeds and their powder are not pungent. Their pungency develops over the course of a few minutes or a few hours when the seeds are soaked in liquid and ground, or the preground seeds are simply moistened. The combination of moisture and cell damage revives the seeds' enzymes and allows them to liberate the pungent compounds from their storage forms. Most prepared mustards are made with acidic liquids—vinegar, wine, fruit juices—which slow the enzymes, but also slow the later

Roman Mustard

Carefully cleanse and sift mustard-seed, then wash it in cold water, and when it has been well cleaned, leave it in the water for two hours. Next take it out and after it has been squeezed in the hands, . . . add pine-kernels, which should be as fresh as possible, and almonds, and carefully crush them together after pouring in vinegar. . . . You will find this mustard not only suitable as a sauce but also pleasing to the eye; for if it is carefully made, it is of an exquisite brilliance.

—Columella, *De re rustica,* 1st century CE

disappearance of the pungent compounds as they gradually react with oxygen and other substances in the mix.

Once the pungency has developed, cooking will drive off and modify the irritant molecules and so reduce the pungency, leaving behind a more generic cabbage-family aroma. Mustard is therefore usually added at the end of the cooking process.

Other Uses for Mustards In addition to their chemical defenses, mustard seeds are about a third protein, a third carbohydrate, and a third oil. When the seeds are ground, the small protein and carbohydrate particles and dissolved mucilage from the seedcoat can coat the surfaces of oil droplets and thus stabilize such sauce emulsions as mayonnaise and vinaigrette (p. 628). The seedcoat of white mustard is especially rich in mucilage (up to 5% of the seed weight), and ground white mustard is used in sausages to help bind the meat particles together.

Mustard oil is a traditional cooking oil in Pakistan and in Northern India, where it lends a distinctive flavor to Bengali fish dishes, pickles, and other preparations. In much of the West, the sale of mustard oil for food use is illegal, for two reasons: it contains large quantities of an unusual fatty acid, erucic acid; and it contains irritating isothiocyanates. Erucic acid causes heart damage in laboratory animals; its significance for human health isn't known. Though our mustard condiments contain the same isothiocyanates as mustard oil, it's possible that daily exposure through foods cooked in the oil could have harmful long-term effects. So far, medical studies are inconclusive. In Asia, it's thought that preheating the oil to the smoking point reduces isothiocyanate content.

Horseradish Horseradish is a west Asian cabbage relative, *Armoracia rusticana,* remarkable for large fleshy white roots rich in sinigrin and its volatile pungent compound. Horseradish pungency develops when the raw root is grated, or when the ground dried root is rehydrated. Horseradish doesn't seem to have been cultivated in Europe until the Middle Ages; today it's used as a relish or dressing for meats and seafood, often in the company of cream to take the edge off its strong flavor.

Wasabi Wasabi is the enlarged stem of an East Asian cabbage relative that also accumulates sinigrin as a chemical defense. *Wasabia japonica* is a native of Japan and Sakhalin Island, where it grows alongside cool mountain streams. Wasabi is now cultivated in several countries and is occasionally available fresh in the West; whole and partly used roots keep for several weeks in the refrigerator.

Most wasabi served in restaurants is in fact ordinary dried horseradish powder, colored green and reconstituted with water. It has a similar pungency, but little else in common with true wasabi. When the fresh stem is grated a few minutes before the meal, it releases more than 20 enzyme-gen-

Dealing with an Overdose of Wasabi or Horseradish

Though a mouthful of food overdosed with chilli can be painful, it's not as startling as too much horseradish or wasabi, whose volatile irritants can quickly get into the airstream and cause a bout of coughing or choking. These reactions can be minimized by remembering to breathe out through the mouth—sparing the nasal passages—and breathing in through the nose, to avoid drawing irritants from the mouth into the lungs.

erated volatiles, some pungent, some oniony, some green, some even sweet.

The Bean Family: Licorice and Fenugreek

Licorice Licorice comes from the roots of *Glycyrrhiza glabra,* a native of southwest Asia. Its English name is a much-altered version of its genus name, which derives from the Greek for "sweet root." The woody roots of this shrub are remarkable for containing a steroid-like chemical, glycyrrhizic acid, that is 50–150 times sweeter than table sugar. The water extract of the roots contains many different compounds, including sugars and amino acids, which undergo flavor- and pigment-producing browning reactions with each other when the extract is concentrated. Licorice extracts are available as dark syrups, blocks, or powders, and are used in various confections, to give color and flavor to dark beers, porter, and stout, and to flavor tobacco for cigars, cigarettes, and chewing. Many licorice candies are flavored with anise-like anethole (p. 414), but licorice root itself has a more complex aroma, with almond and floral notes.

Thanks to its hormone-like chemical structure, glycyrrhizic acid has a number of effects on the human body, some helpful and some not. It helps soothe coughs, but it also can disrupt normal regulation of mineral and blood pressure levels. Licorice is therefore best consumed in moderation and infrequently; daily consumption can sometimes cause a significant rise in blood pressure and other problems.

Fenugreek Fenugreek is the small, hard seed of a bean relative, *Trigonella foenum-graecum,* that's native to southwest Asia and the Mediterranean. Its name comes from the Latin for "Greek hay." Fenugreek is somewhat bitter and has a very distinctive sweet aroma, reminiscent of dry hay as well as maple syrup and caramel, that comes from a chemical called sotolon, which is also an important volatile in molasses, barley malt, coffee, soy sauce, cooked beef, and sherry. The outer cell layer of the fenugreek seed contains a water-soluble storage carbohydrate (galactomannan), so that when the seeds are soaked, they exude a thick, mucilaginous gel that gives a pleasant slipperiness to some Middle-Eastern sauces and condiments (Yemen's *hilbeh*). Fenugreek is a component of various spice mixtures, including Ethiopian berber and some Indian curry powders.

Fenugreek leaves are bitter and slightly aromatic, and are enjoyed as a fresh or dried herb in India and Iran.

Chillis

Chillis, or "chile peppers," the fruits of small shrubs native to South America, are the most widely grown spice in the world. Their active ingredient, the spectacularly pungent chemical capsaicin, protects the seeds of the chilli fruit, and appears to be a chemical repellant aimed specifically at mammals. Birds, which swallow the fruits whole and disperse the seeds widely, are immune to capsaicin; mammals, whose teeth grind up the fruit and destroy the seeds, are pained by it. It's a wonderfully perverse achievement for our mammal species to have fallen in love with this anti-mammalian weapon and spread the chillis much further than any bird ever did!

The success of the chilli has been remarkable. World production and consumption are now some 20 times that of the other major pungent spice, black pepper. It is ubiquitous in Central and South America, Southeast Asia, India, the Middle East, and North Africa. In China the chilli is a major spice in Sichuan and Hunan provinces; in Europe, Hungary has its paprika and Spain its pimenton. In the United States, salsas became more popular than ketchup in the 1980s, thanks to the influence of Mexican restaurants. Mexico remains the most advanced chilli culture, where several different varieties may be blended to obtain a particular flavor, and where the substance of many sauces is contributed by chillis, without the aid of flavorless flours or starches.

Chillis and Capsaicins There are about 25 species of *Capsicum,* most natives of South America, of which five have been domesticated. Most of our common chillis come from one species, *Capsicum annuum,* which was first cultivated in Mexico at least 5,000 years ago. Chillis are hollow fruits, with an outer wall rich in carotenoid pigments that encloses the seeds and the tissue that bears them, a pale, spongy mass called the placenta. (For chillis as vegetables, see p. 331). Their pungent chemicals, the capsaicins, are only synthesized by the surface cells of the placenta, and accumulate in droplets just under the cuticle of the placenta surface. That cuticle can split under the pressure and allow the capsaicin to escape and spread onto the seeds and the inner fruit wall. Some capsaicin also seems to enter the plant's circulation, and can be found in small quantities within the fruit wall and in nearby stems and leaves.

The amount of capsaicin that a chilli contains depends not only on the plant's genetic makeup, but on growing conditions—high temperatures and drought increase production—and on its ripeness. The fruit accumulates capsaicin from pollination until it begins to ripen, when its pungency declines somewhat: so maximum pungency comes around the time that the green fruit begins to change color.

There are several different versions of the capsaicin molecule found in chillis. This may be why different kinds of chillis seem to produce different kinds of pungency—quick and transient, slow and persistent—and to affect different parts of the mouth.

Capsaicin's Effects on the Body The effects of capsaicin on the human body are many and complex. As I write in 2004, the scorecard is fairly positive. Capsaicin does not appear to increase the risk of cancer or stomach ulcers. It affects the body's temperature regulation, making us feel hotter than we actually are, and inducing cooling mechanisms (sweating, increased blood flow in the skin). It increases the body's metabolic rate, so that we burn more energy (and therefore retain less in storage as fat). It may trigger brain signals that make us feel less hungry and more satiated. In sum, it may encourage us to eat less of the meal it's in, and to burn more of the calories that we do eat.

Of course there's also capsaicin's irritating effects, which can be pleasurable in the mouth but not necessarily elsewhere. (This is why "pepper spray" is an effective weapon; it makes breathing and seeing difficult for about an hour.) Capsaicin is potent and oily and hard to wash off surfaces, so small amounts left on fingers can

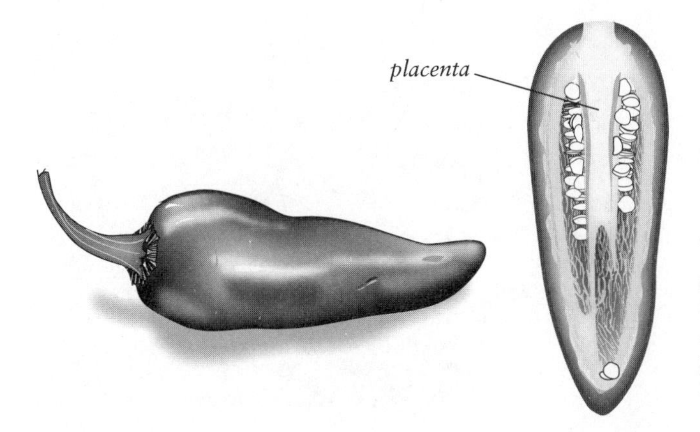

The chilli fruit. The pungent chemical capsaicin is secreted by cells on the surface of the placenta, the pithy tissue that bears the seeds.

end up hours later being rubbed into an eye. Knives, cutting boards, and hands should be thoroughly washed with hot soapy water to avoid this and similar unhappy surprises. On the other hand, capsaicin irritation has found a number of medical applications; for example, when applied to the skin it helps reduce muscle pain by increasing local blood flow.

Controlling Capsaicin Pungency The pungency of any dish that contains chillis is influenced by four main factors: the variety of chilli used, the amount of chilli added, the presence or absence of the capsaicin-rich tissues, and the length of time that the chilli is in contact with the other ingredients. The cook can reduce the pungency of chillis substantially by cutting them in half and carefully dissecting and removing the spongy placental tissue and the seeds.

What about quenching the burn once the mouth is already on fire? The two surest remedies—though they're only temporary—are to get something ice-cold into the mouth, or something solid and rough, rice or crackers or a spoonful of sugar. Cold liquid or ice cools the receptors down below the temperature at which they are activated, and the rough food distracts the nerves with a different kind of signal. Though capsaicin is more soluble in alcohol and oil than it is in water, alcoholic drinks and fatty foods appear to be no more effective than cold or sweetened water at relieving the burn (carbonation adds to the irritation). If all else fails, take comfort in the fact that capsaicin pain generally fades within 15 minutes.

Dried Chillis Dried chillis are much more than a conveniently stable source of pungency and thickening power: they're the source of flavor complexity that is rare even among herbs and spices. The drying process concentrates the contents of the cells in the fruit wall, encouraging them to react with each other and generate dried-fruit, earthy, woody, nutty, and other aromatics. Drying traditionally took several weeks in sun or shade, and in much of the world it still does. Modern machine drying offers more control, and can minimize the loss of light-sensitive pigments and vitamin C, though it brings flavor differences as well. Chillis are sometimes smoke-dried (Mexican chipotles, some Spanish pimentons), which lends a characteristic note.

Other Temperate-Climate Spices

Hops Hops are the dried seed-bearing "cones" of *Humulus lupulus,* a perennial native of the Northern Hemisphere that is a relative of marijuana and hemp. The hop plant was cultivated in the Hallertau region of Germany by the 8th century, and spread to Flanders by the 14th. Though now used

Pepper Terminology

In the United States, pungent capsicum fruits are generally called "peppers," or "hot peppers," terms that stem from the early Spanish identification of capsicum pungency with black-pepper pungency. The native Nahuatl word was *chilli,* which gave rise to Spanish *chile,* and in turn to American *chili* (both a capsicum-flavored stew and the powder used to make it). Chile the country got its name from an entirely unrelated word (Araucanian for "the end of the earth"). Given the many possibilities for confusion, I agree with Alan Davidson and others that we should refer to pungent capsicums with the original and unambiguous Nahuatl name *chilli.*

almost exclusively in beer, they also flavor bread and are made into an herbal tea. Hop aroma depends on the variety, and may include woody, floral, and complex sulfur notes. It's described in more detail in chapter 13.

Mahleb Mahleb or mahaleb is the dried kernel of a small kind of cherry native to Iran, *Prunus mahaleb*. The kernels have a warm aroma vaguely suggestive of bitter almond, and are used to flavor baked goods and sweets in much of the eastern Mediterranean.

Mastic Mastic is a resin exuded from the trunk of a relative of the pistachio, *Pistacia lentiscus*, a tree native to the Eastern Mediterranean that now grows only on the Greek island of Chios. Mastic was chewed like chewing gum (hence its name, from the same root as *masticate*), and is also used to flavor various preparations, from breads and pastries to ice cream, candies,

Chilli Varieties and Pungencies

Here is a list of common chilli varieties and their relative pungencies. Pungency is rated in Scoville units, a measure invented around 1912 by pharmaceutical chemist Wilbur Scoville and later adapted to modern chemical analyses. The original method involved an overnight alcohol extraction of the pepper, and then a tasting of increasing dilutions of the extract until the pungency is barely detectable. The more the extract can be diluted, the more pungent it is, and the higher the Scoville score.

Chilli Variety	Pungency, Scoville Units
Capsicum annuum	
Bell	0–600
New Mexican	500–2,500
Wax	0–40,000
Paprika	0–2,500
Pimento	0
Jalapeno	2,500–10,000
Ancho/ poblano	1,000–1,500
Serrano	10,000–25,000
Cayenne	30,000–50,000
Capsicum chinense	
Habanero, scotch bonnet	80,000–150,000
Capsicum frutescens	
Tabasco	30,000–50,000
Capsicum pubescens	
Rocoto	30,000–60,000
Capsicum baccatum	
Aji	30,000–50,000

and alcohol (ouzo). The main aromatic components of the gum are two terpenes, pine-like pinene and woody myrcene. Myrcene is also the molecule from which the long resin polymers are built. The resin is not very soluble in water, so it must be ground fine and mixed with another powdery ingredient (flour, sugar) to help disperse it evenly in liquid ingredients.

Nigella The small, black, angular seed of *Nigella sativa,* a close Eurasian relative of the common ornamental plant love-in-a-mist, tastes like a milder, more complex version of thyme or oregano, with a hint of caraway. It's used from India through southwest Asia in breads and other dishes.

Saffron Saffron is the world's most expensive spice: a testament not only to the labor required to produce it, but to its unique ability to impart both an unusual flavor and an intense yellow color to foods. It is a part of the flower of a kind of crocus, *Crocus sativus,* which was probably domesticated in or near Greece during the Bronze Age. The saffron crocus was carried eastward to Kashmir before 500 BCE; in medieval times the Arabs took it westward to Spain, and the Crusaders to France and England. (The name comes from the Arabic for "thread.") Today Iran and Spain are the major producers and exporters. They use saffron in their respective rice dishes, pilaf and paella; the French in their fish stew, bouillabaisse; the Italians in risotto milanese; the Indians in biryanis and milk sweets.

The numbers that figure in saffron production are startling. It takes about 70,000 crocus flowers to produce 5 lb/2.25 kg of stigmas, the three dark red ends of the tube ("style") that carries pollen down to the plant's ovary. These 5 pounds in turn dry down to about 1 lb/450 gm of saffron. And because they're so delicate, the stigmas are still harvested and separated from the other flower parts by hand, with nearly 200 hours of labor required for that same 1 pound of dried saffron. Each purple-petaled flower must be harvested on the same day that it begins to open, in late autumn. Once separated, the stigmas are carefully dried, either with a 30-minute toasting over a fire (Spain), or longer times in the sun (Iran), or in a warm room or modern oven.

Saffron Color Saffron's intense color comes from a set of carotenoid pigments (p. 267) that account for 10% or more of the dried spice's weight. The most abundant form, called crocin, is a molecular sandwich of one pigment molecule with a sugar molecule attached at each end. The sugars make the normally oil-soluble pigment into a water-soluble one—and this is why saffron is easily extracted in hot water or milk and works so well as a coloring agent for rice

The saffron crocus. Pure saffron consists of the dried stigmas, the deep red tips that catch pollen grains and send them down the long style to the ovary. Second-quality saffron often includes the pale, relatively flavorless styles.

and other nonfatty foods. Crocin is a powerful colorant, and gives a noticeable tinge to water even at 1 part per million.

Saffron Flavor Saffron flavor is characterized by a notable bitterness and a penetrating, hay-like aroma. It arises largely from another sugar-hydrocarbon combination, picrocrocin, which may be up to 4% of the fresh weight of the stigmas, and is probably a defense against insects and other predators. The combination itself is bitter. When the stigmas dry out and their cell structures are damaged, both the drying heat and an enzyme act on picrocrocin to liberate the hydrocarbon portion, which is a volatile terpene called safranal. Drying the saffron stigmas thus moderates the bitterness and develops the aroma. Several chemical relatives of safranal round out the overall aroma.

Using Saffron Saffron is typically used in small quantities—a few threads, or a "pinch"—and rehydrated in a small amount of warm or hot liquid before being added to a dish, in order to extract both flavor and color. The main pigment is water soluble, but the inclusion of some alcohol or fat in the extraction liquid will dissolve additional fat-soluble carotenoids.

Saffron's color and flavor molecules are readily altered by light and heat, so this valuable spice is best stored in an airtight container in the freezer.

Sumac Sumac is the small, dried, purple-red berry of a shrubby relative (*Rhus coriaria*) of the cashew and mango trees, a native of southwest Asia. Sumac is unusual for being very tart (from malic and other acids), astringent (from abundant tannins, to 4% of its weight), and aromatic, with pine, woody and citrus notes. Sumac is ground and added to a number of savory dishes in the Middle East and North Africa.

A SURVEY OF TROPICAL SPICES

Among the tropical spices, family relationships don't readily translate into flavor relationships. I've therefore listed them all in simple alphabetical order. It's interesting to note, though, that the ginger family includes turmeric, galangal, cardamoms, and grains of paradise; and that allspice and clove are members of the myrtle family, and thus relatives of each other and of two strong-scented fruits, guava and feijoa.

Allspice Allspice is the brown, medium-sized dried berry of a tree of the New World tropics. *Pimenta dioica* is a member of the myrtle family and a relative of the clove. Allspice took on its modern name in the 17th century because it was thought to combine the aromas of several spices, and today it's often described as tasting like a mellow combination of clove, cinnamon, and nutmeg. It is indeed rich in clove's eugenol and related phenolic volatiles, with fresh, sweet, and woody notes (but no cinnamon volatiles). The main producer is Jamaica. The berries are picked when green and at the height of flavor, briefly fermented in heaps,"sweated" in bags to accelerate their drying and browning, then sun-dried for five to six days (or machine-dried). Allspice finds notable use in pickling fish, meats, and vegetables, as well as in pie seasonings.

Annatto Annatto, also known as achiote, is both a flavoring and a colorant. It is the seed of a bush, *Bixa orellana,* native to tropical America, and is much used in various cooked dishes from southern Mexico to northern South America. The bright red-orange pigment bixin is found in the waxy coating of the seeds, and readily changes into a number of chemical variants that are different shades of orange, yellow, and red. Some of these are soluble in water, others in oil; large food manufacturers use annatto extracts to give a vivid color to cheddar-style cheeses, butter, and other products. Annatto seeds are hard, and difficult to grind finely, so they're often heated in a liquid to extract their flavor and color and then are strained out. Commercially ground annatto pastes are also available. The aroma of annatto is dominated by the woody, dry terpene humulone, which is also found in hops.

Cardamom Cardamom is the world's third most expensive spice after saffron and vanilla. It's the seed of a herbaceous plant in the ginger family that is indigenous to the mountains of southwest India, and was grown only there until around 1900. German immigrants then brought it to Guatamala, which is now the largest producer. Cardamom seeds are borne in clusters of fibrous capsules that ripen at different times, so the capsules must be picked by hand one by one, and slightly before full ripeness, when the capsule splits. The word comes from an Arabic root meaning "to warm"; and cardamom has a delicate, warming quality due to two different sets of aromatics, both stored in a layer just below the seed surface: a group of floral, fruity, and sweet terpene compounds (linalool and acetate esters), and more penetrating, eucalyptus-like cineole.

There are two broadly different varieties of cardamom: Malabar, a small, round capsule with a high content of delicate, flowery compounds, and Mysore, a larger, three-angled capsule with mainly pine, woody, and eucalyptus notes. Both are slightly astringent and pungent. Malabar cardamom develops its best flavor after the pods have begun to turn from green to off-white, so it is usually available only in bleached form, after sun-drying or chemical bleaching to make the pod color more uniform. Mysore cardamom is often sold green, its color fixed by a three-hour dose of moderate heat (130ºF/55ºC) before drying.

Cardamom is mentioned along with cinnamon in the Old Testament, but doesn't seem to have reached Europe until the Middle Ages. Today the Nordic countries consume 10% of world trade, mainly in baked goods, while Arab countries take 80% for their cardamom coffee. *Gahwa* is made by boiling together freshly roasted and ground coffee with freshly broken green cardamom pods.

Large cardamom, also called Nepal or greater Indian cardamom, is the seed of a cardamom relative, *Amomum subulatum,* which grows in the eastern Himalayas of North India, Nepal, and Bhutan. (Other species of *Amomum* and *Aframomum* are also used.) The seeds are borne in a reddish pod an inch/2.5 cm long, with a sweet surrounding pulp. Large cardamom has a strong, harsh flavor for two reasons: much of the crop is smoke-dried, and the seeds are rich in the penetrating terpenes cineole and camphor. Large cardamoms are often used in India, west Asia, and China in savory and rice dishes and in pickles.

Cinnamon Cinnamon is the dried inner bark of trees in the tropical Asian genus *Cinnamomum,* a distant relative of the bay laurel. Its inner bark or phloem layer, which carries nutrients from the leaves toward the roots, contains protective oil cells. When the inner bark is cut and peeled from the new growth of these trees, it curls to form the familiar long "quills" or sticks. Cinnamon was one of the first spices to reach the Mediterranean; the ancient Egyptians used it in embalming, and it's mentioned repeatedly in the Old Testament. Asian and Near Eastern peoples have long used cinnamon to flavor meat dishes, and thanks to the influence of the Arabs, medieval European cooks did too. Nowadays most cinnamon goes into sweet dishes and candies.

There are several different species of *Cinnamomum* that provide aromatic bark, but cinnamons fall into two general categories. One is Ceylon or Sri Lankan cinnamon (from *C. verum* or *zeylanicum*), light brown in color, papery and brittle, coiled in a single spiral, and with a mild, delicate cinnamon flavor often described as sweet. The other is the Southeast Asian or Chinese cinnamon, often called cassia, which is typically thick and hard, forming a double spiral, darker in color and much stronger in flavor, bitter and somewhat harsh and burning, as in the American "red-hot" candy. These cinnamons come mainly from China (*C. cassia*), Vietnam (*C. loureirii*), and Indonesia (*C. burmanii*). Cassia types are preferred in most of the world, Sri Lankan types in Latin America. The typical hot, spicy cinnamon aroma comes from a phenolic compound, cinnamaldehyde, of which cassia types have significantly more than Sri Lankan types; the latter are more subtle and complex, with floral and clove notes (linalool, eugenol).

Cloves Cloves are among the most distinctive and strongest of all spices. They're the dried immature flower buds of a tree in the myrtle family, *Syzygium aromaticum,* which is native to a few islands in present-day Indonesia. Cloves were enjoyed in China 2,200 years ago, but weren't much used in European foods until the Middle Ages. Today Indonesia and Madagascar are the biggest producers.

The flower buds of the clove tree are picked just before they open, and then dried for several days. Their distinctiveness results from a high content of the phenolic compound called eugenol, which has a unique aroma that is both somewhat sweet and very penetrating. Clove buds contain the highest concentration of aroma molecules of any spice. They are as much as 17% volatile chemicals by weight, most of this stored just under the surface of the elongated portion, in the flower cap, and in the delicate filaments of the stamens within. The oil is about 85% eugenol. Thanks mainly to eugenol, clove oil is good at suppressing microbes, and it temporarily numbs our nerve endings, properties that have led to its use in mouthwashes and dental products.

In much of the world cloves flavor meat dishes, while Europeans use them mainly in sweets. Cloves are an important element in a number of spice mixes (see box, p. 398). By far their largest role is in the Indonesian flavored cigarette, kretek, which may be 40% shredded clove.

Galangal Galangal is a name given to the underground stem, or rhizome, of two Asian ginger relatives, *Alpinia galanga* or greater galangal, and *Alpinia officinarum* or lesser galangal. The former, sometimes also called Thai ginger, is the more prized and common. Galangal is more austere than ginger, pungent and with overtones of eucalyptus, pine, clove, and camphor, but none of ginger's lemony character. In Thai and other Southeast Asian cuisines it's often combined with lemongrass and many other aromatics. Galangal is also an ingredient in Chartreuse, bitters, and some soft drinks.

Ginger Ginger is the pungent, aromatic rhizome of a herbaceous tropical plant, *Zingiber officinale,* that is distantly related to the banana. It lends its name to a family of about 45 genera that are found throughout the tropics, and that include galangal, grains of paradise, cardamom, and turmeric. The name comes via Latin from the Sanskrit *singabera,* meaning horns or antlers, which the branched rhizomes resemble.

Ginger was domesticated in prehistoric times somewhere in southern Asia, had been brought in dried form to the Mediterranean by classical Greek times, and was one of the most important spices in medieval Europe. The cake known as gingerbread dates from this time; ginger beer and ginger ale from the 19th century, when English taverns sprinkled powdered ginger on their drinks.

To make the dried spice, mature rhizomes are cleaned, scraped to remove most of the skin, sometimes treated with lime or acid to bleach them, and then dried in the sun or a machine. Dried ginger is about 40% starch by weight. Today the main producers of dried ginger are India and China, while Jamaican ginger is considered one of the finest. A surprisingly large fraction of the ginger trade goes to Yemen, where it is added to coffee (as much as 15% of the coffee's weight).

In Asia, and increasingly in the rest of the world, ginger is used fresh. Most fresh ginger in the United States now comes from Hawaii, where the main harvest runs from December to June. Fresh ginger contains a protein-digesting enzyme that can cause problems in gelatin-based preparations (p. 607).

Ginger Aromas Ginger has a remarkable culinary range, flavoring sausages and fish dishes as well as sodas and sweets. It has something of the quality of lemon juice in that it adds a refreshing, bright aroma—from fresh, floral, citrus, woody, and eucalyptus notes—and mild, pepper-like pungency that complements other flavors without dominating them. Gingers from different parts of the world have different qualities. Chinese ginger tends to be mainly pungent; South Indian and Australian gingers have a notable quantity of citral and so a more distinctly lemony aroma; Jamaican ginger is delicate and sweet, African ginger penetrating.

Ginger Pungency Is Variable The pungency of ginger and members of its family comes from the gingerols, chemical relatives of the chilli's capsaicin and black pepper's piperine (p. 394). The gingerols are the least powerful of the group, and the most easily altered by drying and cooking. When ginger is dried, its gingerol molecules lose a small side group of atoms and are transformed into shogaols, which are about twice as pungent: so dried ginger is stronger than fresh. Cooking reduces ginger pungency by transforming some gingerols and shogaols into zingerone, which is only slightly pungent and has a sweet-spicy aroma.

Grains of Paradise Grains of paradise, guinea grains, alligator pepper, and melegueta pepper are all names for the small seeds of *Aframomum melegueta.* This member of the ginger family is a native of West Africa, and was used in Europe from the Middle Ages until the 19th century, when it became a rarity. It's both somewhat pungent from gingerol and relatives (paradols, shogaols), and faintly but pleasantly aromatic, with woody and evergreen notes (humulone and caryophyllene). It is a component of the Moroccan spice mixture ras el hanout, and can serve as an interesting alternative to black pepper.

Mace and Nutmeg Mace and nutmeg have similar aromas and come from the same source: the fruit of a tropical Asian tree, *Myristica fragrans,* which appears to have originated in New Guinea. Along with cloves, nutmeg put the Spice Islands, the Malaccas that are now part of Indonesia, on the maps of European sea powers. The Portuguese and then the Dutch monopolized the nutmeg trade until the 19th century, when the tree was successfully planted in the Caribbean and elsewhere. Nutmeg and mace didn't make much of an impression on European foods until the Middle Ages. Today they provide the characteristic flavoring for doughnuts and eggnog, and are added to hot dogs and other sausages. Nutmeg is also an important element of the classic French béchamel sauce.

Both nutmeg and mace are borne inside the plum- to peach-sized fruits of the tree. When the fruit is ripe, it splits to reveal a shiny, brown-black shell; and entwined around the shell, a narrow, irregular, bright red ribbon. The red ribbon is an aril, a fruit part whose color and sugars attract birds to carry it and the seed away. The aril is the spice called mace, and the seed inside the shell is the nutmeg. The aril is removed from the shell and dried separately. The aroma compounds in nutmeg are concentrated in a layer of oil-containing tissue that weaves through the seed's main body of starchy and fatty storage tissue, which also contains astringent tannins.

Nutmeg and mace have similar but distinct flavors, with mace the gentler and more rounded. Both spices carry fresh, pine, flowery, and citrus notes, but are dominated by woody, warm, somewhat peppery myristicin (also a minor component in fresh dill). Grated nutmeg includes tannic particles of the main seed storage tissue, and is also darker in color than powdered mace. Nutmeg has generally been put to use in sweets and dishes based on cream, milk, and eggs; mace in meat dishes as well as pickles and ketchups. Their flavors tend to become unpleasant with prolonged heat, so they're often grated over a dish at the last minute.

Nutmeg is reputed to have hallucinogenic effects if several grated seeds are consumed at once. Myristicin has been suggested as the active ingredient, but the evidence is scanty.

Black Pepper and Relatives Black pepper was one of the first spices to be traded westward from Asia, and today it remains the preeminent spice in Europe and North America. We think of it as a basic seasoning, like salt, and use its moderate pungency and pleasant aroma to fill out the flavor of many savory dishes, often just before eating them. Pepper is native to the tropical coastal mountains of southwest India, where sea and overland trade with the ancient world began at least 3,500 years ago. It is mentioned in Egyptian papyruses, was well known to the Greeks, and a popular spice in Rome. During this time it was largely gathered from wild forest plants, though sometime before the 7th century the vine was transplanted to the Malay archipelago, Java, and Sumatra. Vasco da Gama discovered the sea route from Europe to southwest India in 1498, and the Portuguese subsequently controlled exports of black pepper for several decades. They were followed by the Dutch and, beginning around 1635, the British, who established pepper plantations. In the 20th century, a number of countries in South America and Africa began producing black pepper. Today India, Indonesia, and Brazil are the major world sources.

Pepper Production Black pepper is the small dried berry of a climbing vine in the genus *Piper,* which includes a number of other spice and herb plants (see box, p. 429). The berries of *Piper nigrum* form on a flower spike a few centimeters long, and take about six months to mature. As berries mature and ripen, their content of pungent piperine continuously increases, while their aromatics reach a peak and then decline. Fully ripe berries may contain less than half of the aroma that they had at the late green stage. The ripe berry skin is red, but turns dark brown to black after harvest thanks to the activity of browning enzymes. The inner seed is largely starch, with some oil, from 3–9% pungent piperine, and 2–3% volatile oil.

Black, White, Green, and Rose-Colored Peppers Pepper berries are processed to make several different versions of the spice.

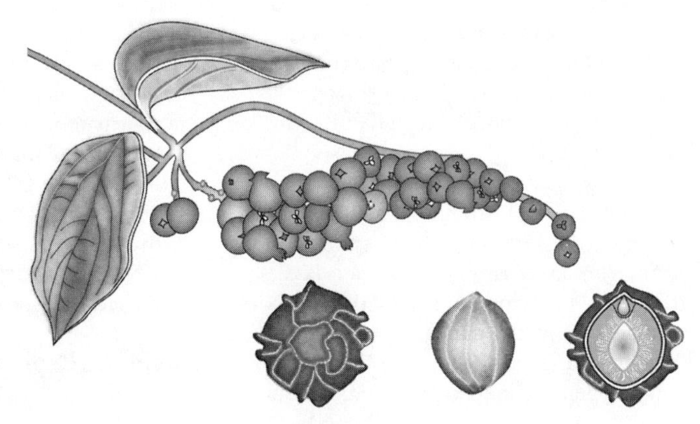

Black and white pepper. Pepper comes from the small fruits of a tropical vine. Black pepper is made by drying the whole fruit; the wrinkled dark outer coating is the dried fleshy fruit layer. White pepper is made by removing the fleshy layer before drying the seed.

- Black pepper, the most common, is made from mature but unripe berries, still green and rich in aromatics. The berry spikes are harvested from the vines and the berries threshed from the spikes. The berries are then blanched for a minute in hot water to clean them and rupture the cells of the fruit layer to speed the work of the browning enzymes. Finally they're sun- or machine-dried for several days, during which the outer fruit layer darkens.
- White pepper consists of the pepper seed only, without the outer fruit layer. It's made from fully ripe berries, which are soaked in water for a week to allow the fruit layer to be degraded by bacteria, then rubbed to remove the fruit layer, and finally dried. White pepper is mainly valued for providing pungency while remaining invisible in light-colored sauces and other preparations. It was developed into a major commercial product in Indonesia, which is still its major producer.
- Green pepper is made from berries harvested a week or more before they would otherwise begin to ripen. The berries are simply preserved by treating with sulfur dioxide and dehydration, by canning or bottling in brine, or by freeze-drying. The flavor depends on the method of preservation, but includes some pungency and pepper aromatics as well as a fresh green-leaf note.
- Pink pepper, or *poivre rose,* is a rarity made by preserving just-ripened red berries in brine and vinegar. (Pink peppercorns are entirely different; see below.)

Pepper Flavor The main pungent compound in pepper is piperine, which is found in the thin fruit layer and the surface layers of the seed. Piperine is about 100 times less pungent than the capsaicin in chillis. The major aroma components in black pepper (the terpenes pinene, sabinene, limonene, caryophyllene, linalool) create an overall impression that is fresh, citrusy, woody, warm, and floral. White pepper is about as pungent as black pepper, but lacks much of its aroma due to the removal of the outer fruit layer. It often has musty and horse-stable notes, probably from the prolonged fermentation of the fruit layer (skatole, cresol).

Pepper is used in the form of whole "corns" in preparations that allow time to extract their flavor: such things as pickles and preserves, and some stocks and sauces. Grinding the peppercorns allows their flavor to be extracted faster for last-minute adjustments. Grinding also frees their aromatics to evaporate, so the most and freshest flavor comes from whole peppercorns ground directly into the preparation. Even whole peppercorns lose much of their aroma after a month in a grinder. Some cooks briefly toast them in a hot pan to enrich their aroma.

Pepper is best stored tightly sealed in the cold and dark. If exposed to light during storage, it loses its pungency because the light energy rearranges piperine to form a nearly tasteless molecule (isochavicine).

Pink Peppercorns Pink peppercorns are fruits of the Brazilian pepper tree, *Schinus terebinthifolius,* which was brought to the southern United States as an ornamental and has become an invasive pest. Its attractive pink fruits were first marketed as a kind of pepper in the 1980s. The tree is in the cashew and mango family, which also includes poison ivy and poison oak, and its brittle, peppercorn-sized fruits contain cardanol, an irritating phenolic compound that limits its usefulness in foods. Pink peppercorns have fresh, pine, citrus, and sweet aroma notes thanks to several terpenes. A close relative from Peru, *S. molle,* is also grown as an ornamental and is called the California pepper tree. Its fruits have a more resinous aroma (thanks to myrcene), with less irritant cardanol.

Sichuan Pepper, Sansho The Chinese spice known as Sichuan pepper and the

Japanese sansho both offer a strange and interesting version of pungency. They come from two small trees in the citrus family sometimes called "prickly ash." Sichuan pepper trees are *Zanthoxylum simulans* or *Z. bungeanum*, and sansho trees are *Zanthoxylum piperitum*. (*Xanthoxylum* is another spelling.) The spices are the small dried fruit rinds, which are aromatic with lemony citronellal and citronellol. The pungent compounds, the sanshools, are members of the same family as piperine from black pepper and capsaicin from chillis. But the sanshools aren't simply pungent. They produce a strange, tingling, buzzing, numbing sensation that is something like the effect of carbonated drinks or of a mild electrical current (touching the terminals of a nine-volt battery to the tongue). Sanshools appear to act on several different kinds of nerve endings at once, induce sensitivity to touch and cold in nerves that are ordinarily nonsensitive, and so perhaps cause a kind of general neurological confusion.

The Chinese and Japanese versions of this spice are different. Chinese Sichuan peppercorns are always toasted, so their citrus affinities are overshadowed by browned, woody notes that go well with meats. Japanese sansho is distinctly lemony, and is used to mask or balance the fattiness of some fish and meats. These spices are almost always used as part of a mixture.

Sandalwood Sandalwood is more familiar in incense than in foods, but the roots and heartwood of the tree *Santalum album* are

Some Relatives of Pepper

There are about a thousand different species in the genus *Piper,* and many relatives of *Piper nigrum* have been also been used as food flavorings, including the herbs hoja santa and lolot (pp. 409, 410). Other notable pepper relatives include the following:

- Long pepper (*Piper longum*). This native of India was probably the first pungent spice after mustard to be appreciated in Europe—the Greeks and Romans preferred it to black pepper—and it gave us our word *pepper* via its Sanskrit name *pippali* (black pepper is *marichi*). Long pepper is so called because it is the entire flower spike with tiny fruits embedded in its surface. It has a somewhat more pungent taste (due to a larger supply of piperine), and a woody aroma. Today it's mainly used in vegetable pickles, though it's also found in some North African spice mixtures. Another plant called long pepper is *Piper retrofractum,* a native of Java and still used in Indonesia and Malaysia. It is said to be more aromatic than Indian long pepper
- Cubeb or tailed pepper (*P. cubeba*). This version of pepper consists of individual berries and their tail-like stems. It is a native of Indonesia, and was used in European cooking in the 17th century; in its home region it still flavors sauces, liquors, lozenges, and cigarettes. In addition to pungency, it has fresh, eucalyptus, woody, spicy, and floral aromatics.
- Ashanti pepper (*P. guineense*). In West Africa this spice lends nutmeg and sassafras notes to various dishes.
- Betel leaf (*P. betle*). The clove-scented leaves of this Asian pepper species have long been wrapped around other ingredients into a bite-sized packet and chewed. The Indian packet, supari, includes lime, the betel nut from the areca palm, and sometimes tobacco.

sometimes used in India to flavor sweets. Its aroma comes mainly from santalol, which has woody, floral, milky, musky qualities.

Star Anise Star anise is the strikingly star-shaped woody fruit of a tree in the magnolia family, *Illicium verum,* a native of South China and Indochina. Its anise flavor comes from the same phenolic chemical, anethole, that flavors the entirely unrelated European anise (p. 414). The fruit itself, which may have six to eight chambers, carries more of the flavor than the seeds, and the unripe fruit is traditionally chewed as a breath sweetener. One traditional and important use of star anise is in Chinese meat dishes simmered in soy sauce; when onions are included, the result is the production of sulfur-phenolic aromatics that intensify the meatiness of the dish.

Tamarind Tamarind is the fibrous, sticky, aromatic, and intensely sour pulp that surrounds the seeds in pods of *Tamarindus indica,* a tree in the bean family native to Africa and Madagascar. The pulp can be extracted by soaking it in water for a few minutes, squeezing the fibrous mass, and straining off the flavored water; tamarind extract is also manufactured and sold as a thick paste. The pulp is about 20% acids, mainly tartaric, 35–50% sugars, and about 30% moisture, and has a complex, savory, roasted aroma thanks to browning reactions that take place on the tree as the pulp becomes concentrated in the hot sun. In much of Asia, tamarind is used to acidify and flavor sweet-sour preserves, sauces, soups, and drinks. Tamarind is also popular in the Middle East, and it's one of the defining ingredients in Worcestershire sauce.

Turmeric Turmeric is the dried underground stem, or rhizome, of a herbaceous tropical plant in the ginger family, *Curcuma longa.* It appears to have been domesticated in prehistoric times in India, probably for its deep yellow pigment (*curcuma* comes from the Sanskrit for "yellow"). Turmeric has long been used to color skin, clothing, and foods for ceremonies surrounding marriage and death. In the United States, the main use of turmeric is to provide color and nonpungent filler in prepared mustards. It's also the major component of most prepared curry powders, making up 25–50% of their weight.

The major pigment in turmeric is a phenolic compound called curcumin, which turns out to be an excellent antioxidant. This may explain why turmeric is considered to have preservative properties; in India fish and other foods are often first dusted with it before cooking, and it goes into many prepared dishes. The color of curcumin is sensitive to pH. In acid conditions it's yellow, while in alkaline conditions it turns orange-red.

To make the spice, turmeric rhizomes are steamed or boiled in slightly alkaline water to set the color and precook the abundant starch, then sun-dried. Turmeric is usually sold preground, though fresh and dried rhizomes can be found in ethnic markets. Turmeric has a woody, dry earth aroma (from mildly aromatic terpenes called turmerone and zingiberene), with slight bitterness and pungency.

Vanilla Vanilla is one of the most popular flavorings in the world. Among the spices it's unique for the richness, depth, and persistence of its flavor. And it's the second most costly, after saffron. So in fact most of the vanilla flavoring consumed in the world today is a synthetic imitation of the original spice.

True vanilla comes from the pod fruit, often called the "bean," of a climbing orchid native to Central and northern South America. There are about 100 species in the tropical genus *Vanilla. V. planifolia* (or *V. fragrans*) was first cultivated by the Totonac Indians along the eastern coast of Mexico near Veracruz, perhaps as long as 1,000 years ago. They sent it north to the Aztecs, who flavored their chocolate drinks with it (p. 695). The first Europeans to taste vanilla were the Spanish, who gave it its name; *vainilla* is the Spanish diminutive for "sheath" or "husk" (from the Latin *vagina*). A 19th-century Belgian botanist, Charles Morren, figured out how to pollinate vanilla flowers by hand, and thus made it possible to produce the spice in regions that lacked the proper pollinating insects. And the French took the vine to the islands off the coast of southeast Africa that now supply much of the world: Madagascar, Réunion, and Comoros, which collectively produce what is called Bourbon vanilla.

Today, Indonesia and Madagascar are the world's largest producers. It's the attentive and extensive labor required to hand-pollinate the vanilla flowers and cure the pods, and the low production of the few regions that cultivate it, that make vanilla so expensive.

Vanilla's rich flavor is the creation of three factors: the pod's wealth of phenolic defensive compounds, preeminently vanillin; a good supply of sugars and amino acids to generate browning-reaction flavors; and the curing process. The plant stores most of its defensive aromatics in inert form by bonding them to a sugar molecule. The active defenses—and aromas—are released when damage to the pod brings the storage forms into contact with bond-breaking enzymes. The key to making good vanilla is thus deliberate damage to the pods, followed by a prolonged drying process that develops and concentrates the flavor, and prevents the pod from spoiling.

Making Vanilla The making of vanilla begins six to nine months after the orchid flowers have been pollinated, with green pods 6–10 inches/15–25 cm long that are just beginning to ripen. On the pod's inner walls, thousands of tiny seeds are embedded in a complex mixture of sugars, fats, amino acids, and phenolic-sugar storage compounds. The enzymes that can liberate the aromatic phenolics from storage are concentrated closer to the outer walls. The first step in curing is to kill the pod so that it doesn't use up its sugars and amino acids, and to damage the pod's cells and allow the phenolic storage compounds to migrate to the liberating enzymes. Both of these goals are accomplished by briefly exposing the pods to high temperatures, either in the sun or in hot water or steam. The cell damage that this causes also allows the browning enzymes (polyphenoloxidases, p. 269) to cluster some phenolic compounds together into colored aggregates, so the pod color changes from green to brown.

Then follow several days during which the pods are alternately exposed to the sun until they're almost too hot to handle, then wrapped in cloth to "sweat" with the residual heat. During this stage, the main flavor components of vanilla—vanillin and related phenolic molecules—are freed from their bondage to sugar molecules. The heat and sunlight also evaporate some of the pod moisture, discourage microbial growth on

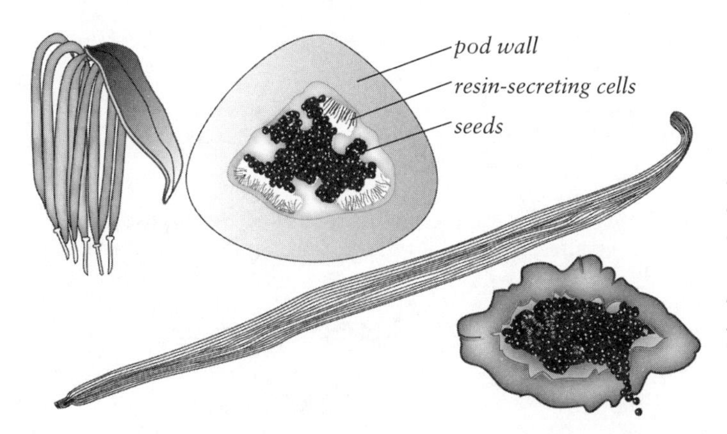

Pods of the vanilla orchid. The fresh pod contains thousands of tiny seeds embedded in a sticky resin of sugars, amino acids, and a storage form of the main vanilla aroma compound, vanillin. The process of drying and curing the pod frees the vanillin and creates additional aroma molecules.

the humid surface, and generate pigments and complex aromas via browning reactions between sugars and amino acids (p. 778). It takes 3 to 5 pounds of fresh pods to produce 1 pound of cured beans.

In the last stage of vanilla processing, the pods are straightened and smoothed by hand, dried for several weeks, then "aged," or stored for some time to develop flavor further (flavor compounds react with oxygen, some heat-resistant enzymes, and each other to form fruity esters and other new notes). In Madagascar, vanilla curing takes 35–40 days, while the Mexican process lasts several months.

The Flavors of Vanilla The cured pod is about 20% water by weight, 20% fiber, 25% sugars, 15% fat, and the remainder amino acids, phenolic compounds, other flavors, and brown pigments. The sugars provide sweetness, free amino acids provide some savoriness, fat richness, and tannins some astringency. The aroma of natural vanilla is complex. More than 200 different volatile compounds have been found in vanilla beans. The principal one, the phenolic compound vanillin, does suggest vanilla on its own, but without the whole spice's richness. Some of the other important vanilla volatiles contribute flavor notes described as woody, floral, green-leaf, tobacco, dried-fruit, clove-like, honey-like, caramel, smoky, earthy, and buttery.

Kinds of Vanilla The different vanilla-producing regions produce beans with broadly different flavors. Bourbon vanilla from Madagascar and neighboring islands is generally considered to be the finest, with the richest, most balanced flavor. Indonesian beans seem lighter, with less vanillin, and sometimes smoky qualities. Mexican beans contain about half the vanillin of Bourbon beans, and have distinctive fruity and winy aromas. Rare Tahitian vanilla beans—from a different species, *V. tahitensis*, also have much less vanillin than Bourbon beans, but carry unique flowery and perfumed notes.

Vanilla Extracts and Flavorings Vanilla extracts are made by chopping whole vanilla beans and repeatedly passing a mixture of alcohol and water over them for several days, then aging the extract to develop a more complex, full flavor. Vanillin and the other flavor components are more soluble in alcohol than water, so the higher the flavor content desired in the extract, the higher the proportion of alcohol necessary to carry it.

Artificial vanilla flavoring contains synthetic vanillin made from various industrial by-products, especially wood lignin, and doesn't have the full, complex, subtle flavor of whole vanilla beans or their extracts. The demand for vanilla flavoring far exceeds the available crop, and natural vanillin costs about 100 times more than synthetic. About 90% of the vanilla flavoring consumed in the United States is artificial; in France, about 50%.

The Virtues of Vanillin

In addition to making vanilla taste like vanilla, vanillin is formed during certain cooking and manufacturing procedures, especially those involving wood fires and wood barrels (pp. 448, 721). It thus contributes to the flavor of such foods as grilled and smoked meats, wines and whiskeys, bread, and boiled peanuts. Vanillin also has several potentially useful biological properties. It is toxic to many microbes, an antioxidant, and inhibits DNA damage.

Cooking with Vanilla Vanilla is used mainly in sweet foods. Almost half of the vanilla flavoring consumed in the United States goes into ice cream, and much of the rest into soft drinks and chocolate. But it also works in savory dishes: lobster and pork are popular examples. Added with a light touch, vanilla can contribute a sense of depth, warmth, roundness, and persistence to almost any food.

The flavor of the whole vanilla bean resides in two different parts of the bean: the sticky, resinous material in which the tiny seeds are embedded, and the fibrous pod wall. The first is easily scraped out of the bean and dispersed in a preparation, while the pod itself must be soaked for some time in order to extract its flavor. Because the volatiles are generally more soluble in fat than in water, the cook can extract more flavor if the extraction liquid includes either alcohol or fat. Prepared vanilla extracts can be dispersed throughout a dish instantly, and are usually best added toward the end of cooking; any period of time spent at a high temperature causes aroma loss.

TEA AND COFFEE

Tea and coffee are the most widely consumed drinks in the world, and their popularity stems from the same source as that of herbs and spices: the plant materials they're made from are crammed with chemical defenses that we have learned to dilute, modify, and love. Tea leaves and coffee beans have one defense in common, and that's caffeine, a bitter alkaloid that has significant effects on our bodies. And they both contain large doses of phenolic compounds. However, they're very different materials. Coffee begins as a seed, a storehouse of protein, carbohydrate, and oil, and is the creation of high heat, a robust epitome of roasted foods and flavors. Tea begins as a new, actively growing leaf, rich in enzymes, and is the delicate creation of those enzymes, carefully captured and preserved by minimal heat and drying. Coffee and tea thus offer two very different experiences of botanical inventiveness and human art.

Caffeine

Caffeine is the most widely consumed behavior-modifying chemical in the world. It is an alkaloid (p. 238) that interferes with a particular signaling system used by many different cells, and therefore has several different effects on the human body. Above all, caffeine stimulates the central nervous system, relieves drowsiness and fatigue, and quickens reaction times. It also increases energy production in muscles and so their capacity for work. It's said to improve mood and mental performance, though recent studies suggest that these may be the result of relieving the initial symptoms of overnight caffeine withdrawal! Less desirably, in high doses it causes restlessness, nervousness, and insomnia. It has complex effects on the heart and arteries, and can produce an abnormally fast heartbeat. There is some evidence that caffeine speeds the loss of calcium from bone, so habitual consumption may contribute to osteoporosis.

Caffeine reaches its maximum levels in the blood between 15 minutes and two hours after consumption, and its levels are reduced by half within three to seven hours. Its effects are more noticeable in people who don't normally consume it. Withdrawal symptoms can be unpleasant, but usually disappear within three days of abstaining.

A chemical relative of caffeine called theophylline is found in tea and is in some respects more potent than caffeine, but tea contains only trace amounts. Though coffee beans are 1–2% caffeine and tea leaves 2–3%, brewed coffee contains more caffeine than brewed tea because a larger weight of coffee is extracted per cup (8–10 grams, vs. 2–5 grams for tea).

Tea, Coffee, and Health

Not so many years ago, both coffee and tea were suspected of contributing to various diseases, including cancers, so they were among the many pleasures to feel guilty about. No longer! Coffee is now recognized as the major source of antioxidant compounds in the American diet (medium roasts have the highest antioxidant activity). Black and especially green teas are also rich in antioxidant and other protective phenolic compounds that appear to reduce damage to arteries and cancer risk.

Certain kinds of brewed coffee do turn out to have an undesirable effect on blood cholesterol levels. Two lipid (fat-like) substances, cafestol and kahweol, raise those levels, though they only get into the coffee when the brewing technique doesn't filter them out. Boiled, plunger-pot, and espresso coffees contain them. The significance of this effect isn't known and may well be small, since the cholesterol raisers are accompanied by a large dose of substances that protect the cholesterol from oxidation and causing damage (p. 255).

Water for Making Tea and Coffee

Brewed tea and coffee are 95–98% water, so their quality is strongly influenced by the quality of the water used to make them. The off-flavors and disinfectant chlorine compounds of most tap waters are largely driven off by boiling. Very hard water, high in calcium and magnesium carbonates, has several undesirable effects: in coffee, these minerals slow flavor extraction, cloud the brew, clog the pipes in espresso machines and reduce the fine espresso foam; in tea, they cause the formation of a surface scum made up of precipitated calcium carbonate and phenolic aggregates. Softened water overextracts both coffee and tea and gives a salty flavor. And very pure distilled water gives a brew best described as flat, with a missing dimension of flavor.

The ideal water has a moderate mineral content, and a pH that is close to neutral, so that the final brew will have a moderately acid pH of around 5, just right to support and balance the other flavors. Some bottled spring waters are suitable (Volvic is used in Hong Kong). Many municipal tap waters

Caffeine Numbers

Daily caffeine consumption in milligrams per capita, 1990s

Norway, Netherlands, Denmark	400
Germany, Austria	300
France	240
Britain	200
United States	170

Caffeine content, milligrams per serving

Brewed coffee	65–175
Espresso	80–115
Tea	50
Cola	40–50
Cocoa	15

are intentionally made alkaline to reduce pipe corrosion, and this can reduce the acidity and liveliness of both tea and dark-roasted coffee (light roasts contribute plenty of their own acid). Alkaline tap water can be corrected by adding tiny pinches of cream of tartar—tartaric acid—until it just begins to have a slightly tart taste.

TEA

Though it has lent its name to many other infusions, *tea*—from the Chinese word *cha*—is a drink prepared from the green leaves of a kind of camellia. Young tea leaves turn out to be as packed with interesting defensive chemicals as any spice. Beginning in southwest China around 2,000 years ago, people learned how to use physical pressure, mild heat, and time to coax a number of different flavors and colors from the tea leaf. Tea became a staple of the Chinese diet around 1000 CE. In 12th-century Japan, Buddhist monks who valued tea as an aid to long hours of study found that tea itself was worthy of their contemplation. They developed the formal tea ceremony, which remains remarkable for the attention it pays to the simplest of preparations, an infusion of leaves in water.

The History of Tea

Tea in China The tea tree, *Camellia sinensis,* is native to Southeast Asia and southern China, and its caffeine-rich, tender young leaves were probably chewed raw long before recorded history. The preparation of tea leaves for infusion in water evolved slowly. There's evidence that by the 3rd century CE the leaves were boiled and then dried for later use, and that by the 8th century they were also stir-fried before drying. These techniques would give green or yellow-green leaves and infusions, and mild but bitter and astringent flavor. More strongly flavored and orange-red teas like modern oolongs were developed around the 17th century, probably beginning with the accidental observation that the leaves develop a distinctive aroma and color when they're allowed to wilt or are pressed before being dried. It was around this time that China began to trade extensively with Europe and Russia, and the new, more complex style of tea conquered England, where consumption rose from 20,000 pounds in 1700 to 20 million in 1800. The strong "black" tea that's most familiar in the West today is a relatively recent invention, the result of intensive pressing; the Chinese developed it in the 1840s specifically for export to the West.

The Spread of Tea Production Until the late 19th century, all tea in world trade was China tea. But when China began to resist Britain's practice of paying for its expensive tea habit with opium, the British intensified tea production in their own colonies, particularly India. For warm regions they cultivated an indigenous variety, *Camellia sinensis* var. *assamica,* or Assam tea, which has more phenolic compounds and caffeine than China tea and produces a stronger, darker black tea. They planted the hardier China types in the Himalayan foothills of Darjeeling and at high elevations in the south. India is now the world's largest tea producer.

Today about three-quarters of the tea produced in the world is black tea. China and Japan still produce and drink more green tea than black.

The Tea Leaf and Its Transformation A fresh tea leaf tastes bitter and astringent, and not much else. This is a reflection of the fact that its major chemical component, even more abundant than its structural materials, is a host of bitter and astringent phenolic substances whose purpose is to make the leaf unattractive to animals. And its aromatic molecules are locked up in nonvolatile combinations with sugar molecules. Green tea retains many of the qualities of the fresh leaf. But the key to making oolong and black teas is encouraging the leaf's own enzymes to transform these austere defensive materials into very different, delightful molecules.

How Tea Enzymes Create Flavor, Color, and Body The period of enzyme activity during tea-making has traditionally been called "fermentation," but it doesn't involve any significant microbial activity. In tea-making, "fermentation" means enzymatic transformation. It occurs when the tea maker presses the leaves to break open their cells, and then allows the leaves to sit for some time while the enzymes do their work.

There are two general kinds of enzymatic transformation in making tea. One is the liberation of a large range of aroma compounds, which in the intact leaf are bound up with sugars and so can't escape into the air. When the cells are crushed, enzymes break the aroma-sugar complex apart. This liberation makes the aroma of oolong and black teas fuller and richer than the aroma of green teas.

The second transformation builds large molecules from small ones, and thereby modifies flavor, color, and body. The small molecules are the tea leaf's abundant supply of three-ring phenolic compounds, which are astringent, bitter, and colorless. The leaf's browning enzyme, polyphenoloxidase, uses oxygen from the air to join the small phenolic molecules together into larger complexes (p. 269). A combination of two phenolics gives a kind of molecule (theaflavin) that's yellow to light copper in color, less bitter but still astringent. Complexes of from three to ten of the original phenolics are orange-red and less astringent (thearubigens). Even larger complexes are brown and not astringent at all. The more the tea leaves are pressed, and the longer they're allowed to sit before the enzymes are killed by heating, the less bitter and astringent and the more colored they become. In oolong teas, about half of the small phenolics have been transformed; in black teas, about 85%.

The red and brown phenolic complexes—and another complex, between double-ring molecules of caffeine and the theaflavins—lend body to brewed tea, because they're large enough to obstruct each other and slow the movement of the water.

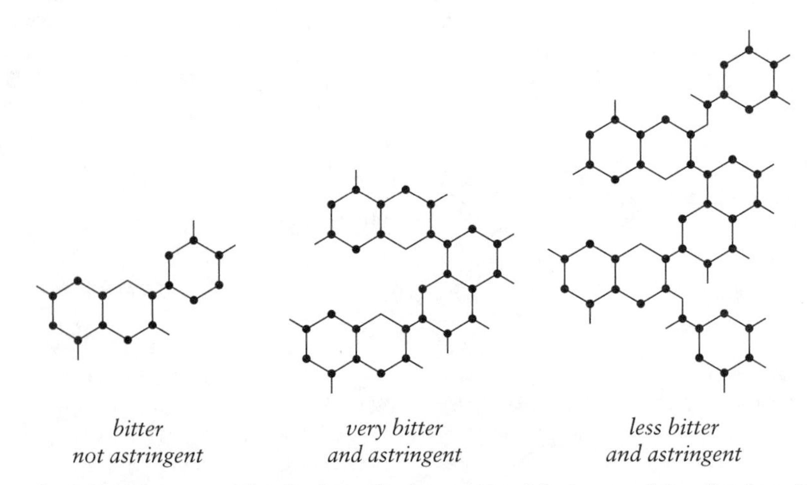

The evolution of tea taste. The fresh tea leaf contains rich stores of simple phenolic compounds (catechin, left) *that are colorless and bitter but not astringent. When the tea leaf is bruised or rolled, leaf enzymes and oxygen combine the simple compounds into larger ones with different colors and tastes. Brief enzyme action produces a yellowish compound (theaflavin,* center) *that is both very bitter and astringent. More extensive enzyme action produces a compound (theaflavin digallate,* right) *that is moderately bitter and astringent. As the phenolic molecules get larger, they get progressively darker and milder.*

Making Tea

The Tea Plant and Its Leaves The best tea is made from the plant's small young shoots and unopened leaf buds, which are the most tender and vulnerable and contain the highest concentrations of chemical defenses and related enzymes. The choice "pluck" is the terminal bud and two adjacent leaves. Most tea is now harvested by machine, and therefore contains a large proportion of older and less flavorful leaves.

Tea Manufacturing The production of tea involves several different steps, some standard and some optional.

- The newly harvested leaves may be allowed to "wither," or sit and wilt for minutes or hours. Withering causes them to shift their metabolism in ways that change their flavor, and to become physically more fragile. The longer the withering, the deeper the flavor and color of the leaves and the brew they make.
- The leaves are almost always "rolled," or pressed to break down the tissue structure and release the cell fluids. If the leaves are rolled while they're still raw, this allows the leaf enzymes and oxygen to transform the cell fluids and generate additional flavor, color, and body.
- The leaves may be heated to inactivate their enzymes and stop the enzymatic production of flavor and color. High dry heat will also generate flavor.
- The leaves are heated to dry them out and preserve them for long keeping.
- The dry leaves are sieved and graded by piece size, which ranges from whole leaves to "dust." The smaller the piece, the faster the extraction of color and flavor.

Major Tea Styles The Chinese developed a half-dozen different styles of tea. Three of them account for most of the tea consumed in the world.

Green Tea Green tea preserves some of the original qualities of the fresh leaf, while heightening them and rounding them out. It's made by cooking the fresh or briefly withered leaves to inactivate their enzymes, then pressing them to release their moisture, and drying them in hot air or on a hot pan. In China, the cooking is done on a hot pan, and this "pan-firing" produces aroma molecules characteristic of roasted foods (pyrazines, pyrroles) and a yellow-green infusion. In Japan, the cooking is done with steam, which preserves more of the grassy flavor and green color in both leaf and tea.

Oolong Tea Oolong tea is made by allowing some modest enzyme transformation of leaf juices. The leaves are withered until they become significantly wilted and weakened. Then they are lightly agitated to bruise the leaf edges, allowed to rest for a

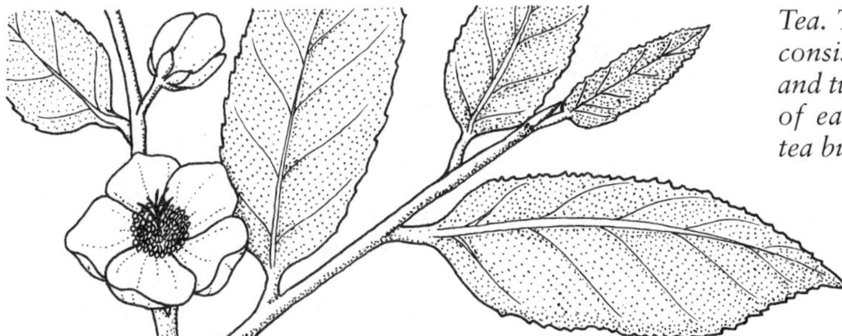

Tea. The choicest pluck consists of the bud tip and two youngest leaves of each branch of the tea bush.

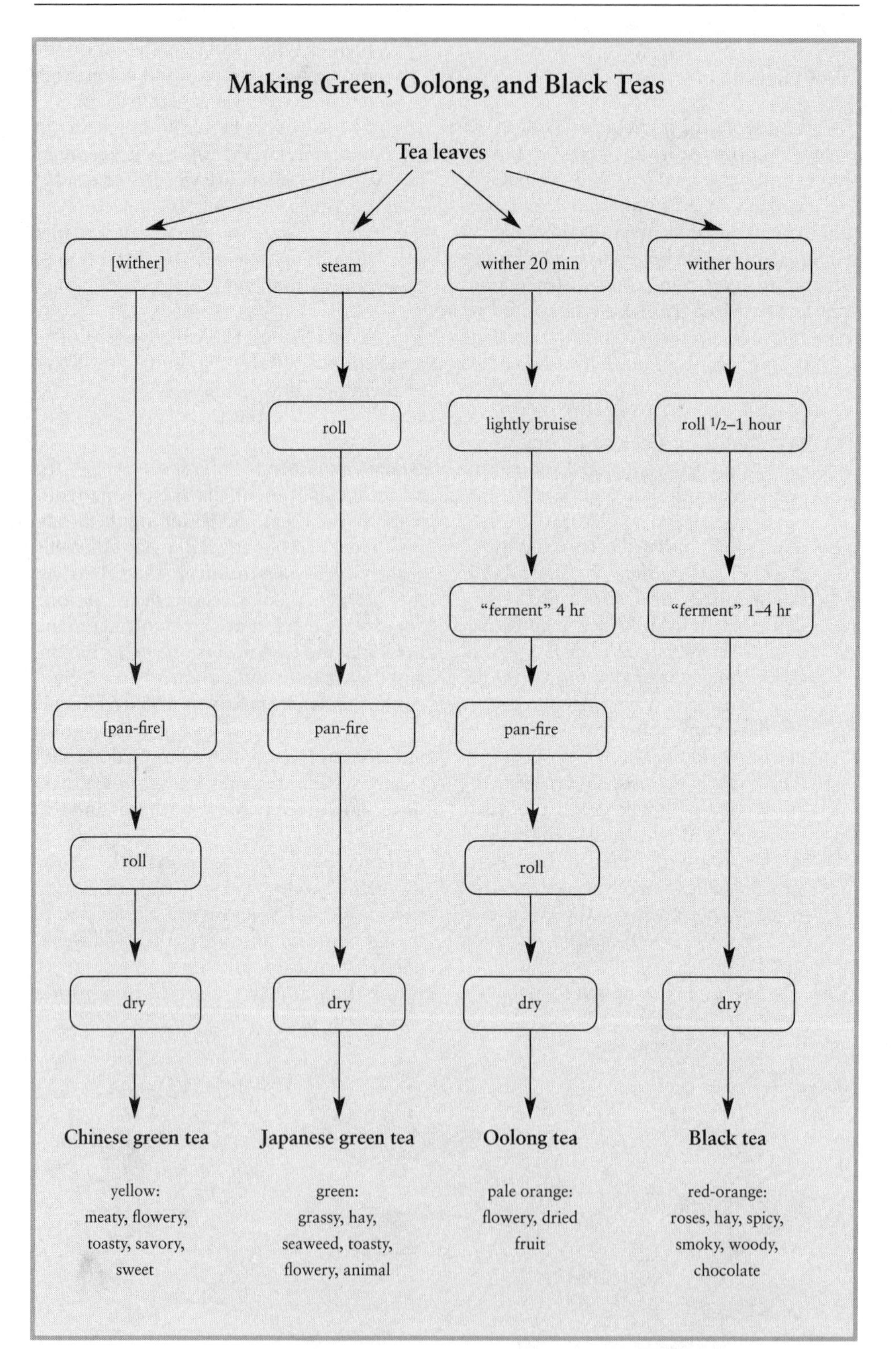
Making Green, Oolong, and Black Teas
Tea leaves
[wither]
steam
wither 20 min
wither hours
roll
lightly bruise
roll 1/2–1 hour
"ferment" 4 hr
"ferment" 1–4 hr
[pan-fire]
pan-fire
pan-fire
roll
roll
dry
dry
dry
dry
Chinese green tea
Japanese green tea
Oolong tea
Black tea
yellow: meaty, flowery, toasty, savory, sweet
green: grassy, hay, seaweed, toasty, flowery, animal
pale orange: flowery, dried fruit
red-orange: roses, hay, spicy, smoky, woody, chocolate

few hours until enzyme action has turned the bruised edges red, pan-fired at a high temperature, rolled, and finally dried gently, at temperatures just below 212°F/100°C. Oolong tea brews to a light amber color with a distinctive fruity aroma.

Black Tea Black tea is made by allowing a profound enzymatic transformation of the leaf fluids. The leaves are withered for hours, rolled repeatedly for as much as an hour, then are allowed to rest for between one and four hours, during which enzyme action turns them a coppery brown and causes them to emit the aroma of apples. Finally, the leaves are air-dried at temperatures around 100°C, and become quite dark.

Tea Flavor The taste of tea, a lively, mouthfilling quality, comes from several different sources. Tea is mildly acid and bitter and contains traces of salt. It's also rich in a unique amino acid, theanine, which is itself sweet and savory, and partly breaks down during manufacturing to savory glutamic acid. Chinese green teas also contain synergizers of savoriness (GMP and IMP, p. 342). Finally, bitter caffeine and astringent phenolics bond to and take the edge off each other and produce the impression of a stimulating but not harsh body. This effect is especially important to the taste of black teas, in which it's called "briskness."

The aromas of different teas are strikingly different. In green teas, early application of heat prevents much enzyme activity in the leaves. Steam heat gives grassy and seaweed, shellfish notes to Japanese green teas (the sea notes from dimethyl sulfide), while pan-firing and drying produce more

Some Prized and Unusual Teas

Here are a number of teas made in unusual ways, with unusual qualities:

- White tea: a Chinese green tea made almost exclusively from buds whose hairs make them look white, withered for two to three days, sometimes steamed, then dried without rolling.
- Pu-erh: a Chinese green tea that is made in the usual way, then moistened and fermented in heaps for some time by a variety of microbes. All of its phenolic contents are converted into nonastringent thearubigens and brown complexes, and it develops a complex, spicy, clove aroma.
- Lapsang souchong: a Chinese black tea, dried over smoky pine fires.
- Scented teas: Chinese teas of various types, scented by being held for 8–12 hours in the same container with flowers, including jasmine, cassia bud, rose, orchid, and gardenia. The packaged tea may include 1–2% flower petals.
- Gyokura and kabesucha: Japanese green teas made from shoots that have been covered with bamboo boxes and almost completely shaded for the two weeks before harvest. They develop a higher content of carotenoid pigments that contribute violet notes to the unique "covered aroma."
- Hoji-cha: Japanese green tea of standard grade that is roasted at high temperatures (360°F/180°C), which triples the volatile content and so boosts flavor.

OPPOSITE: *Making green, oolong, and black teas. Variations in processing produce very different colors and flavors from the same fresh leaves.*

savory, toasted notes in Chinese green teas. In oolong and black teas, enzyme activity liberates floral and fruity aroma molecules from their odorless storage forms, and produces a much richer, stronger aroma (more than 600 volatiles have been identified in black tea).

Cooks exploit tea flavor in a number of different preparations: marinades and cooking liquids, ices and ice creams, in steamed foods, and as a source of aromatic smoke (e.g., Chinese tea-smoked duck).

Keeping and Brewing Tea Well-made tea is fairly stable and can be stored for several months in an airtight container that is kept cool and dark. Tea quality does eventually deteriorate thanks to the effects of oxygen and some residual enzyme activity; aroma and briskness are lost, and the color of black tea infusions becomes less orange-red, more dull brown.

Teas are brewed in various ways in different parts of the world. In the West, a relatively small quantity of black tea leaves—a teaspoon per 6-oz cup/2–5 gm per 180 ml—is brewed once, for several minutes, then discarded. In Asia, a larger quantity of leaves of any tea—as much as a third the volume of the pot—is first rinsed with hot water, then infused briefly several times, with the second and third infusions offering more delicate, subtle flavor balances. The infusion time ranges from 15 seconds to 5 minutes, and depends on two factors. One is leaf size; small particles and their great surface area require less time for their contents to be extracted. The other is water temperature, which in turn varies depending on the kind of tea being brewed. Both oolong and black teas are infused in water close to the boil, and relatively briefly. Green tea is infused longer in much cooler water, 160–110°F/70–45°C, which limits extraction of its still abundant bitter and astringent phenolics, and minimizes damage to its chlorophyll pigment.

In a typical 3–5 minute infusion of black tea, about 40% of the leaf solids are extracted into the water. Caffeine is rapidly extracted, more than three quarters of the total in the first 30 seconds, while the larger phenolic complexes come out much more slowly.

Serving Tea Once tea is properly brewed, the liquid should be separated from the leaves immediately; otherwise extraction continues and the tea gets harsh. All kinds of tea are best drunk fresh; as they stand, their aroma dissipates, and their phenolic components react with dissolved oxygen and each other, changing the color and taste.

Tea is sometimes mixed with milk. When it is, the phenolic compounds immediately bind to the milk proteins, become unavailable to bind to our mouth surfaces and salivary proteins, and the taste becomes much less astringent. It's best to add hot tea to warm milk, rather than vice versa; that way the milk is heated gradually and to a moderate temperature, so it's less likely to curdle.

Lemon juice is sometimes added to tea to bolster its tartness and add the fresh citrus note to its aroma. It also lightens the color of brewed black tea by altering the structure of the red phenolic complexes (the complexes are weak acids themselves, and take up hydrogen ions from the lemon juice). Alkaline brewing water, conversely, tends to produce blood-red infusions from black tea, and can even make green tea red.

Iced Tea Iced tea is the most popular form of tea in the United States; it first caught on at the 1904 World's Fair in steamy St. Louis. It's made by brewing tea with about half-again as much dry tea per cup, to compensate for the later dilution by melting ice. The addition of ice to normally brewed tea tends to make the tea cloudy, due to the formation of particles of a complex between caffeine and theaflavin. The way to avoid this is to brew the initial tea at room or refrigerator temperature, over several hours. This technique extracts less caffeine and theaflavin than brewing in hot water, so the caffeine-theaflavin complexes don't form in sufficient quantities to become visible in the chilled tea.

Coffee

Coffee trees are native to east Africa, and were probably first valued for their sweet cherry-like fruits and for their leaves, which could be made into a kind of tea. Even today an infusion of the dried fruit pulp is enjoyed in Yemen, where the seeds or "beans" were apparently first roasted, ground, and infused in the 14th century. Our word coffee comes from the Arabic *qahwah,* whose own origin is unclear. The coffee tree was taken to south India around 1600, from India to Java around 1700, and from Java (via Amsterdam and Paris) to the French Caribbean shortly thereafter. Today Brazil, Vietnam, and Colombia are the largest exporters of coffee; African countries contribute about a fifth of world production.

The History of Coffee Brewing The original version of brewed roasted coffee beans is the Arab version, which still thrives in the Middle East, Turkey, and Greece. The finely powdered beans are combined with water and sugar in an open pot, the mixture boiled until the pot foams, then settled and boiled to a foam once or twice more, and finally decanted into small cups. This is the coffee that found its way to Europe around 1600; it's concentrated, includes some sediment, and has to be drunk right away or the sediment will increase the already considerable bitterness.

French Refinements The first Western modifications of coffee brewing date from around 1700, when French cooks isolated the solid beans within the liquid by enclosing the grounds in a cloth bag, and thus produced a clearer, less gritty brew. Around 1750, the French came up with the most important advance before espresso: the drip pot, in which hot water was poured onto a bed of grounds and allowed to pass through into a separate chamber. This invention did three things: it kept the temperature of the extracting water below the boil, it limited the contact time between water and ground coffee to a matter of a few minutes, and it produced a sedimentless brew that would keep for a while without getting stronger. The limits on brewing temperature and time meant a less complete extraction of the coffee. This reduced the bitterness and astringency, and allowed the other elements of coffee flavor more prominence, the tartness and aroma that were more appealing to European tastes.

Machine-Age Espresso The 19th century brought the invention of several new brewing methods. There was percolation, or allowing boiling water to rise in a central tube and irrigate a bed of ground coffee. There were plunger pots, which allowed the coffee brewer to steep the grounds, then push the grounds to the bottom with the plunger and pour the beverage off. But the

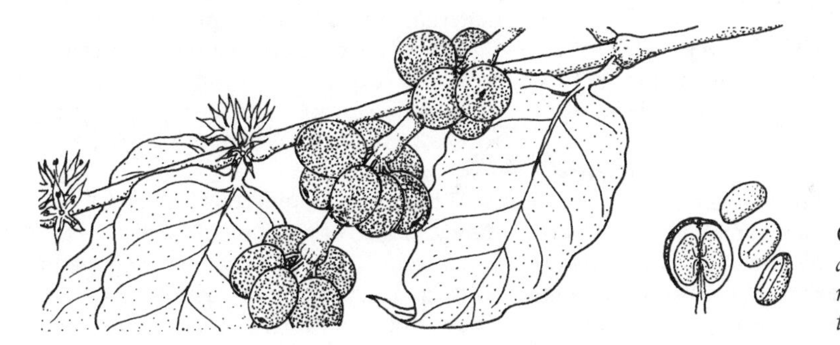

Coffee berries and seeds. Each red berry contains two seeds.

biggest innovation in coffee brewing made its debut at the Paris Exhibition of 1855. That was Italian *espresso,* a word which means something made at the moment it's ordered, rapidly, and for one customer. The way to make coffee fast is to force water through the grounds with high pressure. In the process, the pressure extracts a substantial amount of the coffee bean's oil, and emulsifies it into tiny droplets that create a velvety texture and lingering flavor in the drink. Espresso is an expression of the power of the machine to force the most and the best from a traditional ingredient and make it into something new.

Coffee Beans

Arabica and Robusta Coffees Coffee beans are the seeds of two species of a tropical relative of the gardenia. *Coffea arabica,* a 15 ft/5 m tree that is native to the cool highlands of Ethiopia and the Sudan, produces what are known as "arabica" beans; and *Coffea canephora,* a larger tree native to hotter, more humid West Africa, produces "robusta" beans. About two-thirds of the beans in international trade are arabicas, which develop a more complex and balanced flavor than the robustas. They contain less caffeine (less than 1.5% by weight of the dry bean, vs. 2.5% for robustas), less phenolic material (6.5% vs. 10%), and more oil (16% vs. 10%) and sugar (7% vs. 3.5%). Robusta varieties didn't become prominent until the end of the 19th century, when their disease resistance became important in Indonesia and elsewhere.

Dry and Wet Processing To prepare coffee beans, the ripe coffee berries are picked from the trees, and the seeds cleaned of the fruit pulp by one of two basic methods. In the dry method, the berries are left in the sun to dry, or first piled to ferment for a few days, then spread out in the sun. The fruit is then removed by machine. In the wet method, most of the pulp is rubbed from the seeds by machine, then the remainder is liquefied by a day or two of fermentation by microbes. The seeds are then washed in copious water, dried to about 10% moisture, and the adherent inner "parchment shell" removed by machine. Some sugars and minerals are leached out of wet-processed beans, so they tend to produce coffee with less body and more acidity than dry-processed beans. However they often have more aroma, and tend to be of more uniform quality.

Roasting Raw green coffee beans are as hard as unpopped popcorn, and about as tasty. Roasting transforms them into fragile, easily opened packages of flavor. Most people let the professionals take care of roasting, but it's a fascinating (and smoky) experience to roast coffee at home, as cooks in many countries have long done and still do with equipment ranging from frying pans to popcorn poppers to special roasters.

Coffee beans are roasted to temperatures between 375 and 425°F/190–220°C; the process usually takes between 90 seconds and 15 minutes. As the bean temperature approaches the boiling point of water, the small amounts of moisture inside the cells turn into steam and puff the bean up to half again its original volume. Then at progressively higher temperatures, the proteins, sugars, phenolic materials, and other constituents begin to break into molecular fragments and react with each other, and develop the brown pigments and roasted aromas typical of the Maillard reactions (p. 778). At around 320°F/160°C, these reactions become self-sustaining, like a candle flame, and extreme molecular breakdown generates more water vapor and carbon dioxide gas, whose production rises sharply at 400°F/200°C. If the roasting continues, oil begins to escape from the damaged cells to the bean surface, where it provides a visible gloss.

When the beans have reached the desired degree of roast, the roaster cools the beans immediately with cold air or a water spray to quench the molecular breakdown. The result is a brown, brittle, spongelike bean, with the holes in the sponge filled with carbon dioxide.

The Development of Coffee Flavor The hotter the bean is roasted, the darker it gets, and its color is a good indicator of flavor balance. In the early stages of roasting, sugars are broken down into various acids (formic, acetic, lactic), which together with their own organic acids (citric, malic) give light-brown beans a pronounced tartness. As roasting proceeds, both the acids and astringent phenolic materials (chlorogenic acid) are destroyed, so acidity and astringency decline. However, bitterness increases because some of the browning-reaction products are bitter. And as the bean's color becomes darker than medium brown, the distinctive aromas characteristic of prized beans become overwhelmed by more generic roasted flavors—or, conversely, the flavor deficiencies of second-rate beans become less obvious. Finally, as acids and tannins and soluble carbohydrates decline with dark roasting, so does the brew's fullness of body: there's less there to stimulate our tongue. Medium roasts give the fullest body.

Storing Coffee Once roasted, whole coffee beans keep reasonably well for a couple of weeks at room temperature, or a couple of months in the freezer, before becoming noticeably stale. One reason that whole beans keep as long as they do is that they're filled with carbon dioxide, which helps exclude oxygen from the porous interior. Once the beans have been ground, room-temperature shelf life is only a few days.

Grinding Coffee The key to proper coffee grinding is obtaining a fairly consistent particle size that's appropriate to the brewing method. The smaller the particle size, the greater the surface area of bean exposed to the water, and the faster its contents are extracted. Too great a range of particle sizes makes it hard to control the extraction during brewing. Small particles may be

The Effects of Roasting on Coffee Beans

Weight Loss of Roasted Coffee Beans

Degree of Roast	Weight Loss, %
Cinnamon (375°F/190°C)	12, mostly moisture
Medium	13
City	15
Full city	16, half moisture and half bean solids
French	17
Italian (425°F/220°C)	18–20, mostly bean solids

Composition of Raw and Roasted Coffee Beans, Percent by Weight

	Raw	Roasted
Water	12	4
Protein	10	7
Carbohydrate	47	34
Oil	14	16
Phenolics	6	3
Large complex aggregates that provide color, body	0	25

Coffee Flavor, from the Bean into the Cup

This chart shows the relationships between coffee flavor and the fraction of the coffee bean extracted into the water by various brewing methods. Balanced flavor corresponds to an extraction of around 20% of the coffee solids. The strength of the flavor depends on the relative proportions of coffee and water: espresso is made with a much higher proportion of coffee than other brews.

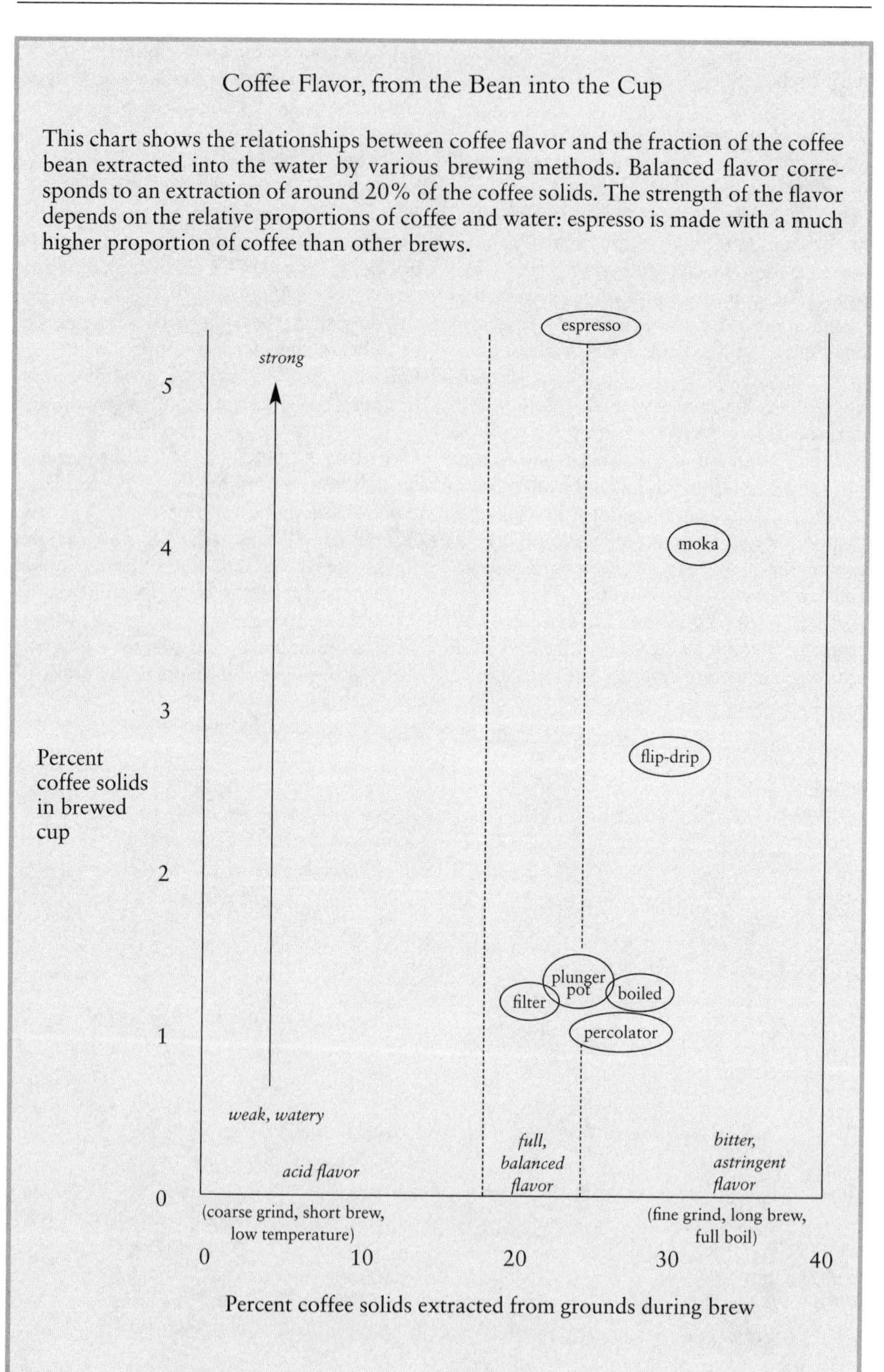

overextracted and large ones underextracted, and the resulting brew can be both bitter and weak. The common propellor grinder smashes all the bean pieces until the machine is stopped, no matter how small the pieces get, so coarse and medium grinds end up containing some fine powder. More expensive burr grinders allow small pieces to escape through grooves in the grinding surfaces, and give a more even particle size.

Brewing Coffee Brewing is the extraction into water of desirable substances from the coffee bean, in amounts that produce a balanced, pleasing drink. These substances include many aroma and taste compounds, as well as browning pigments that provide color (almost a third of the total extract) and cell-wall carbohydrates that provide body (also almost a third). The flavor, color, and body of the finished drink are determined by how much ground coffee is used for a given volume of water, and by what proportion of that coffee is extracted into the water. Inadequate extraction and a watery, acid brew are caused by grinding the beans too coarsely, so that flavor is left inside the particles, by too brief a contact time between coffee and water, or by too low a brewing temperature. Overextraction and a harsh, bitter brew result from an excessively fine grind, or long contact time, or high brewing temperature.

The ideal brewing temperature for any style of coffee is 190–200°F/85–93°C; anything higher extracts bitter compounds too quickly. For a standard cup of American coffee, the usual brewing time ranges from 1 to 3 minutes for a fine grind, to 6 to 8 minutes for a coarse grind.

Brewing Methods There are a number of different methods for brewing coffee. Most of them extract between 20 and 25% of the bean's substance, and produce a cup containing somewhere between 1.3% and 5.5% bean solids by weight. The facing chart places some of the major styles in relation to each other. Standard American filter-drip coffee is the lightest, and Italian espresso the strongest. The initial proportion of coffee to water is 1:15 for American, 1:5 for espresso. One clear lesson from the chart is that it's always best to use too much coffee rather than too little: a strong but balanced cup can be diluted with hot water and remain balanced, but a weak cup can't be improved. This principle can help avoid problems caused by the fact that cup and coffee scoop measures vary, and that scoops themselves are a very approximate measure (one 2-tablespoon/30-ml scoop may deliver anywhere from 8 to 12 gm coffee, depending on grind and packing).

Each brewing method has its drawbacks. Percolators operate at the boil and tend to overextract. Many automatic drip brewers aren't able to deliver near-boiling water, so they brew for a long time to compensate, lose aroma, and extract some bitterness. Manual drip cones give little control over extraction time. The plunger pot leaves tiny suspended particles in the brew that keep releasing bitterness. The Italian stovetop moka pot operates above the boil, at around 230°F/110°C (and 1.5 atmospheres of pressure), and produces a somewhat harsh brew. Overnight extraction in cold water doesn't obtain as many aromatic compounds from the ground coffee as the hot-water methods.

Espresso True espresso is made very quickly, in about 30 seconds. A piston or spring or electrical pump drives 200°F/93°C water through finely ground coffee at 9 atmospheres of pressure. (Inexpensive household machines rely on excessively hot steam, develop far less pressure, and take longer to brew, so the result is relatively thin and harsh.) The proportion of ground coffee is three to four times the amount used in unpressurized brewing, and deposits three to four times the concentration of coffee materials in the brew, creating a substantial, velvety body and intense flavor. These extracted materials include a relatively large amount of coffee oils, which the high pressure forces from the bean particles to form a creamy emul-

Methods of Brewing Coffee

This chart summarizes the important features of some common ways of brewing coffee, and the kinds of brew they produce. The stability of a brew is determined by how many coffee particles remain in it; the more particles, the more bitterness and astringency continue to be extracted in the cup or pot.

	Middle Eastern/ Mediterranean, Boiled	Machine Filter	Manual Filter	Percolator	Plunger Pot (French Press)	Moka	"Espresso" (Steam)	Espresso (Pump)
Coffee grind	Very fine (0.1 mm)	Coarse (1 mm)	Medium (0.5 mm)	Coarse (1 mm)	Coarse (1 mm)	Medium (0.5 mm)	Fine (0.3 mm)	Fine (0.3 mm)
Brew temperature	To 212°F/ 100°C	180–185°F/ 82–85°C	190–200°F/ 87–93°C	212°F/ 100°C	190–195°F/ 87–90°C	230°F/ 110°C	212°F/ 100°C	200°F/ 93°C
Brew time	10–12 min	5–12 min	1–4 min	3–5 min	4–6 min	1–2 min	1–2 min	0.3–0.5 min
Extraction pressure, atmospheres	1	1	1	1(+)	1(+)	1.5	1(+)	9
Flavor	Full but bitter (sweetened)	Light, often bitter	Full	Full, often bitter	Full	Full but bitter	Full but bitter	Very full
Body	Full	Light	Light	Light	Medium	Full	Full	Very full
Stability once brewed	Poor	Good	Good	Good	Poor	Fair	Poor	Poor

sion of tiny droplets, and which contribute to the slow, prolonged release of coffee flavor in the mouth, long after the last sip. Another unique feature of espresso is the *crema,* the remarkably stable, creamy foam that develops from the brew and covers its surface. It's the product of carbon dioxide gas still trapped in the ground coffee, and the mixture of dissolved and suspended carbohydrates, proteins, phenolic materials, and large pigment aggregates, all of which bond in one way or another to each other and hold the bubble walls together. (For the milk foams often served with coffee, see p. 26.)

Serving and Holding Coffee Freshly brewed coffee is best enjoyed immediately—its flavor is evanescent. The ideal drinking temperature is around 140°F/60°C, where a sip won't scald the mouth, and the coffee's full aroma comes out. Because it cools in the cup, coffee is usually held in the pot just below the brewing temperature. High heat accelerates chemical reactions and the escape of volatile molecules, so coffee flavor changes noticeably after less than an hour in the pot; it becomes more acid and less aromatic. Coffee is best kept hot by retaining its original heat in a preheated, insulated, closed container, not on a hot plate that constantly supplies excessive heat from below while heat and aroma escape above.

Coffee Flavor Coffee has one of the most complex flavors of all our foods. At its base is a mouth-filling balance of acidity, bitterness, and astringency. A third or less of the bitterness is due to easily extracted caffeine, the rest to more slowly extracted phenolic compounds and browning pigments. More than 800 aroma compounds have been identified, and they supply notes that are described as nutty, earthy, flowery, fruity, buttery, chocolate-like, cinnamon, tea, honeyed, caramel, bready, roasty, spicy, even winey and gamy. Robusta coffees, with their substantially higher content of phenolic substances than arabicas, develop a characteristic smoky, tarry aroma that is valued in dark roasts (they are also distinctly less acidic than arabicas). Milk and cream reduce the astringency of coffee by providing proteins that bind to the tannic phenolic compounds, but these liquids also bind aroma molecules and weaken the overall coffee flavor.

Decaffeinated Coffee Decaffeinated coffee was invented in Germany around 1908. It's made by soaking green coffee beans with water to dissolve the caffeine, extracting the caffeine from the beans with a solvent (methylene chloride, ethyl acetate), and steaming the beans to evaporate off any remaining solvent. In the "Swiss" or "water" process, water is the only solvent used, the caffeine removed from the water by charcoal filters, and the other water-solubles are then added back to the beans. Some of the organic solvents used in other processes have been suspected of being health hazards even in the tiny traces left in the beans (around 1 part per million). The commonest, methylene chloride, is now thought to be safe. More recently, highly pressurized ("supercritical") and nontoxic carbon dioxide has been used. Where ordinary brewed coffee may contain 60–180 milligrams caffeine per cup, decaffeinated coffee will contain 2–5 mg.

Instant Coffee Instant coffee became commercially practical in Switzerland just before World War II. It's made by brewing ground coffee near the boil to obtain aroma, then a second time at 340°F/170°C and high pressure to maximize the extraction of pigments and body-producing carbohydrates. Water is removed from the two extracts by hot spray-drying or by freeze-drying, which retains more of the volatile aroma compounds and produces a fuller flavor. The two are then blended together and supplemented with aromas captured during the drying stage. Instant coffee crystals contain about 5% moisture, 20% brown pigments, 10% minerals, 7% complex carbohydrate, 8% sugars, 6% acids, and 4% caffeine. As an essentially dry concentrate, instant coffee is a valuable flavoring for baked goods, confections, and ice creams.

WOOD SMOKE AND CHARRED WOOD

Neither wood nor the smoke it gives off is an herb or a spice, strictly speaking. Yet cooks and makers of alcoholic liquids often use burned or burning wood as flavoring agents—in barbecuing meats, in barrel-aging wines and spirits—and some of the flavors they supply are identical to spice flavors: vanilla's vanillin, for example, and clove's eugenol. That's because wood is strengthened with masses of interlinked phenolic units, and high heat breaks these masses apart into smaller volatile phenolics (p. 390).

THE CHEMISTRY OF BURNING WOOD

Charred wood and smoke are products of the incomplete combustion of organic materials in the presence of limited oxygen and at the relatively low temperatures of ordinary burning (below 1,800°F/1,000°C). Complete combustion would produce only odorless water and carbon dioxide.

The Nature of Wood Wood consists of three primary materials: cellulose and hemicellulose, which form the framework and the filler of all plant cell walls, and lignin, a reinforcing material that binds neighboring cell walls together and gives wood its strength. Cellulose and hemicellulose are both aggregates of sugar molecules (pp. 265, 266). Lignin is made of intricately interlocked phenolic molecules—essentially rings of carbon atoms with various additional chemical groups attached—and is one of the most complex natural substances known. The higher the lignin content of a wood, the harder it is and the hotter it burns; its combustion releases 50% more heat than cellulose. Mesquite wood is well-known for its high-temperature fire, which it owes to its 64% lignin content (hickory, a common hardwood, is 18% lignin). Most wood also contains a small amount of protein, enough to support the browning reactions that generate typical roasted flavors (p. 778) at moderately hot temperatures. Evergreens such as pine, fir, and spruce also contain significant amounts of resin, a mixture of compounds related to fats that produce a harsh soot when burned.

How Burning Transforms Wood into Flavor Burning temperatures transform each of the wood components into a characteristic group of compounds (see box, p. 449). The sugars in cellulose and hemicellulose break apart into many of the same molecules found in caramel, with sweet, fruity, flowery, bready aromas. And the interlocked phenolic rings of lignin break apart from each other into a host of smaller, volatile phenolics and other fragments, which have the specific aromas of vanilla and clove as well as a generic spiciness, sweetness, and pungency. Cooks get these volatiles into solid foods, usually meats and fish, by exposing the foods to the smoky vapors emitted by burning wood. Makers of wine and spirits store them in wood barrels whose interiors have been charred; the volatiles are trapped in and just below the barrels' inner surface, and are slowly extracted by the liquid (p. 721).

The flavor that wood smoke imparts to food is determined by several factors. Above all there's the wood. Oak, hickory, and the fruit-tree woods (cherry, apple, pear) produce characteristic and pleasing flavors thanks to their moderate, balanced quantities of the wood components. A second important factor is the combustion temperature, which is partly determined by the wood and its moisture content. Maximum flavor production takes place at relatively low, smoldering temperatures, between 570 and 750°F/300–400°C; at higher temperatures, the flavor molecules are themselves broken down into simpler harsh or flavorless molecules. High-lignin woods burn too hot unless their combustion is slowed by restricted airflow or a high moisture content. When smoking is done by throwing wood chips onto glowing charcoal, the wood chips should be presoaked in water so that they'll cool the coals. Because it's largely pure carbon, charcoal burns mostly smokelessly at temperatures approaching 1,800°F/1,000°C.

Though smoke helps stabilize the flavors of meats and fish, smoke flavor itself is unstable. The desirable phenolic compounds are especially reactive, and largely dissipate in a few weeks or months.

The Toxins in Wood Smoke: Preservatives and Carcinogens In the beginning, smoking was not just a way of giving foods an interesting flavor: it was a way of delaying their spoilage. Wood smoke contains many chemicals that slow the growth of microbes. Among them are formaldehyde, and acetic acid (vinegar) and other organic acids, thanks to which the pH of smoke is a very microbe-unfriendly 2.5. Many of the phenolic compounds in wood smoke are also antimicrobials, and phenol itself is a strong disinfectant. The phenolic compounds are also effective antioxidants, and slow the development of rancid flavors in smoked meats and fish.

In addition to antimicrobial compounds, smoke also contains antihuman compounds, substances that are harmful to our long-term health. Prominent among these are the polycyclic aromatic hydrocarbons, or PAHs, which are proven carcinogens and are formed from all of the wood components in increasing amounts as the temperature is raised. Hot-burning mesquite wood generates double the quantity that hickory wood does. The deposition of PAHs on meat can be minimized by limiting the fire temperature, keeping the meat as far as possible from the fire, and allowing free air circulation to carry soot and other PAH-containing particles away. Commercial smokers use air filters and temperature control for these purposes.

Liquid Smoke

Liquid smoke is essentially smoke-flavored water. Smoke consists of two phases: microscopic droplets that make it visible as a

Wood Components and Smoke Flavors

Wood Component % of dry weight	Combustion Temperature	Combustion By-Products and Their Aromas
Cellulose (cell-wall frame, from glucose) 40–45%	540–610°F 280–320°C	Furans: sweet, bready, floral Lactones: coconut, peach Acetaldehyde: green apple
Hemicellulose (cell-wall filler, from mixed sugars) 20–35%	390–480°F 200–250°C	Acetic acid: vinegar Diacetyl: buttery
Lignin (cell-wall strengthener, from phenolic compounds) 20–40%	750°F 400°C	Guaiacol: smoky, spicy Vanillin: vanilla Phenol: pungent, smoky Isoeugenol: sweet, cloves Syringol: spicy, sausage-like

haze, and an invisible vapor. It turns out that much of the flavor and preservative materials are in the vapor, while the droplets are largely aggregates of tars, resins, and heavier phenolic materials, including the PAHs. PAHs are largely insoluble in water, while most of the flavor and preservative compounds do dissolve to some extent. This difference makes it possible to separate most of the PAHs from the vapors and dissolve the vapors in water. Cooks then use this liquid extract of smoke to flavor foods. Toxicological studies of liquid smoke have found that though it is full of biologically active compounds, the quantities normally used in foods are harmless. Those PAHs that do make it into liquid smoke tend to aggregate and sink over time, so it's best not to shake bottles of liquid smoke before use. Leave the sediment at the bottom.

CHAPTER 9

SEEDS

Grains, Legumes, and Nuts

SEEDS AS FOOD

Seeds are our most durable and concentrated foods. They're rugged lifeboats, designed to carry a plant's offspring to the shore of an uncertain future. Tease apart a whole grain, or bean, or nut, and inside you find a tiny embryonic shoot. At harvest time that shoot had entered suspended animation, ready to survive months of drought or cold while waiting for the right moment to come back to life. The bulk of the tissue that surrounds it is a food supply to nourish this rebirth. It's the distillation of the parent plant's lifework, its gathering of water and nitrogen and minerals from the soil, carbon from the air, and energy from the sun. And as such it's an invaluable resource for us and other creatures of the animal kingdom who are unable to live on soil and sunlight and air. In fact, seeds gave early humans both the nourishment and the inspiration to begin to shape the natural world to their own needs. Ten thousand turbulent years of civilization have unfolded from the seed's pale repose.

The story began when inhabitants of the Middle East, Asia, and Central and South America learned to save some large, easily harvested seeds from wild plants, and sow them in clearings to produce more seeds of a similar kind. It appears that agriculture first arose in the highlands of southeastern Turkey, around the upper reaches of the Tigris and Euphrates rivers, and in the Jordan River valley. The first plants to be brought under human selection there were einkorn and emmer wheat, barley, lentil, pea, bitter vetch, and chickpea: a mixture of seed-bearing cereals and legumes. Gradually the nomadic life of the hunter-gatherer gave way to growing settlements alongside the large grainfields that fed them. The need arose for planning the sowing and the distribution of the harvests, for anticipating seasonal changes before they occurred, for organizing the work, and for keeping records. Some of the earliest known writing and arithmetical systems, dating from at least 5,000 years ago, are devoted to the accounting of grain and livestock. So the culture of the fields encouraged the culture of the mind. At the same time it brought problems, among them a drastic simplification of the hunter-gatherer's varied diet and consequent damage to human health, and the development of a

Seeds of Thought

The development of agriculture had a deep influence on human feeling and thought, on mythology and religion and science, that is hard to capture in a few quotations. The religious historian Mircea Eliade summarized it this way:

> We are used to thinking that the discovery of agriculture made a radical change in the course of human history by ensuring adequate nourishment and thus allowing a tremendous increase in the population. But the discovery of agriculture had decisive results for a quite different reason. . . . Agriculture taught man the fundamental oneness of organic life; and from that revelation sprang the simpler analogies between women and field, between the sexual act and sowing, as well as the most advanced intellectual syntheses: life as rhythmic, death as a return, and so on. These syntheses were essential to man's development, and were possible only after the discovery of agriculture.
>
> —*Patterns in Comparative Religion*, 1958

social hierarchy in which a few benefit from the labor of many.

In the *Odyssey,* Homer called wheat and barley "the marrow of men's bones." It's less obvious to us in the modern industrialized world than it has been through much of human history, but seeds remain the essential food of our species. Grains directly provide the bulk of the caloric intake for much of the world's population, especially in Asia and Africa. The grains and legumes together provide more than two-thirds of the world's dietary protein. Even the industrial countries are fed indirectly by the shiploads of corn, wheat, and soybeans on which their cattle, hogs, and chickens are raised. The fact that the grains come from the grass family adds a layer of significance to the Old Testament prophet Isaiah's admonishment that "All flesh is grass."

As ingredients, seeds have much in common with milk and eggs. All consist of basic nutrients created to nourish the next generation of living things; all are relatively simple and bland in themselves, but have inspired cooks to transform them into some of the most complex and delightful foods we have.

Some Definitions

Seeds Seeds are structures by which plants create a new generation of their kind. They contain an embryonic plant together with a food supply to fuel its germination and early growth. And they include an outer layer that insulates the embryo from the soil and protects it from physical damage and from attack by microbes or animals.

The most important seeds in the kitchen fall into three groups.

Grains, or Cereals These words are near synonyms. The *cereals* (from *Ceres,* the Roman goddess of agriculture) are plants in the grass family, the Gramineae, whose members produce edible and nutritious seeds, the *grains.* But *cereal* is also used to mean their seeds and products made from them—as in "breakfast cereals"—and the plants are sometimes called *grains.* The cereals and other grasses are creatures of the open plain or high-altitude steppe, areas too dry for trees. They live and die in a season or two, and are easily gathered and handled. They grow in densely packed stands that crowd out competition, and produce many small seeds, relying on numbers rather than chemical defenses to ensure that some offspring will survive. These characteristics made the grasses ideal for agriculture. With our help, they have come to cover vast areas of the globe.

Wheat, barley, oats, and rye have been the most important grains in the Middle East and Europe; in Asia, rice; in the New World, maize, or corn; in Africa sorghum and millets. The grains are of special culinary significance because they make possi-

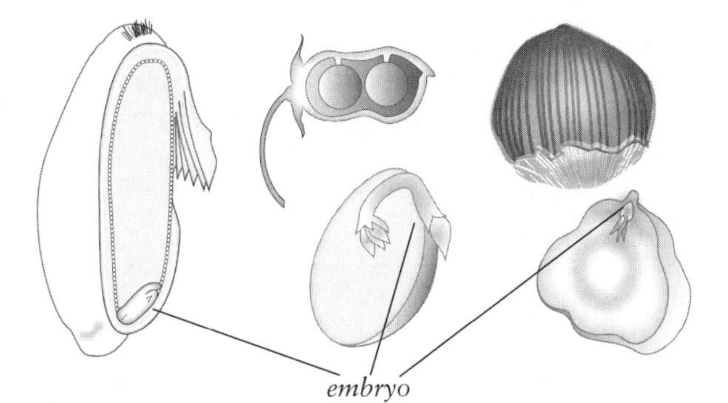

An oat kernel, lentils in their pod, and a hazelnut. All are seeds, and consist of a living embryonic plant together with a food supply to fuel its early growth. In the cereal grains, the food supply is a separate tissue, the endosperm. In the beans and their relatives, and in most nuts, the food supply fills the first two leaves of the embryo, the cotyledons, which are unusually massive and thick.

ble beer and bread, both staples in the human diet for at least 5,000 years.

Legumes The *legumes* (from the Latin *legere,* "to gather") are plants in the bean family, the Leguminosae, whose members bear pods that contain several seeds. The term *legume* is also used to name their seeds. Many legumes are vines that climb on tall grasses and other plants to reach full sunlight, and like the grasses grow, go to seed, and die over a few months. The legumes produce seeds that are especially rich in protein, thanks to their symbiosis with bacteria that live in their roots and feed them with nitrogen from the air. The same symbiosis means that legumes actually enrich the soil they grow in with nitrogen compounds, which is why various legumes have been grown as rotation crops at least since Roman times. Their relatively large seeds are attractive to animals, and it's thought that much of the remarkable diversity in the beans and peas is the result of survival pressures exerted by insects. Legume seeds are camouflaged by colored coats, and protected with an array of several biochemical defenses.

Lentils, broad beans, peas, and chick peas are all native to the Fertile Crescent of the Near East. They were adapted for sprouting and quickly reproducing in the cool, wet season before the summer drought, and were the first substantial foods to ripen in the spring. The soy and mung beans were indigenous to Asia, and peanuts, lima beans, and common beans to the Americas.

Nuts The *nuts* (from an Indo-European root meaning "compressed") come from several different plant families, not just one. They are generally large seeds enclosed in hard shells, and borne on long-lived trees. The seeds are large both to make them attractive to animal dispersers (which bury some for later use and effectively plant the ones they forget), and to give the seedling an adequate food supply for slow, prolonged growth in the partial shade. Most of them store their energy not in starch but in oil, a more compact, concentrated chemical form (p. 121).

Nuts are much less important in the human diet than grains or legumes because nut trees don't begin to bear until years after they're planted, and can't produce as much per acre as the quick-growing grains and legumes. The biggest exception to this rule is the coconut, a staple food in many tropical countries. Another is the peanut, which is a legume with an uncharacteristically oily, tender seed, and which can be grown quickly in massive numbers.

SEEDS AND HEALTH

Our seed foods provide us with many nutritional benefits. To begin with, they're our most important staple sources of energy and protein, and carry the B vitamins that are required for the chemical work of generating energy and building tissue. In fact, they're such a good source of these essential nutrients that cultures have occasionally relied on the grains too heavily, and suffered from dietary deficiencies as a result. The debilitating disease called beriberi plagued rice-eating Asia in the 19th century when milling machines made it easier to remove the inconvenient, unattractive outer bran layer from rice grains—and along with it their thiamin, which the rest of the largely vegetarian diet couldn't make up (meats and fish are rich in thiamin). A different deficiency disease called pellagra struck the rural poor in Europe and the southern United States in the 18th and 19th centuries, when they adopted corn from Central and South America as a staple food, but without the processing method (cooking in alkaline water) that makes its stores of niacin available to the human body.

Beriberi and pellagra led early in the 20th century to the discovery of the vitamins whose deficiencies cause them. Today, even though most people in Asia eat refined rice, and polenta and grits are still not cooked in alkaline water, more balanced diets have made these deficiency diseases far less common.

Valuable Phytochemicals from Seeds

Toward the end of the 20th century, we came to realize that seeds have more to offer us than the basic machinery of life. Epidemiological studies have found a general association between the consumption of whole grains, legumes, and nuts and a reduced risk of various cancers, heart disease, and diabetes. What do these foods provide that refined grains do not? Hundreds or even thousands of chemicals that are concentrated in the outer protective and active layers of the seeds, and that are not found in inner storage tissues, which are mainly depots of starch and protein. Among the chemicals that have been identified and seem likely to be helpful are

- a variety of vitamins, including antioxidant vitamin E and its chemical relatives the tocotrienols
- soluble fiber: soluble but undigestible carbohydrates that slow digestion, moderate blood insulin and blood sugar levels, and reduce cholesterol levels, and provide energy for beneficial intestinal bacteria, which alter their chemical environment, suppress the growth of harmful bacteria, and influence the health of intestinal cells
- insoluble fiber, which speeds passage of food through the digestive system and reduces our absorption of carcinogens and other undesirable molecules
- a variety of phenolic and other defensive compounds, some of which are effective antioxidants, some of which resemble human hormones and may restrain cell growth and thereby the development of cancer

Medical scientists are still in the early stages of identifying and evaluating these substances, but in general it looks as though regular consumption of whole grains, legumes, and nuts can indeed make a real contribution to our long-term health.

Problems Caused by Seeds

Seeds are not perfect foods. Legumes in particular contain defensive chemicals—lectins and protease inhibitors—that can cause malnourishment and other problems. Fortunately, simple cooking disarms these defenses (p. 259). The fava bean contains amino-acid relatives that cause serious anemia in susceptible people (p. 490), but both the bean and the susceptibility are relatively rare. Two other problems are more common.

Seeds Are Common Food Allergens

True food allergies are overreactions of the body's immune system, which mistakes a food component as a sign of invasion by a bacterium or virus and initiates a defense that damages the body. The damage may be mild and manifest itself as discomfort, itchiness, or a rash, or it may be a life-threatening asthma or change in blood pressure or heart rhythm. It's estimated that about 2% of adults in the United States have one or more food allergies, and up to 8% of young children. Allergic reactions to food cause around 200 deaths per year in the United States. Peanuts, soybeans, and tree nuts are among the most common food allergens. The offending components are usually seed proteins, and cooking does not render them less allergenic. Tiny quantities of nut proteins are sufficient to cause reactions, including the levels sometimes found in mechanically extracted nut oils.

Gluten Sensitivity A special form of food allergy is the disease called gluten-sensitive enteropathy, celiac disease, or sprue, in which the body forms defensive antibodies against a portion of the harmless gliadin proteins in wheat, barley, rye, and possibly oats. These defenses end up attacking the nutrient-absorbing cells in the intestine, and therefore cause serious malnourishment. Celiac disease can develop in early childhood or later, and is a lifelong condition. The standard remedy is strict avoidance of all gluten-containing foods. Several grains don't contain gliadin proteins and therefore don't aggravate celiac disease; they are corn, rice, amaranth, buckwheat, millet, quinoa, sorghum, and teff.

Seeds and Food Poisoning

Seeds are generally dry, with only about 10% of their weight coming from water. As a result, they keep well without special treatment; and because we prepare them by thoroughly boiling or roasting them, freshly cooked grains, beans, and nuts generally don't carry bacteria that cause food poisoning. However, moist grain and bean dishes become very hospitable to bacteria as they cool down. Leftovers should be refrigerated promptly and reheated to the boil before serving. Rice dishes are particularly vulnerable to contamination by *Bacillus cereus* and require special care (p. 475).

Even dry seeds aren't entirely immune to contamination and spoilage. Molds, or fungi, are able to grow with relatively little moisture, and can contaminate seed crops both in the field and in storage. Some synthesize deadly toxins that can cause cancer and other diseases (for example, species of *Aspergillus* produce a carcinogen called aflatoxin, and *Fusarium moniliforme* produces another called fumonisin). The presence of fungal toxins in our foods is generally invisible to the consumer, and is monitored by producers and government agencies. They're not now considered a major health risk. But the least sign of mold or other spoilage on grains and nuts means that the food should be discarded.

THE COMPOSITION AND QUALITIES OF SEEDS

Parts of the Seed

All of our food seeds consist of three basic parts: an outer protective coat, a small embryonic portion capable of growing into the mature plant, and a large mass of storage tissue that contains proteins, carbohydrates, and oils to feed the embryo. Each part influences the texture and flavor of the cooked seeds.

The outer protective coat, called the *bran* in grains and the *seed coat* in legumes and nuts, is a dense sheet of tough, fibrous tissue. It's rich in defensive or camouflaging phenolic compounds, including anthocyanin pigments and astringent tannins. And it slows the passage of water into grains and legumes during cooking. It's often removed from grains (especially rice and barley), legumes (notably in Indian dals), and nuts (almonds, chestnuts) to speed the cooking and obtain a more refined appearance, texture, and flavor.

The embryonic portion of legumes and nuts is not of much practical significance, but the germ of the grains is: it contains much of the oil and enzymes in these seeds, and thus is the source of potential flavor, both desirable cooked aromas and undesirable stale ones.

The bulk of the seed is a mass of storage tissue, and its makeup determines the seed's basic texture. The storage cells are filled with particles of concentrated protein, granules of starch, and sometimes with droplets of oil. In some grains, notably barley, oats, and rye, the cell walls are also filled with storage carbohydrates—not starch, but other long sugar chains that like starch can absorb water during cooking. The strength of the cement that holds the storage cells together, and the nature and proportions of the materials they contain, determine the seed's texture. Bean cells and grain cells are filled with solid, hard starch granules and protein bodies; most nut cells are filled with liquid oil, and so are more fragile. Grains retain their shape and some firmness even when we mill away their protective bran envelope and boil them in plenty of water. Beans remain intact as long as we cook them in their seed coats; otherwise they rapidly disintegrate into a puree.

The particular contents of the seed storage cells influence texture and culinary usefulness in a number of ways. So it's worth knowing about the proteins, starches, and oils in some detail.

Seed Proteins: Soluble and Insoluble

Seed proteins are classified by a particular aspect of their chemical behavior, which also determines their behavior during cooking: the kind of liquid in which they dissolve. This may be pure water, water and some salt, water and dilute acid, or alcohol (these types are called "albumins," "globulins," "glutelins," and "prolamins"). Most of the proteins in legumes and nuts are soluble in a salt solution or water alone, so during ordinary cooking in salted water, bean and pea proteins become dispersed in the moisture within the seeds and the cooking liquid surrounding them. By contrast, the main storage proteins in wheat, rice, and other grains are acid-soluble and alcohol-soluble types. In ordinary water, these proteins don't dissolve; instead they bond to each other and clump up into a compact mass. Wheat, rice, corn, and barley kernels develop a chewy consistency in part because their insoluble proteins clump together in the grain during cooking and form a sticky complex with the starch granules.

Seed Starches: Orderly and Disorderly

All the grains and legumes contain a substantial amount of starch, enough that it plays a significant role in the texture of the cooked seeds and their products. It can make one grain variety behave very differently from another variety of the same grain.

Two Kinds of Starch Molecules The parent plant lays down starch molecules in microscopic, solid granules that fill the cells of the seed storage tissue. All starch consists of chains of individual molecules of the sugar called glucose (p. 804). But there are two different kinds of starch molecules in starch granules, and they behave very differently. *Amylose* molecules are made from around 1,000 glucose sugars, and are

The Proportions of Proteins in Seeds

	Water-soluble albumins	Water-salt-soluble globulins	Acid-soluble glutelins	Alcohol-soluble prolamins
Wheat	10	5	40–45	33–45
Barley	10	10	50–55	25–30
Rye	10–45	10–20	25–40	20–40
Oats	10–20	10–55	25–55	10–15
Rice	10	10	75	5
Corn	5	5	35–45	45–55
Beans, peas	10	55–70	15–30	5
Almonds	30	65		

mainly one extended chain, with just a few long branches. *Amylopectin* molecules are made from 5,000 to 20,000 sugars and have hundreds of short branches. Amylose is thus a relatively small, simple molecule that can easily settle into compact, orderly, tightly bonded clusters, while amylopectin is a large, bushy, bulky molecule that doesn't cluster easily or tightly. Both amylose and amylopectin are packed together in the raw starch granule, in proportions that depend on the kind and variety of seed. Legume starch granules are 30% or more amylose, and wheat, barley, maize, and long-grain rice granules are around 20%. Short-grain rice granules contain about 15% amylose, while "sticky" rice starch granules are almost pure amylopectin.

Cooking Separates Starch Molecules and Softens Granules When a seed is cooked in water, the starch granules absorb water molecules, and swell and soften as the water molecules intrude and separate the starch molecules from each other. This granule softening, or *gelation,* takes place in a temperature range that depends on the seed and starch, but is in the region of 140–160°F/60–70°C. (The conversion of solid starch into a starch-water gel is often referred to as "gelatinization," but this is unnecessarily confusing; starch has nothing to do with gelatin.) The tightly ordered clusters of amylose molecules require higher temperatures, more water, and more cooking time to be pulled and kept apart than do the looser clusters of amylopectin molecules. This is why long-grain Chinese rices are made with more water than short-grain Japanese rices.

Cooling Reorganizes Starch Molecules and Firms Granules Once the cooking is finished and the seeds cool down below the gelation temperature, the starch molecules begin to re-form some clusters with pockets of water in between, and the soft, gelated starch granules begin to firm up again. This process is called *retrogradation.* The simpler amylose molecules start bonding to each other again almost immediately, and finish within a few hours at room or refrigerator temperatures. Sprawling, bushy amylopectin molecules take a day or more to reassociate, and form relatively loose, weak clusters. This difference

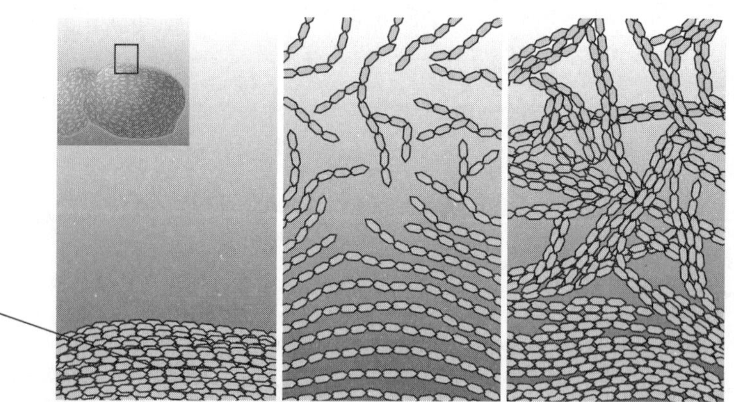

Starch gelation and retrogradation. Starch granules are compact, organized masses of long starch chains (left). *When a starchy cereal is cooked, water penetrates the granule and separates the chains from each other, thus swelling and softening the granule in the process called gelation* (center). *When the cooked cereal cools down, the starch chains slowly rebond to each other in tighter, more organized associations, and the granule becomes firmer and harder, a process called retrogradation* (right).

explains why long-grain rices high in amylose have a firm, springy texture when served right after cooking and get inedibly hard when refrigerated overnight, while short-grain rices low in amylose have a softer, sticky texture and harden much less during overnight refrigeration. The hardness of all leftover grains can be largely remedied simply by reheating and so regelating their starch.

Starch Firming Can Be Useful Reheated grains never get quite as soft as they were when first cooked. This is because during the process of retrogradation, amylose molecules manage to form some clusters that are even more highly organized than the clusters in the original starch granule, crystalline regions that resist breaking even at boiling temperatures. These regions act as reinforcing junctions in the overall network of amylose and amylopectin molecules, and give the granules greater strength and integrity. Cooks take advantage of this strengthening to make bread puddings and starch noodles; parboiled (converted) rice and American breakfast cereals keep their shape because much of their starch has been allowed to retrograde during manufacturing. And it turns out that retrograded starch is good for our bodies! It resists our digestive enzymes and therefore slows the rise in blood sugar following a meal, and feeds desirable bacteria in the large intestine (p. 258).

Seed Oils

Nuts and soybeans are rich in oil, which is kept in the mass of storage tissue in tiny packages called oil bodies. Each is a tiny oil droplet whose surface is covered with two protective materials: phospholipid relatives of lecithin, and proteins called oleosins. The surface coating prevents the oil droplets from pooling together. Seed oil bodies are very similar in size and structure to the fat globules in animal milk. This is why when we eat nuts, they become creamy in the mouth rather than simply greasy. It's also one reason why for a thousand years cooks have made "milks" from almonds, soybeans, and other oil-rich seeds (pp. 494, 504).

Seed Flavors

The most important contributors to the flavors of grains, legumes, and nuts are fragments of the unsaturated fatty acids in the oils and cell membranes, which have individual aromas described as green, fatty, oily, floral, and mushroomy. The outer bran layer grains contains the bulk of the seed's oils and enzymes, and so gives whole grains a stronger flavor, as well as contributing some vanilla and toasted notes from its phenolic compounds. Beans are especially rich in green and mushroom notes. Nuts, which are usually cooked with dry heat, contain products of the browning reactions with typical roasted aromas. The flavors of particular seeds are described below.

HANDLING AND PREPARING SEEDS

Preparations of particular seeds are described in more detailed surveys below. Here are some common aspects of using seeds in the kitchen.

Storing Seeds

Because most of the seeds we eat are designed to survive a dormant, dry period, they are the easiest food ingredients to store. Whole seeds keep well for months in a dry, cool, dark place. Moisture encourages the growth of spoilage microbes, and physical damage, heat and light can accelerate the oxidation of seed oils that leads to stale, rancid aromas and bitter tastes.

The pest that sometimes infests grains, beans, nuts, and flours is the Indian meal moth (*Plodia interpunctella*). It originally came from ears of grain in the field but is now a common inhabitant of our pantries, where its eggs hatch into larvae that consume the seeds and generate unpleasant smells. There's nothing to do with a contaminated batch but discard it. Keeping seeds in separate glass or plastic jars will prevent one batch from contaminating another.

Sprouts

The sprouted seed is a culinary custom of ancient lineage in Asia, but a very recent arrival in the West. Thanks to the sprout, anyone, anytime—even an apartment-dweller in Anchorage in February—can raise a good approximation to fresh vegetables with very little effort. Sprouting often improves a seed's vitamin content and digestibility. And with their nutty flavor and crisp texture, sprouts are simply a nice change from the usual vegetables.

Beans are most commonly sprouted, but many of our food seeds can be sprouted to advantage. Sprouted wheat and barley, for example, develop a sweetness as their enzymes begin to break down stored starch into sugars for the embryonic plant. Sprouts have a nutritional value midway between that of the dry seed, which they just were, and a leafy green vegetable, which they're on the way to becoming. Sprouts are higher in vitamin C and lower in calories than most seeds, and higher than most vegetables in protein (5% versus about 2%) and in the B vitamins and iron.

Cooking Seeds

Seeds are the driest and hardest ingredients that cooks deal with. Most require both moisture and heat to make them edible. Most, but not all: the nuts are generally edible and nutritious fresh out of the shell or after a brief application of dry heat, thanks to their relatively tender cell walls and the cells' content of liquid oil rather than solid starch. But dry grains and legumes are hard and starchy. Hot water softens them by dissolving the strengthening carbohydrates from their cell walls, and moving into the cells to gelate the starch granules and either dissolve or moisten the storage proteins. This makes the seed more nutritious by exposing its nutrients to our digestive enzymes.

There are a few simple facts to remember about cooking grains and legumes in water.

- The outer bran layer or seed coat is designed to control the passage of soil moisture into the embryo and storage tissues during germination. It also slows the passage of cooking water. Seeds that have been milled free of their coats or into small pieces cook much faster than whole seeds.
- Heat penetrates seeds faster than water can, so much of the cooking time is moistening time. Presoaking seeds for a few hours or overnight can cut cooking times by half or more.

Turning Seeds into Meat Substitutes

Vegetarian cooks, particularly Buddhists in China and Japan, have long used grains, beans, and nuts to make foods with the chewy texture and savory flavor of meat. Protein extracts from wheat (gluten or *seitan,* p. 468) and soybean (*yuba,* p. 494) can be manipulated to simulate meat-protein fibers, and fermented to produce savory, meaty flavors. In seed mixtures, whole grains lend chewiness, beans a softer background and some sweetness and complexity of flavor, and nuts both richness and roastiness.

- Most seeds get quite soft when they've absorbed enough liquid to be about 60–70% water by weight. That quantity of water is the equivalent of about 1.7 times the dry weight of the seeds, or about 1.4 times their volume. Recipes generally call for much more water than this to allow for water lost to evaporation during cooking.
- The texture of fully cooked seeds is soft and fragile at cooking temperatures, but firms during cooling. If an intact appearance is important, it's good to let grains and legumes cool down before handling them.

Of course the most important grain and legume foods are made from finely ground flours or extracts. Mix water with ground grains or with starch extracted from beans and the result is a dough or batter, which heat can turn into noodles or flat breads or cakes. Aerate doughs or batters with the help of yeasts or bacteria or chemical leaveners, and the result is raised breads and cakes. Doughs and batters are special materials in their own right, and are described in detail in the next chapter.

Seeds Concentrate Cooking Liquids

Because grains and beans are dry and soak up water, they remove water from the liquid they're cooked in, and therefore effectively concentrate other materials in the liquid. In this way they create a sauce for themselves. When rice or polenta is cooked in milk, for example, the liquid between the grains becomes richer in both milk proteins and fat globules, and so more like cream. When grains are cooked in a meat stock, the stock become more concentrated in gelatin, and so comes to resemble a reduced stock or demiglace.

THE GRAINS, OR CEREALS

Of the approximately 8,000 species in the grass family, only a handful play a significant role in the human diet. Aside from bamboo and sugar cane, these are the cereals. While their grains are very similar in structure and composition, the differences have made for widely divergent culinary histories.

The major Eurasian cereals—wheat, barley, rye, and oats—originally grew wild in extensive stands on the temperate high plains of the Near East. Groups of early humans could harvest enough wheat and barley from these wild fields in a few weeks to sustain themselves for a year. Some 12,000 to 14,000 years ago, the first agriculturalists began to plant and tend wheat and barley seeds selected for their size and the ease with which they could be harvested and used; and farmers gradually spread these crops throughout western and central Asia, Europe, and north Africa. Each cereal had its advantages. Barley was especially hardy, while rye and oats were able to adapt to wet, cold climates, and wheat produced a uniquely elastic paste that could be filled with tiny bubbles and baked into tender raised breads. Around the same time, the inhabitants of tropical and semitropical Asia domesticated rice, with its special ability to grow in wet, hot growing conditions. Somewhat later in warm central and South America arose maize, or corn, whose plants and kernels dwarf those of the other cereals.

Grain Structure and Composition

The edible portion of the cereal plant, commonly called the grain or kernel, is technically a complete fruit whose ovary-derived layer is very thin and dry. Three of the cereals—barley, oats, and rice—bear fruits that are covered by small, tough, leaf-like structures that fuse to form the husk or hull. Bread and durum wheats, rye, and maize bear naked fruits and don't have to be husked before milling.

All the grains have the same basic structure. The fruit tissue consists of a layer of epidermis and several thin inner layers, including the ovary wall; altogether, it's only a few cells thick. Just underneath the seed coat is the *aleurone layer,* only one to four cells thick and yet containing oil, minerals, protein, vitamins, enzymes, and flavor out of proportion to its size. The aleurone layer is the outer layer and only living part of the *endosperm*; the rest is a mass of dead cells that stores most of the carbohydrates and protein, and that takes up most of the grain's volume. Abutting onto the endosperm from one side is the *scutellum,* a single modified leaf that absorbs, digests, and conducts food from the endosperm to the *embryo,* or "germ," which is at the base of the fruit, and which is also well endowed with oil, enzymes, and flavor.

The endosperm (from the Greek: "within the seed") is often the only part of the grain consumed. It consists of storage cells that contain starch granules embedded in a matrix of protein. This matrix is made up of normal cell proteins and membrane materials, and sometimes of spherical bodies of special storage proteins which, squeezed together as the starch granules grow, lose their individual identity and form a monolithic mass. There's generally more starch and less protein per cell near the center of the grain than there is near the surface. This gradient means that the more grains are refined by milling and polishing, the less nutritious they get.

Milling and Refining

People began treating the grains to remove their tough protective layers in prehistoric times. Milling breaks the grains into pieces, and refining sifts away the bran and germ. The very different mechanical properties of endosperm, germ, and bran make this separation possible: the first is easily fragmented, and the others are oily and leathery respectively. The germ and the bran—which in practice include the aleurone layer just underneath it—together account for most of the fiber, oil, and B vitamins contained in the whole grain, as well as some 25% of its protein. Yet these parts of the grain are usually removed entirely or in part from rice and barley grains, and from cornmeal and wheat flours. Why this waste? Refined grains are easier to cook and to chew, and more attractively light in color. And in the case of flours, the high lipid concentrations in the germ and aleurone layer shorten the shelf life of whole-grain flours substantially. The oils are susceptible to oxidation and develop rancid flavors (stale aroma, harsh taste) in a matter of weeks. Today most refined cereals in industrial countries are fortified with B vitamins and iron in order to compensate for the nutrients lost with the bran.

Breakfast Cereals

Apart from breads and pastries, the most common form in which Americans consume grain is probably the breakfast cereal. There are two basic types of breakfast cereals: hot, which require cooking, and ready-to-eat, which are eaten as is, often with some cold milk.

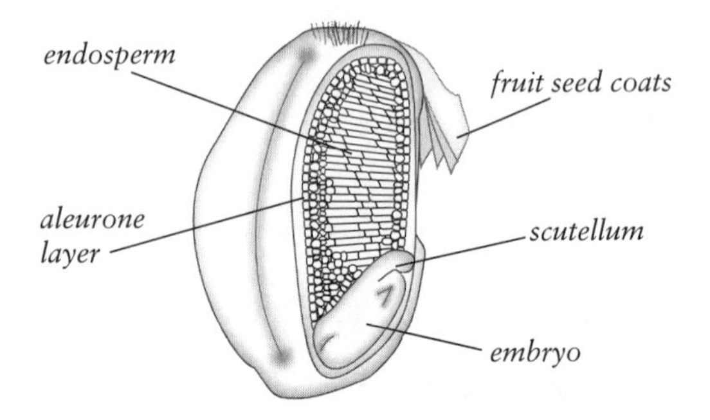

The anatomy of a wheat kernel. It's a miniature but complete fruit, with a dry rather than fleshy ovary wall. The large mass of endosperm cells stores food to nourish the early growth of the embryo or "germ."

Hot Cereals Hot cereals have been eaten since the dawn of civilization in the form of gruels, porridges, and congees. Corn grits, oatmeal, and cream of wheat are modern examples. Cooking the whole or milled cereal in excess of hot water softens the cell walls, gelates the starch grains and leaches starch molecules out, and produces a digestible, bland mush. The only significant improvement brought by the machine age has been a reduction in cooking time, either by grinding the cereal finely enough that it's quickly cooked, or by partly precooking it.

Ready-to-Eat Cereals Ready-to-eat cereals are the more common breakfast cereal by far in the United States. Ironically, the industry that has come under fire for giving children little more than empty calories, a sort of early-morning junk food, began as a "pure" and "scientific" health food, an alternative to the destructive diet of turn-of-the-century America. Its story involves a uniquely American mix of eccentric health reformers, fringe religion, and commercial canniness.

In the middle third of the 19th century, a vegetarian craze arose in opposition to the diet of salt beef and pork, hominy, condiments, and alkali-raised white bread that was prevalent at the time. A pure, plain diet for America was the object, and the issue was not only medical but moral. As Dr. John Harvey Kellogg put it somewhat later in his *Plain Facts for Old and Young,* "A man that lives on pork, fine-flour bread, rich pies and cakes, and condiments, drinks tea and coffee, and uses tobacco, might as well try to fly as to be chaste in thought." Kellogg and his brother Will Keith Kellogg, C. W. Post, and others invented such virtuous preparations as shredded wheat, wheat

The Composition of Grains

Grain composition varies a great deal; these are rough figures, assuming a moisture content of about 10%. Unless otherwise noted, the grains are whole.

Grain	Protein	Carbohydrate	Oil
Wheat	14	67	2
Barley	12	73	2
Barley, pearled	10	78	1
Rye	15	70	3
Oats	17	66	7
Rice, white	7	80	0.5
Rice, brown	8	77	3
Rice, wild	15	75	1
Corn (maize)	10	68	5
Fonio	8	75	3
Millet	13	73	6
Sorghum	12	74	4
Teff	9	77	2
Triticale	13	72	2
Amaranth	18	57	8
Buckwheat	13	72	4
Quinoa	13	69	6

and corn flakes, and Grape Nuts. These precooked cereals did offer a light, simple alternative to the substantial breakfasts of the day, became widely popular, and quickly generated a large, inventive, and profitable industry. Today there are several major varieties of ready-to-eat cereals:

- Muesli is a simple mixture of thinly rolled grains, sugar, dried fruits, and nuts.
- Flakes are made from whole grains (wheat) or grain fragments (corn) that are flavored, steam-cooked, cooled, rolled into thin flakes, and toasted in a drum oven.
- Granola, a term coined by the Kelloggs 100 years ago, is now rolled oats flavored with sweeteners (honey, malt, sugar) and spices, enriched with vegetable oil, toasted and mixed with nuts and/or dried fruit.
- Oven-puffed rice and corn are made by cooking rice grains or corn grits with water and flavorings, partly drying and lightly rolling them, then toasting in an oven that may reach 650°F/340°C,when the remaining moisture evaporates quickly enough to expand the grain structure.
- Puffed rice and wheat are made from whole grains, which are wetted and enclosed in a pressure-cooker "gun" at a temperature of 500–800°F/260–430°C. The steam pressure reaches 14 atmospheres, and is suddenly released, expelling the grains. As the steam within the grains expands from the pressure drop, it expands the grain structure, which then sets as it cools down into a light, porous mass.
- Baked cereals follow in the mold of the 19th century original, C. W. Post's Grape Nuts: dough of some sort is formed, baked, sometimes granulated and rebaked.
- Extruded cereals, usually small crunchy shapes, are made from doughs that are forced at high pressure through small openings, much as dried pastas are made. The pressure and friction generate high temperatures that cook the dough as it's shaped, and the pressure drop as the formed dough exits the extruder can cause it to expand as well.

Grains are still the base for these cereals, but they may actually be outweighed by sugar and other sweeteners. Sucrose is especially favored for its ability to give a frosty or glassy surface to the crisp grain flakes and slow the penetration of milk and resulting sogginess.

Wheat

Wheat was one of the first food plants to be cultivated by humans, and was the most important cereal in the ancient Mediterranean civilizations. After a long hiatus from the Middle Ages to the 19th century,

Food Words: *Cereal, Wheat, Barley, Rye, Oat*

Our word "cereal" comes from the Roman goddess of the fields, Ceres, whose name came in turn from an Indo-European root, *ker,* that meant "to grow"; the same root also led to "create," "increase," and "crescent." *Bhares* was the word for "barley" and also gave us "barn"; *wrughyo* meant "rye." "Wheat" came from the same root that gives us "white" (*kweit*), apparently because its flour was light in color; and "oat" came from *oid,* "to swell."

when hardier but less versatile cereals and potatoes were the principal staple foods, it regained its preeminence in much of Europe. Wheat was brought to America early in the 17th century and had reached the Great Plains by 1855. Compared to other temperate-zone cereals, wheat is a demanding crop. It's susceptible to disease in warm, humid regions and does best in a cool climate, but it can't be grown as far north as rye and oats.

Ancient and Modern Wheats A handful of different wheats have been grown from prehistoric times to the present. Their evolution is fascinating and still somewhat mysterious, and is summarized in the box on p. 466. The simplest wheat and one of the first to be cultivated was *einkorn*, which had the standard genetic endowment of most plants and animals: namely two sets of chromosomes (a "diploid" species). Somewhat less than a million years ago, a chance mating of a wild wheat with a wild goatgrass produced a wheat species with four sets of chromosomes, and this "tetraploid" species gave us the two most important wheats of the ancient Mediterranean world, *emmer* and *durum*. Then, just 8,000 years ago, another unusual mating between a tetraploid wheat species and a goatgrass gave an offspring with six sets of chromosomes: and this offspring gave us our modern *bread wheats*. The extra chromosomes are thought to contribute to the agricultural and culinary diversity found in modern wheats, most importantly the elasticity of the gluten proteins. Today 90% of the wheat grown in the world is hexaploid bread wheat. Most of the remaining 10% is durum wheat, whose main purpose is making pasta (p. 571). The other wheats are still cultivated on a small scale.

Durum Wheat Durum wheat, *T. turgidum durum*, is the most important of the tetraploid wheats. It arose in the Near East and spread to the Mediterranean before Roman times, when it was one of two major wheats. Emmer was better suited to humid climates and had a starchy grain, while durum was better suited to semiarid

World Grain Production

The leading figure for maize is misleading because a large fraction of the maize harvest is used to feed animals and produce industrial chemicals. Some wheat is also used for animal feed, while nearly all rice is eaten directly by humans.

Grain	Whole Grain Production, 2002 *Millions of Metric Tons*
Maize	602
Rice	579
Wheat	568
Barley	132
Sorghum	55
Oats	28
Millets	26
Rye	21
Buckwheat	2

Source: Food and Agriculture Organization, United Nations

conditions and had a glassy grain. Both were used to make breads leavened and unleavened, bulgur, couscous, injera, and other preparations. Southern and central Italy is now the main producer in Europe; India, Turkey, Morocco, Algeria, and the United States and Canada are large producers elsewhere.

Einkorn Wheat Einkorn wheat, *T. monococcum*, was rediscovered in the early 1970s in the Vaucluse region of France and the southern Alps, where it was being grown to make a local porridge. It was probably the first wheat to be cultivated, around 10,000 years ago. It grows best in cool conditions, tends to be rich in yellow carotenoid pigments and is high in protein. However, where the proportions of elastic glutenin and flowing gliadin (p. 521) are 1 to 1 in bread wheat, in einkorn they're 1 to 2. The result is a sticky, fluid gluten that's unsuited to breadmaking.

***Emmer Wheat or* Farro** Emmer wheat, *T. turgidum dicoccum*, was probably the second wheat to be cultivated. It grew in warmer climates than einkorn, and became the most important cultivated form from the Near East through northern Africa and Europe until early Roman times, when it was superseded by durum and bread

The Wheat Family

Relationships in the wheat family are complicated and still the subject of scholarly argument. Here is one plausible version of the family tree. Those wheats whose grains are enclosed in adhering papery husks are noted as "hulled"; all the rest are hullless and are therefore much easier to prepare for cooking or milling. Wheats in common use today are highlighted in bold type.

Wild einkorn (diploid; hulled; *Triticum monococcum boeticum*) → Cultivated einkorn (diploid; hulled; *T. monococcum monococcum*)

A wild wheat, *Triticum urartu* + a goatgrass, *Aegilops speltoides*

↓

Triticum turgidum (tetraploid):
Emmer (hulled; *T. turgidum dicoccum*)
Durum (*T. turgidum durum*)
Khorasan (*T. turgidum turanicum*)
Polish (*T. turgidum polonicum*)
Persian (*T. turgidum carthlicum*)

Triticum turgidum + a goatgrass, *Aegilops tauschii*

↓

Triticum aestivum (hexaploid):
Common, bread wheat (*T. aestivum aestivum*)
Spelt (hulled; *T. aestivum spelta*)
Club (*T. aestivum compactum*)

wheats. But pockets of emmer cultivation survived in parts of Europe, and emmer is now widely available under its Italian name, *farro*. In Tuscany whole farro grains go with beans into a winter soup; the presoaked grains are also made into a risotto-like dish called farrotto.

Kamut Kamut is the registered trademark for an ancient relative of durum wheat, a subspecies of *T. turgidum*. The modern production and commercialization of kamut (Egyptian for "wheat") began after World War II, when seeds said to have been collected in Egypt were planted in Montana. It's characterized by a large grain size and a high protein content, though its gluten is better suited to pasta than to raised breads.

Spelt Spelt, *T. spelta* became known as *Dinkel* in southern Germany, where it has been grown since 4000 BCE. It's often confused with emmer (farro). Spelt is remarkable for its high protein content, as much as 17%. It's still used to make breads and soups. Central Europeans make *Grünkern,* or "green kernel," by gently drying or roasting the green grain and milling it for use in soups and other preparations.

Varieties of Bread and Pasta Wheats

Something on the order of 30,000 varieties of wheat are known, and they're classified into a few different types according to planting schedule and endosperm composition. They're mostly used to make breads, pastries, and pastas, and are described in the next chapter.

Wheat Pigments Most wheat varieties have a reddish-brown bran layer that owes its color to various phenolic compounds and to browning enzymes (p. 269) that assemble them into large colored aggregates. Less common are white wheats, whose bran layer is cream-colored due to a much lower content of phenolic compounds and browning enzymes. White wheats have a less astringent taste and discolor less when some of the bran is included in the flour; they're used to replace ordinary wheats when an especially mild flavor or light color is desired.

The color of durum wheat, its coarse semolina flour, and dry pasta is due mainly to the carotenoid xanthophyll lutein, which can be oxidized to a colorless form by enzymes in the grain and oxygen. This maturation has traditionally been desired in

Protein Content and Quality of Different Wheat Varieties

Gluten quality determines the suitability of a given wheat for particular dishes. Both bread and pasta benefit from a strong, cohesive gluten. Elasticity improves the gas-trapping ability and lightness of bread doughs, but interferes with the rolling of pasta dough into thin sheets.

Wheat Variety	Protein Content, % grain weight	Gluten Quality
Bread	10–15	Strong and elastic
Durum	15	Strong, not very elastic
Einkorn	16	Weak, sticky
Emmer (farro)	17	Moderately strong, not very elastic
Spelt, hard	16	Moderately strong, not very elastic
Spelt, soft	15	Strong, moderately elastic

wheats (remember that the name comes from an ancient root meaning "white"), but is not in durum. Some of the minor wheats are also rich in carotenoid pigments.

Wheat Gluten

Gluten in Wheat Flour Doughs Wheat has long been the West's premier grain primarily because its storage proteins have unique chemical properties. When flour is mixed with water, the gluten proteins bond to each other and form an elastic mass that can expand to accommodate gas bubbles produced by yeast. Without wheat, then, we would not have raised breads, cakes, and pasta as we know them. Gluten quantity and quality vary significantly among different wheats, and determine the uses to which a given type is put.

Gluten as a Separate Ingredient Because they're both cohesive and insoluble in water, the gluten proteins are easily separated from the rest of the flour: you simply make a dough, then knead it in water. The starch and water-soluble substances wash away, and tough, chewy gluten remains. Gluten as a unique food ingredient was discovered by Chinese noodle makers around the 6th century, and by the 11th was known as *mien chin*, or the "muscle of flour." (The Japanese call it *seitan.*) When cooked, concentrated gluten does develop a chewy, slippery texture like that of meats from animal muscle. *Mien chin* became a major ingredient in the vegetarian cooking that developed in Buddhist monasteries; there are recipes dating from the 11th century for imitation venison and jerky, and for fermented gluten. Because gluten contains a high proportion of glutamic acid, fermentation breaks it down into a condiment that was an early version of savory-tasting MSG (p. 342). One of the simplest ways to prepare gluten is to pinch off small bits and deep-fry them; they puff up into light chewy balls that readily absorb the flavor from a sauce. Today gluten is widely available and used to make a variety of vegetarian "meats."

Notable Wheat Preparations Whole grains of wheat, often called wheat berries, are usually sold with their bran fully intact, and can take an hour or more to cook unless presoaked. Farro is now available with part of its bran milled away—similar to partly milled pigmented rice and wild rice—and cooks much more quickly, while retaining the stronger flavor and integrity of separate grains provided by the bran.

Wheat germ is sometimes added to baked goods or other foods; it is a good source of protein (20% by weight), oil (10%), and fiber (13%). Wheat bran is mainly fiber, with about 4% oil. Their oil content makes both bran and germ susceptible to developing stale flavors. They're best stored in the refrigerator.

Bulgur Bulgur or burghul is an ancient preparation of wheat—usually durum—that's still popular in North Africa and the Middle East. It's made by cooking whole grains in water, drying them so that the interior becomes glassy and hard, then moistening them to toughen the outer bran layer, and finally pounding or milling to remove the bran and germ and leave the endosperm in coarse chunks. It's the wheat version of parboiled rice (p. 473). The result is a nutritious form of wheat that keeps indefinitely and cooks relatively quickly. Coarse bulgur (to 3.5 mm across) is used much as rice or couscous is, boiled or steamed to go with a moist dish or made into a pilaf or a salad, while fine bulgur (0.5–2 mm) is made into falafel (deep-fried balls of bulgur and fava-bean flour), and various pudding-like sweets.

Green, or Immature Wheat Green wheat grains have also been enjoyed for their sweetness and unusual flavor. The stalks are cut while the grains are still moist inside, the grains charred over a small straw fire to weaken the husks and add flavor, then eaten fresh or dried for keeping (Turkish *firig*, Arab *frikke*).

Barley

Barley, *Hordeum vulgare,* may have been the first cereal to be domesticated in the grasslands of southwest Asia, where it grew alongside wheat. It has the advantage of a relatively short growing season and a hardy nature; it's grown from the Arctic Circle to the tropical plains of northern India. It was the primary cereal in ancient Babylon, Sumeria, Egypt, and the Mediterranean world, and was grown in the Indus valley civilization of western India long before rice. According to Pliny, barley was the special food of the gladiators, who were called *hordearii,* or "barley eaters"; barley porridge, the original polenta, was made with roasted flaxseed and coriander. In the Middle Ages, and especially in northern Europe, barley and rye were the staple foods of the peasantry, while wheat was reserved for the upper classes. In the medieval Arab world, barley dough was fermented for months to produce a salty condiment, *murri,* that food historian Charles Perry has discovered tastes much like soy sauce.

Today, barley is a minor food in the West; half of production is fed to animals, and a third is used in the form of malt. Elsewhere, barley is made into various staple dishes, including the Tibetan roasted barley flour *tsampa,* often eaten simply moistened with tea; it's an important ingredient in the Japanese fermented soy paste miso; and in Morocco (the largest per capita consumer) and other countries of north Africa and western Asia it's used in soups, porridges, and flat breads. In Ethiopia there are white, black, and purple-grained barleys, some of which are made into drinks. Water simmered with raw or roasted barley has been enjoyed for two thousand years or more, from western Europe to Japan.

The barley grain is notable for containing significant quantities—about 5% each of the grain weight—of two carbohydrates other than starch: the pentosans that give rye flours their stickiness, and the glucans that give oats their gelatinous and cholesterol-lowering qualities (pp. 470, 471). Both are found in the walls of endosperm cells as well as in the bran, and together with barley's water-insoluble proteins they contribute to the distinctively springy texture of the cooked grain. They also cause barley flour to absorb twice the water that wheat flour does.

Pearled Barley There are hull-less barleys, but most food varieties have adherent hulls that are removed as part of the milling process. Barley has more of its grain removed than does rice, the other grain frequently prepared as a whole grain. This is partly because barley bran is brittle and doesn't come off in large flakes, so it can't

Food Words: From Barley Water to *Orgeat, Horchata, Tisane*

The European habit of drinking barley water has mostly faded away, but it lives on in the names of several other beverages or beverage flavorings. The Latin word for barley, *hordeum,* became the French *orge*; *orge mondé,* meaning hulled barley, became *orgemonde* and in the 16th century *orgeat.* Orgeat is still around, but it's now an almond-flavored syrup. *Orgeat* also became the Spanish word *horchata,* which gradually evolved from a barley drink to a drink made with either rice or the chufa or tiger "nut" (p. 308). And *tisane,* the modern French term for an infusion of herbs or flowers? It comes from the Latin *ptisana,* which meant both crushed cleaned barley or the drink made therefrom, which was sometimes flavored with herbs.

be removed during normal milling; and partly because processors eliminate the deep crease in the barley grain to give it a more uniform appearance. The process of "pearling" in a stone mill removes the hull and then portions of the bran. "Pot barley" has lost 7–15% of the grain, but retains the germ and some of the bran, and so more nutrients and flavor. Fine pearled barley has lost the bran, germ, and aleurone and subaleurone layers, a loss of about 33% of the grain's initial weight.

Barley Malt The most important form in which we consume barley is *malt,* a major ingredient in beers and some distilled liquors, and a minor ingredient in many baked goods. Malt is a powder or syrup made from barley grains that are moistened and allowed to germinate, and that become sweet with sugars. Its production and qualities are described below (pp. 679, 743).

Rye

Rye apparently arose in southwest Asia, migrated with domesticated wheat and barley as a weed in the crops of early farmers, reached the coast of the Baltic Sea around 2000 BCE, grew better than the other cereals in the typically poor, acid soil and cool, moist climate, and was domesticated around 1000 BCE. It's exceptionally hardy, and is grown as far north as the Arctic Circle and as high as 12,000 feet/4,000 meters. Up through the last century it was the predominant bread grain for the poor of northern Europe, and even today the taste for rye persists, especially in Scandinavia and eastern Europe. Poland, Germany, and Russia are the leading producers. In Germany, wheat production exceeded rye for the first time only in 1957.

Rye has unusual carbohydrates and proteins, and as a result produces a distinctive kind of bread. It's described in the next chapter (p. 545).

Rye Carbohydrates Rye contains a large quantity, up to 7% of its weight, of carbohydrates called *pentosans* (an old term; the new one is *arabinoxylans*). These are medium-sized aggregates of sugars that have the very useful property of absorbing large amounts of water and producing a

Rye and LSD

In addition to its role as a food, rye has also had an indirect influence on modern medicine and recreational pharmacology. The cool, moist climate in which rye does well is also favorable for the growth of the *ergot* fungus (*Claviceps purpurea*). From the 11th to the 16th centuries, ergot contamination of rye flour was responsible for frequent epidemics of what was called Holy Fire or Saint Anthony's Fire, a disease with two sets of symptoms: progressive gangrene, in which extremities turned black, shrank, and dropped off; and mental derangement. Occasional outbreaks of ergot poisoning from contaminated flour continued well into the 20th century.

Early in the 20th century, chemists isolated from ergot a handful of alkaloids with very different effects: one stimulates the uterine muscle; some are hallucinogens; and some constrict the blood vessels, an action that can cause gangrene, but that also has useful medical applications. All these alkaloids have a basic component in common called lysergic acid. In 1943 the Swiss scientist Albert Hofman discovered the particular variant that would come to such prominence in the 1960s: the hallucinogen lysergic acid diethylamide, or LSD.

thick, viscous, sticky consistency. Thanks to its pentosans, rye flour absorbs eight times its weight in water, while wheat flour absorbs two. Unlike starch, the pentosans don't retrograde and harden after being cooked and cooled. So they provide a soft, moist texture that helps gives rye breads a shelf life of weeks. Rye pentosans also help control appetite; the dried carbohydrates in rye crisps swell in the stomach, thus giving the sensation of fullness, and they are slowly and only partly digested.

Oats

The world produces more oats than rye today, but 95% of the crop is fed to animals. Oats are the grains of *Avena sativa,* a grass that probably originated in southwest Asia and gradually came under cultivation as a companion of wheat and barley. In Greek and Roman times it was considered a weed or a diseased form of wheat. By 1600 it had become an important crop in northern Europe, in whose wet climate it does best; oats require more moisture than any other cereal but rice. Other countries, however, continued to disdain it. Samuel Johnson's *Dictionary* (London, 1755) gives this definition for oats: "A grain, which in England is generally given to horses, but in Scotland supports the people."

Today the United Kingdom and the United States are the largest consumers of food oats. U.S. consumption was boosted in the late 19th century by Ferdinand Schumacher, a German immigrant who developed quick-cooking rolled oats for breakfast, and Henry Crowell, who was the first to turn a cereal from a commodity into a retail brand by packaging oats neatly with cooking instructions, labeling it "Pure," and naming it "Quaker Oats." Oats are now a mainstay in ready-to-eat granolas, mueslis, and manufactured breakfast cereals.

There are several reasons for the relatively minor status of oats. Like barley, oats have no gluten-producing proteins, which means that they can't be made into light raised breads. The kernel has adherent husks that make it difficult to process. Oats contain from two to five times the fat that wheat does, mainly in the bran and endosperm rather than the germ, and also carry large amounts of a fat-digesting enzyme. The combination means that oats have a tendency to become rancid. They require a heat treatment that inactivates the enzyme in order to prevent rapid deterioration during storage.

On the other hand, oats do have several virtues. They're rich in indigestible carbohydrates called beta-glucans, which absorb and hold water, give hot oatmeal its smooth, thick consistency, have a tenderizing, moistening effect in baked goods, and help lower our blood cholesterol levels. The glucans are found mainly in the outer layers of the endosperm under the aleurone layer, and so are especially concentrated in oat bran. Oats also contain a number of phenolic compounds that have antioxidant activity.

Oat Processing Oats are generally used as whole grains, also called groats, because they're much softer than wheat or corn and don't break cleanly into endosperm, germ, and bran. The first stage in their processing is a low-temperature "roasting," which gives the grain much of its characteristic flavor and inactivates the fat-splitting enzyme. (This step also denatures the storage proteins and makes them less soluble, giving the grain greater integrity during cooking.) The whole groats are then processed into various shapes, all of which have the same nutritional value. Steel-cut oats are simply whole groats cut into two to four pieces for faster cooking. Rolled oats are whole kernels that are steamed to make them soft and malleable, then pressed between rollers to make them thin and quick to reabsorb water during cooking or simple soaking (as for muesli). The thinner the oats are rolled, the faster they rehydrate: regular oats are about 0.8 mm thick, "quick-cooking" oats are around 0.4 mm, and "instant" even thinner.

Rice

Rice is the principal food for about half of the world's population, and in such countries as Bangladesh and Cambodia provides nearly three-quarters of the daily energy intake. *Oryza sativa* is a native of the tropical and semitropical Indian subcontinent, northern Indochina, and southern China, and was probably domesticated in several places independently, the short-grain types around 7000 BCE in the Yangtze River valley of south-central China, and long-grain types in Southeast Asia somewhat later. A sister species with a distinctive flavor and red bran, *Oryza glaberrima,* has been grown in west Africa for at least 1,500 years.

Rice found its way from Asia to Europe via Persia, where the Arabs learned to grow and cook it. The Moors first grew large quantities in Spain in the 8th century, then somewhat later in Sicily. The valley of the Po River and the Lombardy plain in northern Italy, the home of risotto, first produced rice in the 15th century. The Spanish and Portuguese introduced rice throughout the Americas in the 16th and 17th centuries. South Carolina was the location of the first commercial American planting in 1685, where the rice-growing expertise of African slaves was important; today most U.S. rice comes from Arkansas and the lower Mississippi region, Texas, and California.

Kinds of Rice There are thought to be more than 100,000 distinct varieties of rice throughout the world. They all fall into one of two traditionally recognized subspecies of *Oryza sativa. Indica* rices are generally grown in lowland tropics and subtropics, accumulate a large amount of amylose starch, and produce a long, firm grain. *Japonica* rices, with upland varieties that do well both in the tropics (Indonesian and Filipino types sometimes known as *javanicas*) and in temperate climates (Japan, Korea, Italy, and California), accumulate substantially less amylose starch than the *indicas,* and produce a shorter, stickier grain. There are also varieties that are intermediate between *indica* and *japonica.* Generally, the higher the amylose content in a rice variety, the more organized and stable the starch granules, and so the more water, heat, and time it takes to cook the grains.

Most rice is milled to remove the bran and most of the germ, and then "polished" with fine wire brushes to grind away the aleurone layer and its oil and enzymes. The result is a very stable refined grain that keeps well for months.

Common categories of rice include the following:

- *Long-grain* rice has an elongated shape, its length four to five times its width. Thanks to its relatively high proportion of amylose (22%), it tends to require the largest proportion of water to rice (1.7 to 1 by weight, 1.4 to 1 by volume), and to produce separate springy grains that become firm as they cool, and distinctly hard if chilled. Most Chinese and Indian rices are long-grain *indicas,* as is most of the rice sold in the United States.
- *Medium-grain* rice is two to three times longer than it is wide, contains less amylose (15–17%) than long-grain rice and requires less water, and develops tender grains that cling to each other. Italian risotto rices and Spanish paella rices are medium-grain *japonicas.*
- *Short-grain* rice is only slightly longer than it is wide, and otherwise similar to medium-grain rice. Short- and medium-grain *japonicas* are the preferred types in north China, Japan, and Korea. They're ideal for sushi because they cling together in small masses and remain tender even when served at room temperature.
- *Sticky* rice, also called waxy, glutinous, or sweet rice, is a short-grain type whose starch is practically all amylopectin. It requires the least water (1 to 1 by weight, 0.8 to 1 by volume) and becomes very clingy and

readily disintegrates when cooked (it's often soaked and then steamed, not boiled). Despite its names, it contains no gluten and isn't sweet, though it's often used to make sweet dishes in Asia. It's the standard rice in Laos and northern Thailand.

- *Aromatic* rices are a distinctive group of mainly long- and medium-grain varieties that accumulate unusually high concentrations of volatile compounds. Indian and Pakistani *basmati* (Urdu for "fragrant," accent on the first syllable), Thai jasmine (an unusual long-grain but low-amylose type), and U.S. Della are well-known aromatic rices.
- *Pigmented* rices have bran layers that are rich in anthocyanin pigments. Red and purple-black colors are the most common. The bran may be left intact, or partly milled away so that only traces of color are left.

Brown Rice Brown rice is unmilled, its bran, germ, and aleurone layers intact. Any kind of rice, whether long-grain, short-grain, or aromatic, may be sold in its brown form. It takes two to three times longer to cook than the milled version of the same variety, and has a chewy texture and a rich aroma, often described as nutty. Thanks to the oil in its bran and germ, it's more susceptible to staling than polished rice, and is best stored in the refrigerator.

Parboiled or Converted Rice For more than 2,000 years, rice producers in India and Pakistan have parboiled nonaromatic varieties before they remove the hull and mill them to white rice. They steep the freshly harvested grain in water, boil or steam it, and then dry it again before hulling and milling. This precooking brings several advantages. It improves the nutritional quality of the milled grain by causing vitamins in the bran and germ to diffuse into the endosperm, and causing the aleurone layer to adhere to the grain. Precooking the starch also hardens the grain and makes its surface less sticky, so when cooked again, parboiled rice produces separate firm

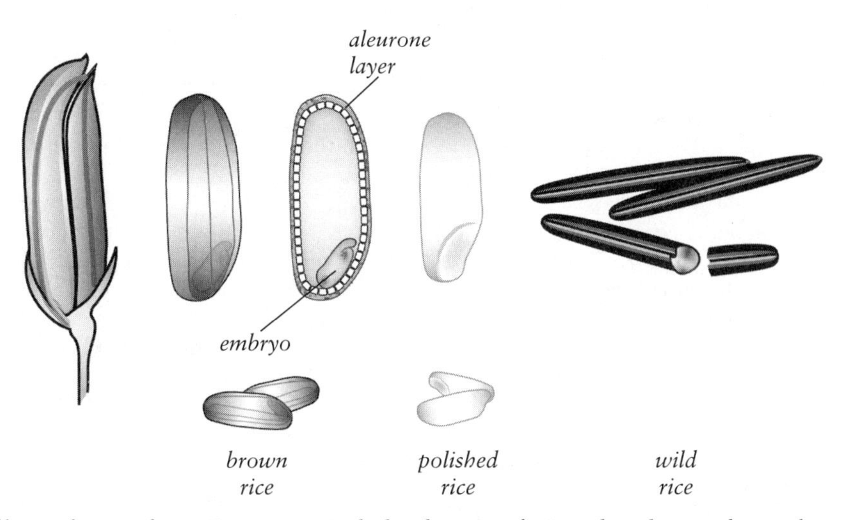

Different forms of rice. Brown rice includes the outer fruit and seed coats that make up the bran, and the embryo and oil- and enzyme-rich aleurone layer. Polished rice is the central mass of endosperm cells, freed from all other parts of the grain; it's mainly starch and protein. Wild rice is the whole grain of a North American grass; it is heated to dry it out and develop flavor, and this processing gives its endosperm a distinctive glassy appearance.

intact grains. Parboiled rice also has a distinctive nutty flavor; the soaking activates enzymes that generate sugars and amino acids that then participate in browning reactions during drying; and partial breakdown of lignin in the attached hull provides vanillin and related compounds. Parboiled rice takes longer to cook than ordinary white rice, a third to half again the time, and its texture is so firm that it can seem coarse.

Quick-Cooking Rice Quick-cooking rice is manufactured by cooking white, brown, or parboiled rice, thus disrupting its cell walls and gelating its starch, then fissuring the grain in order to speed the infiltration of hot water when the consumer cooks it, and finally drying it. The fissuring may be accomplished with dry heat, rolling, microwave treatments, or freeze-drying.

Rice Flavor The flavor of rice depends on the variety and the degree to which it is milled. The outer portions of the rice grain contain more free amino acids, sugars, and minerals, and proportionally less starch. The more a rice kernel is milled, and so the more of its surface is removed, the less flavor and the higher proportion of starch it contains.

The aroma of standard white rices has green, mushroomy, cucumber-like, and "fatty" components (from 6, 8, 9, and 10-carbon aldehydes), as well as a slight popcorn note and floral, corn-like, hay-like, and animal qualities. Brown rices contain these and also small amounts of vanillin and maple-sugar-like sotolon. Aromatic rices are especially rich in the popcorn-like aromatic component (acetylpyrroline), which is also an important element in screwpine leaves (p. 411), and cooked popcorn and bread crust. Because it is volatile and not regenerated during cooking, the popcorn aromatic escapes during cooking, and its concentration declines. This is one reason for presoaking aromatic rices; this step shortens the cooking and minimizes aroma loss.

Cooking with Rice

Many Traditional Methods The cooking of rice is a matter of introducing moisture throughout the grains and heating them enough to gelate and soften the starch granules. Indian cooks boil the rice in an excess of water that's poured off when the rice is done, so the grains end up intact and separate. Chinese and Japanese cooks boil rice with just enough water to moisten and cook it through in a closed pot, which produces a mass of grains that cling together and are easily eaten with chopsticks. Where rice has always been an everyday staff of life, through much of East Asia, it's usually prepared simply in water, and judged by the intactness of its grains and their whiteness, gloss, tenderness, and flavor. Where rice was more unusual and even a luxury, in central Asia, the Middle East, and the Mediterranean, it's often enriched with broths, oils, butter, and other ingredients to make such dishes as pilafs, risottos, and paellas. Iranians, perhaps the most sophisticated rice cooks, make *polo* by partly boiling long-grain rice in excess water, layering it with a variety of cooked meats, vegetables, dried fruits, and nuts, then gently steaming to finish the cooking, and managing the heat so that a brown crust of rice, the prized *tahdig,* forms at the bottom.

Rinsing and Soaking An initial rinsing of the dry rice removes surface starch and thus a source of added stickiness. Some rices, notably basmatis and Japanese varieties, are either soaked in water or allowed to rest for 20–30 minutes after washing; they thus absorb some water, which will speed the subsequent cooking. Brown and wild rices can be treated similarly.

After Cooking: Resting, Reheating Once cooked, rice benefits from a resting period to allow the grains to cool down somewhat and become firmer, so that they aren't as easily broken when scooped from the pot and served. Leftover rice is often hard due to the retrogradation of the starch, which is cured by heating it up to the gelation temperature again. Rice is easily softened by reheating to 160°F/70°C or above, either with a little added water in a pot or in the microwave oven, or fried to make fried rice, rice cakes, or croquettes.

Keeping Rice Safe Cooked rice turns out to be a potential source of food poisoning. Raw rice almost always carries dormant spores of the bacterium *Bacillus cereus,* which produces powerful gastrointestinal toxins. The spores can tolerate high temperatures, and some survive cooking. If cooked rice is left for a few hours at room temperature, the spores germinate, bacteria multiply, and toxins accumulate. Ordinary cooked rice should therefore be served promptly, and leftovers refrigerated to prevent bacterial growth. The rice in Japanese sushi is served at room temperature, but the surface of its cooked grains are coated with a flavorful and antimicrobial mixture of rice vinegar and sugar. Rice salads should be similarly acidified with vinegar, lemon or lime juice.

Some Other Rice Preparations and Products Cultures across the world have found many different inventive uses for rice. Here is a brief sampling.

Rice Flour Rice flour is notable for being around 90% starch, and for having the smallest starch granules of the major cereals, a half to a quarter the size of wheat starch granules. When used to thicken sauces or fillings, it provides an especially fine texture. And thanks to its low protein content, the dry flour absorbs relatively little water. This means that when it's used to make a frying batter for Japanese tempura, rice flour gives a thin consistency with relatively little water, and so the batter readily fries to a crisp, dry texture.

Because rice flour contains no elastic gluten proteins, it can't be used to make raised breads. But the same lack of gluten makes rice flour a useful ingredient for peo-

Risotto: Turning Rice into Its Own Sauce

The risotto of Italy is made with medium-grain varieties that are fairly large and can tolerate the unique cooking method, which abrades and removes starch from the rice surface so that it can thicken the cooking liquid to a creamy consistency.

To make risotto, the rice is cooked through by adding a small amount of hot cooking liquid at a time and stirring the rice until the liquid is absorbed, then repeating until the rice is soft but still has a kernel of chewiness at the center. This time-consuming technique subjects the rice grains to constant friction, and rubs softened endosperm from the surface so that it can become dissolved in the liquid phase (stirring only at the end of cooking breaks the softened grains apart rather than removing the surface layer). In addition, the cooking of small amounts of liquid in an open pan causes much of the moisture to evaporate, which means that more of the cooking liquid must be used, and thus that more of the cooking liquid's flavor becomes concentrated in the dish.

Restaurant cooks prepare risotto to order by cooking the rice well ahead of time the traditional way until it's just short of done, then refrigerating it. This allows some of the cooked starch in the rice to firm (p. 458), giving the grain more resilience than it would have if cooked fully and simply rewarmed. Then just before serving, the chilled rice is reheated and finished with hot broth and enrichments.

ple with gluten intolerance. Bakers make a reasonable approximation of raised bread by supplementing rice flour with xanthan or guar gum or other long-chain carbohydrates, which help bind the dough together and retain the gas bubbles produced by yeasts or chemical leavenings.

Rice Powder Rice powder is a condiment made in Vietnam and Thailand by roasting the grains, then grinding them; it's sprinkled on a variety of dishes just before eating.

Rice Noodles and Rice Paper Despite rice's lack of gluten, noodles and thin sheets can be made from rice-flour doughs (p. 579). Rice paper is used as a wrapper to enclose meat and vegetable preparations, and can be eaten either simply moistened or fried.

Mochi Mochi is the Japanese name for a chewy, almost elastic preparation of sticky rice that may be formed into balls or into thin sheets for wrapping around a filling of some kind. It's made by steaming sticky rice, then pounding it to a paste, or by making a dough from sticky rice flour and kneading it for 30 minutes. The pounding or kneading organizes the bushy amylopectin starch molecules into an intermeshed mass that resists changes in its structure.

Lao Chao Lao chao is a Chinese fermented rice made from sticky rice. The rice is steamed, cooled, made into small cakes with a starter that includes the mold *Aspergillus oryzae* (p. 755) and yeasts, and held at room temperature for two to three days until it becomes soft, sweet, and tart, with a fruity and alcoholic aroma.

Wild Rice Wild rice is not a species of the true tropical rice genus *Oryza*. It's a distant relative, a cool-climate water grass that produces unusually long grains, to three-quarters of an inch (2 cm), with a dark seedcoat and a complex, distinctive flavor. *Zizania palustris* is a native of the upper midwestern Great Lakes region of North America, where it grows in shallow lakes and marshes and was gathered in canoes by the Ojibway and other native peoples. It's the only cereal from North America to have become important as a human food. Wild rice is unusual among the cereals for containing double the usual amount of moisture at maturity, around 40% of the kernel weight. It thus requires more elaborate processing than true rice in order to be stored. It's first matured in moist piles for a week or two, during which immature grains continue to ripen and microbes grow on the grain surfaces, generating flavor and weakening the husks. Then it is parched over a fire to dry the grain, flavor it, and make the husk brittle; and finally it's threshed to remove the husk.

Texture and Flavor Wild rice has a firm, chewy texture thanks to its intact bran layer and the parching process, which gelates and then anneals the starch much as parboiling does for true rice. It takes longer to cook than most grains, sometimes an hour or more, because its starch has been precooked into a hard, glassy mass, and because its bran layers are impregnated with cutins and waxes (p. 262) to resist the absorption of water (in nature, the grains fall into the water and lie dormant for months or even years before germinating). The dark pigmentation may also contribute; it is partly green-black chlorophyll derivatives and partly black phenolic complexes generated by browning enzymes. Producers often slightly abrade the grains to improve their water absorption and shorten cooking time. Cooks can also presoak the grains for hours in warm water.

The flavor of the raw grain has earthy and green, flowery, tea-like notes. Curing amplifies the tea notes (from pyridines) but may add an undesirable mustiness; parching generates browning reactions and a toasted, nutty character (from pyrazines). Different producers use different methods for curing (none, brief, extended) and parching (low or high temperatures, open fires or indirectly heated metal drums), so the flavor of wild rice varies greatly.

Domesticated Wild Rice Relatively little wild rice is still gathered from uncultivated, naturally occurring stands. Today most is grown in artificially flooded paddies, and harvested mechanically after the fields are drained. Cultivated wild rice therefore has more consistently mature, dark seedcoats than the gathered grain. To taste truly wild rice from its native region and savor the differences among small producers, it's necessary to read labels carefully.

Maize, or Corn

Maize, known in the United States as "corn" and among biologists as *Zea mays,* was domesticated in Mexico some 7,000 to 10,000 years ago from a large grass called teosinte (*Zea mexicana*), which grows in open woodlands. Unlike the Old World cereals and the legumes, which human selection altered in relatively minor ways, corn is the result of several drastic changes in the structure of teosinte that concentrated pollen production at the top of the plant and female flower production—and cob and kernel production—along the main stalk. The large size of both plant and fruit made corn agriculture relatively easy, and corn quickly became the basic food plant of many other early American cultures. The Incas of Peru, the Mayas and Aztecs of Mexico, the cliff dwellers of the American Southwest, Mississippi mound builders, and many seminomadic cultures in North and South America depended on corn as a dietary staple. Columbus brought corn back with him to Europe, and within a generation it was being grown throughout southern Europe.

Corn is now the third largest human food crop in the world after wheat and rice, and is the primary nourishment for millions of people in Latin America, Asia, and Africa. In Europe and the United States, where more corn is fed to livestock than to people, it's appreciated for its unique flavor, for the texture and substance it contributes to a variety of boiled, baked and fried foods, and as a snack food. Corn also provides mash for making whiskey, corn starch for thickening sauces and fillings, corn syrup for flavoring and lending viscosity to various sweet preparations, and corn oil. Different parts of the plant are also turned into many industrial products.

Kinds and Colors of Corn There are five general kinds of corn, each characterized by a different endosperm composition. It appears that a high-protein popcorn type was the first corn to be cultivated, but all five were known to native Americans long before the coming of Europeans.

- *Popcorn* and *flint corn* have a relatively large amount of storage protein that surrounds granules of high-amylose starch.
- *Dent corn,* the variety most commonly grown for animal feed and for milled food ingredients (grits, meals, flours), has a localized deposit of

Food Words: *Corn* and *Maize*

The grain that Americans call "corn" was originally known in English as "maize" or "Indian corn." "Maize" comes from the Taino name used in the West Indies, and is the source of the Spanish, Italian, and French names. The word *corn* is a generic term that comes from the same root as *kernel* and *grain* and has the same broad meaning; so "corned" beef is beef cured with grains of salt. "Corn" is also used in different parts of Britain as shorthand for the most important grain of the region. Only in the United States did it come to refer exclusively to maize.

low-amylose, "waxy" starch at the crown of the kernel, which produces a depression, or dent, in the dried kernel.

- *Flour corns,* including the standard varieties of blue corn, are soft and easily ground because their endosperm is a discontinuous and weak combination of relatively little protein, mostly waxy starch, and air pockets. What we call Indian corn today are flour and flint varieties with variegated kernels.
- *Sweet corn,* a popular vegetable in the United States when immature, stores more sugar than starch, and therefore has translucent kernels and loose, wrinkled skins (starch grains reflect light and plump out the kernels in the other types). Most corn-producing countries also eat immature corn, but use the other general-purpose corn types. The native Americans who first developed it apparently enjoyed sweet corn for its full flavor when parched.

The different kinds of corn also come in various colors, some of which were originally selected by native Americans for ceremonial use. The interior is usually either unpigmented and white, or yellow with nutritionally valuable fat-soluble carotenes and xanthophylls (beta carotene, lutein, zeaxanthin). Blue, purple, and red kernels carry water-soluble anthocyanins in their aleurone layer, the nutrient-rich cell layer just under the hull.

Alkaline Treatment: Several Benefits
Corn is unusual among the grains for its large size, and for the thickness and toughness of its outer pericarp, or hull. Early corn eaters developed a special pretreatment to ease the removal of the hull, which is called *nixtamalization* (from an Aztec word): they cooked the kernels in water made alkaline with a variety of substances. The Mayas and Aztecs used ashes or lime; North American tribes, ashes and naturally occurring sodium carbonate deposits; and a contemporary Mayan group burns mussel shells for the same purpose. One of the major glue-like components of plant cells walls, hemicellulose, is especially soluble in alkaline conditions. Nixtamalization softens the hull and partly detaches it from the rest of the kernel so that it can be rubbed off and washed away. It also helps transform the kernels into a cohesive dough for making tortillas and other preparations (see below), and it releases much of their bound niacin so that we can absorb and benefit from it.

Corn Flavor Corn has a distinctive flavor unlike that of any other grain. Popcorn and other dry corn products toasted at a high temperature develop a number of characteristic carbon-ring compounds,

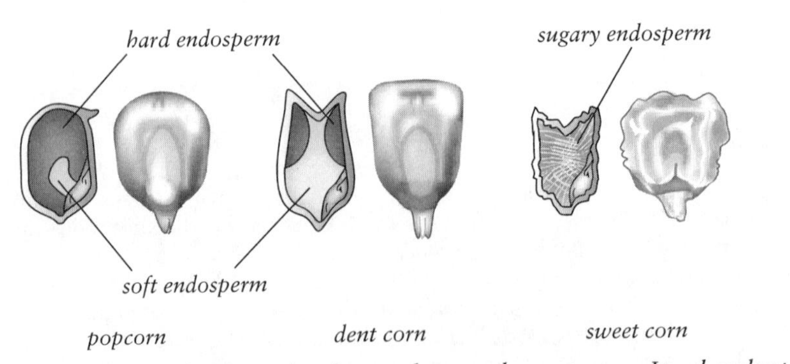

Types of corn. Left to right: *Kernels of pop, dent, and sweet corn. Its abundant hard endosperm helps popcorn contain the steam pressure that eventually explodes it.*

including one that they share with basmati rice (acetylpyrroline). Alkaline processing gives rise to yet another set of distinctive aroma molecules, including one that is a breakdown product of the amino acid tryptophan, and a close chemical and aromatic relative of a characteristic note in concord grapes and wood strawberries (aminoacetophenone, related to the fruits' methyl anthranilate). Masa can also have violet-like and spicy notes (from ionone and vinylguaiacol).

Whole-grain Corns: Hominy, Corn Nuts Common corn ingredients and foods can be divided into two general groups: those based on the whole grain, and those based on ground grain. There's also a basic division between dry, untreated materials and "wet-processed" alkaline-treated materials.

Whole-grain versions of corn are relatively few, with popcorn by far the most common. Hominy consists of whole corn kernels, preferably white, cooked for 20–40 minutes in a solution of lime or lye, then washed of their hulls and excess alkaline solution. Hominy is used in soups (pozole), stews, and side dishes, and has a dense, chewy consistency. Corn Nuts are a familiar snack food made from the largest kernels known, the Cuzco gigante variety from Peru. The kernels are treated with alkali to remove the hulls, soaked for some hours in warm water, fried to develop color, flavor, and a crunchy texture, then flavored.

Popcorn It appears from archaeological remains in Mexico that popping in the embers of a fire may have been the first method for cooking corn. Early explorers described popped corn among the Aztecs, Incas, and North American tribes. In the 19th century, Americans served popcorn as a breakfast cereal, made it into porridges, puddings, and cakes, added it to soups, salads, and main dishes, and mixed it with molasses to make an ancestor of the sweet popcorn ball and Cracker Jack. Popcorn was a popular finger food in the United States in the 1880s, then became associated with movie theatres, and later with watching television in the home. In the 21st century, most of the popcorn sold in supermarkets is packaged for microwave cooking.

How Popcorn Pops Some flint and dent varieties of corn will explode and form a crisp puff, but expand far less than true popping varieties, which are generally

Mud-Fermented Corn

While exploring just east of Lake Huron around 1616, Samuel de Champlain observed what might be called a "fermenting" technique practiced by the Huron Indians. Here is a challenge for the anthropologist: is there a nutritional basis for this recipe, did it just involve the microbial conversion of starch to sugar, or is it the Huron equivalent of the "noble rot?"

> They have another way of eating Indian corn, to prepare which they take it in the ear and put it in water under the mud, leaving it two or three months in that state, until they judge that it is putrid; then they take it out and boil it with meat or fish and then eat it. They also roast it, and it is better this way than boiled, but I assure you that nothing smells so bad as this corn when it comes out of the water all covered with mud; yet the women and children take it and suck it like sugar cane, there being nothing they like better, as they plainly show.

smaller and contain a greater proportion of hard translucent endosperm. Thanks to a denser arrangement of cellulose fibers, the popcorn hull (pericarp) conducts heat several times faster than the hull of ordinary corn; and thanks to both its density and greater thickness, it is several times stronger: so the hull transmits heat more rapidly to the endosperm, and can withstand higher steam pressure from within before giving way.

As the temperature inside the corn kernel reaches and passes the boiling point, the protein matrix and starch granules soften, and moisture in the granules turns into steam. The steam softens the starch even more, and the many thousands of little steam pockets exert a growing pressure against the hull. The softening of starch and protein continues until the internal pressure approaches seven times the external pressure of the atmosphere, at which point the hull breaks open. The sudden pressure drop inside the kernel causes the pockets of steam to expand, and with them expands the soft protein-starch mixture, which puffs up and then stiffens as it cools off, becoming light and crisp. (If the popping is done in a tightly covered pan that offers no escape for the water vapor, the endosperm will retain it and be tough and chewy; the pan lid should be left slightly ajar.)

Popcorn pops best at a temperature around 380°F/190°C, and can be popped in hot oil, in hot air poppers, and in the microwave oven. Different hybrids pop best with different methods. Microwave popcorn bags develop the necessary high temperatures by means of a thin layer of microwave-reflecting mylar.

Dry-milled Corn Foods: Grits, Cornmeal, Corn Flour Most corn is prepared and eaten in ground form, and dry-milled products are ground directly from the stored grain, usually yellow dent corn, without any pretreatment. These days they're generally refined to exclude the hull and germ, an innovation that dates from around 1900 and that made large-scale milling practical. The rarer whole-grain corn meal and flour, sometimes ground between stone wheels, are richer in fiber, flavor, and nutrients, but also stale rapidly thanks to the oils and related substances in the germ, which become oxidized on contact with air.

Grits are relatively coarse corn endosperm particles between 0.6 and 1.2 mm across. They're used to manufacture breakfast cereals, snacks, and beer, and are cooked into a kind of porridge that's especially beloved in the American South. Grits were once made from alkaline-treated hominy, but this is now rare.

Cornmeal is finer than grits, with particles down to 0.2 mm across, absorbs water and cooks faster than grits, and provides a subtler graininess. It's used to make unleavened mush, polenta, and johnnycakes as well as corn breads, muffins, and other baked and fried foods that include some wheat flour and leavening for lightness.

Popcorn Blossoms

Leave it to Henry David Thoreau to see popcorn with fresh eyes. In midwinter of 1842 he wrote in his journal:

> I have been popping corn to-night, which is only a more rapid blossoming of the seed under a greater than July heat. The popped corn is a perfect winter flower, hinting of anemones and houstonias. . . . By my warm hearth sprang these cerealious blossoms; here was the bank where they grew.

Corn flour is the finest-ground form of corn, with particles smaller than 0.2 mm, and is usually mixed with other ingredients to provide flavor in various baked and fried foods.

Wet-milled Corn: Masa, Tortillas, Tamales, Chips Tortillas, tamales, and corn chips are made from corn grains that are milled when wet, after they have undergone the preliminary cooking step called nixtamalization (p. 478). The corn is first cooked in a solution (0.8–5%) of calcium hydroxide, or lime, for a few minutes to an hour, then is left to steep and slowly cool for 8 to 16 hours. During the steeping, the alkalinity softens the hull and cell walls throughout, causes the storage proteins to bond to each other, and breaks apart some of the corn oil into excellent emulsifiers (mono- and diglycerides). After steeping, the soaking solution and softened hulls are washed away, and the kernels, including the germ, are then stone ground to produce the dough-like material called *masa*. Stone grinding cuts the kernels, mashes them, and kneads the mass, mixing together starch, protein, oils, emulsifiers, and cell wall materials, and the lime's molecule-bridging calcium. With further kneading, this combination develops into a cohesive, plastic dough.

The convenience form of masa is masa harina, a flour made by flash-drying freshly made masa into small particles. Because it's made with less water than normal masa and then is dried, masa harina has less masa aroma and an added browned, toasty aroma, and produces a softer texture than fresh masa.

Tortillas, Tamales, and Corn Chips Tortillas are made by forming finely ground masa into thin sheets, then quickly cooking them, traditionally on a hot pan for a minute or two, now in continuous commercial ovens for 20–40 seconds. Tamales are small cakes of masa that enclose a filling, and are traditionally formed in the papery husks of corn ears and steamed. The dough is moistened with a broth and enriched, flavored, and aerated by beating it thoroughly with lard. The lard is semisolid at room temperature and helps lubricate the masa materials and trap air bubbles in a fluffy mass that expands during steaming. And fried chips are made both from tortillas and directly from masa. Tortilla chips are made by deep-frying tortillas, while corn chips are made by forming strips of lower-moisture, coarse masa and then deep-frying them.

Minor Cereals

The following cereals are only occasionally encountered in Europe and the United

Polenta Lore

Polenta, the Italian version of corn meal mush (originally made with barley), has become a popular dish in the United States, and is the subject of much lore. Some cooks make it quickly in the microwave oven, while traditionalists insist on the necessity of slow cooking and constant stirring for an hour or more. Long cooking at the stove does do one useful thing: it develops the corn flavor by the constant application of higher-than-boiling heat to the pot bottom (hence the need to stir to prevent burning) and the exposure to air and drying that takes place at the surface. Busy cooks can develop just as much flavor with less labor by partly covering the pot of just-thickened polenta, putting it into a low oven (250°F/130°C), which heats the bottom and sides in a controlled and even way, and stirring only occasionally.

States, though some are very important in the dry tropics and subtropics.

Fonio Fonio and black fonio are African grasses distantly related to maize and sorghum. *Digitaria exilis* and *D. iburua* were domesticated on the West African savanna around 5000 BCE, and are typical cereals in most respects. The tiny grains are made into porridge and couscous, popped, brewed into beer, and mixed with wheat to make breads.

Millet Millet is the name used for a number of different grains, all of them with very small round seeds, 1–2 mm in diameter (species of *Panicum, Setaria, Pennisetum, Eleusine*). The millets are native to Africa and Asia, and have been cultivated for 6,000 years. They're especially important in arid lands because they have one of the lowest water requirements of any cereal, and will grow in poor soils. The grains are remarkable for their high protein content, from 16 to 22%, and are popped and also made into porridge, breads, malts, and beers.

Sorghum Sorghum (*Sorghum bicolor*) evolved in the steppes and savannas of central and south Africa, was domesticated there around 2000 BCE, and soon after was taken to India and then to China. Thanks to their tolerance of drought and heat, sorghums have become established in most warm countries with marginal croplands. The fruits are small, around 4 mm long and 2 mm wide, and are boiled like rice, popped, and used in many different variations on porridges, flatbreads, couscous, and beers. Sorghum shouldn't be sprouted; as the seed germinates, it produces a protective cyanide-generating system (p. 258).

Teff Teff, *Eragrostis tef,* is the major crop in Ethiopia, but rarely grown elsewhere. Its tiny (1 mm) seeds come in a variety of colors, from dark to red and brown to white, with the pigmented varieties said to have more flavor. Teff is most often made into the spongy flatbread called injera, which unlike most breads stays soft and chewy for several days.

Triticale Triticale is a modern, artificial cross between wheat and rye (*Triticum* x *Secale*), first documented late in the 19th century and grown commercially around 1970. There are many different forms, the most commonly grown a cross between durum wheat and rye. Its grains are generally more similar to wheat than to rye, though the breadmaking qualities of most varieties aren't as good as wheat's. Triticale is now mostly grown for animal feed, and is sometimes sold in health food stores.

Pseudocereals

Amaranth, buckwheat, and quinoa are not members of the grass family and so are not true cereals, but their seeds resemble the cereal grains and are used in similar ways.

Amaranth Amaranth is the tiny seed, just 1–2 mm across, of three species of *Amaranthus* that originated in Mexico and Central and South America, and were cultivated more than 5,000 years ago. (There are also species of amaranth native to the Old World, but they are used exclusively as green vegetables.) Today amaranth supplements other grains in many baked goods, breakfast cereals, and snacks. The Aztec combination of popped seeds and sticky sweetener lives on in the Mexican *alegria* ("joy") and Indian *laddoo.* Amaranth seeds contain substantially more protein and oil than the cereals.

Buckwheat Buckwheat, *Fagopyrum esculentum,* is a plant in the Polygonum family, a relative of rhubarb and sorrel. It's a native of central Asia, was domesticated in China or India relatively recently, around 1,000 years ago, and was brought to northern Europe during the Middle Ages. It tolerates poor growing conditions and matures in a little over two months, so it has long been valued in cold regions with short growing seasons.

Buckwheat kernels are triangular, a sixth to a third of an inch/4–9 mm across, with a dark hull (pericarp). The inner seed is a mass of starchy endosperm surrounding a small embryo and contained in a light green-yellow seed coat. Intact seeds with the hull removed are called groats. Buckwheat is about 80% starch and 14% protein, mostly salt-soluble globulins. It contains about double the oil of most cereals, and this limits the shelf life of groats and flour. The hulled groats are about 0.7% phenolic compounds, some of which give the grain its characteristic astringency. The distinctive aroma of cooked buckwheat has nutty, smoky, green, and slightly fishy notes (due respectively to pyrazines, salicylaldehyde, aldehydes, and pyridines).

Buckwheat flour contains a small amount of mucilage, a complex carbohydrate a bit like amylopectin—it's made up of about 1,500 sugar molecules bonded together into a branched structure. Though a minor component in the flour, the mucilage absorbs water and may provide some of the stickiness that barely holds an all-buckwheat noodle together (p. 577).

Buckwheat is a staple food in parts of China, Korea, and Nepal. In the Himalayan region, buckwheat is used to make *chillare,* a flatbread, as well as fritter-like pakoras and sweets. In northern Italy, it's mixed with wheat to make flat noodles called *pizzoccheri,* and mixed with corn meal in polenta. In Russia it's used to make the small pancakes called blini; and whole groats are toasted to make the nutty-tasting porridge kasha. In Brittany it produces distinctive crêpes. The Japanese make *soba,* buckwheat noodles. In the United States it's most often encountered in pancakes, to which it contributes tenderness and a nutty aroma.

Quinoa Quinoa is a native of northern South America, was domesticated near Lake Titicaca in the Andes around 5000 BCE, and was a staple food of the Incas, second in importance only to the potato. *Chenopodium quinoa* is in the same family as beets and spinach. The grains are small yellow spheres between 1 and 3 mm across. The outer pericarp of many quinoa varieties contains bitter defensive compounds called saponins, which can be removed by brief washing and rubbing in cold water (prolonged soaking deposits saponins within the seed). Quinoa can be cooked like rice or added to soups and other liquid dishes; it's also popped, and is ground and made into a variety of flatbreads.

LEGUMES: BEANS AND PEAS

Beans and peas belong to the third largest family among the flowering plants (after the orchid and daisy families), and the second most important family in the human diet, after the grasses. The distinctive contribution of the legumes is their high content of protein, two to three times that of wheat and rice, which they develop thanks to their symbiosis with certain soil bacteria. Species of *Rhizobium* bacteria invade the roots of legume plants and convert abundant nitrogen in the air into a form that the plant can use directly to make amino acids and thus proteins. Legumes have long been an essential alternative to protein-rich but more costly animal foods, and are especially prominent in the foods of Asia, Central and South America, and the Mediterranean. A remarkable sign of their status in the ancient world is the fact that each of the four major legumes known to Rome lent its name to a prominent Roman family: Fabius comes from the fava bean, Lentulus from the lentil, Piso from the pea, and Cicero—most distinguished of them all—from the chick pea. No other food group has been so honored!

There are about 20 different species of legume cultivated on a large scale (see box, p. 484). The oil crops, soybean and peanut, far outdistance the legumes eaten more or less whole; the oils are used industrially as well as in the kitchen, and soybeans are a major livestock feed in the United States.

Legume Structure and Composition

Legume seeds consist of an embryonic plant surrounded by a protective seed coat. The embryo in turn is made up of two large storage leaves, the cotyledons, together with a tiny stem. The cotyledons provide the bulk of the nourishment, as the endosperm does in the grains. In fact, the cotyledons are actually a transformed endosperm. When pollen joins ovule in the process of fertilization, both an embryo and a primitive nutritional tissue, the endosperm, are formed. In the grains, the endosperm develops along with the embryo and remains the storage organ of

Some Common Beans and Peas

Common Name	Scientific Name
Natives of Europe and Southwest Asia	
Chickpea, garbanzo, bengal gram	*Cicer arietinum*
Lentil, masoor dal	*Lens culinaris*
Pea	*Pisum sativum*
Fava bean, broad bean	*Vicia faba*
Lupine	*Lupinus* species
Alfalfa	*Medicago sativa*
Natives of India and East Asia	
Soybean	*Glycine max*
Mung bean, green/golden gram	*Vigna radiata*
Black gram, urad dal	*Vigna mungo*
Azuki bean	*Vigna acutifolia*
Rice bean	*Vigna umbellata*
Moth bean	*Vigna aconitifolia*
Pigeon pea, red gram	*Cajanus cajan*
Lathyrus, vetch, khesari dal	*Lathyrus sativus*
Hyacinth bean	*Lablab purpureus*
Winged bean	*Psophocarpus tetragonolobus*
Natives of Africa	
Black-eyed pea, cowpea, crowder	*Vigna unguiculata*
Bambara groundnut	*Vigna subterranea*
Natives of Central and South America	
Bean, common bean, haricot, etc.	*Phaseolus vulgaris*
Lima bean, butter bean	*Phaseolus lunatus*
Tepary bean	*Phaseolus acutifolius*
Runner bean	*Phaseolus coccineus*
Peanut	*Arachis hypogaea*

the mature fruit. But in the legumes, the endosperm is absorbed by the embryo, which repackages the nutrients in its cotyledons.

The seed coat is interrupted only at the hilum, the small depression where the seed has been attached to the pod, and where it will absorb water once it's in the ground or the pot. The seed coat may be fairly thin, as in the peanut, but is as much as 15% by weight of the chickpea and 30% of the lupin. The legume seed coat is almost entirely cell-wall carbohydrates, and includes most of the seed's indigestible fiber. The coats of colorful varieties—pinks, reds, black—are rich in anthocyanin pigments and related phenolic compounds, and therefore in antioxidant power.

Most beans and peas are mainly protein and starch (see box, p. 489). The major exceptions are soybeans and peanuts, which are around 25% and 50% oil respectively. Many legumes are several percent sucrose by weight, and noticeably sweet.

Some legume seeds are rich in defensive secondary compounds (p. 258), notably protease inhibitors, lectins, and in the case of tropical lima beans, cyanide-generating compounds (American and European lima varieties generate little or no cyanide). Animals fed a diet of raw beans will actually *lose* weight. All of these potentially toxic compounds are disabled or removed by cooking.

Seed Colors The colors of beans and peas are determined mainly by anthocyanin pigments in the seed coat. Solid reds and blacks generally survive cooking, while mottled patterns become washed out when the water-soluble pigments leak into adjacent nonpigmented areas and into the cooking water. The intensity of color is best maintained by minimizing the amount of cooking water; start the beans in just enough water to cover, and add water only as needed to keep them barely covered. Persistently green peas and dried beans owe their color to chlorophyll.

Pale beans with translucent seed coats sometimes develop a delicate pink color in the small embryonic stem when they're cooked. This is probably a result of the same reaction that causes the reddening of poached quinces and pears (p. 281).

Legumes and Health: The Intriguing Soybean

Beans and peas are generally excellent sources of a number of nutrients, including protein, iron, various B vitamins, folic acid, and starch or oil. Varieties with colored seed coats provide valuable antioxidants. Among all the legumes, however, soybeans appear to have unusual potential for affecting human health. Epidemiological studies have shown that countries in which soybeans are a staple food, notably China and

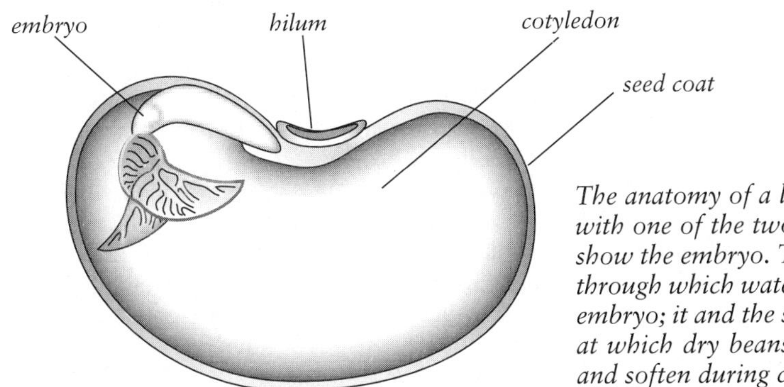

The anatomy of a legume seed. A side view with one of the two cotyledons removed to show the embryo. The hilum is a small pore through which water can pass directly to the embryo; it and the seed coat control the rate at which dry beans and peas absorb water and soften during cooking.

Japan, have significantly lower rates of heart disease and cancer. It may be that soybeans are part of the explanation.

It turns out that soybeans contain storage forms of several phenolic compounds called *isoflavones,* which are liberated by the action of our intestinal bacteria as active compounds (genistein, daidzein, and glycitein) that resemble the human hormone estrogen. The active forms are therefore referred to as "phytoestrogens" (from the Greek *phyton,* "leaf"). Mung beans and other legumes also contain isoflavones, but in much smaller quantities. (Of commonly eaten soy foods, the boiled whole beans contain by far the greatest concentration of isoflavones, about double the amount found in tofu.) Phytoestrogens do appear to have hormone-like and other effects on the human body. There's evidence that they may slow bone loss and the development of prostate cancer and heart disease. However, some studies suggest that phytoestrogens can worsen preexisting breast cancer, and are protective against some cancers only when consumed during adolescence. Our understanding of phytoestrogens is still very incomplete. It's too early to say whether soybeans are more beneficial to human health than any other seed, or whether it's a good idea to eat them often.

Saponins are soap-like defensive compounds that have a water-soluble end and a fat-soluble end, so they can act as emulsifiers and foam stabilizers. They're one of the reasons that a pot of soybeans boils over so readily! Soybeans are a rich source of saponins, which may make up 5% of their total weight, about half of which is in the hulls. Some plant saponins are so strong that they damage our cell membranes. Soy saponins are gentler, and bind to cholesterol so that the body can't absorb it efficiently. Soybeans are also a good source of phytosterols, chemical relatives of cholesterol that also interfere with our absorption of cholesterol and thus lower blood cholesterol levels.

The Problem of Legumes and Flatulence

Several chemical constituents of beans are responsible for an uncomfortable, sometimes embarrassing consequence of eating legumes: the generation of gas in the digestive system.

The Cause: Indigestible Carbohydrates

Everyone produces a mixture of gases from their intestine, about a quart a day, thanks to the growth and metabolism of our resident bacteria. Many legumes, especially soy, navy, and lima beans, cause a sudden increase in bacterial activity and gas production a few hours after they're consumed. This is because they contain large amounts of carbohydrates that human digestive enzymes can't convert into absorbable sugars. These carbohydrates therefore leave the upper intestine unchanged and enter the lower reaches, where our resident bacterial population does the job we are unable to do.

One kind of troublesome carbohydrate is the *oligosaccharides,* molecules consisting of three, four, and five sugar molecules linked together in an unusual way. But the latest research suggests that the oligosaccharides are not the primary source of gas. The cell-wall cements generate just as much carbon dioxide and hydrogen as the oligosaccharides—and beans generally contain about twice as much of these carbohydrates as they do oligosaccharides.

The Cures: Soaking, Long Cooking A commonly used method for reducing the gassiness of beans is to boil them briefly in excess water, let them stand for an hour, then discard the soaking water and start the cooking with fresh water. This does leach out most of the water-soluble oligosaccharides—but it also leaches out significant quantities of water-soluble vitamins, minerals, simple sugars, and seed-coat pigments: that is, nutrients, flavor, color, and antioxidants. That's a high price to pay. An alternative is simple prolonged cooking, which helps by eventually breaking down much of the oligosaccharides and cell-wall cements into digestible single sugars. Oligosaccharides are also consumed by the bean during germination, and consumed by microbes during fermentation: so sprouts, miso, and soy sauce, as well as extracts like bean curd, are less offensive than whole beans.

Bean Flavor

Beans owe their typical beany flavor to a large endowment of the enzyme lipoxygenase, which breaks unsaturated fatty acids into small, odorous molecules. The main components of beaniness are grassy hexanal and hexanol and mushroomy octenol. Lipoxygenase gets its chance to act when the bean cells are damaged and there's enough moisture and oxygen available: for example, when fresh beans are bruised, or dried damaged beans are soaked or brought slowly to a boil. The strong beaniness of soybean products is generally accepted in Asia but objected to in the West, where food scientists have developed techniques to minimize it (see box, p. 494). The aroma of cooked beans also has a distinct sweet note, which comes from lactones, furans, and maltol.

Some beans may be warehoused for years before they find their way to supermarkets or into prepared foods. Prolonged storage causes legumes to lose some typical flavor notes and accumulate stale ones.

Bean Sprouts

Bean sprouts are best known from the cooking of China, where mung bean sprouts became popular in the south and soybean sprouts in the north about 1,000 years ago. Many other legumes are sprouted in Asia and elsewhere, from the tiny alfalfa seed to the massive fava bean. Cooks sometimes trim larger sprouts of their rootlets, primordial leaves, and dense cotyledons, so that the subtle texture and flavor of the stems can be enjoyed without distraction. Sprouts are generally cooked minimally if at all to preserve the delicate flavor and tender but crunchy texture.

Cooking Legumes

Most mature legume seeds are starchy, and require cooking in water to soften the cotyledon cell walls and starch granules. Fresh shell beans are mature but still moist, and so cook fairly quickly, in 10 to 30 minutes. They're also sweeter than the dried beans. Peas, lima beans, cranberry beans, and soybeans (*edamame*) are the legumes most commonly eaten fresh.

Whole dried beans and peas can take an hour or two to cook, much longer than the dried grains. This is due in part to their larger size, but also to the effectiveness of their seed coat at controlling the absorption of water, which is necessary for softening the cell walls and starch. Initially water can enter only through the hilum, the little pore on the curved back of the bean. After 30–60 minutes in cool water (more quickly in hot), the seed coat has become fully hydrated and expanded. From this point on, most of the water flowing into the bean passes across the entire seed coat surface, but the rate of flow is still limited. Legumes whose hulls have been removed—split peas, many Indian *dals*—cook more quickly and disintegrate into a mush.

The Cooking Liquid The quality of cooked beans and the time it takes to cook them depend on the cooking liquid. In vegetable cooking, large volumes of vigorously boiling water minimize enzyme damage to vitamins and pigments by keeping the temperature high when the vegetables are added. Long-cooked legumes are a different story. The greater the volume of cooking water, the more color, flavor, and nutrients are leached out of the beans, and the more they're diluted. So these seeds are best cooked in just enough water for them to soak up *and* to cook in. And though boiling temperatures speed cooking, the turbulence of boiling water can damage the seed coats and cause the beans to disintegrate; lower temperatures (180–200°F/80–93°C) are slower but gentler.

The cooking water's content of dissolved substances also affects cooking times and textures. Hard water with high levels of calcium or magnesium actually reinforces the bean cell walls (p. 282). It can therefore slow the softening of the beans or even prevent them from softening fully. Acidic cooking liquids slow the dissolving of cell-wall hemicelluloses and therefore the softening process, while alkaline cooking water has the reverse effect. Finally, many cooks and cookbooks say that adding salt to the cooking water prevents beans from softening. It does slow the rate at which they absorb water, but they do eventually absorb it and soften. And when beans are presoaked in salted water, they actually cook much faster (below).

Maintaining the Texture of Cooked Beans Three substances slow the softening of beans and therefore make it possible for the cook to simmer beans for hours or reheat them without disintegrating them. *Acids* make the cell-wall hemicelluloses more stable and less dissolvable; *sugar* helps reinforce cell-wall structure and slows the swelling of the starch granules; and *calcium* cross-links and reinforces cell-wall pectins. So such ingredients as molasses—somewhat acid and rich in both sugar and calcium—and acidic tomatoes can preserve bean structure during long cooking or reheating, as for example in baked beans.

Reducing Cooking Times by Presoaking Though beans are an ideal food for slow, easy cooking in a low oven, it's sometimes desirable to cook them faster. In mountainous areas, where high altitude lowers the boiling point, the cooking of dry beans can become an all-day affair.

There are several different ways to reduce the cooking time of beans and peas. The simplest way is to soak dried beans in water before cooking them. This reduces cooking time by 25% or more, and for a very basic reason: heat penetrates a dry seed faster than water does. If beans are cooked directly from the dry state, much of the cooking time is actually spent waiting for water to get to the center. Meanwhile the outer portions of the bean cook more than they need to and may get undesirably fragile.

Briefly Fermented Legumes in India

Though India isn't as fond of fermented foods as many other countries, it did find ways of turning legume and rice gruels into slightly raised pancakes and steamed cakes. In the making of the cakes called *idli,* cooked black gram and rice are ground and mixed to make a thick batter, then allowed to ferment overnight. The same lactic acid bacteria found in fermented milks and creams (*Leuconostoc mesenteroides, Lactobacillus delbrueckii* and *L. lactis, Streptococcus faecalis*) as well as some yeasts (*Geotrichum candidum, Torulopsis* species) feed on the sugars and produce acids, aroma compounds, carbon dioxide gas, and viscous, sticky carbohydrates that thicken the batter and help trap the gas bubbles. The batter is then steamed to produce a spongy, delicately flavored cake. *Dhokla* is a similar preparation made using rice and chickpeas. The *dosa,* a large fried disc, crêpe-like but crisp, is made from a thin fermented batter of rice and black gram. *Papadums,* a familiar side dish in Indian restaurants in the West, are thin wafers of black gram paste that has been allowed to ferment for a few hours before being cut into discs and dried; they're then fried and develop a blistered, delicately brittle texture.

Soaking Times Depend on Temperature Medium-sized beans absorb more than half of their total water capacity in the first two hours of soaking, and plateau at about double their original weight after 10–12 hours. As the soaking temperature goes up, absorption accelerates; and if the beans are first blanched for 1.5 minutes in boiling water, the subsequent water absorption takes only two to three hours in cool water, because the blanching rapidly hydrates the seed coat that controls water movement.

Salt and Baking Soda Speed Cooking Cooking times can be reduced even more by adding various salts to the soaking water. Plain salt at a concentration around 1% (10 g/l, or 2 teaspoons/qt) speeds cooking greatly, apparently because the sodium displaces magnesium from the cell-wall pectins and so makes them more easily dissolved. Baking soda at 0.5% (1 teaspoon/qt) can reduce the cooking time by nearly 75%; it contains sodium and in addition is alkaline, which facilitates the dissolving of the cell-wall hemicelluloses. Of course, added salts affect both the taste and texture of the cooked beans. The alkalinity of baking soda can give an unpleasantly slippery mouth feel and soapy taste. And salt reduces the swelling and gelation of starch granules within beans, which means that it favors a mealy internal texture over a creamy one.

Pressure Cooking Thanks to its temperature of around 250°F/120°C, pressure cooking can cut the cooking time of beans and peas by half or more. Salt-presoaked beans may take just 10 minutes.

Persistently Hard Beans One problem that cooks commonly encounter when cooking dry beans is that some batches take unusually long to soften, or never quite do soften. This may have been caused by growing conditions on the farm, or storage conditions after harvest.

"Hard-seed" is a characteristic found in beans when temperatures are high and humidity and water supply are low during the growing season. The outer seed coat gets very water-resistant, so it takes much longer for water to move into the bean interior. Hard-seed beans are usually smaller than normal beans, so they can sometimes be avoided by picking over the beans and discarding the smallest ones before cooking.

"Hard-to-cook" beans, on the other hand, are normal when harvested, but become resistant to softening when they're stored for a long time—months—at warm temperatures and high humidities. This

The Composition of Dry and Sprouted Legumes

Legume Seed	Water	Protein	Carbohydrate	Oil
Common bean	14	22	61	2
Fava bean	14	25	58	1
Lima bean	14	20	64	2
Mung bean	14	24	60	1
sprout	90	4	7	0.2
Soybean	10	37	34	18
sprout	86	6	6	1
Lentil	14	25	60	1
Chickpea	14	21	61	5
Pea	14	24	60	1

resistance results from a number of changes in bean cell walls and interiors, including the formation of woody lignin, the conversion of phenolic compounds into tannins that cross-link proteins, and the denaturation of storage proteins to form a water-resistant coating around the starch granules. There's no way to reverse these changes and make hard-to-cook beans as soft as regular beans. And there's no way to spot them before cooking. Once cooked, they're likely to be smaller than normal and so may be picked out before serving.

Roasting Though most legumes are cooked in liquid to soften their starch and cell walls, a few are parched in dry heat to create a crisp texture. Peanuts are the most commonly roasted of the legumes, thanks to their nut-like oil content and relatively tender cotyledons. Other beans with lower oil contents, notably soybeans and chickpeas, are also roasted to make a nut-like seed. Because their cotyledons are harder, they're soaked in water first, then roasted. The initial high temperature and moisture soften the cotyledon cell walls and starch granules; continued roasting evaporates most of the water to give a crisp rather than hard texture. The roasting can be done in a hot pan or oven, or—as is done in Asia—in sand that has been heated to 500–600°F/250–300°C. In India, for example, chickpeas are heated to around 180°F/80°C, moistened with water, rested for some hours, then roasted in hot sand so that they puff and the seed coat can be rubbed off.

Characteristics of Some Common Legumes

Fava or Broad Beans The fava bean or broad bean, *Vicia faba,* is the largest of the commonly eaten legumes, and was the only bean known to Europe until the discovery of the New World. It apparently originated in west or central Asia, and was among the earliest domesticated plants. Larger cultivated forms have been found in Mediterranean sites dating to 3000 BCE. There are several sizes, the largest of which seems to have been developed in the Mediterranean region around 500 CE. China is the world's largest producer.

Fava beans are unusual in having a thick, tough seed coat that's often removed from both the meaty cotyledons of the unripe green seeds and from the hard dry seeds. A blanching in alkaline water loosens and softens the coat. In Egypt, the popular dish called *ful medames* is made by boiling the mature beans until soft, then flavoring with salt, lemon juice, oil, and garlic. Mature fava beans are also sprouted, then boiled to make a soup.

Favism Eating fava beans is the cause of a serious disease, favism, in people who have an inherited deficiency of a particular enzyme. Most victims are children who live in the southern Mediterranean and Middle East, or whose ancestors came from that region. When they are exposed to two unusual amino-acid relatives (vicine and convicine) in the beans and in the flower

Food Words: *Bean, Frijol*

The term "bean" that we apply to many different legumes from Eurasia, East Asia, and the Americas originally referred only to the fava bean. The Indo-European root *bhabha* gave us both "fava" and "bean." By Greek and Roman times, the African cowpea or black-eyed pea was also known in the Mediterranean and given the Latin name *phaseolus*—the source of the Spanish *frijol,* and the scientific name for the common beans from the New World.

pollen, their bodies metabolize these chemicals to forms that damage their red blood cells and cause serious, sometimes fatal anemia. The enzyme deficiency also turns out to suppress the growth of the malaria parasite in red blood cells, so it may actually have been an advantageous genetic trait before malaria was brought under control.

Chickpea or Garbanzo Chickpeas are a native of arid southwest Asia, and like the fava bean, pea, and lentil have been cultivated for about 9,000 years. There are two general types, desi and kabuli. Desi are closer to the wild chickpea, with small seeds, a thick, tough seed coat, and a dark color from abundant phenolic compounds. They're the main variety grown in Asia, Iran, Ethiopia, and Mexico. The kabuli type, more common in the Middle East and Mediterranean, is larger, cream-colored, with a thin, light seed coat. There are also varieties with dark green cotyledons. Chickpeas are notable among the legumes for being about 5% oil by weight; most others are 1–2%.

The name comes from the bean's Latin name, *cicer*; in the botanical name, *Cicer arietinum,* the second word means "ram-like" and refers to the seed's resemblance to a ram's head, complete with curling horns. The Spanish *garbanzo* derives from the Greek name. Today, this legume is a frequent ingredient in many Middle Eastern and Indian dishes. Hummus is a chickpea paste flavored with garlic, paprika, and lemon that is popular in the eastern Mediterranean; in parts of Italy, flatbreads are made from chickpea flour. Chickpeas are the most important legume in India, where they're hulled and split to make chana dal, ground into flour for papadums, pakoras, and other fried goods, and are boiled, roasted, and sprouted.

Common Bean, Lima Bean, Tepary Bean The common bean, lima bean, and tepary bean are the important domesticated species of the 30 or so species in the Central American genus *Phaseolus.*

Common Bean The most important species of *Phaseolus* is *P. vulgaris,* or the common bean. The ancestral plant was a native of southwestern Mexico, and the highest consumption of the common bean is still in Latin America. It first came under cultivation about 7,000 years ago, and gradually diffused both north and south, reaching the major continents about 2,000 years ago, and Europe during the age of exploration. The common bean has been developed into many hundreds of varieties of different sizes, shapes, seed-coat colors and color patterns, shininess, and flavors. Most large-seeded varieties (kidney, cranberry, large red, and white) came originally from the Andes, and became established in the American Northeast, Europe, and Africa; smaller-seeded central American types (pinto, black, small red, and white) were concentrated in the American Southwest. In the United States, there are more than a dozen commercial categories based on color and size. Beans are cooked in many ways: simply boiled, made into stews, soups, pastes, cakes, and sweet desserts.

Popping Bean A special kind of common bean is the nuña, or popping bean, which has been cultivated for several thousand years in the high Andes. It can be popped by just 3–4 minutes of high dry heat—a great advantage in the fuel-poor mountains—or in the microwave oven. It doesn't expand nearly as much as popcorn and remains fairly dense, with a powdery texture and nutty flavor.

Lima Bean The use of the common bean in Peru was predated by the larger lima bean—the name derives from Peru's capital—which was native to Central America and domesticated somewhat later than the common bean. Both species were exported to Europe by Spanish explorers. The lima bean was introduced to Africa via the slave trade, and is now the main legume of that continent's tropics. The wild type and some tropical varieties contain potentially toxic quantities of a cyanide defense system, and must be thoroughly cooked to be safe (common commercial varieties are cyanide-free). Lima beans are eaten both fresh and dried.

Tepary Bean Tepary beans, small brown natives of the American southwest, are unusually tolerant of heat and water stress. They're especially rich in protein, iron, calcium, and fiber, and have a distinctive, sweet flavor reminiscent of maple sugar or molasses.

Lentil The lentil is probably the oldest cultivated legume, contemporaneous with wheat and barley and often growing alongside these grasses. Its native ground is arid Southwest Asia, and it's now commonly eaten across Europe and Asia. Most lentils are produced in India and Turkey, with Canada a distant third. The Latin word for lentil, *lens,* gives us our word for a lentil-shaped, or doubly convex, piece of glass (the coinage dates from the 17th century). Lentils contain low levels of antinutritional factors and cook quickly.

Lentils are divided into two groups: varieties with flat and large seeds, 5 mm or more across, and varieties with small, more rounded seeds. Large varieties are more commonly grown, while the small, finer-textured ones include the prized green French lentille du Puy, the black beluga, and the green Spanish pardina. There are varieties with brown, red, black, and green seedcoats; most have yellow cotyledons, though some are red or green. Green seed coats can turn brown with age and during cooking, thanks to the clustering of phenolic compounds into large pigmented complexes (p. 269). Because they are flat and thin, with thin seed coats, cooking water only has to penetrate a millimeter or two from each side, so lentils soften much more quickly than most beans and peas, in an hour or less.

Traditional lentil dishes include Indian masoor dal, whole or hulled and split red lentils cooked into a porridge, and Middle Eastern *koshary* or *mujaddharah,* a mixture of whole lentils and rice.

Pea, Black-eyed Pea, Pigeon Pea

Pea The pea has been cultivated for around 9,000 years and spread quite early from the Middle East to the Mediterranean, India, and China. It's a cool-climate legume that grows during the wet Mediterranean winter and in the spring of temperate countries. It was an important protein source in Europe in the Middle Ages and later, as the old children's rhyme attests: "Pease porridge hot, Pease porridge cold, Pease porridge in the pot, Nine days old." Today, two main varieties are cultivated: a starchy, smooth-coated one that gives us dried and split peas, and a wrinkly type with a higher sugar content, which is usually eaten when immature as a green vegetable. Peas are unusual among legumes in retaining some green chlorophyll in their dry cotyledons; their characteristic flavor comes from a compound related to the aroma compound in green peppers (a methoxy-isobutyl pyrazine).

Black-eyed Pea The so-called black-eyed pea or cowpea is not really a pea, but an African relative of the mung bean that was known to Greece and Rome and brought to the southern United States with the slave trade. It has an eye-like anthocyanin pigmentation around the hilum, and a distinctive aroma. A variety that produces a very long pod and small seeds is the yard-long bean, a common green vegetable in China (p. 336).

Pigeon Pea Pigeon pea is a distant relative of the common bean, native to India, and now grown throughout the tropics. In India it's called toor dal or redgram because the tough seed coat of many varieties is reddish brown, though it's most often hulled and split, and the cotyledons are yellow. It's been cultivated for around 2,000 years, and is made into a simple porridge. Like the other grams, it contains little in the way of antinutritional factors.

Mung Bean, Black Gram, Azuki

The Grams The legume genus *Vigna,* which is native to the Old World, provides the small-seeded "grams" of India and a few other Asian and African seeds. Most of them have the advantages of being small, quick-cooking, and relatively free of antinutritional and discomforting compounds. Green gram, or mung beans, are native to India, spread early on to China, and are now the most widely grown of this group thanks to the popularity of their sprouts. Black gram or urad dal is the most prized of the legumes in India, where it has been cultivated for more than 5,000 years and is eaten whole, split and dehulled, and ground into flour for cakes and breads.

Rice beans are eaten primarily in Thailand and elsewhere in Indochina. The African bambara groundnut resembles the peanut in being borne underground and containing oil, but it isn't nearly as rich as the peanut. In West Africa, the seeds are eaten fresh, canned, boiled, roasted, and made into porridges and cakes.

Azuki Bean Azuki or adzuki (Chinese *chi dou*) is an East Asian species of *Vigna, V. angularis,* about 8 by 5 by 5 mm, and most commonly a deep maroon color, which makes it a favorite ingredient for festive occasions. It was cultivated in Korea and China at least 3,000 years ago, and taken later to Japan; it's now the second most important legume after the soybean in both Korea and Japan. Azuki are a favorite sprouting seed, and are also candied, infused with sugar to make a dessert topping, and used as a base for a hot drink. In Japan most of the azuki crop is made into *an,* a sweet paste composed of equal parts sugar and ground twice-boiled azuki, which are kneaded together.

Lupins Lupins, *lupini* in Italian, come from several different species of *Lupinus* (*albus, angustifolius, luteus*). They're unusual because they contain no starch—they're 30–40% protein, 5–10% oil, and up to 50% soluble but indigestible carbohydrates (soluble fiber, p. 258). Though there are some "sweet" types that require no special processing, many varieties contain bitter and toxic alkaloids and so are soaked in water for up to several days to leach these substances out. They're then boiled until soft, and served in oil, or toasted and salted. A New World species, *L. mutabilis,* grown in the Andes, has a protein content approaching 50% of the dry seed weight.

Soybeans and Their Transformations

Finally, the most versatile legume. Soybeans were domesticated in northern China more than 3,000 years ago, and eventually became a staple food throughout much of Asia; their spread was probably contemporary with and encouraged by the vegetarian doctrine of Buddhism. They were little known in the West until late in the 19th century, but today the United States supplies half of the world production, with China in fourth place after Brazil and Argentina. However, most U.S. soybeans feed not people but livestock, and much of the rest are processed to make cooking oil and a host of industrial materials.

The soybean's many guises have been inspired by both its great virtues and its defects. Soybeans are exceptionally nutritious, with double the protein content of other legumes, a near-ideal balance of amino acids, a rich endowment of oil, and a number of minor constituents that may contribute to our long-term health (p. 485). At the same time, they're pretty unappealing. They contain abundant antinutritional factors and gas-producing oligosaccharides and fiber. When boiled in the usual way, they develop a strong "beany" flavor. And when cooked whole as other beans are, they don't become appealingly creamy; since they contain a negligible amount of starch, their texture remains somewhat firm. The Chinese and others developed two basic ways of making soybeans more palatable: extracting their protein and oil in the form of a milk and then concentrating them in cheese-like curds; and encouraging the growth of microbes that consume undesirable substances and generate an appealing flavor. The results were bean curd and soymilk skins; and soy sauce and miso, tempeh and natto.

Fresh Soybeans One other way to make soybeans more palatable is to eat them before they're fully mature, when they're sweeter, contain lower levels of gassy and antinutritional substances, and have a less pronounced beany flavor. Fresh soybeans, Japanese *edamame* or Chinese *mao dou,* are specialized varieties harvested at 80% maturity, still sweet and crisp and green, then boiled for a few minutes in salted water. Green soybeans are around 15% protein and 10% oil.

Soybean Milk The traditional method for making soymilk is to soak the beans until soft, grind them, and either sieve out the solids and cook the milk (China) or cook the slurry and then sieve out the solids (Japan). The result is a watery fluid filled with soy proteins and microscopic droplets of soy oil. Either method results in a strong soy flavor. The modern method that minimizes enzyme action and soy flavor is to soak the dry beans (an hour at 150ºF/65ºC allows them to absorb their full weight in water without significant cell damage), and then either quickly cook them to 180–212ºF/80–100ºC before grinding, or grind them in that temperature range in a preheated grinder and preheated water.

In the West, soymilk has become a popular alternative to cow's milk, with a roughly similar protein and fat content, but the fat less saturated (soy milk must be fortified with calcium in order to be a good nutritional substitute). But it is dilute, textureless, bland, and not very versatile. The Chinese found two ways to make it more interesting (and to remove the gas-causing oligosaccharides): coagulate the milk into surface skins, or coagulate it into curd.

Soymilk Skin When animal or seed milks are heated in an uncovered pot, a skin of coagulated protein forms on the surface. The skin forms because proteins unfolded by the heat concentrate at the surface, get tangled up with each other, and then lose their moisture to the dry room air. As they dry, they get even more tightly tangled, and form a thin but solid protein sheet, entrap-

The Beany Flavor of Soybeans

The strong aroma of simply cooked soybeans results from two qualities: their high content of polyunsaturated oil, which is especially vulnerable to oxidation, and their highly active oil-breaking enzymes. When the cells of the bean are damaged and their contents mixed together, the enzymes and oxygen break the long carbon chains of the oil into fragments five, six, and eight carbon atoms long. These fragments have aromas reminiscent of grass, paint, cardboard, and rancid fat, and the combination creates a smell usually described as "beany." Some bitter taste and astringency also develop, probably due to free fatty acids or to soy isoflavones that are liberated from their storage form (p. 485).

The key to minimizing the development of beany flavor is to inactivate the beans' enzymes quickly, before they have a chance to attack the oils. This is done by soaking the beans to speed subsequent cooking, and then immersing them in boiling water or pressure-cooking them.

ping oil droplets and developing a fibrous, chewy texture.

Such skins are usually an annoyance, but some cultures make a virtue of them and turn them into a dish. The Indians do this with cow's milk, and for several centuries the Chinese have been using soymilk to make *dou fu pi,* the Japanese *yuba,* which they layer together to form a variety of sweet and savory products, some of them shaped into flowers, fish, birds, even pigs' heads. The skins are also meltingly delicious when eaten just as they're taken from the milk. At some Japanese restaurants, a small pot of soy milk is heated at the table so diners can remove and eat the skins as they form, then add a pinch of salts to the remaining liquid and coagulate it into soft tofu.

Bean Curd, or Tofu Bean curd is curdled soy milk, a concentrated mass of protein and oil formed by coagulating the dissolved proteins with salts that yoke them and the protein-coated oil droplets together. Bean curd was invented in China around 2,000 years ago, was well known by 500 CE, and became a daily food beginning around 1300. Chinese bean curd is traditionally coagulated with calcium sulfate, Japanese bean curd and bean curd from coastal regions of China with what the Japanese call *nigari,* a mixture of magnesium and calcium salts left over when table salt, sodium chloride, is crystallized from seawater.

Making Bean Curd To make bean curd, cooked soy milk is cooled to about 175°F/78°C, then coagulated with calcium or magnesium salts dissolved in a small amount of water. The coagulation takes 8–30 minutes. When the delicate, cloud-like curds have formed, the remaining "whey" is ladled off, or the curd is broken up to release water and drained. The resulting mass is then pressed for 15–25 minutes while still quite hot, around 160°F/70°C, in order to form a cohesive mass that's around 85% water, 8% protein, and 4% oil. In commercial production, the curd is cut into blocks, packaged in water, and the packages pasteurized by immersion in hot water.

Soft or silken tofu, with a custard-like texture, is made by coagulating the soymilk in the package so that the curd remains intact, full of moisture, and delicate.

Freezing Bean Curd Tofu is one of the few foods that can be usefully altered by freezing. When it freezes, the coagulated proteins become even more concentrated, and the solid ice crystals form pockets in the protein network. When the frozen curd

An Early Description of Tofu

One of the earliest European accounts of soybean curd is Friar Domingo Navarrete's, which dates from the 17th century. He called it

> the most usual, common, and cheap sort of food all China abounds in, and which all in that empire eat, from the Emperor to the meanest Chinese; the Emperor and great men as a dainty, the common sort as necessary sustenance. It is called Teu Fu, that is paste of kidney beans. I did not see how they made it. They drew the milk out of the kidney beans, and turning it, make great cakes of it like cheeses, as big as a large sieve, and five or six fingers thick. All the mass is as white as the very snow, to look to nothing can be finer . . . Alone it is insipid, but very good dressed as I say and excellent fried in butter.

thaws, the liquid water flows from the toughened spongy network, especially when the curd is pressed. The sponge is then ready to absorb flavorful cooking liquids, and has a chewier, meatier texture.

Fermented Bean Curd Sufu (*tou fu ru, fu ru*) is soybean curd fermented by molds in the genera *Actinomucor* and *Mucor,* producing the Chinese and vegetarian equivalent of mold-ripened milk cheeses.

Fermented Soybean Products: Soy Sauce, Miso, Tempeh, Natto The great appeal of miso and soy sauce, long-fermented soy products, is their strong, distinctive, and delicious flavor. It develops when microbes break down the bean proteins and other components and transform them into savory substances that then react with each other to generate additional layers of flavor. Tempeh and natto are quick-fermented soy products with their own unusual qualities.

Two-Stage Fermentations Asian mold fermentations generally involve two distinct stages. In the first, dormant green spores of *Aspergillus* molds are mixed with cooked grains or soybeans, which are then kept warm, moist, and well aerated. The spores germinate and develop into a mass of thread-like hyphae, which produce digestive enzymes that break down the food for energy and building blocks. The second stage begins after about two days, when the enzymes are at their peak. The mixture of food and hyphae, called *chhü* in China and *koji* in Japan, is now immersed in a salt brine, often along with more cooked soybeans. In the oxygen-poor brine, the molds die, but their enzymes continue to work. At the same time, microbes that thrive in the absence of oxygen—salt-tolerant lactic acid bacteria and yeasts—grow in the brine, consume some of the building blocks, and contribute their own flavorful by-products to the mixture.

The Origins of Miso and Soy Sauce The first foods that the ancient Chinese fermented in brine were pieces of meat or fish. These were eventually replaced by whole soybeans around the 2nd century BCE. Soy paste became the major condiment around 200 CE and remained so through around 1600, when it was replaced by soy sauce. Soy sauce began as a residue resulting when

Chinese Soy Pastes and Sauces

A number of the condiments used in Chinese cooking as sauces or sauce bases are variations on mold-fermented soybeans, or *chiang*. Their Chinese names reflect this. Some examples are:

- Bean sauce, *yuen-shi chiang,* made from the residue of soy-sauce making, used to make savory sauces
- Bean paste, *to-pan chiang,* essentially a chunky wheat-barley-soy miso, used to make savory sauces
- Hoisin sauce, *ha-hsien chiang,* made from the residue of soy-sauce making, mixed with wheat flour, sugar, vinegar, chilli pepper; served with Peking duck and mu shu pork
- Sweet wheat chiang, *t'inmin chiang,* smooth, soft, brown; made from wheat flour formed into steamed buns or flat sheets, allowed to mold, then brined; used as the base for Peking duck dipping sauce

soy paste was made with excess liquid, but it became more popular than the paste, and by 1000 was prepared for its own sake.

Fermented soy pastes and soy sauce were carried by Buddhist monks to Japan, where sometime around 700 CE a new Japanese name, *miso—mi* meaning flavor—was given to distinctive Japanese versions of the paste. These involved the use of a grain-based koji that provided sweetness, alcohol, finer aromatics, and delicacy. Until the 15th century, Japanese soy sauce was simply the excess fluid, or tamari, ladled from finished soybean miso. By the 17th century, the now-standard formula of roasted cracked wheat and soybeans had been established for making the sauce, and the resulting product given a new name, shoyu. Shoyu began to appear on western tables as an exotic and expensive item by the 17th century.

Miso Miso is used as a soup base, as a seasoning for various dishes, in marinades, and as a medium for pickling vegetables. There are dozens of different varieties.

Miso is made by cooking a grain or legume—usually rice, sometimes barley, sometimes soybean—and fermenting it in shallow trays with koji starter for several days to develop enzymes. The resulting koji is then mixed with ground cooked soybeans, salt (5–15%), and a dose of an earlier batch of miso (to provide bacteria and yeasts). In traditional miso making, the mixture is allowed to ferment (and eventually age) in barrels for months to years at a warm 86–100°F/30–38°C. Various lactic acid bacteria (*Lactobacilli, Pediococci*) and salt-tolerant yeasts (*Zygosaccharomyces, Torulopsis*) break down the seed proteins, carbohydrates, and oils and produce a host of flavor molecules and flavor precursors. Browning reactions generate deeper layers of flavor and color.

Traditionally made miso ends up with a rich, savory, complex flavor dominated by sweet and roasted notes, and sometimes by esters reminiscent of pineapple and other fruits. Modern industrial production cuts the fermentation and aging from months to a few weeks, and compensates for the resulting lack of flavor and color with various additives.

Soy Sauce Soy sauce is made in several different styles today. Broadly speaking, the flavor of traditional soy sauce depends on the proportions of soybeans and wheat. Most Chinese soy sauces, and Japanese tamari, are made primarily or exclusively from soybeans. Japanese soy sauce is generally made from an even mixture of soybeans and wheat, and the starch from the wheat gives it a characteristic sweetness, a higher alcohol content, and more alcohol-

The Delightful Physics of Miso Soup

Miso soup is one of the most common Japanese dishes. It typically includes a dashi broth (p. 238) and small cubes of tofu. As is true of many Japanese preparations, miso soup is a delight to the eye as well as the palate. When the soup is made and poured into the bowl, the miso particles disperse throughout in an even haze. But left undisturbed for a few minutes, the particles gather around the center of the bowl in discrete little clouds that slowly change shape. The clouds mark convection cells, columns in the broth where hot liquid from the bowl bottom rises, is made cooler and so more dense by evaporation at the surface, falls again; is reheated and becomes less dense, rises, and so on. Miso soup enacts at the table the same process that produces towering thunderhead clouds in the summer sky.

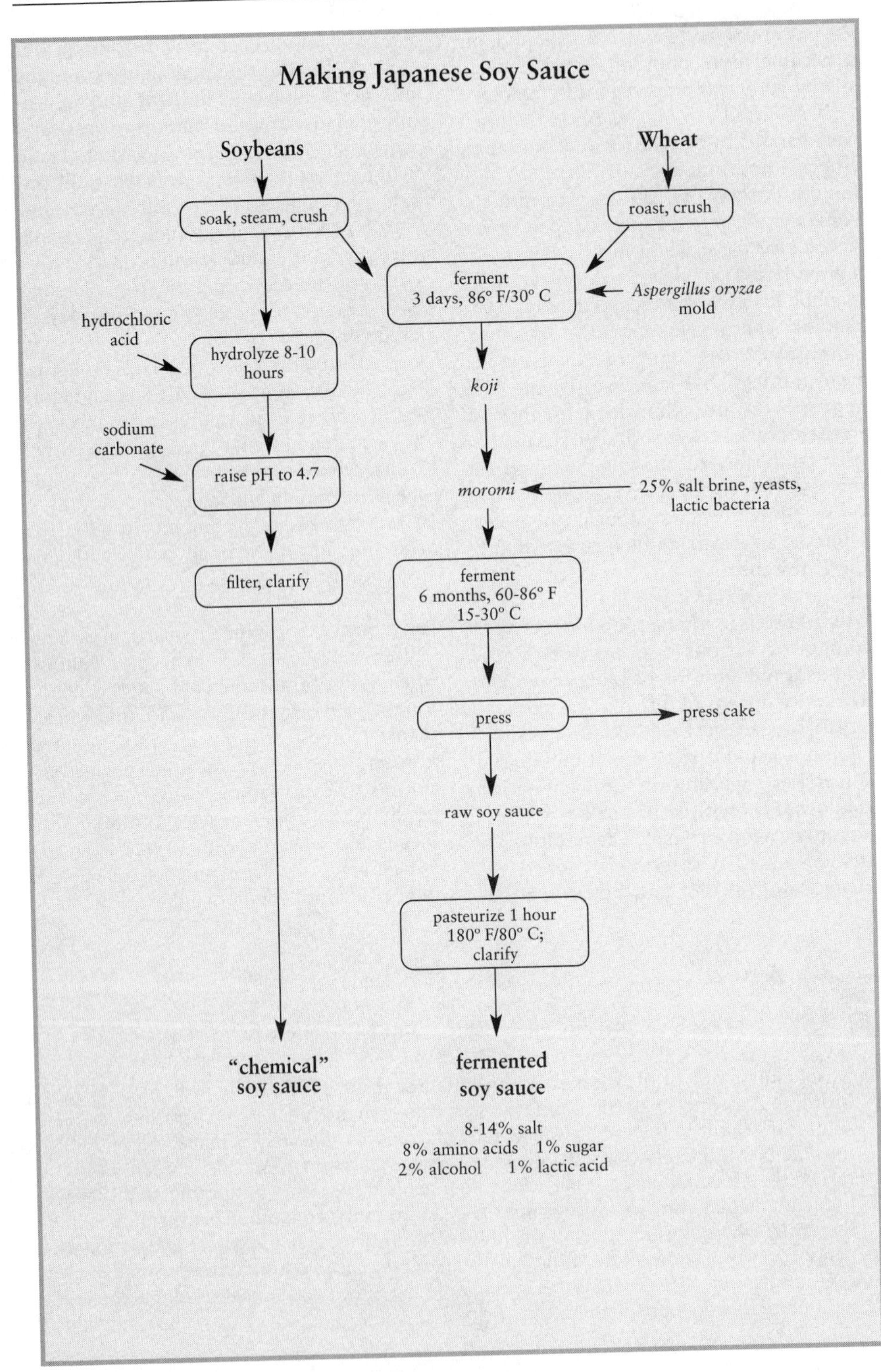
Making Japanese Soy Sauce
Soybeans
Wheat
soak, steam, crush
roast, crush
ferment
3 days, 86° F/30° C
Aspergillus oryzae
mold
hydrochloric
acid
hydrolyze 8-10
hours
koji
sodium
carbonate
raise pH to 4.7
moromi
25% salt brine, yeasts,
lactic bacteria
filter, clarify
ferment
6 months, 60-86° F
15-30° C
press
press cake
raw soy sauce
pasteurize 1 hour
180° F/80° C;
clarify
"chemical"
soy sauce
fermented
soy sauce
8-14% salt
8% amino acids 1% sugar
2% alcohol 1% lactic acid

derived aromatics. *Shiro,* or "white" soy sauce, lighter in color and flavor, is made with more wheat than soybeans.

Japanese Soy Sauce Most of the soy sauce sold in the West is made in Japan or in the Japanese style, which is summarized in the box on p. 498. During the initial brief fermentation, the *Aspergillus* mold produces enzymes that will break down wheat starch into sugars, wheat and soy proteins into amino acids, and seed oils into fatty acids. Then during the longer main fermentation, these enzymes do their work; yeasts produce alcohol and a range of taste and aroma compounds; and bacteria produce lactic, acetic, and other acids, and yet other aromas. Over time, the various enzymes and microbial products also react with each other, the sugars and amino acids forming roasty-smelling pyrazines, acids and alcohols combining to form fruity esters. The high-temperature pasteurization develops yet another layer of flavor by encouraging browning reactions between amino acids and sugars. The result is a liquid that's salty, tart, sweet, savory (from a high concentration of amino acids, mainly glutamic acid), with a rich aroma. Several hundred aroma molecules have been identified in soy sauce, with roasty compounds (furanones and pyrazines), sweet maltol, and a number of meaty sulfur compounds among the more prominent. All in all, soy sauce is a concentrated, mouth-filling liquid, a versatile flavor enhancer for other foods.

Tamari *Tamari* names a kind of Japanese soy sauce closest to its Chinese original: made with little or no wheat, and therefore poor in alcohol and fruity esters derived from it, but with a darker color and richer flavor thanks to the higher concentration of soybean amino acids. Today tamari is sometimes stabilized with added alcohol, which makes its aroma closer to that of ordinary shoyu. Even stronger than true tamari is twice-fermented *saishikomi,* made by making up the mash not with salt water, but with a previous batch of soy sauce.

"Chemical" Soy Sauce Industrial producers have been making nonfermented approximations of soy sauce since the 1920s, when the Japanese first used chemically modified soy protein ("hydrolyzed vegetable protein") as an ingredient. Nowadays, defatted soy meal, the residue of soybean oil production, is broken down—hydrolyzed—into amino acids and sugars with concentrated hydrochloric acid. This caustic mixture is then neutral-

The Original Ketchup

Fermented soy pastes and sauces developed into many different regional variations in Asia. Among them is the Indonesian condiment *kecap,* whose name gave us our term for a sweet-sour tomato condiment. *Kecap* is made by allowing *Aspergillus* mold to grow on cooked soybeans for about a week, brine-fermenting the moldy mass for 2 to 20 weeks, then boiling it for 4 to 5 hours, and filtering off the solids. The salty version is called *kecap asin.* To make sweet *kecap manis,* palm sugar and a variety of spices—among them galangal, makrut lime, fennel, coriander, and garlic—are added to the fermented beans just before the boiling.

OPPOSITE: *Making soy sauce. The more involved and time-consuming fermentation produces a much more flavorful result than the quick chemical production method.*

ized with alkaline sodium carbonate, and flavored and colored with corn syrup, caramel, water, and salt. Such quick "chemical" soy sauce has a very different character from the slow fermented version, and is usually blended with at least some genuine fermented soy sauce to make it palatable.

To make sure you're buying genuine soy sauce, read the label carefully, and avoid products that include added flavorings and colors.

Tempeh Tempeh was invented in Indonesia, and unlike miso and soy sauce is not a salty preserved condiment but an unsalted, quick-fermented, perishable main ingredient. It's made by cooking whole soybeans, forming them into thin layers, and fermenting with a mold, *Rhizopus oligosporus* or *R. oryzae,* for 24 hours at a warm, tropical temperature (85–90°F/30–33°C). The mold grows and produces thread-like hyphae, which penetrate the beans and bind them together, and digest significant amounts of protein and oil to flavorful fragments. Fresh tempeh has a yeasty, mushroomy aroma; when sliced and fried, it develops a nutty, almost meaty flavor.

Natto

Natto has been made in Japan for at least 1,000 years, and is notable for being distinctly alkaline (from the breakdown of amino acids into ammonia) and for developing a sticky, slippery slime that can be drawn with the tip of a chopstick into threads up to 3 ft/1 m long! As with tempeh, no salt is used and the product is perishable. The whole beans are cooked, inoculated with a culture of the bacterium *Bacillus subtilis natto,* and held at about 100°F/40°C for 20 hours. Some bacterial enzymes break down proteins into amino acids and oligosaccharides into simple sugars, while others produce a range of aroma

Some Traditional Fermented Soybean Preparations

Food	Names	Preparation	Qualities
Soy paste, miso	Dou jiang; miso	Beans plus grain fermented with molds, bacteria, yeasts	Rich, savory, salty, sometimes sweet seasoning for many dishes
Soy sauce	Jiang you; shoyu; kecap	Beans plus wheat fermented with molds, bacteria, yeasts	Rich, savory, salty seasoning for many dishes
Black beans, soy nuggets	Dou chi; hamanatto	Beans and wheat flour fermented with mold	Savory, salty ingredient in meat and vegetable dishes
Fermented tofu	Dou fu ru; sufu	Tofu fermented with mold	Cheese-like; a condiment for various dishes
Natto	Na dou; natto	Beans fermented with special bacterium	Soft, distinctive, sticky; eaten with rice or noodles
Tempeh	Tian bei; tempeh	Dehulled beans fermented with special mold	Firm cake, mildly nutty and mushroomy; main ingredient often fried

compounds (buttery diacetyl, various volatile acids, nutty pyrazines), as well as long chains of glutamic acid and long branched chains of sucrose, which form the slimy strings. Natto is served atop rice or noodles, in salads and soups, or cooked with vegetables.

NUTS AND OTHER OIL-RICH SEEDS

From the first, the English word *nut* meant an edible seed surrounded by a hard shell, and this remains the common meaning. Botanists later appropriated the word to refer specifically to one-seeded fruits with a tough, dry fruit layer rather then a fleshy, succulent one. Under this restricted definition, among common nuts only acorns, hazelnuts, beechnuts, and chestnuts qualify as true nuts. The details of anatomy aside, the various seeds that we call nuts differ from grains and legumes in three important ways: they're generally larger, richer in oil, and require little or no cooking to be edible and nourishing. This combination of qualities made nuts an important source of nourishment in prehistoric times. Today, they're especially appreciated for their characteristic rich flavor.

Walnuts, hazelnuts, chestnuts, and pine nuts all have both Old World and New World species, because nut-bearing trees have been around a lot longer than the other food plants, long enough that they existed before North America and Europe split apart some 60 million years ago. Over the last few centuries, humans have spread their prized nut species to nearly every region on the globe with a suitable climate. California has become the largest producer of southwest Asian almonds and walnuts, South American peanuts are grown throughout the subtropics, and Asian coconuts throughout the tropics.

Nut Structures and Qualities

The bulk of most nuts consists of the embryo's swollen storage leaves, or cotyledons, but coconuts and pine nuts are monolithic masses of endosperm, and the Brazil nut is a swollen embryonic stem. Unlike most grains and legumes, nuts are delicious when eaten in their dry, nutrient-concentrated state, somewhat crisped and browned by a quick roasting. Their weak cell walls make them tender, their low starch content prevents them from seeming floury, and their oil gives them a mouth-watering moistness.

An important feature of the nuts is the skin, a protective layer of varying thickness that adheres to the kernel. Chestnut skins are thick and tough, hazelnut skins papery and brittle. The nut skin is usually reddish-brown in color and astringent to the taste. Both qualities are due to the presence of tannins and other phenolic compounds, which may make up a quarter of the skin's dry weight. Many of these

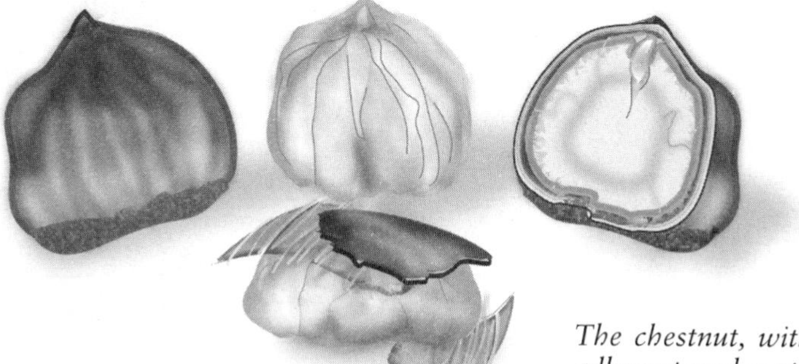

The chestnut, with its shell and tough, adherent seed coat.

same phenolic compounds are effective antioxidants and so nutritionally valuable. But because they're astringent and discolor other ingredients (walnut skins turn breads purple-gray), cooks often remove the skins from nuts.

The Nutritional Value of Nuts

Nuts are very nutritious. After pure fats and oils, they're the richest foods that we eat, averaging around 600 calories (kcal) per quarter-pound/100 gm; by comparison, fatty beef averages 200 calories, and dry starchy grains 350. Nuts can be 50% or more oil, 10–25% protein, and are a good source of several vitamins and minerals and of fiber. Notable among the vitamins is the antioxidant vitamin E, especially concentrated in hazelnuts and almonds, and folic acid, which is thought to be important for cardiovascular health. Most nut oils are made up primarily of monounsaturated fatty acids, and have more polyunsaturates than saturated fats (exceptions are coconuts with a large dose of saturated fat, and walnuts and pecans, which are predominantly polyunsaturated). And nut seed coats are rich in antioxidant phenolic compounds. This cluster of characteristics—a favorable balance of fats, copious antioxidants, and folic acid—may explain why epidemiological surveys have found nut consumption to be associated with a reduced risk of heart disease.

Compositions of Common Nuts and Seeds

The following table gives the major components of nuts and seeds by percentage of the seed's edible weight. Chestnuts and coconut meat are generally sold fresh and so have a relatively high water content.

Nut or Seed	Water	Protein	Oil	Carbohydrates
Almond	5	19	54	20
Brazil	5	14	67	11
Cashew	5	17	46	29
Chestnut	52	3	2	42
Coconut (meat)	51	4	35	9
Flaxseed	9	20	34	36
Hazelnut	6	13	62	17
Macadamia	3	8	72	15
Peanut	6	26	48	19
Pecan	5	8	68	18
Pine nut	6	31	47	12
Pistachio	5	20	54	19
Poppy seed	7	18	45	24
Sesame seed	5	18	50	24
Sunflower	5	24	47	20
Walnut, black	3	21	59	15
Walnut, English	4	15	64	16

Nut Flavor

Nuts provide us with a distinctive, appealing, and versatile set of flavors. Nuttiness is a cluster of qualities: slightly sweet, slightly fatty, slightly roasted or toasted; a delicate flavor, yet with some depth. The nuts' ample endowment of oil is key to their character; the less rich grains take on a pleasant flavor when simply toasted dry, but develop an added dimension when cooked in oil or fat. The quality of nuttiness complements many other foods, whether savory or sweet, all the way from fish to chocolate.

Most nuts contain at least traces of free sugars. Some contain more than a trace and are noticeably sweet; these include chestnuts, cashews, pistachios, and pine nuts.

Handling and Storing Nuts

The same high oil content that makes nuts nutritious and delicious also makes them much more fragile than grains and legumes: that oil readily absorbs odors from the surroundings, and goes rancid when it's split into its component fatty acids and the fatty acids are then fragmented by oxygen and light. The fatty acids have an irritating effect on the mouth, while their fragments have cardboard, paint-like aromas. Walnuts, pecans, cashews, and peanuts are rich in fragile polyunsaturated fats and are especially susceptible to staling. Fat rancidity is favored by bruising, light, heat, and moisture, so it's best to store nuts in opaque containers at cool temperatures. Shell-less kernels are best refrigerated. Because they contain very little water and so don't suffer from the formation of damaging ice crystals, nuts can be frozen for long keeping. Storage containers should be truly air- and odor-tight—glass jars, for example, rather than permeable plastic bags.

Nuts are at their best when they're freshly harvested, usually in late summer and fall (early summer for almonds). Newly harvested nuts are too moist to keep without being vulnerable to molds, so producers dry them with as little heat as possible, usually at 90–100°F/32–38°C. When buying fresh nuts, look for an opaque, off-white interior. Any translucency or darkening is a sign that the cells are damaged, oil has been released, and rancidity is developing.

Cooking Nuts

Unlike most other seed foods, nuts are good simply oven-toasted or fried for a few minutes, which transforms the chewy, pliable, bland, pale seeds into crisp, flavorful, browned morsels. They can also be roasted in the microwave oven. Since nuts are small and dry, frying is generally done at relatively low temperatures for relatively short times, 250–350°F/120–175°C for a few minutes, with lower temperatures and longer times for large nuts (Brazil nuts, macadamias). Doneness should be judged by color and flavor, not texture; heat softens the tissue, which gets crisp as it cools. Stop the cooking just short of the ideal doneness, since nuts continue to cook for some time after they're taken from the heat. Nuts are less brittle when they're warm, so slicing them while warm can give cleaner pieces with fewer flakes and crumbs.

Commercially prepared nuts are often roasted, salted with special flake-shaped particles that have more of a surface to adhere to the nut, then coated with a layer of oil or a protein-emulsifier blend to help retain the salt. Peanuts are salted in the shell by being soaked in brine under a vacuum, which pulls the air from inside the shell and forces the brine in.

Removing Skins Many preparations call for nut skins to be removed so that they don't discolor the dish or add unwanted astringency. Thin skins—those on peanuts and hazelnuts, for example—can often be made brittle enough to rub off by a brief toasting in the oven. The thicker skins of almonds are toughened and loosened by a minute or two in boiling water. Others can often be removed by immersing the nuts in hot water made alkaline with baking soda (3 Tb soda per quart/45 gm per liter), rubbing the softened skins off (alkalinity helps dissolve hemicellulose cement in the cell walls), then reimmersing the nuts in a dilute acid solution to neutralize the slight amount of absorbed alkaline liquid. The color and astringency of hard-to-remove walnut skins can be lightened significantly by a brief boiling in acidified water, which leaches out tannins and bleaches the color of those that remain. Tough chestnut skins are softened by roasting or boiling in the shell, or by a brief period in the microwave oven. They're also simply peeled off like the skin of an apple.

Nut Pastes and Butters All dry and fatty nuts can be ground in a mortar or blender into a butter-like paste, with the oil from ruptured cells coating and lubricating the cell fragments and particles of intact cells. One of the most venerable nut pastes is the Middle Eastern tahini, made from ground sesame seeds; it's used with chickpeas to make hummus, and with eggplant to make baba ghanoush. Throughout the world, nut pastes are added to soups and stews to contribute flavor, richness, and body; almond soups are made in Spain and Turkey, walnut soups in Mexico, coconut soups in Brazil, pecan and peanut soups in the American South.

Nut Oils The oils of a number of nuts are prized for their flavor—walnut oil and coconut oil, for example—and several others as ordinary cooking oils (peanut oil, sunflower seed oil). Oils are extracted from nuts by two different means. "Cold-pressed" or "expeller-pressed" nut oils are made by crushing the nut cells and forcing the oil out with mechanical pressure. The nuts get hot from the pressure and friction, but generally don't exceed the boiling point. Solvent-extracted oils are made by dissolving the oil out of the crushed nuts with a solvent at temperatures around 300°F/150°C, then separating the oil from the solvent. They are more refined than pressed oils, having fewer of the trace compounds that make oils both flavorful and potentially allergenic (p. 455). Cold-pressed oils are generally used as a flavoring, refined oils as cooking oils. Nut oils have a stronger flavor if the nuts are roasted before extraction. Because they often have a large proportion of fragile polyunsaturated fatty acids, they're more vulnerable to oxidation than ordinary vegetable oils, and are best kept in dark bottles in the refrigerator. The leftover solids—nut meal or flour—make a flavorful and nutritious contribution to baked goods.

Nut Milks If nuts are ground while dry, their microscopic oil bodies (p. 459) merge and coalesce to make oil the continuous liquid phase of the paste. But if the raw nuts are first soaked in water, then grinding releases the oil bodies relatively intact into the continuous water phase. When the solid nut particles are strained off, this leaves behind a fluid similar to milk, with

Argan Oil

An unusual nut known in the West almost exclusively for its oil is argan, the seed of a drought-tolerant tree native to Morocco (*Argania spinosa,* a relative of the chicle and miracle fruit trees). The almond-like nuts are removed from their fruits (a process formerly assigned to goats, which eat the fruits and excrete the nuts), shelled and roasted, then ground and pressed. Argan oil has a distinctive, almost meaty aroma.

oil droplets, proteins, sugars, and salts dispersed in water. In medieval Europe, which learned about them from the Arabs, almond milks and creams were both luxurious ingredients and dairy substitutes for fasting days. Today, the most common seed milk is made from coconuts, but it can be made from any oil-rich nut and from soybeans (p. 494).

Modern cooks can use nut milks to make rich and delicious ices, and to enrich sauces and soups. Thanks to the tendency of the nut proteins to coagulate, cooks can thicken nut milks with acid into the equivalent of yogurt, and cook them into a cross between a pudding and custard. Almonds with their high protein content produce the most easily thickened milk. Other nut milks can be boiled to coagulate the proteins into curds, then the curds can be drained of excess fluid, blended until smooth, and heated gently to thicken them further. To impart more flavor to the milk, a small fraction of the nuts can be roasted before grinding.

Characteristics of Some Common Nuts

Almonds Almonds are the world's largest tree-nut crop. They're the seed of a plumlike stone fruit, or drupe; the tree is a very close relative of the plum and peach. There are several dozen wild or minor species, but the cultivated almond, *Prunus amygdalus,* came from western Asia and had been domesticated by the Bronze Age. California is now the largest producer. Thanks to their high content of antioxidant vitamin E and low levels of polyunsaturated fats, almonds have a relatively long shelf life.

Almonds are the main ingredient in marzipan, a paste of sugar and almonds finely ground together and molded and dried into decorative shapes, a Middle Eastern invention that became popular in Europe during the medieval Crusades. Leonardo da Vinci made marzipan sculptures for the Milanese court of Ludovico Sforza in 1470, and wrote that he "observed with pain that [they] gobble up all the sculptures I give them, right to the last morsel." Almond paste is also commonly used as a pastry filling or the base for

Almond Milk and Cream in Medieval Times

Blancmange

Take capons and seeth them, then take them up. Take blanched almonds, grind them, and mix them with the same broth. Cast this milk in a pot. Wash rice, and add, and let it seeth. Then take brawn of capons, tear it small and add. Take lard, sugar and salt, and cast therein. Let it seeth. Then assemble it and decorate it with aniseed confected white or red, and with fried almonds, and serve it forth.

—*The Forme of Cury,* ca. 1390

Cream of Almond Milk

Take almond milk, and boil it, and when it is boiled take it from the fire, and sprinkle on a little vinegar. Then spread it on a cloth, and cast sugar on it, and when it is cold gather it together, and leche [slice] it in dishes, and serve it forth.

—From a medieval manuscript, published in R. Warner, *Antiquitates Culinariae,* 1791

macaroons, cookies whose only other structural ingredient is egg whites.

Why Almonds Don't Taste Like Almond Flavoring Curiously, standard domesticated almonds taste delicately nutty, nothing like the strong and distinctive flavoring called "almond extract." Strong almond flavor is found only in wild or bitter almonds, which are inedibly bitter and toxic. They contain a defensive system that generates deadly and bitter hydrogen cyanide when the kernel is damaged (p. 258). It's estimated that eating a few bitter almonds at a sitting could kill a child. But it turns out that one by-product of cyanide production is benzaldehyde, a volatile molecule that is the essence of wild almond flavor, and that contributes to the aromas of cherry, apricot, plum, and peach. Our safe "sweet" almond varieties lack both the bitterness and the characteristic aroma.

Bitter almonds are generally unavailable in the United States, while in Europe they're used like a spice, added in small numbers to flavor marzipan made from sweet almonds, as well as to amaretti cookies, amaretto liqueur, and other dishes. Apricot and peach kernels are readily available alternative sources of benzaldehyde, though they don't have the intense and otherwise fine flavor of bitter almonds. German cooks make versions of marzipan called *persipan* with apricot and peach kernels.

Brazil Nuts Brazil nuts are unusually large, an inch/2.5 cm or more long, and double the weight of almonds and cashews. They're the seeds of a large tree (*Bertholletia excelsa,* 150 ft/50 m tall, 6 ft/2 m across) native to the Amazon region of South America, where they develop in groups of 8 to 24 inside a hard, coconut-sized shell. South American countries are still the main producers. The pods are gathered only after they fall to the ground. Because they weigh about 5 pounds, they can be lethal missiles, and harvesters must carry shields to protect themselves. The edible portion of the seed is an immensely swollen embryonic stem. Thanks to their size and high oil content, two large Brazil nuts are the caloric equivalent of one egg.

Brazil nuts are notable for containing the

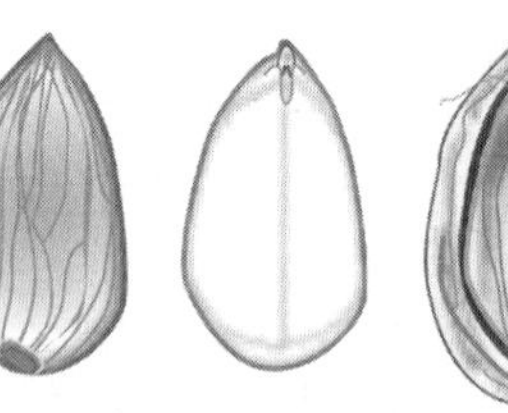
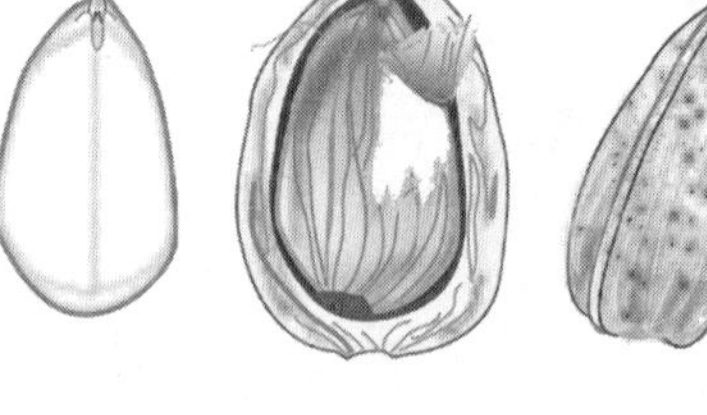

The almond, a close relative of the peach, plum, and cherry, with its stony shell.

Almond Extracts and Imitations

The commonest form of bitter-almond flavor is a bottled extract, which contains aromatic benzaldehyde without the cyanide that accompanies it in the almonds themselves. "Pure" almond extract is derived from bitter almonds, while "natural" extract usually contains benzaldehyde produced from cassia bark (p. 424), and "imitation" extract contains benzaldehyde synthesized from pure chemicals.

highest levels of selenium of any food. Selenium helps to prevent the development of cancer, apparently by several different means, including an antioxidant enzyme and by inducing damaged cells to die. However, an overdose is toxic. The World Health Organization recommends a maximum daily selenium intake corresponding to just a half-oz/14 gm of Brazil nuts.

Cashews Like the Brazil nut, cashews come from the Amazon region, whose natives gave us the name. But the tree was successfully transplanted to India and East Africa by the Portuguese, and today these regions are the world's largest producers. The cashew is second only to the almond in world trade. It's a relative of poison ivy, and that's why we never see cashews for sale in the shell. The shell contains an irritating oil that must be driven off by heating before the seed can be carefully extracted without contamination. In the producing countries, the seed-containing fruit is often discarded in favor of the swollen stem tip or "false fruit" called the cashew apple, which is enjoyed either fresh, cooked, or fermented into an alcoholic drink.

Cashews are unusual among oily nuts in containing a significant amount of starch (around 12% of their weight), which makes them more effective than most nuts at thickening water-based dishes (soups, stews, Indian milk-based desserts).

Chestnuts Chestnuts come from several different species of large trees in the genus *Castanea,* which are found in Europe, Asia, and North America. They're unlike the other common nuts in storing their energy for the future seedling in the form of starch, not oil. Chestnuts are thus usually thoroughly cooked and have a mealy texture. Since prehistory they have been dried, ground into flour, and used in the same way that the starchy cereals are, to make gruels, breads, pastas, cakes, and provide substance in soups. Before the arrival of the potato and corn from the New World, chestnuts were an essential subsistence food in mountainous and marginal agricultural areas of Italy and France. At the opposite extreme, a luxurious chestnut specialty invented in the 17th century is *marrons glacés,* large chestnuts that are cooked, slowly infused over a day or two with a vanilla-flavored syrup, then glazed with a more concentrated syrup.

The American enjoyment of our native chestnut, *Castanea dentata,* was brought to a sad end in the early 20th century, when in the course of a few decades, an imported Asian fungal blight wiped out a tree that used to make up 25% or more of the eastern hardwood forest. Today, the world's leading chestnut producers are China, Korea, Turkey, and Italy.

Because of its high initial moisture content, the chestnut is quite perishable. Chestnuts are best kept covered and refrigerated, and should be eaten fairly quickly. If freshly gathered, however, they should be cured at room temperature for a few days. This improves the flavor by permitting some

Why Brazil Nuts Rise to the Top of the Bowl

A paper published in the 1987 *Physical Review Letters* tried to crack a hard nut: why is it that in a bowl of mixed nuts, the small nuts end up at the bottom and the Brazil nuts on top? The same kind of segregation by size takes place in many different mixtures, from corn flakes to the soil. It turns out that objects in a mixture are pulled downward through gaps in the mixture by gravity—and small gaps are more common than large gaps, so small objects more frequently sink than large ones.

starch to be converted into sugar before the cells' metabolism is slowed down.

Coconuts Coconuts are the largest and most important of all nuts. They are the stone of a drupe, the fruit of *Cocos nucifera,* large (to 100 ft/30 m) tree-like palms that are more closely related to the grasses than to other nut trees. They're thought to have originated in tropical Asia, but their hardy fruits apparently floated to many parts of the world before humans began to transport them. They were largely unknown to Europe until the early Middle Ages. About 20 billion nuts are produced each year, mainly in the Philippines, India, and Indonesia. The word *coconut* comes from the Portuguese *coco,* which means goblin or monkey. The markings on the stem end of the nut can look uncannily like a face. The tiny embryo resides under one of the eyes, through which it grows when it sprouts.

Coconuts consist of a thick fibrous fruit layer, the husk, within which is the seed proper enclosed in a woody shell. The meat and milk constitute the seed's endosperm, which contains enough nutrients and moisture to support the seedling's growth for more than a year. The entire fruit may weight 2–5 lb/1–2 kg, of which about a quarter is meat, 15% free water.

Coconut provides a backbone flavor in many tropical cuisines, from southern India and Southeast Asia to Africa and South America. It's often used in the form of coconut milk, which provides a rich, flavorful liquid in which to cook all kinds of foods, from meats and fish to vegetables and rice. Since the coconut meat can't be roasted intact, its flavor is developed by carefully toasting small flakes or shreds. Unlike other nuts, which provide crunchiness or smooth richness depending on how finely they're broken, coconut has a persistent, chewy texture unless toasted and kept very dry.

The distinctive sweet, rich aroma of coconut is created by derivatives of saturated fatty acids called lactones (octa-, deca-, dodeca-, tetradecalactones)—peaches are also flavored by lactones—while roasting generates more generic nutty notes (from pyrazines, pyrroles, furans).

Coconut Development Coconut fruits are borne and mature year-round. At around four months, the nut fills with liquid; at five, it reaches its full size and begins to develop a jelly-like meat; at seven its shell begins to harden, and it's mature at a year. Immature coconuts, around five to seven months old, offer their own pleasures: a sweet liquid called coconut water (about 2% sugars); and a moist, delicate, gelatinous meat that's mainly water, sugars and other carbohydrates. In the mature coconut of 11–12 months, the liquid has become less sweet and less abundant, and the meat has become firm, fatty, and white. The meat is about 45% water, 35% fat, 10% carbohydrate, 5% protein.

The coconut. This massive seed is borne in a thick, dry husk and contains both solid and liquid endosperm to feed the small embryo, which emerges from one of the three "eyes" at the end of the shell.

Coconut Meat and Milk A good fresh coconut should feel heavy and contain enough liquid to slosh audibly. If coconut meat is pounded in a mortar or ground finely in a blender, it forms a thick paste consisting of microscopic oil droplets and cell debris suspended in water, which makes up about half the volume. Coconut milk is made by mixing the paste with some additional water and straining to remove the solid particles. Left to stand for an hour, this milk separates into a fat-rich cream layer and a thin "skim" layer. Coconut milk can also be made from dried shredded coconut, and is readily available canned.

Coconut Oil For part of the 20th century, coconut oil was the most important vegetable oil in the world. It can be produced in large quantities, is very stable, and has a melting point similar to that of milk fat. But the very quality that makes it stable and versatile also made it appear to be nutritionally undesirable. The fats that make up coconut oil are nearly 90% saturated (15% caprylic and capric, 45% lauric, 18% myristic, 10% palmitic, and just 8% monounsaturated oleic), which means that they raise blood cholesterol levels. During the 1970s and '80s, manufacturers of processed foods therefore replaced coconut oil with less saturated, partly hydrogenated seed oils—which now turn out to contain undesirable trans fatty acids (p. 38).

Given our current and broader understanding of other dietary influences on heart disease (p. 255), there's no reason not to enjoy the coconut's riches as part of a balanced diet that includes plenty of protective fruits, vegetables, and other seeds.

Ginkgo Nuts Gingko nuts are the starchy kernels of *Ginkgo biloba,* the last survivor of a tree family that was prominent during the age of the dinosaurs. The nuts are borne inside fleshy fruits that develop a strong rancid smell when ripe. In Asia, the tree's home, the fruits are fermented in vats of water to soften and remove the pulp, and the seeds are washed, dried, and roasted or boiled, either in-shell or shelled. The kernel has a distinctive but mild flavor.

Hazelnuts Hazelnuts come from a few of the 15 species of mainly bushy trees in the northern-hemisphere genus *Corylus. Corylus avellana* and *C. maxima* are native to temperate Eurasia and were widely exploited in prehistoric times for their nuts and rapidly produced shoots, which were used as walking sticks and a surface for marshy ground. A much taller tree, *C. colurna,* accounts for much of the production in the Black Sea region of Turkey. Another term for the nut is "filbert," which in the United Kingdom is applied to the more elongated varieties, and which may come from St. Philibert's Day in late August, when hazelnuts begin to ripen. The late Roman cookbook of Apicius called for hazelnuts in sauces for birds, boar, and mullet; they're an alternative to almonds in

Coconut "Gelatin"

In addition to products from the seed itself, the coconut palm offers several other distinctive food materials. One of the more unusual is *nata de coco,* or "coconut gelatin," a moist, translucent mass of cellulose produced by a vinegar bacterium (*Acetobacter xylinum*) on the surface of fermenting coconut water. It has little flavor of its own and a uniquely crunchy texture. In the Philippines it's washed of its vinegary home, flavored, packed in sugar syrup, and eaten as a sweet.

Spanish picada and romesco sauces, and an ingredient in the spicy Egyptian spread called dukka and the Italian liqueur frangelico. Hazelnuts remain especially popular in Europe, where Turkey, Italy, and Spain are the main producers. In the United States, nearly all hazelnuts are produced in Oregon.

The distinctive aroma of hazelnuts comes from a compound dubbed filbertone (heptenone), which is present in small quantities in the raw nut, but increases 600- to 800-fold when the nuts are fried or boiled.

Macadamia Nuts Macadamia nuts are newcomers to the world's table. They come from two evergreen tropical trees (*Macadamia tetraphylla* and *M. integrifolia*) native to northeastern Australia, where the aborigines enjoyed them for thousands of years before they were noticed and named by Europeans (for John Macadam, a Scots-born chemist, in 1858). Macadamias were introduced to Hawaii in the 1890s, and became commercially significant there around 1930. Today Australia and Hawaii are the main producers, but their output remains relatively small, and macadamias are therefore among the most expensive nuts. Because their shells are extremely hard, they are sold almost exclusively out-of-shell, often packed in cans or bottles to protect them from air and rancidity. Macadamias have the highest fat content of the tree nuts, and it's mostly monounsaturated (65% oleic acid). Their flavor is mild and delicate.

Peanut This popular nut is not a nut, but the seed of a small leguminous bush, *Arachis hypogaea,* which pushes its thin, woody fruit capsules below ground as they mature. The peanut was domesticated in South America, probably Brazil, around 2000 BCE, and was an important crop in Peru before the time of the Incas. In the 16th century, the Portuguese took it to Africa, India, and Asia, and it soon became a major source of cooking oil in China (peanuts have double the oil content of soybeans). It wasn't until the 19th century that Americans thought of peanuts as anything but animal feed, and not until the early 20th century that the remarkable agricultural scientist George Washington Carver encouraged southern farmers to replace weevil-ravaged cotton with peanuts.

Today India and China are by far the largest peanut producers, with the United States a distant third. Most Asian peanuts are crushed for oil and meal; in the United States most are eaten as food. Peanuts are now prominent in several Asian and African traditions. Pureed, they lend richness, substance, and flavor to sauces and soups. Whole and pureed peanuts are used in Thai and Chinese noodle dishes, in sweet bun fillings, in Indonesian dipping sauces and sambal condiments, and in West African stews, soups, cakes, and confections. A popular snack food in both Asia and the southern United States is peanuts boiled in salted water. When boiled in their shells, peanuts develop a potato-like aroma, with sweet vanilla highlights thanks to the liberation of vanillin from the shell.

In the United States, four varieties are grown for different purposes: large Virginia and small Valencia for nuts sold in the shell, Virginia and small Spanish for mixed nuts and candies, and Runner for baked goods and peanut butter, since its higher content of monounsaturated fat makes it less vulnerable to rancidity.

Peanut Butter The modern version of peanut butter was apparently developed around 1890 in St. Louis or in Battle Creek, Michigan. Commercial peanut butter is made by heating the nuts to an internal temperature around 300°F/150°C to develop flavor, blanching in hot water to remove the skin, and finally grinding them with about 2% salt and up to 6% sugar. The oil can be prevented from separating from the solid peanut particles by adding 3–5% of a hydrogenated shortening that solidifies as the butter cools and forms a host of tiny crystals that hold the very unsaturated, liquid peanut oil in place. Low-fat peanut butter is made by replacing a portion of the peanuts with soy protein and with sugar.

Peanut Flavor Several hundred volatile compounds have been identified in roasted peanuts. The raw seed has a green, bean-like flavor (mainly from green-leaf hexanal and the pyrazine that characterizes peas). The roasted aroma is a composite of several sulfur compounds, a number of generically "nutty" pyrazines, and others, some of which have fruity, flowery, fried, and smoky characters. During storage and staling, the nutty pyrazines disappear and painty, cardboard notes increase.

Peanut Oil Thanks to the productivity of the peanut plant in warm climates, peanut oil is an important cooking oil, especially in Asia. It's made by steaming the peanuts to inactivate enzymes and soften the cellular structure, then pressing them; the oil is then clarified and sometimes refined to remove some of the distinctive flavor and impurities that would lower the smoke point.

Pecans Pecans are the soft, fatty seeds of a very large tree, a distant relative of the walnut that is native to the Mississippi and other river valleys of central North America, and found as far south as Oaxaca. *Carya illinoiensis* is one of about 14 species of hickories, and its nuts among the tastiest and easiest to shell. Wild pecans were enjoyed by the native Americans, and apparently made into a kind of milk that was used for drinking, cooking, and possibly fermenting. The earliest intentional plantings may have been made by the Spanish around 1700 in Mexico, and a few decades later the trees were grown in the eastern British colonies. The first improved varieties were made possible in the 1840s by a Louisiana slave named Antoine, who worked out how to graft wood from superior trees onto seedling stocks. Georgia, Texas, and New Mexico are now the largest producers of pecans.

Compared to the walnut, the pecan is more elongated, the cotyledons thicker and smoother, with a larger proportion of meat to shell. As with walnuts, light-skinned varieties are less astringent than dark-skinned. The distinctive flavor of pecans remains somewhat mysterious. In addition to the generic nutty notes of the pyrazines, one study found a lactone (octalactone) that is also present in coconuts.

Pecans and walnuts are among the nuts with the highest oil and unsaturated fatty acid contents. High oil content generally goes along with fragile texture, which means that the kernels are easily bruised, with seepage of oil to the surface and rapid

Food Words: *Pine, Walnut, Flax, Sesame*

A number of nut names seem simply to name the nuts, without a penumbra of related meanings. This is an indication that almonds and pistachios (from the Greek) and hazelnuts (from the Indo-European) have been basic fare for a very long time. *Pine* comes from an Indo-European root meaning "to swell, to be fat," probably an allusion to the fat-like resin that the tree exudes. *Walnut* is an Old English compound of *wealh,* meaning "Celt" or "foreigner," and *hnutu,* "nut," a reflection of the fact that walnuts were introduced to the British Isles from the east. *Flax* comes from an Indo-European root meaning "to plait," because flax was originally grown for its stem fibers. And *sesame* comes from two words in the ancient Middle Eastern Akkadian language that meant "oil" and "plant."

oxidation and staling. Roasting increases the rate of staling by weakening cells and allowing oil to come into contact with the air. Carefully handled, raw pecans can be stored for years in the freezer.

Pine Nuts Pine nuts are gathered from about a dozen of the 100 species of pines, one of the most familiar evergreen tree families in the Northern Hemisphere. Among the more important sources are the Italian stone pine *Pinus pinea,* the Korean or Chinese pine *P. koraiensis,* and the southwestern U.S. pinyons *P. monophylla* and *P. edulis.* The nuts are borne on the scales of the pine cone, which takes three years to mature. The cones are sun-dried, threshed to shake out the seeds, and the kernels then hulled, nowadays by machine. They have a distinctive, resinous aroma and are rich even for nuts; Asian pine nuts have a higher oil content (78%) than either American or European types (62% and 45% respectively). They're used in many savory and sweet preparations, and pressed to make oil. In Korea, pine pollen is used to make sweets, and Romanians flavor game sauces with the green cones.

Pistachios Pistachios are the seeds of a native of arid western Asia and the Middle East, *Pistacia vera,* a relative of the cashew and the mango. Along with almonds, they have been found at the sites of Middle Eastern settlements dating to 7000 BCE. A close relative, *Pistacia lentiscus,* provides the aromatic gum called mastic (p. 421). Pistachios first became a prominent nut in America in the 1880s, thanks to their popularity among immigrants in New York City. Iran, Turkey, and California are the major producers today.

Pistachios grow in clusters, with a thin, tannin-rich hull around the inner shell and kernel. As the seeds mature, the outer hull turns purple-red and the expanding kernel cracks the inner shell open. Traditionally, the ripe fruits were knocked from the trees and sun-dried, and the hull pigments stained the shell, so the shells were often dyed to make them a uniform red. Today, most California pistachios are hulled before drying, so the shells are their natural pale tan color.

Pistachios are remarkable among the nuts for having green cotyledons. The color comes from chlorophyll, which remains vivid when the trees grow in a relatively cool climate, for example at high elevation, and when the nuts are harvested early, several weeks before full maturity. Pistachios thus offer not only flavor and texture but a contrasting color in pâtés, sausages, and other meat dishes, and in ice creams and sweets. The color is best retained by roasting or otherwise cooking the kernels at low temperatures that minimize chlorophyll damage.

Walnuts Walnuts come from trees in the genus *Juglans,* of which there are around 15 species native to southwestern Asia,

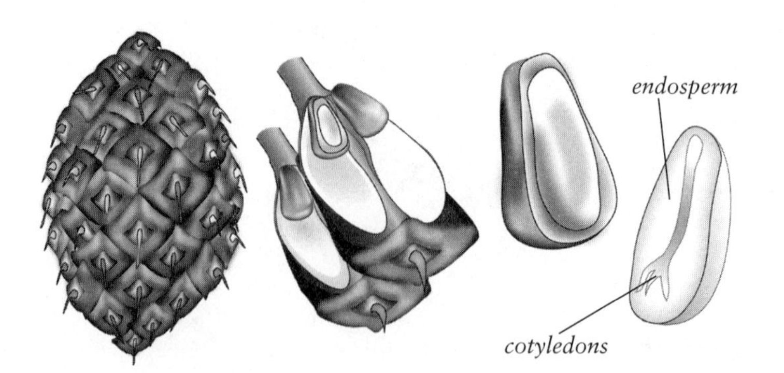

Pine nuts. They are borne on the scales of pine cones, and like the coconut, are mainly endosperm tissue rather than cotyledons.

eastern Asia, and the Americas. The most widely cultivated is the Persian or English walnut, *Juglans regia,* whose seeds have been enjoyed since ancient times in western Asia and Europe, and among tree nuts are second only to almonds in worldwide consumption. In many European languages, the generic term for nut is also the word for walnut. The United States, France, and Italy are the major producers today. Walnuts have long been pressed for their aromatic oil, were once made into milk in Europe and China, and came to provide the rich, flavorful backbone of sauces in Persia (*fesenjan*), Georgia (*satsivi*), and Mexico (*nogado*). In some countries, immature "green" walnuts are harvested in early summer and pickled (England), used to flavor sweetened alcohol (Sicilian *nocino,* French *vin de noix*), or preserved in syrup (the Middle East).

Like its cousins the pecan and hickories, the walnut is the stone of a thin-walled fruit, the edible portion being two lobed, wrinkled cotyledons. Walnuts are exceptionally rich in the omega-3 polyunsaturated linolenic acid, which makes them nutritionally valuable but also especially liable to become rancid; they should be kept in the cold and dark. The aroma of walnuts is created by a complex mixture of molecules derived from the oil (aldehydes, alcohols, and ketones).

Walnut Relatives A North American relative of the Persian walnut, the black walnut (*J. nigra*) is smaller, with a harder shell and a stronger, distinctive flavor. It was once commonly used to make breads, confections, and ice creams, but it's difficult to extract from the shell in large pieces and has been largely ignored. Most still come from wild trees in Missouri. Another American species, the butternut (*J. cinerea*), is even less known, but remarkable for its high protein content—near 30%—and esteemed by enthusiasts as among the tastiest nuts. The Japanese have an indigenous walnut, *J. ailantifolia,* one of whose varieties is the distinctively heart-shaped heartnut.

Characteristics of Other Oil-Rich Seeds

Flaxseed Flaxseed comes from plants native to Eurasia, species of *Linum* and especially *L. usitatissimum,* which have been used for more than 7,000 years as a food and to make linen fiber. The small, tough, reddish-brown seed is about 35% oil and 30% protein, and has a pleasantly nutty flavor and an attractively glossy appearance. Two qualities set it apart from other edible seeds. First, its oil is over half linolenic acid, an "omega-3" fatty acid that the body can convert into the healthful long-chain fatty acids (DHA, EPA) found in seafoods (p. 183). Flax oil (also known as linseed oil, and valued in manufacturing for drying to a tough water-resistant layer) is by far the richest source of omega-3 fatty acids among plant foods. Second, flaxseed is about 30% dietary fiber, a quarter of which is a gum in the seed coat made up of long chains of various sugars. Thanks to the gum, ground flaxseed forms a thick gel when mixed with water, is an effective emulsifier and foam stabilizer, and can improve the volume of baked goods.

Poppy Seeds Poppy seeds come from a west Asian plant, *Papaver somniferum,* that was cultivated by the ancient Sumerians. It's the same plant whose immature seed capsules are cut to collect the latex called opium, a mixture of morphine, heroin, codeine, and other related alkaloid drugs. The seeds are harvested from the capsules after the latex flow has stopped. They may carry traces of opiate alkaloids as well, not enough to have an effect on the body, but enough to cause positive results in drug tests after the consumption of a poppy-flavored cake or pastry.

Poppy seeds are tiny; it takes 3,300 to make a gram, 90,000 an ounce, 1–2 million a pound. The seed is 50% oil by weight. Poppy seeds sometimes have a bitter, peppery taste, the result of damage to the seeds, which mixes oil with enzymes and generates free fatty acids. The striking blue color of some poppy seeds is apparently an optical illusion. Microscopic examination demonstrates that the actual pigment layer of the seed is brown. Two layers above it, however, is a layer of cells containing tiny crystals of calcium oxalate: and the crystals act like tiny prisms, refracting light rays in such a way that blue wavelengths are selectively reflected.

Pumpkin Seeds Pumpkin seeds come from the fruits of the New World native *Cucurbita pepo,* are notable for being deep green with chlorophyll, and for containing no starch, as much as 50% oil, and 35% protein. Pumpkin seeds are eaten widely as a snack and in Mexico are used as a sauce thickener. There are "naked" varieties that lack the usual tough, adherent seed coat and are therefore much easier to work with.

Pumpkin seed oil is a prominent salad oil in central Europe. The oil, containing mainly polyunsaturated linoleic and monounsaturated oleic acids, is intriguingly changeable in color. Pumpkin seeds contain both yellow-orange carotenoid pigments, mainly lutein, and chlorophyll. Oil pressed from raw seeds is green; but when the seed meal is wetted and heated to increase the yield, more carotenoids are extracted than chlorophyll. The result is an oil that looks dark brown in the bottle or bowl from the combination of orange and green pigments; but in a thin layer, for example on a piece of bread dipped into the oil, there are fewer pigment molecules to absorb light, the chlorophyll dominates, and the oil becomes emerald green.

Sesame Seeds Sesame seeds are the seeds of *Sesamum indicum,* a plant of the central African savanna that is now mostly grown in India, China, Mexico, and the Sudan. Sesame seeds are small, with 250–300 per gram and 7,500–9,000 per ounce, come in a variety of colors, from golden to brown, violet, and black, and are about 50% oil by weight. They're usually lightly toasted (250–300°F/120–150°C for 5 minutes) to develop a nutty flavor, which has some sulfur aromatics in common with roasted coffee (furfurylthiol). Sesame seeds are made into the seasoned Middle Eastern paste called tahini, are added to rice balls and made into a tofu-like cake with arrowroot in Japan, and made into a sweet paste in China, as well as decorating a variety of baked goods in Europe and the United States. Sesame oil is also extracted from toasted seeds (360–400°F/180–200°C for 10–30 minutes) and used as a flavoring. The oil is remarkable for its resistance to oxidation and rancidity, which results from high levels of antioxidant phenolic compounds (lignans), some vitamin E, and products of the browning reactions that occur during the more thorough toasting.

Sunflower Seeds The flower of *Helianthus annuus,* the only North American native to become a significant world crop, is a composite of a hundred or more small flowers, each of which produces a small fruit like the "seed" of the strawberry, a single seed contained in a thin hull. The seed is mainly storage cotyledons. The sunflower originated in the American Southwest, was domesticated in Mexico nearly 3,500 years before the arrival of European explorers, and brought to Europe around 1510 as a decorative plant. The first large crops were grown in France and Bavaria in the 18th century to produce vegetable oil. Today, the world's leading producer by far is Russia. Improved Russian oil varieties were grown in North America during World War II, and sunflower is now one of the top annual oil crops worldwide. The eating varieties are larger than the oil types, with decoratively striped hulls that are easily removed. Sunflower seeds are especially rich in phenolic antioxidants and vitamin E.

CHAPTER 10

CEREAL DOUGHS AND BATTERS

Bread, Cakes, Pastry, Pasta

Bread is the most everyday and familiar of foods, the sturdy staff of life on which hundreds of generations have leaned for sustenance. It also represents a truly remarkable discovery, a lively pole on which the young human imagination may well have vaulted forward in insight and inspiration. For our prehistoric ancestors it would have been a startling sign of the natural world's hidden potential for being transformed, and their own ability to shape natural materials to human desires. Bread is nothing like the original grain, loose, hard, chalky, and bland! Simply grinding grains, wetting the particles with water, and dropping the paste on a hot surface, creates a flavorful, puffy mass, crisp outside and moist within. And raised bread is even more startling. Let the paste sit for a couple of days, and it comes alive and grows, inflated from within, and cooks into a bread with a delicately chambered interior that the human hand could never sculpt. Plain parched grains and dense gruels provide just as much nourishment, but bread introduced a new dimension of pleasure and wonder to the mainstays of human life.

So it was bread that became synonymous with food itself in the lands from western Asia through Europe, and took a prominent place in religious and secular rituals (Passover matzoh, Communion bread, wedding cakes). In England, it provided a foundation for naming social relations. "Lord" comes from the Anglo-Saxon *hlaford,* "loaf ward," the master who supplies food; "lady" from *hlaefdige,* "loaf kneader," the person whose retinue pro-

Food Words: *Dough, Bread*

Dough comes from an Indo-European root that meant "to form, to build," and that also gave us the words *figure, fiction,* and *paradise* (a walled garden). This derivation suggests the importance to early peoples of dough's malleability, its clay-like capacity to be shaped by the human hand. (Cooks have long used both clay and dough to make containers for cooking other foods, especially birds, meats, and fish.)

The word *bread* comes from a Germanic root, and originally meant a piece or bit of a loaf, with *loaf* meaning the leavened, baked substance itself. Over time, *loaf* came to mean the intact baked mass, and *bread* took over *loaf*'s original meaning. Otherwise we would now ask for a bread of loaf!

duces what her husband distributes; "companion" and "company" from the late Latin *companio,* or "one who shares bread." The staff of life has also been a mainstay for Western thought.

THE EVOLUTION OF BREAD

Bread's evolution has been influenced by all the elements that go into its making: the grains, the machines for milling them, the microbes and chemicals that leaven the dough, the ovens that bake the loaves, the people who make the bread and eat it. One consistent theme from ancient times has been the prestige of refined and enriched versions of this basic sustenance. Bread has become a product increasingly defined by the use of high-rising bread wheats, the milling of that wheat into a white flour with little of the grain's bran or germ, leavening with ever purer cultures of mild-flavored yeasts, enrichment with ever greater quantities of fat and sugar. In the 20th century we managed to take refinement and enrichment to the extreme, and now have industrial breads with little flavor or texture left in them, and cakes that contain more sugar than flour. In the last couple of decades, bread lovers have led a rediscovery of the pleasures of simple, less refined breads freshly baked in old-fashioned brick ovens, and even supermarket breads are getting more flavorful.

Prehistoric Times

Two prehistoric discoveries laid the foundation for the transformation of grains into breads and noodles, pastries and cakes. The first was that in addition to being cooked into a porridge, pastes of crushed grain and water could also be turned into an interesting solid by cooking them on hot embers or stones: the result was flatbread. The second was that a paste set aside for a few days would ferment and become inflated with gases: and such a paste made a softer, lighter, more flavorful bread, especially when cooked from all sides at once in an enclosed oven.

Flatbreads were a common feature of late Stone Age life in parts of the world where grains were the chief food in the diet; surviving versions include Middle Eastern lavash, Greek pita, Indian roti and chapati, all made mainly from wheat but also other grains, and the Latin American tortilla and North American johnnycake, both made from maize. Such breads were probably first cooked alongside an open fire, then on a griddle stone, and some of them much later in beehive-shaped ovens, which were open at the top and contained both coals and bread; pieces of dough were slapped onto the inside wall.

Bread wheat, the unique species that can make large, light loaves, had evolved by 8000 BCE (p. 465), but the earliest archaeological evidence for leavened breads comes from Egyptian remains of around 4000 BCE. The first raised doughs arose spontaneously, since yeast spores are ubiquitous in the air and on grain surfaces, and they readily infect a moist, nutritious grain paste. Bread makers throughout history have harnessed this natural process by leavening new dough with a leftover piece in which yeast was already growing, but they've also valued less sour starters, especially the frothy residue from brewing beer; yeast production had become a specialized profession in Egypt by 300 BCE. Meanwhile grinding equipment progressed from the mortar and pestle to two flat stones and then, around 800 BCE in Mesopotamia, to stones that could rotate continuously. Continuous milling made feasible the eventual use of animal, water, and wind power, and thus the grinding of grains into very fine flours with little human labor.

Greece and Rome

Leavened loaves of bread arrived fairly late along the northern rim of the Mediterranean. Bread wheat was not grown in Greece until about 400 BCE, and flat barley breads were probably the norm well after. We do know that the Greeks enjoyed breads and cakes flavored with honey, anise, sesame, and fruits, and that they made both whole-grain and partly refined breads. At least from the Greeks on, whiteness in bread was a mark of purity and distinction. Archestratus, a contemporary of Aristotle and author of the *Gastronomia,* a compendious account of ancient Mediterranean eating whose title gave us the word "gastronomy," accorded extravagant praise to a barley bread from the island of Lesbos on just these grounds, calling it "bread so white that it outdoes the ethereal snow in purity. If the celestial gods eat barley bread, no doubt Hermes goes to Eresus to buy it for them."

By late Roman times, wheat bread was a central feature of life, and huge amounts of durum and bread wheats were imported from northern Africa and other parts of the empire to satisfy the public demand. Pliny offers a touching reminder that enriched breads—early cakes and pastries—were great luxuries in turbulent times:

> Some people use eggs or milk in kneading the dough, while even butter has been used by peoples enjoying peace, when attention can be devoted to the varieties of bakers' goods.

The Middle Ages

During the European Middle Ages, bakers were specialists, producing either common brown or luxurious white bread. It wasn't until the 17th century that improvements in milling and in per capita income led to the wide availability of more or less white bread and the dissolution of the brown guild as a separate body. In northern areas, rye, barley, and oats were more common than wheat and were made into coarse, heavy breads. One use of flat bread at this time was the "trencher," a dense, dry, thick slice that served as a plate at medieval meals and then was either eaten or given away to the poor. And pastry was often made to serve as a kind of all-purpose cooking and storage container, a protective and edible wrapping for meat dishes in particular.

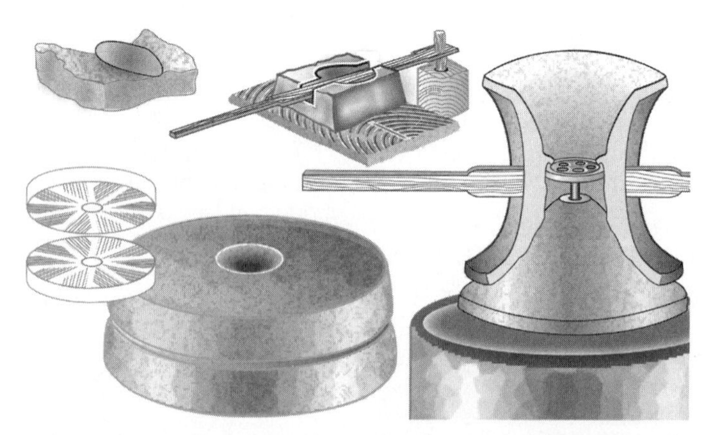

Four stages in the evolution of machines for grinding grain. Clockwise from upper left: *The saddlestone and lever mill were limited by their back-and-forth motion. The hourglass mill, which could be turned continuously in one direction by man or animal, was widely used by Roman times. Flat millstones finally made it possible to harness more elemental forces, and were put to use in water and wind mills. In the modern industrial world, most grain is milled between grooved metal rollers, but some is still stone-ground.*

Early Modern Times

The late medieval period and Renaissance brought notable progress in the art of enriched breads; both puff pastry and choux pastry date from this time. Domestic recipes for bread begin to appear in cookbooks for the emerging middle class, and already look much like modern recipes. English and American cookbooks from the 18th century on contain dozens of recipes for breads, cakes, and cookies. In England around 1800, most bread was still baked in domestic or communal village ovens. But as the Industrial Revolution spread and more of the population moved to crowded city quarters, the bakeries took over an ever increasing share of bread production, and some of them adulterated their flour with whiteners (alum) and fillers (chalk, ground animal bones). The decline of domestic baking was criticized on economic, nutritional, and even moral grounds. The English political journalist William Cobbett wrote in *Cottage Economy* (1821), a tract addressed to the working class, that it is reasonable to buy bread only in cities where space and fuel are in short supply. Otherwise,

> How wasteful, then, and indeed, how shameful, for a labourer's wife to go to the baker's shop . . .
>
> Give me, for a beautiful sight, a neat and smart woman, heating her oven and setting in her bread! And, if the bustle does make the sign of labour glisten on her brow, where is the man that would not kiss that off, rather than lick the plaster from the cheek of a duchess?

The scolding of Cobbett and others failed to reverse the trend. Bread making was one of the most time-consuming and laborious of household tasks, a kiss on the sweaty forehead notwithstanding, and more and more of the work was delegated to the baker.

Innovations in Leavening A new method of leavening made its first appearance in the first American cookbook, Amelia Simmons's 1796 *American Cookery.* Four recipes, two for cookies and two for gingerbread, call for the use of "pearlash," a refined version of potash, which was made by soaking the ash produced when plant materials are burned, draining off the liquid, and drying it down to concentrate the substances dissolved in it. Pearlash is mostly alkaline potassium carbonate, which reacts with acid ingredients in doughs to generate carbon dioxide gas. It was the precursor to baking soda and baking powders, which arrived between 1830 and 1850. These chemical ingredients made it possible to leaven instantly mixtures that living, slow-growing yeasts couldn't very well: such things as fluid cake batters and sweet cookie doughs. Purified commercial yeast cultures for loaf breads, more predictable and less acidic than brewer's yeast, became available from specialist manufacturers around the turn of the 20th century.

Food Words: *Flour*

While the words for ground grain in French, Italian, and Spanish, *farine* and *farina,* come from the Latin for a kind of grain (*far*), the English word "flour" arose in medieval times from "flower," meaning the best part of the ground grain: that is, the portion left after screening out the large particles of germ and bran. To a medieval Englishman, "whole wheat flour" would have been a contradiction in terms!

The Decline and Revival of Traditional Breads

Twentieth-Century Industrialization

The 20th century brought two broad trends to Europe and North America. One was a decline in the per capita consumption of plain bread. As incomes rose, people could afford to eat more meat and more high-sugar, high-fat cakes and pastries. So we now lean less heavily than did our ancestors on the staff of life. The other trend was the industrialization of bread making. Today very little bread is made in the home, and with the exception of countries with a strong tradition of buying fresh bread every day—especially France, Germany, and Italy—most bread is made in large central factories, not in small local bakeries. Mechanical aids to breadmaking, powered mixers and others, began to appear around 1900, and culminated in the 1960s in largely automated factories that produce bread in a fraction of the usual time. These manufacturing systems replace biological dough development, the gradual, hours-long leavening and gluten strengthening of the dough by yeast, with nearly instantaneous, mechanical and chemical dough development. This production method produces breads with a soft, cake-like interior, an uncrusty crust, and an uncharacteristic flavor. They are formulated to remain soft and edible in plastic bags for a week or more. Industrial breads bear little resemblance to traditional breads.

The Return of Flavor and Texture

Europeans and North Americans began to eat significantly more bread in the 1980s than they had the decade before. One reason was the revival of traditional bread making. Small bakeries began to produce bread using less refined grains and grain mixtures, building flavor with long, slow fermentation, and baking small batches in brick ovens that produce a dark, crusty loaf. Another reason was the home cook's rediscovery of the pleasures of baking and eating fresh warm bread. The Japanese invention of the bread machine made it possible for busy home cooks to put all the ingredients into a single chamber, close the lid, and fill the house with the forgotten aroma of fresh-baked bread.

Breads baked by home cooks and artisans account for a small fraction of the overall bread production in England and North America. But their revival demonstrates that people still enjoy the flavors and textures of freshly made traditional breads, and this fact has caught the attention of industrial producers. They have recently developed the "par-baking" system, in which manufacturers ship partly baked and frozen loaves to supermarkets, where they're baked again locally and sold while still crusty and flavorful.

Industrial breads were first "optimized"

Chemical Leavening and the First American Cookie Recipe

Cookies

One pound sugar boiled slowly in half pint water, scum well and cool, add two tea spoons pearl ash dissolved in milk, then two and a half pounds flour, rub in 4 ounces butter, and two large spoons of finely powdered coriander seed, wet with above; make rolls half an inch thick and cut to the shape you please; bake fifteen or twenty minutes in a slack oven—good three weeks.

—Amelia Simmons, *American Cookery,* 1796

to make bread-like products at minimum cost and with maximum shelf life. Finally taste and texture are entering the calculations, and at least some products are improving.

THE BASIC STRUCTURE OF DOUGHS, BATTERS, AND THEIR PRODUCTS

Wheat flour is strange and wonderful stuff! Mix pretty much any other powdery ingredient with water and we get a simple, inert paste. But mix some flour with about half its weight in water, and the combination seems to come alive. At first it forms a cohesive mass that changes its shape reluctantly. With time and kneading, reluctance gives way to liveliness, a bouncy responsiveness to pressure that persists even after the kneader lets go. It's these qualities of cohesiveness and liveliness that set wheat doughs apart from other cereal doughs, and that make possible light, delicate loaves of bread, flaky pastries, and silken pastas.

The various textures of baked goods and pastas are created by the structures of their doughs and batters. Those structures are composed of three basic elements: water, the flour's gluten proteins, and its starch granules. Together, these elements create an integrated, cohesive mass. That cohesiveness is what gives pasta its close-textured silkenness. It's also what makes bread doughs, pastry doughs, and cake batters divisible into microscopically thin but intact sheets. Breads and cakes are light and tender because the protein-starch mass is divided up by millions of tiny bubbles; pastries are flaky and tender because the protein-starch mass is interrupted by hundreds of thin layers of fat.

We call a mixture of flour and water either a *dough* or a *batter,* depending on the relative proportions of the two major ingredients. Generally, doughs contain more flour than water and are stiff enough to be manipulated by hand. All the water is bound to the gluten proteins and to the surfaces of the starch granules, which are embedded in the semisolid gluten-water matrix. Batters, on the other hand, contain more water than flour and are loose enough to pour. Much of the water is free liquid, and both gluten proteins and starch granules are dispersed in it.

The structure of a dough or batter is temporary. When it's cooked, the starch granules absorb water, swell, and create a permanent solid structure from the original, semisolid or liquid one. In the case of breads and cakes, that solid structure is a sponge-like network of starch and protein filled with millions of tiny air pockets. Bakers use the term *crumb* for this network, which constitutes the bulk of the bread or cake. The outer surface, which usually has a dryer, denser texture, is the *crust.*

With this overview in mind, let's look more closely at the structural elements of doughs and batters.

GLUTEN

Chew on a small piece of dough, and it becomes more compact but persists as a gum-like, elastic mass, the residue that the Chinese named "the muscle of flour" and that we call *gluten.* It consists mainly of protein, and includes what may well be the largest protein molecules to be found in the natural world. These remarkable molecules are what give wheat dough its liveliness and make raised breads possible.

Gluten Proteins Form Long Chains That Stick to Each Other Gluten is a complex mixture of certain wheat proteins that can't dissolve in water, but do form associations with water molecules and with each other. When the proteins are dry, they're immobile and inert. When wetted with water, they can change their shape, move relative to each other, and form and break bonds with each other.

Proteins are long, chain-like molecules built up from smaller molecules called amino acids (p. 805). Most of the gluten proteins, the gliadins and the glutenins, are around a thousand amino acids long. The gliadin chains fold onto themselves in a compact mass, and bond only weakly with each other and with the glutenin proteins. The glutenins, however, bond with each other in several ways to form an extensive, tightly knit network.

At each end of the glutenin chain are sulfur-containing amino acids that can form strong sulfur-sulfur bonds with the same amino acids at the ends of other glutenin chains. To do this they require the availability of oxidizing agents—oxygen in the air, certain substances produced by yeasts, or "dough improvers" (p. 529) added by the flour manufacturer or baker. The long, coiled middle stretch of the glutenin molecule consists mainly of amino acids that form weaker, temporary bonds (hydrogen and hydrophobic bonds) with similar amino acids. Glutenin chains thus link up with each other end-to-end to form super-chains a few hundred glutenins long, and coiled stretches along their lengths readily form many temporary bonds with similar stretches along neighboring gluten proteins. The result is an extensive interconnected network of coiled proteins, the *gluten.*

Gluten Plasticity and Elasticity The gluten of the bread wheats is both plastic and elastic; that is, it will change its shape under pressure, yet it resists the pressure and moves back toward its original shape when the pressure is removed. Thanks to this combination of properties, wheat dough can expand to incorporate the carbon dioxide gas produced by yeast, and yet resists enough that its bubble walls won't thin to the breaking point.

Gluten plasticity results from the presence of the gliadin proteins among the glutenins; because they're compact, the gliadins act something like ball bearings, allowing portions of the glutenins to slide past each other without bonding. Elasticity results from the kinked and coiled struc-

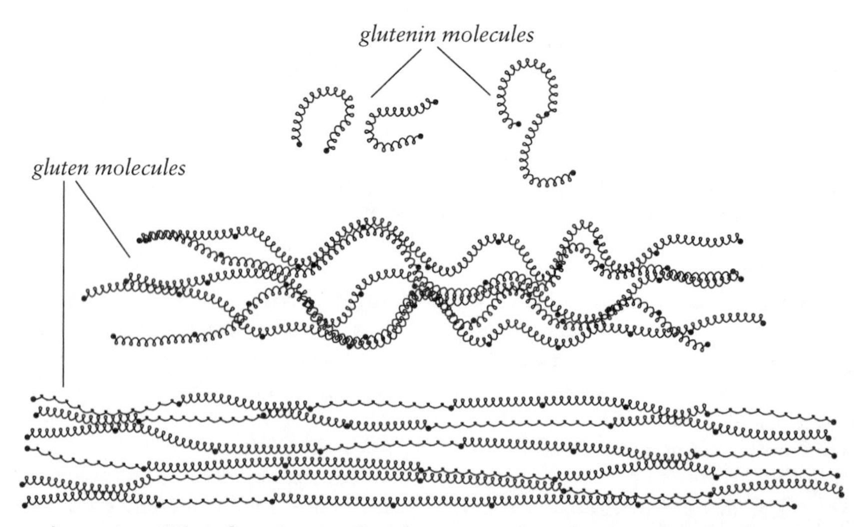

Gluten formation. When flour is mixed with water and made into a dough, glutenin *protein molecules link up end-to-end to form long, composite* gluten *molecules. Dough is elastic because the gluten molecules are coiled and have many kinks in them. When a mass of dough is stretched, the kinks are straightened out, the coils extended, and the proteins get longer* (bottom). *When the stretching tension is released, many of the kinks and coils re-form, the protein mass shortens, and the dough shrinks back toward its original shape.*

ture of the interconnected gluten proteins. Kneading unfolds and aligns the protein molecules, but there are still loops and kinks along their lengths. Stretching the dough straightens out these loops and kinks, but when the pressure is relieved, the molecules tend to revert to their original kinkiness. In addition, the coiled spring-like structure of individual proteins can extend and store some of the energy of stretching, but when the stretching is stopped, the molecules spring back to their compact coiled form. The visible result of these submicroscopic events: the stretched dough creeps back toward its original shape.

Gluten Relaxation Another important characteristic of wheat flour doughs is that their elasticity relaxes with time. An elastic dough that never relaxed could never be formed into the many shapes of raised doughs and pastas! In a well developed dough, the protein molecules have been organized and aligned, and have formed many weak bonds with each other. Because there are so many of them, these bonds hold the proteins firmly in place and resist stretching, so a ball of dough is firm and taut. But because the bonds are weak, the physical tension of the taut ball shape slowly breaks some of them, and the dough structure gradually relaxes into a flatter, more malleable mass.

Controlling Gluten Strength Not all baked goods benefit from a strong, elastic gluten. It's desirable in yeasted breads, bagels, and in puff pastry; but it gives an undesirable toughness to other forms of pastry, to raised cakes, griddle cakes, and cookies. For tender preparations, bakers intentionally limit the development of gluten.

There are a number of ingredients and techniques by which the baker controls the gluten strength and consistency of doughs and batters. They include:

- The kind of flour used. High-protein bread flours produce a strong gluten, low-protein pastry and cake flours a weak one, durum semolina (for pasta) a strong but plastic one.
- The presence in the flour of oxidizing substances—aging and improving agents—which can increase the end-to-end linking of glutenin molecules and thus dough strength (p. 529).
- The water content of the dough, which determines how concentrated the gluten proteins are, and how extensively they can bond to each other. Little water gives an incompletely developed gluten and a crum-

Food Words: *Gluten*

Though Chinese cooks discovered the useful properties of gluten long before anyone else (p. 468), it was two Italian scientists who brought it to the attention of Europe. In a posthumously published manual of 1665 on optics, the Jesuit scholar Francesco Maria Grimaldi noted that durum semolina dough for pasta contains a thick, sticky substance that dries to a hard, brittle one. He named this substance *gluten,* using the Latin word for "glue." *Gluten* in turn came from an Indo-European root *gel-,* which gave rise to a number of words meaning to form into a ball, to make a coagulated lump, to be thick or sticky: these include *cloud, globe, gluteus, clam, cling,* and *clay.* In 1745, Giambattista Beccari studied gluten more carefully and noted its similarity to substances characteristic of animals: that is, he recognized that it is what we now call a protein.

Ingredients That Contribute to the Structure of Doughs, Batters, and Their Products

Ingredient	Kind of Material	Behavior	Main Effects on Structure
Flour			
Glutenin	Protein	Forms interconnected gluten network	Makes dough elastic
Gliadin	Protein	Bonds weakly to glutenin network	Makes dough plastic
Starch	Carbohydrate	Fills gluten network, absorbs water during cooking	Tenderizes dough, sets structure during baking
Water		Allows gluten network to form; dilutes it	Small and large amounts produce tender products
Yeast, leavenings	Live cells, purified chemicals	Produce carbon dioxide gas in doughs and batters	Lighten and tenderize products
Salt	Purified mineral	Tightens gluten network	Makes dough more elastic
Fats, oils, shortenings	Lipids	Weaken gluten network	Tenderize products
Sugar	Carbohydrate	Weakens gluten network, absorbs moisture	Tenderizes products, preserves moistness
Eggs	Proteins; fats and emulsifiers (yolk only)	Proteins coagulate during cooking; fats and emulsifiers weaken gluten network; emulsifiers stabilize bubbles and starch	Supplement gluten structure with tender protein coagulum; tenderize products; slow staling
Milk; buttermilk	Proteins, fats; emulsifiers, acidity	Proteins, fats, emulsifiers, acidity weaken gluten network; emulsifiers stabilize bubbles and starch	Tenderize products; slow staling

bly texture; a lot of water gives a less concentrated gluten and a softer, moister dough and bread.

- Stirring and kneading the flour-water mixture, actions that stretch and organize the gluten proteins into an elastic network.
- Salt, which greatly strengthens the gluten network. The electrically positive sodium and negative chlorine ions cluster around the few charged portions of the glutenin proteins, prevent those charged portions from repelling each other, and so allow the proteins to come closer to each other and bond more extensively.
- Sugar, which at the concentrations typical of raised sweet breads, 10% or more of the flour weight, limits the development of gluten by diluting the flour proteins.
- Fats and oils, which weaken gluten by bonding to the hydrophobic amino acids along the protein chains and so preventing them from bonding to each other.
- Acidity in the dough—as from a sourdough culture—which weakens the gluten network by increasing the number of positively charged amino acids along the protein chains, and increasing the repulsive forces between chains.

Starch

The elastic gluten proteins are essential to the making of raised breads. But proteins account for only about 10% of flour weight, while about 70% is starch. Starch granules serve several functions in doughs and batters. Together with the water they hold on their surfaces, they make up more than half the volume of the dough, interpenetrate the gluten network and break it up, and so tenderize it. In the case of cakes, starch is the major structural material, the gluten being too dispersed in the large amount of water and sugar to contribute solidity. During the baking of bread and cakes, the starch granules absorb water, swell, and set to form the rigid bulk of the walls that surround the bubbles of carbon dioxide. At the same time their swollen rigidity stops the expansion of the bubbles and so forces the water vapor inside to pop the bubbles and escape, turning the foam of separate bubbles into a continuous spongy network of connected holes. If this didn't happen, then at the end of baking the cooling water vapor would contract and cause the bread or cake to collapse.

Gas Bubbles

Gas bubbles are what make leavened doughs and batters light and tender. Breads and cakes are aerated to the point that as much as 80% of their volume is empty space! Gas bubbles interrupt and therefore

Food Words: *Starch*

As far back as the Romans, purified starch has been incorporated into paper to give it body and smooth its surface. In the 14th century, Holland and other northern European countries began stiffening their linen cloth with wheat starch. The word *starch* dates from the 15th century, and comes from a German root that means "to stiffen, to make rigid," which is also what starch does to convert bread dough into bread. The German in turn came from an Indo-European root meaning "stiff"; related words are *stare, stark, stern,* and *starve* (which results in the rigidity of death).

weaken the network of gluten and starch granules, dividing it into millions of very thin, delicate sheets that form the bubble walls.

Bakers use yeasts or chemical leavenings to fill their products with gas bubbles (p. 531). However, these ingredients don't create new bubbles: their carbon dioxide is released into the water phase of the dough or batter, and diffuses into and enlarges whatever tiny bubbles are already there. These primordial bubbles are air bubbles, and are created when the baker first kneads a dough, or creams butter and sugar, or whips eggs. The initial aeration of doughs and batters thus strongly influences the final texture of baked goods. The more bubbles produced during the preparation of a dough or batter, the finer and tenderer the result.

Fats: *Shortening*

Since the early 19th century, the term *shortening* has been used to mean fats or oils that "shorten" a dough, or weaken its structure and thus make the final product more tender or flaky. This role is most evident in pie crusts and puff pastry (p. 561), where layers of solid fat separate thin layers of dough from each other so that they cook into separate layers of pastry. It's less evident but also important in cakes and enriched breads, where fat and oil molecules bond to parts of the gluten protein coils and prevent the proteins from forming a strong gluten. To make a rich bread with a strong gluten (e.g. Italian panettone, p. 546), the baker mixes the flour and water alone, kneads the mix to develop the gluten, and only then works in the fat.

Fats and related substances also play an important but indirect role in the formation of the cooked structure of breads and cakes, where the addition of small quantities significantly increases volume and textural lightness (p. 530).

Dough and Batter Ingredients: Wheat Flours

Though other grains and seeds can also be used, most familiar baked goods and pastas are made from wheat.

Kinds of Wheat

Several kinds of wheat are grown today, each with its own characteristics and uses (see box, p. 527). Most are species of bread wheat, *Triticum aestivum*. Their most important distinguishing characteristic is the content and quality of gluten proteins, with high protein content and strong

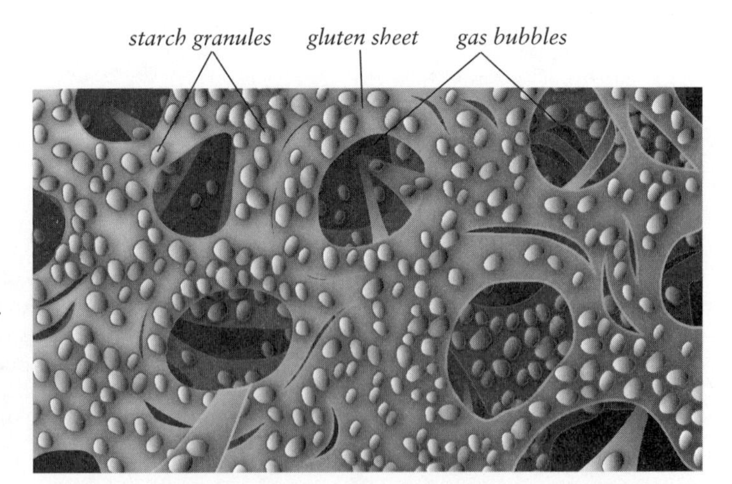

A close-up view of bread dough. The dense mass of gluten and starch is divided and tenderized by gas bubbles.

Doughs and Batters: Representative Compositions

The numbers shown indicate the relative weights of ingredients in doughs and batters, with the weight of flour constant at 100. This chart is meant to give only a general idea of the proportions used in common baked foods; individual recipes vary widely.

	Flour	Total Water	Fat or Oil	Milk Solids	Eggs	Sugar	Salt
			Doughs				
Bread	100	65	3	3	0	5	2
Biscuit	100	70	15	6	0	1	2
Pastry	100	30	65	0	0	1	1
Cookie	100	20	40	3	6	45	1
Pasta	100	25	0	0	5	0	1
Brioche	100	60	45	2	75	3	1
Panettone	100	40	27	1	15*	28	1
			Batters				
Pancake, waffle	100	150–200	20	10	60	10	2
Crêpe, popover	100	230	0	15	60	0	2
Choux	100	200	100	-	130	-	2
Sponge cake	100	75	0	0	100	100	1
Pound cake	100	80	50	4	50	100	2
Layer cake	100	130	40	7	50	130	3
Chiffon cake	100	150	40	0	140	130	2
Angel cake	100	220	0	0	250**	45	3

*yolks only
**whites only

Major Wheat Types

	Protein Content, % by Weight	Use
Hard red spring wheat	13–16.5	Bread flours
Hard red winter wheat	10–13.5	All-purpose flours
Soft red wheat	9–11	All-purpose, pastry flours
Hard white wheat	10–12	Specialty whole-grain flours
Soft white wheat	10–11	Specialty whole-grain flours
Club wheat	8–9	Cake flours
Durum wheat	12–16	Semolina for dried pasta

gluten often coinciding with a hard, glassy, translucent grain interior. *Hard* wheat grains constitute about 75% of the American crop. *Soft* wheats, which make up 20% of the crop, have a lower amount of somewhat weaker gluten proteins. Club wheat is a distinct species, *T. compactum,* whose proteins form an especially weak gluten. Durum wheat is another distinct species (*T. turgidum durum,* p. 465) used mainly to make pasta (p. 571).

In addition to their classification by protein content, North American wheats are named by their growth habit and kernel color. Spring wheats (including durum) are sown in the spring and harvested in the fall, while winter wheats are sown in late fall, live through the winter as a seedling, and are harvested in the summer. The most common wheat varieties are *red,* their seed coat reddish brown with phenolic compounds. *White* wheats, with a much lower phenolic content and a light tan seed coat, are becoming increasingly popular for the light color and less astringent "sweet" taste of their whole-wheat flours and products containing part or all of the bran.

Turning Wheat into Flour

The baking qualities of a particular flour are determined by the wheat from which it's made, and how that wheat is turned into flour.

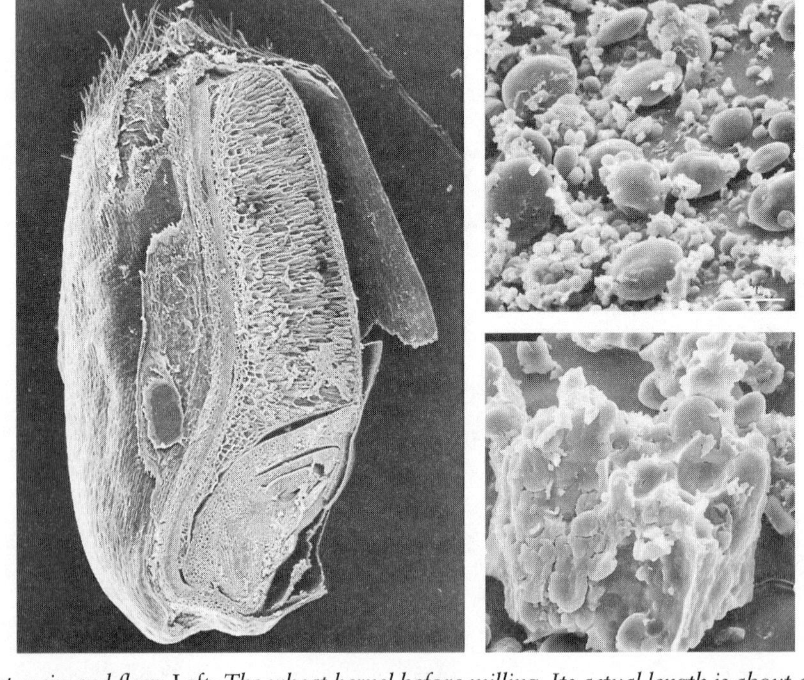

Wheat grain and flour. Left: *The wheat kernel before milling. Its actual length is about a quarter of an inch/6 mm.* Upper right: *Soft wheat flour. The protein in this kind of wheat comes in thin, weak sections interrupted by starch granules and air pockets. When milled, it produces small, fine particles. Soft flour makes a weak gluten and is preferred for tender pastries and cakes.* Lower right: *Hard wheat flour. The protein matrix in hard wheat endosperm is strong enough to break off in chunks during milling. Hard flours make strong gluten and are preferred for most bread making.*

Milling: Conventional and Stone Grinding Milling is the process of breaking the wheat kernel down into small particles, and sifting the particles to make a flour of the desired qualities. Most flours are *refined*: that is, they have been sieved to remove the germ and bran layers from the particles of protein- and starch-rich endosperm. Bran and germ are rich in nutrients and flavor, but they go rancid in a few weeks, and interfere physically and chemically with the formation of a continuous, strong gluten; so whole-grain flours make denser, darker breads and pastries. In conventional milling, grooved metal rollers shear open the grain, squeeze out the germ, and scrape the endosperm away to be ground, sieved, and reground until the particles reach the desired size. Stone grinding, which is much rarer, crushes the whole grain more thoroughly before sieving, so that some of the germ and bran end up in even the refined flours. Stone-ground flour is therefore more flavorful than conventionally milled flour, but also has a shorter shelf-life.

Improving and Bleaching Bakers have known for a long time that freshly milled flour makes a weak gluten, a slack dough, and a dense loaf. As the flour ages for a few weeks in contact with the air, its gluten and baking properties improve. We understand now that oxygen in the air gradually frees the glutenin proteins' end sulfur groups to react with each other and form ever longer gluten chains that give the dough greater elasticity. Beginning around 1900, millers began to save time, space, and money by supplementing freshly milled flour with oxidizing chlorine gas and then with potassium bromate. However, worries about the potential toxicity of bromate residues in the late 1980s led most millers to replace bromate with ascorbic acid (vitamin C) or azodicarbonamide. (Ascorbic acid itself is an antioxidant, but becomes oxidized to dehydroascorbic acid, which in turn oxidizes the gluten proteins.) In Europe, fava bean flour and soy flour have been used as flour improvers; their active fat-oxidizing enzymes, which generate the typical beany flavor, also indirectly cause the oxidation and elongation of gluten proteins.

The traditional air-aging of flour had a visible side effect: the yellowish flour becomes progressively paler as the xanthophyll pigments are oxidized to a colorless form. Once the chemistry of this change was understood, millers began using bleaching agents (azodicarbonamide, peroxide) to whiten flours. Many bakers prefer to use unbleached flours because they have been subjected to less chemical alteration. Bleaching is not allowed in Europe.

Minor Flour Components

The gluten proteins and starch granules in flour account for about 90% of flour weight, and for much of the behavior of

Extraction Rates

The degree to which a flour has been refined is designated by the "extraction rate," or the percentage of the whole grain remaining in the finished flour. Whole wheat flour has an extraction rate of 90%. Most commercial white flours contain between 70 and 72% of the whole grain; French bread flour ranges from 72 to 78%, and so carries more flavor of the whole grain. Home bakers can make their own higher-extraction refined flours by supplementing commercial white flour with a portion of whole wheat flour that they have sifted to remove coarse bran and germ particles.

flour doughs and batters. But some minor components have an important influence.

Fats and Related Molecules Although white flour is only about 1% fats, fat fragments, and phospholipids by weight, these substances are essential to the development of a well-raised bread. There's evidence that some fatty materials can help stabilize the dough bubble walls as they expand and prevent premature rupture and collapse. Others become attached to starch granules and help soften the bread structure and slow staling. Similar ingredients added by the cook or manufacturer can magnify these useful effects (p. 524).

Enzymes Since the flour's normal endowment of sugars is enough to feed yeast cells for only a short period of time, flour manufacturers have long supplemented the ground wheat with malted wheat or barley: grains that have been allowed to sprout and develop the enzymes that break down starch to sugars. Because malt flours give a dark cast to flours and doughs, and because their activity is somewhat variable, manufacturers are increasingly replacing them with enzymes extracted and purified from microscopic molds ("fungal amylase").

The Protein Contents of Common Wheat Flours

These figures are approximate, and assume that the flour contains 12% moisture. The bulk of flour weight, from 70 to 80%, is starch and other seed carbohydrates. High-protein flours absorb substantially more water than low-protein flours, and so will produce stiffer doughs with the same proportion of water.

Flour	Protein Content
Whole wheat, graham	11–15
Durum semolina	13
Bread	12–13
All-purpose (U.S. national brands)	11–12
All-purpose (U.S. regional brands, South, Pacific Northwest)	7.5–9.5
Pastry	8–9
Cake	7–8
0 or 00 (Italian soft wheat)	11–12
Type 55 (French blend of soft and hard wheat)	9–10
English plain	7–10
Vital gluten	70–85

Because different flours have not only different protein contents, but different protein qualities, it's not really possible to turn all-purpose flour into pastry flour or vice versa. However, it's possible to dilute the gluten proteins of a given flour by the addition of cornstarch or another pure starch, or strengthen them by adding powdered vital gluten. To approximate pastry flour with all-purpose, add one part by weight of starch to two parts of all-purpose flour; to approximate all-purpose flour with pastry flour, add one-quarter part of gluten to two parts of pastry flour. (Purified gluten loses a little less than half its strength in the drying process.) With its chlorine-altered starch and fats, cake flour is inimitable.

Kinds of Flour

Though manufacturers and professional bakers can obtain flours from particular wheats, most flours for sale in supermarkets are labeled according to their intended use, with no direct indication of the kind of wheat or wheats they contain—they're usually a blend—or their protein content or quality. Flour compositions can vary significantly from region to region; "all-purpose" flour in much of the United States and Canada has a higher protein content than "all-purpose" flours in the South or Pacific Northwest. Not surprisingly, recipes developed with a particular flour often turn out very differently when made with another, unless care is taken to find a replacement that closely approximates the original. The box on p. 530 lists the compositions of common wheat flours.

Whole wheat flours are high in protein, but a significant fraction of that protein comes from the germ and aleurone layer and does not form gluten; and germ and bran particles interfere with gluten formation. They therefore tend to make dense but flavorful breads. Bread flours are high in strong gluten proteins, and give the lightest, highest, and chewiest loaf breads. Both pastry and cake flours have low levels of weak gluten protein for making tender baked goods. Cake flour is distinctive because it's treated with chlorine dioxide or chlorine gas. This treatment has several effects on the starch granules that are useful in cake making (p. 555), and leaves a trace of hydrochloric acid in the flour, which gives batters and doughs an acid pH and slightly acid taste.

"Self-rising" flours are flours that contain baking powder (1½ teaspoons baking powder per cup flour/5–7 gm per 100 g), and therefore don't require added leavening for the making of quickbreads, pancakes, and other chemically raised foods. "Instant" or "instantized" flours (two brand names are Shake & Blend and Wondra) are low-protein flours whose starch granules have been precooked until they gelate, then dried again. The precooking and drying make it easier for water to penetrate them again during cooking. Instant flours are well suited to tender pastries and last-minute thickening of sauces and gravies.

DOUGH AND BATTER INGREDIENTS: YEASTS AND CHEMICAL LEAVENINGS

Leavenings are the ingredients that fill doughs and batters with bubbles of gas, thus reducing the amount of solid material in a given volume and making the bread or cake less dense, more light and tender.

Yeasts

Humans have been eating raised breads for 6,000 years, but it wasn't until the investigations of Louis Pasteur 150 years ago that we began to understand the nature of the leavening process. The key is the gas-producing metabolism of a particular class of fungus, the yeasts. The word "yeast," how-

> Food Words: *Leavening* and *Yeast*
>
> *Leavening* comes from an Indo-European root meaning "light, having little weight." Related words from the same root include *levity, lever, relieve*, and *lung*. *Yeast* comes from a root word that meant "to seethe, boil, bubble over." This derivation underlines the way in which fermentation seemed to be a kind of cooking of the cereal gruel, a transformation from within.

ever, is as old as the language, and first meant the froth or sediment of a fermenting liquid that could be used to leaven bread.

The yeasts are a group of microscopic single-celled fungi, relatives of the mushrooms. More than 100 different species are known. Some cause human infections, some contribute to food spoilage, but one species in particular—*Saccharomyces cerevisiae,* whose name means "brewer's sugar fungus"—is put to good use in both brewing and baking. For much of human history, yeast was simply recruited from the grain surface or supplied by an earlier piece of dough, or obtained from the surface of beer brewing vats. Today strains especially selected for breadmaking are grown on molasses in industrial fermentation tanks.

Yeast Metabolism Yeasts metabolize sugars for energy, and produce carbon dioxide gas and alcohol as by-products of that metabolism. The overall equation for the conversion that takes place in yeast cells is this:

$$C_6H_{12}O_6 \rightarrow 2C_2H_5OH + 2CO_2$$

(1 molecule of glucose sugar yields 2 molecules of alcohol plus 2 molecules of carbon dioxide)

In making beer and wine, the carbon dioxide escapes from the fermenting liquid, and alcohol accumulates. In making bread, both carbon dioxide and alcohol are trapped by the dough, and both are expelled from the dough by the heat of baking.

In an unsweetened dough, yeasts grow on the single-unit sugars glucose and fructose and on the double-glucose sugar maltose, which enzymes in the flour produce from broken starch granules. A small amount of added table sugar in a dough will increase yeast activity, while a large amount decreases it (see sweet breads, p. 546), as does added salt. Yeast activity is also strongly affected by temperature: the cells grow and produce gas most rapidly at about 95°F/35°C.

In addition to providing carbon dioxide gas to inflate the dough, yeasts release a number of chemicals that affect the dough consistency. The overall effect is to strengthen the gluten and improve its elasticity.

Forms of Baker's Yeast Commercial yeast is sold to home and restaurant cooks in three common forms, each a different genetic strain of *S. cerevisiae* with different traits.

- Cake or compressed yeast is a moist block of fresh yeast cells, direct from the fermentation vat. Its cells are alive, and produce more leavening gas than the other forms. Cake yeast is perishable, must be kept refrigerated, and has a brief shelf life of one to two weeks.
- Active dry yeast, which was introduced in the 1920s, has been removed from the fermentation tank and dried into granules with a protective coat-

An Unusual Chemical Leavening: Hartshorn

The leavening that doesn't involve an acid-base reaction is ammonium salts—ammonium carbonate and/or carbamate—which were once known as "hartshorn" because they were produced by the distillation of deer antlers. (Hartshorn was also a common source of gelatin.) When these compounds are heated to 140°F/60°C, they decompose into two leavening gases, carbon dioxide and ammonia, and don't produce water. They're especially suited to thin, very dry cookies and crackers with a large surface area to release the pungent ammonia during baking.

ing of yeast debris. The yeast cells are dormant and can be stored at room temperature for months. The cook reactivates them by soaking them in warm water, 105–110°F/ 41–43°C, before mixing the dough. At cooler soaking temperatures, the yeast cells recover poorly and release substances that interfere with gluten formation (glutathione).

- Instant dry yeast, an innovation of the 1970s, is dried more quickly than active dry yeast, and in the form of small porous rods that take up water more rapidly than granules. Instant yeast doesn't need to be prehydrated before mixing with other dough ingredients, and produces carbon dioxide more vigorously than active dry yeast.

Baking Powders and Other Chemical Leaveners

Yeast cells produce carbon dioxide slowly, over the course of an hour or more, so the material surrounding them must be elastic enough to contain it for that much time. Weak doughs and runny batters can't hold gas bubbles for more than a few minutes, and are therefore usually raised with a faster-acting gas source. This is the role played by chemical leavenings. These ingredients are concentrated, and small differences in the amount added can cause large variations in the quality of the finished food. Too little leavening will leave it flat and dense, while too much will cause the batter to overexpand and collapse into a coarse structure with a harsh flavor.

Nearly all chemical leavenings exploit a reaction between certain acidic and alkaline compounds that results in the production of carbon dioxide, the same gas produced by yeast. The first chemical leavening was a dried water extract of wood ash—potash, mainly potassium carbonate—which reacts with the lactic acid in a soured dough as follows:

$$2(C_3H_6O_3) + K_2CO_3 \rightarrow 2(KC_3H_5O_3) + H_2O + CO_2$$

(2 molecules of lactic acid plus 1 of potassium carbonate yield 2 molecules of potassium lactate, plus a molecule of water, plus a molecule of carbon dioxide)

The Acid Components of Baking Powders

Some of these acids are available only to manufacturers. Most double-acting supermarket baking powders are a mixture of sodium bicarbonate, MCP, and SAS. Single-acting powders omit the SAS, and the MCP is coated to delay its release artificially.

Leavening Acid	Time of Reaction
Cream of tartar, tartaric acid	Immediately, during mixing
Monocalcium phosphate (MCP)	Immediately, during mixing
Sodium aluminum pyrophosphate (SAPP)	Slow release after mixing
Sodium aluminum sulfate (SAS)	Slow release and heat-activated
Sodium aluminum phosphate (SALP)	Heat-activated, early in cooking (100–104°F/38–40°C)
Dimagnesium phosphate (DMP)	Heat-activated, early in cooking (104–111°F/40–44°C)
Dicalcium phosphate dihydrate (DCPD)	Heat-activated, late in cooking (135–140°F/57–60°C)

Baking Soda The most common alkaline component of chemical leavenings is sodium bicarbonate (or sodium acid carbonate, $NaHCO_3$), usually called baking soda.

Baking soda can be the sole added leavening if the dough or batter contains acids to react with it. Common acid ingredients include sourdough cultures, fermented milks (buttermilk, yogurt), brown sugar and molasses, chocolate, and cocoa (if not dutched, p. 705), as well as fruit juices and vinegar. A general rule of thumb: ½ teaspoon/2 gm baking soda is neutralized by 1 cup/240 ml of fermented milk, or 1 teaspoon/5ml of lemon juice or vinegar, or 1¼ teaspoons/5 gm cream of tartar.

Baking Powders Baking powders are complete leavening systems: they contain both alkaline baking soda and an acid in the form of solid crystals. (The active ingredients are mixed with ground dry starch, which prevents premature reactions in humid air by absorbing moisture, and gives the powder more bulk so that it's easier to measure.) When added to liquid ingredients, the baking soda dissolves almost immediately. If the acid is very soluble, it too will dissolve quickly during mixing and react with the soda to inflate an initial set of gas bubbles. Cream of tartar, for example, releases two-thirds of its leavening potential during two minutes of mixing. If the acid is *not* very soluble, then it will remain in crystal form for a characteristic length of time, or until cooking raises the temperature high enough to dissolve it—and then it reacts with the soda to produce a delayed burst of gas. There are several different acids used in baking powders, each with a different pattern of gas production (see box, p. 533).

Most supermarket baking powders are "double-acting"; that is, they inflate an initial set of gas bubbles upon mixing the powder into the batter, and then a second set during the baking process. Baking powders for restaurant and manufacturing production contain slow-release acids so that leavening power doesn't dissipate while the batter sits before being cooked.

Chemical leavenings can have adverse effects on both flavor and color. Some leavening acids have a distinctly astringent taste (sulfates, pyrophosphates). When acid and base are properly matched, neither is left behind in excess. But when too much soda is added, or when the batter is poorly mixed and not all the powder dissolves, a bitter, soapy, or "chemical" flavor results. Colors are also affected in even slightly alkaline conditions: browning reactions are enhanced, chocolate turns reddish, and blueberries turn green.

BREADS

There are four basic steps in the making of yeast bread. We mix together the flour, water, yeast, and salt; we knead the mixture to develop the gluten network; we give the yeast time to produce carbon dioxide and fill the dough with gas cells; and we bake the dough to set its structure and generate flavor. In practice, each step involves choices that affect the qualities of the finished loaf. There are many ways to make basic bread! The following paragraphs explain some of the more significant choices and their effects. Breads made with special ingredients or methods—sourdoughs, sweet breads, flatbreads—are described later.

The Choice of Ingredients

Bread making begins with the ingredients, especially the flour and the yeasts. Because proportions are important, and the weight of a given volume of flour can vary by as much as 50% depending on whether it has been fluffed up (sifted) or packed down, it's best to weigh ingredients rather than measure them in cups.

Flour The texture and flavor of bread are strongly influenced by the kind of flour used. "Bread flours" are milled from high-protein wheats, require a long kneading period to develop their strong gluten, and produce well-raised loaves with a distinctive, slightly eggy flavor and chewy texture. Lower-protein "all-purpose" flours give breads with a lower maximum volume, more neutral flavor, and less chewy texture, while flours from soft wheats with weak gluten proteins make denser loaves with a tender, cake-like crumb. The more of the outer aleurone, bran, and germ that makes it into the flour, the darker and denser the bread and the stronger the whole-grain flavor. The baker can mix different flours to obtain a particular character. Many artisan breadmakers prefer flours with a moderate protein content, 11–12%, and an extraction rate somewhere between standard white and whole wheat flours.

Water The chemical composition of the water used to make the dough influences the dough's qualities. Distinctly acid water weakens the gluten network, while a somewhat alkaline water strengthens it. Hard water will produce a firmer dough thanks to the cross-linking effects of calcium and magnesium. The proportion of water also influences dough consistency. The standard proportion for a firm bread dough capable of good aeration is 65 parts water to 100 parts all-purpose flour by weight (40% of the combined weight). Less water will produce a firmer, less extensible dough and a denser loaf, while more water produces a soft, less elastic dough and an open-textured loaf. Wet doughs that are barely kneadable—for example the Italian *ciabatta*—may be 80 parts water or more per 100 flour (45%). High-protein flours absorb as much as a third more water than all-purpose flours, so water proportions and corresponding textures also depend on the nature of the flour used.

Salt Though some traditional breads are made without salt, most include it, and not just for a balanced taste. At 1.5–2% of the flour weight, salt tightens the gluten network and improves the volume of the finished loaf. (The tightening is especially evident in the autolyse mixing method, below.) Unrefined sea salts that contain calcium and magnesium impurities may produce the additional gluten strengthening that mineral-rich hard water does. In sourdoughs, salt also helps limit the protein-digesting activity of the souring bacteria, which can otherwise damage the gluten.

Yeast The baker can incorporate yeast in very different forms and proportions. For a simple dough to be fully leavened and baked in a few hours, the standard proportion for cake yeast is 0.5–4% of the flour weight, or 2.5–20 gm per pound/500 gm flour; for dried yeast, approximately half these numbers. If the dough is to be fermented slowly overnight, only 0.25% of flour weight, barely a gram per pound/500 gm is needed. (One gram still contains millions of yeast cells.) As a general rule, the less prepared yeast goes into the dough, and the longer dough is allowed to rise, the better the flavor of the finished bread. This is because the concentrated yeast has its own somewhat harsh flavor, and because

Durum Breads

Durum wheat flour forms an inelastic dough that doesn't rise well, but has nevertheless been used to make dense, distinctively flavored, golden breads in the Mediterranean region for thousands of years. Durum flour absorbs nearly 50% more water than bread flour does, a fact that is part of the reason for the longer shelf life of durum bread.

the process of fermentation generates a variety of desirable flavor compounds (p. 543).

Starters A general method for incorporating yeast into bread dough that maximizes the effective fermentation time and flavor production is the use of *pre-ferments* or *starters,* portions of already fermenting dough or batter that are added to the new mass of flour and water. The starter may be a piece of dough saved from the previous batch, or a stiff dough or runny batter made up with a small amount of fresh yeast and allowed to ferment for some hours, or a culture of "wild" yeasts and bacteria obtained without any commercial yeast at all. This last is called a "sourdough" starter because it includes large numbers of acid-forming bacteria. Starters go by many names—French *poolish,* Italian *biga,* Belgian *desem,* English *sponge*—and develop different qualities that depend on ingredient proportions, fermentation times and temperatures, and other details of their making. Sourdough breads are described on p. 544.

Preparing the Dough: Mixing and Kneading

Mixing The first step in making bread is to mix the ingredients together. The moment flour meets water, several processes begin. Broken starch granules absorb water, and enzymes digest their exposed starch into sugars. The yeast cells feed on the sugars, producing carbon dioxide and alcohol. The glutenin proteins absorb some water and sprawl out into their elongated coils; the coils of neighboring molecules form many weak bonds with each other and thus form the first strands of gluten. We see the dough take on a vaguely fibrous appearance, and feel it cohere to itself. When it's stirred with a spoon, the protein aggregates are drawn together into visible filaments and form what has been vividly described as a "shaggy mass." At the same time, a number of substances in the flour cause breaks in and blocking of the end-to-end bonds of the gluten molecules, and so an initial shortening of the gluten chains. As oxygen from the air and oxidizing compounds from the yeasts enter the dough, the breaking and blocking stop, and the gluten molecules begin to bond end-to-end and form long chains.

Mixing can be done by hand, in a stand mixer, or in a food processor. The processor works in less than a minute, a fraction of the time required for hand or mixer kneading, and therefore offers the advantage of minimizing exposure to air and oxygen, an excess of which bleaches the remaining wheat pigments and alters flavor. The high

Two-Stage Mixing: Autolysis

An alternative to mixing all the dough ingredients at once is the *autolyse* or "autolysis" method championed by a legendary French bread authority, Raymond Calvel, to compensate for some of the disadvantages of rapid industrial production. It has also been adopted by many artisan bakers. Autolysis involves combining only the flour and water and letting them sit for 15–30 minutes before adding the leavening and salt. According to Calvel, this initial preparation gives the starch and the gluten proteins a chance to absorb as much water as possible without the interference of salt, and allows the gluten chains to shorten more (*autolysis* means "self-digestion"). The result is a dough that's easier to manipulate, requires less kneading and therefore less exposure to oxygen, and so better retains the wheat's light golden color and characteristic taste.

energy input heats the dough, which should be allowed to cool before fermentation.

Dough Development: Kneading Once the ingredients have been mixed and the dough is formed, the process of dough development begins. Whether the dough is kneaded by hand or in an electrical mixer, it undergoes a similar kind of physical manipulation: it is stretched, folded over, com-

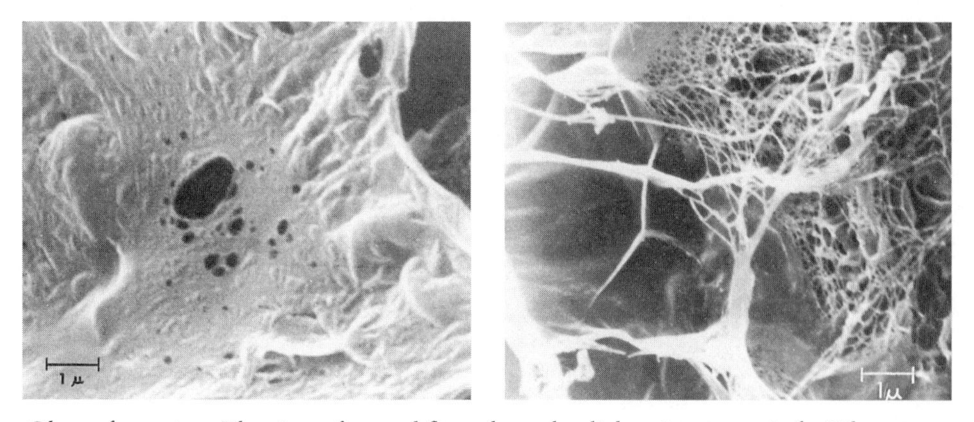

Gluten formation. The view of wetted flour through a light microscope. Left: *When water is first added to flour, the gluten proteins are randomly oriented in a thick fluid.* Right: *As this fluid is stirred, it quickly develops into a tangle of fibers as the glutenin proteins form elongated bundles of molecules.*

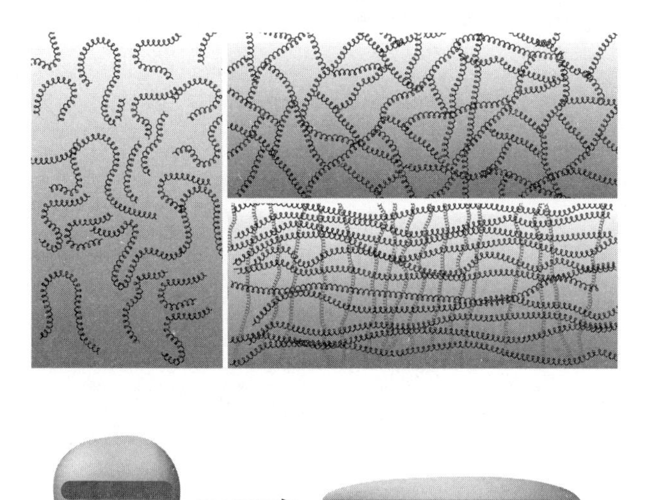

Gluten orientation. When flour is initially mixed with water, the glutenin molecules form a random network of gluten chains. Kneading helps orient the gluten chains in orderly arrays.

Kneading dough. Kneading repeatedly stretches and elongates the gluten, helping to orient the long chains and encourage the side-by-side bonding that contributes to gluten strength.

pressed, stretched, folded, and compressed many times. This manipulation strengthens the gluten network. It unfolds the proteins further, orients them side by side and encourages the development of many weak bonds between neighbors. The glutenin molecules also form strong end-to-end bonds with each other and thus a cohering network of extensive gluten chains. The dough gradually gets stiff, harder to manipulate, and takes on a fine, satiny appearance. (If the dough is worked so hard that many end-to-end bonds start breaking, its overall structure breaks down, and the dough becomes sticky and inelastic. Overdevelopment is a real problem only when kneading is done mechanically.)

Kneading also aerates the dough. As it's repeatedly folded over and compressed, pockets of air are trapped and squeezed under pressure into smaller, more numerous pockets. The more pockets formed during kneading, the finer the texture of the final bread. Most of the air pockets are incorporated as the dough reaches its maximum stiffness.

Some bread recipes call for a bare minimum of kneading. This generally results in fewer and larger air cells, and so a coarse, irregular texture that has its own appeal. The gluten of such doughs is less developed as they begin fermentation, but the rising of the dough continues to develop gluten structure (below), so little-kneaded doughs can eventually rise well to give an airy, tender crumb.

Fermentation, or Rising

Fermentation is the stage during which the dough is set aside for the yeast cells to produce carbon dioxide, which diffuses into the air pockets, slowly inflates them, and thus raises the dough. This gentle stretching action continues the process of gluten orientation and development, as does the oxidizing effect of other yeast by-products, which continue to help the glutenin molecules to link up end-to-end. As a result, even initially wet, barely cohesive doughs become more manageable after fermentation.

Yeasts produce carbon dioxide most rapidly at around 95ºF/35ºC, but they also produce more noticeable quantities of sour and unpleasant-smelling by-products. A fermentation temperature of 80ºF/27ºC is often suggested for a relatively quick rising time of a couple of hours. Lower temperatures may extend fermentation times by an hour or more, and with them the generation of desirable yeast flavors.

The end of the fermentation period is signaled by the dough's volume—it approximately doubles—and by the condition of the gluten matrix. When poked with the finger, fully fermented dough will retain the impression and won't spring back: the gluten has been stretched to the limit of its elasticity. The dough is now gently handled to reconsolidate the gluten, divide the gas pockets, redistribute the yeast cells and their food supply, and even out the temperature and moisture (fermentation generates heat, water, and alcohol). Thanks to the added moisture and to the gluten-interrupting bubbles, fermented doughs feel softer and easier to work than newly kneaded dough.

Food Words: *Knead*

The word *knead* comes from an Indo-European root meaning "to compress into a ball"; related words are *gnocchi, quenelle, knoll,* and *knuckle.*

Doughs made from high-protein flours may be put through a second rising to develop their tougher gluten fully. Either way, the fermented dough is then divided, gently rounded into balls, rested for a few minutes to allow the gluten to relax somewhat, and then molded into loaves. The loaves are then allowed another partial rise, or "proof," to prepare them for the final and dramatic rise during baking.

Retarding the Fermentation Traditional breadmaking can last many hours, and bakers would often have to work through the night in order to sell fresh bread in the morning. In the 1920s, bakers in Vienna began to experiment with breaking the work into two periods, a daytime stint for mixing, fermentation, and molding into loaves, and then an early-morning baking. During the night, the formed loaves were kept in a refrigerated chamber. Cool temperatures slow the activity of microbes substantially; yeasts take 10 times longer to raise bread in the refrigerator than at warm room temperature. Refrigeration of dough is therefore called *retarding*, and the cold chamber a *retarder*. Retarding is now a common practice.

In addition to giving the baker greater flexibility, retarding has useful effects on the dough. Long, slow fermentation allows both yeasts and bacteria more time to generate flavor compounds. Cold dough is stiffer than warm dough, so it's easier to handle without causing a loss of leavening gas. And the cycle of cooling and rewarming redistributes the dough gases (from small bubbles into the water phase, then back out into larger bubbles), and encourages the development of a more open, irregular crumb structure.

Baking

Ovens, Baking Temperatures, and Steam The kind of oven in which bread is baked has an important influence on the qualities of the finished loaf.

Traditional Bread Ovens Until the middle of the 19th century, bread was baked in clay, stone, or brick ovens that were preheated by a wood fire, and that could store a large amount of heat energy. The baker started the fire on the floor of the oven, let it burn for hours, cleaned out the ashes, and then introduced the loaves of dough and closed the oven door. The oven surfaces start out at 700–900°F/350–450°C, the domed roof radiating its stored heat from above, the floor conducting heat directly into the loaves from beneath. As the dough heats it releases steam, which fills the closed chamber and further speeds the transfer of heat to the loaves. Slowly the oven surfaces lose their heat, and the temperature declines during the bake, at the same time that the loaf is browning and therefore becoming more efficient at absorbing heat. The result is a rapid initial heating that encourages the dough to expand, and temperatures high enough to dry the crust well and generate the color and flavors of the browning reactions (p. 778).

Modern Metal Ovens The modern metal oven is certainly easier to bake in than the wood-fired oven, but it isn't as ideally suited to breadmaking. It usually has a maximum cooking temperature of 500°F/250°C. And its thin walls are incapable of storing much heat, so its temperature is maintained by means of gas flames or electrical elements that get red-hot. When these heat sources switch on during baking, the effective temperature temporarily rises well above the target baking temperature, and the bread can be scorched. Because they are vented to allow the escape of combustion gases (carbon dioxide and water), gas ovens don't retain the loaves' steam well during the important early stage. Electric ovens do a better job. Some of the advantages of the traditional stored-heat oven can be obtained from the use of ceramic baking stones or wraparound ceramic oven inserts, which are preheated to the oven's maximum temperature and provide more intensive and even heat during baking.

Steam Steam does several useful things during the first few minutes of baking. It greatly increases the rate of heat transfer from oven to dough. Without steam, the dough surface reaches 195°F/90°C in 4 minutes; with steam, in 1 minute. Steam thus causes a rapid expansion of the gas cells. As the steam condenses onto the dough surface, it forms a film of water that temporarily prevents the loaf surface from drying out into a crust, thus keeping it flexible and elastic so that it doesn't hinder the initial rapid expansion of the loaf, the "oven spring." The overall result is a larger, lighter loaf. In addition, the hot water film gelates starch at the loaf surface into a thin, transparent coating that later dries into an attractively glossy crust.

Professional bakers often inject steam under low pressure into the oven for the first several minutes of baking. In home ovens, spraying water or throwing ice cubes into the hot chamber can produce enough steam to improve the oven spring and crust gloss.

Early Baking: Oven Spring When the bread first enters the oven, heat moves into the bottom of the dough from the oven floor or pan, and into the top from the oven ceiling and the hot air. If steam is present, it provides an initial blast of heat by condensing onto the cold dough surface. Heat then moves from the surface through the dough by two means: slow conduction through the viscous gluten-starch matrix, and much more rapid steam movement through the network of gas bubbles. The better leavened the dough, the faster steam can move through it, and so the faster the loaf cooks.

As the dough heats up, it becomes more fluid, its gas cells expand, and the dough rises. The main cause of this oven spring is the vaporization of alcohol and water into gases that fill the gas cells, and that expand the dough by as much as half its initial volume. Oven spring is usually over after 6–8 minutes of baking.

Mid-Baking: From Foam to Sponge Oven spring stops when the crust becomes firm and stiff enough to resist it, and when the interior of the loaf reaches 155–180°F/68–80°C, the temperature range in which the gluten proteins form strong cross-links with each other and the starch granules absorb water, swell, gelate, and amylose molecules leak out of the granules. Now the walls of the gas cells can no longer stretch to accommodate the rising pressure inside, so the pressure builds and eventually ruptures the walls, turning the structure of the loaf from a closed network of separate gas cells into an open network

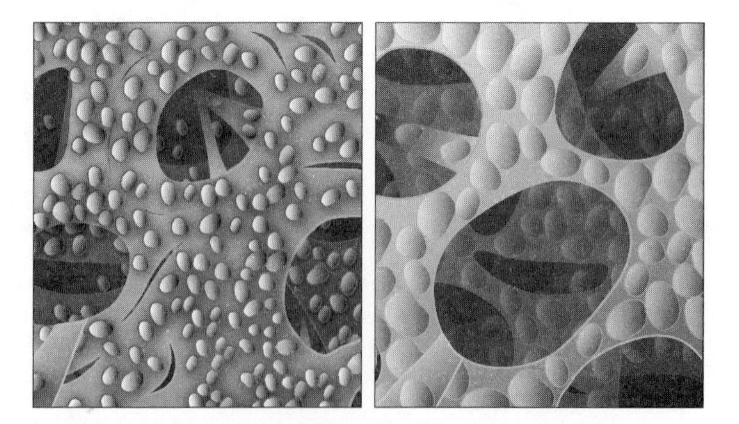

Bread dough before and after baking. As the dough heats up, starch granules absorb moisture from the gluten, swell, and leak some starch molecules, creating reinforcement for the dough walls that surround the gas pockets.

of communicating pores: from an aggregation of little balloons into a sponge through which gases can easily pass. (If the dough were not transformed into a sponge, then cooling would cause each isolated gas cell to shrink, and the loaf would collapse.)

Late Baking: Flavor Development and Cooking Through Baking is continued for some time after the bread center approaches the boiling point. This gelates the starch as thoroughly as possible, thus preventing the center from remaining damp and heavy, and slowing subsequent staling. Continued baking also encourages the surface browning reactions that improve both color and flavor. Though limited to the hot, dry crust, these reactions affect the flavor of the whole loaf because their products diffuse inward. A light-colored loaf will be noticeably less flavorful than a dark one.

Bread is judged to be done when its crust has browned and its inner structure has become fully set. The second condition can be verified indirectly by tapping on the bottom of the loaf. If the interior still contains a continuous gluten mass with embedded bubbles, it will sound and feel heavy and dense. If it has cooked through and become an open sponge, the loaf will sound hollow.

Cooling

Immediately after being removed from the oven, the loaf's outer layer is very dry, around 15% water, and close to 400°F/200°C, while the interior is as moist as the original dough, around 40% water, and around 200°F/93°C. During cooling, these differences partly even themselves out. Moisture diffuses outward, and much of the loaf's moisture loss occurs now. It ranges from 10% to 20% of the dough weight, depending on surface area, with small rolls losing the most and large loaves the least.

As the temperature declines, the starch granules become firmer and so the loaf as a whole becomes easier to slice without tearing. This desirable firming continues over the course of a day or so, and turns out to be the first step in the process called staling.

The Staling Process; Storing and Refreshing Bread

Staling Staling takes place in the days following baking, and seems to involve the loss of moisture: the bread interior gets dry, hard, and crumbly. It turns out that bread will stale even when there's no net loss of moisture from the loaf. This was shown in the landmark study of bread staling in 1852, when the Frenchman Jean-Baptiste Boussingault showed that bread could be hermetically sealed to prevent it from losing water, and yet still go stale. He further showed that staling is reversed by reheating the bread to 140°F/60°C: the temperature, we now know, at which starch gelates.

Staling is now understood to be a man-

Food Words: *Stale*

Though *stale* now suggests a food that is past its prime, old and dried out, it hasn't always had these negative connotations. It is a medieval Teutonic word, and originally meant "to stand" or "to age." It was applied to wines and liquors, which became clarified and stronger in flavor when they were allowed to stand for some time and settle. A kind of settling and strengthening also takes place among the starch molecules in bread, but these have toughening effects that are undesirable, at least for bread to be eaten fresh. Stale toughened bread does have its uses (see box, p. 542).

ifestation of starch retrogradation, the recrystallization, water migration out of the granules, and hardening that take place when cooked starch is then cooled (p. 548). The initial firming of the freshly baked bread loaf, which improves its ability to be sliced, is caused by the retrogradation of the simple straight-chain amylose molecules, and is essentially complete within a day of baking.

The majority of starch molecules, the branched amylopectins within the granule, also retrograde. But thanks to their irregular structure, they form crystalline regions and expel water much more slowly, over the course of several days. This is the process responsible for the undesirable firming in texture *after* the bread has become sliceable. For some reason, both the rate and the extent of staling are lower in lighter, less dense breads.

Certain emulsifying agents have been found to retard staling substantially and for this reason have been added to mass-produced bread doughs for about 50 years. True buttermilk (p. 50) and egg yolks are rich in emusifiers and have the same effect. It's thought that these substances complex with starch or in some other way interfere with water movement, thereby inhibiting recrystallization.

Reheating Reverses Staling As long as much of the water released by the starch granules remains in the surrounding gluten—that is, as long as the loaf isn't too old, or has been wrapped and refrigerated—staling can be reversed by heating the bread above the gelation temperature of wheat starch, 140°F/60°C. Once more the crystalline regions are disrupted, water molecules move in between the starch molecules, and the granules and amylose gels become tender again. This is why toasting sliced bread makes the interior soft, and why a loaf of bread can be refreshed by heating it in the oven.

Storing Bread: Avoid the Refrigerator Staling proceeds most rapidly at temperatures just above freezing, and very slowly below freezing. In one experiment, bread stored in the refrigerator at 46°F/7°C staled as much in one day as bread held at 86°F/30°C did in six days. If you're going to use bread in a day or two, then store it at room temperature in a breadbox or paper bag, which reduces moisture loss while allowing the crust to remain somewhat crisp. If you need to keep bread for several days or more, then wrap it well in plastic or foil and freeze it. Refrigerate bread (well wrapped) only if you're going to toast or otherwise reheat it.

Bread Spoilage Compared to many foods, bread contains relatively little water, and so it often dries out before it becomes infected by spoilage microbes. Keeping bread at room temperature in a plastic bag

The Virtues of Stale Bread

Cooks have long known that stale bread is a very useful ingredient in its own right. It is more robust than fresh bread, and retains its sponge-like structure in wet preparations that would cause fresh bread to disintegrate, such dishes as bread salads, bread puddings, and *pain perdu*. Similarly, bread crumbs retain their individual identity when wetted, and can serve as tender binding agent in stuffings, panades, and breadings for frying. The source of dry bread's structural integrity is its starch. When it retrogrades, it forms some regions that are extremely ordered and stable and that hold the rest of the starch network strongly together (p. 458).

allows moisture from the staling starch granules to collect on the bread surfaces and encourages the growth of potentially toxic molds, especially blue-green species of *Aspergillus* and *Penicillium*, gray-white *Mucor* species, and red *Monilia sitophila*.

Bread Flavor

The incomparable flavor of simple wheat bread has three sources: the flavor of wheat flour, the products of yeast and bacterial fermentation, and the reactions caused by oven heat during baking. The aroma of low-extraction white flour is dominated by vanilla, spicy, metallic, and fatty notes (from vanillin, a furanone, and fatty aldehydes), while whole-meal flour is richer in most of these and in addition has cucumber, fried, "sweaty," and honey notes (from other fatty aldehydes and alcohols and phenylacetic acid). Yeast fermentation generates a "yeasty" character, a large part of which comes from fruity esters and eggy sulfur compounds. Baking contributes the toasty products of browning reactions. Starters add general complexity and a distinctive sour note from acetic and other organic acids.

Mass-Produced Breads

The manufacture of commercial breads bears little resemblance to the process described above. Ordinary mixing, kneading, and fermentation require several hours of work and waiting from the bread maker. In bread factories, high-powered mechanical dough developers and chemical maturing agents (oxidizers) can produce a "ripe" dough, with good aeration and gluten structure, in four minutes. Yeast is added to such doughs mainly as flavoring. The formed loaves are proofed briefly and then baked as they move through a tunnel-like metal oven. These breads tend to have a very fine, cakelike texture, because machines are far more efficient at aerating dough than are hands or stand mixers. The

A Scientific Definition of Bread Quality

Raymond Calvel is an eminent figure in the world of baking, a researcher and teacher who made great contributions to the understanding and improvement of bread quality in postwar France. His definition of high quality in good French bread doesn't necessarily apply in detail to other bread styles, but it shows how much there is to appreciate in a well-made loaf.

> A good bread—a real quality loaf— . . . will have a creamy white crumb. The proper creamy-white color of the crumb shows that the dough oxidation during mixing has not been excessive. It also presages the distinctive aroma and taste that are a subtle blend of the scent of wheat flour—that of wheat germ oil, along with the delicate hint of hazelnut aroma that comes from the germ. All of these are combined with the heady smell that comes from alcoholic dough fermentation, along with the discreet aromas that are the results of caramelization and crust baking. . . . the grain of French bread should be open, marked here and there by large gas cells. These should be thin-walled cells, with a lightly pearlescent appearance. This unique structure, resulting from the combination of numerous factors including the level of dough maturation and the loaf forming method, is basic to the eating qualities, flavor, and gustatory appeal of French bread.
>
> —*The Taste of Bread*, transl. R. L. Wirtz.

flavor of manufactured bread can sometimes be marked by such unpleasant aroma compounds as sour, sweat-like isovaleric and isobutyric acids, which are produced by flour and yeast enzymes in unbalanced amounts during intensive mixing and high-temperature proofing.

Special Kinds of Loaf Breads: Sourdough, Rye, Sweet, Gluten-Free

Bakers make distinctive variations on the basic loaf bread from a variety of grains and other ingredients. Here are brief descriptions of some of them.

Sourdough Breads Sourdough breads get their name from the fact that both the dough and bread are acid. The acidity, along with other distinctive flavor components, is produced by bacteria that grow in the dough along with various yeasts. The bacteria often include some of the same lactic acid bacteria that make milk into yogurt and buttermilk (p. 44). The leavening for this kind of bread begins as a "wild" starter, a mixture of whatever microbes happened to be on the grain and in the air and other ingredients when flour was mixed with water. The mixture of yeasts and bacteria is then perpetuated by saving a portion of the dough to leaven the next batch of bread.

The first breads probably resembled modern sourdoughs, and bread in much of the world is made with sourdough starters that give distinctive regional flavors. The bacteria somehow delay starch retrogradation and staling, and the acids they produce make the bread resistant to spoilage microbes: so sourdough breads are especially flavorful and keep well. Because browning reactions are slowed in acid conditions, sourdough breads tend to be lighter in color than straight yeast breads, and their flavor less toasty.

It isn't easy to make good bread with sourdough cultures. There are two reasons for this. One is that the bacteria grow faster than the yeasts, almost always outnumber them by factors of a hundred or a thousand, and inhibit the yeasts' gas production: so sourdoughs often don't rise very well. The other is that acid conditions and bacterial protein-digesting enzymes weaken the dough gluten, which makes it less elastic and the resulting bread more dense.

Guidelines for Working with Sourdoughs The key to successful baking with sourdough starters is to limit bacterial growth and acidification, and encourage a healthy yeast population. In general, this means keeping sourdough starters relatively cool, and "refreshing" them frequently by adding new flour and water and aerating

Frozen and Par-baked Doughs and Breads

Bread dough can be frozen, thawed, and baked into bread, but freezing kills a large proportion of the yeast cells, which means less leavening power, a slower rise, and the spread of yeast chemicals that weaken gluten. Sweet rich doughs turn out to freeze the best.

The best stage at which to freeze bread dough is after the dough has risen and baked for 70 to 80% of its usual baking time. This frozen "par-baked" bread can be thawed and finished with just a few minutes in a hot oven. Yeast survival is no longer important, because the yeast cells have done their leavening and are killed during the initial bake.

them vigorously. Here are rules of thumb to keep in mind.

Both yeasts and bacteria grow fastest in liquid starters, which allow the microbes easier access to nutrients; in a semisolid dough they grow more slowly and require less frequent attention. Because growing microbes consume nutrients rapidly, and produce acid and other growth-inhibiting substances, starters need to be divided and refreshed frequently, two or more times per day. Adding new water and flour dilutes the accumulated acids and other growth inhibitors, and provides a fresh supply of food. Aerating the starter—whisking a liquid one, or kneading a doughy one—supplies the oxygen that yeasts require to build cell membranes for new cells. The more frequently the starter is divided and refreshed, the better the yeasts will be able to grow, and the more leavening power the starter will have. Starters should be incorporated into a dough when they're actively growing and at their bubbliest. While bacteria thrive at warm temperatures, 86–96°F/30–35°C, yeasts in an acid environment grow better at a cooler 68–78°F/20–25°C; so both starters and rising doughs should be kept relatively cool.

Finally, sourdoughs should be well salted. Salt limits bacterial protein-digesting enzymes, and tightens the vulnerable gluten.

Rye Breads Though a minor grain compared to wheat, rye is still found in many breads in Germany and elsewhere in northern Europe and Scandinavia. Most rye breads baked today are made from mixtures of rye and wheat flours, with rye providing its distinctive, full flavor and wheat the rising power of gluten. Rye proteins simply don't form an elastic network like gluten, apparently because the glutenin molecules can't link up end-to-end into long chains. Rye has another major bread-making liability: it tends to begin sprouting before harvest, so its starch-digesting enzymes are active during baking and break down the other major source of dough structure. Nevertheless bakers in northern Europe found a way of making a unique raised bread from pure rye flour.

Pumpernickel True pumpernickel was apparently born during a famine in Westphalia in the 16th century. It starts with coarse whole-grain rye flour and a several-stage sourdough fermentation; the acidity helps limit starch breakdown and also makes the dough more elastic. The dough manages to retain some carbon dioxide bubbles thanks to its high content of gummy cell-wall materials called pentosans (p. 470). The fermented rye dough is baked in a pan at a low oven temperature, or steamed, and for a very long time: 16 to 24 hours. The loaf develops only a thin crust, and turns a dark chocolate-brown and takes on a rich flavor thanks to the long cooking time and high concentration of

Food Words: *Pumpernickel, Bagel, Pretzel, Brioche, Panettone, Pandoro*

Three of these bread names are German in origin, three from Romance languages. *Pumpernickel* comes from Westphalian dialect words for the devil (St. Nick) and for "fart": this is a high-fiber bread. *Bagel* comes via Yiddish from a German root meaning "ring," and *pretzel* directly from a German word of Latin origin meaning "little bracelet," so both are named for their shape. *Brioche* is French, its root apparently *broyer*, meaning to grind or knead. It first appears in the 15th century to name breads enriched with butter but not yet with eggs. *Panettone* and *pandoro* are 19th-century Italian coinages meaning "grand bread" and "golden bread."

free sugars and amino acids, which undergo browning reactions. Because the abundant starch-digesting enzymes are active for a long time during the slow baking, pumpernickel may end up very sweet, with a sugar content of 20%.

The distinctive, complex flavor of rye bread comes largely from the grain itself, which has mushroom, potato, and green notes (from octenone, methional, nonenal). The traditional sourdough fermentation adds malty, vanilla, fried, buttery, sweaty, and vinegar notes.

Sweet and Rich Breads: Brioche, Panettone, Pandoro Bread doughs that contain substantial amounts of fat and/or sugar pose special challenges to the baker. Both fat and sugar slow gluten development and weaken it, sugar because it binds up water molecules and interrupts the gluten-water network, fat because it bonds to fat-loving portions of the gluten chains and prevents them from bonding to each other. Rich doughs are therefore relatively soft and fragile. Bakers often build them by holding back the fat and sugar and kneading these in only after developing the gluten network, and then bake the doughs in containers that support their weight and prevent them from sagging and flattening. Large amounts of sugar slow the growth of yeast by dehydrating the cells, so sweet doughs are often made with more yeast than ordinary breads, and they may take longer to rise. Sugar also makes sweet doughs prone to begin browning early in the baking, so they're usually baked at a relatively low oven temperature to prevent the surface from browning before the interior has set.

French brioche dough is especially rich in butter and eggs. It's often retarded (chilled, p. 539) for 6–18 hours to stiffen it, then rolled out and briefly rested. This makes the dough easier to handle and form before its final rise. Italian panettone and pandoro are remarkable holiday breads that are enriched with large quantities of sugar, egg yolks, and butter, but that keep well because they are built from a sourdough that starts with a naturally leavened sponge.

Gluten-Free Breads People whose immune systems are intolerant of gluten must avoid wheat and its close relatives, and therefore can't eat ordinary bread, where gluten plays a major role in texture. A reasonable facsimile of raised bread can be made with gluten-free flours or starches—rice flour, for example—that are supplemented with xanthan gum and emulsifiers. The gum, which is secreted by a bacterium and purified from industrial-scale fermenters, provides a modest gluten-like elasticity, while the emulsifiers stabilize the gas bubbles and slow the diffusion of carbon dioxide from the dough during baking.

Other Breads: Flatbreads, Bagels, Steamed Breads, Quick Breads, Doughnuts

Light oven-baked loaves are the standard form of bread in Europe and North America, but there are many other versions of

Milk in Bread

Both fresh and powdered milk are sometimes included in doughs for their flavor and nutritiousness, but they can weaken the gluten of bread dough and produce a dense loaf. The culprit appears to be a whey protein, which can be inactivated by scalding the milk—bringing it just to the boil—before use. (The milk must be cooled before mixing to avoid precooking the flour and damaging the yeast.)

the staff of life. Here are brief descriptions of some of them.

Flatbreads Thin flatbreads were the original breads, and are still a major source of nourishment in many countries throughout the world. The essential characteristic of flatbreads is that they cook very quickly, in as little as two minutes, on a simple hot surface, whether a pan, an oven floor or wall, or a mass of hot pebbles. The heat is often very high—pizza ovens can run at 900°F/450°C—and this means that tiny air-pockets in the dough are puffed up by rapidly vaporizing steam, essentially leavening the dough without the necessity of fermentation (though many flatbreads are made with leavened doughs). This puffing, and the breads' thinness, make them tender; and since neither requires a strong gluten, flatbreads can be made from all kinds of grains. Despite the short baking time, the high temperatures develop a delicious toasted flavor across the extensive surface of flatbreads.

Flatbreads often puff to an impressive if temporary volume, and the central cavity of pita and similar breads is used as a pocket for filling with other foods. Puffing occurs when the two bread surfaces have set in the heat and become tougher than the inner layer, where steam accumulates and eventually tears the tender interior, forcing the two surfaces apart. When puffing is undesirable—in crackers, for example, which would become too fragile—the sheeted dough is "docked," or pricked in a regular pattern with a pointed utensil—a fork, or a special stamp—to form dense gluten nodes that resist puffing.

Pretzels Pretzels are unusual for their woven shape, dark brown crust, and unusual flavor. Like crackers, they're made from a stiff yeast dough of soft wheat flour. In manufacturing, the formed dough is sprayed for 10–15 seconds with a hot 1% solution of alkaline sodium hydroxide (lye) or sodium carbonate. The heat and moisture combine to gelate the surface starch. The dough is then salted and baked for about five minutes in a very hot oven. The starch gel hardens to a shiny finish and thanks to the alkaline conditions created by the lye, browning-reaction pigments and flavor compounds rapidly accumulate. (The lye reacts with carbon dioxide in the oven to form a harmless edible carbonate.) The final step is a long, slow bake to dry the whole pretzel out. The pretzel is crisp but fragile thanks to tiny airy bubbles and ungelated starch granules throughout, and it has a distinctive flavor from its alkaline-browned surface.

Soft and homemade pretzels may be allowed to rise before being boiled briefly in a solution of baking soda and then baked for 10 or 15 minutes in a hot oven.

Bagels The bagel is a relatively small, ring-shaped bread that arose in Eastern Europe, and was introduced to the United States by immigrants to New York in the early 20th century.

Traditionally, the bagel had a shiny, thick, chewy crust and a dense interior; after its popularity grew in the late 20th century, many bakers began to make it larger and softer. Bagels are made with strong-gluten flour, which is made into a very stiff dough (a standard bread dough has 65 parts water to 100 flour; bagel dough has only 45 to 50). Traditional bagels are made by forming the dough, allowing it to rise somewhat (an 18-hour retardation gives a good crumb), immersing it in boiling water for 1.5–3 minutes on both sides to expand the interior and develop a thick crust, and then baking it. In the modern method, which is simpler to automate and takes a fraction of the usual time, the formed dough is steamed and then baked, with no slow rise and no immersion in boiling water. The steaming puffs the dough up more than rising and boiling do, and produces a thinner crust. The result is a lighter, softer ring.

Asian Steamed Breads The Chinese have been making and eating steamed breads and buns for around 2,000 years. Asian

Some Flatbreads of the World

Country	Bread	Qualities
	Unleavened	
Israel	Matzoh	Very thin, cracker-like
Armenia	Lavash	Paper-thin, often dried and rehydrated
Italy (Sardinia)	Parchment bread, carta di musica	Semolina flour, very thin
Norway	Lefse	Flour & potatoes, often with butter, cream
Scandinavia	Various rye, oat, barley flatbreads	Many dry
Scotland	Bannocks	Oat cakes
Tibet	Barley bread	From roasted barley flour, tsampa
China	Shaobing	Flour, water, lard, folded and rolled, layered
	Baobing	Hot-water dough, rolled very thin for wrappers
India	Chapati	Whole-wheat, dry-roasted on pan
	Phulka	Chapati cooked, then puffed directly on coals
	Paratha	Folded with ghee, rolled, layered
	Puri, golegappa, lucchi	Deep-fried, puffed
Mexico	Tortillas	Wheat flour or maize
	Leavened	
Iran	Sangak	Whole wheat, baked on hot pebbles
Italy	Focaccia	Moderate thickness
	Pizza	Thin, cooked in very hot oven (to 900°F/450°C)
Egypt	Baladi	Pocket bread
Ethiopia	Injera	Soured teff batter, bubbly and soft
India	Naan	Dough enriched with yogurt, baked in tandoor
United States	Soda cracker	Sourdough neutralized by baking soda
	English muffin	Small diameter, thick
	Pretzel	Thin knotted cylinder

breads are generally small, round, very white, with a smooth, shiny surface and thin skin, and a moist, springy texture that may be chewy (*mantou*) or tender and fluffy (*bao*). They are generally made from soft wheats with moderate gluten content and strength. The relatively stiff dough is fermented, rolled out several times, then cut, formed, proofed, and steamed for 10–20 minutes.

Quickbreads: Biscuits, Biscotti, Scones

Quickbreads are appropriately named in two ways: they are quick to prepare, being leavened with rapid-acting chemicals and mixed briefly to minimize gluten development; and they should be quickly eaten, because they stale rapidly. Batter breads are moister, richer, and keep longer (p. 554).

The term *biscuit* is an ambiguous one. It comes from the French for "twice-cooked," and originally referred to breads and pastries that were baked until dry and hard. The Italian hard cookies called *biscotti* remain true to this heritage; they're lean doughs leavened with baking powder, baked in flattish loaves, then cut crosswise into thin pieces and rebaked at a low oven temperature to dry them out. French *biscuits* proper, and English *biskets,* were long-keeping sweets, small bread-like loaves made from foamed egg whites, flour, and sugar. To this day in England, the word is used for little sweet dry cakes, what Americans call *cookies*. Modern French *biscuits* are dryish cakes made from egg foams, usually moistened with a flavored syrup or cream.

Biscuits became something entirely different in America, and early in its history (see box below). American biscuits contain no sugar and often no eggs, are made from a moist dough of milk or buttermilk, flour, pieces of solid fat, and baking soda, and are cooked briefly into a soft, tender morsel. There are two styles: one with a crusty, irregular top and tender interior; another with a flat top and flaky interior. The first is made with minimal handling to avoid gluten development, the second with just enough folding and kneading to develop a structure with alternating layers of dough and fat. The simple recipe and short cooking mean that the flavor of the flour itself is prominent.

English scones are similar to American biscuits in their simplicity, basic composi-

Early American Biscuits

Despite their name, these American biscuits were cooked only once, and were rich and moist, not dry.

Biscuit

One pound flour, one ounce butter, one egg, wet with milk and break while oven is heating; and in same proportion.

Butter Biscuit

One pint each milk and emptins [liquid yeast], laid into flour, in sponge; next morning add one pound butter melted, not hot, and knead into as much flour as will with another pint of warmed milk, be of a sufficient consistence to make soft—some melt the butter in the milk.

—Amelia Simmons, *American Cookery,* 1796

tion, and floury taste. Irish soda bread is made with soft whole-wheat flour and without fat.

Doughnuts and Fritters Doughnuts and fritters are essentially pieces of bread or pastry dough that are fried in oil rather than baked. Doughnuts have a moist interior and little or no crust, while fritters are usually fried until crisp.

The word *doughnut* was coined in the United States in the 19th century to name what the Dutch called *olykoeks,* portions of fried sweetened dough. Their great popularity flowered in the 1920s, when machinery simplified the handling of the soft, sticky doughs, which are rich in sugar, fat, and sometimes eggs. There are two main styles: yeasted doughnuts are light and fluffy, while cake doughnuts, leavened with baking powder, are denser. Light, yeasted doughnuts ride on the oil surface and must be turned, which leaves a white band around their circumference where the oil surface cooks the dough less thoroughly. Doughnuts are fried at a moderate temperature, originally in lard and now usually in a hydrogenated vegetable shortening, which solidifies when the doughnut cools to provide a dry rather than oily surface.

THIN BATTER FOODS: CRÊPES, POPOVERS, GRIDDLE CAKES, CREAM PUFF PASTRY

Batter Foods

The difference between doughs and batters is reflected in their names. *Dough* comes from a root meaning "to form," while *batter* comes from a root meaning "to beat." Doughs are firm enough to develop and sculpt by hand. Batters are too fluid and elusive to hold, so we contain them in a bowl, mix them by battering them repeatedly from within—by stirring—and cook them in a container to give them form and solidity.

Batters are fluid because they include two to four times more water than doughs. The water disperses the gluten proteins so widely that they form only a very loose, fluid network. When we cook a batter, the starch granules absorb much of the water, swell, gelate, leak amylose, stick to each other, and thus turn the fluid into a solid but tender, moist structure. The gluten proteins play a secondary structural role, providing an underlying cohesiveness that prevents the food from being crumbly. But if the gluten is overdeveloped, it makes the food elastic and chewy. Batters often contain eggs, and the egg proteins also contribute a nonelastic solidity when they coagulate in the cooking heat. Fluid batters can't retain much of the gas slowly evolved by yeast, and are usually leavened either chemically, or else mechanically, by beating air into the batter or its components.

Most batter products are meant to be delicate and tender. Tenderness can be encouraged in several different ways.

- The concentration of gluten proteins in the batter is reduced by the use of pastry flour, or low- or no-gluten flours (buckwheat, rice, oat), or all-purpose flour mixed with cornstarch or other pure starches.
- Gluten development is minimized by minimal stirring of the ingredients.
- Replacing milk or water with soured dairy products, notably buttermilk and yogurt, helps produce an especially tender texture. This is mainly an effect of their thick consistency, which means that it takes less flour to make the batter properly thick. A spoonful of the finished batter therefore contains less flour, less starch and gluten, and cooks to a more delicate structure.
- Leavening the batter with gas bubbles not only divides it up into innumerable thin sheets surrounding the gas bubbles, it makes the batter more viscous (as in sauce foams, p. 595), which again means that less flour is needed to thicken it.

It's useful to divide batter foods into two groups. Thin batters are both thin in consistency and cooked in the form of small, thin, free-standing cakes. Quickbreads and cakes are made from thicker batters, and are cooked in pans in larger, deeper masses.

Crêpes

Crêpes and their relatives (Eastern European blintzes and palaschinki), thin unleavened pancakes that are cooked on a shallow pan and folded over a filling of some kind, have been made for a thousand years from a simple batter of flour, milk and/or water, and eggs. Their delicacy comes from their thinness. The batter is carefully mixed to minimize gluten formation, allowed to stand for an hour or more to allow the proteins and damaged starch to absorb water and air bubbles to rise and escape, and then cooked for just a couple of minutes per side. In France, the milk in crêpe batter is sometimes partly replaced with beer, and wheat flour with buckwheat, especially in Brittany.

Popovers

Popovers are an American version of English Yorkshire pudding, which is cooked in the fat rendered from a beef roast. The batter is almost identical to crêpe batter, but a different cooking method transforms it into a large air pocket surrounded by a thin layer of pastry. Popover batter is vigorously beaten to incorporate air and cooked immediately, before the air bubbles have a chance to escape. The batter is poured into a preheated, liberally oiled pan and set in a hot oven. The surfaces of the batter set almost immediately. The air bubbles within the batter are trapped, expand with the rising temperature, coalesce into one large bubble, and the liquid batter balloons around it and sets into a thin blister. When cooked in a pan with many cups, popovers rise unevenly, because the cups around the outside heat faster than the ones in the middle of the pan.

Griddle Cakes: Pancakes and Crumpets

Griddle cakes are made from a more floury, viscous batter than crêpes, popovers, and choux pastry, and can retain gas cells for the few minutes that it takes to cook them; so they rise on the hot pan surface and develop an aerated, tender structure. Pancakes may be leavened (and flavored) with yeast, or with whipped egg whites folded into the batter, or chemical leavening, or some combination of these. Russian blinis sometimes include beer, which may contribute effervescence. The batter is poured onto the griddle and allowed to cook until bubbles begin to rise and break on the upper surface; the cook then flips the cake to set the second side before the leavening gas escapes.

Crumpets are an English invention, small, flat, yeast-raised cakes with a distinctively pale, cratered top surface. They're made from a somewhat thick pancake batter that's allowed to become bubbly from yeast activity, then poured into ring molds to a depth of about 0.75 in/2 cm, cooked very slowly until bubbles break and set on the surface, then unmolded and turned to cook briefly on the hole-y side.

Food Words: *Crêpe*

The French word *crêpe* comes from the Latin for "curled, wavy," and probably refers to the curling of the edge as it dries during cooking.

Griddle Cakes: Waffles and Wafers

Waffles and wafers have two things in common: the root word for their names, and the unique method by which they're cooked. Their flour-water mixture is spread in a thin layer and pressed between two heated and embossed metal plates, which spread them even thinner, conduct heat into them rapidly, and imprint them with an attractive and often useful pattern. The usual square indentations increase the area of crisp, browned surface and collect the butter, syrup, and other enrichments that are often added on top. The French version, the *gaufre,* goes back to medieval times, when street vendors would make them to order and serve them hot on religious feast days.

Today the difference between wafers and waffles is a matter of texture. Wafers are thin, dry, crisp, and when high in sugar are dense, almost hard. The most familiar wafer is the ice cream cone; there are also French cookies called *gaufres* that are similar to very thin, crisp *tuiles.* Waffles came to the United States from Holland in the 18th century and are thicker, lighter, and more delicate thanks to leavening by either yeast or a baking powder, which interrupts the cooked structure with gas bubbles. They're served fresh and hot, their honeycomb structure filled with butter or syrup.

Modern waffle recipes are often essentially a lean pancake batter cooked in a waffle iron instead of on a griddle, and they often produce a disappointing result that is more leathery than crisp. Crispness requires a high proportion of fat, sugar, or both: otherwise the batter essentially steams rather than fries, the flour proteins and starch absorb too much softening water, and the surface ends up tough.

Cream Puff Pastry, Pâte à Choux

Choux is the French word for "cabbage," and choux pastry forms little irregular cabbage-like balls that are hollow inside like popovers. Unlike popovers, choux pastry becomes firm and crisp when baked. It provides the classic container for cream fillings in such pastries as cream puffs (profiteroles) and éclairs, and also makes such savory bites as cheese-flavored gougères and deep-fried beignets, whose lightness inspired the name *pets de nonne,* "nun's farts."

Choux paste was apparently invented in late medieval times, and is prepared in a very distinctive way. It's a cross between a batter and a dough, and is cooked twice: once to prepare the paste itself, and once to transform the paste into hollow puffs. A large amount of water and some fat are brought to the boil in a pan, the flour is added, and the mixture stirred and cooked

An Early French Waffle Recipe

Waffles of Milk or Cream

Put a litron [13 oz/375 g] of flour in a bowl, break in two or three eggs and mix together while adding some cream or milk and a pinch of salt. Add a piece the size of two eggs of freshly made cream cheese, or simply soft cheese from whole milk, and a quarteron [4 oz/125 g] of melted butter. If you only add a half quarteron of butter, that is enough provided that you add a half quarteron of good beef marrow crushed very small.

Mix all this together, and when it is well bound, one should put the waffle irons on the fire, and make the waffles. These waffles should be eaten while they're hot.

—La Varenne, *Le Cuisinier françois,* 1651

over low heat until it forms a cohesive ball of dough. Several eggs are then beaten sequentially into the dough until it becomes very soft, almost a batter. This paste is then formed into balls or other shapes and baked in a hot oven or deep-fried. As with the popover, the surface sets while the interior is still nearly liquid, so the trapped air coalesces and expands into one large bubble.

Frying Batters

A number of foods, especially seafood, poultry, and vegetables, are sometimes coated with a layer of flour batter before deep-frying (or sometimes baking). A good batter adheres well to the food, and fries into a crust that has a long-lasting crunchiness and that readily breaks apart in the mouth without an oily residue. Problematic batters fall away during frying or produce a crust that's greasy, chewy, and tough, or soft and mushy.

Batters include some kind of flour, a liquid that might be water, milk, or beer, sometimes a chemical leavening to provide gas bubbles and lightness, and often eggs, whose proteins promote adherence to the food and allow the use of less flour. Of all the ingredients, the flour has the largest influence on batter quality. Too much can produce a tough, bready coating; too little and it will be fragile. The gluten proteins in ordinary wheat flour are valuable for the clinginess they provide, but they form elastic gluten and absorb both moisture and fat, and so are responsible for both chewiness and oiliness in the fried crust. For these reasons, moderate-protein flours make better batters than bread flour, and some batters are made from other grains, or from a mixture of wheat flour and other flours or starches. Rice proteins don't form gluten and absorb less moisture and fat, so batters that contain a substantial proportion of rice flour fry crisper and drier. Similarly, corn flour improves crispness because its relatively large particles are less absorbant, and its proteins dilute wheat gluten and reduce the chewiness of the crust. Adding some pure corn starch also reduces the proportion and influence of wheat gluten proteins. Root flours and starches don't work well in batters because their starch granules gelate and disintegrate at relatively low temperatures, do so early in frying, and produce a soft crust that gets soggy quickly.

Batters adhere better to moist foods when the foods are predipped in dry particles, whether seasoned flour or breadcrumbs; the dry particles stick to the moist surface, and the moist batter then clings to the rough surface created by the particles. Batters are more likely to produce

The Logic of Cream Puff Pastry

The technique for preparing cream puff pastry may seem tediously elaborate, but it's a brilliant invention. It produces an especially rich and moist paste that the cook can shape and cook into a hollow, crisp vessel for other ingredients. Cooking the flour with water and fat tenderizes the gluten proteins, preventing them from developing elasticity, and it swells and gelates the starch to turn what would normally be a batter into a dough. The subsequent addition of raw eggs contributes the richness of the yolks and the cohesive, structure-building proteins of the whites, and thins the dough into a near-batter so that air pockets in the interior will be able to move and coalesce during cooking. During the baking, the fat helps crisp and flavor the outer surface. And both eggs and fat contribute to a structure that resists moisture and stays crisp while holding the cream filling.

crisp, tender crusts if they're prepared just before frying, with cold liquid and little mixing to minimize water absorption and gluten development (see Japanese tempura, p. 214). If a batter stands for a long time before the food is dipped and fried, many of its air and gas bubbles can leak out, and some of the remaining chemical leavening reacts too early; the result is a dense casing rather than a light puff.

THICK BATTER FOODS: BATTER BREADS AND CAKES

Batter Breads and Muffins

Batter breads and muffins are moister, usually sweeter versions of quickbreads (p. 549). They're leavened with baking powder or soda, and often contain moderate amounts of egg and fat in addition to sugar. They develop a dense, moist texture that accommodates nuts, dried fruits, and even such fresh fruits and vegetables as apples, blueberries, carrots and zucchini, whose moistness readily blends into the moistness of the crumb. Potatoes and bananas can be mashed to become part of the batter itself.

Muffin batters generally contain less sugar, eggs, and fat than quickbreads. The ingredients are mixed together just enough to dampen the solids, and the mix baked in small portions rather than a large loaf. Well-made muffins have an even, open, tender interior. They stale quickly because the small proportion of fat is dispersed unevenly by the minimal mixing and can't protect much of the starch. Overmixing produces a less tender, finer interior with occasional coarse tunnels, which develop when the overly elastic batter traps the leavening gas in large pockets.

Green Blueberries and Blue Walnuts Sometimes the solid ingredients folded into bread and muffin batters turn disconcerting colors: blueberries, carrots, and sunflower seeds may go green, and walnuts blue. This happens when the mix contains too much baking soda, or when the soda isn't evenly mixed in the batter, so that there are concentrated alkaline pockets. Because the anthocyanin and related pigments in fruits, vegetables, and nuts are sensitive to pH, and their normal surroundings are acidic, alkaline batters cause their colors to change (p. 281). Small brown spots on the surface of the finished bread or muffin are also a sign of incomplete mixing; the browning reactions proceed faster in more alkaline portions of the batter.

Cakes

The essence of most cakes is sweetness and richness. A cake is a web of flour, eggs, sugar, and butter (or shortening), a delicate structure that readily disintegrates in the mouth and fills it with easeful flavor. Cakes often contain more sugar and fat than they do flour! And they serve as a base for even sweeter and richer custards, creams, icings, jams, syrups, chocolate, and liqueurs. As suits their luxurious nature, they're often elaborately shaped and decorated.

A cake's structure is created mainly by flour starch and by egg proteins. The tender, melt-in-the-mouth texture comes from gas bubbles, which subdivide the batter into fragile sheets, and from the sugar and fat, which interfere with gluten formation and egg protein coagulation, and interrupt the network of gelated starch. The sugar and fat can compromise lightness if they weaken the cake structure so much that it can't support its own weight. Of course dense, heavy cakes can be delicious in their own way. Flourless chocolate cakes, nut cakes, and fruit cakes are examples.

Traditional Cakes: Limited Sweetness and Hard Work Well into the 20th century, risen cakes were typified by the English pound cake or French *quatre quarts*, "four quarters," which contain equal weights of the four major ingredients: structure-building flour and eggs, and structure-weakening butter and sugar. These proportions push the flour's starch and the eggs' proteins to their limit for holding the fat and sugar in a tender, light scaffolding; more butter or sugar collapses the scaffolding and makes dense, heavy cakes. And because cake batter must be filled with many small bubbles without the help of yeasts, which generate gas too slowly for the batter to hold them, traditional cake making was hard work. In 1857, Miss Leslie described a technique by which the cook could beat eggs "for an hour without fatigue" and then added: "But to stir butter and sugar is the hardest part of cake making. Have this done by a manservant." Fannie Farmer warned in 1896 that "A cake can be made fine grained only with long beating."

Modern American Cakes: Help from Modified Fats and Flours Beginning around 1910, several innovations in oil and flour processing led to major changes in American cakes. The first innovation made it possible to leaven cakes with much less work. The hydrogenation of liquid vegetable oils to make solid fats allowed manufacturers to produce specialized shortenings with the ideal properties for incorporating air quickly at room temperature (p. 557). Modern cake shortenings also contain tiny bubbles of nitrogen that provide preformed gas cells for leavening, and emulsifiers that help stabilize the gas cells during mixing and baking, and disperse the fat in droplets that won't deflate the gas cells.

The second major innovation was the development of specialized cake flour, a soft, low-protein flour that is very finely milled and strongly bleached with chlorine dioxide or chlorine gas. The chlorine treatment turns out to cause the starch granules to absorb water and swell more readily in high-sugar batters, and produce a stronger starch gel. It also causes fats to bind more readily to the starch granule surface, which may help disperse the fat phase more evenly. In combination with the new shortening and with double-acting baking powders, cake flour allowed U.S. food manufacturers to develop "high-ratio" packaged cake mixes, in which the sugar can outweigh the flour by as much as 40%. The texture of the cakes they make is distinctively light and moist, fine and velvety.

Thanks to these qualities and to the convenience of premeasured ingredients, packaged cake mixes were a great success: just 10 years after their major introduction following World War II, they accounted for half of all cakes baked in U.S. homes. The very sweet, tender, moist, light cake became the American standard; and hydrogenated shortening and chlorinated flour became

An Early English Pound Cake Recipe

In the centuries before electric mixers and preleavened shortenings, filling a dense cake batter with many fine air bubbles was hard, prolonged work!

To Make a Pound Cake

> Take a pound of butter, beat it in an earthen pan, with your hand one way, till it is like a fine thick cream; then have ready twelve eggs, but half the whites, beat them well, and beat them up with the butter, a pound of flour beat in it, and a pound of sugar, and a few caraways; beat it all well together for an hour with your hand, or a great wooden spoon. Butter a pan, and put it in and bake it an hour in a quick oven.
>
> —Hannah Glasse, *The Art of Cookery Made Plain and Easy*, 1747

standard kitchen supplies for cakes made "from scratch."

The Disadvantages of Modified Fats and Flours Hydrogenated vegetable shortenings and chlorinated flour are very useful, but have drawbacks that lead some bakers to avoid them. Hydrogenated shortening does not have the flavor that butter does, and has the more serious disadvantage of containing high levels of trans fatty acids (10–35% compared to butter's 3–4%; see p. 38). Chlorinated flour has a distinctive taste that some bakers dislike (others find that it enhances cake aroma). And the chlorine ends up in fat-like flour molecules that accumulate in animal bodies. There's no evidence that this accumulation is harmful, but the European Union and the United Kingdom consider the safety of chlorinated flour unproven, and forbid its use. The U.S. FDA and the World Health Organization consider chlorinated flour a safe ingredient for human consumption.

Manufacturers are addressing some of these problems and uncertainties. For example, the effects of flour chlorination can be approximated by heat treatment, and vegetable oils can be hardened without the production of trans fatty acids. So it's likely that cooks will eventually be able to make high-ratio cakes with less questionable ingredients.

Cake Ingredients Cakes are generally made with flour, eggs, sugar, and either butter or shortening. Eggs are 75% water and may provide all the moisture in a recipe; or various dairy products—milk, buttermilk, sour cream—may be included to provide moisture as well as richness and flavor. Because the sugar is used to incorporate air into the mix, the preferred form is finely granulated ("Extra-Fine," "Superfine") so as to maximize the numbers of small sharp edges that will cut into the fat or eggs. Because they're filled with bubbles in the mixing process, cake recipes usually call either for no chemical leavening, or for less than other batter recipes do.

Flours, Starches, Cocoa Cake bakers use low-protein pastry or cake flours to minimize the toughening that comes with gluten formation. They're not really interchangeable; cake flours are both chlorinated and milled into very small particles to produce a fine, velvety texture. Cooks who prefer not to use cake flours can approximate their protein content and increase fineness by adding starch to all-purpose or pastry flours. Corn starch is the most commonly available starch in the United States; potato and arrowroot starches lack cornstarch's cereal flavor and gelate at lower temperatures, which can reduce cooking times and produce a moister cake. Some cakes are

Standard Cake Proportions and Qualities

	Flour	Eggs	Fat	Sugar	Qualities
Pound cake	100	100	100	100	Moist, soft, rich
Butter cake	100	40	45	100	Moist, soft
Génoise	100	150–200	20–40	100	Light, springy, somewhat dry
Biscuit	100	150–220	0	100	Light, springy, dry
Sponge cake	100	225	0	155	Light, springy, sweet
Angel food cake	100	350 (whites)	0	260	Light, springy, very sweet
Chiffon	100	200	50	135	Light, moist

made with pure starch or starchy chestnut flour only, and no wheat flour at all.

In chocolate cakes, cocoa powder takes on some of the water-absorbing and structural duties of flour; it's around 50% carbohydrate, including starch, and 20% nongluten protein. Cocoa powders may be "natural" and acidic or "dutched" and alkaline (p. 705), a difference that affects both leavening and flavor balance; cake recipes should specify which kind to use, and the baker should not substitute one for the other. If chocolate rather than cocoa is used, it must be melted and carefully incorporated into the fat or eggs. Different chocolates contain widely varying proportions of cocoa fat, cocoa solids, and sugar (p. 704), so again, bakers and recipes should be clear about what kind of chocolate is to be used.

Fats In the standard method for making pound and layer cakes, the cook beats fat and sugar together to incorporate air bubbles, until the mixture reaches the fluffy consistency of whipped cream. Solid fats retain air bubbles thanks to their semisolid consistency: the air carried along by the sugar crystals and beater becomes immobilized in the mixture of crystalline and liquid fat. Butter is the traditional cake fat, and is still the fat of choice for bakers who value flavor more than lightness of texture.

But modern vegetable shortenings do a better job of incorporating air bubbles into the cake batter. Animal fats—butter and lard—tend to form large fat crystals that collect large air pockets, which rise in thin batter and escape. Vegetable shortenings are made to contain small fat crystals that trap small air bubbles, and these bubbles stay in the batter. Manufacturers also fill shortenings with preformed bubbles of nitrogen (around 10% of the volume), and bubble-stabilizing emulsifiers (up to 3% of the shortening weight). Butter is best aerated at a relatively cool 65°F/18°C, while shortening creams most effectively at a warm room temperature, between 75 and 80°F/24–27°C.

Fat Replacers The moistening and tenderizing effects of fat—but not its aerating abilities—can be imitated by some concentrated fruit purees, notably prune, apple, apricot, and pear. Their high levels of viscous plant carbohydrates, mainly pectins and hemicelluloses, bind water and also interrupt the gluten and starch networks.

Techniques for Aerating Cakes

- Fat-sugar aeration: Sugar beaten into butter or shortening; other ingredients then folded in
  - Pound cake, French *quatre quarts,* American butter and layer cakes, fruit cake
- Egg-sugar aeration: Sugar beaten with whole eggs, or separated yolks or whites; other ingredients then folded in
  - Whole eggs: *Génoise*
  - Yolks and whites aerated separately: French *biscuit,* Black Forest cake
  - Yolks only: Sponge cake
  - Whites only: Angel food cake; chiffon cake; flourless meringue and *dacquoise*
- All-ingredients aeration: Flour, eggs, sugar, shortening all beaten together
  - Commercial cake mixes
- No aeration: Ingredients stirred together with minimal incorporation of air
  - Dissolved-sugar cakes: *pain d'épices,* spice cake

So these fruit purees can be used to replace some of the fat in cake recipes. The result is usually moist and tender, but also denser than a full-fat cake.

Mixing Cake Batters In cake making, the mixing step doesn't just combine the ingredients into a homogeneous batter: it has the critical purpose of incorporating air bubbles into the batter, and thereby strongly influencing the final texture of the cake. The various ways of aerating the batter help define families of cakes (see box, p. 557). They involve beating the sugar and/or the flour into the fat, the eggs, or all the liquid ingredients. The fine solid particles carry tiny air pockets on their surfaces, and the particles and beating utensils carry those pockets into the fat or liquid. Flour is often added only after the foam is formed, and then by gently folding it in, not beating, to avoid popping a large fraction of the bubbles, and to avoid developing gluten. (For folding as a mixing technique, see p. 112.) Mixing the dry flour and fat together also prevents the gluten proteins from bonding strongly to each other.

Preleavened shortening and electric mixers have helped to turn cake making into a far less onerous task than it once was, but the mixing stage can still take 15 minutes or more.

Bakers often modify or combine elements of these techniques. In the "pastry-blend" method, the flour, sometimes with the sugar, is creamed with the fat, then the liquid ingredients are added and mixed long enough to augment the initial aeration. Another alternative is a combination of the fat and egg aerations: some of the sugar is used to aerate the fat, some the eggs, and the two foams are then combined.

Baking Cakes Cake baking can be divided into three stages: expansion, setting, and browning. During the first stage, the batter

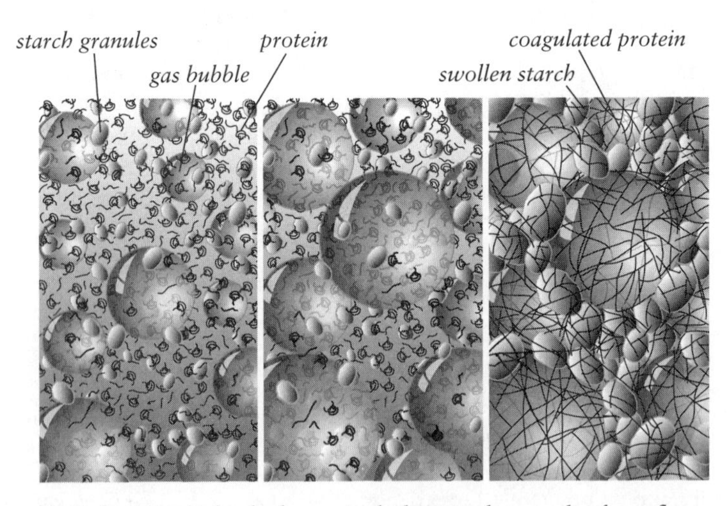

Baking a cake. Left: *A typical cake batter includes starch granules from flour, egg proteins that coagulate when heated, and gas bubbles incorporated during mixing, all swimming in a syrup of water and sugar. (Most cakes include some form of fat, which is not shown here for the sake of clarity.)* Center: *When the mix is heated, the gas bubbles expand, causing the mix to rise. At the same time, the proteins begin to unfold and the starch granules begin to absorb water and swell.* Right: *At the end of baking, the fluid batter has set into a porous solid, thanks to the continued swelling and gelation of the starch granules and the coagulation of the egg proteins.*

expands to its full volume. As the batter temperature rises, the gases in the air cells expand, chemical leavening releases carbon dioxide, and beginning around 140ºF/60ºC, water vapor begins to form and expand the air cells even further. During the second stage of cake baking, the risen batter is set into its permanent shape by the oven heat. Beginning around 180ºF/80ºC, the egg proteins coagulate, and starch granules absorb water, swell, and gelate. The actual setting temperature depends strongly on the proportion of sugar, which delays both protein coagulation and starch swelling; in a high-ratio cake, the starch may not gelate until close to 212ºF/100ºC. In the last stage, batter solidification is completed, flavor-enhancing browning reactions take place in the now-dried surface, and the cake often shrinks slightly, an indication that it should be taken out of the oven. Another test of doneness is to probe the center with a toothpick or wire cake tester, which should come out clean of any batter or crumb particles.

Cakes are generally baked at moderate oven temperatures, 350 to 375ºF/175–190ºC. Below this range, the batter sets slowly, expanding gas cells can coalesce to produce a coarse, heavy texture, and the upper surface sinks. Above this range, the outer portions of the batter set before the inside has finished expanding, which produces a peaked, volcano-like surface, and the surface browns excessively.

Cake Pans By affecting the rate and distribution of heating, cake pans can have an important influence on their contents. The ideal pan size is that which matches the final volume of the cake, which is usually 50–100% greater than the initial batter volume. Doughnut-shaped tube pans, with the hole at their center, provide a greater surface area and speed the penetration of heat into the batter. Bright surfaces reflect radiant heat, transmit heat poorly to the food they contain, and slow the baking process. A dull metal pan or a glass one (which also transmits radiant heat well) will cook a cake as much as 20% faster than a shiny pan, while a black surface tends to absorb heat quickly and cause rapid surface browning. Recent innovations in nonmetal baking containers include flexible silicone molds and paper molds,

Baking Cakes at High Altitudes

Cake recipes that work well at sea level can produce dry, dense disasters in kitchens at high altitudes. The reason for this is that the low air pressure in the mountains allows water to boil at a lower temperature than the 212ºF/100ºC characteristic of sea level. The drops in pressure and boiling point have several effects on a cake in the oven. The batter starts to lose moisture at lower temperatures, and dries out more rapidly. The air bubbles and leavening expand faster at temperatures below the setting temperature, and the protein and starch set and stabilize that structure slowly, because the batter temperature doesn't get as hot. So a cake baked in the mountains tends to end up dry, coarse, and flat.

A cake recipe developed at sea level must be adjusted in order for it to work at several thousand feet. The loss of moisture can be compensated for by the addition of extra liquid. The overexpansion of the gas cells can be reduced by reducing the amount of leavening. And the structure-stabilizing elements can be set earlier by reducing the levels of sugar and fat and increasing the eggs and/or flour. Increasing the oven temperature also speeds the protein coagulation and starch gelation that set the structure.

larger and stiffer and more elegant versions of muffin and cupcake papers.

Cooling and Storing Cakes Most cakes require a cooling period before they're removed from their pans or otherwise handled. Their structures are quite delicate when still hot, but become firm as the starch molecules begin to settle back into close, orderly associations with each other. Pound and butter cakes are fairly robust, their structure coming mainly from gelated starch, and can be removed from their pans after just 10–20 minutes. The sweeter egg-aerated cakes are held up largely by the coagulated egg proteins, which form a more gas-tight film around the gas cells than starch does, and therefore shrink as the gas within cools and contracts. The result can be a collapsed cake. To avoid this, angel, sponge, and chiffon cakes are cooked in tube pans that can be inverted and suspended over a bottle to cool. Gravity keeps the cake structure stretched to its maximum volume while the walls of the gas cells firm and develop cracks that allow the pressure inside and outside the cells to equalize.

Cakes keep for several days at room temperature, and they can be refrigerated or frozen. They stale more slowly than bread, thanks to the presence of emulsifiers and their high proportions of moisture, fat, and moisture-retaining sugar.

PASTRIES

Pastries bear little family resemblance to cakes or breads or pastas. They're a very different expression of the nature of the wheat grain. In making other dough and batter foods, we use water to fuse the particles of wheat flour into an integrated mass of gluten and starch granules, and further knit that mass together with cooking. By contrast, pastry is an expression of the fragmentary, discontinuous, particulate qualities of wheat flour. We use just enough water to make a cohesive dough from the flour, and work in large amounts of fat to coat and separate flour particles and dough regions from each other. Cooking gelates half or less of the water-deprived starch, and produces a dry mass that readily crumbles or flakes in the mouth, releasing the fat's complementary moist richness.

Many pastries are not prepared and eaten on their own as other dough and batter foods are. Instead they serve as a contrasting container for a moist filling, whether savory (quiche, pâté, meat pies, vegetable tarts) or sweet (fruit pies and tarts, creams, custards). The container may be open, as in tarts and open-faced pies, a closed double-crust pie, or fully enclosed turnovers—samosas, empanadas, pasties, pierogi, piroshki. We also use the term

Varieties of Pastry in Elizabethan Times

Beginning in the Middle Ages, one of the primary purposes of pastry was to contain and help preserve preparations of meat. Meat brought to a simmer inside a thick, durable crust would essentially be pasteurized and protected against contamination by microbes in the air, so in a cool place it would keep for many days. Other dishes to be eaten freshly baked would be made with a more delicate pastry. As Gervase Markham wrote around 1615 in *The English Housewife*,

> our English housewife must be skilful in pastry, and know how and in what manner to bake all sorts of meat, and what paste is fit for every meat, and how to handle and compound such pastes.

"pastry" for what are essentially sweet breads whose structure is divided by layers of fat. Croissants and Danish pastries are really bread-pastry hybrids.

Pastry making flowered in the Mediterranean region in the late Middle Ages and early Renaissance, when puff and cream puff pastries first appeared. By the time of La Varenne in the 17th century, both crumbly and puff pastries were standard preparations. The bread-pastry hybrids are a more recent invention from the late 19th and 20th centuries.

Pastry Styles

There are several different styles of pastry, each with a different texture that is created by the kinds of particles into which they come apart when chewed.

- *Crumbly pastries*—short pastry, pâte brisée—come apart into small, irregular particles.
- *Flaky pastries*—American pie crusts—come apart in small, irregular, thin flakes.
- *Laminated pastries*—puff pastry, phyllo, strudel—are constructed of large, separate, very thin layers that shatter in the mouth into small, delicate shards.
- *Laminated breads*—croissants, Danish pastries—combine the layering of the laminated pastries with the soft chewiness of bread.

These varied structures and textures depend on two key elements: the way the fat is incorporated into the dough, and the development of the flour gluten. Pastry makers work fat in so that it either isolates very small dough particles from each other, isolates larger masses or even whole sheets of dough from each other, or both. And pastry cooks carefully control gluten development to avoid making a dough that's hard to shape and a pastry that's tough and chewy instead of tender and delicate.

Pastry Ingredients

Flours Pastries are made from several different kinds of flour. A crumbly texture, which requires minimal gluten development, is best obtained with a pastry flour moderately low in protein; some protein is necessary to provide continuity in the

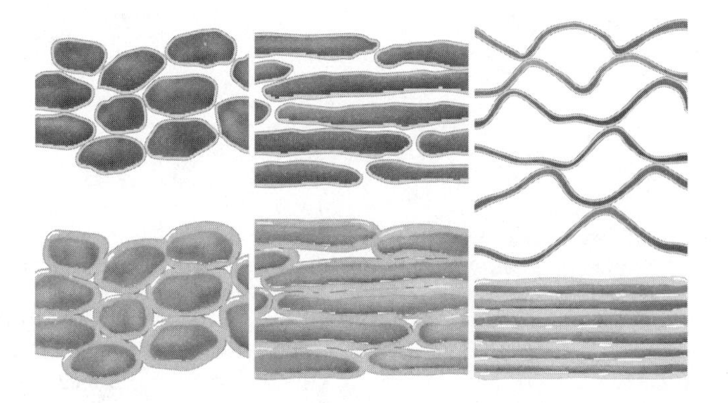

Pastry structures (uncooked doughs shown at bottom; cooked pastries at top). The key to pastry structure is the distribution of the fat, here shown as a light layer surrounding darker masses of dough. Left: *In crumbly pastries, fat coats and separates small particles of dough.* Center: *In flaky pastries, fat coats and separates flattened pieces of dough.* Right: *In laminated pastries, fat coats and separates extended, thin sheets of dough. The sheets in laminated pastries are so light that cooking steams them apart into a light, airy structure.*

dough particles, or the pastry comes out chalky rather than crumbly. Flakiness and the laminated structure of puff pastry depend on controlled gluten development, and can be achieved with pastry flour or with flour of a higher protein content, the equivalent of U.S. national all-purpose flours (11–12%). Highly stretched strudel and phyllo can benefit from the very high protein content of bread flours and the strong gluten they form.

Fats Much of the flavor of pastry—and much of the pleasure—comes from its fat, which may be a third or more of its weight. But pastry makers often choose a fat that has little or no flavor. This is because the fat must have the necessary consistency for producing the desired texture. Broadly speaking, any fat or oil can be worked finely into flour to make crumbly pastry, while flaky and laminated pastries require fats that are solid but malleable at cool room temperature: namely butter, lard, or vegetable shortening. Of these, shortenings are the easiest to work with, and produce the best textures.

Fat Consistency: Butter and Lard Are Demanding At any given temperature, solid fats have different consistencies that depend on what fraction of their molecules is in solid crystals, and what fraction is liquid. Above about 25% solids, fat is too hard and brittle to roll into an even layer. Below about 15% solids, fat is too soft to work; it sticks to the dough, doesn't hold its shape, and leaks liquid oil. The ideal fat for flaky and laminated pastries is therefore one that has between 15 and 25% solids at kitchen temperature, and at the temperatures that the pastry dough reaches as it's mixed and formed. It turns out that butter has the right consistency for making pastry in a relatively narrow temperature range, between 58 and 68°F/15–20°C. Lard is properly workable at only slightly warmer temperatures, up to 75°F/25°C. Our flavorful natural fats easily get too soft in the kitchen to make good pastry. This is why pastry makers often prechill ingredients and utensils, work on a cold marble surface that keeps the ingredients cool during the mixing and rolling out, and value assistants with constitutionally cold hands.

Fat Consistency: Shortenings Are Forgiving Manufacturers of vegetable shortenings control the consistency of their products by controlling how much of the base oil's unsaturated fat is hydrogenated (p. 801). Standard cake shortening has the desirable 15–25% solids over a temperature range triple that of butter, from 53 to

Food Words: *Pastry, Pasta, Pâté, Pie*

The English word *pastry,* Italian *pasta,* and French *pâte* and *pâté* all go back to a suggestive group of ancient Greek words having to do with small particles and fine textures: they variously referred to powder, salt, barley porridge, cake, and an embroidered veil. A later Latin derivative, *pasta,* was applied to flour that had been wetted to a paste, then dried; it led to Italian *pasta* and to *pâte* meaning "dough." *Pâté* is a medieval French word that was given originally to a chopped meat preparation enclosed in a dough, but eventually came to name the meat preparation itself, with or without enclosure. *Pie* was the near equivalent of the original *pâté* in medieval English, and meant a dish of any sort—meat, fish, vegetable, fruit—enclosed in pastry. The word had less to do with doughs than with odds and ends: it came from *magpie,* a bird with variegated coloring that collects miscellaneous objects for its nest.

85°F/12–30°C. It's therefore much easier to make flaky pastry with shortening than with butter. Because laminated pastries and breads are especially tricky to make, professionals and manufacturers often use shortenings that have been formulated specifically for their production. Danish margarines are workable up to 95°F/35°C, and puff-pastry margarines to 115°F /46°C: they don't melt until well into the baking process! However, high melting points have an important drawback: they mean that the fat remains solid at mouth temperature. Where butter and lard melt in the mouth and release luscious flavor, manufactured pastry shortenings can leave a pasty or waxy residue in the mouth, and have no true flavor of their own (they're often flavored with milk solids).

Water in Pastry Fats An important difference between butter and either lard or shortening is that butter is about 15% water by weight, and therefore doesn't separate dough layers as thoroughly as the pure fats do; water droplets in the fat can glue adjacent layers together. Pastry makers generally prefer European-style butters, which contain less water than standard American butter (p. 35). However, some water is useful for producing steam that pushes apart the dough layers of laminated pastries. Manufacturers formulate puff-pastry margarine with about 10% water.

Other Ingredients Water is essential for binding the flour particles into a dough, and the water content is especially critical in pastries because there is so little. Pastry cooks say that as little as ½ teaspoon/3 ml variation in water in 1 cup/120 gm flour can make the difference between a crumbly texture and a tough one. Eggs are often used to provide richness and added cohesiveness to crumbly pastries, and of course also contribute water. Various dairy products, including milk, cream, sour cream, crème fraîche, and cream cheese may replace some or all of the water, and at the same time provide flavor and fat as well as sugars and proteins for the browning reactions. Salt is added mainly for flavor, though it does have a tightening effect on gluten.

Cooking Pastries

Baking Pans Two portions of the same pastry dough, baked in the same oven but in different kinds of pan, will cook differently. Shiny pans reflect much of the oven's radiant heat (p. 782) away from the crust and so are slow cooking. Black pans absorb most of the radiant heat and conduct it to the crust, and clear glass allows it to pass right through and heat the crust directly. Thin metal pans can't hold much heat in themselves and so tend to slow heating and produce uneven browning. Heavier gauge metal pans and ceramic plates and molds can accumulate oven heat, get hotter than thin pans, and transmit the heat more evenly to the pastry.

Baking Apart from bready croissants and Danish, most pastry doughs contain very little water, not nearly enough to gelate all the starch granules. Cooking therefore partly gelates the starch and dries the gluten network well, and produces a firm, crunchy or crisp texture and a golden brown exterior. Pastry crusts in particular are cooked at relatively high oven temperatures so that the dough heats through and sets quickly. Slow heating just melts the pastry dough's fat, and the protein-starch network slumps before the starch gets hot enough to absorb water from the gluten and set the structure.

The filling in open pies and tarts blocks oven heat from reaching the pastry surface directly, and can prevent the crust from cooking through, so that it ends up pale and soggy rather than brown and dry. This problem is prevented by precooking the crust on its own (baking it "blind," or empty, often with dry beans or ceramic pie weights to support the dough and prevent slumping). A crisper bottom crust also results from higher oven temperatures and from putting the container on the lowest rack or directly on the oven floor. Crispness can be preserved under a moist filling by sealing the crust surface during precooking with a moisture-resistant layer of egg yolk or white, or afterwards with cooked-down jams or jellies, or chocolate, or a moisture-absorbing layer of compatible crumbs.

Crumbly Pastries: Short Pastry, Pâte Brisée

Crumbly but firm pastries are especially prominent in French cooking, where thin but robust crusts support quiches, various savory pies, and fruit tarts. Where American pie crusts are too tender to support themselves and are served from the pan, French tarts are almost always removed from the pan and stand on their own. In the standard French version of crumbly pastry, pâte brisée, coarse pieces of butter and egg yolks are placed in the midst of the proper amount of flour, and the liquid and solids gently worked together with the fingers to form a rough dough. The dough is then kneaded by pushing it into and along the work surface with the heel of the hand, an action that disperses the butter finely into the dough. The butter separates small flour aggregates from each other and prevents them from forming a continuous, tough mass, while the egg yolks provide moisture, fats, and proteins that will coagulate during cooking and help hold the flour aggregates together. The butter may be replaced by vegetable oils, poultry fats—chicken, duck, goose—and lard and beef tallow, depending on the nature of the filling. The dough is allowed to rest in the refrigerator to firm its consistency for the subsequent rolling out and shaping.

Pâte sucrée and pâte sablée—"sugar pastry" and "sandy pastry"—are versions of crumbly pastry made with sugar. The large proportion of sugar in pâte sablée gives a distinctly grainy character to the pastry.

One simple way to make crumbly pastry crusts is to start with premade crumbs, bread or cookie crumbs moistened with fat and simply pressed into the pan before a quick baking.

Flaky Pastries: American Pie Pastry

The methods for making American-style pie dough produce a crust that is both tender and flaky. They disperse some of the fat

Early American Pie Pastries

American pie pastry is notable for having some of its fat rubbed into the flour to tenderize, and some rolled in for flakiness. The first American recipes—Amelia Simmons's "Puff Pastes for Tarts"—are notable for their terseness and variety. She gives several; here are three.

> No. 1. Rub one pound of butter into one pound of flour, whip 2 whites and add with cold water and one yolk; make into paste, roll in six or seven times one pound of butter, flouring it each roll. This is good for any small thing.
>
> No. 3. To any quantity of flour, rub in three fourths of its weight in butter (twelve eggs to a peck), rub in one third or half, and roll in the rest.
>
> No. 8. Rub in one and half pound of suet to six pounds of flour, and a spoon full of salt, wet with cream, roll in, in six or eight times, two and half pounds of butter—good for a chicken or meat pie.
>
> —*American Cookery,* 1796

evenly into the dough, separating small particles from each other, and some coarsely, separating different layers of the dough from each other. There are various ways to accomplish this. One is to work the fat into the dry flour in two different stages, the first time finely, the second in pea-sized pieces. Another is to add the fat all at once, and use the fingers to fragment and gently rub the chunks down to pea size; the rubbing does the fine dispersion. (This method works better with shortening than with butter, which warm fingers can soften excessively.) A small amount of cold water, 2–4 tablespoons per cup/15–30 gm per 100 gm flour, is then added and the mixture manipulated very briefly, just until the water is absorbed and a dough forms.

The dough is rested in the refrigerator to rechill the fat and let the water become more evenly distributed, and then is rolled out. The rolling stretches the dough and thus develops some gluten, and flattens the fat chunks into thin sheets. The combination creates the layered texture. The rolled dough is then rested to allow the gluten sheets to relax, and shaped with minimal stretching; otherwise the gluten may rebound and the crust shrink during baking. In the oven, the sheets of fat, trapped air, and steam from the dough water (and the water in any butter) all help to separate the dough into layers and give it a flaky texture.

Shortening and lard generally produce more tender and flaky crusts than butter, which melts into the dough at a lower temperature and whose water can cause dough particles and flakes to stick to each other.

Laminated Pastries: Puff Pastry, Pâte Feuilleté

According to the food historian Charles Perry, puff and sheet pastries appear to have been invented by the Arabs and the Turks respectively, sometime around 1500. Though the aim in both is to produce many layers of very thin pastry, they involve two very different techniques.

Making Puff Pastry Preparing puff pastry dough is elaborate and time consuming. There are several different ways to construct the dough-fat sandwich, and several different ways to make the folds; here for simplicity I'll describe the standard one.

The cook first mixes pastry flour with ice water to make a moderately moist initial dough, with about 50 parts water per 100 parts flour. Sometimes butter and/or lemon

Early Recipes for Laminated Pastry: Early English "Puff Paste"

Gervase Markham's recipe for "puff paste" is a cross between laminated and sheet pastries.

> Now for the making of puff paste of the best kind, you shall take the finest wheat flour after it hath been a little baked in a pot in the oven, and blend it well with eggs, whites and yolks all together, after the paste is well kneaded, roll out a part thereof as thin as you please, and then spread cold sweet butter over the same, then upon the same butter roll another leaf of the paste as before; and spread it with butter also; and thus roll leaf upon leaf with butter between till it be as thick as you think good: and with it either cover any baked meat, or make paste for venison, Florentine, tart or what dish else you please and so bake it.
>
> —*The English Housewife*, 1615

juices are included to weaken the gluten and make the dough more easily shaped. The mixing is done with minimal manipulation to minimize gluten development, which the later rollings-out accomplish. The dough is shaped into a square.

Now the fat, traditionally butter and weighing about half the initial dough weight is pounded with a rolling pin until it warms up to about 60°F/15°C and becomes pliable, its consistency matching the consistency of the dough. (Firmer fat would tear the dough, softer fat would be squeezed out during the later rolling. Shortenings, with their low water content, produce a lighter and crisper puff pastry, though a less flavorful one.) The fat is formed into a flat piece, placed on the dough square, and the combination repeatedly folded onto itself and rolled out, with turns to vary the direction of rolling and rests in the refrigerator to give the fat a chance to resolidify and the gluten to relax. The sequence of turning, rolling, folding, and refrigerating is repeated several times, for a total of six "turns." With each rolling out, the gluten becomes more developed, and the dough more elastic and difficult to shape.

The result of this work, which takes several hours, is a dough made up of 729 layers of moistened flour separated by 728 layers of fat. (The term *millefeuille,* or "thousand-leaf," is applied to a pastry made by stacking two baked pieces of puff pastry, with a layer of pastry cream in between.) The dough is rested for at least an hour after the final turn, and then is rolled out for baking to a thickness of about a quarter of an inch/6 mm. This means that each layer in the dough is microscopically thin, around a thousandth of an inch or a hundredth of a millimeter. This is much thinner than paper thin, about the diameter of an individual starch granule.The dough must be cut with a very sharp knife; a dull blade will press the many layers together at the edge and restrain their expansion. When puff pastry is baked in a very hot oven, the expanding air and water vapor puff the separate layers apart from each other and cause the volume to increase by four or more times.

Quick Puff Pastry "Quick" puff pastry, also known as "flaky pastry" (England), "American" puff pastry, or demi-feuilleté, is a shortcut hybrid of true puff pastry and American flaky pie pastry. Again there are many versions. Usually some or all of the fat is cut coarsely into the flour as for pie pastry, cold water added to make a dough, any remaining fat sandwiched with the dough, and the dough then folded and rolled out two or three times, with periods of refrigeration to rechill the fat and relax the gluten.

Even quick puff pastry dough takes a couple of hours to make. Fortunately these doughs freeze well and are commercially available in frozen form.

Sheet Pastries: Phyllo, Strudel

Unlike puff pastry doughs, sheet pastry doughs are prepared one layer at a time, and are assembled into pastries with a few dozen layers just before cooking. Charles Perry speculates that phyllo pastry was invented in Istanbul in the time of the early

Food Words: *Phyllo, Strudel*

Phyllo is the Greek ancestor of French *feuille* and like it means "leaf." *Strudel* reflects the unusual rolled form of this version of sheet pastry: it is German for "eddy" or "whirlpool."

Ottoman empire around 1500; it's now used to make the Eastern Mediterranean honey-nut sweet baklava, savory turnovers (Turkish boreks), and many savory pies (Greek spanakopita and others). During the period when the Ottoman Turks ruled parts of eastern Europe, the phyllo leaf was adopted in Hungary as *retes* and in Austria as *strudel.*

Phyllo dough is prepared by making a stiff flour-water dough (about 40 parts water to 100 flour) with a little salt and often some tenderizing acid or oil. The dough is thoroughly kneaded to develop the gluten, rested overnight, and then stretched out either in a single mass, or in small balls that are rolled out into a thin disc, sprinkled with starch, and rolled out again. The dough eventually gets thin enough to become translucent, around 5 thousandths of an inch/0.1 mm thick. This is so thin that the silken dough quickly dries out and becomes brittle, so it's brushed with oil or melted butter to keep it supple until it's cut, stacked into a many-layered pastry, and baked.

The variant of phyllo called strudel is made somewhat differently. The initial dough is wetter, 55–70 parts water per 100 flour, and contains a small amount of fat and often whole egg. The dough is kneaded, rested, rolled fairly thin, rested again, and then gradually stretched with the hands into one large sheet, which is then used as a wrapper to roll around a variety of savory and sweet preparations.

Both phyllo and strudel doughs are especially tricky to make, and are available refrigerated and frozen.

Pastry-Bread Hybrids: Croissants, Danish Pastries

Croissants and Danish pastries are made in very much the same way that puff pastry is. Because the doughs for croissants and Danish pastries are essentially bread doughs, both moister and softer than puff dough, they are easily torn by cold, hard fat. The proper consistency of butter or margarine is therefore especially important in making croissants and Danish pastries.

Croissants According to Raymond Calvel, croissants first made a splash at the 1889 Paris World's Fair, where they were one of many kinds of *Wienerbrod,* or Vienna goods brought from the city that specialized in rich, sweet pastries. The original croissants were enriched yeast-raised breads shaped into a crescent. It wasn't until the 1920s that Parisian bakers had the idea of forming them from a laminated dough, thus creating a marvelous pastry that is both flaky and moistly, richly, tenderly bready.

Croissants are made by preparing a firm but malleable dough with minimal kneading from flour, milk, and yeast; the proportion of liquid is 50–70 parts to 100 flour. Some butter may be added to the dough during mixing to make the dough more extensible and easily rolled out. In earlier times, the dough was allowed an initial rise of six to seven hours; today it's only around one hour. The more time allowed for fermentation, the fuller the flavor and the lighter the finished pastry. The risen dough is deflated and chilled, then rolled out, covered with a layer of butter or pastry margarine, and repeatedly folded, rolled out, and chilled as puff pastry is, for a total of four to six turns. The finished dough is then rolled out to around a quarter of an inch/6 mm thick, cut into triangles, the triangles rolled up into tapered cylinders and allowed a final rise of about an hour at a temperature cool enough to prevent the fat from melting. When baked, the outer layers of the dough expand and dry out to form flaky, puff pastry–like sheets, while the inner layers remain moist and bake into exquisitely delicate sheets of bread, translucent and pebbled with tiny bubbles.

Danish Pastries What Americans call "Danish" pastries also originated as Vienna goods, but were introduced to the United States via Copenhagen. In the 19th century, Danish bakers took a basic Viennese enriched bread dough and added even more layering butter, thus making a lighter, crisper pastry than the original. They also used the dough to surround a variety of fillings, notably *remonce* (butter creamed with sugar and often including some form of almonds). Danish pastries are made in essentially the same way as croissants. The initial dough is moister and softer, includes sugar and also whole eggs, so it's sweeter, richer, and distinctively yellow, and it isn't given an initial rising. Often more butter or margarine is used for the laminations, and the dough may only be turned three times, so the layers are fewer and thicker. Danish pastry dough is often used as a container for sweet or rich fillings, or rolled out, covered with a combination of nuts, raisins or flavored sugar, rolled up, and cut into spiral cross-sections. Once the final pastry is formed, it's allowed to rise until about doubled in volume (again, at temperatures that keep the shortening solid), and then is baked.

Tender Savory Pastry: Hot-Water Pastry, Pâte à Pâté

Hot-water pastry, or pâté pastry, is unlike the other pastries. Its original purpose in medieval times was to provide a sturdy container for meat dishes meant to be kept for some time (p. 560). Today it's used to enclose meat pâtés, to make meat pies, and sometimes as an alternative to puff pastry that surrounds the tenderloin in beef Wellington and the salmon in coulibiac. It is easily rolled out and formed into a container, able to retain juices released during cooking, yet tender to both knife and tooth. It's made with a relatively large amount of water—50 parts per 100 flour—and about 35 parts lard. The water and lard are heated together near the boil, and the flour is then added, the mixture stirred just until it forms a homogeneous mass, then rested. The large proportion of fat limits gluten development, thus providing tenderness, and also acts as a kind of water repellant, thus providing a barrier against cooking juices. The precooking swells and gelates some of the flour starch, which takes up water and gives the dough a thick, workable consistency in place of an elastic gluten structure.

COOKIES

Common cookies are simple pleasures, but the microcosm of all cookies is a *summa* of the baker's art. Cookies include sweet bite-sized baked goods of all sorts: crumbly and laminated pastries, wafers, butter and sponge cakes, biscuits, meringues, nut pastes. The term comes from the medieval Dutch for "little cake." The French equivalent is *petits fours,* or "little oven goods," and the German *klein Gebäck* means much the same. Their miniature size and the numerous possibilities for shaping, decorating, and flavoring have resulted in a great diversity of cookies, many of them developed by the French and named in the same spirit that gave us Italian pastas called butterflies, little worms, and priest-stranglers: hence cat's tongues, Russian cigarettes, eyeglasses, and Nero's ears.

Cookie Ingredients and Textures

Most cookies are both sweet and rich, with substantial proportions of sugar and fat. They're also tender, thanks to ingredients, proportions, and mixing techniques that minimize the formation of a gluten network. But then they may be moist or dry, crumbly or flaky or crisp or chewy. The diversity of textures arises from a handful of ingredients, and from the proportions and methods of combining them.

Flour Most cookies are made with pastry or all-purpose flour, but both bread flour and cake flour produce doughs and batters that spread less (thanks respectively to more gluten and more absorbant starch). A high proportion of flour to water, as in shortbread and pastry-dough cookies, limits both gluten development and starch gelation—as little as 20% of the starch in some dry cookies is gelated—and produces a crumbly texture. A high proportion of water to flour, as in batter-based cookies, dilutes gluten proteins, allows extensive starch gelation, and produces either a soft, cakelike texture or a crisp, crunchy one, depending on the method and how much moisture is baked out of the cookie. For doughs that need to hold their shape during baking—those rolled out and stamped with a cookie cutter—a high flour content and some gluten development are necessary. The baker gives fluid batters some solidity by chilling them, and then shapes them by extruding them through a pastry pipe or setting them in molds.

A coarser but more fragile backbone can be created by replacing some or all of the flour with ground nuts, as in classic macaroons made only with egg whites, sugar, and almonds.

Sugar Sugar makes several contributions to cookie structure and texture. When creamed with the fat, or beaten with egg, it introduces air bubbles into the mix and lightens the texture. It competes with the flour starch for water, and raises the starch gelation temperature nearly to the boiling point: so it adds hardness and crispness. A large proportion of pure table sugar, sucrose, contributes to hardness in another way. The proportion of sugar in some cookie doughs is so high that only about half the sugar dissolves in the limited amount of moisture. When the dough heats up during baking, more sugar can dissolve, and the added liquid causes the cookie to soften and spread. Then when the cookie cools, some of the sugar recrystallizes, and the initially soft cookie develops a distinctive snap—a process that may take a day or two. Other forms of sugar—honey, molasses, corn syrup—tend to absorb water rather than crystallize (chapter 12), so when heated they form a syrup that permeates the cookie, helps it to spread, and firms as it cools, making it moist and chewy.

Eggs Eggs generally provide most of the water in a cookie mix, as well as proteins that help bind the flour particles together and coagulate during baking to add solidity. The fat and emulsifiers in the yolk enrich and moisten. The higher the proportion of whole eggs or yolks in a recipe, the more cake-like the texture.

Fat Fat provides richness, moistness, and suppleness. When it melts during cooking, it lubricates the solid particles of flour and sugar and encourages the cookie to spread and thin—a quality that is sometimes desirable, sometimes not. Because butter melts at a lower temperature than margarine or shortening, it gives cookies more time to spread before the protein and starch set. Butter is about 15% water, and is the main or only source of moisture in such low-egg recipes as shortbread and tea cookies.

Leavening Leavening, whether tiny bubbles of air or of carbon dioxide, helps tenderize cookies, and encourages them to puff. Many cookies are leavened only with air bubbles incorporated when the sugar is creamed with the fat, or beaten with the eggs. Some are supplemented with chemical leavenings. Alkaline baking soda may be used when the dough includes such acid ingredients as honey, brown sugar, and cake flour.

Making and Keeping Cookies

There are as many ways to prepare cookies as there are ways to produce cakes and pastries—and then some. The standard American categories are:

- *Drop cookies*, formed from a soft dough that is portioned by spoonfuls onto the baking sheet, where they spread out during baking. Chocolate chip and oatmeal cookies are examples.
- *Cut-out cookies*, formed from a stiffer dough that holds its shape. The dough is rolled out and por-

tioned with a cookie cutter; baking sets the cookies in their original shape. Sugar cookies and butter cookies are examples.

- *Hand-shaped cookies*, formed from batters that are stiffened by chilling and then carefully piped or molded for baking. Examples are ladyfingers and madeleines.
- *Bar cookies*, shaped after baking, not before. They're cut from the thin cake-like mass produced when the cookie batter is baked in a shallow pan. Date and nut bars and brownies are examples.
- *Icebox cookies*, formed by slicing cross-sections from a premade cylinder of dough kept in the refrigerator for use when needed. Many cookie doughs can be treated this way.

Thanks to their small size, thinness, and high sugar content, cookies quickly brown at oven temperatures. Their bottoms and

Some Cookie Doughs and Batters: Ingredients and Typical Proportions

Cookie	Flour	Total Water	Eggs	Butter	Sugar	Chemical Leavening
Shortbread (crumbly)	100	15	—	100	33	—
Biscotti (crunchy after drying)	100	35	45	—	60	yes
Chocolate chip (cakey)	100	38	33	85	100	yes
Tuiles, wafers (crisp)	100	80	80	50	135	—
Pastry doughs						
Short: Sablés (crumbly)	100	25	22	50	50	—
Puff: Palmiers (flaky)	100	35	—	75	(topping)	—
Creamed-butter batters						
Tea cookies (crumbly)	100	25	18 (whites)	70	45	—
Lady wafers, cat's tongue (delicate, crisp)	100	90	100	100	100	—
Russian cigarette (thin, crisp)	100	180	180 (whites)	140	180	—
Spongecake batters						
Ladyfingers (light, dry)	100	150	200	—	100	
Madeleines (soft, moist, cakey)	100	145	170	110	110	(yes)

edges may get too dark while the centers finish cooking, a problem that can be minimized by lowering the oven temperature and using light baking sheets that reflect radiant heat, rather than dark ones that absorb it. Slight underbaking helps produce a moister, chewier texture. Immediately after baking, many cookies are soft and malleable, a fact that allows the cook to shape thin wafer cookies into flowerlike cups, rolled cylinders, and arched tiles that then stiffen as they cool.

With their low water content, cookies are especially prone to losing their texture during storage. Crisp, dry cookies absorb moisture from the air and get soft; moist, chewy cookies lose moisture and become dry. Cookies are therefore best stored in an airtight container. With their low moisture and high sugar levels, they are not very hospitable to microbes, and keep well.

PASTA, NOODLES, AND DUMPLINGS

One of the simplest preparations of cereal flour gave us one of the most popular foods in the world: pasta. The word is Italian for "paste" or "dough,"and pasta is nothing more than wheat flour and water combined to make a clay-like mass, formed into small pieces, and boiled in water until cooked through—not baked, as are nearly all other doughs. *Noodle* comes from the German word for the same preparation, and generally refers to pasta-like preparations made outside the Italian tradition. The keys to pasta's appeal are its moist, fine, satisfyingly substantial texture and its neutral flavor, which makes it a good partner for a broad range of other ingredients.

Two cultures in the world have thoroughly explored the possibilities of boiled grain paste: Italy and China. Their discoveries were different, and complementary. In Italy, the availability of high-gluten durum wheat led to the development of a sturdy, protein-rich pasta, one that can be dried and stored indefinitely, one that readily lent itself to industrial manufacturing, and that can be formed into hundreds of fanciful shapes. The Italians also refined the art of making fresh pastas from soft wheat flours, and evolved an entire branch of cooking based on pasta as the principal ingredient, its combination of substance and tenderness providing the foundation for flavorful sauces—usually just enough to coat the surfaces—and fillings. In China, which had soft, low-gluten wheats, cooks concentrated on simple long noodles and thin wrappers, prepared them fresh and by hand, sometimes with great panache and just moments before cooking, and served the soft, slippery results almost exclusively in large amounts of thin broth. More remarkably, Chinese cooks found ways to make noodles from many different materials, including other grains and even pure, protein-free starch from beans and root vegetables.

The History of Pasta and Noodles

It's a story often told, and often refuted, that the medieval traveler Marco Polo found noodles in China and introduced them to Italy. A recent book by Silvano Serventi and Françoise Sabban has set the record straight in authoritative and fascinating detail. China was indeed the first country to develop the art of noodle making, but there were pastas in the Mediterranean world long before Marco Polo.

Noodles in China Despite the fact that wheat was grown in the Mediterranean region long before it arrived in China, the northern Chinese appear to have been the first to develop the art of noodle making, sometime before 200 BCE. Around 300 CE, Shu Xi wrote an ode to wheat products (*bing*) that names several kinds of noodles and dumplings, describes how they're made, and suggests their luxurious qualities: poets frequently likened their appearance and texture to the qualities of silk (see box, p. 572). In 544, an agricultural treatise called *Important Arts for the People's Welfare* devoted an entire chapter to dough products. These included not only several different shapes of wheat noodle, most made by mixing flour with meat broth and one made with egg, but also noodles made from rice flour and even from pure starch (p. 579).

China also invented filled pasta, the original ravioli, in which the dough surrounds and encloses a mass of other ingredients. Both small, thin-skinned, delicate *hundun* or wontons, now usually served in southern soups, and the thicker *chiao-tzu* or pot sticker, often steamed and fried in the north, are mentioned in written records before 700 CE, and archaeologists have found well-preserved specimens that date to the 9th century. Over the next few hundred years, recipes describe making thin noodles by slicing a rolled dough sheet and by repeatedly pulling and folding a dough rope; and the doughs themselves are made with a variety of liquids, including radish and leaf juices, vegetable purees, juice pressed from raw shrimp (which makes the noodles pink), and sheep's blood.

Noodles—*mian* or *mein*—and filled dumplings began in the north as luxury foods for the ruling class. They gradually became staples of the working class, with dumplings retaining the suggestion of prosperity, and spread to the south around the 12th century. Noodles made their way to Japan by the 7th or 8th century, where several kinds of *men* evolved (p. 578).

Pasta in the Middle East and Mediterranean Far to the west of China, in the homeland of wheat, the earliest indications of pasta-like preparations come in the 6th century. A 9th century Syrian text gives the Arabic name *itriya* to a preparation of semolina dough shaped into strings and dried. In 11th-century Paris, mention is made of vermicelli, or "little worms." In

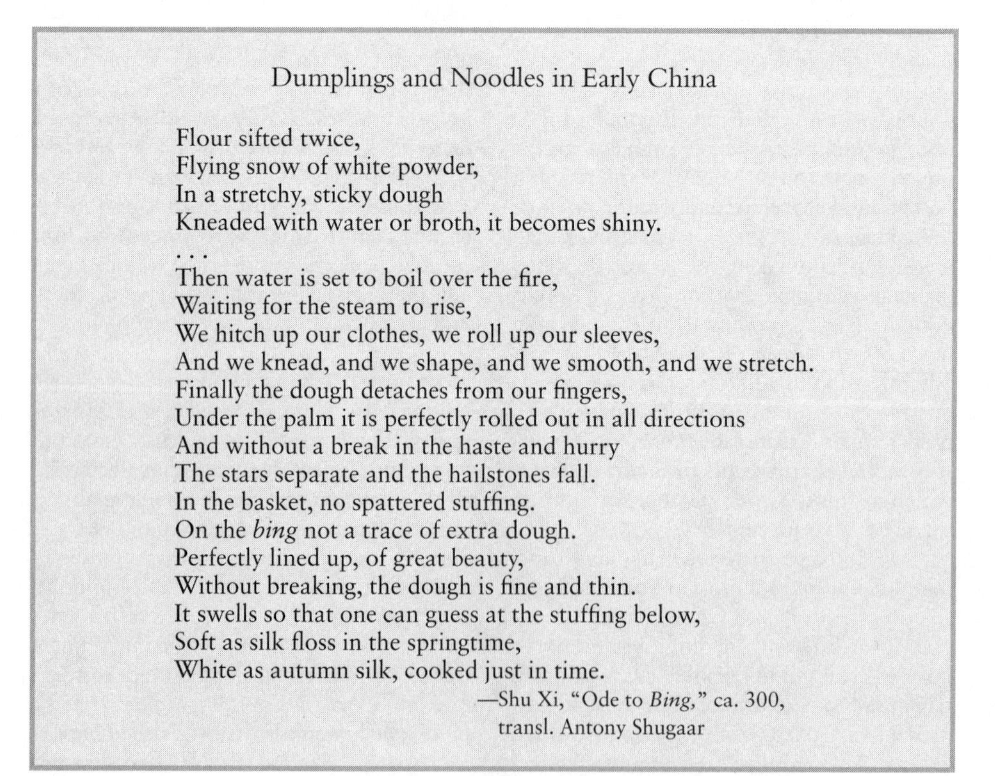

Dumplings and Noodles in Early China

Flour sifted twice,
Flying snow of white powder,
In a stretchy, sticky dough
Kneaded with water or broth, it becomes shiny.
. . .
Then water is set to boil over the fire,
Waiting for the steam to rise,
We hitch up our clothes, we roll up our sleeves,
And we knead, and we shape, and we smooth, and we stretch.
Finally the dough detaches from our fingers,
Under the palm it is perfectly rolled out in all directions
And without a break in the haste and hurry
The stars separate and the hailstones fall.
In the basket, no spattered stuffing.
On the *bing* not a trace of extra dough.
Perfectly lined up, of great beauty,
Without breaking, the dough is fine and thin.
It swells so that one can guess at the stuffing below,
Soft as silk floss in the springtime,
White as autumn silk, cooked just in time.

—Shu Xi, "Ode to *Bing*," ca. 300,
transl. Antony Shugaar

the 12th century—around 200 years before Marco Polo's travels—the Arab geographer Idrisi reported that the Sicilians made thread-like *itriya* and exported them. The Italian term *macaroni* first appeared in the 13th century and was applied to various shapes, from flat to lumpy. Medieval cooks made some pastas from fermented doughs; they cooked pasta for an hour or more until it was very moist and soft; they frequently paired it with cheese, and used it to wrap around fillings.

The postmedieval evolution of pasta took place largely in Italy. Pasta makers formed guilds and made fresh types from soft wheat flour throughout Italy, dried types from durum semolina in the south and in Sicily. Italian cooks developed the distinctive preparation style called *pastasciutta* or "dry pasta," pasta served as the main component of the dish, moistened with sauce but not drowning in it or dispersed in a soup or stew. With its ideal climate for drying raw noodles, a tricky process that took one to four weeks, Naples became the center of durum pasta manufacturing.

Thanks to the mechanization of dough kneading and extrusion, by the 18th century durum pasta had become street food in Naples, and common in much of Italy. Perhaps because street vendors minimized cooking and open-air consumers enjoyed chewing on something substantial, it was in Naples that people began to prefer pasta cooked for minutes rather than hours, so that it retains some firmness. This practice spread to the rest of the country in the late 19th century, and the term *al dente,* or cooked "to the tooth," appeared after World War I. Subsequent decades brought effective artificial drying and the machinery and understanding necessary to turn pasta making from a batch-by-batch process to a continuous one. Dried durum pasta is now made on an industrial scale in many countries. In addition, modern heat treatments and vacuum-packing have also made it possible to keep fresh pasta for several weeks in the refrigerator.

Recent decades have brought a revival in Italy of small-scale manufacturing using selected wheat varieties, old-fashioned extrusion dies that produce a rough, sauce-holding surface, and longer-time, low-temperature drying that is said to produce a finer flavor.

Making Pasta and Noodle Doughs

Basic Ingredients and Methods The aim in making pasta and noodle dough is to transform dry flour particles into a cohesive mass that is malleable enough to be shaped into thin strips, but strong enough to stay intact when boiled. With wheat flours, the cohesiveness is provided by the gluten pro-

Pasta, Cheese, and Wine Without End

By the time of the great storyteller Giovanni Boccaccio, who died in 1375, pasta was familiar enough in Italy to be part of a glutton's paradise:

> in a country called Bengodi . . . there was a mountain made entirely of grated Parmesan cheese, on which lived people who did nothing but make macaroni and ravioli and cook them in capon broth. And then they threw them down, and the more of them you took, the more you had. And nearby ran a rivulet of white wine whose better was never drunk, and without a drop of water in it.
>
> —*Decameron*, Day 8, Tale 3

teins. Durum wheats have the advantage of a high gluten content, and a gluten that is less elastic than bread-wheat gluten and so easier to roll out. Water normally makes up around 30% of the dough weight, compared to 40% or more for bread doughs.

After the ingredients are mixed and briefly kneaded into a homogenous but stiff mass, pasta dough is rested to allow the flour particles to absorb the water and the gluten network to develop. With time the dough becomes noticeably easier to work, and the finished noodles end up with a cohesive consistency rather than a crumbly one. The dough is then rolled gently and repeatedly to form an ever thinner sheet. This gradual sheeting presses out air bubbles that weaken the dough structure, and organizes the gluten network, compressing and aligning the protein fibers, but also spreading them out so that the dough becomes more easily stretched without snapping back.

Egg Pasta and Fresh Pasta Noodles made from standard bread wheat and eggs are preferred in much of northern Europe, and most fresh pastas sold in the United States are of this type. Eggs perform two functions in noodles. One is to enhance color and richness. Here the yolk is the primary factor, and yolks alone can be used; their fat content also makes the dough more delicate and the noodles tender. The second function is to provide additional protein for moderate-protein flours used in both home and industrial production. The egg white proteins make the dough and noodles more cohesive and firm, reduce the gelation and leaking of starch granules, and reduce cooking losses. In U.S. commercial noodles, dried egg is added at 5–10% of the weight of the flour. In Italy, Alsace, Germany, and in specialty and homemade noodles in the United States, fresh eggs are used, and in larger proportions. They may be the only source of water in the dough. Some pastas from the Piedmont region in northwest Italy contain as many as 18 yolks per pound of flour/40 yolks per kg.

To make egg pasta, the ingredients are mixed in proportions that give a stiff dough; the dough is kneaded until smooth, allowed to rest and relax, rolled out, and then cut into the desired shapes. The fresh pasta is perishable, and if made with eggs carries a slight possibility of salmonella contamination; it should be cooked immediately or wrapped and refrigerated. Prolonged drying at kitchen temperature may allow microbes to multiply to hazardous levels. Fresh pasta cooks quickly, in a few seconds or minutes depending on thickness.

Dried Durum Pasta The standard Italian pastas, and Italian-style pastas from around the world, are made from durum wheat, with its distinctive flavor, attractive yellow color, and abundant gluten proteins. Durum pastas are seldom made with eggs. Their gluten proteins give the dried noodles a hard, glassy interior; during cooking they limit the loss of dissolved proteins and gelated starch, and make a firm noodle.

Making Durum Pasta Dough and Shapes Durum pasta is made from *semolina,* which is milled durum endosperm with a characteristically coarse particle size, 0.15–0.5 mm across, thanks to the hard nature of durum endosperm (finer grinding causes excessive damage to starch granules). Flat pasta shapes are punched from out of a sheet of dough. Long noodles and short thick ones are formed by extruding the dough through the holes of a die at high pressure. The movement, pressure, and heat of extrusion change the structure of the dough by shearing the protein network apart, mixing it more intimately with starch granules that have been partly gelated by the heat and pressure, and allowing broken protein bonds to re-form and stabilize the new network. Noodles extruded through modern low-friction Teflon dies end up with a glossier, smoother surface, with fewer pores and cracks through which hot water can leak in and dissolved starch can leak out. They generally lose less starch to the cooking water, absorb less cooking water, and therefore have a firmer texture than the same noodle extruded through a traditional bronze die. Proponents of traditional dies prefer the rougher surface, which they say better retains the sauce in the finished dish.

Drying Durum Pasta Before the invention of mechanical driers, manufacturers held the new pasta at ambient temperatures and humidities for days or weeks. Early industrial driers operated at 100–140°F/40–60°C and took about a day. Modern drying takes only two to five hours and involves rapid predrying at or above 185°F/84°C, and then a more extended phase of drying and resting periods. The modern high-temperature method rapidly inactivates enzymes that can destroy the yellow xanthophyll pigments and cause brown discoloration, and it cross-links some of the gluten protein and produces a firmer, less sticky cooked noodle. However, proponents of slow drying say that high heat also damages flavor.

Cooking Pasta and Noodles

When pasta is cooked in water, the protein network and starch granules absorb water and expand, the outer protein layer is ruptured, and the dissolving starch escapes into the cooking water. Deeper within the noodle there's less water available, so the starch granules aren't completely disrupted: the center of the noodle therefore stays more intact than the surface. Cooking pasta *al dente* means stopping the cooking when the center of the noodle still remains slightly underdone and offers some resistance to chewing; at this point, the noodle surface is 80–90% water, the center 40–60% (somewhat moister than freshly baked bread). Pasta is sometimes cooked just short of this point and then finished in the sauce that will dress it.

Cooking Water It's generally recommended that pasta be cooked in 10 or more times its weight of vigorously boiling water (around 5 quarts or liters water per

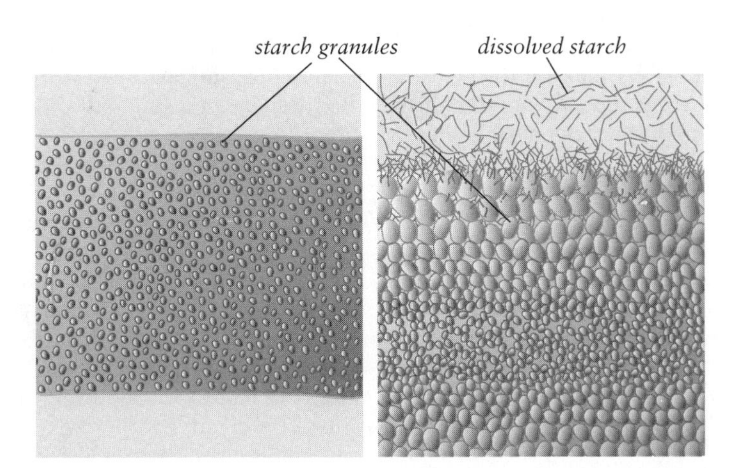

Cooking pasta. Left: *Uncooked pasta dough consists of raw starch granules embedded in a matrix of gluten protein.* Right: *When pasta is cooked in water, starch granules at and near the noodle surface absorb water, swell, soften, and release some dissolved starch into the cooking water. In pasta done* al dente, *hot water has penetrated to the center of the noodle, but the starch granules there have absorbed relatively little, and the starch-gluten matrix remains firm.*

pound/500 gm). This allows for the pasta's absorption of 1.6–1.8 times its weight, and leaves plenty to dilute the starch that escapes during cooking, and to separate the noodles from each other so that they cook evenly and without sticking. Hard water—water that is alkaline and contains calcium and magnesium ions—increases both cooking losses and stickiness in noodles (it probably weakens the protein-starch film at the noodle surface, and the ions act as a glue to bond noodle surfaces to each other). Most city tap water has been made alkaline to reduce pipe corrosion, so pasta cooking water can often be improved by adding some form of acid (lemon juice, cream of tartar, citric acid) to adjust the pH to a slightly acidic 6.

Stickiness Noodles stick to each other during cooking when they're allowed to rest close to each other just after they're added to the cooking water. Their dry surfaces absorb the small amount of water between them so there's none left for lubrication, and the partly gelated surface starch glues the noodles together. Sticking can be minimized by constantly stirring the noodles for the first few minutes of cooking, or by adding a spoonful or two of oil to the pot and then lifting the noodles through the water surface a few times to lubricate them. Salt in the cooking water not only flavors the noodles, but limits starch gelation and so reduces cooking losses and stickiness.

Stickiness after cooking is caused by surface starch that dries out and cools down after the noodles have been drained, and develops a gluey consistency. It can be minimized by rinsing the drained noodles, or moistening them with some sauce, cooled cooking water, oil, or butter.

Couscous, Dumplings, Spätzle, Gnocchi

Couscous Couscous is an elegantly simple pasta that appears to have been invented by the Berber peoples of northern Algeria and Morocco between the 11th and 13th centuries. It remains a staple dish in North Africa, the Middle East, and Sicily. In its traditional form, couscous is made by sprinkling salted water into a bowl containing whole wheat flour, then stirring with the fingers to form little bits of dough. The bits are rubbed between the hands and sieved to obtain granules of uniform size, usually 1–3 mm in diameter. There is no kneading and therefore no gluten development, so this gentle technique can be and is applied to many other grains. Couscous granules are small enough that they can be cooked not in a large excess of water but in steam (traditionally over the fragrant stew that it will accompany), which allows them to develop a uniquely light, delicate texture. Couscous works best with thin sauces that spread easily over the large surface area of the small granules.

"Israeli" or "large" couscous is actually an extruded pasta invented in Israel in the 1950s. It's made from a dough of hard wheat flour formed into balls a few millimeters in diameter and lightly toasted in an oven to add depth to the flavor. It's cooked and served in the same ways as pasta and rice.

Dumplings and Spätzle Western dumplings and Spätzle (a word in a Bavarian dialect meaning "clod, clump," not "sparrow" as is often said) are essentially coarse, informal portions of dough or batter that are dropped into a pot of boiling water and cooked through, and served as is in a stew or braise or sautéed to accompany a meat dish. Unlike pasta doughs, dumpling doughs are minimally kneaded to maximize tenderness, and benefit from the inclusion of tiny air pockets, which provide lightness. The progress of cooking is judged by the position of the dumpling in the pot; when it rises to the top, it's considered almost done, given another minute or so, and then scooped out. This tendency to rise with cooking is due to the expansion of the dough's air pockets, which fill with vaporized water as the dumpling interior approaches the boiling point and make the dough less dense than the surrounding water.

Gnocchi Gnocchi—the word is Italian and means "lumps"—got their start in the 1300s as ordinary dumplings made from bread crumbs or flour (Roman gnocchi are still made by baking squares of a cooked dough of milk and semolina). But with the arrival of the New World's potato, Italian cooks transformed gnocchi into a form of dumpling with an unusually light texture. The starchy potato flesh became the main, tender ingredient, with just enough flour added to absorb moisture and provide gluten to hold it together into a formable dough. Eggs are sometimes added to provide additional binding and yolky richness, though they also add a springy quality. Old potatoes, and mealy rather than waxy varieties, are preferred for their lower water and higher starch contents, which means that less flour is needed to make the dough, so less gluten forms and the dumpling is more tender. The potatoes are cooked, peeled, and riced immediately to allow as much moisture as possible to evaporate; then cooled or even chilled, and kneaded into a dough with just as much flour as necessary, usually less than 1 cup/120 gm per lb/500 gm potatoes. The dough is formed into a thin rope and cut into small pieces, the pieces shaped, and then boiled in water until they rise to the top of the pot. Gnocchi can also be made by replacing the potato with other starchy vegetables or with ricotta cheese.

Asian Wheat Noodles and Dumplings

Two very different families of noodles are made in Asia. Starch noodles are described below. Asian wheat noodles—Chinese *mian*—bear some resemblance to European pastas made from bread wheat. They're typically made from low- or moderate-protein flours, and are formed not by extrusion but by sheeting and cutting or by stretching. The most spectacular form of noodle production is that of Shanghai's hand-pulled noodles, *la mian,* for which the maker starts with a thick rope of dough, swings, twists, and stretches it to arms' length, brings the ends together to make the one strand into two—and repeats the stretching and folding as many as eleven times to make up to 4,096 thin noodles! Asian noodles are both elastic and soft, their texture created by both their weak gluten and by amylopectin-rich starch granules. Salt, usually at around 2% of the noodle weight, is an important ingredient in Asian noodles. It tightens the

Soba: Japanese Buckwheat Noodles

Buckwheat noodles were made in northern China in the 14th century, and had become a popular food in Japan by around 1600. It's difficult to make noodles exclusively with buckwheat flour because the buckwheat proteins do not form a cohesive gluten. Japanese *soba* noodles may be from 10% to 90% buckwheat, the remainder wheat. They're traditionally made from freshly milled flour, which is mixed very quickly with the water and worked until the water is evenly absorbed and the dough firm and smooth. Salt is omitted because it interferes with the proteins and mucilage that help bind the dough (p. 483). The dough is rested, then rolled out to about 3 mm thick and rested again, then cut into fine noodles. The noodles are cooked fresh, and when done, are washed and firmed in a container of ice water, drained, and served either in a hot broth or cold, accompanied by a dipping sauce.

gluten network and stabilizes the starch granules, keeping them intact even as they absorb water and swell.

Chinese Wheat Noodles and Dumplings

White and Yellow Noodles Salted white noodles arose in northern China and are now most widely known in their Japanese version, udon (below). Yellow noodles, which are made with alkaline salts, appear to have originated in southeast China sometime before 1600, and then spread with Chinese migrants to Indonesia, Malaysia, and Thailand. The yellowness of the traditional noodles (modern ones are sometimes colored with egg yolks) is caused by phenolic compounds in the flour called flavones, which are normally colorless but become yellow in alkaline conditions. The flavones are especially concentrated in the bran and germ, so less refined flours develop a deeper color. Because they're based on harder wheats, southern yellow noodles have a firmer texture than white salted noodles, and alkalinity (pH 9–11, the equivalent of old egg whites) increases this firmness. The alkaline salts (sodium and potassium carbonates at 0.5–1% of noodle weight) also cause the noodles to take longer to cook and absorb more water, and they contribute a characteristic aroma and taste.

Dumplings The Chinese version of filled pasta is thin sheets of wheat-flour dough enclosing seasoned morsels of meat, shellfish, or vegetables. Some doughs are simply made from flour and water, but robust potsticker dough is made by boiling part of the water before adding the flour, so that some of the starch is gelated and contributes to the dough's cohesiveness. The formed dumplings may be steamed, boiled, fried, or deep-fried.

Japanese Wheat Noodles The standard thick Japanese noodles (2–4 mm in diameter), called *udon,* are descendents of the Chinese white salted noodle. They're white and soft and made from soft wheat flour, water, and salt. *Ra-men* noodles are light yellow and somewhat stiff, and are made from hard wheat flour, water, and alkaline salts (*kansui*). Very thin noodles (around 1 mm) are called *so-men.* Japanese noodles are usually cooked in water of pH 5.5–6, which is often adjusted by adding some acid. After cooking, the noodles are drained and washed and cooled in running water, which causes the surface starch to set into a moist, slippery, nonsticky layer.

Tapioca

Tapioca pearls, which are widely used to absorb moisture and flavor, thicken puddings and pie fillings, and nowadays to provide chewy "bubbles" in teas and other drinks, are translucent, glossy, and elastic, and based on the same principle as the starch noodle. They are spheres 1–6 mm across made up of tapioca starch granules held together by a matrix of gelatinized tapioca starch (about 17% amylose). A wet mass of the starch granules (40–50% water by weight) is broken up into coarse grains, and the grains then fed into rotating pans, where they roll around and gradually agglomerate into little balls. They're then steamed until a little more than half the starch is gelated, mostly in the outer layer, and then are dried, so that a firm retrograded starch matrix forms. When cooked in liquid, they soak up water and the rest of their starch gelates while the retrograded matrix maintains their structure.

The Japanese instant version of Chinese-style noodles, *ra-men,* was born in 1958. They're manufactured by making thin, quickly rehydrated noodles, then steaming them, frying them at 280°F/140°C, and air-drying at 180°F/80°C.

Asian Starch and Rice Noodles

All the pastas we've looked at so far are held together by the gluten proteins of wheat flour. Starch and rice noodles contain no gluten whatsoever. Starch noodles in particular are a remarkable, even startling invention: unlike all other noodles, they're translucent. They're often called glass or cellophane noodles, and in Japan are given the lovely name *harusame,* "spring rain" noodles.

Starch Noodles Dried noodles made out of pure starch—usually from mung beans (China), rice (Japan), or sweet potato—are prized for several qualities: their clarity and glossy brilliance, their slippery, firm texture, and their readiness for eating after just a few minutes of soaking in hot liquid, whether plain hot water or a soup or braised dish.

The firmest noodles are made from starches high in the straight-chain amylose form (p. 457). Where ordinary long-grain rice is 21–23% amylose, special noodle rices are 30–36%, and mung-bean starch is 35–40% amylose. Starch noodles are made by first cooking a small amount of dry starch with water to make a sticky paste that will bind the rest of the starch into a cohesive dough. The paste is mixed with the rest of the dry starch and more water to make a dough with 35–45% moisture, and the dough is then extruded through small holes in a metal plate to form noodles. The noodles are immediately boiled to gelate all the starch and form a continuous network of starch molecules throughout, and then are drained and held at the ambient temperature or chilled for 12–48 hours before being air-dried. During the holding period, the gelated starch molecules fall into a more orderly arrangement, or retrograde (p. 458). The smaller amylose molecules cluster together to form junctions in the network, crystalline regions that resist disruption even by boiling temperatures. The dried noodles are thus firm and strong, but the less orderly parts of the network readily absorb hot liquid and swell to become tender without the need for active cooking.

Starch noodles are translucent because they're a uniform mixture of starch and water, with no particles of insoluble protein or intact starch granules to scatter light rays.

Rice Noodles and Wrappers Like starch noodles, rice noodles are held together by amylose, not gluten; but because they contain protein and cell-wall particles that scatter light, they're opaque rather than translucent. Rice noodles are made by soaking high-amylose rice in water, grinding it into a paste, cooking the paste so that much but not all of the starch is gelated, kneading the paste into a dough and extruding it to form noodles, steaming the noodles to finish the gelation process, cooling and holding for 12 hours or more, and drying with hot air or by frying them in oil. Again, the holding and drying cause starch retrogradation and the formation of a structure that stands up to rehydration in hot water. Fresh rice noodles, *chow fun,* need no rehydration before being stir-fried.

Rice papers, *banh trang* in Vietnamese, are thin, parchment-like discs that are used as wrappers for southeast Asian versions of the spring roll. They're made by soaking and grinding rice, soaking it again, pounding it into a paste and spreading it into a thin layer, steaming it and then drying it. Rice papers are rehydrated briefly in lukewarm water, then used immediately as wrappers that can be eaten fresh or fried.

CHAPTER 11

SAUCES

Sauces are liquids that accompany the primary ingredient in a dish. Their purpose is to enhance the flavor of that ingredient—a portion of meat or fish or grain or vegetable—either by deepening and broadening its own intrinsic flavor, or by providing a contrast or complement to it. While the meat or grain or vegetable is always more or less itself, a sauce can be anything the cook wants it to be, and makes the dish a richer, more various, more satisfying composition. Sauces help the cook feed our perpetual hunger for stimulating sensations, for the pleasures of taste and smell, touch and sight. Sauces are distillations of desire.

The word *sauce* comes from an ancient root word meaning "salt," which is the original concentrated flavoring, pure mineral crystals from the sea (p. 639). Our primary foods—animal flesh, grains and breads and pastas, starchy vegetables—are pretty bland, and cooks have found or invented a vast range of ingredients with which to make them more flavorful. The simplest are *seasonings* provided by nature: salt, pungent black pepper and chillis, sour juices of unripe fruits, sweet honey and sugar, distinctively aromatic herbs and spices. More complex are prepared *condiments,* many of them foods preserved and transformed by fermentation: sour and aromatic vinegar, salty and savory soy sauce and fish sauce, salty and sour pickles, pungent and sour mustard, sweet and sour and fruity ketchup. And then there are *sauces,* the ultimate composed flavorings. The cook conceives and prepares sauces for particular dishes, and can give them any flavor. They always include seasonings, sometimes condiments, and sometimes artfully intensified flavors of the primary foods themselves, or of other foods, or of the cooking process.

In addition to their heightened flavor, sauces give tactile pleasure by the way they move in the mouth. Cooks construct sauces to have a consistency somewhere between the resistant solidity of animal or plant tissues and the elusive thinness of water. This is the consistency of luscious ripe fruit that melts in the mouth and seems to feed us willingly, and of the fats that give a persistent, moist fullness to animal flesh and to cream and butter. The fluidity of a sauce allows it to coat the solid food evenly and lend it a pleasing moistness, while the substantial, lingering quality helps the sauce cling to the food and to our tongue and palate as well, prolonging the experience of its flavor and providing a sensation of richness.

A last pleasure that a sauce can provide is an attractive appearance. Many sauces are nondescript, but others have the vibrant color of their parent fruit or vegetable, or the depth of tone that comes with roasting and long cooking. Some have an attractive sheen, and some are intriguingly transparent. The visual beauty of a sauce is a sign of the care with which it was made, a suggestion of intensity and clarity of flavor and of presence on the tongue: an anticipation of pleasures to come.

There are several basic ways of making sauces. Many of them involve disrupting organized plant and animal tissues and freeing the juices that carry their flavor. Once extracted from their source, the juices can be combined with other flavorful materials, and then often benefit from thickening to help them linger on the food and in the mouth. The cook thickens juices by filling them with a variety of large molecules or particles that obstruct the flow of the water molecules. Most of this chapter deals with different thickening methods and their applications.

Sauces are closely related to two other basic preparations. *Soups* are also liquid foods of various consistencies, and may differ from sauces only in being somewhat less concentrated in flavor, so that they can be eaten as a food in themselves, not an accent. And *jellies* are thickened liquids with enough gelatin in them to set at room temperature, thus becoming a temporarily solid food that melts into a sauce in the mouth.

THE HISTORY OF SAUCES IN EUROPE

Europe is just one of several regions in the world that have evolved sauces with broad appeal in modern times. Many sauces are now popular far from their birthplaces, among them Chinese soy-based sauces, Indian sauces thickened and flavored with spices, and Mexican salsas and chilli-thickened moles. But it was in Europe, more precisely in France, that generations of cooks developed sauce making into a systematic art, and made it the heart of a national cuisine that became an international standard.

Ancient Times

Our first real knowledge of sauce-like preparations in Europe comes from Roman times. A Latin poem from around 25 CE describes a peasant farmer making a spread of pounded herbs, cheese, oil, and vinegar—an ancestor of *pesto genovese*—that gave a pungent, salty, aromatic savor to his flatbread (see box, p. 583).

A few centuries later, the Latin recipe book attributed to Apicius makes it clear that sauces played an essential part in the dining of the Roman elite. More than a quarter of the nearly 500 recipes are for sauces, the term for which was *ius*, the ancestor of our "juice." Most contained at least a half dozen herbs and spices, as well as vinegar and/or honey, and some form of the fermented fish sauce *garum* (p. 235), which provided saltiness, savoriness, and a distinctive aroma (much as anchovies do today). And they were thickened in a variety of ways: with the pounded flavorings themselves; with pounded nuts or rice; with pounded liver or sea urchins; with pounded bread, pieces of pastry, and with pure wheat starch itself; with egg yolks, both raw and

---

Harmonizing Flavors in Ancient China

The addition, intensification, and blending of flavors that characterize good sauce making are central to the art of cooking, and have been considered such for at least 2,000 years. Here is an ancient Chinese description of the process that centers on the making of a stew or soup, a preparation in which the solid food both provides part of the sauce and cooks in the sauce.

> In the business of harmonious blending, one must make use of the sweet, sour, bitter, pungent and salty. Whether things are to be added earlier or later and in what amounts—their balancing is very subtle and each thing has its own characteristic. The transformation which occurs in the cauldron is quintessential and wondrous, subtle and delicate. The mouth cannot express it in words; the mind cannot fix upon an analogy. It is like the subtlety of archery and horsemanship, the transformation of Yin and Yang, or the revolution of the four seasons. Thus [the food] is long-lasting yet does not spoil; thoroughly cooked yet not mushy; sweet yet not cloying; sour yet not corrosive; salty yet not deadening; pungent yet not acrid; mild yet not insipid; oily-smooth yet not greasy.
>
> —attributed to the chef I Yin in the *Lü Shih Chhun Chhiu* (*Master Lü's Spring and Autumn Annals*), 239 BCE, transl. Donald Harper and H. T. Huang

cooked. The sauce maker's most important tool was clearly the mortar, but the sea urchins, eggs, and starch are early versions of more refined thickening methods.

The Middle Ages: Refinement and Concentration

We don't know much about cooking in Europe between the time of Apicius and the 14th century, the period from which a number of manuscript recipe collections survive. In some respects, sauce making hadn't changed much. Medieval sauces often contained many spices, the mortar and pestle still pounded ingredients—now including meats and vegetables—and most of the Roman thickeners were still used. Bread was most common, toasted to provide additional color and flavor, while pure

Sauce Recipes from Ancient Rome

. . . the bulb [of garlic] with the leaves he kept and dipped in water, then dropped into the round hollow stone. On it he sprinkled some grains of salt, and as the salt dissolved added hard cheese, then heaped on the herbs he had gathered [parsley, rue, coriander], and with his left hand wedged the mortar into his shaggy groin; his right hand first mashed the pungent garlic with the pestle, then pounded everything so as to mix the juices evenly. Round and round went his hand; gradually the original ingredients lost their own properties and one color emerged from several, not wholly green, since the milky fragments held out, nor shining milky white, being variegated by all the herbs. . . . he poured in some drops of olive oil and on top added a tiny drop of pungent vinegar, and once again mixed and thoroughly remixed the mass. Finally with two fingers he wiped round the whole mortar and brought together the parts into a single ball so as to produce a *moretum,* perfect in appearance as in name.

—*Moretum,* transl. E.J. Kenney

White Sauce for Boiled Foods

Pepper, *liquamen* [fish sauce], wine, rue, onion, pine nuts, spiced wine, a few pieces of bread cut up to thicken, oil.

For Stuffed Squid

Pepper, lovage, coriander, celery-seed, egg yolk, honey, vinegar, *liquamen,* wine, and oil. You will thicken it [by heating].

Pastry-Milk Chicken

Cook the chicken in *liquamen,* oil, and wine, to which you add a bundle of coriander and onion. Then when it is done, lift it from its sauce and put into a new pan some milk, a little salt and honey, and very little water. Set it by a slow fire to warm, crumble some pastry, and add it gradually, stirring carefully so that it doesn't burn. Add the chicken, whole or cut up, turn out on a dish, and pour over the following sauce: pepper, lovage, oregano, honey, a little grape syrup, and cooking liquid. Mix. Bring it to the boil in a pan. When it boils, thicken with starch, and serve.

—Apicius

starch was no longer used, and cream and butter still weren't.

New Flavors, New Clarity, and Jellies
But there were some important differences, and genuine progress. Fish sauce had disappeared, its place taken by vinegar and unripe grape juice, or *verjus*. Thanks in part to the Crusades, which brought Europeans to the Middle East and into contact with Arab trade and traditions, many local Mediterranean flavorings had been displaced by exotic imports from Asia, especially cinnamon, ginger, and grains of paradise; and the nut of choice for thickening was now the almond. The mortar was joined by a second indispensable utensil: the cloth sieve or strainer (French *étamine* or *tamis*) through which sauces were passed to remove coarse particles of spice and thickener and produce a finer consistency. Cooks had discovered the principle of thickening meat broths by concentration—by boiling off unwanted water—and

Refinements in Medieval Sauce Making

These recipes from more than 500 years ago show the great care with which medieval cooks made sauces and jellies. The broth recipe is remarkable for its exact descriptions of consistency and stirring time off the heat to prevent curdling.

Fish or Meat Jelly

Cook [the fish or meat] in wine, verjuice, and vinegar . . . then grind ginger, cinnamon, cloves, grains of paradise, long pepper, and infuse this in your bouillon, strain it, and put it to boil with your meat; then take bay leaves, spikenard, galingale, and mace, and tie them in your bolting [flour-sieving] cloth, without washing it, along with the residue of the other spices, and put this to boil with your meat; keep the pot covered while it is on the fire, and when it is off the fire keep skimming it until the preparation is served up; and when it is cooked, strain your bouillon into a clean wooden vessel and let it sit. Set your meat on a clean cloth; if it is fish, skin and clean it and throw your skins into your bouillon until it has been strained for the last time. Make certain that your bouillon is clear and clean and do not wait for it to cool before straining it. Set out your meat in bowls, and afterwards put your bouillon back on the fire in a bright clean vessel and boil it constantly skimming, and pour it boiling over your meat; and on your plates or bowls in which you have put your meat and broth sprinkle ground cassia buds and mace, and put your plates in a cool place to set. Anyone making jelly cannot let himself fall asleep. . . .

—Taillevent, *Le Viandier,* ca. 1375, transl. Terence Scully

A Fine Thick Broth

For ten servings, get three egg yolks per serving, good verjuice, good meat broth, a little saffron and fine spices; mix everything together, strain it and put it into a pot on the coals, stirring constantly until it coats the spoon; and so take it off the fire, stirring constantly for the length of two *Our Fathers*; then dish it out, putting mild spices on top. . . .

—*The Neapolitan Recipe Collection,* ca. 1475, transl. Terence Scully

so developed both the consommé and the solid jelly, part of whose value was the way it could coat cooked meat or fish and protect it from the air and spoilage. The transparency of clear jellies in turn led by the 15th century to an improved strainer for removing the tiniest particles from them: a protein "fabric" of whipped egg whites that clarified the liquid from within.

Sauce Terminology Another important development during the Middle Ages was the elaboration of a new vocabulary for sauces and other flavorful fluids, and a more systematic approach to them. The Roman term *ius* was replaced by derivatives of the Latin *salsus,* meaning "salted": *sauce* in France, *salsa* in Italy and Spain. In French, *jus* came to mean meat juices; *bouillon* was a stock produced by simmering meat in water; *coulis* was a thickened meat preparation that gave flavor and body to sauces, to *potages*—substantial soups—and other prepared dishes. The French *soupe* was the equivalent of the English *sop,* a flavorful liquid imbuing a piece or pieces of bread. A number of manuscripts divide their recipes into categories: there are uncooked sauces, cooked sauces, sauces in which to cook meat, and others with which to serve meats, thin and thick potages, and so on. And the English word *gravy* appears, derived apparently but mysteriously from the French *grané.* The latter, whose name derives from the Latin *granatus,* "made with grains, grainy," was a kind of stew made with meat and meat juices, and not a separate mixture of spices and liquid.

French Sauces from the 17th Century

In the recipe books of La Varenne and Pierre de Lune, we can find a hollandaise-like "fragrant sauce," the cream-like emulsion still called *beurre blanc* or "white butter," and the thin *court bouillon* ("short-cooked bouillon") traditionally used for poaching and serving fish. Notice the simplicity of flavoring compared to the medieval dishes.

Asparagus in Fragrant Sauce

Choose the largest, scrape the bottoms and wash, then cook in water, salt well, and don't let them cook too much. When cooked, put them to drain, make a sauce with good fresh butter, a little vinegar, salt, and nutmeg, and an egg yolk to bind the sauce; take care that it doesn't curdle; and serve the asparagus garnished as you like.

—La Varenne, *Le Cuisinier françois,* 1651

Trout in Court Bouillon

Cook your trout with water, vinegar, a packet [of chive, thyme, cloves, chervil, parsley, sometimes a piece of lard, all tied with a string], parsley, salt, bay, pepper, lemon, and serve the same way.

Perches in Beurre Blanc

Cook them with wine, verjus, water, salt, cloves, bay; remove the scales and serve with a thickened sauce that you make with butter, vinegar, nutmeg, slices of lemon; it should be well thickened.

—Pierre de Lune, *Le Cuisinier,* 1656

Early Modern Sauces: Meat Essences, Emulsions

It's in the three centuries between 1400 and 1700 that the sauces of our own time have their roots. Recipes call for fewer spices and a lighter hand with them; vinegar and verjus begin to give way to lemon juice; coarse bread and almond thickeners are replaced by flour and by butter and egg emulsions (see box, p. 585). And in France, meat broths become the central element of fine cooking. This is the era in which experimental science began to flourish, and some influential French cooks conceived of themselves as the chemists—or alchemists—of meat. Around 1750, François Marin echoed the Chinese description of flavor harmony from 2,000 years before, but with some telling twists (see box below).

Both Marin and I Yin speak of harmony and balance. But the Chinese cauldron brings together sweet, sour, bitter, salty, and pungent ingredients, while the French pot contains only meat juices, and generates complexity and harmony by concentrating them. Marin said that "Good taste has forbidden the burning juices and caustic ragouts of the *ancienne cuisine,*" with their Asian spices and abundant vinegar and verjus. Meat bouillon was now "the soul of cooking." The meat's juices are its essence, and the cook extracts them, concentrates them, and then uses them to imbue other foods with their flavor and nourishment. The purpose of a sauce is not to add new flavors to a food, but to deepen its flavor and integrate it with the underlying flavor of the other dishes.

Many of these preparations required prodigious amounts of flesh, the solid part of which did not appear in the final dish. A small amount of consommé, for example, was made with 2 lb/1 kg each of beef and veal, two partridges, a hen, and some ham. This meat was first cooked with some bouillon—itself already a meat extract—until the liquid bouillon and meat juices evaporated, and the meat began to stick to the pan and caramelize. Then yet more bouillon was added along with some vegetables, the mixture cooked for four hours, and strained to produce a liquid "yellow like gold, mild, smooth, and cordial."

The Flowering of French Sauces Marin called his collection of *bouillons, potages, jus, consommés, restaurants* ("restoring" soups), *coulis,* and *sauces* "the foundation of cooking," and said that by adopting a systematic approach to them, even a bourgeois family with limited resources would be "able to imagine an infinity of sauces and different stews." French cookbooks soon began to include dozens of different soups and sauces, and several of the classic sauces were soon developed and named. Among these were alternatives to the meat-juice preparations, including two egg-emulsified sauces, hollandaise and mayonnaise, and the economical béchamel, the basic, neutral white sauce of milk, but-

François Marin on Cooking as a Chemical Art

Modern cooking is a species of chemistry. The science of the cook today is to break down, digest, and distill meats into their quintessence, to take their light and nourishing juices, mix and confound them together, in such a way that none dominates and all can be tasted; finally, to give them that unity which painters give their colors, and render them homogeneous enough that their different flavors result only in a fine and piquant taste; in, if I may say it, a harmony of all tastes joined together. . . .

—*Dons de Comus*, 1750

ter, and flour. But the great majority of sauces were made from meat, and meat juices were the underlying, unifying element in French cooking.

THE CLASSIC FRENCH SYSTEM: CARÊME AND ESCOFFIER

In 1789 came the French Revolution. The great houses of France were much reduced, and their cooks no longer had unlimited help and resources. Some lost their positions, and survived by opening the first fine restaurants. The culinary impact of these upheavals was assessed by the renowned chef Antonin Carême (1784–1833). In the "Preliminary Discourse" to his *Maître d'Hôtel français,* he noted that the "splendor of the old cuisine" was made possible by the lavish expenditures of the master on personnel and materials. After the Revolution, cooks lucky enough to retain a position

> were thus obliged, for want of help, to simplify the work in order to be able to serve dinner, and then to do a great deal with very little. Necessity brought emulation; talent made up for everything, and experience, that mother of all perfection, brought important improvements to modern cuisine, making it at the same time both healthier and simpler.

Restaurants too brought improvements; "in order to flatter the public taste," the commercial chefs had to come up with novel, ever more "elegant" and "exquisite" preparations. So social revolution became a new motivating force for culinary progress.

Sauce Families Carême made a number of contributions to this progress, and perhaps the most notable involved the sauces. His idea, set forth in *The Art of French Cooking in the 19th Century,* was to organize the infinity of possibilities that Marin foresaw, and thereby help cooks realize them. He classified the sauces of the time into four families, each headed by a basic or leading sauce, and each expandable by playing variations on that basic theme. Only one of the leading sauces, espagnole, was based on expensive, highly concentrated meat extract; both velouté and allemande used unreduced stock, and béchamel used milk. Many of these sauces were thickened with flour, which is much more economical than reduced meat bouillon.This approach suited the limits and needs of postrevolutionary cuisine.The parent sauces could be prepared in advance, with the novel but minor modifications and seasonings to be done at the last minute on the day of the meal. As Raymond Sokolov puts it in his guide to the classic sauces, *The Saucier's Apprentice,* these sauces were conceived as "convenience foods at the highest level."

Less than a century after Carême, the great compilation of classic French cuisine, Auguste Escoffier's *Guide Culinaire* (1902), lists nearly 200 different sauces, not including dessert sauces. And Escoffier attributed the eminence of French cooking directly to its sauces. "The sauces represent the *partie capitale* of the cuisine. It is they which have created and maintained to this day the universal preponderance of French cuisine."

Of course this flavoring system was the creation of the line of professional cooks going back to medieval times. Alongside it there developed a more modest domestic tradition, which is accomplished in its own way. Disinclined to the labor and expense of long-simmered stocks and sauces, middle-class home cooks refined other methods: for example, making a broth from the trimmings of a roast, using the broth to dissolve the flavorful crust from the roasting pan, and boiling this relatively small amount of liquid to reduce and thicken it, or binding it with cream or flour.

SAUCES IN ITALY AND ENGLAND

Purees and Meat Juices From the Middle Ages through the 16th century, Italian court cooking was as innovative as French cooking, and sometimes more so. Yet it

The Classic French Sauce Families

Carême's original classification of the sauces has undergone various modifications, as have the ingredients in many of the derived sauces. Here is one modern version of the family tree that shows a number of the more familiar derived sauces. Stocks and roux are brown if the meat, vegetables, or flour are browned at relatively high heat before liquid is added; otherwise they are yellow or white, and lighter in flavor as well.

Basic Sauce: Brown, or Espagnole, made with brown stock (beef, veal), brown roux, tomatoes

Sauce	Ingredients
Bordelaise ("from Bordeaux")	Red wine, shallots
Diable ("devil")	White wine, shallots, cayenne
Lyonnaise ("from Lyon")	White wine, onion
Madeira	Madeira wine
Périgueux (village in Perigord region)	Madeira wine, truffles
Piquante	White wine, vinegar, gherkins, capers
Poivrade ("peppered")	Vinegar, peppercorns
Red wine sauces	Red wine
Robert	White wine, onion, mustard

Basic Sauce: Velouté ("velvety"), made with white stock (veal, poultry, fish), yellow roux

Sauce	Ingredients
Allemande ("German")	Egg yolks, mushrooms
White Bordelaise	White wine, shallots
Ravigote ("invigorated")	White wine, vinegar
Suprême	Poultry stock, cream, butter

Basic Sauce: Béchamel (a gourmand), made with milk, white roux

Sauce	Ingredients
Crème	Cream
Mornay (a family)	Cheese, fish or poultry stock
Soubise (army commander)	Onion puree

Basic Sauce: Hollandaise ("from Holland"), made with butter, eggs, lemon juice, or vinegar

Sauce	Ingredients
Mousseline (light cloth)	Whipped cream
Béarnaise ("from Béarn")	White wine, vinegar, shallots, tarragon

Basic Sauce: Mayonnaise (uncertain etymology), made with vegetable oil, eggs, vinegar, or lemon juice

Sauce	Ingredients
Rémoulade (twice ground)	Gherkins, capers, mustard, anchovy paste

stagnated in the 17th century, according to historian Claudio Benporat, as part of a general political and cultural decline caused by an absence of strong Italian leaders and the influence of other European powers on the several Italian courts. The sauces that have come to be known as distinctively Italian are mainly domestic and relatively unrefined in character, based not so much on essences as on whole materials: the purees of tomato fruits and basil leaves, for example. The basic Italian meat sauce, or sugo, is made in the manner of Marin's 18th-century consommé: meat is slowly cooked to liberate its juices, which are allowed to cook down and brown on the pan bottom; then meat broth is used to redissolve the browned residues, and allowed to concentrate and itself brown: and the process repeated to produce a concentrated flavor. The meat is not discarded, but becomes part of the sauce. Not only Italy but much of the Mediterranean region, including southern France, has been less interested in extracting meat essences than in highlighting and combining flavors.

Sauces in England: Gravies and Condiments According to an 18th-century bon mot attributed to Domenico Caracciolli, with implicit contrast to France: "England has sixty religions and one sauce"—that one sauce being melted butter! And the sharp-toothed Alberto Denti di Pirajno begins the chapter on sauces in his *Educated Gastronome* (Venice, 1950) with these pointed sentences:

> Doctor Johnson defined a sauce as something which is eaten with food in order to improve its flavor. It would be difficult to believe that a man of the intelligence and culture of Dr. Johnson . . . had expressed himself in these terms, if we did not know that Dr. Johnson was English. Even today his compatriots, incapable of giving any flavor to their food, call on sauces to furnish their dishes that which their dishes do not have. This explains the sauces, the jellies and prepared extracts, the bottled sauces, the *chutneys,* the *ketchups* which populate the tables of this unfortunate people.

England's culinary standards were not formed at the Court or in the noble houses; they remained grounded in the domestic habits and economies of the countryside. English cooks ridiculed French cooks for their essences and quintessences. The French gastronome Brillat-Savarin (1755–1826) tells the story of the prince of Soubise being presented with a request from his chef for 50 hams, to be used at one supper party. Accused of thievery, the chef responds that all this meat is essential for the sauces to be made: "Command me, and I can put these fifty hams which seem to bother you into a glass bottle no bigger than your thumb!" The prince is astonished, and won over, by this assertion of the cook's power to concentrate flavor. By contrast, in her popular 18th-century cookbook, the English writer Hannah Glasse gives several French sauce recipes that require more meat than the meal they will accompany, and then remarks on "the Folly of these fine French Cooks" in running up such huge expenses for so little. Glasse's principal sauce is "gravy," made by browning some meat, carrots, onions, several herbs and spices, shaking in some flour, adding water, and stewing. In the 19th century, similarly homely anchovy, oyster, parsley, egg, caper, and butter sauces were popular.

And the Worcestershire sauces and chutneys and ketchups that Denti di Pirajno mocked? These condiments had become a part of English cooking in the 17th century thanks to the commercial activities of the East India Company, which brought back Asian soy and fish sauces—including Indonesian *kecap* (p. 499)—and pickled fruits and vegetables, all preserved foods with intensified flavors. Many of these preparations are rich in savory amino acids, and the English imitations were often made with similarly savory mushrooms and anchovies. Our familiar tomato ketchup is a sweetened version of salty, vinegary, spicy tomato preserves. So an English contemporary of Carême's, William Kitchiner, includes a recipe for béchamel in his recipe book, but also presents "Wow Wow Sauce," which contains parsley, pickled cucumbers or walnuts, butter, flour, broth, vinegar, catsup, and mustard. These strongly flavored concoctions were quick and easy to use, and were evidently enjoyed for their strong contrast to the flavor of the foods they accompanied, not for subtle enhancement.

Modern Sauces: Nouvelle and Post-Nouvelle

The 20th Century: Nouvelle Cuisine Back in the 18th century, François Marin and his colleagues described their bouillon-based cooking as *nouvelle cuisine,* or the "new cooking." In the hands of Carême and Escoffier, that *nouvelle cuisine* was augmented with a few new sauces and became classic French cooking, the standard throughout the western world for fine dining. In time, the classic system became increasingly rigid and predictable, with most chefs essentially preparing the same standard dishes from the same precooked sauce bases. The 20th century brought a new *nouvelle cuisine,* along with the New Novel and the New Wave in cinema. In the 1960s a number of well established French chefs, including Paul Bocuse, Michel Guérard, the Troisgros family, and Alain Chapel, led the way in rethinking the French tradition. They asserted the chef's creative role and the virtues of simplicity, economy, and freshness. Foods were no longer to be distilled into their essences, but were to be presented intact, as themselves.

In 1976, the journalists Henri Gault and Christian Millau published the Ten Commandments of *nouvelle cuisine,* of which the seventh was: "You shall eliminate brown and white sauces." The new cooks still thought that, in the words of Michel Guérard, "the great sauces of France must be described as the cornerstones of cuisine," but they used them more selectively and with restraint. Lighter-flavored veal, chicken, and fish stocks served as poaching and braising liquids; reductions of these were used to give depth to last-minute pan sauces; and sauces in general were thickened less with flour and starch, more often with cream, butter, yogurt and fresh cheese, vegetable purees, and with bubbly foams.

Post-Nouvelle: Diverse and Innovative Sauces At the opening of the 21st century, the classic brown and white sauces have become scarce, so much so that perhaps we're ready to appreciate their virtues again. Those restaurant and home cooks who do serve time-consuming meat stocks and reductions seldom make them from scratch; these products are well suited to manufacture on an industrial scale, and good versions are available in frozen form. The rich cream and butter sauces popularized by the *nouvelle cuisine* have become less common; simpler broths, reduced pan deglazings, and vinaigrettes more so. Thanks to the international scope of modern cooking, restaurant diners encounter a wider range of sauces than ever before. Many of them are contrasting purees made from fruits, vegetables, nuts, and spices, or else thinner soy- and fish-based Asian dipping sauces; these are attractive to restaurateurs because they require less time, labor, and often less skill than the classic French sauces. Similarly, home cooks are now likely to buy time-saving and versatile bottled sauces and dressings. And a few inventive chefs are experimenting with unusual tools and materials—among them liquid nitrogen, high-powered pulverizers, thickeners derived from seaweeds and microbes—to make new forms of suspensions, emulsions, foams, and jellies.

The subtleness and delicacy described by I Yin and François Marin are not especially prominent among contemporary sauces. On the other hand, never before in history have we had so many distillations of desire from which to choose!

THE SCIENCE OF SAUCES: FLAVOR AND CONSISTENCY

FLAVOR IN SAUCES: TASTE AND SMELL

The primary purpose of a sauce is to provide flavor in the form of a liquid with a pleasing consistency. It's much easier to generalize about consistency, how it is created, and how it can go wrong, than to generalize about flavor. There are many thousands of different flavor molecules; they can be combined in untold numbers of ways, and different people perceive them differently. Still, it's useful to keep a few basic facts about flavor in mind when constructing a sauce.

The Nature of Flavor Flavor is mainly a combination of two different sensations, taste and smell. Taste is perceived on the tongue, and comes in five different sensations: saltiness, sweetness, sourness, savoriness, bitterness. The molecules that we taste—salt, sugars, sour acids, savory amino acids, bitter alkaloids—are all easily soluble in water. (The astringent sensation caused by tea and red wine is a form of touch, and the "hot" pungency of mustard is a form of pain. They are not true tastes, but we also perceive them on the tongue and their causes are also water-soluble molecules.) Smell is perceived in the upper nasal region, and comes in thousands of different aromas that we usually describe by the foods they remind us of, fruity or flowery or spicy or herbaceous or meaty. The molecules that we smell are more soluble in fat than in water, and tend to escape from water into the air, where our smell detectors can sniff them.

It can be useful to think of taste as the backbone of a flavor, and smell as its fleshing out. Taste alone is what we experience when we take some food in the mouth and pinch our nostrils shut; smell alone is what we experience when we sniff some food without putting it in the mouth. Neither is fully satisfying on its own. And recent research has shown that taste sensations affect our smell sensations. In a sweet food, the presence of sugar enhances our perception of aromas, and in savory foods, the presence of salt has the same effect.

The Spectrum of Sauce Flavors When considered as carriers of flavor, sauces form a broad spectrum. At one end are simple mixtures that provide a pleasing contrast to the food itself, or add a flavor that it lacks. Melted butter offers a subtle richness, vinaigrette salad dressings and mayonnaise a tart richness, salsas tartness and pungency. At the other end of the spectrum are complex flavor mixtures that fill the mouth and nose with sensations, and provide a rich background into which the flavor of the food itself blends. Among these are the meat-based sauces of the French tradition, whose complexity comes largely from the extraction and concentration of savory amino acids and other taste molecules, and from the generation of meaty aromas by way of the browning reactions between amino acids and sugars (p. 778). Chinese braising liquids based on soy sauce are similarly complex thanks to the cooking and fermentation of the soybeans (p. 496), while the spice blends of India and Thailand and the moles of Mexico typically combine a half dozen or more strongly aromatic and pungent ingredients.

Improving Sauce Flavor Perhaps the most common problem with sauce flavor is that there doesn't seem to be enough of it, or that "there's something missing" in it. Perfecting the flavor of any dish is an art that depends on the perceptiveness and skill of the cook, but there are two basic principles that can help anyone analyze and improve a sauce's flavor.

- Sauces are an accompaniment to a primary food, are eaten in small amounts compared to the primary food, and therefore need to have a concentrated flavor. A spoonful of sauce alone should taste too strong, so that a little sauce on a piece of meat or pasta will taste just right. Thickening agents tend to reduce the flavor of a sauce (p. 596), so it's important to check and adjust the flavor after thickening.

- A satisfying sauce offers stimulation to most of our chemical senses. A sauce that doesn't seem quite right is probably deficient in one or more tastes, or doesn't carry enough aroma. The cook can taste the sauce actively for its saltiness, sweetness, acidity, savoriness, and aroma, and then try to correct the deficiencies while maintaining the overall balance of flavors.

Sauce Consistency

Though the main point of sauces is their flavor, we also enjoy them for their consistency, their feeling in the mouth. And problems with consistency—with the sauce's physical structure—are far more likely than flavor problems to make a sauce unusable. Curdled or congealed or separated sauces are not pleasant to look at or to feel in the mouth. So it's good to understand the physical structures of common sauces, how they're put together and how they're ruined.

Food Dispersions: Mixtures That Create Texture The base ingredient in nearly all flavorful food liquids is water. That's because foods themselves are mostly water. Meat juices, vegetable and fruit purees are all obviously watery; cream and mayonnaise and the hot egg sauces less obviously so, but they too are built on water. In each of these preparations, water is the *continuous phase*: the material that bathes all the other components, the material in which all the other components swim. (The only common exceptions are some vinaigrettes and butter and nut butters, in which fat is the continuous phase.) Those other components are the *dispersed phase*. The task of giving sauces a desirable consistency is a matter of making the continuous, base phase of water seem less watery, more substantial. The way this is done is to add some nonwatery substance—a dispersed phase—to the water. This substance may be particles of plant or animal tissue, or various molecules, or droplets of oil, or even bubbles of air. And how do the added substances make the water seem more substantial? By obstructing the free movement of the water molecules.

Obstructing the Movement of Water Molecules Individual water molecules are small—just three atoms, H_2O. Left to themselves, they're very mobile: so water is runny and flows as easily as a stream. (Oil molecules, by contrast, have three chains stuck together, each 14 to 20 atoms long, so they drag against each other and move more slowly. This is why oil is more viscous than water.) But intersperse solid particles or long, tangly molecules, or oil droplets, or air bubbles among the water molecules, and the water molecules can move only a small distance before they collide with one of these foreign, less mobile substances.

Food Words: *Liaison*

To name both the act of thickening and the agents of thickening, early French cooks used the word *liaison,* which meant a close connection or bond, whether physical, political, or amorous. When the English got around to borrowing the word in the 17th century, it was the culinary application that came first; military and romantic liaisons didn't arrive until the 19th century.

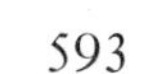

They're then able to make only slow progress, so they flow more reluctantly.

Thickening agents in saucemaking are just such obstructing agents. Cooks have traditionally thought of them as binding agents, and this view makes its own kind of sense. The dispersed materials essentially divide the liquid into many small, local masses: and by dividing, they organize and collect it and give it a kind of coherence that it lacked beforehand. Some thickening agents also literally bind water molecules to themselves and so take them out of circulation altogether, and this too has the effect of reducing the fluidity of the continuous phase.

In addition to giving watery fluids a thicker consistency, the substances in the dispersed phase can give them textures of various kinds. Solid particles may make them grainy or smooth, depending on the particle size; oil droplets make them seem creamy; dispersed molecules with a tendency to adhere to each other may make them seem sticky or slimy; air bubbles make them seem light and evanescent.

There are four common ways of thickening watery food juices. Each produces a different kind of physical system, and lends different qualities to the finished sauce.

Cloudy Suspensions: Thickening with Particles Most of our raw ingredients—vegetables, fruits, herbs, meats—are plant or animal tissues built from microscopic cells that are filled with watery fluids. The cells are contained within walls, membranes, or thin sheets of connective tissue. (Dry seeds and spices contain no juices, but are still made up of solid cells and cell walls.) When any of these foods is broken apart into small pieces by being ground in a mortar or pulverized in a blender, they are turned inside out, so the fluids form a continuous phase that contains fragments of the solid cell walls and connective tissue. These fragments obstruct and bind the water molecules, and thus thicken the consistency of the mixture. Such a mixture of a fluid and solid particles is called a *suspension*: the particles are suspended in the fluid. Sauces made from pureed foods are suspensions.

The texture of a suspension depends on the size of its particles. The smaller the particles, the less noticeable they are to the tongue, and the smoother the texture. Also, the smaller the particles are, the more of them there are to do the obstructing and the more surface area they have to take up a layer of water molecules: and so the thicker the consistency they produce. Suspensions are always opaque, because the solid particles are large enough to block the passage of light rays and either absorb them or bounce them back toward their source. Because the particles and water are very different materials, suspensions tend

Thickening a liquid with food particles. In a suspension, microscopic chunks of plant or animal tissue are suspended in liquid, and give the impression of thickness by interfering with the liquid's flow.

to settle and separate into thin fluid and concentrated particles. Cooks work to prevent separation by reducing the volume of the continuous phase (draining off or boiling away excess water), or by augmenting the dispersed phase (adding starch or other long molecules or fat droplets).

Nut butters and chocolate are suspensions of solid seed particles not in water, but in oils and fats.

Clear Dispersions and Gels: Thickening with Molecules A single microscopic fragment of a tomato cell wall or muscle fiber is built up of many thousands of submicroscopic molecules. Not all of the large molecules in those fragments can be teased away from each other so that they are individually dispersed in water. But those that can be extracted in this way—starch, pectins, such proteins as gelatin—are very useful thickening agents. Because single molecules are so much smaller and lighter than intact starch granules and cell fragments, they don't settle out and separate. And they are too small and too widely separated to block the passage of light rays: so unlike suspensions, molecular dispersions are usually translucent and glassy-looking. In general, the longer the molecule, the better it is at obstructing water movement, because long molecules more readily get tangled up in each other. So a small quantity of long amylose starch molecules will do the same thickening job as a large quantity of short amylopectin (p. 611), and long gelatin molecules thicken more efficiently than short ones. Thickening with molecules often requires heat, either to liberate the molecules from the larger structures—starch molecules from their granules, gelatin molecules from meat connective tissue—or to shake out compactly folded molecules—egg proteins—into their long, extended, tangly form.

Solid Dispersions: Jellies When the water phase of a food fluid has enough thickening molecules dissolved in it, and the fluid is left undisturbed and allowed to cool, those molecules can bond to each other and form a loose but continuous tangle or network that permeates the fluid, with the water immobilized in pockets between the network molecules. Such a network thickens the fluid to the point that it becomes a very moist solid, or a *gel*. It's possible to make a solid—if wobbly—jelly that is 99% water and just 1% gelatin. If the gel is made from dissolved molecules, then it will be translu-

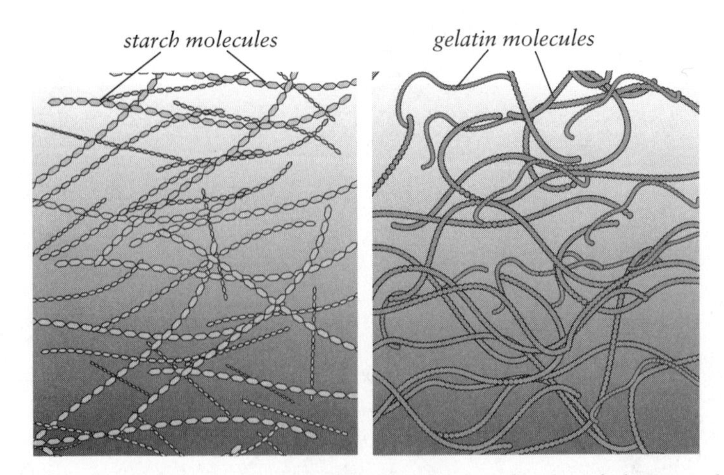

Thickening a liquid with long food molecules. Dissolved molecules of plant starch or animal gelatin get tangled up with each other and impede the flow of the liquid.

cent, like the dispersion from which it is formed. Familiar examples are savory jellies made from gelatin and sweet jellies made from fruit pectin. If the solution also contains particles—the remains of starch granules, for example—then the jelly will be opaque.

Emulsions: Thickening with Droplets Thanks to their very different structures and properties, water molecules and oil molecules don't mix evenly with each other (p. 797). Neither can dissolve in the other. If we use a whisk or blender to force a small portion of oil to mix into a larger one of water, the two form a milky, thick fluid. Both the milkiness and the thickness are caused by small droplets of oil, which block light rays and the free movement of water molecules. The oil droplets thus behave much as the solid particles in a suspension do. Such a mixture of two incompatible liquids, with droplets of one liquid dispersed in a continuous phase of the other, is called an *emulsion*. The term comes from the Latin word for "milk," which is just such a mixture (p. 17).

Emulsifiers In addition to the two incompatible liquids, a successful emulsion requires a third ingredient: an *emulsifier*. An emulsifier is a substance of some kind that coats the oil droplets and prevents them from coalescing with each other. Several different materials can serve this function, including proteins, cell-wall fragments, and a group of hybrid molecules (for example, egg-yolk lecithin) that have an oil-like end and a water-soluble end (p. 802). To make an emulsified sauce, we add oil to a mixture of water and emulsifiers (egg yolk, ground herbs or spices), and break the oil up into microscopic droplets, which the emulsifiers immediately coat and stabilize. Or we can begin with a premade emulsion. Cream is an especially robust and versatile base for emulsified sauces.

Foams: Thickening with Bubbles At first it seems surprising that a fluid can be thickened by adding air to it. Air is the opposite of substantial! Yet think of the foams on an espresso coffee or a glass of beer: they all have enough body to hold their shape when scooped with a spoon. Similarly, a pancake batter gets noticeably thicker if you stir the chemical leavening in last. In a fluid, air bubbles have much the same effect as solid particles: they interrupt the mass of water molecules and

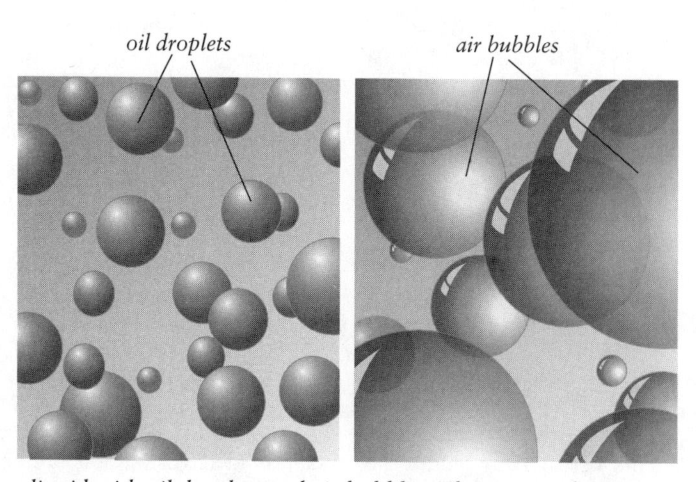

Thickening a liquid with oil droplets and air bubbles. These tiny spheres act much as solid food particles do, interfering with the flow of the liquid surrounding them.

obstruct the water's flow from one place to another. The disadvantage of foams is that they are fragile and evanescent. The force of gravity unceasingly drains fluid from the bubble walls, and when the walls get just a few molecules thick, they break, the bubbles pop, and the foam collapses. This outcome can be delayed in a couple of ways. The cook can thicken the fluid with truly substantial particles or molecules (oil droplets, egg proteins) to slow its drainage from the bubble cell walls, or include emulsifiers (egg-yolk lecithin) that stabilize the bubble structure itself. On the other hand, the very delicacy and evanescence of unreinforced foams is a part of their appeal. Such foams must be prepared at the last minute and savored as they disappear.

Real Sauces: Multiple Thickeners The sauces that cooks actually make are seldom simple suspensions, molecular dispersions, emulsions, or foams. They're usually a combination of two or more. Purees usually contain both suspended particles and dispersed molecules, starch-thickened sauces contain both dispersed molecules and remnants of the granules, emulsified sauces include proteins and particles from milk or eggs or spices. Cooks often thicken and enrich sauces of all kinds at the last minute by melting a piece of butter into it or stirring in a spoonful of cream, thus making them in part a milkfat emulsion. Such complexity of the dispersed phase may well make sauce texture more subtle and intriguing.

The Influence of Consistency on Flavor

Thickeners Reduce Flavor Intensity In general, the components of a sauce that create its consistency have little or no flavor of their own. They therefore only dilute whatever flavors the sauce has. Thickening agents also actively reduce the effectiveness of the flavor molecules in the sauce. They bind some of those molecules so that our palate never senses them, and they slow their movement from the sauce into our taste buds or nasal passages. Because aroma molecules tend to be more fat soluble than water soluble, fat in a sauce hangs onto aroma molecules and decreases aromatic intensity. Amylose starch molecules trap aroma molecules (the aroma molecules in turn make the starch molecules more likely to bond to each other into light-scattering, milky aggregates). And wheat flour binds more sodium than pure starches, so flour-thickened preparations require more added salt than starch-thickened sauces.

As a general rule, then, a thin sauce will have a more intense and immediate flavor than the same sauce with thickeners added. But the thickened sauce will release its flavor more gradually and persistently. Each effect has its uses.

Many sauces can be thickened not just by adding thickeners, but by removing some of the continuous phase—boiling off water—so that the thickeners already present in the sauce become more concentrated. This technique doesn't diminish flavor, because whatever flavor the sauce's particles and molecules can bind have already been bound. And in fact it can intensify flavor, because the concentration of flavor molecules may increase just as the thickeners' concentration does.

The Importance of Salt Recent research has uncovered intriguing indications that thickeners reduce our perception of aroma in part because they reduce our perception of saltiness. Various long-chain carbohydrates, including starch, first reduce the apparent saltiness of the sauce, either by binding sodium ions to themselves or by adding another sensation (viscosity) for the brain to attend to. Then this reduced saltiness reduces the apparent aroma intensity—despite the fact that the same number of aroma molecules are flowing out of the sauce and across the smell receptors in our nose. The practical significance of this finding is that thickening a sauce with flour or starch diminishes its overall flavor, and that both taste *and* aroma can be restored to some extent by the simple addition of more salt.

SAUCES THICKENED WITH GELATIN AND OTHER PROTEINS

If we gently heat a piece of meat or fish alone in a pan, it releases flavorful juices. Normally we make the pan hot enough to evaporate the water the moment it comes out, so that the flavor molecules become concentrated on the meat and pan surfaces, and react with each other to form brown pigments and a host of new flavor molecules (p. 778). But if the juices remain juices, they constitute a very basic sauce, a product of the meat that can be added back to moisten and flavor the mass of coagulated muscle protein from which they've been squeezed. The problem is that the meat or fish only gives up a small amount of juice compared to the solid mass. To satisfy fully our appetite for those juices, cooks have invented methods for making meat and fish sauces for their own sake, and in any quantity. The main thickening agent in these sauces is gelatin, an unusual protein that cooking releases from the meat and fish. Cooks also use other animal proteins to thicken sauces, but their behavior is very different and more problematic, as we'll see (p. 603).

The Uniqueness of Gelatin

Gelatin is a protein, but it's unlike the other proteins that the cook works with. Nearly all food proteins respond to the heat of cooking by unfolding, bonding permanently to each other, and coagulating into a firm, solid mass. It turns out that gelatin molecules can't easily form permanent bonds with each other, due to their particular chemical makeup. So heat simply causes them to shake loose from the weak, temporary bonds that hold them together, and disperses them in water. Because gelatin molecules are very long and get tangled up with each other, they give the mixture a definite body, and can even set it into a solid gel (p. 605). However, gelatin is relatively inefficient at thickening. Its molecules are very flexible, while those of starch and other carbohydrates are rigid and better at interfering with the movement of water and each other. This is one reason why gelatin-thickened sauces are usually augmented with starch. A sauce that contains only gelatin requires a large concentration, 10% or more, to have real weight. But at that concentration, the sauce is quick to congeal on a cooling plate, and it can also cause the teeth to stick together (gelatin makes an excellent glue!).

Gelatin Comes from Collagen Free gelatin molecules don't exist in meat and fish. They're woven tightly together to form the fibrous connective-tissue protein called collagen (p. 130), which gives mechanical strength to muscles, tendons, skin, and bones. Single gelatin molecules are chains of around 1,000 amino acids. Thanks to the repeating pattern of their amino acids, three gelatin molecules naturally fit alongside each other and form weak, reversible bonds that arrange the three molecules in the form of a triple helix. Many triple helixes then become cross-linked to each other to form the strong, rope-like fibers of collagen.

Cooks generate gelatin from collagen by using heat to dismantle the collagen fibers. For the muscles of land animals, it takes a temperature of around 140°F/60°C to agitate the muscle molecules enough to break the weak bonds of the triple helix. The orderly structure of the collagen fibers then collapses and the fibers shrink, thus squeezing juices from the muscle fibers. Some of the juices bathe the fibers, and single gelatin molecules or small aggregates may disperse into the juice. The higher the meat temperature goes, the more gelatin becomes dispersed. However, many of the collagen fibers remain intact thanks to the strong cross-linking bonds. The older the animal and the more work its muscles do, the more strongly cross-linked its collagen fibers are.

Extracting Gelatin and Flavor from Meats

The muscles that make up meat are mainly water and the protein fibers that do the work of contraction, which are not dispersable in water. The soluble and dispersable materials in muscle include about 1% by weight of collagen, 5% other cell proteins, 2% amino acids and other savory molecules, 1% sugars and other carbohydrates, and 1% minerals, mainly phosphorus and potassium. Bones are around 20% collagen, pig skin around 30%, and cartilaginous veal knuckles up to 40%. Bones and skin are thus much better sources of gelatin and thickening power than meat. However, they carry only a small fraction of the other soluble molecules that provide flavor. To make sauces with good meat flavor, it's meat that must be extracted, not bones or skin.

When meat is thoroughly cooked, it releases about 40% of its weight in juice, and the flow of juice pretty much ends when the tissue reaches 160°F/70°C. Most of the juice is water, and the rest the soluble molecules carried in the water. If meat is cooked in water, then gelatin can be freed from the connective tissue and extracted over a long period of time. When cooks make stocks, extraction times range from less than an hour for fish, to a few hours for chicken or veal stocks, to a day for beef. Optimum extraction times depend on the size of the bones and meat pieces, and on the age of the animal; the more cross-linked collagen of a steer takes longer to free than the collagen from a veal calf. At long extraction times, the gelatin molecules that have already been dissolved are gradually broken down into smaller pieces that are less efficient thickeners.

Meat Stocks and Sauces

There are several general strategies for making meat and fish sauces. The simplest of them center on the juices produced when the meat for the final dish is cooked, which can be flavored and/or thickened at the last minute with purees, emulsions, or a starch-based mixture. In the more versatile system developed by French cooks, one begins by making a water extract of meat and bones ahead of time, and then uses that stock to cook the final dish, or concentrates it to make intensely flavored, full-bodied sauces. These stocks and concentrates used to be the heart of restaurant cooking. They're less important now, but still represent the state of the art in meat sauces.

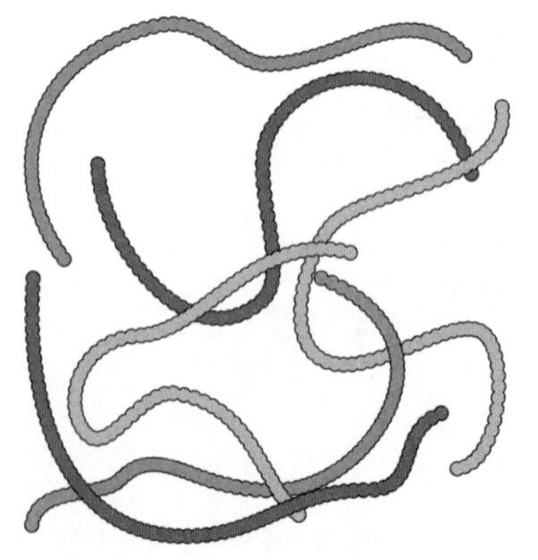

Collagen and gelatin. Collagen molecules (left) *contribute mechanical strength to connective tissue and bone in animal muscles. They are made up of three individual protein chains wound closely together into a helix to make a rope-like fiber. When heated in water, the individual protein chains come apart* (right) *and dissolve into the water. The unwound, separate chains are what we call gelatin.*

The Choice of Ingredients The aim in making meat stock is to produce a full-flavored liquid with enough gelatin that it will also become full-bodied when reduced. Meat is an expensive ingredient, an excellent source of flavor, and a modest source of gelatin. Bones and skins are less expensive, poor sources of flavor, but excellent sources of gelatin. So the most flavorful and expensive stocks are made with meat, the fullest bodied and cheapest with bones and pork skin, and everyday stocks with some of each. Beef and chicken stocks taste distinctly of their respective meats, while veal bones and meat are valued for their more neutral character, as well as their higher yield of soluble gelatin. Cartilaginous veal knuckles and feet give especially large amounts. Typically, the meat and bones are cooked in between one and two times their weight in water (1–2 quarts or liters per 2 lb/1 kg solids), and yield about half their weight in stock, thanks to gradual evaporation during cooking. The smaller the pieces into which they're cut, the more quickly their contents can be extracted in the water.

In order to round out the flavor of a stock, cooks usually cook the meat and bones along with aromatic vegetables—celery, carrots, onions—a packet of herbs, and sometimes wine. Carrots and onions contribute sweetness as well as aroma, wine tartness and savoriness. Salt is never added at this stage, because the meats and vegetables release some, and it becomes concentrated as the stock reduces.

Cooking the Stock A classic meat stock should be as clear as possible, so that it can be made into soup broths and aspics that will be attractive to the eye. Many of the details of stock making have to do with removing impurities, especially the soluble cell proteins that coagulate into unsightly gray particles.

The bones and often meat as well (and skin, if any) are first washed thoroughly. To make a light stock, they are then put in a pot of cold water that is brought to the boil; they're then removed from the pot and rinsed. This blanching step removes surface impurities and coagulates surface proteins on the bones and meat so that they won't cloud the cooking liquid. To make a dark stock for brown sauces, the bones and meat are first roasted in a hot oven to produce color and a more intense roasted-meat flavor with the Maillard reactions between proteins and carbohydrates. This process also coagulates the surface proteins and makes blanching unnecessary.

The Importance of a Cold Start and Uncovered, Slow Heating After the blanching or browning, the meat solids are started in an uncovered pot of cold water, which the cook brings slowly to a gentle simmer and keeps there, regularly skimming off the fat and scum that accumulate at the surface. The cold start and slow heating allow the soluble proteins to escape the solids and coagulate slowly, forming large aggregates that either rise to the surface

Food Words: *Stock, Broth*

The word *stock* as it's applied in the kitchen reflects the professional cook's approach to sauce making. It derives from an old Germanic root meaning "tree trunk," and has more than 60 related meanings revolving around the idea of basic materials, sources, and supplies. It's thus the culinary application of a very general term, and was first used in the 18th century. Much more specific and ancient is *broth*, which goes back to 1000 CE and a Germanic root *bru* meaning "to prepare by boiling" and the material so prepared, both it and the boiling liquid. *Bouillon* and *brew* are related terms.

and are easily skimmed off, or settle onto the sides and bottom. A hot start produces many separate and tiny protein particles that remain suspended and cloud the stock; and a boil churns particles and fat droplets into a cloudy suspension and emulsion. The pot is left uncovered for several reasons. Because this allows water to evaporate and cool the surface, it makes it less likely that the stock will boil. It also dehydrates the surface scum, which becomes more insoluble and easier to skim. And it starts the process of concentration that will give the stock a more intense flavor.

Single and Double Stocks After the scum has mostly stopped forming, the vegetables, herbs, and wine are added and the cooking is continued at a gentle simmer until most of the flavor and gelatin have been extracted from the solids. The liquid is strained through cheesecloth or a metal strainer without pressing on the solids, which would extract cloudy particles. It's then thoroughly chilled, and the solidified fat removed from the surface. (If the cook doesn't have the time to chill the stock, he can soak away much of the fat from the surface with cloth or paper towels or specially designed plastic blotters.) The stock is now ready to use as an ingredient, to make braised and stewed meats and meat soups, or as a savory cooking liquid for vegetables; or it may be reduced for use in a sauce. The cook may also use stock to extract a new batch of meat and bones and produce the especially flavorful, highly prized—and expensive—double stock. (Double stock can in turn be combined with more fresh meat and bones to make a triple stock.)

Because a standard kitchen extraction of eight hours releases only about 20% of the gelatin in beef bones, the bones may be extracted for a second time, for a total of up to 24 hours. The resulting liquid can then be used to start the next fresh extraction of meat and bones.

Concentrating Meat Stocks: *Glace* and *Demi-glace* Slowly simmered until it's reduced to a tenth its original volume, stock becomes *glace de viande,* literally "meat ice" or "meat glass," which cools to a stiff, clear jelly. Glace has a thick, syrupy, sticky consistency thanks to its high gelatin content, about 25%, an intensely savory taste thanks to the concentrated amino acids, and a rounded, mellow, but somewhat flat aroma thanks to the long hours during which volatile molecules have been boiled off or reacted with each other. Meat glace is used in small quantities to lend flavor and body to sauces. Intermediate between stock and glace is *demi-glace* or "half-*glace,*" which is stock simmered down to 25–40% of its original volume, often with some tomato puree or paste to add flavor and color, and with some flour or starch to supplement its lower gelatin content (10–15%). The tomato particles and flour

Concentrating Stock and Flavor to Finish a Dish

An alternative to cooking stock down in bulk is to reduce it in small quantities to augment the pan juices of a roast or sauté. Once the meat is cooked and its juices concentrated and browned on the pan bottom, the cook can repeatedly add a small quantity of stock to the pan and cook it down until its solids begin to brown, then dissolve the successive brownings in a final dose of stock to make the liquid sauce. The high pan temperature helps break down the gelatin molecules into shorter lengths, so the resulting sauce is less sticky and congeals more slowly than it would if the gelatin were intact.

gluten proteins cloud the stock and are removed by skimming the stock as it reduces, and then by a final straining. The starch in demi-glace, around 3–5% of its final weight, is largely an economy measure—it gives a greater thickness with less stock reduction and loss of volume to evaporation—but it also has the advantage of sparing some of the stock's flavor from being boiled off, and avoiding the sticky consistency of very concentrated gelatin.

Demi-glace is the base for many classic French brown sauces, which are given particular flavors and nuances with the addition of various other ingredients (meats, vegetables, herbs, wine) and final enriching thickeners (butter, cream). Because they're versatile but tedious to prepare, demi-glace and glace are manufactured and widely available in frozen form.

Consommé and Clarification with Egg Whites One of the most remarkable soups is consommé, an intensely flavored, amber-colored, clear liquid with a distinct but delicate body. (The name comes from the French for "to consume," "to use up," and referred to the medieval practice of cooking the meat broth down until it reached the right consistency.) It is made by preparing a basic stock mainly from meat, not flavor-poor bones or skin, and then clarifying it while simultaneously extracting a second batch of meat and vegetables. It's a kind of double stock made expressly for soup; as much as a pound/0.5 kg of meat may go into producing one serving.

The clarification of consommé is accomplished by stirring finely chopped meat and vegetables into the cold stock along with several lightly whisked egg whites. The mixture is then brought slowly to the simmer, and kept there for around an hour. As the stock heats up, the abundant egg white proteins begin to coagulate into a fine cheesecloth-like network, and essentially strain the liquid from within. Soluble proteins from the fresh batch of meat help produce large protein particles that are easily trapped by the egg-white network. Gradually the protein mesh rises to the top of the pot to form a "raft," which continues to collect particles brought to the surface by convection in the liquid. When the cooking is done, the raft is skimmed off and any remaining particles are removed by a final straining. The resulting liquid is very clear. Clarification with egg whites does remove both flavor molecules and some gelatin from the stock, which is why the cook supplies fresh meat and vegetables during the clarification.

Commercial Meat Extracts and Sauce Bases

These days many restaurants and home cooks rely on commercial meat extracts and bases for making their sauces and soups. The pioneer of mass-produced meat extracts was Justus von Liebig, inventor of the mistaken theory that searing meat seals in the juices (p. 161), who was motivated by the equally mistaken belief that the soluble substances in meat contain most

Chinese Meat Stock: Clarifying Without Eggs

The proteins in egg whites are especially effective at removing tiny protein and other particles from meat stocks, but meat proteins themselves can also do the job. Chinese cooks make clear meat broths by cooking the meat and bones of chicken and pork in water, and then clarifying the liquid twice with batches of finely chopped chicken meat, which are gently simmered for 10 minutes and then carefully strained off.

of its nutritional value. However, they do contain much of its savory flavor. Today, meat extracts are made by simmering meat scraps and/or bones in water, then clarifying the stock and evaporating off more than 90% of the water. The initial stock is more than 90% water and 3–4% dissolved meat components; the finished extract is a viscous material that is about 20% water, 50% amino acids, peptides, gelatin, and related molecules, 20% minerals, mainly phosphorus and potassium, and 5% salt. (There are also less concentrated fluid extracts, and solid bouillon cubes that have various natural and artificial flavors added.) Because gelatin would make such concentrated material too thick to work with, manufacturers intentionally break it down into smaller molecules by extending the initial cooking by several hours, and by pressure-cooking the clarified stock (at around 275°F/135°C for 6–8 minutes; this step also coagulates the remaining soluble proteins). In order to limit browning reactions and keep the extract light in both color and roasted flavor, much of the water evaporation is done at temperatures below 170°F/75°C.

Manufacturers now also produce more conventional sauce bases with their gelatin intact. These are often sold in the form of demi-glace or glace de viande.

Cooks can improve the flavor of commercial meat extracts and canned broths by cooking them briefly with herbs and/or diced aromatic vegetables. This fills out the extract's aroma, which is generically meaty to begin with and depleted during the concentration process.

Fish and Shellfish Stocks and Sauces

Like mammals and birds, fish have bones and skin rich in connective tissue. But thanks to the cold environment in which fish bodies function (p. 189), their collagen differs from mammal and bird collagens. Fish collagen is less cross-linked, and so melts and dissolves at much lower temperatures. The collagen and gelatin of warm-water fish like tilapia melt at around 77°F/25°C, that of cold-water cod around 50°F/10°C. So we can extract fish gelatin at cooking temperatures far below the boil, and in relatively short times. The collagen of

Wine in Sauce Making

Wines enter into the making of a variety of sauces, and sometimes are the main ingredient, as in the Burgundian *sauce meurette* (red wine reduced by half with meat and vegetables, then thickened with flour and butter). They contribute several flavor elements, including the tartness of their acids, the sweetness of any residual sugar, the savoriness of succinic acid, and their distinctive aromas. The aromas are modified by cooking, while the tartness, sweetness, and savoriness are not, and become concentrated if the wine is cooked long enough to reduce it. The alcohol in wine can seem harsh when warm, so the wine is usually cooked enough to evaporate much of it. Gentle simmering is said to produce a finer flavor than a fast boil. The tannins in red wine can be a problem, especially when a bottle of wine is reduced down to a few syrupy tablespoons: the tannins become concentrated and unbearably astringent. This outcome can be avoided by cooking the wine down with protein-rich ingredients, including finely chopped meat or a gelatinous stock reduction. The tannins bind to the proteins in those ingredients instead of to the proteins in our mouth (just as tea tannins bind to milk proteins), and so lose their astringent effect.

squid and octopus is more cross-linked than fish collagen, so these molluscs require more prolonged heating at 180°F/80°C to give up much of their gelatin. Most cooks recommend cooking fish stocks for less than an hour so as to avoid making the stock cloudy and chalky with calcium salts from the disintegrating bones. Another reason for short and gentle extractions is that fish gelatins are relatively fragile and more readily break down into small pieces when cooked. And because they associate more loosely with each other, they form delicate gels that melt far below mouth temperature, at 70°F/20°C and lower.

Because fish flavor deteriorates quickly, it's important that fish stock or *fumet* be made with very fresh ingredients. Whole fish, bones, and skin should be thoroughly cleaned and rinsed, and the blood-rich, very perishable gills discarded. Cooks often briefly cook the ingredients in butter to develop the flavor. A gelatin-bodied sauce can be made from the cooking liquid of poached or steamed fish, since even brief cooking will extract flavor and gelatin into the liquid. The traditional cooking liquid for fish is a *court bouillon* or "quick bouillon" made by briefly cooking water, salt, wine, and aromatics together (p. 215).

The bony shells of crustaceans don't contain collagen, so cooking them in water won't give body to the extract. In fact crustacean shells are normally extracted in butter or oil, since their pigments and flavors are more soluble in fat than in water (p. 220).

Other Protein Thickeners

Gelatin is the easiest, most forgiving protein any cook deals with. Heat it up with water and its molecules let go of each other and become dispersed among the water molecules; cool it and they rebond to each other; heat it again and they disperse again. Nearly all other proteins in animals and plants behave in exactly the opposite way: heat causes them to unfold from their normally compact shape, become entangled, and form strong bonds with each other, so that they coagulate permanently and irreversibly into a firm solid. Thus liquid eggs solidify, pliable muscle tissue becomes stiff meat, and milk curdles. Of course a solid piece of coagulated protein can't be a sauce.

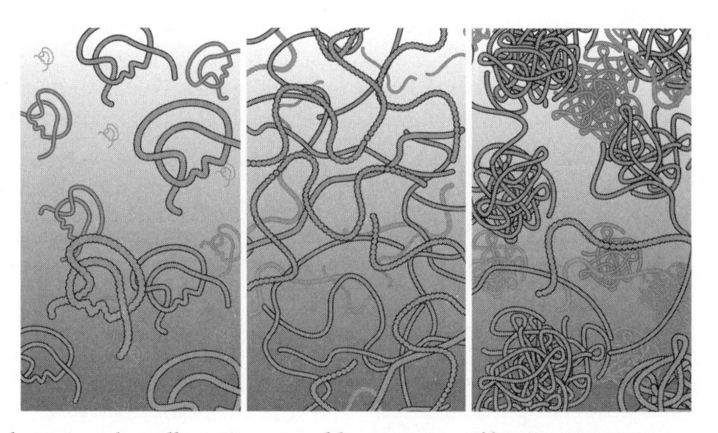

Protein thickening and curdling. Two possible outcomes of heating egg proteins, which start out folded in compact shapes (left). *If conditions favor their unfolding, they form a loose network of long chains* (center) *and thicken the sauce. If heated excessively, the chains aggregate and coagulate in compact clumps* (right) *that give the sauce a curdled consistency and appearance.*

But it's possible to control protein coagulation so that it can give body to sauces.

Careful Temperature Control Cooks first make the flavorful but thin liquid that will be the bulk of the sauce and then add a source of finely suspended proteins. An example is the fricassee, in which the liquid is the stock in which chicken or another meat has been cooked, and the protein source is egg yolks. The mixture is then heated gently. At the point that the proteins unfold and begin to tangle—but before they form strong bonds—the sauce thickens noticeably: it clings to a spoon rather than running off. The attentive cook immediately takes the sauce from the heat and stirs, thus preventing the proteins from forming very many strong bonds, until the sauce cools enough to prevent further bonding. If the sauce gets too hot and the proteins do form strong bonds, they clot together into dense particles, and the sauce becomes grainy and thins out again. Most animal proteins coagulate beginning around 140°F/60°C, but this critical point can vary, so there's no substitute for careful monitoring of the sauce's consistency. Once the sauce has thickened, careful straining can remove the few particles that may have formed.

In all protein-thickened sauces, the cook must take care when mixing the hot sauce with the cool thickener. It's always safest to stir some of the sauce into the thickener, thus heating the thickener gently and diluting it, and then add that mixture to the rest of the sauce. If the thickener goes directly into the sauce, then some of the thickener will get instantly overheated and coagulate into grainy particles. Cooks sometimes work pastes of liver or shellfish organs into butter and then chill the mix. When a chunk of the mix is added to the sauce, the butter melts and slowly releases the thickener into the sauce, while making it somewhat harder for the thickener proteins to bond to each other and coagulate. The inclusion of flour or starch can protect sauce proteins from coagulating; the long starch molecules get in the way of the proteins and prevent them from forming many strong bonds with each other.

If you do overcook a protein-thickened sauce and it separates into thin liquid and grainy particles, you can salvage it by remixing the sauce in a blender, straining it of any remaining coarse particles, and if necessary rethickening it with whatever materials are handy (egg yolks, flour, or starch).

Egg Yolks Egg yolks are the most efficient protein thickeners, in part because they are so concentrated: egg yolks are only 50% water and 16% protein. They are also the handiest, being a common, inexpensive ingredient, and their proteins are already finely dispersed in a rich, creamy fluid. They're mainly used to thicken light-colored white sauces, blanquettes, and fricassees. Yolk-thickened sauces can be brought to the boil as long as they're also partly thickened by starch.

Sabayon sauces are also partly thickened by the coagulation of yolk proteins (p. 639).

Liver Liver is a flavorful thickener, but has the disadvantage of requiring disintegration before it can be used. The coagulable proteins are concentrated inside its cells, so the cook must break the cells open by pounding the tissue, and then strain away the particles of connective tissue that hold the cells together.

Blood Blood is the traditional thickening agent in *coq au vin,* the French rooster in wine sauce, and in braises of game animals (civets). It's about 80% water and 17% protein, and consists of two phases: the various cells, including the red cells colored by hemoglobin, and the fluid plasma in which the cells float. The plasma makes up about two thirds of cattle and pig blood and contains dispersed proteins, about 7% by weight. Albumin is the protein that causes blood to thicken when heated above 167°F/75°C .

Shellfish Organs The liver and eggs of crustaceans, and the sexual tissues of sea urchins have the same advantages and disadvantages as liver, and thicken and coagulate at much lower temperatures. They should be added carefully to a sauce that has first been allowed to cool well below the boil.

Cheese and Yogurt These cultured milk products differ from the other protein thickeners in that their casein proteins have already been coagulated by enzyme activity and/or acidity. They're thus unable to develop a new thickness by being heated with a sauce. Instead, they lend their own thickness as they're mixed into the sauce. They're best subjected to only moderate heat, since temperatures approaching the boil can cause curdling. Yogurt is a more effective thickener if it has been drained of its watery whey. The best cheeses for thickening have a creamy consistency themselves, an indication that the protein network has been broken down into small, easily dispersed pieces; more intact casein fibers can form stringy aggregates (p. 65). Most cheeses are a concentrated source of fat, emulsified droplets of which also contribute body.

Almond Milk This water extract of soaked ground almonds contains a significant amount of protein that causes the liquid to thicken when heated or acidified (p. 504).

SOLID SAUCES: GELATIN JELLIES AND CARBOHYDRATE JELLIES

When a meat or fish stock is allowed to cool to room temperature, it may set into a fragile solid, or gel. This behavior can be undesirable, for example when it causes some sauce to congeal on the plate. But cooks also exploit it to make delightful jellies, a sort of solid sauce. A gel forms when the gelatin concentration is sufficiently high, around 1% or more of the stock's total weight. At these concentrations, there are enough gelatin molecules in the stock that their long chains can overlap with each other to form a continuous network throughout the stock. As the hot stock cools down to the melting temperature of gelatin, around 100°F/40°C, the extended gelatin chains begin to assume the coiled shape that they had in the original triple helix of the collagen fibers (p. 597). And when coils on different molecules approach each other,

Protein-Thickened Sauces and Health

Sauces thickened with proteins are very nutritious, and microbes can multiply rapidly in them. They're best held either above 140°F/60°C or below 40°F/5°C to prevent the growth of bacteria that can cause food poisoning. When cooling large quantities of meat stock, the cook should divide the stock into small portions so that their temperature will fall rapidly through the potentially dangerous temperature range.

Like well-browned meats themselves, meat stocks and sauces whose flavor comes from browned pan juices or from long reduction carry small quantities of chemicals called heterocyclic amines. HCAs are known to damage DNA and therefore may contribute to the development of cancer (p. 124). We don't yet know whether the levels found in meats and sauces pose a significant risk. Vegetables in the cabbage family contain chemicals that prevent HCAs from damaging DNA, so it may be that other foods in a well-balanced diet protect us from the toxic effects of HCAs.

they nest closely alongside each other and bond to form new double and triple helixes. These reassembled collagen junctions give some rigidity to the network of gelatin molecules, and they and the water molecules they surround can no longer flow freely: so the liquid turns into a solid. A 1% gelatin gel is fragile and quivery and breaks easily when handled; the more familiar and robust dessert jellies made with commercial gelatin are usually 3% gelatin or more. The higher the proportion of gelatin, the more firm and rubbery the gel is.

Jellies are remarkable in two ways. At their best they are translucent, glistening, beautiful on their own or as settings for the foods embedded in them. And the temperature at which the gelatin junctions are shaken apart is right around body temperature: so gelatin gels melt effortlessly in the mouth to a full-bodied fluid. They bathe the mouth in sauce. No other thickener has this quality.

Jelly Consistency

The firmness or strength of a gelatin gel, and therefore its tolerance of handling and its texture in the mouth, depends on several factors: the gelatin molecules themselves, the presence of other ingredients, and the way in which the mixture is cooled.

Gelatin Quality and Concentration The most important influence on the texture of a jelly is the concentration and quality of its gelatin. Gelatin is a highly variable material. Even manufactured gelatin (below) is only 60–70% intact, full-length gelatin molecules; the remainder consists of smaller pieces that are less efficient thickeners. The gelatin in a stock is especially unpredictable, since meat and bones vary in their collagen content, and long cooking causes progressive breakdown of the gelatin chains.The best way to assess gel strength is to cool a spoonful of the liquid in a bowl resting in ice water, see if the liquid sets, and how firm the gel is. A liquid lacking in firmness can be reduced further to concentrate the gelatin, or it can be supplemented with a small amount of pure gelatin.

Additional Ingredients Other common ingredients have various effects on gel strength when included in a jelly.

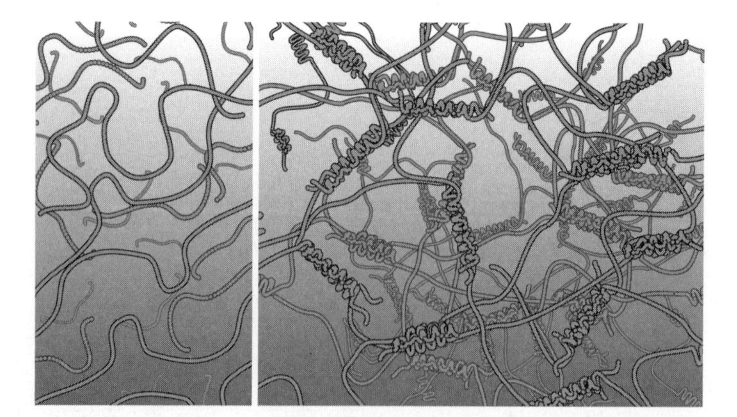

How gelatin turns a liquid into a solid. When the gelatin solution is hot (left), *the water and protein molecules are in constant, forceful movement. As the solution cools and the molecules move more gently, the proteins naturally begin to form little regions of collagen-like helical association* (right). *These "junctions" gradually form a continuous meshwork of gelatin molecules that traps the liquid in its interstices, preventing any noticeable flow. The solution has become a solid gel.*

- Salt lowers gel strength by interfering with gelatin bonding.
- Sugars (except for fructose) increase gel strength by attracting water molecules away from the gelatin molecules.
- Milk increases gel strength.
- Alcohol increases gel strength until it becomes 30 to 50% of the gel, when it will cause the gelatin to precipitate into solid particles.
- Acids—vinegar, fruit juices, wine—with a pH below 4 produce a weaker jelly by increasing repulsive electrical charges on the gelatin molecules.

The gel-weakening effects of salt and acids can be compensated for by increasing the gelatin concentration.

Both strongly acid ingredients and the tannins in tea or red wine can cloud a jelly, the acids by precipitating proteins in a meat or fish stock into tiny particles, and the tannins by binding to and precipitating the gelatin molecules themselves. These ingredients are best cooked briefly with the gelatin solution so that it can be strained or clarified before setting.

A number of fruits—papaya, pineapple, melons, and kiwi among them—contain protein-digesting enzymes that break gelatin chains into short pieces and thus prevent them from setting into a gel at all. They and their juices can be made into a jelly only after a brief cooking to inactivate the enzymes.

Cooling Temperatures The temperature at which the gel forms and ages affects its texture. When "snap-chilled" in the refrigerator, the gelatin molecules are immobilized in place and bond to each other quickly and randomly, so the bonds and the structure of the network are relatively weak. When allowed to set slowly at room temperature, the gelatin molecules have time to move around and form more regular helix junctions, so once it forms, the network is more firm and stable. In practice, jellies should be set in the refrigerator to minimize the growth of bacteria. Gelatin bonds continue to form slowly in the solid jelly, so snap-chilled jellies become as firm as slow-chilled jellies after a few days.

Jellies from Meat and Fish: Aspics

Meat and fish jellies go back to the Middle Ages (p. 584), and are still delightful showpieces. They're made much as consommé is, ideally from a flavorful meat stock—often cooked with a veal foot to provide enough gelatin—or from a double fish stock. The stocks are clarified with egg whites and chopped meat or fish, then filled and flavored just before they set. Aspics should be firm enough to be cut as necessary, but quivery and tender in the mouth, not rubbery. When made to coat a terrine or whole portion of meat, or to bind chopped meat together, they must be firmer, around 10–15% gelatin, so that they don't run off the food or crumble. Fish jellies and aspics are especially delicate due to the low melting temperature of fish gelatin; they and their plates should be kept distinctly cold to prevent premature melting. A homely ver-

Food Words: *Gel, Gelatin, Jelly*

Gel and *jelly,* words for a fragile solid that is largely water, and *gelatin,* the name of the protein that can gel water into a solid, all come from an Indo-European root meaning "cold" or "to freeze." The jelly maker freezes liquids with molecules instead of icy temperatures.

sion of the meat aspic is *boeuf à la mode,* a pot roast braised in stock and wine along with a veal foot, then sliced and embedded in the strained jelly made by the cooking liquid. *Chauds-froids* are meat or fish jellies that include cream.

Other Jellies and Gelées; Manufactured Gelatins

The first jellies were meat and fish dishes, but cooks soon began to use animal gelatins to set other ingredients into pleasing solids, especially creams and fruit juices, and prepared gelatin became a standard ingredient for the pastry cook, who also uses it to give a melting firmness to some mousses, whipped creams, and pastry creams. The most familiar jellies in the United States today, both made from manufactured gelatin powders, are sweet, fruit-flavored, fluorescently colored desserts, and "shooters" fortified with vodka and other spirits. More refined preparations, often named by the French *gelée,* take advantage of the fact that other ingredients can be added at the last minute when the mix is barely warm and about to set, so fresh and delicate flavors can be preserved in the jelly: such things as champagne or the "water" from a seeded tomato.

Gelatin Production Most manufactured gelatin in the United States and Europe comes from pigskin, though some is also made from cattle skins and from bones. Industrial extraction is far more efficient and gentler on the gelatin chains than kitchen extractions. The pigskins are soaked in dilute acid for 18–24 hours to break the collagen's cross-linking bonds, and then are extracted in several changes in water, beginning at just 130°F/55°C, and ending around 195°F/90°C. The low-temperature extracts contain the most intact gelatin molecules, produce the strongest gels, and are the lightest in color; higher temperatures damage more gelatin chains and cause a yellow discoloration. The extracts are then filtered, purified, their pH adjusted to 5.5, evaporated, sterilized, and dried into sheets or granules that are 85–90% gelatin, 8–15% water, 1–2% salts, and 1% glucose. Gelatin quality is sometimes indicated by a "Bloom" number (named for Oscar Bloom, inventor of the measuring device), with high numbers (250) indicating high gelling power.

Types of Gelatin Gelatin is sold in several different forms. Granulated gelatin and sheet gelatin are given an initial soaking in cold water so that the solid gelatin network can absorb moisture and dissolve readily when warm liquid is added. If added to the warm liquid directly, the outer layers of the solid granules can become gluey and stick neighboring granules together, though even these clusters eventually disperse. Sheets with their small surface area introduce less air into the liquid, which can be an advantage when the cook wants great clarity in the jelly. There is also an "instant" gelatin that is manufactured by drying the extract rapidly before the gelatin chains can form junctions, so it disperses directly in warm liquid. And hydrolyzed gelatins have been intention-

Gelatin Doesn't Strengthen Nails or Hair

Though it's widely believed that gelatin supplements strengthen both nails and hair, there is no good evidence that this is true. Nails and hair are made of a very different protein called keratin, and gelatin has no advantage over other protein sources in supplying the building blocks for keratin production.

ally broken into chains too short to form a gel; they're used in food manufacturing as an emulsifying agent (p. 627).

The standard proportion advised by dessert gelatin packages is one 7-gm package per cup/240 ml, or about a 3% solution; 2% and 1% solutions are progressively more tender.

Carbohydrate Gelling Agents: Agar, Carrageenan, Alginates

Gelatin isn't the only ingredient that cooks have at their disposal for turning a flavorful liquid into an intriguing solid. Starch gels give us various pie fillings and the candy called Turkish Delight, and pectin gels the many fruit jellies and jams (p. 296). Along the seacoasts of the world, cooks found long ago that various seaweeds release a viscous substance into hot water that forms a gel when the water cools. These substances are not proteins like gelatin, but unusual carbohydrates with some unusual and useful properties. Food manufacturers use them to make gels and to stabilize emulsions (cream and ice cream, for example).

Agar Agar, a shortened version of the Malay *agar agar,* is a mixture of several different carbohydrates and other materials that has long been extracted from several genera of red algae (p. 341). It's now manufactured by boiling the seaweeds, filtering the liquid, and freeze-drying it in the form of sticks or strands, which are readily available in Asian groceries. The solid pieces of agar can be eaten uncooked as a chewy ingredient in cold salads, soaked and cut into bite-sized pieces. In China agar is made into an unflavored gel that's sliced and served in a complex sauce; it's also used to gel flavorful mixtures of fruit juice and sugar, and stews of meats, fish, or vegetables. In Japan agar is made into jellied sweets.

Agar forms gels at even lower concentrations than gelatin does, less than 1% by weight. An agar jelly is somewhat opaque, and has a more crumbly texture than a gelatin jelly. To make an agar jelly, the dried agar is soaked in cold water, then heated to the boil to fully dissolve the carbohydrate chains, mixed with the other ingredients, and the mixture strained and cooled until it sets, at around 110°F/38°C. But where a

Gelatinous Delicacies: Tendons, Fins, and Nests

The Chinese are great admirers of gelatinous textures, the semi-solid stickiness of long-cooked gelatin-rich connective tissue, and make soups from several ingredients that in the West are hardly considered to be edible. Beef tendons are one example; they are essentially pure connective tissue, and when simmered for hours develop a texture that is simultaneously gelatinous and crunchy. Shark fins are a delicacy that are dried after being taken from that cartilaginous fish, then rehydrated, simmered in several changes of water to remove off-flavors, and then simmered in broth.

Most unusual of all are the nests of cave-dwelling birds in the swallow family, swiftlets of the genus *Collocalia,* which are found throughout Southeast and South Asia. The males build their nests up from strands of their saliva, which stick to the cave walls and dry to form a small but strong cup. The harvested nests are soaked in cold water to rinse out impurities and to let them absorb water and swell. They're then simmered in broth, and enjoyed for their semisolid, gelatinous consistency, which is due not to gelatin itself, but to salivary proteins called mucins, which are related to the mucins in egg white (p. 77).

gelatin gel sets and remelts at around the same temperature, an agar gel only melts again when its temperature reaches 185°F/85°C. So an agar gel won't melt in the mouth; it must be chewed into particles. On the other hand, it will remain solid on hot days, and can even be served hot. Modern cooks have used this property to disperse small agar-gelled morsels of contrasting flavor into a hot dish.

Carrageenan, Alginates, Gellan Experimentally minded cooks are exploring a number of other unusual carbohydrate gelling agents, some traditional and some not. *Carrageenan,* from certain red algae (p. 341), has long been used in China to gel stews and flavored liquids, and in Ireland to make a kind of milk pudding. Purified fractions of crude carrageenan produce gels with a range of textures, from brittle to elastic. *Alginates* come from a number of brown seaweeds, and form gels only in the presence of calcium (in milk and cream, for example). Inventive cooks have taken advantage of this to make small flavored spheres and threads: they prepare a calcium-free alginate solution of the desired flavor and color, and then drip or inject it into a calcium solution, where it immediately gels. *Gellan,* an industrial discovery, is a carbohydrate secreted by a bacterium, and in the presence of salts or acid forms very clear gels that release their flavor well.

SAUCES THICKENED WITH FLOUR AND STARCH

Many sauces, from long-simmered classic French brown sauces to last-minute gravies, owe at least part of their consistency to the substance called *starch.* Unlike the other thickening agents, starch is a major component of our daily diet. It's the molecule in which most plants store the energy they generate from photosynthesis, and provides about three-quarters of the calories for the earth's human population, mainly in the form of grains and root vegetables. It's the least expensive and most versatile thickener the cook has to work with, a worthy adjunct to gelatin and fat. The cook can choose among several different kinds of starch, each with its own qualities.

Agar: From Pudding to Petri Dish

Solid gels made from agar have long been a standard tool in the study of microbes. Scientists make them up to include various nutrients, and then grow colonies of microbes on their surface. Agar gels have several important advantages over the original growth medium, which was gelatin. Very few bacteria can digest the unusual agar carbohydrates, so agar gels remain intact and the bacterial colonies separate, while many bacteria digest proteins and can quickly liquefy a gelatin gel into a useless soup. And agar gels remain solid at the ideal temperatures for bacterial growth, often around 100°F/38°C, a temperature at which gelatin begins to melt.

How did microbiologists come to use agar? In the late 19th century, Lina Hesse, the American wife of a German scientist, recalled the advice of family friends who had lived in Asia, and made agar jellies and puddings that stayed solid in the summer heat of Dresden. Her husband relayed his wife's suggestion to his boss, the pioneering microbiologist Robert Koch, who then used agar to isolate the bacterium that causes tuberculosis.

The Nature of Starch

Starch molecules are long chains of thousands of glucose sugar molecules linked up together. There are two kinds of starch molecules: long straight chains called amylose, and short, branched, bushy chains called amylopectin. Plants deposit starch molecules in microscopic solid granules. The size, shape, amylose and amylopectin contents, and cooking qualities of the starch granules vary from species to species.

Linear Amylose and Bushy Amylopectin The shapes of amylose and amylopectin molecules have a direct effect on their ability to thicken a sauce. The straight amylose chains coil up into long helical structures when dissolved in water, but they retain their basically linear shape. Their elongation makes it very likely that one chain will knock into another or into a granule: each sweeps through a relatively large volume of liquid. By contrast, the branched shape of amylopectin makes for a compact target and therefore a molecule less likely to collide with others; and even if it does collide, it's less likely to get tangled up and slow the motion of other molecules and granules in the vicinity. A small number of very long amylose molecules, then, will do the job of more but shorter amylose molecules, and of many more bushy amylopectins. For this reason, the cook can obtain the same degree of thickening from a smaller amount of long-amylose potato starch than from moderate-amylose wheat and corn starches.

Swelling and Gelation What makes starch so useful is its behavior in hot water. Mix some flour or cornstarch into cold

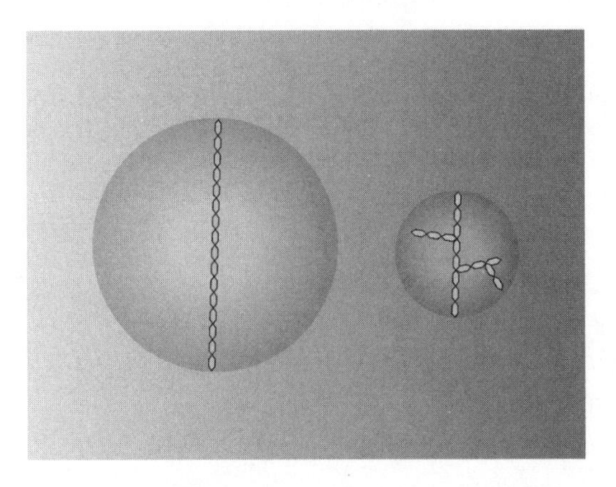

Two kinds of starch. Starch molecules are chains made up of hundreds or thousands of glucose molecules bonded together. They take two forms: straight chains of amylose (left), *and branched chains of amylopectin* (right). *A long amylose chain moves around in a larger volume than the more compact amylopectin containing the same number of glucose molecules, and is more likely to tangle with other chains. Amylose is therefore a more effective thickener than amylopectin.*

Pure Starch

Starch has been separated from the proteins and other materials in grains since ancient times. The Romans called it *amylum,* which meant "not ground at the mill." They made it by grinding wheat in a mortar and then soaking the flour for days, during which bacteria grew and digested the grain's cell walls and gluten proteins while leaving the dense, solid starch grains intact. They reground the dregs, and then pressed them through fine linen, which retained the small grains. The starch grains were dried in the sun, and then either cooked in milk or used to thicken sauces (p. 583).

water, and nothing much happens. The starch granules slowly absorb a limited amount of water, about 30% of their own weight, and they simply sink to the bottom of the pot and sit there. But when the water gets hot enough, the energy of its molecules is sufficient to disrupt the weaker regions of the granule. The granules then absorb more water and swell up, thereby putting greater and greater stress on the more organized, stronger granule regions. Within a certain range of temperatures characteristic of each starch source but usually beginning around 120–140°F/50–60°C, the granules suddenly lose their organized structure, absorb a great deal of water, and become amorphous networks of starch and water intermingled. This temperature is called the *gelation range,* because the granules become individual gels, or water-containing meshworks of long molecules. This range can be recognized by the fact that the initially cloudy suspension of granules suddenly becomes more translucent. The individual starch molecules become less closely packed together and don't deflect as many light rays, and so the mixture becomes clearer.

Thickening: The Granules Leak Starch Depending on how concentrated the starch granules are to begin with, the starch-water mixture may noticeably thicken at various points during their swelling and gelation. Most sauces are rather dilute (less than 5% starch by weight) and thicken during gelation, when the mixture begins to become translucent. They reach their greatest thickness after the gelated granules begin to leak amylose and amylopectin molecules into the surrounding liquid. The long amylose molecules form something like a three-dimensional fishnet that not only entraps pockets of water, but blocks the movements of the whale-like, water-swollen starch granules.

Thinning: The Granules Break Once it reaches its thickest consistency, the starch-water mixture will slowly thin out again. There are three different things that the cook may do that encourage thinning: heating for a long period of time after thickening occurs, heating all the way to the boil, and vigorous stirring. All of these have the same effect: they shatter the swollen and fragile granules into very small fragments. While

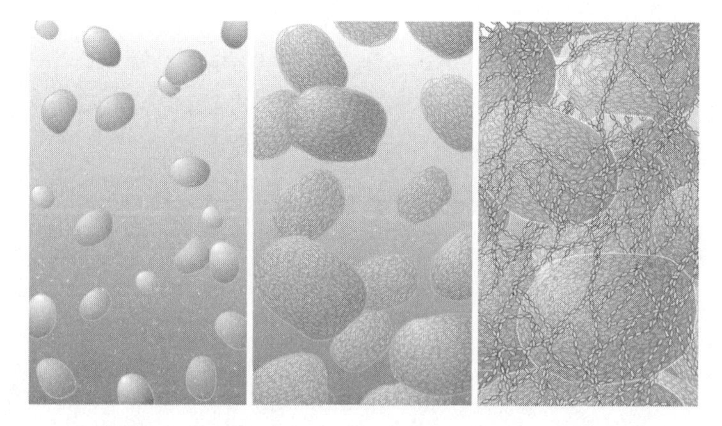

Thickening a sauce with starch. Uncooked starch granules offer little obstruction to the flow of the surrounding liquid (left). *As the sauce heats up and the temperature reaches the gelation range, the granules absorb water and swell, and the sauce consistency begins to thicken* (center). *As cooking continues and the temperature approaches the boil, the granules swell even more and leak starch chains into the liquid* (right). *It's at this stage that the sauce reaches its maximum thickness.*

this does mean that even more amylose is released into the water, it also means that there are many fewer large bodies to get caught in the amylose tangle. In other words, the amount of netting increases, the mesh grows finer, but at the same time the big whales become small minnows. This thinning effect is especially striking in the case of very thick pastes, less obvious in normal sauces. If the granules are few and far between to begin with, their disintegration is less noticeable. This thinning is accompanied by a greater refinement of texture, as the starch particles disappear and only indetectably small molecules remain.

Some of the thinning of long-simmered starch-based sauces is caused by the gradual breakdown of the starch molecules themselves into smaller fragments. Acidity accelerates this breakdown.

Cooling, Further Thickening, and Congealing Once the starch in a sauce has gelated, its amylose has leaked out, and the cook judges the sauce to be properly cooked, he stops the cooking, and the temperature of the sauce begins to fall. As the mixture cools down, the water and starch molecules move with less and less energy, and at a certain point the force of the temporary bonds among them begins to hold the molecules together longer than they are kept apart by random collisions. Gradually, the longer amylose molecules form stable bonds among themselves, the kind of bonds that held them together in the granule initially. Water molecules settle in the pockets between starch chains. As a result, the liquid mixture gets progressively thicker. If the amylose molecules are concentrated enough, and the temperature falls far enough, the liquid mixture congeals into a solid gel, just as a gelatin solution settles into a jelly. (Bushy amylopectin molecules take much longer to bond to each other, so low-amylose starches are slow to congeal.) This is the way in which pie fillings, puddings, and similar solid but moist starch concoctions are made.

Judge Sauce Consistency at Serving Temperatures It's important for the cook to anticipate this cooling and thickening. We create and evaluate most sauces on the stove at high temperatures, around 200°F/93°C, but when they're poured in a

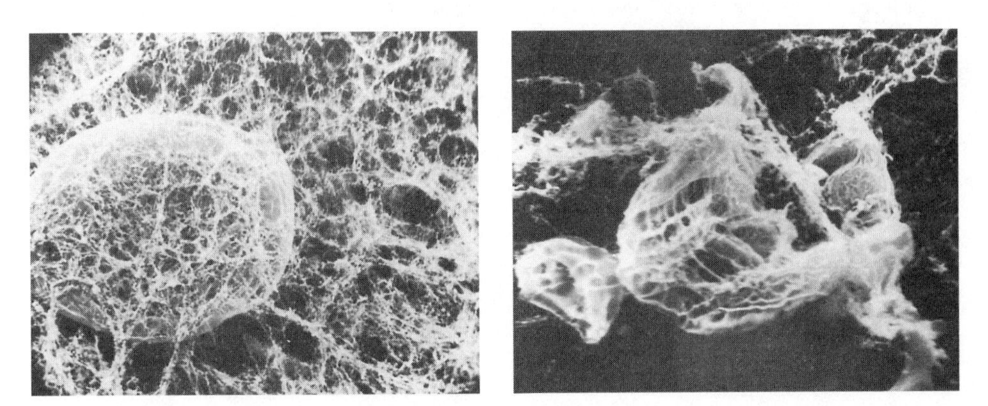

Starch in sauce making. A swollen granule of potato starch caught in a meshwork of molecules freed from it and other granules (left). *A starch-thickened sauce is thickest at this stage, when both starch granules and molecules block the movement of water. A granule of wheat starch that has lost nearly all of its starch molecules to the surrounding liquid* (right). *As the granules in a starch-thickened sauce disintegrate, they no longer get caught in the mesh of free starch, and the sauce thins out.*

thin layer onto food and served, they immediately begin to cool and thicken. However thick a sauce is in the pan, it's going to be thicker when the diner actually tastes it, and it may even congeal on the plate. So sauces should be thinner at the stove than they're meant to be at the table. (Minimizing the amount of thickener will also reduce the extent to which the sauce's flavor is muted.) The best way to predict the final texture of a sauce is to pour a spoonful into a cool dish and then sample it.

Different Starches and Their Qualities

Cooks have several different forms of starch to choose among for thickening sauces, each with its own particular qualities. They fall into two families: starches from grains, including flour and cornstarch, and starches from tubers and roots, including potato starch and arrowroot. Less commonly seen except on the ingredient labels of processed foods is sago starch, from the stem pith of a Pacific palm (*Metroxylon sagu*).

Grain Starches Starches from grains tend to share several characteristics. Their starch granules are medium-sized, and contain small but significant amounts of lipids (fats, fatty acids, phospholipids) and protein. These impurities somehow give the starch granules some structural stability, which means that it takes a higher temperature to gelate them; and they lend a cloudiness and distinct "cereal" flavor to starch-water mixtures. Light that passes right through a gelated mesh of pure starch and water is scattered by tiny starch-lipid or starch-protein complexes, producing a milky, impenetrable appearance. Grain starches contain a high proportion of moderately long amylose molecules that readily form a network with each other, and so make sauces that quickly thicken and congeal when cooled.

Wheat Flour Wheat flour is made by grinding wheat grains and sieving the bran and germ from the starch-rich endosperm (p. 528). Wheat flour is only about 75% starch, and includes about 10% by weight of protein, mainly the insoluble gluten proteins. It's therefore a less efficient thickener than pure cornstarch or potato starch; it takes more flour to obtain the same consistency. A common rule of thumb is to use 1.5 times as much flour as starch. Flour has a distinct wheat flavor that cooks often transform by precooking the flour before adding it to a sauce (p. 617). The suspended particles of gluten protein make flour-based sauces especially opaque and give their surface a matte appearance, unless the sauce is cooked for hours and skimmed to remove the gluten.

Cornstarch Cornstarch is practically pure starch and so a more efficient thickener than flour. Cornstarch is manufactured by soaking the whole maize grain, milling it coarsely to remove the germ and hull, and grinding, sieving, and centrifuging the remainder to separate the seed proteins. The resulting starch is washed, dried, and reground into a fine powder consisting of single granules or small aggregates. During this wet processing, the starch granules absorb odors and develop their own when their traces of lipids are oxidized, so cornstarch has a distinctive flavor unlike that of wheat flour, which is milled dry.

Rice Starch Rice starch is seldom seen in Western markets. Its granules have the smallest average size of the starches, and produce an especially fine texture in the early stages of thickening.

Tuber and Root Starches Compared to the starches from dry grains, the starches from moist underground storage organs come in the form of larger granules that retain more water molecules, cook faster, and release starch at lower temperatures. They contain less amylose, but their amylose chains are up to four times longer than cereal amyloses. Root and tuber starches contain a fraction of the lipids and proteins that are associated with cereal starches, which makes them more readily gelated—lipids delay gelation by stabilizing granule structure—and gives them less pronounced flavors. These starches leave their sauces with a translucent, glossy appearance. The properties of root starches suit them for last-minute corrections of sauce consistency: less of them is required to lend a given thickness, they thicken quickly, and don't need precooking to improve their flavor.

Potato Starch Potato starch was the first commercially important refined starch and is still an important food starch in Europe. It is unusual for several characteristics. Its granules are very large, up to a tenth of a millimeter across, and its amylose molecules are very long. This combination gives potato starch an initial thickening power far greater than that of any other starch. The long amylose chains tangle with each other and with the giant granules to block easy movement of the sauce fluid. This entanglement also creates long aggregates of amylose and granules that can give the impression of stringiness. And the large swollen granules give a noticeable initial graininess to sauces. However the granules are fragile, and readily fragment into finer particles; so having reached its thickest and grainiest, the consistency of a potato-starch sauce rapidly gets both finer and thinner. Potato starch is also unusual for having a large number of attached phosphate groups, which carry a weak electric charge and cause the starch chains to repel each other. This repulsion helps keep the starch chains evenly dispersed in a sauce, and contributes to the thickness and clarity of the dispersion and its low tendency to congeal into a gel on cooling.

Tapioca Tapioca, derived from the root of a tropical plant known as manioc or cassava (*Manihot esculenta,* p. 305), is a root starch used mostly in puddings. It tends to form unpleasantly stringy associations in water and so is usually made into large pregelatinized pearls (p. 578), which are then cooked only long enough to be softened. Because tapioca keeps well in the ground and is processed into starch within days of harvest, it doesn't develop the strong aromas found in wheat and corn starches or in potato starch, which is typically extracted from long-stored, second-quality tubers. Tapioca starch is especially prized for its neutral flavor.

Properties of Some Common Thickening Starches Cooked in Water

	Gelation Temperature	Maximum Thickness	Consistency	Stability to Prolonged Cooking	Appearance	Flavor
Wheat	126–185°F 52–85°C	+	Short	Good	Opaque	Strong
Corn	144–180°F 62–80°C	++	Short	Moderate	Opaque	Strong
Potato	136–150°F 58–65°C	+++++	Stringy	Poor	Clear	Moderate
Tapioca	126–150°F 52–65°C	+++	Stringy	Poor	Clear	Neutral
Arrowroot	140–187°F 60–86°C	+++	Stringy	Good	Clear	Neutral

Arrowroot Arrowroot starch as it's known in the West is refined from the roots of a West Indian plant (*Maranta arundinacea*). Arrowroot starch has smaller granules than potato or tapioca starches, produces a less stringy consistency, and doesn't thin out as much on prolonged cooking. Its gelation temperature is higher than the other root starches, more like the range for cornstarch. A number of other plants and their starches are also called arrowroot in Asia and Australia (species of *Tacca, Hutchenia, Canna*).

Root Starches in China In China, starch was originally extracted from millet and water chestnuts. Nowadays most Chinese sauces are thickened with corn, potato, or sweet potato starch—all plants from the New World. Other Asian sources of starch are yams, ginger, lotus, and the tuber of the kudzu vine (*Pueraria*).

Modified Starches Food manufacturers have not been content with the starches available in nature, mainly because the consistency they create isn't stable throughout the cycle of production, distribution, storage, and use by the consumer. They've therefore engineered a variety of starches that are more stable. Plant breeders have developed so-called "waxy" varieties of corn whose seeds contain little or no amylose and are nearly all amylopectin, which doesn't form networks as readily as amylose. Waxy starches therefore make sauces and gels that resist congealing and separation into a firm solid phase and watery residue, a problem to which high-amylose starches are prone.

Ingredient manufacturers also use physical and chemical treatments to modify the starch molecules from standard plant varieties. They precook and dry starches in various ways to produce powders or granules that readily absorb cold water or disperse in and thicken liquids without requiring cooking. And they alter them with chemicals—cross-linking chains to each other, or oxidizing them, or substituting fat-soluble side groups along the chain—to make them less prone to breakdown during cooking, to make them more effective emulsion stabilizers, and to give them other qualities that "native" starches don't normally have. Such starches are listed on ingredient labels as "modified starch."

The Influence of Other Ingredients on Starch Sauces

Flavorings: Salt, Sugar, Acid Starch and water are the basis for a sauce's structure, and most other ingredients have only secondary effects on that structure. Salt, acid, and sugar are frequently added for their contributions to flavor. Salt slightly lowers the gelation temperature of starch, while sugars increase it. Acids in the form of wine or vinegar encourage the fragmentation of starch chains into much shorter lengths, so that starch granules gelate and disintegrate at lower temperatures, and the final product is less viscous for a given amount of starch. Root starches are noticeably affected by moderate acidity (a pH lower than 5), while grain starches can withstand the acidities typical of yogurt and many fruits (pH 4). Gentle and brief cooking will minimize acid breakdown.

Proteins and Fats Two other materials are commonly found in sauces and have some influence on their texture. Flour is about 10% protein by weight, and much of this fraction is insoluble gluten. Gluten aggregations probably get caught in the starch network and so slightly increase the viscosity of the solution, though the pure starches are generally more powerful thickeners overall. Sauces based on concentrated meat stocks also contain a good deal of gelatin, but gelatin and starch seem not to affect each other's behavior.

Finally, fats are usually present in the form of butter, oil, or the drippings from a roast. They do not mix with water or water-soluble compounds, but they do slow the penetration of water into starch granules. Fat does contribute the sensations of smoothness and moistness to a sauce, and when used to precook the flour in a roux, it coats the flour particles, prevents them from clumping together in the water, and so safeguards against lumps.

Incorporating Starch into Sauces

In order to thicken a sauce with starch, the cook must get the starch into the sauce. Very basic, but not so simple! If you add flour or starch directly to a hot sauce, it lumps up and never disperses evenly: the moment they hit the hot liquid, the clumps of starch granules develop a partly gelated, sticky surface that seals the dry granules inside and prevents them from dispersing.

Slurries, Beurre Manié, Floured Meat

Cooks use four methods for incorporating starch into a sauce. The first is to mix the starch with some cold water, so that the granules are wetted and separated before they encounter gelation temperatures. The starch-water slurry can then be added directly to the sauce. A second method is to separate the starch or flour particles not with water, but with fat. *Beurre manié,* or "kneaded butter," is flour worked into a paste with its weight in butter. When a piece of the paste is added to a hot sauce in need of last-minute thickening, the butter melts and gradually releases greased starch particles into the liquid, where their swelling and gelation are slowed by the water-repelling surface layer.

A third method for getting starch into a sauce is to introduce it early in the cooking, not late. Many stews and fricassees are made by dusting pieces of meat with flour, then sautéing the pieces, and only then adding a cooking liquid that will become the sauce. In this way the starch has already been dispersed over the large surface area of the meat pieces, and it has been precoated with the sautéing fat, which prevents clumping when the liquid is added.

Roux The fourth method for getting starch into a sauce, and one that has been developed into a minor art of its own, is to preheat the starch separately in fat to make what the French call a *roux,* from the word for "red." The basic principle works with any form of starch and any fat or oil. In the traditional French system, the cook carefully heats equal weights of flour and butter in a pan to one of three consecutive endpoints: the mixture has had the moisture cooked out of it, but the flour remains white; the flour develops a light yellow color; or the flour develops a distinctly brown color.

Improvements in Flavor, Color, and Dispersability In addition to coating the flour particles with fat and making them easier to disperse in a hot liquid, roux making has three other useful effects on the flour. First, it cooks out the raw cereal flavor and develops a rounded, toasty flavor that becomes more pronounced and intense as the color darkens. Second, the color itself—the product of the same browning reactions between carbohydrates and proteins that produce the toasty flavor—can lend some depth to the color of the sauce.

Finally, the heat causes some of the starch chains to split, and then to form new bonds with each other. This generally means that long chains and branches are broken down into smaller pieces that then form short branches on other molecules. The short, branched molecules are less efficient at thickening liquids than the long chains, but they're also slower to bond to each other and form a continuous network as the liquid cools. The sauce is therefore less prone to congeal on the plate. The darker the roux, the more starch chains are modified in this way, and so the more roux is required to create a given thickness. It takes more of a dark brown roux than a light one to thicken a given amount of liquid. (The industrial version of roux making to make a starch more dispersable and stable to cooling is called *dextrinization,* and involves heating dry starch together with some dilute acid or alkali to 375°F/190°C.)

Outside of France, roux are especially

prominent in the cooking of New Orleans, where flour is cooked to a number of different stages from pale to chocolate-brown, and where cooks may use several roux in a single gumbo or stew to lend their distinct layers of flavor.

Starch in Classic French Sauces

In the code formalized by Auguste Escoffier in 1902, there are three leading mother sauces that are thickened in part with flour: the stock-based brown and white sauces, or espagnole and velouté; and the milk-based béchamel. Each of these relies on a distinctive combination of roux and liquid. Brown sauce consists of a stock made from browned vegetables, meat, and bones, then reduced after thickening with a roux that is cooked until the flour browns as well. White sauce uses a stock made from *un*browned meat, vegetables, and bones, and is bound with a pale yellow roux. Béchamel combines milk with a roux that is not allowed to change color at all. From these three parent sauces, the cook can produce scores of offspring simply by finishing the sauce with different seasonings and enrichments.

Once the roux has been added to the stock, the mixture is allowed to simmer for quite a while—two hours for velouté, and up to ten in the case of brown sauce. During this time, the flavor is concentrated as water evaporates, and the starch granules dissolve and disperse among the gelatin molecules, with a very smooth texture the result. Brown sauce is cooked for the longest time because it's meant to be quite clear to the eye, and this requires that the gluten proteins coagulate and be carried to the surface, where they and the tomato solids can be skimmed off.

Escoffier said that a sauce should have three characteristics: a "decided" taste, a texture that's smooth and light without being runny, and a glossy appearance. The taste is a matter of making fine stocks and being judicious in seasoning, while the consistency and appearance depend on how the thickening is accomplished. Generally, long and patient simmering is necessary, so that there will be little or no vestige of granular structure left to the starch, and the

The First Printed Recipe for *Roux*

It was long thought that the first recipes for roux appeared in late 17th-century French cookbooks, but here is one of two German recipes from 150 years before La Varenne. They suggest that this version of the starch thickener was developed in late medieval times.

How to Cook a Wild Boar's Head, Also How to Prepare a Sauce for It

A wild boar's head should be boiled well in water and, when it is done, laid on a grate and basted with wine, then it will be thought to have been cooked in wine. Afterwards make a black or yellow sauce with it. First, when you would make a black sauce, you should heat up a little fat and brown a small spoonful of wheat flour in the fat and after that put good wine into it and good cherry syrup, so that it becomes black, and sugar, ginger, pepper, cloves and cinnamon, grapes, raisins and finely chopped almonds. And taste it, however it seems good to you, make it so.

—*Das Kochbuch der Sabina Welserin,* 1533, transl. Valoise Armstrong

insoluble gluten proteins will be caught up in the surface scum and so removed from the sauce. Gelatin contributes some body to the stock-based sauces, but the starch is what gives them most of their viscosity. After reduction, the concentration of starch in these sauces is around 5%, the gelatin concentration probably about half that.

Milk-Based Sauces: Béchamel and Boiled Sauces based on milk rather than stock are of course much easier to make, and more forgiving; because they're already milky, the cook doesn't have to worry about long simmering to clarify them. The classic starch-thickened milk sauce is béchamel, whose only other ingredients are seasonings and the butter in which the starch is precooked for a couple of minutes. Once the milk has been added to the roux, the sauce is simmered for 30–60 minutes with occasional skimming of the skin of milk and flour proteins that forms at the surface. Starch is more effective at thickening milk than it is meat stocks, apparently because it bonds both to the milk proteins and the fat globules and so recruits these weighty ingredients into its flow-slowing network. Thanks to its pleasant but neutral flavor, béchamel is a versatile sauce that can be imbued with many flavors and served with many main ingredients. It's also made in several thicknesses for a variety of purposes. Thick preparations (6% flour by weight) serve as the base for soufflés, somewhat thinner ones as a moistening and enrichment for gratins.

In the "boiled dressing" often used in the United States to moisten coleslaw and other robust salads, flour not only thickens the milk and/or cream, but also helps prevent the vinegar from curdling the milk and egg-yolk proteins into coarse particles.

Gravy

We come now to the homely Anglo-American cousin of French sauces, the starch-thickened gravy typically made to accompany a roast. This is a last-minute sauce that's put together just before serving, and consists of the roast's juices, extended with additional liquid, and thickened with flour. The drippings from the roast, both fat and browned solids, give the gravy its flavor and color. First the fat is poured off and reserved, and the pan is "deglazed": the browned solids are lifted from the roasting pan with a small amount of water, wine, beer, or stock. The liquid dissolves the browning-reaction products that have stuck to the pan and so takes up their especially rich flavors. The deglazing liquid is poured off and reserved separately. Now some of the fat is returned to the pan with an equal

Escoffier on Future Roux

Though he was a traditionalist in many ways, Escoffier openly looked forward to the day when pure starch would replace flour as the thickener in stock-based sauces.

> Indeed, if [starch] is absolutely necessary to give the mellowness and velvetiness to the sauce, it is much simpler to give it pure, which permits one to bring it to the point in as little time as possible, and to avoid a too prolonged sojourn on the fire. It is therefore infinitely probable that before long starch, fecula, or arrowroot obtained in a state of absolute purity will replace flour in the roux.

Today's proponents of the classic sauces, however, generally remain loyal to flour.

volume of flour, and the flour cooked until it has lost its raw aroma. The deglazing liquid is added, around a cup/250 ml for every 1–2 tablespoons/10–20 gm flour. The mixture is cooked until it thickens, a matter of a few minutes.

Because they're made at the last minute, gravies are not cooked long enough to cause the disintegration of the starch granules, and therefore generally have a slightly coarse texture, even when lump-free. This gives gravies a character very different from that of the suave sauce: hearty, and when they are extremely thick, almost bready. The cook can obtain a smoother consistency by making an initial preparation from the flour and a fraction of the deglazing liquid, heating the mixture until the starch granules gelate and crowd up against each other to form a thick paste, and whisking the paste vigorously to smash the granules into each other and break them up into finer pieces. This paste is then mixed with the rest of the deglazing liquid and simmered until it's evenly dispersed and the liquid reaches the desired consistency.

SAUCES THICKENED WITH PLANT PARTICLES: PUREES

Some of the most delicious sauces we eat, including tomato sauces and applesauce, are made simply by crushing fruits and vegetables. Crushing, or pureeing, frees the juices from the cells of the fruit or vegetable, and breaks the cell walls into fragments that become suspended in the juices and block their flow, so giving them some thickness. Crushed nuts and spices have no juices of their own, but they thicken a liquid to which they're added by absorbing some of its water and providing dry cell particles that obstruct the liquid's flow.

Until recently, most purees of plant tissue would have been made by cooking the tissue to soften it, and then either grinding it in a mortar or forcing it through a fine sieve. Raw purees could only be made from fruits softened by ripening, or from brittle nuts. Today's cooks can use powerful machines—blenders, food processors—to puree any fruit or vegetable or seed with ease, whether they're raw or cooked.

Plant Particles: Coarse and Inefficient Thickeners

Compared to the other ways of thickening, simple pureeing tends to produce a coarse sauce that more readily separates into solid particles and thin fluid. The solid fragments of plant cell walls are clumps of many thousands of carbohydrate and protein molecules. If those molecules were dispersed separately and finely throughout the fluid—as gelatin or starch molecules are in other sauces—they would bind many more water molecules, get tangled up in each other, and be far too small for the tongue to detect as particles. But plant-cell fragments range from 0.01 to 1 millimeter across; they give a grainy impression on the tongue and they're far less efficient than individual mol-

Food Words: *Puree*

The word *puree,* meaning thoroughly crushed fruits, vegetables, or animal tissue, comes ultimately from the Latin *purus,* meaning "pure." England borrowed a form of the French descendent, the verb *purer,* which had both a general meaning, "to purify," and a specific one: to drain excess water from beans and peas left to soak. The beans and peas would go on to be cooked into a mush, and the consistency of that mush appears to be the prototype of other purees.

ecules at binding water or interfering with fluid flow. And because the fragments are usually denser than the cell fluids, they end up sinking and separating from the fluids. Heating without stirring tends to speed this separation, because the free water is able to flow and rise from the bottom of the pot through the thicker particle phase, and accumulate above it.

Some sauces and related preparations aren't meant to be suave and smooth; instead the cook leaves some pieces of tissue intact to highlight the texture of the fruit or vegetable itself. Mexican tomato and tomatillo salsas, unstrained cranberry sauce, and applesauce are familiar examples.

Refining the Texture of Purees Cooks can refine the basic coarseness of purees by modifying either the solid plant particles or the fluid that surrounds them.

Making the Plant Particles Smaller There are several ways to make the plant particles as fine as possible.

- The pureeing process itself is a physical crushing or shearing that breaks the plant tissue into pieces and liberates thickening molecules from them. Blenders and mortars are the most effective tools for this; food processors slice rather than crush. Producing a fine puree can take some time even in a blender, several minutes or more.
- Straining through a sieve or cheesecloth removes the large particles, and forcing the puree through a fine mesh breaks large particles into smaller ones.
- Heating softens cell walls so that they'll break into smaller pieces, and shakes loose long-chain carbohydrates from the cell walls and gets them into the watery phase, where they can act as separate starch and gelatin molecules do.
- Freezing a puree and then thawing it causes ice crystals to damage cell walls, which can help liberate more pectin and hemicellulose molecules into the liquid.

Preventing Separation The consistency of a puree is also improved by reducing the

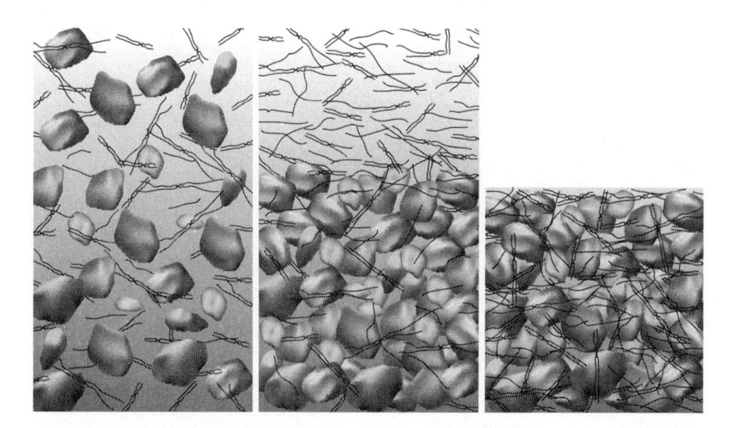

A fruit or vegetable puree. Grinding plant tissue turns it inside out, freeing the cell fluids and breaking the cell walls and other structures into small particles. A puree is a mixture of plant particles and molecules floating in water (left). *If left to stand, most purees will separate, with the larger particles settling to the bottom* (center). *This separation can be prevented, and the puree consistency thickened, by cooking the puree down and evaporating the excess water* (right).

amount of water in the continuous phase. The simplest way to do this is to cook the whole puree down, simmering gently, until the separate thin phase disappears. Another way that better preserves the puree's fresh flavor is to drain the thin fluid off the solids and either discard it, or cook it down separately and then add it back. Or the cook can remove some of the fruit's or vegetable's water before crushing it, for example by partly drying halved tomatoes in the oven.

The binding power of the puree particles themselves can be supplemented by adding some other thickener, dry spices or nuts for example, or flour or starch.

Fruit and Vegetable Purees

Any fruit or vegetable can be turned into a sauce by crushing it. Here are brief observations about some of the more commonly pureed foods.

Raw Purees: Fruits Raw purees are generally made from fruits, whose ripening enzymes often break down their cell walls from within, and thus allow their intact flesh to turn into a puree in the mouth. Raspberries, strawberries, melons, mangoes, and bananas are examples of such naturally soft fruits. The flavor of a raw puree is usually accentuated by the addition of sugar, lemon juice, and aromatic herbs or spices. But that flavor is fragile and changeable. Pureeing mixes the cell contents with each other and with oxygen in the air, so enzyme action and oxidation begin immediately (see below for the effects in cooked purees of tomato, a botanical fruit). The best way to minimize this change is to chill the puree, which slows all chemical reactions.

Raw Purees: Pesto The Italian puree of basil leaves, *pesto genovese,* also contains olive oil and so is partly an emulsion as well. Pesto takes its name from the same root that gives us *pestle,* and the basil leaves and garlic were traditionally ground with a pestle and mortar. Because this takes some time and effort, modern cooks usually prepare pesto in a blender or food processor. The choice of appliance and how it's used influence both consistency and flavor. The crushing and shearing action of the pestle, the shearing action of the blender, and the slicing action of the processor all produce different proportions of intact and broken cells. The more thoroughly the cells are broken, the more their contents are exposed to each other and to the air, and the more their flavor evolves. A coarse pesto will have a flavor most like the flavor of fresh basil leaves.

Cooked Purees: Vegetables, Applesauce Most vegetable purees are made by first cooking the vegetable to soften its tissues, break apart the cells, and free their thickening molecules. Some that develop an especially suave smoothness have cell walls rich in soluble pectin, which escapes from the softened wall fragments during pureeing. These vegetables include carrots, cauliflower, and capsicum peppers; more than 75% of the cell-wall solids in capsicum puree is pectin. Many root and tuber vegetables (though not carrots) contain starch granules, which when cooked absorb much of the water in the vegetable and make it less watery. However, such vegetables are best crushed gently, without breaking open the cells.Thorough pureeing that liberates the gelated starch turns the vegetable into a super-thick potato gravy, gluey and stringy.

Even though fruits are presoftened by ripening, cooks often heat them to improve their texture, flavor, and storage life. One of the most popular cooked fruit sauces is applesauce, which is meant to have a certain coarseness and yet not seem grainy. The cells of different varieties have different tendencies to adhere to each other, and that tendency can change with time in storage. Most of the soft varieties used to make sauce produce finer-grained purees with time, while the Macintosh produces coarser ones.

Tomato Sauce: The Importance of Enzymes and Temperatures The most familiar vegetable puree in the West, and perhaps in the world, is tomato sauce and paste. The solids in tomatoes are about two-thirds flavorful sugars and organic acids, and 20% cell-wall carbohydrates that have some thickening power (10% cellulose, and 5% each pectin and hemicelluloses). In the United States, commercial tomato purees may include all the water in the original tomatoes, or just a third. Tomato paste is tomato puree cooked down so that it contains less than a fifth of the water of the raw vegetable. Tomato paste is thus a concentrated source of flavor, color, and thickening power. (It's also an effective emulsion stabilizer; see p. 628.)

There are several variables in the preparation of purees that can affect their final texture and flavor. Food scientists have shown this most clearly for mass-produced tomato puree. The general lessons are also relevant to the preparation of purees from other fruits and vegetables.

Tomato Enzymes and Consistency The final consistency of a tomato puree depends not just on how much water has been removed, but also on how long the puree spends at either moderate or high temperatures. Ripe tomatoes have very active enzymes whose job is to break down pectin and cellulose molecules in the fruit cell walls, and so give the fruit its soft, fragile texture. When the tomatoes are first crushed, the enzymes and their target molecules are thoroughly mixed together, and the enzymes start breaking down the cell-wall structures. If the raw puree is held at room temperature for a while, or heated to a temperature below the denaturation temperature of the pectin enzymes, around 180°F/80°C, then the enzymes will break down a lot of the cell-wall reinforcements, and these liberated molecules will give a noticeably thicker consistency to the puree.

However, when the puree is then heated to remove water and concentrate it, the high temperatures break the already enzyme-damaged molecules into smaller pieces that are less efficient thickeners, and the paste requires that much greater reduction to obtain the desired thickness. If instead the raw puree is cooked quickly close to the boil, the result is a thicker sauce that requires less subsequent reduction. The pectin and cellulose enzymes are denatured and become inactive, while at the same time the cell walls are disrupted by the heat. The cell-wall pectins that escape into the fluid phase during the concentration heating are longer molecules and more efficient thickeners.

Tomato Enzymes and Flavor In addition to tomato enzymes that affect texture, there are enzymes that affect flavor: and in the case of flavor, some initial enzyme activity can be desirable. The fresh, "green"-smelling molecules (hexanal and hexanol, p. 274) that are an important element in ripe tomato flavor are generated by the action of enzymes on fatty acids

The Thickening Components of Tomatoes

	Total Solids, % by weight	Pectin and Hemicellulose Content, % by weight
Raw tomatoes	5–10	0.5–1.0
Canned tomato puree	8–24	0.8–2.4
Canned tomato paste	40	4

when the fruit tissue is crushed, either in the mouth or in the pot. Rapid cooking to the boil minimizes this fresh flavor element, while allowing the raw puree to sit at room temperature—in a Mexican salsa, for example—or only slowly heating it, will cause the accumulation of these flavor molecules in the puree. A method that home cooks sometimes use is to halve or quarter tomatoes, then bake them in a slow oven to remove water, and finally cook them relatively quickly into a sauce. This technique minimizes the mixing of enzymes and targets, so cells stay relatively intact, and relatively little green aroma develops.

Then there's the traditional Italian preparation called *estratto*, which begins with fresh tomatoes cooked down to some extent, and then mixes them with some olive oil and spreads the paste on boards to dry down further in the sun. This is often described as a relatively "gentle" process compared to cooking, and it probably does spare some damage to the pectin molecules. But in fact it subjects a number of sensitive molecules—including the antioxidant tomato pigment lycopene and unsaturated fatty acids in the olive oil—to powerful and damaging ultraviolet light, which gives *estratto* a distinctively strong, cooked flavor.

Nuts and Spices as Thickeners

Among seeds and other dry plant materials, only oily nuts can be made into a sauce base on their own. When such nuts are ground into "butters," the oil provides the fluid continuous phase that lubricates the particles of cell walls and proteins. But most of the time, nuts are mixed with other ingredients, including liquids, so they become part of a complex suspension and help thicken both with their dry particles and with their oil, which becomes emulsified into tiny droplets. Almonds have long served this purpose in the Middle East and Mediterranean in such sauces as romesco (with red peppers, tomatoes, and olive oil) and picada (garlic, parsley, oil), and the coconut in tropical Asia, where it's pounded along with spices and herbs to become part of the sauce for cooked meats, fish, and vegetables.

Nuts and other finely ground seeds and spices help thicken liquid sauces thanks to their very dryness, which allows their particles to absorb water from the sauce and thus reduce the amount of liquid that needs to be filled with particles. At the same time, the particles themselves swell and become larger obstacles to the liquid's flow. Dry spices such as turmeric, cumin, and cinnamon are both flavorings and thickeners in Indian sauces, and coriander is especially effective thanks to its fibrous, absorbant seed coat. Dried chilli peppers, ground nuts, and spices thicken Mexican mole sauces. Powdered versions of dried chilli pepper are prominent in Spanish and Hungarian sauces (pimenton, paprika); mustard is also widely used. Some spices also release efficient thickening molecules into the liquid. Fenugreek seeds exude a gum that gives a gelatinous consistency to the Yemeni sauce hilbeh; and the dried leaves of the sassafras tree, ground to make filé powder, release carbohydrates that give Louisiana gumbos a slightly stringy viscosity.

Complex Mixtures: Indian Curries, Mexican Moles

The most complex and sophisticated puree sauces are made in Asia and Mexico. The sauce or "gravy" for many Indian and Thai dishes begins with finely ground plant tissues—onions, ginger, garlic in northern India, coconut in southern India and in Thailand—and a number of different spices and herbs. These ingredients are then fried in hot oil until much of the moisture has boiled off, and the plant solids are sufficiently concentrated that the sauce clings to itself and the oil separates. The frying also cooks the sauce, eliminating raw flavors and developing new ones. The sauce is then slightly thinned with some water, and the main ingredient cooked in it. Mexican mole sauces are prepared in much the same way, except that the foundation ingredient is usually rehydrated dried chillis; pumpkin and other seeds are another major element. Thanks to the high pectin content of the chillis, moles have a more suave, finer consistency than the Asian purees. But both are marvels of mouth-filling pleasure.

SAUCES THICKENED WITH DROPLETS OF OIL OR WATER: EMULSIONS

The sauces we've examined so far are liquids thickened with a fine dispersion of solid materials: protein molecules, starch granules and molecules, particles of plant tissue and cell-wall molecules. A very different thickening method is to fill the water-based liquid with droplets of oil, which are much more massive and slow-moving than individual molecules of water, impede their motion, and so create a thick and creamy consistency in the mixture as a whole. Such a dispersion of one liquid in another is called an *emulsion.* The word comes from the Latin for "to milk out," and referred originally to the milky fluids that can be pressed from nuts and other plant tissues. Milk, cream, and egg yolks are natural emulsions, while sauce emulsions include mayonnaise, hollandaise sauce, beurre blanc, and oil-and-vinegar salad dressings. Modern chefs have applied the basic idea to the thickening of all kinds of liquids, and often actually describe the result on the menu as an *emulsion,* a word that lingers on the tongue longer than *sauce* does.

Emulsified sauces offer a special challenge to the cook: unlike sauces thickened with solids, emulsions are basically unstable. Whisk oil and a little vinegar together in a bowl, and the vinegar forms droplets in the oil: but they soon sink and coalesce, and in a few minutes the two liquids have separated again. Cooks not only have to form the emulsion, they also have to prevent the emulsion from being undone by the basic incompatibility of the two liquids.

THE NATURE OF EMULSIONS

An emulsion can only be made from two liquids that don't dissolve in each other, and therefore retain their distinct identities

The Relative Proportions of Fat and Water in Common Food Emulsions

Food	Parts Fat to 100 Parts Water
Fat-in-Water Emulsions	
Whole milk	5
Half-and-half	15
Light cream	25
Heavy cream	70
Heavy cream reduced by a third	160
Egg yolk	65
Mayonnaise	400
Water-in-Fat Emulsions	
Butter	550
Vinaigrette	300

even when mixed. The molecules of water and alcohol, for example, mix freely and so can't form an emulsion. In addition to sauces, cosmetic creams, floor and furniture waxes, some paints, asphalt, and crude oil are all emulsions of water and oil.

Two Liquids: Continuous and Divided The two liquids in an emulsion can be thought of as the container and the contained: one liquid is broken up into separate droplets, and these droplets are contained in and surrounded by the intact mass of the other liquid. In the usual shorthand, an "oil-in-water" emulsion is one in which oil is dispersed in a continuous water phase; "water-in-oil" names the reverse situation. The dispersed liquid takes the form of tiny droplets, between a ten-thousandth and a tenth of a millimeter across. The droplets are large enough to deflect light rays from their normal path through the surrounding liquid, and give emulsions their characteristically milky appearance.

The more droplets that are crowded into the continuous phase, the more they get in the water's and each other's way, and the more viscous the emulsion is. In light cream, the fat droplets take up about 20% of the total volume and water 80%; in heavy cream, the droplets are about 40% of the volume; and in stiff, semisolid mayonnaise, oil droplets occupy nearly 80% of the volume. If the cook works more of the dispersed liquid into the emulsion, then it gets thicker; if he adds more of the continuous liquid, then there's more space between droplets, and the emulsion becomes thinner. Clearly it's important to keep in mind which phase is which.

Because nearly all emulsified sauces are oil-in-water systems, I'll assume in most of the following discussion that the continuous phase is water, the dispersed phase oil.

Forming Emulsions: Overcoming the Force of Surface Tension It takes work to make an emulsion. We all know from experience that when we pour water and oil into the same bowl, they form two separate layers: one doesn't just turn into tiny droplets and invade the other. The reason for this behavior is that when liquids can't mix for chemical reasons, they spontaneously arrange themselves in a way that minimizes their contact with each other. They form a single large mass, which exposes less surface area to the other liquid than does the same total mass broken into pieces. This tendency of liquids to mini-

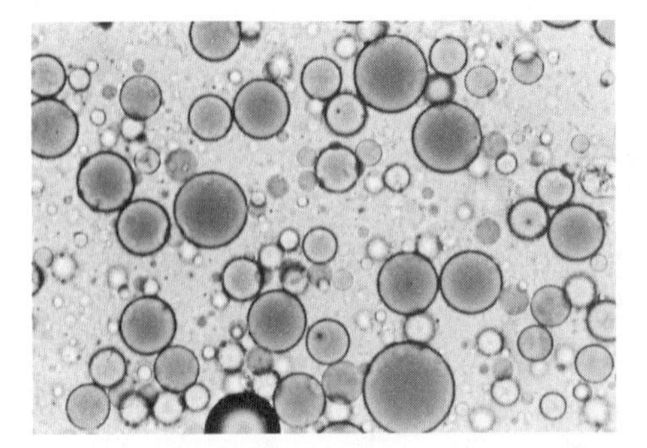

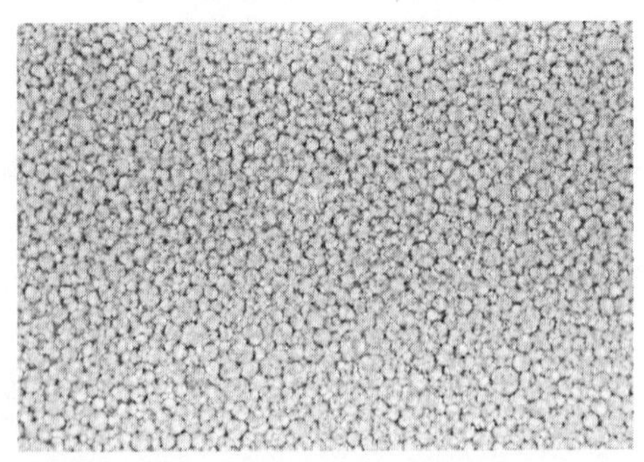

Mayonnaise formation. Two stages in making mayonnaise as seen through a light microscope. One tablespoon/15 ml of oil beaten into 1 egg yolk plus water gives a sparse emulsion of coarse, unevenly sized oil droplets (left). *Eight tablespoons/120 ml of oil give a tightly packed, semisolid emulsion of small droplets* (right). *The yolk emulsifiers and stabilizing proteins must be effective enough to withstand considerable physical pressure in order to prevent the oil droplets from coalescing into a separate layer.*

mize their surface area is an expression of the force called *surface tension.*

Making Billions of Droplets from One Tablespoon It's on account of surface tension, then, that the cook must pour energy into the liquid to be dispersed. To make a sauce, its natural monolithic arrangement must be shattered. And seriously shattered: when you beat a single tablespoon/15 ml of oil into a mayonnaise, you break it up into about 30 billion separate droplets! Serious whisking by hand or in a kitchen mixer provides enough shearing force to make droplets as small as 3 thousandths of a millimeter across. A blender can get them somewhat smaller, and a powerful industrial homogenizer can reduce them to less than one thousandth of a millimeter. The size of the droplets matters, because smaller droplets are less likely to coalesce with each other and break the sauce into two separate phases again. They also produce a thicker, finer consistency, and seem more flavorful because they have a larger surface area from which aroma molecules can escape and reach our nose.

Two factors make it easier for the cook to generate small droplets. One is the thickness of the continuous phase, which drags harder on the droplets and transfers more shearing force to them from the whisk. Shake a little oil in a bottle of water, and the oil droplets are coarse and quickly coalesce; shake a little water in the more viscous oil, and the water gets broken into a persistent cloud of small droplets. It's helpful, then, to start with as viscous a part of the continuous phase as possible, and dilute it with any other ingredients after the emulsion has formed.

The second factor that makes it easier to produce small droplets in an emulsion is the presence of emulsifiers.

Emulsifiers: Lecithin and Proteins Emulsifiers are molecules that lower the surface tension of one liquid dispersed in

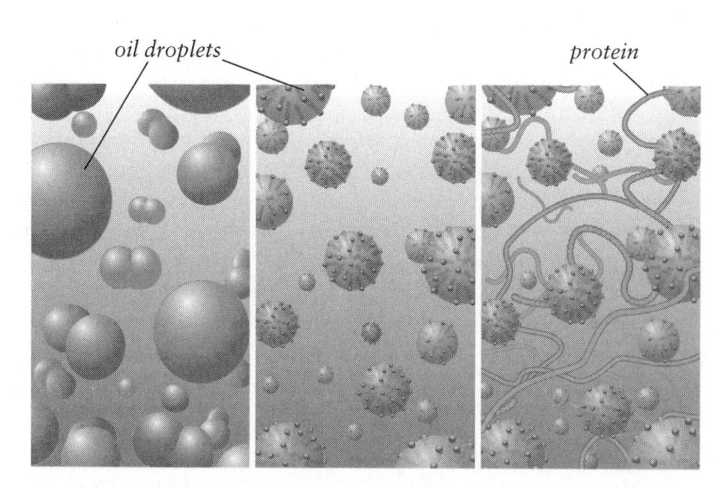

Unstable and stable emulsions. Oil and water are incompatible substances; they can't mix evenly with each other. When oil is whisked into water, the resulting oil droplets tend to coalesce with each other and separate into a layer on top of the water (left). *Emulsifiers are molecules with a fat-compatible tail and water-compatible head (p. 802). They embed their long tails in the fat droplets, leaving their electrically-charged heads projecting into the surrounding water. Coated in this way, the droplets repel each other instead of coalescing* (center). *Large water-soluble molecules, including starch and proteins, help stabilize emulsions by blocking the fat droplets from each other* (right).

another, and therefore make it easier to make small droplets and a fine, creamy emulsion. They do so by coating the surface of the droplets, and shielding the droplet surface from the continuous liquid. Emulsifiers are therefore a true liaison: they must be partly soluble in each of the two mutually incompatible liquids. They manage this by having two different regions on the same molecule, one soluble in water and the other in fat.

There are two general kinds of molecules that can act as emulsifiers. One kind is typified by the egg phosopholipid lecithin. These are relatively small molecules with a fat-like tail that buries itself in the fat phase and an electrically-charged head that is attracted to water molecules (p. 802). The other kind of emulsifier is the proteins, which are much larger molecules, long chains of amino acids that have a number of different fat-compatible and water-compatible regions. The yolk proteins in eggs and the casein proteins in milk and cream are the best protein emulsifiers.

Stabilizers: Proteins, Starch, Plant Particles Emulsifiers make it easier for the cook to prepare an emulsion, but they don't necessarily result in a stable emulsion. Once formed, the droplets may be so crowded that they bump into each other or are forced up against each other, and the force of surface tension may pull them together and cause them to coalesce again. Fortunately, there are many kinds of molecules and particles that can help stabilize an emulsion once it's formed. They all have in common the property of getting in the way, so that two approaching droplets encounter the stabilizers rather than each other. Large, bulky molecules like proteins do this well, as do starch, pectins, and gums, and particles of pulverized plant tissue. Ground white mustard seed is especially effective thanks both to its particles and to a gum that it releases when wetted. Tomato paste contains a considerable amount of protein (around 3%) as well as cell particles, and is a useful emulsifier and stabilizer.

Guidelines for Successful Emulsified Sauces

Forming Emulsions Emulsions have always been considered fickle concoctions, by chemists as well as cooks. One chemist wrote in 1921 that contemporary books on pharmacy were "filled with elaborate details as to the making of emulsions," and recorded two such details: "If one starts stirring to the right, one must continue stirring to the right, or no emulsion will be formed. Some books go so far as to say that a left-handed man cannot make an emulsion, but that seems a little absurd." The worry is always that at some point the emulsion may break and separate into blobs of oil and water again.This can happen, but it's almost always because the cook has made one of three mistakes: he has added the liquid to be dispersed too quickly to the continuous liquid, or added too much of the dispersed liquid, or

Bouillabaisse, an Emulsified Soup

Bouillabaisse is a Provençal fish soup that takes advantage of gelatin's thickening and emulsifying properties. It's made by cooking a variety of whole fish and fish parts, some of them bony and gelatin-producing rather than meaty, in an aromatic broth with some olive oil. The soup is finished at a vigorous boil, which breaks the oil into tiny droplets and coats them with a stabilizing layer of gelatin. The consistency is thus a combination of gelatin's viscosity and the enriching creaminess of the emulsified oil droplets.

allowed the sauce to get either too hot or too cold.

There are several basic rules that apply to the making of any emulsified sauce:

- The first materials into the bowl are the continuous phase—usually the water-based ingredient—and at least some emulsifying and stabilizing ingredients. The dispersed phase is always added to the continuous phase, not the other way around: otherwise it can't be dispersed!
- The dispersed phase should be added very gradually to begin with, a small spoonful at a time, while the cook whisks or blends the mixture vigorously. Only after an emulsion has formed and developed some viscosity should the oil be added more rapidly.
- The proportions of the two phases must be kept in balance. For most emulsified sauces, the volume of the dispersed phase shouldn't exceed three times the volume of the continuous phase. If the droplets are crowded so closely together that they are in continuous contact, then they're more likely to pool together. When the consistency of an emulsion becomes stiff, this is a sign that the cook should add more of the continuous phase to give the droplets more room.

Starting Slowly There's a simple reason for starting the emulsion slowly and carefully, with small amounts of the dispersed phase. In the early mixing, when little or no oil has yet been emulsified, it's easy for large droplets to avoid the churning action of the whisk and collect at the surface. If a large volume of oil is added before the previous one has been fully emulsified, then the bowl may end up with more unemulsified oil than water. The oil then becomes the continuous phase, the normally continuous water becomes dispersed in it, and the result is an inside-out emulsion, oily and runny. By whisking in the first portion of oil in small doses, the cook makes sure to produce and maintain a growing population of small droplets. Then when the rest of the oil is incorporated more rapidly into the already well-emulsified system, the existing droplets work as a kind of mill, automatically breaking down the incoming oil into particles of their own size. In the last stages of sauce making the cook's whisk need not break up the oil drops directly, but has the easier job of mixing the new oil with the sauce, distributing it evenly to all parts of the droplet "mill."

Using and Storing Emulsified Sauces

Once emulsified sauces have been successfully made, there are two basic rules for using them.

- The sauce must not get too hot. At high temperatures, the molecules and droplets in a sauce are moving very energetically, and the droplets may collide hard enough to coalesce. Temperatures above 140°F/60°C also cause the proteins in egg-emulsified sauces to coagulate, so they're no longer able to protect the droplets. And a cooked sauce that is held before serving on gentle heat may lose enough water by evaporation that the dispersed fat droplets become overcrowded. So cooked emulsions should be made and held at warm, rather than hot, temperatures and should not be spooned onto a piece of food still sizzling from the pan.
- The sauce must not get too cold. At low temperatures, surface tension increases, making it more likely that neighboring droplets will coalesce. Butterfat solidifies at room temperature, and some oils do so in the refrigerator. The resulting sharp-edged fat crystals rupture the layer of emulsifier on the droplets, so that they coalesce and separate when stirred or warmed. Refrigerated emulsions often need to be reemulsified before use. (Manufactured mayonnaise is made with

oils that remain liquid at refrigerator temperatures.)

Rescuing a Separated Sauce When an emulsified sauce breaks and the droplets of the dispersed phase puddle together, there are two ways to reemulsify it. One is simply to throw the sauce in a blender and use its mechanical power to break the dispersed phase apart again. This generally works for sauces that still have plenty of intact emulsifier and stabilizer molecules in the sauce, but not for cooked egg sauces that have been overheated and their proteins coagulated. The second and more reliable technique is to start with a small amount of the continuous phase, perhaps supplemented with an egg yolk and its wealth of emulsifiers and stabilizers, and carefully beat the broken sauce back into it. If proteins in the initial sauce had coagulated from overheating, the lumps should be strained out before reemulsifying; otherwise the rescue process may leave the protein particles too small to strain out, but large enough to leave a grainy impression in the mouth.

Cream and Butter Sauces

Cream and butter don't need to be made into sauces—they themselves are sauces! In fact they're prototypes for sauces in general, with their lingering, mouthfilling consistency and rich but delicate flavor. A ramekin of melted butter in which to dip a morsel of lobster or an artichoke leaf, a pour of cream over fresh berries or pastry—these are wonderful combinations. But cream and butter are versatile ingredients, and cooks have found other ways to exploit them in saucemaking.

Milk and Cream Emulsions Cream owes its versatility to its origins in milk. Milk is a complex dispersion whose continuous phase is water, and whose dispersed phases are milkfat in the form of microscopic droplets, or globules, and protein particles in the form of casein aggregates (p. 19). The droplets are coated with a thin membrane of emulsifiers, both lecithin-like phospholipids and certain proteins; and other non-casein proteins float free in the water. Both the globule membranes and the various proteins are tolerant of heat: so plain milk and cream can be boiled hard without the fat globules coalescing and separating, or the proteins coagulating and curdling.

Whole milk is only about 4% fat, so its fat globules are too few and far between to block the flow of the water phase and give much of an impression of thickness. Cream is a portion of milk in which the fat globules have been concentrated and crowded: light cream is around 18% fat, and heavy or whipping cream around 38%. In addition to its fat supply, cream offers proteins and emulsifying molecules that can help stabilize other, more fragile emulsions (beurre blanc).

Heavy Cream Resists Curdling The casein proteins in milk and cream are stable to boiling temperatures, but they're sensitive to acidity, and the combination of heat and acid will cause them to curdle. Many sauces include flavorful acid ingredients: sauté pans are often deglazed with wine, for example. This means that most milk and cream products, including light cream and sour cream, can't actually be cooked to make a sauce; they must be added as a last-minute enrichment. The exceptions are heavy cream and crème fraîche, which contain so little casein that its curdling simply isn't noticeable (p. 29).

Reduced Cream When heavy cream is added to another liquid to enrich and thicken it—to a meat sauce, or deglazing liquid, or a vegetable puree—then of course its fat globules are diluted and its consistency thins down. In order to make cream a more effective thickener, cooks concentrate it even further by boiling off water from the continuous phase. When the volume of cream is reduced by a third, the globule concentration reaches 55% and the consistency is like that of a light starch-thickened sauce; when reduced by half, the globules take up 75% of the volume and the consistency is very thick, almost semi-solid. Stirred into a thinner liquid, these reduced creams have enough fat globules to fill it and lend a substantial body. Cream reduction and thickening can also be done at the last minute, for example after a sauté pan has been deglazed; the cook adds cream to the deglazing liquid and boils the mixture until it reaches the desired consistency.

Crème Fraîche in Sauce Making Reduced creams have several disadvantages. They take time and attention to prepare, develop a cooked flavor, and are very rich, sometimes overly so for the balance of a given sauce. A useful alternative to reduced creams is crème fraîche, a version of heavy cream whose consistency has been thickened not by boiling down, but by fermentation (p. 49). The acid produced by lactic bacteria causes the casein proteins in the water phase to cluster together and form a network that immobilizes the water. Some strains of bacteria also secrete long carbohydrate molecules that further thicken the water phase and act as stabilizers. Used in place of reduced cream, crème fraîche requires no preparation, is less rich, and has a fresher flavor. Thanks to its low protein content, it tolerates temperatures that would curdle sour cream.

Butter Like its parent material, cream, butter is an emulsion: but it's one of the few food emulsions in which the continuous phase is fat, not water. In fact, butter is made by "inverting" the fat-in-water cream emulsion to produce a water-in-fat emulsion (p. 33). The continuous fat phase of butter, together with some intact fat globules that survived churning, takes up about 80% of its volume, and the dispersed water droplets about 15%. When it melts, the heavier water droplets sink to the bottom and form a separate layer. The consistency of melted butter, then, is the consistency of the butterfat itself, which thanks to its long fat molecules is naturally more slow-flowing and viscous than water. So melted butter, whole or clarified ("drawn") to remove the water phase, makes a simple and delicious sauce. Cooks also heat whole butter until the water boils off and the milk solids turn brown, which gives the fat a nutty aroma. The French *beurre noisette* and *beurre noir*, or "hazelnut" and "black" butters, are such browned butters, often made into a temporary emulsion with lemon juice and vinegar respectively.

Compound and Whipped Butters There are other ways to take advantage of butter's semisolid consistency and background richness. One is to make a "compound butter" by incorporating pounded herbs, spices, shellfish eggs or livers, or other ingredients; another is to whip softened butter with a flavorful liquid into a combined emulsion and foam. Pieces or dollops of these flavored butters can then be melted into a rich, flavorful coating atop a piece of meat or fish, or on some vegetables or pasta, or they can be swirled into an otherwise finished sauce.

Turning Butter Back into Cream: Enriching Sauces with Butter Butter is remarkable for being a convertible emulsion. This offspring of cream can be turned back into cream! Its convertibility is what makes butter so useful as a finishing enrichment for many sauces, including simple pan deglazings, and it's what makes possible the sauce called beurre blanc, literally "white butter." There's only one requirement for converting butter into the equivalent of cream with 80% fat: the process must start in a small amount of water. If you melt butter on its own, the fat phase remains the continuous phase, and the water droplets settle out of it. But if you melt butter in some water, then you're starting with water as the continuous phase. As the fat molecules are released into the water, they're surrounded by water—and by the substances contained in the butter's own water droplets, which merge right into the cooking water. The droplets contain milk proteins and remnants of the emulsifier membranes that coated the fat globules in the original cream. And those protein and phospholipid remnants reassemble themselves onto the fat as it melts into the water, coating and protecting separate fat droplets and forming the fat-in-water emulsion. However, the droplet coatings in this reconstituted cream are sparser and more fragile than the original fat-globule membranes, and will begin to leak fat if heated close to 140°F/60°C.

Any water-based sauce can thus be thickened and enriched simply by swirling a pat of butter into it at the end. This is especially handy in the last-minute thickening of pan juices, which don't have the benefit of containing much gelatin or any starch. Incorporating one volume of butter into three volumes of deglazing liquid—off the heat, to avoid damaging the fragile droplet coatings—will produce a consistency (and fat content) approximating that of light cream.

In purees and starch-thickened sauces, a small amount of butter (or cream) lubricates the solid thickeners and lends a smoother consistency. Because these sauces are rich in emulsion-stabilizing molecules and particles, they can be heated to the boil without causing the reconstituted fat droplets to separate.

Beurre Blanc The French sauce *beurre blanc* probably evolved from the practice of enriching cooking liquids with butter. It's made by preparing a flavorful reduction of vinegar and/or wine, then whisking pieces of butter into the reduction. Each piece of butter carries all the ingredients necessary for a new portion of sauce, so the cook can whisk in one piece of butter, or 100. The proportions are entirely up to the cook's taste and needs. The consistency of beurre blanc is like that of thick cream, and can be made somewhat thicker by adding water-free clarified butter once the initial emulsion has been formed. The phospholipids and proteins carried in the butter's water are capable of emulsifying two to three times the butterfat in which they're embedded.

Beurre blanc will begin to separate and leak butterfat if its temperature rises above about 135°F/58°C. However, the phospholipid emulsifiers can tolerate heat and re-form a protective layer. An overheated sauce can usually be restored with a small amount of cool water and brisk whisking. The addition of a spoonful of cream supplies more emulsifying materials and can make a beurre blanc more stable. Most damaging to beurre blanc is letting it cool below body temperature. The butterfat solidifies and forms crystals around 85°F/30°C, and the crystals poke through the thin membrane of emulsifiers and fuse with each other, forming a continuous network of fat that separates when the sauce is reheated. Ideally, beurre blanc should be kept at around 125°F/52°C. Because water will evaporate at this temperature and may overconcentrate the fat phase, it's a good idea to add a little water periodically if the sauce has to be held for any time.

Beurre Monté A preparation closely related to beurre blanc is *beurre monté*, "worked up" or "prepared" butter, which is simply an unflavored beurre blanc made with an initial dose of water rather than vinegar or wine. Beurre monté is used among other things as a poaching medium. Thanks to the relatively low heat conductivity and heat capacity of fat compared to water, it cooks delicate fish and meats more gradually than does a broth at the same temperature.

Eggs as Emulsifiers

As we've already seen, cooks can use egg yolks to thicken all kinds of hot sauces. The yolk proteins unfold and bond to each other when heated, and so form a liquid-immobilizing network (p. 604). Egg yolks are also very effective emulsifiers, and for a simple reason: they themselves are a concentrated and complex emulsion of fat in water, and are therefore filled with emulsifying molecules and molecule aggregates.

Emulsifying Particles and Proteins Of the various yolk components, two in particular provide most of the emulsifying power. One is the low-density lipoproteins or LDLs (the same LDLs that circulate in our blood and whose levels are measured in blood tests because they carry potentially artery-blocking cholesterol). LDLs are particles made up of emulsifying proteins, phospholipids, and cholesterol, all surrounding a core of fat molecules. The intact LDL particles appear to be more effective emulsifiers than any of their components. The other major emulsifying particles are the larger yolk granules, which contain both LDLs and HDLs (the "good-cholesterol" high-density lipoproteins are even more effective emulsifiers than LDL) as well as dispersed emulsifying protein, phosvitin. Yolk granules are so large that they can't cover a droplet surface very well, but when they're exposed to moderate concentrations of salt they fall apart into their separate LDLs, HDLs, and proteins, and these are very effective indeed.

Using Eggs to Emulsify Sauces As emulsifiers, egg yolks are most effective when they're raw, and when they're warm. Fresh out of the refrigerator, the various yolk particles move only sluggishly and don't coat the fat droplets as quickly and completely. When yolks are cooked, the proteins unfold and coagulate, thus ending their usefulness as flexible surface coatings. Hard-cooked yolks are sometimes used instead of raw yolks to make emulsified sauces; their disadvantage is that the proteins have been coagulated in place and phospholipids probably trapped in the coagulated particles, so they have far less emulsifying power, and the yolk texture can give a subtle graininess.

And egg whites? They're a less concentrated source of protein, and designed for a fat-free, watery environment, and therefore of little help in coating fat droplets. However, the white proteins provide some viscosity thanks to their large size and loose associations with each other, so they have some value as emulsion stabilizers.

Cold Egg Sauces: Mayonnaise

Mayonnaise is an emulsion of oil droplets suspended in a base composed of egg yolk, lemon juice or vinegar, water, and often mustard, which provides both flavor and stabilizing particles and carbohydrates (p. 417). It's the sauce most tightly packed

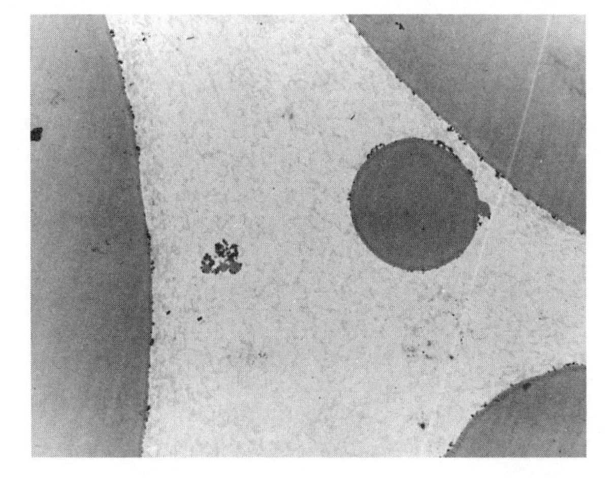

Oil droplets in mayonnaise. A view through an electron microscope. Protein and emulsifier molecules and aggregates, all from egg yolk, are present between the large droplets and on their surfaces, and help prevent them from coalescing.

with oil droplets—as much as 80% of its volume is oil—and is usually dense and too stiff to pour. It can be thinned and flavored with various water-based liquids, including purees and stocks, or it can enrich such liquids the way cream does; it can also be aerated with the addition of whipped cream or egg whites. As a room-temperature preparation, mayonnaise is generally served with cold dishes of various sorts. But thanks to the yolk proteins, it also reacts usefully to heat. It lends body and richness when added to thin broths and briefly cooked; and when layered onto fish or vegetables and broiled, it moderates the heat, puffs up and sets into a rich coating.

Traditionally, mayonnaise is made with raw egg yolks, and therefore carries a slight risk of salmonella infection. Manufacturers use pasteurized yolks, and cooks concerned about salmonella can now find pasteurized eggs in supermarkets. Both vinegar and extra-virgin olive oil kill bacteria, but mayonnaise is best treated as a highly perishable food that should be served immediately or kept refrigerated.

Making Mayonnaise All of the ingredients for making mayonnaise should be at room temperature; warmth speeds the transfer of emulsifiers from the yolk particles to the oil droplet surfaces. The simplest method is to mix together everything but the oil—egg yolks, lemon juice or vinegar, salt, mustard—and then whisk in the oil, slowly at first and more rapidly as the emulsion thickens. However, the cook can produce more stable small droplets by whisking a portion of the oil into just the yolks and salt to start, and then adding the remaining ingredients when the emulsion gets stiff and needs to be thinned. The salt causes the yolk granules to fall apart into its component particles, which makes the yolks become both more clear and more viscous. If left undiluted, this viscosity will help break the oil into smaller droplets.

Though cookbooks often say that the ratio of oil to egg yolk is critical, that one yolk can only emulsify a half-cup or cup of oil, this just isn't true. A single yolk can emulsify a dozen cups of oil or more. What *is* critical is the ratio of oil to water: there must be enough of the continuous phase for the growing population of oil droplets to fit into. For every volume of oil added, the cook should provide about a third of that volume in the combination of yolks, lemon juice, vinegar, water, or some other water-based liquid.

A Sensitive Sauce Because mayonnaise is chock-full of oil, so much so that the droplets press up against each other, its emulsion is easily damaged by extremes of cold, heat, and agitation. It will tend to leak oil in near-freezing refrigerators and on hot rather than warm food. These prob-

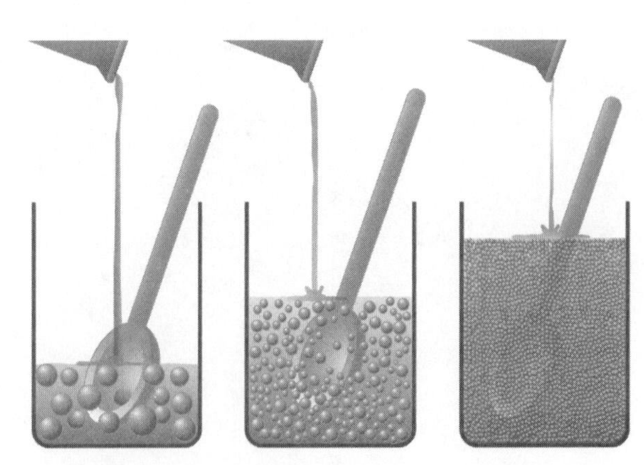

Making mayonnaise. The cook begins with a small volume of the water phase—mostly egg yolk—and slowly beats oil into droplets in this base (left). *As more oil is incorporated, the mixture becomes thicker and the oil is broken into smaller droplets* (center). *When the sauce is done, as much as 80% of its volume is occupied by oil droplets, and its consistency is semisolid* (right).

lems are ameliorated in manufactured mayonnaise by the addition of stabilizers, usually long carbohydrate or protein molecules that fill the spaces between droplets. American bottled "salad dressing" is a very stable hybrid of mayonnaise and a boiled white sauce made with water instead of milk. The texture of such modified sauces, however, is noticeably different from the dense, creamy original. Refrigerated mayonnaise should be handled gently, since some oil may have crystallized and escaped from their droplets. If so, stir gently to re-emulsify it, perhaps with the addition of a few drops of water.

Hot Egg Sauces: Hollandaise and Béarnaise

The classic hot egg sauces, hollandaise and béarnaise and their offspring, are egg-emulsified butter sauces. They are similar to mayonnaise in many respects, but of course must be hot to keep the butter fluid. Their dispersed fat phase is usually a smaller proportion of the sauce, between one- and two-thirds of the total volume. Hollandaise and béarnaise differ mainly in seasoning; hollandaise is only lightly flavored with lemon juice, while béarnaise begins with a tart and aromatic reduction of wine, vinegar, tarragon, and shallots.

Heat Thickens—and Curdles The consistency of the hot egg sauces depends on two factors. One is the form and amount in which the butter is incorporated. Whole butter is about 15% water, so each addition thins the egg phase and the sauce as a whole; clarified butter is all butterfat, and thickens the sauce with every addition. The second influence on consistency is the degree to which the egg yolks are heated and thickened. The main trick in making these sauces is to heat the egg yolks enough to obtain the desired thickness, but not so much that the yolk proteins coagulate into little solid curds and the sauce separates. This happens at around 160–170°F/70–77°C. A double boiler or a saucepan resting above a larger pan of simmering water will guarantee a gentle and even heat but will also slow the cooking; for this reason, some cooks prefer the riskier but rapid direct heat of a burner. Heating the yolks with the acidic reduction also minimizes curdling; if the pH is around 4.5, the equivalent of yogurt's acidity, the yolks can be

Olive Oil Can Make Crazy Mayonnaise

Mayonnaise can be made with any kind of oil. One popular choice, unrefined extra virgin olive oil, often produces an unstable mayonnaise, one that forms properly, but then separates just an hour or two later. Ironically, the likely troublemakers are molecules with emulsifying abilities: oil molecules that have been broken into fragments that have a fat-like tail and water-soluble head, just like lecithin (p. 802). They are concentrated in the oil, and when the cook breaks the oil into droplets, they move to the droplet surface, where they end up pushing the bulkier, more effective egg emulsifiers off the droplet surfaces. Because the droplets are crowded tightly together, this causes the droplets to coalesce and form puddles of oil.

This delayed disintegration of olive-oil mayonnaise is well known in Italy, where the sauce is said to "go crazy" (*impazzire*). Old and improperly stored oils are most likely to have suffered damage to their oil molecules and therefore to cause problems in mayonnaise. Two ways to avoid crazy mayonnaise are to use refined olive oil, and to use extra-virgin oil as a flavoring, with the bulk of the oil being any flavorless refined oil.

safely heated to 195°F/90°C. (The acid causes the proteins to repel each other, so that they unfold before bonding to each other and form an extended network rather than dense curds.) Cooks concerned about salmonella should make sure the yolks are cooked at least to 160°F/70°C, or else should use pasteurized eggs.

Making Hollandaise and Béarnaise

There are at least five different ways of making hollandaise and béarnaise, each with its advantages and disadvantages.

- Cook the egg and water-based ingredients first to a thick consistency, then whisk in pats of whole butter to emulsify the butterfat and thin the continuous phase. This is Carême's method, and is the trickiest because the small volume of the initial egg mixture is easily overcooked.
- Warm the yolks and water-based ingredients, whisk in either whole or clarified butter, then cook the mixture until it reaches the desired consistency. This is Escoffier's method, and has the advantage that the cook can control the final consistency directly, and by heating the entire volume of sauce.
- Put all of the ingredients for the sauce in a cold saucepan, turn the heat on low, and start stirring. The butter gradually melts and releases itself into the egg phase as both heat up together, and the cook then continues to heat the formed sauce until it reaches the desired consistency. This is the simplest method.
- Don't cook the yolks at all; just warm them and the water-based ingredients above the melting point of butter, then whisk clarified butter in until the crowding of droplets creates the desired consistency. This is essentially a butter mayonnaise, and eliminates the possibility of overcooking the yolks.
- Make the butter-sauce version of a sabayon (p. 639). Whisk the egg yolks and some water while heating them until they form an airy foam, and then gently incorporate melted or clarified butter and the lemon juice or acid reduction. This version is of course much lighter, and is also made with less butter per yolk.

It's possible to make hot egg sauces with fats and oils other than butter, and to fla-

Alternative Oil Emulsions

These days we think of mayonnaise exclusively as an egg-emulsified sauce, but this hasn't always been the case, and there are a number of other ways to form and stabilize a flavorful oil emulsion. In 1828, perhaps a few decades after the supposed invention of mayonnaise, the great chef and sauce-systematizer Antonin Carême gave three recipes for *magnonnaise blanche,* only one of which includes egg yolks. The others are made with a ladleful of starchy velouté or béchamel sauce, and with a gelatinous reduced extract of veal meat and bones. In these versions, gelatin and milk proteins (in the béchamel) are emulsifiers, and starch is a stabilizer. Some versions of the herb-flavored Italian *salsa verde,* "green sauce," emulsify olive oil with a hard-boiled yolk and bread. The Provençal *aïoli* and Greek *skorthaliá* are emulsified with a combination of pounded garlic and cooked potato; garlic and bread are also used, as are fresh cheeses. None of these ingredients is as effective at emulsifying and stabilizing as a raw egg yolk, so they will emulsify less oil and the sauces will tend to leak some free oil.

vor the water phase with meat reductions or vegetable purees.

Holding and Salvaging Hot Egg Sauces Butter sauces need to be kept warm to prevent the butter from solidifying, and are best held at around 145°F/63°C to discourage the growth of bacteria. Because the egg proteins slowly continue to bond to each other at this temperature, the cook should stir the sauce occasionally to keep it from thickening. The container should be covered to prevent the sauce's moisture from evaporating and overcrowding the fat droplets, and to prevent the formation of a protein skin on the surface.

Curdled egg sauces can be rescued by straining out the solid bits of protein, keeping the whole mess warm, beginning with another warm egg yolk and one tablespoon/15 ml water, and slowly whisking the sauce into the new yolk. The same technique will revive a sauce that has been refrigerated and so had its butterfat crystallized; the crystals melt to form fat puddles when the sauce is simply rewarmed.

VINAIGRETTES

A Water-in-Oil Emulsion The most commonly and easily made emulsified sauce is the simple oil-and-vinegar salad dressing known as *vinaigrette,* from the French word for "vinegar." Vinaigrette does a good job of clinging to lettuce leaves and other vegetables, and lending a refreshing tart counterpoint to their taste. The standard proportions for a vinaigrette are 3 parts oil to 1 vinegar, similar to the proportions in mayonnaise, but the preparation is much simpler. The liquids and other flavorings—salt, pepper, herbs—are often simply shaken into a cloudy, temporary emulsion at the last minute, then poured onto and mixed with the salad. When made in this casual way, a vinaigrette is the odd sauce out: instead of being oil droplets dispersed in water, it's water (vinegar) droplets dispersed in oil. Without the help of an emulsifier, one part of water simply cannot accommodate three parts of oil, so the more voluminous phase, the oil, becomes the continuous phase.

There are good reasons for making oil the continuous phase of a vinaigrette, and for not worrying about the stability of the emulsion. Where many sauces are served under or atop large pieces of food, oil-and-vinegar emulsions are used almost exclusively as salad dressings, whose role is to provide a very fine and even coat for the extensive surface area of lettuce leaves and cut vegetables. A thin, mobile sauce is more effective at this than a thick, creamy one, and oil adheres to the vegetable surfaces better than the water-based vinegar, whose high surface tension causes it to bead up rather than leave a film. And because the

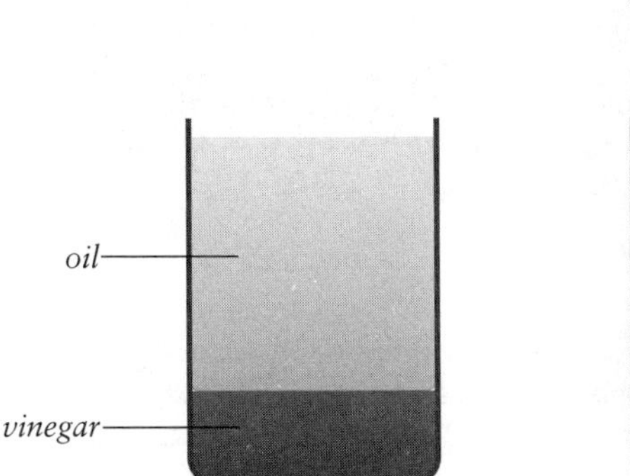

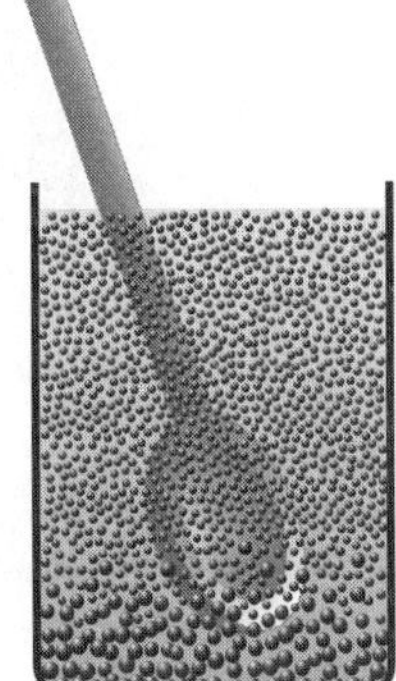

Making a vinaigrette dressing. The proportion of oil to the water phase in a vinaigrette is similar to the proportion in mayonnaise, but in a vinaigrette the water is the phase dispersed in droplets, and the oil is the continuous phase. This emulsion is much less crowded with droplets, and accordingly a vinaigrette is more fluid than mayonnaise.

sauce is so spread out, it doesn't matter as much that the dispersed droplets be carefully stabilized. Because water and oil are antagonists, the salad fixings should be well dried before they're tossed with vinaigrette; surfaces wet with water will repel the oil.

Untraditional Vinaigrettes Nowadays the term *vinaigrette* is used very broadly to mean almost any kind of emulsified sauce enlivened with vinegar, whether water-in-oil or oil-in-water, cold or hot, destined for salads or vegetables or meats or fish. You can make an oil-in-water version simply by changing the proportions: reducing the oil content and diluting the vinegar with other watery ingredients to provide more of the continuous phase without excessive acidity. Creamy but thin oil-in-water vinaigrettes can spread and cling reasonably well, and have the advantage over a classic vinaigrette of being slower to discolor and wilt lettuce leaves. (Oil seeps through breaks in the waxy leaf cuticle and spreads into the leaf interior, where it displaces air and causes the leaf to darken and its structure to collapse.)

Inventive cooks now make vinaigrettes with a variety of fats, including flavorsome olive and nut oils, neutral vegetable and seed oils, melted butter, and even hot meat and poultry fats (pork, duck); the water phase may contain vegetable or fruit juices or purees, meat juices or stock reductions; and the droplets may be emulsified or stabilized by thorough pulverizing to a small size in a blender, or with pounded herbs or spices, vegetable purees, mustard, gelatin, or cream. Today's vinaigrette is a very versatile kind of sauce!

Bottled salad dressings that look like vinaigrettes are generally stabilized and given body with starch or carbohydrate gums, which in low-fat versions can produce a slimy consistency.

SAUCES THICKENED WITH BUBBLES: FOAMS

Like emulsions, foams are a dispersion of one fluid in another. In the case of foams, the fluid is not a liquid, but a gas, and the dispersed particles are not droplets, but bubbles. Still, the bubbles do the same thing that droplets do in a sauce: they get in the way of water molecules in the sauce, prevent them from flowing easily, and thus give the sauce as a whole a thicker body. At the same time, they provide two unique characteristics: a large surface area in contact with air that can enhance the release of aromas to the nose; and a light insubstantiality and evanescence that offers a refreshing contrast to the texture of nearly any food they accompany.

There's one classic foam sauce, the sabayon, which is made by cooking and whipping egg yolks at the same time to form a stable mass of bubbles. And both whipped cream and whipped egg whites can be folded with their bubbles into any water-based sauce. But cooks nowadays make foams from all kinds of water-based liquids and semisolids that contain dissolved or suspended or structure-stabilizing molecules of some kind. The Catalan chef Ferran Adrià pioneered this development, with foams of—among other ingredients—cod, shellfish, foie gras, asparagus, potatoes, raspberries, and cheese. Cooking broths and their reductions, protein- and starch-thickened sauces, juices, purees, and emulsified sauces can all be lightened by incorporating bubbles into them. And it's a quick last-minute preparation: just agitate some of the liquid until it froths, then scoop off the bubble-rich portion, add it to the food, and serve.

Making and Stabilizing Foams

There are several different ways of getting bubbles into a liquid and stabilizing them. Whipping with a whisk or hand-held blender introduces air by agitating the liquid surface; the foaming wands on espresso machines shoot steam, a mixture of water vapor and air; and foaming devices for whipped cream and seltzer water mix a stream of pressurized carbon dioxide or nitrous oxide with the liquid. Any dissolved or suspended molecules in the liquid collect at the interface of the air and liquid and give the bubble wall some solid reinforcement.

However, the reinforcement will be momentary and the bubbles short-lived unless the molecules can form a stable layer at the interface. This is exactly what emulsifiers like lecithin and proteins do, and for the same reason that they stabilize oil droplets in emulsions: they have a water-soluble portion that rests in the bubble wall, and a water-insoluble portion that rests in the air. Because the bubbles in a typical foam are between 0.1 and 1 millimeter across, much larger than most emulsion droplets, they require very little emulsifier to cover their surface area, typically just 0.1% of the liquid weight (1 gram per quart or liter).

Stabilizing Foams A liquid that is even modestly supplied with proteins or yolk phospholipids will form an impressive mass of bubbles, solid enough to stand up without flowing or even slumping. However, the foam may still collapse within a minute or two. Air and water have very different densities, so when the foam is left to stand on its own, the air bubbles rise while gravity pulls the liquid in their walls in the opposite direction. This means that liquid drains from the bubble walls, which also lose water to evaporation. Eventually, the foam at the surface becomes dry, around 95% air and just 5% liquid, the bubble walls become too thin, fail, and the bubbles pop.

This instability of the foam as a whole can be prevented by the same materials that stabilize the emulsified sauces: namely materials that interfere with the free movement of water molecules, and thereby slow the drainage and thinning of the bubble walls. Foam stabilizers include the microscopic particles in purees, proteins, thickening carbohydrates like starch, pectin, and gums—and even emulsified fat. Free fat or oil is a foam killer, because the fat spreads at the interface with the air—it's chemically more compatible with air than with water—and prevents emulsifiers from settling at the interface and stabilizing it. However, if the fat is emulsified—for example in an egg yolk or yolk-based sauce—then it remains dispersed in the water phase, and its droplets only interfere with the flow of liquid from the bubble walls.

Heat-Stabilized Foams: Sabayons Both the method and the name of the French sabayon derive from the Italian *zabaglione*, a sweet, winy foam of egg yolks (p. 113). Though rich in proteins and phospholipids, egg yolks don't foam well on their own because they don't contain enough water. Add water and beat and they foam prodigiously but temporarily; heat while beating and the yolk proteins unfold and bond to each other into a thickening, stabilizing network. This is how sabayons are made, with the water replaced by a flavorful liquid of some some sort, a broth or juice or puree for example. The hot egg-emulsified butter sauces can be made in the style of a sabayon, with the butter folded in gently at the end so as not to pop too many of the foam bubbles. (The butter doesn't need to be beaten in because the foam has created a large surface area over which the butter can spread and stay suspended, much as a vinaigrette is spread over lettuce leaves.) The proteins in aerated yolks thicken around 120°F/50°C, and may coagulate and separate if heated much above that, so many cooks prepare sabayons over a pot of hot water rather than over direct stovetop heat.

SALT

The word *sauce* comes via Latin from an ancient root word meaning "salt," the primordial condiment that was prepared by the earth billions of years before early humans learned to enliven their foods with it. Salt is an important flavoring, but also much more than that, and is an ingredient in nearly every preparation described in this book. The relevant chapters explain its role in the making of such foods as cheese, cured meats and fish, pickled vegetables, boiled vegetables, soy sauce, and bread. Here are a few pages about salt itself.

The Virtues of Salt Salt is like no other substance we eat. Sodium chloride is a simple, inorganic mineral: it comes not from plants or animals or microbes, but from the oceans, and ultimately from the rocks that erode into them. It's an essential nutrient, a chemical that our bodies can't do without. It's the only natural source of one of our handful of basic tastes, and we therefore add it to most of our foods to fill out their flavor. Salt is also a taste enhancer and taste modifier: it strengthens the impression of aromas that accompany it, and it suppresses the sensation of bitterness. It's one of the very few ingredients that we keep in pure form at the table, to be added to individual taste as we eat.

In addition to *sauces* and *salads,* somewhat bitter leaves dressed to make them more palatable, another food named for salt is *sausage,* one of the preparations in which salt is more than just a flavoring. Thanks to its basic chemical nature, salt can alter other ingredients in useful ways. Sodium chloride dissolves in water into separate single atoms that carry electrical charge—positively charged sodium ions and negatively charged chloride ions. These atoms are smaller and more mobile than any molecule, and therefore readily penetrate our foods, where they react in useful ways with proteins and with plant cell walls. And because a concentrated solution of any kind draws water out of living cells by osmosis—water in the less concentrated cell fluid moves out of the cell to relieve the imbalance—the presence of sufficient salt in a food discourages the growth of spoilage bacteria while allowing harmless flavor-producing (and salt-tolerant) bacteria to grow. It thus preserves the food and improves it at the same time.

Salt is a remarkable ingredient. No wonder that people from earliest times have found it indispensable, that it's embedded in everyday words and sayings (*salary,* from the Roman practice of paying soldiers in salt; *worth his salt*; *salt of the earth*), and that it has been the occasion for governmental monopolies and taxes and popular revolts against them, from revolutionary France to Gandhi's 1930 salt march to Dandi.

Salt Production

People have been gathering crystalline salt since prehistoric times, both from the seacoasts and from inland salt deposits. The rock-salt deposits, some of which are hundreds of millions of years old, are masses of sodium chloride that crystallized when ancient seas were isolated by rising land masses and evaporated, and their beds then covered over by later geological processes. Until the 19th century, salt was produced mainly for the preservation and flavoring of foods. Nowadays large amounts are used in industrial manufacturing of all kinds, as well as in the de-icing of winter roads, and salt production itself has been industrialized. Most rock salt is now mined by solution, or pumping water into the deposits to dissolve the salt, then evaporating the brine down in vacuum chambers to form solid crystals. While some sea salt is still produced by gradual solar evaporation from open-air salt pans in sufficiently warm, dry regions, much is now produced by more rapid vacuum evaporation.

Removing Bitter Minerals Salt comes from seawater, and seawater contains significant quantities of several bitter minerals, the chloride and sulfate salts of magnesium and calcium. Producers have a couple of ways of dealing with these minerals.

They can remove them from rock salt by dissolving the salt, then adding sodium hydroxide and carbon dioxide to the brine to precipitate magnesium and calcium. They can remove them from seawater by the same means, or else by slow and gradual concentration in open-air pans, during which the calcium salts become insoluble, crystallize, and settle before the sodium chloride does, and so can be separated. The sodium chloride in turn crystallizes before the magnesium salts, whose slight residue on the crystal surfaces can then be washed off in new brine.

Crystal Shapes These days both edible rock salt and sea salts are produced from brines by evaporating the water away. The evaporation process determines the kinds of salt crystals produced. If the brine becomes concentrated rapidly in a closed tank and crystallization takes place throughout the brine, then many small, regular cubic crystals are formed: the familiar granulated salt of the salt shaker. However, if the evaporation proceeds slowly and at least partly in an open container or sea-side pool, so that crystallization occurs primarily at the brine surface, then the salt solidifies into fragile, hollow, pyramid-shaped flakes, a useful shape for sticking to the surfaces of baked goods, and for dissolving rapidly. To be preserved, the flakes must be scooped off the surface before they settle and sink into the brine, where they fill in and become the large, coarse crystal often seen in minimally processed sea salts.

Once collected and dried, both granular and flake salts can be rolled, compacted, and crushed to make various particle sizes and shapes.

Kinds of Salt

Worldwide, about half of all salt production comes from the sea, and about half from salt mines; in the United States, 95% is mined. Depending on how they've been processed, edible salts range from 98 to 99.7% sodium chloride, with the lower figures typical of table salts treated with anticaking additives.

Granulated Table Salt Granulated table salts come in the form of small, regular, cubic crystals, are the densest salts, and take the longest to dissolve. Standard table salt is often supplemented with additives, as much as 2% of the total weight, that prevent the crystal surfaces from absorbing moisture and sticking to each other. These additives include aluminum and silicon compounds of sodium and calcium, silicon dioxide—the material of glass and ceramics (p. 788)—and magnesium carbonate. Other compounds called humectants may be added to keep *these* additives from excessive drying and caking. Most anticaking additives do not dissolve as readily as salt, and cloud the brines for pickled vegetables, so specialized pickling salts omit them. These additives may also contribute slight undesirable tastes of their own.

Iodized Salt Many granulated table salts and some sea salts are fortified with potassium iodide to help prevent devastating iodide deficiency (below). This practice began in the United States in 1924. Because iodide is sensitive to acidity, manufacturers usually supplement iodized salt with stabilizing traces of sodium carbonate or thiosulfate and sugar. When dissolved in chlorinated tap water, iodized salt can develop a distinct seaweed-like iodine odor, the result of a reaction between the iodide and chlorine compounds.

Flake Salt Flake salts come in flat, extended particles rather than compact, dense granules. Flake salts are produced by surface evaporation of the mother brine, or by mechanically rolling granulated salts. Maldon sea salt from the south coast of England includes individual hollow-pyramid crystals measuring as much as a half-inch/1 cm across. The large particles of flake salts and minimally processed sea salts are easier to measure and add by the pinch. Sprinkled onto a food at the last minute, flake salt provides a crunchy texture and a burst of flavor.The flat crystals don't pack together as compactly as cubic crystals, so a given volume measure of flake salt weighs less than the same measure of granulated salt.

Kosher Salt Kosher salt is salt used for the koshering process, the preparation of meats according to Jewish dietary laws (p. 143). It comes in coarse particles, often flakes, and is sprinkled on the freshly butchered meat for the purpose of drawing out blood. Because it's meant to remove impurities, the salt itself is not iodized. Many cooks like to use kosher salt in general cooking for its relative purity and ease of dispensing by hand.

Unrefined Sea Salt Unrefined sea salts are produced in the way that agricultural crops are: their beds are managed and tended, the salt is harvested when ready, and minimally processed. The tending consists of a slow progressive concentration of the seawater, and can take as much as five years. In most places the freshly harvested salt is washed of its surface impurities before drying. Unrefined versions are not systematically washed of their coating of minor minerals, algae and a few salt-tolerant bacteria. They therefore carry traces of magnesium chloride and sulfate and calcium sulfate, as well as particles of clay and other sediments that give the crystals a dull gray cast (unrefined French salts are called *sel gris,* "gray salt"). Because taste and aroma compounds are often detectable in minute concentrations, and these salts include both organic and mineral impurities, it's possible that they would have a more complex flavor than refined salts, though that complexity would be overwhelmed by any food to which the salt is added.

Fleur de sel Fleur de sel, literally "flower of salt," meaning the finest and most delicate, is a special product of the sea-salt beds of west-central France. It consists of the crystals that form and accumulate at the surface of the salt pans when the humidity and breezes are right; they're gently raked off the surface before they have a chance to fall below the surface, where the ordinary gray sea salt accumulates. Fleur de sel forms delicate flakes, doesn't carry the particles of sediment that darken and dull the gray salt, but is said to carry traces of algae and other materials that contribute a characteristic aroma. This is possible, since the interface between water and air is where aroma molecules and other fatty materials would concentrate; but to date the aroma of sea salts has not been much studied. Thanks to the labor required to make it, fleur de sel is expensive, and is used as a last-minute condiment rather than as a cooking salt.

Flavored and Colored Salts In addition to providing its own saltiness, salt is sometimes turned into a carrier for other flavors and for decorative colors. Examples of flavored salts include celery salt with ground celery seeds, garlic salt with dehydrated garlic granules, and the smoked and roasted salts found in Wales, Denmark, and Korea. The "black salt" of India, more of a gray-pink when ground, is an unrefined mixture of minerals with a sulfurous smell. Black and red Hawaiian salts are made by mixing ordinary sea salt with finely ground lava, clay, or coral.

Salt and the Body

Salt and Blood Pressure Sodium and chloride ions are essential components of the system that keeps our general body chemistry in working balance. They mostly remain in the fluid that surrounds all our cells, the *plasma,* the fluid portion of the blood, where they balance the potassium and other ions inside the cells. It's estimated that we need something like 1 gram of salt per day, a requirement that goes up with physical activity since we lose body fluids and minerals in sweat. Thanks to its presence in nearly all manufactured foods, the average daily salt intake in the United States is around ten times the requirement.

Medical scientists have long suspected

that constant excessive salt intake results in an excessive volume of plasma being contained in our blood vessels, and therefore causes high blood pressure, which damages the blood vessels and increases the risk of heart disease and stroke. However, low-salt diets have been found to lower high blood pressure only modestly, and only in some people. And low-salt diets have surprising side effects of their own, including undesirable increases in blood cholesterol levels. At this time, it appears that the most beneficial nonmedical influences on blood pressure are general dietary balance—more vegetables, fruits, and seeds rich in potassium, calcium, and other minerals—together with physical exercise that conditions the whole cardiovascular system.

Effects on Kidneys, Bones, and the Digestive System Excess sodium is absorbed from the blood and excreted by the kidneys, which help regulate many body systems. High sodium levels thus have the potential for having indirect effects on those systems. There's evidence that they can cause loss of bone calcium and thus increase our dietary calcium requirement, as well as worsening chronic kidney disease.

Though our bodies have ways of diluting and excreting excessive doses of salt, eating salty foods exposes the surfaces of our digestive system to potentially cell-damaging concentrations. There is evidence from China and elsewhere in Asia that diets high in salt increase the risk of several cancers of the digestive system.

Iodized Salt Some salts do carry an undisputed health benefit. Iodized salts include trace amounts of potassium iodide, and thus are a source of a mineral that's essential for proper functioning of the thyroid gland, which regulates the body's heat production, protein metabolism, and development of the nervous system. Iodine is a chemical relative of chlorine and readily found in ocean fish, seaweeds, and crops and animals raised near the seacoast. Iodine deficiency was once common in inland areas, and is still a significant problem in rural China. It causes both physical and mental impairment, especially in children.

Salt to Taste: Salt Preference Both the sensitivity to salt and the preference for saltiness in foods vary a great deal from person to person. They depend on several factors, including inherited differences in the numbers and effectiveness of taste receptors on the tongue, general health, age, and experience. Most young adults can identify as salty a water solution with

The Physical Properties of Salt

Salt generally remains a solid in the kitchen unless it's dissolved. Room-temperature water can dissolve around 35% of its weight in salt, to give a saturated solution of 26% salt that boils at around 228°F/109°C at sea level.

The particle size of salt crystals determines how fast they will dissolve, a fact that can make a big difference when adding salt to a low-moisture food, for example to a bread dough that has been made by the autolysis method (p. 536). Flake salts may dissolve four to five times faster than granulated salt, and finely ground salt nearly 20 times faster.

Solid salt crystals melt at 1,600°F/800°C, and evaporate at around 3,000°F/1,500°C, temperatures reached in wood fires and glowing coals, which can vaporize salt and deposit a thin film on foods above them.

0.05% salt, or 1 teaspoon in 10 quarts/liters, while people older than sixty years generally detect saltiness only at double that concentration. Many manufactured soups, which many people experience as moderately to very salty, are around 1% salt (10 grams, or 2 teaspoons per quart/liter), approximately the same concentration as our blood plasma. Some may be 3% salt, which is the average salinity of seawater.

It appears that the basic liking for saltiness is innate in humans, no doubt because salt is an essential nutrient. The preference for a certain *level* of saltiness is learned through repeated eating experiences and the expectations they create in us. Preferences can be changed by constant exposure to different salt levels, which changes expectations. But this takes time, usually two to four months.

CHAPTER 12

SUGARS, CHOCOLATE, AND CONFECTIONERY

Ordinary sugar is an extraordinary food. Sugar is pure sensation, crystallized pleasure. All human beings share an innate liking for its sweetness, which we first experience in mother's milk, and which is the taste of the energy that fuels all life. Thanks to this deep appeal, sugar and sugar-rich foods are now among the most popular and widely consumed of all foods. In centuries past, when sugar was rare and expensive, they were luxuries reserved for the wealthy and for the climax of the meal. Today sugar is cheap, and manufactured sweets have become everyday, casual pleasures, affordable and entertaining morsels. Some are soothing classics, cream and sugar cooked into rich brown caramels, or clear sugar tinted to look like a shard of stained glass. And others are provocative novelties with glaringly unnatural colors, whimsical shapes, hidden pockets of hissing gas, and burningly excessive doses of acidity or spice.

In the kitchen, sugar is a versatile ingredient. Because sweetness is one of a small handful of basic taste sensations, cooks add sugar to dishes of all kinds to fill out and balance their flavor. Sugar interferes usefully with the coagulation of proteins, and so tenderizes the gluten network of baked goods and the albumen network of custards and creams. If we heat sugar enough to break its molecules apart, it generates both appealing colors and an increasing complexity of flavor: no longer just sweetness, but acidity, bitterness, and a full, rich aroma. And sugar is a sculptural material. Provide it with some moisture and high heat, and we can coax from it a broad range of shapeable consistencies, creamy and chewy and brittle and rock hard.

The story of sugar is not all sweetness and light. Its appeal was a destructive force in the history of Africa and the Americas, whose peoples were enslaved to satisfy the European hunger for it. And today, by displacing more nourishing foods from our diet, sugar contributes indirectly to several modern diseases of affluence. Like most good things in life, it's best enjoyed in moderation. And like that other good thing, fat, it's easy to consume a lot of sugar in manufactured foods without realizing it.

Chocolate, the cooked, sculptable paste of a South American tree seed, has been married to sugar ever since its arrival in Europe nearly 500 years ago, and is in some respects sugar's complement. Where sugar is a single molecule purified from complex plant fluids, chocolate is a mixture of hundreds of different molecules produced by fermenting and roasting a plain bland seed. It's one of the most complex flavors we experience, and yet it lacks and is completed by basic, simple sweetness.

Gathering honey in prehistoric times. This rock painting, found in the Spider Cave at Valencia, Spain, dates back to about 8000 BCE *and appears to show two people raiding a wild beehive. The leader* (enlarged at right) *may be carrying a basket for the honeycomb. Artificial hives and the domestication of bees are known from about 2500* BCE *in Egypt. (Redrawn from H. Ransome,* The Sacred Bee, *1937.)*

THE HISTORY OF SUGARS AND CONFECTIONERY

Before Sugar: Honey

After mother's milk, the first significant source of sweetness in human experience must have been fruits. Some warm-climate fruits like the date can approach a sugar content of 60%, and even temperate fruits become very sweet when they dry out. But the most concentrated natural source of sweetness is honey, the stored food of certain species of bees, which reaches 80% sugars. It's clear from a remarkable painting in the Spider Cave of Valencia that humans have gone out of their way to collect honey for at least 10,000 years. The "domestication" of bees probably goes back 4,000 years, judging by Egyptian hieroglyphs that show clay hives.

However our ancestors obtained it, honey came to represent pleasure and fulfillment to them, and is a prominent metaphor in some of the earliest literature we know. A love poem inscribed 4,000 years ago on a Sumerian clay tablet describes a bridegroom as "honeysweet," the bride's caress as "more savory than honey," and their bedchamber as "honey-filled." In the Old Testament, the promised land is pictured several times as a land flowing with milk and honey, a metaphor of delightful plenty that is itself used figuratively in the Song of Songs, where another bridegroom chants, "Thy lips, O my spouse, drop as the honeycomb: honey and milk are under thy tongue . . ."

Honey remained an important ingredient in both the food and culture of classical Greece and Rome. The Greeks offered it in ceremonies to the dead and the gods, and priestesses of the goddesses Demeter, Artemis, and Rhea were called *melissai*: the Greek *melissa,* like the Hebrew *deborah,* means "bee." The prestige of honey was due in part to its mysterious origins and to a belief that it was a little bit of heaven fallen to earth. The Roman natural historian Pliny speculated in entertaining detail on honey's nature.

> Honey comes out of the air . . . At early dawn the leaves of trees are found bedewed with honey . . . Whether this is the perspiration of the sky or a sort of saliva of the stars, or the moisture of the air purging itself, nevertheless it brings with it the great pleasure of its heavenly nature.

It was more than 1,000 years before the true roles of flower and bee in the creation of honey were uncovered (p. 663). In fact, honey making is the natural model for all human sugar production. We too take sweet juices from plants and separate the sugars from the water. Palm trees in South Asia, maple and birch trees in northern forests, agave plants and maize stalks in the Americas: all these have provided the sweet

Sweet Manna

In the Old Testament book of Exodus, God fed the exiled Israelites with *manna,* which is described as "like coriander seed, white; and the taste of it was like wafers made with honey." Today this term is used for the sugar-rich secretion of certain trees and also certain insects. In the Middle East, the tamarisk tree produces enough manna that Bedouin nomads can collect several pounds in a morning, and go on to make halvah with it. The sugar alcohol *mannitol* (p. 662) owes its name to the fact that it was first found in and extracted from manna.

juices. But none of them has been as generous as sugarcane.

Sugar: Beginnings in Asia

Europe barely knew what we now consider ordinary table sugar until around 1100, and it was a luxury until 1700. Our first major source of sucrose was the sugar cane, *Saccharum officinarum,* a 20-foot-tall member of the grass family with an unusually high sucrose content—about 15%—in its fluids. Sugar cane originated in New Guinea in the South Pacific and was carried by prehistoric human migration into Asia. Sometime before 500 BCE, people in India developed the technology of making unrefined, "raw" sugar by pressing out the cane juice and boiling it down into a dark mass of syrup-coated crystals. By 350 BCE, Indian cooks were combining this dark *gur* with wheat, barley, and rice flours and with sesame seeds to make a variety of shaped confections, some of them fried. A couple of centuries later, Indian medical texts distinguished among a number of different syrups and sugars from cane, including crystals from which the dark coating had been washed. These were the first refined white sugars.

Early Confectionery in Southwest Asia

Around the 6th century CE, both the cane and sugar-making technology were carried westward from the delta of the Indus River to the head of the Persian Gulf and the delta of the Tigris and Euphrates rivers, where the Persians made sugar a prized ingredient in their cooking. One modern survival of this esteem is the sprinkling of large sugar crystals over a dish called "jeweled rice." Islamic Arabs conquered Persia in the 7th century and took the cane to northern Africa, Syria, and eventually Spain

Pulled Sugar and Almond Confection in 13th-century Baghdad

Medieval Arab cooks were among the first to explore sugar's remarkable sculptural qualities, as these early examples of pulled sugar and marzipan show.

Dry Halwa

Take sugar, dissolve in water, and boil until set: then remove from the dish, and pour onto a soft surface to cool. Take an iron stake with a smooth head and plant it in the mass, then pull up the sugar, stretching it with the hands and drawing it up the stake all the time, until it becomes white: then throw once more onto the surface. Knead in pistachios, and cut into strips and triangles. If desired, it may be colored, either with saffron or with vermilion.

Faludhaj

Take a pint of sugar and one-third of a pint of almonds and grind both together fine, then scent with camphor. Take one-third of a pint of sugar, and dissolve in an ounce of rose-water over a slow fire, then remove. When cooled, throw in the ground sugar and almonds, and knead. If the mixture needs strengthening, add more sugar and almonds. Make into middling pieces, melons, triangles, etc. Then lay on a dish and serve.

—*Kitab al Tabikh,* transl. A. J. Arberry

and Sicily. Arab cooks combined sugar with almonds to make marzipan paste, cooked it down with sesame seeds and other ingredients to make chewy halvah, made great use of sugar in syrups aromatized with rose petals and orange blossoms, and were pioneers in confectionery and in sugar sculpture. There are records of a 10th-century feast in Egypt that was adorned with sugar models of trees, animals, and castles!

In Europe: A Spice and Medicine

Western Europeans first encountered sugar during their Crusades to the Holy Land in the 11th century. Shortly thereafter Venice became the hub of the sugar trade from Arab countries to the West, and the first large shipment to England that we know of came in 1319. At first, Europeans treated sugar the way they treated pepper, ginger, and other exotic imports, as a flavoring and a medicine. In medieval Europe, sugar was used in two general sorts of preparations: preserved fruits and flowers, and small medicinal morsels. Sweets, or candy, began not as little entertaining treats but as "confections" (from the Latin *conficere,* "to put together," "to prepare") composed by the apothecaries, or druggists, to balance the body's principles. Sugar served several medicinal purposes. Its sweetness covered the bitterness of some drugs and made all preparations more pleasant. Its meltability and stickiness made it a good vehicle for mixing and carrying other ingredients. The solidity of a fused mass of sugar meant that it could release its medicine slowly and gradually. And its own supposed effect on the body—encouraging both heat and moisture—was thought to balance the effects of other foods and enhance the digestive process. A number of soothing medicinal sweets remain popular to this day, including lozenges, pastilles, and comfits.

Confectionery for Pleasure

It's thought that the first nonmedical confection in Europe may have been made around 1200 by a French druggist who coated almonds with sugar. Medieval recipes from the French and English courts call for sugar to be added to fish and fowl sauces, to ham, and to various fruit and cream-egg desserts. Chaucer's Tale of Sir Topas, a 14th-century parody of the chivalric romance, included sugar in a list of "royal spicery," along with gingerbread, licorice, and cumin. By the 15th century, wealthy Europeans had come to appreciate the purely pleasurable virtues of sugar and its ability to complement the flavors of many foods. The Vatican librarian Platina wrote around 1475 that sugar was being produced in Crete and Sicily as well as India and Arabia, and added,

> The ancients used sugar only in medicines, and for this reason make no mention of sugar in their foods. They certainly missed out on a great delight, since nothing given us to eat is so flavorless that sugar cannot make it savory. . . . By melting it, we make almonds . . . pine nuts, hazelnuts, coriander, anise, cinnamon, and many

Food Words: *Sugar* and *Candy*

Our language bears the traces of sugar's passage from India through the Middle East to Europe. The English word *sugar* comes from the Arabic imitation of the Sanskrit *sharkara,* meaning gravel or small chunks of material; *candy* from the Arabic version of the Sanskrit for sugar itself, *khandakah.*

> other foods into beautiful things. The quality of sugar then almost crosses over into the qualities of the things to which it clings in the confection.

Advances in Confectionery In the 15th and 16th centuries, confectionery became more of an art, done with greater sophistication and intended more and more to delight the eye. Molten sugar was now spun into delicate threads and pulled to develop a satiny sheen, and confectioners began to develop ways of determining the different states of a sugar syrup and their appropriateness to different preparations. By the 17th century, court confectioners were making whole table settings and massive decorations out of sugar, hard sugar candies had become common, and cooks had developed systems for marking the syrup concentrations suitable for different confections—ancestors of today's thread-ball-crack scale (see box, p. 651).

A Pleasure for All

Sugar became more widely available in the 18th century, when whole cookbooks were devoted to confectionery. England developed an especially strong sugar habit, and consumed large amounts in the tea and jams that fueled the working class. The per capita consumption rose from 4 pounds/2 kg a year in 1700 to 12 pounds/5 kg in 1780. By contrast, the French limited their use of sugar mainly to preserves and to desserts. In the 19th century, the growing production of sugar from beets, and the development of machines that automated the cooking, manipulation, and shaping of sugar preparations, brought inexpensive candies for all and encouraged an inventiveness that continues to this day. It's in the 19th century that familiar modern candies and chocolates were invented, and the control of crystallization was refined. *Taffy* or *toffee,* from the Creole for a mixture of sugar and molasses, and *nougat,* from the vulgar Latin for "nut cake," entered the language early in the century; *fondant,* from the French for "melting," the basic material of fudge and all semisoft or creamy centers, was developed around 1850. Most candy today is a variation of some kind on bonbons, taffy, and fondant.

The Rise of the Sugar Industry The 18th-century explosion in European sugar consumption was made possible by colonial rule in the West Indies and the enslavement of millions of Africans. Columbus carried the cane to Hispaniola (now Haiti and the Dominican Republic) on his second voyage

Sugar as Disguise

The medicinal origins of confections live on in expressions that we use today. While "honey" is almost invariably a term of praise, "sugar" is often ambivalent. Sugary words, a sugary personality, suggest a certain calculation and artificiality. And the idea of "sugaring over" something, the deception of hiding something distasteful in a sweet shell, would seem to be taken directly from the druggist's confections. As early as 1400, the phrase, "Gall in his breast and sugar in his face" was used, and Shakespeare has Hamlet say to Ophelia,

> 'Tis too much prov'd, that with devotion's visage
> And pious action we do sugar o'er
> The devil himself. (III.i)

in 1493. By about 1550, the Spanish and Portuguese had occupied many Caribbean islands and the coasts of western Africa, Brazil, Mexico, and were producing sugar in significant quantities; English, French, and Dutch colonists followed in the next century. By 1700, some 10,000 Africans were being traded via the Portuguese colony São Tomé to the Americas every year. The sugar industry was not the only force behind the great expansion of slavery, but it probably was the major force and helped ease its introduction into the southern American colonies and the cotton plantations. According to one estimate, fully two-thirds of the 20 million Africans enslaved in the Americas worked on sugar plantations. The intricate trade in sugar, slaves, rum, and manufactured goods made major ports out of the hitherto minor cities of Bristol and Liverpool in England, and Newport, Rhode Island. And the huge fortunes made by plantation owners helped finance the opening stages of the Industrial Revolution.

In the 18th century, just when it seemed at its strongest, the West Indian sugar industry began a rapid decline. The horrors of slavery gave rise to abolition movements, especially in Britain. Slaves staged revolts, and received some support from the very countries that had carried them to the plantations. One by one, through the mid-19th century, European countries outlawed slavery in the colonies.

The Development of Beet Sugar The severest blow to West Indian sugar was the development of an alternative to the sugar cane that could grow in northern climates. In 1747, a Prussian chemist, Andreas Marggraf, showed that by using brandy to extract the juice of the white beet (*Beta vulgaris,* var. *altissima*), a common European vegetable, he could isolate crystals that were identical to those purified from sugar cane, and in comparable quantities. Marggraf foresaw a kind of cottage industry by which individual farmers could sat-

Stages of Sugar Cooking in the 17th Century

This early system for recognizing the concentration of boiled sugar syrups comes from *Le Confiturier françois.* Then as now, the confectioner needed tough fingers.

Cookings of Sugar

The first is to the ribbon. It is reached when the syrup begins to thicken, so that in taking it with the finger and putting it on the thumb, it doesn't flow, and remains round as a pea.

Cooked to the pearl. The second cooking is reached when, in taking the syrup with the finger and putting it on the thumb, and opening the fingers, it forms a small thread . . . .

Cooked to the feather. This cooking has many different names. . . . It is recognized by placing a spatula in the syrup, and shaking the syrup in the air; the syrup flies away as if dry feathers without stickiness. . . . This cooking is the one for preserves and tablets.

Cooking to the burning smell. This cooking is recognized when one dips the finger in cool water, then in the sugar, and when putting the finger back into the cool water, the sugar breaks neatly like a glass without stickiness . . . . This cooking is for the large citron *biscuit*, for caramel, and pulled sugar, or *penide*, and this is the last cooking of the sugar.

isfy their own needs for sugar, but this never came about, and many years passed before the idea escaped the laboratory. In 1811, the Emperor Napoleon officially set the goal of freeing France from dependence on the English colonies for various commodities, and in 1812 personally awarded a medal to Benjamin Delessert, who had developed a working sugar-beet factory. In the next year, 300 such factories sprang up. A treaty resuming trade between France and England was signed in 1814, making West Indian sugar available once again, and the fledgling industry crashed as suddenly as it had begun. But it rose again in the 1840s and has flourished ever since.

Sugar in Modern Times

At present, beet sugar accounts for about 30% of the sucrose produced in the world. Russia, Germany, and the United States are the major beet growers, with California, Colorado, and Utah the leading states. The Caribbean is now a minor source of cane sugar, its role having been assumed by India and Brazil. Florida, Hawaii, Louisiana, and Texas also produce sugar cane. Spurred by the demand of an increasingly populous and affluent West, world sugar production increased sevenfold between 1900 and 1964, a rate matched by no other major crop in history. And thanks to the development of methods for making sweeteners from corn, an even less expensive source, sugar has never been cheaper or more abundant in our diet. This is not necessarily good for our long-term health (p. 657), and one of the major developments in 20th-century food manufacturing has been the development of ingredients that mimic the flavor and physical characteristics of sugar without having adverse effects on body weight and the regulation of blood sugar (p. 659).

THE NATURE OF SUGARS

Ordinary sugar is one member of a group of many chemicals, all of which are given

Recipes for Caramel, Pulled Sugar, and Sugar Ham in the 17th Century

Caramelle

Make some sugar cooked to the burning smell, take it off the fire, put in a little amber, rub a stone of marble or plate with oil of sweet almonds, throw your caramel on in little pieces as if preserves, and take them up with a spoon.

Twisted Sugar

Make some sugar cooked to the burning smell; take from the fire and throw it on a marble stone that you have rubbed with sweet-almond oil; rub your hands also, and work it well, have iron hooks to pull and draw out, and dress as a wreathed marzipan.

Slices of Ham

Make some sugar cooked to the feather, put it in three containers; in one put some lemon juice, in another some roses of Provence, and in the other some powdered cochenille, or pomegranate juice or powdered barberry. Make a layer of the white on some paper, two layers of red, continue until the sugar has the thickness of a ham, and cut it by the slice in the form of a slice of ham.

—*Le Confiturier françois*

the general name *sugars*. All sugars are made from just three kinds of atoms, carbon, hydrogen, and oxygen, with the carbon atoms providing a kind of backbone to which the other atoms are attached. Some sugars are simple molecules, while others are made from two or more simple sugars joined together. Glucose and fructose are simple *monosaccharides*, while table sugar, or sucrose, is a *disaccharide* made up of one glucose and one fructose joined together.

Living things put the sugars to two primary uses. The first is the storage of chemical energy. All life depends on sugars for the energy that fuels the activity of cells. This is why we have taste receptors that register the presence of sugars, and why our brain attaches pleasure to that sensation: sweetness is the sign of a food that can help supply our need for calories. The second major role for sugars is to provide building blocks for physical structures, especially in plants. The cellulose, hemicellulose, and pectin that give bulk and strength to plant cell walls are long chains of various sugars. The simple physical bulk of sugar is also useful to the cook, who can construct from it a variety of interesting textures.

One chemical characteristic of sugars is especially important in the kitchen. Sugars have a strong affinity for water, so they readily dissolve in water, and form temporary but strong bonds to water molecules in their vicinity. Sugars therefore retain moisture in baked goods, keep frozen desserts from solidifying into a solid block of ice, form a sticky matrix that holds food particles together in such things as marzipan and granola bars, maintain a moist, glossy appearance in glazes, and help preserve fruits by drawing moisture out of spoilage microbes and preventing their growth.

Kinds of Sugar

The cook works with just a handful of the many different sugars in nature. All of them are sweet, but each has its distinctive qualities.

Glucose Glucose, also called *dextrose,* is a simple sugar, and the most common sugar from which living cells directly extract chemical energy. Glucose is found in many fruits and in honey, but always in a mix-

Sweets Around the World

Sugar is universally popular, but different cultures have made different uses of it. Here are examples of sweets that are characteristic of some nations and regions.

India	Reduced-milk sweets, deep-fried batters in syrup, halvah (pastes of sugar, wheat, or chickpea flour, fruits, vegetables)
Middle East	Halvah (pastes of sugar syrup and semolina, sesame), pastries in syrup (baklava), marzipan
Greece	Spoon fruits, pastries in syrup
France	Caramel, nougat, dragées
England, United States	Novelty candies
Scandinavia	Licorice
Mexico	Dulce de leche (reduced milk), penuche (brown-sugar fudge)
Japan	Agar jelly candies, bean-paste candies, sweet-rice mochi, tea ceremony sweets

ture with other sugars. It's the building block from which starch chains are constructed. Cooks encounter it most often as the sweet substance in corn syrup, which is made by breaking starch down into individual glucose molecules and small glucose chains (p. 677). A chain of two glucoses is called *maltose*. Compared to table sugar, or sucrose, glucose is less sweet, less soluble in water, and produces a thinner solution. It melts and begins to caramelize at around 300°F/150°C.

Fructose Fructose, also called *levulose*, has exactly the same chemical formula as glucose, but the atoms are arranged in a different structure. Like glucose, fructose is found in fruits and honey, and certain corn syrups are treated with enzymes to convert their glucose into fructose. It's also sold in pure crystalline form. Fructose is the sweetest of the common sugars, the most soluble in water (4 parts will dissolve in 1 part room-temperature water), and absorbs and retains water most effectively. Our bodies metabolize fructose more slowly than glucose and sucrose, so it causes a slower rise in blood glucose levels, a quality that makes it preferable to other sugars for diabetics. Fructose melts and begins to caramelize at a much lower temperature than the other sugars do, just above the boiling point of water at 220°F/105°C.

The fructose molecule exists in several different shapes when dissolved in water, and the different shapes have different effects on our sweet receptors. The sweetest shape, a six-corner ring, predominates in cold, somewhat acid solutions; in warm or hot conditions, this shape shifts to less sweet five-corner rings. The apparent sweetness of fructose is cut nearly in half at 140°F/60°C. Neither glucose nor sucrose changes so drastically. Fructose is thus a useful substitute for table sugar in cold drinks, where it can provide the same sweetness with half the concentration and a calorie savings approaching 50%. In hot coffee, however, its sweetness drops to the level of table sugar.

Sucrose Sucrose is the scientific name for table sugar. It is a composite molecule made of one molecule each of glucose and fructose. Green plants produce sucrose in the process of photosynthesis, and we extract it from the stalks of sugar cane and the storage stems of sugar beets. Of all the common sugars, it has the most useful combination of properties. It is the second sweetest, after fructose, but is alone in having a pleasant taste even at the very high concentrations

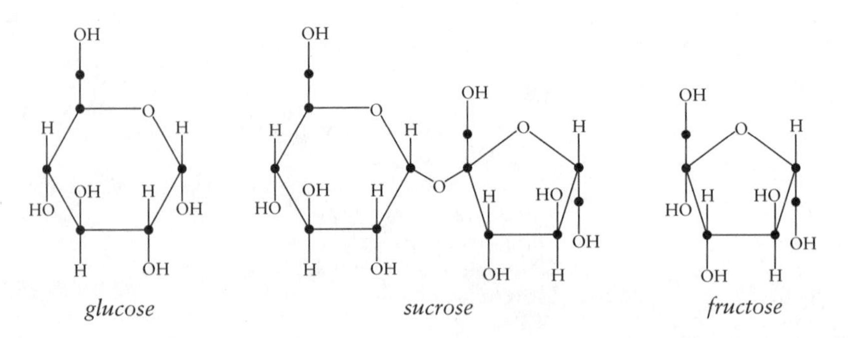

Common sugars. Carbon atoms are shown as dots. Glucose and fructose have the same chemical formula, $C_6H_{12}O_6$*, but different chemical structures, and different degrees of sweetness. A given concentration of fructose tastes much sweeter than the same concentration of glucose. Table sugar, or sucrose, is a combination of glucose and fructose (a molecule of water is released when the two sugars bond to make sucrose).*

found in candies and preserves; other sugars can seem harsh. Sucrose is also the second most soluble sugar—two parts can dissolve in one part of room-temperature water—and it produces the greatest viscosity, or thickness, in a water solution. Sucrose begins to melt around 320ºF/160ºC, and caramelizes at around 340ºF/170ºC.

When a solution of sucrose is heated in the presence of some acid, it breaks apart into its two subsugars. Certain enzymes will do the same thing. Breaking sucrose into glucose and fructose is often referred to as *inversion,* and the resulting mixture is called *invert sugar* or *invert syrup.* ("Inversion" refers to a difference in optical properties between sucrose and a mixture of its components parts.) Invert syrups are about 75% glucose and fructose, 25% sucrose. Invert sugar only exists as a syrup, since the fructose component won't fully crystallize in the presence of glucose and sucrose. Sucrose inversion and invert sugars are useful in candy making because they help limit the extent of sucrose crystallization (p. 685).

Lactose Lactose is the sugar found in milk. It is a composite of two simple sugars, glucose and galactose. Cooks seldom encounter it in pure form. Because it's much less sweet than table sugar, manufacturers use it much as they do the sugar alcohols (p. 662), more for its physical bulk than for its sweetness.

The Complexities of Sweetness

There's more to the sweetness of sugars than the sensation of sweetness pure and simple. Sweetness helps mask or balance both sourness and bitterness from other ingredients. And flavor chemists have shown that it has a strong enhancing effect on our perception of food aromas, perhaps by signaling the brain that the food is a good energy source and therefore deserves special attention.

Different sugars give different impressions of sweetness. Sucrose takes some time to be detected on the tongue, and its sweetness lingers. By comparison, the sweetness of fructose registers quickly and strongly, but it also fades quickly. And corn syrup is slow to taste sweet, peaks at about half the intensity of sucrose, and lingers even longer than sucrose. The quick action of fructose is said to enhance certain other flavors in foods, especially fruitiness, tart-

The Composition and Relative Sweetnesses of Different Sugars

Sugar sweetness is designated by comparison to the sweetness of table sugar, which is assigned a value of 100.

Sugar	Composition	Sweetness
Fructose		120
Glucose		70
Sucrose (table sugar)		100
Maltose		45
Lactose		40
Corn syrup	Glucose, maltose	30–50
High-fructose corn syrup	Fructose, maltose	80–90
Invert sugar syrup	Glucose, fructose, sucrose	95

ness, and spiciness, by allowing us to perceive them clearly without the masking effect of residual sweetness.

Crystallization

Sugars are wonderfully robust materials! Unlike proteins that easily denature and coagulate, unlike fats that are damaged by air and heat and go rancid, unlike starch chains that break apart into smaller chains of glucose molecules, sugars themselves are small and stable molecules. They mix easily with water, tolerate the heat of boiling, and when sufficiently concentrated in water, they readily bond to each other and collect themselves into pure, solid masses, or crystals. This tendency to form crystals is the means by which we obtain pure sugar from plant juices, and it's the way that we make many kinds of candies. Sugar crystallization is described in detail on p. 682.

Caramelization

Caramelization is the name given to the chemical reactions that occur when any sugar is heated to the point that its molecules begin to break apart. This destruction triggers a remarkable cascade of chemical creation. From a single kind of molecule in the form of colorless, odorless, simply sweet crystals, the cook generates hundreds of new and different compounds, some of them small fragments that are sour or bitter, or intensely aromatic, others large aggregates with no flavor but a deep brown color. The more the sugar is cooked, the less sugar and sweetness remain, and the darker and more bitter it gets.

Though caramel is most often made with table sugar, its sucrose molecules actually break apart into their glucose and fructose components before they begin to fragment and recombine into new molecules. Glucose and fructose are "reducing sugars," meaning that they have reactive atoms that perform the opposite of oxidation (they donate electrons to other molecules). A sucrose molecule is made from one glucose and one fructose joined by their reducing atoms, so sucrose has no reducing atoms free to react with other molecules, and is therefore less reactive than glucose and fructose. This is why sucrose requires a higher temperature for caramelization (340°F/170°C) than glucose (300°F/150°C) and especially fructose (220°F/105°C).

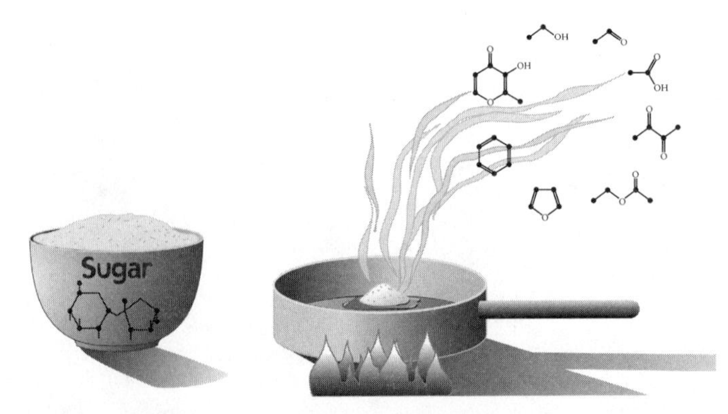

The flavors of caramelization. Heat transforms table sugar, a sweet, odorless, single kind of molecule, into hundreds of different molecules that generate a complex flavor and rich brown color. A few aromatic examples (clockwise from top left)*: alcohol, sherry-like acetaldehyde, vinegary acetic acid, buttery diacetyl, fruity ethyl acetate, nutty furan, solvent-like benzene, and toasty maltol.*

Making Caramel The usual technique for making caramel is to mix table sugar with some water, then heat until the water has boiled off and the molten sugar colors. Why add water if the first thing you do is boil it off? Water makes it possible to cook the sugar over high heat from the very beginning without the danger of burning it. In addition, the presence of water prolongs the period during which the syrup is cooked, gives these reactions more time to proceed, and develops a stronger flavor than heating the sugar on its own very quickly. And water enhances the conversion of sucrose into its glucose and fructose components. Cooking the syrup in the microwave oven has been found to produce a somewhat different spectrum of flavors than ordinary stovetop cooking.

Once caramelization and color and flavor generation begin, the overall set of reactions actually gives off heat, and can run away and burn the sugar if it's not carefully controlled. It's helpful to have a bowl of cold water ready to cool the pan down as soon as the caramel is done. Excessive caramelization turns the syrup very dark, bitter, and viscous or even solid.

The Flavor of Caramelized Sugar The aroma of a simple caramelized sugar has several different notes, among them buttery and milky (from diacetyl), fruity (esters and lactones), flowery, sweet, rum-like, and roasted. As the reactions proceed, the taste of the mixture becomes less sweet as more of the original sugar is destroyed, with more pronounced acidity and eventually bitterness and an irritating, burning sensation. Some of the chemical products in caramel are effective antioxidants and can help protect food flavors from damage during storage.

When sugars are cooked with ingredients that include proteins or amino acids—milk or cream, for example—then in addition to true caramelization, some of the sugars participate with the proteins and amino acids in the Maillard browning reactions (p. 778), which produce a larger range of compounds and a richer aroma.

Sugars and Health

"Empty Calories" In one sense, sugars are highly nourishing. Pure sugars are pure energy. After fats and oils, they're the most concentrated source of calories we have. The problem is that most people in the developed world consume more energy than they need to fuel their activity, and less than they need of hundreds of other nutrients and plant substances that contribute to long-term health (p. 253). To the extent that sugar-rich foods displace more broadly nourishing foods from our diet, they are detrimental to human health, a source of calories "empty" of any other nutritional value, and a major contributor to the modern epidemic of obesity and

Food Words: *Caramel*

Our word for browned sugar may come from its resemblance in color to straw. *Caramel* first appears in French in the 17th century as a borrowing via Spanish from the Portuguese *caramel,* which meant both the elongated sugar loaf and "icicle," perhaps because they shared a similar shape and sparkly appearance.The Portuguese in turn seems to derive from the Latin *calamus,* meaning "reed." The Greek *kalamos* meant "straw," and the original Indo-European root meant "grass." The Italian *calamari,* "squid," comes from the same root! Perhaps the common element is the brown color of dry grass, partly refined sugar, cooked sugar syrup, and camouflaging squid skin.

associated health problems, including diabetes (p. 659).

People in the developed world, particularly in the United States, consume large amounts of refined sugars. Adults in the United States get about 20% of their calories from refined sugars, children between 20% and 40%. Most of this sugar intake comes not from candies and confections, but from soft drinks. Significant amounts of sugar also find their way into most processed foods, including many savory sauces, dressings, meats, and baked goods. The total sugar content in processed foods is often unclear from the ingredients list, where different sugars can be listed separately as sucrose, dextrose, levulose, fructose, corn syrup, high-fructose corn syrup, etc.

Sugars and Tooth Decay It has been common knowledge for thousands of years that sweet foods encourage tooth decay. In the Greek book of *Problems* attributed to Aristotle, the question is asked, "Why do figs, which are soft and sweet, destroy the teeth?" Nearly 2,000 years later, as sugar cane was being established in the West Indies, a German visitor to the English court named Paul Hentzner described Queen Elizabeth I as she appeared in 1598:

> Next came the Queen, in the Sixty-fifth Year of her Age, as we were told, very majestic; her Face oblong, fair, but wrinkled; her Eyes small, yet black and pleasant; her Nose a little hooked; her Lips narrow, and her Teeth black; (a defect the English seem subject to, from their too great use of Sugar) . . .

We now know that certain kinds of *Streptococcus* bacteria colonize the mouth and cling to undisturbed surfaces, where they live on food residues, converting sugars into sticky "plaque" carbohydrates that anchor and protect them, and into defensive acids that eat away at tooth enamel and so cause decay. Clearly, the more food there is for the bacteria, the more active they will be, and hard sugar candies that slowly dissolve in the mouth provide a feast for them. But pure sugar is not the only culprit in tooth decay. Starchy foods like bread, cereals, pasta, and potato chips are also harmful because they stick to the teeth and then are broken down into sugars by enzymes in the saliva. A few other foods, notably chocolate, cocoa, and licorice extract among candy ingredients, as well as coffee, tea, beer, and some cheeses, actually inhibit decay-causing bacteria. There's evidence that phenolic compounds interfere with the adhesion of bacteria to the teeth. The sugar alcohols in low-calorie candies (p. 662) are generally not metabolized by

Caramel Food Colorings

Cooks have been confecting caramel candies and syrups for many centuries, and have been making "burnt" sugar for its brown color since ancient times. The commercial production of caramel syrups as food colorings began in Europe and the United States in the middle of the 19th century. They're now the most common food coloring, and provide the deep brown of colas, root beers and other soft drinks, spirits, candies, and many prepared foods. In addition to color, the pigment molecules also have some antioxidant activity that helps preserve flavor. Caramel colors were originally produced by heating sugar syrup in an open pan. With time, closed vacuum pans were introduced to control color development more finely, and manufacturers began to add various chemicals to obtain pigments with good dispersing or emulsifying properties.

bacteria in the mouth and don't contribute to tooth decay.

Food Sugars and Blood Sugar: The Problem of Diabetes Some foods rich in sugars can contribute to the disruption of the body's system for controlling its own sugar levels. Glucose is the body's primary form of chemical energy, so it's distributed to all cells via the blood. On the other hand, glucose is a reactive molecule, and excess quantities can damage the circulatory system, eyes, kidneys, and nervous system. So the body tightly regulates blood glucose levels, and does so with the hormone insulin. Diabetes is a disease in which the insulin system is unable to control blood glucose adequately. And a high intake of some food sugars overloads the blood with glucose and puts stress on the insulin system. This is dangerous for people who suffer from diabetes. The foods that raise blood glucose levels the most are foods rich in glucose itself, including such starchy foods as potatoes and rice that our enzymes digest into glucose. Table sugar, a combination of glucose and fructose, has a somewhat smaller effect, and fructose itself has a much smaller effect, since it must be metabolized in the liver before the body can use it for energy. One valuable property of many sugar substitutes is that they do not raise blood sugar levels.

Sugar Substitutes

Sugars combine several useful qualities in one ingredient: energy, sweetness, substance, moisture binding, and the ability to caramelize. The problem with this versatility is that each quality comes with the others. And sometimes we want just one or two alone: the pleasure of sweetness without the calories or stress on the body's system for regulating blood sugar levels, for example, or the substance without the sweetness, or substance and sweetness without the tendency to brown when cooked. Manufacturers have therefore developed ingredients that offer some but not all of the properties of sugars. Many of these ingredients were originally discovered in plants; a few are entirely artificial. Inventive cooks are now experimenting with some to make candy-like savory foods and other novelties.

There are two main kinds of sugar substitutes. The first includes various carbohydrates that provide bulk without being as digestible as the sugars. They therefore don't raise blood sugar levels as quickly, and supply fewer calories. The second is high-intensity sweeteners: molecules that provide the sensation of sweetness without supplying many calories, usually because they are hundreds or thousands of times sweeter than sugar, and are used in tiny quantities. Low- and no-calorie sweets

The Glycemic Index ofVarious Sugars and Foods

The "glycemic index" is a measure of how much a given food raises blood glucose levels. The glycemic index of glucose itself is set at 100.

Sugar	Glycemic Index	Sugar	Glycemic Index
Maltose	110	Table sugar	90
Glucose	100	Banana	60
Potatoes	95	Fruit preserves	55
White rice	95	Fructose	20
Honey	90		

Some Sugar Substitutes and Their Qualities

In this table, the sweetness of table sugar is represented as 100. A sweetness of 50 means that the substance is half as sweet as table sugar; a sweetness of 500 means that it is 5 times sweeter. The sugar alcohols and corn syrups with sweetnesses less than 1 are mainly useful for providing bulk and viscosity with reduced calories and effects on blood sugar. The intensive sweeteners, with sweetnesses greater than 100, provide taste with reduced calories and effects on blood sugar. Even those sugar substitutes that were originally found in nature are now manufactured by chemical modification of a natural or synthetic starting material.

Ingredient	Relative Sweetness	Original Source	Date of Commercialization	Notable Qualities
Polydextrose (Litesse)	0	Glucose (modified)	1980s	Produces high viscosities
Corn syrup	40	Starch	1860s	
Trehalose	50	Honey, mushrooms, yeasts	2000s?	
Sugar alcohols:				
Lactitol	40	Lactose, modified	1980s	
Isomalt (Palatinit)	50	Sucrose, modified	1980s	Less prone than sugar to crystallize or absorb moisture
Sorbitol	60	Fruits	1980s	Cooling; absorbs moisture
Erythritol	70	Fruits, fermentation	2000s?	
Mannitol	70	Mushrooms, algae	1980s	Cooling
Maltitol	90	Maltose, modified	1980s	
Xylitol	100	Fruits, vegetables	1960s	Especially cooling
Tagatose	90	Heated milk	2000s?	

Ingredient	Relative Sweetness	Original Source	Date of Commercialization	Notable Qualities
Sucrose	100	Sugar cane & beet	Traditional	
High-fructose corn syrup	100	Starch	1970s	
Fructose crystals	120–170	Fruits, honey	1970s	
Cyclamate	3,000	Synthetic	1950s	Banned in U.S., allowed in Europe
Glycyrrhizin	5,000–10,000	Licorice root	Traditional	
Aspartame	18,000	Amino acids (modified)	1970s	Not stable at cooking temperatures
Acesulfame K	20,000	Synthetic	1980s	Stable at cooking temperatures
Saccharin	30,000	Synthetic	1880s	Stable at cooking temperatures
Stevioside	30,000	South American plant	1970s	
Sucralose	60,000	Sucrose + chlorine	1990s	Stable at cooking temperatures
Neohesperidin dihydrochalcone	180,000	Citrus fruits (modified)	1990s	
Alitame	200,000	Amino acids (modified)	1990s	
Thaumatin	200,000–300,000	African plant	1980s	
Neotame	800,000	Aspartame (modified)	2000s?	

are made by combining these two kinds of ingredients, whose qualities are summarized in the chart on pp. 660–661.

Bulking Ingredients: Sugar Alcohols The most common ingredients that provide sugar-like bulk are the sugar alcohols, or polyols—chemicals whose names end in *-itol*—which are essentially sugars with one corner of their molecule modified (for example, sorbitol is derived in this way from glucose). Small amounts of some sugar alcohols—sorbitol, mannitol—are found in many fruits and plant parts. Because the human body is designed to make use of sugars, not sugar alcohols, we absorb only a fraction of these molecules from food, and use that fraction inefficiently: so they cause only a slow rise in blood insulin levels. The rest are metabolized by the microbes in our intestines, and we obtain their energy indirectly. All told, sugar alcohols provide 50–75% of the caloric value of sugar.

Sugar alcohols don't have the chemical structure (aldehyde group) that initiates the browning reactions with each other and with amino acids, so they have the sometimes useful property of being resistant to discoloration and flavor changes when heated to make confections.

Intensive Sweeteners Though most of the intensive sweeteners that we consume today were synthesized in industrial laboratories, a number of them occur in nature and have been enjoyed for centuries. Glycyrrhizin or glycyrrhizic acid, a compound found in licorice root, is 50–100 times sweeter than sucrose, and is the reason that licorice was first made into a sweet by extracting the root in hot water, then boiling down the extract. The sweetness of the extract builds slowly in the mouth and lingers. And the leaves of a South American plant commonly known as stevia, *Stevia rebaudiana,* have been used for centuries in its homeland to sweeten maté tea. Its active ingredient, stevioside, is available in a purified powdered form. Neither it nor the plant has been approved by the U.S. FDA as a food additive, so they're sold as dietary supplements.

Intensive sweeteners often have some flavor qualities that make them imperfect replacements for table sugar. For example, saccharin has a metallic aftertaste and can seem bitter; stevioside has a woody aftertaste. Many are slower than table sugar to trigger the sensation of sweetness, and their taste persists longer after swallowing. The relative sweetness of these sweeteners actually goes down as their individual concentration goes up, while combining them produces a synergistic effect. So manufacturers often use two or more to minimize their odd qualities and maximize their taste intensity.

Aspartame, a synthetic combination of two amino acids, is the most widely used

Modern Licorice

Today licorice is seldom used as a sweetener. The root of the licorice plant is extracted with ammonia to produce an ammonium salt of the sweet-tasting glycyrrhizic acid. The extract is much more expensive than molasses (the source of blackness in traditional licorice candies), sugar, gelatin, starch, and other ingredients in licorice candy, so it's used mainly as an aromatic flavoring. Licorice is especially popular in Denmark, where it's strangely combined in candies with salt and with ammonia. Glycyrrhizin also has effects on the hormone system that controls blood pressure and volume, and so in large doses can cause high blood pressure and swelling.

noncaloric sweetener. It is 180–200 times sweeter than table sugar, so that though it carries the same number of calories in a given weight, much smaller amounts are needed. Aspartame's disadvantage is that it is broken down by heat and by acidity and therefore can't be used in cooked preparations.

Sweetness Inhibitors Not only are there artificial sweeteners: there are also substances that block us from experiencing the sweetness of sugars. These taste inhibitors are useful for reducing the sweetness of a preparation whose texture depends on a high sugar concentration. Lactisole (tradename Cypha) is a phenolic compound found in small quantities in roasted coffee, patented as a flavor modifier in 1985, and used in confectionery and snacks. In very small amounts it reduces the apparent sweetness of sugar by two-thirds.

SUGARS AND SYRUPS

HONEY

Honey was the most important sweetener in Europe until the 16th century, when cane sugar and its more neutral sweetness became more widely available. Germany and the Slavic countries were leading producers in the meantime, and honey wine or *mead* (from the Sanskrit word for "honey") was a great favorite in both central Europe and Scandinavia. Honey is now valued as an alternative to sugar, a premade syrup with many distinctive flavors to offer.

The Honeybee While the New World certainly knew and enjoyed honey before the arrival of European explorers, North America did not. The bees native to the New World, species of the genera *Melipona* and *Trigona,* are exclusively tropical. They also differ from the European honeybees in being stingless and in collecting fluids not just from flowers, but also from fruits, resins, and even carrion and excrement—sources that make for unhealthful honeys as well as rich and strange flavors. European colonization brought a fundamental change to North America by introducing, around 1625, the bee that produces practically all the honey in the world today, *Apis mellifera.*

Bees are social insects that have evolved along with nectar-producing flowering plants. The two organisms help each other out: plants provide the insects with food, and insects carry cross-fertilizing pollen from one flower to another. Honey is the form in which flower nectar is stored in the hive. It appears from the fossil record that bees have been around for some 50 million years, their social organization for half that time. *Apis,* the principal honey-producing genus, originated in India. *Apis mellifera,* the honey bee proper, evolved in subtropical Africa and now inhabits the whole of the Northern Hemisphere up to the Arctic Circle.

How Bees Make Honey

Nectar The principal raw material of honey is the nectar collected from flowers, which produce it in order to attract pollinating insects and birds. Secondary sources include nectaries elsewhere on the plant and honeydew, the secretions of a particular group of bugs. The chemical composition of nectar varies widely, but its major ingredient by far is sugars. Some nectars are mostly sucrose, some are evenly divided among sucrose, glucose, and fructose, and some (sage and tupelo) are mostly fructose. A few nectars are harmless to bees but poisonous to humans, and so generate toxic honeys. Honey from the Pontic region of eastern Turkey was notorious in ancient Greece and Rome; a local species of rhododendron carries "grayanotoxins," which interfere with both lung and heart action.

The most important sources of nectar are the flowers of plants in the bean family, especially clover, and in the lettuce family, a large group that includes the sunflower, dandelion, and thistles. Though most honey is made from a mixture of nectars from different flowers, some 300 different "monofloral" honeys are produced in the world, with citrus, chestnut, buckwheat, and lavender honeys especially valued for their distinctive tastes. Some honeys, chestnut and buckwheat in particular, are much darker than others, thanks in part to the higher protein content in their nectars, which reacts with the sugars to produce dark pigments as well as a toasted aroma.

Gathering Nectar The bee gathers nectar from a flower by inserting its long proboscis down into the nectary. In the process, its hairy body picks up pollen from the flower's anthers. The nectar passes through the bee's esophagus into the honey sac, a storage tank that holds the nectar until the bee returns to the hive. Certain glands secrete enzymes into the sac, and these work to break down starch into smaller chains of sugars and sucrose into its constituent glucose and fructose molecules.

A few remarkable figures are worth quoting. A strong hive contains one mature queen, a few hundred male drones, and some 20,000 female workers. For every pound of honey taken to market, eight pounds are used by the hive in its everyday activities. The total flight path required for a bee to gather enough nectar for this pound of surplus honey has been estimated at three orbits around the earth. The average bee forages within one mile of the hive, makes up to 25 round trips each day, and carries a load of around 0.002 of an ounce, or 0.06 grams—approximately half its weight. With its light chassis, a bee would get about 7 million miles to a gallon (3 million km per liter) of honey. In a lifetime of

The Advance of the Bee in North America

We're lucky to have a near-contemporary description of the honey bee's movement across North America. In 1832, Washington Irving toured what is now the Oklahoma region and published his observations in *A Tour on the Prairies*. The ninth chapter describes a "Bee-hunt," the practice of finding honey in the wild by following bees back to their hive.

> It is surprising in what countless swarms the bees have overspread the Far West within but a moderate number of years. The Indians consider them the harbinger of the white man, as the buffalo is of the red man; and say that, in proportion as the bee advances, the Indian and buffalo retire. We are always accustomed to associate the hum of the bee-hive with the farm-house and flower-garden, and to consider those industrious little animals as connected with the busy haunts of man, and I am told that the wild bee is seldom to be met with at any great distance from the frontier. They have been the heralds of civilization, steadfastly preceding it as it advanced from the Atlantic borders, and some of the ancient settlers of the West pretend to give the very year when the honey-bee first crossed the Mississippi. The Indians with surprise found the mouldering trees of their forests suddenly teeming with ambrosial sweets, and nothing, I am told, can exceed the greedy relish with which they banquet for the first time upon this unbought luxury of the wilderness.

For those of us who buy our luxury in jars, this initial sense of wonder is worth reimagining.

gathering, a bee contributes only a small fraction of an ounce of honey to the hive.

Transforming Nectar into Honey In the hive, the bees concentrate the nectar to the point that it will resist bacteria and molds and so keep until it is needed. "House bees" pump the nectar in and out of themselves for 15 or 20 minutes, repeatedly forming a thin droplet under their proboscises from which water can evaporate, until the water content of the nectar has dropped to 50 or 40%. The bees then deposit the concentrated nectar in a thin film on the honeycomb, which is a waxy network of hexagonal cylinders about 0.20 inch/5 mm across, built up from the secretions of the wax glands of young workers. Here, with workers keeping the hive air in continuous motion by fanning their wings, the nectar loses more moisture, until it's less than 20% water. This process, known as "ripening," takes about three weeks. The bees then fill the honeycomb cells to capacity with fully ripe honey and cap them with a layer of wax.

The ripening of honey involves both evaporation and the continuing work of bee enzymes. One important enzyme converts the sucrose almost entirely to glucose and fructose, because a mixture of single-unit sugars is more soluble in water than the equivalent amount of its parent sucrose, and so can be more highly concentrated without crystallizing. Another enzyme oxidizes some glucose to form gluconic acid and peroxides. Gluconic acid lowers the honey's pH to about 3.9 and makes it less hospitable to microbes, and the peroxides also act as an antiseptic. In addition to these and other enzyme activities, the various components of ripening honey react with each other and cause gradual changes in color and flavor. Hundreds of different substances have been identified in honey, including more than 20 different sugars,

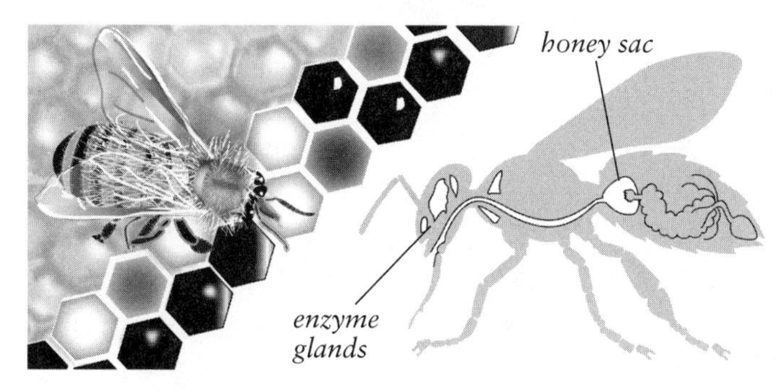

Honeycomb, and the anatomy of the worker bee. Worker bees hold freshly gathered nectar in the honey sac, together with enzymes from various glands, until they return to the hive.

Sweet Ants

Among the more unusual sweeteners are the honey or honeypot ants (species of *Melophorus, Camponotus, Myrmecocystus*) of Australia, Mexico, and the American southwest. Their colonies include a caste of workers whose role it is to store nectar and honeydew in their abdomens, which swell to the size of a pea or grape and become translucent. Honey ants are enjoyed by pinching off the abdomens and popping them directly into the mouth, or by folding them into tortillas.

savory amino acids, and a variety of antioxidant phenolic compounds and enzymes.

Processing Honey Some honey is sold in its beeswax honeycomb, but producers extract most of their honey from the comb and then treat it to extend its shelf life. They remove the honeycomb from the hive and spin it in a centrifuge to separate liquid honey from solid wax. They then generally heat the honey to around 155°F/68°C to destroy sugar-fermenting yeasts, strain it to remove pieces of wax and debris, sometimes blend it with other honeys, and finally filter it under pressure to remove pollen grains and very small air bubbles that would cloud the liquid. The honey may be packaged as a liquid at this stage, or else crystallized to form a spreadable paste, or "cream," that doesn't run and drip the way liquid honey does. Though it seems solid, 85% of cream honey remains in its liquid form, dispersed around the 15% that has solidified into tiny crystals of glucose.

Because all sugars become increasingly soluble as the temperature rises, cream honey softens and begins to melt into liquid honey when it's warmed above about 80°F/26°C. By the same token, liquid honey that has granulated during storage can be reliquefied with gentle heat.

Storing Honey Honey is one of our more stable foods, but unlike table sugar it can spoil. This is because it contains some moisture and absorbs more from the air whenever the relative humidity exceeds 60%. Sugar-tolerant yeasts can grow on the honey and produce off-flavors. It's therefore best to store honey in a moisture-tight container.

Thanks to its high concentration of sugars and the presence of some amino acids and proteins, honey is prone to undesirable, flavor-flattening browning reactions, not just when heated, but also when stored for a long time at room temperature. If you use honey infrequently, it's best to keep it at temperatures below 50°F/15°C. Liquid honey will slowly granulate in the refrigerator, and cream honey will get somewhat coarser.

Honey Flavor The most delightful quality of honeys is their flavors, which make them into natural sauce-like condiments. All honeys share a sweet taste base that is slightly tart and savory as well, and a complex aroma that has several different elements: caramel, vanilla, fruity (esters), floral (aldehydes), buttery (diacetyl), sweet-spicy (sotolon, p. 418). Then honeys made from single nectars add their own distinctive notes. Buckwheat honey is malty (methylbutanal); chestnut honey carries the distinct note of corn tortillas (aminoacetophenone, with both floral and animal elements); citrus and lavender honeys are citrusy and herbal but share a grapy note (methyl anthranilate); linden honey includes a mixture of mint, thyme, oregano, and tarragon aromatics.

The Composition of a Typical Honey

	% by weight		% by weight
Water	17	Other disaccharides	7
Fructose	38	Higher sugars	1.5
Glucose	31	Acids	0.6
Sucrose	1.5	Minerals	0.2

Honey in Cooking Unlike sugar, which is often a hidden ingredient in processed foods, honey is a very visible sweetener; most of it is added to foods by individual consumers. With its syrup-like viscosity, glossiness, and range of brown shades, it makes an attractive topping for pastries and other foods. It is the characteristic sweetener in such pastries as baklava and lebkuchen, such confections as nougat and torrone, halvah and pasteli, and in such liqueurs as Benedictine, Drambuie, and Irish Mist. Although honey wine, or mead, has all but disappeared, honey beer is popular in Africa. Americans use honey in many baked goods for a variety of reasons. It can be substituted for sugar—1 measure of honey is considered the sweetening equivalent of 1.25–1.5 measures of sugar, although the amount of added liquid must be decreased because honey does contain some water. Because it is more hygroscopic, or water attracting, than table sugar, honey will keep breads and cakes moister than sugar will, losing water to the air more slowly, and even absorbing it on humid days. Thanks to its antioxidant phenolic compounds, it slows the development of stale flavors in baked goods and warmed-over flavors in meats. Bakers can use its acidity to react with baking soda and leaven quickbreads. And its reactive reducing sugars accelerate desirable browning reactions and the development of flavor and color in the crusts of baked goods, in marinades and glazes, and other preparations.

Honey and Health; Infant Botulism Though honey has not been refined the way table sugar is and is chemically complex, it is no wonder food. Its vitamin content is negligible; bees get most of theirs from pollen. Its antibacterial properties, which led early physicians to use it to dress wounds, are due largely to hydrogen peroxide, one of the products of glucose-oxidizing enzyme and a substance well known and long employed in medicine. And honey should not be fed to children less than a year old. It often carries the seed-like dormant spores of the botulism bacterium (*Clostridium botulinum*), which are able to germinate in immature digestive systems. Infant botulism can cause difficulty in breathing and paralysis.

Tree Syrups and Sugars: Maple, Birch, Palm

When bees make honey, they perform two basic tasks: they remove a very dilute solution of sugar from plants, and then evaporate off most of the water. What the bees evolved to do instinctively and with their own muscles and enzymes, humans have learned to do with the help of tools and fire. We make syrups and sugars by extracting dilute juices from plants, and then boiling off most or all of the water. Of the man-made sweets, tree syrups and sugars are most like honey in that they retain nearly all the original contents of the sap, and are not refined to the extent that cane and beet sugars are.

Food Words: *Honey*

Though we think of the essence of honey as sweetness, the English word arises from its color. *Honey* comes from an Indo-European root meaning "yellow." The Indo-Europeans of course enjoyed honey and had a name for it. The modern descendants of that root, *melit-*, include *molasses, marmalade, mellifluous,* and *mousse* (via the Latin *mulsus,* "honey-sweet").

Maple Syrup and Sugar Long before Europeans introduced the honeybee, the natives of North America had developed their own delicious concentrated sweets. Several Indian tribes, notably the Algonquins, Iroquois, and Ojibways, had well-established myths about and terminologies for maple sugaring by the time that European explorers encountered them. Thanks to a remarkable document, we have some idea of how ingenious they were at extracting and concentrating the tree sap (see box below). All they needed was a tomahawk to cut into the tree, a wood chip to keep the wound open, sheets of elm bark for containers, and cold nights to freeze the water into pure ice crystals that could then be removed from the ever more concentrated sap.

Maple sugar was an important part of the native Americans' diet, worked into bear fat, or mixed with corn meal to make a light, compact provision for journeys. For the colonists, maple sugar was cheaper and more available than the heavily taxed cane sugar from the West Indies. Even after the Revolution, many Americans found a moral reason for preferring maple sugar to cane; cane sugar was produced largely with slave labor. Toward the end of the nineteenth century, cane and beet sugar became so cheap that the demand for maple sugar declined steeply. Today the production of maple syrup is a cottage industry concentrated in the eastern Canadian provinces, especially Quebec, and in the American Northeast.

The Sap Run The maple family originated in China or Japan and numbers some 100 species throughout the Northern Hemisphere. Of the four North American species good for sugaring, the hard or rock maple, *Acer saccharum,* produces sap of greater quality and in greater quantity than the others, and accounts for most of the syrup produced today. In the spring, sap is collected from the first major thaw until the leaf buds burst, at which point the tree fluids begin to carry substances that give the syrup a harsh flavor. The sap run is improved by four conditions: a severe winter that freezes the roots, snow cover that keeps the roots cold in the spring, extreme variations in temperatures from day to night, and good exposure to the sun. The northeastern states and eastern Canadian

Maple Sugaring Without Metal or Fire

In 1755, a young colonist was captured and "adopted" by a small group of natives in the region that is now Ohio. In 1799 he published his story in *An Account of the Remarkable Occurrences in the Life and Travels of Col. James Smith,* which includes several descriptions of how the Indians made maple sugar. Here's the most ingenious method.

> We had no large kettles with us this year, and the squaws made the frost, in some measure, supply the place of fire, in making sugar. Their large bark vessels, for holding the stock-water, they made broad and shallow; and as the weather is very cold here, it frequently freezes at night in sugar time; and the ice they break and cast out of the vessels. I asked them if they were not throwing away the sugar? they said no; it was water they were casting away, sugar did not freeze and there was scarcely any in that ice. . . . I observed that after several times freezing, the water that remained in the vessel, changed its colour and became brown and very sweet.

provinces meet these needs most consistently.

Sap does run in other trees in early spring, and some of them—birch, hickory, and elm, for example—have been tapped for sugar. But maples produce more and sweeter sap than any other tree, thanks to an intricate physical mechanism by which the tree forces sugars from the previous growing season out of storage in the trunk and into the outer, actively growing zone, the cambium.

Syrup Production From colonial times to the 20th century, sugar producers collected the sap by punching a small hole in the maple tree, inserting a wooden or metal spout into the cambium, and hanging a bucket into which the sap dripped. This picturesque collection method has mostly given way to systems of plastic taps and tubing, which carry the sap from many trees to a central holding tank. Over a six-week season, the taps remove around 10% of a tree's sugar stores, in an average of 5 to 15 gallons/20–60 liters per tree (some give as much as 80 gallons). It takes around 40 parts of sap to make 1 part syrup. The sap contains around 3% sucrose at the beginning of the season, half that at the end; so late-season sap must be boiled longer and is therefore darker and stronger-flavored. Today, many producers use energy-efficient reverse osmosis devices to remove about 75% of the sap water without heat, then boil the concentrated sap to develop its flavor and obtain the desired sugar concentration. They aim for a temperature around 7°F/4°C above the boiling point of water, the equivalent of a syrup that's around 65% sugars.

The Flavors of Maple Syrups The final composition of maple syrup is approximately 62% sucrose, 34% water, 3% glucose and fructose, and 0.5% malic and other acids, and traces of amino acids. The characteristic flavor of the syrup includes sweetness from the sugars, a slight tartness from the acids, and a range of aroma notes, including vanilla from vanillin (a common wood by-product) and various products of sugar caramelization and browning reactions between the sugars and amino acids. The longer and hotter the syrup is boiled, the darker the color and the heavier the taste. Maple syrups are graded according to color, flavor, and sugar content, with grade A assigned to the lighter, more delicately flavored, sometimes less concentrated syrups that are poured directly onto foods. Grades B and C are stronger in caramel flavor and are more often used for cooking, for example in baked goods and meat glazes. Because true maple syrup is expensive, many supermarket syrups contain little or none, and are artificially flavored.

Maple Sugar Maple sugar is made by concentrating the syrup's sucrose to the point that it will crystallize when the syrup cools. This point is marked by a boiling temperature of 25–40°F/14–25°C above the boiling point of water, or 237–250°F/114–125°C at sea level. Left to itself, the syrup will form coarse crystals thinly coated with the remainder of the brown, flavorful syrup. Maple cream, a malleable mixture of very fine crystals in a small amount of dispersed syrup, is made by cooling the syrup very rapidly to about 70°F/21°C by immersing the pan in baths of iced water, and then beating it continuously until it becomes very stiff. This mass is then gently rewarmed until it becomes smooth and semisoft.

Birch Syrup The inhabitants of far northern parts of the globe, including Alaska and Scandinavia, have long made a sweet syrup from the sap of birch trees, various species of *Betula* that are the dominant forest trees in northern latitudes. Birch sap runs for two to three weeks in early spring. It is much more dilute than maple sap, around 1% sugars, mainly an even mixture of glucose and fructose. It takes around 100 parts of sap to make 1 of syrup, both because there's less sugar to begin with, and because a mixture of glucose and fructose is thinner than the equivalent amount of sucrose; producers therefore aim for a final sugar concentration of 70–75%. Thanks to the different sugars and their reactions, the syrup is reddish brown and has a more caramel-like flavor than maple syrup; the level of vanillin is lower, too.

Palm Syrup and Sugar; Agave Syrup Among sugar-giving trees, certain tropical palms are by far the most generous. The Asian sugar palm *(Borassus flabellifer)* can be tapped for up to half the year, and yields 15–25 quarts/liters per day of a sap that may be 12% sucrose! Individual trees can give 10–80 pounds of raw sugar every year. Coconut, date, sago, and oil palms are less productive, but still far more so than maples and birches. The sap is collected either from the flowering stalks at the top of the tree, or from taps in the trunk, and then is boiled down either to a syrup called palm honey, or to a crystallized mass, which in India is known as *gur* (Hindi) or *jaggery* (English, via Portuguese from the Sanskrit *sharkara*). These same words are also used for unrefined cane sugars. Unrefined palm sugar has a distinctive, winey aroma that contributes to the flavor of Indian, Thai, Burmese, and other South Asian and African cuisines. Some palm sugar is refined to make more neutral white sugar.

Agave syrup is produced from the sap of various species of agave, desert plants native to the New World that are related to the cactus family. The sugars in agave syrup are about 70% fructose and 20% glucose, so this syrup tastes sweeter than most.

Table Sugar: Cane and Beet Sugars and Syrups

The processing of cane and beet sugar is much more complicated than the production of honey and maple and palm sugars, and for one basic reason. Bees and tree tappers begin with an isolated plant fluid that contains little else besides water and sugar. But the raw material for table sugar is the crushed whole stem of the cane, or the whole root of the beet. Cane and beet juices include many substances—proteins, complex carbohydrates, tannins, pigments—that not only interfere with the sweet taste themselves, but decompose into even less palatable chemicals at the high temperatures necessary for the concentration process. Cane and beet sugar must therefore be separated from these impurities.

Preindustrial Sugar Refining From the late Middle Ages until the 19th century, when machinery changed nearly every sort of manufacturing, the treatment of sugar followed the same basic procedure. There were four separate stages:

- clarifying the cane juice
- boiling it down into a thick syrup to concentrate and crystallize the sucrose
- draining the impurity-laden syrup from the solid crystals
- washing the remaining syrup from the crystals

The cane stalks were first crushed and pressed, and the resulting juice was cleared of many organic impurities by heating it with lime and a substance such as egg white or animal blood, which would coagulate and trap the coarse impurities in a scum that could be skimmed off. The remaining liquid was then boiled down in a series of shallow pans until it had lost nearly all of its water, and poured into cone-shaped clay molds a foot or two long with a capacity of 5 to 30 lb/2–14 kg. There it was cooled, stirred, and allowed to crystallize into "raw sugar," a dense mass of sucrose crystals coated with a thin layer of syrup containing other sugars, minerals, and various dissolved impurities. The clay cones were left to stand inverted for a few days, during which time the syrup film, or *molasses,* would run off through a small hole in the tip. In the final phase, a fine wet clay was packed over the wide end of the cone, and its moisture was allowed to percolate through the solid block of sugar crystals for eight to ten days. This washing, which could be repeated several times, would remove most of the remaining molasses, though the resulting sugar was generally yellowish.

Modern Sugar Refining Today, sugar is produced by somewhat different means. Because most sugarcane has been grown in colonies or developing countries, and sugar refining requires expensive machinery, cane sugar production came to be divided into two stages: the crystallization of raw, unrefined sugar in factories near the plantations; and refining into white sugar in industrial countries that are the major consumers. Sugar beets, on the other hand, are a temperate crop, grown mainly in Europe and North America, so they are processed all the way to refined sugar in a single factory. Harvested sugarcane is very perishable and must be processed immediately; sugar beets may be stored for weeks to months before they are processed into sugar.

Sugar production requires two basic kinds of work: crushing the cane to collect the juice, and then boiling off the juice's water. The crushing is hard physical labor, and the boiling requires large amounts of heat. In the Caribbean, these needs were filled by slave labor and deforestation. Three 19th-century innovations helped make sugar a less costly pleasure: the application of steam power to the crushing; the vacuum pan, which boils the syrup at reduced pressure and so at a lower, gentler temperature; and the multiple evaporator, which recycles the heat of one evaporation stage to heat the next.

The initial clarification of cane and beet juice is now accomplished without eggs or blood; heat and lime are generally used to coagulate and remove proteins and other impurities. Rather than waiting for gravity to draw off molasses, refiners use centrifuges, which spin the raw sugar as a salad spinner spins greens, forcing the liquid off the crystals in minutes rather than weeks. The sucrose is whitened by the technique of decolorization, in which granular carbon—a material like activated charcoal that can absorb undesirable molecules on its large surface area—is added to the centrifuged, redissolved sugar. After it absorbs the last remaining impurities, the charcoal is filtered out. The final crystallization process is carefully controlled to give individual sugar crystals of uniform size. Our table sugar is an astonishingly pure 99.85% sucrose.

Impurities in White Sugar It turns out that the tiny fraction of impurities in table sugar can make a noticeable difference in its color and flavor. Make a concentrated syrup from just water and sugar and it will have a yellow, sometimes hazy cast, thanks to large carbohydrate and pigment molecules that either get trapped between sucrose molecules as they crystallize, or

From Sugarloaf to Sugar Cube

Until the late 19th century, sugar was sold in the conical masses formed by the draining molds. These masses were called *loaves*: hence the name "Sugarloaf" that has been given to various hills and mountains for their supposed resemblance. In 1872, a onetime grocer's assistant named Henry Tate, who had worked his way to the top of a Liverpool sugar refinery, was shown an invention that cut up sugarloaves into small pieces for household use. Tate patented the device, went into production, and in a short time made a fortune with "Tate's Sugar Cube." He became a philanthropist and built the National Gallery of British Art, better known as the Tate Gallery, which he filled with his own collection.

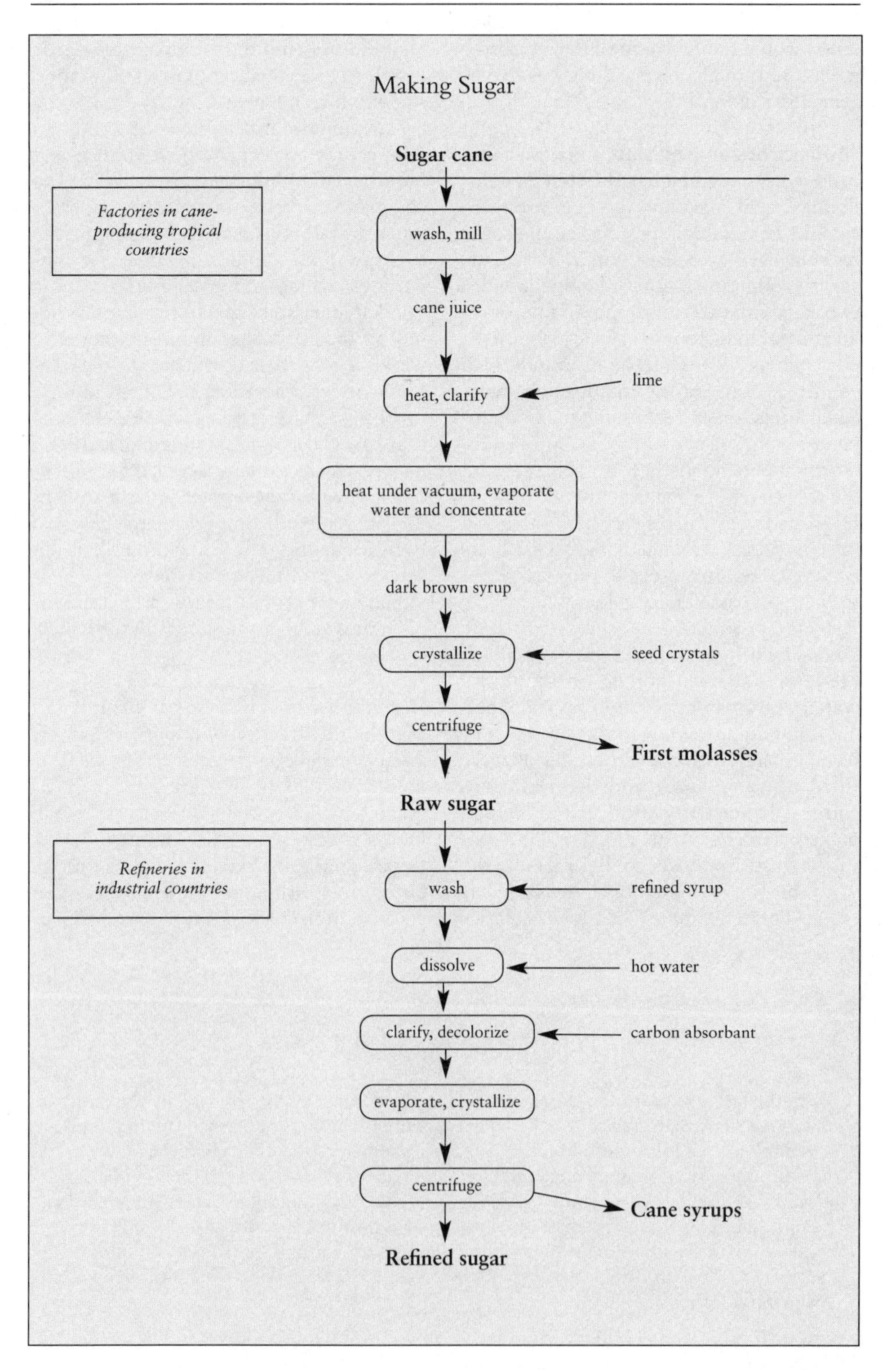
Making Sugar
Sugar cane
Factories in cane-producing tropical countries
wash, mill
cane juice
heat, clarify
lime
heat under vacuum, evaporate water and concentrate
dark brown syrup
crystallize
seed crystals
centrifuge
First molasses
Raw sugar
Refineries in industrial countries
wash
refined syrup
dissolve
hot water
clarify, decolorize
carbon absorbant
evaporate, crystallize
centrifuge
Cane syrups
Refined sugar

remain stuck to the crystal surface. Beet sugar in particular sometimes carries earthy, rancid off-odors. Where sugarcane grows above the ground and is so perishable that it is processed immediately after harvest, the beet grows underground and may be stored for weeks or months between harvest and processing, during which time it can be tainted by soil bacteria and molds that remain on its surface. In addition, beet sugar sometimes carries traces of defensive chemicals called saponins, which resemble soaps. These are known to cause the development of a scum in syrups, and may also be responsible for the poor baking performance sometimes attributed to beet sugar. (This reputation may be an undeserved legacy of the early 20th century, when refining techniques weren't as effective and the quality of beet sugar often didn't measure up to that of cane sugar.)

Kinds of White Sugar White sugar comes in a number of different forms, which differ mainly in the size of the crystals. They go by many different names. Ordinary table sugar, used for general cooking and dissolving in drinks, is midsized. Coarser crystals are mainly used for decorating baked goods and confections, and for that reason are specially treated to produce a sparkling, crystal-clear appearance. They are made from exceptionally pure batches of sucrose, with the least possible residue of the impurities that give ordinary sugar solutions a yellowish look. They're even washed with alcohol to remove sucrose dust on their surfaces. When a cook wants to make the

Forms of White Sugar: Names and Dimensions

The lengths listed below are the approximate largest dimension of the whole or powdered crystals. Our tongues sense particles larger than about 0.02 mm as gritty.

Large-grain sugars: 1–2 mm
- Coarse
- Sanding
- Pearl

Standard granulated table sugar: 0.3–0.5 mm

Fine granulated sugars: 0.1–0.3 mm
- Fruit
- Baker's special
- Caster
- Superfine, ultrafine

Powdered sugars: 0.01–0.1 mm
- Confectioner's
- Powdered
- Fondant
- Icing

OPPOSITE: *Making cane sugar. The initial processing of sugar cane into raw sugar is carried out in tropical and sub-tropical cane-growing countries; most of the subsequent refining of raw sugar into white sugar in consuming countries. Beet sugar is made in essentially the same way, except that most sugar beets are grown in temperate industrialized countries and processed there, and beet molasses and syrups are not palatable.*

whitest possible fondant, or the clearest possible syrup, it's best to use these coarse or "sanding" sugars.

At the finer end of the scale, there are a number of sugars with smaller particles than table sugar. Extra-fine, baker's special, and English caster sugars all offer more crystalline surfaces that can introduce air into fat during the creaming stage of making cakes (p. 556). "Powdered" sugars have been ground into even smaller particles, some small enough that they offer no roughness to the tongue, and can be made directly into very smooth icings, frostings, and fillings. Powdered sugars contain around 3% starch by weight to absorb moisture and prevent caking, and therefore have a slightly floury taste and feeling on the tongue.

Brown Sugars Brown sugars are sucrose crystals that are coated with a layer of dark syrup from one or another stage of sugar refining, and therefore have a more complex flavor than pure sucrose. There are several basic kinds of brown sugars.

Factory Brown Sugars "Factory" brown sugars were originally produced during the initial processing of the cane juice into unrefined sugar. These include *demerara*, *turbinado*, and *muscovado* sugars. Demerara (named after a region in Guyana) came from the first crystallization stage of light cane juice, and took the form of sticky, large, yellow-gold crystals. Turbinado was raw sugar partly washed of its molasses coat during the centrifugation, also yellow-gold and large but not as sticky as demerara. Muscovado was the product of the final crystallization from the dark mother liquor (p. 675); it was brown, small-grained, sticky, and strong-flavored.

Refinery Brown Sugars Today, these evocative names of factory sugars are often applied to a different product, brown sugars produced at the refinery using raw sugar as the starting material, not the cane juice. All ordinary brown sugar is also made in this way. There are two ways to make refinery brown sugars: redissolving the raw sugar in a syrup of some kind and then recrystallizing it, so that it retains some of the syrup on its crystal surfaces; or refining the raw sugar all the way to pure white sugar, and then coating or "painting" its surfaces with a thin film of syrup or molasses.

The basic difference between factory and refinery brown sugars is that true factory sugars retain more of the flavor of the

The Compositions of White and Brown Sugars

"Soft" brown sugars retain a coating of the syrup from which they were crystallized; "coated" sugars are white sugars that have had a thin film of brown syrup added after they've been crystallized and washed.

Sugar	Sucrose	Glucose + Fructose	Other Organic Material	Minerals	Water
White Sugar	99.85	0.05	0.02	0.03	0.05
Brown Sugars					
Soft	85–93	1.5–4.5	2–4.5	1–2	2–3.5
Coated	90–96	2–5	1–3	0.3–1	1–2.5

original cane juice, including green, fresh, and vegetable-ocean aromas (from hexanol, acetaldehyde, and dimethyl sulfide). Both kinds have an important vinegar aroma (from acetic acid), as well as caramel and buttery notes (the buttery one from diacetyl, indeed found in butter), and salty and bitter tastes (from minerals). Refinery brown sugars also develop what's described as a licorice aroma from the long, slow heating of the syrups.

Whole Sugars It's still possible to taste what might be called whole sugar, crystalline sugar still enveloped in the cooked cane juice from which it formed. This is the sugar sold in Indian groceries as *jaggery* or *gur,* and in Latin American shops as *piloncillo, papelon,* or *panela.* The flavor is highly variable, and ranges from mild caramel to strong molasses.

Using Brown Sugars Brown sugar is soft and clingy because its molasses film—whose glucose and fructose are more hygroscopic than sucrose—contains a significant amount of water. Of course, if brown sugar is left exposed to dry air, it will lose its moisture through evaporation and become hard and lumpy. It can be kept moist by storing it in an airtight container, and resoftened by closing it up with a damp towel or piece of apple from which it can absorb moisture. Because brown sugar tends to trap air pockets between groups of adhering crystals, it should be packed down before its volume is measured.

Molasses and Cane Syrups

Molasses Molasses, which is called *treacle* in the United Kingdom, is generally defined as the syrup left over in cane sugar processing after the readily crystallizable sucrose has been removed from the boiled juice. (There is such a thing as beet molasses, but it has a strong, unpleasant odor, and so is used to feed animals and industrial fermentation microbes.) In order to extract as much sucrose as possible from cane juices, crystallization is performed in several different steps, each of which results in a different grade of molasses. "First" molasses is the product of centrifuging off the raw sugar crystals, and still contains some sucrose. It is then mixed with some uncrystallized sugar syrup, recrystallized, and recentrifuged. The resulting "second" molasses is even more concentrated in impurities than the first. Repeating this process once more yields "third,"or final, or "blackstrap" molasses (from the Dutch *stroop* for "syrup"). The brown-black color of final molasses is due to the extreme caramelization of the remaining sugars and to chemical reactions induced by the high temperatures reached during the repeated boilings. These reactions, together with the high concentration of minerals, give final molasses a harsh flavor that makes it generally unfit for direct human consumption, although it's sometimes sold blended with corn syrup. A small amount is also used in tobacco curing.

Food Words: *Molasses, Treacle*

Molasses comes from the late Latin word *mellaceus,* which meant "like honey." The English term *treacle* comes via the French *triacle* from the Latin *theriaca,* meaning antidotes against poison. Medieval pharmacists used sugar syrups to compound their drugs, and came to refer to the syrups by a term for the remedies. Today, *treacle* can mean both dark, strong molasses or the lighter, more delicate refiner's syrups.

Kinds of Molasses First and second molasses have been used in foods for many years, and for a long time were the only form of sugar available to slaves and the poor of the rural South, usually bleached with sulfur dioxide and strongly sulfurous to the taste. Today, most molasses available to consumers are actually blends of molasses and syrups from various stages throughout the sugar-making process. They range from mild to pungent and bitter, from golden brown to brown-black. The darker the molasses, the more its sugars have been transformed by caramelization and browning reactions, and so the less sweet and more bitter it is. Light molasses may be 35% sucrose and 35% invert sugars, and 2% minerals; blackstrap molasses may be 35% sucrose, 20% invert sugars, and 10% minerals.

Molasses in Cooking The flavor of cane molasses is complex, with woody and green notes as well as sweet, caramel, buttery ones. Its complexity has made it a popular background flavor in many foods; popcorn balls, gingerbread, licorice, barbeque sauces, and baked beans are examples. Cane molasses is usually but unpredictably acidic; its pH varies between 5 and neutral 7, so it can sometimes react with baking soda and produce leavening carbon dioxide in baked goods. Thanks to its invert sugars, it helps retain moisture in foods. And a variety of components contribute to a general antioxidant capacity, which helps slow the development of off-flavors.

Cane and Sorghum Syrups Cane syrups may be produced directly from cane juice at sugar factories, or from raw sugar at refineries. They generally contain a combination of sucrose (25–30%) and invert sugars (50%), are golden to medium brown in color, and have a mild flavor with caramel, butterscotch, and leafy aromas. Louisiana cane syrups were traditionally made from whole cane juice, concentrated and clarified. The same basic product, with about half of its sucrose inverted by acid or enzymes, is now sometimes called "high-test molasses." It has been heated less, and

Fruit Syrups: Ancient Saba, Modern Fruit Sweeteners

In Europe, the original sweet syrups were made not from cane, but from grapes. Italian *saba* is grape juice cooked down to a concentrated, viscous syrup. It contains about equal amounts of glucose and fructose, and in addition has a distinct tartness due to the simultaneous concentration of the grape acids. In the 16th century, Nostradamus described making various sweet preparations with saba, and noted that "in places where there is neither sugar nor honey, the sovereign sun produces and nourishes other fruits which . . . come to satisfy our sensuous desire. . . ."

Manufactured fruit syrups are a relatively recent version of the traditional syrups. They're made from batches of various fruits, including apples, pears, and grapes, that are in surplus, damaged, or otherwise not suitable for other uses. Both aroma and color are removed from the juice, which is concentrated to about 75% sugars, mainly glucose and fructose due to the action of the fruit acids on sucrose. The acids are also concentrated, so the pH of the syrup is around 4. Food manufacturers value these fruit syrups in part because they can be identified appealingly as "fruit sweeteners" on the label, rather than as sugar or corn syrup. They may also contain significant amounts of pectin and other cell-wall carbohydrates that help stabilize emulsions and reduce crystal size in frozen preparations.

so has a more aromatic, less bitter flavor than true molasses. "Golden syrup" is a refinery syrup made from raw sugar, filtered through charcoal to give it a characteristic light, crystal-clear appearance and delicate flavor. Cane syrups offer more character (though also a more intense sweetness) than corn syrup in such dishes as pecan pie.

Sorghum syrup is made in small quantities in the American South and Midwest from the stalk juice of sweet sorghum, specialized varieties of a cereal plant normally grown for its grain (*Sorghum bicolor*, p. 482). Sorghum syrup is mainly sucrose, and has a distinctive pungency.

Corn Syrups, Glucose and Fructose Syrups, Malt Syrup

Sugars from Starch We come now to a source of sugar that is relatively new, but today rivals cane and beet sugars in commercial importance. In 1811, a Russian chemist, K. S. Kirchof, found that if he heated potato starch in the presence of sulfuric acid, the starch was transformed into sweet crystals and a viscous syrup. A few years later, he discovered that malted barley had the same effect as the acid (and thereby laid the foundations for a scientific understanding of beer brewing). We now know that starch consists of long chains of glucose molecules, and that both acids and certain plant, animal, and microbial enzymes will break these long chains down into smaller pieces and eventually into individual glucose molecules. The sugars make the syrup sweet, and the remaining fragments of glucose chains give the solution a thick, viscous consistency. In the United States, the acid technique was used to produce syrup from potato starch in the 1840s, and from corn starch beginning in the 1860s.

High-Fructose Corn Syrups The 1960s brought the invention of fructose syrups. These start out as plain corn or potato syrups, but an additional enzyme process converts some of the glucose sugars into fructose, which is much sweeter and therefore gives the syrups a higher sweetening power. The solids in standard high-fructose corn syrup are around 53% glucose and 42% fructose, and provide the same sweetness as the syrup's equivalent weight in table sugar. Because high-fructose syrups are relatively cheap, soft-drink manufacturers began to replace cane and beet sugars with them in the 1980s, and Americans began to consume more corn syrups than cane and beet sugar. Today they're a very important sweetener in food manufacturing.

Making Corn Syrups To make corn syrups, manufacturers extract starch granules from the kernels of common dent corn (p. 477), and then treat them with acid and/or with microbial or malt enzymes to develop a sweet syrup that is then clarified, decolorized, and evaporated to the desired con-

Fructose Crystals

Crystalline fructose has been commercially available for only a few decades. Fructose is so hygroscopic, or water-absorbing, that it's hard to get it to crystallize from a water solution. It's now made by mixing high-fructose corn syrup with alcohol, in which fructose is much less soluble. If fructose crystals are sprinkled onto a food as decoration, they'll quickly disappear into a thin, sticky syrup as they absorb moisture from the food and air and dissolve.

centration. Nowadays, enzymes from the easily cultured molds *Aspergillus oryzae* (also used in Japan to break rice starch down into fermentable sugars for *sake*) and *A. niger* are used almost exclusively. In Europe, potato and wheat starch are the main sources for making what is called "glucose" or "glucose syrup," which is essentially the same as American corn syrup.

The Properties and Uses of Corn Syrups Among the usual sweeteners available to the cook, corn syrups are alone in providing long carbohydrate molecules that get tangled up with each other and slow down the motion of all molecules in the syrup, thus giving it a thicker consistency than any but the most concentrated sucrose syrups. It's largely these long tangly molecules that have made corn syrup increasingly important in confectionery and other prepared foods. Because the tangling interferes with molecular motion, it also has the valuable effect of preventing other sugars in candy from crystallizing and producing a grainy texture. All molecules in the syrup are flowing very slowly, and the sucrose crystal faces keep getting covered with chains that can't become part of the crystal. (The same behavior helps minimize the size of ice crystals in ice cream and fruit ices, thus encouraging a smooth, creamy consistency.) Another consequence of corn syrup's viscosity is that it imparts a thick, chewy texture to foods. And because it includes glucose, a water-binding sugar that is less sweet than table sugar, corn syrup helps prevent moisture loss and prolongs the storage life of various foods without the cloying sweetness that honey or sucrose syrup imparts. Finally, all corn syrups are somewhat acid, with a pH between 3.5 and 5.5, so in baked goods they can react with baking soda to produce carbon dioxide and thus contribute to leavening.

Grades of Corn Syrup Corn syrup is an especially versatile ingredient in food manufacturing because its sweetness and viscosity can be varied simply by controlling the thoroughness of the enzymatic digestion of starch into sugars. The most common consumer grade of corn syrup is about 20% water, 14% glucose, 11% maltose, and 55% longer glucose chains. It is only

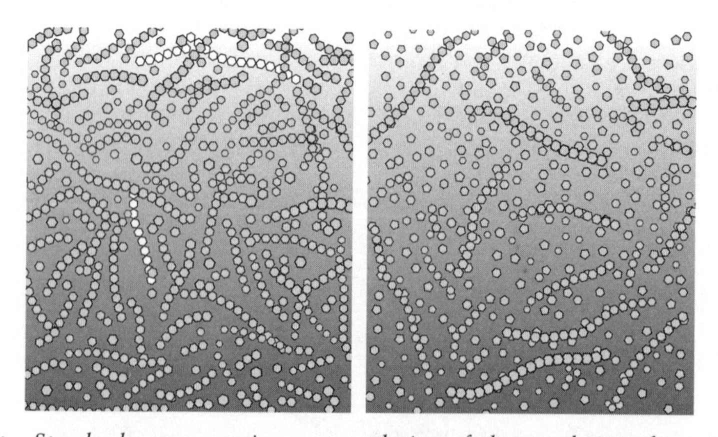

Corn syrups. Standard corn syrup is a water solution of glucose chains of varying lengths (left). *One- and two-unit sugars taste sweet, while taste-free longer chains make the syrup viscous. By controlling the relative populations of different chains, the manufacturer can tailor the syrup's balance of sweetening and thickening powers. High-fructose corn syrup* (right) *has been treated with an enzyme to convert a portion of the single glucose molecules (small hexagons) into fructose molecules (small pentagons), which taste sweeter.*

moderately sweet, and fairly viscous. Several others grades are available to manufacturers:

- Maltodextrins are syrups that contain less than 20% glucose plus maltose, and are used mainly to give viscosity and body with little sweetness and moisture absorption.
- High-fructose corn syrups are around 75% fructose plus glucose, and give an overall sweetness around that of table sugar. They and high-glucose syrups help develop color and retain moisture in baked goods.
- High-maltose syrups are valuable in ice creams, and some confections, where lowered freezing points or interference with crystallization are desired but sweetness is not; maltose is less sweet than either table sugar or glucose. In baked goods, maltose feeds yeasts and improves leavening.

Malt Syrup and Extract Malt syrup is made from a combination of germinated cereal grains, preeminently barley, and ordinary cooked grains. It's among the most ancient and versatile of sweetening agents, and was the predecessor of modern-day high-tech corn syrups. Along with honey, malt syrup was the primary sweetener in China for 2,000 years, until around 1000 CE; it's still made in both China and Korea. Malt syrup had the advantage that it could be made in households from readily available and easily stored materials, the same whole grains that were grown as staple foods, including wheat, rice, and sorghum. It was therefore a far more affordable sweetener than cane sugar.

There are three stages to making malt syrup. First a portion of whole grain is *malted*: soaked in water and allowed to germinate partly, then dried again by means of carefully controlled heating (p. 744). The germinating embryo produces enzymes that will digest the grain's starch into sugars to fuel its growth; barley is preferred in malting because it produces unusually copious and active enzymes. Drying preserves these enzymes, and also develops color and flavor by means of browning reactions. In the second stage, the malted grain is mixed with some water and with unmalted but cooked grains—rice, wheat, barley—and the malt enzymes digest the cooked starch granules to produce a sweet slurry. In the final stage, the slurry is extracted with additional water, and the liquid is boiled down to concentrate it. The result is a concentrated syrup of maltose, glucose, and some longer glucose chains. Malt syrup is therefore much less sweet than a similarly viscous sucrose syrup. In Asia it is used to provide color and gloss in savory dishes—for example, it's painted onto the skin of Peking duck—as well as in confections.

The Composition of Malt Extract

	% of Malt Extract by Weight
Water	20%
Protein	5
Minerals	1
Total sugars	60
Glucose	7–10
Maltose (double-glucose chains)	40
Maltotriose (triple-glucose chains)	10–15
Longer glucose chains	25–30

Malt syrup has a relatively mild malt aroma because the malted barley is a small fraction of the grain mixture. If the malted barley is soaked on its own, without any added cooked grains, then the malt flavor is much stronger. Such a preparation is usually called "malt extract." It is frequently used in baking to provide maltose and glucose for yeast growth and moisture retention (p. 530). In the United States, malted milk and malt balls are made from a mixture of barley malt and powdered milk.

SUGAR CANDIES AND CONFECTIONERY

All sugar candies, whether brittle or creamy or chewy, are essentially mixtures of two ingredients: sugar and water. Cooks manage to create very different textures from the same materials by varying the relative proportions of sugar and water, and the physical arrangement of the sugar molecules. They control the proportions as they cook the sugar syrup, and they control the physical arrangement as they cool it. Depending on how hot the syrup gets, how quickly it cools, and how much it's stirred, it can solidify into coarse sugar crystals, fine sugar crystals, or a monolithic crystal-free mass. To a large extent, the art of the confectioner depends on the science of crystallization.

Setting the Sugar Concentration: Cooking the Syrup

The first factor that influences candy texture is the concentration of sugar in the cooked syrup. Confectioners have found from long experience that certain syrup concentrations are best for making certain kinds of candy. Generally, the more water the syrup contains, the softer the final product will be. So the cook must know how to make and recognize particular syrup concentrations. This turns out to be pretty simple. When we dissolve sugar or salt in water, the boiling point of the solution becomes higher than the boiling point of pure water (see p. 785). This increase in the boiling point depends predictably on the amount of material dissolved: the more dissolved molecules in the water, the higher the boiling point. So the boiling point of a solu-

Frostings, Icings, and Glazes

Frostings, icings, and glazes are sweet coatings for cakes and other baked goods. In addition to being tasty and decorative, they protect the food underneath from drying out. These preparations began in the 17th century as plain syrup glazes, and gradually evolved into more elaborate forms. Today, glazes are glossy, thin, dense coatings made with a combination of powdered sugar, a small amount of water, corn syrup, and sometimes fat (butter, cream). The corn syrup and fat prevent the sugar from forming coarse crystals, and the corn syrup provides a moisture-attracting liquid phase to fill the space between sugar particles and create a smooth, glass-like surface. A warm fondant (around 100°F/38°C) poured over the cake or pastry produces a similar effect. Simple frostings are made by whipping sugar and air into a solid fat—butter, cream cheese, or vegetable shortening—to make a sweet, creamy, light mass. The sugar particles must be small enough not to make the frosting seem grainy, so fine grades of powdered sugar are the usual choice. Cooked frostings and icings include eggs or flour and owe their body in part to the egg proteins or flour starch. Because the sugar dissolves during the cooking, its particle size is unimportant.

tion is an indicator of the concentration of the dissolved material. The graph in the box below shows, for example, that a sugar syrup that boils at 250°F/125°C is about 90% sugar by weight.

Cooking the Syrup Raises the Sugar Concentration As a sugar solution boils, water molecules evaporate from the liquid phase into the air, while the sugar molecules stay behind. The sugar molecules therefore account for a larger and larger proportion of all the molecules in the solution. So as it boils, the syrup gets more and more concentrated: and this in turn causes its boiling point to continue to rise. In order to make a syrup of a given sugar concentration, all the candy maker has to do is heat a mixture of sugar and water until it boils, and then keep it at the boil and watch its temperature. At 235°F/113°C, or about 85% sugar, the cook can stop the concentration process and make fudge; at 270°F/132°C, or 90%, taffy; at 300°F/149°C and above, nearing 100% sugar, brittles and hard candies.

The Cold-Water Test Although it was invented 400 years ago by Sanctorius, the thermometer has been a common household appliance for only a few decades. Beginning in the 16th century and continuing to this day, confectioners have used a more direct means of sampling the syrup's fitness for different candies: they scoop out a small amount, cool it quickly, and note its behavior. Thin syrups will simply form a thread in the air. Somewhat more concentrated syrups form a ball when dropped into cold water, and the ball will be soft

Syrup Boiling Points Depend on Sugar Concentration

The boiling point of a sugar solution increases as the concentration of sugar increases. This graph shows the relationship between boiling point and sugar concentration at sea level.

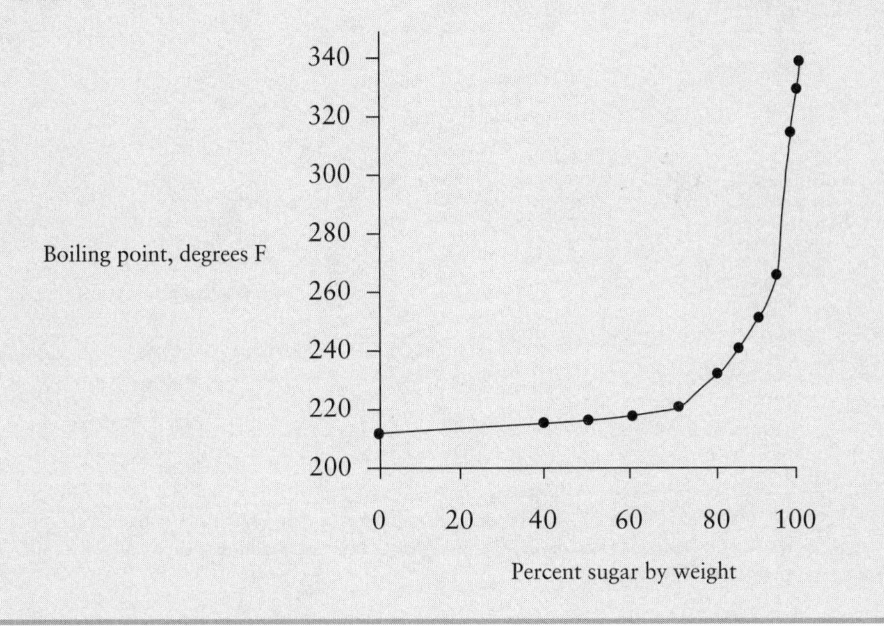

and malleable between the fingers; as the concentration increases, the cooled ball becomes harder. The most concentrated syrups make a cracking sound as they turn into hard, brittle threads. Each of these stages indicates a particular temperature range and suitability for a particular kind of candy (see box below).

The Heating Rate Accelerates During Cooking As we cook a sugar syrup, most of the heat goes into the work of evaporating water molecules from the syrup, and less into actually raising the temperature of the syrup; so the syrup temperature rises only gradually. But as the sugar concentration passes 80%, there's so little water left that both the temperature of the syrup and its boiling point rise more rapidly. As the concentration approaches 100%, the temperature rises very fast, and can easily overshoot the desired range and brown or scorch the sugar. To avoid this, the cook should reduce the heat toward the end of cooking and keep a careful eye on the syrup temperature.

Setting the Sugar Structure: Cooling and Crystallization

The final texture of a candy is determined by the way in which the sugar molecules in the cooked syrup cool and settle into a solid structure. If the sugar forms a few large crystals, then the candy texture will be coarse and grainy. If it forms many millions

Confections and the Sugar Syrups They're Made from

Sugar confections are made from syrups with particular sugar concentrations. This chart lists some common confections and two distinguishing qualities of their syrups.

Syrup Behavior in Cold-Water Test	Syrup Boiling Point* °F/°C	Confection
Thread	215–235/102–113	Syrups, preserves
Soft ball	235–240/113–116	Fondant, fudge
Firm ball	245–250/118–121	Caramel candies
Hard ball	240–265/121–130	Marshmallows, nougat
Soft crack	270–290/132–143	Taffy
Hard crack	300–310/149–154	Butterscotch, brittle
	320–335/160–168	Hard candies, toffee
	340/170	Light caramel for syrups, color, and flavor
	355–360/180–182	Spun sugar, sugar cages; medium caramel
	375–380/188–190	Dark caramel
	410/205	Black caramel

*Above 330°F/165°C, the sugar syrup is more than 99% sucrose. It no longer boils, but begins to break down and caramelize. Boiling points depend on elevation. For each 1,000 feet/305 meters above sea level, subtract 2°F/1°C from every boiling point listed.

of microscopic crystals that are lubricated by just the right amount of syrup, then the candy will be smooth and creamy. And if it forms no crystals at all, then it will be a hard, monolithic mass. The trickiest stage of candy making thus comes *after* the cooking, when the syrup cools from 250–350°F/ 120–175°C down to room temperature. The rate of cooling, the movement of the syrup, and the presence of the smallest particles of dust or sugar can have drastic effects on the candy's structure and texture.

How Sugar Crystals Form Sugar molecules have a natural tendency to bond to each other in orderly arrays and form dense solid masses, or crystals. When sugar crystals are dissolved in water to make a syrup, the water molecules overcome that tendency by forming their own bonds with the sugar molecules, surrounding and separating them from each other. If the dissolved sugar molecules in a syrup get too crowded for the water molecules to keep the sugars apart from each other, the sugars will begin to bond to each other again and form crystals. When the tendency of a dissolved substance to bond to itself is exactly balanced by the water's ability to prevent this bonding, the solution is called *saturated.*

The saturation point depends on temperature. The rapidly moving water molecules in a hot sugar solution can keep more sugar molecules dissolved than the sluggish water molecules in a cold solution can. The moment that a hot and saturated solution begins to cool, it becomes *super*saturated. That is, it temporarily contains more dissolved sugar than it normally could at that temperature. And once the solution has become supersaturated, the smallest disturbance will induce sugar crystals to form and grow. As the sugar molecules gather into solid crystals, they leave the solution around them less concentrated. When the solution reaches the sugar concentration appropriate for its new temperature, the sugar crystals stop forming and growing. The sugar is now in two different states: some remains dissolved in the syrup, and some is packed in the solid crystals surrounded by the syrup.

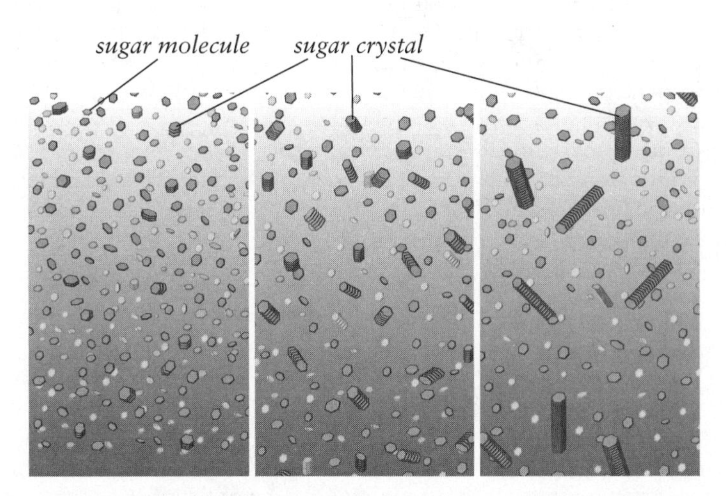

The growth of sugar crystals as a hot syrup cools. Left: *Crystals are tightly organized, solid clusters of molecules.* Center: *When conditions favor the formation of crystal seeds, the dissolved sugar molecules can join many seeds, the resulting crystals are small, and the candy texture is fine.* Right: *When conditions limit the formation of crystal seeds, the dissolved sugar molecules can join only a few seeds, the resulting crystals are large, and the candy texture is coarse.*

There are two steps in sugar crystallization: the formation of crystal "seeds," and the growth of those seeds into mature, full-sized crystals. Seed formation determines how many crystals will form, and crystal growth determines how large they get. Both steps affect the final texture of a candy.

Particles, Temperature, and Stirring Influence Crystallization The crystal "seed" is an initial surface to which sugar molecules can attach themselves and accumulate in a solid mass. The seed can be a few sugar molecules that happen to come together during random movements in the syrup. Stirring and agitation have the effect of bumping solution molecules together more often than they otherwise would, and thereby encourage the formation of crystal seeds. Other things can also serve as seeds in a cooling syrup and initiate crystallization. Among the more common are the tiny crystals that form when the syrup spatters on the side of the pan or dries off on a spoon, and that then are stirred back into the syrup. Dust particles and even tiny air bubbles can also act as crystal seeds. A metal spoon can induce crystallization by conducting heat away from local areas of the syrup, cooling them and so leaving them super-supersaturated. Experienced candy makers therefore prevent premature crystallization by using wooden spoons, avoiding agitation of the syrup once it's cooked and begins to cool, and carefully removing dried syrup spatter from the pan walls with a moist brush.

Controlling Crystal Size and Candy Texture The cook has to worry about premature crystallization because candy texture is affected by the syrup temperature at which crystallization begins. Generally, hot syrups produce coarse crystals, and cool syrups produce fine crystals. Here's the logic. Because more sugar molecules will arrive at the crystal surface during a given time in a hot syrup with fast-moving molecules than in a cold, lethargic one, crystals grow more rapidly in hot syrups. At the same time, because stable crystal *seeds* are *less* likely to form at higher temperatures—an aggregate of a few sugar molecules is more easily knocked apart in fast-moving surroundings—the total number of crystals formed in a hot syrup will be lower. Put these two trends together, and we see that when a hot syrup begins to crystallize, it will produce fewer and larger crystals than a cool one, and therefore a coarse texture. This is why recipes for fudge or fondant, candies with a fine, creamy texture, call for the syrup to be cooled drastically—from 235°F/113°C down to around 110°F/43°C—before the cook initiates crystallization by stirring.

Stirring Makes Smaller Crystals Crystal size and texture are also influenced by stirring. We've seen that agitation favors the

Rules for Creating Fine-Textured Candies

In order to produce many small sugar crystals from a syrup, the candy maker should

- include some corn syrup in the recipe to interfere with crystal formation
- remove dried syrup from the pan interior before cooling the syrup
- allow the syrup to cool before initiating crystallization
- avoid agitating the syrup while it cools
- when the syrup is cool, agitate continuously and vigorously for as long as the syrup is workable

formation of crystal seeds by pushing sugar molecules into each other. A syrup that is stirred infrequently will develop only a few crystals, while one that is kept in motion continuously will produce great numbers. And the more crystals there are in a syrup, all competing for the remaining free molecules, the fewer free molecules there are to go around, and so the smaller the average size of each crystal. The more a syrup is stirred, then, the finer the consistency of the final candy. This is the justification for wearing your arm out when making fudge: the moment you let up, the formation of seeds slows down, the crystals you've made up to that point begin to grow in size, and the candy gets coarse and grainy.

Preventing Crystal Formation: Making Sugar into a Glass Candy makers produce an entirely different structure and texture when they cool a syrup so rapidly that the sugar molecules stop moving before they have a chance to form any crystals at all. This is how transparent hard candies are made. If the water content of the cooked syrup is just 1 or 2%, then it's essentially molten sugar with a trace of water dispersed in it. The syrup is very viscous, and if it cools quickly, the sucrose molecules never have a chance to settle into orderly crystals. Instead, they just set in place in a disorganized mass. Such an amorphous, noncrystalline material is called a *glass*. Ordinary window and table glass is a noncrystalline version of silicon dioxide. Like this mineral glass, sugar glass is brittle and transparent (and often stands in for its harder and more dangerous cousin in the movies and on stage!). Glasses are transparent because individual sugar molecules are too small to deflect light when they're randomly arranged. Crystalline solids appear opaque because even tiny crystals are solid masses of many molecules, and their surfaces are big enough to deflect light.

Limiting Crystal Growth with Interfering Agents In practice, it's not easy to control or prevent the crystallization of pure sucrose syrups, and candy makers have long relied on other ingredients that interfere with and therefore limit crystal formation and growth. These interfering agents help the cook prepare clear noncrystalline hard candies and fine-textured creams, fudges, and other soft candies.

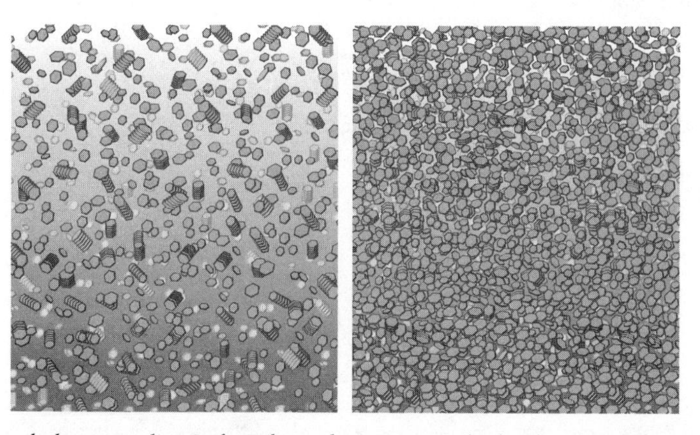

Crystalline and glassy candies. Left: *When a hot syrup cools slowly enough for the molecules to cluster together, they form tightly organized crystals.* Right: *When a very concentrated syrup cools quickly and traps the sugar molecules in place before they can cluster, they solidify into a disorganized, noncrystalline glass.*

Invert Sugar The original interfering agents were glucose and fructose, or "invert sugar" (p. 655). When heated along with a small amount of acid (often cream of tartar), sucrose is broken down into its two components, glucose and fructose. Glucose and fructose interfere with sucrose crystallization by bonding temporarily to the crystal surface and blocking the way of sucrose molecules. Honey is a natural source of invert sugar, and "invert syrup" is an artificial preparation of a glucose-fructose mixture. Thanks to their fructose content, both honey and invert syrup readily caramelize and can cause undesirable browning in some sweets. Acid-inverted syrups brown less because their acidity slows caramelization.

Corn Syrup Because acid treatment of sucrose is somewhat unpredictable, most modern confectioners instead use corn syrup, which is an especially effective inhibitor of crystallization, and doesn't readily caramelize. The assorted long glucose chains form a tangle that impedes the motion of both sugar and water molecules and makes it more difficult for the sucrose to find a crystal onto which to fit. The glucose and maltose molecules interfere in the same way that invert sugar does. Corn syrup also provides body and chewiness, is less sweet than sugars, and has the advantage for manufacturers of being less expensive than crystalline sugar.

Other Candy Ingredients Confectioners add a number of other ingredients to the basic sugar syrup for candies to modify taste and texture. All interfere with sucrose crystallization to some extent and so tend to encourage finer crystals.

Milk Proteins and Fat Milk proteins thicken candy body and, because they brown easily, add a rich flavor to caramels and fudge. The casein proteins contribute to a chewy body, whey proteins to browning and flavor development, and both help emulsify and stabilize butterfat droplets. Butterfat lends smoothness and moistness to butterscotch, caramel, toffee, and fudge, and reduces the tendency of chewy candies to stick to the teeth. Because milk proteins curdle in acid conditions, and caramelization and browning reactions generate acids, candies that include milk solids are sometimes neutralized with baking soda. The reaction between acids and baking soda generates bubbles of carbon dioxide, so such candies may be filled with small bubbles that give them a more fragile texture, less chewy or hard or clinging.

Gelling Agents Confectioners also give a firmer body to certain candies with a number of ingredients that bond to each other and to water to form solid but moist gels. These ingredients include gelatin, egg white, grain starches and flours, pectin, and plant gums. Gelatin and pectin in particular are used to make gummy and jelly candies, often in combination. Gelatin provides a tough chewiness, while pectin makes a more tender gel. Gum tragacanth, a carbohydrate from a West Asian shrub in the bean family (*Astragalus*), has been used

Candy Colors

Many candies are intensely colored to strike the eye as strongly as the taste buds. The pigments in such candies are generally synthesized from petroleum by-products, and are much more intense and stable than natural colorings. Iridescent effects are produced with a combination of thin plates of mica (potassium aluminum silicate) and either titanium dioxide or ferric oxide (mineral pigments).

for centuries to make the sugar dough from which lozenges are cut and dried.

Acids Many candies include an acid ingredient to balance the overwhelming sweetness. Jelly beans, for example, have a tart surface. These flavoring acids are added after the syrup has cooled down, so as to avoid excessive inversion of the sucrose into glucose and fructose. Different acids are said to have different taste profiles. Citric and tartaric acids give a rapid impression of acidity, while malic, lactic, and fumaric acids are slower to register on the tongue.

Kinds of Candies

It's convenient to divide sugar confections into three groups: noncrystalline candies, crystalline candies, and candies whose texture is modified with gums, gels, and pastes. In practice these groups overlap: there are crystalline and noncrystalline versions of caramels, hard candies, nougat, sugar work, and so on. Here are brief descriptions of the principal candies made today.

Noncrystalline Candies: Hard Candies, Brittles, Caramel and Taffy, Sugar Work

Hard Candies Hard candies are the simplest noncrystalline candies; they include hard drops, clear mints, butterscotch, bonbons, lollipops, and so on. Hard candy is made by boiling the syrup high enough that the final solid will contain only 1 or 2% moisture, then pouring the syrup onto a surface and cooling it down, kneading in

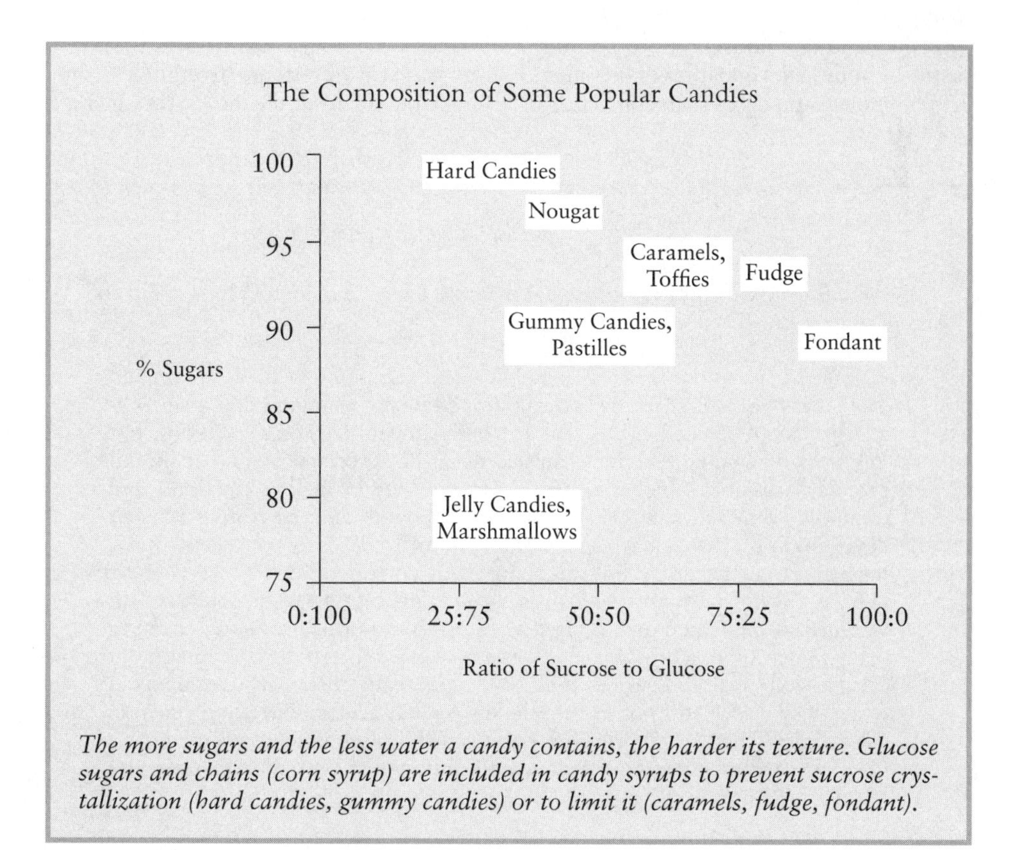

The more sugars and the less water a candy contains, the harder its texture. Glucose sugars and chains (corn syrup) are included in candy syrups to prevent sucrose crystallization (hard candies, gummy candies) or to limit it (caramels, fudge, fondant).

colors and flavors while it's still malleable, and shaping it. The very high sugar concentration makes this syrup liable to form crystals at the slightest excuse, so a substantial proportion of corn syrup is used to prevent this and produce a clear sugar glass. The high cooking temperatures also encourage caramelization and a yellow-brown discoloration, which are not desirable in these candies; they're often manufactured under reduced pressure, which allows them to reach the proper sugar concentration at a lower temperature.

Intentionally Crystalline Hard Candies The development of crystals is considered a defect in many hard candies, and results from too little interfering corn syrup, or the introduction of seed crystals from the pan sides, or too much moisture in the syrup. But some hard candies are intentionally manipulated to form tiny crystals, which give the candy a "short," more crumbly texture. Candy canes and after-dinner mints are common examples of such confections. An opaque but satin- or silk-like sheen results when the cooled but malleable syrup is repeatedly pulled and folded back onto itself. This working incorporates some air bubbles, and these in turn encourage the formation of tiny sucrose crystals. Both bubbles and crystals interrupt the candy structure, giving it a crisp, light quality and making it easier to break between the teeth. (See "Sugar Work" below.)

Cotton Candy Cotton candy or candy floss is a very different kind of hard candy, filaments of sugar glass so fine that they have the consistency of a cotton ball and dissolve away the moment they touch the moist mouth. Cotton candy is made in a special machine that melts the sugar and forces it through tiny spinnerets into the air, where it instantly solidifies into threads. It was introduced at the 1904 World's Fair in St. Louis.

Brittles Brittles are also cooked to a very low moisture content, around 2%, but unlike the other hard candies, they include

Caramel, Caramels, Caramelization

These very similar words mean somewhat different things, and aren't always used consistently.

- *Caramelization* is the cooking of a plain sugar syrup until it turns brown and aromatic. It is similar to the browning or Maillard reactions that give color and aroma to roasted meats, baked goods, and other complex foods, but unlike the browning reactions it proceeds in the absence of amino acids and proteins. It requires higher temperatures than the browning reactions, and produces a different mixture of aromatic compounds and therefore a different flavor (p. 777). Cooks have spoken of "caramelized" or "carmelized" meats for better than a century, but this is not really correct.
- *Caramel* is first of all the brown, sweet, aromatic syrup produced in caramelization, which may be used as coloring and/or flavoring ingredient in many preparations. But cooks use the same word to mean the combination of caramelized sugar and various milk products, ideally cream, which are mixed while the sugar is still hot so that the milk solids are browned and generate color and aroma as well. This kind of caramel is often used as a sauce.
- *Caramels* are solid candies made from a mixture of caramelized sugar and cream.

butter and milk solids, and usually pieces of nuts. They're thus opaque with fat droplets and protein particles, and brown in color thanks to extensive browning reactions between sugars and proteins. Baking soda is often added to brittle syrups after they're cooked, for several reasons: alkaline conditions favor browning reactions, help neutralize some of the acids produced thereby, and the bubbles of carbon dioxide that result from this neutralization become trapped in the candy, giving it a lighter texture. The original French *praline* was a brittle made with almonds. (The modern New Orleans praline is soft and chewy, more like a caramel, and contains New-World pecans instead of almonds.)

Caramels, Toffees, and Taffies Caramels and their relatives are generally noncrystalline candies that contain milkfat and milk solids, usually in the concentrated form of sweetened condensed milk. (Cheap versions are made with milk powder and vegetable shortening.) They are chewy rather than hard, and wonderfully mouthwatering because chewing liberates droplets of butterfat from the sugar mass. Their chewiness comes from a lower cooking temperature and so a higher moisture content than hard candies, a large proportion of corn syrup, and the presence of milk casein proteins. The characteristic caramel flavor develops from the milk ingredients and reactions between these and the syrup sugars during the cooking. In Britain, butter for toffee was often stored to develop some rancidity (from free butyric acid), which produced a desirably stronger dairy flavor in the finished candy. (American chocolate manufacturers have done much the same thing; see box, p. 703). The higher the fat content, the less these candies stick to the teeth.

Caramels are cooked to the lowest temperature of the noncrystalline candies, have the highest moisture content, and are the softest. Toffees and taffies contain less butter and milk solids—taffies sometimes none at all—and are cooked 50°F hotter than caramels, so they're more firm. Taffies are often pulled to produce an aerated, finely crystalline, less dense, less chewy version. Caramels made with dairy products owe some of their flavor to caramelized sugar, but they include flavors from the Maillard reactions. Like the terminology, caramelized sugar and dairy flavors blend easily with each other. This may be in part because one of the important products of sugar caramelization is diacetyl, an aromatic chemical that provides the pronounced buttery aroma of cultured butter (p. 35). Caramel has a rich, complex flavor and consistency, viscous and sticky and creamy all at once, that works well with most sweets and fruits, with coffee and chocolate, and even with salt: the prized caramels of Brittany are made with a notable dose of sea salt.

Sugar Work The most spectacular sugar preparations are those that take advantage of sugar's similarity to glass: its transparency and capacity to be sculpted, blown, and drawn out into countless shapes. "Sugar work," as such preparations are called, goes back at least 500 years. A "nest of silken threads," probably similar to our spun sugar, was made from malt syrup for the Chinese Imperial household before 1600; and in 17th-century Italy, various banquet decorations, including dishes, were made from sugar. In Japan, there is a traditional street entertainment called "sweet candy craft," *amezaiku,* in which the performers sculpt flowers, animals, and other shapes while people watch.

The basic material for sugar work is molten sucrose mixed with a large portion of glucose and fructose to help prevent crystallization. The glucose and fructose may be added in the form of corn syrup, or the pure sugars, or they may be formed from the sucrose itself during the cooking of the syrup with added acid (cream of tartar). The sugar mixture is heated until it reaches 315–330°F/157–166°C, at which point there is practically no water left. Any residual water can cause crystallization and milkiness by making it easier for the sucrose molecules to move around and nest together. At somewhat higher temperatures, the sugar begins to caramelize and turn yellow-brown, which is undesirable for much sugar work but encouraged for spun sugar and sugar cages, which are made by drizzling the hot syrup in threads over a solid form or a wooden rack, where they harden almost instantly. For more elaborate sugar work, the entire sugar mass is cooled to around 130–120°F/55–50°C, a range in which it has a pliable, doughy consistency. Now it can be handled and formed, blown like glass into hollow spheres and other shapes, and kept workable with a heat lamp. Though pastry chefs with seasoned fingertips can sculpt sugar barehanded, many use thin latex gloves in order to avoid transferring moisture and skin oils from their fingers.

One of the more striking forms of sugar work is pulled sugar, which develops a lovely delicate satin-like opacity. The cook pulls a piece of the sugar mixture into a long rope, then folds and twists it onto itself and pulls again. By repeating this action many times, he forms the sugar mixture into many fine, partly crystalline strands separated by columns of air, a combination that becomes a solid fabric of shiny threads.

Crystalline Candies: Rock Candy, Fondant, Fudge, Panned Candies, Lozenges

About the only candy in which large, coarse crystals are valued is rock candy, a vivid demonstration of crystal growth. Simply cook a syrup to the hard ball stage, then pour into a small glass, with a toothpick to serve as a removable foundation, and let it sit for a few days. The resulting crystals can be preserved by washing the encrusted stick briefly under cold water, shaking off the excess, and letting it dry.

Fondant and Fudge Fondant and fudge are the two most common finely crystalline candies, whose nature is to dissolve to a creamy consistency on the tongue. The name *fondant* comes from the French *fondre,* meaning "to melt," and fondant is the base for what are called candy "creams," the flavored, moist, melt-in-the-mouth interiors of filled chocolates and other candies. It also serves as an icing for cakes and pastries; it can be rolled out and molded onto a cake, or warmed or thinned until runny and poured into a thin layer. Fudge is essentially fondant made with added milk, fat, and sometimes chocolate solids (it can also be thought of as a crystallized version of caramel). Penuche is fondant made with brown sugar (some New Orleans pralines are penuche that includes pecans).

Fondant and fudge are made with the help of corn syrup, which favors the production of small crystals. After the syrup has been boiled and then cooled to 130–100°F/54–38°C, the cook beats it continuously for about 15 minutes, until crystallization is complete.

The texture of these candies depends on how much water they're left with. If the syrup has become especially concentrated, the texture will be dry and crumbly, the appearance dull; if it's less cooked or absorbs moisture from the air during cooling and beating, it will be soft, even runny, the appearance glossy thanks to the abundance of syrup between crystals. Small variations in water content—just 1 or 2%—make a noticeable difference. Fudge is more complex than fondant, its syrup carrying milk solids and fat droplets as well as sugar crystals.

Panned Candies These are the modern version of the medieval *dragées*: flavorful nuts or spices coated in sugar. There are two basic ways to coat candies in a pan. In hard panning, the nut or spice or other center is rolled around a hot pan and periodically sprayed with a concentrated sucrose syrup, whose moisture evaporates and leaves behind tightly interlocked, hard layers of crystals, just 0.01–0.02 mm thick. In soft panning, most often applied to jellybeans, the jelly candy is rolled around in a cool pan with a glucose syrup and powdered sugar. Instead of crystallizing, the syrup is absorbed by the powder, and excess moisture is dried off. Soft-panned layers are thicker and less crystalline.

Lozenges Lozenges are among the oldest and simplest of confections—they require no high-temperature cooking. They're made by preparing a binding agent in water—gum tragacanth is standard, though gelatin also works—and then making a "dough" by adding finely ground icing sugar and flavoring. The dough is then rolled out, cut into pieces, and dried. Lozenges have a crumbly texture.

Aerated Candies: Marshmallow, Nougat Candies with a light, chewy texture are made by combining a sugar syrup with an ingredient that forms a stable foam. Egg whites, gelatin, and soy protein are the most common foaming agents. Usually they and interfering agents prevent the syrup from crystallizing, but some aerated candies are made crystalline by combining a fine fondant with the foam.

Marshmallows Marshmallows were first made in France from the gummy root juice of the marsh mallow (*Althaea officinalis*), a weedy relative of the hollyhock; the confection was called *pâte de Guimauve*. The juice was mixed with eggs and sugar and then beaten to a foam. Today, marshmallows are made by combining a viscous protein solution, usually gelatin, with a sugar syrup concentrated to about the caramel stage, and whipping the mixture to incorporate air bubbles. The protein molecules collect in the bubble walls, and this reinforcement, together with the viscosity of the syrup, stabilizes the foam structure. The gelatin accounts for 2–3% of the mixture, and produces a somewhat elastic texture. Marshmallows made with egg whites are lighter and softer.

Nougat Nougat is a traditional sugar candy made in Provence that contains nuts and is aerated with egg-white foam. Italian *torrone* and Spanish *turron* are similar. It's a cross between a meringue and a candy, and is made by preparing a meringue and then streaming hot, concentrated sugar syrup into it while continuing to beat. It can be either soft and chewy or hard and crunchy depending on the degree to which the sugar syrup is cooked and the proportion of sugar syrup to egg white. Honey is often an ingredient.

Flashy Candies: Lightning in the Mouth

Mix together crystals of table sugar and essence of wintergreen and you get something startling: a candy that seems to give off sparks when you eat it! When highly orderly sucrose crystals are fractured between the teeth, the sudden split leaves an imbalance of electrical charge between the two pieces: there are more electrons on one side than the other. The electrons then jump the gap to the more positively charged piece. En route, they collide with nitrogen molecules in the air, which then discharge the sudden jolt of kinetic energy in the form of light energy. The same kinds of electron jumping and colliding produce lightning strikes between electrically charged clouds and the earth. Of course, sugar crystals give off a much weaker glow than true lightning. And much of that glow is in the invisible ultraviolet part of the light spectrum. Here's where the wintergreen plays a role. Its aromatic essence, methyl salicylate, is fluorescent: it absorbs the invisible ultraviolet rays and re-radiates them in the visible part of the light spectrum. It thus amplifies the dimmer sucrose glow to the point that, in a dark room, we can see blue flashes when the candy is crushed.

Chewy Jelly and Paste Candies; Marzipan A number of different candies are made by incorporating a sugar syrup into a solution of starch, gelatin, pectin, or plant gums, and then allowing the mixture to solidify into a dense, chewy mass. In Japan and elsewhere in Asia, sweets are often gelled with the seaweed extract agar (p. 609), which is effective in unusually small amounts (as little as 0.1% of the mix).

Turkish Delight Turkish delight, or *lokum rahat* in Turkish, is one of the most venerable of this kind, having been made in the Middle East and the Balkans for centuries. It is thickened with starch (around 4%), translucent, and traditionally flavored with essence of rose.

Licorice Licorice is usually made with wheat flour and molasses, around 30% and 60% of the mix respectively, with licorice extract around 5%; it's dense and opaque, like its flavor. The licorice is often complemented with anise, and in Scandinavian countries there's a curious pairing of licorice with ammonia—in foods, an aroma usually encountered only in overripe cheeses!

Jelly Beans and Gummy Candies These favorites are made with approximately equal weights of sucrose and corn syrup and a mixture of gelatin and pectin. The gelatin may be between 5 and 15% of the candy weight, and by itself produces an increasingly elastic, even rubbery texture; pectin at around 1% introduces a complex microstructure into the candy, gives a shorter, more crumbly texture and also causes candy tastes and aromas to seem more intense. Gelatin is degraded in high heat, so a concentrated solution is added to the sugar syrup after it has been cooked and mostly cooled. These candies are relatively moist, being about 15% water.

Marzipan Marzipan is essentially a paste of sugar and almonds, has been made in the Middle East and Mediterranean region for centuries, and is especially prized as a sculptural material; it's shaped and colored to resemble fruits and vegetables, animals, people, and many other objects. The solid phase in nut pastes like marzipan is provided by finely granulated sugar and the particles of nut proteins and carbohydrates. It can be made by cooking almonds and

Fizzy and Crackling Candies

Candies that fizz and crackle in the mouth were developed in the 19th century by embedding the equivalent of baking powder in a very low-moisture sugar syrup as it cools and hardens. Remember that baking powder is a mixture of an acid together with alkaline baking soda. When the two components are moistened together in a batter, they react to produce carbon dioxide gas. Similarly, when the dry crystals of citric or malic acid and sodium bicarbonate in a candy are moistened together in the mouth, they react and form bubbles of carbon dioxide that provide the sensation of tartness and prickly foaminess.

A 20th-century industrial twist on this idea produced Pop or Space Rocks, which instantly burst and then disappear in the mouth. A scientist at General Foods found that he could supercharge a concentrated sugar syrup with carbon dioxide gas, then chill it down quickly and under pressure to lock the gas in the solidified candy. When the candy is depressurized much of the gas escapes, but some remains. And when the candy dissolves in the mouth's moisture, the gas bursts out with a startling crackle. Some chefs use these gasified candies as a source of unexpected sensations; they embed them in dishes that are sufficiently dry or cold not to dissolve them prematurely.

syrup together and then cooling and crystallizing the mixture; or ground almonds can be mixed with a premade fondant and powdered sugar. Egg white or gelatin is sometimes added to improve the binding.

Chewing Gum

This quintessentially American confection has ancient roots. Humans have chewed on gums, resins, and latexes secreted by various plants for thousands of years. The Greeks named the resin of a kind of pistachio tree with their word for "to grind the teeth together, to chew": that was mastic (p. 421), whose root also gives us "masticate." Europeans and North Americans chewed the relatively harsh resin of spruce trees; and the Maya chewed chicle, the latex of the sapodilla tree (*Achras sapote*), ten centuries before it was commercialized in New York City. The idea of mixing gum with sugar goes back to the early Arab sugar traders, who used the exudation of certain kinds of acacia, a substance now known as gum arabic. It and gum tragacanth are slightly soluble and eventually dissolve when chewed; they were used in early medicine as carriers that would release drugs slowly. This is still one of the purposes of chewing gum, which are to release a pleasant flavor for some time while giving the jaw muscles something to do and stimulating a cleansing flow of saliva.

Gum in America The history of modern chewing gums begins in 1869, when a New York inventor by the name of Thomas Adams was introduced to chicle from Central and South America. Chicle is a latex, a milky, water-based plant fluid that carries tiny droplets of long, coiled carbon-hydrogen chains. These chains have the property of being elastic: they uncoil and stretch out when pulled, but snap back when released. The best known of these latex substances is rubber. Adams got the idea of using the chicle as a gum base, and patented chicle gum in 1871. With sugar and sassafras or licorice flavorings, it quickly caught on. By 1900, entrepreneurs with such names as Fleer and Wrigley had developed gumballs and peppermint and spearmint flavors, and in 1928 a Fleer employee perfected bubble gum by developing a very elastic latex mixture from longer hydrocarbon polymers.

Modern Synthetic Gums Today, chewing gum is made mostly of synthetic polymers, especially styrene-butadiene rubber—also found in auto tires—and polyvinyl acetate—in adhesives and paints—though some brands still contain chicle or jelutong, a natural latex from the Far East. The crude gum base is first filtered, dried, and then cooked in water until syrupy. Powdered sugar and corn syrup are mixed in, then flavorings and softeners—vegetable oil derivatives that make the gum easier to chew—and the material is cooled, kneaded to an even, smooth texture, cut, rolled thin, and cut again into strips, and packaged. The final product is about 60% sugar, 20% corn syrup, and 20% gum materials. Sugar-free gums are made using sugar alcohols and intensive sweeteners (p. 659).

Candy Storage and Spoilage

Because of their generally low water content and concentrated sugars, which draw moisture out of living cells, candies are seldom spoiled by the growth of bacteria or molds. Their flavor can be degraded, however, by the oxidation and consequent rancidity of added fats, whether in milk solids or butter. This process can be slowed down by refrigeration or freezing, but cold storage encourages another problem called "sugar bloom." Changes in temperature can cause moisture from the air to condense on the candy surface, and some sugar will dissolve into the liquid. When the moisture evaporates again or is drawn deeper into the candy, the surface sugar crystallizes into a rough, white coating. Airtight wrapping will prevent sugar bloom.

CHOCOLATE

Chocolate is one of our most remarkable foods. It is made from the astringent, bitter, and otherwise bland seeds of a tropical tree, yet its flavor is exceptionally rich, complex, and versatile, the product of both fermentation and roasting. Its consistency is like no other food's: hard and dry at room temperature, melting and creamy in the warmth of the mouth. It can be sculpted into almost any shape, and its surface can be made as glossy as glass. And chocolate is one of the few examples of a food whose full potential was first revealed in industrial manufacturing. The chocolate that we know and love, a dense, smooth, sweet solid, has existed for only a tiny fraction of chocolate's full history.

The History of Chocolate

An Exotic Drink The story of chocolate begins in the New World with the cacao tree, which probably evolved in the river valleys of equatorial South America. The tree bears large, tough seed pods that also contain a sweet, moist pulp, and early peoples may have carried the pods into Central America and southern Mexico as a portable source of energy and moisture. It appears that the first people to cultivate the tree were the Olmecs of the southern Gulf coast of Mexico. They in turn introduced it sometime before 600 BCE to the Maya, who produced it in the tropical Yucatan peninsula and Central America, and traded it to the Aztecs in the cool and arid north. The Aztecs roasted and ground cacao seeds and made them into a drink that was served in religious ceremonies and associated with human blood. The seeds were valuable enough to serve as a form of currency. The first Europeans to see the cacao bean were probably the crew of Columbus's fourth voyage in 1502, who brought some back to Spain. In 1519 one of Cortez's lieutenants, Bernal Diaz del Castillo, saw the Aztec emperor Montezuma at table and in passing described the prepared drink:

> Fruit of all the kinds that the country produced were laid before him; he ate very little, but from time to time a liquor prepared from cocoa, and of an aphrodisiac nature, as we were told, was presented to him in golden cups. . . . I observed a number of jars, above fifty, brought in, filled with foaming chocolate, of which he took some. . . .

One of the first detailed accounts of the original chocolate comes from the *History of the New World* (1564) by the Milanese Girolamo Benzoni, who traveled in Central America. He remarked that the region had made two unique contributions to the world: Indian fowls," or turkeys, and "cavacate," or the cacao bean.

> **Food Words:** ***Cocoa, Chocolate***
>
> The word *cocoa* comes via the Spanish *cacao,* which in turn came via the Maya and Aztec from a probable Olmec word *kakawa* coined 3,000 years ago. *Chocolate* has a more complicated history. The Aztec (Nahuatl) word for cocoa-water was *cacahuatl,* but the early Spanish coined *chocolate* for themselves. According to historians Michael and Sophie Coe, they may have done so to distinguish the hot Maya version that they preferred from the cold Aztec one—in the Yucatan, "hot" was *chocol*; the Aztec for "water" *atl.*

They pick out the kernels and lay them on the mats to dry; then when they wish for the beverage, they roast them in an earthen pan over the fire, and grind them with the stones which they use for preparing bread. Finally, they put the paste into cups . . . and mixing it gradually with water, sometimes adding a little of their spice, they drink it, though it seems more suited for pigs than men.

. . . The flavor is somewhat bitter, but it satisfies and refreshes the body without intoxicating: the Indians esteem it above everything, wherever they are accustomed to it.

Benzoni and other visitors reported that the Maya and Aztecs flavored their chocolate drinks with a number of different ingredients, including aromatic flowers, vanilla, chilli, wild honey, and achiote (p. 423). The Europeans then began to add their own flavorings, among them sugar, cinnamon, cloves, anise, almonds, hazelnuts, vanilla, orange-flower water, and musk. According to the English Jesuit Thomas Gage, they dried the cocoa beans and spices, ground them up and mixed them together, and heated them to melt the cocoa butter and form a paste. Then they scraped the paste onto a large leaf or piece of paper, allowed it to solidify, and then peeled it off as a large tablet. According to Gage, there were several ways of preparing chocolate, both hot and cold.

The one most used in Mexico is to take it hot with *atole* [a maize gruel], dissolving a tablet in hot water, and then stirring and beating it in the cup with a molinet, and when it is well stirred to a scum or froth, then to fill the cup with hot *atole,* and so drink it sup by sup.

The first European "factories" for making the spiced chocolate paste were built in Spain around 1580, and within 70 years chocolate had found its way into Italy, France, and England. These countries purged the drink of most added flavorings except sugar and vanilla. At first, vendors of lemonade sold it in Paris; coffeehouses—themselves an innovation—served it London. But by the late 17th century, chocolate houses were thriving in London as a kind of specialty coffeehouse. The idea of making hot chocolate from milk seems to have arisen in these places.

Early Chocolate Confections For a couple of centuries, Europe knew chocolate almost exclusively as a beverage. The use of the cacao bean in confectionery was quite limited. The Englishman Henry Stubbe noted in his treatise on chocolate, *The Indian Nectar* (1662), that in Spain and the Spanish colonies "there is another way of taking it made into *Lozenges,* or shaped into Almonds," and that people were aware of what we now know to be the effects of the caffeine in chocolate: "The Cacao-nut being made into Confects, being eaten at night, makes Men to wake all night-long: and is therefore good for Souldiers, that are upon the Guard." Cookbooks of the 18th century generally included a handful of recipes that call for chocolate, among them dragées, marzipans, and biscuits, creams and ices and mousses. There are some remarkable Italian recipes for lasagna sauced with almonds, walnuts, anchovies, and chocolate, for liver with chocolate, and polenta with chocolate. And in the 18th-century French *Encyclopédie,* we find that *chocolat* was commonly sold as a half-cocoa, half-sugar cake flavored with some vanilla and cinnamon, and was not so much a delightful confection as an emergency meal—perhaps the first instant breakfast!

When one is in a hurry to leave one's lodgings, or when during travel one does not have the time to make it into a drink, one can eat a tablet of one ounce and drink a cup [of water] on top of that, and let the stomach churn to dissolve this impromptu breakfast.

Even in the middle of the 19th century, the English compendium *Gunter's Modern*

Confectioner devoted only four pages out of 220 to chocolate recipes.

Dutch and English Innovations: Cocoa Powder and Eating Chocolates The main reason for this lack of interest in solid chocolate was probably the coarse, crumbly texture of the chocolate paste. The suave confections that are so popular today were made possible by several innovations, the first of which came in 1828. Conrad van Houten, whose family ran a chocolate business in Amsterdam, was trying to find a way to make chocolate less oily so that the drink would be less heavy and filling. The weight of the cacao bean is better than half fat, or "cocoa butter." Van Houten developed a screw press that removed most of the cocoa butter from the ground bean—in itself, not a novelty—and then sold the defatted cocoa powder, which carries nearly all the flavor, for making hot chocolate. Cocoa powder was a long-lasting success, though recently there has been a revival of interest in richer versions of hot chocolate full of cocoa butter.

At first, the pure cocoa butter extracted by Van Houten's screw press was a mere by-product. But it turned out to be the key to the development of modern chocolate candy. This was cocoa butter that could be *added* to a paste of ordinary ground cocoa beans and sugar to provide a rich, melting matrix for the dry particles, and make the paste seem less pasty. The first solid "eating chocolate" was introduced by the English firm of Fry and Sons in 1847, and soon inspired many imitations throughout Europe and the United States.

Swiss Innovations: Milk Chocolate and Refined Texture In 1917, Alice Bradley's *Candy Cook Book* devoted an entire chapter to "assorted Chocolates," and noted that"more than one hundred different chocolates may be found in the price lists of some manufacturers." The South American bean had come of age as a major ingredient in confectionery.

Two technical developments had helped expand chocolate's appeal. In 1876, a Swiss confectioner named Daniel Peter used the new dried milk powder produced by his countryman Henri Nestlé to make the first solid milk chocolate. Not only do milk flavors blend well with chocolate, but the milk powder dilutes the strong chocolate flavor, and milk proteins reduce its astringency and make the taste milder. Today, most chocolate is now consumed in the form of milk chocolate. Then in 1878, a Swiss manufacturer named Rudolphe Lindt invented the conche, a machine which ground cacao beans, sugar, and milk powder slowly for hours and even days, and developed a much finer consistency than had been possible before. This is the consistency that we now take for granted in even the most ordinary chocolates.

Having contributed so much to the evolution of modern chocolate, the Swiss are understandably the world's champion chocolate eaters, and have been so for a long time. At about an ounce/30 gm per day, Switzerland's per capita consumption is nearly double that of the United States.

Making Chocolate

The transformation of the fresh cacao bean into a finished chocolate is an intriguing collaboration between the tremendous potential of the natural world and human ingenuity at finding nourishment and pleasure in the most unpromising materials. Right out of the pod, the bean is astringent, bitter, and essentially aroma-less. Cacao farmers and chocolate manufacturers develop its potential in several distinct processing steps:

- Farmers ferment the mass of beans and pulp in order to generate the precursors to chocolate flavor.
- Manufacturers roast the fermented beans to transform flavor precursors into flavors.
- Manufacturers grind the beans, add sugar, and then physically work the mixture to refine its flavor and create a silken texture.

The Cacao Bean The cacao tree, named *Theobroma cacao* by Linnaeus—*theobroma* is Greek for "food of the gods"—is a broad-leaved evergreen that grows between 20° north and south of the equator, and reaches about 20 feet/7 m in height. It produces fruits in the form of fibrous pods from 6 to 10 inches/15–25 cm long, 3 or 4 inches/7.5–10 cm in diameter, and containing 20 to 40 seeds, or "beans," each about an inch/2.5 cm long, embedded in a sweet-tart pulp.

Varieties There are a number of different cacao varieties that fall into three botanical groups: the Criollos, Forasteros, and Trinitarios. Criollo trees produce relatively mild beans with some of the finest, most delicate flavors, reminiscent of flowers and tea. Unfortunately, they are also disease-prone, low-yielding trees, and so provide less than 5% of the world crop. High-yielding, robust Forasteros provide most of the world's cacao crop in the form of full-flavored "bulk" beans. Trinitarios are hybrids of a Criollo and Forastero, and have intermediate qualities.

West Africa (Ivory Coast and Ghana) now accounts for more than half of world cacao production, and Indonesia also outproduces Brazil, the largest producer in cacao's original homeland.

Storage and Defensive Cells Cacao beans consist mainly of the embryo's storage leaves, or cotyledons (p. 453), and contain two distinct groups of cells. Around 80% of the cells are storage depots of protein and of fat, or cocoa butter, nutrients that will feed the seedling as it germinates and develops on the shady floor of the tropical forest. The other 20% are defensive cells meant to deter the many forest animals and microbes from feasting on the seed and its nutrients. These cells are visible in the cotyledons as purplish dots, and contain astringent phenolic compounds, their chemical relatives the anthocyanin pigments, and two bitter alkaloids, theobromine and caffeine. The beans are moist, around 65% water. The composition of the dried fermented beans is shown in the box on p. 698.

Fermentation and Drying The first important step in the development of chocolate flavor is the least controlled and predictable. Fermentation takes place where cacao is grown, on thousands of small farms and larger plantations, and may be done carefully or casually or not at all, depending on the resources and skill of the farmer. The quality of cacao beans thus varies tremendously, from unfermented to badly overfermented and even moldy. The

Cacao pods contain many large seeds that are covered with a sweet pulp. The seeds consist mainly of the embryo's tightly folded food-storage cotyledons, which are speckled with purple defensive cells rich in alkaloids and astringent phenolic compounds.

first challenge for the chocolate manufacturer is to find good-quality, fully fermented beans.

Soon after the cacao pods are harvested, workers break them open and pile the beans and sugary pulp together at the ambient tropical temperature. Microbes immediately begin growing on the sugars and other nutrients in the pulp. A proper fermentation lasts from two to eight days, and generally has three phases. In the first, yeasts predominate, converting sugars to alcohol and metabolizing some of the pulp acids. As the yeasts use up the oxygen trapped in the pile, they are succeeded by lactic acid bacteria, many of which are the same species found in fermented dairy products and vegetables. When workers turn the mass of beans and pulp to aerate it, the lactic bacteria are succeeded by acetic acid bacteria, the makers of vinegar, which consume the yeasts' alcohol and convert it into acetic acid.

Fermentation Transforms the Beans Cacao fermentation is a fermentation of the pulp, not the beans, but it transforms the beans as well. The acetic acid produced by the vinegar bacteria penetrates into the beans and etches holes in cells as it does so, spilling the contents of the cells together and allowing them to react with each other. The astringent phenolic substances mix with proteins, oxygen, and each other, and form complexes that are much less astringent. Most important, the beans' own digestive enzymes mix with the storage proteins and sucrose sugar and break them down into their building blocks—amino acids and simple sugars—which are much more reactive than their parent molecules, and will produce more aromatic molecules during the roasting process. Finally, the perforated beans soak up some flavor molecules from the fermenting pulp, including sugars and acids, fruity and flowery and winey notes. So a properly conducted fermentation converts the astringent but bland beans into vessels laden with desirable flavors and flavor precursors.

Drying Once fermentation is complete, the cacao farmers dry the beans, often just by spreading them out on a flat surface in the sun. Drying can take several days, and if not done carefully can allow undesirable bacteria and molds to grow both on and within the beans and taint them with undesirable flavors.

Once dried to about 7% moisture, the beans are resistant to further microbial spoilage. They're then cleaned, bagged, and shipped to manufacturers all over the world.

Roasting Dried fermented cacao beans are less astringent and more flavorful than unfermented beans, but their flavor is still unbalanced and undeveloped, and often dominated by vinegary acetic acid. After selecting, sorting, and blending the dried beans, the chocolate manufacturer roasts

The Composition of Dried Fermented Cacao Beans

	% by weight		% by weight
Water	5	Sugars	1
Cacao butter	54	Phenolic compounds	6
Protein and amino acids	12	Minerals	3
Starch	6	Theobromine	1.2
Fiber	11	Caffeine	0.2

them to develop their flavor. The time and temperature vary according to whether the beans are to be roasted whole, in their thin shell, or as the cracked inner kernels, the *nibs*, or as nibs that have been ground into small, quickly heated particles. Whole beans take 30–60 minutes at 250–320°F/120–160°C. This is a much gentler treatment than coffee beans require, thanks to the abundance of reactive amino acids and sugars that readily participate in Maillard browning to generate flavor (p. 778). In fact, gentle roasting helps preserve some of the flavors that are intrinsic to the beans or developed during fermentation.

Grinding and Refining After roasting, the beans are cracked open and the nibs are separated from the shells. The nibs are then passed between several sets of steel rollers, and are transformed from solid chunks of plant tissue into a thick, dark fluid called cocoa liquor. This grinding stage has two purposes: to break the bean cells open and release their stores of cocoa butter; and to break the cells down into particles too small for the tongue to detect as separate, gritty grains. Because the nibs are around 55% cocoa butter, this fat becomes the continuous phase, and the solid fragments of the cells—mainly protein, fiber, and starch—are suspended in the fat. The final grinding, or refining, brings the particle size down to 0.02–0.03 mm. Swiss and German chocolates have traditionally been ground smoother than English and American.

Further treatment of the cocoa liquor varies according to the manufacturer's needs. To make cocoa powder and cocoa butter, the liquor is pressed through a fine filter that retains the cocoa particles while allowing the butter to flow through. The compacted cake of cocoa particles is then made into cocoa powder (below), while the butter becomes an important ingredient in all kinds of manufactured chocolate.

Conching Pure cocoa liquor has a concentrated chocolate taste, and may be hardened and packaged as is for use in baked goods. But its flavor is relatively rough, bitter, and astringent and acidic. To make it into something not only edible but delicious, manufacturers add a few other ingredients: sugar for dark chocolate, sugar and dry milk solids for milk chocolate, some vanilla (the whole bean, or an extract, or artificial vanillin), and a supplement of pure cocoa butter. And they subject the mixture to an extended agitation called *conching*, a process named after the shell-like shape of the first machines. Conches rub and smear the mixture of cocoa liquor, sugar, and milk solids against a solid surface. The combination of friction and supplemental heat raises the temperature of the mass to 115–175°F/45–80°C (milk chocolate is kept at 110–135°F/43–57°C). Depending on the machine and manufacturer, conching may last for 8 to 36 hours.

Refining Texture and Flavor The original conche was a mechanized version of the Mayan stone grinding slab: a heavy granite roller moved back and forth over a granite bed, both mixing the ingredients together and grinding the still somewhat coarse particles to a finer size. Today the various solids are ground to the proper dimensions before conching, which now serves two main functions. First, it breaks up small aggregates of the solid particles, separates them from each other, and coats all of them evenly with cocoa butter, so that when the finished chocolate melts, it flows smoothly. Second, conching greatly improves the flavor of the chocolate, not by heightening it, but by mellowing it. The aeration and moderate heat causes as much as 80% of the volatile aromatic compounds (and excess moisture) to evaporate out of the chocolate liquor. Fortunately, many of these are undesirable volatiles, including various acids and aldehydes; acidity steadily declines during conching. At the same time, a number of desirable volatiles are augmented by the heat and mixing, notably those with roasted, caramel, and malty aromas (pyrazines, furaneol, maltol).

Both cocoa butter and a small amount

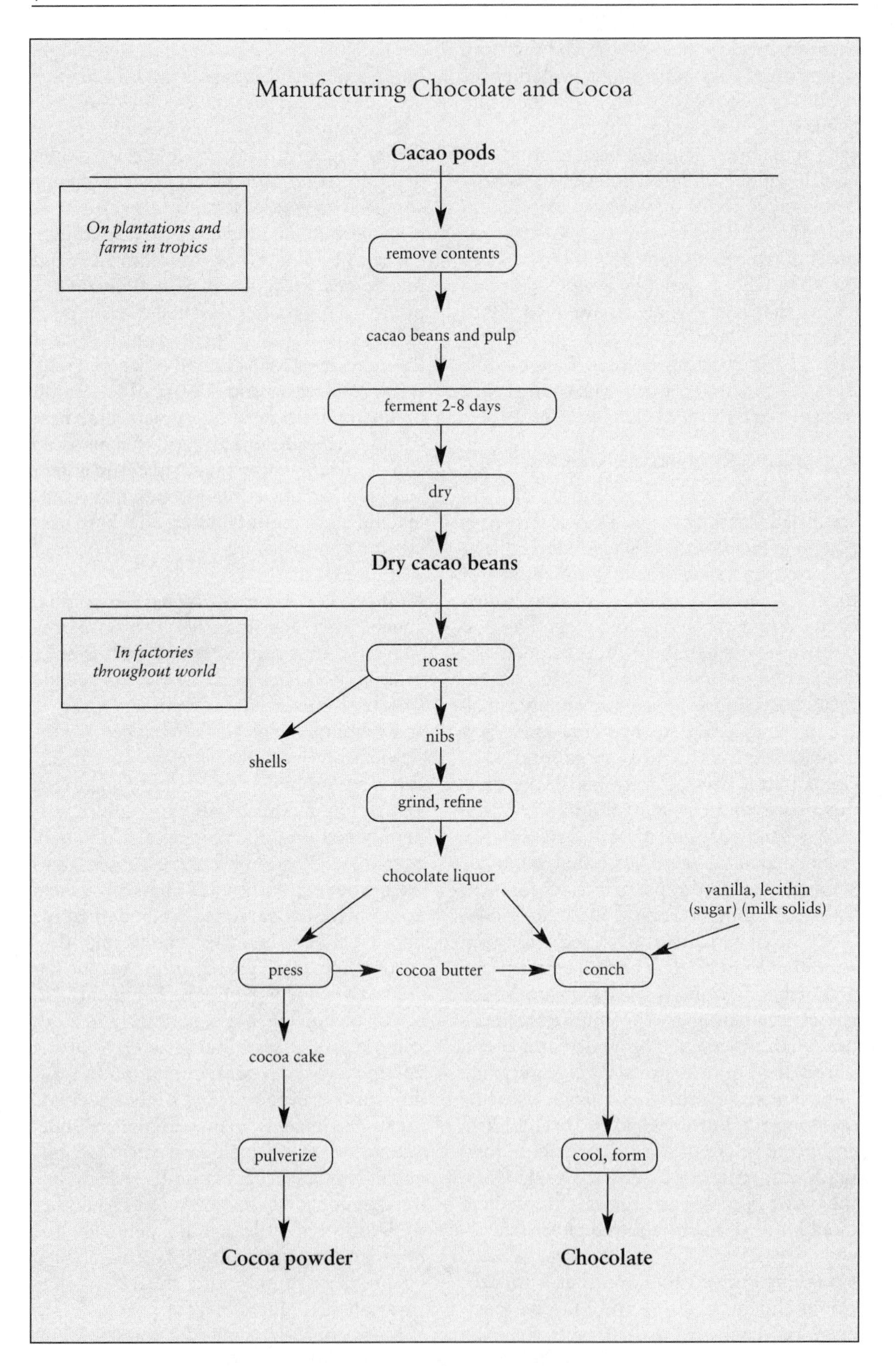
Manufacturing Chocolate and Cocoa
Cacao pods
On plantations and farms in tropics
remove contents
cacao beans and pulp
ferment 2-8 days
dry
Dry cacao beans
In factories throughout world
roast
shells
nibs
grind, refine
chocolate liquor
vanilla, lecithin (sugar) (milk solids)
press
cocoa butter
conch
cocoa cake
pulverize
cool, form
Cocoa powder
Chocolate

of the emulsifier lecithin (p. 802) are added to the chocolate mass toward the end of conching. The additional cocoa butter is necessary to provide sufficient lubrication for all the added sugar particles to make the mixture creamily fluid rather than pasty when it melts. The higher the ratio of sugar to ground nibs, the more added cocoa butter is required. Lecithin, whose use in chocolate dates to the 1930s, coats the sugar particles with the fat-like ends of its molecules and helps reduce the amount of cocoa butter needed to lubricate the particles; one part of lecithin replaces 10 parts of butter. It typically makes up 0.3–0.5% of chocolate weight.

Cooling and Solidifying After conching, dark chocolate is essentially a warm fluid mass of cocoa butter that contains suspended particles of the original cacao beans and of sugar. Milk chocolate also contains butterfat, milk proteins, and lactose, and proportionally less cacao bean solids.

The last step in manufacturing chocolate is to cool the fluid chocolate to room temperature and form the familiar solid bars. It turns out that this transition from fluid to solid is a tricky one. To obtain stable cocoa butter crystals and a glossy, snappy chocolate, manufacturers carefully cool and then rewarm the liquid chocolate to particular temperatures before portioning it into molds, where it finally cools to room temperature and solidifies.

Cooks often melt manufactured chocolate in order to give it a special shape or to coat other foods. If they want it to resolidify with its original gloss and snap, then they must repeat in the kitchen this cycle of warming up and cooling down, or *tempering* (p. 709).

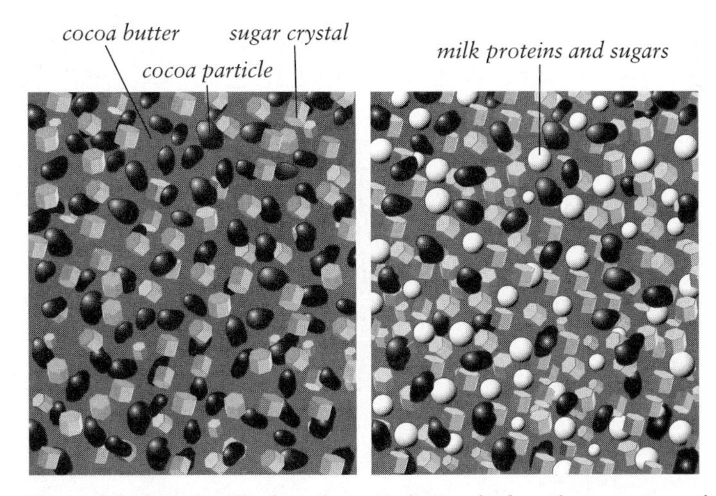

The composition of dark and milk chocolates. Left: *Dark chocolate consists of cacao-bean particles and sugar crystals embedded in a base of cocoa butter.* Right: *In milk chocolate, a significant proportion of cacao-bean particles is replaced with particles of dried milk protein and sugar.*

OPPOSITE: *Making chocolate. As is true of cane sugar, chocolate is processed in two stages, the first in the tropical cacao-growing countries and the second in manufacturing plants throughout the world.*

The Special Qualities of Chocolate

Consistency and Appearance: The Creations of Cocoa Butter The remarkable appearance and consistency of chocolate are a direct expression of the physical qualities of cocoa butter, the part of chocolate that surrounds the solid particles of cacao bean and holds them together. When carefully prepared, chocolate has a silken or glassy surface, is hard and not greasy at room temperature, breaks with a delightful snap, yet melts to a smooth creaminess in the mouth. These are very unlike the qualities of any other food, and are a consequence of the structure of cacao fat molecules, which are mostly saturated and unusually regular (most of them are constructed from just three kinds of fatty acids). This structure means that the fat molecules are capable of forming a dense network of compact, stable crystals, with little liquid fat left over to ooze out between the crystals.

However, this special network only develops when the fat crystallization is carefully controlled. Cocoa butter can solidify into six different kinds of fat crystals! Only two are stable kinds that produce a glossy, dry, hard chocolate; the other four are unstable kinds that produce a looser, less organized network, with more liquid fat, and crystals whose fat molecules readily detach and ooze away. When chocolate melts and then resolidifies in an uncontrolled way—for example, when it's temporarily left too close to a hot stove, or in a hot car—it's the unstable crystals that predominate, and they produce a greasy, soft, mottled chocolate. To rescue its original consistency, such chocolate must be tempered.

Chocolate Flavor Chocolate has one of the richest and most complex flavors of any food. In addition to its slight acidity, pronounced bitterness and astringency, and the sweetness of its added sugar, chemists have detected more than 600 different kinds of volatile molecules in chocolate. While a handful of these may account for the basic roasted quality, many others contribute to its depth and wide range. The

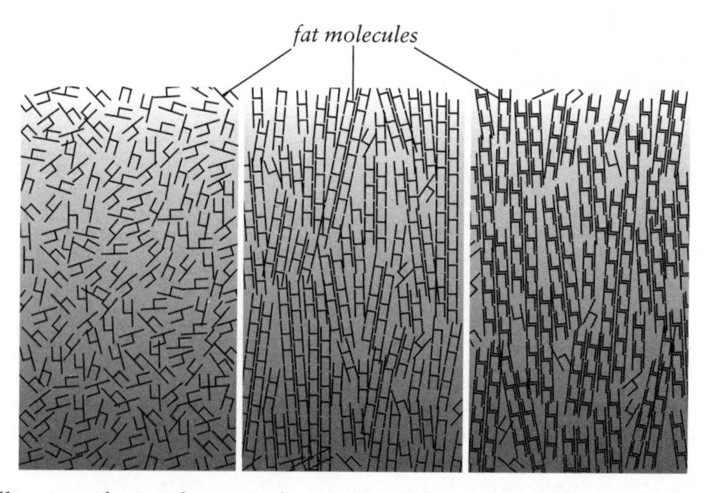

The crystallization of cocoa butter. Left: *In melted chocolate, the fat molecules (p. 798) of cocoa butter are in constant random motion.* Center: *When chocolate cools in an uncontrolled way, the fat molecules form loosely packed, unstable crystals, and the chocolate is soft and greasy.* Right: *When chocolate is carefully cooled, its fat molecules form tightly packed, stable crystals, and the chocolate is snappy and dry.*

richness of chocolate flavor arises from two factors. One is the cacao bean's intrinsic flavor potential, its combination of sugars and proteins, and the enzymes that break these down into the building blocks of flavor. The second factor is the complexity of chocolate's preparation, which combines the chemical creativity of microbes and of high heat.

Among the flavors that an attentive taster can detect in chocolate are these:

- From the bean itself, astringency and bitterness (phenolic compounds, theobromine)
- From the fermented pulp, the flavors of fruits and wine and sherry and vinegar (acids, esters; alcohols; acetaldehyde; acetic acid)
- From the self-digested bean, almond and dairy and flowery notes (benzaldehyde; diacetyl and methyl ketones; linalool)
- From roasting and the browning reactions, roasted, nutty, sweet, earthy, flowery, and spicy notes (pyrazines and thiazoles; phenyls; phenylalkanals; dienals), as well as a more pronounced bitterness (diketopiperazines)
- From added sugar and vanilla, sweetness and the warm character of the spice
- From added milk solids, caramel and butterscotch and cooked-milk and cheese notes

Chocolate made from poorly fermented or handled beans can have a variety of unpleasant aromas, among them rubbery, burned, smoky, hammy, fishy, moldy, cardboard, and rancid notes.

Some confectioners add a small amount of salt to their products, especially milk chocolates. Saltiness is the one basic taste sensation missing from simply sweetened chocolate, and adding it is said to give the overall flavor a certain bite and clarity.

The Kinds of Chocolate

Manufacturers produce a wide range of different chocolates, some meant for eating as is, some meant for cooking or confectionery, some for all three. They fall into several general categories.

- Mass-produced, inexpensive chocolates are made from ordinary beans that are processed in largely automated plants, and contain the minimum amount of cocoa solids and

The Different Flavors of Milk Chocolate

The milk chocolates made in Europe, England, and the United States have traditionally had distinct flavors. In continental Europe, where it was invented, milk chocolate is made using dried whole milk powder, which has a relatively fresh milk flavor. In England, the preference has been to mix liquid milk with sugar, concentrate the mixture to 90% solids, mix it with chocolate liquor, and finish drying it into a material called "chocolate crumb." The milk proteins and sugars undergo browning reactions during the concentration and drying and produce a special cooked-milk, caramelized flavor that isn't obtained by ordinary drying. And in the United States, large manufacturers have long encouraged their milk fat to undergo some breakdown by fat-digesting enzymes. This breakdown develops a slight note of rancidity, whose cheesy, animal overtones blend well in their own way with chocolate flavor and make a positive contribution to the complexity of flavor.

cocoa butter and the maximum amounts of sugar and milk solids. Their flavor is mild and unremarkable.

- "Fine" expensive chocolates are made from beans selected for their excellent flavor potential, often processed in small batches to optimize flavor development, and contain far more than the minimum amount of cocoa solids and cocoa butter. Their flavor is stronger and more complex.
- Dark chocolate contains cocoa solids, cocoa butter, and sugar, but no milk solids. It is manufactured in a range of compositions, from sugarless *bitter*, to *bittersweet*, to *sweet*. Some manufacturers now label their premium chocolates with their content of cocoa beans: "70% chocolate" is 70% by weight cocoa butter and cocoa solids, and about 30% sugar; "62% chocolate" is about 38% sugar (there are also small amounts of lecithin and vanilla). The higher the proportion of cocoa solids, the more intense the chocolate flavor, including its bitterness and astringency. Strong chocolates deliver more flavor to cream, egg, and flour mixtures, whose proteins bind to the phenolic substances and reduce the apparent astringency.
- Milk chocolate is the most popular form of chocolate, and the mildest. It contains milk solids and a large proportion of sugar, which together usually outweigh the combination of cocoa solids and cocoa butter. Thanks to its relatively low cocoa butter content, milk chocolate tends to be softer and less snappy than bittersweet chocolate.
- Couverture chocolate (from the French for "to cover") is dark or milk chocolate formulated to flow easily when melted, and therefore to work well for forming thin, delicate chocolate coatings. This means adding more cocoa butter than usual to pro-

The Composition of Some Kinds of Chocolate

Chocolates vary widely in composition, especially among "bittersweet" and "semisweet" versions. The figures below, which are given in percent of the chocolate weight, are very approximate, but useful for broad comparisons.

	Minimum cocoa solids + added cocoa butter, U.S.	Cocoa liquor	Added cocoa butter	Sugar	Milk solids	Total fat	Total carbo-hydrates	Protein
Unsweetened		99	0	0	0	53	30	13
Bittersweet/ semisweet	35	70–35	0–15	30–50	0	25–38	45–65	4–6
Sweet (dark)	15	15	20	60	0	32	72	2
Milk	10	10	20	50	15	30	60	8
Unsweetened cocoa powder						20	40	15

vide plenty of room for the cocoa and sugar particles to move past each other. Most couvertures are 31–38% fat.

- “White chocolate” is chocolate-less chocolate: it contains no cocoa particles whatsoever, and therefore has little or no chocolate flavor. White chocolate was invented around 1930, and is a mixture of purified, usually deodorized cocoa butter, milk solids, and sugar. It does offer a valuable decorative contrast to ordinary chocolate.

Some manufacturers are now packaging nibs, or small pieces of the roasted beans, which offer crunchy particles of intense flavor. Whole roasted beans can sometimes be found in Latin markets.

Storing Chocolate; Fat Bloom The best storage temperature for chocolate is a constant 60–65°F/15–18°C, without fluctuations that would encourage the melting and recrystallization of the cocoa butter fats. Sometimes stored chocolate will develop a white, powdery-looking coating on its surface. This “fat bloom” is cocoa butter that has melted out of unstable crystals, migrated to the surface, and formed new crystals there. Fat bloom is normally prevented by proper tempering of the chocolate in the first place. Its development can be slowed down by the addition to the melted chocolate of some clarified butter, which makes the mix of fats more random and so retards the formation of crystals.

Thanks to its abundant antioxidant molecules and chemically stable saturated fats, chocolate has a remarkably long shelf life. It keeps for many months at room temperature. White chocolate, which lacks the antioxidants in the cocoa solids, has a room-temperature shelf life of only a few weeks; after that, or sooner if it’s exposed to bright light, its fats are damaged and it develops a stale, rancid flavor.

Cocoa Powder Manufacturers produce cocoa powder from the cakes of roasted cocoa bean particles left behind when they extract cocoa butter (p. 699). The particles remain coated with a thin layer of cocoa butter; the fat content of the powder ranges from 8 to 26%. The solid particles of the cacao bean are the source of chocolate’s flavor and color. Cocoa is therefore the most concentrated version of chocolate, and a valuable ingredient in its own right. Natural cocoa powder has a strong chocolate taste and pronounced astringency and bitterness. It’s also distinctly acid, with a pH around 5.

“Dutched” or Alkalized Cocoa In Europe and sometimes the United States, cocoa powder is produced from cocoa beans that have been treated with an alkaline substance, potassium carbonate. This treatment, sometimes called “dutching” because its inventor was the Dutch chocolate pioneer Conrad van Houten, raises the cocoa pH to a neutral 7 or alkaline 8. The application of an alkaline material to the beans either before or after roasting has a strong influence on their general chemical com-

Chocolate Cools the Mouth

Well-made chocolate has an unusual and refreshing characteristic for such a rich food: as it melts, it cools the mouth. This happens because its stable fat crystals melt in a very narrow temperature range, and just below body temperature. The phase change from solid to liquid absorbs much of the mouth’s heat energy and leaves little to raise the temperature of the chocolate, which therefore feels persistently cool.

position. In addition to adding a distinctly alkaline taste (like that of baking soda), alkaline treatment reduces the levels of roasty, caramel-like molecules (pyrazines, thiazoles, pyrones, furaneol) and of astringent, bitter phenolics, which now bond to each other to form flavorless dark pigments. The result is a cocoa powder with a milder flavor and darker color. Dutched cocoa can be produced in shades running from light brown to nearly black; the darker the color, the milder the flavor.

Cocoas in Baking It's important for bakers to be aware of the difference between "natural" and alkalized cocoa powders. Some recipes rely on acidic natural cocoa to react with baking soda and generate leavening carbon dioxide. If the same recipe is made with an alkalized cocoa, the reaction won't take place, no carbon dioxide will be generated, and the taste will be alkaline and soapy.

Instant Cocoa So-called "instant" cocoas for hot chocolate include lecithin, an emulsifier that helps separate the particles so that they mix readily with water. Sugar is frequently added to instant cocoa mix and may account for up to 70% of its weight.

Chocolate and Cocoa as Ingredients

Chocolate and cocoa are versatile ingredients. They're incorporated into many mixtures of ingredients, and not just sweets; savory Mexican mole sauces and some European meat stews and sauces borrow depth and complexity from them. Chocolate and cocoa provide flavor, richness, and structure-building capacity; their dry particles contain both starch and protein, absorb moisture, and contribute thickness and solidity to baked goods, soufflés, fillings and icings. Flourless cakes can be made with chocolate or cocoa as the starchy and fatty ingredients, eggs as the moistening and setting agent. In a chocolate mousse, the foam structure provided by the whipped eggs is reinforced by both the dry particles and the gradually crystallizing cocoa butter.

Of course, chocolate can be presented as is, part of a pastry construction for example, or melted over a preparation and then hardened to provide a coating. It's when we melt and cool it for coating or molding that it requires the most care (p. 708). Otherwise, keeping in mind a few facts about chocolate will prevent most problems.

Working with Chocolate Dark chocolate is a fully cooked, fully developed ingredient in its own right, robust and forgiving. Remember that it has been roasted and then heated again to fairly high temperatures in the conche, and that it's a relatively simple physical mixture of cocoa and sugar particles in fat. The most that a cook needs to do to it is melt it to perhaps 120°F/50°C, but it can be heated to 200°F/93°C and then some without suffering disastrous effects. It won't separate, and it won't burn unless it's left over direct stovetop heat or in a microwave oven without stirring. It can be melted and solidified repeatedly if necessary.

Because they contain more milk solids than they do cocoa solids, milk chocolate and white "chocolate" are less robust than dark chocolate and are best melted gently.

Melting Chocolate Chocolate can be successfully melted in several different ways: quickly, over direct stove heat, with care and constant stirring to avoid burning; more slowly, but with less attention; in a bowl set over a pan of hot water, from 100°F/38°C to the simmer (the hotter the water, the faster it melts); in the microwave oven, with frequent interruptions for stirring and checking the temperature. Because chocolate is a poor heat conductor, it's best to chop it into small pieces or process it into crumbs to speed its melting, or its blending with hot ingredients.

Chocolate and Moisture The one vulnerable aspect of chocolate is its extreme dryness, and the vast number of tiny sugar and cacao particles whose surfaces attract moisture. If a small amount of water is stirred into molten chocolate, the chocolate will seize up into stiff paste. It seems perverse that adding liquid to a liquid produces a solid: but the small amount of water acts as a kind of glue, wetting the many millions of sugar and cocoa particles just enough to make patches of syrup that stick the particles together and separate them from the liquid cocoa butter. It's important, then, either to keep chocolate completely dry, or to add enough liquid to dissolve the sugar into a syrup, not just wet it. It's therefore best to add solid chocolate to hot liquid ingredients, or pour the hot liquid all at once onto the chocolate, rather than add the liquid gradually to molten chocolate. Seized chocolate can be salvaged by adding more warm liquid until the paste turns into a thick fluid.

Different Chocolates Are Not Interchangeable Both recipe writers and cooks need to be as precise as possible about the kinds of chocolate they use. Different chocolates have very different proportions of cocoa butter, cocoa particles, and sugar. The proportions of cocoa particles and sugar are especially important when chocolate is combined with wet ingredients. Sugar dissolves into syrup, thereby increasing the volume of a preparation's liquid phase and contributing fluidity, while cocoa particles absorb moisture, decrease the volume of the liquid phase, and reduce fluidity. A recipe developed for sweet chocolate may fail badly if it's made with a 70% premium bittersweet chocolate, which has far more drying cocoa particles and far less syrup-making sugar.

Ganache Once of the simplest and most familiar of chocolate preparations is ganache, a mixture of chocolate and cream that can be infused with many other flavors, whipped to lighten its richness, or further enriched with butter. It serves as a filling for chocolate truffles and pastries, and as both filling and topping for cakes. The dessert called pot de crème, made by melting some chocolate into about twice its weight of cream, is essentially a ganache served on its own.

Ganache Structure A soft ganache is made with approximately equal weights of cream and chocolate. A firm ganache, more suitable for holding a shape and with a stronger chocolate flavor, is made with two parts chocolate for every one of cream. To make ganache, the cream is scalded and the chocolate melted into it to form a complex combination of an emulsion and a suspension (p. 818). The continuous phase of this mixture, the portion that permeates it, is a syrup made from the cream's water and the chocolate's sugar. Suspended in the syrup are the milk fat globules from the cream, and cocoa butter droplets and solid cocoa particles from the chocolate.

Food Words: *Ganache*

The word *ganache* is French, and before it was applied to a mixture of chocolate and cream, it meant "cushion." The confectioner's ganache is indeed a kind of melting cushion for the mouth, soft and plush. Ganache seems to have been invented in France or Switzerland in the middle of the 19th century. Chocolate truffles, morsels of ganache shaped into rough balls and coated either with cocoa powder or a thin layer of hard chocolate, were a simple homemade sweet until well into the 20th century, when they became fashionable luxuries.

In an even mixture of cream and chocolate, there's an abundance of the syrup phase to hold the fat and particles; but in a firm, high-chocolate mixture, there's less syrup, and proportionally more cocoa particles that slowly absorb moisture from the syrup and reduce its volume even further. With chocolates high in cocoa solids, the cocoa particles can eventually absorb so much moisture that they swell and stick to each other. The water-deprived emulsion then fails, allowing the fat globules and droplets to coalesce, and the fat to separate from the swollen particles. This is why high-chocolate ganaches are often unstable and coarsen with time.

Maturing Ganache Many confectioners let ganache mixtures stand at a cool room temperature overnight before working with them. This gradual cooling allows the cocoa butter to crystallize so that when the ganache is shaped or eaten, it softens and melts more slowly. Ganache refrigerated immediately after making hardens without forming many crystals, and becomes soft and greasy when it warms.

Thanks to the initial scalding of the cream and the chocolate's sugar content, moisture-absorbing cocoa particles, and abundant microbe-unfriendly phenolic compounds, ganache has a surprisingly long shelf life of a week or more at room temperature.

Tempered Chocolate for Coating and Molding

Like sugar, chocolate can be shaped to please the eye. Pastry cooks and confectioners make thin chocolate sheets by brushing melted chocolate onto a surface, then letting the chocolate set completely, and stamping or cutting it into shapes, or nudging it into a ruffle. Melted chocolate can be painted onto plant leaves, allowed to harden into the leaves' mirror images, then gently peeled off. It can be squeezed through a pastry bag and tip to form a myriad of lines, drops, and filled shapes.

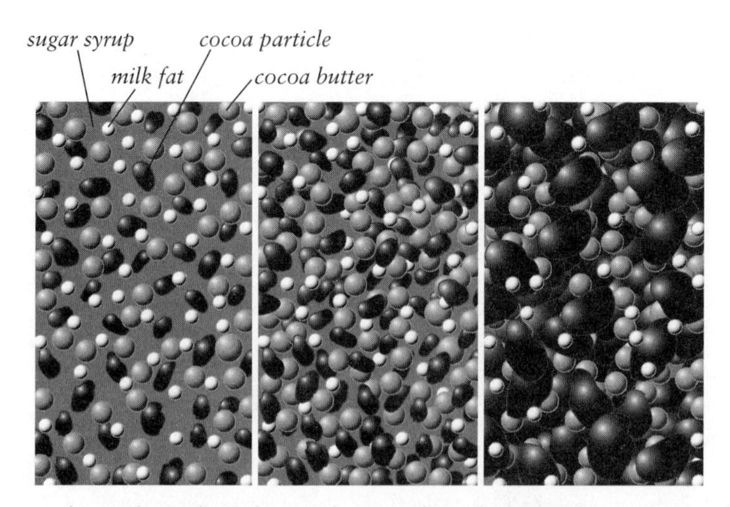

The structure of ganache. Left: *Soft ganache is made with an equal proportion of chocolate and cream, with cocoa particles and droplets of cocoa and milk fat surrounded by a syrup of the chocolate's sugar and the cream's water.* Center: *A firm ganache, made with more chocolate than cream, contains proportionally more dry cocoa particles and less water.* Right: *With time, the cocoa particles in a firm ganache absorb water from the syrup and swell. This can crowd the fat droplets so tightly that they coalesce and the ganache separates.*

And of course it can be used to line molds and produce shapes from hollow spheres to Easter bunnies.

Chocolate lovers often melt chocolate and then use it as a coating for cookies or strawberries or handmade truffles. This can be done easily and casually, the chocolate simply warmed until it melts and then used immediately, the results sometimes chilled in the refrigerator to speed their solidification. Chocolate handled in this way will taste fine, but it's likely to look dull and mottled, and to be soft instead of snappy. This is because the chocolate cooled down so quickly that its cocoa fat solidified into a loose, weak network of unstable crystals instead of the dense, hard network of stable crystals. If appearance and consistency matter, as they do to professional cooks and confectioners, then the cook must *temper* the melted chocolate, or prime it with desirable stable crystals of cocoa fat, just as the manufacturer did before forming it into bars.

Tempering Chocolate The tempering process consists of three basic steps: heating the chocolate to thoroughly melt all of its fat crystals, cooling it somewhat to form a new set of starter crystals, and carefully heating it again to melt the unstable crystals, so that only desirable stable crystals remain. The stable starter crystals will then direct the development of the dense, hard crystal network when the chocolate finally cools and solidifies.

Unstable cocoa butter crystals are crystals that melt relatively easily, which means at relatively cool temperatures, between about 59 and 82°F/15–28°C. The desirable stable crystals (sometimes referred to as "beta" or "beta prime" or "Form V" crystals) melt only at warmer temperatures, between 89 and 93°F/32–34°C. The temperature range in which a particular kind of crystal melts is also the range in which it *forms* as the chocolate cools. Unstable crystals therefore form when molten chocolate is cooled rapidly, so that the stable crystal types—the ones that begin to form at warmer temperatures—don't have time to gather most of the fat molecules to themselves before the unstable crystals begin to form. Stable crystals predominate in melted chocolate when the cook carefully holds it at temperatures *above* the melting point of the unstable crystals, but *below* the melting point of the stable crystals. This tempering range is 88–90°F/31–32°C for dark chocolate, somewhat lower for milk and white chocolates thanks to their mixture of cocoa and milk fats.

Tempering Methods There are several different ways to obtain melted chocolate that is in temper. All of them require an accurate

Temperatures for Tempering Different Kinds of Chocolate

The ideal temperatures for preparing milk and white chocolates depend on the formulation of the particular chocolate, and are best obtained from the manufacturer. This chart gives figures generally used in the chocolate industry.

Kind of Chocolate	Melting Temperature	Cooling Temperature	Tempering Range
Dark	113–122°F/45–50°C	82–84°F/28–29°C	88–90°F/31–32°C
Milk	104–113°F/40–45°C	80–82°F/27–28°C	86–88°F/30–31°C
White	104°F/40°C	74–76°F/24–25°C	80–82°F/27–28°C

thermometer, a gentle heat source (often a pot of hot water over which the bowl of chocolate can be held), and the cook's full attention. And all of them end with the chocolate at a temperature where stable crystals can form and unstable crystals can't.

Of the two common methods for tempering chocolate, one creates the stable crystals from scratch, while the other uses a small amount of tempered chocolate to "seed" the melted chocolate with stable crystals.

- To temper the chocolate from scratch, heat it to 120°F/50°C to melt all crystals, and cool it down to around 105°F/40°C. Then either stir the chocolate as it cools further, until it thickens noticeably (an indication of crystal formation), or pour a portion onto a cool surface and scrape and mix it until it thickens, and return it to the bowl. Then carefully raise the temperature of the chocolate to the tempering range, 88–90°F/31–32°C, and stir to melt any unstable crystals that might have formed during the stirring or scraping.
- To seed melted chocolate with stable crystals, chop and hold in reserve a portion of solid tempered chocolate. Heat the chocolate to be tempered to 120°F/50°C to melt all crystals, and cool it to 95–100°F/35–38°C, just above the temperature range in which stable crystals form. Then stir in the solid portion with its stable crystals, while keeping the temperature in the tempering range, 88–90°F/31–32°C.

No matter how chocolate is tempered, its temperature must be held in the tempering range until it is used. If allowed to cool, it will begin to solidify prematurely, won't flow evenly, and produces an uneven consistency and appearance.

Melting Tempered Chocolate While Maintaining It in Temper It's also possible to obtain tempered melted chocolate without actually doing the tempering. Nearly all manufactured chocolate is sold in tempered

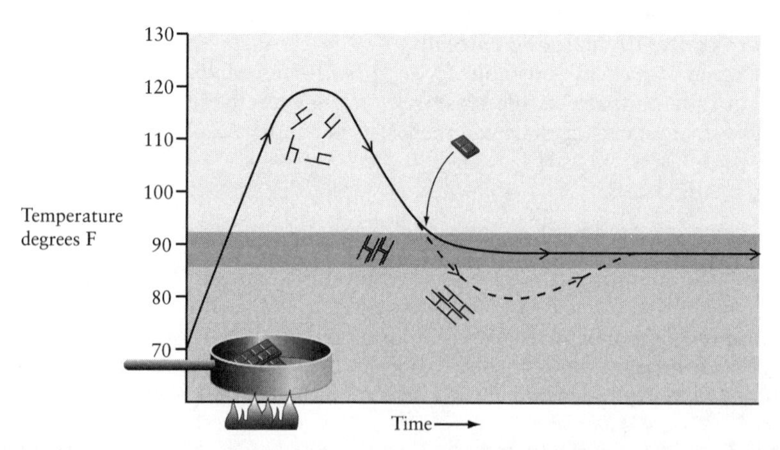

Tempering chocolate. To make chocolate with stable fat crystals, the cook first heats the chocolate to melt all the crystals. In one method, he then cools the chocolate to the temperature range in which only stable crystals can form, adds a portion of tempered chocolate to provide stable crystal seeds, and keeps the mixture warm until it's used for molding or coating. In a second method (dotted line), *the cook allows the molten chocolate to cool below the stable-crystal temperature and form a mixture of crystal types, then warms it to melt the unstable crystals while retaining the stable ones.*

form. A cook using new, well-made chocolate can warm it carefully and directly to the tempering range, 88–90°F/31–32°C, so that it melts but still retains some of its desirable fat crystals. This is easily done by stirring the finely chopped chocolate in a bowl over a pot filled with water at 90–95°F/32–34°C. If for some reason the chocolate is overheated, so that it loses all of its fat crystals, or if the cook is using previously melted and resolidified chocolate with a mixture of crystals, then it's necessary to temper the chocolate with one of the methods described above.

The Art of Tempering Though an accurate thermometer and careful temperature control are necessary for successful tempering, they aren't sufficient. The art of tempering lies in recognizing when the chocolate has accumulated *enough* stable crystals to form a dense, hard network as it cools. Insufficient tempering time, or insufficient stirring, produce too few stable crystal seeds and undertempered chocolate, which will form some unstable crystals when it cools. Too much stirring or time produce too many or too large stable crystals and overtempered chocolate, in which individual crystals predominate over the joined network. Overtempered chocolate is stable, but it can seem coarse, crumbly rather than snappy, dull in appearance, and waxy in the mouth.

Testing for Temper Molten chocolate can be tested for its temper by placing a small, thin portion on a room-temperature surface, a plate or piece of foil. Properly tempered chocolate solidifies in a few minutes to a clean, silky-surfaced mass; the side in contact with the cool surface is shiny. Chocolate out of temper takes many minutes to harden, and has an irregular powdery or grainy appearance.

Working with Tempered Chocolate

Once chocolate has been tempered, it must be handled so that it stays in temper. It should be kept warm, in the tempering range of 88–90°F/31–32°C. When shaped, it should be poured into molds or coated onto fillings that are neither so cold that they cause the cocoa butter to solidify quickly and unstably, nor so warm that they melt the stable crystal seeds in the chocolate. Confectioners recommend a temperature around 77°F/25°C. Similarly, the room temperature should be moderate, neither chilly nor hot.

It turns out that tempered chocolate shrinks by about 2% in each dimension as it solidifies, because the fat molecules in the stable crystals are more densely packed than they were in liquid form. This shrinkage is helpful in making molded chocolates, because the chocolate pulls away from the mold as it hardens. But it can

Specialty Coatings

Ordinary chocolate is not well suited for certain kinds of coatings, including those on ice cream and other frozen ingredients, and candies that are eaten in the heat of the summer or the tropics. For these kinds of products, manufacturers have developed replacements for cocoa butter that don't require tempering to look good, break with a snap, and remain hard at high temperatures. Some closely resemble cocoa butter and can be mixed with chocolate, while others are very different, not compatible with chocolate, and must be flavored with low-fat cocoa. Among the former are fats purified from a number of tropical nuts (palm, shea, illipe, sal); among the latter are "lauric fats" derived from coconut and palm oils. Coatings made with these ingredients are often called "nontempering" chocolates.

cause the thin coating on a candy or truffle to crack, especially if the filling is cold and expands slightly when coated with the warm chocolate. The snappy hardness of tempered chocolate takes several days to develop fully as the crystal network continues to grow and become stronger.

Modeling Chocolate "Modeling" or "molding" chocolate is a version made expressly for shaping into decorations. It's made by mixing molten chocolate with a third to half its weight of corn syrup and sugar, then kneading the mixture into a pliable mass. The resulting "chocolate" is now a concentrated sugar syrup that is filled and thickened with cocoa particles and droplets of cocoa butter. The pieces stiffen as the syrup phase loses moisture to the air and to the dry cocoa particles.

Chocolate and Health

Fats and Antioxidants Cocoa beans, like all seeds, are rich in nutrients that support the plant embryo until it develops leaves and roots. They're especially rich in saturated fats, which are notorious for contributing to raised blood cholesterol levels and therefore to the risk of heart disease. However, much of the saturated fat in cocoa butter is a particular fatty acid that the body immediately converts into an unsaturated one (stearic acid is converted to oleic acid). So chocolate is not thought to pose a risk to the heart. In fact, it may well be beneficial. Cocoa particles are a tremendously rich source of antioxidant phenolic compounds, which account for 8% of the weight of cocoa powder. The higher the cocoa solids content of a chocolate or candy, the higher its antioxidant content. Any added sugar, milk products, or cocoa butter simply dilute the cocoa solids and their phenolics. The dutching process also reduces the levels of desirable phenolics in cocoa powder, and the milk proteins in milk chocolate appear to bind to the same molecules and prevent us from absorbing them.

Caffeine and Theobromine Chocolate contains two related alkaloids, theobromine and caffeine, in the ratio of about 10 to 1. Theobromine is a weaker stimulant of the nervous system than caffeine is (p. 433); its main effect seems to be a diuretic one. (However it is quite toxic to dogs, who can suffer serious poisoning from chocolate candies.) A 1-oz /30 gm piece of unsweetened chocolate contains around 30 mg of caffeine, around a third the dose in a cup of coffee; sweetened and milk chocolates contain substantially less. Cocoa powder has around 20 mg caffeine per tablespoon/10 g.

Cravings for Chocolate Because many people, especially women, experience cravings for chocolate that border on the symptoms of addiction, it has been thought that chocolate might contain psychoactive chemicals. Chocolate does turn out to contain both "cannabinoid" chemicals—chemicals similar to the active ingredient in marijuana—as well as other molecules that cause brain cells to accumulate cannabinoid chemicals. But these are present in extremely small amounts that probably have no practical significance. Similarly, chocolate contains phenylethylamine, a naturally occurring body chemical that has amphetamine-like effects—but then so do sausages and other fermented foods. In fact there is good experimental evidence that chocolate does not contain any drug-like substances capable of inducing a true addiction. Psychologists have shown that chocolate cravings can be satisfied by imitations that have no real chocolate in them, while these cravings are not satisfied by capsules of genuine cocoa powder or chocolate that are swallowed without tasting. It appears to be the sensory experience of eating chocolate, no more and no less, that is powerfully appealing.

CHAPTER 13

WINE, BEER, AND DISTILLED SPIRITS

Like all good foods, wine, beer, and spirits nourish and satisfy the body. What sets them apart is the very direct way in which they touch the mind. They contain alcohol, which is both a source of energy and a drug. In moderate amounts, alcohol causes us to feel and express emotions of all kinds—happiness, conviviality, sadness, anger—with more freedom. In large amounts, it's a narcotic: it numbs feeling and clouds thought. Alcoholic drinks thus offer various degrees of release from our usual state of mind. Small wonder that they were once considered an earthly version of the nectar of the gods, foods that give mortals a taste of being carefree masters of life!

Humankind has always had a thirst for alcohol, and now satisfies it with mass-produced drinks that offer an inexpensive respite from the world and its cares. But some wines and beers and spirits are among the most finely crafted foods there are, the best that the world and care have to offer. Their flavor can be so rich, balanced, dynamic, and persistent that they touch the mind not with release from the world, but with a heightened attentiveness and connection to it.

Wine, beer, and spirits are the creation of microscopic yeasts, which break food sugars down into alcohol molecules. Alcohol is a volatile substance whose own aroma is relatively diffuse. It has the effect of lending a new dimension to the flavor of grapes and grains, a kind of open stage on which the food's own volatile molecules can appear. Yeasts are also prodigious flavor chemists, so during the fermentation they fill that stage with dozens of new aromas. The winemaker or brewer then directs the transformation of this teeming, unruly cast into a balanced, harmonious ensemble.

Though they share this basic nature, wine and beer and spirits are very different foods. Wine begins with fruits that are fragrant and sweet with sugars, and therefore ready-made to ferment into an aromatic drink—but only during the few days of the year when they're ripe. Grapes and wine are a gift of nature, a form of grace, which the winemaker must accept when they're given, and can leave largely to themselves to realize their innate potential for producing flavor. Beer and rice alcohols, by contrast, are the expression of everyday human effort and ingenuity. They're made from sugarless, aroma-less dry grains, the charmless but dependable staff of life. Brewers transform grains into something fermentable and aromatic by sprouting them or cultivating molds on them for days, and cooking them for hours. They can do this at any time of the year, anywhere in the world. Beer is thus our universal alcohol, com-

The Drink of Happiness

Nearly 4,000 years ago, a Sumerian poet put these words into the mouth of the goddess Inanna, who ruled both heaven and earth and was as delighted as any mortal by the experience of drinking beer. Ninkasi was the goddess of beer. (I've omitted the poem's many repetitions.)

May Ninkasi live together with you!
Let her pour for you beer [and] wine,
Let [the pouring] of the sweet liquor resound
pleasantly for you!
In the . . . reed buckets there is sweet beer,
I will make cupbearers, boys, [and]
brewers stand by,
While I circle around the abundance of beer,
While I feel wonderful, I feel wonderful,
Drinking beer, in a blissful mood,
Drinking liquor, feeling exhilarated,
With joy in the heart [and] a happy liver—
While my heart full of joy,
[And] [my] happy liver I cover with a
garment fit for a queen!
The heart of Inanna is happy again,
The heart of the queen of heaven is happy again!

—Transl. Miguel Civil

fortably local and everyday and ordinary, yet sometimes extraordinary. And distilled spirits are the heart of wine and beer, concentrates of their volatile and aromatic content, and drinks of unmatched intensity.

The pleasure of tasting a good beer or wine or spirit grows with the recognition that its flavor is the expression of many natural, cultural, and personal particulars: a place and its traditions, certain plants and the soil they grew in, a year and its weather, the course of fermentation and maturation, the taste and skills of the maker. Their rich natural and human parentage explains why alcohols are so absorbingly diverse, and why a thoughtful sip can momentarily fill us with the world and delight.

THE NATURE OF ALCOHOL

Alcohol molecules are made in many living cells as a by-product of breaking down sugar molecules for their chemical energy. Most cells then break down the alcohol molecules to extract their energy content too. The great exception to this rule is certain yeasts, which excrete alcohol into their surroundings. Like the lactic acid in cheeses and pickled vegetables, like the powerful aromas in herbs and spices, the alcohol in wine and beer is a defensive chemical weapon, which the yeasts deploy to protect themselves against competition from other microbes. Alcohol is toxic to living cells. Even the yeasts that make it for us can only tolerate a certain amount. The pleasant feeling that it gives us is a manifestation of the fact that it's disrupting the normal function of our brain cells.

Yeasts and Alcoholic Fermentation

Yeasts are a group of about 160 species of single-celled microscopic molds. Not all are useful: some cause the spoilage of fruits and vegetables, some cause human disease (for example, the yeast infection of *Candida albicans*). Most of the yeasts used in making bread and alcoholic drinks are members of the genus *Saccharomyces,* whose name means "sugar fungus." We cultivate them for the same reason that we use particular bacteria to sour milk: they make foods resistant to infection by other microbes, and produce substances that are mainly pleasant to us. Essential to the yeasts' production of alcohol is their ability to survive on very little oxygen, which most living cells use to burn fuel molecules for energy, leaving behind only carbon dioxide and water. In the absence of oxygen, the fuel

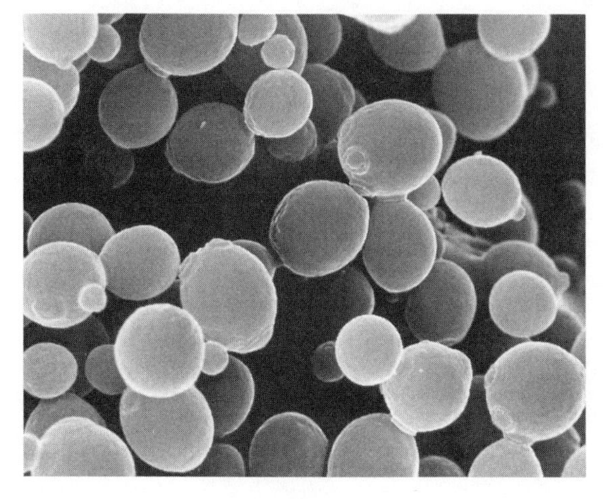

Yeast. Cells of brewer's yeast, Saccharomyces cerevisiae, *as seen through an electron microscope. Each is about 0.005 mm in diameter. The cell at the upper right center is in the process of reproducing, and bears the scars of previous buddings.*

can be broken down only partly. The overall equation for the production of energy from glucose without oxygen goes like this:

$$C_6H_{12}O_6 \rightarrow 2CH_3CH_2OH + 2CO_2 + \text{energy}$$
Glucose → alcohol + carbon dioxide + energy

Yeasts introduce a variety of other compounds into the grape juice or grain mash that contributes characteristic flavors. For example, they produce savory succinic acid, and transform amino acids in the liquid into "higher," or longer-chain alcohols; they combine alcohols with acids to make fruity-smelling esters; they produce sulfur compounds reminiscent of cooked vegetables, coffee, and toast. And when a yeast cell dies, its enzymatic machinery digests the cell and releases its contents into the liquid, where they continue to generate flavor. Because growing yeast cells synthesize proteins and B vitamins, they can actually make a fruit juice or cereal mash more nutritious than it was when fresh.

The Qualities of Alcohol

In chemistry, the term *alcohol* is applied to a large family of substances with a similar molecular structure. Our everyday word *alcohol* refers to one particular member of this family, which chemists call *ethyl alcohol*, or *ethanol*. In this chapter I'll use *alcohol* in its common sense, but I'll also refer to "higher" alcohols, or molecules in the alcohol family with more atoms than ethanol has.

Physical and Chemical Qualities Pure alcohol is a clear, colorless liquid. The alcohol molecule is a small one, CH_3CH_2OH, whose backbone is just two carbon atoms. One end of the alcohol molecule, the CH_3, resembles fats and oils, while the OH group at the other end is two-thirds of a water molecule. Alcohol is therefore a versatile liquid. It mixes easily with water, but also with fatty substances, including cell membranes, which it excels at penetrating, and aroma molecules and carotenoid pigments,

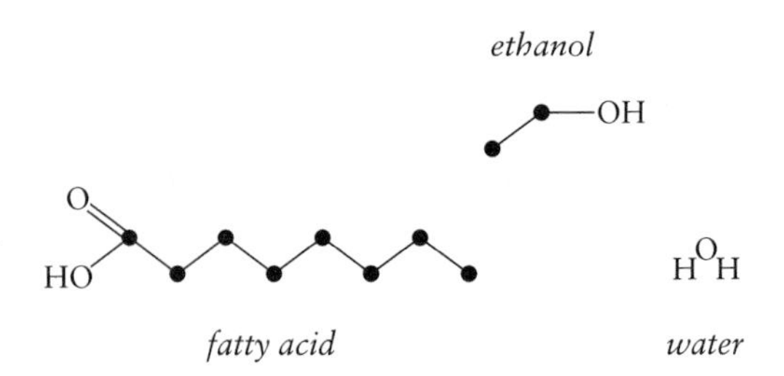

Ethanol, or common alcohol. The versatile ethanol molecule has one end that resembles the fatty-acid carbon chain of fats and oils, and one end that resembles water.

Alcoholic Fermentation Helped Form Modern Biology

The mystery of fermentation attracted some of the best and most headstrong scientists of the 19th century, including Justus von Liebig and Louis Pasteur, and helped give rise to the science of microbiology. The first microorganisms to be isolated in pure cultures were beer and wine yeasts prepared in the laboratory of the Carlsberg Brewery in Copenhagen around 1880. And scientists coined the word *enzyme,* denoting the remarkable protein molecules that living cells use to transform other molecules, from the Greek words for "in yeast," where sugar is transformed into alcohol.

which it excels at extracting from cells. The higher alcohols, which yeasts also produce in small quantities and which become concentrated in distilled spirits, have a longer fat-like end to their molecules (p. 762), and behave more like fats. They lend an oily, viscous quality to whiskies and other spirits. They also tend to concentrate in the membranes of our cells, and therefore are more irritating and more potent narcotics than simple alcohol.

Several of alcohol's physical properties have important consequences for the cook and food lover.

- Alcohol is more volatile than water, more easily evaporated and brought to the boil. Its low boiling point, 176°F/78°C, is what makes it possible to distill alcohol into a much stronger solution than wine or beer.
- Alcohol is flammable, which makes possible spectacular flaming dishes fueled by brandy or rum. The food doesn't get scorched because the heat of combustion is fully absorbed by the vaporization of the spirits' water.
- Alcohol has a much lower freezing point than water, –173°F/–114°C. This makes it possible to concentrate alcoholic liquids in winter cold or the freezer (see box, p. 761).
- A given volume of alcohol weighs 80% as much as the same volume of water, so a mixture of alcohol and water is lighter than pure water. This helps makes possible layered cocktails (see box, p. 770).

Alcohol and Flavor We experience the presence of alcohol in a food through our senses of taste, smell, and touch. The alcohol molecule bears some resemblance to a sugar molecule, and indeed it has a slightly sweet taste. At high concentrations, those typical of distilled spirits and even some strong wines, alcohol is irritating, and produces a pungent, "hot" sensation in the mouth, as well as in the nose. As a volatile chemical, alcohol has its own distinctive aroma, which we experience at its purest in unflavored grain alcohol or vodka. Its chemical compatibility with other aroma compounds means that concentrated alcohol tends to bind aromas in foods and drinks and inhibit their release into the air. But at very low concentrations, around 1% or less, alcohol actually enhances the release of fruity esters and other aroma molecules into the air. This is one reason that wine, vodka, and other alcohols are valuable ingredients in general cooking, provided that the proportion is small or the alcohol mostly removed by long cooking.

The Tears of Strong Wine and Spirits

Regular drinkers of strong wines and spirits have probably mused upon the odd phenomenon known as "tears" or "legs," films of liquid on the inside of the glass that seem to be in slow but constant movement up and down. These moving films are created by the dynamic nature of alcohol-water mixtures. Alcohol lowers the forces of attraction between water molecules in wine or spirits; but at the edge of the surface, alcohol evaporates from the water, the water bonds more strongly to itself and to the glass, and the decreasingly alcoholic water pulls itself up the side of the glass until gravity wins out and it falls back in a droplet. The higher the alcoholic content of the liquid, and the easier it is for alcohol to evaporate—warm temperatures and wide-mouthed shallow glasses are most favorable—the more pronounced the tears and legs are.

Effects on Living Things One consequence of alcohol's chemical versatility is that it readily penetrates the membranes of living cells, which are made in part of fat-like molecules. When it does so, it disturbs the action of the membrane proteins. A high enough concentration of alcohol will cause such a disturbance that this critical boundary between cell and environment fails, and the cell dies. The yeasts that produce alcohol can tolerate a concentration of about 20%, and most other microbes are killed by much less. When the solution also contains acid or sugar, as in wines, alcohol is an even more effective microbial poison. This is why, unlike beer and wine, distilled spirits and such alcohol-fortified wines as sherry and port don't spoil after they're opened.

Our own pleasant inebriation when we drink alcohol is in part a symptom of mild membrane and protein disturbance throughout our nervous system.

Alcohol as a Drug: Intoxication

Alcohol is a *drug*: it alters the operation of the various tissues into which it diffuses. We value it most for its influence on the central nervous system, where it acts as a narcotic. The fact that it seems to stimulate more animated, excited behavior than usual is actually a symptom of its depressant effect on the higher functions of the brain, those that normally control our behavior with various kinds of inhibition. As more alcohol reaches the brain, it interferes with more basic processes: memory, concentration, and thinking in general; muscular coordination, speech, vision. With regard to the idea that alcohol is an aphrodisiac, modern investigators continue to cite the authority of the Porter in Shakespeare's *Macbeth*, who says of drink that "Lechery, sir, it provokes, and unprovokes: it provokes the desire, but it takes away the performance."

The degree to which someone is intoxicated depends on the concentration of alcohol in the cells. Once alcohol is absorbed from the digestive tract, the blood rapidly distributes it to all body fluids, and it readily diffuses into and across membranes to penetrate all cells. Large people can therefore drink more than small people without being drunker: they have a greater volume of body fluids and cells in which to dilute the alcohol. Impaired coordination and impulsive behavior usually appear when the concentration of alcohol in the blood is

The Source of Happiness and Oblivion

Observers of the human condition have long remarked on the ways in which alcohol helps people deal with that condition. Here are two of the earliest and simplest formulations, from the Ayurvedic tradition of India, and from the Old Testament.

> Wine is the foremost of all things that lead to cheerfulness.
> Abuse of wine is the foremost of all causes that lead to loss of intelligence and memory.
>
> —*Charaka-Samhita*, ca. 400 BCE

> Give strong drink unto him that is ready to perish, and wine unto them that be of heavy hearts. Let him drink, and forget his poverty, and remember his misery no more.
>
> —Proverbs, ca. 500 BCE

0.02–0.03%. Falling-over drunkenness is the result at 0.15%, and 0.4% can be fatal.

As drugs go, alcohol is a relatively weak one. It takes grams of pure alcohol, not milligrams, to have noticeable effects, and this allows us to enjoy moderate amounts of wine and beer without harming ourselves. But like other narcotic drugs, alcohol can be addictive, and the habitual consumption of large quantities is destructive. It has been the cause of widespread misery and premature death for thousands of years, and it still is. Alcohol and the molecule to which it's first metabolized, acetaldehyde, disrupt many systems and organs in the body. Their constant presence can therefore cause a broad range of serious and even fatal diseases.

How the Body Metabolizes Alcohol

Our bodies eliminate alcohol by breaking it down in a series of chemical reactions and using the energy freed by those reactions. Alcohol's chemical structure has similarities to both sugar and fat, and it has a nutritional value between the two, around 7 calories per gram (sugar has 4 calories per gram, fat 9). It provides around 5% of the calories in the American diet, much more among heavy drinkers.

Alcohol is broken down and converted into energy in two organs, the stomach and the liver. The "first-pass" metabolism of alcohol in the stomach consumes a portion before it gets to the small intestine and then into the blood. That portion is around 30% in men, but only 10% in women. Men therefore experience a slower rise in blood alcohol when they drink, and can drink more before they feel its effects. And there are strong genetic influences on how well individuals are able to handle alcohol.

Overall, the body can metabolize around 10–15 grams of alcohol per hour, the equivalent of one standard-sized drink every 60–90 minutes. The level of alcohol in the blood reaches a maximum 30–60 minutes after consumption. Foods, and especially fats and oils, delay the passage of the stomach's contents into the small intestine, giving the stomach enzymes more time to work, slowing the rise in blood alcohol, and reducing its peak to about half of what it reaches on an empty stomach. On the other hand, aspirin interferes with the stomach's alcohol metabolism and so causes a quicker rise in blood alcohol levels. The carbon dioxide bubbles in sparkling wines and beer cause the same accelerated rise by as yet unknown means.

The Hangover Then there's the misery of the hangover, the general feeling of illness

The Benefits of Moderate Drinking

One consistent finding from several decades' worth of studies is that people who regularly consume the equivalent of one or two alcoholic drinks per day die less often from heart disease and stroke. (Higher consumption is associated with higher death rates from cancer and accidents.) Alcohol itself raises the levels of desirable HDL cholesterol and lowers the levels of blood factors that induce clotting and thus contribute to blockages. And red wine and dark beer are good sources of antioxidant phenolic compounds (p. 255). Wine phenolics also cause arteries to widen and reduce the tendency of red blood cells to stick together, and a few of these compounds, notably *resveratrol* and its relatives, inhibit an enzyme (cyclooxygenase) that's associated with damaging inflammation reactions and the development of arthritis and certain cancers.

that we wake up with the morning after we've had too much alcohol. The folk remedies for this affliction are many and ancient. In medieval times, the medical school of Salerno was already recommending the hair of the dog:

Si nocturna tibi noceat potatio vini,
Hoc tu mane bibas iterum, et fuerit medicina.
If an evening of wine does you in,
More the next morning will be medicine.

The hangover is in part a mild withdrawal syndrome. The night before, the body adjusted to a high concentration of alcohol and related narcotic chemicals, but by morning the drug is going or gone. Hypersensitivity to sound and light, for example, may be a leftover compensation for the general depression of the nervous system. The logic of the morning-after drink is simple but insidious: it restores many of the conditions to which the body had become accustomed, as well as lightly anesthetizing it. But this only postpones the body's true recovery from intoxication.

Only a few of the different symptoms that constitute a hangover can be directly treated. The dry mouth and headache can be due to the dehydration that alcohol causes, so that drinking liquids may relieve them. Alcohol can also cause a headache by enlarging the cranial blood vessels; the caffeine in coffee and tea has the opposite effect, and may bring some relief.

Cooking with Alcohol

Cooks use wines, beers, and distilled spirits as ingredients in a broad range of dishes, from savory soups and sauces and stews to sweet creams and cakes, soufflés and sorbets. They contribute distinctive flavors, often including acidity, sweetness, and savoriness (from glutamic and succinic acids), and the aromatic dimension provided by alcohol and other volatile substances. Some qualities can be a challenge for the cook to work with, including the astringency of red wines (p. 737) and the bitterness of most beers. The alcohol itself also provides a third kind of liquid—in addition to water and oil—into which flavor and color molecules can be extracted and dissolved, as well as reactive molecules that can combine with other substances in the food to generate new aromas and greater depth of flavor. While large amounts of alcohol tend to trap other volatile molecules in the food, small traces boost their volatility and so intensify aroma.

At the same time that alcohol itself can be an asset for the cook, it can also be a liability. Alcohol has its own pungent, slightly medicinal qualities, and these qualities are heightened and can become harsh in hot foods. Cooks may therefore simmer or boil sauces for some time to evaporate off as much alcohol as possible. In the showy preparation called the *flambé,* from the French for "to flame," they ignite the heated vapors of spirits and high-alcohol wines into flickering, ghostly blue flames to burn off the alcohol and give a lightly singed flavor to a dish. However, none of these techniques leave a food free of alcohol. Experiments have shown that long-simmered stews retain about 5% of the alcohol initially added, briefly cooked dishes from 10 to 50%, and flambés as much as 75%.

Alcoholic Liquids and Wood Barrels

The great good fortune of wine and beer is that microbes can "spoil" fruit juice and gruel into something both delicious and pleasantly inebriating. A few centuries ago, winemakers and distillers discovered another remarkable piece of good luck: simply storing wine, spirits, and vinegars in wood barrels turns out to give them a new and complementary dimension of flavor.

Oak and Its Qualities Though chestnut and cedar have been used in Europe and redwood in the United States, most barrels for aging wines and spirits are made from oak. Oak heartwood, the older inner wood, is a mass of dead cells that supports the outer living layers. The heartwood cells are filled with compounds that deter boring insects. These are mainly tannins, but they include such aromatic compounds as clove-like eugenol, vanilla-like vanillin, and oaky "oak lactones," relatives of the characteristic aromatics of coconut and peach. From 90 to 95% of the heartwood solids are cell-wall molecules, cellulose, hemicellulose, and lignin. These are mostly insoluble, but the lignins can be partly broken down and extracted by strong alcohol, and all can be transformed into new aromatic molecules when the wood is heated during barrel making (p. 449).

Coopers rely mainly on two European oak species (*Quercus robur* and *Q. sessilis*), and ten North American species, the most important being the white oak (*Q. alba*). The European species are mostly made into wine barrels, American oak into barrels for aging distilled spirits. American oak tends to have lower levels of extractable tannins and higher levels of the oak lactones and vanillin.

Making Barrels: Forming and Cooking In order to make barrels, the cooper splits the heartwood into pieces, dries them, and forms them into thin, elongated staves, which are then roughly hooped together and heated to make them more pliable and easily bent into the final barrel shape. In Europe, the barrel interior is heated with a small brazier of burning wood scraps to 400°F/200°C. Once the softened staves have been tightly hooped into their final positions, the interior is "toasted" further at 300–400°F/150–200°C, for 5 to 20 minutes, depending on the degree of cooking desired: less for wine barrels, more for spirits. In the United States, the heat treatment for whiskey barrels is more extreme. The hooped staves are first steamed to soften them, and then the barrel interior is charred with an open gas burner for from 15 to 45 seconds.

Barrel Flavors Several things happen when alcoholic liquids are stored in new barrels. First, the liquid extracts soluble materials that contribute color and flavor, including tannins, oak and clove and vanilla aromas, and the sugars, browning-reaction products, and smoky volatiles formed when the barrel was heated. In the charred American barrels used for whiskey, the carbonized surface acts something like an activated charcoal absorbant, removing some materials from the whiskey and so accelerating the maturation of its flavor. Gaps and pores in the wood allow the liquid to absorb limited amounts of oxygen. And the rich chemical brew of wine or spirits, wood components, and oxygen slowly undergoes innumerable reactions and evolves toward a harmonious equilibrium.

New oak barrels give a pronounced flavor to liquids stored in them, one that can

Barrel Fermentation

Some wines and vinegars are fermented in the barrel as well as matured there after fermentation, and develop a distinctive barrel-fermented flavor. One unusual component of that flavor, produced by the action of yeast enzymes on compounds found in toasted oak, is a sulfur-containing chemical whose aroma is reminiscent of roasted coffee and roasted meat (furfurylthiol).

overwhelm the inherent qualities of delicate wines. The producer can control the contribution of the wood by limiting the aging time in new barrels, or by working with used barrels, which have already had much of their flavor components extracted.

Alternatives to Barrels Oak barrels are expensive, so only relatively expensive wines and spirits are aged in them. There are other ways of getting oak flavor into less expensive products. *Boisés,* extracts made by boiling wood chips in water, are a traditional finishing additive in French brandies, including Cognac and Armagnac. In recent years, large-volume winemakers have begun putting barrel staves, oak chips, and even sawdust into wines while they mature in containers made of steel and other inert materials.

WINE

The juice of the grape is just one of the naturally sweet liquids with which our ancestors learned to make alcoholic drinks. Perhaps just as ancient as grape wine is *koumiss,* the fermented mare's milk of the central Asian nomads. One Greek word for wine, *methu,* came from the Indo-European word for fermented honey water, whose name in English is *mead.* The Romans fermented dates and figs. And before they tasted wine, the inhabitants of northern Europe drank apple juice fermented into cider.

But the grape turned out to be uniquely suited to the development of a diverse family of alcoholic drinks. The grapevine is a highly productive plant that can adapt to a wide range of soils and climates. Its fruits retain large amounts of an unusual acid, tartaric acid, which few microbes can metabolize, and which favors the growth of yeasts. The grapes ripen with enough sugar that the yeasts' alcohol production can suppress the growth of nearly all other microbes. And they offer striking colors and a variety of flavors.

Thanks largely to these qualities, grapes are the world's largest fruit crop, with about 70% of the annual production used to make wine. France, Italy, and Spain are the world's largest wine producers and exporters.

THE HISTORY OF WINE

The evolution of wine is long and fascinating, and ongoing. Here are a few highlights.

Ancient Times: Aged Wines and Connoisseurship As I write, the earliest evidence we have for wine made from grapes, residues at the bottom of a pot found in western Iran, dates from around 6000 BCE. From 3000 BCE on, wine was a prominent part of trade in western Asia and Egypt. Wild grapes and the first wines were red, but the Egyptians had a color mutant of the grape plant and made white wines from it. They would ferment grape juice in large clay jars. The contents of the jars were eventually sampled and graded, and the jars marked, stoppered, and sealed with mud. The airtight containers allowed wine to be aged for years. Many wine amphoras found in the tombs of the pharaohs carry labels with the date of production, the region in which the wine was made, sometimes a brief description and the name of the winemaker. Wine connoisseurship is ancient!

Greece and Rome Phoenician and Greek traders introduced the cultivated vine throughout the Mediterranean basin, where the Greeks developed the cult of Dionysos, god of vegetation, the vine, and the temporary release from ordinary life that wine made possible. By Homer's time, about 700 BCE, wine had become a standard beverage in Greece, one that was made strong, watered down before drinking, and graded in quality for freeman and slave. The culture of the vine was not established in Italy until about 200 BCE, but it took hold so well that the Greeks took to calling southern Italy *Oenotria,* "land of the grape."

Over the next couple of centuries, Rome advanced the art of winemaking considerably. Pliny devoted a full book of his *Natural History* to the grape. He noted that there were now an infinite number of varieties, that the same grape could produce very different wines in different places, and named Italy, Greece, Egypt, and Gaul (France) as admired sources. Like the Egyptians, the Romans had airtight amphoras that allowed them to age wine for years without spoiling. The Greeks and Romans also preserved and flavored wines with tree resins or the pitch refined from them, salt, and spices.

It was in Roman times that wooden casks—an innovation of northern Europe—arrived along the Mediterranean as an alternative to clay amphoras. During subsequent centuries, they became the standard wine vessel, and amphoras disappeared. Casks had the advantage of being lighter and less fragile, but the disadvantage of not being airtight. This meant that wines could only be stored in them for a handful of years before they became overoxidized and unpleasant to drink. Excellent aged wines therefore disappeared along with the amphora, and only reappeared after more than a thousand years with the invention of the cork-stoppered bottle (p. 724).

The Spread of Winemaking in Europe; the Rise of France After the fall of Rome around the 5th century CE, Christian monasteries advanced the arts of viticulture and winemaking in Europe. Local rulers endowed them with tracts of land, which they then cleared of forest and reclaimed from swamps, bringing systematic, organized agriculture to sparsely settled regions, and the grape to northern France and Germany. Wine was required for the sacrament of Communion, and it and beer were made for daily consumption, to serve guests, and to sell. It was in the Middle Ages that the wines of Burgundy became famous.

From the late Middle Ages on, France slowly became the preeminent source of wine in Europe. By the 1600s the wines of France, and especially Bordeaux, which had the advantage being a port, were important exports to England and Holland. Meanwhile Italy fell behind, a victim of political and economic circumstance. Until the middle of the 19th century it was not a nation but a collection of city-states, each with protective tariffs and little of the international trade that brought competition and improvement to the wine regions of France. Most of the wine was consumed locally, and the grapevines grown not in vineyards but in sharecroppers' plots, between rows of food plants or trained on trees.

Food Words: *Wine, Vine, Grape*

Our language bears witness to the fact that from the very earliest times, people thought of the grapevine not as the source of edible fruit, but as the source of wine. Our words *vine* and *wine* come from the same root word, and that word meant the fermented juice of the vine's fruit. This root is so ancient that it predates the divergence of Indo-European from other prehistoric languages of western Asia. The words for the fruit itself, on the other hand, are different in different languages. The English word *grape* appears to come from an Indo-European root meaning "curved" or "crooked," probably referring to the curved blade of the knife used to harvest grape bunches, or to the shape of the bunch stem. *Grapple* and *crumpet* are related words.

New Wines and New Containers Early modern times brought the invention of several wonderful variants on plain fermented grape juice, and important improvements in wine storage. Sometime before 1600, Spanish winemakers found that they could both stabilize and give a new character to wines by fortifying them with brandy; the result was sherry. Around 1650, Hungarian winemakers managed to make deliciously concentrated and very sweet Tokaji wine from grapes infected by an otherwise destructive fungus, which came to be known as the "noble rot." This was the forerunner of French Sauternes and similar sweet German wines. At about the same time, English importers of white wine from the Champagne region east of Paris discovered that they could make the wine delightfully bubbly by transferring it from barrel to bottles before it had finished fermenting. And a few decades later, the English developed port in the effort to stabilize strong red wines during their sea journey from Portugal. The shippers added distilled alcohol to the wines to prevent spoilage, and thus discovered the pleasures of fortified sweet red wines.

Bottles and Corks The 17th and 18th centuries brought two major innovations that once again made it possible to age wines for many years, a possibility that had disappeared when the impermeable amphora was replaced by wood barrels. These momentous developments were slim bottles and cork stoppers! The English discovery of sparkling Champagne depended on the fact that they had begun plugging bottle necks with compressible gas-tight cork instead of cloth, and that they had especially strong bottles that could withstand the inner pressure (the glass strength came from manufacturing with hot coal fires rather than wood fires). And during the 18th century, the wine bottle gradually evolved from a short, stout flask to the familiar elongated bottle. The bulky bottles were only used to convey the wine from barrel to table or to hold it for a day or two. When bottles had slimmed down enough that they could lie on their sides, their contents wetting the cork and preventing it from shrinking and admitting air, then wine could be stored in them for many years without spoiling, and sometimes with great improvements in flavor.

Pasteur and the Beginnings of a Scientific Understanding of Wine In 1863 the French Emperor Louis Napoleon asked the great chemist Louis Pasteur to study the "maladies" of wine. Three years later, Pasteur published the landmark *Etudes sur le vin*. Pasteur and others had already demonstrated that yeast is a living mass of microbes, and thus had made it possible to begin to identify and control the kinds of microbes that both make wine and spoil it. But Pasteur was the first to analyze the development of wine, to discover the central role of oxygen, and show why both barrel and bottle were indispensable for making good wine, the barrel for providing oxygen to the young wine to help mature it, and the bottle for excluding oxygen from the mature wine to help preserve it.

> In my view, it is oxygen which *makes* wine; it is by its influence that wine ages; it is oxygen which modifies the harsh principles of new wine and makes the bad taste disappear . . .
>
> It is necessary to aerate the wine slowly to age it, but the oxidation must not be pushed too far. It weakens the wine too much, wears it out, and removes from red wine nearly all its color. There exists a period . . . during which the wine must pass from a permeable container [the barrel] to one nearly impermeable [the bottle].

Scientific Approaches to Making Wine Pasteur planted the seed of a scientific approach to winemaking. That seed soon took root in both France and the United States. In the 1880s, the University of Bordeaux and the University of California established institutes of oenology. The Bordeaux group focused on understanding and improving traditional French methods for producing fine wines, and discovered the nature of the malolactic fermentation (p. 730). The California institute moved from Berkeley to Davis in 1928, and studied how best to build a wine industry in the absence of a local tradition, including determining what grape varieties were best suited to various climatic conditions. Today, thanks to this and similar work in a number of countries, and to the general modernization of winemaking, more good wine is being made in more parts of the world than ever before.

Traditional and Industrial Wines There's now a spectrum of approaches among winemakers, and so a spectrum of wines from which we can choose. At one end is the relatively straightforward approach found in traditional winemaking regions: the grapes are grown in a place and with methods that maximize wine quality; they're simply crushed, fermented, the new wine matured for some time, and bottled. At the other end of the spectrum are advanced manufacturing processes that treat grapes and wine like other industrial materials. These aim to approximate the qualities of the traditionally produced wine by nontraditional means that are less labor intensive and less expensive. The grapes themselves need not be coaxed to an ideal ripeness because the winemaker can use various separation technologies to adjust their content of water, sugar, acid, alcohol, and other components. The effects of barrel and bottle aging can be simulated inexpensively and rapidly by means of oak chips or sawdust, and the bubbling of pure oxygen through wine stored in huge steel tanks.

Industrial wines are marvels of reverse engineering, and often taste good, clean, and without obvious faults. Wine made on a small scale with minimal manipulation is less predictable in its quality, but this is because it is more distinctive, an expression of grapes that were grown in a particular place and year, and transformed by a particular winemaker. Such wine is more expensive than industrial wine, sometimes much better, and usually more interesting.

Wine Grapes

Grapes provide the substance of wine, and therefore determine many of its qualities. Their most important components are

- Sugars, which the yeasts feed on and convert into alcohol. Wine grapes are generally harvested with 20–30% sugar, mainly glucose and fructose.
- Acids, mainly tartaric and some malic, which help prevent the growth of undesirable microbes during fermentation, and are a major component of wine flavor.
- Tannins and related phenolic compounds, which contribute an astringent feeling and thereby a body and weightiness to wine (p. 737).
- Pigment molecules that provide color, and sometimes contribute to astringency as well. Red grapes contain anthocyanin pigments (p. 267), mainly in the skins. "White" grapes lack anthocyanins; their yellowish color comes from a different group of phenolic compounds, the flavonols.
- Aroma compounds, which may be generically grapey, or distinctive of a particular grape variety. Many aromatics are chemically bound to other molecules, often sugars, and so aren't evident in the raw fruit; during winemaking, fruit and yeast enzymes liberate the aromatics and so make them available for us to enjoy.

Grape Varieties and Clones The grapevine evolved with the ability to regenerate itself and grow vigorously in the spring. It's easily propagated by cuttings, and readily lends itself to creating identical versions, or *clones* of a given plant. And it's a variable species, one that offers many differences in growth habit, requirements for water and temperature, and fruit composition. For several millennia, and until around 1800, grapes were mostly cultivated and made into wine throughout western Asia and Europe by small groups of people essentially isolated from each other and living in different environments. So there developed a large number of distinctive grape varieties, each selected by particular people for characteristics they found desirable.

Today it's estimated that there are around 15,000 different varieties of the Eurasian grape, *Vitis vinifera.* A single variety—Pinot Noir, for example, or Cabernet Sauvignon—may exist in the form of several hundred different clones, each a somewhat different version of that variety. Some varieties have very distinctive aromas; others are more subtle or even anonymous and therefore allow the aromas of the fermentation and aging more prominence. The term *noble* is applied to varieties that produce wines with the potential for developing great complexity over many years in the bottle; these include the French Cabernet Sauvignon, Pinot Noir, and Chardonnay, the Italian Nebbiolo and Sangiovese, and the German Riesling.

The Influence of Growing Conditions; Vintage and "*Terroir*"

Pampered Vines Don't Make the Best Wines As Pliny observed 2,000 years ago, "the same vine has a different value in different places." The quality of grapes, and of the wine made from them, is influenced by the conditions in which the grapes grow and mature. To produce a decent wine, the grapes must ripen to an adequate sweetness, and so the vine must get enough sun, warmth, minerals, and water. On the other hand, abundant water produces watery fruit, abundant soil nitrogen produces excess foliage that shades the fruit and gives it odd flavors, and abundant sun and warmth produce fruit with plenty of sugar but reduced acidity and aroma compounds, and thus a strong but flat wine.

Vintage Wines The grapes that make the best wines seem to be produced in a narrow range of conditions—barely adequate water, minerals and light and heat—that encourage complete but slow, gradual ripening. Those conditions may or may not be realized in a given year. Hence the significance for many wines of the *vintage,* the particular year in which the grapes are grown and harvested. Some years yield better wines than others.

Terroir In recent times, much has been said and written about the importance in winemaking of *terroir*: the influence on a wine of the particular place in which the grapes were grown. The French word includes the entire physical environment of the vine-

Hybrid and American Wine Grapes

The Eurasian wine grape *Vitis vinifera* has a number of sister species in North America with which it can breed, and over the centuries plant breeders have produced a number of different Euro-American hybrids. These were generally denigrated by European connoisseurs and bureaucrats for their untypical flavors, but the better of them, and American varieties themselves, are coming to be appreciated for their own qualities. They include grapes based on the mainly northeastern *Vitis labrusca* (Concord, floral Catawba, strawberry-like Ives), midwestern *Vitis aestivalis* (Norton, Cynthiana), southeastern *Vitis rotundifolia* (floral-citrus Scuppernong), and on complex parentage (Chambourcin, developed in the Loire region of France).

yard: the soil, its structure and mineral content; the water held in the soil; the vineyard's elevation, slope, and orientation; and the microclimate, the regime of temperature, sunlight, humidity, and rainfall. Each of these aspects can vary over small distances, from one vineyard to the next; and each can affect the growth of the vine and the development of its fruit, sometimes in indirect ways. For example, sloping ground and certain kinds of soil encourage water to drain away from the roots, and absorb and release the sun's heat to the vine in different ways. A south-facing slope can increase the exposure to autumn sunlight by 50% over a planting on level ground, and thus extend the growing season and accumulation of flavor compounds.

The wine connoisseur enjoys detecting and marveling at the expression of *terroir* in wines, the palpable differences in the wines made from neighboring vineyards. The winemaker, on the other hand, generally tries to manage and minimize the effects of less than ideal *terroirs* and vintages. There's nothing new about this effort to make the best of things. The French have been adding sugar to their fermenting grapes for centuries to compensate for incomplete ripening. What's new nowadays is the degree to which the grape composition can be manipulated after harvest, so that the wine becomes less the product of a particular place and year, and more the product of modern fermentation technology.

Making Wine

The making of a basic table wine can be divided into three stages. In the first, the ripe grapes are crushed to free their juice. In the second, the grape juice is fermented by sugar-consuming, alcohol-producing yeasts into new wine. The third stage is the aging or maturing of the new wine. This is a period during which the chemical constituents of the grape and the products of fermentation react with each other and with oxygen to form a relatively stable ensemble of flavor molecules.

Crushing Grapes to Make the Must

Crushing extracts from the grape the liquid that will become wine. This step therefore determines to a large extent the final wine's composition and potential qualities.

The substances important to wine quality are not evenly distributed in the grape. The stems contain bitter-tasting resins and are usually separated from the grapes as they are crushed. The skin holds much of the fruit's phenolic compounds, both pigments and tannins, as well as most of the acid and the many compounds that give the grape its characteristic aroma. Like the stem, the seeds at the center are full of tannins, oils, and resins, and care is taken not to break them open during the pressing.

As a mass of grapes is crushed in a mechanical press, the first juice to come out, the *free run*, is primarily from the middle of the pulp, and is the clearest, purest

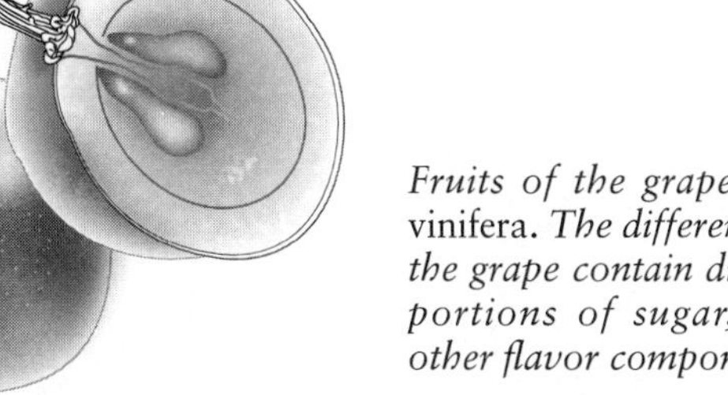

Fruits of the grape vine, Vitis vinifera. *The different regions of the grape contain different proportions of sugar, acid, and other flavor components.*

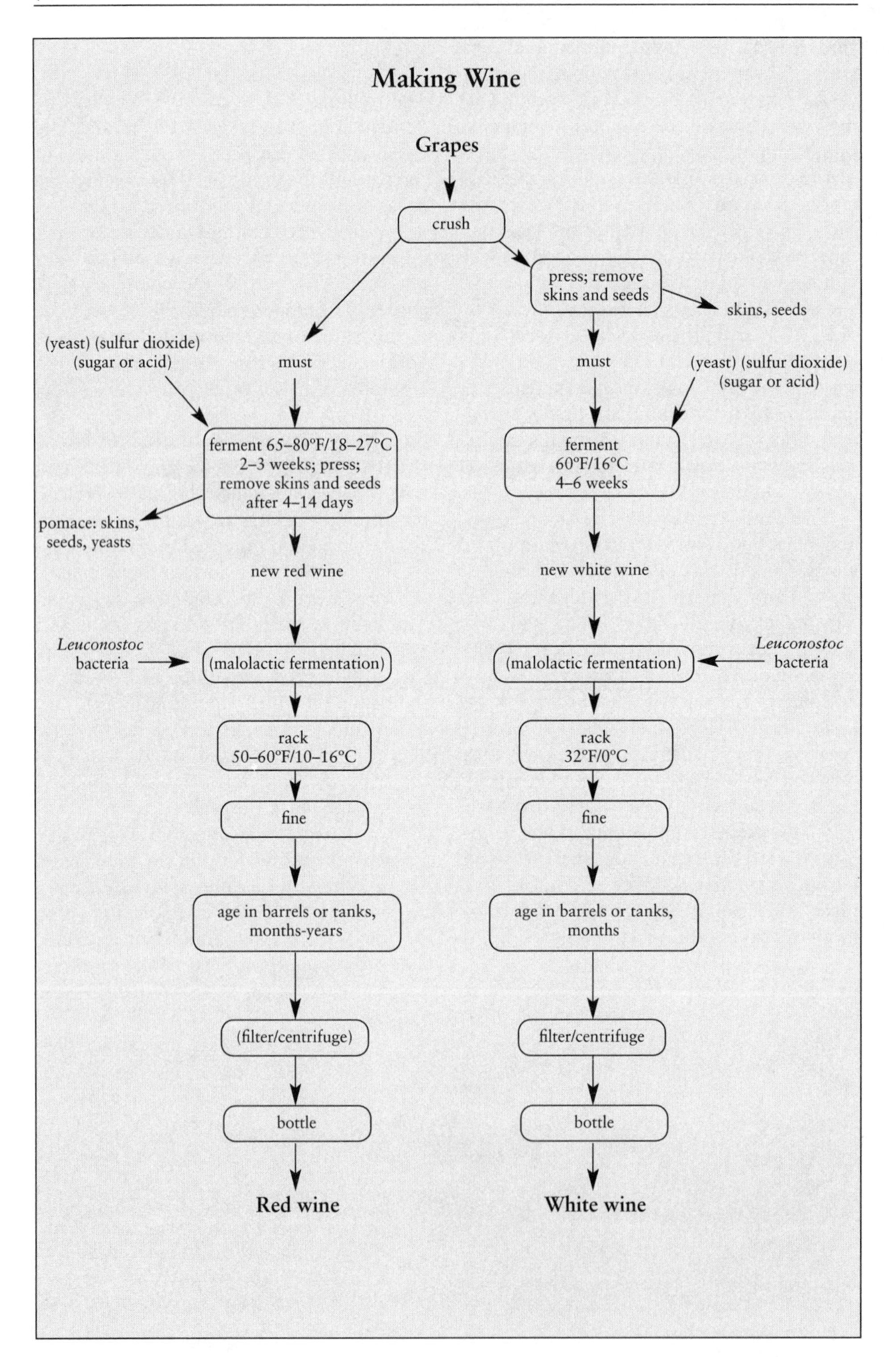
Making Wine
Grapes
crush
press; remove skins and seeds
skins, seeds
(yeast) (sulfur dioxide) (sugar or acid)
must
must
(yeast) (sulfur dioxide) (sugar or acid)
ferment 65–80°F/18–27°C 2–3 weeks; press; remove skins and seeds after 4–14 days
ferment 60°F/16°C 4–6 weeks
pomace: skins, seeds, yeasts
new red wine
new white wine
Leuconostoc bacteria
(malolactic fermentation)
(malolactic fermentation)
Leuconostoc bacteria
rack 50–60°F/10–16°C
rack 32°F/0°C
fine
fine
age in barrels or tanks, months-years
age in barrels or tanks, months
(filter/centrifuge)
filter/centrifuge
bottle
bottle
Red wine
White wine

essence of the grape, sweet and largely tannin free. As mechanical pressure is applied, juices from just under the skin and around the seeds augment the free run with a more complex character. The extent of pressing will have an important influence on the character of the final wine. The liquid portion, called the *must,* is 70 to 85% water, 12 to 27% sugars, mainly glucose and fructose, and about 1% acids.

After the Crush In the case of white wines, the must is left in contact with the skins for a few hours and removed with gentle pressure before fermentation. It thus picks up little tannic material or pigmentation. Rosé musts and red wine musts are partly fermented in contact with the red skins. The longer the must is in contact with skin and seeds, and the harder it is pressed, the deeper the color (whether yellow or red) and the more astringent the taste.

Before beginning the fermentation, the winemaker usually adds two substances to the must. One, sulfur dioxide, suppresses the growth of undesirable wild yeasts and bacteria, and prevents the oxidation of both flavor and pigment molecules (the same treatment is given to many dried fruits, and for the same reason). Though this treatment may sound antiseptically modern, it is centuries old. One of the natural by-products of fermentation that is increased by sulfuring is sulfites, sulfur compounds that can induce an allergic reaction in sensitive people.

The second additive is either sugar or acid, and it is used to correct the balance between these two substances. Grapes that ripen in a cool climate can lack the sugar to produce enough alcohol for a stable wine; grapes that ripen in a hot climate metabolize some of their acids and can produce a flat-tasting wine. French winemakers usually add sugar; California winemakers often add tartaric acid.

The Alcoholic Fermentation

The Fermentation Yeasts Fermentation can begin with or without the addition of a starter culture of yeast. The winemaker can choose among many different strains of *Saccharomyces,* or allow the fermentation to begin spontaneously with "wild" yeasts from the grape skins (species of *Kloeckera, Candida, Pichia, Hansenula,* and others). These are always eventually displaced by *Saccharomyces cerevisiae,* which has a greater tolerance for alcohol, but they do contribute flavor compounds to the finished wine.

The primary job of the yeast is to convert sugar to alcohol, but it also produces various volatile, aromatic molecules that the grape itself cannot supply. Prominent among them are the longer-chain alcohols, and esters, a class of compounds that combine an acid with an alcohol or phenol. Both yeast and grape enzymes and the acid conditions also liberate aromatic molecules from the nonvolatile sugar complexes in which some are stored in the grape, so fermentation also brings out the grape's own flavor potential.

Temperatures and Times The winemaker varies the conditions of fermentation according to the particular kind of wine being made. In the case of delicate white wines, the must is fermented for four to six weeks at about 60°F/16°C. With more robust red wines, the must is fermented at a temperature between 65 and 80°F/18–27°C in contact with the skins, to extract pigments, tannins, and flavor. This phase may last for 4 to 14 days (less if heat or a carbon dioxide treatment is applied). Then the must is separated from the skins and fermented again for a total of two to three weeks. One of the most critical variables

OPPOSITE: *Making red and white wines. White wines are fermented at a lower temperature and without the grape skins and seeds; they're also clarified at a lower temperature, in part so that they won't cloud when served chilled.*

during fermentation is temperature. The lower the temperature, the slower and longer the fermentation, and the more aromatic molecules accumulate.

The main fermentation is considered complete when essentially all the sugar in the must has been converted into alcohol. A wine with no residual sugar is called *dry.* Sweet wines are made by stopping the fermentation before all the sugar has been consumed, or more commonly, by adding some reserved sweet grape juice to the dry wine after its yeast has been removed.

Malolactic Fermentation Winemakers sometimes allow or even induce a second bacterial fermentation in the new wine after the main yeast fermentation. The bacterium *Leuconostoc oenos* consumes the wine's malic acid and converts it into lactic acid, which is less strong and sour. This "malolactic" fermentation thus reduces the apparent tartness of the wine. It also produces a number of distinctive aroma compounds, among them buttery diacetyl. (A relative of *L. oenos, L. mesenteroides,* contributes the same compound to cultured butter itself!) Some winemakers work to prevent a spontaneous malolactic fermentation from developing, so that they can retain the sharpness and flavor of the original wine.

Maturation Once fermentation is complete, the new wine is drained out of the fermentation tanks to begin the work of clarification and aging, in which the cloudy, rough-tasting liquid develops into a clear, smooth one.

Racking and Fining The solid particles of grape and yeasts are cleared from the wine by the process of racking: letting the yeast cells and other large particles settle, carefully drawing the wine off the sediment into a new container, and repeating the process every few months. Interesting exceptions to this rule are wines intentionally aged for months or even years *sur lie,* or "on the lees," in contact with the yeast sediment, whose cells slowly break apart and contribute more flavor and body to the wine. Champagne and Muscadet are two wines that are aged on the lees.

The cool racking temperatures—less than 60°F/16°C for red wines, around 32°F/0°C for whites—reduce the solubility of all the dissolved solids, and cause the wine to cloud up with a fine precipitate of various complexes of proteins, carbohydrates, and tannins. Late in the racking procedure, the wine may be *fined*: that is, a substance will be added to the wine that attracts these suspended particles to itself, and then settles to the bottom, carrying them with it. Gelatin, egg white, bentonite clay, and synthetic materials are used. Any particles remaining in the wine after racking and fining can be removed by passing it through a

Brettanomyces, the Controversial Barrel Yeast

Some wines, including classic reds from Burgundy and Bordeaux, develop striking and unusual aroma notes suggestive of the barnyard or horse stable. Oenologists have recently discovered the major source of these aromas to be a group of yeasts in the genus *Brettanomyces,* which readily colonize wine barrels. At low levels, their unusual aromas may suggest tobacco leaf; they also include smoky, medicinal, clove-like, and musty notes (the barnyard notes come from ethyl phenols and isovaleric and isobutyric acids). Some wine lovers consider the "brett" aroma a defect caused by contamination and inadequate winery sanitation, while others appreciate it as an intriguing contribution to the variety and complexity of wine flavors.

centrifuge or filter. Winemakers may choose to limit or omit the fining and filtration steps, since they remove some flavor and body along with the undesirable particles.

Barrel Aging New wine has a raw flavor and a strong, simple, fruity aroma. As the wine rests after fermentation, a host of chemical reactions slowly proceeds, and results in the development of balance and complexity in the flavor. If the wine is being held in a new wood barrel, it also absorbs various substances from the wood that either provide flavor directly—for example, vanilla-like vanillin and the coconut-woody oak lactones—or that modify the wine's own flavor molecules. In traditional winemaking, the months during which the wine is racked and shifted from container to container are a time when the wine's chemical evolution is directed by periodic exposure to the air. In the presence of oxygen, the tannins, anthocyanin pigments, and other phenolic compounds react with each other to form large complexes, so the wine's astringency and bitterness decline. Some of the molecules that provide aroma break apart or react with oxygen and each other to form a new suite of aromas, so fruity, floral notes fade in favor of a more subdued general "wineyness." White and light red wines are generally bottled young, after 6–12 months, with a fairly fresh, fruity bouquet, while astringent dark reds may require a year or two to develop and smooth out.

Most wines are made by blending two or more different varieties, and this important test of the winemaker's art occurs just before bottling. The final wine may then be filtered to remove any remaining microbes and haze, and given a final dose of sulfur dioxide to prevent microbial growth during storage. It may also be pasteurized. This practice is not limited to inexpensive wines. The Burgundy wines of Louis Latour are flash-heated for 2–3 seconds at 165ºF/72ºC, a treatment that is said not to have a detrimental effect on the wine's continuing flavor development.

Bottle Aging After a period of a few months to two years in barrels or tanks that allow controlled contact with oxygen, the wine goes into impermeable glass bottles. For the last two centuries, the standard stopper for wine bottles has been cork, which is made from the bark of a species of oak. Because cork can be the source of off-flavors, some wine producers are now using metal and plastic stoppers (see box below).

Corks, Cork Taints, Corked Wine

Cork is the outer protective layer of an evergreen oak, *Quercus suber,* that is native to the western Mediterranean. Where most tree bark is fibrous, cork is composed of tiny air cells. Nearly 60% of the cell wall in cork is made up of *suberin,* a complex waxy substance similar to the cutin that coats many fruits, and this makes cork water-resistant and long-lived.

Cork is a natural, organic material, and as such can be infected by molds and bacteria. Molds produce moldy, earthy, mushroomy, and smoky smells. And certain bacteria can act on phenolic compounds in the cork and traces of chlorine disinfectant to produce trichloroanisole, an especially unpleasant and potent molecule that smells like a dank cellar. It's estimated that cork taints spoil from 1 to 5 percent of the wine bottles stoppered with cork. The problem of "corked" wine has led wine producers to experiment with alternative stoppers, including metal caps and stoppers made from foamed plastic.

Wine continues to be affected by oxidation long after it leaves the cask. It picks up some air when it is filled into the bottle, and the bottle is sealed with a small space between wine and cork. So while oxidation is greatly slowed in the bottle, it does continue, though it may be outweighed by a different set of reactions, "reductive" rather than oxidative. The chemical changes that occur are not well understood, but include the ongoing liberation of aromatic molecules from nonaromatic complexes, and aggregation reactions among tannins and pigments that further lower astringency and cause a shift in pigment hues, usually toward the brown.

White wines and light-red rosés benefit from about a year of bottle aging, during which time the aroma develops and the amount of free, odorous sulfur dioxide decreases. Many red wines improve greatly after a year or two in the bottle, and some may develop for decades. All wines have a finite life span, and their quality eventually declines. White wines develop overtones of honey, hay, wood, and chemical solvents; red wines lose much of their aroma and become flat-tasting and more plainly and harshly alcoholic.

Special Wines

In the last few pages, I've described the general method for making dry table wines, which usually accompany meals. Sparkling, sweet, and fortified wines are often sipped on their own. Here is a brief account of their special qualities and how they are produced.

Sparkling Wines: Champagne and Others Sparkling wines delight by emitting bubbles that catch the light and prick the tongue. The bubbles come from the wine's considerable dissolved reserves of carbon dioxide gas, a by-product of yeast metabolism that ordinarily escapes into the air from the surface of the fermenting wine. To make a sparkling wine, the wine is confined under pressure—either in the bottle or in special tanks—so that the carbon dioxide can't escape as it's produced, and instead comes to saturate the liquid. A bottle of Champagne holds a gas pressure of 3–4 atmospheres, somewhat higher than the pressure in car tires, and contains about six times its volume in carbon dioxide!

When we remove the cork and thus relieve the pressure, the excess carbon dioxide leaves the solution in the form of gas bubbles. The bubbles form wherever the liquid comes into contact with a microscopic air pocket into which the dissolved carbon dioxide can diffuse. In the glass, the bubbles form on scratches and other surface imperfections. The refreshingly sharp prickle in the mouth comes from the irritating dose of carbonic acid that the bubbles deliver as they redissolve into the unsaturated layer of saliva.

Many countries have their own versions of sparking wine, which range from the carefully crafted to the mass-produced. The best-known example of a sparkling wine is Champagne, which strictly speaking is the wine made in the region of that name east of Paris, and accounts for less than a tenth of the world's production of sparkling wine. From the late 17th century to the late 19th, Champagne evolved to become the most refined expression of this style. The French invented the method of inducing a bubble-making second fermentation in the bottle, and this *méthode champenoise* has become a worldwide benchmark for making fine sparkling wines.

Making Champagne The first stage in making Champagne is to produce a base wine, which is made primarily from Pinot Noir and/or Chardonnay grapes. Next comes the secondary fermentation, which must be carried out in a closed container in order to retain the gas. Sugar is added to the dry base wine as food for the yeast. The wine, sugar, and yeast are put into individual bottles, corked, clamped, and kept at about 55°F/13°C.

Though the secondary fermentation is

usually complete after about two months, the wine is left to age in contact with the yeast sediment for anywhere from a few months to several years. During this time, most of the yeast cells die, fall apart, and release their contents into the wine, giving it a distinctive, complex flavor with toasted, roasted, nutty, coffee, even meaty notes (due in large part to complex sulfur compounds). In addition to flavor, yeast proteins and carbohydrates will stabilize bubbles when they form in the glass, and help produce the very fine bubbles typical of Champagne. After aging on the yeast, the sediment is removed and the bottle topped off with additional wine, and finished with a small amount of aged wine mixed with sugar and brandy. The bottle is then recorked.

Making Other Sparkling Wines The traditional Champagne process is labor intensive, time consuming, and expensive. More affordable and less complex sparkling wines are made all over the world in a number of different ways. One is simply to minimize or eliminate aging on the yeast sediment. Others involve fermenting the base wine for the second time not in individual bottles, but in large tanks, or not fermenting again at all: the cheapest sparkling wines are simply carbonated as soft drinks are, with tanks of pressurized carbon dioxide.

Sweet Wines Table wines are generally fermented until they are *dry*: that is, until the yeast consumes essentially all the grape's sugars and converts them into alcohol. Sweet or dessert wines with 10–20% "residual" sugar are made in several different ways:

- An ordinary dry wine is sweetened with some unfermented grape juice, and the combination treated to prevent further fermentation with a dose of sulfur dioxide, or filtration that removes all yeast and bacteria from the wine.
- Grapes are dried on the vine or after picking to concentrate their sugars to 35% or more of the grape weight. This leaves residual sugar in the wine when the yeasts reach their maximum alcohol level and the fermentation stops. German *Trockenbeerenauslese* and Italian *recioto* wines are examples.
- Grapes are left on the vine past the first frost and picked when frozen (or frozen artificially), and then gently pressed while cold to separate the concentrated juice from the ice crystals. The concentrated juice ferments into a stable wine with residual sugar. German *Eiswein* dates from around 1800.
- The grapes are allowed to become infected with "noble rot," the mold *Botrytis cinerea,* which dehydrates the grapes, concentrates their sugars, and transforms their flavor and con-

Enjoying Sparkling Wines

In order to appreciate their sparkle, sparkling wines are best served very cold, around 40°F/5°C, in tall, narrow glasses that allow their rising bubbles to be admired for several seconds. Carbon dioxide is more soluble in water at low temperatures, so the bubbles will be smaller and longer-lasting in cold wine. Because soaps, fats, and oils cause bubbles to collapse (p. 639), bubbliness is reduced when our lips deposit lipstick or oils from food on the glass, or when the glass has been incompletely rinsed and carries traces of dish soap.

sistency. This method originated in the Tokaji region of Hungary around 1650, and was adopted in the German Rheingau by 1750 and in the Sauternes region of Bordeaux around 1800.

The Noble Rot: Tokaji, Sauternes, and Others The noble rot (French *pourriture noble,* German *Edelfäule*) *Botrytis cinerea* is also known as bunch rot, and it is mainly a destructive disease of grapes and other fruits. It becomes noble only in the right climatic conditions, when the initial infection in humid weather is followed by a dry period that limits the infection. In this situation, the mold does several useful things. It perforates the skin of the grape, thus allowing it to lose moisture and become concentrated during the subsequent dry period; it metabolizes some of the tartaric acid at the same time that it consumes some of the grape's sugars, so the balance between sweetness and acidity doesn't suffer; it produces glycerol, which lends the eventual wine an incomparably dense body; and it synthesizes a number of pleasing aroma compounds, notably the maple-sugar-like sotolon, mushroomy octenol, and a number of terpenes. The honeyed flavor of these wines can develop in the bottle for decades.

Fortified Wines *Fortified* wines are so called because the strength of the base wine is boosted by the addition of distilled spirits to 18–20% alcohol, a level that prevents spoilage by vinegar bacteria and other microbes. Fortification appears to have begun in the sherry-producing region of Spain sometime before 1600. Winemakers take advantage of the stability of fortified wines by exposing them to the air for months or years. They thus make a virtue of the normally undesirable oxidative changes in flavor that come with keeping leftover wine. Most fortified wines keep well for weeks in an opened bottle or decanter.

Madeira Beginning in the 15th century, Portuguese ships embarking on long voyages to the Indies would pick up barrels of ordinary wine on the Portuguese island of Madeira. Sailors and producers soon found that the long barrel aging in extreme temperatures and with constant agitation produced an unusual but attractive wine. By 1700, ships were sailing to the East Indies and back just to age the barrels of Madeira stored on board; by 1800, the wine was

Styles of Port

Today there are several distinct styles of port, the most common of which are these:

- Vintage port is made from the best grapes in especially good years, and is barrel-aged for two years and bottle-aged unfiltered for a minimum of 10 years, often many decades longer. It's dark and fruity, must be decanted off its considerable sediment, and must be drunk within a few days after the bottle is opened.
- Tawny port, so named for its brown color (the result of precipitation of the red pigments), is typically barrel-aged for 10 years before being filtered and bottled. It is much more oxidized than vintage port of the same age, and can be kept in an opened bottle or decanter for weeks.
- Ruby port is an intermediate product, aged for three years in barrels before filtration and bottling.

being fortified with brandy and hot-aged on the island. Today, the base wine, which can be white or red, is fortified, sometimes sweetened, then artificially heated to a temperature around 120°F/50°C, where it's held for three months before slowly cooling down again. It's then aged in barrels in a sherry-like *solera* system (below) before bottling. There are several different styles of Madeira, from sweet to nearly dry.

Port The name *port* was originally the English term for any Portuguese wine. The addition of brandy was introduced in the 18th century as a way of guaranteeing that the wines would get to England in drinkable condition, and it resulted in the development of an unusual group of sweet red wines. Port is made by stopping the fermentation of the base red wine while about half the grape sugar is left, and fortifying it with distilled spirits to give an alcohol content around 20%. The wine is then aged in barrel and finally in bottle for anywhere from two to 50 years. Older ports are characterized by the maple-like compound sotolon and other sweetly aromatic compounds, likely products of browning reactions, which are also found in botrytized wines and sherries.

Sherry Sherry is a fortified, oxidized white wine that was developed in the Spanish port city Jerez de la Frontera, whose name was Anglicized to "sherry" around 1600. True sherry gets its distinctive flavor from the *solera* system of maturing wine, which was developed early in the 19th century. The solera is a series of casks, each initially containing the fortified new wine of a particular year, but not completely filled, so a significant area of wine surface is in direct contact with the air. The wine therefore develops a characteristic intense, oxidized flavor. As the cask contents evaporate and become more concentrated, each is replenished with wine from the next younger cask. The final wine is bottled from the casks containing the oldest wines, and thus is a blend of wines from many different vintages and degrees of development.

There are a variety of rapid industrial methods for making sherry-like wines. The fortified base wines may be heated to develop the flavor, or cultured with a "submerged flor": the wine and flor yeasts (see box) are kept in large tanks and agitated and aerated.

Vermouth Modern-day vermouth derives from a medicinal wine made in 18th-century Italy, which the Germans named *Vermut* after its main ingredient, wormwood (see box, p. 771). Today it's essentially a flavored wine fortified to about 18% alcohol, used mainly in mixed drinks

Styles of Sherry

True sherry from the Jerez region of Spain is made in several different ways to produce different styles of wine.

- *Fino* is the lightest, least fortified, and least oxidized. Its surface in the solera is protected from the air by a layer of unusual yeasts called the *flor*.
- *Amontillado* is essentially *fino* sherry that did not develop or retain a *flor* in the solera, and so is more oxidized, darker, and heavier.
- *Oloroso* sherries (from the Spanish for "fragrant," "perfumed") are made from heavier, more strongly fortified base wines that don't develop a *flor*, reach 24% alcohol, and become dark brown and concentrated.

and in cooking. Vermouth is made in both Italy and France from a neutral white wine flavored with dozens of herbs and spices, and sometimes sweetened (up to 16% sugar). The French usually extract the flavorings in the wine itself, while the Italians extract or co-distill them with the fortification alcohol. Once fortified, the wine is aged for several months.

Storing and Serving Wine

Wine Storage Wines are sensitive liquids, and require some care in order to keep well and even improve during storage. They're best kept in some version of the traditional cellar: a moderately humid, dark, cool place. Bottles are stored on their sides, so that the wine wets the cork and prevents it from drying out, shrinking, and allowing air in. Moderate humidity keeps the outer portion of the cork from shrinking, and constant temperature prevents volume and pressure changes in the liquid and air inside the bottle, which can cause air and wine movement in the space between bottle and cork. Darkness minimizes the penetration of high-energy light into sparkling and other white wines, where it can cause a sulfury off-aroma similar to that found in light-struck beer and milk (pp. 749, 21). And low temperatures, between 50 and 60°F/10–15°C, slow the wine's development, so that it remains complex and interesting for the longest possible time.

Serving Temperatures Different kinds of wine taste their best at different serving temperatures. The colder a wine, the less tart, sweet, and aromatic it seems. Intrinsically tart and mildly aromatic wines, usually light white and rosé wines, are best served cold, 42–55°F/5–13°C. Less tart, more aromatic red wines are more full-flavored at 60–68°F/16–20°C. The strongly alcoholic, richly aromatic port is said to taste best at 65–72°F/18–22°C. Complex white wines may be served at higher temperatures than their light cousins; similarly, many light red wines are better at cooler temperatures.

"Breathing" and Aeration Wines can sometimes be improved just before serving by a period of aeration or "breathing." Such a treatment allows volatile substances in the wine to escape into the air, and it allows oxygen from the air to enter the wine, where it reacts with volatile and other molecules and changes the wine's aroma. No significant aeration occurs when a wine is simply uncorked and left to sit in the open bottle. The most effective way to aerate a wine is to pour it, and into a broad, shallow decanter that continues to expose a large surface area to the air. Aeration can improve a wine's aroma by accelerating the escape of some off-odors (for example, excess sulfur dioxide in some white wines), and by providing a kind of accelerated aging to young, undeveloped red wines. But it allows desirable aromas to escape as well, and can undo the complexity of a mature wine that has developed slowly over years in the bottle.

Wine also absorbs oxygen when it's poured into the glass and as it rests there, and its aroma often evolves noticeably between the first sip and the last. To discover this dynamic quality and follow its course is one of the pleasures of drinking wine.

The Alcohol Content of Wines

In the United States, the approximate alcohol content of a wine is listed on its label. There is a total of 3% leeway allowed, so a wine labeled as containing 12% alcohol by volume may contain anywhere from 10.5% to 13.5%.

Keeping Leftovers The key to preserving the quality of leftover wine is to minimize chemical change. Lowering the wine's temperature slows all chemical activity, and simple refrigeration works well for white wines, which usually keep well simply by being recorked and refrigerated. However, chilling causes dissolved substances in more complex red wines to precipitate into solid particles, and this causes irreversible changes in flavor. Leftover red wine is best handled by a minimizing its exposure to flavor-altering oxygen, which can be done by means of inexpensive devices that pull the air out of a partly empty bottle, or by replacing the air with inert nitrogen gas, or by gently pouring a partial bottle into smaller bottles that can be filled all the way to the top—though the act of pouring itself introduces some air into the wine.

Enjoying Wine

For those who love it, wine can be endlessly fascinating. The varieties of grape, the place they were grown in, the weather that year, the yeasts that ferment them, the skills of the winemaker in handling them, the years they spend in oak or in glass: all these factors and more affect what we taste in a sip of wine. And there's a lot to taste in that sip, because wine has one of the most complex flavors of all our foods. Wine connoisseurs have developed an elaborate vocabulary to try to capture and describe these fugitive sensations, one that may seem forbiddingly complicated and fanciful. Many of us most of the time would be content with the five *F*'s proposed 800 years ago in the Regimen of Health for the School of Salerno:

Si bona vina cupis, quinque haec laudantur in illis:
Fortia, formosa, et fragrantia, frigida, frisca.
If you desire good wines, these five things are praised in them:
Strength, beauty, and fragrance, coolness, freshness.

On the other hand, we can learn to taste much more in a sip of wine, and get more pleasure from it, if we know something about what's in that sip, about the kinds of substances that can and do contribute to the flavor of wine (see box, p. 738).

Clarity and Color The appearance of a wine can give some important clues about how it will taste. If the wine is cloudy and the particles don't settle with a few hours' standing, it has probably undergone an unintended bacterial fermentation in the bottle, and its flavor is likely to be off. Tiny crystals (which *do* settle) are usually salts of excess tartaric or oxalic acid, and are not signs of spoilage; in fact they indicate a good level of acidity. "White" wines actually range in color from straw yellow to deep amber. The darker the color, the older the wine—the yellow pigments turn brownish when oxidized—and the more mature the flavor. Most red wines retain a deep, ruby-like color for some years, along with a fruity character in the flavor. As they age, the anthocyanin pigments complex with some of the tannins and precipitate, leaving more of the brownish tannins visible. The wine develops an amber or tawny tint, which goes along with its less fruity, more complex flavor.

Feeling and Taste in the Mouth When we experience a sip of wine in the mouth, the senses of both touch and taste come into play.

Astringency The *feel* of a wine is largely a matter of its astringency and viscosity. Astringency—the word comes from the Latin for "to bind together"—is the sensation we have when the tannins in wine "tan" the lubricating proteins in our saliva the way they do leather: they cross-link the proteins and form little aggregates that make the saliva feel rough rather than slick. This dry, constricting feeling, together with the smoothness and viscosity caused by the presence of alcohol and other extracted components, and in sweet wines sugar, cre-

Some Aromas and Molecules in Wines

Here are examples of the molecules and aromas that chemists have found in wines and that contribute significantly to wine flavor.

Aroma Quality	Wine	Chemical
Fruits: Apple, pear	Many wines	Ethyl esters
Banana, pineapple	Many wines	Acetate esters
Strawberry	Concord grape wine	Furaneol
Guava, grapefruit, passion fruit	Sauvignon Blanc, Champagne	Sulfur compounds
Citrus fruits	Riesling, Muscat	Terpenes
Apple	Sherries	Acetaldehyde
Flowers: Violets	Pinot Noir, Cabernet Sauvignon	Ionone
Citrus, lavender	Muscat	Linalool
Rose	Gewürztraminer	Geraniol
Rose	Sake	Phenethyl alcohol
Rose, Citrus	Riesling	Nerol
Wood: Oak	Barrel-aged wines	Lactones
Nuts: Almond	Barrel-aged wines	Benzaldehyde
Vegetables: Bell pepper, green peas	Cabernet Sauvignon, Sauvignon Blanc	Methoxyisobutylpyrazines
Grass, tea	Many wines	Norisoprenoids
Asparagus, cooked vegetables	Many wines	Dimethyl sulfide
Spices: Vanilla	Barrel-aged wines	Vanillin
Clove	Barrel-aged red wines	Ethyl, vinyl guaiacol
Tobacco	Barrel-aged red wines	Ethyl, vinyl guaiacol
Earthiness: Mushrooms	Botrytized wines	Octenol
Stone	Cabernet Sauvignon, Sauvignon Blanc	Sulfur compound
Smoke, Tar	Many red wines	Ethyl phenol, ethyl guaiacol, vinyl guaiacol
Sweet, caramel: Maple syrup, fenugreek	Sherry, port	Sotolon
Butter	Many white wines	Diacetyl
Roasted: Coffee, toasted brioche	Champagne	Sulfur compounds
Grilled meats	Sauvignon Blanc	Sulfur compounds
Animals: Leather, horse, stable	Many red wines	Ethyl phenol, ethyl guaiacol, vinyl guaiacol
Cat	Sauvignon Blanc	Sulfur compounds
Solvent: Kerosene	Riesling	TDN (trimethyldihydronaphthalene)
Nail polish remover	Many wines	Ethyl acetate

ate the impression of the wine's body, of substance and volume. In strong young red wines, the tannins can be palpable enough that "chewy" seems a good description. In excess, they are drying and harsh.

Taste The *taste* of a wine is mostly a matter of its sourness, or a balance between sour and sweet, and a savory quality that has been attributed to succinic acid and other products of yeast metabolism. Phenolic compounds can sometimes contribute a slight bitterness. The acid content of a wine is important in preventing it from tasting bland or flat; it's sometimes said to provide the "backbone" for the wine's overall flavor. White wines are usually around 0.85% acid, red wines 0.55%. Wines that are fermented dry, with no residual sugar, may still have a slight sweetness thanks to the alcohol and glycerol, a sugar-like molecule produced by the yeasts. Fructose and glucose are the predominant sugars in grapes, and they begin to provide a noticeable sweetness when left in wines at levels around 1%. Sweet dessert wines may contain more than 10% sugars. In strong wines, alcohol itself can dominate other sensations with its pungent harshness.

Wine Aroma If acidity is the backbone of a wine, viscosity and astringency its body, then aroma is its life, its animating spirit. Though they account for only about one part in a thousand of wine's weight, the volatile molecules that can escape the liquid and ascend into the nose are what fill out its flavor, and make wine much more than tart alcoholic water.

An Ever-Changing Microcosm A given wine contains several hundred different kinds of volatile molecules, and those molecules have many different kinds of odors. In fact they run the gamut of our olfactory world. Some of the same molecules are also found in temperate and tropical fruits, flowers, leaves, wood, spices, animal scents, cooked foods of all kinds, even fuel tanks and nail polish remover. That's why wine can be so evocative and yet so hard to describe: at its best, it offers a kind of sensory microcosm. And that little world of molecules is a dynamic one. It evolves over months and years in the bottle, by the minute in the glass, and in the mouth with every passing second. The vocabulary of wine tasting thus amounts to a catalogue of things in the world that can be smelled, and whose smell can be recognized, however fleetingly, in an attentive sip.

A few of the aromatic substances in wine are contributed directly by particular varieties of grape, mainly the flowery terpenes of some white grapes and unusual sulfur compounds in the Cabernet Sauvignon family. But the primary creators of wine aroma are the yeasts, which apparently generate most of the volatile molecules as incidental by-products of their metabolism and growth. The yeasts and 400 generations of winemakers, who noticed and cultivated those incidental pleasures, made a tart alcoholic liquid into something much more stimulating.

BEER

Wine and beer are made from very different raw materials: wine from fruits, beer from grains, usually barley. Unlike grapes, which accumulate sugars in order to attract animals, the grains are filled with starch to provide energy for the growing embryo and seedling. Yeasts can't exploit starch directly, and this means that before they can be fermented, grains must be treated to break down their starch to sugars. While it's true that grapes are much more easily fermented—yeasts begin to flourish in the sweet juice as soon as they break open—grain has several advantages as a material for producing alcohol. It's quicker and easier to grow than the grapevine, much more productive in a given acreage, can be stored for many months before being fermented, and it can be made into beer any day of the year, not just at harvest time. Of course grains bring a very different flavor to beer than grapes do to wine; it's the flavor of the grasses, of bread, and of cooking, which is essential to the beer-making process.

The Evolution of Beer

Three Ways to Sweeten Starchy Grains

Our ingenious prehistoric ancestors discovered no fewer than three different ways to turn grains into alcohol! The key to each was enzymes that convert the grain starch into fermentable sugars. Because every enzyme molecule can do its starch-splitting operation perhaps a million times, a small quantity of the enzyme source can digest a large quantity of starch into fermentable sugars. Inca women found the enzymes in their own saliva: they made *chicha* by chewing on ground corn, then mixing that corn with cooked corn. In the Far East, brewers found the enzymes in a mold, *Aspergillus oryzae,* which readily grew on cooked rice (p. 754). This preparation, called the *chhü* in China, *koji* in Japan, was then mixed with a fresh batch of cooked rice. In the Near East, the grain itself supplied the enzyme. Brewers soaked the grain in water and allowed it to germinate for several days, then heated the ground seedling with ungerminated grain. This technique, called *malting,* is the one most widely used today to make beer.

Beer in Ancient Times Malting is much like the making of sprouts from beans and other seeds, and may have begun in the sprouting of grains simply to make them softer, moister, and sweeter. There's clear evidence that barley and wheat beers were being brewed in Egypt, Babylon, and Sumeria by the third millennium BCE, and that somewhere between a third and half of the barley crop in Mesopotamia was reserved for brewing. We know that brewers preserved the malted grain, or malt, by baking it into a flat bread, then soaked the bread in water to make beer.

The knowledge of beer making seems to have passed from the Middle East through western Europe to the north, where in a

Some Beer-like Brewed Drinks

This chapter concentrates on standard beers brewed with barley malt, but there are many other ways to make an alcoholic drink from starchy foods. Here are a few examples.

Name of Drink	Region	Main Ingredient
Chicha	South America	Boiled maize, chewed to contribute saliva enzymes
Manioc beer	South America and elsewhere	Boiled manioc root, chewed to contribute saliva enzymes
Millet, sorghum, rice beers	Africa, Asia	Millet, sorghum, rice
Boza/bouza	Southwest Asia, North Africa	Bread made from malted millet, wheat
Pombe ya ndizi	Kenya	Bananas and malted millet
Kvass	Russia	Rye bread
Roggenbier	Germany	Malted rye

climate too cold for the vine, beer became the usual beverage. (Among the nomadic tribes of northern Europe and central Asia who did not even cultivate grain, milk was fermented into the drinks called *kefir* and *koumiss.*) To this day, beer remains the national beverage of Germany, Belgium, Holland, and Britain.

Wherever both have been available, beer has been the drink of the common people and wine the drink of the rich. The raw material for beer, grain, is cheaper than grapes and its fermentation is less tricky and drawn out. To the Greeks and Romans, beer remained an imitation wine made by barbarians who did not cultivate the grape. Pliny described it as a cunning if unnatural invention:

> The nations of the West also have their own intoxicant, made from grain soaked in water. There are a number of ways of making it in the various provinces of Gaul and Spain. . . . Alas, what wonderful ingenuity vice possesses! A method has actually been discovered for making even water intoxicated.

Germany: Hops and Lagering In the centuries following the fall of Rome, beer continued to be an important beverage in much of Europe. Monasteries brewed it for themselves and for nearby settlements. By the 9th century, alehouses had become common in England, with individual keepers brewing their own. Until 1200, the English government considered ale to be a food, and did not tax it.

It was in medieval Germany that two great innovations made beer largely what it is today: brewers preserved and flavored it with hops, and began to ferment it slowly in the cold to make mild-flavored *lager.*

Hops The earliest brewers probably added herbs and spices to beer, both to give it flavor and to delay the development of off-flavors from oxidation and the growth of spoilage microbes. In early Europe this mixture, called *gruit* in German, included bog-myrtle, rosemary, yarrow, and other herbs. Coriander was also sometimes used, juniper in Norway, and sweet gale (*Myrica gale*) especially in Denmark and Scandinavia. It was around 900 that hops, the resinous cones of the vine *Humulus lupulus,* a relative of marijuana, came into use in Bavaria. Thanks to its pleasant taste and effectiveness in delaying spoilage, it had largely replaced *gruit* and other herbs by the end of the 14th century. In 1574, Reginald Scot noted in *A Perfite Platforme of a Hoppe Garden* that the advantages of hops were overwhelming: "If your ale may endure a fortnight, your beer through the benefit of the hops, shall continue a month, and what grace it yieldeth to the taste, all men may judge that have sense in their mouths." Still, it was not until about 1700 that English ale was hopped as a matter of course.

Lager From the times of Egypt and Sumeria to the Middle Ages, brewers made beer without much control over the temperature of fermentation, and with yeasts that grew at the surface of the liquid. The beer fermented in a few days, and it was consumed

Food Words: *Malt*

Our ancestors probably began soaking and sprouting grains because it was an easy way to make them soft enough to eat as is, and quicker to cook. In fact our word for soaked, partly sprouted cereal grain comes from an Indo-European root meaning "soft." Words that are related to *malt* include *melt, mollusc* (p. 226), and *mollify.*

within days or weeks. Sometime around 1400, in the foothills of the Bavarian Alps, there evolved a new kind of beer. It was fermented in cool caves over the period of a week or more, and with a special yeast that grew below the surface of the liquid. Then it was packed in ice for several months before it was drawn off the yeast sediment for drinking. The cool, slow fermentation gave the beer a distinctive, relatively mild flavor, and the cold temperature and long settling time produced a sparkling-clear appearance. This *lager* beer (from the German *lagern,* "to store," "to lay down") remained distinctly Bavarian until the 1840s, when the special yeast and techniques were taken to Pilsen, Czechoslovakia, to Copenhagen, and to the United States, and became the prototype of most modern beers. England and Belgium are the only major producers that still brew most of their beer in the original way, at warm temperatures and with top-fermenting yeasts.

England: Bottles and Bubbles, Specialty Malts The English were late to accept hops, but pioneered in the making of bottled beer. Ordinary ale—the original English word for beer—was fermented in an open tank, and like wine it lost all its carbon dioxide to the air: the bubbles simply rose to the surface and burst. Some residual yeast might grow while the liquid was stored in a barrel, but it would lose its light gassiness as soon as the barrel was tapped. Sometime around 1600, it was discovered that ale kept in a corked bottle would become bubbly. Quite early on, the discovery was attributed to Alexander Nowell, dean of St. Paul's Cathedral. Thomas Fuller, in his 1662 *History of the Worthies of England,* wrote:

> Without offense it may be remembered, that leaving a bottle of ale, when fishing, in the grass, [Nowell] found it some days after, no bottle, but a gun, such the sound at the opening thereof: and this is believed (casualty is mother of more invention than industry) the original of bottled ale in England.

By 1700, glass-bottled ale sealed with cork and thread had become popular, along with sparkling Champagne (p. 724). But both were largely novelties. Most beer was drunk flat, or close to it, from barrels. Centuries later, with the development of airtight kegs, of carbonation, and the increasing tendency to drink beer at home instead of at the tavern, bubbly beer became the rule.

Specialty Malts The 18th and 19th centuries were an innovative time in Britain, and it was early in this period that many of today's familiar British brewing names—Bass, Guinness, and others—got their start. By 1750, the greater control that coke and coal heat gave the maltster made gently dried pale malts possible, and thereby pale ales. And in 1817, "patent malt" was developed. This was malted barley roasted very

Food Words: *Ale* and *Beer; Brew*

The original English word for a fermented barley drink was not *beer,* but *ale.* It apparently derives from the effects of alcohol; the Indo-European root of *ale* had to do with intoxication, magic, and sorcery, and may be related to a root meaning "to wander, to be in exile." The alternative name, *beer,* comes via Latin from a much more prosaic connection: its root is the word for "to drink." *Brew* is related to *bread, broil, braise,* and *ferment*; they all come from an Indo-European root meaning "to boil, to bubble, to effervesce."

dark, and used in small amounts only to adjust the color and flavor of ales and beers, not to provide fermentable sugars. Patent and pale malts made it possible to produce a range of dark beers with a combination of light, largely fermentable malt and very dark coloring malt. This was the beginning of porter and stout as we know them today: darker and heavier than ordinary brews, but much lighter and less caloric than they were 200 years ago.

Beer in America The U.S. preference for light, even characterless brews would seem to be the result of climate and history. Heavy beer is less refreshing when the summers get as hot as ours do. And the original British colonists seem to have been more interested in making whiskey than beer (p. 760). We had no strong national tradition in the matter of beer, so the way was clear for later German immigrants to set the taste around 1840, when someone—perhaps one John Wagner near Philadelphia—introduced the newly available lager yeast and technique, and the distinctive brew caught on.

Both Milwaukee and St. Louis quickly became centers of lager brewing: in the former, Pabst, Miller, and Schlitz; in the latter, Anheuser and Busch; and Stroh in Detroit all got their starts in the 1850s and 1860s, and Coors in Denver in the 1870s. Several of these names and their light, Pilsner-style beers remain dominant today, while the stronger traditional brews of England and Germany appeal to a relatively small number of beer lovers. The only indigenous American style of beer is "steam beer," a rare relic of the California Gold Rush. Without the large supply of ice necessary to make lager beer, San Francisco brewers used the yeast and techniques appropriate to cool bottom fermentation, but brewed at top-fermentation temperatures. The result: a full-flavored and gassy beer that gave off a lot of foam when the keg was tapped.

Beer Today Today, the countries with the largest per capita consumption of beer are mainly traditional European beer producers: Germany, the Czech Republic, Belgium, and Britain and its former colony Australia. In the United States, beer accounts for more than three quarters of the alcoholic drinks consumed. Most American beer remains bland and uniform, produced by a handful of large companies in largely automated factory-like breweries. The 1970s brought a revival of interest in more flavorful alternatives, and a flourishing of "microbreweries" making specialty beers in small quantities, brewpubs that both brew and serve beer, and home brewing. Some of these small enterprises have grown with their success, and the giant brewers are now making microbrew lookalikes. And it's now possible to find beers from all over the world in liquor stores and supermarkets. These are good times for exploring the many different styles of beer and ale.

Brewing Ingredients: Malt

Beer begins with barley. Other grains—oats, wheat, corn, millet, sorghum—have also been used, but barley has become the grain of choice because it's the best at generating starch-digesting enzymes.

Malting The first step in converting barley grain into malt is to steep the dry grain in cool water and then allow it to germinate for several days at around 65°F/18°C. The embryo restarts its biochemical machinery and produces various enzymes, including some that break down the barley cell walls, and others that break down the starch and proteins inside the cells of the food-storage tissue, the endosperm. These enzymes then diffuse from the embryo into the endosperm, where they work together to dissolve away the cell walls, penetrate the cells, and digest some of the starch granules and protein bodies inside. The embryo also secretes the hormone gibberellin, which stimulates the aleurone cells to produce digestive enzymes as well.

The maltster's aim is to maximize the breakdown of the endosperm cell walls and the grain's production of starch- and protein-digesting enzymes. The cell walls have been adequately weakened by the time that the growing tip of the embryo reaches the end of the kernel, some five to nine days after the grain is first soaked. If the maltster is going to make a pale malt, then he keeps starch digestion to a minimum and malts for a shorter time; for a darker malt that will benefit from more sugars for the browning reactions, he malts for a longer time, and may finish by holding the moist barley at 140–180°F/60–80°C to maximize the action of the starch-digesting, sugar-producing enzymes.

Kilning Once the barley reaches the desired balance of enzymes and sugars, the maltster fixes that balance by drying and heating it in a kiln. The dehydration and heat kill the embryo, and they also generate color and flavor. To make malts with high enzyme activities, the maltster dries the barley gently, over about 24 hours, and brings the temperature slowly up to around 180°F/80°C. Such a malt is pale, and makes a light-colored, light-flavored brew. To make malts that have little enzyme activity but are rich in color and flavor, he kilns the barley at a high temperature, 300–360°F/150–180°C, to encourage browning reactions. Dark malts develop flavors that range from toasted to caramelized to sharp, astringent, and smoky. Brewers have many different kinds of malt to choose from—their names include pale or lager, ale, crystal, amber, brown, caramel, chocolate, and black—and often mix two or more malts in a single brew to obtain a particular combination of flavor, color, and enzyme power.

Once the malt has been kilned, the dried kernels can be stored for several months until they're needed, when they're ground into a coarse powder. They're also made into malt extracts for both commercial and home brewers: the malted barley is soaked in hot water to remove its carbohydrates, enzymes, color, and flavor, and the liquid then concentrated to a syrup or a dry powder (p. 679).

Brewing Ingredients: Hops

Hops are the female flowers, or "cones," of a Eurasian-American vine, *Humulus lupulus,* which bear small resin and aromatic oil glands near the base of their floral leaves, or bracts. They are an essential flavoring ingredient in beers. There are now several dozen brewing varieties, most of them European or European-American hybrids. Hops are picked off the plant when mature,

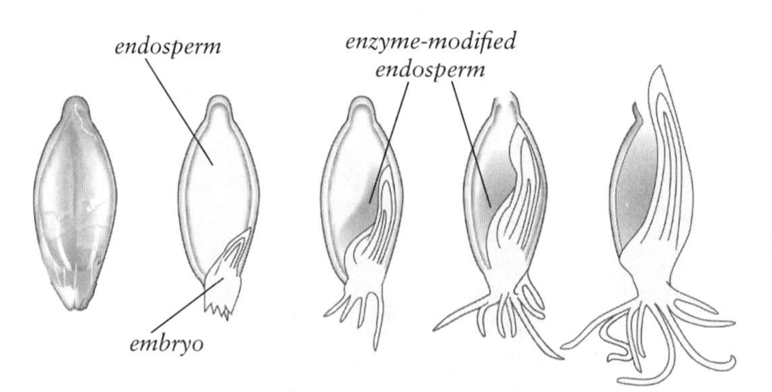

Four stages in the malting process. As the barley kernel germinates, it generates digestive enzymes, weakens cell walls, and begins the process of converting starch into fermentable sugars. The shading indicates the progress of cell-wall weakening and starch digestion. Malting is stopped when the growing shoot just reaches the tip of the kernel.

dried, sometimes powdered and formed into pellets, and then stored until needed. They're added to the brew liquid at the rate of about 0.5 to 5 grams per quart/liter, with the low figure typical of bland commercial brews, the high figure of flavorful microbrews and traditional Pilsners.

Bitterness and Aroma Hops provide two different elements: bitterness from phenolic "alpha acids" in their resins, and aroma from their essential oil. Some hop varieties contribute a dependable level of bitterness, while others are prized for their aroma.The important bittering compounds are the alpha acids humulone and lupulone. In their native form they're not very soluble in water, but prolonged boiling transforms them into soluble structures that flavor the beer effectively. (Brewers sometimes use hop extracts that have been pretreated to produce the more soluble alpha acids.) Because boiling evaporates away many of the volatile aroma compounds, another dose of hops is sometimes added to the brew after the boiling, specifically to add aroma. The aroma of ordinary hops is characterized by the terpene myrcene, which is also found in bay leaf and verbena, and is woody and resinous. "Noble" hop varieties are dominated by humulene, which is more delicate, and often contain pine and citrus notes from other terpenes (pinene, limonene, citral). The American variety "Cascades" has a distinctive floweriness (due to linalool and geraniol).

Brewing Beer

The brewing of beer takes place in several stages.

- *Mashing*: ground barley malt is soaked in hot water. This revives the barley enzymes, which break starch into sugar chains and sugars, and protein into amino acids. The result is a sweet, brown liquid called the *wort.*
- *Boiling*: hops are added to the wort, and the two are boiled together. This treatment extracts the hop resins that flavor the beer, inactivates the enzymes, kills any microbes present, deepens the color of the wort, and concentrates it.
- *Fermentation*: yeasts are added to the cooled wort and allowed to consume sugars and produce alcohol until the desired levels of each are reached.
- *Conditioning*: the new beer is held for some time to purge it of off-flavors, clear it of yeasts and other materials that give it a cloudy appearance, and develop carbonation.

Here are some details of each stage.

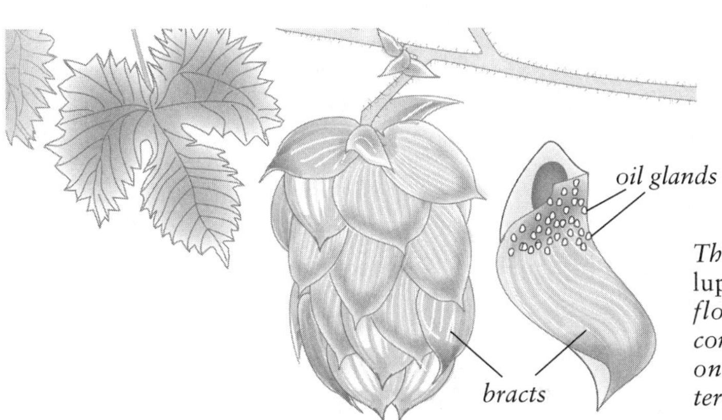

The hop vine, Humulus lupulus, *and its female flower structure, the cone, with a close-up of one of the cone's clustered leaves, or bracts.*

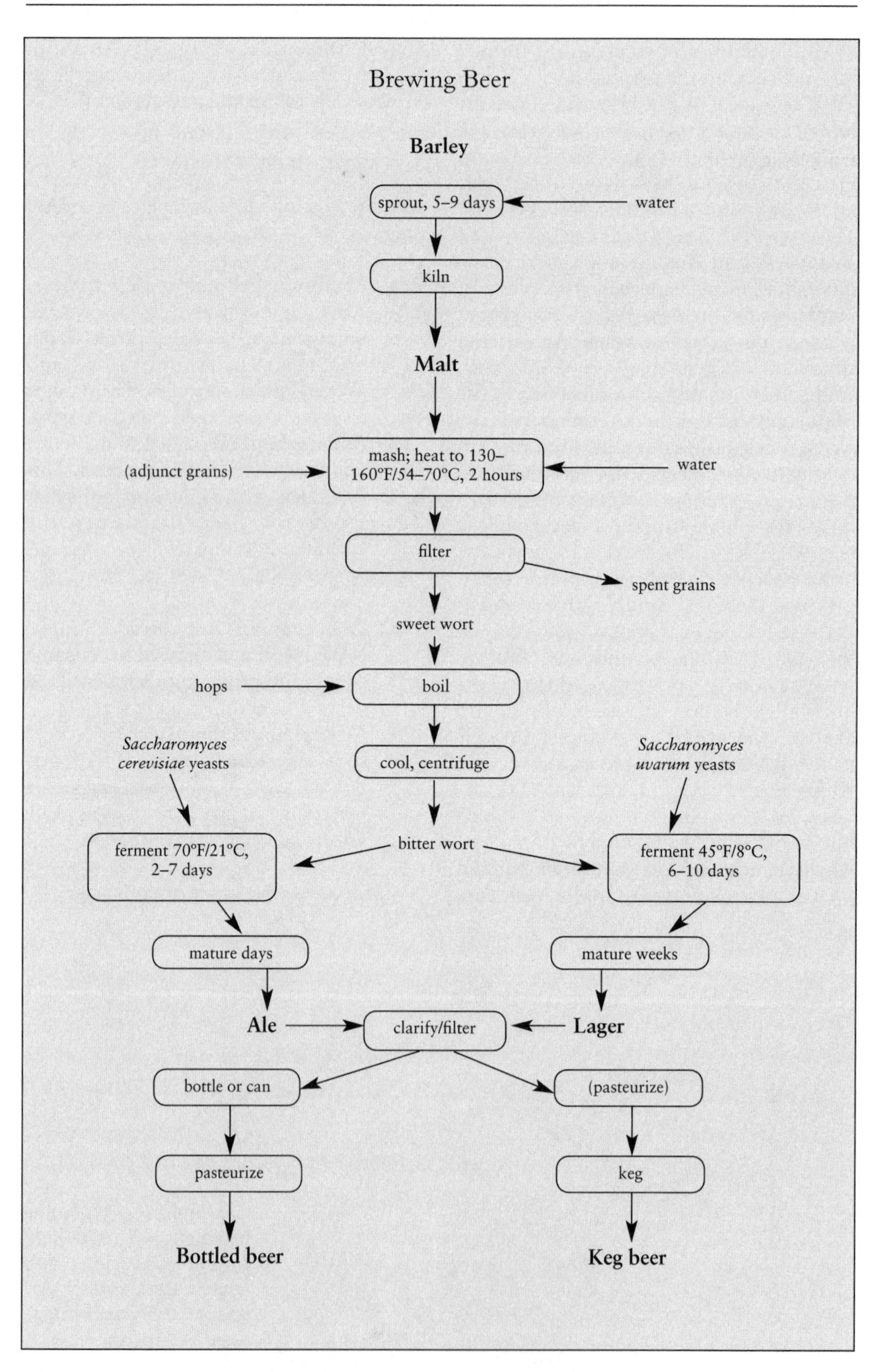
Brewing Beer
Barley
sprout, 5–9 days
water
kiln
Malt
mash; heat to 130–160°F/54–70°C, 2 hours
(adjunct grains)
water
filter
spent grains
sweet wort
hops
boil
cool, centrifuge
Saccharomyces cerevisiae yeasts
Saccharomyces uvarum yeasts
bitter wort
ferment 70°F/21°C, 2–7 days
ferment 45°F/8°C, 6–10 days
mature days
mature weeks
Ale
clarify/filter
Lager
bottle or can
(pasteurize)
pasteurize
keg
Bottled beer
Keg beer

Mashing In the stage known as mashing, the coarsely ground malt is soaked in water at between 130 and 160°F/54–70°C for a couple of hours. Typical proportions are around eight parts water per one part malt. Mashing is completed by running the wort off the solid remains of the malt, which are then rinsed with hot water—"sparged"—to remove some remaining extractable materials before being discarded.

Mashing accomplishes several purposes. Above all, it gelates the starch granules and allows the barley's enzymes to break down long starch molecules into shorter sugar chains and small fermentable sugars, and proteins into foam-stabilizing amino-acid chains and fermentable single amino acids. And it extracts all these substances, along with color and flavor substances, from the grain particles and into the water.

Because the different enzymes work fastest at different temperatures, the brewer can adjust the ratio of fermentable sugars to sugar chains, and amino acids to amino-acid chains, by varying the temperature and time of mashing. By this means he controls the beer's final body, and the stability of its foam. Fully 85% of the carbohydrate in malt is starch. In the liquid wort, 70% or more is in the form of various sugars, mainly the two-glucose sugar called maltose. Most of the remaining carbohydrates, 5 to 25% of the dissolved solids, are the so-called *dextrins,* or sugar chains of from four glucose units to a few hundred, which get tangled up with each other, impede the movement of the water, and so provide a full-bodied consistency to the wort and beer. The dextrins and amino-acid chains will also slow the draining of fluid from the bubble walls of the beer foam, and so contribute to its stability in the glass.

Cereal Adjuncts Making the wort with nothing but barley malt and hot water is the standard method in Germany, and in many U.S. microbreweries. In most large breweries in the United States and elsewhere, unmalted "adjunct" sources of carbohydrate—ground or flaked rice, corn, wheat, barley, even sugar—are commonly added to the liquid to lower the amount of malt needed, and so the brewer's production costs. Unlike malt, they contribute little or no flavor of their own. They're therefore mostly limited to pale, mild brews like standard American lagers, which may start with almost as much adjunct grain as malt.

Water Water is the main ingredient in beer, so its quality has a definite influence on beer quality. Though modern brewers can tailor the mineral content of their water to the kind of beer they're making, early brewers tailored their beers in part to make the best of the local waters. The sulfate-rich water of Burton-on-Trent gave English pale ales a bitterness that limited the use of hops, while the mild water of Pilsen encouraged Czech brewers to add large amounts of bitter and aromatic hops. The alkaline, carbonate-rich waters of Munich, southern England, and Dublin can balance the acidity of dark malts that normally extract too much astringent material from barley husks, and encouraged the development of dark German beers and British porters and stouts.

Boiling the Wort Once the liquid wort has been drawn off the grain solids, the brewer runs it into a large metal tank, adds hops, and boils it vigorously for up to 90 minutes. Boiling converts the insoluble hop alpha acids into their soluble form and develops the beer's bitterness, and inactivates the barley enzymes and so fixes the carbohydrate mix—a certain portion of sugar for the yeasts to convert into alcohol,

OPPOSITE: *Making beer. Beer is made in two basic ways. Ales are fermented in less than a week at a warm temperature and matured for days, while lagers are fermented for more than a week at a cold temperature and matured for weeks.*

a certain portion of dextrins for the beer's body. It sterilizes the wort so that the brewing yeasts won't have any competition during fermentation, and it concentrates the wort by evaporating off some of its water. Boiling deepens the wort's color by encouraging browning reactions, mainly between the sugar maltose and the amino acid proline. And it begins the process of clarifying the brew by coagulating large proteins and causing them to bind with tannins from the barley hulls, form large masses, and precipitate out of the solution. When boiling is finished, the wort is strained, then cooled and aerated.

Fermentation With boiling, the brewer has finished transforming the bland barley grain into a rich, sweet liquid. Now the yeast cells transform this liquid into beer, which is far less sweet, but more complex in flavor.

There are two basic methods for fermenting beer, and they produce distinctive results. One is rapid fermentation at a high temperature with ale yeasts (strains of *Saccharomyces cerevisiae*) that clump together, trap the carbon dioxide gas that they produce, and rise to the wort surface. The other is slow fermentation at a low temperature with lager yeasts (*Saccharomyces uvarum* or *carlsbergensis*) that remain submerged in the wort and fall to the bottom when fermentation is over. These are often called "top" and "bottom" fermentations.

Top fermentation is usually carried out at between 64 and 77°F/18–25°C and takes two to seven days, during which the yeasty foam is skimmed off several times. Because the yeast layer at the top has a good supply of oxygen and is inevitably contaminated by other airborne microbes, including lactic-acid bacteria, top-fermented beers are often relatively acidic and strong in flavor. Bottom fermentation goes on at distinctly lower temperatures, 43 to 50°F/6–10°C, takes six to ten days, and produces a milder flavor. Bottom fermentation is the standard technique in the United States. Because warm temperatures encourage yeasts to generate particular aroma compounds (esters, volatile phenols), top fermentation produces fruity, spicy aromas; cold, slow fermentation produces crisp beers with a dry, bready flavor.

Conditioning The treatment of beer after fermentation varies according to the type of fermentation that has taken place: brief for fast top fermentation, prolonged for slow bottom fermentation.

Top-fermented beer is cleared of yeast and then run into a tank or cask for conditioning. The green beer, as it's called fresh from fermentation, contains little carbon dioxide, has a sulfurous, harsh flavor, and is

The Different Flavor of Beer on Tap

Bottled and canned beers and ales are generally cold stabilized and pasteurized (at 140–160°F/60–70°C) to survive extreme temperatures during shipping and storage, while keg beers, which are kept refrigerated continuously, may not be. This is why bottle and keg versions of the same beer can taste very different. However, even keg beer is a world apart from the traditional *cask-conditioned* beer. Keg beer has been cleared of all its yeast before the keg is filled, while in cask conditioning, the new beer and the yeast that will help mature it are sealed together in the cask. Cask-conditioned beer is thus in contact with yeast until the moment it's dispensed, and its flavor reflects this. Cask beer is fragile and has a drinkable life of about a month, compared with three months for keg beer.

hazy with the detritus of dead yeast cells. In conditioning, a secondary fermentation is induced by adding to the green beer either a small amount of yeast and some sugar or fresh wort, or some actively fermenting wort (this is called *Kräusening*). Inside the closed cask or tank, the liquid traps and absorbs the carbon dioxide produced. Undesirable odors can be forced out of the beer by opening the container briefly and allowing some gas to escape. These traditional techniques are sometimes replaced by simply pumping pure carbon dioxide into the beer—carbonating it. Some hops or hop extract may also be added at this point to augment aroma, bitterness, or both. A few days of cooling and the use of a "fining" agent—isinglass (fish gelatin), clay, and vegetable gums are common—precipitate suspended proteins and tannins that might later form a haze when the beer is chilled for drinking; this is called "cold stabilization." The beer is then centrifuged to remove any remaining yeast and precipitate, filtered, packaged, and usually pasteurized.

Lagering The conditioning process for bottom-fermented beer is somewhat different. The original Bavarian lager was packed in ice and allowed to rest in contact with its yeast dregs for several months. The yeast slowly produced carbon dioxide, which helped purge the beer of sulfury off-odors. Today, some traditional lagers are still aged for several months; but because storage has the economic disadvantage of tying up money and materials, the tendency is to lager the green beer at temperatures just above freezing for two to three weeks. Carbon dioxide may be pumped in to purge undesired aromas; and centrifuges, filters, and additives help clarify the beer. As a replacement for wooden casks, some beech or hazelwood chips may be thrown into the tank for flavor.

Additives More than 50 additives are permitted in American beer, including preservatives, foaming agents (usually vegetable gums), and enzymes—similar to meat tenderizers—that break down proteins into smaller molecules that are less likely to cloud the brew. Some companies avoid the use of preservatives, and usually advertise this fact on the label.

The Finished Beer In the end, brewing has transformed the dry, tasteless barley grain into a bubbly, bitter, acidic liquid (its pH is about 4) that is 90% water, 1 to 6% alcohol, and between 2 and 10% carbohydrates, mainly the long-chain dextrins that provide body.

Storing and Serving Beer

In contrast to wines, with their higher alcohol and antioxidant contents, most beers do not improve with age, and are at their best fresh from the brewery. Oxidation causes the gradual development of a stale, cardboard aroma (from nonenal, a fatty-acid fragment) and harshness on the tongue (from hop phenolic substances). Browning reactions cause other undesirable changes. Top-fermented ales develop a solvent-like note. Staling is slowed at low temperatures, so when possible beer should be stored in the cold. Britain does make "laying-down beers," and Belgium *bières de garde*, which start out fermentation with a very high soluble carbohydrate content and continue to ferment slowly in the bottle, the continuous production of carbon dioxide and other substances helping to prevent oxidation and staling. They end with alcohol levels of 8% or more, and improve for a year or two.

Keep Beer in the Dark Beer should also be kept away from bright light, especially sunlight, and especially if it has been bottled in clear or green glass: otherwise it will develop a strong sulfurous odor. A cup of beer at a picnic can go skunky in a few minutes; bottled beer in a fluorescent-lit display case may deteriorate in a few days. It turns out that light in the blue-green to ultraviolet parts of the light spectrum reacts with one of the hop acids to form an unstable free radical, which in turn reacts with sulfur compounds to form a close relative of chemicals in the skunk's defensive arsenal. Brown glass can absorb blue-green wavelengths before they get to the beer inside, but green bottles don't. As a result, green-bottled German and Dutch beers are often sulfurous, and many consumers now expect this! One American brewer with trademark clear bottles developed a modified hop extract that's free of the vulnerable hop acid, and this prevents its beer from going skunky.

Serving Beer In the United States, beer is often drunk ice-cold and straight from the can or bottle. This is fine for a light, thirst-quenching beer, but doesn't do justice to beers designed to have some character. The colder any food is, the less full its flavor will seem. Lager beers are usually best served somewhat warmer than refrigerator temperature, around 50°F/10 C, while top-fermented ales are served at a cool room temperature, from 50 to 60°F/10–15°C. Beers worth savoring are poured into a glass, where some of the carbon dioxide gas can escape and moderate their prickliness, and where their color and head of foam can be appreciated.

Beer Foam: The "Head" Beer is not the only intrinsically bubbly liquid we enjoy, but it's the only one whose bubbles we expect to persist long enough to form a "head" of foam atop the glass. Beer lovers even value the ability of the foam to cling to the glass as the liquid level drops, a quality known as *lacing* (or, in more impressive German, *Schaumhaftvermögen*). There are many factors that influence foaming, from the amount of carbon dioxide dissolved in the beer to the way the beer is released from keg or can. Here are some of the most interesting.

Grain Proteins Stabilize the Head Foam stability depends on the presence in the bubble walls of emulsifier molecules with water-loving and water-avoiding ends (p. 802); the water-avoiding ends project into the gas while the water-loving ends stay in the liquid, and thus reinforce the gas-liquid interface. In beer, these molecules are mostly medium-sized proteins that come from the malt or from cereal adjuncts, whose proteins are more intact than malt's and significantly improve head stability. Hop acids also contribute to foam stability, and become concentrated enough in the foam to make it noticeably more bitter than the liquid beneath. Cool-fermented lagers generally give more persistent foams than warm-fermented ales because the latter contain more foam-destabilizing higher alcohols from yeast metabolism (p. 762).

Nitrogen Makes Creamy Foams In the last decade, many beers have come to be endowed with an especially fine, creamy head that used to be largely limited to stouts. The creamy head comes from an artificial dose of nitrogen gas that may be injected into beer at the brewery, or in the bar or pub by the tap that delivers beer from the keg, or by a small device inside an individual beer can. Nitrogen is less soluble in water than carbon dioxide, so its bubbles are slower to lose gas to the surrounding liquid, and slower to coarsen and deflate. Nitrogen bubbles remain small, and persist. They also don't carry the tart prickliness of carbon dioxide, which becomes carbonic acid when it dissolves in beer and on the surface of our tongue.

Foam in the Glass An initially vigorous pouring action develops the head of foam with a small, easily controlled portion of the beer. Once the foam is of the desired thickness, the rest of the beer can be poured in gently along the side of the glass, avoiding aeration and nucleation of new bubbles. The glass itself should be clean of any residues of oil or soap, which interfere with foaming. (These molecules have water-avoiding ends that pull the similar ends of the bubble-stabilizing proteins out of the bubbles.) By the same token, if a newly poured beer threatens to foam over, it can often be stopped in its tracks by touching the rim with a finger or lip, which carry traces of oil.

Kinds and Qualities of Beer

Beers are a wonderfully diverse group of drinks, and a good beer can be a mouth-filling experience, one that rewards slow savoring. There are several qualities worth appreciating:

- Color, which can range from pale yellow to impenetrable brown-black, and comes from the kinds of malt used
- Body, or weightiness in the mouth, which comes from the long remnants of starch molecules in the malt
- Astringency, from malt phenolic compounds
- Prickly freshness, from dissolved carbon dioxide
- Taste, which may include saltiness from the water, sweetness from unfermented malt sugars, acidity from roasted malt and from fermentation microbes, bitterness from hops and from dark-roasted malts, savoriness from malt amino acids
- Aroma, from woody, floral, citrusy hops; malty, caramel, and even smoky malt; and from the yeasts and other microbes, which can produce notes that seem fruity (apple, pear, banana, citrus), flowery (rose), buttery, spicy (clove), and even horsey or stable-like (p. 738)

Ales develop a characteristic tartness and fruitiness from their diverse group of fermentation microbes. Lagers have a more subdued aroma, part of whose foundation is cooked-corn-like DMS (dimethyl sulfide), which comes from a precursor in the lightly roasted malt and is produced while the wort is boiled and then cooled. But there's tremendous variation in the flavors of these basic beers. Rich and somewhat sweet brews—porter, stout, barley wine—can even go well with desserts.

Beers Low in Calories, Alcohol, and Beer Flavor

Nowadays there are versions of beer for people who like beer but don't want to consume alcohol, or want to consume alcohol but reduce their intake of calories. A standard 12 oz/360 ml container of American lager contains about 14 grams of alcohol and 11 grams of carbohydrate, for a total of about 140 calories. Low-calorie "light" or "dry" beers have 100–110 calories, a savings produced by using a lower proportion of malt and adjuncts to water, and then adding enzymes that digest more of the carbohydrates into fermentable sugars. The fermentation then produces a brew with only a slightly lower alcohol content, but about half the sugar chains—and very little body.

"Nonalcoholic beers" can be made by modifying the fermentation so that the yeasts produce little alcohol (very low temperatures, abundant oxygen), or by removing the alcohol from normally fermented beer using a molecular version of sieving called "reverse osmosis." The lowest-alcohol malt product is "malta," a popular drink in the Caribbean made by bottling a full-fledged wort without any fermentation at all. It is dense and sweet.

Then there are "malt beverages," which have the alcohol and calorie content of beer, but taste nothing like beer: they're more like soft drinks. In these products, the only purpose of the malt is to generate sugars for fermentation into alcohol; neither it nor the yeast contributes any flavor.

Some Styles and Qualities of Beer

Beer Style	Alcohol Content, Percent by Volume	Unusual Ingredients	Qualities
Pale lager: European	4–6		Malty, bitter and spicy/floral from hops
American/International	3.5–5	Unmalted grains	Little malt or hop aroma or bitterness; cooked-corn, green apple notes
Dark lager: European	4.3–5.6		Malty, somewhat sweet
American	4–5	Unmalted grains, caramel coloring	Little malt or hop aroma; cooked-corn note; some sweetness
Bock	6–12		Malty, caramel, somewhat sweet
Pale ale: English	3–6.2		Balanced malt and hop aromas, fruity, moderately bitter
Belgian	4–5.6	Spices	Spicy, fruity, moderately bitter
American	4–5.7		
India	5–7.8		Strong hop aroma and bitterness
Brown ale	3.5–6		Somewhat sweet, nutty, fruity
Porter	3.8–6	Dark malts	Malty, roasted coffee/chocolate notes, somewhat sweet
Stout	3–6	Dark malts, roasted unmalted barley	Like porter but less sweet, more bitter
Imperial stout	8–12	Dark malts, roasted unmalted barley	Like stout, but stronger (originally for export to Russia)
Wheat beer: Bavarian	2.8–5.6	Wheat malt, special yeasts	Wheat, grain, tartness, banana and clove notes
Berlin	2.8–3.6	*Lactobacillus* culture	Wheat, slightly fruity, sour
Belgian	4.2–5.5	Unmalted wheat, spices, bitter orange peel, special yeasts, *Lactobacillus* culture	Wheat, spice, citrus, tartness
American	3.7–5.5	Normal yeasts	Wheat, grain, light hop aromas, light bitterness
Belgian lambics		Unmalted wheat, aged hops, wild yeasts and bacteria	
Faro	4.7–5.8	Spices, sugar	Spicy, sweet
Gueuze	4.7–5.8	Blends of several ages	Tart, fruity, complex
Fruit	4.7–5.8	Cherry, raspberry, other fruits	Tart, fruity, complex
Barley wine	8–12+		Malty, fruity, full-bodied

Adapted from "Guide to Beer Styles," Beer Judge Certification Program 2001, and other sources.

In addition to the many variations on the two themes of beer and ale, there are two kinds of beer that are worth special mention for their distinctiveness.

Wheat Beers German wheat beers differ from the usual Bavarian brew in four ways. First, a large fraction of the barley malt is replaced by wheat malt, which carries more protein, produces a more foamy and hazy brew, and lightens the typical malt flavor. Second, wheat beers are top-fermented like ales, and so develop more tartness and fruitiness. Third, the culture often includes an unusual yeast (*Torulaspora*) that produces aroma compounds not usually found in beer. These volatile phenols (vinyl guaiacol, p. 738) may suggest cloves and similar spices, but also a medicinal quality like that of plastic bandages, or an animal quality reminiscent of the barnyard or stable. Finally, some wheat beers are not fully clarified, and retain some of their yeast, which gives them a cloudy appearance and yeasty flavor. German wheat beers may be called *Weizen* for "wheat," *Hefe-weizen* for "yeast-wheat," or *Weissen* for "white," referring to their cloudy appearance.

Some American breweries now produce wheat beers on the German model, but usually without the phenol-producing yeast; they are mild, tart, and cloudy.

Belgian Lambic Beers The brewers of Belgium have been more inventive than any others. They allow many different microbes to participate in the fermentation; they ferment some beers for years, either continuously or by restarting them; they flavor their beers with spices and herbs, and even re-ferment them with fresh fruits to make a hybrid beer and fruit wine. They generally used aged hops, which are less harmful to the unusual brewing microbes, less bitter, and higher in drying, slightly astringent, wine-like tannins.

The most unusual Belgian beers are the *lambics*. The hallmark of brewing traditional lambic is a spontaneous and months-long fermentation of the wort in wood barrels. Once the wort has been boiled, it is cooled in a broad open tank, where it picks up microbes from the ambient air. The cool wort is then poured into wooden casks that contribute microbes from previous batches, and ferments in the casks for 6 to 24 months. The fermentation proceeds in four stages: an initial growth of wild yeasts (*Kloeckera* and others) and various bacteria (*Enterobacter* and others) that takes 10–15 days and produces acetic acid and vegetable aromas; the main alcohol-producing growth of *Saccharomyces* yeasts, which dominate for several months; at 6 to 8 months, the acid-producing growth of lactic and acetic bacteria (*Pediococcus, Acetobacter*); and finally the growth of *Brettanomyces* yeasts, which produce a range of fruity, spicy, smoky, and animal aromas (see box, p. 730). The resulting brew may then be blended with other lambics and aged to make *gueuze,* with a wine-like acidity and complexity; or blended with some ordinary top-fermented ale and flavored with sugar and coriander to make *faro*; or re-fermented in the barrel for four to six months with fresh whole cherries or raspberries, to make *kriek* and *framboise.*

ASIAN RICE ALCOHOLS: CHINESE *CHIU* AND JAPANESE *SAKE*

SWEET MOLDY GRAINS

The peoples of eastern Asia developed their own distinctive form of alcohol, one that the rest of the world is coming to appreciate more and more. It's not exactly a wine, because it's fermented from starchy grains, mainly rice. But it's not exactly a beer either, because the grain starch is not digested into fermentable sugars by grain enzymes. Instead, Asian brewers use a mold to supply the starch-digesting enzymes, and the mold digests the grain starch at the same time that the yeasts are converting the sugars into alcohol. The resulting liquid can reach an alcohol concentration of 20%, far stronger than Western beers and wines. Chinese *chiu* and Japanese *sake* don't have the grapey fruitiness or acidity of wine, nor the malt or hop characters of beer. Because it's made from only the starchy heart of the rice grain, sake is perhaps the purest expression of the flavor of fermentation itself, surprisingly fruity and flowery even though no fruit or flower has come near it.

Why and how did Asians come up with this alternative to sprouting grains? The historian H. T. Huang suggests that the key was their reliance on small, fragile millet and rice grains, which unlike barley and wheat are easily and usually cooked whole. Huang speculates that leftover cooked grains were frequently left sitting out long enough to get moldy; and because there are air spaces in a mass of grains, the oxygen-requiring molds would have grown well and digested starch throughout the mass. People eventually noticed that moldy rice tasted sweet and smelled alcoholic. Sometime before the 3rd century BCE, these simple observations led to a regular technique for producing alcoholic liquids. By 500 CE, a Chinese source lists nine different mold preparations and 37 different alcoholic products.

Today, few people outside of China have heard of *chiu,* but many millions have heard of its Japanese counterpart, *sake* (pronounced "sa-kay"). Rice cultivation and probably *chiu* production were brought to Japan from the Asian mainland around 300 BCE. Over the following centuries, Japanese brewers so refined *chiu* that it became something distinctive.

Starch-Digesting Molds

Though modern industrial production has brought a number of common shortcuts and simplifications, Chinese and Japanese brewers have traditionally used very different preparations for breaking rice starch down into fermentable sugars.

Chinese *Chhü*: Several Molds and Yeasts The ancient Chinese preparation, *chhü,* is usually made from wheat or rice, and includes several different kinds of mold as well as the yeasts that will eventually produce the alcohol. Some of the wheat may be roasted or left raw, but most is steamed, coarsely ground, shaped into cakes, and then left to mold in incubation rooms for several weeks. Species of *Aspergillus* grow on the outside, and species of *Rhizopus* and *Mucor* on the inside. *Aspergillus* is the same kind of mold used to digest soybeans to make soy sauce and *Rhizopus* is the major mold in soybean *tempeh* (p. 496, 500), while *Mucor* is important in some kinds of aged cheeses. All of them accumulate starch- and protein-digesting enzymes, and generate trace by-products that contribute flavor. Once the grain cakes have been well permeated with microbes, they're dried for storage. When

Pasteurization Before Pasteur

Unlike European wines and beers, *chiu* was usually served warm or hot. Perhaps because they noticed that heated leftovers kept better than the original batch, by 1000 CE the Chinese had developed the practice of steaming containers of newly fermented *chiu* to slow its deterioration. In the 16th century, Japanese brewers refined this method by lowering the heating temperature to 140–150°F/60–65°C, which is high enough to kill most enzymes and microbes, but does less damage to the flavor of the sake. Asian brewers were thus "pasteurizing" their alcohols centuries before Louis Pasteur suggested gently heating wine and milk to kill spoilage microbes.

needed for *chiu* production, they're soaked in water for several days to reactivate the microbes and their enzymes.

Japanese *Koji* and *Moto*: One Mold, Separate Yeast The Japanese *koji,* by contrast, is made fresh for each particular sake brewing, is based only on polished, unground rice, and is inoculated with a selected culture of *Aspergillus oryzae* alone, with no other molds. The mold preparation for sake therefore doesn't provide the complexity of flavor that the Chinese preparation does, with its roasted wheat, variety of microbes, and period of drying.

Because the koji contains no yeasts, the Japanese system requires a separate source of yeast. The traditional yeast preparation, the *moto,* is made by allowing a mixture of koji and cooked rice gruel to sour spontaneously with a mixed population of bacteria, mainly lactic acid producers (*Lactobacillus sake, Leuconostoc mesenteroides*, and others) that contribute tart and savory tastes and some aroma. A pure yeast culture is then added and allowed to multiply. Because this microbe-soured moto takes more than a month to mature, it has been largely replaced by the simple addition of organic acids to the moto mash, or by the addition of acids and concentrated yeasts directly to the main fermentation. These time-saving methods tend to produce lighter, less complex sakes.

Brewing Rice Alcohols

Simultaneous, Stepwise Fermentation Chinese and Japanese brewing methods differ in important details, but they also share several important features. The starch-digesting molds and alcohol-producing yeasts are added to the cooked rice gruel together, and work simultaneously. Unlike the making of beer, where a liquid is extracted from the grain and only the liquid is fermented, the thick gruel of cooked rice is fermented whole. And the rice is introduced gradually into the fermentation, not all at once: new portions of cooked rice and water are added to the vat at intervals during the fermentation, which lasts from two weeks to several months. All of these practices apparently contribute to the yeasts' ability to continue producing alcohol to high concentrations. When rice is added toward the end of fermentation, some sugar remains unmetabolized by the yeasts, and the resulting alcohol is sweet.

Chinese Practice: Ordinary Rice and High Temperatures Traditional Chinese brewing begins with soaking the mold preparation in water for several days, and then proceeds with the periodic addition of ordinary cooked rice over the course of an initial fermentation that may last one or two weeks at a temperature around 85°F/30°C. At the end of this phase, the mash is often divided into smaller containers and held at cooler temperatures for weeks or months. The liquid is then pressed from the solids, filtered, adjusted with water and colored with caramel, pasteurized at 190–200°F/85–90°C for 5–10 minutes, matured for several months, then filtered and packaged. The high-temperature pasteurization helps develop the finished flavor.

Japanese Practice: Polished Rice and Low Temperatures Chinese brewers use rice that has been milled to remove about 10% of the grain, only slightly more than is removed to make ordinary white rice for cooking (p. 472). In Japan, however, the rice for anything above the standard grade of sake must be milled to remove a minimum of 30% of the grain, and the highest grades of sake are made with rice that has been polished down to 50% or less of its original weight. The center of the rice grain is the portion that contains the most starch and the least protein or oil, so the more the outer layers of the rice are ground away, the simpler and purer the remaining grain, and the less grain flavor it contributes to the final liquid.

Sake is also fermented at significantly lower temperatures than Chinese rice alco-

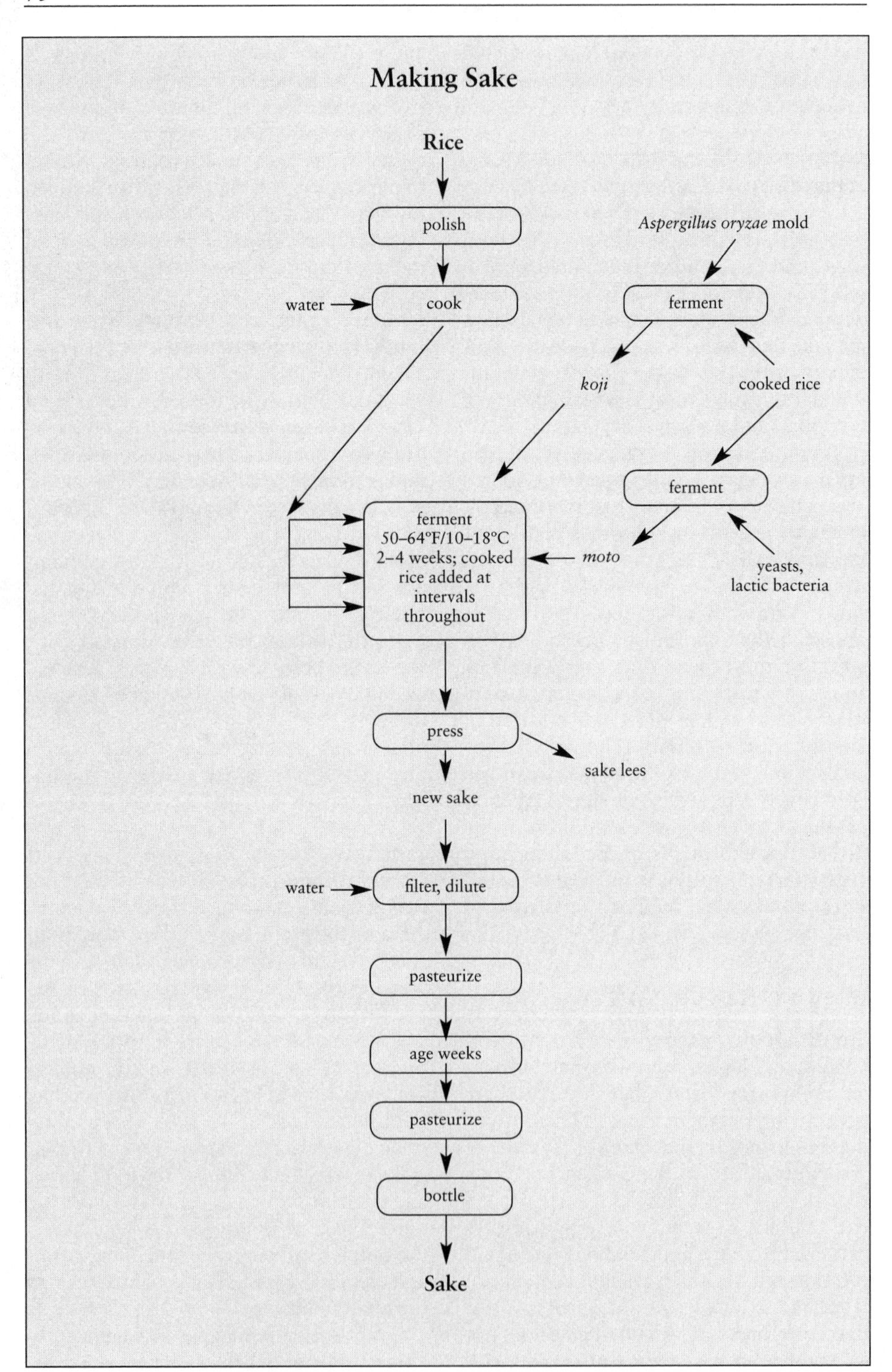
Making Sake
Rice
polish
Aspergillus oryzae mold
water
cook
ferment
koji
cooked rice
ferment
ferment
50–64°F/10–18°C
2–4 weeks; cooked
rice added at
intervals
throughout
moto
yeasts,
lactic bacteria
press
sake lees
new sake
water
filter, dilute
pasteurize
age weeks
pasteurize
bottle
Sake

hols. Beginning in the 18th century, most sake brewing was reserved for the winter months, and this remains largely the case today. The upper limit for sake brewing is around 64°F/18°C, and brewers of the highest grades will keep the temperature at a distinctly chilly 50°F/10°C. In these conditions, the fermentation takes about a month instead of two to three weeks, and the mash accumulates two to five times the normal quantity of aroma compounds, notably the esters that provide apple, banana, and other fruity notes.

Once the sake fermentation is complete, the liquid is pressed from the solids, then filtered, diluted with water to 15–16% alcohol, and held for some weeks to allow the flavor to mellow. It's also pasteurized (at 140–150°F/60–65°C) after filtering and again before bottling to denature any remaining enzymes, one of which otherwise slowly generates a particularly unpleasant volatile (sweaty isovaleraldehyde).

Varieties of Sake There is a broad range of different grades and kinds of sake. Both the cheapest and the standard grades are made by adding substantial amounts of pure alcohol to the mash just before pressing. This became standard industrial practice during the war years because it greatly increases the yield from a given quantity of rice. Sugar and various organic acids can also be added to these grades. At the other end of the scale, there are premium versions made with nothing but rice, water, and microbes, painstakingly cultured in the traditional way. The box below gives examples of some kinds worth seeking out.

Though much sake is drunk warm as Chinese rice alcohols are, connoisseurs pre-

Some Kinds of Sake

Sake made to be savored is usually of the grade *ginjo,* or "special," in which pure alcohol is the only allowed additive and at least 40% of the rice grain has been removed. *Junmaishu* is sake made only with rice and water. Some interesting specialty sakes include the following:

Genmaishu	Brewed with brown rice
Genshu	Undiluted, and so around 20% alcohol
Kimoto	"Live" moto, the yeast preparation soured slowly by bacteria and not instantly with pure acids
Namazake	"Live" sake because unpasteurized and so containing active enzymes, to be kept refrigerated and drunk soon
Orizake and Nigorizake	Cloudy sakes that include lees, yeast cells, and other fine particles from the mash
Shizuku	"Droplets" sake, made by allowing the liquid to drain by gravity from the mash rather than pressing
Taruzake	"Keg" sake, aged in cedar barrels

OPPOSITE: *Making sake. One of the unusual features of sake fermentation is the repeated addition of cooked rice to the fermenting mash over the course of several weeks.*

fer to chill finer examples. In general, sake is less tart and more delicately flavored than wine. Savory amino acids are an important element. Its aroma varies a great deal depending on how it was made, and features the biochemical artistry of the yeasts. Fruity esters and flowery complex alcohols are usually prominent.

Sake Is Fragile Sake and its delicate flavors are vulnerable to alteration by exposure to both light and high temperatures. It's best drunk as young as possible. The clear and blue bottles in which it's usually packaged offer little protection, so sake should be kept in the cool and dark of the refrigerator, and once opened should be consumed quickly.

DISTILLED SPIRITS

Distilled spirits are the concentrated essence of wine and beer. They're the product of a basic chemical fact: different substances boil at different temperatures. The boiling point of alcohol is about 173°F/78°C, well below water's 212°F/100°C. This means that if a mixture of water and alcohol is heated, more of the alcohol than the water will end up in the vapor. That vapor can then be cooled and condensed back into a liquid that has a higher alcoholic content than the original beer or wine.

Distilled spirits were first valued, and still are, for their high alcohol content. But there's much more to them than their intoxicating power. Like alcohol, the substances that give wine and beer their aroma are also volatile: so the same process that concentrates alcohol also concentrates aroma. Distilled alcohols are some of the most intensely flavorful foods we have.

The History of Distilled Spirits

The Discovery of Distillation High concentrations of alcohol are toxic to all living things, including the yeasts that produce it. Brewing yeasts can't tolerate more than about 20%. So stronger drink can only be made by physically concentrating the alcohol in fermented liquids. The key to discovering distilled alcohol would have been two observations: that the vapors of a heated liquid can be recaptured by condensing them on a cool surface, and that the vapors of heated wine or beer are more strongly alcoholic than the original liquid.

The practice of distillation itself appears to be very old. There's evidence that the Mesopotamians were concentrating the essential oils of aromatic plants more than

Japanese Cooking Alcohols: Mirin and Sake Lees

Mirin is a sweet Japanese cooking alcohol. It's made by combining cooked polished rice, koji, and *shochu,* a distilled spirit made from a low-grade sake. The alcohol inhibits any further alcoholic fermentation. Instead, during two months at a warm 77–86°F/25–30°C, the koji mold and enzymes slowly convert the rice starch into glucose. The full-bodied liquid is drawn off and clarified, and ends up at around 14% alcohol and from 10% to as much as 45% sugar. Industrial imitations are made from grain alcohol, sugar, and flavorings.

The solids left after pressing and filtering the sake mash are called *sake kasu,* or sake lees. They include starch, proteins, the cell walls of rice, yeasts, and molds, and some acids, alcohol, and enzymes. These sake lees are much used in Japanese cooking, especially in the making of vegetable pickles, marinades for fish, and soups.

5,000 years ago, using a simple heated pot and a lid onto which vapors condensed and could be collected. And Aristotle noted in the 4th century BCE in his *Meteorology* that "Sea water, when it is converted into vapor, becomes drinkable, nor does it form sea water when it condenses again." Concentrated alcohol may have been discovered for the first time in ancient China. Archaeological finds and written documents suggest that Chinese alchemists were distilling small amounts of concentrated alcohol from grain preparations around 2,000 years ago. A privileged few were drinking it before the 10th century, and by the 13th it was a commercial product.

Spirits and Waters of Life In Europe, significant quantities of distilled alcohol were produced around 1100 at the medical school in Salerno, Italy, where it developed its reputation as a uniquely valuable medicine. Two hundred years later, the Catalan scholar Arnaud of Villanova dubbed the active principle of wine *aqua vitae,* the "water of life," a term that lives on in Scandinavia (*aquavit*), in France (*eau de vie*), and in English: *whisky* is the anglicized version of the Gaelic for "water of life," *uisge beatha* or *usquebaugh,* which is what Irish and Scots monks called their distilled barley beer. Throughout the Old World, alchemists thought of distilled alcohol as a uniquely powerful substance, the *quintessence* or fifth element that was as fundamental as earth, water, air, and fire. The first printed book devoted to distillation, Hieronymus Brunschwygk's *Liber de arte distillandi* (1500), explained that the process achieves

> the separation of the gross from the subtle and the subtle from the gross, the breakable and destructible from the indestructible, the material from the immaterial, so as to make the body more spiritual, the unlovely lovely, to make the spiritual lighter by its subtlety, to penetrate with its concealed virtues and force into the human body to do its healing duty.

It's this connection between distillation and the pure and ethereal that gives us our synonym for distilled alcohol, *spirits.*

From Medicine to Pleasure and Drug of Oblivion For several centuries after its discovery, *aqua vitae* was produced in apothecaries and monasteries and prescribed as a *cordial,* a medicine to stimulate the circulation (the word comes from the Latin for "heart"). It seems to have been liberated from the pharmacy and drunk for pleasure in the 15th century, when the terms *Bernewyn* and *brannten Wein,* ancestors of our word *brandy* that meant "burning" or "burnt" wine, appear in German laws about public drunkenness. This is also when winemakers in the Armagnac region of southwest France began to distill their wine into spoilage-resistant brandy for shipping to northern Europe. Gin, a whisky-like medicinal concoction from rye, with juniper added for its flavor and diuretic effect, was first formulated in 16th-century Holland. The renowned brandy of France's Cognac, just to the north of Bordeaux, arose around 1620. Rum was first made from molasses in the English West Indies around 1630, and monastic liqueurs

Food Words: *Distill*

The word *distill* comes from the Latin *destillare,* "to drip." It thus captures the moment at which the barely visible vapors from a hot liquid condense and rematerialize on a cool surface.

like Benedictine and Chartreuse date from about 1650 on.

Over the next couple of centuries, the drinkability of spirits improved as distillers learned how to refine their composition. First came double distillation, in which a wine or beer is distilled, and the distillate then distilled a second time; then in the late 18th and early 19th centuries came ingenious French and English column stills, which produce alcohols of greater purity in one continuous process. The growing availability and drinkability of distilled liquors meant that addiction became a serious problem, particularly among the urban populations of the Industrial Revolution. In England the principal scourge was cheap gin, which the average Londoner in the late 18th century consumed at the rate of nearly a pint/400 ml a day "to seek relief in the temporary oblivion of his own misery," as Charles Dickens later wrote in *Sketches by Boz*. Government control of production and social progress later moderated the problem of alcohol addiction, but hasn't eliminated it.

Whiskey in America Distilled alcohol was so popular in North America that it gave us an enduring legacy: the Internal Revenue Service! In the early days of the colonies and then the United States, molasses was more plentiful than barley, and rum more common than beer. Rye and barley spirits were also being distilled in the northern colonies by 1700, and Kentucky corn whiskey by 1780. After the Revolutionary War, the new American government tried to raise revenues for its war debts by taxing distillation, and in 1794 the largely Scots-Irish region of western Pennsylvania rose in the short-lived Whiskey Rebellion. When President Washington called out federal troops to put it down, the rebellion went underground and "moonshining" became entrenched, especially in the poor hills of the South where the small amount of corn that could be grown would fetch a better price if fermented and distilled. This evasion led the federal government to form the Office of Internal Revenue in 1862. Sixty years later, the national taste for hard liquor was an important stimulus to the temperance movement that culminated in Prohibition.

Recent Times: The Rise of the Cocktail

It was in the 19th century that mixtures of distilled and other alcohols, or cocktails, became fashionable before-dinner drinks in Europe and the Americas. This development led to a mind-numbing explosion of inventiveness: bartenders' manuals now list hundreds of different named cocktails. The origins of the preeminent cocktail, the martini (gin and vermouth), are disputed; it may have been invented several times in different places. The gin and tonic comes from British India, where gin helped make antimalarial quinine water more palatable. In the United States, one of the first famous mixed drinks was the sazerac of New Orleans (brandy and bitters), while Winston Churchill's mother is said to have

Food Words: *Alcohol*

Our word *alcohol* comes from medieval Arab alchemy, which strongly influenced Western science and gave it several other important terms, including *chemistry, alkali,* and *algebra.* To the Arabs, *al kohl* was the dark powder of the metal antimony, which women used to darken their eyelids. By a process of generalization, it came to mean any fine powder, and then for the essence of any material. *Alcohol* was first used to mean the essence of wine itself by the 16th-century German alchemist Paracelsus.

incited the creation of the manhattan (whiskey, vermouth, bitters) at a New York club. Prohibition and harsh "bathtub gin" slowed further progress from 1920 to 1934. In the 1950s, mixologists discovered the value of vodka as a largely flavorless alcohol, and the appeal of sweet-tart fruit juices and sweet liqueurs. Over the next few decades they concocted such broadly popular drinks as the mai tai, piña colada, screwdriver, daiquiri, margarita, and tequila sunrise. In the 1970s, vodka dethroned whiskey as America's best-selling spirit.

The late 20th century brought a modest revival of interest in the more austere classic cocktails, and in fine distilled spirits of all kinds, mixed with nothing more than water.

Making Distilled Alcohols

All distilled alcohols are made in basically the same way.

- Fruits, grains, or other sources of carbohydrates are fermented with yeasts to make a liquid with a moderate alcohol content, from 5 to 12% by volume.
- This liquid is heated in a chamber that collects the alcohol- and aroma-rich vapors as they escape from the boiling liquid, and then passes them across cooler metal surfaces, where the vapors condense and are collected as a separate liquid.
- The concentrated alcoholic liquid is then modified in various ways for consumption. It may be flavored with herbs or spices, or aged in wood barrels. The alcohol content is usually adjusted with the addition of water before it's bottled for sale.

The Distilling Process The essential principle of distillation is a simple one: both alcohol and aromatic substances are more volatile in water than water itself, so they evaporate in disproportionate amounts from wine and beer and become concentrated in the vapor. But it's not a simple matter to make a delicious distilled alcohol, or even a drinkable one. Yeast fermentation produces thousands of volatile substances, and not all of them are desirable. Some are unpleasant, and others, notably methanol, are dangerously toxic.

Selecting Desirable Volatiles Distillers must therefore control the composition of the distilled liquid. They do this by subdividing

Concentration by Freezing

Distillation is the most common way of making concentrated alcohols, but it's not the only way. Freezing also concentrates the alcohol in fermented liquids, by causing the water to form a mass of solid crystals from which an alcohol-enriched fluid can be drained. (Alcohol freezes at –173°F/–114°C, far below water's freezing point of 32°F/0°C.) In the 17th century, Francis Bacon noted Paracelsus's claim that "if a glass of wine be set upon a terrace in bitter frost, it will leave some liquor unfrozen in the center of the glass, which excelleth *spiritus vini* [spirit of wine] drawn by fire." The nomads of central Asia apparently applied "freezing-out" to their alcoholic mare's milk, *koumiss,* and European settlers in North America made apple brandy—applejack—in the same way. Freezing-out produces a different kind of concentrated alcohol. There's no heating step that alters the aroma, and unlike distillation it retains and concentrates the sugars, savory amino acids, and other nonvolatile substances that contributed to the original liquid's taste and body.

the vapor into fractions that are more and less volatile, and collecting mainly the fraction that is richest in alcohol. The fraction more volatile than alcohol, often called the "heads" or "foreshots" because it evaporates earlier than alcohol, includes toxic methanol, or wood alcohol, and acetone. The fraction that's less volatile than alcohol, the "tails" or "feints," includes a host of aromatic substances that are desirable. Among these "congeners" (substances that accompany alcohol) are esters, terpenes, and volatile phenolics, along with some substances that are desirable in limited amounts. The most notable of the latter are the "higher" alcohols, whose long, fat-like chains can give spirits a full, almost oily body, but also contribute a pronounced harsh flavor and unpleasant aftereffects. They're often called *fusel oils.* (*Fusel* is the German for "rotgut.") A small dose of fusel oils gives a distilled alcohol character; too much makes it unpleasant.

Purity and Flavor The best indication of how strongly flavored a distilled alcohol will be is the percentage of alcohol in the liquid immediately after distillation, before it's further treated by aging and/or dilution with water to its final strength (see box, p. 765). The higher the alcohol content to which it's distilled, the purer a mixture of alcohol and water it is, the lower the proportion of fusel oils and other aromatics, and so the more neutral the flavor. Vodkas are usually distilled to 90% alcohol or more; brandies and flavorful malt and corn whiskies to 60–80%. Crudely distilled moonshine is only 20–30% alcohol, and therefore harsh and even hazardous.

Pot or Batch Distillation: Selecting Volatiles by Time There are two ways for distillers to separate the vapor into undesirable heads, somewhat desirable tails, and the desirable main run of alcohol. The original way, and the way that is still used for many of the finest liquors, is separation in a simple pot still by *time.* It can take 12 hours or more for a batch of beer or wine to be heated close to the boil and then distilled. The very volatile head vapors come off first, followed by the main alcohol-rich run, and then the less volatile fusel-oil tails. So the distiller can divert the initial vapors, collect the desirable main run in a different

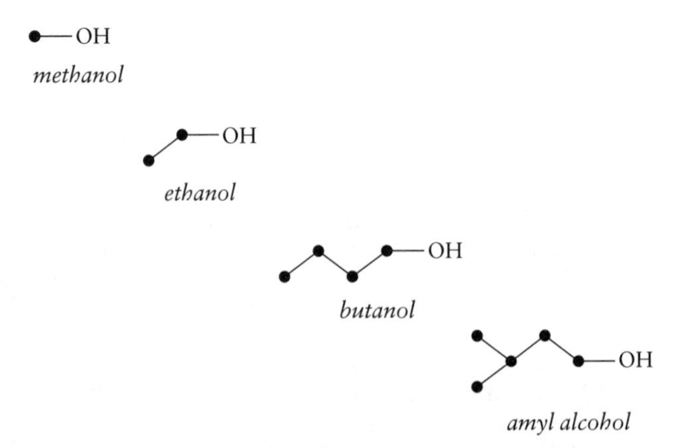

The structures of several different alcohols. Methanol is a poison because our bodies convert it into formic acid, which accumulates and damages the eyes and brain. Ethanol is the main alcohol produced by yeasts. Butyl and amyl alcohols are two of the "higher," or longer-chain alcohols. When concentrated by distillation, they contribute an oily consistency to spirits thanks to their fat-like hydrocarbon tails.

container, and then divert the late vapors again. In practice, distillers repeat the pot distillation, the first pass giving spirits with 20–30% alcohol, and the second 50–70%.

Continuous Distillation: Selecting Volatiles by Position The second way in which distillers can separate the desirable volatiles from the rest is by their *position* in a column still, an elongated chamber developed by French and British distillers during the Industrial Revolution. In a column still, the starting wine or beer is fed into the column from the top, and the column is heated from the bottom with steam. The bottom of the column is therefore the hottest region, the top the coolest. Methanol and other low-boiling substances remain vaporized throughout all but the very top of the column, while fusel oils and other aromatics with high boiling points will condense on collection plates at hotter positions toward the bottom of the column, and alcohol will condense—and can be collected—at an intermediate point. The advantage of the column still is that it can be operated continuously and without the necessity of close monitoring; the disadvantage is that it offers less opportunity than the pot still for the distiller to control the composition of the distillate. When two or more columns are run together in series, they're capable of producing a neutral distillate that is 90–95% alcohol.

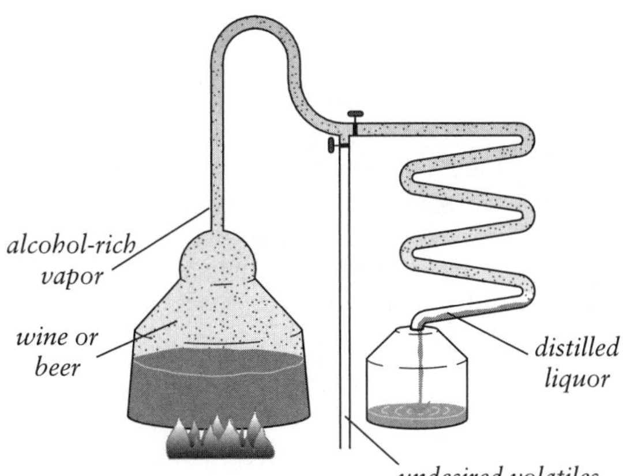

Pot distillation. As wine or beer is gradually heated, the composition of its vapors changes, with very volatile substances evaporating first, less volatile substances later. The distiller diverts early and late vapors with their undesirable volatiles, and collects the "main run," rich in alcohol and desirable aromas.

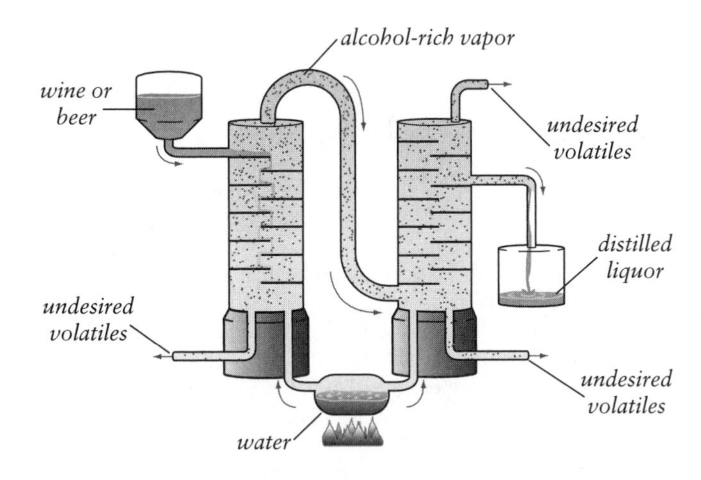

Continuous distillation in a column still. The plates in each column are hottest at the steam input and coolest at the other end. Substances with low boiling points, including alcohol, are concentrated in the vapor that leaves the first column and rises in the second, and the alcohol-rich fraction is collected at a particular position in the second column.

Maturation and Aging Fresh from the still, distilled liquors are as colorless as water, or "white." They're also rough and harsh, so all are matured for weeks or months to allow the various components to react with each other, form new combinations, and become less irritating. From this point, the spirits are handled differently according to the kind of product they're meant to become. "White" spirits, including vodka and eaux de vie made from fruits, are not aged; they may be flavored, then adjusted to the proper alcohol content by the addition of water, and bottled. "Brown" spirits, including brandies and whiskies, are so called because they're aged in wood barrels, from which they derive a characteristic tawny color and complexity of flavor. (Some brown spirits may be colored with caramel instead.) Spirits may be barrel-aged for anything from a few months to decades, during which their flavor changes considerably.

The extraction, absorption, and oxidation processes that take place during barrel aging result in the development of a mellow, rich flavor (p. 720). And the barrel allows both water and alcohol to evaporate from the spirits, thus concentrating the remaining substances. A barrel may lose several percent of its volume per year; that portion is called "the angels' share," and it may approach half the barrel volume after 15 years.

Final Adjustments When spirits are judged ready for bottling, they're usually blended to obtain a consistent flavor, and diluted with water to the desired final alcohol content, in the neighborhood of 40%. Small quantities of other ingredients may be added to fine-tune the flavor and color; these include caramel coloring, sugar, a water extract made by boiling wood chips (the *boisé* of Cognac and Armagnac), and wine or sherry (blended U.S. and Canadian whiskeys).

Chill-Filtering Many spirits are *chill-filtered*: chilled to below the freezing point of water, and then filtered to remove the cloudy material that forms. The substances that form the cloud are poorly soluble fusel oils and volatile fatty acids from the original spirits, and a variety of similar substances extracted from the barrel. Their removal prevents the spirits from clouding when the drinker chills them or dilutes them with water, but it also removes some flavor and body, so some producers choose not to chill-filter. Clouding does not occur in spirits with more than about 46% alcohol, so such undiluted "cask-strength" spirits are often not chill-filtered. (Some spirits cloud spectacularly; see p. 771).

Serving and Enjoying Spirits

Crystal Decanters Can Be Hazardous High-alcohol spirits are biologically and

Concentrated Alcohol: *Proof*

The term *proof* is sometimes used to designate the alcoholic content of distilled liquors. In the United States, the proof designation is just about double the percentage by volume of alcohol, so that 100 proof, for example, designates 50% alcohol. (The proof number is slightly more than double the percentage, because alcohol causes a volume of water to contract on mixing.) The term *proof* comes from a 17th-century test for proving the quality of spirits that involved moistening gunpowder with the spirits, and then putting a flame to it. If the gunpowder burned slowly, the spirits were at proof; if it spluttered or burst into flame, they were under or over proof respectively.

chemically stable and can be kept for years without spoiling. One traditional and decorative way of storing them has been the decanter made of glass crystal, which derives its weight and appearance from the element lead. Unfortunately, lead is powerfully toxic to the nervous system, and readily leaches from crystal into spirits and other acidic liquids. Old decanters that have been used many times have been preextracted and are safe to use; new ones should either be pretreated to remove lead from the inner surfaces, or only used for serving, not storing.

The Flavors of Spirits Spirits are served at temperatures ranging from ice cold (Swedish *aquavit*) to steaming hot (Calvados). To appreciate nuances of flavor, they're best served at room temperature, and if necessary warmed in the hands. Their aroma is intense, so much so that it can be just as enjoyable to sniff as to sip; Scotch lovers call this *nosing*. At distilled strengths, alcohol has an irritating and then numbing effect on the nose that is accentuated at high temperatures. To reduce the interference of alcohol and bring out more delicate aro-

Some Popular Distilled Spirits

The alcohol content after distillation is an indicator of how much flavor is carried over from the base wine or beer to the spirits. The higher the alcohol content, the lower the content of other aromatics and the more neutral the flavor.

Spirit	Original Material	Alcohol Content after Distillation (percent)	Aging
Brandy	Grapes	Up to 95	Oak barrels
Armagnac		52–65	Oak barrels
Cognac		70	Oak barrels
Grappa, marc	Grape pomace	70	None
Calvados	Apples	70	Oak barrels
Eaux de vie	Various fruits	70	None
Whiskies and whiskeys			
Scotch malt	Malted barley	70	Used oak barrels
Grain	Grains, malted barley	95	Used oak barrels
Irish	Grains, malted barley	80	Used oak barrels
Bourbon	Corn, malted barley	62–65	Charred new oak barrels
Canadian	Grains, malted barley	90	Used oak barrels
Gin	Grains, malted barley	95	None
Vodka	Grains, potatoes, malted barley	95	None
Rum	Molasses	70–90	None/oak barrels
Tequila	Agave	55	None/oak barrels

mas, connoisseurs often dilute whiskies with good-quality water to 30% or 20% alcohol. Different kinds of spirits have very different flavors, which derive from the original ingredient—grape or grain—from the yeasts and fermentation, from the prolonged heat of distillation, and from contact with wood and the passage of time. Spirits with a high fusel oil content have an unctuous quality in the mouth, while more neutral spirits give a cleansing, drying effect. The aromas of spirits often persist in the mouth long after the liquid itself has been swallowed.

Kinds of Spirits

Distilled spirits are made all over the world from all kinds of alcoholic liquids. Here are brief descriptions of the more prominent.

Brandies Brandies are spirits distilled from grape wine. The two classic brandies are Cognac and Armagnac, the first named for a town and the second for a region in southwestern France, each not far from Bordeaux. Both are made from neutral white grapes (mainly Ugni blanc) that are casually fermented into wine, and the wines distilled between harvest and mid-spring (the best brandies are distilled first; as the wine sits, it loses esters and develops volatile acidity and off-aromas). Cognac is double-distilled from the wine with its yeast lees to an alcohol content of about 70%, most Armagnac single-distilled without yeast in a traditional column still to about 55%. Each is then aged in new French oak barrels for a minimum of six months; some Cognacs are aged for 60 years or more. Before bottling, each is diluted to about 40% alcohol and may be adjusted with sugar, oak extract, and caramel. Cognac has a fruity, flowery character thanks to the distillation of esters from the wine yeasts. Armagnac is relatively rough and complex thanks to its higher content of volatile acids; it's said to have a prune-like aroma. With long aging, both develop a prized *rancio* ("rancid") character from the transformation of fatty acids into methyl ketones, which also provide the distinctive aroma of blue cheese (p. 62).

Less renowned brandies are made elsewhere in France and throughout the world in a variety of ways, from the industrial to the artisanal. Especially interesting are brandies distilled from more distinctive grape varieties than the purposely neutral Ugni blanc.

Eaux de vie, Fruit Alcohols, White Alcohols These are various terms that are less confusing than their synonym "fruit brandy": they name spirits that are distilled from fermented fresh fruits other than grapes. Unlike true "burned wines," which offer a complicated, transformed winey-ness, eaux de vie capture and concentrate the distinctive essence of the fruits from which they're made, so they can be savored almost pure rather than in their flesh. France, Germany, Italy, and Switzerland are especially noted for their fine fruit alcohols. Some popular examples are apple (Calvados), pear (Poire Williams), cherry (Kirsch), plum (Slivovitz, Mirabelle, Quetsch), and raspberry (Framboise); less widely known are apricot (French Abricot), figs (North African and Middle Eastern *Boukha*), and watermelon (Russian *Kislav*).

A single bottle of eau de vie may represent from 10 to 30 lb/4.5–13.5 kg of the fruit. Fruit alcohols are generally double-distilled to about 70% alcohol and are not aged in barrels—hence their lack of color—because their point is to concentrate the fruit's own flavor into an intense, full, but pure essence. One prominent exception to this rule is Calvados, an apple eau de vie that is distilled in Brittany from a blend of varieties, some too sour or bitter for eating. The apples are slowly fermented into cider over the course of several cool weeks in the autumn, and the cider is then distilled in either pot or column stills, depending on the district. The distillate is then matured in old barrels for a minimum of two years. Slivovitz, a Balkan plum alcohol, is also barrel-aged.

Whiskies and Whiskeys Whiskies (United Kingdom) and whiskeys (elsewhere) are spirits that have been distilled from fermented grains, mainly barley, maize, rye, and wheat, and then aged in barrels. The term comes from a barley distillate of medieval Britain, but is now applied to largely maize distillates in the United States and Canada, and mixed grain distillates in many countries.

Scotch and Irish Whiskies There are three kinds of Scotch whisky. One, *malt whisky,* is made in the Highlands and islands entirely from malted barley. It's distilled twice in pot stills to about 70% alcohol, and has a strong, distinctive flavor. Another, *grain whisky,* is less flavorful and less costly; it is made in the lowlands from various cereals and just a small portion (10–15%) of malted barley to convert their starches into sugars, and distilled in a continuous still to a neutral 95% alcohol. The third and most common is a blend of malt and grain whiskies, with grain whisky accounting for 40–70%. Such blending began in the 1860s for economic reasons, and turned out to produce a milder, more widely appealing drink just in time to replace brandy when the insect scourge phylloxera devastated European vineyards in the 1870s and 1880s. This is when Scotch developed its international reputation. Today, Scotch connoisseurs prize the distinctive "single-malt" whiskies produced by the few remaining small distillers of all-malt whisky.

Whisky makers produce beer, omitting the hops, and then distill it, yeasts and all. The distillate is aged in used oak barrels for a minimum of three years, then diluted with water to around 40% alcohol, and is usually chill-filtered. Scotch whisky owes its special flavor largely to the barley malt. Malt whiskies from Scotland's west coast have a unique, smoky flavor that comes from the use of peat fire for drying the malt, and peaty water for mashing the grain before fermentation. Peat, the mat of decaying and decayed vegetation that once was the cheapest fuel available in swampy areas of Britain, contributes volatile organic molecules to the brew that find their way into the distillate.

Most Irish whisky is made from a mixture of about 40% malted and 60% unmalted barley. For this reason, and because it is pot-distilled twice and then again in a column still, Irish whisky is milder than malt Scotch and even some Scotch blends.

American and Canadian Whiskeys North American whiskeys are produced mainly from the New World's indigenous grain, maize. The most prominent corn whiskey is bourbon, which is named for a county in Kentucky where maize grew well in colonial times, and where there was abundant water for both mashing the grains and cooling the distillate.

Food Words: *Aperitif, Digestif*

These French words describe two functions attributed to concentrated alcohols in the Middle Ages, ideas that live on in both the words and drinking habits. *Aperitif* comes from an Indo-European root meaning "to uncover, to open," and is a drink to be had before a meal in order to open our system to the nourishment to come. *Digestif* comes from an ancient root meaning "to act, to do," and names a drink for the end of the meal that will stimulate our system to assimilate the meal's nourishment. Research has found that alcohol does indeed stimulate the secretion of digestive hormones in the stomach.

Bourbon is made from a mash that's usually 70–80% maize, 10–15% malted barley to digest the starch, and the remainder rye or wheat. After fermenting for two to four days, the whole mash, grain residues and yeast included, is distilled in a column and then a kind of continuous pot still to 60–80% alcohol. The distillate is aged for at least two years in new, charred American oak barrels, which give bourbon a deeper color and stronger vanilla note than Scotch whiskies have. Summer temperatures that can reach 125°F/53°C in the warehouses modify and accelerate the chemical reactions of aging. Bourbons are generally chill-filtered; in fact this technique was invented by the Tennessee whiskey maker George Dickel around 1870. Unlike French brandies and Canadian whiskeys, bourbon cannot be colored, sweetened, or flavored; the only additive allowed is water.

Canadian whiskeys are among the mildest and most delicate of the spirits made from grains. They are a blend of a light-flavored column-distilled grain whiskey with small amounts of stronger whiskeys; they can also include wines, rum, and brandy, up to 9% of the blend. They're aged for a minimum of three years in used oak casks.

Gins There are two principal styles of *distilled* gin made today, English and Dutch, as well as cheaper gin that cannot be called distilled because its flavorings are simply added to neutral alcohol.

The traditional Dutch production method is to distill a fermented mixture of malt, corn, and rye two or three times in pot stills at low proof: that is, the distillate contains a fair amount of congeners, and resembles a light whisky. Then this distillate is distilled one last time, to a minimum of 37.5% alcohol, along with juniper berries and other spices and herbs, whose aromatic molecules end up in the final gin.

English-style, or "dry" gin, begins with neutral 96% alcohol produced from grain or molasses by other distilleries. This flavorless liquid is then diluted with water and redistilled in a pot with juniper and other flavorings. Juniper is required for the product to be called gin, and most gins also contain coriander. The other ingredients may include citrus peel and a great variety of spices. This distillate is diluted before bottling to 37.5 to 47% alcohol.

The primary aromas in gin come from the terpene aromatics (p. 390) in the spices and herbs, especially notes of pine, citrus, flowers, and wood (pinene, limonene, linalool, myrcene). Dutch gin is generally enjoyed on its own, while beginning in the 1890s, English dry gin inspired many cocktails and tall mixed drinks, including the martini, gimlet, and gin and tonic.

Rums Rum got its start in the early 17th century as a by-product of sugar making in the West Indies. Yeasts and other microbes readily grew in the leftover molasses and wash waters, the yeasts producing alcohol and the bacteria all kinds of aromatic substances, many of them not pleasant. From this mixed material, primitive distillation equipment and methods produced a strong, harsh liquid that was given mainly to slaves and sailors, and traded to Africa for more slaves. Controlled fermentations and improvements in distilling technology brought more drinkable rums in the 18th and 19th centuries.

There are now two distinct styles of rum. The modern light style is made by fermenting a molasses solution with a pure yeast culture for 12–20 hours, then distilling it to about 95% alcohol in a continuous still, aging it for a few months to eliminate roughness in the flavor, and diluting and bottling it at around 43% alcohol. Some light rums are given a brief time in barrels, but then are passed over charcoal to remove the color and some of the flavor.

Traditional Rums Traditional rums are made very differently, and have a much stronger flavor and darker color. Most come from Jamaica and the French-speaking Caribbean (Martinique, Guadeloupe). They were once fermented for up to two weeks with a spontaneous group of microbes, and often by adding the already strong-flavored lees of one fermentation to the next vat. Today, most traditional rums are fermented for a day or two with mixed microbial cultures dominated by an unusual yeast (*Schizosaccharomyces*) that excels in ester production. They're then pot-distilled to a much lower alcohol content, and therefore end up with four to five times the quantity of aroma compounds that light rum has. Finally they're aged in used American whiskey casks, where they get most of their color. Caramel can be added to deepen the color and flavor, a procedure that seems appropriate since rum is made from sugar in the first place.

Rums as Ingredients Rums are delicious on their own, but it's their aptitude for other foods that accounts for much of their popularity. Light rums go well with tart-sweet fruits and are the base for a number of tropical cocktails, including piña coladas and daiquiris. Medium and dark rums are a useful ingredient in sweets of all kinds thanks to their full caramel flavor.

Vodkas Vodka was first distilled in Russia in medieval times and for medical purposes, and became a popular drink in the 16th century. Its name means "little water." It has traditionally been made from the cheapest source of starch available, usually grain, but sometimes potatoes and sugarbeets. The source is unimportant, since the fermented base is distilled to eliminate most aromatics, and the remainder is removed by filtration through powdered charcoal to produce a smooth, neutral flavor. The essentially pure mixture of alcohol and water is then diluted with water to the desired strength, a minimum of around 38%, and bottled without aging.

Vodka was scarcely known in the United States until the 1950s, when it was discovered as an ideal alcohol for blending with fruit and other flavors in cocktails and tall mixed drinks. Recent years have brought vodkas flavored with citrus and other fruits, with chillis, and with barrel aging.

Grappa, Marc These are the Italian and French names for spirits distilled from *pomace,* the residue of grape skins and pulp, seeds and stems left behind when wine grapes are pressed. These drinks were born from frugality, as a way of getting the most out of the grapes. The solid remains still have juice, sugar, and flavor in them, so with some water and another period of fermentation, they generate alcohol and flavors that can then be concentrated by distillation, leaving behind the harsh astringency and bitterness. Pomace distillates were very much a by-product, usually distilled just once and often without diverting the heads and tails, and were bottled as is: so they were strong and harsh, something to warm and stimulate the vineyard workers, but not something to savor. In the last few decades, producers have been distilling more selectively and sometimes aging the results to make a fine drink.

Tequila and Mezcal These spirits are distilled from the carbohydrate-rich heart of certain Mexican species of the agave, a succulent plant in the Amaryllis family that resembles a cactus. Tequila is made mainly by large distilleries around the northerly city of Jalisco from the blue agave, *Agave tequilana,* while the more rustic mezcal is made by small producers around central Oaxaca from the maguey, *Agave angustifolia.*

The agave stores its energy in the sugar fructose and the long fructose chains called inulin (p. 805). Because humans lack an enzyme for digesting inulin, people have learned to cook inulin-rich foods for a long time at a low temperature, a treatment that breaks the chains into their component sugars, and also develops an intense and characteristic browned flavor. Tequila makers steam the inulin-rich agave hearts, which may weight 20–100 lb/9–45 kg, while mezcal producers roast them in large charcoal-fired pit ovens and generate smoky aromas that carry over into the spirits. The cooked, sweet hearts are then mashed with water and fermented, and the resulting alcoholic liquid distilled. Tequila distillation is industrial; mezcal is double-distilled, first in small clay pots, then in a larger metal pot still. Most tequila and mezcal is bottled without aging.

Tequila and mezcal have distinctive flavors that include roasty aromas, but also flowery ones (linalool, damascenone, phenylethyl alcohol), and vanilla (vanillin).

Flavored Alcohols: Bitters and Liqueurs

Alcohol's split chemical personality, its resemblance to fats as well as water, makes it an excellent solvent for other volatile, aromatic molecules. It does a good job of extracting and holding flavors from solid ingredients. Herbs, spices, nuts, flowers, fruits: all these and more are soaked in alcohol, or distilled along with alcohol, to make a host of flavored liquids. Gin is the best known of these. Most of the others fall into two families: bitters, which are just that, and liqueurs, which are sweetened to varying degrees with sugar.

Bitters Bitters are modern descendents of medicinal herbal brews that were first made with wine. Purely bitter ingredients include angostura (*Galipea cusparia*), a South American relative of the citrus family, Chinese rhubarb root, and gentian (*Gentiana* species); plant materials that are both bitter and aromatic include wormwood, chamomile, bitter orange peel, saffron, bitter almond, and myrrh (*Commifera molmol*). Most bitter alcohols are complex mixtures.They may be made by macerating the plant material or by distilling it along with the source of alcohol. Among the bitters commonly used today are Angostura and Peychaud's bitters, condiment-like

Some Examples of Flavored Alcohols

Flowers: Sambuca (elderflower), Gul (rose)
Spices: Anisette (anise), Pimento (allspice)
Nuts: Amaretto (almond); Frangelico (hazel); nocino (green walnuts)
Coffee: Kahlúa, Tia Maria
Chocolate: Crème de cacao
Fruits: Cointreau, Curaçao, Grand Marnier, Triple Sec (orange); Midori (melon); Cassis (black currants); limoncello (lemon); sloe "gin" (plum)
Herbs: Benedictine, Chartreuse, Jaegermeister, Crème de menthe, peppermint schnapps

Layering Liqueurs

The added sugar that sweetens liqueurs also contributes to their body and density. And because different liqueurs have different proportions of light alcohol and dense sugar, they have densities different enough to allow the mixologist to form distinct layers in the glass, with the densest liqueurs at the bottom (red grenadine, brown Kahlúa) and the lightest at the top (amber Cointreau, green Chartreuse). When the liqueurs have different colors and complementary flavors, this can produce a pleasant novelty drink. Fruit juices and syrups can also play a part in such constructions. Eventually, adjacent liquids will diffuse into each other and the layers disappear.

19th-century formulations that are added to mixed drinks and foods as a flavor accent, and such drinkable aperitifs and digestifs as Campari (unusually sweet) and Fernet Branca.

Liqueurs Liqueurs are a distilled alcohol sweetened with sugar and flavored with herbs, spices, nuts, or fruits. The flavoring agents may be extracted by soaking in the distilled alcohol, or they may themselves be distilled along with the alcohol. Most liqueurs have a neutral grain alcohol as their base, but there are a few whose base is a brandy or whisky. Examples are Grand Marnier, Cognac plus orange peel; Drambuie, Scotch whisky plus honey plus herbs; and Southern Comfort, bourbon whiskey plus peach brandy and peaches. Some liqueurs include stabilized cream.

Anise and Caraway Alcohols These spirits get their dominant flavor from the seeds of plants in the carrot family, and may be either sweet or dry. Anise is especially popular; there are French, Greek, Turkish, and Lebanese versions among others (*pernod* and *anisette, ouzo, raki, araq*). Caraway seeds flavor dry Scandinavian aquavits and the sweet German *Kümmel.* When clear anise alcohols are diluted with clear liquid water or ice cubes that melt, the mixture becomes surprisingly cloudy. This is because the aromatic terpene molecules are insoluble in water, and soluble in alcohol only when the alcohol is highly concentrated. As the alcohol becomes diluted, the terpenes separate from the continuous liquid into little water-avoiding droplets, and these scatter light like the fat globules in milk.

VINEGAR

Vinegar is alcohol's fate, the natural sequel to an alcoholic fermentation. Alcohol makes a liquid more resistant to spoilage because most microbes can't tolerate it. But there are a few important and ubiquitous exceptions: bacteria that can use oxygen to metabolize alcohol and extract energy from it. In the process they convert it to acetic acid, which is a far more potent antimicrobial agent than alcohol, and came to be one of the most effective preservatives of ancient and modern times. Alcoholic wine thus becomes pungently acidic wine: in French, *vin aigre.*

Absinthe

The most notorious herbal alcohol is absinthe, a green-tinged, anise-flavored liqueur whose main ingredient is parts of the wormwood plant, *Artemisia absinthium.* Wormwood has a harsh, bitter flavor and carries an aromatic compound, thujone, that in high doses is toxic not only to intestinal parasites and insects—hence the plant's name—but also to the human nervous system, muscles, and kidneys. Absinthe was hugely popular in late 19th-century France, and *l'heure verte,* "the green hour" of the afternoon when people dripped water through a sugar cube into the absinthe and caused it to cloud up, was depicted by a number of Impressionist painters and by the young Picasso. Absinthe developed a reputation for inducing convulsions and insanity, and was therefore outlawed in many countries around 1910, and replaced by simpler anise-flavored spirits. Whatever toxicity wormwood had for the heavy drinker was probably exacerbated by absinthe's high alcohol content, around 68% and nearly double the strength of most spirits. Absinthe remains legal in a number of countries and has recently enjoyed a modest and moderate revival.

An Ancient Ingredient

Because fermented plant juices naturally turn sour with acetic acid, our ancestors discovered wine and vinegar together. In fact, a major challenge in winemaking was to delay this souring by limiting the wine's exposure to the air. The Babylonians were making vinegar from date wine, raisin wine, and beer around 4000 BCE. They flavored their vinegar with herbs and spices, used it to pickle vegetables and meats, and added it to water to make it safe to drink. The Romans mixed vinegar and water to make an ordinary drink called *posca,* pickled vegetables in vinegar and brine, and judging by the late Roman recipe book of Apicius, often enjoyed vinegar in combination with honey. Pliny said that "no other sauce serves so well to season food or to heighten a flavor." In the Philippines there developed a tradition of serving a variety of uncooked fish, meats, and vegetables in vinegar made from palm sap and tropical fruits. And the Chinese evolved dark, complex vinegars from rice, wheat, and other grains, which are sometimes roasted before fermentation.

For millennia, vinegar was made simply by allowing partly filled containers of wine and other alcoholic liquids to sour, an unpredictable process that took weeks or months. The first system for more rapid production, a bed of grapevine twigs over which the wine was regularly poured to aerate it, was invented in France in the 17th century. In the 18th a Dutch scientist, Hermann Boerhaave, introduced the continuous trickling of wine over an aerating bed. In the 19th century, Louis Pasteur demonstrated the essential roles of microbes and oxygen in the traditional Orléans process (p. 773). Modern methods for growing baker's yeast and producing penicillin were adapted to vinegar manufacture after World War II, and now produce finished vinegar in a day or two.

The Virtues of Acetic Acid

Acetic acid contributes two different flavor elements to foods. One is its acidity on the tongue, and the other is its characteristic aroma in the nose, which can intensify to a kind of startling pungency, particularly when the vinegar is heated. The vinegar molecule can exist in two forms: as the intact molecule, and broken into its main portion and a free hydrogen ion. The hydrogen ion gives the main impression of acidity, while only the intact molecule is volatile and can escape from the vinegar or food, travel through the air, and reach the nose. Both the intact and "dissociated" forms coexist side by side, in proportions that are determined by their chemical surroundings. If the food is already acidic—thanks to the presence of tartaric acid in wine vinegar, for example—then less of the acetic acid dissociates, more is intact and volatile, and the vinegar aroma is stronger.

Acetic acid is an especially effective preserving agent. A solution as weak as 0.1%—the equivalent of a teaspoon of standard-strength vinegar in a cup of water/5 ml in 250 ml—will inhibit the growth of many microbes.

Acetic acid has a higher boiling point

Food Words: *Vinegar, Acid, Acetic*

Though it doesn't look or sound related, the word *vinegar* comes from the same root as both *acid* and *acetic*: the Indo-European *ak-*, meaning "sharp." (The *aigre* in *vinaigre* was originally the Latin *acer.*) *Edge, acute, acrid, ester,* and *oxygen* are related words, oxygen because its presence was once thought necessary to make an acid.

than water, 236°F/118°C. This means that vinegar will get more concentrated if it's boiled. Because half of its molecule is more fat-like than water-like, it is a better solvent than water for many chemical relatives of fats, including the aroma compounds in herbs and spices. This is why cooks flavor vinegars by steeping herbs and spices in them, and why vinegar can help remove greasy films from various surfaces.

The Acetic Fermentation

It takes three ingredients to make vinegar: an alcoholic liquid, oxygen, and bacteria of the genus *Acetobacter* or *Gluconobacter*, mainly *A. pasteurianus* and *A. aceti*. These bacteria are among the few microbes that are able to use alcohol as an energy source. Their metabolism of alcohol leaves behind two by-products, acetic acid and water.

$$CH_3CH_2OH + O_2 \rightarrow CH_3COOH + H_2O$$

Alcohol + oxygen → acetic acid + water

Acetic acid bacteria require oxygen, and so live on the surface of the fermenting liquid, where with other microbes they form a film sometimes called the "mother." Especially thick films are created by *Acetobacter xylinum*, which secretes a form of cellulose. (Such mats are sometimes cultivated and eaten for themselves; see p. 509.) Acetobacteria thrive in warm conditions, so vinegar fermentations are often carried out at relatively high temperatures, from 82 to 104°F/28–40°C.

The concentration of alcohol in the starting liquid affects the acetic fermentation and the stability of the resulting vinegar. An alcohol concentration around 5% will produce a vinegar that is around 4% acetic acid, which is strong enough to prevent the vinegar solution itself from spoiling. Above 5% alcohol, the resulting vinegar will be stronger in acetic acid and so more stable, but the fermentation proceeds more slowly because the high alcohol levels inhibit the activity of the bacteria. For this reason, and to minimize residual alcohol in the finished vinegar, wines of 10–12% alcohol are usually diluted with water before acetic fermentation. However this also dilutes the wine's flavorful components; so patient vinegar makers may still choose to ferment their wine straight.

Vinegar Production

There are three standard ways of producing vinegar in the West.

The Orléans Process The simplest, oldest, and slowest method was perfected in the Middle Ages in the French city of Orléans, where spoiled barrels of Bordeaux and Burgundy wine on their way to Paris were identified and salvaged as vinegar. In the Orléans process, wood barrels are partly filled with diluted wine, inoculated with a mother from a previous batch, and allowed to ferment. Periodically, some vinegar is drawn off and replaced by new wine. This method is slow, because the transformation of alcohol to acetic acid is limited to the wine surface exposed to the air. But the slow fermentation leaves time for reactions among the alcohol, acetic acid, and other molecules, and produces the finest flavor. When optimized, this process can yield a barrel full of vinegar in two months.

Streamlined Trickling and Submerged Cultures In the second, "trickling" method, the wine is poured repeatedly over

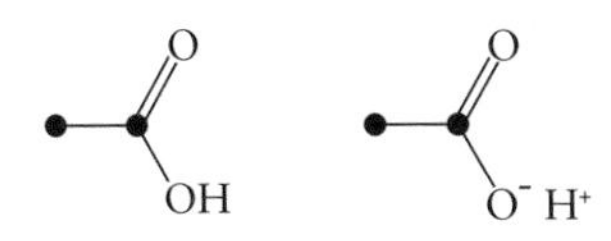

acetic acid

The intact acetic acid molecule, and the acid dissociated into its acetate and hydrogen ions. Only the intact molecule is volatile and detectable in the nose by its distinctive smell. Adding vinegar to an alkaline ingredient—egg whites or baking soda, for example—causes the acetic acid molecules to dissociate, and diminishes its aroma.

a porous, air-rich matrix—wood shavings, or a synthetic material—onto which the acetic bacteria cling. This greatly increases the effective surface area of the wine, and regularly exposes all parts of the liquid to both oxygen and bacteria. The result is a quick fermentation that takes only a few days. Finally, there is the "submerged culture" method, in which free-swimming bacteria are supplied oxygen in the form of air that is bubbled through the tank. This industrial method converts the liquid's alcohol into acetic acid in 24–48 hours.

After Fermentation After fermentation, nearly all vinegars are pasteurized at 150–160°F/65–70°C to kill remaining bacteria of all kinds, but especially the acetobacteria themselves, which respond to the disappearance of the alcohol by metabolizing acetic acid to water and carbon dioxide and thus weakening the vinegar. Most vinegars are aged for a few months, a period in which their flavor becomes less harsh and more mellow, thanks in part to the combination of acetic and other acids with various compounds to form new, less pungent, often aromatic substances.

Common Kinds of Vinegar

Cooks can choose among several different kinds of vinegar. Though all have the basic aroma and pungency of acetic acid, each is distinctive, because they're made with different starting materials, and may or may not be matured in wood.

Wine Vinegars Wine vinegars are made from a base of yeast-fermented grape juice. They therefore have a winey character from the aromatic and savory by-products of the yeast fermentation. Interestingly prominent in wine and cider vinegars are buttery aroma compounds (diacetyl, butyric acid). Balsamic and sherry vinegars are specialized versions of wine vinegar (see pp. 775–776).

Cider Vinegars Cider vinegar is made from a base of yeast-fermented apple juice. It therefore includes some of the characteristic aroma components of apples, and others that are especially accentuated in apple fermentation; these include the volatile phenols that give animal and stable aromas to grape wines (ethyl guaiacol and ethyl phenol, p. 738). Apples are rich in malic acid, so cider vinegars undergo a malolactic fermentation (p. 730) that may augment aroma while softening acidity. Thanks to its pulp and tannin content, cider vinegar often becomes cloudy with tannin-protein complexes.

Fruit Vinegars Fruit vinegars may simply be ordinary vinegars flavored by contact

Making Vinegars in the Kitchen

Cooks can easily make their own vinegars from leftover wine or from fruits of their own choosing. A few guidelines will improve the odds of getting a good result. Sweet liquids will become alcoholic and sour spontaneously, but "wild" microbes may produce off-flavors. This possibility is minimized by starting with a cultured yeast and a vinegar "mother" from an active vinegar crock or commercial source. The warmer the temperature (up to around 85°F/30°C) and the larger the surface area exposed to the air, the faster the liquid will acetify. Fruits with less than about 10% sugar in their juices will produce less than 5% alcohol and thus less than 4% acetic acid in the final vinegar, which will be prone to spoiling. Such fruits should be supplemented with table sugar, which boosts the subsequent alcohol and acetic acid levels.

with fresh fruit, including apples, but they're also made by fermenting the fresh fruit juices. Pineapple and coconut vinegars are examples. Fruit vinegars are interesting for their expression of the fruit's flavor through the alcoholic and acetic fermentations.

Malt Vinegars Malt vinegar is essentially made from unhopped beer: that is, from cereal grains and sprouted barley. It has overtones of barley malt. This was the standard form of vinegar in beer-drinking Britain, where it was originally called *alegar.*

Asian Vinegars Asian rice and grain vinegars are made from grains whose starch is broken down to sugars by means of a mold culture rather than sprouted grains (p. 753). Chinese vinegars can be especially flavorful and savory because they're made from whole, sometimes roasted grains, fermented in continuous contact with the grain solids, and often aged in contact with the molds, yeasts, and bacteria, all of which release amino and other organic acids and other flavor compounds into the vinegar.

White Vinegars White vinegar is among the purest sources of acetic acid. It's made by acetic fermentation of pure alcohol that has been either distilled or synthesized from natural gas, and is not aged in or softened by contact with wood. It contains few or none of the aromatic and savory byproducts of the alcoholic fermentation. In the United States, more white vinegar is made than any other kind. It's used mainly in the manufacture of pickles, salad dressings, and mustards.

Distilled Vinegars Distilled vinegar in the United States is white vinegar made with distilled alcohol; in the United Kingdom, it's vinegar that is made by acetic fermentation of unhopped beer, and then distilled to concentrate the acetic acid.

Vinegar Strength When developing and following recipes in which vinegar is a prominent ingredient, cooks should take care to note not only the kind of vinegar, but also the strength, which is usually indicated on the label. In the United States, most industrially produced vinegars are adjusted to 5% acetic acid, but many wine vinegars are 7% or even stronger. Mild Japanese rice vinegars, by contrast, may be 4% (the U.S. minimum), black Chinese vinegars as little as 2%. A spoonful may thus provide half as much acetic acid as expected, or twice as much, depending on the vinegars that are called for and actually used.

Balsamic Vinegar

True balsamic vinegar, *aceto balsamico,* is a vinegar like no other: almost black in color, syrupy, sweet, remarkably complex in flavor, and remarkably expensive, all thanks to decades-long fermentation, aging, and concentration in wood casks. It has been made in the northern Italian state of Emilia-Romagna since medieval times. Individual households produced their own as a kind of general-purpose, soothing tonic, or balsam. It wasn't until the 1980s that the rest of the world discovered balsamic vinegar, a discovery that fostered the

Using Balsamic Vinegar

Traditional balsamic vinegar is applied by the drop to a variety of dishes, from salads and grilled meats and fish to fruits and cheese. Mass-produced versions are added in larger quantities to lend depth of flavor to soups and stews, and to make mellower vinaigrette dressings than plain wine vinegar does.

development of less elaborate and less costly approximations. The label term *tradizionale,* "traditional," is reserved for the original version.

Making Traditional Balsamic Vinegar

Traditional balsamic vinegar begins with wine grapes: white Trebbiano, red Lambrusco, and a number of other varieties are used. Their juice is boiled until the volume is reduced by about a third. Boiling removes enough water to concentrate the juice to around 40% dissolved sugars and acids, and begins the sequence of browning reactions between sugars and proteins that generate both rich flavor and color (p. 778). The juice is then placed in the first of a sequence of progressively smaller barrels, often made from a variety of woods (oak, chestnut, cherry, juniper), which are kept in an attic or other location where they're exposed to the variations and extremes of the local climate. In summer heat, the concentrated sugars and amino acids react with each other to produce aroma molecules more commonly found in roasted and browned foods, and the fermentation products and by-products react with each other to form a heady mixture. As evaporation continues to remove water and concentrate the must (about 10% of the barrel disappears each year), each barrel is replenished with must from the next younger barrel. Finished vinegar, whose average age must be a minimum of 12 years, is removed from the oldest barrel. According to one estimate, it takes about 70 lb/36 kg of grapes to make 1 cup/250 ml of traditional balsamic vinegar.

Notice that there's no initial alcoholic fermentation before the acetification begins. Instead, a mixed culture of yeasts and bacteria simultaneously converts a portion of the abundant grape sugars into alcohol, and that alcohol into acetic acid. These conversions proceed slowly, over the course of several years, because the high concentration of grape sugars and acids inhibits the growth of all microbes. The alcoholic fermentation is carried out by unusual yeasts, *Zygosaccharomyces bailii* or *bisporus,* that are adapted to surviving in environments high in sugars and in acetic acid. At the same time that the two fermentations take place, so do the processes of maturation and aging.

In the end, traditional balsamic vinegar may contain anywhere from 20 to 70% unfermented sugars, about 8% acetic and 4% tartaric, malic, and other nonvolatile acids, an aroma-enhancing 1% alcohol, and up to 12% glycerol, a product of the yeast fermentation that contributes to the velvety viscosity.

The "condiment" grades of balsamic vinegar are made much more rapidly than the traditional grade, and are far less concentrated and fine-flavored. The better mass-produced vinegars include some cooked-down grape must and young balsamic vinegar, and are aged for a year or so. Cheap balsamic vinegars are no more than ordinary wine vinegar colored with caramel and sweetened with sugar.

Sherry Vinegar

A style of vinegar that lies somewhere between ordinary wine vinegar and balsamic vinegar is the solera-aged sherry vinegar of Spain. This starts from the young sherry wine, which contains no residual sugar. Like sherry wines and balsamic vinegars, sherry vinegar is blended with older batches and matured for years or decades in a series of partly-filled barrels. The concentration by evaporation, and extended contact with microbes and wood, leave sherry vinegar with high levels of savory amino acids and organic acids, and viscous glycerol. In old soleras, the acetic acid concentration can reach 10% and more. Sherry vinegar isn't as dark and savory as balsamic vinegar, but is noticeably more intense and nutty than other wine vinegars.

CHAPTER 14

COOKING METHODS AND UTENSIL MATERIALS

Each of the basic methods of cooking, from grilling over a fire to irradiating in a microwave oven, has its own particular influence on food. This chapter briefly explains how these methods work, and describes the properties of the various metal and ceramic utensils we use to heat foods.

First, though, it's good to consider an important transformation that foods undergo when they're subjected to sufficient heat, no matter what the cooking method. The browning reactions come up in every chapter this book. They have remarkable effects on both the flavor and appearance of a host of foods, from condensed milk to grilled meats, from chocolate to beer.

BROWNING REACTIONS AND FLAVOR

While the chemical changes caused by moderate heat modify or intensify flavors that are intrinsic to a food, the browning reactions produce new flavors, flavors that are characteristic of the cooking process. These reactions are named for the typical colors that they also create, which may actually range from yellow to red to black, depending on the conditions.

CARAMELIZATION

The simplest browning reaction is the caramelization of sugar, and it's not simple at all (p. 656). When we heat plain table sugar, essentially just molecules of sucrose, it first melts into a thick syrup, then slowly changes color, becoming light yellow and progressively deepening to a dark brown. At the same time, its flavor, initially sweet and odorless, develops acidity, some bitterness, and a rich aroma. The chemical reactions involved in this transformation are many, and they result in the formation of hundreds of different reaction products, among them sour organic acids, sweet and bitter derivatives, many fragrant volatile molecules, and brown-colored polymers. It's a remarkable change, and a fortunate one: it contributes to the pleasures of many candies and other sweets.

THE MAILLARD REACTIONS

Even more fortunate and complex are the reactions responsible for the cooked color and flavor of bread crusts, chocolate, coffee beans, dark beers, and roasted meats, all foods that are *not* primarily sugar. These are known as the Maillard reactions, after Louis Camille Maillard, a French physician, who discovered and described them around 1910. The sequence begins with the reaction of a carbohydrate molecule (a free sugar or one bound up in starch; glucose and fructose are more reactive than table sugar) and an amino acid (free or part of a protein chain). An unstable intermediate structure is formed, and this then undergoes further changes, producing hundreds of different by-products. Again, a brown coloration and full, intense flavor result. Maillard flavors are more complex and meaty than caramelized flavors, because the involvement of the amino acids adds nitrogen and sulfur atoms to the mix of carbon, hydrogen, and oxygen, and produces new families of molecules and new aromatic dimensions (see illustration below, and box, p. 779).

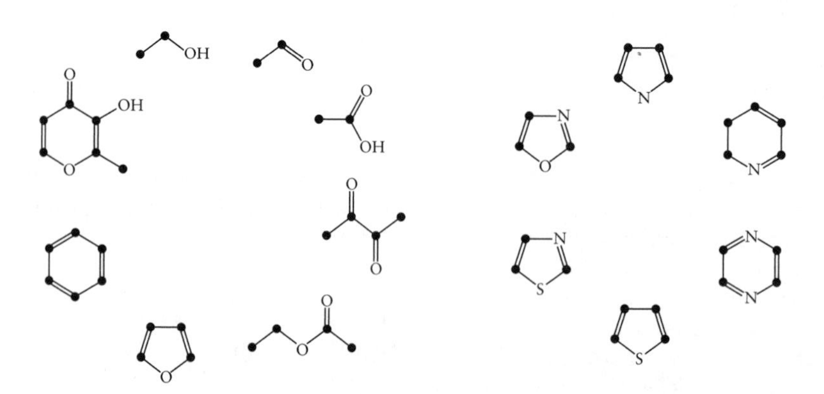

Representative aroma molecules produced by caramelization (left; see p. 656) *and by the Maillard reactions between carbohydrates and amino acids* (right). *Amino acids contribute nitrogen and sulfur atoms to produce the distinctive central rings of* (clockwise from top) *pyrroles, pyridines, pyrazines, thiophenes, thiazoles, and oxazoles. Each ring can be augmented with other structures attached to the carbon atoms. Maillard products have a range of qualities, from leafy and floral to earthy and meaty.*

High Temperatures and Dry Cooking Methods

Both caramelization and Maillard browning proceed at a rapid rate only at relatively high temperatures. Caramelization in table sugar becomes noticeable at around 330°F/165°C, Maillard browning perhaps 100°F/50°C below that. Large amounts of energy are required to force the initial molecular interactions. The practical consequence of this is that most foods brown only on the outside and during the application of dry heat. The temperature of water can't rise above 212°F/100°C until it is vaporized (unless it's under high pressure in a pressure cooker). So foods that are cooked in hot water or steamed, and the moist interiors of meats and vegetables, will never exceed 212°F. But the outer surfaces of foods cooked in oil or in an oven quickly dehydrate and reach the temperature of their surroundings, perhaps 300 to 500°F/159–260°C. So it is that foods cooked by "moist" techniques—boiling, steaming, braising—are generally pale and mild compared to the same foods cooked by "dry" methods—grilling, baking, frying. This is a useful rule to keep in mind. For example, one key to a rich-tasting stew is to brown the meat, vegetables, and flour quite well by frying them before adding any liquid. On the other hand, if you want to emphasize the intrinsic flavors of the foods, avoid the high temperatures that create the intense but less individualized browning flavors.

Slow Browning in Moist Foods

There are exceptions to the rule that browning reactions require temperatures above the boil. Alkaline conditions, concentrated solutions of carbohydrates and amino acids, and prolonged cooking times can all generate Maillard colors and aromas in moist foods. For example, alkaline egg whites, rich in protein, with a trace of glucose, but 90% water, will become tan-colored when simmered for 12 hours. The base liquid for brewing beer, a water extract of barley malt that contains reactive sugars and amino acids from the germinated grains, deepens in color and flavor with several hours of boiling. Watery meat or chicken stock will do the same as it's boiled down to make a concentrated demiglace. Persimmon pudding turns nearly black thanks to its combination of reactive glucose, alkaline baking soda, and hours of cooking; balsamic vinegar turns nearly black over the course of years!

Some of the Flavors Produced by Caramelization and Maillard Browning

Caramelization 330°F/165°C and above	**Maillard Reactions 250°F/120°C and above**
Sweet (sucrose, other sugars)	Savory (peptides, amino acids)
Sour (acetic acid)	Floral (oxazoles)
Bitter (complex molecules)	Onions, meatiness (sulfur compounds)
Fruity (esters)	Green vegetables (pyridines, pyrazines)
Sherry-like (acetaldehyde)	Chocolate (pyrazines)
Butterscotch (diacetyl)	Potato, earthy (pyrazines)
Caramel (maltol)	
Nutty (furans)	Plus caramelization flavors

Drawbacks of the Browning Reactions

Browning reactions do have some drawbacks. First, many dehydrated fruits are prone to gradual browning over weeks or months at room temperature, because the carbohydrates and amine-containing molecules are especially concentrated (browning caused by enzymes can also be a factor). Small amounts of sulfur dioxide are commonly added to these foods to block these unwanted changes in color and taste. Second, the nutritional value of the foods is slightly reduced because amino acids are altered or destroyed.

Finally, there's evidence that some products of the browning reactions can damage DNA and may cause cancer. In 2002, Swedish researchers found worrisome levels of acrylamide, a known carcinogen in rats, in potato chips, french fries, and other starchy fried foods, apparently the product of reactions between sugars and the amino acid asparagine. The health significance of this and similar findings remains unclear. The ubiquity of browned foods, both today and through thousands of years of history, would suggest that they do not constitute a major threat to public health. And other browning-reaction products have been found to protect against DNA damage! But it's probably prudent to make charred meats and fried snacks occasional pleasures, not everyday ones.

FORMS OF HEAT TRANSFER

Cooking can be defined in a general way as the transformation of raw foods into something different. Most often, we transform foods by heating them—by transferring energy from a heat source into the foods, so that the food molecules move faster and faster, collide harder and harder, and react to form new structures and flavors. Our various cooking methods—boiling, broiling, baking, frying, and so on—achieve their various effects by employing very different materials as the medium through which the heat moves, and by drawing on different forms of heat transfer. There are three ways to transfer heat, and an acquaintance with them will help us understand how particular cooking techniques affect foods the way they do.

Conduction: Direct Contact

When thermal energy is exchanged from one particle to a nearby one by means of a collision or a movement that induces movement (for example, through electrical attraction or repulsion), the process is called *conduction*. Though it's the most straightforward means of heat transfer in matter, conduction takes different forms in different materials. For example, metals are usually good conductors of heat because, while their atoms are fixed in a lattice-like structure, some of their electrons are very loosely held and tend to form a free-moving "fluid" or "gas" in the solid that can carry energy from one region to another. This same electron mobility makes metals good electrical conductors. But in nonmetallic solids like ceramics, conduction is more mysterious. It seems that heat is propagated not by the movement of energetic electrons—in solids of ionic- or covalent-bonded compounds, the electrons are not free to move—but by the vibration of individual molecules or a portion of the lattice, which is transferred to neighboring areas. This transfer of vibration is a much slower and less efficient process than electron movement, and nonmetals are therefore usually referred to as thermal or electrical *insulators,* rather than conductors. Liquids and gases, because their molecules are relatively far apart, are very poor conductors.

The conductivity of a material determines its behavior on the stove. The better the conductor, the faster a pan heats up and cools off, and the more evenly heat is distributed across the pan bottom. Uneven heating creates hot spots that can burn foods: during frying, for example, or the boiling down of a puree or sauce.

Conduction Within a Food Heat also travels from the outside to the center of a solid piece of food—a piece of meat or fish or vegetable—by means of conduction. Because the cellular structure of foods impedes the movement of heat energy, foods behave more like insulators than like metals, and heat up relatively slowly. One of the keys to good cooking is knowing how to heat a food to the desired doneness at its center without overheating its outer regions. This is not a simple task, because different kinds of foods heat through at different rates. One of the most important variables is the thickness of the food. Though common sense might suggest that a piece of meat one inch thick would take twice as long to cook through as a half-inch piece, it turns out that it takes somewhere between twice and four times longer, depending on the overall shape: less for a compact chop or chunk, more for a broad steak or fillet. There's no absolutely reliable way to predict how long it will take heat to move from the food surface to its center, so the best rule is to check the doneness frequently.

Convection: Movement in Fluids

In the form of heat transfer called *convection,* heat is transferred by the movement of molecules in a fluid from a warm region to a cooler one. The fluid may be a liquid such as water, or it may be the air or other gases. Convection is a process that combines conduction and mixing: energetic molecules move from one point in space to another, and then collide with slower particles. Convection is an influential phenomenon, contributing as it does to winds, storms, ocean currents, the heating of our homes, and the boiling of water on the stove. It occurs because air and water take up more space—become less dense—when their molecules absorb energy and move faster, and so they rise when they heat up and sink again as they cool off.

Radiation: The Pure Energy of Radiant Heat and Microwaves

We all know that the earth is warmed by the sun. How does solar energy reach us across millions of miles of nearly empty space, where there's nothing there to conduct or convect? The answer is thermal *radiation,* a process that does not require direct physical contact between heat source and object. All matter emits thermal radiation all the time, though normally we can detect it only when something is very hot. The warmth we feel from sunlight or a stove burner comes from thermal radiation. It's emitted by atoms and molecules which, having absorbed energy, release it again not in the form of faster movement, but as waves of pure energy.

Radiant Heat Is Invisible "Infrared" Radiation As unlikely as it may seem, radiated heat is close kin to radio waves, microwaves, visible light, and X rays. Each of these phenomena is a part of the *electromagnetic spectrum,* waves of varying energies created by the movement of electrically charged particles, often electrons within atoms. Such movement creates electrical and magnetic fields that radiate, or spread out, as waves. And conversely, when such energetic waves hit other atoms, they cause increased movement in those atoms. One of the first to recognize that heat radiation is related to light was the English oboist and astronomer William Herschel, who noticed in 1800 that if a thermometer was moved from one end of a prism-produced light spectrum to the other, the highest temperatures would register below the red band, where no light was visible. Because of its position in the spectrum, heat radiation is called *infrared* (*infra* is Latin for "below").

Different Kinds of Radiation Carry Different Amounts of Energy Different kinds of radiation carry different energies, and the energy of a given kind of radiation determines the kind of effect that it will have.

- At the bottom end of the scale, radio waves are so weak that they can only cause increased movement in free electrons. This is why metal antennas and their mobile electrons are necessary to transmit and receive such radiation.
- Next come microwaves, which are energetic enough to set polar molecules like water moving faster. (*Microwave* refers to the fact that their wavelength is shorter than radio wavelengths.) Since most foods are mostly water molecules, microwave radiation is an effective means of cooking.
- Then there's heat radiation, the cook's standard energy source, which causes the increased movement of nonpolar molecules—including carbohydrates, proteins, and fats—as well as polar water.
- Visible and ultraviolet light is capable of altering the orbits of electrons bound in molecules, and so can initiate chemical reactions that cause damage to pigments and fats and the development of stale, rancid flavors. Visible and ultraviolet rays from the sun can ruin the flavor of milk and beer, and ultraviolet rays can burn our skin, damage our DNA, and cause cancer.
- X and gamma rays penetrate matter and *ionize* it, or strip electrons from its molecules. Along with controlled beams of certain subatomic particles, they damage DNA and kill microbes, and are used to "cold-pasteurize" and sterilize some foods.

Useful Heat Radiation Is Generated by High Temperatures Because all molecules are vibrating to some extent, everything around us is emitting at least some infrared radiation all the time. The hotter an object gets, the more energy it radiates in higher regions of the spectrum. So it is that glowing metal is hotter than metal that does not radiate visible light, and that yellow-hot metal is hotter than red-hot. It turns out that the rate of infrared radiation is relatively low below about 1,800°F/980°C, or the point at which objects begin to glow visibly red. Cooking by radiation is thus a slow process except at very high cooking temperatures, those characteristic of grilling and broiling near glowing coals, electrical elements, or gas flames. At typical baking and frying temperatures, conduction and convection tend to be more significant than infrared radiation. But as the oven temperature goes up, the proportion of heat contributed by the radiating oven walls goes up with it. The cook can control this contribution by moving the food close to the walls or ceiling to increase it, or shielding the food with reflective foil, which reduces it.

BASIC METHODS OF HEATING FOODS

Pure examples of the three different forms of heat transfer are seldom found in every-

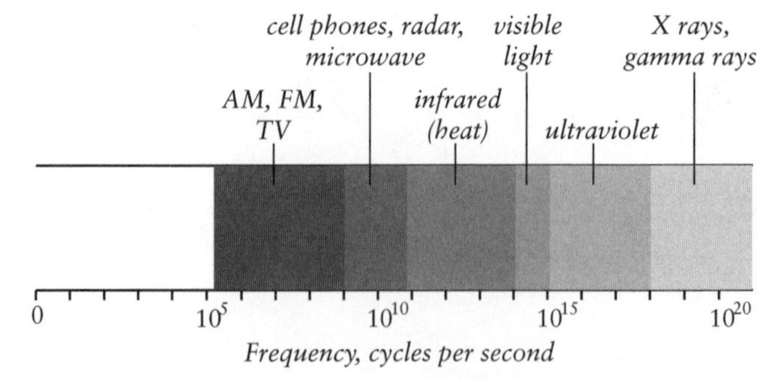

The spectrum of electromagnetic radiation. We use both microwave and infrared radiation to cook our foods. (The scale employs a standard scientific abbreviation for large numbers; 10^5 means a 1 followed by 5 zeroes, or 100,000.)

day life. All hot utensils radiate heat to some degree, and cooks usually work with combinations of solid containers that conduct and fluids that circulate. As simple an operation as heating a pan of water on the stove involves radiation and conduction from an electrical element (radiation and convection from a gas flame), conduction through the pan, and convection in the water. Still, one kind of heat transfer usually predominates in a given cooking technique and, together with the cooking medium, has a distinctive influence on foods.

Grilling and Broiling: Infrared Radiation

Grilling and broiling are the modern, controlled versions of the oldest culinary technique, roasting over an open fire or glowing coals. In grilling, the heat source is below the food; in broiling, above. Though air convection contributes some heat, especially as the distance between heat source and food is increased, broiling is largely a matter of infrared radiation. The heat sources used in these techniques all emit visible light and so are also intense radiators of infrared energy. Glowing coals or the nickel-chrome alloys used in electrical appliances reach about 2,000°F/1,100°C, and a gas flame is closer to 3,000°F/1,600°C. The walls of an oven, by contrast, rarely exceed 500°F/250°C. The total amount of energy radiated by a hot object is proportional to the fourth power of the absolute temperature, so that a coal or metal rod at 2,000°F is radiating more than 40 times as much energy as the equivalent area of oven wall at 500°F.

This tremendous amount of heat is at once the great advantage and the principal challenge of grilling and broiling. On one hand, it makes possible a rapid and thorough browning of the surface, and so produces intense flavors. On the other, there's a huge disparity between the rate of heat radiation at the surface and the rate of heat conduction within the food. This is why it's so easy to end up with a steak that's charred on the outside and cold at the center.

The key to grilling and broiling is to position the food far enough from the heat source to match the browning rate with the inner conduction rate, or to brown the surface well with intense heat, and then

Induction Cooking

An innovative version of heating with electromagnetic radiation is induction heating. It's an alternative to the stovetop burner or electrical element, and heats the pot that then heats the food. In induction heating, the heating element, under a ceramic cooktop surface, is a wire coil through which a rapidly alternating electrical current flows (between 25,000 and 40,000 cycles per second). The current causes the coil to generate a magnetic field that extends some distance from the coil, and that alternates at the same rate. If a pot made from a magnetic material—cast iron, steel, stainless steel of the proper crystal structure (ferritic)—is placed near the coil, then the alternating magnetic field *induces* an alternating electrical current in the pot. That is, it causes electrons to move in the pot, and that movement rapidly generates heat.

Induction heating has two notable advantages over burners and radiant elements. Like microwave heating, it's more efficient, because all the energy goes into the object to be heated, not into the surrounding air. And only the pot and its contents get very hot. The ceramic surface above the induction coil is heated only indirectly by the pot, because its electrons aren't free to be moved by the magnetic field.

move the food to finish cooking through with a more distant or weaker heat source. This might be a spot on the grill with fewer coals below, or a moderate oven.

Baking: Air Convection and Radiation

When we bake a food, we surround it with a hot enclosure, the oven, and rely on a combination of radiation from the walls and hot-air convection to heat the food. Baking easily dehydrates the surface of foods, and so will brown them well provided the oven temperature is high enough. Typical baking temperatures are well above the boiling point, from 300 to 500°F/150–250°C), and yet baking is nowhere near as efficient a means of heat transfer as is boiling. A potato can be boiled in less time than it takes to be baked at a much hotter temperature. This is so because neither radiation nor air convection at 500°F transfers heat very rapidly to food. Oven air is less than a thousandth as dense as water, so the collisions between hot molecules and food are much less frequent in the oven than in the pot (this is why we can reach into a hot oven without immediately burning our hand). *Convection ovens* increase the rate of heat transfer by using fans to force more air movement, and significantly reduce baking times.

Because baking requires a fairly sophisticated container, it was probably a late addition to the culinary repertoire. The earliest ovens seem to have accompanied the refinement of bread making around 3000 BCE in Egypt; they were hollow cones of clay that contained a layer of coals, with the bread stuck onto an inside wall. As a relatively compact metal box easily installed in individual homes, the modern oven dates from the late 19th century. Before then, most meat cooking was done over the fire.

Boiling and Simmering: Water Convection

In boiling and its lower-temperature versions, simmering and poaching, food is heated by the convection currents in hot water. The maximum temperature possible is the boiling point, 212°F/100°C at sea level, which is usually not high enough for these "moist" cooking methods to trigger browning reactions. Despite the relatively low cooking temperature, boiling is a very efficient process. The entire surface of the food is in contact with the cooking medium, and water is dense enough that its molecules constantly collide with the food and rapidly impart their energy to it.

As a cooking technique, boiling probably followed roasting and preceded baking. It requires containers that are both water- and fireproof, and so probably had to await the development of pottery, around 10,000 years ago.

The Boiling Point: A Reliable Landmark It isn't always easy for the cook to recognize and maintain a particular cooking temperature, and reproduce the same temperature reliably. Thermostats, thermometers, and our senses are all fallible. So one of the great advantages of water as a cooking medium is that its boiling point is constant—212°F/100°C at sea level—and it's instantly recognizable. The sure sign of boiling water is bubbling. Why? When the water in a pan is heated near boiling, molecules at the bottom, where the pan is hottest, vaporize and become steam, and form regions that are less dense than the surrounding liquid. (The small bubbles that form very early on are pockets of air that had been dissolved in the cold water but became less soluble as the temperature rose.) Because all the pan heat at the boil goes into vaporizing the liquid water, the temperature of the water itself stays the same (p. 816). It's only slightly higher at a full, rolling boil than in a gently bubbling pot, and will not get any higher until the phase change from liquid to gas has been completed.

The Boiling Point Depends on Elevation The boiling point of water is constant given a constant physical environment, but it varies from place to place and even in the same place. The boiling point of any liquid depends on the atmospheric pressure bearing down on its surface: the higher the pressure, the more energy it takes for liquid molecules to escape the surface and become a gas, and so the higher the temperature at which the liquid boils. Every 1,000 feet/305 meters in elevation above sea level lowers the boiling point about 2°F below the standard 212°F (or 1°C below 100°C). And food takes longer to cook at 200° than it does at 212°. Even a low-pressure weather front can lower the boiling point, or a high-pressure front raise it, by as much as a degree or two.

Pressure Cooking: Raising the Boiling Point The same principal is put to use to *speed* cooking in the pressure cooker. This appliance reduces cooking times by trapping the steam that escapes from boiling water, thereby increasing the pressure on the liquid, and so raising its boiling point—and maximum temperature—to about 250°F/120°C. This is the equivalent of boiling water in an open pan at the bottom of a pit 19,000 feet/5,800 meters *below* sea level.

The pressure cooker was invented by the French physician Denis Papin in the 17th century.

The Boiling Point Is Increased by Dissolved Sugar and Salt When salt, sugar, or any other water-soluble substance is added to pure water, the boiling point of the resulting solution becomes higher than the boiling point of water, and the freezing point lower than water's freezing point. Both effects are due to the fact that the water molecules are diluted by the dissolved particles, which interfere with the water molecules as they change phase from liquid to gas or liquid to solid. In the case of the boiling point, the solution contains sugar molecules or salt ions that also absorb heat energy, but cannot themselves turn into a gas. So at water's normal boiling point, there is a smaller proportion of molecules with enough energy to escape from the liquid and form a bubble of vapor, and the cook has to add more energy than usual in order to get those bubbles to form. The boiling point and freezing point rise and fall predictably as the concentration of dissolved sugar or salt increases, a fact that is handy for making both sugar candies and ice creams.

It's true that adding salt to water raises its boiling point, and so speeds cooking. However, it takes one ounce of salt in a quart of water—around the salinity of the ocean—to raise the boiling point a negligible 1°F. A Denverite who wanted to boil water at the same temperature as someone in Boston would have to add more than half a pound of salt to that quart of liquid (225 grams to a liter).

Cooking Below the Boil Though the boil is a handy temperature landmark, it's not necessarily the best temperature at which to cook foods in water. Fish and many meats develop an ideal texture at temperatures around 140°F/60°C. If they're cooked in boiling water, which is 70°F hotter, then the outer portions of the food overcook and dry out while the interior heats through. Lower water temperatures reduce this overcooking, though they also prolong cooking times. A water temperature of 180°F/80°C, verified by thermometer, offers a good compromise between gentle and efficient cooking.

Steaming: Heating by Vapor Condensation and Convection

Though it's less dense than liquid water and so makes less frequent contact with the food, steam compensates for this loss in efficiency with a gain in energy. It takes a large amount of energy to turn liquid water into a gas, and conversely gaseous water releases that same large amount of energy when it condenses onto a cooler object. So molecules of steam don't just impart their energy of motion to the food; they impart their energy of vaporization also. This means that steaming does an especially quick job of bringing the surface of the food up to the boiling point, and an effective job of keeping it there.

Pan-Frying and Sautéing: Conduction

Frying and sautéing are methods that heat foods for the most part by conduction from a hot, oiled pan, with temperatures between 350 and 450°F/175–225°C that encourage Maillard browning and flavor development. The fat or oil has several roles to play: it brings the uneven surface of the food into uniform contact with the heat source, it lubricates and prevents sticking, and it supplies some flavor. As is true in broiling, the trick in frying is to prevent the outside from overcooking before the inside is done. The surface is quickly dehydrated by the high temperatures—odd as it sounds, frying in oil is a "dry" technique—while the interior remains largely water and never exceeds 212°F/100°C. In order to reduce the disparity between outer and inner cooking times, we generally fry only thin cuts of food. It's also common practice to fry meats at a high initial temperature—to sear them—in order to accomplish the browning, and then to reduce the heat while the interior heats through. Yet another way to avoid overcooking the outer portions of the food is to coat it in another material that develops pleasant flavors when fried, and acts as a kind of insulation to protect the inner food from direct contact with high heat. Breadings and batters are such insulators.

How far back frying goes is hard to tell. The rules for sacrifice in Leviticus 2, which dates from about 600 BCE, distinguish between bread baked in an oven and cooked "on the griddle" or "in the pan." Pliny, in the 1st century CE, records a prescription for spleen disease that calls for eggs steeped in vinegar and then fried in oil. And by Chaucer's time, the 14th century, frying was common enough to serve as a colorful metaphor. The Wife of Bath says of her fourth husband

> That in his owene grece I made
> hym frye
> For angre, and for verray jalousye.
> By God! in erthe I was his
> purgatorie,
> For which I hope his soule be in
> glorie.

Deep Frying: Oil Convection

Deep frying differs from pan frying by employing enough oil to immerse the food altogether. As a technique, it resembles boiling more than pan frying, with the essential difference that the oil is heated far above the boiling point of water, and so will dehydrate the food surface and brown it.

Microwaving: Microwave Radiation

Microwave ovens transfer heat via electromagnetic radiation, but with waves that carry only a ten-thousandth the energy of infrared radiation from glowing coals. This shift makes for a unique heating effect. Whereas infrared waves are energetic enough to increase the vibratory movement of nearly all molecules, microwaves tend to affect only polar molecules (p. 793), whose electrical imbalance gives the radiation a kind of handle with which to move them. So foods that contain water are heated directly and rapidly by microwaves. But the oven air, composed of nonpolar nitrogen, oxygen, and hydrogen molecules, and nonpolar container materials like glass, stoneware, and plastic (made of hydrocarbon chains), are unaffected by the microwaves; the food heats them as it heats up.

Here's how a microwave oven works. A transmitter, very much like a radio transmitter, sets up an electromagnetic field in the oven which reverses its polarity some 2 or 5 billion times every second. (It operates at a frequency of either 915 or 2,450 million cycles per second, compared to wall socket currents at 60 cycles, and FM radio signals at some 100 million cycles per second.) Polar water molecules in the food are pulled by the field to orient themselves with it, but because the field is constantly changing, the molecules oscillate back and forth with it. The water transmits this motion to neighboring molecules by knocking into them, and the temperature of the food as a whole quickly rises.

Metal foil and utensils can be put in the microwave oven with moist foods without causing problems, provided they're reasonably large and kept some distance from the walls and each other to prevent arcing. Fine metallic decoration on china will spark and suffer damage. Foil is useful for partly shielding some foods from the radiation, the thin edges of fish fillets, for example.

Microwave ovens are a recent invention. In 1945 Dr. Percy Spencer, a scientist working for Raytheon in Waltham, Massachusetts, filed a patent for the use of microwaves in cooking after he had successfully popped corn with them. This kind of radiation had already been used in diathermy, or deep heat treatment for patients with arthritis, as well as in communications and navigation. Microwave ovens became a popular appliance in the 1970s.

Advantages and Disadvantages of Microwaves Microwave radiation has one great advantage over infrared: the fact that it cooks food much faster. Microwaves can penetrate food to a depth of about an inch/2.5 cm, while infrared energy is almost entirely absorbed at the surface. Because heat radiation can travel to the center of foods only by the slow process of conduction, it's easily beaten by microwaves, with their substantially deeper reach. This reach, along with the microwaves' concentration on heating the food and not its surroundings, results in a very efficient use of energy.

Microwave cooking has several disadvantages. One is that, in the case of meats, speedy heating can cause greater fluid loss and so a drier texture, and makes it more difficult to control doneness. This can be partly overcome by pulsing the oven on and off to slow the heating. Another problem is that microwaves cannot brown many foods unless they essentially dehydrate them, since the food surface gets no warmer than the interior. Thin, radiation-concentrating metal sheets in special microwavable food packaging can help heat the food surface to the point that it browns.

UTENSIL MATERIALS

Finally, a brief discussion of the materials from which we make our pots and pans. We generally want two basic properties in a utensil. Its surface should be chemically unreactive so that it won't change the taste or edibility of food. And it should conduct heat evenly and efficiently, so that local hot spots won't develop and burn the contents. No single material provides both properties.

The Different Behaviors of Metals and Ceramics

As we've seen, heat conduction in a solid proceeds either by the movement of energetic electrons, or by vibration in crystal structures. A material whose electrons are mobile enough to conduct heat well is also likely to give up those electrons to other atoms at its surface: in other words, good conductors like metals are usually chemically reactive. By the same token, inert compounds are poor conductors. Ceramics are stable, unreactive mixtures of compounds (magnesium and aluminum oxides, silicon dioxide) whose covalent bonds hold electrons tightly. They therefore transmit heat slowly by means of inefficient vibrations. If subjected to the direct and intense heat of the stovetop, ceramics can't distribute the energy evenly. Hot areas expand while cooler areas do not, mechanical stresses build up, and the utensil cracks or shatters. This is why ceramics are generally used only in the oven, where they encounter only moderate and diffuse heat, or are applied in thin coatings on the surface of metals, so that the metals can do the job of distributing the heat evenly.

Spontaneous Ceramic Coatings on Metals It turns out that most of the metals commonly used in kitchen utensils naturally cover themselves with a very thin layer of ceramic material. Metallic electrons are mobile, and oxygen is electron-hungry. When metal is exposed to the air, the surface atoms undergo a spontaneous reaction with atmospheric oxygen to form a very stable metal oxide compound. (The discoloration on silver and copper that we call *tarnish* is a metal-sulfur compound; the sulfur comes mainly from air pollution.) These oxide films are both unreactive and fairly tough. Aluminum oxide, when it occurs in crystals rather than on pans, makes up the abrasive called corundum, and is also the principal material of rubies and sapphires (the gem colors come from chromium and titanium impurities). The problem is that these natural coatings are only a few molecules thick, and are easily scratched through or worn away during cooking.

Metallurgists have found two ways to take advantage of metal oxidation at the pan surface. The film over aluminum can be made up to a thousandth of an inch/ 0.03 mm thick, and so fairly impervious, by a chemical treatment. And iron can be protected by mixing it with other metals that form a tougher oxide surface and so produce stainless steel (p. 791).

Here are brief descriptions of the materials from which most kitchen utensils are made today, and their particular advantages and disadvantages.

Ceramics

Earthenware, Stoneware, Glass Ceramics are varying mixtures of a number of different compounds, notably the oxides of silicon, aluminum, and magnesium. *Glass* is a particular variety of ceramic whose composition is more regular, and usually includes a preponderance of silica (silicon dioxide). Until fairly recently, these materials were made from naturally occurring mineral aggregates: the word *ceramic* comes from the Greek for "potter's clay." The molding and drying of simple clay pottery, or *earthenware,* dates from about 9,000 years ago, or about the time that plants and animals were first domesticated. Less porous and coarse than earthenware, and much stronger, is *stoneware,* which contains enough silica and is fired at a high enough temperature that it vitrifies, or becomes partly glass. The Chinese invented this refinement sometime before 1500 BCE. *Porcelain* is a white but translucent stoneware made by mixing kaolin, a very light clay, with a silicate mineral, and firing at high kiln temperatures; it dates from the T'ang Dynasty (618–907 CE). This fine ceramic was introduced to Europe with the tea trade in the 17th century, and in England was first called "Chinaware," and then simply "China." The first glass containers were not molded or blown, but laboriously sculpted from blocks, and date from 4,000 years ago in the Near East.

The Qualities of Ceramic Pots The outstanding characteristic of ceramic materials is chemical stability: they are unreactive, resist corrosion, and don't affect the flavor or other qualities of foods. (One exception to this rule is the fact that clays and glazes sometimes contain lead, which is a nerve poison, and which can be leached out into acidic foods. Imported ceramic containers made with high-lead clays or glazes still occasionally cause cases of lead poisoning.) Ceramic pots tend to be used only in slow, uniform cooking processes, especially oven baking and braising, because direct high heat can shatter them. Heat-resistant forms of glass incorporate an oxide of boron that has the effect of reducing thermal expansion by a factor of about 3, and for this reason are less affected by thermal shock, though they're still not immune.

Enamelware In utensils called *enamelware,* powdered glass is fused into a thin layer onto the surface of iron or steel utensils. This was first done to cast iron early in the 19th century, and today enameled metal is widely used in the dairy, chemical, and brewing industries, as well as on bathtubs. In kitchen utensils, the metal diffuses the direct heat evenly, the ceramic layer is thin enough that it can expand and contract uniformly, and it protects the food from direct contact with the metal. Enamelware is reasonably durable, though it still requires some care: the ceramic layer can be chipped or damaged by quenching a hot pan in cold water.

The Advantages of Poor Conductivity

The poor conductivity of ceramic materials is an advantage if the cook needs to keep food hot. Good conductors like copper and aluminum quickly give up heat to their surroundings, while ceramics retain it well. Similarly, ovens with ceramic (brick) walls are unparalleled for the evenness of their heating. The walls slowly absorb and store large quantities of energy while the oven is heated up, and then release it when the food is placed inside. Modern metal ovens can't store much heat and so must cycle their heating elements on and off. This causes large temperature fluctuations, and can scorch breads and other foods that are baked at high temperatures.

Aluminum

Aluminum has been used in pots and pans for barely a century, despite the fact that it's the most abundant metal in the earth's crust. It is never found in nature in the pure state, and a good method for separating the metal from its ore wasn't developed until 1890. In cookware, it is usually alloyed with small amounts of manganese and sometimes copper. Aluminum's prime advantages are its relatively low cost, a heat conductivity second only to copper's, and a low density that makes it lightweight and easily handled. Its ubiquitous presence in the form of foil wrappings and beer and soft drink cans testifies to its usefulness. But because unanodized aluminum develops only a thin oxide layer, reactive food mole-

Nonstick Coatings and Silicone "Pans"

The materials for nonstick coatings were developed around the middle of the 20th century by industrial chemists, and nonstick utensils were introduced in the 1960s. Teflon and its relatives are long chains of carbon atoms with fluorine atoms projecting from the backbone. They produce a plastic-like material with a smooth, slippery surface, and are as inert as ceramics at moderate cooking temperatures. Above about 500°F/250°C, however, they decompose into a number of noxious and toxic gases. Nonstick utensils therefore need to be used with care to avoid overheating. The coatings have the additional disadvantage of being easily scratched, and food sticks to the scratches.

Beginning in the 1980s, flexible nonstick sheets and containers made of silicone have been used by bakers to line metal baking sheets or replace molded metal pans. Silicone is also a long-chain molecule, with a backbone of alternating silicon and oxygen atoms, and small fat-like carbon chains projecting from it. The backbone gives the material its flexibility, and the hydrophobic projections make the surface behave like a permanently well-oiled pan surface. Food-grade silicones decompose at temperatures above about 480°F/240°C, so like nonstick pans, silicone bakeware must be used with some caution.

cules—acids, alkalis, the hydrogen sulfide evolved by cooked eggs—will easily penetrate to the metal surface, and a variety of aluminum oxide and hydroxide complexes, some of them gray or black, are formed. These can mar light-colored foods. Today, most aluminum utensils are either given a nonstick coating or are *anodized,* a process that involves making the metal the positive pole (anode) in a solution of sulfuric acid, and so forcing the oxidation of its surface to make a thick protective oxide layer.

Copper

Copper is unique among the common metals because it can be found naturally in the metallic state. For this reason it was the first metal to be used in tool making, about 10,000 years ago. In the kitchen, it is prized for its unmatched conductivity, which makes fast and even heating a simple matter. But copper is also relatively expensive, since its conductivity has made it the preferred material for millions of miles of electrical circuitry. It is troublesome to keep polished, because it has a high affinity for oxygen *and* sulfur, and forms a greenish coating when exposed to air. Most important, copper cookware can be harmful. Its oxide coating is sometimes porous and powdery, and copper ions are easily leached into food solutions. Copper ions can have useful effects: they stabilize foamed egg whites (p. 102), and the green color of cooked vegetables is improved by their presence. But the human body can excrete copper in only limited amounts, and excessive intake may cause gastrointestinal problems and, in more extreme cases, liver damage. No one will be poisoned by the occasional meringue whipped in a copper bowl, but bare copper isn't a good candidate for everyday cooking. To overcome this major drawback, manufacturers line copper utensils with stainless steel or, more traditionally, with tin. Tin has its own limitations (p. 791).

Iron and Steel

Iron was a relatively late discovery because it exists in the earth's crust primarily in the form of oxides, and had to be encountered in its pure form by accident, perhaps when a fire was built on an outcropping of ore. Iron artifacts have been found that date from 3000 BCE, though the Iron Age, when the metal came into regular use without replacing copper and bronze (a copper-tin alloy) in preeminence, is said to begin around 1200 BCE. *Cast iron* is alloyed with about 3% carbon to harden the metal, and also contains some silicon; *carbon steel* contains less carbon, and is heat-treated to obtain a less brittle, tougher alloy that can be formed into thinner pans. The chief attractions of cast iron and carbon steel in kitchen work are their cheapness and safety. Excess iron is readily eliminated from the body, and most people can actually benefit from additional dietary iron. Their greatest disadvantage is a tendency to corrode, though this can be avoided by regular seasoning (below) and gentle cleaning. Like aluminum, iron and carbon steel can discolor foods. And iron turns out to be a poorer conductor of heat than copper or aluminum. But exactly for this reason, and because it's denser than aluminum, a cast iron pan will absorb more heat and hold it longer than a similar aluminum pan. Thick cast iron pans provide steady, even heat.

"Seasoning" Cast Iron and Carbon Steel Cooks who appreciate cast iron and carbon steel pans improve their easily corroded surface by building up an artificial protective layer. They "season" them by coating them with cooking oil and heating them for several hours. The oil penetrates into the pores and fissures of the metal, sealing it from the attack of air and water. And the combination of heat, metal, and air oxidizes the fatty acid chains and encourages them to bond to each other ("polymerize") to form a dense, hard, dry layer (just as linseed and other "drying oils" do on wood and on paintings). Highly unsaturated oils—soy oil, corn oil—are especially prone to oxidation and polymerizing. To avoid removing the protective oil layer, cooks carefully clean seasoned cast iron pans with mild soaps and a dissolving abrasive like salt, rather then with detergents and scouring pads.

Stainless Steel

The important exception to the rule that metals form protective surface coatings is iron, which rusts in the presence of air and moisture. The orange complex of ferric oxide and water ($Fe_2O_3 \cdot H_2O$) is a loose powder rather than a continuous film, and so does not protect the metal surface from further contact with the air. Unless it's protected by some other means, iron metal will corrode continuously (this is why pure iron is not found in nature). Efforts to make this cheap and abundant element more resistant to rusting resulted in the 19th century in the development of *stainless steel,* an iron-carbon alloy that—in cookware—is formulated with about 18% chromium and 8–10% nickel. Chrome is synonymous with bright and permanent shininess because chromium is extremely prone to oxidation and naturally forms a thick protective oxide coat. In the stainless steel mixture, oxygen reacts preferentially with the chromium atoms at the surface, and the iron never gets the opportunity to rust.

This chemical stability is bought at a price. Stainless steel is more expensive than cast iron and carbon steel, and it's an even poorer heat conductor. The addition of large numbers of foreign atoms apparently interferes with electron movement by causing structural and electrical irregularities in the metal. The transfer of heat in a stainless pan can be evened out by coating the underside of the pan with copper, or by inserting a copper or aluminum plate in the pan bottom, or by making the pan out of two or more layers, with a good conductor just under the surface. Of course these refinements add further to the cost of the utensil. Still, these hybrids are the closest thing we have to the ideal chemically inert but thermally responsive pan.

Tin

Tin was probably first used in combination with copper to make the mechanically tougher alloy called bronze. Today tin is generally found only as a nontoxic, unreactive lining in copper utensils. This limited role is the result of two inconvenient properties: a low melting point, 450°F/230°C, that can be reached in some cooking procedures, and a softness that makes the metal very susceptible to wear. The tin alloy called *pewter,* which used to contain some lead and now is made with 7% antimony and 2% copper, is not much used today.

CHAPTER 15

THE FOUR BASIC FOOD MOLECULES

This chapter describes the characters of the four chemical protagonists in foods and the cooking process, the molecules referred to constantly in the first fourteen chapters.

- *Water* is the major component of nearly all foods—and of ourselves! It's also a medium in which we heat foods in order to change their flavor, texture, and stability. One particular property of water solutions, their acidity or alkalinity, is a source of flavor, and has an important influence on the behavior of the other food molecules.
- *Fats, oils,* and their chemical relatives are water's antagonists. Like water, they're a component of living things and of foods, and they're also a cooking medium. But their chemical nature is very different—so different that they can't mix with water. Living things put this incompatibility to work by using fatty materials to contain the watery contents of cells. Cooks put this quality to work when they fry foods to crisp and brown them, and when they thicken sauces with microscopic but intact fat droplets. Fats also carry aromas, and produce them.
- *Carbohydrates,* the specialty of plants, include sugars, starch, cellulose, and pectic substances. They generally mix freely with water. Sugars give many of our foods flavor, while starch and the cell-wall carbo-

hydrates provide bulk and texture.

- *Proteins* are the sensitive food molecules, and are especially characteristic of foods from animals: milk and eggs, meat and fish. Their shapes and behavior are drastically changed by heat, acid, salt, and even air. Cheeses, custards, cured and cooked meats, and raised breads all owe their textures to altered proteins.

WATER

Water is our most familiar chemical companion. It's the smallest and simplest of the basic food molecules, just three atoms: H_2O, two hydrogens and an oxygen. And its significance is hard to overstate. Leaving aside the fact that it shapes the earth's continents and climate, all life, including our own, exists in a water solution: a legacy of life's origin billions of years ago in the oceans. Our bodies are 60% water by weight; raw meat is about 75% water, and fruits and vegetables up to 95%.

Water Clings Strongly to Itself

The important properties of ordinary water can be understood as different manifestations of one fact. Each water molecule is electrically unsymmetrical, or *polar*: it has a positive end and a negative end. This is because the oxygen atom exerts a stronger pull than the hydrogen atoms on the electrons they share, and because the hydrogen atoms project from one side of the oxygen to form a kind of V shape: so there's an oxygen end and a hydrogen end to the water molecule, and the oxygen end is more negative than the hydrogen end. This polarity means that the negative oxygen on one water molecule feels an electrical attraction to the positive hydrogens on other water molecules. When this attraction brings the two molecules closer to each other and holds them there, it's called a *hydrogen bond.* The molecules in ice and liquid water are participating in from one to four hydrogen bonds at any given moment. However, the motion of the molecules in the liquid is forceful enough to overcome the strength of hydrogen bonds and break them: so the hydrogen bonds in liquid water are fleeting, and are constantly being formed and broken.

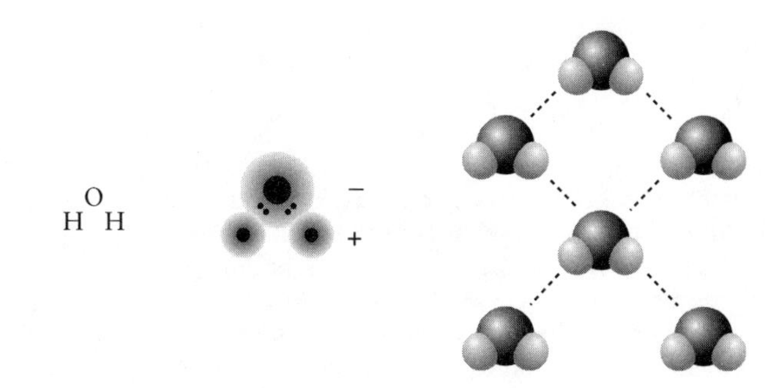

Water molecules. Here are three different ways of representing a molecule of water, which is formed from one oxygen and two hydrogen atoms. Because the oxygen atom exerts a stronger pull on the electrons (small dots) it shares with the hydrogen atoms, the water molecule is electrically unsymmetrical. The separation of positive and negative centers of charge leads to the formation of weak bonds between oppositely charged centers on different molecules. These weak bonds between molecules, shown here by dashed lines, are called hydrogen bonds.

This natural tendency of water molecules to form bonds with each other has a number of effects in life and in the kitchen.

Water Is Good at Dissolving Other Substances

Water forms hydrogen bonds not only with itself, but with other substances that have at least some electrical polarity, some unevenness in the distribution of positive and negative electrical charges. Of the other major food molecules, which are much larger and more complex than water, both carbohydrates and proteins have polar regions. Water molecules are attracted to these regions and cluster around them. When they do this, they effectively surround the larger molecules and separate them from each other. If they do this more or less completely, so that each molecule is mostly surrounded by a cloud of water molecules, then that substance has *dissolved* in the water.

Water and Heat: From Ice to Steam

The hydrogen bonds among its molecules have a strong effect on how water absorbs and transmits heat. At low temperatures, water exists as solid ice, its molecules immobilized in organized crystals. As it warms up, it first melts to become liquid water; and then the liquid water is vaporized to form steam. Each phase is affected by hydrogen bonding.

Ice Damages Cells Normally, the solid phase of a given substance is denser than the liquid phase. As the molecules' attraction for each other becomes stronger than their movements, the molecules settle into a compact arrangement determined by their geometry. In solid water, however, the molecular packing is dictated by the requirement for even distribution of hydrogen bonds. The result is a solid with *more* space between molecules than the liquid phase has, by a factor of about one-eleventh. It's because water expands when it freezes that water pipes burst when the heat fails in winter; that bottles of beer put in the freezer for a quick chill and then forgotten will pop open; that containers of leftover soup or sauce will shatter in the freezer if they're too full for the liquid to expand freely. And it's why raw plant and animal tissues are damaged when they're frozen and leak liquid when thawed. During freezing, the expanding ice crystals rup-

Hard Water: Dissolved Minerals

Water is so good at dissolving other substances that apart from distilled water, it's seldom found in anything like pure form. Tap water is quite variable in composition, depending on its ultimate source (well, lake, river) and its municipal treatment (chlorination, fluoridation, and so on). Two common minerals in tap water are carbonate (CO_3) and sulfate (SO_4) salts of calcium and magnesium. Calcium and magnesium ions are troublesome because they react with soaps to form insoluble scums, and because they leave crusty precipitates on showerheads and teapots. Such so-called hard water can also affect the color and texture of vegetables, and the consistency of bread dough (pp. 282, 535). Hard water can be softened either city-wide or in the home, usually by one of two methods: precipitating the calcium and magnesium by adding lime, or using an ion-exchange mechanism to replace the calcium and magnesium with sodium. Distilled water, which is produced by boiling ordinary water and collecting the condensed steam, is fairly free of impurities.

ture cell membranes and walls, which then lose internal fluids when the crystals melt.

Liquid Water Is Slow to Heat Up Again thanks to the hydrogen bonding between water molecules, liquid water has a high *specific heat,* the amount of energy required to raise its temperature by a given amount. That is, water absorbs a lot of energy before its temperature rises. For example, it takes 10 times the energy to heat an ounce of water 1° as it does to heat an ounce of iron 1°. In the time that it takes to get an iron pan too hot to handle on the stove, water will have gotten only tepid. Before the heat energy added to the water can cause its molecules to move faster and its temperature to rise, some of the energy must first break the hydrogen bonds so that the molecules are *free* to move faster.

The basic consequence of this characteristic is that a body of water—our body, or a pot of water, or an ocean—can absorb a lot of heat without itself quickly becoming hot. In the kitchen, it means that a covered pan of water will take more than twice as long as a pan of oil to heat up to a given temperature; and conversely, it will hold that temperature longer after the heat is removed.

Liquid Water Absorbs a Lot of Heat as It Vaporizes into Steam Hydrogen bonding also gives water an unusually high "latent heat of vaporization," or the amount of energy that water absorbs without a rise in temperature as it changes from a liquid to a gas. This is how sweating cools us: as the water on the skin of our overheated body evaporates, it absorbs large amounts of energy and carries it away into the air. Ancient cultures used the same principle to cool their drinking water and wine, storing them in porous clay vessels that evaporate moisture continuously. Cooks take advantage of it when they bake delicate preparations like custards gently by partly immersing the containers in an open water bath, or oven-roast meats slowly at low temperatures, or simmer stock in an open pot. In each case, evaporation removes energy from the food or its surroundings and causes it to cook more gently.

Steam Releases a Lot of Heat When It Condenses into Water Conversely, when water vapor hits a cool surface and condenses into liquid water, it gives up that same high heat of vaporization. This is why steam is such an effective and quick way of cooking foods compared with plain air—also a gas—at the same temperature. We can put a hand into an oven at 212°F/100°C and hold it there for some time before it gets uncomfortably warm; but a steaming pot will scald us in a second or two. In bread baking, an initial blast of steam increases the dough's expansion, or oven spring, and produces a lighter loaf.

Water and Acidity: The pH Scale

Acids and Bases Despite the fact that the molecular formula for water is H_2O, even absolutely pure water contains other combinations of oxygen and hydrogen. Chemical bonds are continually being formed and broken in matter, and water is no exception. It tends to "dissociate" to a slight extent, with a hydrogen occasionally breaking off from one molecule and rebonding to a nearby intact water molecule. This leaves one negatively charged OH combination, and a positively charged H_3O. Under normal conditions, a very small number of molecules exist in the dissociated state, something on the order of two ten-millionths of a percent. This is a small number but a significant one, because the presence of relatively mobile hydrogen ions, which are the basic units of positive charge (protons), can have drastic effects on other molecules in solution. A structure that is stable with a few protons around may be unstable when many protons are in the vicinity. So significant is the proton concentration that humans have a specialized taste sensation to estimate it: sourness. Our term for the class of chemical compounds that release protons into solutions, *acids,* derives from the Latin *acere,* meaning to taste sour. We call the complementary chemical group that accepts protons and neutralizes them, *bases* or *alkalis.*

The properties of acids and bases affect us continually in our daily life. Practically every food we eat, from steak to coffee to oranges, is at least slightly acidic. And the degree of acidity of the cooking medium can have great influence on such characteristics as the color of fruits and vegetables and the texture of meat and egg proteins. Some measure of acidity would clearly be quite useful. A simple scale has been devised to provide just that.

The pH Scale The standard measure of proton activity in solutions is *pH,* a term suggested by the Danish chemist S. P. L. Sørenson in 1909. It's essentially a more convenient version of the minuscule percentages of molecules involved (for some details, see box below). The pH scale runs from 0 to 14. The pH of neutral, pure water, with equal numbers of protons and OH ions, is set at 7. A pH lower than 7 indicates a greater concentration of protons and so an acidic solution, while a pH above 7 indicates a greater prevalence of proton-*accepting* groups, and so a basic solution. Here's a list of common solutions and their usual pH.

Liquid	pH
Human gastric juice	1.3–3.0
Lemon juice	2.1
Orange juice	3.0
Yogurt	4.5
Black coffee	5.0
Milk	6.9
Egg white	7.6–9.5
Baking soda in water	8.4
Household ammonia	11.9

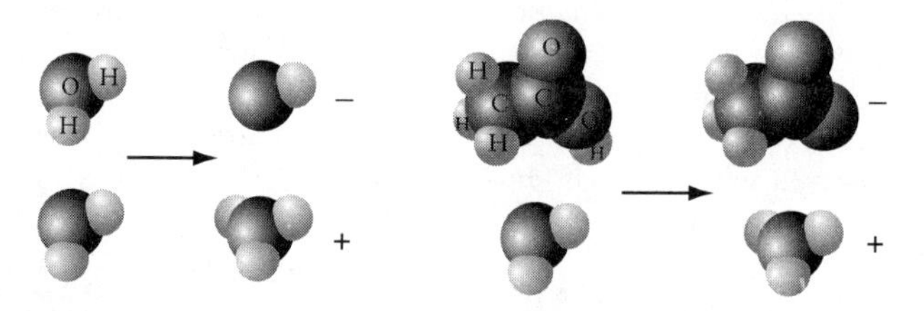

Acids. Acids are molecules that release reactive hydrogen ions, or protons, in water, where neutral water molecules pick them up and become positively charged. The acids themselves become negatively charged. Left: *Water itself is a weak acid.* Right: *Acetic acid.*

The Definition of pH

The pH of a solution is defined as "the negative logarithm of the hydrogen ion concentration expressed in moles per liter." The logarithm of a number is the exponent, or power, to which 10 must be raised in order to obtain the number. For example, the hydrogen ion concentration in pure water is 10^{-7} moles per liter, so the pH of pure water is 7. Larger concentrations are described by smaller negative exponents, so a more acidic solution will have a pH lower than 7, and a less acidic, more basic solution will have a pH higher than 7. Each increment of 1 in pH signifies an increase or decrease in proton concentration by a factor of 10; so there are 1,000 times the number of hydrogen ions in a solution of pH 5 as there are in a solution of pH 8.

FATS, OILS, AND RELATIVES: LIPIDS

LIPIDS DON'T MIX WITH WATER

Fats and oils are members of a large chemical family called the *lipids*, a term that comes from the Greek for "fat." Fats and oils are invaluable in the kitchen: they provide flavor and a pleasurable and persistent smoothness; they tenderize many foods by permeating and weakening their structure; they're a cooking medium that allows us to heat foods well above the boiling point of water, thus drying out the food surface to produce a crisp texture and rich flavor. Many of these qualities reflect a basic property of the lipids: they are chemically unlike water, and largely incompatible with it. And thanks to this quality, they have played an essential role in the function of all living cells from the very beginnings of life. Because they don't mix with water, lipids are well suited to the job of forming boundaries—membranes—between watery cells. This function is performed mainly by phospholipids similar to lecithin (p. 802), molecules that cooks also use to form membranes around tiny oil droplets. Fats and oils themselves are created and stored by animals and plants as a concentrated, compact form of chemical energy, packing twice the calories as the same weight of either sugar or starch.

In addition to fats, oils, and phospholipids, the lipid family includes beta-carotene and similar plant pigments, vitamin E, cholesterol, and waxes. These are all molecules made by living things that consist mainly of chains of carbon atoms, with hydrogen atoms projecting from the chain. Each carbon atom can form four bonds with other atoms, so a given carbon atom in the chain is usually bonded to two carbon atoms, one on each side, and two hydrogens.

This carbon-chain structure has one overriding consequence: lipids can't dissolve in water. They are "hydrophobic" or "water-fearing" substances. The reason for this is that carbon and hydrogen atoms pull with a similar force on their shared electrons. So unlike the oxygen-hydrogen bond, the carbon-hydrogen bond is not polar, and the hydrocarbon chain as a whole is nonpolar. When polar water and nonpolar lipids are mixed together, the polar water molecules form hydrogen bonds with each other, the long lipid chains form a weaker kind of bond with each other (van der Waals bonds, p. 814), and the two substances segregate themselves. Oils minimize the surface at which they contact water by coalescing into large blobs, and resist being divided into smaller droplets.

Thanks to their chemical relatedness, different lipids can dissolve in each other. This is why the carotenoid pigments—the beta-carotene in carrots, the lycopene in tomatoes—and intact chlorophyll, whose molecule has a lipid tail, color cooking fats much more intensely than they do cooking water.

Lipids share two other characteristics. One is their clingy, viscous, oily consistency, which results from the many weak bonds formed between their long carbon-hydrogen molecules. And those same molecules are so bulky that all natural fats, solid or liquid, float on water. Water is a denser substance due to its extensive hydrogen bonding, which packs its small molecules more tightly together.

THE STRUCTURE OF FATS

Fats and oils are members of the same class of chemical compounds, the *triglycerides*. They differ from each other only in their melting points: oils are liquid at room temperature, fats solid. Rather than use the technical *triglyceride* to denote these compounds, I'll use *fats* as the generic term. Oils are liquid fats. These are invaluable ingredients in cooking. Their clingy viscosity provides a moist, rich quality to many foods, and their high boiling point makes them an ideal cooking medium for the production of intense browning-reaction flavors (p. 778).

Glycerol and Fatty Acids Though they contain traces of other lipids, natural fats and oils are triglycerides, a combination of three *fatty acid* molecules with one molecule of *glycerol.* Glycerol is a short 3-carbon chain that acts as a common frame to which three fatty acids can attach themselves. The fatty acids are so named because they consist of a long hydrocarbon chain with one end that has an oxygen-hydrogen group and that can release the hydrogen as a proton. It's the acidic group of the fatty acid that binds to the glycerol frame to construct a glyceride: glycerol plus one fatty acid makes a monoglyceride, glycerol plus two fatty acids makes a diglyceride, and glycerol plus three fatty acids makes a triglyceride. Before it bonds to the glycerol frame, the acidic end of the fatty acid is polar, like water, and so it gives the free fatty acid a partial ability to form hydrogen bonds with water.

Fatty acid chains can be from 4 to about 35 carbons long, though the most common in foods are from 14 to 20. The properties of a given triglyceride molecule depend on the structure of its three fatty acids and their relative positions on the glycerol frame. And the properties of a fat depend on the particular mixture of triglycerides it contains.

Saturated and Unsaturated Fats, Hydrogenation, and Trans Fatty Acids

The Meaning of Saturation The terms "saturated" and "unsaturated" fats are familiar from nutrition labels and ongoing discussions of diet and health, but their meaning is seldom explained. A *saturated* lipid is one whose carbon chain is saturated—filled to capacity—with hydrogen atoms: there are no double bonds between carbon atoms, so each carbon within the

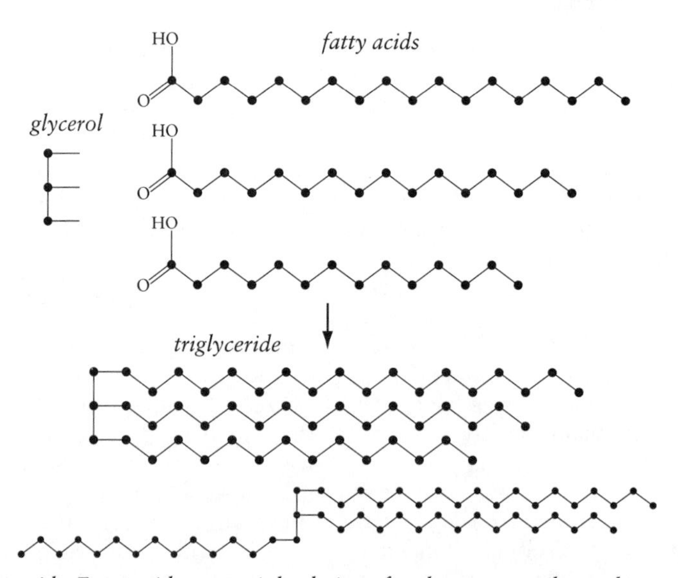

Fats and fatty acids. Fatty acids are mainly chains of carbon atoms, shown here as black dots. (Each carbon atom has two hydrogen atoms projecting from it; the hydrogen atoms are not shown.) A fat molecule is a triglyceride, *which is formed from one molecule of glycerol and three fatty acids. The acidic heads of the fatty acids are capped and neutralized by the glycerol, so the triglyceride as a whole no longer has a polar, water-compatible end. The fatty-acid chains can rotate around the glycerol head to form chair-like arrangements* (bottom).

chain is bonded to two hydrogen atoms. An *unsaturated* lipid has one or more double bonds between carbon atoms along its backbone. The double-bonded carbons therefore have only one bond left for a hydrogen atom. A fat molecule with more than one double bond is called *polyunsaturated.*

Fat Saturation and Consistency Saturation matters in the behavior of fats because double bonds significantly alter the geometry and the regularity of the fatty-acid chain, and so its chemical and physical properties. A saturated fatty acid is very regular and can stretch out completely straight. But because a double bond between carbon atoms distorts the usual bonding angles, it has the effect of adding a kink to the chain. Two or more kinks can make it curl.

A group of identical and regular molecules fits more neatly and closely together than different and irregular molecules. Fats composed of straight-chain saturated fatty acids fall into an ordered solid structure—the process has been described as "zippering"—more readily than do kinked unsaturated fats. Animal fats are about half saturated and half unsaturated, and solid at room temperature, while vegetable fats are about 85% unsaturated, and are liquid oils in the kitchen. Even among the animal fats, beef and lamb fats are noticeably harder than pork or poultry fats, because more of their triglycerides are saturated.

Double bonds are not the only factor in determining the melting point of fats. Short-chain fatty acids are not as readily "zippered" together as the longer chains, and so tend to lower the melting point of fats. And the more variety in the structures of their fatty acids, the more likely the mixture of triglycerides will be an oil.

Fat Saturation and Rancidity Saturated fats are also more stable, slower to become

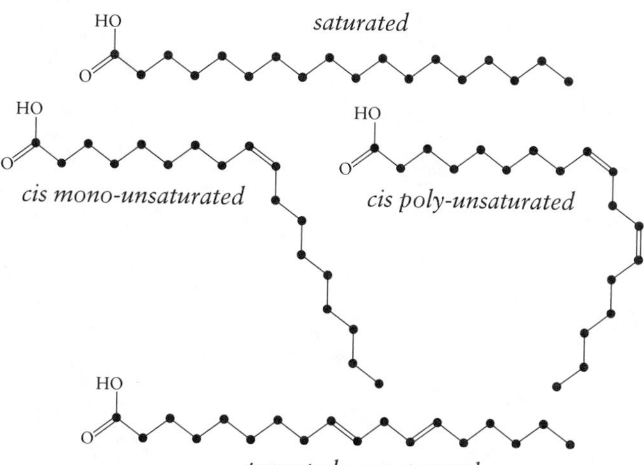

Saturated and unsaturated fatty acids. An unsaturated fatty acid has one or more double bonds along its carbon chain, and a rigid kink at that point in the chain. The structural irregularity caused by the double bond makes it more difficult for these molecules to solidify into compact crystals, so at a given temperature, unsaturated fats are softer than saturated fats. In the hydrogenation of vegetable oils to make them harder, some cis-unsaturated fatty acids are converted to trans-unsaturated fatty acids, which are less kinked and behave more like a saturated fatty acid, both in cooking and in the body.

rancid than unsaturated fats. The double bond of an unsaturated fat opens a space unprotected by hydrogen atoms on one side of the chain. This exposes the carbon atoms to reactive molecules that can break the chain and produce small volatile fragments. Atmospheric oxygen is just such a reactive molecule, and is one of the major causes of flavor deterioration in foods containing fats. Water and metal atoms from other food ingredients also help fragment fats and cause rancidity. The more unsaturated the fat, the more prone it is to deterioration. Beef has a longer shelf life than chicken, pork, or lamb because its fat is more saturated and so more stable.

Some small volatile fragments of unsaturated lipids actually have desirable and distinctive aromas. The typical aroma of crushed green leaves and of cucumber both come from fragments of membrane phospholipids generated not just by oxygen, but by special plant enzymes. And the characteristic aroma of deep-fried foods comes

Saturated and Unsaturated Fatty Acids in Foods and Cooking Fats

Proportions of fatty acids are given as a percentage of the total fatty-acid content.

Fat or Oil	Saturated Fatty Acids	Monounsaturated Fatty Acids	Polyunsaturated Fatty Acids
Butter	62	29	4
Beef	50	42	4
Lamb	47	42	4
Pork	40	45	11
Chicken	30	45	21
Coconut oil	86	6	2
Palm kernel oil	81	11	2
Palm oil	49	37	9
Cocoa butter	60	35	2
Vegetable shortening	31	51	14
Cottonseed oil	26	18	50
Stick margarine	19	59	18
Tub margarine	17	47	31
Peanut oil	17	46	32
Soybean oil	14	23	58
Olive oil	13	74	8
Corn oil	13	24	59
Sunflower seed oil	13	24	59
Grapeseed oil	11	16	68
Canola oil	7	55	33
Safflower oil	9	12	75
Walnut oil	9	16	70

in part from particular fatty-acid fragments created at high temperatures.

Hydrogenation: Altering Fat Saturation For more than a century now, manufacturers have been making solid, fat-like shortenings and margarines from liquid seed oils to obtain both the desired texture and improved keeping qualities. There are several ways to do this, the simplest and most common being to saturate the unsaturated fatty acids artificially. This process is called *hydrogenation,* because it adds hydrogen atoms to the unsaturated chains. A small amount of nickel is added to the oil as a catalyst, and the mixture is then exposed to hydrogen gas at high temperature and pressure. After the fat has absorbed the desired amount of hydrogen, the nickel is filtered out.

Trans Fatty Acids It turns out that the hydrogenation process straightens a certain proportion of the kinks in unsaturated fatty acids not by adding hydrogen atoms to them, but by rearranging the double bond, twisting it so that its bend is less extreme. These molecules remain chemically unsaturated—the double bond between two carbons remains—but they have been transformed from an acutely irregular *cis* geometry to a more regular *trans* structure (see illustration, p. 799). *Cis* is Latin for "on this side of," and *trans* for "across from"; the terms describe the positions of neighboring hydrogen atoms on the double bond between carbon atoms. Because the trans fatty acids are less kinked, more like a saturated fat chain in structure, they make it easier for the fat to crystallize and so make it firmer. They also make the fatty acid less prone to attack by oxygen, so it's more stable. Unfortunately, trans fatty acids also resemble saturated fats in raising blood cholesterol levels, which can contribute to the development of heart disease (p. 38). Manufacturers will soon be required to list the trans fatty acid content of their foods, and they're beginning to implement other processing techniques that harden fat consistency without creating trans fatty acids.

Fats and Heat

Most fats do not have sharply defined melting points. Instead, they soften gradually over a broad temperature range. As the temperature rises, the different kinds of fat molecules melt at different points and slowly weaken the whole structure. (An interesting exception to this rule is cocoa

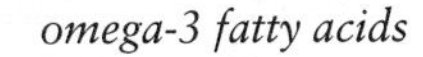

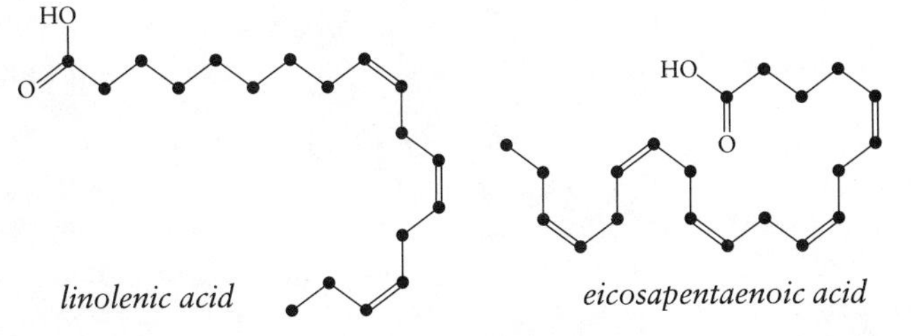

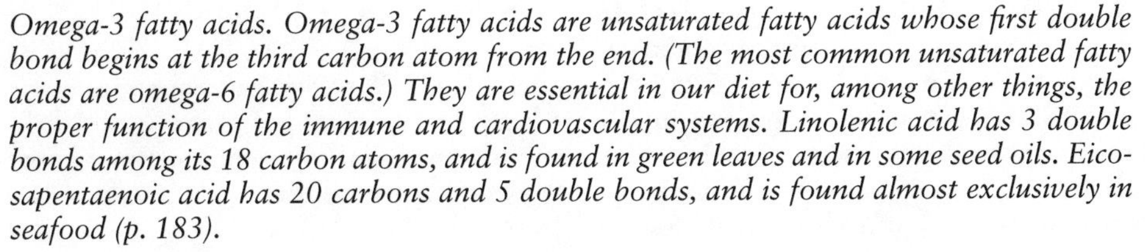
Omega-3 fatty acids. Omega-3 fatty acids are unsaturated fatty acids whose first double bond begins at the third carbon atom from the end. (The most common unsaturated fatty acids are omega-6 fatty acids.) They are essential in our diet for, among other things, the proper function of the immune and cardiovascular systems. Linolenic acid has 3 double bonds among its 18 carbon atoms, and is found in green leaves and in some seed oils. Eicosapentaenoic acid has 20 carbons and 5 double bonds, and is found almost exclusively in seafood (p. 183).

butter, p. 705). This behavior is especially important in making pastries and cakes, and it's what makes butter spreadable at room temperature.

Melted fats do eventually change from a liquid to a gas: but only at very high temperatures, from 500° to 750°F/260–400°C. This high boiling point, far above water's, is the indirect result of the fats' large molecular size. While they can't form hydrogen bonds, the carbon chains of fats do form weaker bonds with each other (p. 814). Because fat molecules are capable of forming so many bonds along their lengthy hydrocarbon chains, the individually weak interactions have a large net effect: it takes a lot of heat energy to knock the molecules apart from each other.

The Smoke Point Most fats begin to decompose at temperatures well below their boiling points, and may even spontaneously ignite on the stovetop if their fumes come into contact with the gas flame. These facts limit the maximum useful temperature of cooking fats. The characteristic temperature at which a fat breaks down into visible gaseous products is called the *smoke point*. Not only are the smoky fumes obnoxious, but the other materials that remain in the liquid, including chemically active free fatty acids, tend to ruin the flavor of the food being cooked.

The smoke point depends on the initial free fatty acid content of the fat: the lower the free fatty acid content, the more stable the fat, and the higher the smoke point. Free fatty acid levels are generally lower in vegetable oils than in animal fats, lower in refined oils than unrefined ones, and lower in fresh fats and oils than in old ones. Fresh refined vegetable oils begin to smoke around 450°F/230°C, animal fats around 375°F/190°C. Fats that contain other substances, such as emulsifiers, preservatives, and in the case of butter, proteins and carbohydrates, will smoke at lower temperatures than pure fats. Fat breakdown during deep frying can be slowed by using a tall, narrow pan and so reducing the area of contact between fat and atmosphere. The smoke point of a deep-frying fat is lowered every time it's used, since some breakdown is inevitable even at moderate temperatures, and trouble-making particles of food are always left behind.

Emulsifiers: Phospholipids, Lecithin, Monoglycerides

Some very useful chemical relatives of the true fats, the triglycerides, are the diglycerides and monoglycerides. These molecules act as *emulsifiers* to make fine, cream-like mixtures of fat and water—such sauces as mayonnaise and hollandaise—even though fat and water don't normally mix with each other. The most prominent natural emulsifiers are the diglyceride *phospholipids* in egg yolks, the most abundant of which is *lecithin* (it makes up about a third of the yolk lipids). Diglycerides have only two fatty-acid chains attached to the glycerol frame, and monoglycerides just one, with the remaining positions on the frame being occupied by small polar groups of atoms. These molecules are thus water-soluble at the head, and fat-soluble at the tail. In cell membranes, the phospholipids assemble themselves in two layers, with one set of polar heads facing the watery interior, the other set the watery exterior, and the tails of both sets mingling in between. When the cook whisks some fat into a water-based liquid that contains emulsifiers—oil into egg yolks, for example—the fat forms tiny droplets that would normally coalesce and separate again. But the emulsifier tails become dissolved in the droplets, and the electrically charged heads project from the droplets and shield the droplets from each other. The emulsion of fat droplets is now stable.

These "surface-active" molecules have many other applications as well. For example, monoglycerides have been used for decades in the baking business because they help retard staling, apparently by complexing with amylose and blocking starch retrogradation.

CARBOHYDRATES

The name for this large group of molecules comes from the early idea that they were made up of carbon and water. They are indeed made up of carbon, hydrogen, and oxygen atoms, though the oxygen and hydrogen are not found as intact water complexes within the molecules. Carbohydrates are produced by all plants and animals for the purpose of storing chemical energy, and by plants to make a supporting skeleton for its cells. Simple sugars and starch are energy stores, while pectins, cellulose, and other cell-wall carbohydrates are the plant's structural materials.

SUGARS

Sugars are the simplest carbohydrates. There are many different kinds of sugar molecules, each distinguished by the number of carbon atoms it contains, and then by the particular arrangement it assumes. Five-carbon sugars are especially important to all life because two of them, ribose and deoxyribose, form the backbones of ribonucleic acid (RNA) and deoxyribonucleic acid (DNA), the carriers of the genetic code. And the 6-carbon sugar glucose is the molecule from which most living things obtain the energy to run the biochemical machinery of their cells. Sugars are such an important nutrient that we have a special sense designed specifically to detect them. Sugars taste sweet, and sweetness is a nearly universal source of pleasure. It's the essence of the dishes we serve at the end of the meal, as well as of candies and confections. Sugars and their properties are described in detail in chapter 12.

OLIGOSACCHARIDES

The oligosaccharides ("several-unit sugars") raffinose, stachyose, and verbascose are 3-, 4-, and 5-ring sugars, respectively, all too large to trigger our sweet detectors, so they're tasteless. They're commonly found in the seeds and other organs of plants, where they make up part of the energy sup-

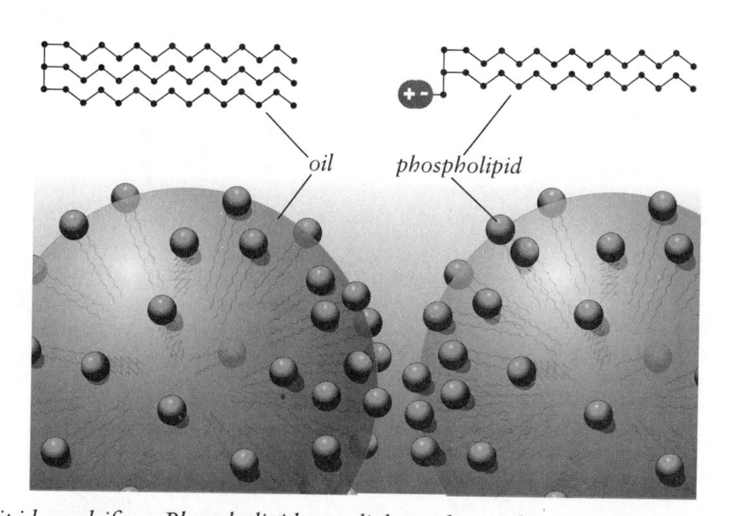

Phospholipid emulsifiers. Phospholipids are diglycerides, and are excellent emulsifiers, molecules that make possible a stable mixture of oil and water. Unlike the triglycerides of fat and oil, they have a polar, water-compatible head. Such emulsifiers bury their fatty-acid tails in oil droplets, while their water-compatible, electrically charged heads project from the surface and block the droplets from contacting each other and coalescing.

ply. These sugars all affect our digestive system, thanks to the fact that we don't have digestive enzymes capable of breaking them down into single sugars that can be absorbed by the intestine. As a result, the oligosaccharides are not digested and pass intact into the colon, where various bacteria do digest them, producing large quantities of carbon dioxide and other gases in the process (p. 486).

POLYSACCHARIDES: STARCH, PECTINS, GUMS

Polysaccharides, which include starch and cellulose, are sugar *polymers,* or molecules composed of numerous individual sugar units, as many as several thousand. Usually only one or a very few kinds of sugars are found in a given polysaccharide. Polysaccharides are classified according to the overall characteristics of the large molecules: a general size range, an average composition, and a common set of properties. Like the sugars of which they're composed, polysaccharides contain many exposed oxygen and hydrogen atoms, so they can form hydrogen bonds and absorb water. However, they may or may not dissolve in water, depending on the attractive forces among the polymers themselves.

Starch By far the most important polysaccharide for the cook is *starch,* the compact, unreactive polymer in which plants store their supply of sugar. Starch is simply a chain of glucose sugars. Plants produce starch in two different configurations: a completely linear chain called *amylose,* and a highly branched form called *amylopectin,* each of which may contain thousands of glucose units. Starch molecules are deposited together in a series of concentric layers to form solid microscopic granules. When starchy plant tissue is cooked in water, the granules absorb water, swell, and release starch molecules; when cooled again, the starch molecules rebond to each other and can form a moist but solid gel. Various aspects of starch—the way it determines the texture of cooked rice, its formation into pure starch noodles, its role in breads, pastries, and sauces—are described in detail in chapters 9–11.

Glycogen Glycogen, or "animal starch," is an animal carbohydrate similar to amylopectin, though more highly branched. It's

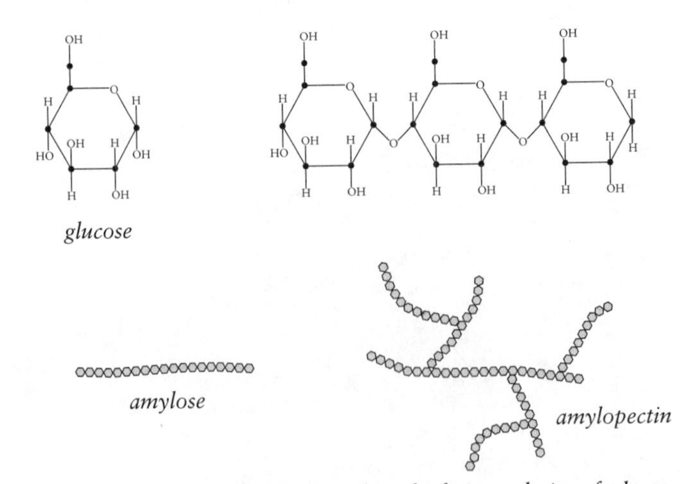

A sugar, glucose, and a polysaccharide, starch, which is a chain of glucose molecules. Plants produce two broadly different forms of starch: simple long chains called amylose, and highly branched chains called amylopectin.

a fairly minor component of animal tissue and so of meats, although its concentration at the time of slaughter will affect the ultimate pH of the meat, and thereby its texture (p. 142).

Cellulose Cellulose is, like amylose, a linear plant polysaccharide made up solely of glucose sugars. Yet thanks to a minor difference in the way the sugars are linked to each other, the two compounds have very different properties: cooking dissolves starch granules but leaves cellulose fibers intact; most animals can digest starch, but not cellulose. Cellulose is a structural support that's laid down in cell walls in the form of tiny fibers analogous to steel reinforcing bars, and it's made to be durable. Few animals can digest cellulose, and hay-eating cattle and wood-eating termites can do so only because their guts are populated by cellulose-digesting bacteria. To other animals, including ourselves, cellulose is indigestible fiber (which has its own value; see p. 258).

Hemicelluloses and Pectic Substances These polysaccharides (made from a variety of sugars, including galactose, xylose, arabinose) are found together with cellulose in the plant cell walls. If the cellulose fibrils are the reinforcing bars in the cell walls, the amorphous hemicelluloses and pectic substances are a sort of jelly-like cement in which the bars are embedded. Their significance for the cook is that, unlike cellulose, they are partly soluble in water, and therefore contribute to the softening of cooked vegetables and fruits. Pectin is abundant enough to be extracted from citrus fruits and apples and used to thicken fruit syrups into jams and jellies. These carbohydrates are described in detail in chapter 5.

Inulin Inulin is a polymer of fructose sugars, from a handful to hundreds per molecule. Inulin is a form of energy storage and a source of antifreeze (sugars lower the freezing point of a water solution) in members of the onion and lettuce families, notably garlic and the sunchoke. Like the oligosaccharides, inulin is not digestible, and so feeds bacteria in our large intestine and generates gas.

Plant Gums There are a number of other plant carbohydrates that cooks and manufacturers have found useful for thickening and gelling liquid foods, helping to stabilize emulsions, and producing smoother consistencies in frozen goods and candies. Like the cell-wall cements, they're generally complex polymers of several different sugars or related carbohydrates. They include:

- Agarose, alginates, and carrageenans, cell-wall polymers from various seaweeds
- Gum arabic, which exudes from cuts in various species of *Acacia* trees
- Gum tragacanth, an exudate from various species of *Astralagus* shrubs
- Guar gum, from seeds of a shrub in the bean family (*Cyamopsis tetragonobola*)
- Locust-bean gum, from seeds of the carob tree, *Ceratonia siliqua*
- Xanthan gum and gellan, polysaccharides produced by certain bacteria in industrial fermentation

PROTEINS

Of all the major food molecules, proteins are the most challenging and mercurial. The others, water and fats and carbohydrates, are pretty stable and staid. But expose proteins to a little heat, or acid, or salt, or air, and their behavior changes drastically. This changeability reflects their biological mission. Carbohydrates and fats are mainly passive forms of stored energy, or structural materials. But proteins are the active machinery of life. They assemble all the molecules that make a cell, themselves included, and tear them down as well; they move molecules from one place in the cell to another; in the form of muscle fibers, they move whole animals. They're at the heart of all organic activity, growth, and movement. So it's the nature of proteins to be active and sensitive. When we cook foods that contain them, we take advantage of their dynamic nature to make new structures and consistencies.

Amino Acids and Peptides

Like starch and cellulose, proteins are large polymers of smaller molecular units. The smaller units are called *amino acids*. They consist of between 10 and 40 atoms, mainly carbon, hydrogen, and oxygen, with at least one nitrogen atom in an *amine* group—NH_2—that gives the amino acids their family name. A couple of amino acids include sulfur atoms. There are about 20 different kinds of amino acids that occur in significant quantities in food. Particular protein molecules are dozens to hundreds of amino acids long, and often contain many of the 20 different kinds. Short chains of amino acids are called *peptides*.

Amino Acids and Peptides Contribute Flavor Three aspects of amino acids are especially important to the cook. First, amino acids participate in the browning reactions that generate flavor at high cooking temperatures (p. 778). Second, many single amino acids and short peptides have tastes of their own, and in foods where proteins have been partly broken down—aged cheeses, cured hams, soy sauce—these tastes can contribute to the overall flavor. Most tasty amino acids are either sweet or bitter to some degree, and a number of peptides are also bitter. But glutamic acid, better known in its concentrated commercial form MSG (monosodium glutamate), and some peptides have a unique taste that is designated by such words as *savory*, *brothy*, and *umami* (Japanese for "delicious"). They lend an added dimension of flavor to foods that are rich in them, including tomatoes and certain seaweeds as well as salt-cured and fermented products. When heated, sulfur-containing amino acids break down and contribute eggy, meaty aroma notes.

Amino Acids Influence Protein Behavior The third important characteristic of amino acids is that they have a variety of chemical natures, and these influence the structure and behavior of the protein they're a part of. Some amino acids have portions resembling water and can form hydrogen bonds with other molecules, including water. Some have short carbon chains or carbon rings that resemble fats, and can form van der Waals bonds with other similar molecules. And some, especially those that include a sulfur atom, are especially reactive, and can form strong covalent bonds with other molecules, including other sulfur-containing amino acids. This means that a single protein has many different chemical environments along its chain: parts that attract water molecules, parts that avoid water molecules, and parts that are ready to form strong bonds with similar parts on other proteins, or on other parts of the same protein.

Protein Structure

Proteins are formed by linking the amine nitrogen of one amino acid with a carbon atom on another amino acid, and then repeating this "peptide bond" to make a chain dozens or hundreds of amino acids long. The carbon-nitrogen backbone of the protein molecule forms a sort of zigzag pattern, with the "side groups"—the other atoms on each amino acid—sticking out to the sides.

The Protein Helix One effect of the peptide bond is a certain kind of regularity that causes the molecule as a whole to twist and form a spiral, or helix. Very few proteins exist as a simple regular helix, but those that do tend to join together in strong fibers. These include connective-tissue collagen in meat, an important factor in its tenderness, and the source of gelatin (pp. 130, 597).

Protein Folds The other influence on protein structure is the side groups of its amino acids. Because the protein chain is so long, it can bend back on itself and bring together amino acids that are some distance along the chain from each other. Amino acids with similar side groups can then bond to each other in various ways, including via hydrogen bonds, van der Waals bonds, ionic bonds (p. 813), and strong covalent bonds (especially between sulfur atoms). This bonding is what gives a particular protein molecule the characteristic shape that allows it to carry out its particular job. The weak, temporary nature of the hydrogen and hydrophobic bonds

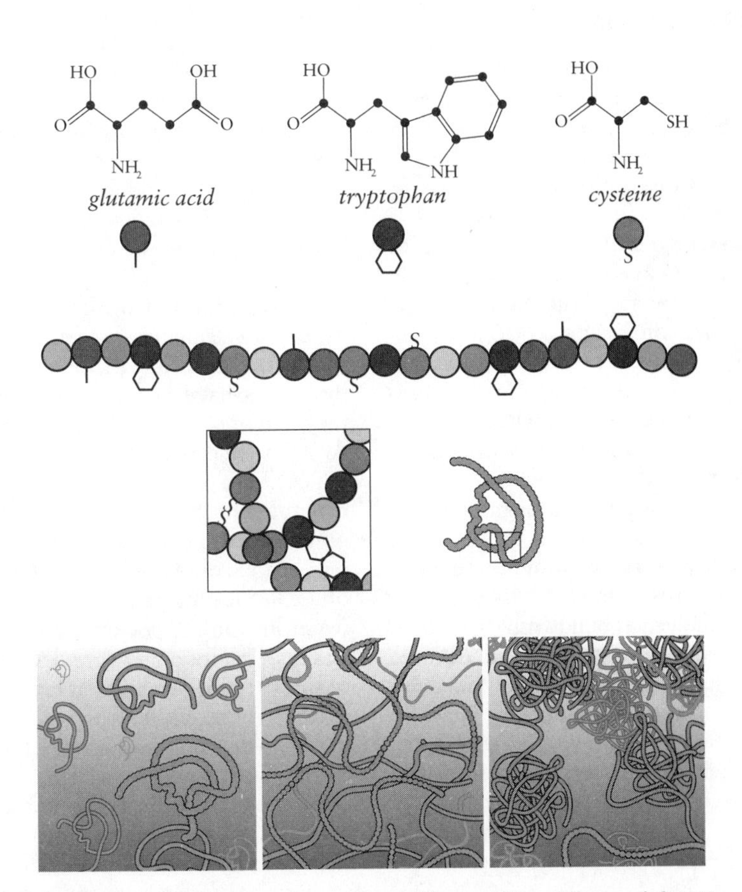

Amino acids and proteins, denaturation and coagulation. Top: *Three of the 20-odd amino acids important in food. Each amino acid has a common end including an amino (NH_2) group, by which amino acids bond to each other into long chains called proteins, and a variable end or "side group" that can form different kinds of bonds with other amino acids.* Center: *A chain of amino acids shown schematically, with some of the side groups projecting from the chain. The amino acid chain can fold back on itself, and some of the side groups form bonds with each other to hold the chain in a folded shape.* Bottom: *Heating and other cooking processes can break the fold-stabilizing bonds and cause the long chains to unfold, or* denature (left, center). *Eventually the exposed side groups form new bonds between different protein chains, and the proteins* coagulate, *or form a permanently bonded solid mass* (right).

allows it to change its shape as it works. The overall shape of a protein can range from a long, extended, mostly helical molecule with a few kinks or loops, to compact, elaborately folded molecules that are called "globular" proteins. Collagen is an example of a helical protein, and the various proteins in eggs are mainly globular.

Proteins in Water

In living systems and in most foods, protein molecules are surrounded by water. Because all proteins are capable to some extent of hydrogen bonding, they absorb and hold at least some water, although the amounts vary greatly according to the kinds of side groups present and the overall structure of the molecule. Water molecules can be held "inside" the protein, along the backbone, and "outside," on polar side groups.

Whether or not a protein is *soluble* in water depends on the strength of the bonds between molecules, and on whether water can separate the molecules from each other by hydrogen bonding. The wheat proteins that form gluten when flour is mixed with water are a kind of protein that absorbs considerable amounts of water but doesn't dissolve, because many fat-like groups along their molecules bond with each other, hold the proteins together, and exclude water. Similarly, the proteins that make up the contracting muscle fibers in meat are held together by ionic and other bonds. On the other hand, many of the proteins in milk and eggs are quite soluble.

Protein Denaturation

A very important characteristic of proteins is their susceptibility to *denaturation,* or the undoing of their natural structure by chemical or physical means. This change involves breaking the bonds that maintain the molecule's folded shape. (The strong backbone bonds are broken only in extreme conditions or with the help of enzymes.) Denaturation is not a change in composition, only a change in structure. But structure determines behavior, and denatured proteins behave very differently from their originals.

Proteins can be denatured in many ways: by exposing them to heat—usually to somewhere between 140–180ºF/60–80ºC—or to high acidity, or to air bubbles, or to a combination of these. In each case, the unusual chemical or physical conditions—increased molecular agitation, or lots of reactive protons, or the drastic difference between the air bubble and the liquid wall that surrounds it—breaks many of the bonds between amino acid side groups that hold the protein molecule in its specific folded shape. The long proteins therefore unfold, exposing many more of their reactive side groups to the watery environment.

Protein Coagulation There are several general consequences of denaturation that follow for most food proteins. Because the molecules have been extended in length, they're more likely to bump into each other. And because their side groups are now exposed and available for forming bonds, denatured proteins begin to bond with each other, or *coagulate.* This happens throughout the food, and results in the development of a continuous network of proteins, with water held in the pockets between protein strands. The food therefore develops a kind of thickness or density that can be delicate and delightful, as in a barely set custard or perfectly cooked piece of fish. However, if cooking or other denaturing conditions continue, given the extreme physical or chemical environment that caused the proteins to denature in the first place, only the stronger bonds can form and survive, which means that the proteins bond together more and more tightly, densely, and irreversibly. And as they do so, they squeeze the pockets of water out from between them. The custard gets dense and a watery fluid separates from the solid portion; the fish gets tough and dry.

The details of protein denaturation and coagulation in any given food are intricate and fascinating. For example, acidity and salts can cause egg proteins to cluster together even before they begin to unfold, and thus affect the consistency of scrambled eggs and custards. Such details are noted in the descriptions of particular foods.

ENZYMES

There's a particular group of proteins that are important to the cook not so much for their direct contribution to food texture and consistency, but for the way they change the other components of the food they're in. These proteins are the *enzymes.* Enzymes are biological catalysts: that is, they increase the rate of specific chemical reactions that otherwise would occur only very slowly, if at all. Enzymes thus cause chemical change. Some enzymes build molecules up, or modify them; some break molecules down. Human digestive enzymes, for example, break proteins into individual amino acids, and starch into individual glucose units. A singe enzyme molecule can catalyze as many as a million reactions per second.

Enzymes matter to the cook because foods contain enzymes that once did important work for the plant or animal when it was alive, but that can now harm the food by changing its color, texture, taste, or nutritiousness. Enzymes help turn green chlorophyll in vegetables dull olive, cause cut fruits to turn brown and oxidize their vitamin C, and turn fish flesh mushy. And bacterial spoilage is largely a matter of bacterial enzymes breaking the food down for the bacteria's own use. With a few exceptions—the tenderizing of meat by its own internal enzymes, the firming of some vegetables before further cooking, and fermentations in general—the cook wants to prevent enzymatic activity in food. Storing foods at low temperatures delays spoilage in part because it slows the growth of spoilage microbes, but also because it slows the activity of the food's own enzymes.

Cooking Accelerates Enzyme Action Before Stopping It Because the activity of an enzyme depends on its structure, any change in that structure will destroy its effectiveness. So cooking foods sufficiently will denature and inactivate any enzymes they may contain. One vivid example of this principle is the behavior of raw and cooked pineapple in gelatin. Pineapples and certain other fruits contain an enzyme that breaks proteins down into small fragments. If raw pineapple is combined with gelatin to make a jelly, the enzyme digests the gelatin molecules and liquifies the jelly. But canned pineapple has been heated enough to denature the enzyme, and makes a firm gelatin jelly.

There's a complication, though. The reactivity of most chemicals increases with increasing temperature. The rule of thumb is that reactivity doubles with each rise of 20ºF/10ºC. The same tendency goes for enzymes, up to a range in which they begin to denature, become less effective, and finally become completely inactive. This means that cooking gives enzymes a chance to do their damage more and more quickly as the temperature rises, and only stops them once they reach their denaturation temperature. In general, the best rule is to heat foods as rapidly as possible, thereby minimizing the period during which the enzymes are at their optimum temperatures, and to get them all the way to the boiling point. Conversely, desirable enzyme action—meat tenderizing, for example—can be maximized by slow, gradual heating to denaturing temperatures.

APPENDIX

A CHEMISTRY PRIMER

Atoms, Molecules, Energy

Cooking is applied chemistry, and the basic concepts of chemistry—molecules, energy, heat, reactions—are keys to a clearer understanding of what our foods are and how we transform them. A casual acquaintance with these concepts is enough to follow most of the explanations in this book. For readers who'd like to get to know them better, here's a brief review.

ATOMS, MOLECULES, AND CHEMICAL BONDS

It was the ancient Greeks who gave us the idea of *atoms,* fundamental and invisibly small particles of matter, and also the word *atom,* which means "uncuttable," "indivisible." Greek philosophers proposed that there are just four basic kinds of particles in the world—atoms of earth, air, water, and fire—and that all material things, our bodies and our foods and everything else, are built from these primary particles. The modern scientific view of matter's invisible innards is more complicated, but also more precise and illuminating.

Atoms and Molecules

All matter on earth is a mixture of around 100 pure substances, which we call the *elements*: hydrogen, oxygen, nitrogen, carbon, and so on. An atom is the smallest particle into which an element can be subdivided without losing its characteristic properties. Atoms are very small indeed: several million would fit into the period at the end of this sentence. All atoms are made up of smaller "subatomic" particles, *electrons, protons,* and *neutrons.* The different properties of the elements arise from the varying combinations of subatomic particles that make up their atoms, and in particular their quotas of protons and electrons. Hydrogen atoms contain one proton and one electron; oxygen carries 8 of each, and iron 26.

When two or more atoms bond together, which they do by sharing electrons with each other, they form a *molecule* (from the Latin for "mass," "bulk"). The molecule is to a chemical compound what the atom is to an element: the smallest unit that retains the properties of the original material. Most matter on earth, including food, is a mixture of different chemical compounds.

Protons and Electrons Carry Positive and Negative Electrical Charges There's one primary driving force behind all the chemical activity that makes life and cooking possible, and that's the electrical attraction between protons and electrons. Protons carry a positive electrical charge, and electrons an exactly balancing negative charge. (The neutral neutron carries no charge.) Opposite electrical charges attract each other; similar electrical charges repel each other. In each atom, protons in the central nucleus attract a cloud of electrons that orbit constantly at various distances from the nucleus. Stable forms of the elements are electrically neutral, which means that their atoms contain equal numbers of protons and electrons.

(If like charges repel each other and opposite charges attract, then why is it that the protons in the nucleus don't push each other away and the orbiting electrons fall straight into the nucleus? It turns out that there are forces besides electricity at work in the atom. The protons and neutrons are bound together by very strong nuclear forces, while it's the nature of electrons to be in continual motion. So protons and electrons are always attracted to each other and move in response to the other's presence, but they never consummate their attraction.)

Electrical Imbalance, Reactions, and Oxidation

The electrons in atoms are arranged around the nucleus in orbits that determine how strongly any particular electron is held there. Some electrons are held close and tightly to the nucleus, while others range far away and are held more weakly. The behavior of the outermost electrons largely determines the chemical behavior of the elements. For example, the elements classified as metals—copper, aluminum, iron—hold their outermost electrons very weakly, and easily give them away to the atoms of other elements—oxygen, chlorine—that are hungrier for electrons, and that tend to grab up any that are loosely held. This imbalance in electrical pulls among different elements is the basis of most *chemical reactions*. Reactions are encounters among atoms and molecules that result in the loss, gain, or sharing of electrons, and thus

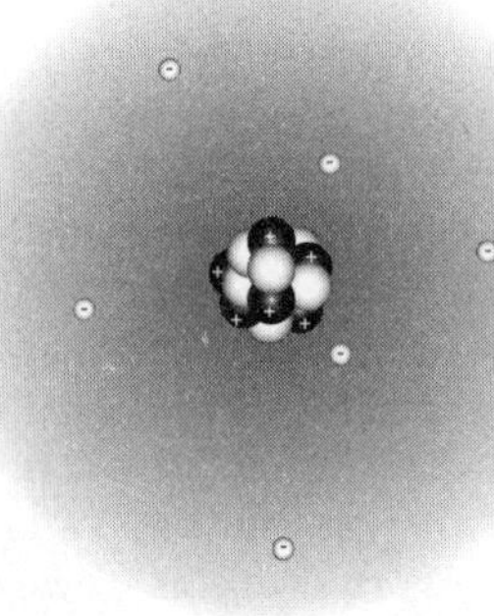

An atom of carbon. Carbon has six protons and six neutrons in its nucleus, and six electrons orbiting around the nucleus.

changes in the properties of the atoms and molecules involved.

Of all the electron-grabbing elements, the most important is oxygen, so much so that chemists use the term *oxidation* to name the general chemical activity of grabbing electrons from other atoms, even if a chlorine atom is doing the grabbing. Oxidation is very important in the kitchen, because oxygen is always present in the air, and readily robs electrons from the carbon-hydrogen chains of fats, oils, and aroma molecules. This initial oxidation triggers a cascade of further oxidations and other reactions that end up breaking the original large fat molecules into small, strong-smelling fragments. *Antioxidant* substances—for example, phenolic compounds found in many foods made from plants—prevent this breakdown by giving oxygen the electrons it wants without starting a reaction cascade, thus sparing the fat molecules from oxidation.

Electrical Imbalance and Chemical Bonds

Electron hunger is also the basis for the *chemical bond,* an interaction between atoms or molecules that holds them together, either loosely or tightly, momentarily or indefinitely. There are several different kinds of chemical bonds that are important in the kitchen, as they are throughout nature.

Ionic Bonds; Salt One kind of chemical bond is the *ionic bond,* in which one atom completely captures the electron(s) of another, so great is the difference between their electron hungers. Chemical compounds held together by ionic bonds don't simply dissolve in water; they come apart into separate *ions,* or atoms that are electrically charged because they either carry extra electrons or gave up some of their electrons. (The term was coined by the pioneer of electricity, Michael Faraday, from the Greek word for "going," to name those electrically charged particles that move when an electrical field is set up in a water solution.) Salt, our most common seasoning, is a compound of sodium and chlorine held together with ionic bonds. In a solid crystal of pure salt, positively charged sodium ions alternate with negatively charged chloride ions, the sodiums having lost their electrons to the chlorines. Because several positive sodium ions are always in a

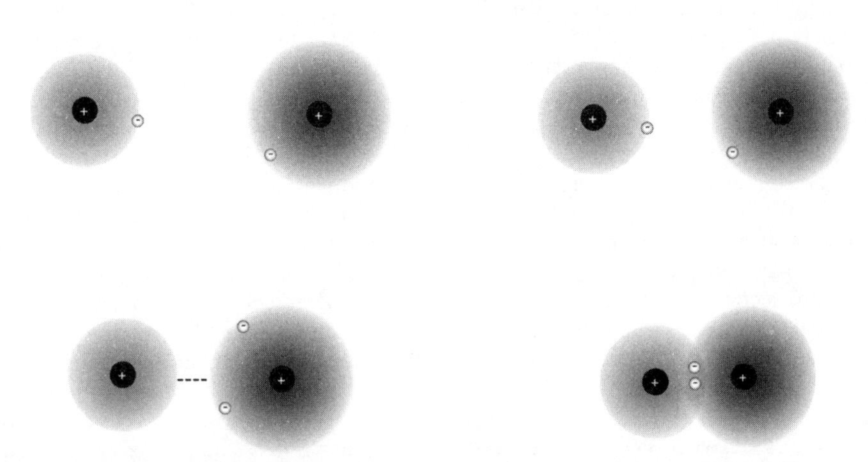

Ionic and covalent bonds. Left: *An ionic bond results when an atom completely captures one or more electrons of another atom, and the two atoms experience an attractive force (dotted line) due to their opposite electrical charges.* Right: *In the covalent bond, atoms share electrons, and thereby form stable combinations called molecules.*

state of attraction to several negative chloride ions, we can't really speak of individual molecules of salt, with one particular sodium atom bonded to a particular chlorine atom. In water, salt dissolves into separate positive sodium ions and negative chloride ions.

Strong Bonds That Make Molecules A second kind of chemical bond, called *covalent* (from the Latin, "of equal power"), produces stable *molecules*. When two atoms have roughly similar affinities for electrons, they will *share* them rather than gain or lose them entirely. In order for sharing to occur, the electron clouds of two atoms must overlap, and this condition results in a fixed arrangement in space between two particular atoms, which thus form a stable combined structure. The bonding geometry determines the overall shape of the molecule, and molecular shape in turn defines the ways in which one molecule can react with others.

The elements most important to life on earth—hydrogen, oxygen, carbon, nitrogen, phosphorus, sulfur—all tend to form covalent bonds, and these make possible the complex, stable assemblages that constitute our bodies and our foods. The most familiar pure chemical compounds in the kitchen are water, a covalent combination of two hydrogen atoms and an oxygen; and sucrose, or table sugar, a combination of carbon, oxygen, and hydrogen atoms. Covalent bonds are generally strong and stable at room temperature: that is, they're not broken in significant numbers unless subjected to heat or to reactive chemicals, including enzymes. Unlike salt, which dissolves into electrically charged ions, covalently bonded molecules that can dissolve in water generally do so as intact, electrically neutral molecules.

Weak Bonds Between Polar Molecules: Water A third kind of chemical bond, about a tenth as strong and stable as covalent bonds, is the *hydrogen* bond. The hydrogen bond is one of several "weak" bonds that do not form molecules, but do make temporary links between different molecules, or between different parts of one large molecule. Weak bonds come about because most covalent bonds leave at least a slight electrical imbalance among the participating atoms. Consider water, whose chemical formula is H_2O. The oxygen atom has a greater hunger for electrons than the two hydrogen atoms, and so the shared electrons are held closer to the oxygen than to the hydrogens. As a result, there's an overall negative charge in the vicinity of the oxygen, and an overall positive charge around the hydrogen atoms. This unequal distribution of charge, together with the geometry of the covalent bonds, results in a molecule with a positive end and a negative end. Such a molecule is called *polar* because it has two separate centers, or poles, of charge.

A hydrogen bond results from the attraction between oppositely charged ends of polar molecules (or portions of molecules). This kind of bond is important because it's very common in materials that contain water, because it brings different kinds of molecules into close association, and because it's weak enough that these molecular associations can change rapidly at room temperature. Many of the chemical interactions in plant and animal cells occur via hydrogen bonds.

Very Weak Bonds Between Nonpolar Molecules: Fats and Oils A fourth kind of chemical bond is very weak indeed, between a hundredth and a ten-thousandth as strong as a molecule-making covalent bond. These *van der Waals* bonds, named after the Dutch chemist who first described them, are the kind of flickering electrical attraction that even nonpolar molecules can feel for each other, thanks to brief fluctuations in their structures. Where electrically polar water is held together as a liquid by hydrogen bonds, nonpolar fat molecules are held together as a liquid and given their appealingly thick consistency by van der Waals bonds. Though these bonds are indeed weak, their effect can add up to a significant force: fat molecules are long chains and include dozens of carbon atoms, so each fat molecule can interact with many more other molecules than a small water molecule can.

ENERGY

Energy Causes Change

The paragraphs directly above describe various bonds as "weak" and "strong," easily or not so easily formed and broken. The idea of bond strength is useful because most cooking is a matter of the systematic breaking of certain chemical bonds and the formation of others. The key to the behavior of chemical bonds is *energy.* The word is a Greek compound of "in" and "force" or "activity," and now has as its standard definition "the capacity for doing work," or "the exertion of a force across a distance." Most simply, energy is that property of physical systems that makes possible *change.* A system with little energy is largely unchanging. Conversely, the more energy available to an object, the more likely that object is to be changed, or to change its surroundings. Our kitchens are organized around this principle. Stoves and ovens change the qualities of food by pouring heat energy into it, while the refrigerator preserves food by removing heat and thus slowing down the chemical changes that constitute spoilage.

Atoms and molecules can absorb or release energy in several different forms, two of which are important in the kitchen.

The Nature of Heat: Molecular Movement

One kind of energy is the energy of motion, or *kinetic energy.* Atoms and molecules can move from one place to another; or spin in place, or vibrate, and all of these changes in position or orientation require energy. *Heat* is a manifestation of a material's kinetic energy, and *temperature* is a measure of that energy: the higher the temperature of a food or pan, the hotter it is, the faster its molecules are moving and colliding with each other. And simple movement is the key to transforming molecules and foods. As molecules move faster and more forcefully, their motions and collisions begin to overcome the electrical forces holding them together. This frees some atoms to find new partners and rearrange themselves in new molecules. Heat thus encourages chemical reactions and chemical change.

Bond Energy

The second kind of energy that's important in the kitchen is the energy of the chemical bonds that hold molecules together. When two or more atoms become a molecule by sharing electrons and bonding with each other, they're pulled together by an electrical force. So in the process of forming the bond, some of their electrical energy is transformed into energy of motion. And the stronger the electrical

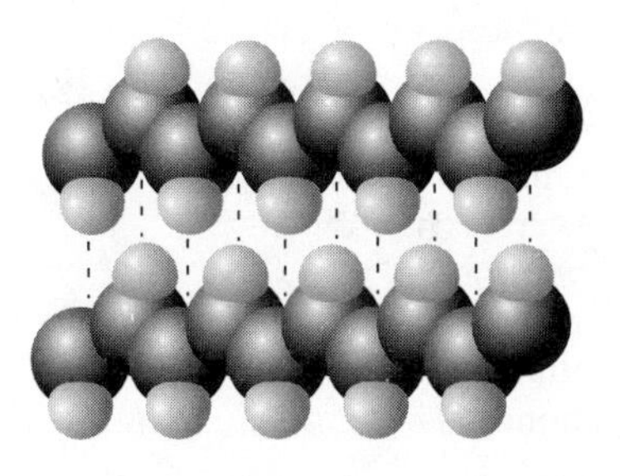

Van der Waals bonds. Thanks to fluctuations in the positions of their shared electrons, even the nonpolar chains of carbon and hydrogen atoms in fats experience weakly attractive electrical forces (dotted lines).

force, the more rapidly they accelerate toward each other. The stronger the bond, the more energy is released—lost—from the molecule in the form of motion. Strong bonds, then, "contain" less energy than weak bonds. This is another way of saying that they are more stable, less susceptible to change, than weak bonds.

Bond strength is defined as the amount of energy released from the participating atoms when they form the bond. This is the same as the amount of energy required to break that bond once it's formed. When the atoms in a molecule are heated up so they move with the same kinetic energy that they had released when they bonded to each other, then those bonds begin to break apart, and the molecule begins to react and change.

The strong covalent bonds typical of our major food molecules—proteins, carbohydrates, fats—are broken by about 100 times the average kinetic energy of molecules at room temperature. This means that they break very rarely at room temperature, and don't change at a significant rate until we heat them. The weaker, temporary hydrogen and van der Waals bonds *between* molecules are constantly being broken and re-formed at room temperature, and this welter of activity increases as the temperature rises. This is why fats melt and become thinner in consistency as we heat them: the energy of their motion increasingly overpowers the forces attracting them to each other.

THE PHASES OF MATTER

In our everyday life, we encounter matter in three different states, or *phases* (the word comes from the Greek for "appearance" or "manifestation"). These states are the solid phase, the liquid phase, and the gas phase. The temperatures at which a material melts—changes from solid to liquid—and boils—changes from liquid to gas—are determined by the bonding forces among the atoms or molecules. The stronger the bonds, the more energy needed to overcome them, and so the higher the temperature at which the material shifts from one phase to another. During that shift, all the heat added to the material goes into completing the phase change. The temperature of a solid-liquid mix will remain fairly constant until all the solid has melted. Similarly, the temperature of a pot of boiling water on a high flame remains constant—at the boiling point—until all the liquid water has been turned into steam.

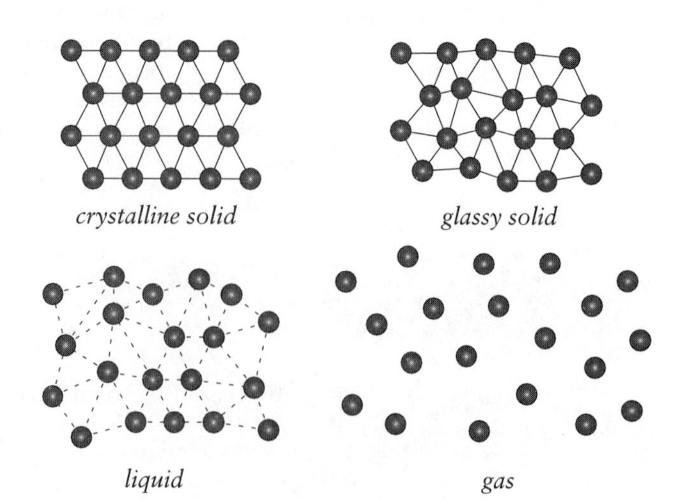

States of matter. Crystalline solids such as salt and sugar are made up of atoms or molecules bonded together in highly ordered, regular arrays. Amorphous solids, such as hard candies and glass, are masses of atoms or molecules that have bonded to each other in a random arrangement. Liquids are a loosely bonded, fluid mass of atoms or molecules, while a gas is a fluid and dispersed group of atoms or molecules.

Solids

At low temperatures, atomic motion is limited to rotation and vibration, and the immobilized atoms or molecules bond tightly to each other in solid, closely packed, well-defined structures. Such structures define the solid phase. In a crystalline solid—salt, sugar, tempered chocolate—the particles are arranged in a regular, repeating array, while in amorphous solids—boiled candies, glass—they are randomly oriented. Large, irregular molecules like proteins and starch often form both highly ordered, crystalline regions and disordered amorphous regions in the same chunk of material. Ionic bonds, hydrogen bonds, and van der Waals bonds may be involved in holding the particles of a solid together.

Liquids

At a temperature that is characteristic of each solid substance, the rotation and vibration of individual molecules in that substance becomes forceful enough that the electrical forces holding them in place are overpowered. The fixed structure then breaks up, leaving the molecules free to move from one place to another. However, most of the molecules are still moving slowly enough that they remain influenced by the forces that once immobilized them, and so they remain loosely associated with each other. They're free to move, but move together. This fluid but cohesive phase is a *liquid.*

Gases

If the temperature continues to rise and the molecules move with a kinetic energy high enough that they can break away from each other's influence completely and move freely into the air, the substance become a different kind of fluid, a *gas*. The most familiar transition from the liquid phase to the gas phase is boiling, in which we transform liquid water into bubbles of water vapor, or steam. Less obvious to the eye, because it's so gradual, is the evaporation of water at temperatures below the boiling point. The molecules in a liquid move with a wide range of kinetic energies, and a small portion of the molecules in room-temperature water are moving fast enough to escape from the surface and move into the air.

In fact, water molecules can even escape as a gas from solid ice! This direct transformation of a solid into a gas is called *sublimation,* and is the cause of that deterioration in foods known as "freezer burn," in which crystalline water evaporates into the freezer's cold, dry air. Freeze-drying is a controlled version of the same process.

Many Food Molecules Can't Change Phase

Most of the molecules that the cook works with can't simply change from one phase to another when heated. Instead, they react to form entirely different kinds of molecules. This is because food molecules are large, and form so many weak bonds between molecules that they're in fact held very strongly together. It takes as much energy to break them apart from each other as it does to break the molecules themselves apart: and so rather than melting or evaporating, the molecules become transformed. For example, sugar will melt from a solid into a liquid, but rather than then vaporize into a gas as water does, it breaks apart and forms hundreds of new compounds: a process we call caramelization. Fats and oils melt, but break down and smoke before they begin to boil. Starch, which is a long chain of sugar molecules joined together, won't even melt: it and proteins, also very large molecules, begin to break down as solids.

Mixtures of Phases: Solutions, Suspensions, Emulsions, Gels, Foams

Cooks seldom deal with pure chemical compounds or even single phases. Foods are mixtures of different molecules, different phases, and even different kinds of mixtures! Here are brief definitions of mixtures that are important in the kitchen.

- A *solution* is a material in which individual ions or molecules are dispersed in a liquid. Salt brines and sugar syrups are simple culinary examples.
- A *suspension* is a material in which a substance is dispersed in a liquid in clusters or particles consisting of many molecules. Nonfat milk is a suspension of milk-protein particles in water. Suspensions are usually cloudy because the clusters are large enough to deflect light rays (individual dissolved molecules are too small to do so, so solutions are clear).
- An *emulsion* is a special kind of suspension, one in which the dispersed substance is a liquid that can't mix evenly with the containing liquid. Cream is an emulsion of butterfat in water, and an oil-and-vinegar dressing is usually an emulsion of vinegar in oil.
- A *gel* is a dispersion of water in a solid: the molecules of the solid form a sponge-like network, and the pockets of water are trapped in the network. Examples are savory or sweet jellies made with gelatin, and fruit jellies made with pectin.
- A *foam* is a dispersion of gas bubbles in a liquid or solid. Soufflés, bread, and the head on a glass of beer are all foams.

SELECTED REFERENCES

The scientific and historical literatures on food are vast, as is the literature of cooking itself! The following list is a selection of the sources that I've referred to in the course of writing this book. It provides documentation for the more important facts and ideas, leads for finding more detailed information, and gives credit to researchers and translators. I begin by listing works that I've relied on throughout the book. Then follow bibliographies by chapter. Each is divided into two sections: first, books and articles intended for the general reader; and second, more specialized and technical studies.

GENERAL SOURCES

Books on Food and Cooking

Behr, E. *The Artful Eater.* New York: Atlantic Monthly, 1992.

Child, J., and S. Beck. *Mastering the Art of French Cooking.* 2 vols. New York: Knopf, 1961, 1970.

Davidson, A. *The Oxford Companion to Food.* Oxford: Oxford Univ. Press, 1999.

Kamman, M. *The New Making of a Cook.* New York: Morrow, 1997.

Keller, T., S. Heller, and M. Ruhlman. *The French Laundry Cookbook.* New York: Artisan, 1999.

Mariani, J. *The Dictionary of American Food and Drink.* New York: Hearst, 1994.

Robuchon, J. et al., eds. *Larousse gastronomique.* Paris: Larousse, 1996.

Steingarten, J. *It Must've Been Something I Ate.* New York: Knopf, 2002.

———. *The Man Who Ate Everything.* New York: Knopf, 1998.

Stobart, T. *The Cook's Encyclopedia.* London: Papermac, 1982.

Weinzweig, A. *Zingerman's Guide to Good Eating.* Boston: Houghton Mifflin, 2003.

Willan, A. *La Varenne Pratique.* New York: Crown, 1989.

Word Meanings and Etymologies

Battaglia, S., ed. *Grande dizionario della lingua italiana.* 21 vols. Turin: Unione tipografico-editrice torinese, 1961–2002.

Bloch, O. *Dictionnaire étymologique de la langue française.* 5th ed. Paris: Presses universitaires, 1968.

Oxford English Dictionary. 2nd ed. 20 vols. Oxford: Clarendon, 1989.

Watkins, C. *The American Heritage Dictionary of Indo-European Roots.* 2nd ed. Boston: Houghton Mifflin, 2000.

Writings on Food Science for the General Reader

Barham, P. *The Science of Cooking.* Berlin: Springer-Verlag, 2001.

Corriher, S. *CookWise.* New York: Morrow, 1997.

Kurti, N. The physicist in the kitchen. *Proceedings of the Royal Institution* 42 (1969): 451–67.

McGee, H. *The Curious Cook.* San Francisco: North Point, 1990.

This, H. *Révélations gastronomiques.* Paris: Belin, 1995.

This, H. *Les Secrets de la casserole.* Paris: Belin, 1993.

Books on Major Regional Cuisines

Achaya, K.T. *A Historical Dictionary of Indian Food.* New Delhi: Oxford Univ. Press, 1998.

———. *Indian Food: A Historical Companion.* Delhi: Oxford Univ. Press, 1994.

Anderson, E.N. *The Food of China.* New Haven: Yale Univ. Press, 1988.

Artusi, P. *La Scienza in cucina e l'arte di mangiar bene.* 1891 and later eds. Florence: Giunti Marzocco, 1960.
Bertolli, P. *Cooking by Hand.* New York: Clarkson Potter, 2003.
Bugialli, G. *The Fine Art of Italian Cooking.* New York: Times Books, 1977.
Chang, K.C., ed. *Food in Chinese Culture.* New Haven: Yale Univ. Press, 1977.
Cost, B. *Bruce Cost's Asian Ingredients.* New York: Morrow, 1988.
Ellison, J.A., ed. and trans. *The Great Scandinavian Cook Book.* New York: Crown, 1967.
Escoffier, A. *Guide Culinaire,* 1903 and later editions. Translated by H.L. Cracknell and R.J. Kaufmann as *Escoffier: The Complete Guide to the Art of Modern Cooking.* New York: Wiley, 1983.
Hazan, M. *Essentials of Classic Italian Cooking.* New York: Knopf, 1992.
Hosking, R. *A Dictionary of Japanese Food.* Boston: Tuttle, 1997.
Kennedy, D. *The Cuisines of Mexico.* New York: Harper and Row, 1972.
Lo, K. *The Encyclopedia of Chinese Cooking.* New York: Bristol Park Books, 1990.
Mesfin, D.J. *Exotic Ethiopian Cooking.* Falls Church, VA: Ethiopian Cookbook Enterprises, 1993.
Roden, C. *The New Book of Middle Eastern Food.* New York: Knopf, 2000.
St.-Ange, E. *La Bonne cuisine de Mme E. Saint-Ange.* Paris: Larousse, 1927.
Shaida, M. *The Legendary Cuisine of Persia.* Henley-on-Thames: Lieuse, 1992.
Simoons, F.J. *Food in China.* Boca Raton: CRC, 1991.
Toomre, J., trans. and ed. *Classic Russian Cooking: Elena Molokhovets'* A Gift to Young Housewives. Bloomington: Indiana Univ. Press, 1992.
Tsuji, S. *Japanese Cooking: A Simple Art.* Tokyo: Kodansha, 1980.

Books About the History of Food

Benporat, C. *Storia della gastronomia italiana.* Milan: Mursia, 1990.
Coe, S. *America's First Cuisines.* Austin: Univ. of Texas Press, 1994.
Dalby, A. *Siren Feasts: A History of Food and Gastronomy in Greece.* London: Routledge, 1996.
Darby, W.J. et al. *Food: The Gift of Osiris.* 2 vols. New York: Academic, 1977. Food in ancient Egypt.
Flandrin, J.L. *Chronique de Platine.* Paris: Odile Jacob, 1992.
Grigg, D.B. *The Agricultural Systems of the World: An Evolutionary Approach.* Cambridge: Cambridge Univ. Press, 1974.
Huang, H.T., and J. Needham. *Science and Civilisation in China.* Vol. 6, part V: *Fermentations and Food Science.* Cambridge: Cambridge Univ. Press, 2000.
Kiple, K.F., and K.C. Ornelas, eds. *The Cambridge World History of Food.* 2 vols. Cambridge: Cambridge Univ. Press, 2000.
Peterson, T.S. *Acquired Taste: The French Origins of Modern Cooking.* Ithaca: Cornell Univ. Press, 1994.
Redon, O. et al. *The Medieval Kitchen.* Trans. E. Schneider. Chicago: Univ. of Chicago Press, 1998.
Rodinson, M., A.J. Arberry, and C. Perry. *Medieval Arab Cookery.* Totnes, Devon: Prospect Books, 2001.
Scully, T. *The Art of Cookery in the Middle Ages.* Rochester, NY: Boydell, 1995.
Singer, C.E. et al. *A History of Technology.* 7 vols. Oxford: Clarendon, 1954–78.
Thibaut-Comelade, E. *La table médiévale des Catalans.* Montpellier: Presses du Languedoc, 2001.
Toussaint-Samat, M. *History of Food.* Trans. Anthea Bell. Oxford: Blackwell, 1992.
Trager, J. *The Food Chronology.* New York: Holt, 1995.
Wheaton, B.K. *Savoring the Past: The French Kitchen and Table from 1300 to 1789.* Philadelphia: Univ. of Penn. Press, 1983.
Wilson, C.A. *Food and Drink in Britain.* Harmondsworth: Penguin, 1984.

Historical Sources

Anthimus. *On the Observation of Foods.* Trans. M. Grant. Totnes, Devon: Prospect Books, 1996.
Apicius, M.G. *De re coquinaria: L'Art culinaire.* J. André, ed. Paris: C. Klincksieck, 1965. Edited and translated by B. Flower and E. Rosenbaum as *The Roman Cookery Book.* London: Harrap, 1958.
Brillat-Savarin, J. A. *La Physiologie du goût.* Paris, 1825. Translated by M.F.K. Fisher as *The Physiology of Taste.* New York: Harcourt Brace Jovanovich, 1978.
Cato, M.P. *On Agriculture.* Trans. W.D. Hooper. Cambridge, MA: Harvard Univ. Press, 1934.
Columella, L.J.M. *On Agriculture.* 3 vols. Trans. H.B. Ash. Cambridge, MA: Harvard Univ. Press, 1941–55.
Grewe, R. and C.B. Hieatt, eds. *Libellus De Arte Coquinaria.* Tempe, AZ: Arizona Center for Medieval and Renaissance Studies, 2001.
Hieatt, C.B. and S. Butler. *Curye on Inglysch.* London: Oxford Univ. Press, 1985.
La Varenne, F.P. de. *Le Cuisinier françois.* 1651. Reprint, Paris: Montalba, 1983.
Platina. *De honesta voluptate et valetudine.* Ed. and trans. by M.E. Milham as *On Right Pleasure and Good Health.* Tempe, AZ: Renaissance Soc. America, 1998.

Pliny the Elder. *Natural History.* 10 vols. Trans. H Rackham et al. Cambridge, MA: Harvard Univ. Press, 1938–62.
Scully, T., ed. and trans. *The Neapolitan Recipe Collection.* Ann Arbor: Univ. of Michigan Press, 2000.
———, ed. and trans. *The Viandier of Taillevent.* Ottawa: Univ. of Ottawa Press, 1988.
———, ed. and trans. *The Vivendier.* Totnes, Devon: Prospect Books, 1997.
Warner, R. *Antiquitates culinariae.* London: 1791; Reprint, London: Prospect Books, n.d.

Encyclopedias of Food Science and Technology (referred to below as "Caballero" and "Macrae")

Caballero, B. et al., eds. *Encyclopedia of Food Sciences and Nutrition.* 10 vols. Amsterdam: Academic, 2003. [2nd ed. of Macrae et al.]
Macrae, R. et al., eds. *Encyclopaedia of Food Science, Food Technology, and Nutrition.* 8 vols. London: Academic, 1993.

Comprehensive Writings on Food Chemistry, Microbiology, Botany, Physiology

Ang, C.Y.W. et al., eds. *Asian Foods: Science and Technology.* Lancaster, PA: Technomic, 1999.
Ashurst, P.R. *Food Flavorings.* Gaithersburg, MD: Aspen, 1999.
Belitz, H.D., and W. Grosch. *Food Chemistry.* 2nd English ed. Berlin: Springer, 1999.
Campbell-Platt, G. *Fermented Foods of the World.* London: Butterworth, 1987.
Charley, H. *Food Science.* 2nd ed. New York: Wiley, 1982.
Coultate, T.P. *Food: The Chemistry of Its Components.* 2nd ed. Cambridge: Royal Society of Chemistry, 1989.
Doyle, M.P. et al., eds. *Food Microbiology.* 2nd ed. Washington, DC: American Society of Microbiology, 2001.
Facciola, S. *Cornucopia II: A Source Book of Edible Plants.* Vista, CA: Kampong, 1998.
Fennema, O., ed. *Food Chemistry.* 3rd ed. New York: Dekker, 1996.
Ho, C.T. et al. Flavor chemistry of Chinese foods. *Food Reviews International* 5 (1989): 253–87.
Maarse, H., ed. *Volatile Compounds in Foods and Beverages.* New York: Dekker, 1991.
Maincent, M. *Technologie culinaire.* Paris: BPI, 1995.
Paul, P.C., and H.H. Palmer, eds. *Food Theory and Applications.* New York: Wiley, 1972.
Penfield, M.P., and A.M. Campbell. *Experimental Food Science.* 3rd ed. San Diego, CA: Academic, 1990.
Silverthorn, D.U. et al. *Human Physiology.* Upper Saddle River, NJ: Prentice Hall, 2001.
Smartt, J., and N. W. Simmonds, eds. *Evolution of Crop Plants.* 2nd ed. Harlow, Essex: Longman, 1995.
Steinkraus, K.H., ed. *Handbook of Indigenous Fermented Foods.* 2nd ed. New York: Dekker, 1996.

CHAPTER 1: MILK AND DAIRY PRODUCTS

Brown, N.W. *India and Indology.* Delhi: Motilal Banarsidass, 1978.
Brunet, P., ed. *Histoire et géographie des fromages.* Caen: Université de Caen, 1987.
Calvino, I. *Mr. Palomar.* Trans. W. Weaver. San Diego. CA: Harcourt Brace Jovanovich, 1985.
Grant, A.J., trans. *Early Lives of Charlemagne.* London: Chatto and Windus, 1922.
Macdonnell, A.A. *A Vedic Reader for Students.* Oxford: Oxford Univ. Press, 1917.
Masui, K., and T. Yamada. *French Cheeses.* New York: Dorling Kindersley, 1996.
O'Flaherty, W.D., ed. and trans. *The Rig Veda.* Harmondsworth: Penguin, 1981.
Polo, M. *Travels* (ca. 1300). Trans. W. Marsden. New York: Dutton, 1908.
Rance, P. *The French Cheese Book.* London: Macmillan, 1989.
———. *The Great British Cheese Book.* London: Macmillan, 1982.

Blackburn, D.G. et al. The origins of lactation and the evolution of milk. *Mammal Review* 19 (1989): 1–26.
Bodyfelt, F.W. et al. *The Sensory Evaluation of Dairy Products.* New York: Van Nostrand Reinhold, 1988.
Buchin, S. et al. Influence of pasteurization and fat composition of milk on the volatile compounds and flavor characteristics of a semi-hard cheese. *J Dairy Sci.* 81 (1998): 3097–108.
Curioni, P.M.G., and J.O. Bosset. Key odorants in various cheese types as determined by gas chromatography-olfactometry. *International Dairy J* 12 (2002): 959–84.
Dupont, J., and P.J. White. "Margarine." In Macrae, 2880–95.
Durham, W. H. *Coevolution: Genes, Culture, and Human Diversity.* Stanford, CA: Stanford Univ. Press, 1991.
Fox, P.F., ed. *Cheese: Chemistry, Physics, Microbiology.* 2 vols. London: Elsevier, 1987.
Garg, S.K., and B.N. Johri. Rennet: Current trends and future research. *Food Reviews International* 10 (1994): 313–55.
Gunderson, H.L. *Mammalogy.* New York: McGraw-Hill, 1976.
Jensen, R.G., ed. *Handbook of Milk Composition.* San Diego, CA: Academic, 1995.

Juskevich, J.C., and C.G. Guyer. Bovine growth hormone: Human food safety evaluation. *Science* 249 (1990): 875–84.
Kosikowski, F.V., and V.V. Mistry. *Cheese and Fermented Milk Foods.* 3rd ed. Westport, CT: F.V. Kosikowski LLC, 1997.
Kurmann, J.A. et al. *Encyclopedia of Fermented Fresh Milk Products.* New York: Van Nostrand Reinhold, 1992.
Mahias, M.C. Milk and its transmutations in Indian society. *Food and Foodways* 2 (1988): 265–88.
Marshall, R.T., and W.S. Arbuckle. *Ice Cream.* 5th ed. New York: Chapman and Hall, 1996.
Miller, M.J.S. et al. Casein: A milk protein with diverse biologic consequences. *Proc Society Experimental Biol Medicine* 195 (1990): 143–59.
Muhlbauer, R.C. et al. Various selected vegetables, fruits, mushrooms and red wine residue inhibit bone resorption in rats. *J Nutrition* 133 (2003): 3592–97.
Queiroz Macedo, I. et al. Caseinolytic specificity of cardosin, an aspartic protease from the cardoon: Action on bovine casein and comparison with chymosin. *J Agric Food Chem.* 44 (1996): 42–47.
Reid, G. et al. Potential uses of probiotics in clinical practice. *Clinical and Microbiological Reviews* 16 (2003): 658–72.
Robinson, R.K., ed. *Modern Dairy Technology.* 2 vols. London: Chapman and Hall, 1993.
Schmidt, G.H. et al. *Principles of Dairy Science.* 2nd ed. Englewood Cliffs, NJ: Prentice Hall, 1988.
Scott, R. *Cheesemaking Practice.* London: Applied Science, 1981.
Stanley, D.W. et al. Texture-structure relationships in foamed dairy emulsions. *Food Research International* 29 (1996): 1–13.
Starr, M.P. et al., eds. *The Prokaryotes: A Handbook on Habitats, Isolation, and Identification of Bacteria.* 2 vols. Berlin: Springer-Verlag, 1981.
Stini, W.A. Osteoporosis in biocultural perspective. *Annual Reviews of Anthropology* 24 (1995): 397–421.
Suarez, F.L. et al. Diet, genetics, and lactose intolerance. *Food Technology* 51 (1997): 74–76.
Tamime, A.Y., and R.K. Robinson. *Yoghurt: Science and Technology.* 2nd ed. Cambridge, UK: Woodhead, 1999.
Virgili, R. et al. Sensory-chemical relationships in Parmigiano-reggiano cheese. *Lebensmittel-Wissenschaft und Technologie* 27 (1994): 491–95.
The Water Buffalo. Rome: U.N. Food and Agriculture Organization, 1977.
Wheelock, V. *Raw Milk and Cheese Production: A Critical Evaluation of Scientific Research.* Skipton, UK: V. Wheelock Associates, 1997.

CHAPTER 2: EGGS

Davidson, A., J. Davidson, and J. Lang. Origin of crême brulée. *Petits propos culinaires* 31 (1989): 61–63.
Healy, B., and P. Bugat. *The French Cookie Book.* New York: Morrow, 1994.
Hume, R.E. *The Thirteen Principal Upanishads Translated from the Sanskrit.* Oxford: Oxford Univ. Press, 1921.
Radhakrishnan, S. *The Principal Upanisads.* Atlantic Highlands, NJ: Humanities, 1992.
Smith, P., and C. Daniel. *The Chicken Book.* Boston: Little Brown, 1975.
Wolfert, P. *Couscous and Other Good Foods from Morocco.* New York: Harper and Row, 1973.

Board, R.G., and R. Fuller, eds. *Microbiology of the Avian Egg.* London: Chapman and Hall, 1994.
Burley, R.W., and D.V. Vadehra. *The Avian Egg: Chemistry and Biology.* New York: Wiley, 1989.
Chang, C.M. et al. Microstructure of egg yolk. *J Food Sci.* 42 (1977): 1193–1200.
Gosset, P.O., and R.C. Baker. Prevention of gray-green discoloration in cooked liquid whole eggs. *J Food Sci.* 46 (1981): 328–31.
Jänicke, O. *Zur Verbreitungsgeschichte und Etymologie des fr.* meringue. *Zeitschrift für romanischen Philologie* 84 (1968): 558–71.
Jiang, Y. et al. Egg phosphatidylcholine decreases the lymphatic absorption of cholesterol in rats. *J Nutrition* 131 (2001): 2358–63.
Maga, J.A. Egg and egg product flavor. *J Agric Food Chem.* 30 (1982): 9–14.
McGee, H. On long-cooked eggs. *Petits propos culinaires* 50 (1995): 46–50.
McGee, H. J., S.R. Long, and W.R. Briggs. Why whip egg whites in copper bowls? *Nature* 308 (1984): 667–68.
Packard, G.C., and M.J. Packard. Evolution of the cleidoic egg among reptilian ancestors of birds. *American Zoologist* 20 (1980): 351–62.
Perry, M.M., and A.B. Gilbert. The structure of yellow yolk in the domestic fowl. *J Ultrastructural Res.* 90 (1985): 313–22.
Stadelman, W.J., and O.J. Cotterill. *Egg Science and Technology.* 3rd ed. Westport, CT: AVI, 1986.
Su, H.P., and C.W. Lin. A new process for preparing transparent alkalised duck egg and its quality. *J Sci Food Agric.* 61 (1993): 117–20.
Wang, J., and D.Y.C. Fung. Alkaline-fermented foods: A review with emphasis on pidan fermentation. *CRC Critical Revs in Microbiology* 22 (1996): 101–38.
Wilson, A.J., ed. *Foams: Physics, Chemistry and Structure.* London: Springer-Verlag, 1989.
Woodward, S.A., and O.J. Cotterill. Texture and

microstructure of cooked whole egg yolks and heat-formed gels of stirred egg yolk. *J Food Sci.* 52 (1987): 63–67.

———. Texture profile analysis, expressed serum, and microstructure of heat-formed egg yolk gels. *J Food Sci.* 52 (1987): 68–74.

CHAPTER 3: MEAT

Cronon, W. *Nature's Metropolis.* New York: Norton, 1991.

Kinsella, J., and D.T. Harvey. *Professional Charcuterie.* New York: Wiley, 1996.

Paillat, M., ed. *Le Mangeur et l'animal.* Paris: Autrement, 1997.

Rhodes, V.J. How the marking of beef grades was obtained. *J Farm Economics* 42 (1960): 133–49.

Serventi, S. *La grande histoire du foie gras.* Paris: Flammarion, 1993.

Woodard, A. et al. *Commercial and Ornamental Game Bird Breeders Handbook.* Surrey, BC: Hancock House, 1993.

Abs, M., ed. *Physiology and Behavior of the Pigeon.* London: Academic, 1983.

Ahn, D.U., and A.J. Maurer. Poultry meat color: Heme-complex-forming ligands and color of cooked turkey breast meat. *Poultry Science* 69 (1990): 1769–74.

Bailey, A.J., ed. *Recent Advances in the Chemistry of Meat.* London: Royal Society of Chemistry, 1984.

Bechtel, P.J., ed. *Muscle as Food.* Orlando, FL: Academic, 1986.

Campbell-Platt, G., and P.E. Cook, eds. *Fermented Meats.* London: Blackie, 1995.

Carrapiso, A.I. et al. Characterization of the most odor-active compounds of Iberian ham headspace. *J Agric Food Chem.* 50 (2002): 1996–2000.

Cornforth, D.P. et al. Carbon monoxide, nitric oxide, and nitrogen dioxide levels in gas ovens related to surface pinking of cooked beef and turkey. *J Agric Food Chem.* 46 (1998): 255–61.

Food Standards Agency, U.K. *Review of BSE Controls.* 2000, http://www.bsereview.org.uk.

Gault, N.F.S., "Marinaded meat." In *Developments in Meat Science*, edited by R. Lawrie, 5, 191–246. London: Applied Science, 1991.

Jones, K.W., and R.W. Mandigo. Effects of chopping temperature on the microstructure of meat emulsions. *J Food Sci.* 47 (1982): 1930–35.

Lawrie, R.A. *Meat Science.* 5th ed. Oxford: Pergamon, 1991.

Lijinsky, W. N-nitroso compounds in the diet. *Mutation Research* 443 (1999): 129–38.

Maga, J.A. *Smoke in Food Processing.* Boca Raton, FL: CRC, 1988.

———. Pink discoloration in cooked white meat. *Food Reviews International* 10 (1994): 273–386.

Mason, I.L., ed. *Evolution of Domesticated Animals.* London: Longman, 1984.

McGee, H., J. McInerny, and A. Harrus. The virtual cook: Modeling heat transfer in the kitchen. *Physics Today* (November 1999): 30–36.

Melton, S. Effects of feeds on flavor of red meat: A review. *J Animal Sci.* 68 (1990): 4421–35.

Milton, K. A hypothesis to explain the role of meat-eating in human evolution. *Evolutionary Anthropology* 8 (1999): 11–21.

Morgan Jones, S.D., ed. *Quality Grading of Carcasses of Meat Animals.* Boca Raton, FL: CRC, 1995.

Morita, H. et al. Red pigment of Parma ham and bacterial influence on its formation. *J Food Sci.* 61 (1996): 1021–23.

Oreskovich, D.C. et al. Marinade pH affects textural properties of beef. *J Food Sci.* 57 (1992): 305–11.

Pearson, A.M., and T.R. Dutson. *Edible Meat Byproducts.* London: Elsevier, 1988.

Pinotti, A. et al. Diffusion of nitrite and nitrate salts in pork tissue in the presence of sodium chloride. *J Food Sci.* 67 (2002): 2165–71.

Rosser, B.W.C., and J.C. George. The avian pectoralis: Biochemical characterization and distribution of muscle fiber types. *Canadian J Zoology* 64 (1986): 1174–85.

Rousset-Akrim, S. et al. Influence of preparation on sensory characteristics and fat cooking loss of goose foie gras. *Sciences des aliments* 15 (1995): 151–65.

Salichon, M.R. et al. Composition des 3 types de foie gras: Oie, canard mulard et canard de barbarie. *Annales Zootechnologie* 43 (1994): 213–20.

Saveur, B. Les critères et facteurs de la qualité des poulets Label Rouge. *INRA Productions Animales* 10 (1997): 219–26.

Skog, K.I. et al. Carcinogenic heterocyclic amines in model systems and cooked foods: A review on formation, occurrence, and intake. *Food and Chemical Toxicology* 36 (1998): 879–96.

Solyakov, A. et al. Heterocyclic amines in process flavours, process flavour ingredients, bouillon concentrates and a pan residue. *Food and Chemical Toxicology* 37 (1999): 1–11.

Suzuki, A. et al. Distribution of myofiber types in thigh muscles of chickens. *Journal of Morphology* 185 (1985): 145–54.

Varnam, A.H., and J.P. Sutherland. *Meat and Meat Products: Technology, Chemistry, and Microbiology.* London: Chapman and Hall, 1995.

Wilding, P. et al. Salt-induced swelling of meat. *Meat Science* 18 (1986): 55–75.

Wilson, D.E. et al. Relationship between chemical

percentage intramuscular fat and USDA marbling score. A.S. Leaflet R1529. Iowa State University: 1998.
Young, O.A. et al. Pastoral and species flavour in lambs raised on pasture, lucerne or maize. *J Sci Food Agric.* 83 (2003): 93–104.

CHAPTER 4: FISH AND SHELLFISH

Alejandro, R. *The Philippine Cookbook.* New York: Putnam, 1982.
Bliss, D. *Shrimps, Lobsters, and Crabs.* New York: Columbia Univ. Press, 1982.
Davidson, A. *Mediterranean Seafood.* 2nd ed. London: Allan Lane, 1981.
———. *North Atlantic Seafood.* New York: Viking, 1979.
Kurlansky, M. *Cod.* New York: Walker, 1997.
McClane, A.J. *The Encyclopedia of Fish Cookery.* New York: Holt Rinehart Winston, 1977.
McGee, H. "The buoyant, slippery lipids of the snake mackerels and orange roughy." In *Fish: Foods from the Waters*, edited by H. Walker, 205–9. Totnes, UK: Prospect Books, 1998.
Peterson, J. *Fish and Shellfish.* New York: Morrow, 1996.
Riddervold, A. *Lutefisk, Rakefisk and Herring in Norwegian Tradition.* Oslo: Novus, 1990.

Ahmed, F.E. Review: Assessing and managing risk due to consumption of seafood contaminated with microorganisms, parasites, and natural toxins in the US. *Int J Food Sci. and Technology* 27 (1992): 243–60.
Borgstrom, G., ed. *Fish as Food.* 4 vols. New York: Academic, 1961–65.
Chambers, E., and A. Robel. Sensory characteristics of selected species of freshwater fish in retail distribution. *J Food Sci.* 58 (1993): 508–12.
Chattopadhyay, P. et al. "Fish." In Macrae, 1826–87.
Doré, I. *Fish and Shellfish Quality Assessment.* New York: Van Nostrand Reinhold, 1991.
Flick, G.J., and R.E. Martin, eds. *Advances in Seafood Biochemistry.* Lancaster, PA: Technomic, 1992.
Funk, C.D. Prostaglandins and leukotrienes: Advances in eicosanoid biology. *Science* 294 (2001): 1871–75.
Gomez-Guillen, M.C. et al. Autolysis and protease inhibition effects on dynamic viscoelastic properties during thermal gelation of squid muscle. *J Food Sci.* 67 (2002): 2491–96.
Gosling, E. *The Mussel Mytilus: Ecology, Physiology, Genetics and Culture.* Amsterdam: Elsevier, 1992.
Haard, N.F., and B.K. Simon. *Seafood Enzymes.* New York: Dekker, 2000.
Hall, G.M., ed. *Fish Processing Technology.* 2nd ed. New York: VCH, 1992.
Halstead, B.W. *Poisonous and Venomous Marine Animals of the World.* 2nd rev. ed. Princeton, NJ: Darwin, 1988.
Hatae, K. et al. Role of muscle fibers in contributing firmness of cooked fish. *J Food Sci.* 55 (1990): 693–96.
Iversen, E.S. et al. *Shrimp Capture and Culture Fisheries of the United States.* Cambridge, MA: Fishing News, 1993.
Jones, D.A. et al. "Shellfish." In Macrae, 4084–118.
Kobayashi, T. et al. Strictly anaerobic halophiles isolated from canned Swedish fermented herring. *International J Food Microbiology* 54 (2000): 81–89.
Korringa, P. *Farming the Cupped Oysters of the Genus Crassostrea.* Amsterdam: Elsevier, 1976.
Kugino, M., and K. Kugino. Microstructural and rheological properties of cooked squid mantle. *J Food Sci.* 59 (1994): 792–96.
Lindsay, R. "Flavour of Fish." In *Seafoods: Chemistry, Processing, Technology, and Quality,* edited by F. Shahidi and J.R. Botta, 74–84. London: Blackie, 1994.
Love, R.M. *The Food Fishes: Their Intrinsic Variation and Practical Implications.* London: Farrand, 1988.
Mantel, L.H., ed. *Biology of Crustacea.* Vol. 5, *Internal Anatomy and Physiological Regulation*; vol. 9, *Integument, Pigments, and Hormonal Processes.* New York: Academic, 1983; Orlando, FL: Academic, 1985.
Martin, R.E. et al., eds. *Chemistry and Biochemistry of Marine Food Products.* Westport, CT: AVI, 1982.
Morita, K. et al. Comparison of aroma characteristics of 16 fish species by sensory evaluation and gas chromatographic analysis. *J Sci Food Agric.* 83 (2003): 289–97.
Moyle, P.B., and J.J. Cech. *Fishes: An Introduction to Ichthyology.* 4th ed. Upper Saddle River, NJ: Prentice Hall, 2000.
Nelson, J.S. *Fishes of the World.* 3rd ed. New York: Wiley, 1994
Ò Foighil, D. et al. Mitochondrial cytochrome oxidase I gene sequences support an Asian origin for the Portuguese oyster *Crassostrea angulata. Marine Biology* 131 (1998): 497–503.
Ofstad, R. et al. Liquid holding capacity and structural changes during heating of fish muscle. *Food Microstructure* 12 (1993): 163–74.
Oshima, T. Anisakiasis: Is the sushi bar guilty? *Parasitology Today* 3 (2) (1987): 44–48.
Pennarun, A.L. et al. Identification and origin of the character-impact compounds of raw oyster *Crassostrea gigas. J Sci Food Agric.* 82 (2002): 1652–60.
Royce, W.F. *Introduction to the Practice of Fishery Science.* San Diego, CA: Academic, 1994.
Shimizu, Y. et al. Species variation in the gel-forming [and disintegrating] characteristics of fish

meat paste. *Bulletin Jap Soc Scientific Fisheries* 47 (1981): 95–104.
Shumway, S. E., ed. *Scallops: Biology, Ecology, and Aquaculture*. Amsterdam: Elsevier, 1991.
Sikorski, Z.E. et al., eds. *Seafood Proteins*. New York: Chapman and Hall, 1994.
Sternin, V., and I. Doré. *Caviar: The Resource Book*. Moscow and Stanwood, WA: Cultura, 1993.
Tanikawa, E. *Marine Products in Japan*. Tokyo: Koseisha-Koseikaku, 1971.
Taylor, R.G. et al. Salmon fillet texture is determined by myofiber-myofiber and myofiber-myocommata attachment. *J Food Sci.* 67 (2002): 2067–71.
Triqui, R., and G.A. Reineccius. Flavor development in the ripening of anchovy. *J Agric Food Chem.* 43 (1995): 453–58.
Ward, D. R., and C. Hackney. *Microbiology of Marine Food Products*. New York: Van Nostrand Reinhold, 1991.
Whitfield, F.B. Flavour of prawns and lobsters. *Food Reviews International* 6 (1990): 505–19.
Wilbur, K.M., ed. *The Mollusca*. 12 vols. New York: Academic, 1983.

CHAPTER 5: EDIBLE PLANTS

Harlan, J.R. *Crops and Man*. Madison, WI: Am. Soc. Agronomy, 1992.
Heiser, C.B. *Seed to Civilization*. Cambridge, MA: Harvard Univ. Press, 1990.
Thoreau, H.D. "Wild Apples" (1862). In H.D. Thoreau, *Wild Apples and Other Natural History Essays*, ed. W. Rossi. Athens, GA: Univ. of Georgia Press, 2002.
Wilson, C. A. *The Book of Marmalade*. New York: St. Martin's, 1985.

Bidlack, W.R. et al., eds. *Phytochemicals: A New Paradigm*. Lancaster, PA: Technomic, 1998.
Borchers, A.T. et al. Mushrooms, tumors, and immunity. *Proc Society Experimental Biol Medicine* 221 (1999): 281–93.
Buchanan, B.B. et al., eds. *Biochemistry and Molecular Biology of Plants*. Rockville, MD: Am. Society of Plant Physiologists, 2000.
Coulombe, R.A. "Toxicants, natural." In *Wiley Encyclopedia of Food Science and Technology*. Edited by F.J. Francis, 2nd ed., 4 vols, 2336–54. New York: Wiley, 2000.
Daschel, M.A. et al. Microbial ecology of fermenting plant materials. *FEMS Microbiological Revs.* 46 (1987): 357–67.
Dewanto, V. et al. Thermal processing enhances the nutritional value of tomatoes by increasing total antioxidant activity. *J Agric Food Chem.* 50 (2002): 3010–14.
Dominy, N.J., and P.W. Lucas. Importance of trichromic vision to primates. *Nature* 410 (2001): 363–66.
Elson, C.E. et al. Isoprenoid-mediated inhibition of mevalonate synthesis: Potential application to cancer. *Proc Society Experimental Biol Medicine* 221 (1999): 294–305.
Francis, F.J. Anthocyanins and betalains: Composition and applications. *Cereal Foods World* 45 (2000): 208–13.
Gross, J. *Pigments in Vegetables: Chlorophylls and Carotenoids*. New York: Van Nostrand Reinhold, 1991.
Karlson-Stiber, C., and H. Persson. Cytotoxic fungi: an overview. *Toxicon* 42 (2003): 339–49.
Larsen, C.S. Biological changes in human populations with agriculture. *Annual Reviews Anthropology* 24 (1995): 185–213.
Luck, G. et al. Polyphenols, astringency, and proline-rich proteins. *Phytochemistry* 37 (1994): 357–71.
Muhlbauer, R.C. et al. Various selected vegetables, fruits, mushrooms and red wine residue inhibit bone resorption in rats. *J Nutrition* 133 (2003): 3592–97.
Santos-Buelga, C., and A. Scalbert. Proanthocyanidins and tannin-like compounds—nature, occurrence, dietary intake and effects on nutrition and health. *J Sci Food Agric.* 80 (2000): 1094–1117.
Smith, D., and D. O'Beirne. "Jams and preserves." In Macrae, 2612–21.
Vincent, J.E.V. Fracture properties of plants. *Advances in Botanical Research* 17 (1990): 235–87.
Vinson, J.A. et al. Phenol antioxidant quantity and quality in foods: Vegetables. *J Agric Food Chem.* 46 (1998): 3630–34.
———. Phenol antioxidant quantity and quality in foods: Fruits. *J Agric Food Chem.* 49 (2001): 5315–21.
Walter, R.H., ed. *The Chemistry and Technology of Pectin*. San Diego, CA: Academic, 1991.
Tomás-Barberán, F.A., and R.J. Robins, eds. *Phytochemistry of Fruit and Vegetables*. New York: Oxford Univ. Press, 1997.
Waldron, K.W. et al. New approaches to understanding and controlling cell separation in relation to fruit and vegetable texture. *Trends Food Sci Technology* 8 (1997): 213–21.

CHAPTER 6: A SURVEY OF COMMON VEGETABLES

Arora, D. *Mushrooms Demystified*. 2nd ed. Berkeley, CA: Ten Speed, 1986.
Chapman, V.J. *Seaweeds and Their Uses*. 3rd ed. New York: Chapman and Hall, 1980.
Dunlop, F. *Land of Plenty*. New York: Morrow, 2003.
Fortner, H.J. *The Limu Eater: A Cookbook of Hawaiian Seafood*. Honolulu: Univ. of Hawaii, 1978.

Olivier, J.M. et al. *Truffe et trufficulture.* Perigueux: FANLAC, 1996.

Phillips, R., and M. Rix. *The Random House Book of Vegetables.* New York: Random House, 1993.

Schneider, E. *Uncommon Fruits and Vegetables.* New York: Harper and Row, 1986.

———. *Vegetables from Amaranth to Zucchini.* New York: Morrow, 2001.

Alasalvar, C. et al. Comparison of volatiles . . . and sensory quality of different colored carrot varieties. *J Agric Food Chem.* 49 (2001): 1410–16.

Andersson, A. et al. Effect of preheating on potato texture. *CRC Critical Revs Food Sci Nutrition* 34 (1994): 229–51.

Aparicio, R. et al., "Biochemistry and chemistry of volatile compounds affecting consumers' attitudes towards virgin olive oil." In *Flavour and Fragrance Chemistry,* edited by V. Lanzotti and O. Tagliatela-Scarfati, 3–14. Amsterdam: Kluwer, 2000.

Bates, D.M. et. al., eds. *Biology and Utilization of the Cucurbitaceae.* Ithaca, NY: Comstock, 1990.

Block, E. Organosulfur chemistry of the genus *Allium. Angewandte Chemie,* International Edition 31 (1992): 1135–78.

Buttery, R.G. et al. Studies on flavor volatiles of some sweet corn products. *J Agric Food Chem.* 42 (1994): 791–95.

Duckham, S.C. et al. Effect of cultivar and storage time in the volatile flavor components of baked potato. *J Agric Food Chem.* 50 (2002): 5640–48.

Fenwick, G.R., and A.B. Hanley. The genus *Allium. CRC Critical Reviews in Food Sci Nutrition* 22 (1985): 199–271, 273–377.

Fukomoto, L.R. et al. Effect of wash water temperature and chlorination on phenolic metabolism and browning of stored iceberg lettuce photosynthetic and vascular tissues. *J Agric Food Chem.* 50 (2002): 4503–11.

Gomez-Campo, C., ed. *Biology of Brassica Coenospecies.* Amsterdam: Elsevier, 1999.

Heywood, V.H. Relationships and evolution in the *Daucus carota* complex. *Israel J Botany* 32 (1983): 51–65.

Hurtado, M.C. et al. Changes in cell wall pectins accompanying tomato paste manufacture. *J Agric Food Chem.* 50 (2002): 273–78.

Jirovetz, L. et al. Aroma compound analysis of *Eruca sativa* SPME headspace leaf samples using GC, GC-MS, and olfactometry. *J Agric Food Chem.* 50 (2002): 4643–46.

Kozukue, N., and M. Friedman. Tomatine, chlorophyll, ß-carotene and lycopene content in tomatoes during growth and maturation. *J Sci Food Agric.* 83 (2003): 195–200.

Lipton, W.J. Postharvest biology of fresh asparagus. *Horticultural Reviews* 12 (1990): 69–155.

Lu, Z. et al. Effects of fruit size on fresh cucumber composition . . . . *J Food Sci.* 67 (2002): 2934–39.

Mau, J.-L. et al. 1-octen-3-ol in the cultivated mushroom . . . *J Food Sci.* 57 (1992): 704–6.

McDonald, R.E. et al. Bagging chopped lettuce in selected permeability films. *HortScience* 25 (1990): 671–73.

Mithen, R.F. et al. The nutritional significance, biosynthesis and bioavailability of glucosinolates in human foods. *J Sci Food Agric.* 80 (2000): 967–84.

Mottur, G.P. A scientific look at potato chips. *Cereal Foods World* 34 (1989): 620–26.

Noble, P.S., ed. *Cacti: Biology and Uses.* Berkeley: Univ. of Calif. Press, 2001.

Oruna-Concha, M.J. et al. Comparison of the volatile components of two cultivars of potato cooked by boiling, conventional baking, and microwave baking. *J Sci Food Agric.* 82 (2002): 1080–87.

Petersen, M.A. et al. Identification of compounds contributing to boiled potato off-flavour (POF). *Lebensmittel-Wissenschaft und Technologie* 32 (1999): 32–39.

Pacioni, G. et al. Insects attracted by Tuber: A chemical explanation. *Mycological Res.* 95 (1991): 1359–63.

Rodger, G. Mycoprotein—a meat alternative new to the U.S. *Food Technology* 55 (7) (2001): 36–41.

Rouseff, R.L., ed. *Bitterness in Foods and Beverages.* Amsterdam: Elsevier, 1990.

Smith, D.S. et al. *Processing Vegetables: Science and Technology.* Lancaster, PA: Technomic, 1997.

Suarez, F. et al. Difference of mouth versus gut as site of origin of odiferous breath gases after garlic ingestion. *American J Physiology* 276 (1999): G425–30.

Takahashi, H. et al. Identification of volatile compounds of kombu and their odor description. *Nippon Shokuhin Kagaku Kaishi* 49 (2002): 228–37.

Talou, T. et al. "Flavor profiling of 12 edible European truffles." In *Food Flavors and Chemistry,* edited by A.M. Spanier et al. London: Royal Society of Chemistry, 2000.

Tanikawa, E. *Marine Products in Japan.* Tokyo: Koseisha-Koseikaku, 1971.

Terrell, E.E., and L.R. Batra. *Zizania latifolia* and *Ustilago esculenta,* a grass-fungus association. *Economic Botany* 36 (1982): 274–85.

Valverde, M.E. et al. Huitlacoche as a food source—biology, composition, and production. *CRC Critical Revs Food Sci Nutrition* 35 (1995): 191–229.

Van Buren, J.P. et al. Effects of salts and pH on

heating-related softening of snap beans. *J Food Sci.* 55 (1990): 1312–14.

Walter, W.M. Effect of curing on sensory properties and carbohydrate composition of baked sweet potato. *J Food Sci.* 52 (1987): 1026–29.

CHAPTER 7: A SURVEY OF COMMON FRUITS

Foust, C.W. *Rhubarb.* Princeton, NJ: Princeton Univ. Press, 1992.

Grigson, J. *Jane Grigson's Fruit Book.* New York: Atheneum, 1982.

Morgan, J., and A. Richards. *The Book of Apples.* London: Ebury, 1993.

Saunt, J. *Citrus Varieties of the World.* Norwich, UK: Sinclair, 1990.

Schneider, E. *Uncommon Fruits and Vegetables.* New York: Harper and Row, 1986.

Arnold, J. Watermelon packs a powerful lycopene punch. *Agricultural Research* (June 2002): 12–13.

Arthey, D., and P.R. Ashurst. *Fruit Processing.* 2nd ed. Gaithersburg, MD: Aspen, 2001.

Buettner, A., and P. Schieberle. Evaluation of aroma differences between hand-squeezed juices from Valencia late and navel oranges . . . . *J Agric Food Chem.* 49 (2001): 2387–94.

Dawson, D. M. et al. Cell wall changes in nectarines. *Plant Physiology* 100 (1992): 1203–10.

Hulme, A.C., ed. *The Biochemistry of Fruits and Their Products.* 2 vols. London: Academic, 1970–71.

Janick, J., and J.N. Moore, eds. *Advances in Fruit Breeding.* West Lafayette, IN: Purdue Univ. Press, 1975.

Lamikanra, O., and O.A. Richard. Effect of storage on some volatile aroma compounds in fresh-cut cantaloupe melon. *J Agric Food Chem.* 50 (2002): 4043–47.

Lota, M.L. et al. Volatile components of peel and leaf oils of lemon and lime species. *J Agric Food Chem.* 50 (2002): 796–805.

Mithra, S.K. *Postharvest Physiology and Storage of Tropical and Subtropical Fruits.* Wallingford, UK: CAB, 1997.

Morton, I.D., and A.J. Macleod, eds. *Food Flavours C: Flavours of Fruits.* Amsterdam: Elsevier, 1990.

Nagy, S. et al., eds. *Fruits of Tropical and Subtropical Origin.* Lake Alfred, FL: Florida Science Source, 1990.

Somogyi, L.P. et al. *Processing Fruits: Science and Technology.* Vol 1. Lancaster, PA: Technomic, 1996.

Wilhelm, S. The garden strawberry: A study of its origin. *American Scientist* 62 (1974): 264–71.

Wyllie, S.G. et al. "Key aroma compounds in melons." In *Fruit Flavors,* edited by R.L. Rouseff and M.M. Leahy, 248–57. Washington, DC: American Chemical Society, 1995.

CHAPTER 8: FLAVORINGS FROM PLANTS

Dalby, A. *Dangerous Tastes: The Story of Spices.* Berkeley: Univ. of Calif. Press, 2000.

Knox, K. and J.S. Huffaker. *Coffee Basics.* New York: Wiley, 1997.

Koran. Trans. N.J. Dawood. London: Penguin, 1974.

Kummer, C. *The Joy of Coffee.* Shelburne, VT: Chapters, 1995.

Man, R., and R. Weir. *The Compleat Mustard.* London: Constable, 1988.

Ortiz, E.L. *The Encyclopedia of Herbs, Spices, and Flavorings.* New York: Dorling Kindersley, 1992.

Peterson, T.S. *Acquired Taste: The French Origins of Modern Cooking.* Ithaca: Cornell Univ. Press, 1994.

Staples, G. *Ethnic Culinary Herbs: A Guide to Identification and Cultivation in Hawaii.* Honolulu: Univ. of Hawaii Press, 1999.

Stobart, T. *Herbs, Spices, and Flavorings.* Woodstock, NY: Overlook, 1982.

Bryant, B.P., and I. Mezine. Alkylamides that produce tingling paraesthesia activate tactile and thermal trigeminal neurons. *Brain Research* 842 (1999): 452–60.

Caterina, M.J., and D. Julius. The vanilloid receptor. *Annual Rev Neuroscience* 24 (2001): 487–517.

Chadwick, C.I. et al. The botany, uses, and production of *Wasabia japonica. Economic Botany* 47 (1993): 113–35.

Charalambous, G., ed. *Spices, Herbs, and Edible Fungi.* Amsterdam: Elsevier, 1994.

Charles, D.J. et al. "Essential oil content and chemical composition of finocchio fennel." In *New Crops,* edited by J. Janick and J.E. Simon, 570–73. New York: Wiley, 1993.

Clarke, R.J., and O.G. Vizthum. *Coffee: Recent Developments.* Oxford: Blackwell, 2001.

Clarke, R.J., and R. Macrae, eds. *Coffee.* 6 vols. Vol. 2: Technology. London: Elsevier, 1985.

Dalla Rosa, M. et al. Changes in coffee brews in relation to storage temperature. *J Sci Food Agric.* 50 (1990): 227–35.

del Castillo, M.D. et al. Effect of roasting on the antioxidant activity of coffee brews. *J Agric Food Chem.* 50 (2002): 3698–703.

Dignum, M.J.W. et al. Vanilla production. *Food Revs International* 17 (2001): 199–219.

Hiltunen, R., and Y. Holm, eds. *Basil.* Amsterdam: Harwood, 1999.

Illy, A., and R. Viani, eds. *Espresso Coffee: The Chemistry of Quality.* San Diego, CA: Academic, 1995.

Jagella, T., and W. Grosch. Flavour and off-flavour compounds of black and white pepper II [black pepper]. *Eur J Food Research and Technology* 209 (1999): 22–26.

———. Flavour and off-flavour compounds of black and white pepper III [white pepper]. *Eur J Food Research and Technology* 209 (1999): 27–31.

Jordt, S.E. et al. Mustard oils and cannabinoids excite sensory nerve fibers through the TRP channel ANKTM1. *Nature* 427 (2004): 260–65.

Kintzios, S.E., ed. *Sage.* Amsterdam: Harwood, 2000.

Maga, J. A. *Smoke in Food Processing.* Boca Raton, FL: CRC, 1988.

McGee, H. In victu veritas. *Nature* 392 (1998): 649–50.

Nasrawi, C.W., and R.M. Pangborn. Temporal effectiveness of mouth-rinsing on capsaicin mouth-burn. *Physiology and Behavior* 47 (1990): 617–23.

Nemeth, E., ed. *Caraway.* Amsterdam: Harwood, 1998.

Noleau, E. et al. Volatile compounds in leek and asafoetida. *J of Essential Oil Research* 3 (1991): 241–56.

Peter, K.V., ed. *Handbook of Herbs and Spices.* Cambridge, UK: Woodhead, 2001.

Prescott, J. et al. Effects of oral chemical irritation on tastes and flavors in frequent and infrequent users of chili. *Physiology and Behavior* 58 (1995): 1117–27.

Rozin, P., and D. Schiller. The nature and acquisition of a preference for chili peppers by humans. *Motivation and Emotion* 4 (1980): 77–101.

Shimoda, M. et al. Comparison of volatile compounds among different grades of green tea and their relations to odor attributes. *J Agric Food Chem.* 43 (1995): 1621–25.

Sivetz, M., and N.W. Desrosier. *Coffee Technology.* Westport, CT: AVI, 1979.

Takeoka, G. "Volatile constituents of asafoetida." In *Aroma Active Constituents of Foods,* 33–44. Oxford: Oxford Univ. Press, 2001.

Taucher, J. et al. Analysis of compounds in human breath after ingestion of garlic using proton-transfer-reaction mass spectrometry. *J Agric Food Chem.* 44 (1996): 3778–82.

Werker, E. et al. Glandular hairs and essential oil in developing leaves of [basil]. *Annals of Botany* 71 (1993): 43–50.

Winterhalter P., and M. Straubinger. Saffron—renewed interest in an ancient spice. *Food Revs International* 16 (2000): 39–59.

Yamanishi, T., ed. Special issue on tea. *Food Revs International* 11 (1995), no. 3.

Yu, H.C. et al., eds. *Perilla.* Amsterdam: Harwood, 1997.

Zamski, E. et al. Ultrastructure of capsaicinoid-secreting cells in pungent and nonpungent red pepper (*Capsicum annuum* L.) cultivars. *Botanical Gazette* 148 (1987): 1–6.

CHAPTER 9: SEEDS

Champlain, S., ed. *The Voyages, 1619.* Translated by H.H. Langton and W.F. Ganong. *The Works of Samuel Champlain,* vol. 3. Toronto: Champlain Society, 1929.

Eliade, M. *Patterns in Comparative Religion.* Trans. R. Sheed. New York: Sheed and Ward, 1958.

Fussell, B. *The Story of Corn.* New York: Knopf, 1992.

National Research Council. *Lost Crops of Africa.* Vol. 1, *Grains.* Washington, DC: National Academy Press, 1996.

Rosengarten, F.J. *The Book of Edible Nuts.* New York: Walker, 1984.

Shurtleff, W., and A. Aoyagi. *The Book of Miso.* New York: Ballantine, 1981.

———. *The Book of Tofu.* New York: Ballantine, 1979.

Thoreau, H.D. "Journal, Jan. 3, 1842." In *The Writings of Henry David Thoreau: Journal I, 1837–46,* edited by B. Torrey. New York: AMS, 1968.

Bakshi, A.S., and R.P. Singh. Kinetics of water diffusion and starch gelatinization during rice parboiling. *J Food Sci.* 45 (1980): 1387–92.

Bernath, J., ed. *Poppy.* Amsterdam: Harwood, 1998.

Bett-Garber, K.L. et al. Categorizing rice cultivars based on cluster analysis of amylose content, protein content and sensory attributes. *Cereal Chemistry* 78 (2001): 551–58.

Bhattacharjee, P. et al. Basmati rice: A review. *International J Food Sci Technology* 37 (2002): 1–12.

Bushuk, W. *Rye: Production, Chemistry, and Technology.* 2nd ed. St. Paul, MN: Am. Assoc. of Cereal Chemists, 2001.

Cassidy, A. Potential risks and benefits of phytoestrogen-rich diets. *International J Vitamin Nutrition Research* 73 (2003): 120–26.

Fast, R.B., and E.F. Caldwell, eds. *Breakfast Cereals and How They Are Made.* 2nd ed. St. Paul, MN: Am. Assoc. Cereal Chemists, 2000.

Fischer, K.H., and W. Grosch. Untersuchungen zum Leguminosenaroma roher Erdnusse. *Lebensmittel-Wissenschaft und Technologie* 15 (1982): 173–76.

Fujimura, T., and M. Kugimiya. Gelatinization of starches inside cotyledon cells of kidney beans. *Starch* 46 (1994): 374–78.

Glaszmann, J.C. Isozymes and classification of Asian rice varieties. *Theoretical and Applied Genetics* 74 (1987): 21–30.

Granito, M. et al. Identification of gas-producing components in different varieties of *Phaseolus vulgaris* by in vitro fermentation. *J Sci Food Agric.* 81 (2001): 543–50.

Hahn, D.M. et al. Light and scanning electron microscope studies on dry beans. *J Food Sci.* 42 (1977): 1208–12.

Hallauer, A.R., ed. *Specialty Corns.* 2nd ed. Boca Raton, FL: CRC, 2001.

Harries, H.C. "Coconut Palm." In Macrae, 1098–1104.

Hickenbottom, J.W. Processing, types, and uses of barley malt extracts and syrups. *Cereal Foods World* 41 (1996): 788–90.

Huang, S. et al. Genes encoding plastid acetyl-Co-A carboxylase . . . and the evolutionary history of wheat. *Proceedings of the National Academy of Sciences* 99 (2002): 8133–38.

Jezussek, M. et al. Comparison of key aroma compounds in cooked brown rice varieties. . . . *J Agric Food Chem.* 50 (2002): 1101–5.

Khush, G.S. Origin, dispersal, cultivation, and variation of rice. *Plant Molecular Biology* 35 (1997): 25–34.

Kimber, I., and R.J. Dearman. Factors affecting the development of food allergies. *Proceedings Nutrition Society* 61 (2002): 435–39.

Lentz, D.L. et al. Prehistoric sunflower (*Helianthus annuus* L.) domestication in Mexico. *Economic Botany* 55 (2001): 370–76.

Lin, S.H. Water uptake and gelatinization of white rice. *Lebensmittel-Wissenschaft und Technologie* 26 (1993): 276–78.

Liu, K. *Soybeans: Chemistry, Technology, and Utilization.* Gaithersburg, MD: Aspen, 1999.

———. Storage proteins and hard-to-cook phenomenon in legume seeds. *Food Technology* 51 (1997): 58–61.

Lumpkin, T.A., and D.C. McClary. *Azuki Bean: Botany, Production, and Uses.* Wallingford, UK: CAB, 1994.

MacGregor, A.W., and R.S. Bhatty, eds. *Barley: Chemistry and Technology.* St. Paul, MN: Am. Assoc. of Cereal Chemists, 1993.

Marshall, H.G., and M.E. Sorrells, eds. *Oat Science and Technology.* Madison, WI: American Society of Agronomy, 1992.

O'Donnell, A.U., and S.E. Fleming. Influence of frequent and longterm consumption of legume seeds on excretion of intestinal gases. *American J of Clinical Nutrition* 40 (1984): 48–57.

Oelke, E.A. et al. Wild rice. *Cereal Foods World* 42 (1997): 234–47.

Paredes-Lopez, O., ed. *Amaranth: Biology, Chemistry, Technology.* Boca Raton, FL: CRC, 1994.

Pattee, H.E., and H.T. Stalker, eds. *Advances in Peanut Science.* Stillwater, OK: American Peanut Research and Education Assoc., 1995.

Rockland, L.B., and F.T. Jones. Scanning electron microscope studies on dry beans. *J Food Sci.* 39 (1974): 342–46.

Rosato, A. D. et al. Why the Brazil nuts are on top: Size segregation of particulate matter by shaking. *Physical Review Letters* 58 (1987): 1038–42.

Salunkhe, D.K. et al. *Postharvest Biotechnology of Food Legumes.* Boca Raton, FL: CRC, 1985.

———. *World Oilseeds: Chemistry, Technology, and Utilization.* New York: Van Nostrand Reinhold, 1992.

Santerre, C.R. *Pecan Technology.* New York: Chapman and Hall, 1994.

Shan, L. et al. Structural basis for gluten intolerance in celiac sprue. *Science* 297 (2002): 2275–79.

Smartt, J. *Grain Legumes.* Cambridge: Cambridge Univ. Press, 1990.

Smith, C.W., and R.A. Frederiksen, eds. *Sorghum: Origin, History, Technology, and Production.* New York: Wiley, 2000.

Sobolev, V.S. Vanillin content in boiled peanuts. *J Agric Food Chem.* 49 (2001): 3725–27.

van Schoonhoven, A., and O. Voysest, eds. *Common Beans: Research for Crop Improvement.* Wallingford, UK: CAB, 1991.

Wang, J., and D.Y.C. Fung. Alkaline-fermented foods: A review with emphasis on pidan fermentation. *CRC Critical Revs in Microbiology* 22 (1996): 101–38.

Williams, J.T., ed. *Cereals and Pseudocereals.* London: Chapman and Hall, 1995.

Woodruff, J.G. *Coconuts: Production, Processing, Products.* 2nd ed. Westport, CT: AVI, 1979.

———. *Tree Nuts.* 2nd ed. Westport, CT: AVI, 1979.

Wrigley, C. The lupin—the grain with no starch. *Cereal Foods World* 48 (2003): 30–31.

CHAPTER 10: CEREAL DOUGHS AND BATTERS

Beranbaum, R.L. *The Cake Bible.* New York: Morrow, 1988.

———. *The Pie and Pastry Bible.* New York: Scribner, 1998.

David, E. *English Bread and Yeast Cookery.* London: Penguin 1977.

Friberg, B. *The Professional Pastry Chef.* 3rd ed. New York: Van Nostrand Reinhold, 1996.

Glezer, M. *Artisan Baking.* New York: Artisan, 2000.

Healy, B., and P. Bugat. *The Art of the Cake.* New York: Morrow, 1999.

———. *The French Cookie Book.* New York: Morrow, 1994.

Perry, C. "Couscous and its cousins." In *Medieval Arab Cookery,* edited by M. Rodinson et al., 233–38. Totnes, UK: Prospect Books, 2001.

———. Puff Paste Is Spanish. *Petits propos culinaires* 17 (1984): 57–61.

———. "The taste for layered bread among the nomadic Turks and the Central Asian origins of baklava." In *Culinary Cultures of the Middle East,* edited by R. Tapper and S. Zubaida, 87–92. London: I.B. Tauris, 1994.

Serventi, S., and F. Sabban. *Pasta: The Story of a Universal Food.* Trans. A. Shugaar. New York: Columbia Univ. Press, 2002.

Siesby, B. The Turkish crescent and the Danish pastry. *Petits propos culinaires* 30 (1988): 7–10.

Udesky, J. *The Book of Soba.* Tokyo: Kodansha, 1995.

Wolfert, P. *Couscous and Other Good Food from Morocco.* New York: Harper and Row, 1973.

Barsby, T.L. et al., eds. *Starch: Advances in Structure and Function.* Cambridge: Royal Society of Chemistry, 2001.

Bath, D.E., and R.C. Hoseney. A laboratory-scale bagel-making procedure. *Cereal Chemistry* 71 (1994): 403–8.

Bernardin, J.E., and D.D. Kasarda. The microstructure of wheat protein fibrils. *Cereal Chemistry* 50 (1973): 735–45.

Bhattacharya, M. et al. Physicochemical properties related to quality of rice noodles. *Cereal Chemistry.* 76 (1999): 861–67.

Blanshard, J.M.V. et al., eds. *Chemistry and Physics of Baking.* London: Royal Society of Chemistry, 1986.

Brooker, B.E. The stabilisation of air in cake batters—the role of fat. *Food Microstructure* 12 (1993): 285–96.

Calvel, R. *The Taste of Bread.* Trans. R.L. Wirtz. Gaithersburg, MD: Aspen, 2001.

Czerny, M., and P. Schieberle. Important aroma compounds in freshly ground wholemeal and white wheat flour: Identification and quantitative changes during sourdough fermentation. *J Agric Food Chem.* 50 (2002): 6835–40.

Dexter, J.E. et al. Scanning electron microscopy of cooked spaghetti. *Cereal Chemistry* 55 (1978): 23–30.

Eliasson, A.C., and K. Larsson. *Cereals in Breadmaking: A Molecular Colloidal Approach.* New York: Dekker, 1993.

Fabriani, G., and C. Lintas, eds. *Durum Wheat: Chemistry and Technology.* St. Paul, MN: Am. Assoc. Cereal Chemists, 1988.

Fik, M., and K. Surowka. Effect of prebaking and frozen storage on the sensory quality and instrumental texture of bread. *J Sci Food Agric.* 82 (2002): 1268–75.

Frazier, P.J. et al., eds. *Starch: Structure and Functionality.* Cambridge: Royal Society of Chemistry, 1997.

Heidolph, B.B. Designing chemical leavening systems. *Cereal Foods World* 41 (1996): 118–26.

Hoseney, R.C. "Physical chemistry of bread dough." In *Physical Chemistry of Foods*, edited by H.G. Schwartzberg and R.W. Hartel. New York: Dekker, 1992.

Hoseney, R.C., and P.A. Seib. Structural differences in hard and soft wheats. *Bakers Digest* 47 (1973): 26–28.

Kruger, J.E. et al. *Pasta and Noodle Technology.* St. Paul, MN: Am. Assoc. Cereal Chemists, 1996.

Loewe, R. Role of ingredients in batter systems. *Cereal Foods World* 38 (1993): 673–77.

Matsunaga, K. et al. Influence of physicochemical properties of starch on crispness of tempura fried batter. *Cereal Chemistry* 80 (2003): 339–45.

O'Brien, R.D. *Fats and Oils: Formulating and Processing for Applications.* Lancaster, PA: Technomic, 1998.

Pomeranz, Y., ed. *Wheat: Chemistry and Technology.* 2 vols. St. Paul, MN: Am. Assoc. Cereal Chemists, 1988.

Schieberle, P., and W. Grosch. Potent odorants of rye bread crust—differences from the crumb and from wheat bread crust. *Zeitschrift für Lebensmittel-Untersuchung und-Forschung* 198 (1994): 292–96.

Sluimer, I.P. Principles of dough retarding. *Bakers Digest* 55, no. 4 (1981): 6–10.

Stear, C.A. *Handbook of Breadmaking Technology.* London: Elsevier, 1990.

Tester, R.F., and S.J.J. Debon. Annealing of starch—a review. *Int J Biological Macromolecules* 27 (2000): 1–12.

Thiele, C. et al. Contribution of sourdough lactobacilli, yeast, and cereal enzymes to the generation of amino acids in dough relevant for bread flavor. *Cereal Chemistry* 79 (2002): 45–51.

Weiss, T.J. *Food Oils and Their Uses.* 2nd ed. Westport, CT: AVI, 1983.

Zweifel, C. et al. Influence of high-temperature drying on structural and textural properties of durum wheat pasta. *Cereal Chemistry* 80 (2003): 159–67.

CHAPTER 11: SAUCES

Armstrong, V., trans. *Cookbook of Sabina Welserin.* 1553. www.daviddfriedman.com/Medieval/Cookbooks/Sabrina_Welserin.html

Brears, P. Transparent pleasures—the story of the jelly. *Petits propos culinaires* 53: 8–19 and 54 (1996): 25–37.

Harper, D. Gastronomy in ancient China—cooking for the Sage King. *Parabola* 9, no. 4 (1984): 38–47.

Kenney, E.J., trans. *The Ploughman's Lunch: Moretum.* Bristol: Bristol Classical Press, 1984.

Kurlansky, M. *Salt: A World History.* New York: Walker, 2002.

Mennell, S. *Lettre d'un pâtissier anglois, et autres contributions à une polémique gastronomique du XVIIIème siècle.* Exeter: Univ. of Exeter Press, 1981.

Mortimer, P. Koch's colonies and the culinary con-

tribution of Fanny Hesse. *Microbiology Today* 28 (2001): 136–37.
Peterson, J. *Sauces: Classical and Contemporary Sauce Making*. New York: Van Nostrand Reinhold, 1991.
Rao, H. et al. Institutional change in Toque Ville: Nouvelle cuisine as an identity movement in French gastronomy. *American Journal of Sociology* 108 (2003): 795–843.
Sokolov, R. *The Saucier's Apprentice*. New York: Knopf, 1983.

Augustin, J. et al. Alcohol retention in food preparation. *J Am Dietetic Assoc.* 92 (1992): 486–88.
Chang, C.M. et al. Electron microscopy of mayonnaise. *Canadian Institute of Food Science and Technology Journal* 5 (1972): 134–37.
Cook, D.J. et al. Effect of hydrocolloid thickeners on the perception of savory flavors. *J Agric Food Chem.* 51 (2003): 3067–72.
Dickinson, E., and J.M. Rodriguez Patino, eds. *Food Emulsions and Foams*. Cambridge: Royal Society of Chemistry, 1999.
Frazier, P.J. et al., eds. *Starch: Structure and Functionality.* Cambridge: Royal Society of Chemistry, 1997.
Gudmundsson, M. Rheological properties of fish gelatins. *J Food Science* 67 (2002): 2172–76.
Harris, P., ed. *Food Gels*. London: Elsevier, 1990.
Hoover, R. Composition, molecular structure, and physicochemical properties of tuber and root starches: A review. *Carbohydrate Polymers* 45 (2001): 253–67.
Leuenberger, B.H. Investigation of viscosity and gelation properties of different mammalian and fish gelatins. *Food Hydrocolloids* 5 (1991): 353–62.
Martinez Padilla, L., and J. Hardy. "Rheological study of interactions among wheat flour milk proteins and lipids of bechamel sauce." In *Food Colloids,* edited by R.D. Bee et al., 395–99. Cambridge: Royal Society of Chemistry, 1989.
Miller, B.S. et al. A pictorial explanation for the increase in viscosity of a heated wheat starch-water suspension. *Cereal Chemistry* 50 (1973): 271–80.
Niman, S. Using one of the oldest food ingredients—salt. *Cereal Foods World* 41 (1996): 729–31.
Oda, M. et al. Study on food components: The structure of N-linked asialo carbohydrate from the edible bird's nest built by *Collocalia fuciphaga*. *J Agric Food Chem.* 46 (1998): 3047–53.
Pearson, A.M., and T.R. Dutson. *Edible Meat Byproducts*. London: Elsevier, 1988.
Sayaslan, A. et al. Volatile compounds in five starches. *Cereal Chemistry* 77 (2000): 248–53.
Solyakov, A. et al. Heterocyclic amines in process flavors, process flavor ingredients, bouillon concentrates and a pan residue. *Food and Chemical Toxicology* 37 (1999): 1–11.
Thebaudin, J.Y. et al. Rheology of starch pastes from starches of different origins: Applications to starch-based sauces. *Lebensmittel-Wissenschaft und Technologie* 31 (1998): 354–60.
Walstra, P., and I. Smulders. Making emulsions and foams: An overview. In *Food Colloids*, edited by E. Dickinson and B. Bergenstahl, 367–81. Cambridge, UK: Royal Society of Chemistry, 1997.
Ward, A.G., and A. Courts, eds. *Science and Technology of Gelatin*. New York: Academic, 1977.
Weel, K.G.C. et al. Flavor release and perception of flavored whey protein gels: Perception is determined by texture rather than release. *J Agric Food Chem.* 50 (2002): 5149–55.
Westphal, G. et al. "Sodium chloride." In *Ullmann's Encyclopedia of Industrial Chemistry,* A24: 317–39. Weinheim: VCH, 1993.
Whistler, R.L., and J.N. BeMiller, eds. *Industrial Gums*. 3rd ed. San Diego, CA: Academic, 1993.
Whistler, R.L. et al., eds. *Starch: Chemistry and Technology.* 2nd ed. Orlando, FL: Academic, 1984.

CHAPTER 12: SUGARS, CHOCOLATE, AND CONFECTIONERY

Alper, J. Crazy candies. *ChemMatters*. October 11, 1993.
Benzoni, G. *History of the New World* (1565). Trans. W.H. Smyth. London: Hakluyt Society, 1857.
Beranbaum, R.L. Rose's sugar bible. *Food Arts* (April 2000).
Coe, S.D., and M.D. Coe. *The True History of Chocolate*. London: Thames and Hudson, 1996.
Gage, T. *The English-American: His Travail by Sea and Land,* 1648. Ed. J.E.S. Thompson. Norman: Univ. of Oklahoma Press, 1958.
Presilla, M. *The New Taste of Chocolate*. Berkeley, CA: Ten Speed, 2001.
Richardson, T. *Sweets: A History of Candy.* New York: Bloomsbury, 2002.
Teubner, C. *The Chocolate Bible*. New York: Penguin Studio, 1997.

Alexander, R.J. *Sweeteners: Nutritive*. St. Paul, MN: Eagan, 1997.
Baikow, V.E. *Manufacture and Refining of Raw Cane Sugar.* Amsterdam: Elsevier, 1982.
Beckett, S.T. *The Science of Chocolate*. Cambridge: Royal Society of Chemistry, 2000.
Beckett, S.T., ed. *Industrial Chocolate Manufacture and Use*. 3rd ed. Oxford: Blackwell, 1999.
Birch, G.G., and K.J. Parker. *Sugar: Science and Technology.* London: Applied Science, 1979.

Blackburn, F. *Sugar-cane.* London: Longman, 1984.
Clarke, M.A. "Syrups." In Macrae, 5711–16.
Edwards, W.P. *Science of Sugar Confectionery.* Cambridge: Royal Society of Chemistry, 2000.
Galloway, J.H. *The Sugar Cane Industry: An Historical Geography From Its Origins to 1914.* Cambridge: Cambridge Univ. Press, 1989.
Godshall, M.A. et al. Sensory properties of white beet sugars. *International Sugar J* 97 (1995): 296–300.
Harris, N. et al. *A Formulary of Candy Products.* New York: Chemical Publishing Co., 1991.
Harris, P., ed. *Food Gels.* London: Elsevier, 1990.
Hickenbottom, J.W. Processing, types, and uses of barley malt extracts and syrups. *Cereal Foods World* 41 (1996): 788–90.
Hurst, W.J. et al. Cacao usage by the earliest Maya civilization. *Nature* 418 (2002): 289.
Jackson, E.B., ed. *Sugar Confectionery Manufacture.* New York: Van Nostrand Reinhold, 1990.
Kroh, L.W. Caramelisation in food and beverages. *Food Chemistry* 51 (1994): 373–79.
Michener, W., and P. Rozin. Pharmacological versus sensory factors in the satiation of chocolate craving. *Physiology and Behavior* 56 (1994): 419–22.
Minifie, B. *Chocolate, Cocoa, and Confectionery: Science and Technology.* 3rd ed. New York: Van Nostrand Reinhold, 1989.
Nabors, L.O., ed. *Alternative Sweeteners.* 3rd ed. New York: Dekker, 2001.
Pennington, N.L., and C.W. Baker. *Sugar: A User's Guide to Sucrose.* New York: Van Nostrand Reinhold, 1990.
Sweeting, L.M. Experiments at home: Wintergreen candy and other triboluminescent materials. 1998, http://www.towson.edu/~sweeting/wg/candywww.htm.
Taylor, C.N. Truffles and pralines. *The Manufacturing Confectioner* (May 1997), 90–94.
Vinson, J.A. et al. Phenol antioxidant quantity and quality in foods: Cocoa, dark chocolate, and milk chocolate. *J Agric Food Chem.* 47 (1999): 4821–24.
Whistler, R.L., and J.N. BeMiller, eds. *Industrial Gums.* 3rd ed. San Diego, CA: Academic, 1993.
Whistler, R.L. et al., eds. *Starch: Chemistry and Technology.* 2nd ed. Orlando: Academic, 1984.
Winston, M. *The Biology of the Honey Bee.* Cambridge, MA: Harvard Univ. Press, 1987.

CHAPTER 13: WINE, BEER, AND DISTILLED SPIRITS

Brode, B. et al. *Beer Judge Certification Program: Guide to Beer Styles for Home Brew Beer Competitions.* Hayward, CA: BJCP, 2001.
Civil, M. Modern brewers re-create an ancient beer. http://oi.uchicago.edu/OI/IS/CIVIL/NN_FAL91/NN_Fal91.html
Harper, P. *The Insider's Guide to Saké.* Tokyo: Kodansha, 1998.
Jackson, M. *Great Beer Guide.* New York: Dorling Kindersley, 2000.
Johnson, H. *Vintage: The Story of Wine.* New York: Simon and Schuster, 1989.
Johnson, H., and J. Robinson. *The World Atlas of Wine.* 5th ed. London: Mitchell Beazley, 2001.
Kramer, M. *Making Sense of Wine.* 2nd ed. Philadelphia: Running Press, 2003.
McGovern, P.E. et al., eds. *The Origins and Ancient History of Wine.* Amsterdam: Gordon and Breach, 1996.
Papazian, C. *The Home Brewer's Companion.* New York: Avon, 1994.
Robinson, J. *The Oxford Companion to Wine.* Oxford: Oxford Univ. Press, 1994.
Waymack, M.H., and J.F. Harris. *The Book of Classic American Whiskeys.* Chicago: Open Court, 1995.
Wilson, J.E. *Terroir: The Role of Geology, Climate, and Culture in the Making of French Wines.* Berkeley: Univ. of California Press, 1998.

Adams, M.R. "Vinegar." In *Microbiology of Fermented Foods*, edited by B.J.B. Wood, 2 vols. Vol. 1, 1–45. New York: Elsevier, 1985.
Augustin, J. et al. Alcohol retention in food preparation. *J American Dietetic Assoc.* 92 (1992): 486–88.
Aylott, R.I., and E.G. Hernandez. "Gin." In Caballero, 2889–98.
Bakalinsky, A.T., and M.H. Penner. "Alcohol." In Caballero, 107–28.
Bertrand, A., and R. Cantagrel. "Brandy and Cognac." In Caballero, 584–605.
Blanchard, L. et al. Formation of furfurylthiol exhibiting a strong coffee aroma during oak barrel fermentation from furfural released by toasted staves. *J Agric Food Chem.* 49 (2001): 4833–35.
Cocchi, M. et al. Determination of carboxylic acids in vinegars and in aceto balsamico tradizionale di Modena by HPLC and GC methods. *J Agric Food Chem.* 50 (2002): 5255–61.
Conner, H.A., and R.J. Allgeier. Vinegar: Its History and Development. *Advances in Applied Microbiology* 20 (1976): 81–133.
Conner, J.M. et al. Release of distillate flavor compounds in Scotch malt whisky. *J Sci Food Agric.* 79 (1999): 1015–20.
De Keersmaecker, J. The mystery of lambic beer. *Scientific American* (August 1996), 74–78.
Ebeler, S. Analytical chemistry: Unlocking the secrets of wine flavor. *Food Reviews International* 17 (2001): 45–64.
Fahrasmane, L., and A. Parfait. "Rum." In Caballero, 5021–27.
Fix, G. *Principles of Brewing Science.* Boulder, CO: Brewers Publications, 1989.

Fleming, M. et al. "Ethanol." In *Goodman and Gilman's The Pharmacological Basis of Therapeutics,* edited by L.S. Goodman et al., 10th ed. 429–45. New York: McGraw-Hill, 2001.
Harris, R., and D.H. West. "Caribbean rum: Its manufacture and quality." In *Chemistry and Processing of Sugarbeet and Sugarcane,* edited by M.A. Clarke and M.A. Godshall, 313–40. Amsterdam: Elsevier, 1988.
Hayman, C.F. "Vodka." In Caballero, 6068–69.
Jackson, R.J. *Wine Tasting: A Professional Handbook.* San Diego, CA: Academic, 2002.
Jackson. R.S. *Wine Science.* 2nd ed. San Diego, CA: Academic, 2000.
Lavigne, V. et al. Identification and determination of sulfur compounds responsible for "grilled" aroma in wines. *Science des Aliments* 18 (1998): 175–91.
Ledauphin, J. et al. Chemical and sensorial aroma characterization of freshly distilled Calvados. *J Agric Food Chem.* 51 (2003): 433–42.
Licker, J.L. et al. "What is 'Brett' *(Brettanomyces)* flavor? A preliminary investigation." In *Chemistry of Wine Flavor,* edited by A.L. Waterhouse and S.E. Ebeler, 96–115. Washington, DC: American Chemical Society, 1998.
Mosedale, J.R., and J.L. Puech. "Barrels: wine, spirits, and other beverages." In Caballero, 393–402.
Neve, R.A. *Hops.* London: Chapman and Hall, 1991.
Noble, A.C., and G.F. Bursick. The contribution of glycerol to perceived viscosity and sweetness in white wine. *Am J Enology and Viticulture* 35 (1984): 110–12.
Olson, R.W. et al. Absinthe and γ-aminobutyric acid receptors. *Proceedings of the National Academy of Sciences* 97 (2000): 4417–18.
Peynaud, E. *The Taste of Wine.* London: Macdonald, 1987.
Piggott, J.R., and Conner, J.M. "Whisky, whiskey, and bourbon." In Caballero, 6171–83.
Swings, J. "The genera *Acetobacter* and *Gluconobacter.*" In *The Prokaryotes,* edited by A. Balows et al., 2nd ed. Vol. 3, 2268–86. New York: Springer, 1992.
Verachtert, H., and R. De Mot, eds. *Yeast: Biotechnology and Biocatalysis.* New York: Dekker, 1990.
Wiese, J.G. et al. The alcohol hangover. *Annals of Internal Medicine* 132 (2000): 897–902.

CHAPTER 14: COOKING METHODS AND UTENSIL MATERIALS

Fennema, O., ed. *Food Chemistry.* 3rd ed. New York: Dekker, 1996.
Hallström, B. et al. *Heat Transfer and Food Products.* London: Elsevier, 1990.
McGee, H. From raw to cooked: The transformation of flavor. In *The Curious Cook: More Kitchen Science and Lore,* 297–313. San Francisco: North Point, 1990.
McGee, H., J. McInerny, and A. Harrus. The virtual cook: modeling heat transfer in the kitchen. *Physics Today* (November 1999): 30–36.
Scientific American. Special issue on "Materials." September 1967.

CHAPTER 15: THE FOUR BASIC FOOD MOLECULES

Barham, P. *The Science of Cooking.* Berlin: Springer-Verlag, 2001.
Fennema, O., ed. *Food Chemistry.* 3rd ed. New York: Dekker, 1996.
Penfield, M.P., and A.M. Campbell. *Experimental Food Science.* 3rd ed. San Diego, CA: Academic, 1990.

APPENDIX: A CHEMISTRY PRIMER

Hill, J.W., and D.K. Kolb. *Chemistry for Changing Times.* 8th ed. Upper Saddle River, NJ: Prentice Hall, 1998.
Snyder, C.H. *The Extraordinary Chemistry of Ordinary Things.* New York: Wiley, 1992.

INDEX

Principal page references to topics are in **boldface;** page references to illustrations and boxed material are in *italics.*

PERMISSIONS

Page 31: Micrographs of whipped cream courtesy of H.D. Goff and A.K. Smith, University of Guelph.

Page 60: Diagram of cheesemaking, created by Soyoung Scanlan, Andante Dairy, and reprinted by permission.

Page 76: Micrograph of egg yolk from C.M. Chang, W.D. Powrie, and O. Fennema, Microstructure of egg yolk. *Journal of Food Science* 42 (1977): 1193–1200. Reprinted with permission.

Pages 114 and 584: Excerpts from *The Viandier of Taillevent,* translated and edited by Terence Scully. Copyright © 1988 by the University of Ottawa Press (Web site: www.uopress.uottawa.ca). Reprinted with permission.

Page 119: "The Rage of Achilles" by Homer, from *The Iliad* by Homer, translated by Robert Fagles, copyright © 1990 by Robert Fagles. Used by permission of Viking Penguin, a division of Penguin Group (USA) Inc.

Pages 120 and 626: Micrographs of meat fibers and oil droplets from Palmer, Helen Hanson; Osman, Elizabeth; Campbell, Ada Marie; Bowers, Jane; Drahn, Marcia; Palumbo, Mary; Jacobson, Marion; Charley, Helen G.; Berkeley, Selma; *Food Theory and Applications,* 1st edition, copyright © 1986. Reprinted by permission of Pearson Education, Inc., Upper Saddle River, NJ.

Page 528: Micrograph of wheat grain courtesy of Ann Hirsch. Flour micrographs from R.C. Hoseney and P.A. Seib, Structural differences in hard and soft wheats. *Bakers Digest* 47 (1973): 26–28. Reprinted by permission.

Page 537: Micrographs of gluten from J.E. Bernardin and D.D. Kasarda, The microstructure of wheat protein fibrils. *Cereal Chemistry* 40 (1973): 735–45. Reprinted by permission.

Page 572: Excerpt from "Ode to *Bing*" from *Pasta: The Story of a Universal Food* by Silvano Serventi and Françoise Sabban, translated by Antony Shugaar. Copyright © 2002 Columbia University Press. Reprinted with permission of the publisher.

Page 583: Excerpt from *Moretum,* translated by E.J. Kenney. Copyright © 1984 E.J. Kenney. Reprinted with permission of Gerald Duckworth and Co., Ltd.

Page 614: Micrographs of starch granules from B.S. Miller, R.I. Derby, and H.B. Trimbo, A pictorial explanation for the increase in viscosity of a heated wheat starch-water suspension. *Cereal Chemistry* 50 (1973): 271–80. Reprinted by permission.

Page 633: Micrograph of oil droplets from C.M. Chang, W.D. Powrie, and O. Fennema, Electron microscopy of mayonnaise. *Canadian Institute of Food Science and Technology Journal* 5 (1972): 134–37. Reprinted by permission.

Page 714: Excerpt from Hymn to Ninkasi, English translation by Miguel Civil. http://oi.uchicago.edu/OI/IS/CIVIL/NN_FAL91/NN_Fal91_hymn.html. Copyright © 2002 Oriental Institute, University of Chicago. Reprinted courtesy of the Oriental Institute of the University of Chicago.

Page 715: Micrograph of yeast courtesy of Alastain Pringle, Research Director, Anheuser Busch Inc.